BIOPOLYMERS
AT INTERFACES

SURFACTANT SCIENCE SERIES

1. Nonionic Surfactants, *edited by Martin J. Schick* (see also Volumes 19, 23, and 60)
2. Solvent Properties of Surfactant Solutions, *edited by Kozo Shinoda* (see Volume 55)
3. Surfactant Biodegradation, *R. D. Swisher* (see Volume 18)
4. Cationic Surfactants, *edited by Eric Jungermann* (see also Volumes 34, 37, and 53)
5. Detergency: Theory and Test Methods (in three parts), *edited by W. G. Cutler and R. C. Davis* (see also Volume 20)
6. Emulsions and Emulsion Technology (in three parts), *edited by Kenneth J. Lissant*
7. Anionic Surfactants (in two parts), *edited by Warner M. Linfield* (see Volume 56)
8. Anionic Surfactants: Chemical Analysis, *edited by John Cross*
9. Stabilization of Colloidal Dispersions by Polymer Adsorption, *Tatsuo Sato and Richard Ruch*
10. Anionic Surfactants: Biochemistry, Toxicology, Dermatology, *edited by Christian Gloxhuber* (see Volume 43)
11. Anionic Surfactants: Physical Chemistry of Surfactant Action, *edited by E. H. Lucassen-Reynders*
12. Amphoteric Surfactants, *edited by B. R. Bluestein and Clifford L. Hilton* (see Volume 59)
13. Demulsification: Industrial Applications, *Kenneth J. Lissant*
14. Surfactants in Textile Processing, *Arved Datyner*
15. Electrical Phenomena at Interfaces: Fundamentals, Measurements, and Applications, *edited by Ayao Kitahara and Akira Watanabe*
16. Surfactants in Cosmetics, *edited by Martin M. Rieger* (see Volume 68)
17. Interfacial Phenomena: Equilibrium and Dynamic Effects, *Clarence A. Miller and P. Neogi*
18. Surfactant Biodegradation: Second Edition, Revised and Expanded, *R. D. Swisher*
19. Nonionic Surfactants: Chemical Analysis, *edited by John Cross*
20. Detergency: Theory and Technology, *edited by W. Gale Cutler and Erik Kissa*
21. Interfacial Phenomena in Apolar Media, *edited by Hans-Friedrich Eicke and Geoffrey D. Parfitt*
22. Surfactant Solutions: New Methods of Investigation, *edited by Raoul Zana*
23. Nonionic Surfactants: Physical Chemistry, *edited by Martin J. Schick*
24. Microemulsion Systems, *edited by Henri L. Rosano and Marc Clausse*
25. Biosurfactants and Biotechnology, *edited by Naim Kosaric, W. L. Cairns, and Neil C. C. Gray*
26. Surfactants in Emerging Technologies, *edited by Milton J. Rosen*
27. Reagents in Mineral Technology, *edited by P. Somasundaran and Brij M. Moudgil*
28. Surfactants in Chemical/Process Engineering, *edited by Darsh T. Wasan, Martin E. Ginn, and Dinesh O. Shah*
29. Thin Liquid Films, *edited by I. B. Ivanov*
30. Microemulsions and Related Systems: Formulation, Solvency, and Physical Properties, *edited by Maurice Bourrel and Robert S. Schechter*
31. Crystallization and Polymorphism of Fats and Fatty Acids, *edited by Nissim Garti and Kiyotaka Sato*
32. Interfacial Phenomena in Coal Technology, *edited by Gregory D. Botsaris and Yuli M. Glazman*
33. Surfactant-Based Separation Processes, *edited by John F. Scamehorn and Jeffrey H. Harwell*
34. Cationic Surfactants: Organic Chemistry, *edited by James M. Richmond*
35. Alkylene Oxides and Their Polymers, *F. E. Bailey, Jr., and Joseph V. Koleske*
36. Interfacial Phenomena in Petroleum Recovery, *edited by Norman R. Morrow*
37. Cationic Surfactants: Physical Chemistry, *edited by Donn N. Rubingh and Paul M. Holland*
38. Kinetics and Catalysis in Microheterogeneous Systems, *edited by M. Grätzel and K. Kalyanasundaram*
39. Interfacial Phenomena in Biological Systems, *edited by Max Bender*
40. Analysis of Surfactants, *Thomas M. Schmitt* (see Volume 96)

41. Light Scattering by Liquid Surfaces and Complementary Techniques, *edited by Dominique Langevin*

42. Polymeric Surfactants, *Irja Piirma*

43. Anionic Surfactants: Biochemistry, Toxicology, Dermatology. Second Edition, Revised and Expanded, *edited by Christian Gloxhuber and Klaus Künstler*

44. Organized Solutions: Surfactants in Science and Technology, *edited by Stig E. Friberg and Björn Lindman*

45. Defoaming: Theory and Industrial Applications, *edited by P. R. Garrett*

46. Mixed Surfactant Systems, *edited by Keizo Ogino and Masahiko Abe*

47. Coagulation and Flocculation: Theory and Applications, *edited by Bohuslav Dobiáš*

48. Biosurfactants: Production • Properties • Applications, *edited by Naim Kosaric*

49. Wettability, *edited by John C. Berg*

50. Fluorinated Surfactants: Synthesis • Properties • Applications, *Erik Kissa*

51. Surface and Colloid Chemistry in Advanced Ceramics Processing, *edited by Robert J. Pugh and Lennart Bergström*

52. Technological Applications of Dispersions, *edited by Robert B. McKay*

53. Cationic Surfactants: Analytical and Biological Evaluation, *edited by John Cross and Edward J. Singer*

54. Surfactants in Agrochemicals, *Tharwat F. Tadros*

55. Solubilization in Surfactant Aggregates, *edited by Sherril D. Christian and John F. Scamehorn*

56. Anionic Surfactants: Organic Chemistry, *edited by Helmut W. Stache*

57. Foams: Theory, Measurements, and Applications, *edited by Robert K. Prud'homme and Saad A. Khan*

58. The Preparation of Dispersions in Liquids, *H. N. Stein*

59. Amphoteric Surfactants: Second Edition, *edited by Eric G. Lomax*

60. Nonionic Surfactants: Polyoxyalkylene Block Copolymers, *edited by Vaughn M. Nace*

61. Emulsions and Emulsion Stability, *edited by Johan Sjöblom*

62. Vesicles, *edited by Morton Rosoff*

63. Applied Surface Thermodynamics, *edited by A. W. Neumann and Jan K. Spelt*

64. Surfactants in Solution, *edited by Arun K. Chattopadhyay and K. L. Mittal*

65. Detergents in the Environment, *edited by Milan Johann Schwuger*

66. Industrial Applications of Microemulsions, *edited by Conxita Solans and Hironobu Kunieda*

67. Liquid Detergents, *edited by Kuo-Yann Lai*

68. Surfactants in Cosmetics: Second Edition, Revised and Expanded, *edited by Martin M. Rieger and Linda D. Rhein*

69. Enzymes in Detergency, *edited by Jan H. van Ee, Onno Misset, and Erik J. Baas*

70. Structure–Performance Relationships in Surfactants, *edited by Kunio Esumi and Minoru Ueno*

71. Powdered Detergents, *edited by Michael S. Showell*

72. Nonionic Surfactants: Organic Chemistry, *edited by Nico M. van Os*

73. Anionic Surfactants: Analytical Chemistry, Second Edition, Revised and Expanded, *edited by John Cross*

74. Novel Surfactants: Preparation, Applications, and Biodegradability, *edited by Krister Holmberg*

75. Biopolymers at Interfaces, *edited by Martin Malmsten*

76. Electrical Phenomena at Interfaces: Fundamentals, Measurements, and Applications, Second Edition, Revised and Expanded, *edited by Hiroyuki Ohshima and Kunio Furusawa*

77. Polymer-Surfactant Systems, *edited by Jan C. T. Kwak*

78. Surfaces of Nanoparticles and Porous Materials, *edited by James A. Schwarz and Cristian I. Contescu*

79. Surface Chemistry and Electrochemistry of Membranes, *edited by Torben Smith Sørensen*

80. Interfacial Phenomena in Chromatography, *edited by Emile Pefferkorn*

81. Solid–Liquid Dispersions, *Bohuslav Dobiáš, Xueping Qiu, and Wolfgang von Rybinski*
82. Handbook of Detergents, *editor in chief: Uri Zoller*
    Part A: Properties, *edited by Guy Broze*
83. Modern Characterization Methods of Surfactant Systems, *edited by Bernard P. Binks*
84. Dispersions: Characterization, Testing, and Measurement, *Erik Kissa*
85. Interfacial Forces and Fields: Theory and Applications, *edited by Jyh-Ping Hsu*
86. Silicone Surfactants, *edited by Randal M. Hill*
87. Surface Characterization Methods: Principles, Techniques, and Applications, *edited by Andrew J. Milling*
88. Interfacial Dynamics, *edited by Nikola Kallay*
89. Computational Methods in Surface and Colloid Science, *edited by Małgorzata Borówko*
90. Adsorption on Silica Surfaces, *edited by Eugène Papirer*
91. Nonionic Surfactants: Alkyl Polyglucosides, *edited by Dieter Balzer and Harald Lüders*
92. Fine Particles: Synthesis, Characterization, and Mechanisms of Growth, *edited by Tadao Sugimoto*
93. Thermal Behavior of Dispersed Systems, *edited by Nissim Garti*
94. Surface Characteristics of Fibers and Textiles, *edited by Christopher M. Pastore and Paul Kiekens*
95. Liquid Interfaces in Chemical, Biological, and Pharmaceutical Applications, *edited by Alexander G. Volkov*
96. Analysis of Surfactants: Second Edition, Revised and Expanded, *Thomas M. Schmitt*
97. Fluorinated Surfactants and Repellents: Second Edition, Revised and Expanded, *Erik Kissa*
98. Detergency of Specialty Surfactants, *edited by Floyd E. Friedli*
99. Physical Chemistry of Polyelectrolytes, *edited by Tsetska Radeva*
100. Reactions and Synthesis in Surfactant Systems, *edited by John Texter*
101. Protein-Based Surfactants: Synthesis, Physicochemical Properties, and Applications, *edited by Ifendu A. Nnanna and Jiding Xia*
102. Chemical Properties of Material Surfaces, *Marek Kosmulski*
103. Oxide Surfaces, *edited by James A. Wingrave*
104. Polymers in Particulate Systems: Properties and Applications, *edited by Vincent A. Hackley, P. Somasundaran, and Jennifer A. Lewis*
105. Colloid and Surface Properties of Clays and Related Minerals, *Rossman F. Giese and Carel J. van Oss*
106. Interfacial Electrokinetics and Electrophoresis, *edited by Ángel V. Delgado*
107. Adsorption: Theory, Modeling, and Analysis, *edited by József Tóth*
108. Interfacial Applications in Environmental Engineering, *edited by Mark A. Keane*
109. Adsorption and Aggregation of Surfactants in Solution, *edited by K. L. Mittal and Dinesh O. Shah*
110. Biopolymers at Interfaces: Second Edition, Revised and Expanded, *edited by Martin Malmsten*

ADDITIONAL VOLUMES IN PREPARATION

Biomolecular Films: Design, Function, and Applications, *edited by James F. Rusling*

Structure–Performance Relationships in Surfactants: Second Edition, Revised and Expanded, *edited by Kunio Esumi and Minoru Ueno*

BIOPOLYMERS AT INTERFACES

Second Edition, Revised and Expanded

edited by

Martin Malmsten

Institute for Surface Chemistry and
Royal Institute of Technology
Stockholm, Sweden

CRC Press
Taylor & Francis Group
Boca Raton London New York

CRC Press is an imprint of the
Taylor & Francis Group, an **informa** business

First published 2003 by Marcel Dekker, Inc.

Published 2019 by CRC Press
Taylor & Francis Group
6000 Broken Sound Parkway NW, Suite 300
Boca Raton, FL 33487-2742

First issued in paperback 2019

No claim to original U.S. Government works

ISBN 13: 978-0-367-44684-0 (pbk)
ISBN 13: 978-0-8247-0863-4 (hbk)

Visit the Taylor & Francis Web site at
http://www.taylorandfrancis.com

and the CRC Press Web site at
http://www.crcpress.com

Library of Congress Cataloging-in-Publication Data
A catalog record for this book is available from the Library of Congress.

Preface

It has been a few years since the first edition of *Biopolymers at Interfaces* was published. Issues relating to the interfacial behavior of biopolymers and notably proteins are as relevant as ever both to understanding biophysical processes and to controlling numerous industrial applications. In recent years, there has been continued progress in our understanding of the basics of biopolymer interfacial behavior, and new experimental methods are being developed and applied for further investigation. Furthermore, the role of biopolymer interfacial behavior in numerous applications is progressively becoming better understood, and at the same time new and exciting application areas are emerging.

The second edition of *Biopolymers at Interfaces* addresses many of these issues. For example, the interesting development of biopolymer-based multilayer structures (with potential applications in, e.g., biosensors) is discussed in the chapters by Schaaf and Voegel (Chapter 13) and Ariga and Lvov (Chapter 14). Also, the progress made in the understanding of protein behavior at the air–water interface, facilitated by developments in experimental techniques, is highlighted and detailed examples are given by Britt et al. (Chapter 16) and Horne and Rodriguez Patino (Chapter 30).

Experimental developments are also reviewed by Fromell et al. (calorimetry, Chapter 20), Malmsten (ellipsometry and reflectometry, Chapter 21), Sinner et al. (SPR, Chapter 22), Lu (neutron reflection, Chapter 23), and Griesser et al. (XPS, ToF-SIMS, MALDI-MS, Chapter 24). The section covering different experimental techniques has thus been significantly expanded compared with the first edition.

Finally, some fascinating recent developments concerning protein interfacial behavior in microfabricated total analysis systems and microarrays are discussed by Dérand and Malmsten (Chapter 28), and the importance of protein interfacial behavior in the oral cavity (e.g., in relation to biological function) is highlighted by Arnebrant (Chapter 29). Biological function of interfacial protein layers is also the focus of the chapter by Horbett (Chapter 15).

I would like to thank all the contributors for their excellent support during the preparation of this edition.

Martin Malmsten

Contents

Preface   *iii*

Contributors   *ix*

Part I.   Concepts

1. Macromolecular Adsorption: A Brief Introduction   1
    Martinus A. Cohen Stuart

2. Driving Forces for Protein Adsorption at Solid Surfaces   21
    Willem Norde

3. Thermodynamics of Adsorption of Amino Acids, Small Peptides,
    and Nucleic Acid Components on Silica Adsorbents   45
    Vladimir A. Basiuk

4. Quantitative Modeling of Protein Adsorption   71
    Charles M. Roth and Abraham M. Lenhoff

5. Interfacial Behavior of Protein Mutants and Variants   95
    Martin Malmsten, Thomas Arnebrant, and Peter Billsten

6. Orientation and Activity of Immobilized Antibodies   115
    James N. Herron, Hsu-kun Wang, Věra Janatová, Jacob D. Durtschi,
    Karin Caldwell, Douglas A. Christensen, I-Nan Chang, and
    Shao-Chie Huang

7. Interactions Between Surfaces Coated with Carbohydrates, Glycolipids,
    and Glycoproteins   165
    Per M. Claesson

8. Protein Adsorption Kinetics   199
    Jeremy J. Ramsden

9. Mobility of Biomolecules at Interfaces   221
    Robert D. Tilton

10. Protein Adsorption in Relation to Solution Association
    and Aggregation   259
    Tommy Nylander

11. Mechanism of Interfacial Exchange Phenomena for Proteins
    Adsorbed at Solid–Liquid Interfaces   295
    Vincent Ball, Pierre Schaaf, and Jean-Claude Voegel

12. Interactions Between Proteins and Surfactants at Solid Interfaces   321
    Marie Wahlgren, Camilla A.-C. Karlsson, and Stefan Welin-Klintström

13. Toward Functionalized Polyelectrolyte Biofilms   345
    Pierre Schaaf and Jean-Claude Voegel

14. Self-Assembly of Functional Protein Multilayers: From Planar Films to
    Microtemplate Encapsulation   367
    Katsuhiko Ariga and Yuri M. Lvov

15. Biological Activity of Adsorbed Proteins   393
    Thomas A. Horbett

16. Protein Interactions with Monolayers at the Air–Water Interface   415
    David W. Britt, G. Jogikalmath, and Vladimir Hlady

Part II.   Methods

17. Local and Global Optical Spectroscopic Probes of Adsorbed Proteins   435
    Vladimir Hlady and Jos Buijs

18. Proteins on Surfaces: Methodologies for Surface Preparation
    and Engineering Protein Function   467
    Krishnan K. Chittur

19. Studies on the Conformation of Adsorbed Proteins with the Use of
    Nanoparticle Technology   497
    Peter Billsten, Uno Carlsson, and Hans Elwing

20. Scanning Calorimetry in Probing the Structural Stability of Proteins
    at Interfaces   517
    Karin Fromell, Shao-Chie Huang, and Karin Caldwell

21. Ellipsometry and Reflectometry for Studying Protein Adsorption   539
    Martin Malmsten

22. Surface Plasmon Resonance Spectroscopies for Protein Binding
    Studies at Functionalized Surfaces   583
    Eva-Kathrin Sinner, Kazutoshi Kobayashi, Thomas Lehmann,
    Thomas Neumann, Birgit Prein, Jürgen Rühe, Fang Yu,
    and Wolfgang Knoll

23. Neutron Reflection Study of Protein Adsorption at the Solid–Solution
    Interface   609
    J. R. Lu

Contents

24. XPS, ToF-SIMS, and MALDI-MS for Characterizing Adsorbed
    Protein Films   641
    Hans J. Griesser, Sally L. McArthur, Matthew S. Wagner,
    David G. Castner, Peter Kingshott, and Keith M. McLean

Part III. Applications

25. Interaction of Proteins with Polymeric Synthetic Membranes   671
    Georges Belfort and Andrew L. Zydney

26. Protein Adsorption in Intravenous Drug Delivery   711
    Martin Malmsten

27. Control of Protein Adsorption in Solid-Phase Diagnostics
    and Therapeutics   741
    Krister Holmberg and Gerard Quash

28. Protein Interfacial Behavior in Microfabricated Analysis Systems
    and Microarrays   773
    Helene Dérand and Martin Malmsten

29. Protein Adsorption in the Oral Environment   811
    Thomas Arnebrant

30. Adsorbed Biopolymers: Behavior in Food Applications   857
    David S. Horne and J. M. Rodriguez Patino

Index   *901*

Contributors

Katsuhiko Ariga   ERATO Nanospace Project, Japan Science and Technology Corporation, Tokyo, Japan

Thomas Arnebrant   Pharmaceuticals and Food Section, Institute for Surface Chemistry, Stockholm, and Department of Prosthetic Dentistry, Malmö University, Malmö, Sweden

Vincent Ball[*]   Department of Biophysical Chemistry, Biocenter, University of Basel, Basel, Switzerland

Vladimir A. Basiuk   Instituto de Ciencias Nucleares, Universidad Nacional Autónoma de México, Mexico City, Mexico

Georges Belfort   Howard P. Isermann Chemical Engineering Department, Rensselaer Polytechnic Institute, Troy, New York, U.S.A.

Peter Billsten[†]   Department of Applied Physics, IFM, Linköping University, Linköping, Sweden

David W. Britt   Department of Biological Engineering, Utah State University, Logan, Utah, U.S.A.

Jos Buijs   Ångstrom Laboratory, Division of Ion Physics, Uppsala University, Uppsala, Sweden

Karin Caldwell   Center for Surface Biotechnology, Uppsala University, Uppsala, Sweden

---

Current affiliation:
[*] Institute of Chemistry, LIMBO, Strasbourg, France.
 Analysis and Formulations, Astra Draco AB, Lund, Sweden.

Uno Carlsson   Department of Chemistry, IFM, Linköping University, Linköping, Sweden

David G. Castner   Departments of Bioengineering and Chemical Engineering, University of Washington, Seattle, Washington, U.S.A.

I-Nan Chang   Biochemistry Division, Development Center for Biotechnology, Taipei, Taiwan, Republic of China

Krishnan K. Chittur   Department of Chemical Engineering, University of Alabama in Huntsville, Huntsville, Alabama, U.S.A.

Douglas A. Christensen   Department of Bioengineering, University of Utah, Salt Lake City, Utah, U.S.A.

Per M. Claesson   Department of Chemistry, Surface Chemistry, Royal Institute of Technology, and Institute for Surface Chemistry, Stockholm, Sweden

Martinus A. Cohen Stuart   Department of Physical and Colloidal Chemistry, Wageningen Agricultural University, Wageningen, The Netherlands

Helene Dérand   Gyros AB, Uppsala, Sweden

Jacob D. Durtschi   Department of Bioengineering, University of Utah, Salt Lake City, Utah, U.S.A.

Hans Elwing   Department of General and Marine Microbiology, Göteborg University, Göteborg, Sweden

Karin Fromell   Center for Surface Biotechnology, Uppsala University, Uppsala, Sweden

Hans J. Griesser   Ian Wark Research Institute, University of South Australia, Mawson Lakes, Australia

James N. Herron   Department of Material Science and Engineering, University of Utah, Salt Lake City, Utah, U.S.A.

Vladimir Hlady   Department of Bioengineering, University of Utah, Salt Lake City, Utah, U.S.A.

Krister Holmberg   Department of Applied Surface Chemistry, Chalmers University of Technology, Göteborg, Sweden

Thomas A. Horbett   Departments of Bioengineering and Chemical Engineering, University of Washington, Seattle, Washington, U.S.A.

David S. Horne   Food Quality Group, Hannah Research Institute, Ayr, Scotland

Shao-Chie Huang   Diagnostic Products Corporation, Los Angeles, California, U.S.A.

Věra Janatová*   Department of Material Science and Engineering, University of Utah, Salt Lake City, Utah, U.S.A.

G. Jogikalmath   Department of Material Science and Engineering, University of Utah, Salt Lake City, Utah, U.S.A.

Camilla A.-C. Karlsson   Department of Food Engineering, University of Lund, Lund, Sweden

Peter Kingshott   The Danish Polymer Center, Risø National Laboratory, Roskilde, Denmark

Wolfgang Knoll   Materials Science, Max-Planck Institute for Polymer Research, Mainz, Germany

Kazutoshi Kobayashi   Research and Development Center, Hitachi Chemical Company, Ltd., Ibaraki, Japan

Thomas Lehmann   New Applications Business Team, Wacker Polymer Systems, Burghausen, Germany

Abraham M. Lenhoff   Department of Chemical Engineering, University of Delaware, Newark, Delaware, U.S.A.

J. R. Lu   Department of Physics, University of Manchester Institute of Science and Technology, Manchester, England

Yuri M. Lvov   Institute for Micromanufacturing, Louisiana Tech University, Ruston, Louisiana, U.S.A.

Martin Malmsten   Institute for Surface Chemistry, and Department of Chemistry, Surface Chemistry, Royal Institute of Technology, Stockholm, Sweden

Sally L. McArthur   Department of Bioengineering, University of Washington, Seattle, Washington, U.S.A.

Keith M. McLean   Department of Molecular Science, Commonwealth Scientific and Industrial Research Organisation, Clayton, Australia

Thomas Neumann   Max-Planck Institute for Polymer Research, Mainz, Germany

Willem Norde   Department of Physical and Colloid Chemistry, Wageningen Agricultural University, Wageningen, The Netherlands

---

Present location: Prague, Czech Republic.

Tommy Nylander   Department of Physical Chemistry 1, Center for Chemistry and Chemical Engineering, University of Lund, Lund, Sweden

Birgit Prein   Department of Biochemistry, Graz University of Technology, Graz, Austria

Gerard Quash   Laboratory of Immunochemistry, Faculty of Medicine Lyon-Sud, Oullins, France

Jeremy J. Ramsden   Institute of Experimental Medicine, Hungarian Academy of Sciences, Budapest, Hungary

J. M. Rodriguez Patino   Department of Chemical Engineering, University of Seville, Seville, Spain

Charles M. Roth   Center for Engineering in Medicine, Massachusetts General Hospital, Harvard Medical School, and Shriners Burns Hospital, Boston, Massachusetts, U.S.A.

Jürgen Rühe   Institute for Microsystems Technology, University of Freiburg, Freiburg, Germany

Pierre Schaaf   Institut Charles Sadron (CNRS–ULP), Strasbourg, France

Eva-Kathrin Sinner   Membrane Biochemistry, Max-Planck Institute for Biochemistry, Martinsried, Germany

Robert D. Tilton   Colloids, Polymers, and Surfaces Program, Department of Chemical Engineering, Carnegie Mellon University, Pittsburgh, Pennsylvania, U.S.A.

Jean-Claude Voegel   Faculté de Chirurgie Dentaire, INSERM—Unité 424, Strasbourg, France

Matthew S. Wagner   Department of Chemical Engineering, University of Washington, Seattle, Washington, U.S.A.

Marie Wahlgren   Department of Food Technology, University of Lund, Lund, Sweden

Stefan Welin-Klintström   Department of Physics and Measurement Technology, Linköping University, Linköping, Sweden

Hsu-kun Wang   Department of Material Science and Engineering, University of Utah, Salt Lake City, Utah, U.S.A.

Fang Yu   Max-Planck Institute for Polymer Research, Mainz, Germany

Andrew L. Zydney   Department of Chemical Engineering, University of Delaware, Newark, Delaware, U.S.A.

1

Macromolecular Adsorption: A Brief Introduction

MARTINUS A. COHEN STUART Wageningen Agricultural University, Wageningen, The Netherlands

I. BIOLOGICAL MACROMOLECULES

The primary structure of many biological macromolecules is simply that of a linear chain of monomers. In this respect, they do not differ from most synthetic polymers. Behind the chemical simplicity, however, rather complex and specific behavior may be hidden. For example, the secondary and tertiary structures of proteins often play an essential role in their specific biological function. For such molecules it may seem impossible to develop a universal description as is now well established for synthetic polymers.

However, one may wonder whether this has been attempted. Some of the concepts fruitfully introduced in polymer physics may be quite relevant for understanding the behavior of, say, globular proteins. It is the purpose of this chapter to discuss polymers at interfaces from a physical chemist's point of view, highlighting processes and central concepts rather than specific systems. Effects of conformational entropy, internal cohesion, electrostatic interactions, and intermolecular interactions on adsorbed layers will be considered, and processes like transport, attachment, unfolding, and trapping will be discussed.

As said, most biological polymers are chemically more complex than homopolymers and have a copolymer character that cannot be ignored. However, considering the entire plenitude of phenomena that various classes of copolymers can display would by far exceed the limits of this introductory presentation. We will therefore not explicitly discuss copolymers.

II. MACROMOLECULES ARE SOFT PARTICLES

Around the bonds that connect the atoms in the main chain of a linear polymer, some rotation is usually possible. This gives the polymer a certain degree of flexibility, which is usually enough to allow the molecule to assume a large number of shapes or *conformations*. With this large number of conformations, a conformation entropy is associated. In the absence of external forces, the chain will wiggle because of thermal motion and assume an average, approximately spherical *coil* shape with a characteristic size R dependent on the number of monomers N per chain (i.e., the length of the chain). This dependence can be generally expressed as a power law: $R \sim N^v$. The conformational entropy gives the coil a certain elastic resistance to deformations such as squeezing and stretching. This resilience is rather weak; the molecules can be considered "soft spheres."

The characteristic size R of the coil depends strongly on the solvent [1,2], but it can always be written as a power law of the chain length N: $R \sim N^\alpha$. *Good solvents* tend to keep the individual monomers separate. In such a solvent, most shapes permitted by bond rotation are energetically equivalent. If the monomers would not occupy any volume (i.e., the chain is infinitely thin, this is called an *ideal chain*, analogous to ideal gases), one could describe the chain in terms of a random walk (diffusion-like) process. However, because monomers do have a volume, walks that lead to overlap between monomers must be excluded. The proper description is therefore that of a *self-avoiding walk*. The characteristic size R of such a structure is known to scale as $N^{3/5}$ ($\alpha = 3/5$), and such molecules are denoted as *swollen coils*. The monomer–monomer repulsion also ensures that the molecules do not cluster but form a homogeneous solution. At higher concentration, however, this repulsion is not enough to keep the coils apart; as soon as they fill the entire solution volume, they begin to interpenetrate, thus forming a transient network. This situation is now commonly referred to as a *semidilute solution*.

In less good solvents, there is, effectively, an attraction between the monomers. As long as this attraction is very weak, the coil may somewhat contract, but it remains swollen. At a certain point, the attraction becomes strong enough to compensate the effect of the excluded volume of the monomers. This is called the θ point or θ temperature. The behavior of the coil around the θ point is equivalent to that of an ideal chain, for which random walk statistics apply and one can prove $R \sim N^{1/2}$ ($\alpha = 1/2$); this is called a Gaussian coil because the density of monomers inside the coil is a Gaussian function of the distance to the center [2].

Should the solvent become even worse, the entropy can no longer maintain the open, dilute structure of the coil, and the molecule collapses to a compact *globule*. Inside this globule, the monomer density is essentially constant and therefore the mass must scale as R^3 so that $R \sim N^{1/3}$ ($v = 1/3$). Of course, monomer–monomer attraction will not only occur between monomers within the same chain, but also between monomers on different chains. Hence, in poor solvents there is a concentration where the molecules accumulate into a dense phase and we have phase separation.

As a parameter that describes the strength of the monomer interaction it is customary to use the so-called Flory–Huggins parameter χ, which is essentially an excess Gibbs energy of monomer–solvent contacts (with respect to monomer–monomer and solvent–solvent contacts) normalized by kT. In good solvents, χ is close to zero; at the θ point it equals 0.5; and for poor solvents, $\chi > 0.5$. In many texts, the parameter $v = 1 - 2\chi$ is used. It measures the strength of the monomer–monomer interaction relative to that at the θ point and is called the *excluded volume* parameter. Its value is zero at the θ point, positive in better solvents, and negative in worse solvents (Fig. 1).

In poor solvents, where the molecules collapse into globules, deformation is not only counteracted by the conformational entropy, but also by the fact that the number of (unfavorable) monomer–solvent contacts increases at the expense of (fa-

(a)   (b)   (c)

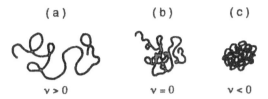

$v > 0$ $v = 0$ $v < 0$

FIG. 1   Polymer molecules in solution under (a) good, (b) θ, and (c) poor solvent conditions.

vorable) monomer–monomer contacts, just as for a droplet of an inmiscible liquid in some liquid medium. One may therefore assign a certain *interfacial tension* γ to the molecules of order kTv^2. This means that collapsed macromolecules are also deformable, but to an extent dictated by the surface tension (v^2) rather than the conformational entropy. in other words, they become "harder" as $|v|$ increases [3].

As is well known, protein molecules are rather compact. This compact structure is stabilized by hydrophobic attractions between lipophilic amino acids which tend to accumulate inside the globular structure of the molecule because water is a poor solvent for them. In this sense, there is a clear parallel with simple homopolymers in a poor solvent. As one approaches the temperature where the hydrophobic interactions become too weak to keep the globule together, the particles become "softer" (get lower surface tension), and unfolding becomes increasingly easier. Therefore one can consider the denaturation temperature of the protein as a kind of θ temperature. Of course, the important difference between proteins and simple polymers in a poor solvent is that the former are copolymers; they have sufficient hydrophilic monomers on their external surface to suppress the accumulation of molecules into a macroscopic phase. A detailed discussion of protein structure can be found in Chapter 2.

III.   ADSORPTION AND PARTICLE DEFORMATION

Adsorption of a polymer occurs whenever its monomers are attracted sufficiently strongly by a surface. Therefore, on the one hand, the molecules attempt to maximize their favorable monomer–surface interactions by spreading out to a large extent. For each polymer–surface contact established, a solvent molecule is detached and the net Gibbs energy of contact formation is therefore $u_p - u_s$. For convenience, the parameter $\chi_s = (u_p - u_s)/kT$ has been introduced as a measure of adsorption affinity. On the other hand, there is a penalty in the form of the resistance to deformation. For the Gaussian or swollen coil, this resistance is weak. In the case of weak attraction, the polymer is deformed into a "pancake" of thickness D, where D scales as $(\varepsilon/kT)^{-\alpha/(1-\alpha)}$, where ε is the adsorption energy per monomer [1]. For the Gaussian coil, one may argue that the Gibbs energy increase due to entropy loss ($-TS$) upon adsorption is, per monomer unit, of order $\ln(\omega^b/\omega^s)$, where ω^b and ω^s are the orientational degrees of freedom per monomer in the bulk (b) and in the surface (s), respectively. This entropic contribution is small, typically around 0.5 kT, and must be balanced by an equal adsorption energy ε for appreciable adsorption to occur. Values of 1 kT for ε are quite common; this is enough to completely squeeze the chains into an essentially flat structure (Fig. 2a) [4].

(a)   (b) 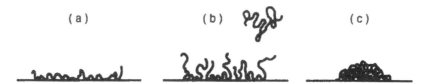   (c)

FIG. 2 Polymers on a surface: (a,b) good solvent conditions; (c) poor solvent conditions.

Hence, already at moderate adsorption strength per monomer, large polymers can make an appreciable number of contacts with the surface and will, therefore, adhere tenaciously. As a consequence, the surface will be fully covered before any free polymer in the equilibrium solution is detected. In other words, the adsorption has a strongly *high-affinity* character. When the molecules start to crowd on the surface, they begin to interpenetrate, thus forming a more or less homogeneous layer on the surface with a high monomer concentration (Fig. 2b). Since in good solvents, the monomers repel each other, the Gibbs energy of adsorption increases in this stage, eventually bringing the system to equilibrium. The homogeneity of the adsorbed layer allows us to consider it as a polymer solution with a concentration which only varies normal to the surface, the so-called density profile [4]. In Section V I explain how one can calculate this profile theoretically from known solution properties of the polymer and from the adsorption energy.

For the globular polymer molecule (Fig. 2c), deformation leads to exposure of more monomers to the solvent. In other words, work must be done against the interfacial tension. Therefore, globular polymers adsorb in a way analogous to the spreading of liquid droplets [5]. The driving force is, again, given by the low polymer–substrate interfacial tension (proportional to χ_s), but the resistance is a measure of the monomer–monomer attraction as given by χ. Large deformation will occur as soon as $\chi_s = \chi^2$. For a strongly deformed polymer, the monomer–monomer contacts are so few that one can no longer speak of a surface tension; this occurs when the thickness becomes of order ξ_t, the thermal blob size given by $-1/v$ [3,5].

IV. FORMATION OF AN ADSORBED LAYER

When a bare surface is exposed to a solution of adsorbing macromolecules, an adsorbed layer will form. Each adsorbing molecule must pass through the following steps:

1. Transport toward the surface of convection and diffusion
2. Attachment
3. Spreading

In the following, we shall assume that the formation of the polymer layer takes place under well-defined hydrodynamic conditions, so that the first step (transport) can be treated simply by solving the convective diffusion equation. Analysis shows that most of the process takes place under steady-state conditions where a fixed concentration gradient drives the process. This leads to an adsorption rate [6]

$$\frac{d\Gamma}{dt} = k(c^b - c^s) \qquad (1)$$

where c^s is the concentration of polymer in the immediate neighborhood of the surface, often referred to as the subsurface concentration, c^b is the bulk concentration, and k is a rate coefficient depending on diffusion coefficient and hydrodynamic conditions. Expressions for k in various flow geometries have been given by Adamczyk et al. [7]. The value of the *subsurface concentration* c^s depends on what happens near the surface, i.e., on the attachment process. In general, one can treat this as a first-order reaction. For the forward rate,

$$\left.\frac{d\Gamma}{dt}\right|_+ = Kc^s \qquad (2)$$

Of course, the rate constant K will increase with increasing coverage Γ, because fewer and fewer surface sites are available for the polymer to attach to. However, even on a bare surface, it may be lowered by some repulsive barrier (e.g., electrostatic repulsion).

As the coverage increases, so does the rate of *desorption*. Eventually, the system should come to its equilibrium condition, and the rates of adsorption and desorption should become equal. This implies for the net attachment rate, i.e., the sum of forward and backward rates:

$$\frac{d\Gamma}{dt} = K(c^s - c^{eq}) \qquad (3)$$

where c^{eq} is the equilibrium concentration corresponding to the value of Γ at some moment during adsorption. The function $c^{eq}(\Gamma)$ is the inverse adsorption isotherm. Combining Eqs. (1), (2), and (3), one obtains

$$\frac{d\Gamma}{dt} = \frac{c^b - c^{eq}}{1/k + 1/K} \qquad (4)$$

Provided the attachment barrier is negligible ($K^{-1} = 0$) and the adsorbed layer entirely relaxed, one can use this equation to deduce the adsorption rate from the equilibrium isotherm and vice versa.

For example, the high-affinity adsorption isotherm so characteristic for most polymers implies that $c^{eq} \simeq 0$ up to almost saturation, so that $d\Gamma/dt = kc^b$, and Γ must increase linearly in time with a slope k up to almost saturation; this is easily verified in a well-controlled experiment [8]. Since the rate of adsorption cannot increase beyond kc^b, this is called the limiting flux J_0. Lower rates are indicative for adsorption barriers. One can also use the equation to describe desorption into pure solvent by putting $c^b = 0$ so that $d\Gamma/dt = -kc^{eq}(\Gamma)$ [9].

In order to develop this further and predict (starting from the same assumptions) adsorption and desorption rates, the function $c^{eq}(\Gamma)$ and, hence, an equilibrium adsorption theory is required. This will be considered in the following section.

V. THE HOMOGENEOUS LAYER IN EQUILIBRIUM

Developing a general theory for an arbitrary kind of polymer on an arbitrary surface is of course a hopeless task. We therefore shall have to limit ourselves to a tractable case, which nevertheless brings out essential features. We shall consider an ideally planar and homogeneous surface in contact with a solution of a simple, uncharged homopolymer. The polymer is homodisperse: all the polymer molecules have the same molar mass. (This limitation can be overcome, however [10].) The monomer units interact with sites on the surface with a net energy $-kT\chi_s$, which may be negative (attraction) or positive (repulsion); see Section III. As pointed out, polymers in θ or good solvents form a layer of interpenetrated molecules, the density ϕ of which varies only normal to the surface. We now shall try to calculate this density profile $\phi(z)$.

One particularly transparent way to arrive at a theoretical description is the mean-field approach. On a crowded surface with many polymers, there are many interactions between the chains. In any statistical theory of adsorption, all these interactions have to be considered. However, this is obviously very complicated. In a mean-field theory, these interactions are replaced by interactions between a single chain and an average environment, called the *field*, that originates from the presence of all the other chains. Hence, each monomer at a distance z from the surface just "feels" a potential (Gibbs energy) given by the average density at and around z. This reduces the problem to one of a chain in an external, z-dependent field. Of course, the strength of the field at z is a function of the monomer density at z. To a reasonable approximation, one can ignore the interactions of the chain with itself. We then have the case of a Gaussian chain in an external field, which can be treated with random walk statistics and leads to an expression for the density profile and, hence, to a field. Of course, this field must be the same as that which was initially imposed; it must be self-consistent, hence the term *self-consistent field theory* [4,11,12].

The first step in the calculation of ϕ at position z is to realize that a monomer finding itself at z [so that it contributes to $\phi(z)$] must be somewhere in a polymer chain. It therefore has to be the end point of *two* random walks (in the field), starting from either end of the chain, and ending exactly at z. Hence, we can write $\phi(z)$ in terms of the product of (1) the end point probabilities $g(z, s)$ and $g(z, N - s)$ (with respect to the same probabilities in the solution far from the wall) for the two walks that make up a chain of length N [namely, s steps from one end and $(N - s)$ steps from the other] and (2) the density of s-monomers in the bulk solution ϕ^b/N [4]:

$$\phi(z, s) = \frac{\phi^b}{N} \, g(z, s)g(z, N - s) \tag{5}$$

We now have to find the function $g(z, s)$ by considering the random walk in a field problem. This is equivalent to solving the diffusion equation

$$\lambda \frac{d^2g}{dz^2} = ug + \frac{dg}{ds} \tag{6}$$

where λ is the fraction of bonds normal to the interface and ug represents the effect of the external field u on g. Equations of this type occur frequently. One example is the motion of elementary particles as described by Schrödinger's equation. The general solution is written as

$$g(z, s) = \sum_i g_i(z)^{\varepsilon_i^s} \tag{7}$$

where ε_i is the eigenvalue of the ith term. Note that the solution is factorized into

z- and s-dependent parts. Short random walks will be rather different from long ones, and therefore $g(z, s)$ and $g(z, N - s)$ are also different for small values of s or $N - s$, i.e., near the end points. However, as the walks become longer, they will visit the same z again and again, and eventually the z-dependence of g becomes more or less independent of s. In terms of the Schrödinger equation, we expect stationary solutions for long times and a spectrum of well-separated stationary eigenfunctions. The assumption that is therefore usually made at this point is that only the first term in the series of Eq. (7) is important; this is equivalent to supposing that the system is in its "ground state" (this is called the ground state approximation, or GSA) [4]. Hence,

$$g(z, s) = g(z)e^{\varepsilon^s} \tag{8}$$

Here, ε can be seen as the free energy per monomer in the ground state. Using this we can rewrite Eq. (5) as

$$\phi(z, s) = \frac{\phi^b}{N} g^2 e^{\varepsilon N} \tag{9}$$

Summing this over s gives

$$\phi(z) = N\phi(z, s) = \phi^b g^2 e^{\varepsilon N} \tag{10}$$

We note that in the solution far from the wall we must have $g(z) = 1$, giving

$$\varepsilon = -\frac{\ln(\phi_b)}{N} \tag{11}$$

The "potential" u can be considered as the extra reversible work needed to bring one monomer from the solution to the adsorbed layer, which in good solvents is well approximated by $\phi(z)$. Inserting this into Eq. (6) one obtains, for the case of an attracting wall [and using the boundary condition that $\phi(z)$ has a negative slope at the wall],

$$\phi(z) = \frac{\varepsilon}{\sinh^2(z + D/d)} \tag{12}$$

where D is a constant related to the adsorption energy per monomer, and d equals $\sqrt{\lambda/\varepsilon}$ [13]. For the adsorption isotherm in the region of the semiplateau, one deduces by integration:

$$\Gamma = \frac{2\lambda}{d} \left\{ \coth \left(\frac{D}{d - 1} \right) \right\} \approx 2\lambda \left(\frac{1}{D} - \frac{1}{d} \right) \tag{13}$$

Note that the adsorbed amount is linear in ε, hence to $[\ln(\phi_b)/N]^{1/2}$; it is not really constant but increases very slowly with the polymer concentration. This modest increase in Γ comes from the growth of the dilute periphery of the layer with concentration and with molecular weight. Of course, the approximations introduced can be avoided if one solves the basic equations numerically. Then end effects and short chains are no longer neglected. This has been done in great detail by Scheutjens and Fleer [11].

Further analysis has shown that good agreement between numerical results and analytical theory can be obtained when the expansion Eq. (7) is not truncated after the first term but after the second term; there is a second function $g_2(z)$ which has almost the same eigenvalue as g_1 but describes walks that *do not touch the wall*. In quantum mechanical terms, the spectrum is degenerate. Just as g_1^2 corresponds to chains which have at least two attached monomers (thereby forming loops), g_2^2 relates to conformations that do not touch the wall (free chains), and the cross product $g_1 g_2$ to chains that start on the surface and end in the solution (tails). Taking these three contributions into account, excellent agreement between numerical and analytical solutions can be established [4,13–15].

VI. ADSORPTION AND DESORPTION RATES

The function $\Gamma(c^{eq})$ being available, we can invert it to $c^{eq}(\Gamma)$ and insert it into Eq. (4). One sees immediately that, provided k is a constant and other resistances do not occur ($K^{-1} = 0$), we obtain a straightforward differential equation for $\Gamma(t)$. As pointed out, Γ is predicted to increase proportionally to t, the slope being given by k. As soon as the semiplateau is reached, c^{eq} increases very rapidly (roughly as $e^{M(\Gamma - \Gamma_{ov})}$, where Γ_{ov} is the coverage where molecules start to overlap), $c^b - c^{eq}$ drops rapidly, and $d\Gamma/dt$ becomes very small. Hence $\Gamma(t)$ has a second part where it rises very slowly (Fig. 3).

In the same way, we can analyze the case of desorption. Now, $c^b = 0$ is imposed. Initially Γ decreases, as c^{eq} has the finite value c^0, but as Γ decreases, c^{eq} must become very small because of the high-affinity character of $\Gamma(c)$ (Fig. 3). Hence, $d\Gamma/dt$ also becomes very small. We can now use Eq. (12); to a very good approximation, $\Gamma \sim \log c$. If this is the case, one can show that Γ must decay logarithmically [9]:

$$\Gamma = \Gamma_0 \left[1 - p \ln\left(1 + \frac{t}{\tau} \right) \right] \tag{14}$$

where p is a small coefficient of order $1/N$ and τ equals p/kc^0 (Fig. 4). This result teaches us that per decade in time a constant (small) amount $p\Gamma_0$ is desorbed, say 1% of the initial value. Hence, the time needed for desorption increases exponentially with the amount to be desorbed. It follows that, in a practical sense, long chains desorb too slowly for any rate to be measured, which is in agreement with the experimental observation that adsorbed polymer layers cannot simply be removed

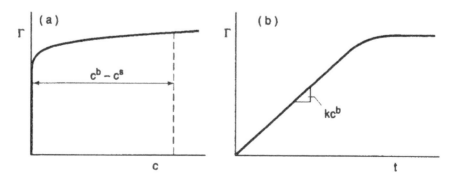

FIG. 3   (a) Adsorption isotherm $\Gamma(c)$; (b) corresponding kinetic curve $\Gamma(t)$.

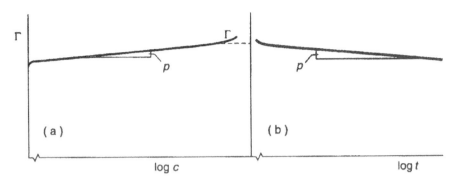

FIG. 4 (a) Adsorption isotherm $\Gamma(\log c)$; (b) corresponding desorption curve. Note that the slope of $\Gamma(\log c)$ is the same as the one for $\Gamma(\log t)$ but with a minus sign.

by rinsing with the solvent they were prepared from [9]. Yet one should realize that this does not imply that densities and conformations at the surface cannot relax!

VII. STRUCTURE OF ADSORBED LAYERS

As pointed out, isolated random coil polymers tend to become strongly compressed normal to the surface. However, as soon as crowding occurs, the driving force for the spreading decreases because entirely free surface is no longer available. More polymer can adsorb, but the molecules must protrude into the solution and form loops and loose ends ("tails"). Hence, the dilute periphery of the adsorbed layer grows in the normal direction.

The protrusions, in particular the tails, have a pronounced effect on solvent flow along the surface. Despite their very low density they are able to suppress flow velocities near the surface very strongly, just as a few trees on an open field break the wind [4,16]. As a consequence, the surface apparently bears a considerable stagnant layer of solvent, the thickness of which is called the *hydrodynamic layer thickness* δ_h. Measurements and calculations show that δ_h is very small at low coverage (chains spread out) but increases enormously as soon as saturation is approached, and that the maximum layer thickness observed becomes strongly dependent on molecular weight. This is clearly shown by the (theoretical) curves given in Fig. 5 [16]. In some systems, particularly for good solvents, a dependency of Γ on molecular weight could not be detected, yet δ increased strongly with increasing M; the few dominating tails hardly contribute to the adsorbed mass [17]. This corresponds quite well with what we see in Fig. 5 for good solvents: δ may increase very strongly with r while θ is virtually constant.

Turning this around, one concludes that very minor desorption of molecules from a saturated layer must show up in a strong decrease of δ_h; this idea has been exploited to verify the desorption rate equation [9]. Data are shown in Fig. 6 for *static* thickness measurements (at finite c^b) and for the *dynamic case* ($c^b = 0$).

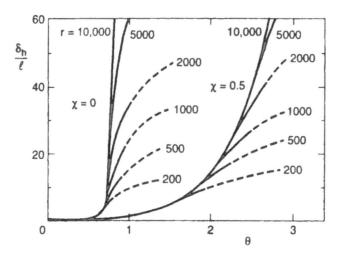

FIG. 5 Hydrodynamic layer thickness δ_h as a function of coverage θ for various molar masses (here indicated by the number of chain segments r). The curves on the left are for $\chi = 0$, those on the right for $\chi = 0.5$, a θ solvent. These curves are obtained by a theoretical based on flow perturbation in a (theoretically calculated) density profile [4].

VIII.   POLYMER MIXTURES: PREFERENCE AND EXCHANGE

It is often stated that polymers adsorb irreversibly, implying that a molecule, once adsorbed, cannot leave the surface anymore. This is a very misleading idea. It has been observed for many systems that an adsorbed polymer molecule can readily desorb if another molecule (even when identical to the leaving one) takes its place

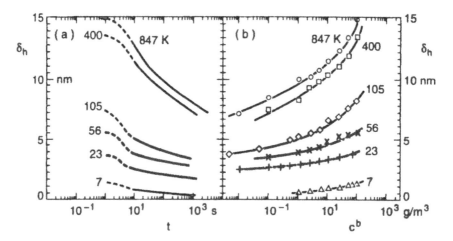

FIG. 6   (a) Decrease of layer thickness as a function of log t during a kinetic desorption experiment (the layer is continuously flushed with pure solvent); (b) increase of layer thickness as a function of log c in a static thickness measurement (fixed c for each point). Note the similarity in slope and shape between $\delta_h(\log t)$ and $\delta_h(\log c)$, just as between $\Gamma(\log t)$ and $\Gamma(\log c)$ in Fig. 4.

so that Γ does not need to decrease. In other words, *exchange* with molecules in solution occurs frequently. This may have various consequences.

When the solution contains more than one solute component, exchange may lead to a change in the composition of the adsorbed layer. Most commercial polymers are far from pure components, as they contain a large variation in chain lengths so that one may anticipate shifts in the average chain length on the surface. Such *polydispersity effects* have indeed been observed in many systems [4,18,19]. In dilute solutions, there is a clear driving force for preferential adsorption of long chains. Each long chain (of length N) that adsorbs at the expense of an equal mass of short chains (of length n) liberates on average N/n chains, which leads to a small net gain in translational entropy of order $kT(N/n - 1)$. Since exchange can even be observed in the absence of a net driving force, we should not be surprised to see exchange occur between chemically identical short and long chains. For an entire distribution of chain lengths (as is the usual case) one can show that the surface chain length shifts gradually to higher values as the polymer concentration in solution is increased; the amount of available surface area as compared to the solution volume is then a crucial parameter, determining the final distribution of each fraction in the polymer sample over the solution and the surface [10,20]. How this exchange process proceeds and what its rate is are discussed in the following section.

Chemically different polymer molecules often have an even more pronounced tendency for exchange. In this case the driving force is a lowering of the Gibbs energy of the interfacial contact. Since the adsorbed mass in equilibrium depends only logarithmically on the bulk concentration of a component, whereas the polymer-surface interaction is linear in the surface composition, a large concentration difference is needed to offset a difference in surface energy [4,21]. Therefore, simple experiments of sequential adsorption may quickly tell which of a given monomer adsorbs more strongly.

The limiting case of exchange between chemically different species is displacement of a polymer by a small molecule. This may even occur when the small displacer molecule adsorbs just a bit stronger than the monomer unit but is present in large amounts [22–24]. A simple case is the adsorption of a polymer from its own "monomer." Theory predicts that this will not occur as there is no change in surface contact energy, hence no driving force for spreading and adsorption. Diluting the displacer will eventually restore the situation of adsorption; this happens at a sharply defined *critical displacer concentration*. Thermodynamic analysis allows us to determine the adsorption energy per monomer unit from this critical displacer concentration [22,23].

Occurrence of exchange is also well documented for proteins. When a solution of a protein is exposed to a surface, adsorbed molecules may again spontaneously exchange with molecules from the solution. It has been found for some cases that as a result of adsorption followed by desorption one finds molecules in the solution which have undergone irreversible changes in conformation. Exchange between different proteins occurs frequently in mixtures and is known as the *Vroman effect* [25].

IX.   SPREADING: EXPERIMENTAL EVIDENCE

In Section III we argued that macromolecular adsorption is always accompanied by deformation or *spreading* of the molecules. In our analysis of the rate of the adsorption process we totally ignored this spreading process. In other words, we considered adsorption processes where spreading occurs very fast as compared to the experimental time scale. As a result, no information on the rate of the spreading process could be obtained. What if this is not so? Qualitatively, one consequence is immediately seen.

If the molecules need a certain time τ_s to spread, we can compare that to the time τ_d required to deposit molecules to the surface. If the deposition is slow ($\tau_s < \tau_d$), each molecule can unfold and spread before it is surrounded by other molecules. The adsorbed amount will correspond to a thin layer and therefore be low. In the reverse case—fast supply ($\tau_s > \tau_d$)—the molecules are enclosed by neighbors before they had time to unfold. The layer will therefore be thicker and Γ higher. The two cases are sketched in Fig. 7.

In order to develop this more quantitatively, we need to specify the occupied area as a function of time. Let us suppose that one single time scale τ_s dominates the spreading process, i.e.,

$$a = a_0 + \alpha(1 - e^{-t/\tau_s}) \tag{15}$$

where a_0 is the size of the molecule upon first attachment and α denotes the extra surface occupied at full spreading. In addition, we need to specify the relation between the rate of attachment and the coverage. We simply suppose that this rate is proportional to the unoccupied area β. We now have two parallel processes, supply and spreading, which are coupled by the instantaneous value of β. This leads to a differential equation for the adsorbed mass versus time, which now becomes a function of the limiting flux J_0 [26,27]. If no desorption occurs (so that spreading stops as soon as $\beta = 0$), one gets a family of $\Gamma(t)$ curves with increasing saturation plateaus as J_0 increases. This allows us to determine the spreading rate constant. An experimental example is given in Fig. 8 for the adsorption of the protein savinase [27].

Desorption may also occur, in which case some molecules may increase their occupied area at the expense of others; this could be called "competitive spreading." If this occurs, one expects that at high supply rates many molecules are packed on the surface. As they begin to spread, some molecules are "kicked off" the surface and the coverage decreases spontaneously; the $\Gamma(t)$ curve has an overshoot. At lower supply rates, the spreading process can keep up with the supply, and the overshoot disappears. Such behavior is seen, e.g., for the protein savinase on silica (Fig. 9) [28].

Effects as discussed here occur frequently for proteins but not for most of the swollen polymers, although one case for a polyelectrolyte has been reported. This seems to indicate that the spreading rate is strongly decreased by the high density in the protein globules, probably as a result of strong friction within the molecule

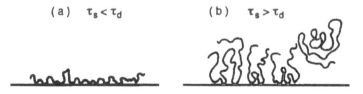

(a) $\tau_s < \tau_d$ (b) $\tau_s > \tau_d$

FIG. 7   Adsorbed layers obtained at (a) fast and (b) slow spreading.

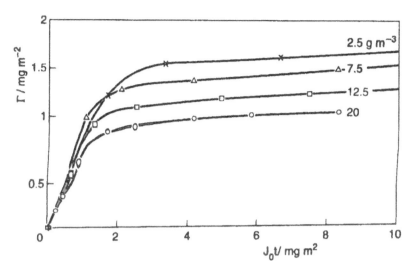

FIG. 8 Adsorption kinetics of savinase at various concentrations, hence various supply rates (limiting flux) $J_0 = kc^b$. The time axis has been renormalized by J_0 to make the initial slopes all equal to unity [27].

as it is being deformed. Swollen polymers do not have much internal friction. However, their spreading rate may be slowed down because monomers must detach and reattach to the surface, and this requires an activation energy which is equivalent to friction with the substrate. Alternatively, competition with surrounding molecules may play a role.

That surrounding molecules do so is deduced from exchange experiments between two polymeric species where the detected rate of desorption of one species reveals the rate of spreading of the other (Fig. 10) [29,30]. Most observations have shown that this process has two stages: (1) the invading chain attaches rapidly to the surface and (2) the displaced polymer molecules leave the surface more slowly. In fact, in many cases one can stop the supply of displacing polymer without inter-

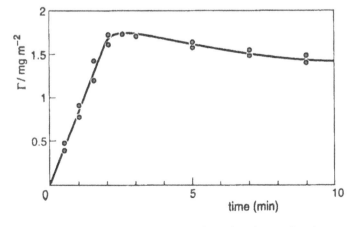

FIG. 9 Kinetic curve for the adsorption of savinase, showing an overshoot [28].

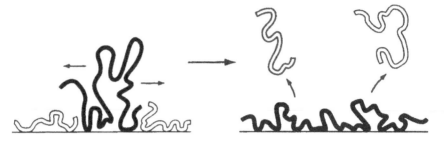

FIG. 10   Desorption of molecules induced by a spreading process; the detected rate of desorption reveals the rate of spreading.

rupting the desorption process; the displacer is already "stored" on the surface. In our view, desorption then follows as a result of competitive spreading, where the invading chain increases its occupied area at the expense of the leaving species. Detailed studies on several systems have shown that the rate of competitive spreading depends on (1) stiffness of the invading chain and (2) its net adsorption energy. This seems plausible, as both increase the friction between the polymer and the substrate.

X. TRAPPING

Granick [31] has emphasized the importance of topological effects in dense layers of flexible polymers. Just as in concentrated solutions, long chains tend to become entangled, and this hampers translational diffusion very strongly. Hence, instead of displacement by spreading (as described above) an incoming polymer can also *trap* chains that are already on the surface. The trapped chain must escape from "under" the invading chain which may involve a kind of slithering motion (reptation) that could be very slow indeed.

This trapping phenomenon has been investigated in more detail with a system consisting of two rather similar polysaccharides. These were first sequentially supplied to the surface in order to allow displacement and/or trapping to occur. Then the solvent was changed such that the first polymer was encouraged to leave the substrate. Desorption was allowed to take place. Finally, the composition of the adsorbed layer was analyzed. The experiment was repeated for several lengths of the displacer polymer. Surprisingly, both displacement and trapping were observed. When the second polymer was shorter than the first one, displacement did not occur, but substantial trapping was seen. Increasing the displacer's chain length led to substantial displacement (up to about 80%) and much less trapping (Fig. 11) [32].

XI. POLYELECTROLYTES AND POLYAMPHOLYTES

So far we have not said anything about electrostatic effects. Yet many water-soluble macromolecules, particularly those of biological origin, carry charges. A simple case is that of a water-soluble homopolymer with a fixed fraction of equally charged monomer. If these changes arise from strong dissociating groups, they will remain

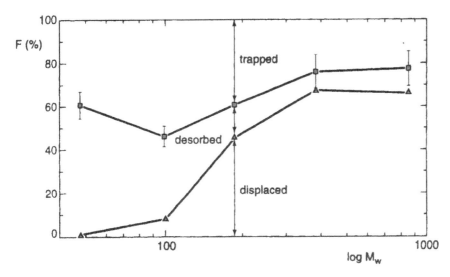

FIG. 11 Fraction of initially adsorbed polymer that is displaced, desorbed, and trapped as a function of displacer/trapper chain length. System: carboxypullulan adsorbed on polystyrene from water, trapped by uncharged pullulan [32].

charged over the entire pH range; this is called a *quenched polyelectrolyte*. The opposite case, where the charge arises from reversible proton transfer and is thus pH dependent, is the *annealed* case [33]. Another case of importance, certainly in the context of biological matter, is that of polyampholytes, which carry both positive and negative groups.

Polyelectrolytes tend to swell due to essentially two electrostatic effects: the chains become locally more rigid and the effective excluded volume increases. Very long polymers will still have a swollen coil shape, but short ones may eventually become rodlike. Polymers in poor solvents, which would be collapsed in their neutral state, may swell dramatically as soon as they acquire enough charges; there are well-documented cases for annealed polyelectrolytes which behave in this way (e.g., poly-methacrylic acid) [34].

Polyampholytes behave differently. If their net charge is around zero, they tend to contract, because the formation of ion pairs between positive and negative groups leads again to an increase of entropy which drives the association. Synthetic poly-ampholytes therefore tend to be poorly soluble. Adding salt may improve the solu-bility and narrow the pH range in which there is phase separation ("salting in") because the entropy gain of the ions becomes less. Hence, it may be misleading to consider only the *net* charge on a molecule. At larger distances (of order κ^{-1}, the *Debye length*) the local variations in charge become "invisible" and only the net charge is important.

Surfaces may be classified in roughly the same way. Many inorganic surfaces consist of metal (hydr)oxides and can be considered as annealed surfaces that get their charge (density σ_s) from reversible proton transfer reactions. Various insoluble salts behave likewise, but they get their charge from other adsorbing ions. Most of these annealed surfaces are amphoteric and have a point where their net charge is zero; this is the point of zero charge (pzc) [35]. Quenched surfaces are obtained if strongly dissociating groups are chemically attached to a surface; a typical example is the surface of polymer latex particles with sulfate groups. The behavior of poly-electrolytes and polyampholytes is complicated because both the electrostatic and the nonelectrostatic (short range) interactions may contribute, and all sorts of patterns may arise.

In order to appreciate the role of electrostatics, we first discuss *pure electro-sorption* of a simple polyelectrolyte on a surface with opposite, quenched charge, from a solution of very low ionic strength [4]. In its neutral state, the polymer experiences no attraction from this surface, and adsorption does not take place. A small amount of charge suffices to induce substantial adsorption up to a point where the charge on the surface and that on the polymer are just equal. Beyond that *charge compensation* point, only little more charged polymer can adsorb (for entropic reasons the net charge of surface plus polymer need not be zero) up to a certain extent of overcompensation [36,37]. Beyond this value of σ_p the adsorption process can be considered as a kind of ion exchange process, where the macroion takes the place of many small counterions at the surface [38]. Clearly, the driving force for this process is the large entropy gain (at low ionic strength) due to the liberation of the small ions. We should therefore expect that the adsorbed amount comes close to charge stoichiometry. Hence, if the charge density σ_p on the polymer increases further, less and less polymer is needed to reach the same overcompensation; the adsorption decreases with increasing σ_p. A curve of Γ versus σ_p is obtained with a pronounced maximum located at the compensation point (Fig. 12) [38]. This pattern is rather general and occurs for many polyelectrolytes. If there are additional, non-electrostatic interactions contributing to the adsorption, the adsorbed amounts at low σ_p will be increased, particularly those around $\sigma_p = 0$.

An experimental case of pure electrosorption (for an annealed polyelectrolyte) has been reported (Fig. 13) [39]. In this case, the amphoteric nature of the annealed surface played a crucial role. On the basis of net charge, electrosorption would not be expected to occur beyond the pzc of the surface, where polymer and surface carry the same charge, because there would be no net attraction between the polymer and

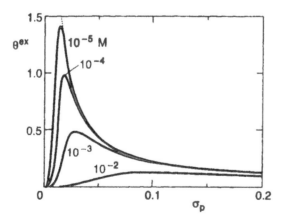

FIG. 12 Adsorption θ^{ex} of a polyelectrolyte on a surface with fixed charge as a function of polymer charge density σ_p [38].

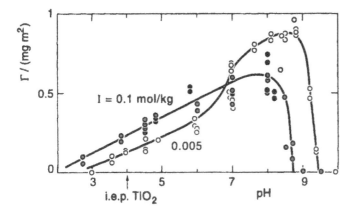

FIG. 13   Adsorption versus pH of PDMAEMA on TiO$_2$ at two different ionic strengths [39].

the surface. Yet, adsorption occurs well into that range. This can only mean that the polymer is able to find and attach to sites of opposite charge; it sees the heterogeneity of the surface. That nonelectrostatic forces might be responsible for this effect could be excluded because the polymer was unable to adsorb in its uncharged form.

When we have electrosorption of a quenched polyelectrolyte on an annealed surface, the polymer structure remains unaffected by pH. The system then behaves as a capacitor with a fixed capacitance so that charge and potential are linearly related as are (by Nernst's law) potential and pH. As a result, the adsorption is simply linear in pH [39]. In many cases, both polymer and surface can adjust their charge. Provided the pzc of the surface is well separated from the pK of the polyelectrolyte, we can observe a combined pattern [40].

At this point, we should consider the question to what extent polyelectrolyte adsorption is reversible. Adsorption of polyelectrolytes is a self-killing process; the accumulation of charged polymer leads to a high electrostatic potential which will repel new incoming molecules. Even if these molecules could anchor under strong short-range interactions (complexation, hydrophobic interaction), they may not be able to reach the surface. If this occurs, the adsorbed amount is kinetically limited, rather than the outcome of a free energy balance. Lowering σ_p (pH shift) or adding salt may then lower the kinetic barrier and promote adsorption, but subsequently restoring the original conditions does *not* restore the corresponding adsorption. In other words, adsorption hysteresis occurs under cycling the pH or the ionic strength. This is analogous to the aggregation of charged colloidal particles, which can be very effectively enhanced by adding salt or bringing the system into the pzc but often cannot be undone by restoring the conditions under which the dispersion was stable. Kinetic barriers due to electrostatic repulsion for the flexible polyelectrolytes were recently considered theoretically [41]; examples of experimental systems are numerous in the literature. We give one example in Fig. 14.

Polyampholytes are attracted to almost any charged surface. As with the amphoteric surface, they can find (short-range) electrostatic attraction even when the net surface charge is repulsive. This has been shown theoretically and experimentally for synthetic polyampholytes [42,43]. Should nonelectrostatic interactions also be present, then adsorption is even more likely to occur, unless a very strong barrier

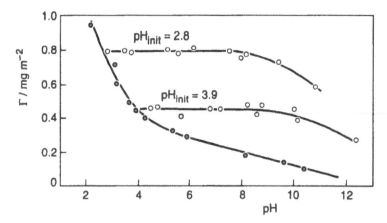

FIG. 14 Polyelectrolyte adsorption hysteresis upon cycling pH: adsorption of carboxy methyl cellulose on titanium dioxide (TiO_2, rutile) particles from aqueous electrolyte solution. Closed circles: adsorption was measured at fixed pH; open circles: adsorption took place at low pH (2.8 and 3.9, respectively) but was determined after increasing the pH [40].

prohibits the attachment of the polymer. Hence, proteins in their pzc are likely to adsorb to almost any substrate and certainly to hydrophobic substrates.

Adding electrolyte will reduce all effects due to charges and therefore we should expect to eventually obtain a case of pseudoneutrality [4]. How adsorption depends on ionic strength therefore depends in the first place on the level of adsorption in the absence of any charge. If this is high, one may anticipate salt to enhance adsorption; if this is low or zero, then one must expect lower adsorption [4,37]. In some subtle cases there is a weak maximum because screening of lateral interactions occurs before extensive ion competition for surface sites comes into play [39,44].

XII. CONCLUDING REMARKS

The adsorption process of a macromolecule is a complex process in which transport in solution, attachment barriers, and shape relaxations together determine the final outcome. In particular, shape relaxations are difficult to study directly, but recent research has shed much light on the factors that control the rate of these surface processes. With the help of these insights, it may soon become feasible to construct a theory for the adsorption of biopolymers.

REFERENCES

1. P. G. de Gennes, *Scaling Concepts in Polymer Physics*. Cornell University Press, Ithaca, NY, 1979.
2. H. Yamakawa, *Modern Theory of Polymer Solutions*. Harper & Row, 1971.
3. I. M. Lifschitz, A. Yu. Grosberg, and A. R. Khoklov, Structure of a polymer globule formed by saturating bonds. *Zh. Exp. Teor. Fyz. 71*:1634–1643 (1976).

Macromolecular Adsorption 19

4. G. J. Fleer, M. A. Cohen Stuart, J. M. H. M. Scheutjens, T. Cosgrove, and B. Vincent, *Polymers at Interfaces*. Chapman & Hall, London, 1993.
5. A. Johner and J. F. Joanny, Polymer adsorption in a poor solvent. J. Phys. II *1*:181–194 (1991).
6. M. A. Cohen Stuart and G. J. Fleer, Adsorbed polymers in nonequilibrium situations. Ann. Rev. Mat. Sci. *26*:463–500 (1996).
7. Z. Adamczyk, T. Dabros, J. Czarnecki, and T. G. M. van de Ven, Particle transfer to solid surfaces. Adv. Colloid Interface Sci. *19*:183–252 (1983).
8. J. C. Dijt, M. A. Cohen Stuart, J. E. Hofman, and G. J. Fleer, Kinetics of polymer adsorption in stagnation point flow. Colloids Surfaces *51*:141–158 (1990).
9. J. C. Dijt, M. A. Cohen Stuart, and G. J. Fleer, Kinetics of polymer adsorption and desorption in capillary flow. Macromolecules *25*:5416–5423 (1992).
10. G. J. Fleer, Multicomponent polymer adsorption. A useful approximation based upon ground-state solutions. Colloids Surf. A: Physicochem. Eng. Aspects *104*:271–284 (1995).
11. J. M. H. M. Scheutjens and G. J. Fleer, Statistical theory of the adsorption of chain molecules. I: Partition function, segment density distribution and adsorption isotherms. J. Phys. Chem. *83*:1619–1635 (1979); II: Train, loop, tail size distribution *84*:178–190 (1980).
12. P. G. de Gennes, Polymer solutions near an interface. I: Adsorption and depletion layers. Macromolecules *14*:1637–1644 (1981); II: Interaction between two plates carrying adsorbed polymer layers *15*:492–500 (1982).
13. G. J. Fleer, Ground state description of the adsorption of homodisperse and polydisperse polymers. Macromol. Symp. *113*:177–196 (1997).
14. J. M. H. M. Scheutjens, G. J. Fleer, and M. A. Cohen Stuart, End effects in polymer adsorption: a tale of tails. Colloids Surf. *21*:285–306 (1986).
15. A. N. Semenov, J. Bonet-Avalos, A. Johner, and J. F. Joanny, Adsorption of polymer solutions onto a flat surface. Macromolecules *29*:2179–2196 (1996).
16. M. A. Cohen Stuart, F. H. W. H. Waajen, T. Cosgrove, T. L. Crowley, and B. Vincent, Hydrodynamic thickness of adsorbed polymer layers. Macromolecules *17*:1825–1830 (1984).
17. C. W. Hoogendam, C. J. W. Peters, A. de Keizer, M. A. Cohen Stuart, and B. H. Bijsterbosch, Layer thickness of adsorbed CMC on hematite. J. Colloid Interface Sci. (submitted).
18. M. A. Cohen Stuart, J. M. H. M. Scheutjens, and G. J. Fleer, Polydispersity effects and the interpretation of polymer adsorption isotherms. J. Poly. Sci. Poly. Phys. Ed. *18*:559–573 (1980).
19. J. M. H. Scheutjens and G. J. Fleer, Some implications of recent polymer adsorption theory, in *The Effect of Polymers on Dispersion Properties* (Th. F. Tadros, ed.). Academic Press, London, 1982, pp. 145–168.
20. S. P. F. M. Roefs, J. M. H. M. Scheutjens, and F. A. M. Leermakers, Adsorption theory for polydisperse polymers. Macromolecules *27*:4810–4816 (1994).
21. D. C. Leonhardt, H. E. Johnson, and S. Granick, Adsorption isotope effect for protio- and deuteropolystyrene at a single surface. Macromolecules *23*:685 (1990).
22. M. A. Cohen Stuart, G. J. Fleer, and J. M. H. M. Scheutjens, Displacement of polymers. I: Theory. Segmental adsorption energy from polymer desorption in binary solvents. J. Colloid Interface Sci. *97*:515–525 (1984).
23. G. P. van der Beek, M. A. Cohen Stuart, G. J. Fleer, and J. E. Hofman, A chromatographic method for the determination of segmental adsorption energies of polymers. Polystyrene on silica. Langmuir *5*:1180–1186 (1989).
24. G. P. van der Beek, M. A. Cohen Stuart, G. J. Fleer, and J. E. Hofman, Segmental adsorption energies for polymers on silica and alumina. Macromolecules *24*:6600–6611 (1991).

25. L. Vroman and A. L. Adams. Adsorption of proteins out of plasma and solutions in narrow spaces. J. Colloid Interface Sci. *111*:391 (1986).

26. E. Pefferkorn and A. Elaissari, Adsorption–desorption processes in charged polymer–colloid systems. Structural relaxation of adsorbed macromolecules. J. Colloid Interface Sci. *138*:187 (1990).

27. M. C. P. van Eijk and M. A. Cohen Stuart, Polymer adsorption kinetics: effects of supply rate. Langmuir *13*:5447 (1997).

28. J. Buijs, P. A. W. van den Berg, J. W. Th. Lichtenbelt, W. Norde, and J. Lyklema, Adsorption dynamics of IgG and its F(ab')$_2$ and Fc fragments studied by reflectometry. J. Colloid Interface Sci. *178*:594–605 (1996); M. C. L. Maste, Proteolytic stability in colloid systems. Ph.D. thesis, Wageningen University, 1996.

29. J. C. Dijt, M. A. Cohen Stuart, and G. J. Fleer, Surface exchange kinetics of chemically different polymers. Macromolecules *27*:3229–3237 (1994).

30. H. E. Johnson and S. Granick, Exchange kinetics between the adsorbed state and free solution: PMMA in CCl$_4$. Macromolecules *23*:3367–3374 (1990).

31. H. E. Johnson, J. F. Douglas, and S. Granick, Regimes of polymer adsorption desorption kinetics. Phys Rev. Lett. *70*:3267 (1990).

32. A. Krabi and M. A. Cohen Stuart, Sequential adsorption of polymer—displacement or trapping? Macromolecules (1990).

33. I. Borukhov, D. Andelman, and H. Orland, Polymer solutions between charged surfaces, Europhys. Lett. *32*:499 (1995).

34. J. C. Leyte and M. Mandel. Potentiometric behavior of poly(methacrylic acid). J. Pol. Sci. A2:1879–1891 (1964).

35. J. Lyklema, *Fundamentals of Interface and Colloid Science*, Vol. II. Academic Press, London, 1995, Chap. 3.

36. M. R. Böhmer, O. A. Evers, and J. M. H. M. Scheutjens, Weak polyelectolytes between two surfaces: adsorption and stabilization. Macromolecules *23*:2288–2301 (1990).

37. H. G. M. van de Steeg, M. A. Cohen Stuart, A. de Keizer, and B. H. Bijsterbosch, Polyelectrolyte adsorption: a subtle balance of forces. Langmuir *8*:2538–2546 (1992).

38. M. A. Cohen Stuart, Polyelectrolytes on solid surfaces, in *Short and Long Chains at Interfaces*, Proc. 30th Rencontres de Moriond (J. Daillant, P. Guenoun, C. Marques, P. Muller, and J. Trân Thanh Vân, eds.). Editions Frontières, 1995, pp. 3–12.

39. N. G. Hoogeveen, M. A. Cohen Stuart, and G. J. Fleer, Polyelectrolyte adsorption on oxides. I: Kinetics and adsorbed amounts. J. Colloid Interface Sci. *182*:133–145 (1996).

40. C. W. Hoogendam, A. de Keizer, M. A. Cohen Stuart, B. H. Bijsterbosch, and J. G. Batelaan, Adsorption mechanics of carboxymethylcelluloses on mineral surfaces. Langmuir (submitted).

41. M. A. Cohen Stuart, C. W. Hoogendam, and A. de Keizer, Kinetics of polyelectrolyte adsorption. J. Phys. Condensed Matter (1997).

42. J. F. Joanny, Adsorption of a polyampholyte chain. J. Phys. France II *4*:1281–1288 (1994).

43. S. Neyret, L. Ouali, F. Candau, and E. Pefferkorn, Adsorption of polyampholytes on PS latex. Effect on the colloid stability. J. Colloid Interface Sci. *176*:86–94 (1995).

44. B. C. Bonekamp, Adsorption of polylysins at solid–liquid interfaces. Ph.D. thesis, Wageningen University, 1984.

2

Driving Forces for Protein Adsorption at Solid Surfaces

WILLEM NORDE   Wageningen Agricultural University, Wageningen, The Netherlands

I. INTRODUCTION

Proteins are copolymers of some 22 different amino acids of varying hydrophobicity (Fig. 1). As a consequence, proteins are more or less amphiphilic and, therefore, usually highly surface active. Moreover, a number of amino acid residues in the side groups along the polypeptide chain contain positive or negative charges. This makes the protein an amphoteric polyelectrolyte.

Based on the spatial organization in protein molecules the following distinctions may be made:

1. Molecules that are highly solvated and flexible, resulting in a disordered (randomly) coiled structure. This group comprises some proteins of which the natural function is nutritional, such as the caseins in milk and glutelins in wheat grains. Furthermore, unfolding of ordered proteins, for instance by heat treatment or by adding a denaturant, often leads to a loosely coiled structure.
2. Molecules that have adopted a regular structure (e.g., helices and pleated sheets), the so-called fibrillar proteins. Fibrillar proteins are usually found in connective tissue and they are often insoluble in water.
3. Molecules that contain different structural elements, i.e., helices, pleated sheets, and parts that are unordered, which are folded up into a compact, densely packed structure: the globular proteins. By way of example, computer graphic images of the globular protein bovine pancreas ribonuclease are shown in Fig. 2. Although the globular proteins represent only a minority of the available protein mass, they make up by far the greatest proportion of protein species. These proteins have evolved to fulfill specific functions. This applies to enzymes, transport proteins, and immunoproteins (antibodies). The biological functioning of a given protein is directly related to its specific three-dimensional structure.

With respect to practical applications and implications the adsorption of the globular proteins is most relevant. Examples can be found in biomedical engineering, biosensors, immunological test systems, immobilized-enzyme bioreactors, biofouling of

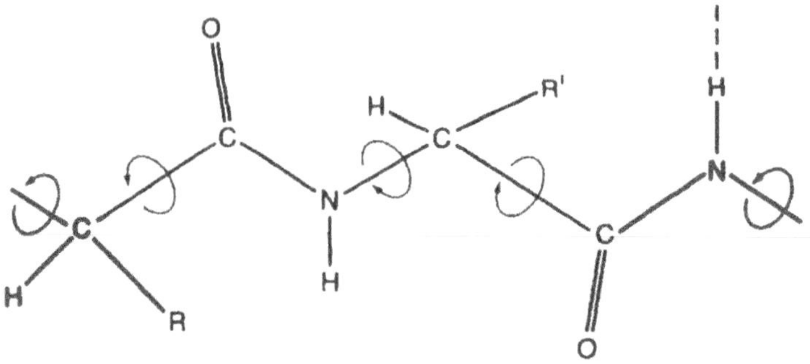

FIG. 1 Schematic representation of two peptide units in a polypeptide chain. Two of the three backbone bonds in the peptide unit are free to rotate; the other one is fixed. The amino acid side groups R, R', ... may be polar or apolar and may contain ionized groups.

processing equipment, various forms of chromatography, and many others. Therefore, this chapter focuses on the adsorption of globular proteins.

The theoretical understanding of the adsorption of (randomly) coiled flexible polymers, including polyelectrolytes, has greatly advanced over the last few decades. However, because of their intricate highly specific structures, a general theory for globular protein adsorption has not been developed yet. At best some general principles may be indicated.

In the next section, the theoretical trends, which as a rule are experimentally verified, for the adsorption of uncharged and charged flexible polymers are briefly summarized. A more detailed discussion on this subject is given by Martinus A. Cohen Stuart in Chapter 1. These trends may serve as a starting point in the discussion of globular protein adsorption in subsequent sections.

II. TRENDS IN THE ADSORPTION BEHAVIOR OF UNCHARGED AND CHARGED FLEXIBLE POLYMERS

Flexible polymers in solution possess a high conformational entropy resulting from the various states each of the many segments in the polymer chain can adopt. The expansion of a polymer coil is determined by the quality of the solvent. The better the solvent, the more expanded the coil is and the higher its conformational entropy. Adsorption leads to a reduction of this conformational entropy. Hence, adsorption takes place only if the loss in conformational entropy is compensated by sufficient attraction between polymer segments and the surface. The critical Gibbs energy for adsorption to occur spontaneously is typically a few tenths of a kT unit per segment. Even if the Gibbs energy of adsorption is only slightly higher than the critical value, the whole polymer molecule adsorbs tenaciously and, apparently, irreversibly. This is because the contribution from each adsorbing segment adds to the Gibbs energy of adsorption of the whole polymer molecule. The resulting high affinity between the polymer and the surface is reflected in the shape of the adsorption isotherm (where the adsorbed mass Γ is plotted against the polymer concentration in solution after adsorption c_p); the initial part of the isotherm merges with the Γ-axis because

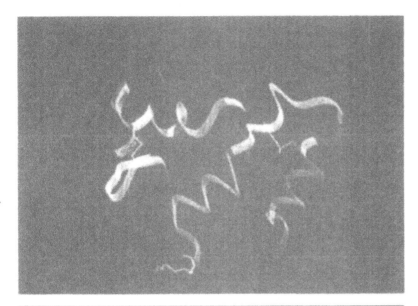

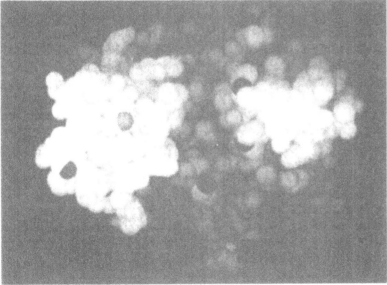

FIG. 2 Computer graphic images of bovine pancreas ribonuclease, showing (top) the polypeptide backbone made up of α-helices (spirals seen for example in the lower right of the molecule), β-sheets (upper left) and "unordered" parts, and (bottom) a space-filling model showing the compact packing of a globular protein molecule.

at low polymer supply all of the polymer is adsorbed until the surface is saturated (Fig. 3).

Figure 4 depicts how the segments of an adsorbed flexible polymer molecule may be distributed over trains, loops, and tails. Trains account for the attached segments; they are rarely very long and they do not completely cover the entire sorbent

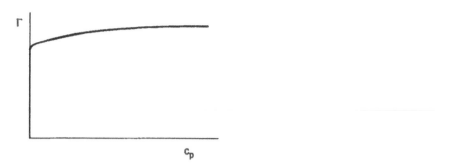

FIG. 3 High-affinity adsorption isotherm, typical for polymer adsorption.

surface, leaving about 20–30% of the surface uncovered. Loops account for most of the adsorbed mass. Their occurrence limits the reduction in conformational entropy. Their extension is determined by the solvent quality. A high loop density is tolerated close to the surface only if the solvent is relatively poor. For a poor solvent the adsorbed mass in the (pseudo-) plateau region of the isotherm typically is in the range of 2–4 mg m^{-2}; for a good solvent it amounts to 0.5–1.0 mg m^{-2}. For entropic reasons tails usually extend far in the solution and, therefore, play a dominant role in steric interaction. The experimental thickness of the adsorbed polymer layer depends on the method of determination. For instance, some methods, e.g., ellipsometry and reflectometry, determine an optical thickness ($\approx$ average loop extension), whereas other methods, such as dynamic light scattering and viscometry, yield the hydrodynamic thickness ($\approx$ tail extension). Invariably, the tails extend further from the sorbent surface than the average of the loops.

Based on these considerations the adsorption behavior of flexible polyelectrolytes may be predicted. Because of the charge they carry, polyelectrolytes are strongly expanded in aqueous solution; in other words water is an excellent solvent for flexible polyelectrolytes. As a consequence the formation of loops is strongly suppressed. Hence, polyelectrolytes tend to adsorb in thin layers up to only a few tenths of a milligram per square meter. As with uncharged polymers, polyelectrolytes also require some critical attractive interaction with the sorbent to become adsorbed. In addition to an electrostatic contribution, the Gibbs energy of adsorption may also

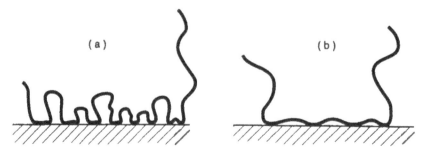

FIG. 4 Conformations of a flexible polymer adsorbed on a surface: (a) poor solvent and (b) good solvent.

comprise a nonelectrostatic component. Depending on the charge signs of the poly-electrolyte and the surface, the electrostatic contribution is attractive or repulsive; it may or it may not outweigh the nonelectrostatic contribution. If too strongly repulsive, it prevents the polyelectrolyte from adsorbing. For sake of comparison, the electric part of the Gibbs energy of a monovalent ionic group in an electric field is ca. 1 kT for every 25 mV, and the Gibbs energy of dehydration of a $—CH_2—$ group is ca. 1.5 kT, both at room temperature. Hence, polyelectrolytes with some hydro-carbon groups in their chain may readily adsorb on a hydrophobic surface against an unfavorable electric potential.

In distinction to uncharged polymers, the adsorption of polyelectrolytes is very sensitive to indifferent low-molecular-weight electrolytes. These electrolytes exert a dual effect: (1) They screen the intramolecular repulsion between charged segments, which manifests itself in water becoming a poorer solvent. Therefore, the addition of salt promotes the formation of loops and hence results in more adsorbed mass per unit sorbent surface. (2) They also screen electrostatic interactions between a polymer segment and the sorbent surface. Attachment of the segment to the sorbent surface is promoted/opposed by electrolyte if this interaction is repulsive/attractive. Along similar lines the influence of the pH on the adsorption of polyelectrolytes containing weak ionic groups (e.g., carboxyl and/or amino groups) may be explained. At a pH where such a polyelectrolyte is uncharged it adsorbs in a relatively thick layer; the adsorbed amount is then high and independent of ionic strength. However, at a pH where the ionic groups are fully charged the adsorbed amount is low, but it increases with increasing salt concentration.

For similar reasons amphiphilic polyelectrolytes show maximum adsorption at their isoelectric point. This maximum is less pronounced at higher salt concentrations.

III. STRUCTURE AND STABILITY OF GLOBULAR PROTEINS IN AQUEOUS SOLUTION

Unlike the flexible polymers discussed in the foregoing section, globular proteins in aqueous solution acquire compact, ordered conformations in which the atomic packing density, expressed in volume fraction, reaches values of 0.70–0.80. In such a compact conformation the rotational freedom along the polypeptide chain is severely restricted, implying a low conformational entropy. The compact structure is possible only if interactions within the protein molecule and interactions between the protein molecule and its environment are sufficiently favorable to compensate for the low conformational entropy. These interactions are of different nature, i.e., Coulomb, hydrophobic, hydrogen bonding, and interactions between fixed and/or induced dipoles.

Protein adsorption research is often focused on structural rearrangements in the protein molecules, not only because of its relevance to the biological functioning of the molecules, but also because of the significant role such rearrangements play in the mechanism of the adsorption process. Knowledge of the factors that determine the protein structure helps to understand the behavior of proteins at interfaces. The major factors will be briefly discussed.

A. Conformational Entropy of Proteins

Most globular proteins contain a significant amount (~40–80%) of ordered structural elements, notably α-helices and/or β-sheets. These structures are stabilized by hydrogen bonds between peptide units in the polypeptide backbone (Fig. 5). These intramolecular hydrogen bonds reduce the rotational mobility of the bonds in the polypeptide chain and, as a consequence, the conformational entropy. The dense packing of the polypeptide backbone and of the side groups within the interior of the folded molecule will further reduce the conformational entropy. Thus, for a protein molecule consisting of 100 amino acid residues (corresponding to a molar mass of ca. 10,000 Da) the conformational entropy decrease upon folding the polypeptide chain to a compact conformation containing ca. 50% ordered secondary structure (α-helix and/or β-sheet) is estimated to be several hundred joules per degree Kelvin, corresponding to a Gibbs energy increase of up to a few hundred kilojoules per mole at 300 K [1].

B. Hydrophobic Interaction

Dehydration of nonpolar components in an aqueous environment results in an increase of the entropy of the water molecules released from those components and, therefore, in a lowering of the Gibbs energy of the system. This favorable hydrophobic dehydration causes apolar parts of the polypeptide in water to associate. The relevance of hydrophobic dehydration for protein folding was first recognized by Kauzmann [2], and it is now considered the primary driving force for the folding process.

To estimate the contribution from hydrophobic interaction to the stabilization of a compact structure, the hydrophobicities of the constituting amino acids must be known. These hydrophobicities may be assessed by partitioning the amino acids between water and a nonpolar solvent. It has thus been established that the Gibbs energy of dehydration decreases by 9.2 kJ mol^{-1} after reducing the hydrophobic water-accessible surface area by 1 nm^2 [3]. Then, assuming that ca. 60% of the protein's interior consists of apolar amino acid residues [4], the contribution from hydrophobic dehydration to the Gibbs energy of stabilization of a compact globular molecule of 10,000 Da molar mass is about 500 kJ mol^{-1} at room temperature.

C. Coulomb Interaction

Most of the charged amino acid residues in a protein molecule are located in its aqueous periphery. Model calculations where the protein molecule is approximated by a sphere with its surface exhibiting evenly, but discretely, distributed positive and negative charges yields attractive Coulomb interaction in the isoelectric region and repulsive interaction remote from the isoelectric point [5–7]. Hence, near the isoelectric point Coulomb interaction is expected to favor a compact conformation, and at more extreme pH values a more expanded structure is promoted. It should, however, be realized that the Coulomb effects are largely dependent on the distribution of the charged residues on the protein molecule. For instance, ion pairs on the protein surface stabilize the compact conformation [8]. Furthermore, the electrolyte concentration largely affects the distance over which charged groups interact.

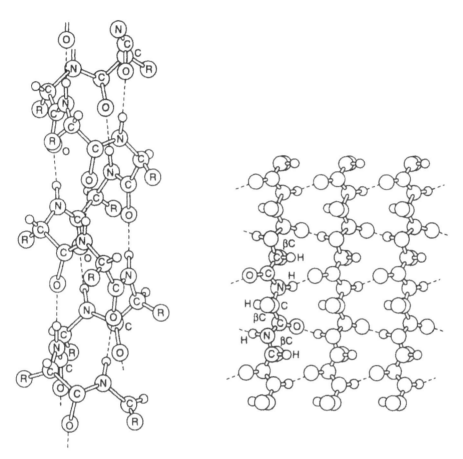

FIG. 5 Ordering of polypeptide chains into an α-helix structure (left) and a parallel β-pleated sheet (right).

If charged residues are present in the interior of a globular protein molecule, they usually occur as ion pairs. Unfolding of the protein would imply rupture of the electrostatically favorable ion pairs but also leads to favorable hydration of both ionic groups. Because of these compensating effects, ion pairing in the apolar, non-aqueous interior of the protein does not make a major contribution to the stabilization of a compact protein structure taking the unfolded polypeptide chain as in the reference state [8].

Ionization of residues originally buried in the low-dielectric interior of the globular protein molecule in the nonionized form (e.g., histidine and tyrosine) may be a significant driving force for unfolding [9,10].

D. Lifshitz–van der Waals Interactions

Lifshitz–van der Waals interactions originate from interactions between fixed and/ or induced dipoles. They are very sensitive to the separation distance r between the dipoles, varying as r^{6}. Upon folding the polypeptide chain into a compact structure, dipolar interactions between the protein and water are disrupted, but on the other hand dipolar interactions inside the protein molecule and between water molecules are newly formed. The overall effect of dipolar interactions on protein stability is not clear, although it is generally assumed that because of the relatively high packing density, dipolar interactions tend to promote a compact structure [11,12].

E. Hydrogen Bonds

Most of the hydrogen bonds in globular proteins are those between amide and carbonyl groups of the polypeptide backbone. In α-helices and β-sheets such intrachain hydrogen bonds enforce each other because they are aligned more or less parallel to one another (Fig. 5). The number of hydrogen bonds involving amino acid side groups is, as a rule, relatively small.

As with dipolar interactions, folding of the polypeptide chain implies loss of hydrogen bonds between protein and water and formation of intramolecular hydrogen bonds in the protein and among water molecules. The net effect on the stabilization of the protein structure remains unclear. Studies using model components [13–15] indicate that two peptide–water interactions on the one hand are more favorable than peptide–peptide and water–water interactions on the other. This would mean that hydrogen bonding does not stabilize a compact protein structure. However, if due to other types of interaction (e.g., hydrophobic bonding) the polypeptide chain and side chains that are capable of forming hydrogen bonds are forced in the apolar interior of a compact protein, the formation of intramolecular hydrogen bonds would largely stabilize that structure.

F. Bond Lengths and Bond Angles

It is probable that in a tightly packed compact conformation not all lengths and angles of the covalent bonds attain the most favorable values. Indeed, energy-minimization calculations point to distortion of covalent bonds in (crystallographic) globular proteins that significantly opposes the folded conformation, possibly up to several kilojoules per mole [1,16].

The various contributions that determine the protein structure in aqueous solution are summarized in Table 1. Hydrophobic interaction and conformational entropy of the protein are the major factors. Because of their opposing effects, which are of comparable magnitude, the folded globular conformation is thermodynamically only marginally stable. Typically, the Gibbs energy of stabilization of the globular structure (relative to the unfolded structure) is in the range of some tens of kilojoules per mole, corresponding to, e.g., the energy required to rupture a few hydrogen bonds in an apolar environment.

Thus, although conformational entropy and hydrophobic dehydration dominate protein folding, none of the other factors are unimportant. As a consequence, (small) disturbances in the environment, e.g., changes in temperature, pH, ionic strength, etc., and the introduction of an interface, may induce perturbations in the protein structure.

By way of example, Fig. 6 shows the Gibbs energy and the heat-induced unfolding of α-lactalbumin, together with its enthalpic and entropic components. It is

TABLE 1 Contributions to the Gibbs Energy G of the Native (N) Globular Protein Structure Relative to the Unfolded Denatured (D) Structure

Type of interaction	Contribution to G_N/G_D	Remarks
Rotational mobility along the polypeptide chain	$\gg 0$	Folding, in particular into secondary structures, reduces the conformational entropy of the polypeptide chain
Hydrophobic dehydration	$\ll 0$	Entropy increase of water released from contact with hydrophobic components
Coulomb	>0 or <0	Depending on the pH relative to the isoelectric point of the protein-sorbent complex
Lifshitz–van der Waals interactions	$\leqslant 0$	Dispersion interactions slightly favor the folded structure because of dense atomic packing in the globular form
Hydrogen bonds	$\geqslant 0$ (?)	Loss of peptide–water bonds may not be fully compensated by peptide–peptide and water–water bonds
Bond lengths and bond angles	>0	Some bonds are more or less under stress in the folded structure

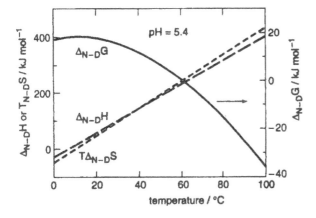

FIG. 6 Changes in the Gibbs energy, enthalpy, and entropy for the heat-induced denaturation of α-lactalbumin. Ionic strength is 0.05 M KCl.

clear that the small change in Gibbs energy is the result of substantial enthalpy–
entropy compensation.

IV. ADSORPTION OF PROTEINS FROM AQUEOUS SOLUTION ONTO SOLID SURFACES

Proteins are polymers. Their adsorption behavior shares some features with those for
the flexible polymers, summarized in Section II and discussed more extensively in
Chapter 1. First, an adsorbing protein molecule forms numerous contacts with the
sorbent surface [15,17], usually leading to a high-affinity adsorption isotherm (Fig.
3). As a consequence, the rate of desorption upon diluting the system is often below
the detection limit. However, adsorbed proteins may readily be exchanged against
other surface-active solutes being either of the same or of another kind [18–22]. For
instance, surface-active components of relatively low molecular weight may effectively strip off proteins from surfaces [23–25]. Furthermore, the sequential adsorption of different proteins from complex biological fluids is well documented [26,27].
Second, the adsorption pattern of proteins is typical for a polyampholyte; the amount
adsorbed generally is at a maximum around the isoelectric point of the protein/
sorbent complex, i.e., at conditions where the charge on the protein and the sorbent
just compensate each other. Also, similar to flexible polymers, proteins may tend to
spread over the sorbent surface. However, due to their relatively strong internal
cohesion the spreading rate of proteins is far slower than that of flexible polymers.
If the rate of spreading is comparable to the rate of deposition at the surface, the
extent of spreading (= conformation change) decreases with increasing flux toward
the surface and, as a consequence, the adsorbed mass will be larger. Such behavior
has been observed for proteins [28,29], whereas under usual experimental conditions
the spreading of flexible polymers is much faster than the transport so that the adsorbed amounts are essentially independent of the flux [30]. Another, even more
basic, difference in the adsorption behavior between globular proteins and flexible
polymers concerns the change in the conformation entropy. In contrast to flexible
polymers, globular proteins are highly ordered, low-entropy structures. Upon adsorption this structure may (partly) break down, thereby increasing the conformational entropy of the protein. Hence, where attachment to a surface always leads to
a loss of conformational entropy of a flexible polymer [31], protein adsorption may
result in increased conformational entropy. This entropy gain may be sufficiently
large to cause spontaneous adsorption of the protein under otherwise adverse conditions, i.e., at a hydrophilic, electrostatically repelling surface [32,33].

In forthcoming sections the primary contributions to protein adsorption on a
smooth, rigid surface will be discussed in more detail. (The influence of water-soluble
oligomers or polymers preadsorbed or grafted on the sorbent surface will not be
considered.) The major contributions originate from (1) redistribution of charged
groups (ions) when the electrical double layers around the protein molecule and the
sorbent surface overlap, (2) dispersion interaction between the protein and the sorbent, (3) changes in the hydration of the sorbent surface and of the protein molecule,
and (4) structural rearrangements in the protein. Although these factors will be discussed more or less individually, it may be clear that their actions are interdependent,
being either synergistic or antagonistic. For instance, the structural flexibility of an

adsorbed protein molecule strongly affects the electrostatic and hydrophobic inter-action between the protein and the sorbent.

A. Redistribution of Charged Groups Resulting from Overlapping Electrical Double Layers

In an aqueous environment both the protein molecule and the sorbent surface are, as a rule, electrically charged. The charge on the protein arises from association or dissociation of acidic and basic amino acid residues. Depending on the type of sorbent material, its surface charge stems from the dissociation/association of (strong) acidic or basic groups (e.g., sulfonated, carboxylated, and aminated polymer latexes, oxide surfaces, and surfaces of biological materials) or from the binding of ions other than protons (e.g., silver halide surfaces). Other surfaces may not contain charged groups, such as those of polyethylene oxide, Teflon, etc.

Both the protein and the sorbent are surrounded by counterions and co-ions. A fraction of these ions may be specifically absorbed to the protein and sorbent surface, whereas the remaining ions are diffusely distributed in the solution. The charge on the surface together with its compensating surrounding charge is referred to as the electrical double layer. Figure 7 gives a schematic representation of the electrical double layer according to the Gouy–Stern model [34]. The electrical part of the Gibbs energy of an electrical double layer G_{el} equals the isothermal, reversible work of charging it [35,36]:

$$G_{el} = \int_0^{\sigma_0} \phi_0' \, d\sigma_0' \tag{1}$$

where ϕ_0' and σ_0' are the variable surface potential and surface charge density, respectively, during the charging process.

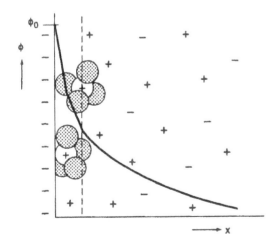

FIG. 7 Schematic representation of the Gouy–Stern model of an electrical double layer, indicating the distribution of the counterions ($+$) and co-ions ($-$), compensating the negative surface charge and the resulting potential decay.

Solving Eq. (1) requires knowledge of $\phi_0'(\sigma_0')$. This functionality follows from combining the Poisson equation [Eq. (2)] with an expression for the charge distribution in the electrical double layer. At any point in the electrical double layer the electrostatic potential $\phi(x)$ is related to the volume charge density $\rho(x)$ by

$$\nabla^2\phi(x) = -\frac{\rho(x)}{\varepsilon\varepsilon_0} \tag{2}$$

where $\varepsilon\varepsilon_0$ is the dielectric permittivity of the medium considered, and ε_0 that of free space. To evaluate $\phi(x)$, assumptions have to be made concerning the distribution of charge across the electrical double layer. For instance, for charged surfaces the Gouy–Stern model (Fig. 7) is usually adopted, where the surface charge is located at $x = 0$ and where the countercharge is diffusely distributed in the solution, separated from the surface by a charge-free layer having a thickness equal to the radius of a hydrated counterion. Then, taking into account the proper boundary conditions, which for a planar surface are

$$\left(\frac{\partial\phi}{\partial x}\right)_{x=r} = \frac{\sigma_{x=r}}{\varepsilon_0\varepsilon_{x=r}} \tag{3}$$

where $x = r$ refers to any value of x in the electrical double layer, Eq. (2) can be solved and $\phi_0'(\sigma_0')$ can be evaluated. In turn, G_{el} can be computed using Eq. (1).

When the protein and the surface approach each other, their electrical double layers overlap, giving rise to electrostatic interaction (Fig. 8). The resulting change in Gibbs energy, $\Delta_{ads}G_{el}$, is the difference between G_{el} after and before adsorption. It follows that Eq. (1) has to be applied three times, i.e., for the protein-covered sorbent surface, for the bare sorbent surface, and for the protein molecule in solution. According to this procedure, various attempts have been made to calculate $\Delta_{ads}G_{el}$. Ståhlberg and Jönsson [37,38] thus compared the interaction for constant charge densities at the surfaces with that for constant potentials. Their data on protein retention in ion exchange chromatography could be reasonably well described by the constant-potential approach. It points to a charge regulation when the protein comes

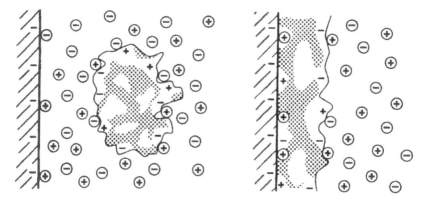

FIG. 8 Schematic representation of a protein–sorbent surface system before and after adsorption. The charged groups on the surface and the protein molecule are indicated by $+/-$, and the low-molecular-weight ions are indicated by $\oplus/\ominus$. The shaded areas represent hydrophobic regions.

close to the stationary phase. A similar approach has been undertaken by Lenhoff and coworkers [39–41]. They emphasized the effect of the heterogeneous charge distribution in the protein molecule. Proteins having a relatively large dipole moment, such as ribonuclease [39] and chymotrypsinogen A [40,41], are expected to adsorb in a preferred orientation, whereas the less dipolar lysozyme [40,41] approaches the sorbent surface in a more random orientation. Similarly, Roush et al. [42] computed a preferred orientation of cytochrome b5 on an anion exchange surface. A more detailed electrostatic analysis was undertaken by Noinville et al. [43], who computed interaction energies for all atoms in a protein/sorbent system. In this way they predicted a preferred orientation of α-lactalbumin and lysozyme at a poly(vinyl imidazole) surface.

The values obtained for $\Delta_{ads}G_{el}$ in the simulation studies mentioned depend on parameters such as charge densities (or potentials) at the surfaces of the protein and the sorbent, the separation distance between these two components, the size and the orientation of the protein molecule in the electric field, the ionic strength, and the dielectric constant of the surrounding medium. Thus, in an aqueous environment of 0.1 M ionic strength and at a separation between the sorbent and the protein molecule of, say, 1 nm the value for $\Delta_{ads}G_{el}$ typically varies from a few RT up to a few tens of RT per mol of protein.

At close "atomic" contact between protein and sorbent the aforementioned approaches may not be satisfactory. In particular, they may not properly account for charge regulation. Experimental and theoretical analyses show that under many conditions the Debye length is smaller than the dimensions of the protein molecule. For instance, in a 0.1 M solution of a 1:1 electrolyte in water the Debye length is ca. 1 nm, whereas the thickness of an adsorbed protein layer usually is at least a few nanometers. In that case, the protein layer shields the sorbent–protein contact region from the solution. It implies that the sorbent surface charge should be largely compensated by countercharge in the protein–sorbent contact region. The extent of charge neutralization was predicted by Norde and Lyklema [44,45] based on a three-layer model for the adsorbed protein. This model, depicted in Fig. 9, assumes coverage of the sorbent surface by a compact protein layer (cf. Section IV.D). All sorbent surface charge is located at $x = 0$. The inner region 1, $0 < x \leq m$, contains a fraction of the adsorbed protein charge and any ions trapped between the protein of the sorbent surface. The thickness of region 1 is of the order of the diameter of a hydrated ion, which is in the range of a few tenths of a nanometer. The extension of the outer region 3, $p \leq x \leq d$, is assumed to be comparable to the distance over which charged groups (including their hydration layer) on the protein surface protrude into the aqueous medium; for this distance 0.7 nm is taken. Analogous to the interiors of native state globular proteins, the central region 2, $m < x < p$, is considered to be void of isolated charged groups. The thickness of this region follows from measured thicknesses (see Section IV.D) of adsorbed protein layers corrected for the thicknesses assumed for regions 1 and 3. Because of the requirement of overall electroneutrality,

$$\sigma_0 + \sigma_1 + \sigma_2 + \sigma_3 + \sigma_d = 0 \tag{4}$$

where the indices refer to the sorbent surface, the three regions of the adsorbed layer, and the diffuse part of the electrical double layer, respectively. Based on the assumptions that σ_0 is located at $x = 0$, that σ_1, σ_2 (= 0), and σ_3 are distributed

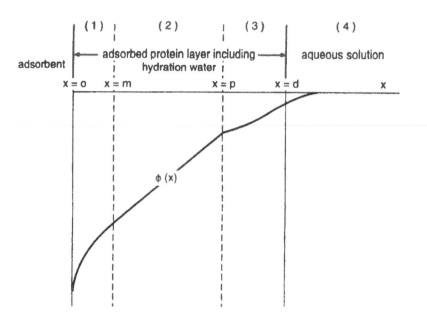

FIG. 9 Model for the adsorbed protein–sorbent interface, in which the course of the electrostatic potential is indicated. For details refer to the text.

homogeneously over regions 1, 2, and 3 and that σ_d is exponentially distributed according to the Gouy–Stern model (Fig. 7), Norde and Lyklema [45] derived expressions for $\phi(x)$ across the adsorbed layer and within the aqueous solution. By applying Eq. (2) a qualitative representation of $\phi(x)$ is shown in Fig. 9.

For all possible values of σ_0 (derived from titration data for the bare sorbent surface [46]) and σ_d (derived from electrokinetic data [47]), $\phi(x)$ shows a strong dependence on the assumed division of charge between regions 1 and 3. Since region 1 has a relatively low dielectric permittivity, any net charge in the contact zone between the protein and the sorbent leads to a high electrostatic potential and is therefore highly unfavorable. Realistically, it is not to be expected that $\phi(x)$ attains values exceeding a few hundred millivolts. Hence, any mismatch of protein and sorbent charge in region 1 must be compensated by low-molecular-weight ions to make this region nearly electrically neutral. An example of charge compensation predicted by the model is represented by the curve in Fig. 10; it shows a strong dependence on the electrical states of the protein and the sorbent. Here, the number of counterions taken up was estimated from calculated values of σ_1, which follow from reasonable estimates for ϕ_m (e.g., -100 mV) and the relation [45]

$$\frac{d\phi_m}{d\sigma_1} = \frac{p - m}{\varepsilon_0 \varepsilon_2} + \frac{d - p}{2\varepsilon_0 \varepsilon_3} \tag{5}$$

Equation (5) indicates that ϕ_m is highly sensitive to changes in σ_1 and, hence, to the number of coadsorbed ions.

The tendency of charge neutralization is experimentally confirmed by the shift in the proton titration curves of proteins upon adsorption at a charged surface [48–51]. By way of example, proton titration curves of α-lactalbumin in solution and

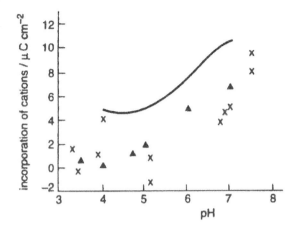

FIG. 10 Charge compensation in desorbed layers of human serum albumin on a negatively charged polystyrene surface ($\sigma_0 = -15.5$ μC cm^{-2}) as predicted from the model (——). Experimentally determined incorporation of cations in the adsorbed layer from 0.02 M $BaCl_2$ solution (×) or 0.02 M $MnCl_2$ solution (▲) at 25°C.

adsorbed on both positively and negatively charged polystyrene latexes are presented in Fig. 11. Charge adjustments may also occur on the sorbent surface. This has been clearly demonstrated by Fraaije for the adsorption of bovine serum albumin on silver iodide crystals [20]. Apart from adjustments on the protein and, possibly, the sorbent surface, the charge density in the protein–sorbent contact region may be further regulated by the transfer of indifferent electrolyte between that region and the solution. Indications for such transfer were inferred from electrokinetic data [44,49], and direct experimental evidence was obtained by tracing radiolabeled ions [52]. Experimental data on the uptake of ions are included in Fig. 10. It is clear that the number of coadsorbed cations tends to increase with increasing pH, i.e., with increasing charge antagonism between the protein and the sorbent surface. Moreover,

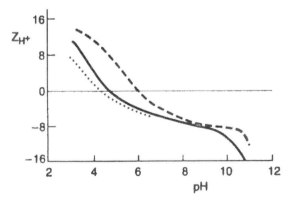

FIG. 11 Proton titration curve for α-lactalbumin in solution (——) and adsorbed on positively (· · · ·) and negatively (– – – –) charged polystyrene surfaces. Ionic strength 0.05 M; 25°C.

on the basis of phenomenological (thermodynamic) linkage relations the partitioning of ions between the adsorbed protein layer and the solution can be derived from the electrolyte dependency of the protein adsorption isotherms [53].

So we conclude that by active participation of small ions in the overall adsorption process, adverse conditions of charge on the protein and the sorbent surface may be eliminated. As a consequence of this charge regulation the absolute value of $\Delta_{ads}G_{el}$ does not exceed a few tens of RT, and it is not very sensitive to the charge on the protein (or, for that matter, the pH) and on the sorbent surface. Its sign and value primarily depend on the charge distributions and the dielectric constants of the electrical double layers before and after the adsorption process, respectively [45,54].

In addition to charge regulation, the transfer of ions from an aqueous to a nonaqueous protein layer includes a chemical effect as well. Because the protein-aceous environment is a poorer "solvent" for the low-molecular-weight ions compared to water, the chemical effect opposes protein adsorption [34,54]. Hence, if a protein is repelled from a like-charged surface, it is due to the unfavorable chemical effect that would result from the incorporation of ions needed to regulate the charge in the protein–sorbent contact region. Indeed, maximum protein adsorption affinity has been found under conditions where the charge on the protein just compensates the charge on the sorbent surface so that no additional ions are needed [20].

Finally, it should be realized that all these approaches, except the one by Noinville et al. [43], are based on continuum models. Each of the components, i.e., the sorbent surface, the protein molecule, and the solvent, are treated as continuous media having their own, constant dielectric permittivity, and in each of the phases the charge is smeared out according to the assumed distribution. However, in reality, in particular at high ionic strength, the distances between the individual charged groups on the protein and on the sorbent surface may exceed the Debye length, so a discrete-charge model would be more appropriate. Furthermore, as a result of the polar heterogeneity of most protein molecules, different areas in the adsorbed layer may have different permittivities. Including these specific effects in the analysis would not only require exact knowledge of the three-dimensional protein structure, but also detailed information concerning the orientation and structural perturbation of the protein upon adsorption. For most systems such detailed information is not available.

B. Dispersion Interaction

The most advanced computation of dispersion interaction between macroscopic bodies is based on quantum electrodynamics of continuous media, as presented in the Lifshitz theory [55]. A more approximate, simpler treatment is that in which the interaction energy is obtained by pairwise summation of London–van der Waals energies between all molecules of the interacting bodies. This approach has been elaborated by Hamaker [56] and de Boer [57]. The Hamaker–de Boer approximation may deviate by 10–30% with respect to the absolute magnitude [34]. However, since the (adsorbed) protein molecules and, sometimes, the sorbent material are often not sufficiently well defined to allow for quantum-dynamic computations, it is reasonable to utilize the Hamaker–de Boer theory for these systems. Then, for a sphere interacting with a planar surface the contribution from van der Waals interaction to the Gibbs energy of adsorption, $\Delta_{ads}G_{vdW}$, is given by

$$\Delta_{ads}G_{vdW} = -\frac{A_{132}}{6}\left(\frac{a}{h} + \frac{a}{h + 2a} + \ln\frac{h}{h + 2a}\right) \tag{6}$$

where A_{132} is the Hamaker constant for the interaction between body 1 (the sorbent) and body 2 (the protein) across a medium 3 (the aqueous solution) and where a is the radius of the sphere and h the distance of closest approach between the sphere and the planar surface. For $h \ll a$,

$$\Delta_{ads}G_{vdW} = -\frac{A_{132}a}{6h} \tag{6a}$$

Values for Hamaker constants (of the individual components) are given in Refs. 11, 34, 58–60. The Hamaker constant A_{132} for the system can be obtained from the individual ones according to the following rules:

$$A_{132} = (A_1^{1/2} - A_3^{1/2})(A_2^{1/2} - A_3^{1/2}) \tag{7}$$

and

$$A_{132} = (A_{131}A_{232})^{1/2} \tag{8}$$

In aqueous media, usually $A_1 > A_3$ and $A_2 > A_3$, so $A_{132} > 0$ and hence $\Delta_{ads}G_{vdW} < 0$, which implies attraction.

The Hamaker constant for interaction across water is ca. 6.6×10^{-21} J for proteins [11], $1–3 \times 10^{-19}$ J for metals [34] and $4–12 \times 10^{-21}$ J for synthetic polymers, e.g., polystyrene and Teflon [34]. According to Eq. (6a), $\Delta_{ads}G_{vdW}$ increases linearly with increasing dimensions of the protein molecule and it drops off steeply (hyperbollically) with increasing separation distance between the protein and the sorbent. For example, for a spherical protein molecule of radius 3 nm at a distance of 0.1 nm from the sorbent surface, $\Delta_{ads}G_{vdW}$ at room temperature is calculated to be 1–4 RT/mol in case of a synthetic polymer surface and 6–11 RT at metal surfaces. The values given here may not be better than a qualitative indication for the magnitude of van der Waals interaction in protein adsorption. Obtaining more accurate values is hampered by the irregular geometry of the protein molecule and, possibly, the sorbent surface. Furthermore, structural perturbation in the adsorbed protein may affect the Hamaker constant to an unknown extent.

C. Hydration Changes

When the surfaces of the protein molecule and the sorbent are hydrophilic, their hydration is favorable. Therefore, if protein adsorption occurs it is likely that some hydration water is retained between the sorbent surface and the adsorbed layer. However, when (one of) the contacting surfaces are (is) hydrophobic, dehydration would simulate the adsorption process.

1. Protein Hydrophobicity

Protein molecules contain both polar and apolar parts. As discussed in Section III.B, in globular proteins in an aqueous environment the apolar residues tend to be buried in the interior of the molecule where they are shielded from contact with water. However, due to other interactions that are participating in determining the protein structure and to geometrical constraints, as a rule not all apolar parts are hidden in the interior and not all the polar parts are exposed at the aqueous periphery of the protein molecule. For relatively small proteins (e.g., lysozyme, α-lactalbumin, ribonuclease, cytochrome c, etc.) having a molar mass of ca. 15,000 Da, apolar atoms occupy about 40–50% of the water-accessible surface area [4]. For larger proteins, which have a smaller surface/volume ratio the apolar fraction of the surface is usually less. Furthermore, water-soluble, nonaggregating proteins show a more or less even distribution of the polar and apolar residues over their surface so that no pronounced hydrophobic patches are present.

Several studies [31,61,62] confirm that the hydrophobicity of the protein influences its adsorption. However, not only the hydrophobicity of the protein surface, but the overall hydrophobicity of the protein may be relevant for the adsorption behavior. The overall hydrophobicity influences the extent of structural perturbation upon adsorption, which, in turn, affects the adsorption affinity (cf. Section IV.D). As a consequence, it is virtually impossible to establish the influence of protein hydrophobicity, as such, on the adsorption process.

2. Sorbent Hydrophobicity

As for the protein, the impact on adsorption by the sorbent hydrophobicity is difficult to establish experimentally because a variation in the hydrophobicity usually involves a variation in the electrostatic potential (or charge) as well. Experiments using hydrophobicity gradients [63] are probably best for using this matter in more detail.

The contribution from dehydration of a component to the Gibbs energy of adsorption $\Delta_{ads}G_{hydr}$ may be estimated from partition coefficients of (model) compounds in water/nonaqueous two-phase systems (e.g., Ref. 64). It has thus been established that, at room temperature, dehydration of most hydrophobic surfaces is predominantly entropically determined, involving effects in the range of 20–50 μJ $K^{-1} m^{-2}$; this corresponds to a lowering of the Gibbs energy with 5–12 mJ m^{-2}. For a protein of molar mass 15,000 Da that adsorbs ca. 1 mg m^{-2}, it corresponds to $\Delta_{ads}G_{hydr}$ ranging between -30 and -75 RT/mol of adsorbed protein. It demonstrates that hydrophobic dehydration easily overrules effects from the redistribution of charged groups and from dispersion interaction (cf. Sections IV.A and IV.B).

D. Rearrangements in the Protein Structure

As discussed in Section III, the three-dimensional structure of a globular protein molecule in aqueous solution is only marginally stable, and interaction with an interface may induce structural perturbations in the protein. However, unlike flexible polymers globular proteins do not adsorb in a loosely structured train loop tail–like conformation. The general observation that the thickness of an adsorbed protein layer is comparable with the dimensions of the native molecule in solution [65–68] points to a compact structure of the adsorbed protein molecules, even if they have undergone structural rearrangements (Fig. 8). In Section IV.A it has been explained that the formation of such a compact layer requires coadsorption of low-molecular-weight ions in the contact region between the protein and the sorbent to prevent a high electrostatic potential in that region. An alternative way to avoid the development of a high potential would be the unfolding of the protein into an open and loose structure that is freely penetrable for water and electrolyte, as depicted in Fig. 4. In such a layer the dielectric permittivity would not differ too much from that of the bulk solution. Because of the general observation that globular proteins do not form such loose adsorbed structures, it is concluded that the unfavorable chemical effect of ion incorporation (Section IV.A) is less unfavorable than the exposure of apolar residues of the protein to water, as would occur upon extensive unfolding.

When a protein molecule arrives at the sorbent surface, at one side of the molecule the aqueous environment is replaced by the sorbent material. As a consequence, intramolecular hydrophobic interaction becomes less important as a structure-stabilizing factor; i.e., apolar parts that tend to be buried in the interior of the dissolved molecule may become exposed to the sorbent surface without making contact with water. Because hydrophobic interaction between amino acid side groups in the protein's interior support the formation of secondary structures as α-helices and β-sheets, a reduction of this interaction tends to destabilize such structures. Decrease in α-helix and/or β-sheet content is indeed expected to occur only if peptide units released from these structures can form hydrogen bonds with the sorbent surface, as is the case for oxides (e.g., glass, silica, and metal oxides) or with residual water molecules remaining at the sorbent surface. Then the decrease in secondary structure may lead to an increased conformational entropy of the protein, contributing to $\Delta_{ads}G$ with a few tens of RT per mol of protein [23,69–71] (cf. Section III.A).

If in the nonaqueous protein–sorbent contact region it is not possible for the peptide units to form hydrogen bonds with the sorbent surface, as is the case for hydrophobic surfaces, adsorption may induce formation of extra intramolecular peptide–peptide hydrogen bonds, thereby promoting the formation of α-helices and/or β-sheets [71,72]. Hence, whether adsorption on a hydrophobic surface results in an increased or decreased order in protein structure depends on the subtle balance between energetically favorable interactions and the protein's conformational entropy.

V. CONCLUSIONS

Protein adsorption is a complex process that is controlled by a number of subprocesses, the major ones being (1) electrostatic interactions between the protein and the sorbent surface, giving rise to coadsorption of small ions, (2) dispersion interaction, (3) changes in the state of hydration of the sorbent surface and parts of the protein molecule, and (4) structural rearrangements in the protein. These subprocesses are not independent of each other. For instance, hydrophobic bonding between the protein and the surface requires close contact between the two components, which may be optimized by structural rearrangements in the protein. This may involve decreased or increased flexibility along the polypeptide chain. At hydrophilic surfaces, increased conformational freedom is anticipated and indeed experimentally found, which in turn improves the ability of the protein to form ion pairs with oppositely charged groups on the sorbent surface. The synergistic and antagonistic effects of these interactions are indicated in Scheme 1. Based on these considerations it is to be expected that all proteins adsorb on hydrophobic surfaces, even under electrostatically adverse conditions. With respect to their adsorption behavior on hydrophilic surfaces, distinction may be made between structurally stable ("hard") and labile ("soft") proteins. The hard proteins adsorb on hydrophilic surfaces only if they are electrostatically attracted. The soft proteins are more prone to undergo

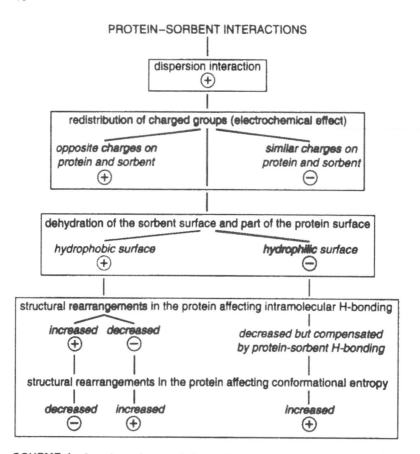

SCHEME 1 Interdependency of the major subprocesses that are involved in the overall protein adsorption process. Adsorption promotion is denoted by + and adsorption opposition by −.

structural rearrangements when they adsorb, and the ensuing gain in conformational entropy may be sufficiently large to cause adsorption on a hydrophilic electrostatically repelling surface.

REFERENCES

1. T. E. Creighton, *Proteins: Structures and Molecular Properties*, 2nd ed. W. H. Freeman, New York, 1993, Chap. 5.
2. W. Kauzmann, Some factors in the interpretation of protein denaturation. Adv. Protein Chem. *14*:1–63 (1959).
3. E. M. Richards, Areas, volumes, packing and protein structure. Ann. Rev. Biophysics. Bioeng. *6*:151–176 (1977).
4. B. Lee and F. M. Richards, The interpretation of protein structures: estimation of static accessibility. J. Mol. Biol. *55*:379–400 (1971).
5. C. Tanford and J. G. Kirkwood, Theory of protein titration curves. I. General equations for impenetrable spheres. J. Am. Chem. Soc. *79*:5333–5339 (1957).

6. C. Tanford and J. G. Kirkwood, Theory of protein titration curves. II. Calculations for simple models at low ionic strength. J. Am. Chem. Soc. *79*:5340–5347 (1957).

7. J. B. Matthew and F. R. N. Gurd, Calculation of electrostatic interactions in proteins. Methods Enzymol. *130*:413–453 (1986).

8. D. J. Barlow and J. M. Thornton, Ion pairs in proteins. J. Mol. Biol. *168*:867–885 (1983).

9. F. Franks and D. Eagland, The role of solvent interactions in protein conformation. Crit. Rev. Biochem. *3*:165–219 (1975).

10. J. A. Schelman, Solvent denaturation. Biopolymers *17*:1305–1322 (1978).

11. S. Nir, van der Waals interactions between surfaces of biological interest. Prog. Surf. Sci. *8*:1–58 (1977).

12. C. A. Haynes, K. Tamura, H. R. Körfer, H. W. Blanch, and J. M. Prausnitz, Thermodynamic properties of aqueous α-chymotrypsin solutions from membrane osmometry measurements. J. Phys. Chem. *96*:905–912 (1991).

13. G. C. Kresheck and I. M. Klotz, The thermodynamics of transfer of amides from an apolar to an aqueous solution. Biochemistry *8*:8–11 (1969).

14. C. Tanford, Protein denaturation. Part C. Theoretical models for the mechanism of denaturation. Adv. Protein Chem. *24*:1–95 (1970).

15. W. Norde, Adsorption of proteins from solution at the solid–liquid interface. Adv. Colloid Interface Sci. *25*:267–340 (1986).

16. M. Levitt, A simplified representation of protein conformations for rapid simulation of protein folding. J. Mol. Biol. *104*:59–107 (1976).

17. B. W. Morrissey and R. R. Stromberg, The conformation of adsorbed blood proteins by infrared bound fraction measurements. J. Colloid Interface Sci. *46*:152–164 (1974).

18. J. L. Brash and Q. M. Samak, Dynamics of interactions between human albumin and polyethylene surface. J. Colloid Interface Sci. *65*:495–504 (1978).

19. B. M. C. Chan and J. L. Brash, Conformational change in fibrinogen desorbed from glass surface. J. Colloid Interface Sci. *84*:263–265 (1981).

20. J. G. E. M. Fraaije, Interfacial thermodynamics and electrochemistry of protein partitioning in two-phase systems. Ph.D. thesis, Wageningen Agricultural University, The Netherlands, 1987.

21. V. Ball, A. Bentaleb, J. Hemmerle, J.-C. Voegel, and P. Schaff, Dynamic aspects of protein adsorption onto titanium surfaces: mechanism of desorption into buffer and release in the presence of proteins in the bulk. Langmuir *12*:1614–1621 (1996).

22. P. R. Van Tassel, P. Viot, and G. Tarjus, A kinetic model of partially reversible protein adsorption. J. Chem. Phys. *106*:761–770 (1997).

23. W. Norde and J. P. Favier, Structure of adsorbed and desorbed proteins. Colloids Surf. *64*:87–93 (1992).

24. W. Norde and A. C. I. Anusiem, Adsorption, desorption and re-adsorption of proteins on solid surfaces. Colloids Surf. *66*:299–306 (1992).

25. M. C. Wahlgren and T. Arnebrant, Interaction of cetyltrimethylammonium bromide and sodium dodecyl sulfate with β-lactoglobulin and lysozyme at solid surfaces. J. Colloid Interface Sci. *142*:503–511 (1991).

26. L. Vroman and A. L. Adams, Adsorption of proteins out of pasma and solutions in narrow spaces. J. Colloid Interface Sci. *111*:391–402 (1986).

27. T. A. Horbett, Adsorption of proteins from plasma to a series of hydrophilic–hydrophobic copolymers. II. Compositional analysis with the prelabeled technique. J. Biomed. Mater. Res. *15*:673–695 (1981).

28. J. J. Ramsden, Concentration of protein deposition kinetics, Phys. Rev. Lett. *71*:295–298 (1993).

29. M. C. P. van Eijk and M. A. Cohen Stuart, Polymer adsorption kinetics: effects of supply rate. Langmuir *13*:5447–5450 (1997).

30. J. C. Dijt, M. A. Cohen Stuart, and G. J. Fleer, Reflectometry as a tool for adsorption studies. Adv. Colloid Interface Sci. *50*:79–101 (1994).

31. J. M. H. M. Scheutjens and G. J. Fleer, Statistical theory of the adsorption of interacting chain molecules. 1. Partition function, segment density distribution, and adsorption isotherms. J. Phys. Chem. *83*:1619–1635 (1979).

32. T. Arai and W. Norde, The behavior of some model proteins at solid–liquid interfaces. 1. Adsorption from single protein solutions. Colloids Surf. *57*:1–16 (1990).

33. C. A. Haynes and W. Norde, Globular proteins at solid–liquid interfaces. Colloids Surf. B: Biointerfaces 2:517–566 (1994).

34. J. Lyklema, *Fundamentals of Interface and Colloid Science*. Volume I: *Fundamentals*. Academic Press, London, 1991.

35. E. J. W. Verwey and J. T. G. Overbeek, *The Theory of the Stability of Lyophobic Colloids*. Elsevier, Amsterdam, 1948.

36. A. J. Babchin, Y. Gur, and I. J. Lin, Repulsive interface forces in overlapping electric double layers in electrolyte solutions. Adv. Colloid Interface Sci. 9:105–141 (1978).

37. J. Ståhlberg, U. Appelgren, and B. Jönsson, Electrostatic interactions between a charged sphere and a charged planar surface in an electrolyte solution. J. Colloid Interface Sci. *176*:397–407 (1995).

38. J. Ståhlberg and B. Jönsson, Influence of charge regulation in electrostatic interaction chromatography of proteins. Anal. Chem. *68*:1536–1544 (1996).

39. B. J. Yoon and A. M. Lenhoff, Computation of the electrostatic interaction energy between a protein and a charged surface. J. Phys. Chem. *96*:3130–3134 (1992).

40. C. M. Roth and A. M. Lenhoff, Electrostatic and van der Waals contributions to protein adsorption: computation of equilibrium constants. Langmuir 9:962–972 (1993).

41. C. M. Roth and A. M. Lenhoff, Electrostatic and van der Waals contributions to protein adsorption: comparison of theory and experiment. Langmuir *11*:3500–3509 (1995).

42. D. J. Roush, D. S. Gill, and R. C. Willson, Electrostatic potentials and electrostatic interaction energies of rat cytochrome b_5 and a simulated anion-exchange adsorbent surface. Biophys. J. *66*:1290–1300 (1994).

43. V. Noinville, C. Vidal-Madjar, and B. Sébille, Modeling of protein adsorption on polymer surfaces. Computation of adsorption potential. J. Phys. Chem. *99*:1516–1522 (1995).

44. W. Norde and J. Lyklema, The adsorption of human plasma albumin and bovine pancreas ribonuclease at negatively charged polystyrene surfaces. IV. The charge distribution in the adsorbed state. J. Colloid Interface Sci. *66*:285–294 (1978).

45. W. Norde and J. Lyklema, Thermodynamics of protein adsorption. Theory with special reference to the adsorption of human plasma albumin and bovine pancreas ribonuclease at polystyrene surfaces. J. Colloid Interface Sci. *71*:350–366 (1979).

46. K. Furusawa, W. Norde, and J. Lyklema, A method for preparing surfactant-free polystyrene latices of high surface charge. Kolloid-Z Z Polymere *250*:908–909 (1972).

47. W. Norde and J. Lyklema, The adsorption of human plasma albumin and bovine pancreas ribonuclease at negatively charged polystyrene surfaces. III. Electrophoresis. J. Colloid Interface Sci. *66*:277–284 (1978).

48. W. Norde and J. Lyklema, The adsorption of human plasma albumin and bovine pancreas ribonuclease at negatively charged polystyrene surfaces. II. Hydrogen ion titrations. J. Colloid Interface Sci. *66*:266–276 (1978).

49. C. A. Haynes, E. Sliwinski, and W. Norde, Structural and electrostatic properties of globular proteins at a polystyrene–water interface. J. Colloid Interface Sci. *164*:394–409 (1994).

50. F. Galisteo and W. Norde, Protein adsorption at the AgI–water interface. J. Colloid Interface Sci. *172*:502–509 (1995).

51. F. Galisteo and W. Norde, Adsorption of lysozyme and α-lactalbumin on poly(styrene sulphonate) latices. II. Proton titrations. Colloids Surf. B: Biointerfaces 4:389–400 (1995).

52. P. van Dulm, W. Norde, and J. Lyklema, Ion participation in protein adsorption at solid surfaces. J. Colloid Interface Sci. *82*:77–82 (1981).

53. J. G. E. M. Fraaije, R. M. Murris, W. Norde, and J. Lyklema, Interfacial thermodynamics of protein adsorption, ion co-adsorption and ion binding in solution. I. Phenomenological linkage relations for ion exchange in lysozyme chromatography and titration in solution. Biophys. Chem. *40*:303–315 (1991).

54. W. Norde and J. Lyklema, Why proteins prefer interfaces. J. Biomater. Sci. Polymer Edn. *2*:183–202 (1991).

55. L. D. Landau and E. M. Lifshitz, *Electrodynamics of Continuous Media*, 2nd ed. Pergamon, Oxford, 1963.

56. H. C. Hamaker, The London–van der Waals attraction between spherical particles. Physica. *4*:1058–1072 (1937).

57. J. H. de Boer, Influence of van der Waals forces and primary bonds in binding energy, strength, and orientation, with special reference to some artificial resins. Trans. Faraday Soc. *32*:10–38 (1936).

58. J. Visser, Hamaker constants. Comparison between Hamaker constants and Lifshitz–van der Waals constants. Adv. Colloid Interface Sci. *3*:331–363 (1972).

59. T. Afshar-Rad, A. I. Baily, P. F. Luckham, W. MacNaughtan, and D. Chapman, Forces between proteins and model polypeptides adsorbed on mica surfaces. Biochem. Biophys. Acta. *915*:101–111 (1987).

60. J. N. Israelachvili, *Intermolecular and Surface Forces*, 2nd ed. Academic Press, New York, 1992.

61. F. E. Regnier, The role of protein structure in chromatographic behavior. Science *238*: 319–323 (1987).

62. A. A. Gorbunov, A. Y. Lukyanov, V. A. Pasechnik, and A. V. Vakrushev, Computer simulation of protein adsorption and chromatography. J. Chromatog. *365*:205–212 (1986).

63. H. Elwing, A. Askendal, U. Nilsson, and I. Lundström, A wettability gradient method for studies of macromolecular interactions at the liquid–solid interface. J. Colloid Interface Sci. *91*:248–255 (1987).

64. G. Némethy, H. A. Scheraga, Structure of water and hydrophobic bonding in proteins. J. Chem. Phys. *36*:3401–3417 (1962).

65. N. de Baillou, P. Dejardin, A. Schmitt, and J. L. Brash, Fibrinogen dimensions at an interface: variations with bulk concentration, temperature and pH. J. Colloid Interface Sci. *100*:167–174 (1984).

66. P. Schaaf and P. Dejardin, Structural changes within an adsorbed fibrinogen layer during the adsorption process: a study by scanning angle reflectometry. Colloids Surf. *31*:89–103 (1988).

67. E. Blomberg, P. M. Claesson, and R. D. Tilton, Short-range interaction between adsorbed layers of human serum albumin. J. Colloid Interface Sci. *166*:427–436 (1994).

68. E. Blomberg, P. M. Claesson, J. C. Fröberg, and R. D. Tilton, Interaction between adsorbed layers of lysozyme studied with the surface force technique. Langmuir *10*: 2325–2334 (1994).

69. A. Kondo, F. Mukarami, and K. Higashitani, Circular dichroism studies on conformational changes in protein molecules upon adsorption on ultrafine polystyrene particles. Biotechnol. Bioeng. *40*:889–894 (1992).

70. A. Kondo, S. Oku, and K. Higashitani, Structural changes in protein molecules adsorbed on ultrafine silica particles. J. Colloid Interface Sci. *143*:214–221 (1991).

71. T. Zoungrana, G. H. Findenegg, and W. Norde, Structure, stability, and activity of adsorbed enzymes. J. Colloid Interface Sci. *190*:437–448 (1997).

72. M. C. L. Maste, W. Norde, and A. J. W. G. Visser, Adsorption-induced conformational changes in the serine proteinase savinase: a tryptophan fluorescence and circular dichroism study. J. Colloid Interface Sci. *196*:224–230 (1997).

3
Thermodynamics of Adsorption of Amino Acids, Small Peptides, and Nucleic Acid Components on Silica Adsorbents

VLADIMIR A. BASIUK   Instituto de Ciencias Nucleares, Universidad Nacional Autónoma de México, Mexico City, Mexico

I. INTRODUCTION

Adsorption of amino acids, peptides, and nucleic acid constituents on silica surfaces is a very important aspect of biomolecular adsorption. One cannot imagine the modern practice of their analysis and separation, so widely used in biotechnology, medicine, diagnostics, etc., without high-performance liquid chromatographic (HPLC) methods employing silica stationary phases (see, for example, Refs. 1–3). The rational design of new biocompatible materials and drug delivery systems also needs a detailed knowledge on biomolecule adsorption interactions. Furthermore, studying the adsorption of the biopolymers' building blocks and the development of structure–property relationships can give insight to the interfacial behavior of the biopolymers themselves.

A key point in the characterization of adsorption processes on solid–liquid interfaces is determination of the corresponding thermodynamic parameters: Gibbs free energy ($\Delta G°$), enthalpy ($\Delta H°$), and entropy ($\Delta S°$) changes during adsorption. Now adsorption from solutions on solid surfaces is a highly advanced theory accounting for different possible types of adsorption isotherms and allowing a detailed quantitative description of various particular systems to be performed [4–8]. In practice, however, "a reasonable balance between theoretical foundations and derivations, on the one hand, and results, on the other hand" is necessary [5]. In this regard, thermodynamics of adsorption of amino acids, small peptides, and nucleic acid constituents on silica surfaces is especially illustrative because, as is demonstrated in this chapter, so far it has been described using only the simplest experimental approaches and the simplest numeric interrelations. In particular, the case of adsorption was always considered from very diluted aqueous solutions; this fact can be readily explained by the dominating role of the works on HPLC separations in the whole body of publications related to biomolecular adsorption. Due to the same circumstances, particular types of adsorption isotherm and real surface nonuniformity are also neglected.

II. EXPERIMENTAL APPROACHES TO DETERMINE ADSORPTION THERMODYNAMIC CONSTANTS

A. Static Approach

The traditional approach to the determination of adsorption thermodynamic constants is based on measuring adsorption isotherms under static conditions. The adsorption equilibrium constant (or distribution coefficient) K can be relatively easily found from a slope of the isotherm initial rectilinear part; this then enables calculation of the free energy change during adsorption $\Delta G°$:

$$\Delta G° = -RT \ln K \tag{1}$$

Measuring the isotherms under several different temperatures and plotting $\ln K$ versus $1/T$ (Van't Hoff plots) enables us to derive the entropy and enthalpy changes $\Delta S°$ and $\Delta H°$.

One should note, however, that for solutes which are adsorbed more weakly, the results will be less accurate. Moreover, measuring the adsorption isotherms themselves is not always possible. This is the case when adsorption is very slight, e.g., the equilibrium constants are close to one, especially if $K < 1$. The latter case means that molecule concentration in the solid phase appears less than in the bulk liquid phase (negative excess adsorption), and it is impossible to detect any concentration changes in the solution.

B. Dynamic Approach

The dynamic (liquid chromatographic, usually HPLC) approach is based on measuring a compound's retention value k' on a chromatographic column [9–11]. The equilibrium constant in this case is equated to

$$K = \frac{k'}{\phi} \tag{2}$$

where ϕ is the column phase ratio and k' is the capacity factor for a given solute,

$$k' = \frac{(t_i - t_0)}{t_0} = \frac{(V_i - V_0)}{V_0} \tag{3}$$

where t_i and V_i are retention time and retention volume of the solute i; t_0 and V_0 are retention time (usually called the *dead time*) and retention volume (the *dead*, or *void, volume* equivalent to the total volume of eluent in the column) for a compound which is not retained (or not adsorbed). The simplest definition for ϕ is the ratio of volumes of stationary (V_s) to mobile (V_m) phases in the chromatographic column [10]. A more strict definition of the phase ratio accounts for the mass (m) and specific surface area S of the stationary phase [12]:

$$\phi = \frac{mS}{V_m} = \frac{V_s\rho S}{V_m} \tag{4}$$

where ρ is the density of the stationary phase. The free energy change during adsorption $\Delta G°$ is then calculated according to Eq. (1). Performing the chromatographic procedure under different temperatures allows us to derive the corresponding $\Delta S°$ and $\Delta H°$ values from Van't Hoff plots.

Like the static approach, this method has its drawbacks as well. The most serious problem is that in practice a column dead volume V_0 (or t_0) and phase ratio are relatively difficult to determine precisely [11,13]. An additional problem appears when solutes chromatographed possess relatively low capacity factors k', for instance, when $k' < 1$ precise measurement is difficult; as is shown subsequently, this is the case for most amino acids. Thus it seems quite reasonable to consider some thermodynamic parameters derived from HPLC experiments as "questionable" [11]. Realizing these problems, it is preferable to discuss not *determination*, but only *estimation* of these parameters.

Nevertheless, when the same HPLC column (and the same chromatographic system as a whole) is employed for a series of the measurements, errors due to the dead volume and phase ratio are the same for all solutes studied. As a result, any graphic interrelation between thermodynamic parameters and solute molecular properties, e.g., linear trends for the plots $\Delta G°$ versus number of aliphatic carbon atoms found for amino acids and peptides (see Sections III.A–C), are maintained but are shifted along the ordinate, e.g., the $\Delta G°$ axis, depending on these errors. Even if the latter might appear significant, one can discuss comparative adsorption behavior of solutes. For instance, regardless of the errors due to incorrect measurements of the dead volume and phase ratio, a solute retained for a longer time in a chromatographic column will have a higher $-\Delta G°$ (or lower $\Delta G°$) value. Suppose two compounds A and B are chromatographed on the same column and have the capacity factors k'_A and k'_B, respectively. Their adsorption equilibrium constants, K_A and K_B, depend on the same ϕ value, which can be completely excluded if we consider not corresponding $\Delta G°_A$ and $\Delta G°_B$ by themselves, but their difference $\delta(\Delta G°_{AB})$, i.e., a *relative* free energy change:

$$\delta(\Delta G°_{AB}) = \Delta G°_A - \Delta G°_B = -RT \ln \frac{K_A}{K_B} = -RT \ln \frac{k'_A}{k'_B} = -RT \ln \alpha_{A/B} \qquad (5)$$

Equation (5) includes one more parameter, $\alpha_{A/B}$, which is the selectivity of resolution of A and B. Conventionally, the compound retained longer is designed as A, i.e., $k'_A > k'_B$, in order to obtain $\alpha_{A/B} > 1$. Such description is indispensable in the HPLC resolution of enantiomeric compounds to characterize their chiral discrimination (see, for example, Ref. 14).

Values of $\Delta S°$ are also affected by the errors due to incorrect measurements of the dead volume and phase ratio, whereas $\Delta H°$ values found from slopes of the Van't Hoff plots are not and thus appear to be the most reliable thermodynamic parameters derived from HPLC retention data.

III. THERMODYNAMICS OF ADSORPTION OF SMALL BIOMOLECULES

A. Amino Acids

Despite obvious implications of amino acid adsorption on silica adsorbents for their liquid chromatographic analysis, and continuously growing number of publications on the HPLC of amino acids, there are not many works focused on the thermodynamic aspects. Nevertheless, to get a general idea on their comparative behavior, even barely chromatographic data can be useful. A good example is the paper by

Molnár and Horváth [15]. The authors found that under isocratic elution (i.e., the solvent composition is maintained the same during the chromatographic procedure) of a homologous series of α-amino acids having n-alkyl side chains on a column with octadecyl silica, the relationship between the log k' values and the carbon number of the side chain is linear (Fig 1). The capacity factors of the species with undissociated carboxylic groups (at pH 0.2) are higher than those of the about half dissociated species (at pH 2.1), indicating that on octadecyl-silica surface the former are adsorbed stronger than the latter. Admitting that both chromatographic retention on nonpolar stationary phases in aqueous media and partitioning between organic solvents and aqueous solutions have a similar physicochemical basis, the authors compared the partitioning coefficients, P, for protein amino acids in octanol–water and the retention values obtained for octyl silica (pH 6.7) and found another linear relationship, shown in Fig. 2.

An attempt to establish a direct relationship between the structure of amino acids and the free energy changes has been undertaken for the case of adsorption on bare silica from purely aqueous solutions using the HPLC approach [16–18]. The choice of pure water as the adsorption medium excluded any interference of the

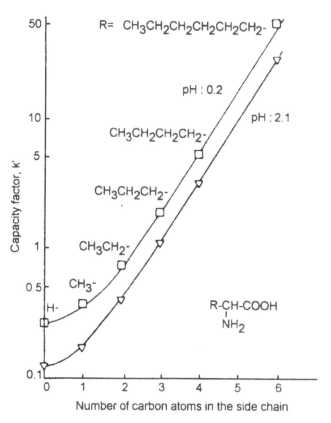

FIG. 1 Plots of log k' against the side chain carbon number for the homologous series of α-amino acids. Stationary phase, LiChrosorb RP-18 octadecyl silica; eluents: 0.5 M HClO$_4$, pH 0.2, and 0.1 M phosphate buffer, pH 2.1; temperature 70°C. (From Ref. 15.)

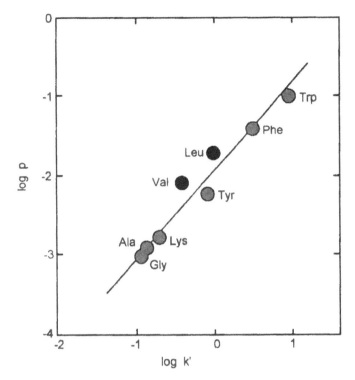

FIG. 2 Plot of log P versus log k' for the series of α-amino acids. Stationary phase, LiChrosorb RP-8 octyl silica; eluent: 0.1 M phosphate buffer, pH 6.7, temperature 70°C. (From Ref. 15.)

components of buffer solutions and simplified extremely the adsorption system. As it was found from the experimental retention values using Eqs. (1) and (2), for the overwhelming majority of α-amino acids the values of $\Delta G°$ were positive (ranging from 130 J mol^{-1} for Phe to 3640 J mol^{-1} for Asp; for Pro, -400 J mol^{-1}) and K < 1, indicating that the molecular concentration in the solid phase is lower than in the bulk liquid phase. As was mentioned in Section II.A, the traditional static approach cannot be successful for thermodynamic characterization of such adsorption systems.

Plotting the $\Delta G°$ values versus the number of aliphatic carbon atoms in the molecules revealed a linear interrelation for n-alkyl (aliphatic bifunctional) α-amino acids, similar to that reported by Molnár and Horváth [15]; this is shown in Fig. 3. From the slope, an increment in $\Delta G°$ for each aliphatic C atom has been obtained to be about -300 J mol^{-1}. In addition, amino acid adsorbability is considerably influenced by heteroatoms and other nonaliphatic moieties in the α-substituent. Imidazole nucleus (for His) and carboxylic groups (Asp and Glu) cause the sharpest increase in $\Delta G°$ values (and corresponding decrease in adsorbability); amide (Asn and Gln) and alcohol functions (Ser and Thr) also reduce adsorbability but to a much lesser extent. The sulfur atom noticeably decreases $\Delta G°$ in the case of Cys (as compared to Ala) but only slightly contributes in the case of Met (as compared to Val). The $\Delta G°$ decrease can also be due to phenyl nucleus (Phe versus Ala), but

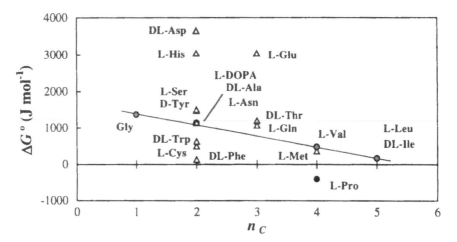

FIG. 3 Plot of $\Delta G°$ versus the number of aliphatic carbon atoms (n_C) in the amino acid molecules for the adsorption on bare silica from water. Aliphatic bifunctional amino acids and proline (●) and other amino acids (Δ). Temperature 19°C. (From Ref. 18.)

only if the nucleus does not contain hydroxy substituents, as in Tyr and 3,4-dioxy-phenylalanine (DOPA). Proline is adsorbed much stronger than its open-chain analog valine, which is apparently explained by a cyclic structure of Pro.

We attempted a comparison with some other molecular and thermodynamic properties of amino acids [17]. For example, the relationship between $\Delta G°$ and apparent molar volume showed a good fit to a straight line. Plotting the heats and free energies of amino acid formation versus $\Delta G°$ also revealed trends close to linear, but the fits were much worse.

Some idea of the accuracy of the dynamic determination of K and $\Delta G°$ values has been provided from the following estimates [17]. The error of recording the retention volumes typically did not exceed 2%. Taking a solute with the average retention volume V_i of 200 μL as an example, due to this error it can vary from 196 to 204 μL. Correspondingly, the under- and overestimation of the K values gives 0.70 and 0.76 (4% deviation from the average value of 0.73 calculated for V_i of 200 μL); for the free energy changes, the values of 870 and 670 J mol^{-1} are obtained (12 to 14% deviation from the average $\Delta G°$ value of 760 J mol^{-1}), respectively. The problem of precise determination of the dead volume and phase ratio still remained an open issue.

A more careful approach is to avoid any estimates of the free energy changes and focus on the enthalpy change determination from Van't Hoff plots. Uddin et al. [19] did this to study Gln, Met, Phe, and Trp adsorption on bare silica surface from water at 20, 37, and 55°C. They used a liquid chromatographic technique as well, though it was not HPLC. Nevertheless, evident advantage of their experimental setup was the use of the method of disturbance peaks, perhaps the most precise method to determine the column dead volume [13]. Thus, the K values presented should be highly reliable and should give rise to reliable corresponding $\Delta G°$ values. In Table 1 the latter are presented for 20°C along with the enthalpy changes derived from the Van't Hoff plots (Fig. 4a). As shown, both the $\Delta G°$ and $\Delta H°$ values increase in the

TABLE 1 Comparison of the Free Energy (for 20°C) and Enthalpy Changes for the Adsorption of Phe, Met, Trp, and Gln on Bare Silica from Water Obtained from Dynamic Measurements by Two Different Groups

Amino acid	ΔG°_{20} (J mol^{-1})		ΔH° (J mol^{-1})	
	From Ref. 19[a]	From Ref. 20	From Ref. 19	From Ref. 20
Gln	460	1040	−2470	−2150
Met	−830	760	−3190	−9450
Phe	−1530	280	−6030	−11620
Trp	−490	910	−2910	−8110

[a]Calculated from the reported K values.

same order of Gln > Trp > Met > Phe. This, as well as the fact that the K values decrease (Fig. 4a) and ΔG° values correspondingly increase with temperature, indicates that the adsorption process is enthalpy driven.

It is interesting to compare these results with our recently published data for the same adsorption systems, also obtained using the dynamic (but HPLC) approach [20]. An important difference in the experimental setup was that we used a microcolumn of 64 × 2 mm I.D., whereas Uddin et al. [19] used a bigger column, 100 × 11 mm I.D. Figure 4 demonstrates that the two sets of the Van't Hoff plots are very similar though shifted along the ln K axis. While in Fig. 4a the plots are situated basically above the abscissa, our series appears completely below it. Consequently, ΔG° values in the latter case are all positive. Considerable discrepancies are found also for the ΔH° values (Table 1). One should note, however, that both the ΔG° and ΔH° values increase in the same order, Gln > Trp > Met > Phe, as in the work by Uddin et al. Revalidation of all the quantitative results would by highly desirable.

In addition to Gln, Met, Phe, and Trp, we performed the estimates of ΔH° for many other α-amino acids as well as attempted to determine complementary ΔS°

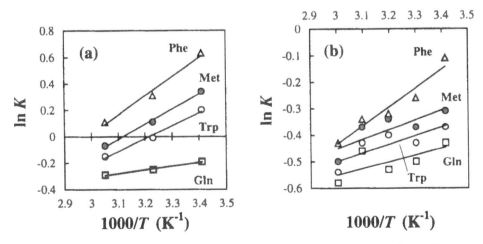

FIG. 4 Van't Hoff plots for the adsorption of Phe, Met, Trp, and Gln on bare silica from water. [(a) From Ref. 19. (b) From Ref. 20.]

values [20]. As a whole, a significant scatter was observed for the points in the Van't Hoff plots (Fig. 4b exemplifies this). In most cases, the correlation coefficients [20] exceeded 0.75, reaching 0.97 (Ala and Phe). However, for Ser and Asn their values appear unsatisfactory: 0.31 and 0.38, respectively. Therefore the estimates of $\Delta H°$ and $\Delta S°$ for these two amino acids cannot be considered reliable at all.

The $\Delta G°$ values increase with temperature in all cases due to enthalpy-driven adsorption. All the $\Delta H°$ and $\Delta S°$ values are negative. The $\Delta H°$ values vary from -550 J mol^{-1} for Ser to -6030 J mol^{-1} for Phe; the $\Delta S°$ values, from -6.6 J mol^{-1} K^{-1} for Asn to -24.0 J mol^{-1} K^{-1} for Asp. The latter data suggest the adsorption process to be unfavorable.

An interesting and useful way to analyze the thermodynamic data derived from the HPLC measurements is the approach based on enthalpy–entropy compensation. This phenomenon manifests itself in a linear interrelation between the free energy changes and the corresponding enthalpy changes for intrinsically similar classes of solutes [21,22]. If such interrelation is observed, the related equilibrium process is said to be isoequilibrium one. In other words, compounds which exhibit compensation behavior are considered to be adsorbed according to an essentially identical mechanism. Conventionally, compensation behavior is verified by linearity of the plots of $\Delta H°$ versus $\Delta S°$ [21,22]. However, under such conditions the linearity sometimes arises not only from the compensation itself, but also from statistical effects due to possible errors associated with the determination of enthalpy changes [23,24]. On the other hand, if the plots of $\Delta H°$ versus $\Delta G°$ are used, the statistical effects are minimized and the linearity is indicative of real compensation behavior.

We also analyzed our data from this point of view, plotting $\Delta H°$ versus $\Delta S°$ and $\Delta G°$ [20]. For both interrelations, the points did not completely fit onto straight lines (Figs. 5 and 6), apparently due to experimental errors. Nevertheless, the trends toward linearity were evident. The points for amino dicarboxylic acids, Glu and Asp, are separate from those for the majority of amino acids; this was clearly seen, es-

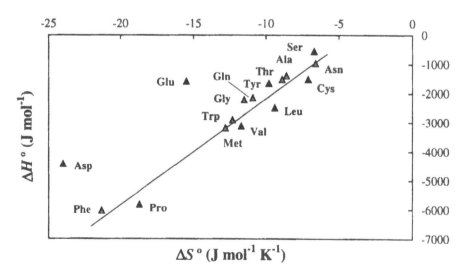

FIG. 5 Plot of $\Delta H°$ versus $\Delta S°$ for the adsorption of amino acids on bare silica from water. (From Ref. 20.)

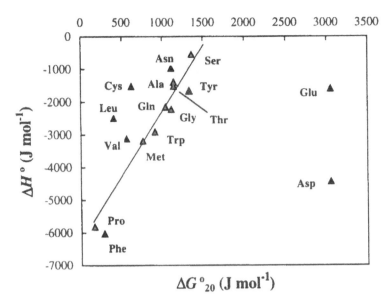

FIG. 6 Plot of $\Delta H°$ versus $\Delta G°$ (at 20°C) for the adsorption of amino acids on bare silica from water. (From Ref. 20.)

pecially for the plot of $\Delta H°$ versus $\Delta G°_{20}$ (Fig. 6). This may be due to the presence of a second carboxylic group in the molecules, resulting in certain differences between a mechanism of Glu and Asp adsorption and that for other amino acids.

Some suggestions on the nature of these differences can be made based on the available data on ion equilibria in amino acid aqueous solutions. Accounting for the dual nature of amino acid molecules, such equilibria can be discussed in terms of acid and base pK_i values as well as corresponding isoelectric points (pI). Use of the latter is more appropriate in the present case, since the pI values reflect overall charge of the molecules. It is known that Glu and Asp have isoelectric points at 3.22 and 2.77, respectively; whereas pI values for other amino acids mentioned in Figs. 5 and 6 vary from 5.41 (Asn) to 6.30 (Pro) [25]. Thus, none of them exceeded the almost neutral pH values of the diluted aqueous solutions. Under such conditions, an equilibrium exists between the zwitterionic form and the form having a deprotonated amino group, i.e., $NH_3^+CHRCOO^-$ and $NH_2CHRCOO^-$, respectively. For Glu and Asp, which possess much lower isoelectric points, the concentration of the zwitterions is lower, and the concentration of negatively charged species is correspondingly higher than for other amino acids, resulting in the observed deviation from the common linear trend. To elucidate whether the adsorption equilibrium correlates with the ionic equilibria in the solutions, we tested the graphic interrelation between our experimental ln K values and amino acid isoelectric points. As a result, a general linear regularity was obtained (Fig. 7), similar to the plots for $\Delta H°$ versus $\Delta S°$ and $\Delta G°$. Lower ln K values correspond to higher pI values, i.e., amino acids having higher isoelectric points are, as a rule, adsorbed on bare silica more strongly. This is evidence for major contribution of electrostatic interactions to the adsorption process and against hydrophobic interactions, since no direct correlation between our equilibrium data and the amino acid hydrophobicity indices has been found [20].

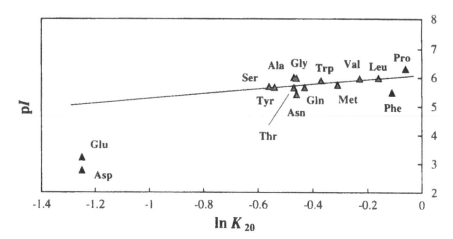

FIG. 7 Interrelation between ln K (at 20°C; bare silica) and isoelectric points of amino acids, pI. (From Ref. 20.)

The latter fact seems quite natural for the case of bare silica; its surface contains ionizable ≡Si—OH groups and no hydrophobic hydrocarbon moieties. Nevertheless, the complementary adsorption studies on a hydrophobic octadecyl silica, widely used in HPLC, demonstrated that the close-to-linear interrelation between the equilibrium constants and pI is observed in this case as well [20]. What could well be expected here is a linear correlation between the equilibrium data and the amino acid hydrophobicity indices because amino acids are retained on C_{18} phases according basically to a hydrophobic mechanism. To verify this, we plotted the hydrophobicity coefficients (16 sets compiled and developed by Wilce et al. [26]) versus the ln K values. In most cases the plotting gave rather random distribution of points; however, in a few cases linear trends were found (Fig. 8).

Another important difference, as compared to the adsorption on bare silica, was that at ambient temperature, for example, hydrophobic amino acids (Val, Nva, Leu, Ile, Nle, Tyr, Phe, Trp, and Met) have $K > 1$, i.e., the solute concentration in the adsorbed phase is higher than that in the bulk liquid phase (positive excess adsorption). The $\Delta G°$ values cover the range of −8400 (Trp) to 3220 J mol^{-1} (Glu). Similarly to the case of bare silica, K values decrease and $\Delta G°$ values increase with temperature. Statistical treatment of the Van't Hoff plots revealed a very good fit of the points to straight lines for all amino acids; the correlation coefficients varied from 0.97 (Phe and Trp) to 1.00 (Asp and Met) [20]. Thus, the reliability of the thermodynamic estimates in the case of octadecyl silica is much higher than in the case of bare silica.

The $\Delta H°$ values were found to range from −6310 J mol^{-1} for Asp to −31360 J mol^{-1} for Trp, being negative for all amino acids, indicating enthalpy-driven adsorption. The $\Delta S°$ values, as a whole, were much more negative compared to those for adsorption on bare silica—the lowest value was derived for Trp, −79.7 J mol^{-1} K^{-1}, and Asp had the highest one, −32.2 J mol^{-1} K^{-1}. The latter data suggest that the adsorption process is unfavorable, as in the case of bare silica. Analyzing the thermodynamic data for possible compensation behavior, we found that points in the plot of $\Delta H°$ versus $\Delta G°$ fit to a straight line better than those for amino acid ad-

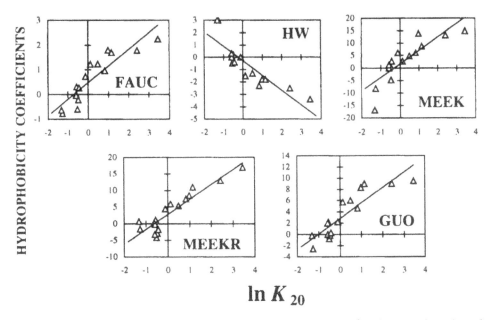

FIG. 8 Interrelations between the experimental ln K values (at 20°C) for the adsorption of amino acids on octadecyl silica from water and some hydrophobicity coefficients. Hydrophobicity increases as hydrophobicity coefficient increases for FAUC, MEEK, MEEKR, and GUO, or as hydrophobicity coefficient decreases for HW. For abbreviations and references, see Ref. 26. (From Ref. 20.)

sorption on bare silica. Regarding the amino dicarboxylic acids Glu and Asp, the corresponding points were separate from the general regularity in the plot $\Delta H°$ versus $\Delta S°$, whereas they fit the linear trend of $\Delta H°$ versus $\Delta G°$. Thus we concluded that the adsorption behavior of amino acids on octadecyl silica is more uniform than that on bare silica, i.e., all the amino acids studied are adsorbed according to an essentially identical mechanism.

All the preceding thermodynamic data refer to the amino acid adsorption on bare and octadecyl silicas from pure water and by no means are directly applicable to other solvent systems. Both pH variations and even minor additives of other compounds (and all the more components of conventional buffers) can drastically change the adsorption characteristics. This provides almost unlimited opportunities for optimizing the chromatographic resolution of amino acids and related compounds by shifting the ion equilibria, introducing ion-pairing reagents, surfactants, etc. (see, for example, Ref. 27).

One of the most interesting HPLC developments afforded in this way was the resolution of amino acid enantiomers on conventional C_{18} silicas dynamically modified with substituted amino acids containing a long (C_7–C_{18}) linear hydrocarbon radical (e.g., [28]). Such compounds are strongly adsorbed and not washed away by aqueous solvents. After complexation of the adsorbed molecules with copper(II) ions, the resulting stationary phase can be used for an efficient resolution of amino acid racemates. In our studies [29], one of the stationary phases used for this purpose was C_{18} silica dynamically modified with *N*-octyl-L-proline. Eluting racemic amino acids (Val, Nva, Leu, Nle, Phe, Tyr, and Met) with water containing 10^{-4}–10^{-3} M $CuSO_4$, we were able to afford their enantioseparation with the selectivity parameter $\alpha_{D/L}$ [Eq. (5)] ranging from 1.62 (Met, 10^{-3} M $CuSO_4$) to 2.51 (Leu, 10^{-4} M $CuSO_4$). According to the same Eq. (5), it is easy to calculate the corresponding relative free energy changes $\delta(\Delta G^{\circ}_{D/L})$: ca. -1170 to -2240 J mol^{-1}, respectively. A plethora of similar thermodynamic estimates can be derived from the published selectivity values; however, they are considered of secondary importance in HPLC resolution of enantiomers due to merely practical orientation of such works. On the contrary, the chromatographic approach to the determination of relative free energies of interaction between hydrophobic and amphiphilic amino acid side chains, applied by Pochapsky and Gopen [30] to a series of amino acid N-acetyl C-(N'-methyl)amides, can appear useful just from a fundmental point of view, e.g., for studying the thermodynamics of protein folding.

As one can see, a lot of information on the thermodynamics of amino acid adsorption on silicas can be obtained by using the dynamic method. And what is the situation with the traditional batch technique? I am aware of no publications reporting on the amino acid adsorption thermodynamics from batch-measured isotherms. The isotherms alone, however, have been presented for several amino acids and modified silicas, which enables rough estimates of the free energy changes. In particular, Kubota et al. [31] studied the adsorption of glycine, leucine, histidine, and lysine on silica chemically modified with zirconium phosphate and aminobenzenesulfonic groups:

$$\equiv Si-(CH_2)_3-N\!=\!N\!\!-\!\!\left\langle\!\!\bigcirc\!\!\right\rangle\!\!-SO_3H$$

The authors focused on the maximum adsorption capacities for amino acids and did not pay special attention to the initial parts of the isotherms, so that now it is difficult to determine their slopes more or less precisely. Based on the available data, one can say that for the silica modified with aminobenzenesulfonic groups (Fig. 9) the ΔG° values increase in the series Gly < His < Leu < Lys (adsorbability from the diluted solution increases in the reverse order), whereas the maximum adsorption capacity increases in the order Leu < Lys < His < Gly; i.e., there is no direct correlation between the two parameters. For the adsorption on the silica modified with zirconium phosphate groups (Fig. 10), the orders of increasing the maximum adsorption capacity and adsorbability from the diluted solution do not coincide as well, His < Lys < Gly = Leu and Lys < His < Gly < Leu, respectively. On average, the ΔG° values vary in the range of -14.2 to -9.1 kJ mol^{-1}.

B. Linear Peptides

Logically, peptides should be adsorbed stronger than the amino acids constituting them: the longer the peptide chain, the stronger should be the adsorption. Such a cumulative effect ultimately results in a high adsorbability of proteins.

For the simplest case of linear dipeptides adsorbed on bare silica from water, the free energy changes have been estimated from HPLC retention data [17,18]. Most dipeptides, similar to amino acids, have positive ΔG° values and $K < 1$; the exceptions are L-Val-L-Val and Gly-L-Leu (Fig. 11). For the dipeptides derived solely from

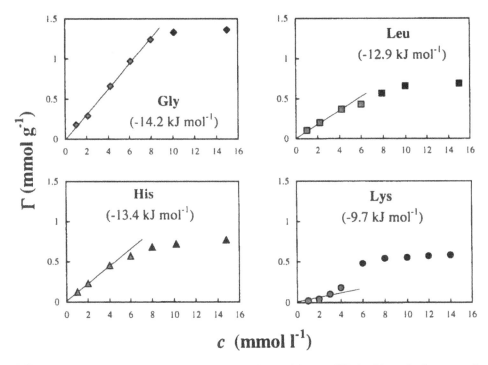

FIG. 9 Adsorption isotherms of the amino acids on silica modified with aminobenzenesulfonic groups and the $\Delta G°$ values derived from the isotherm initial parts. Temperature 25°C; pH 4. (From Ref. 31.)

aliphatic bifunctional amino acids, the dependence of $\Delta G°$ on the number of aliphatic C atoms does not fit well onto a straight line (Fig. 11, thick line). However, if we classify these dipeptides into those derived from Gly (Gly-Gly, Gly-DL-Ala, Gly-L-Val, and Gly-L-Leu) and homodipeptides (again Gly-Gly, Ala-Ala diasteromers, and L-Val-L-Val), two separate linear trends can be obtained (Fig. 11, lower and upper fine lines, respectively). The increments in $\Delta G°$ for each aliphatic C atom are about -390 and -260 J mol^{-1}, respectively, i.e., adsorbability increases more sharply in the first series of dipeptides. A linear interrelation was also found between the $\Delta G°$ values and apparent molar volumes for the dipeptides Gly-Gly, Gly-DL-Ala, and Gly-L-Leu [17].

Similar to the case of amino acids (Section III.A), the presence of heteroatoms in the α-substituent substantially influences the adsorption characteristic of dipeptides [17,18]. The highest $\Delta G°$ value of 1170 J mol^{-1} was found for L-His-L-Leu, containing imidazole ring. The amide grouping of Asn moiety also decreases adsorbability, whereas indolyl and phenyl nuclei increase it. In the series of glycine dipeptides (Gly-DL-Ala, Gly-DL-Asn, and Gly-DL-Phe), the change of Ala to Asn residue increases $\Delta G°$ by 320 J mol^{-1}; the change of Ala to Phe lowers it by 720 J mol^{-1}. A qualitatively similar picture can be found for the alanine dipeptides (DL-Ala-DL-Ala, DL-Ala-DL-Asn and DL-Ala-DL-Trp).

An important aspect of the comparative characterization of amino acid and peptide adsorption behavior is the influence of peptide chain length. From compar-

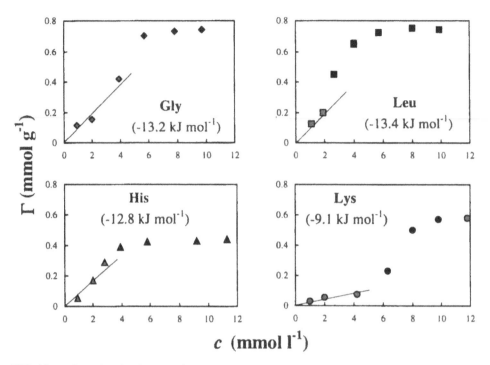

FIG. 10 Adsorption isotherms of the amino acids on silica modified with zirconium phosphate groups and the $\Delta G°$ values derived from the isotherm initial parts. Temperature 25°C; pH 4. (From Ref. 31.)

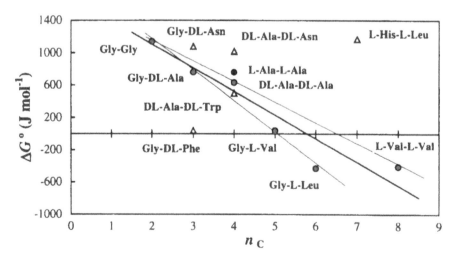

FIG. 11 Plot of $\Delta G°$ versus the number of aliphatic carbon atoms (n_C) in the linear dipeptide molecules for the adsorption on bare silica from water. Dipeptides derived from solely aliphatic bifunctional (●) and other amino acids (△). Lines, see text. Temperature 19°C. (From Ref. 18.)

ison of the data for amino acids and related peptides, it is clearly seen that the latter
have better adsorbability (Figs. 3 and 11). In particular, the difference in the $\Delta G°$
values for Gly and Gly-Gly is 230 J mol^{-1}; for DL-Ala and DL-Ala-DL-Ala, 510 J
mol^{-1}; and for L-Val and L-Val-L-Val, 890 J mol^{-1}. These values are the increments
in $\Delta G°$ for each amino acid residue, and they increase as the number n_C in amino
acid residue increases. If we attribute the increments to the corresponding n_C values,
we obtain approximately the same values for the cited three pairs of solutes: 230,
255, and 220 J mol^{-1} (this result is rather close to that for the series of aliphatic
bifunctional amino acids).

Of special importance (e.g., for understanding the behavior of higher-molecular-weight peptides and proteins) is the question of how the adsorption characteristics
change upon lengthening the peptide chain. For example, when studying the factors
influencing retention of peptides in HPLC, Meek and Rosetti [32] found no general
effect of the chain length. However, the authors considered a large number (100) of
peptides with a random sequence and amino acid composition. For a valid comparison closely related compounds should be chosen, as was done by Molnár and Horváth [15]. Considering the series of L-Ala to its hexamer chromatographed at pH 0.2
and 2.1 on a C_{18} silica, they plotted log k' versus the number of Ala residues to yield
straight lines (Fig. 12). The slopes of the straight lines were much smaller than the
slopes obtained with aliphatic bifunctional amino acids (Fig. 1) under identical chromatographic conditions. The log k' increment of a methylene group of the aliphatic

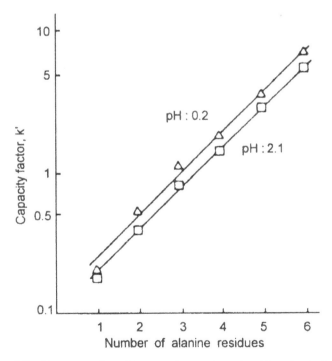

FIG. 12 Plots of log k' versus the number of residues in alanine oligomers. Stationary phase,
LiChrosorb RP-18 octadecyl-silica; eluents: 0.5 M HClO$_4$, pH 0.2, and 0.1 M phosphate buffer,
pH 2.1; temperature 70°C. (From Ref. 15.)

side chain was roughly twice as large as the log k' increment of the Ala residue. For phenylalanine and its oligomers, a similar linear dependence of log k' on the number of Phe residues can be obtained for up to (at least) Phe_5 [33].

A direct linear interrelation between the free energy changes and the number of amino acid residues was found for the adsorption of glycine and its oligomers (up to Gly_4) on bare silica from water (Fig. 13a) [17,18]. The slope gives the $\Delta G°$ increment of -220 J mol^{-1} per one Gly residue. Even Gly_4 appears still to have a positive $\Delta G°$ value (700 J mol^{-1}); extrapolation indicates that only the homopeptides beginning with Gly_8 will have negative $\Delta G°$ values (positive excess adsorption) in this system.

How long can such linear trends be maintained upon lengthening the peptide chain? For the case of reversed-phase HPLC, several research groups reported the linear relationships for peptides larger than eight amino acid residues (see Ref. [34] and references therein). Substantially longer peptides do not obey this regularity. This is generally assumed to be due to stabilization of secondary and tertiary structures in the oligo- and polypeptides, which remove some amino acid residues from contact with the silica surface. Obviously, the peptide chain folding should depend strongly both on the chemistry of silica surface and solvent composition, and no general predictions can be made so far. Even the simplest systems with the shortest peptides can exhibit deviations from the linear interrelation. This can best be exemplified by the retention of the same glycine series on the same bare silica considered in Fig. 13a, but using mixed acetate buffer–acetonitrile solvents instead of pure water [18]. As is seen from Figure 13b, a linear trend is no longer found for the plot of log k' (and consequently $\Delta G°$) versus n_{Gly}.

Short peptide chain folding/unfolding in chromatographic systems is a relatively fast process compared to that for long peptides and proteins. While the former are always detected as a single peak, slow kinetics of long peptide and protein unfolding upon adsorption on the stationary phase results in split peaks corresponding to two stable confirmations [35,36]; one is more stable in the solution, whereas the other is more stable on the surface. The single compound thus can be characterized by two different $\Delta G°$ values; nevertheless, in the present context only the value

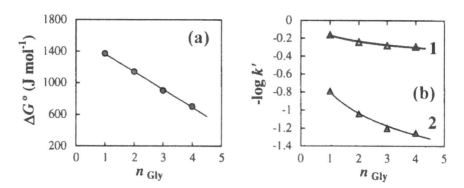

FIG. 13   (a) Plot of $\Delta G°$ versus the number of glycine residues (n_{Gly}) for the adsorption on bare silica from water. Temperature 19°C. (b) Plot of log k' versus n_{Gly}. Mobile phase: acetate buffer solution (pH 5.21)–acetonitrile with volume ratio of (1) 45:55 and (2) 25:75. Temperature 20°C. (From Ref. 18.)

which corresponds to the surface-stable conformation (equilibrium adsorption conditions) is relevant.

All the thermodynamic data discussed relate to the adsorption systems where the eluent composition is maintained the same during the chromatographic process (isocratic elution). For gradient elution, which is of special importance for modern HPLC, other thermodynamic treatment is necessary. One of the basic relationships used for the description in linear gradient elution can be expressed as follows:

$$\log \bar{k} = \log k_0 - S\bar{\Psi} \tag{6}$$

where $\bar{\Psi}$ is the value of the volume fraction of the organic solvent (or the median organic mole fraction) as the solute band passes the center of the column, $\bar{k}$ is the median capacity factor, k_0 is the extrapolated value of $\bar{k}$ for $\bar{\Psi} = 0$, and S is a parameter related to the magnitude of contact surface area between the solute and the stationary phase [36–40]. The median capacity factor $\bar{k}$ thus substitutes k' that is used in the case of isocratic elution.

One example of the thermodynamic characterization of gradient elution systems was presented in a paper by Purcel et al. [37] where oligopeptides (up to nine amino acid residues) related to human growth hormone were chromatographed on octadecyl silica stationary phases. Negative ΔH° values (Table 2) derived from the Van't Hoff plots indicate that heat is released upon the adsorption. At the same time, the ΔS° values found were in some cases negative (-32.8 to -122.2 J mol^{-1} K^{-1} for the open-chain (α- and β-linked peptides) and in some cases positive (2.6 J mol^{-1} K^{-1} for Phe$_5$ and 3.8 J mol^{-1} K^{-1} for the imide, both of which exist as helices in solution), depending on flexibility of the solute molecules. The negative ΔS° values are indicative of an increased ordering of the overall system, and suggest that the open-chain peptides exist in more flexible conformation in solution than on stationary phase, whereas the helical structure is more rigid and constrained [37].

Note, that for the β-linked Leu-Ser-Arg-Leu-Phe-Glu-Asn-Ala-Gly, two sets of the thermodynamic parameters (Table 2) have been derived from the Van't Hoff plot, log k versus $1/T$, having two parts of different slopes which correspond to two different conformations at 5–65°C and 65–75°C. This peptide is composed of nine amino acid residues, and conformational transitions became even more expressed upon lengthening the chain. This is conventionally demonstrated by the plots of log

TABLE 2 Thermodynamic Data for Peptides Chromatographed on the C$_{18}$ Column Using Gradient Elution

Peptide	ΔH° (kJ mol^{-1})	ΔS° (J mol^{-1} K^{-1})
Leu-Ser-Arg-Leu-Phe-Asp-Asn-Ala (imide)	-3.3 ± 0.6	3.8 ± 0.6
Leu-Phe-Asp-Asn-Ala-Gly (α)	-8.7 ± 0.3	-39.7 ± 1.2
Leu-Ser-Arg-Leu-Phe-Glu-Asn-Ala-Gly (β)	-35.4 ± 8.9[a]	-122.2 ± 30.6[a]
	-6.1 ± 1.0[b]	-32.8 ± 5.5[b]
Phe$_5$	-3.3 ± 0.6	2.6 ± 1.1

[a] 65–75°C.
[b] 5–65°C.
Source: Ref. 37.

k_0 (or alternatively S, the contact area parameter) versus temperature (°C) which, however, cannot be used to derive $\Delta H°$ and $\Delta S°$. The small peptide models N-acetylphenylalanine ethyl ester and N-acetyltryptophanamide exhibit a uniform decrease in log k_0 with increase in temperature (Fig. 14a) and therefore no structural perturbations. Phe$_5$, which exists as a helix in solution and has greater conformational flexibility, displays minor fluctuations with temperature (Fig. 14a). For bombesin, composed of 14 amino acid residues, a transition in the range of 5–25°C, corresponding to two different interactive structures, is observed on the C_{18} stationary phase. (On the C_4 phase such transition proceeds at much higher temperatures, as shown in Fig. 14b.) A behavior essentially similar to that of bombesin was found for bovine insulin A chain (21 amino acid residues), whereas the B chain (30 amino acid residues) demonstrated multistep conformational changes within the whole temperature range of 5–85°C remnant of the behavior of the parent bovine insulin (Fig. 14c). Thus the thermodynamic characterization for the adsorption of flexible peptide molecules displaying conformational transitions should be very complicated since each conformation has its own $\Delta H°$ and $\Delta S°$ values. A detailed thermodynamic treatment of such a nonlinear Van't Hoff behavior on hydrophobic silicas was exemplified for biologically active peptides bombesin, β-endorphin, and glucagon, as well as for some synthetic peptides, in Refs. 38, 41, and 42.

C. Cyclic Peptides

Scarce data are available so far on the adsorption thermodynamics of cyclic peptides. The only example studied using the HPLC technique was a series of cyclic dipeptides (or piperazine-2,5-diones) derived from α-alkyl amino acids adsorbed on bare silica from pure water [17,18]. Contrary to amino acid and linear dipeptide adsorption under the same conditions, most piperazinediones exhibit values of $K > 1$ and $\Delta G°$ < 0 (positive excess adsorption); the exceptions are cyclo-Gly$_2$, cyclo-Gly-DL-Ala, and diastereomeric cyclo-Ala$_2$ ($K < 1$ and $\Delta G° > 0$), as well as cyclo-Aib$_2$ ($K = 1$ and $\Delta G° = 0$). Plotting $\Delta G°$ versus the number of aliphatic carbon atoms (n_C), we obtained a linear trend (Fig. 15) similar to those shown for aliphatic bifunctional amino acids (Fig. 3) and related dipeptides (Fig 11). In the present case, the increment in $\Delta G°$ for one aliphatic carbon atom is ca. -510 J mol^{-1}, i.e., adsorbability increases more significantly when increasing the size of the α-substituent than that for amino acids and linear dipeptides (see Sections III.A and B). Comparison of the $\Delta G°$ values found for piperazinediones and related linear compounds shows that the cyclization of the latter into the former results in $\Delta G°$ decrease by -70 J mol^{-1} for Gly-Gly $\rightarrow$ cyclo-Gly$_2$; by -180 J mol^{-1} for DL-Ala-DL-Ala $\rightarrow$ cyclo-(DL-Ala)$_2$; and by -990 J mol^{-1} for L-Val-L-Val $\rightarrow$ cyclo-(L-Val)$_2$. In other words, amino and carboxylic groups reduce adsorbability of amino acids and related peptides on bare silica not only being present in the α-substituent: this conclusion can be applied to the terminal groups as well.

D. Nucleic Acid Bases

Despite the fact that many papers have been published on HPLC resolution of purine and pyrimidine nucleic acid bases, they have not focused on the thermodynamic aspects. Much retention data have been generated (e.g., Refs. 43 and 44) using

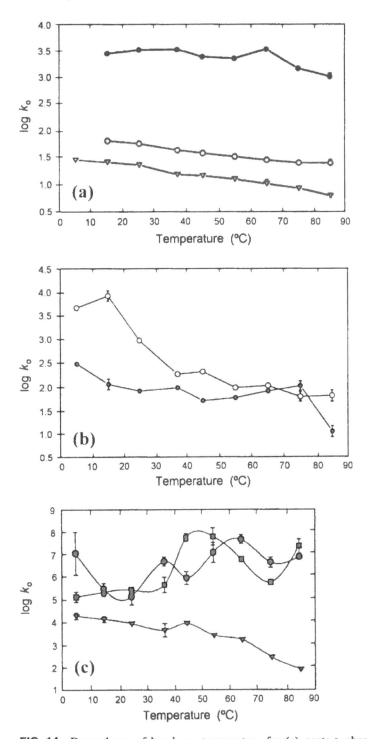

FIG. 14 Dependence of log k_0 on temperature for (a) penta-L-phenylalanine (●), N-acetyl-phenylalanine ethyl ester (○) and N-acetyltryptophanamide (▽) chromatographed on C_{18} silica; (b) bombesin chromatographed on C_{18} (○) and C_4 (●) silica; (c) bovine insulin (●) and its A-(▼) and B-chain (■) chromatographed on C_{18} silica. For the details of gradient elution, see Refs. 36 and 38. (From Refs. 36 and 38.)

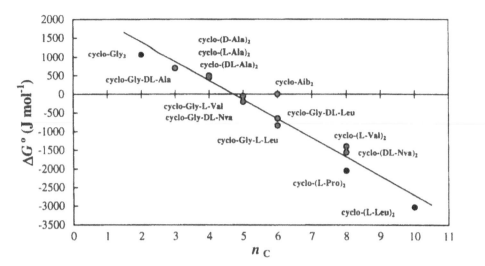

FIG. 15 Plot of $\Delta G°$ versus the number of aliphatic carbon atoms (n_C) in the piperazine-2,5-dione molecules for the adsorption on bare silica from water. Temperature 19°C. (From Ref. 18.)

various silica stationary phases (bare silica, C_2, C_8, and C_{18}, of different surface and porosity properties) and solvent systems; however, they were not treated to derive the thermodynamic characteristics. Some qualitative comments on the relative behavior of the solutes were provided. In particular, in the normal-phase chromatography (bare silica and mixed dichloromethane/methanol/aqueous eluents) [43], contrary to the reversed-phase mode, methyl substituents generally decrease the retention parameters; thus the adsorbability also decreases ($\Delta G°$ increases). For the reversed-phase sorbents, the retention time is longer (and adsorption is correspondingly higher) on C_{18} than on C_8; at the same time no significant differences have been found comparing the C_2 and C_8 materials [44].

Apparently, the brief communication by Kazakevich and El'tekov [45] remains after more than a decade the only work containing direct thermodynamic estimates for nucleic base adsorption. Using the chromatographic method, the $\Delta H°$ values have been determined for the adsorption of adenine, thymine, uracil, and cytosine on phenyl silica from solution of acetonitrile in water (20 vol%): 32.2, 15.9, 4.2, and 6.7 kJ mol^{-1}, respectively.

Driving forces for the purine adsorption on C_{18} and CN silicas from acetonitrile–water solutions were analyzed (in terms of chromatographic retention only, without thermodynamic estimates) by El'tekova and El'tekov [46], taking the parent purine, adenine, guanine, and xanthine as test adsorbates. Electrostatic, dispersion, dipole–dipole, and hydrophobic forces all contribute to the adsorption process. In particular, the dispersion interactions contribute more significantly on the hydrophobic C_{18} surface than on the CN silica, giving rise to 1.5 to 2.5-fold increase in the capacity factors. On the other hand, the presence of nitrile groups on the latter silica increases the contribution from dipole–dipole interactions. The purine structure has a pronounced effect on the retention characteristics as well.

E.   Nucleosides

The situation outlined in the first paragraph of the previous section can be also applied to nucleosides. From the results of the HPLC retention studies one can conclude that ribonucleosides are systematically adsorbed stronger than their parent bases and the corresponding 2'-deoxyribonucleosides when chromatographed in the normal-phase mode [43], although no direct thermodynamic data have been presented.

Some results have been afforded using the static (batch) approach [17,47]. Figure 16 shows that the isotherms for adenosine and inosine adsorption on bare silica from water have essentially the same shape (Langmuir-type), with about the same maximum adsorption capacity (ca. 45 nmol m^{-2}) and $\Delta G°$ values (-14.2 and -13.5 kJ mol^{-1}, respectively). Since the evident structural difference, i.e., the amino group in adenosine against the oxo group in inosine, does not influence noticeably the adsorption properties, one can assume that the purine nuclei do not contact the silica surface and, therefore, the carbohydrate moiety is responsible for the adsorption process. This assumption can be supported by analogous results obtained for the adenosine and inosine adsorption on γ-aminopropyl silica from water [47], though the corresponding $\Delta G°$ values were found to be -12.9 and -11.5 kJ mol^{-1}, respectively (Fig. 17), i.e., a little higher than in the case of bare silica.

F.   Nucleotides

In the same works [17,47], the adsorption isotherms were measured for adenosine-5'-triphosphate and uridine-5'-triphosphate on bare and γ-aminopropyl silica from water. On bare silica, the nucleotides had lower values of maximal adsorption capacity (31 and 35 nmol m^{-2} for UTP and ATP, respectively) than the nucleosides (Fig. 16). This fact can be explained by the substantially bigger size of the nucleoside triposphate anions compared to that for the nucleosides by themselves, and their mutual electrostatic repulsion in the adsorbed phase. The calculated free energy changes are -11.9 kJ mol^{-1} (ATP) and -13.9 kJ mol^{-1} (UTP).

For the case of γ-aminopropyl silica (Fig. 17), the results are quite different. The maximum adsorption capacity for ATP, 140 nmol m^{-2}, despite the difference in the molecular size, is almost four times higher than the corresponding value for the parent nucleoside, 36 nmol m^{-2}. A big difference in their $\Delta G°$ values was also found: -17.6 versus -12.9 kJ mol^{-1}, respectively. Strong electrostatic attraction of the triphosphate anions to the surface aminopropyl groups is the most obvious reason for both phenomena. One could expect that UTP, due to a smaller molecular size, would have even higher adsorption capacity than ATP has. Actually the reverse effect is observed, which is hard to explain. As regards $\Delta G°$, the value found for UTP is -18.2 kJ mol^{-1}, i.e., slightly lower than for ATP.

One should note that studying these adsorption systems with nucleosides and nucleotides using the HPLC approach was inconvenient. The compounds were strongly retained on the bare silica column when eluted with pure water and gave very diffuse, poorly recognizable peaks [17,47]. Nevertheless, for buffer solutions with a higher ionic strength and organic additives typically used in HPLC (see, e.g., Ref. 44 and references therein), the dynamic method should be indispensable.

Nucleotides, and especially the diphosphates and triphosphates, are rather "fragile" compounds and require certain precautions in their handling. It is best to

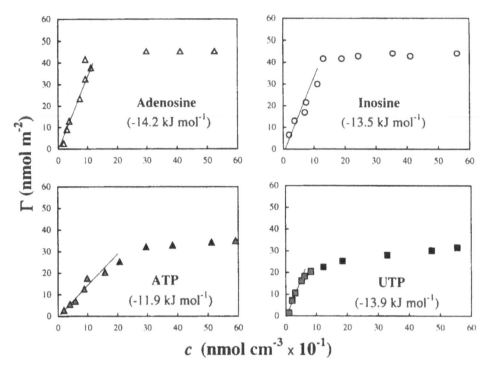

FIG. 16 $\Delta G°$ values determined from the isotherm initial parts for adsorption of adenosine, inosine, adenosine-5'-triphosphate, and uridine-5'-triphosphate on bare silica from water. Detection: UV at 270 nm. Temperature 15°C. (From Ref. 17.)

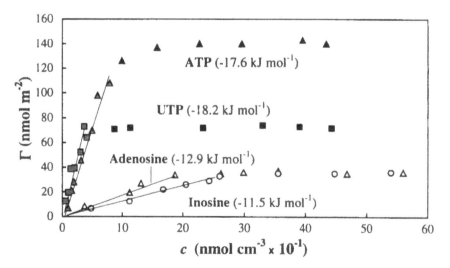

FIG. 17 $\Delta G°$ values determined from the isotherm initial parts for adsorption of adenosine, inosine, adenosine-5'-triphosphate, and uridine-5'-triphosphate on γ-aminopropyl silica from water. Detection: UV at 270 nm. Temperature 15°C. (From Ref. 47.)

buffer their solutions (unless unbuffered systems are of special interest) and to not expose the latter for a long time at even ambient temperatures. Otherwise one cannot be sure that after awhile the triphosphate remains triphosphate, and not monophosphate or diphosphate. What can result is clear: the adsorption isotherm and other characteristics will relate to the decomposition product(s). Take ATP as an example. The hydrolysis of its phosphate linkages can produce ADP or AMP and one or two phosphate anions. If UV absorption is used to determine equilibrium concentrations of the nucleotide in solution, an isotherm remnant of the ATP and adenosine isotherms should be obtained; adenine nuclei are detected, the total amount of which in the system remains the same. But if phosphate moieties are measured, the equilibrium concentration might appear to be overestimated by up to three times (at the extreme, one AMP molecule plus two phosphate anions). In particular, the method of phosphate ionometry was used in Ref. 48 to study the adsorption of biochemically significant phosphates on bare silica from unbuffered solutions. Apparently only the hydrolysis of phosphate linkages was the reason that ATP and ADP were found to be adsorbed very weakly, practically not adsorbed at all, strikingly differing from AMP (Fig. 18; for easy comparison, the scale has been adapted to that used in Figs. 16 and 17).

No direct data are available on the thermodynamic behavior of oligonucleotides, beginning with dinucleotides. From the HPLC retention data and simple chromatograms presented elsewhere (e.g., Refs. 49 and 50), one can expect essentially similar regularities to those found for oligopeptides (Section III.B), at least to some limit of the chain length. Results of a recent study of the adsorption of fluorescently tagged oligonucleotide 5′-GTC AAG GCT GCC CAA TTT GAG-3′, which encodes a region of the human FcgRIIA gene, on γ-aminopropyl–derivatized porous glass from phosphate buffer of pH 7.4 [51] suggest that this oligonucleotide (21 nucleotides long) still has a rigid rod structure, lying upon adsorption along its long axis parallel to the surface. The adsorption isotherm (Fig. 19) is a Langmuir isotherm, as in the case of the triphosphates (Figs. 16 and 17). At the same time, a free energy change calculated for the oligonucleotide (-10.4 kJ mol^{-1}; Fig. 19) appears to be much higher than the $\Delta G°$ values for ATP and UTP adsorption on the same γ-aminopropyl surface (Fig. 17) due to the strong effect of the phosphate buffer. In

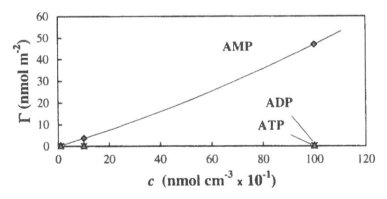

FIG. 18 Isotherms of AMP, ADP, and ATP adsorption on bare silica from unbuffered aqueous solutions as measured by phosphate ionometry. (From Ref. 48.)

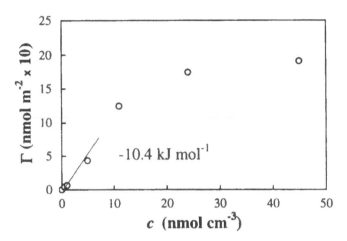

FIG. 19 Adsorption isotherm of fluorescently tagged oligonucleotide 5'-GTC AAG GCT GCC CAA TTT GAG-3' on γ-aminopropyl–derivatized porous glass from phosphate buffer of pH 7.4. (From Ref. 51.)

this particular case the adsorption was reversible due to high ionic strength. On the other hand, one can expect irreversible adsorption of the same oligonucleotide from solutions of low ionic strength (e.g., water as the simplest case) due to the formation of polyphosphate salts with the surface γ-aminopropyl groups. Adsorption reversibility should necessarily be verified prior to determining equilibrium constants from the initial isotherm slopes and then deriving any thermodynamic data, since in the case of irreversible binding the conventional Langmuir model cannot be applied (e.g., for long-chain DNA adsorption [52]).

IV. CONCLUSIONS AND FUTURE PROSPECTS

It is clearly seen that the main body of thermodynamic data for biomolecular adsorption on silica has been afforded for amino acid adsorption on the surface of chromatographic sorbents. This is quite understandable taking into account the role of HPLC in the modern practice of amino acid analysis and separation. Much thermodynamic data have been derived (basically for the adsorption on bare and octadecyl silica from water). As an example of structure–property relationships, dependence of the Gibbs free energy on the amino acid structure (carbon number) has been shown. Also, compensation behavior has been shown to be applicable for amino acid series adsorption on bare and octadecyl silica. As a whole, the HPLC approach has proved to be the most useful to study thermodynamics of amino acid adsorption on silicas; some data, nevertheless, can be afforded from adsorption isotherms obtained from batch measurements. At the same time, some descrepancies can be revealed in the thermodynamic data presented by different groups, thus indicating the necessity of their revalidation.

Thermodynamic characterization of small peptide adsorption is so far at a less advanced level. Perhaps the most important generalization remains a linear dependence of the retention values, equilibrium constants and free energy changes on the number of amino acid residues in homopeptides, which is maintained to a certain limit of peptide chain length. Naturally, there is a great similarity in adsorption behavior of amino acids and related peptides, which can be found in the dependence of $\Delta G°$ on the carbon number of peptide molecules.

Regarding nucleic acid components, only scarce data have been reported, derived mostly from the batch-measured adsorption isotherms. There is much to do in this area; of special interest would be systematic studies on how adsorption characteristics change from pyrimidine and purine bases to related nucleosides, nucleotides, and further to oligonucleotides.

To conclude, studying the relative behavior within a class of compounds can be much more useful to understand the interfacial behavior of biomonomers and biopolymers than attempts to estimate particular thermodynamic values for a given compound, which are often insufficiently precise and meaningful only for a given adsorption system.

REFERENCES

1. A. Fallon, R. F. G. Booth, L. D. Bell, *Applications of HPLC in Biochemistry*. Elsevier, Amsterdam, 1987.
2. C. T. Mant and R. S. Hodges, eds., *High Performance Liquid Chromatography of Peptides and Proteins: Separation, Analysis and Conformation*. CRC Press, Boca Raton, FL, 1991.
3. E. D. Katz, ed., *High Performance Liquid Chromatography: Principles and Methods in Biotechnology*. Wiley, Chichester, 1996.
4. G. D. Parfitt, C. H. Rochester, eds., *Adsorption from Solution at the Solid/Liquid Interface*. Academic Press, London, 1983.
5. H.-H. Kohler. In: B. Dobias, ed., *Coagulation and Flocculation*. Marcel Dekker, New York, 1993, pp. 1–36.
6. J. Torrent, F. Sanz. J. Electroanal. Chem. *286*:207–215 (1990).
7. C. H. Giles, T. H. MacEwan, S. N. Nakhwa, D. Smith. J. Chem. Soc. 3973–3993 (1960).
8. C. H. Giles, D. Smith, A. Huitson. J. Colloid Interface Sci. *47*:755–765 (1974).
9. R. P. W. Scott. In: P. R. Brown, R. A. Hartwick, eds., *High Performance Liquid Chromatography*. Wiley, New York, 1989, pp. 117–143.
10. B. Karger. In: J. J. Kirkland, ed., *Modern Practice of Liquid Chromatography*. Wiley, New York, 1971, Chap 1.
11. K. S. Yun, C. Zhu, J. F. Parcher. Anal. Chem. *67*:613–619 (1995).
12. N. A. Eltekova, D. Berek, I. Novak. Zh. Fiz. Khim. (Russian) *63*:2675–2678 (1989).
13. Y. V. Kazakevich, H. M. NcNair. J. Chromatogr. Sci. *31*:317–322 (1993).
14. S. Ahuja, ed., *Chiral Separations by Liquid Chromatography*. American Chemical Society, Washington, D.C, 1991.
15. I. Molnár, C. Horváth. J. Chromatogr. *142*:623–640 (1977).
16. V. A. Basiuk, T. Yu. Gromovoy, E. G. Khil'chevskaya. Polish J. Chem. *68*:777–781 (1994).
17. V. A. Basiuk, T. Yu. Gromovoy, E. G. Khil'chevskaya. Origins Life Evol. Biosphere *25*: 375–393 (1995).
18. V. A. Basiuk, T. Yu. Gromovoy. Adsorption *2*:145–152 (1996).
19. M. S. Uddin, K. Hidajat, C.-B. Ching. Ind. Eng. Chem. Res. *29*:647–651 (1990).
20. V. A. Basiuk, T. Yu. Gromovoy. Colloids Surfaces A *118*:127–140 (1996).
21. W. Melander, D. E. Campbell, C. Horváth. J. Chromatogr. *158*:215–225 (1978).

22. A. Vailaya, C. Horváth, J. Phys. Chem. *100*:2447–2455 (1996).
23. R. R. Krug, W. G. Hunter, R. A. Grieger. J. Phys. Chem. *80*:2335–2341 (1976).
24. R. R. Krug, W. G. Hunter, R. G. Grieger. J. Phys. Chem. *80*:2341–2351 (1976).
25. J. P. Greenstein, M. Winitz. *Chemistry of the Amino Acids,* Vol 1. RE Kreiger Publishing, Malabar, 1986, p. 486.
26. M. C. J. Wilce, M.-I. Aguilar, M. T. W. Hearn. Anal. Chem. *67*:1210–1219 (1995).
27. A. H. Rodgers, M. G. Khaledi. Anal. Chem. *66*:327–334 (1994).
28. V. A. Davankov, A. S. Bochkov, Yu. P. Belov. J. Chromatogr. *218*:547–557 (1981).
29. V. A. Basiuk, A. A. Chuiko. J. Chromatogr. *521*:29–42 (1990).
30. T. C. Pochapsky, Q. Gopen. Protein Sci. *1*:786–795 (1992).
31. L. T. Kubota, A. Gambero, A. S. Santos, J. M. Granjeiro. J. Colloid. Interface Sci. *183*: 453–457 (1996).
32. J. L. Meek, Z. L. Rossetti. J. Chromatogr. *211*:15–28 (1981).
33. M. T. W. Hearn, B. Grego. J. Chromatogr. *296*:309–319 (1984).
34. C. T. Mant, T. W. L. Burke, J. A. Black, R. S. Hodges. J. Chromatogr. *458*:193–205 (1988).
35. K. Benedek. J. Chromatogr. *646*:91–98 (1993).
36. A. W. Purcell, M. I. Aguilar, M. T. W. Hearn. J. Chromatogr. A *711*:71–79 (1995).
37. A. W. Purcell, M. I. Aguilar, M. T. W. Hearn. J. Chromatogr. *476*:125–133 (1989).
38. A. W. Purcell, M. I. Aguilar, M. T. W. Hearn. J. Chromatogr. *593*:103–117 (1992).
39. A. W. Purcell, M.-I. Aguilar, M. T. W. Hearn. Anal. Chem. *65*:3038–3047 (1993).
40. B. de Collongue-Poyet, C. Vidal-Madjar, B. Sebille, K. K. Unger. J. Chromatogr. B *664*: 155–161 (1995).
41. A. W. Purcell, G. L. Zhao, M. I. Aguilar, M. T. W. Hearn. J. Chromatogr. *852*:43–57 (1999).
42. R. I. Boysen, Y. Wang, H. H. Keah, M. T. W. Hearn. Biophys. Chem. *77*:79–97 (1999).
43. M. Ryba, J. Beránek. J. Chromatogr. *211*:337–346 (1981).
44. A. Rizzi, H. R. M. Lang. J. Chromatogr. *331*:33–45 (1985).
45. Yu. V. Kazakevich, Yu. A. El'tekov. Russ. J. Phys. Chem. *61*:233–234 (1987).
46. N. A. El'tekova, Yu. A. El'tekov. Russ. J. Phys. Chem. *74*:1324–1329 (2000).
47. V. A. Basiuk, T. Yu. Gromovoy, E. G. Khil'chevskaya. Polish J. Chem. *69*:127–131 (1995).
48. K. Hamdani, K. L. Cheng. Colloids Surfaces *63*:29–31 (1992).
49. M. Colpan, D. Riesner. J. Chromatogr. *296*:339–353 (1984).
50. R. Bischoff, L. W. McLaughlin. J. Chromatogr. *296*:329–337 (1984).
51. V. Chan, D. J. Graves, P. Fortina, S. E. McKenzie. Langmuir *13*:320–329 (1997).
52. K. A. Melzak, C. S. Sherwood, R. F. B. Turner, C. A. Haynes. J. Colloid Interface Sci. *181*:635–644 (1996).

4

Quantitative Modeling of Protein Adsorption

CHARLES M. ROTH Massachusetts General Hospital, Harvard Medical School, and Shriners Burns Hospital, Boston, Massachusetts, U.S.A.

ABRAHAM M. LENHOFF University of Delaware, Newark, Delaware, U.S.A.

I. INTRODUCTION

The general principles that have been deduced from experimental observations of protein adsorption are covered in several major review articles [1–3]. Our focus here is more specific, namely, on the development of a *quantitative* understanding of the mechanistic bases driving proteins to accumulate at solid–liquid interfaces and on the ability to predict extent of adsorption from physicochemical properties of the protein, aqueous solvent, and sorbent. Such a capability would allow rational approaches to be developed toward the solution of problems involving protein adsorption. For example, a detailed understanding of how the properties of protein and surface are manifested in their adsorption behavior would aid in the design or selection of materials that selectively adsorb proteins onto micropatterned surfaces for tissue engineering and biosensor applications. For protein chromatography, a quantitative understanding of the extent of adsorption as a function of protein and adsorbent properties and solvent strength would improve the process of selecting adsorbents and conditions most amenable to particular protein separations.

One key factor that distinguishes protein adsorption to biomaterials from adsorption to chromatographic stationary phases is the extent to which the adsorption is reversible. The adsorption of blood proteins, which often possess rather flexible structures, to the polymeric substrates usually used as biomaterials has been found to be irreversible in many cases; often conformational change occurring over long contact times is associated with it [1–3]. On the other hand, the ability to elute proteins from a chromatographic column, particularly in isocratic operation, demonstrates that protein adsorption in these systems is essentially reversible. Because the current level of quantitative understanding regarding protein adsorption is not very advanced, we focus here on systems of the latter type. The understanding of protein adsorption that does exist currently is in the form of basic thermodynamic concepts, such as a change in the Gibbs free energy upon adsorption [4]; dealing with irreversible adsorption within a thermodynamic framework is beyond the scope of present capabilities. In practical terms, this excludes adsorption in which conformational change and/or strong hydrophobic interactions are significant. Nonetheless, understanding of the mechanisms driving protein adsorption will be necessary in understanding these more complex situations as well.

Here, we focus on aqueous systems for which protein adsorption occurs as the net result of protein molecular interactions with the adsorbent relative to those of the protein and sorbent with their solvent. Electrostatic interactions are very important in this case, as many of the residues of a protein are ionized at any given pH. Dispersion, or van der Waals, forces are almost always attractive and so form a weak but nonnegligible contribution to adsorption. Solvation and steric forces may be significant as well. These different contributions are discussed in more detail in Section II.

Given the contribution of each of these mechanisms to the interactions that govern adsorption, one needs a framework to combine the interaction energies or forces that are computed with information regarding the formation and structure of the adsorbed layer. Two classes of approach exist: one in which a thermodynamic framework utilizes configurational averaging to estimate bulk properties and a second in which the adsorption process is simulated from the interaction potentials (forces and energies). In either case it is possible to obtain descriptions of adsorption that can be related to protein and adsorbent properties and can thus be applied with some predictive capability. The descriptions sought are limited to ones of the adsorbed amount, which is usually of principal experimental interest. In Section III, these mechanistic models for predicting adsorption behavior are reviewed.

Early models for protein adsorption did not account for structural aspects of the protein or sorbent, but instead treated the protein as a monofunctional chemical entity and the sorbent as a collection of chemical sites. In this approach, isotherms are generated via mass action equations from the proposed reaction, and slight variations result from altering the description of the reaction stoichiometry. However, for mechanistic understanding and a priori prediction of adsorption behavior, more detailed models accounting for structural aspects of the protein and/or surface are required.

Protein structures and their associated behavior are quite complex. The amino acids that comprise proteins span a wide range of chemical functions, and a particular protein is likely to contain each of the 20 major amino acids somewhere within it. Therefore, a folded protein is likely to have different regions that are hydrophilic or hydrophobic; positively charged, negatively charged, or neutral; internal or exposed to solvent; and functional or structural. Because of this heterogeneity at the amino acid residue level, chemical treatments of protein adsorption are inadequate.

With recent advances in computational capabilities, incorporation of atomic information for both proteins and surfaces into mechanistic models of protein adsorption has begun to be tractable in some cases. Molecular mechanics computations and molecular dynamics simulations have become commonplace in modeling smaller molecules in aqueous solvents, but their application to macromolecules, including proteins, is still limited by computational resources. Nonetheless, atomistic models of proteins and polymer surfaces have been used in a few cases to calculate adsorption energies [5–7] and to investigate dynamic [8] and conformational [9] effects involving the adsorption of proteins onto polymer surfaces.

These models have provided some insight into the relative contributions of various types of interactions (e.g., electrostatic, dispersion, solvation) to protein–polymer interactions, but the problem of simulating entire protein molecules and their adsorbents with sufficient water and ions to be realistic remains too demanding for extensive study. Consequently, a number of approaches have been applied in which the protein molecule is treated atomistically, but continuum approximations are made for the solvent and/or adsorbent. Noinville et al. [10] used atomistic descriptions of protein and poly(vinylimidazole) polymer, the force field parameters of the latter determined via quantum mechanical computations. Solvent was included by modifying the intermolecular potentials (AMBER force field [11]) by a distance-dependent dielectric constant and by an exponential screening term that accounts for the double-layer screening with increasing ionic strength. Others have also described protein and polymer atomistically, using interatomic potentials designed to account for the presence of intervening water [5]. The solvation energy absent from such a description was added by a fragment method based on aqueous–organic transfer free energies of constituent amino acids [6,7]. This approach has been further extended to account for the presence of grafted polymer chains on the adsorbent substrate [12].

Continuum, or colloidal, methods represent an alternative to atomistic approaches that are particularly advantageous in their ability to deal with the problems of solvent and electrolyte. The particle size over which a colloidal description is appropriate ranges from about a nanometer to a micrometer. Most proteins fall toward the lower end of this range but are certainly within it. Since the sizes are small, protein interaction energies are relatively small as well, and consequently proteins can be sensitive to minor changes in environment. Choosing a colloidal approach engenders a few assumptions. First, treating a protein molecule as a rigid particle makes it virtually impossible to account for conformational change in the adsorption process. While protein lability can be an important driving force for adsorption [13], predicting conformational changes is beyond the range of current capabilities. Second, colloidal approaches deal primarily with nonspecific interactions. Specificity is included only with respect to strengthening of dispersion interactions via complementarity of shape. For example, it is not possible to assess, from the point of view of particle–surface interactions, changes in the hydrogen-bonding state of protein, surface, or solvent. Third, all materials involved in the adsorption process are treated as continua. Consequently, properties of multifunctional adsorbents are taken into account only in an average or statistical sense. Other solutes are generally treated as part of the solvent medium, particularly ions, which are treated as point charges and lumped into the Debye screening parameter, which depends on the ion valence and concentration but not on its specific chemical type. Overall, the implication is that a colloidal description is good for describing the effect of nonspecific interactions on bulk association but is not useful for specific effects, strongly hydrophobic sorbents, or adsorption driven by conformational change.

In the remainder of this chapter, we emphasize a molecular understanding of the governing interactions as a fundamental basis for analyzing and manipulating the bulk adsorption and chromatographic behavior of proteins. These interactions can be described from either atomistic or colloidal perspectives, and the choice of formulation depends on the extent of structural information and computational resources available. We review quantitative approaches to describing the molecular interactions involved in protein adsorption in both atomistic and colloidal frameworks. Thereafter, we proceed to use these methods as the basis for calculating adsorption equilibrium, and isotherms. A suitable thermodynamic framework allows a quantitative model for interactions at the molecular (or, from the colloidal view, particulate) level to be used to make predictions or interpretations regarding bulk behavior, which can be tested experimentally.

II. PROTEIN–SURFACE MOLECULAR INTERACTIONS

A. Electrostatics

At the atomic level, electrostatics are treated pairwise, with the potential energy U_{ij} of each charge pair following from Coulomb's law. Thus, the electrical energy U_{elec} of a configuration of atoms is

$$U_{elec} = \sum_{i<j} \frac{q_i q_j}{\varepsilon r} \tag{1}$$

where q_i and q_j are the charges on any two atoms i and j, ε is the dielectric permittivity of the intervening medium, and r is the distance between charge centers. The source of charges within the protein usually consists not just of the fixed charges resulting from titratable amino acid residues, but also of partial charges of the polypeptide backbone and side chains. The electrical energy contains contributions from both intramolecular and intermolecular Coulombic interactions; under the assumption of a rigid body, the intramolecular interactions can be omitted from the summation. Intermolecular interactions in aqueous media are strongly mediated by the presence of water and ions. In a dynamic simulation, these can be included explicitly; however, in a static calculation, approximations for their effects on interatomic charge interactions are introduced. The presence of water is treated by inclusion of the solvent dielectric constant between each pair of atoms in Eq. (1); sometimes a distance-dependent value is used. The presence of electrolyte can be approximated by scaling the interactions in Eq. (1) by an exponentially decaying function of distance, with the Debye length as the decay length.

The problems of intervening solvent and electrolyte are often not suitably addressed with the aforementioned approximations, and as a result continuum treatments for electrostatics have been developed. These generally build on the cavity dielectric model of Kirkwood [14], in which the protein is treated as a low-dielectric body immersed in a continuous electrolyte solution. The interior of the protein is governed by the Poisson equation:

$$\nabla^2 \phi = -\frac{\rho}{\varepsilon} \tag{2}$$

which relates the electrical potential ϕ in a region of dielectric permittivity ε to the distribution of charges ρ in that region. For interactions in an electrolyte solution, the free ions are assumed to follow a Boltzmann distribution, leading to the Poisson–Boltzmann equation (PBE):

$$\nabla^2 \phi = -\frac{1}{\varepsilon} \sum_i \rho_i^o z_i e \exp\left(-\frac{z_i e \phi}{kT}\right) \tag{3}$$

where k is the Boltzmann constant, T the temperature, z_i the valency of ion species

i, e the magnitude of the electronic charge, and ρ_i^o the bulk concentration of species i. The linearized form of Eq. (3) (LPBE) is used frequently, and while linearization is strictly valid only for very low potentials, surprisingly good agreement between energies computed using the PBE and the LPBE has been found for a variety of double-layer calculations [15–18]. Solution of these coupled equations provides the electrostatic potential distribution throughout space, from which the electrostatic interaction energy is calculated by integration of the potential changes at each charge site [19,20] using an expression of the form

$$F = \frac{1}{2} \int_{\partial P} \sigma \phi^e \, d\mathbf{A} + \frac{1}{2} \sum_{k=1}^{n} q_k \phi^i(\mathbf{x}_k) \tag{4}$$

Here, the superscripts e and i denote exterior and interior potentials, σ is the charge density on a uniformly charged planar adsorbent surface, $\mathbf{x}_k$ are the locations of the n charges, and ∂P represents the planar surface.

For a geometrically complex particle such as a protein, the governing electrostatic equations must be solved numerically. A number of programs for solution of these equations in and around protein molecules have been developed. Early implementations relied almost exclusively on the finite difference technique [21,22], but alternatives such as finite elements [23] and the boundary element method [18,20, 24–27] have been developed. While these programs are generally rather computer intensive, one calculation is able to give a wealth of information in terms of the electrostatic potential distribution everywhere in space. The growing literature on this subject has addressed many of the key issues in the model, including the boundary condition at the protein–solvent interface, intramolecular interactions among charged groups (pK_a shifts) [28], and the dielectric constant of the protein [20,30].

For the continuum methods described, the electrostatic potential distribution and the free energy of interaction are both useful in explaining adsorption behavior. The electrostatic potential distribution alone has provided some qualitative explanations of protein adsorption orientation and layer structure. The potential distribution around ribonuclease A, along with its interaction energy, has been found to favor its adsorption to a negatively charged surface in a side-on orientation with its active site facing toward the surface [31]. Furthermore, since the opposite side of the ribonuclease molecule features a negative potential, an adjustment to an end-on orientation for close-packed adsorbed molecules could be stabilized by favorable electrostatic interactions. This proposed mechanism was used to explain observed structural changes in adsorbed layers of ribonuclease A [31] based on surface forces measurements and enzymatic activity changes. Explicit inclusion of a charged surface in electrostatic calculations [20] corroborated these conditions. In a similar manner, Haggerty and Lenhoff [32] used the electrostatic potential distribution around lysozyme in solution to propose a mechanism for electrostatically favored ordering of protein molecules adsorbed at a surface. While the potential distribution does not allow direct prediction of adsorption extent or adsorbed layer structure, these studies highlight the important role of electrostatics in protein adsorption and complement the quantitative models discussed later.

Molecular detail in finding the free energy of interaction is also informative. One observation is that a striking dependence on orientation is possible, e.g., that some orientations result in favorable energies of interaction while other orientations exhibit unfavorable energies of interaction for proteins and surfaces of opposite net charge. This can be seen in Fig. 1, where the electrostatic free energies as a function of orientation are presented for chymotrypsinogen A (net charge +4 at pH 7) interacting with a uniformly charged anionic surface. A range of interaction energies exists, but in most orientations, chymotrypsinogen A is attracted to the negatively charged surface; such behavior is consistent with its positive net charge. Analogous behavior has been reported for the interaction of rat cytochrome b_5 at pH 8, where it is negatively charged, with an array of model anion exchange ligands represented atomistically [33]. However, the calculated attractive energies in this system are enormous (up to -58 kT), presumably due to the large protein net charge (-9.4 e) and surface charge density (10 $\mu C/cm^2$) employed.

The orientational dependence of the free energy of interaction has also been investigated in a system where the protein and sorbent are both positively charged. Ribonuclease A and cytochrome C have been observed to adsorb to anion exchange (positively charged) surfaces under conditions at which both proteins are positively charged [34]; this behavior has been ascribed to patches of negative electrostatic potential around the generally positively charged proteins. Asthagiri and Lenhoff [35] have computed interaction energies for these proteins in orientations most likely to produce this patch-controlled behavior. When the sorbent was modeled as a surface with uniform charge density, electrostatics computations were unable to reproduce the favorable interaction energies necessary for patch-controlled adsorption. However, when the sorbent charges were modeled as an array of discrete charges or as isolated low-dielectric charged spheres (approximating the topography of polymer chain end groups), favorable interaction energies did result from the model.

Such a pronounced orientational dependence of the interaction energy is not a universal phenomenon. For lysozyme, the dependence of electrostatic energy on orientation (Fig. 2) is different from that for ribonuclease or chymotrypsinogen. Under the same conditions as presented for chymotrypsinogen A, lysozyme exhibits some dependence on orientation, with values ranging from -1.6 to -4.0 kT, but in all cases the electrostatic interaction is attractive. An explanation is provided by the fact

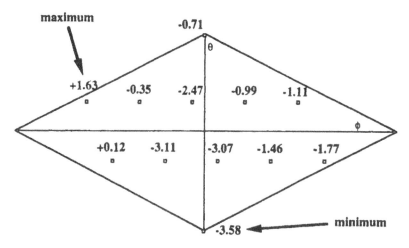

FIG. 1 Chymotrypsinogen A–surface electrostatic interaction energies as a function of orientation (ϕ, θ). Conditions: gap = 7.66 Å, ionic strength = 0.1 M, $\sigma = -4.6$ $\mu C/cm^2$.

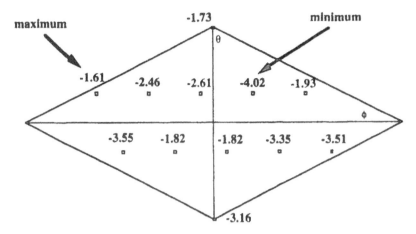

FIG. 2 Lysozyme–surface electrostatic interaction energies as a function of orientation (ϕ, θ). Conditions: gap = 7.66 Å, pH 7, ionic strength = 0.1 M, $\sigma = -4.6$ $\mu C/cm^2$.

that the molecular charges of lysozyme are unusually evenly distributed [36]. One measure of the charge distribution is the dipole moment, which for lysozyme is significantly less than that for chymotrypsinogen A (approximately 70 versus 520 Debye at pH 7).

The electrostatic interaction energy provides a means to assess the importance, relative to other intermolecular forces, of electrostatic interactions in protein adsorption, and often approximate determinations are adequate because of the high computational demand of protein molecular electrostatics computations. The sphere–plate geometry is most convenient, albeit a numerical solution is nontheless required. Exact enumeration of electrostatic potentials and energies for a sphere and plate are useful in capturing the essential physics of colloidal particle–surface electrostatic interactions. The main drawback in application to proteins is the neglect of the heterogeneity of the charge distribution within a protein molecule that is likely to favor certain configurations in protein–surface interactions.

An intermediate representation of protein electrostatics can be made by incorporating the dipole moment of the charge distribution into the sphere model [37]. For proteins with a relatively symmetric charge distribution, such as lysozyme, the addition of the dipole moment is a relatively small perturbation and the interaction energies have a fairly narrow distribution, with the most favorable orientation being that in which the positive end of the dipole faces the surface (Fig. 3). This is in fairly good agreement with the protein computations of Fig. 2, but the sphere–dipole model underestimates the maximum free energy somewhat. For proteins with an asymmetric charge distribution, such as chymotrypsinogen A, the addition of the dipole moment to the charge distribution is a large perturbation, with the result that the distribution of energies is quite large and actually overestimates the range observed for the protein as a whole (Fig. 4). Nonetheless, incorporation of the dipole moment is relatively simple and seems to improve the accuracy of the characterization of electrostatic energies. Higher-order moments of the charge distribution could be incorporated as well. An alternative approach has been developed by Grant and Saville [38], who represented lysozyme as a sphere with the charge distribution

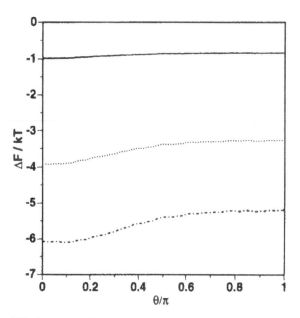

FIG. 3 Interaction energies for lysozyme, represented as a sphere with charges mimicking the protein's monopole and dipole moments, as a function of dipole orientation, indicated by the angle θ between the dipole axis and the surface normal. Conditions: $R = 15.6$ Å, pH 7 (net charge = +8, dipole moment = 72 D), ionic strength = 0.1 M, $\sigma = -4.6$ μC/cm^2. (———) gap = 21.98 Å; ($\cdots$) gap = 7.66 Å; (—·—·—·) gap = 2.15 Å.

smeared onto a finite set of patches on its surface. This model also is able to produce orientation-dependent energies that match those calculated from detailed protein electrostatics semiquantitatively.

For some applications, an approximate analytical solution to the sphere-plate electrostatics equations is used. The most commonly employed approximation to the double-layer interaction energy in colloid science is that provided by the linear superposition approximation (LSA). The LSA represents the potential at any point between two charged bodies as the sum of the unperturbed potentials of the interacting bodies. This approximation is essentially correct when the bodies are far apart, but is less valid as the double layers overlap more extensively. Furthermore, the scalings of energy with respect to distance, surface charge density, and to some extent ionic strength predicted by the LSA hold for the models employing structural details, given a particular protein-surface orientation [39]. An analytical LSA result for a sphere with a nonuniform charge distribution interacting with a charged plate has also been developed [97].

B. van der Waals Interactions

van der Waals forces arise from fluctuations in atomic dipoles and the polarizability of constituent atoms within the given materials that are interacting. For an atomistic description of macromolecules, the interaction is described by the Lennard–Jones potential

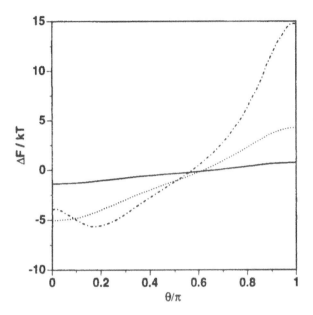

FIG. 4 Interaction energies for chymotrypsinogen A, represented as a sphere with charges mimicking the protein's monopole and dipole moments, as a function of dipole orientation, indicated by the angle θ between the dipole axis and the surface normal. Conditions: $R = 19.3$ Å, pH 7 (net charge = +4, dipole moment = 516 D), ionic strength = 0.1 M, $\sigma = -4.6$ $\mu C/cm^2$. (——) gap = 21.98 Å; ($\cdots$) gap = 7.66 Å; (—·—·—·) gap = 2.15 Å.

$$U_{ij} = \sum_{i<j} \left(\frac{B_{ij}}{r_{ij}^{12}} - \frac{C_{ij}}{r_{ij}^6} \right) \tag{5}$$

where C_{ij} is a constant dependent mainly on the polarizabilities of the two atoms or groups (London dispersion force), but also on their dipole moments (Keesom and Debye forces). The first term in Eq. (5), parameterized by its constant B_{ij}, is the Born repulsion term, discussed in more detail in Section II.D. The total van der Waals energy of a discrete collection of atoms is formed by summation of the second term in Eq. (5) over all pairs of atoms ij, although multibody effects, such as those captured by the Lifshitz theory discussed subsequently, are not properly accounted for.

Atomistic descriptions are suitable for very short range interactions, but are very demanding computationally for the same reason as discussed previously for electrostatic effects, namely, the need to include explicit water molecules. Thus, at somewhat longer range a continuum colloidal approach is usually adopted. The treatment of van der Waals interactions in colloids is usually based on the approach of Hamaker [40], who described the total interaction of two bodies by integrating the characteristic $1/r^6$ dependence of gaseous dispersion interactions over the volume of the interacting bodies. A major consequence of this approach is that the van der Waals interaction energy depends on the product of a material property, named the Hamaker constant A, and a function of geometry. This separation is a direct consequence of the assumption of pairwise additivity inherent in the Hamaker approach. The Lifshitz theory [41] is a rigorous alternative, based on quantum electrodynamics, that accounts for the many-body interactions within interacting continuous bodies. Calculations for the only two tractable geometries—sphere/sphere and plane/plane —suggest that the Hamaker approach accounts correctly for the geometric dependence at close separations [42,43].

The Hamaker integration has been performed for a variety of geometries [44]. The result most often applied to protein adsorption models is that for a sphere and a plate,

$$\Delta F_{vdW} = -\frac{A_{132}}{6}\left[\frac{R}{z} + \frac{R}{2R+z} + \ln\left(\frac{z}{2R+z}\right)\right]$$

(6)

where R is the sphere radius and z the separation distance between the sphere and the plate. It can be readily seen that the van der Waals interaction energy increases with the size of the spherical particle and decays with increasing separation distance. Mathematically, the interaction becomes infinitely attractive as the sphere and plate touch ($z = 0$), but short-range repulsive forces prevent this from occurring in practice, and the validity of the continuum representation is questionable here in any event. The van der Waals interaction is effective only at close range, but in that range it can be quite significant. The consequence for protein adsorption is that van der Waals interactions are probably less important in drawing a protein to a surface than in binding it once it arrives at the surface.

Because of the complex structure of a protein molecule, most analyses of van der Waals interactions involving proteins assume the simplified geometry of a sphere [37,45] or even a plate [46,47]. The effect of protein geometry on van der Waals interactions has been investigated [48]. It was found that the interaction of proteins with planar surfaces is generally disfavored relative to that of spheres and planes because the roughness of the protein surface prevents most of its volume from being close enough to the surface to interact with it [48]. Certain orientations, however, are able to bring a fraction of the protein volume comparable to that of an equivalent sphere near to the surface. The converse, however, is also true: when two bodies have complementary shapes, their van der Waals attraction is enhanced [48]. For application to protein adsorption, the sorbent is frequently not well characterized in terms of molecular topography, and the exact value of the van der Waals contribution is difficult to ascertain.

In order to apply the approach of Hamaker quantitatively, it is necessary to estimate the material property describing the mutual polarizability of protein and surface—the Hamaker constant. The rigorous framework of Lifshitz theory [4] allows the Hamaker constant to be estimated from the frequency-dependent dielectric spectra of the materials involved (including solvent). In particular, the adsorption spectra, along with estimates of the static dielectric constant and the refractive index in the visible region for each of the interacting materials as well as the intervening solvent, are sufficient to obtain a fairly accurate value of the Hamaker constant [49,50].

Relatively little work has been devoted to obtaining the dielectric data required for accurate determination of Hamaker constants, especially for studying proteins. The only complete set of dielectric data is that of Inagaki et al. [51], who measured the dielectric response of bovine serum albumin (BSA) over a wide frequency range. Since most proteins are composed of the same 20 amino acids, it can reasonably be assumed that the BSA data are representative of proteins in general [48]. Using these data, the Hamaker constant for two proteins interacting through water is found to be about 3.1 kT [48]. In aqueous electrolyte solution, the zero-frequency contribution is partially screened. The result is a reduction in the Hamaker constant for two bodies with an intervening aqueous electrolyte of about 0.75 kT; this reduction is independent of the nature of the materials, as long as their static dielectric constants are significantly less than that of water. For the analysis of protein adsorption, Hamaker constants for the interaction of BSA with a selection of materials have been evaluated [48]. The Hamaker constant for protein interactions with other proteins is higher than that for protein interactions with most of the materials on which they are likely to adsorb [48,52].

C. Solvation Forces

Solvation, or hydration, forces are thought to arise from perturbations in the molecular structure of water in the vicinity of dissolved solutes or surfaces that can be either hydrophobic or hydrophilic in nature. We concentrate here on hydrophobic phenomena, which are of major importance for many protein adsorption situations. By coming together, two hydrophobic forces can reduce the total surface area exposed to solvent. The details of solvation interactions are still incompletely understood, and the term "hydrophobic effect" is probably an overused catch phrase. Nonetheless, there is experimental evidence implicating solvent-induced forces [53,54].

Since the amount of protein adsorbed generally increases with surface hydrophobicity [2], with hydrophobicity usually measured by water contact angle, it is sometimes stated that hydrophobic interactions provide the primary driving force for protein adsorption [55]. While this may well be the case for many hydrophobic surfaces such as those used in biomaterials studies, it is probably an overgeneralization. Nonetheless, the hydrophobic driving force is certainly an important one, and the poor understanding of its nature is a major factor in much of the uncertainty and controversy in protein adsorption studies.

In view of the molecular origins of the hydrophobic effect, much effort has been expended on calculations and simulations involving water structure at the molecular level. In molecular dynamics, no hydrophobic force is necessary to produce solvation effects; they are a natural consequence of water as a solvent. Descriptions of solvation effects at the colloidal level are more difficult to model accurately, but are straightforward to implement into quantitative models of protein adsorption. There exist a number of macroscopic theories regarding solvation effects [56,57]. These have in large part been linked to what is commonly called hydrophobic hydration, which is generally characterized for a given species by the free energy of transferring that species from a nonpolar solvent to water. For hydrocarbons, the free energy is positive, reflecting the preference for the nonpolar solvent. The transfer free energy is correlated with the surface area of the solute, with approximately 85–180 J/mol Å^2 being contributed for nonpolar amino acid side chains [58]. More detailed breakdowns into atomic [59] or larger fragments [60] have also been suggested. Indeed, the latter approach has been used to estimate the hydrophobic contribution to the adsorption energy for proteins at polymer surfaces [6]. The hydrophobic interaction energy ΔF_{hphob} in this approach is computed by a summation of terms:

$$\Delta F_{hphob} = \sum_k RT \ln P_k \qquad\qquad (7)$$

where P_k is a partition coefficient for the atom or fragment. The validity of applying partition coefficients measured for amino acids to fragments of a protein remains unclear, however. Furthermore, the dependence on separation distance must be added by an empirical exponential decay function [7].

Hydrophobic interactions have been measured in surface forces apparatus experiments [61,62], but the applicability of the results to protein interactions is questionable for at least two reasons. First, the force measurements are based on extended surfaces, and the variation with curvature remains to be clarified. Second, the heterogeneity of the protein surface complicates characterization of surface hydrophobicity.

D. Steric Forces

Steric forces are described by repulsive potentials that arise from near overlap between different moieties, be they atoms, molecules, or parts of macromolecules (e.g., segments or chains). They can manifest themselves in various forms for these different moieties [62], the most obvious being the overlap of the outer electron clouds of pairs of atoms as they approach each other, often referred to as Born repulsion. This situation (and the more complex one involving the approach of colloidal particles) can be modeled by the simple and intuitive hard-sphere potential, according to which there is no interaction until contact is reached, at which point the interaction potential jumps to infinity. A more accurate approach makes use of a slightly "softer" potential, for example, with an exponential or a power-law dependence, such as the Born repulsion term in the Lennard–Jones potential [Eq. (5)].

Steric repulsion can occur in several other forms and for more complex reasons [62]. Some of these phenomena owe their existence to the presence of solvent and may, thus, also be categorized as solvation forces. However, for proteins and other macromolecules the thermal fluctuations of surface groups may also play a role, with their suppression as solutes approach each other (or a surface) giving rise to an entropic repulsive force. The significance for proteins of contributions such as this one will depend on the mobility of surface chains and groups, and hence on the stability of the folded structure, but quantitative characterization remains problematic.

III. RELATION TO BULK ADSORPTION MEASUREMENTS

While a number of insights regarding adsorption behavior have been gained from analysis of the configuration-dependent interaction energies discussed in the previous section, the energies alone are insufficient to make quantitative predictions regarding adsorption equilibrium, kinetics or adsorbed layer structure. Here we address the incorporation of molecular interactions into frameworks that allow such predictions to be formed and evaluated. Equilibrium extents of adsorption are most amenable to computation, whereas nonequilibrium behavior, including kinetic aspects and irreversible adsorption, is less so.

Limitations in the quantitative understanding of the effects involved and in computational capabilities preclude full prediction of adsorption behavior in many situations. For example, molecular dynamics calculations applied to proteins at surfaces [8,9] can predict conformational changes within the constraints of their respective assumptions, but the configurational exploration required to predict adsorption equilibrium is not yet feasible. Thus the discussion below is limited to a relatively restricted set of conditions. In particular, the adsorbing molecules are assumed to be rigid and to retain their conformation during adsorption, an assumption that is likely to be satisfied mainly for "hard" globular proteins adsorbing on surfaces of limited hydrophobicity. Cases that are notable for not satisfying these assumptions include many widely studied experimental systems involving blood proteins such as albumin and fibrinogen adsorbing on polymetric biomaterials. On the other hand, systems in which long-range electrostatic forces are dominant are most amenable to mechanistic modeling, despite the fact that the adsorbed protein layers are expected to neutralize the adsorbent surface (e.g., lysozyme on mica [63]) at least partially.

Here we consider first protein–surface interactions, addressing primarily the issues of adsorption reversibility and bulk–surface equilibrium. We then expand the discussion to include protein–protein interactions, which dictate ultimate surface coverage and adsorbed layer structure.

A. Protein–Surface Interactions

An analysis of protein–surface interactions is most usefully carried out at low surface coverages, where the confounding effect of interactions among adsorbate molecules is absent. The analysis is then usually performed in terms of equilibrium thermodynamics, specifically an adsorption equilibrium constant $K_{eq} = C_s/C_b$, where C_s and C_b denote surface and bulk concentrations, respectively; since these concentrations are expressed per unit area and volume, respectively, K_{eq} as defined here has units of length. Thus, K_{eq} is a Henry's law [64] constant, i.e., one in which the surface concentration is low enough that the adsorbed protein may be considered to be at infinite "dilution."

The link between the experimentally accessible K_{eq} and the model calculations described previously is through the free energy of interaction, which is also often a quantity more easily interpretable in physical terms: the meaning of an adsorption equilibrium constant of 100 nm might not be clear, but a binding energy of 100 kJ/mol is likely to be more so. The most common expression relating free energy to the equilibrium constant is

$$\Delta G = -RT \ln K_{eq} \tag{8}$$

where ΔG is the Gibbs free energy change between adsorbed and free protein. Since K_{eq} is not dimensionless, ΔG should in principle be expressed relative to that for a standard state, but in this case there is no clear way to define a standard state. The difficulty can be concealed when adjustable parameters are included in the free-energy model. It is also possible to scale K_{eq} by a characteristic length, e.g., a specific surface area (phase ratio) as used in chromatographic analysis, but incorporation of the geometric factor in the thermodynamic analysis is inconsistent with the dependence of adsorption equilibrium on the bulk concentration, rather than amount, of protein.

An alternative approach that both reconciles these difficulties and probably better reflects the physical situation is one that abandons the notion of a single adsorbed state in favor of a distribution of states in which protein is localized in the vicinity of the sorbent as a consequence of their favorable interaction. The extent to which the protein is localized depends directly on the free energy for a particular configuration (i.e., separation distance and mutual orientation) of protein and sorbent; specifically, the protein concentration should follow a Boltzmann distribution with respect to interaction energy.

This idea of localization is essentially the Gibbs surface excess notion of adsorption and has been applied previously to gas adsorption [65]. The adsorbed amount is given as the total surface excess, the number of moles of solute actually present relative to that in the absence of the wall. This leads to the equilibrium constant, which for a homogeneous adsorbing surface is written as

$$K_{eq} = \int_{z_0}^{x} dz \int_{\Omega} (e^{-\Delta F(\Omega, z)/kT} - 1)\, d\Omega \tag{9}$$

where z is the solute–interface separation distance, z_0 is a cutoff distance, and Ω refers to orientational space, which must be included for nonspherically symmetric solute molecules such as proteins. Within this formalism, the equilibrium constant can be estimated from any model for the configuration-dependent total interaction energy $\Delta F(\Omega, z)$, which is typically computed as a Helmholtz free energy but is equivalent to the Gibbs free energy, as volume changes in the adsorption process are negligible.

The assumption of equilibrium on which the discussion here has been based implies that adsorption is reversible. In many systems adsorption appears to be irreversible, but this may be a function of the experimental observation time. Typically, adsorption is diffusion limited with a characteristic time much shorter than the time required for desorption. Measurement of protein–surface equilibrium in nominally irreversible adsorption at low coverage may in fact be possible with long time measurements at extremely low bulk protein concentrations, where the mass transfer limitation is increased and as a result the rate of adsorption more closely matches the low desorption rate.

Models of protein–surface equilibrium have been applied to a wide variety of adsorption data, even for high coverages. The amount of detail in such models has also varied, but in view of the uncertainties associated with several of the contributions discussed in Section II, more elaborate descriptions are not necessarily more successful at providing predictive mechanistic descriptions. They are, however, often capable of describing experimental trends with the aid of adjustable parameters, and in such situations an extensive data set, obtained under a range of conditions, is essential to test the model adequately. An example of such an approach is the model of Norde and Lyklema [66,67], in which the protein, adsorbate, solvent, and electrolyte are incorporated, and the effects accounted for include electrostatics, hydration, structural rearrangements, and transfer of hydrogen and other ions. However, there are uncertainties in the description of each mechanism.

Models with a narrower focus allow less latitude, but obviously at the price of neglecting some effects. The most widely used have been those applied to electrostatically dominated systems, especially for correlating ion exchange chromatographic data, where adsorption must be reversible for operation to be successful. The earliest such model was a chemical one, the stoichiometric displacement model

(SDM) of Boardman and Partridge [68], in which adsorption occurred as a protein displaced one or more ions on the sorbent in a strict ion exchange. The basic idea has been revised to include the concentration of ions in solution and the concept of distinct binding sites on the protein [69,70], but has remained quite popular for the correlation of ion exchange data. The essential result of this model is a linear dependence of log K_{eq} on the logarithm of ionic strength, with the (negative) slope given by the net charge or number of binding sites on the protein, depending on interpretation. In either case, the slope parameter does not depend upon sorbent properties and so is not useful for scale-up or even the prediction of elution order.

The SDM has been extended to account for behavior at high ionic strengths, where adsorption increases for reasons that are thought to be related to sequestration by electrolyte of the water molecules necessary to hydrate the protein. Solvophobic theory, based on the formation of a cavity to accept a solute, is able to correlate this salting out of protein [71,72], but again predictive capabilities are limited.

Colloidal models are more amenable to a priori specification of at least protein properties such as size and charge, and in some cases to adsorbent properties as well. The application of such a model to describing retention in ion exchange chromatography was initially based on the colloidal interaction of two plates, for which the LPBE has an analytical solution [73,74]. Despite its geometric simplicity, this model successfully correlates ion exchange data both at low ionic strengths, where the retention is dominated by electrostatic attraction, and at higher ionic strengths.

Two other interesting points emerged from this work. First, although log K_{eq} is predicted by the model to be proportional to the inverse square root of ionic strength [46,47,74], as opposed to the inverse first power of ionic strength predicted by the SDM, the model adequately fitted data previously shown [70] to be described well by the SDM. Second, within the context of this model, the increase in retention at high ionic strengths is predicted to result from the screening of electrostatic repulsion that is due to oppositely charged, but greatly mismatched, surfaces. This is very different from the solvophobic mechanism previously discussed [70–72], but requires fitting several adjustable parameters, including unrealistically high surface charge densities.

A higher level of realism in colloidal modeling is that in which the protein molecule is described as a sphere of radius R and net charge Q [37]. Electrostatic and van der Walls interaction energies are calculated as described in Sections II.A and II.B, and an equilibrium constant is obtained as in Eq. (9). The adsorbent, too, is characterized in terms of physically meaningful quantitative properties, namely, the surface charge density σ and the Hamaker constant A characterizing the material propensity for van der Walls interactions with proteins. In general, the model predicts that the charges on the protein and on the surface determine the steepness of the change in equilibrium constant with increasing ionic strength, whereas the Hamaker constant affects the level of the equilibrium constant [37].

The great benefit of this approach is that most of the parameter values can be specified without allowing them to be adjustable. The size of a protein molecule is estimated easily from its molecular weight. The charge on a protein or surface can be obtained from titration experiments, although this may be difficult for the sorbent if its specific area is low. For a protein, in fact, the primary sequence is usually sufficient to estimate the charge, as the titration curve is often not significantly affected by its local environment such as ionic strength [75]. The Hamakar constant is the most difficult parameter to evaluate, but the Lifshitz theory provides a means to obtain at least an estimate, and Hamaker constants derived from calculation and experiment are available for a number of systems involving proteins [48,76–79].

Such models can be extended from a description of the protein as a sphere to one accounting more realistically for the geometry, based on crystallographic information [80] or on assumed confirmational forms [81]. The extension is, of course, more straightforward conceptually than it is in practice, but this barrier should diminish as computational capabilities continue to improve.

Several of these modeling efforts have been used directly to compare with adsorption data obtained by ion exchange chromatography [82], fluorescence spectroscopy [80], and solution depletion [81]. Comparisons of calculations with adsorption trends, e.g., effects of mutations, have also been described [10,33]. Trends in K_{eq} of multiple proteins and peptides with respect to ionic strength, and to a fair extent the quantitative values are adequately described by the models. Perhaps most remarkable, however, is how well grossly simplified models are able to describe the same adsorption behavior, e.g., with the protein or peptide represented as a point charge [81], plate [46,47,74], or sphere [80,82]. This robustness is probably a reflection of the fact that even simpler models can capture the dominant effect when electrostatics are globally attractive, with an implicit or explicit adjustable parameter aiding in "tuning" the actual values of adsorption constants. Trends such as those with varying ionic strength are then more easily reproduced. When the adsorption behavior is qualitatively less easily predictable, e.g., when electrostatics are generally repulsive but patch-controlled adsorption occurs, results are more sensitive to modeling details [35].

Although the adsorbent physicochemical properties, namely, the surface charge density and the Hamaker constant for interactions with proteins, are well-defined physical quantities, there is uncertainty in how meaningful they are in characterizing protein–surface interactions. Among the sources of uncertainty are the effects of discreteness of charge, the adsorbent geometry (to which van der Waals interactions are especially sensitive [48]), and neglect of other effects such as hydration-related ones. Thus the adsorbent properties included in the model may be surrogates for other effects in addition to those that they are nominally intended to capture. In particular, descriptions of van der Waals interactions, being of short range, may also account to some extent for other short-range interactions such as solvation interactions; these have in common their strong dependence on the size of the interacting areas on the protein and surface. This view gives rise to a simpler but more pragmatic formulation in which the protein electrostatics are represented by a sphere with net charge, while all other effects are captured in a short-range interaction energy that increases with protein size [82].

B. Protein–Protein Interactions

At higher coverages than those considered in Section III.A, the Henry's law limit implied by the use of an equilibrium constant K_{eq} is no longer observed, and the finite amount of surface area available comes into play in the absence of multilayer adsorption. The result is then that C_s no longer increases linearly with C_b, but instead that a plot of C_s versus C_b typically adopts the convex-upward shape characteristic of adsorption isotherms of many kinds.

Adsorption behavior of this kind is most frequently described using a Langmuir adsorption isotherm, despite the widespread recognition that most of the assumptions underlying this model are not satisfied for protein adsorption. The Langmuir formulation assumes adsorption to occur on discrete sites, to be reversible, and interactions among sites to be negligible, and the hyperbolic form of the isotherm is then a consequence of the finite number of sites available. For adsorption of proteins, all of these assumptions become questionable. First, the issue of reversibility has been mentioned earlier and is discussed again subsequently. Next, the adsorbate molecules are so much larger than the surface lattice dimensions that the adsorbent is more reasonably considered to be a continuum than comprising discrete sites. Thus, the finite number of sites in the Langmuir model should be replaced by consideration of the finite adsorbent area.

More generally, this excluded-area feature may be thought of in terms of steric interactions among adsorbate molecules, and in this context it is also necessary to consider protein–protein energetic interactions on the surface. The most obvious energetic interactions are long-range electrostatics, which will generally be repulsive in view of the like charge carried by the protein molecules. Shorter-range interactions become more important as complete monolayer coverage is approached, but their effects are more difficult to assess. Because such attractive contributions as van der Waals and hydrophobic interactions are strongly dependent on molecular complementarity [48], strong attraction is not as likely as would be expected for general colloidal particles at short range. Beyond just the energetics of the interactions is also the possibility, even for conformationally rigid molecules, of orientational adaptations as the coverage increases [31,83].

Because of these and other effects, the fits of the Langmuir equation to protein adsorption data often show systematic discrepancies. In particular, experimental isotherms are often "softer" than can be fitted by the Langmuir form, i.e., the approach to a plateau is less steep than predicted, and the plateau region may show a continued gradual increase in C_s, with coverage still not exceeding theoretical monolayer levels as estimated from simple geometrical packing arguments. Such experimental isotherms can be adequately described by various alternative models. For instance, the steric mass action (SMA) model [84] is an extension of the SDM discussed earlier for ion exchange chromatography, and although it is not explicitly formulated in terms of excluded area, the essential features are analogous. The resulting isotherms are thus the convex-upward ones widely observed in practice.

For more mechanistically based models of high-coverage adsorption, however, the formulation and solution are complicated by the many-body nature of the problem, specifically the need to know the relative positions of the molecules involved, i.e., the arrangement of molecules on the surface. The configuration directly affects the protein–protein energetics, and there is generally a trade-off between this, generally repulsive, contribution, and the attractive protein–surface energetics. In addition, the questions raised earlier of whether adsorption is reversible and whether the adlayer is an equilibrium phase must be considered. To illustrate some of the considerations involved, we examine various examples of irreversible adsorption first.

Several possibilities arise even in the absence of protein–protein energetic interactions, as can be seen by exploring the predicted adsorption behavior when the proteins are modeled as simple hard spheres. From simple geometric arguments, close-packed hexagonal and square layers have coverages of 90.7 and 78.5% respectively. However, such close packing may be unattainable. For the limiting case in which the spheres approach the surface sequentially and are adsorbed irreversibly, if there is no overlap with previously adsorbed spheres, the process is known as random sequential adsorption (RSA), and the maximum coverage, the so-called jamming limit, is about 54.7% [85,86]. If the RSA model is modified to allow sterically excluded particles to "roll over" the blocking particles in order to reach the surface (ballistic deposition), the ultimate coverage is about 61% [87,88]. Thus, the randomness of the locations at which particles adsorb results in a considerable additional fraction of uncovered surface.

When there are energetic interactions, especially repulsive electrostatics, between the protein molecules, the picture becomes more complicated. The RSA model has been modified to include an energetic threshold for rejection of a potential adsorption event [89–91], leading to jamming limits lower than the usual 54.7%. Although this model captures observed experimental behavior, a more likely explanation of the physics involved is that repulsion by adsorbed molecules presents a kinetic barrier precluding additional adsorption [91]. In either case, an increase in salt concentration reduces protein–protein repulsion, leading to a predicted increase in adsorbed amount. For relatively small globular proteins, however, the repulsion may be too weak for the modified RSA models to be valid.

The stated bases for estimating surface coverage are premised on the assumption that adsorption is irreversible. One consequence of this is that the coverage is independent of the bulk protein concentration, i.e., any "isotherm" would rise infinitely steeply to the plateau. The issue of reversibility is also paramount if adsorbed particles are able to diffuse on the surface: irreversible adsorption would ultimately lead to a close-packed surface layer. Thus, if particles are able to move on the surface, it is reversible adsorption that is usually of interest, with the limiting behavior at low coverage represented by the equilibrium constant K_{eq} as discussed in the previous section. Configurational issues are crucial here; statistical mechanical approaches are possible for simple intermolecular potentials [92], but simulations are required for more realistic representations.

As the coverage increases, the energy gain due to adsorption (protein–surface interaction) becomes offset by a penalty due to protein–protein interaction. This penalty is present even for hard spheres devoid of energetic interactions with one another, which give rise to a surface pressure if they are mobile. Such interactions can be accounted for by treating the adsorbate layer as a two-dimensional fluid [93]. As Fig. 5 shows, the resulting isotherm is appreciably "softer" than its Langmuir counterpart; the figure also shows, for comparison, the coverages predicted by the various models invoking irreversible adsorption. That the abscissa is in terms of $K_{eq}C_b$ illustrates nicely the central importance of the protein–surface interactions, offset by the increasing significance of protein–protein interactions as the coverage increases. However, the form of the abscissa somewhat obscures some important features of the predicted isotherms. First, the discrepancies between the Langmuir and the softer isotherm are exaggerated by the logarithmic axis. Second, scaling by K_{eq} can result in conspicuous changes in isotherms plotted on the more conventional linear abscissa C_b. Specifically, for small K_{eq} the isotherms rise rather gradually, and a clearly identifiable plateau is not seen, whereas for high K_{eq} a steeper (high-affinity) isotherm and fairly flat plateau result. Thus for electrostatically driven adsorption,

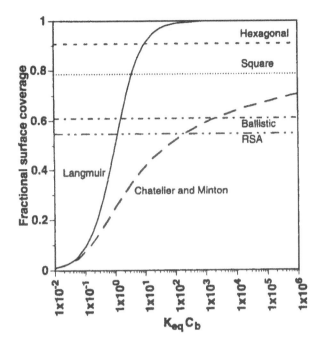

FIG. 5 Fractional surface coverages predicted by various models of adsorption of spheres.

the reduction in K_{eq} with increasing salt concentration is predicted to give rise to a monotonic decrease in the adsorbed amount at a given protein concentration. This is in fact what is usually seen under conditions such as those encountered in ion exchange chromatography, but it is opposite to the trend described here resulting from screening of protein–protein interactions, e.g., in RSA.

The ideal array concept can also be modified to include energetic interactions in describing equilibrium adsorption [94]. Accounting for electrostatic and van der Waals interactions leads to isotherms with a generally "soft" convex-upward form, with the same trend with increasing salt as described in the previous paragraph, i.e., screening of protein–surface interactions is more important than screening of protein–protein interactions. However, a direct comparison with Fig. 5 is difficult because of the different scaling involved. Beyond the general isotherm form, the existence of multiple-valued isotherms was predicted. This result, which is qualitatively similar to previous predictions for different models of adsorption, suggests possible explanations for apparent anomalies in protein adsorption isotherms, e.g., steps and kinks.

The general case in which reversibility, mobility, and randomness are all possible can be undertaken only by simulation methods such as Brownian dynamics [95,96], with tractability generally requiring use of pairwise additive analytical approximations for evaluating potentials. The grand canonical Brownian dynamics scheme [96] accounts directly for all interactions, configurational issues, and adsorption dynamics. For a small globular protein such as lysozyme at pH 7, it predicts a range of adsorption behaviors under electrostatically controlled conditions. At very low salt concentrations, the predicted isotherms show high-affinity behavior, with the plateau amounts increasing with ionic strength; i.e., repulsive protein–protein interactions dominate. At higher ionic strengths, however, the trend predicted by Chatelier and Minton [93] and Johnson et al. [94] is recovered.

The number of degrees of freedom in the case of high surface coverage leaves this as an extremely complex situation, even when the individual molecules are treated in an idealized fashion. Allowing for the anisotropy of protein molecules adds complexity of a different kind, possibly leading to counterintuitive behavior, e.g., favorable intermolecular electrostatic interactions even for molecules with a nonzero net charge [32]. Even simple packing arguments are complicated by the different sizes of the protein "footprint" in different orientations.

IV. SUMMARY

The primary thrust of this chapter is the quantitative determination of protein–surface interaction energies, and the implications of such information for experimentally accessible quantities, e.g., adsorption isotherms. The calculation of interaction energies is feasible for systems dominated by electrostatic and van der Waals interactions, with both rigorous approaches and simpler approximations available. Such calculations are, however, predicated on the availability of protein structural information and surface characteristics, as well as on the assumption that both the protein and the surface are fairly rigid. When these requirements are satisfied, adsorption equilibrium can be predicted quite well in absolute terms, and key trends are captured very well. In particular, these calculations provide a meaningful solid fundamental framework within which to develop rational methods for manipulating adsorption.

Although these capabilities are valuable, many challenges remain, and in view of the complexity, even qualitative, that is involved with some of them, it is not clear what solutions may emerge. Probably the most important outstanding features are characterization and modeling of hydrophobic interactions and understanding the cause and effect of protein conformational changes during adsorption.

REFERENCES

1. J. D. Andrade and V. Hlady, Protein adsorption and materials biocompatability: a tutorial review and suggested hypotheses. Adv. Polym. Sci. 79:1–63 (1986).
2. W. Norde, Adsorption of proteins from solution at the solid–liquid interface. Adv. Coll. Interf. Sci. 25:267–340 (1986).
3. C. A. Haynes and W. Norde, Globular proteins at solid/liquid interfaces. Colloids Surf. B: Biointerfaces 2:517–566 (1994).
4. W. Norde and J. Lyklema, Why proteins prefer interfaces. J. Biomater. Sci. Polym. Edn. 2:183–202 (1991).
5. D. R. Lu and K. Park, Protein adsorption on polymer surfaces: calculation of adsorption energies. J. Biomater. Sci. Edn. 1:243–260 (1990).
6. D. R. Lu, S. J. Lee, and K. Park, Calculation of solvation interaction energies for protein adsorption onto polymer surfaces. J. Biomater. Sci. Polym. Edn. 3:127–147 (1990).
7. S. J. Lee and K. Park, Protein interactions with surfaces: separation distance-dependent interaction energies. J. Vac. Sci. Technol. A 12:2949–2955 (1994).

8. K. Lim and J. N. Herron, Molecular simulation of protein–PEG interaction. In *Poly (Ethylene Glycol) Chemistry: Biotechnical and Biomedical Applications* (J. M. Harris, ed.). Plenum Press, New York, 1992, pp. 29–57.

9. D. J. Tobias, W. Mar, J. K. Blasie, and M. L. Klein, Molecular dynamics simulations of a protein on hydrophobic and hydrophilic surfaces. Biophys. J. *71*:2933–2941 (1996).

10. V. Noinville, C. Vidal-Madjar, and B. Sébille, Modeling of protein adsorption on polymer surfaces. Computation of adsorption potential. J. Phys. Chem. *99*:1516–1522 (1995).

11. S. J. Weiner, P. A. Kollman, D. A. Case, U. C. Singh, C. Ghio, G. Alagona, S. Profeta, and P. Weiner, A new force field for molecular mechanical simulation of nucleic acids and proteins. J. Am. Chem. Soc. *106*:765–784 (1984).

12. I. Szleifer, Protein adsorption on surfaces with grafted polymers: a theoretical approach. Biophys. J. *72*:595–612 (1997).

13. T. Arai and W. Norde, The behavior of some model proteins at solid–liquid interfaces. I. Adsorption from single protein solution. Colloids Surf. *51*:1–15 (1990).

14. J. G. Kirkwood, Theory of solutions of molecules containing widely separated charges with special application to zwitterions. J. Chem. Phys. *2*:351–361 (1934).

15. R. Hogg, T. W. Healy, and D. W. Fuerstenau, Mutual coagulation of colloidal dispersions. Trans. Faraday Soc. *62*:1638–1651 (1966).

16. J. T. G. Overbeek, The role of energy and entropy in the electrical double layer. Colloids Surf. *51*:61–75 (1990).

17. S. A. Palkar and A. M. Lenhoff, Energetic and entropic contributions to the interaction of unequal spherical double layers. J. Coll. Interf. Sci. *165*:177–194 (1994).

18. H.-X. Zhou, Macromolecular electrostatic interaction energy within the nonlinear Poisson–Boltzmann equation. J. Chem. Phys. *100*:3152–3162 (1994).

19. K. A. Sharp and B. Honig, Electrostatic interactions in macromolecules: theory and applications. Ann. Rev. Biophys. Chem. *19*:301–332 (1990).

20. B. J. Yoon and A. M. Lenhoff, Computation of the electrostatic interaction energy between a protein and a charged surface. J. Phys. Chem. *96*:3130–3134 (1992).

21. M. K. Gilson, K. A. Sharp, and B. Honig, Calculating the electrostatic potential of molecules in solution: method and error assessment. J. Comp. Chem. *9*:327–335 (1988).

22. J. Warwicker and H. C. Watson, Calculation of the electric potential in the active site cleft due to α-helix dipoles. J. Mol. Biol. *157*:671–679 (1982).

23. W. H. Orttung, Direct solution of the Poisson equation for biomolecules of arbitrary shape, polarizability density, and charge distribution. Ann. NY Acad. Sci. *303*:22–37 (1977).

24. R. J. Zauhar and R. S. Morgan, A new method for computing the macromolecular electric potential. J. Mol. Biol. *186*:815–820 (1985).

25. A. H. Juffer, E. F. F. Botta, B. A. M. van Keulen, A. van der Plug, and J. C. Berendsen, The electric potential of a macromolecule in a solvent: a fundamental approach. J. Comp. Phys. *97*:144–171 (1991).

26. B. J. Yoon and A. M. Lenhoff, A boundary element method for molecular electrostatics with electrolyte effects. J. Comp. Chem. *11*:1080–1086 (1990).

27. H.-X. Zhou, Boundary element solution of macromolecular electrostatics: interaction energy between two proteins. Biophys. J. *65*:955–963 (1993).

28. D. Bashford and M. Karplus, Multiple site titration curves of proteins: an analysis of exact and approximate methods for their calculation. J. Phys. Chem. *95*:9556–9561 (1991).

29. M. K. Gilson and B. H. Honig, Energetics of charge–charge interactions in proteins. Proteins Struct. Funct. Genet. *3*:32–52 (1988).

30. L. Haggerty and A. M. Lenhoff, Relation of protein electrostatics computations to ion-exchange and electrophoretic behavior. J. Phys. Chem. *95*:1472–1477 (1991).

31. C. S. Lee and G. Belfort, Changing activity of ribonuclease A during adsorption: a molecular explanation. Proc. Natl. Acad. Sci. USA 86:8392–8396 (1989).

32. L. Haggerty and A. M. Lenhoff, Analysis of ordered arrays of adsorbed lysozyme by scanning tunneling microscopy. Biophys. J. 64:886–895 (1993).

33. D. J. Roush, D. S. Gill, and R. C. Willson, Electrostatic potentials and electrostatic interaction energies of rat cytochrome b$_5$ and a simulated anion-exchange adsorbent surface. Biophys. J. 66:1290–1300 (1994).

34. V. Lesins and E. Ruckenstein, Patch controlled attractive electrostatic interactions between similarly charged proteins and adsorbents. Colloid. Polym. Sci. 266:1187–1190 (1988).

35. D. Asthagiri and A. M. Lenhoff, Influence of structural details in modelling electrostatically driven protein adsorption. Langmuir 13:6761–6768 (1997).

36. D. J. Barlow and J. M. Thornton, The distribution of charged groups in proteins. Biopolymers 25:1717–1733 (1986).

37. C. M. Roth and A. M. Lenhoff, Electrostatic and van der Waals contributions to protein adsorption: computation of equilibrium constants. Langmuir 9:962–972 (1993).

38. M. L. Grant and D. A. Saville, Colloidal interactions in protein crystal growth. J. Phys. Chem. 98:10358–10367 (1994).

39. C. M. Roth, Electrostatic and van der Waals contributions to protein adsorption. PhD dissertation, University of Delaware, Newark, DE, 1994.

40. H. C. Hamaker, The London–van der Waals attraction between spherical particles. Physica 4:1058–1072 (1937).

41. I. E. Dzyaloshinskii, E. M. Lifshitz, and L. P. Pitaevskii, The general theory of van der Waals forces. Adv. Phys. 10:165–209 (1961).

42. D. J. Mitchell and B. W. Ninham, van der Waals forces between two spheres. J. Chem. Phys. 56:1117–1126 (1972).

43. J. E. Kiefer, V. A. Parsegian, and G. H. Weiss, An easily calculable approximation for the many-body van der Waals attraction between two equal spheres. J. Coll. Interf. Sci. 57:580–582 (1976).

44. J. Lyklema, *Fundamentals of Interface and Colloid Science.* Volume 1: *Fundamentals.* Academic Press, London, 1991.

45. E. Ruckenstein and D. C. Prieve, Adsorption and desorption of particles and their chromatographic separation. AIChE J. 22:276–283 (1976).

46. J. Ståhlberg, B. Jönsson, and C. Horváth, Combined effect of coulombic and van der Waals interactions in the chromatography of proteins. Anal. Chem. 64:3118–3124 (1992).

47. J. Ståhlberg and B. Jönsson, Influence of charge regulation in electrostatic interaction chromatography of proteins. Anal. Chem. 68:1536–1544 (1996).

48. C. M. Roth, B. L. Neal, and A. M. Lenhoff, Van der Waals interactions involving proteins. Biophys. J. 70:977–987 (1996).

49. D. B. Hough and L. R. White, The calculation of Hamaker constants from Lifshitz theory with applications to wetting phenomena. Adv. Coll. Interf. Sci. 14:3–41 (1980).

50. V. A. Parsegian, Long range van der Waals forces. In *Physical Chemistry: Enriching Topics from Colloid and Surface Science* (H. van Olphen and K. J. Mysels, eds.). Theorex, La Jolla, CA, 1975, pp. 27–72.

51. T. Inagaki, R. N. Hamm, E. T. Arakawa, and R. D. Birkhoff, Optical property of bovine plasma albumin between 2 and 82 eV. Biopolymers 14:839–845 (1975).

52. S. Nir, van der Waals interactions between surfaces of biological interest. Progr. Surf. Sci. 8:1–58 (1976).

53. J. Israelachvili, Solvation forces and liquid structure as probed by direct force measurements. Acc. Chem. Res. 20:415–421 (1987).

54. R. P. Rand, N. Fuller, V. A. Parsegian, and D. C. Rau, Variation in hydration forces between neutral phospholipid bilayers: evidence for hydration attraction. Biochemistry 27:7711–7722 (1988).

55. R. D. Tilton, C. R. Robertson, and A. P. Gast, Manipulation of hydrophobic interactions in protein adsorption. Langmuir 7:2710–2718 (1991).

56. A. Ben-Naim, *Hydrophobic Interactions*. Plenum Press, New York, 1990.

57. C. Tanford, *The Hydrophobic Effect*. Wiley, New York, 1980.

58. T. E. Creighton, *Proteins: Structures and Molecular Properties*. 2nd ed. Freeman, New York, 1993.

59. D. Eisenberg and A. D. McLachlan, Solvation energy in protein folding and binding. Nature 319:199–203 (1986).

60. D. J. Abraham and A. J. Leo, Extension of the fragment method to calculate amino acid zwitterion and side chain partition coefficients. Proteins 2:130–152 (1987).

61. J. Israelachvili and R. Pashley, The hydrophobic interaction is long range, decaying exponentially with distance. Nature 300:341–342 (1982).

62. J. N. Israelachvili, *Intermolecular and Surface Forces*. 2nd ed. Academic Press, New York, 1992.

63. E. Blomberg, P. M. Claesson, J. C. Froberg, and R. D. Tilton, The interaction between adsorbed layers of lysozyme studied with the surface force technique. Langmuir 10: 2325–2334 (1994).

64. S. I. Sandler, *Chemical and Engineering Thermodynamics*. 2nd ed. Wiley, New York, 1989.

65. J. A. Barker and D. H. Everett, High temperature adsorption and the determination of the surface area of solids. Trans. Faraday Soc. 58:1608–1623 (1962).

66. W. Norde and J. Lyklema, The adsorption of human plasma albumin and bovine pancreas ribonuclease at negatively charged polystyrene surfaces: IV. The charge distribution in the adsorbed state. J. Coll. Interf. Sci. 66:285–294 (1978).

67. W. Norde and J. Lyklema, Thermodynamics of protein adsorption: theory with special reference to the adsorption of human plasma albumin and bovine pancreas ribonuclease at polystyrene surfaces. J. Coll. Interf. Sci. 71:350–366 (1979).

68. N. K. Boardman and S. M. Partridge, Separation of neutral proteins on ion-exchange resins. Biochem. J. 59:543–552 (1955).

69. W. Kopaciewicz, M. A. Rounds, J. Fausnaugh, and F. E. Regnier, Retention model for high-performance ion exchange chromatography. J. Chromatogr. 266:3–21 (1983).

70. W. R. Melander, Z. El Rassi, and C. Horváth, Interplay of hydrophobic and electrostatic interactions in biopolymer chromatography: effect of salts on the retention of proteins. J. Chromatogr. 469:3–27 (1989).

71. W. Melander and C. Horváth, Salt effects on hydrophobic interactions in precipitation and chromatography of proteins: an interpretation of the lyotropic series. Arch. Biochem. Biophys. 183:200–215 (1977).

72. W. R. Melander, D. Corradini, and C. Horváth, Salt-mediated retention of proteins in hydrophobic-interaction chromatography: application of solvophobic theory. J. Chromatogr. 317:67–85 (1984).

73. V. A. Parsegian and D. Gingell, On the electrostatic interaction across a salt solution between two bodies bearing unequal charges. Biophys. J. 12:1192–1204 (1972).

74. J. Ståhlberg, B. Jönsson, and C. Horváth, Theory for electrostatic interaction chromatography of proteins. Anal. Chem. 63:1867–1874 (1991).

75. Y. Nozaki and C. Tanford, Examination of titration behavior. In *Methods in Enzymology*. Vol. XI: *Enzyme Structure* (C. Hirs, ed.). Academic Press, New York, 1967, pp. 715–734.

76. V. A. Parsegian and S. L. Brenner, The role of long range forces in ordered arrays of tobacco mosaic virus. Nature 259:632–635 (1976).

77. T. Afshar-Rad, A. I. Bailey, P. F. Luckham, W. MacNaughton, and D. Chapman, Forces between protein and model polypeptides adsorbed on mica surfaces. Biochim. Biophys. Acta *915*:101–111 (1987).

78. C. A. Helm, W. Knoll, and J. N. Israelachvili, Measurements of ligand–receptor interactions. Proc. Natl. Acad. Sci. USA *88*:8169–8173 (1991).

79. D. E. Leckband, F.-J. Schmitt, J. N. Israelachvili, and W. Knoll, Direct measurements of specific and nonspecific protein interactions. Biochemistry *33*:4611–4624 (1994).

80. C. M. Roth and A. M. Lenhoff, Electrostatic and van der Waals contributions to protein adsorption: comparison of theory and experiment. Langmuir *11*:3500–3509 (1995).

81. N. Ben-Tal, B. Honig, R. M. Peitzsch, G. Denisov, and S. McLaughlin, Binding of small basic peptides to membranes containing acidic lipids: theoretical models and experimental results. Biophys. J. *71*:561–575 (1996).

82. C. M. Roth, K. K. Unger, and A. M. Lenhoff, A mechanistic model of retention in protein ion-exchange chromatography. J. Chromatogr. A *726*:45–56 (1996).

83. J. L. Robeson and R. D. Tilton, Spontaneous reconfiguration of adsorbed lysozyme layers observed by total internal reflection fluorescence with a pH-sensitive fluorophore. Langmuir *12*:6104–6113 (1996).

84. C. A. Brooks and S. M. Cramer, Steric mass-action ion exchange: displacement profiles and induced salt gradients. AIChE J. *38*:1969–1978 (1992).

85. J. Feder, Random sequential adsorption. J. Theor. Biol. *87*:237–254 (1980).

86. J. Feder and I. Giaever, Adsorption of ferritin. J. Coll. Interf. Sci. *78*:144–154 (1980).

87. R. Jullien and P. Meakin, Random sequential adsorption with restructuring in two dimensions. J. Phys. A: Math. Gen. *25*:L189–L194 (1992).

88. H. S. Choi, J. Talbot, G. Tarjus, and P. Viot, First-layer formation in ballistic deposition of spherical particles: kinetics and structure. J. Chem. Phys. *99*:9296–9303 (1993).

89. Z. Adamczyk, M. Zembala, B. Siwek, and P. Warszynski, Structure and ordering in localized adsorption of particles. J. Coll. Interf. Sci. *140*:123–137 (1990).

90. Z. Adamczyk, B. Siwek, M. Zembala, and P. Belouschek, Kinetics of localized adsorption of colloid particles. Adv. Coll. Interf. Sci. *48*:151–280 (1994).

91. M. R. Oberholzer, J. M. Stankovich, S. L. Carnie, D. Y. Chan, and A. M. Lenoff, 2-D and 3-D interactions in random sequential adsorption of charged particles. J. Coll. Interf. Sci. *194*:138–153 (1997).

92. D. Nicholson and N. G. Parsonage, *Computer Simulation and the Statistical Mechanics of Adsorption.* Academic Press, London, 1982.

93. R. C. Chatelier and A. P. Minton, Adsorption of globular proteins on locally planar surfaces: models for the effect of excluded surface area and aggregation of adsorbed protein on adsorption equilibria. Biophys. J. *71*:2367–2374 (1996).

94. C. A. Johnson, P. Wu, and A. M. Lenhoff, Electrostatic and van der Waals contributions to protein adsorption: 2. Modelling of ordered arrays. Langmuir *10*:3705–3713 (1994).

95. E. Dickinson, Brownian dynamics with hydrodynamic interactions: the application to protein diffusional problems. Chem. Soc. Rev. *14*:421–455 (1985).

96. M. R. Oberholzer, N. J. Wagner, and A. M. Lenhoff, Grand canonical Brownian dynamics simulation of colloidal adsorption. J. Chem. Phys. *107*:9157–9167 (1997).

97. J. E. Sader and A. M. Lenoff, J. Coll. Interf. Sci. (1998).

5
Interfacial Behavior of Protein Mutants and Variants

MARTIN MALMSTEN Institute for Surface Chemistry and Royal Institute of Technology, Stockholm, Sweden

THOMAS ARNEBRANT Institute for Surface Chemistry, Stockholm, and Malmö University, Malmö, Sweden

PETER BILLSTEN* Linköping University, Linköping, Sweden

I. INTRODUCTION

The interfacial behavior of proteins is of importance in biochemical and biophysical processes, such as complement activation, thrombus formation, initial stages of arteriosclerosis, dental pellicle formation, etc., and a host of biomedical applications, such as intravenous drug delivery, solid-phase diagnostics, extracorporeal therapy, biomaterials, biosensors, biotechnical separation methods, biofouling, dental implants, etc. Not surprisingly, therefore, substantial work has been devoted to understanding the behavior of proteins at interfaces and the interplay between factors determining this interfacial behavior.

Despite considerable progress in the understanding of these processes, regarding, e.g., the relative importance of electrostatic effects, conformational stability, etc., and of adsorption kinetics and interfacial exchange phenomena, an undeniable fact is that the current understanding of protein interfacial behavior lags that of the interfacial behavior of homopolymers, copolymers, polyelectrolytes, and polyampholytes [1-3]. Naturally, this is a consequence of the higher complexity of proteins as compared to these simpler macromolecules. However, it is also possible that the lack of awareness among scientists concerned with protein adsorption about recent progress in the field of polymer adsorption is partly to blame for the current situation. Although it is important to remember that proteins are more complex than simpler polymers, they are still subject to the same thermodynamics, and therefore at least some general features should be analogous to those displayed by simpler systems. Indeed, this view is stressed repeatedly throughout this book.

One of the complexities of proteins is their rather unique structure, which to some extent has hindered systematic investigation of the effects of e.g., molecular weight, protein–surface interactions, structural stability, etc., on the protein adsorption, since more often than not such investigations have involved the comparison of proteins differing in more than one respect. For example, when investigating the

---

Current affiliation: Analysis and Formulations, Astra Draco AB, Lund, Sweden.

effect of the molecular weight, typically totally different proteins have been com-pared, which means that not only the molecular weight, but also, e.g., the protein structure and structural stability, as well as the protein isoelectric point, net charge, and charge distribution are different, precluding a conclusive analysis.

Although there are some naturally occurring proteins existing in several quite similar variants which can be, and have been, used for this type of investigation, it is mainly after the recent developments in protein engineering that they have really become feasible. It is the aim of the present chapter to try to exemplify how inves-tigations with both naturally occurring protein variants and protein mutants may yield information on the effects of protein molecular weight, structural stability, and pro-tein–surface and protein–protein interactions on the interfacial behavior of proteins. Note, however, that it is not our aim to make a complete inventory of reported studies with protein variants and mutants. For convenience, we have subdivided the different types of mutants (variants) rather arbitrarily and nonstringently into three categories —stability mutants, (surface) interaction mutants, and association mutants, the name indicating the main effect of the mutation (variation). Where appropriate, compari-sons will be made between the findings for the protein mutants and variants on one hand and those for simpler macromolecules, e.g., homopolymers, copolymers, and polyelectrolytes, on the other.

II. INTERFACIAL BEHAVIOR OF NATURALLY OCCURRING PROTEIN VARIANTS

One natural starting point for systematic studies of the effects of, e.g., molecular weight, conformational stability, and protein–surface and protein–protein interac-tions for the interfacial behavior of proteins is to investigate the adsorption prop-erties of naturally occurring variants of the same protein, and there are several papers where this approach has been used. For example, in an early study Horseley et al. investigated the adsorption of hen egg white and human milk lysozyme at various surfaces with total internal reflection fluorescence spectroscopy (TIRF), and found that these proteins indeed behave differently at interfaces [4]. A more recent study by Xu and Damodaran with those and other variants has shown that these differences exist also at the air–water interface [5]. The latter investigation also clearly indicated that the adsorption at this interface increases with the degree of protein denaturation (see subsequent discussion). Furthermore, as will be discussed in more detail, the β-lactoglobulin variants A and B, displaying different self-assembly behavior, were investigated by Elofsson et al. [6–9]. Yet another example of a study of the interfacial behavior of naturally occurring protein variants is that by Elbaum et al., in which the surface activity of hemoglobin S and other human hemoglobin variants was investigated, and in which the variants hemoglobin S and hemoglobin C_{Harlem}, both containing $\beta6Glu \rightarrow Val$ substitutions, were found to be more surface active than hemoglobin A, hemoglobin C, and hemoglobin Korle Bu ($\beta73Asp \rightarrow Asn$) [10]. This was interpreted as being related to the ligated state of the β' chain in the tetramer. However, as for the other examples given above, with the exception of the β-lactoglobulin variants in the studies by Elofsson et al., the mechanistic interpretation of the differences in interfacial behavior of these variants is somewhat difficult.

III. INTERFACIAL BEHAVIOR OF STABILITY MUTANTS

It has been found for a range of proteins that their tendency for adsorption at various surfaces increases on approaching their thermal denaturation. For example, this behavior has been observed for a series of lysozyme variants by Xu and Damodaran [5] and for β-lactoglobulin by Elofsson et al. [9] and Arnebrant et al. [11]. Analogously, proteins capable of undergoing interfacial conformational change are frequently found to adsorb at surfaces even when electrostatics counteracts adsorption [2,3]. There could, in principle, be several reasons for this. One of these could be that on undergoing conformational changes amino acids typically hidden in the protein interior may be made accessible to interaction with both the surface and the solvent. Since these residues typically are hydrophobic, the overall solvency of the protein could decrease. This could be expected to lead to an increased surface activity due to both direct solubility effects and indirect effects due to solvency-induced protein self-assembly, analogous to the performance of simpler systems, such as homo- and copolymers [1], and to the maximum adsorption frequently found for proteins at their isoelectric point [2,3]. Furthermore, it has been observed that the adsorption of proteins at various interfaces may be entropically driven, and it has been suggested that one origin of this increase in the entropy is the loss of ordered structure, e.g., α-helices, on adsorption, which in fact has been observed in numerous experimental systems [2,3] (see following discussion). However, other mechanisms for the observed entropy gain on adsorption can also be envisaged, including, e.g., the release of hydration water and/or counterions for both the protein and the surface on adsorption.

Protein mutants offer interesting possibilities for the investigation of the effects of the conformational stability on the adsorption of proteins, since structure mutants are fairly straightforwardly achievable by protein engineering at an essentially fixed molecular weight and unchanged composition of residues at the protein surface [12]. Not surprisingly, therefore, structural stability investigations constitute one of the most frequent of the different types of studies possible with protein mutants. Until today several different systems have been investigated, including T4 lysozyme [12–17], tryptophane synthase α-subunits [18], and human carbonic anhydrase II [19,20].

One of the first systems investigated regarding the effects of protein structural stability on its interfacial behavior was tryptophan synthase α-subunits, which is available in numerous different mutants ranging in denaturation free energy between about 5 and 10 kcal/mol (at pH 9), which can be further controlled by varying pH. As found by Kato and Yutani, these have very different surface activity [18]. More precisely, a clear correlation was found between the protein structural stability and its surface activity, since the air–water surface tension was found to vary in an essentially linear manner with the free energy of denaturation. Since the surface and interfacial tension are crucial parameters for foam formation and emulsification, also the latter were found to depend on the protein structural stability in a similar manner —i.e., the less stable the protein, the better the emulsification and the better the foam stability. However, only a limited amount of information on the mechanisms of the increased interfacial activity with a decreasing protein structural stability was obtained from these experiments.

Analogous to the tryptophan synthase α-subunits, bacteriophage T4 lysozyme is available in a range of conformational stabilities, achieved primarily through point mutations in the Ile3 position [12]. Also the interfacial behavior of these mutants has been investigated. For example, McGuire et al. investigated the adsorption of a series of T4 lysozyme stability mutants at silica and methylated silica surfaces with ellipsometry, and found some trends for the equilibrium adsorbed amount at both these surfaces to follow the protein structural stability (Table 1) [13]. A somewhat more straightforward relationship appeared to exist between the degree of removal ("elutability") of the adsorbed protein by a cationic surfactant and the structural stability, indicating the less stable proteins undergo larger interfacial conformational changes. (Subsequent investigations showed that removal by an ionic surfactant occurred at a lower surfactant concentration for the less stable mutant than for the wild-type protein. From the ionic strength dependence it was inferred that this was due to an earlier onset of the cooperative binding of the surfactant to the less stable, i.e., more structurally altered, protein [39].) A similar conclusion was reached from an analysis of the adsorption kinetics displayed by these mutants [17]. Later, this was shown directly by Billsten et al. for T4 lysozyme at silica, using circular dichroism (CD) [15]. As seen in Fig. 1, the helical content after adsorption at silica of three proteins with different structural stability but with essentially identical helical content in solution decreases with the protein structural stability. Note that the kinetics of interfacial conformational change are also affected by the protein structural stability and increase with decreasing stability (Fig. 1b). Indeed, for the less stable mutants the interfacial conformational changes are of such a scale that they may even be detected by methods monitoring the overall adsorbed layer thickness. Thus, Fröberg et al. investigated the adsorption of wild-type lysozyme and the mutant Ile3 → Trp, where the latter is 2.8 kcal/mol less stable than the former, and found that after adsorption at mica the adsorbed layer thickness for the stability mutant was much lower (15–17 Å) than that of the wild type (45–50 Å) (Fig. 2) [16]. It was also observed that the short-range attraction on separation between the adsorbed protein

TABLE 1 Amount of T4 Lysozyme Mutants Adsorbed at Silica and Methylated Silica after (a) Adsorption from 0.01 M Sodium Phosphate Buffer, pH 7.0, for 30 min, (b) Rinsing with Buffer for 30 min after the Initial Adsorption, and (c) Addition of the Cationic Surfactant DTAB for 45 min, Followed by Another Rinsing with Buffer

Protein	$\Delta\Delta dG$ (kcal/mol)	Γ (mg/m$_2$)			Elutability resistance
		(a)	(b)	(c)	
Ile3 → Cys	+1.2	2.55*	1.63	0.45	0.28
		1.59	1.32	0.22	0.17
Wild type	±0	2.84	2.38	0.84	0.35
		3.28	2.94	0.65	0.22
Ile3 → Ser	−1.2	4.29	3.05	0.87	0.29
		3.80	2.17	0.66	0.30
Ile3 → Trp	−2.8	2.24	1.90	1.11	0.58
		2.19	2.16	0.62	0.29

*First and second values refer to results obtained for silica and methylated silica, respectively.
Source: Ref. 13.

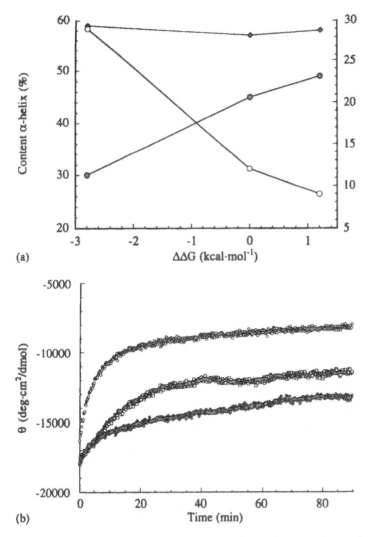

FIG. 1 (a) Content of α-helix in adsorbed (filled circles) and nonadsorbed (filled diamonds) T4 lysozyme mutants of different structural stability calculated from CD spectra. Shown also is the loss in α-helix content on adsorption (open circles). (b) Molecular ellipticity at 222 nm as a functon of time for the wild type (squares), Ile3 $\rightarrow$ Trp (circles), and Ile3 $\rightarrow$ Cys (diamonds) adsorbing at silica particles. (Data from Ref. 15.)

layer and pure mica was an order of magnitude larger for the conformationally altered mutant in comparison to the wild type (not shown), indicating a strong effect on the protein–surface interaction. Hence, it seems clear that structural changes do occur in these systems on adsorption to an extent depending on the protein structural stability.

Another system which has been investigated in relation to the effect of structural stability on the protein interfacial behavior is human carbonic anhydrase II. In this case, a series of variants of different structural stability can be obtained by

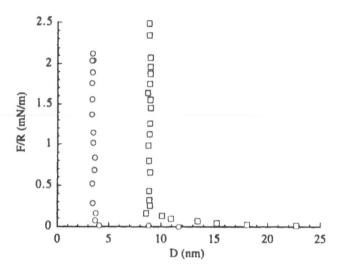

FIG. 2 Force normalized by curvature for wild-type lysozyme (squares) and the Ile3 → Trp
T4 lysozyme mutant (circles) adsorbed at mica from water. Only forces measured on approach
are shown. (Data from Ref. 16.)

truncation of the N terminus with a certain number of amino acids [19]. Although
this system suffers from the molecular weight not being entirely constant for the
different stability variants, it has the nice feature of being straightforwardly quantifiable regarding biological activity. It has been found that a decreased activity is
reached at lower concentrations of a denaturing cosolute (GuHCl) for the truncated
proteins as compared to the intact proteins, which was interpreted as being a consequence of the destabilization of the native structure relative to an intermediate state
[19]. Using a series of human carbonic anhydrase II proteins, Billsten et al. investigated the adsorption of this protein at silica using a range of techniques, including
CD and fluorescence, and found that as the structural stability of the protein was
decreased, the content of ordered structure in the protein after adsorption was reduced, in line with the findings for T4 lysozyme [20]. Furthermore, the active site
structure was changed on adsorption for the truncated proteins but not for the intact
protein (Fig. 3). Although the interfacial enzymatic activity was not determined, these
findings seem to indicate an inverse correlation between enzymatic activity and interfacial conformational changes for this system. This is in line with the frequently
observed loss in biological activity on adsorption, e.g., in relation to protein and
peptide drugs. Note, however, that the loss in activity and native structure on adsorption by no means is universal. In fact, there are systems which actually depend
on these events. Examples of this are lipases, which may have a lower activity in
bulk solution than after adsorption at an interface [21,22, and references therein].

    Although the existence of a correlation between structural stability in solution,
structural loss on adsorption, and interfacial activity seems clear in numerous systems, the origin of these effects is still somewhat unclear. In particular, the relative
importance of exposure of hydrophobic groups, affecting the protein–surface, protein–protein, and protein–solvent interactions, entropy gain related to conformational
freedom and translational entropy related to hydration water and counterions for both

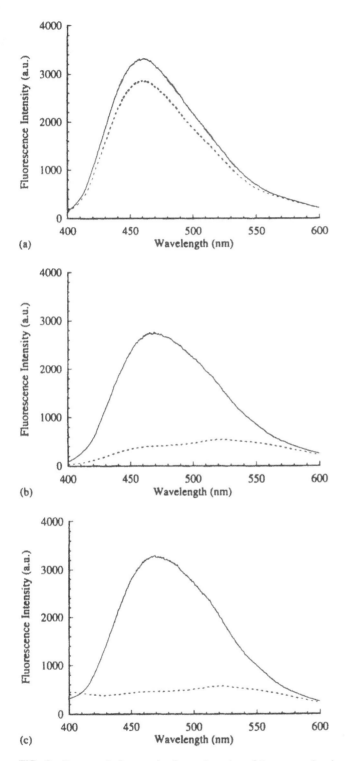

FIG. 3 Structural changes in the active site of human carbonic anhydrase II as observed by monitoring the fluorescence of an extrinsic active site probe, DNSA. Shown are the results obtained for the proteins in solution (solid lines) and for the proteins after 24 h equilibration with silica particles (dashed lines) for (a) the pseudo–wild-type protein, (b) Trunc 5 (i.e., four amino acids truncated), and (c) Trunc 17 (i.e., 16 amino acids truncated). (Data from Ref. 20.)

the protein and the surface, is unclear at present. As indicated, McGuire et al. have in several studies compared the behavior at silica and methylated silica, and found that both the adsorption and the interfacial conformation depend on the protein structural stability for both these surfaces, but primarily so for the hydrophilic and negatively charged silica surface [13,17]. Furthermore, as already discussed, significant interfacial conformational changes were found by Billsten et al. [15] and Fröberg et al. [16] for hydrophilic silica and mica surfaces, respectively. This could be interpreted to indicate that the effects on interfacial conformational change of the protein-surface interaction is of minor importance, since the protein–surface interaction is not likely to be favored by a conformational change and exposure of hydrophobic residues in the case of silica and mica. This, however, is contradicted by the adhesion data from the surface force measurements by Fröberg et al. (see above). On the other hand, Su et al. concluded that the limiting adsorption of lysozyme at silica was dependent on the protein–protein interactions [23]. As the protein stability decreases, structural adaptations become more accessible, which may therefore affect the adsorption. This is to some extent analogous to the oligomerization-dependent adsorption extensively discussed in Chapter 12. Regarding the conformational and translational entropy gain processes on adsorption, further systematic studies are in our mind required to reach an improved understanding as to their relative importance.

IV. INTERFACIAL BEHAVIOR OF INTERACTION MUTANTS

When a homopolymer adsorbs at an interface, it typically loses translational and conformational entropy, since it is not free to move as extensively as in the bulk solution and since the presence of the interface dramatically reduces the number of possible conformations [1]. Consequently, the adsorption of homopolymers is enthalpically driven, and in the absence of a polymer–surface attractive interaction the polymer is depleted from the interface. In the presence of an attraction outbalancing the entropic loss the polymer adsorbs to an extent increasing with the polymer-surface interaction until high adsorption energies, when the adsorption becomes essentially independent of the polymer–surface attraction. Due to the typically large number of monomers in a polymer, even a weak monomer–surface interaction results in a high overall adsorption driving force, and the adsorption is therefore highly cooperative.

For proteins, the situation is somewhat more complex, since the adsorption of a protein may result in a conformational entropy gain due to loss in ordered structure. As discussed, it has been concluded that the adsorption of proteins may in some cases be entropically driven, e.g., due to the conformational entropy gain on reduction in the content of ordered conformations or due to gain in entropy related to protein and surface counterions and/or hydrating water molecules. Nevertheless, also for these systems the adsorption could be expected to be favored by an attractive protein–surface interaction. Examples showing this include the higher adsorption for a number of (negatively charged) proteins at hydrophilic and positively charged surfaces than at similarly hydrophilic and negatively charged surfaces, as well as the typically higher adsorption at the latter of similarly charged hydrophilic and hydrophobic surfaces (cf silica and methylated silica) [2,3,24,25].

In order to illustrate the importance of protein–surface and protein–protein interactions for the adsorption of proteins, Fig. 4 shows the adsorption at methylated silica of proteoheparan sulfate, a glycoprotein consisting of a hydrophobic peptide domain and hydrophilic and highly negatively charged polysaccharide domains [26]. Naturally, the former promotes adsorption at the hydrophobic surface, whereas the adsorption is opposed by electrostatic interactions due to the polysaccharide domains, originating, e.g., from an image charge repulsion between the protein charges and the low-dielectric-constant surface, from a direct electrostatic interaction between the similarly charged protein and surface, and from an electrostatic protein–protein interaction [1,26,27]. Clearly, the adsorption will be determined by the balance in these interactions. From Fig. 4, one can see that at low electrolyte concentration, a limited adsorption occurs at methylated silica, indicating that the repulsive interactions do not totally dominate. For silica, on the other hand, no adsorption was observed. This could be expected, since no hydrophobic adsorption driving force exists in this case. When $CaCl_2$ is added to the system, the adsorption at methylated silica increases notably, which is a consequence of the increased screening of the electrostatic interactions, making the hydrophobic attraction relatively more important. However, also at high electrolyte concentration, no adsorption was observed at methylated silica for the polysaccharide side chains in the absence of the hydrophobic peptide domain [28]. Clearly, in this case the system is effectively below the critical adsorption energy, and the repulsive protein–surface interaction dominates.

Although the example indicates the importance of protein–surface and protein–protein interactions for adsorption, the results are not entirely conclusive. In particular, the molecular weight of the polysaccharide side chains is much lower than that of the intact proteoheparan sulfate. Since the adsorption is expected to increase with increasing molecular weight, especially at highly electrostatically screened conditions [1], molecular weight effects cannot be excluded.

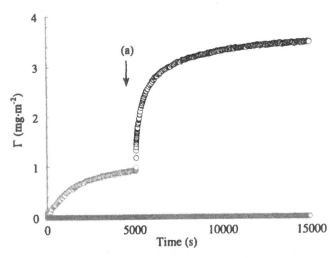

FIG. 4 Amount of proteoheparan sulfate adsorbed at methylated silica as a function of time (open circles). At zero time and point (a), proteoheparan sulfate and 1.25 mM $CaCl_2$ were added, respectively. Illustrated also is the (null) adsorption of the heparan sulfate side chains. (Data from Ref. 26.)

The structure of proteins in general is governed by a balance of effects, e.g., electrostatic, hydrophobic, and van der Waals interactions; hydrogen bonding; solvation; and entropic effects due to conformational restrictions [2,3,27]. This balance may be quite delicate and thus easily perturbed by changing the solution condition (e.g., pH or electrolyte concentration), temperature, addition of cosolutes (e.g., surfactants, urea, and guanidinium hydrochloride, GuHCl), the presence of an interface (see above), and, notably, mutations. Since pertubations in protein structure and/or structural stability may affect the protein interfacial behavior (discussed above), it is important when investigating the effects of interactions on protein adsorption that the interaction be changed essentially without affecting the protein structure and structural stability.

In one of the first attempts to probe these effects, McGuire et al. investigated the effects of net charge and charge location on the adsorption and surfactant-induced elutability of bacteriophage T4 lysozyme at silica and methylated silica with ellipsometry [14]. In this investigation, a series of proteins ranging in net charge from +5 to +9 were obtained by point mutation in either the C- or the N-terminal lobe. Unfortunately, no clear correlation was observed between the protein charge, on one hand, and adsorption and elutability, on the other. Instead, the location of the mutation was found to be of importance. At silica, mutants allowing close proximity of positive charges of the protein and the surface negative charges were found to have a decreased elutability, whereas at methylated silica a decreased elutability was observed for mutations favoring a hydrophobic interaction between the protein and the surface. However, the point mutation approach used did not allow the structural stability to be preserved between the different proteins. Instead, the denaturation free energy spanned the range -1.5 kcal/mol $\leq \Delta\Delta G \leq +0.5$ kcal/mol compared to the wild type. Since the structural stability has been shown to be of importance for the adsorption of this and other proteins, effects related to this parameter cannot be excluded in the interpretation of the data.

In order to investigate the importance of the protein–surface interaction for protein adsorption and interfacial activity, Wannerberger and Arnebrant studied the adsorption, e.g., of the wild type and a mutant of lipase from *Humicola lanuginosa*, where the latter was modified from the former by an Asp96 → Leu mutation [21,22]. Thus, the mutant is more hydrophobic than the wild type. As can be seen in Table 2, the higher hydrophobicity of the mutant resulted in a higher adsorbed amount at methylated silica compared to the wild type. The effect of the surface hydrophobicity on the adsorption of the two proteins was investigated over a range of contact angles ($<10°$–$90°$). For both proteins, the adsorbed amount was found to decrease with decreasing contact angle. For silica, with a contact angle of less than $10°$, a very low adsorbed amount (<0.10 mg/m^2) was observed for both proteins. The latter finding indicates that the higher adsorption of the mutant at methylated silica really is due to a larger hydrophobic protein–surface attraction and not to solvency effects. Considering this, the finding that the nondesorbable fraction after surfactant addition and subsequent rinsing was larger for the mutant than for the wild type [21] is quite expected. In parallel to the adsorption measurements, Wannerberger and Arnebrant monitored the interfacial enzymatic activity with respect to the hydrolysis of a water-soluble substrate (*p*-nitrophenylacetate, PNPA) and, as can be seen in Table 2, the specific activity of the mutant is lower than that of the wild type. The latter finding may possibly be explained by the stronger hydrophobic interaction between the pro-

TABLE 2 Adsorbed Amount (Γ; mg/m^2), Total Enzymatic Activity (A_{tot}; nmol/ min), Reference Activity (A_{ref}; nmol/min),[a] and Specific Activity (A_{spec}; nmol, m^2/min, mg)[b] of Lipase Adsorbed from a 345 nM Solution to Methylated Silica (MeSi) of Different Wettability

	Γ	A_{tot}	A_{ref}	A_{spec}
Wild-type				
Solution	—	—	23 ± 2	—
MeSi/90°	1.83 ± 0.15	74 ± 7	23 ± 2	29 ± 5
MeSi/80°	1.68 ± 0.26	73 ± 2	23 ± 2	30 ± 3
MeSi/75°	1.20 ± 0.00	68 ± 2	23 ± 2	38 ± 2
Mutant				
Solution	—	—	56 ± 0	—
MeSi/90°	2.18 ± 0.02	75 ± 6	56 ± 0	9 ± 3
MeSi/80°	1.97 ± 0.08	87 ± 3	56 ± 0	16 ± 1
MeSi/75°	1.92 ± 0.03	87 ± 8	56 ± 0	16 ± 4

[a]The reference activity is the contribution from the cuvette walls, Teflon tubings, and magnetic stirrer bar.
[b]The specific activity is the total activity with the reference activity subtracted, divided by the adsorbed amount after rinsing.
Source: Ref. 21.

tein and the surface, causing the mutant to adsorb with its more hydrophobic active site surroundings oriented toward the surface in order to maximize the attractive interaction. This is further supported by the finding of an increasing specific activity with a decreasing surface contact angle, allowing orientations other than those with the active site surrounding directed toward the surface to a higher extent.

Malmsten and Veide studied the effects of insertions of hydrophobic peptide stretches in a hydrophilic and structurally stable protein called Z, a synthetic protein corresponding to the IgG binding domain of staphylococcal protein A [29]. By performing insertions of the peptide stretches T$_n$ [(AlaTrpTrpPro)$_n$] and I$_n$ [(AlaIle-IlePro)$_n$] in the C terminus of Z or its dimer ZZ, the number of inserted hydrophobic amino acids could be varied from 0 to 6 without any detectable change in structure or structural stability, as inferred from CD and GuHCl titrations (Fig. 5) [30–33]. Furthermore, due to the Z and ZZ proteins being much larger than the peptide insertions, these measurements were performed at an essentially constant molecular weight. In parallel, measurements of the adsorption of the inserted peptide stretches per se as well as of oligo amino acids were performed.

It was found that as the length of the hydrophobic stretch increases, so does the adsorption at hydrophobic methylated silica surfaces (Fig. 6a). This is a consequence of the increasingly attractive interactions between the hydrophobic peptide stretch and the hydrophobic surface, since no adsorption was found irrespective of peptide stretch length at hydrophilic and negatively charged (silica) and hydrophilic and uncharged [poly(ethylene oxide) modified] surfaces. The increase was found for the Ile- and Trp-containing peptide stretches and for the Z and the ZZ proteins. Also for the inserted peptide stretches alone and the oligo amino acids, the adsorption at methylated silica was found to increase with the length of the molecules, whereas no adsorption was detected at the hydrophilic surfaces.

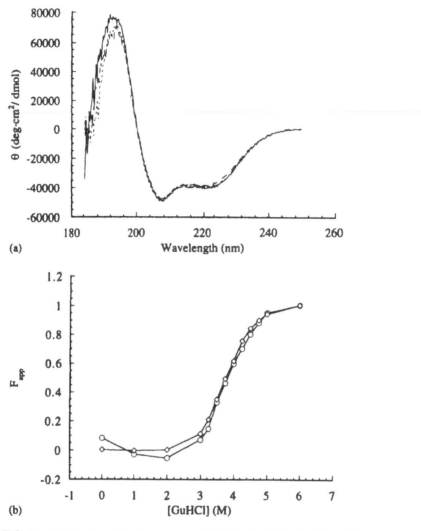

FIG. 5 (a) Circular dichroism spectra for ZT_0 (solid line), ZT_1 (dashed line), ZT_2 (dotted line), and ZT_3 (dashed-dotted line). (b) GuHCl titration for ZT_0 (diamonds) and ZT_2 (circles). (Data from Ref. 32.)

It is interesting to compare these results with those found for simpler systems, i.e., homopolymers and copolymers. As described earlier these lose conformational entropy at close proximity to an interface, and for the adsorption to occur this entropy loss must be compensated by an enthalpic adsorption driving force. At a given polymer–surface interaction (neglecting end group effects), the adsorption becomes more favorable with increasing molecular weight, since the entropy loss on adsorption (per segment) decreases with increasing polymer molecular weight. Therefore, the adsorbed amount typically increases with molecular weight. Although this effect is less pronounced for polyelectrolytes, especially at low salt concentration, it occurs also for these systems at high electrolyte concentrations, and one could therefore expect

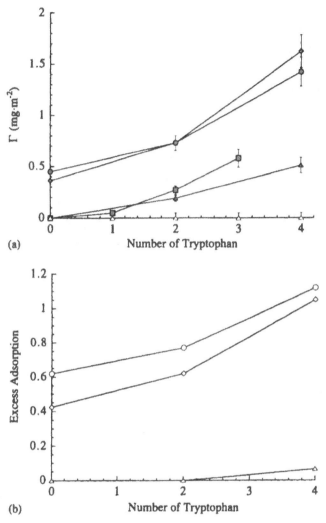

FIG. 6 (a) Ellipsometrically determined adsorbed amount of ZZT_n (circles), ZT_n (diamonds), T_n (triangles), and oligo-Trp (squares) versus the number of Trp in the protein/peptide at methylated silica (filled symbols) and silica and poly(ethylene oxide)–modified silica (all open symbols). (b) Excess adsorption [$\Gamma_{ex} = \Sigma_i (\Phi_i - \Phi_b)$, where Φ_i and Φ_b are the protein volume fractions in layer i and bulk solution, respectively] for ZZT_n (circles), ZT_n (diamonds), and T_n (triangles) at methylated silica. (Data from Ref. 29.)

it to occur at least to some extent also for proteins and peptides at physiological electrolyte concentrations. Consequently, the increasing adsorption with the molecular size found for the inserted peptides and the oligo amino acids is in line with these findings for simpler systems.

For the Z and ZZ proteins, the peptide insertions occur at essentially constant molecular weight and overall protein structure and structural stability, the only essential effect of the insertions being an increasingly long hydrophobic anchor attached at the protein. It is therefore interesting to compare the findings for these systems with those for block copolymers with one weakly adsorbing block and one strongly adsorbing block in a selective solvent at fixed total molecular weight. Thus, for such a system, the adsorbed amount is found to increase with an increasing length of the adsorbing block, at least to a limit corresponding to a fraction of preferentially adsorbing blocks typically larger than those used in the Z and ZZ proteins [1].

Furthermore, although the adsorbed amount of the Z protein is lower than that of the ZZ protein, presumably due to a lower molecular weight, a given number of hydrophobic amino acids inserted has a bigger effect for the adsorption of the former protein, which could be expected since each hydrophobic Trp or Ile residue is balanced by fewer hydrophilic amino acids for the Z than for the ZZ proteins, and therefore the effect of each hydrophobic residue will be larger for the corresponding Z protein. Clearly, these proteins appear to behave similarly to the much simpler block copolymer systems, and, in fact, these effects could be modelled successfully with a lattice mean-field theory for heterogeneous systems [34] (Fig. 6b).

The combined effects of hydrophobic and electrostatic interaction on the adsorption of the ZZ proteins were investigated previously [35]. In this study, the adsorption of proteins containing T_n stretches was compared to the adsorption of proteins containing stretches of N_n [(AlaTrpTrpAspPro)$_n$] and P_n [(AlaTrpTrp-LysPro)$_n$], i.e., identical except for the additional Asp and Lys negative and positive charges, respectively. As for the ZZT$_n$ proteins, neither the ZZP$_n$ nor the ZZN$_n$ proteins were found to adsorb at the hydrophilic and negatively charged silica surfaces (Fig. 7a). This is expected for the latter proteins, and somewhat less expected for the former. However, the net charge of the ZZ protein at the conditions used in the measurements was −5, and also for ZZP$_1$ the protein thus has a net negative charge (of −2) [33]. Although the measurements were performed at highly electrostatically screening conditions $\kappa_D^{-1} \approx 4$ Å), the charge of the ZZ domain outbalances the attractive interaction due to the peptide stretch for this surface (Fig. 7a,b).

Going from silica to methylated silica the electrostatic charge of the surface is essentially unchanged (electrostatic potential about −45 mV for both surfaces at 1 mM NaCl, pH 7.0) [35] at the same time as the contact angle with water in air increases from less than 10° to about 95° [25]. Thus the interaction between the P_n stretches and the surface is expected to be more attractive for methylated silica than for silica, as seen also from the higher adsorbed amount at the former surface (Fig. 7b). This is the case also for the T_n and N_n stretches. In the latter case, the hydrophobic attraction between the Trp amino acids and the surface is counteracted by an electrostatic repulsion between the similarly charged Asp and dissociated silanol surface groups, and therefore the adsorption of N_n at methylated silica is somewhat smaller than that of T_n.

Also for the modified ZZ proteins, this interplay between hydrophobic and electrostatic interactions may be observed. Thus, for all stretch lengths investigated, the adsorption at methylated silica increases in the presence of an additional electrostatic attraction (ZZP$_n$) and decreases in the presence of an additional electrostatic repulsion (ZZN$_n$). As discussed previously, this behavior is analogous to that of much simpler systems, such as copolymers and polyelectrolytes. Thus, for a polyelectrolyte, the effective interaction with the surface (χ_s^{eff}) is given by

$$\chi_s^{\text{eff}} = \chi_s - \frac{\tau e \alpha \sigma_d}{\varepsilon \kappa kT} \tag{1}$$

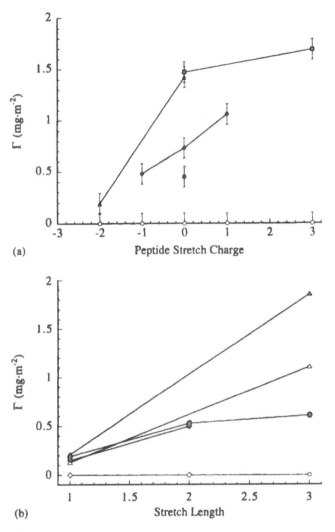

(a)

(b)

FIG. 7 (a) Effects of the charge of the inserted peptide stretch on the adsorption of ZZ at methylated silica ($\zeta = -45$ mV) at a stretch length of 3 (squares), 2 (triangles), 1 (diamonds), and 0 (filled circle). For comparison, results obtained for silica are also shown (all open circles). (b) Effect of the stretch length on the saturation adsorption at silica (open symbols) and methylated silica (filled symbols) of T_n (circles), N_n (diamonds), and P_n (triangles). The adsorption was performed from 0.5 M potassium phosphate, pH 7.0. (Data from Ref. 35.)

where χ_s is the nonelectrostatic polyelectrolyte–surface interaction, τe the charge per segment, α the dissociation constant, σ_d the Stern charge density, κ the inverse Debye length, ε the dielectric permittivity, k the Boltzmann constant, and T the temperature [1]. Therefore, χ_s^{eff} decreases (increases) with an increasing (decreasing) segment and/or surface charge density for similarly charged polyelectrolyte and surface. For oppositely charged polyelectrolyte and surface the electrostatic interaction provides an additional polymer–surface attraction. Considering this, the modulation of the adsorption for both the ZZ proteins and the Trp-containing peptide stretches in the presence of Asp and Lys charges (Fig. 7a,b) are in line with this behavior, as is the high adsorption observed to the cationic diaminocyclohexane plasma polymer surface, irrespective of the length and nature of the stretch insertion (not shown) [35].

The effects of the protein–surface interaction on the kinetics of protein adsorption have also been investigated for mutants of the soluble core tryptic fragment of cytochrome b5. Thus, using an optical waveguide technique at well-defined hydrodynamic conditions, Ramsden et al. investigated effects of glutamic acid to glutamine point mutations in positions 15 or 48 [36]. Since the soluble core of cytochrome b5 is a quite stable molecule, these mutations could be achieved at essentially constant structural stability. From the experimental data, it is clear that changing the location of a single charge in this protein drastically affects the adsorption kinetics. Furthermore, from a quantitative analysis of the adsorption kinetics of the two different mutants at a range of surfaces, it was concluded that the adsorption could largely be interpreted in terms of the protein–surface interactions.

V. INTERFACIAL BEHAVIOR OF ASSOCIATION MUTANTS

As discussed, homopolymers are typically found to adsorb more extensively with increasing molecular weight, which is a consequence of a decreasing entropy loss on adsorption per segment. Although this effect is less pronounced for charged macromolecules, it is still significant for polyelectrolytes at high electrolyte concentrations [1]. Therefore, also for proteins at high electrolyte concentrations, e.g., physiological conditions, the adsorption could be expected to increase with the protein size, although this effect could in some systems be obscured by the solvation water and counterion contributions to the total entropy, as well as by a conformational entropy gain on adsorption due to loss in content of ordered structure. In order to probe these effects, it is advantageous to utilize self-assembling systems, since for this type of system the molecular weight effects can be investigated at an essentially constant structural stability and, at least in certain cases, a nearly constant set of interactions.

A number of naturally occurring self-assembling protein systems have been investigated regarding their interfacial behavior. This is the case, e.g., for bovine serum albumin (BSA) and human serum albumin (HSA). The adsorption of BSA monomers and dimers has been investigated, e.g., by Okubo et al. [37], who found the dimer to adsorb preferentially over the monomer for a range of surfaces and a range of protein concentrations. Similar findings have been obtained for HSA by Lensen et al. [38]. For a more extensive discussion on this, see Chapter 15.

Furthermore, as discussed in much greater detail in Chapter 12, insulin self-associates to form a range of oligomers. In a series of investigations, Nylander, Arnebrant, and others have studied the adsorption of this protein in relation to its self-association. As discussed by Nylander, self-assembly induced, e.g., by nonspecific electrostatic screening or specific cations generally favors adsorption (Chapter 12). Of particular interest to the present chapter is that investigations have been performed with a mutated insulin not undergoing self-assembly. As can be seen from Table 1 in Chapter 12, the mutated and the wild-type insulin adsorb to a similar extent at methylated silica in the absence of Zn^{2+}, where the absorption is believed to occur mainly through monomers. On the other hand, at silica, where insulin adsorbs mainly in its hexameric form, no adsorption was found for the monomeric mutant. Although the oligomerization dependence could be expected to be at least partly due to a molecular weight effect, this system is somewhat more complex, since the monomeric form is more hydrophobic than the oligomeric forms, and therefore both the protein–surface and "protein–protein" interaction, as well as general solvency effects, should be considered in the interpretation of these results.

Another protein undergoing self-assembly is β-lactoglobulin. A nice feature of this protein is that it exists in several different variants showing different tendencies for self-assembly, with essentially identical structure and structural stability [6–9]. Analogous to the behavior of insulin, HSA, and BSA, the adsorption of the two β-lactoglobulin variants A and B was found to display a shift in the concentration-dependent adsorption corresponding to the difference in dissociation constant for the two proteins (see Fig. 8, Chapter 12) [6]. Hence, it is clear that the dimer adsorbs preferentially over the monomer, analogous to the behavior of BSA and HSA, and to the molecular-weight dependence of the adsorption of simpler macromolecules, such as homopolymers, copolymers, and polyelectrolytes [1]. The effects of self-assembly on protein interfacial behavior are discussed more extensively in Chapter 10.

VI. CONCLUDING REMARKS

Both naturally occurring protein variants and mutated proteins from genetic engineering offer interesting possibilities for systematic investigations of the mechanisms of protein adsorption, exemplified in this chapter by studies of the effects of the protein molecular weight, structural stability, and protein–surface and protein–protein interactions for the interfacial behavior. However, it is essential that the amino acid variations are performed in such a way that only one of these parameters is changed at a time if systematic information is to be obtained as to the relative importance of these and other effects. To this end, it is crucial to consider how the variations should be achieved, using either point mutations in the protein interior or exterior, insertion of peptide stretches, truncations, etc. Finally, a close analogy appears to exist between protein adsorption, on one hand, and the interfacial behavior of simpler macromolecules, such as homopolymers, copolymers, and polyelectrolytes, on the other.

ACKNOWLEDGMENTS

This work was financed by the Foundation for Surface Chemistry, Sweden, the Swedish National Board for Industrial and Technical Development (NUTEK) (MM), the Swedish Research Council for Engineering Sciences (TFR) (PB and TA), and the Swedish Medical Research Council (MFR) (TA). Tommy Nylander, Joseph McGuire, Svend Havelund, Marie Wahlgren, Ulla Elofsson, Kristin Wannerberger, Johan Fröberg, Per Claesson, and Hans Elwing are thanked for fruitful discussions on this subject over the years.

REFERENCES

1. G. J. Fleer, M. A. Cohen Stuart, J. M. H. M. Scheutjens, T. Cosgrove, and B. Vincent, *Polymers at Interfaces*. Chapman & Hall, London, 1993.
2. W. Norde, Adsorption of proteins from solution at the solid–liquid interface. Adv. Colloid. Interface Sci. *25*:267–340 (1986).
3. C. A. Haynes and W. Norde, Globular proteins at solid/liquid interfaces. Colloids Surf. B *2*:517–566 (1994).
4. D. Horsley, J. Herron, V. Hlady, and J. D. Andrade, Human and hen lysozyme adsorption: a comparative study using total internal reflection fluorescence spectroscopy and molecular graphics. In *Protein at Interfaces: Physicochemical and Biochemical Studies*. (J. L. Brash, T. A. Horbett, eds.). ACS Symposium, Washington DC, 1987, pp. 290–305.
5. S. Xu and S. Damodaran, Comparative adsorption of native and denatured egg-white, human, and T4 phage lysozymes at the air–water interface. J. Colloid. Interface Sci. *159*:124–133 (1993).
6. U. M. Elofsson, M. A. Paulsson, and T. Arnebrant, Adsorption of β-lactoglobulin A and B in relation to self-association: effect of concentration and pH. Langmuir *13*:1695–1700 (1997).
7. U. M. Elofsson, M. A., Paulsson, and T. Arnebrant, Adsorption of β-lactoglobulin A and B: effect of ionic strength and phosphate ions. Colloids Surf. B *8*:163–169 (1997).
8. M. Wahlgren and U. Elofsson, Simple models for adsorption kinetics and their correlation to the adsorption of β-lactoglobulin A and B. J. Colloid. Interface Sci. *188*:121–129 (1997).
9. U. M. Elofsson, M. A. Paulsson, P. Sellers, and T. Arnebrant, Adsorption during heat treatment related to the thermal unfolding/aggregation of β-lactoglobulins A and B. J. Colloid. Interface Sci. *183*:408–415 (1996).
10. D. Elbaum, J. Harrington, E. F. Roth, and R. L. Nagel, Surface activity of hemoglobin S and other human hemoglobin variants. Biochim. Biophys. Acta *427*:57–69 (1976).
11. T. Arnebrant, K. Barton, and T. Nylander, Adsorption of α-lactalbumin and β-lactoglobulin on metal surfaces versus temperature. J Colloid. Interface Sci. *119*:383–390 (1987).
12. M. Matsumura, W. J. Becktel, and B. W. Matthews, Hydrophobic stabilization in T4 lysozyme determined directly by multiple substitutions of Ile 3. Nature *334*:406–410 (1988).
13. J. McGuire, M. C. Wahlgren, and T. Arnebrant, Structural stability effects on the adsorption and dodecyltrimethylammonium bromide–mediated elutability of bacteriophage T4 lysozyme at silica surfaces. J. Colloid. Interface Sci. *170*:182–192 (1995).
14. J. McGuire, M. C. Wahlgren, and T. Arnebrant, The influence of net charge and charge location on the adsorption and dodecyltrimethylammonium bromide–mediated elutability of bacteriophage T4 lysozyme at silica surfaces. J. Colloid. Interface Sci. *170*:193–202 (1995).
15. P. Billsten, M. Wahlgren, T. Arnebrant, J. M. McGuire, and H. Elwing, Structural changes of T4 lysozyme upon adsorption to silica nanoparticles measured by circular dichroism. J. Colloid. Interface Sci. *175*:77–82 (1995).
16. J. C. Fröberg, T. Arnebrant, J. McGuire, and P. M. Claesson, Effects of structural stability on the characteristics of adsorbed layers of T4 lysozyme. Langmuir *14*:456–462 (1998).
17. B. Singla, V. Krisdhasima, and J. McGuire, Adsorption kinetics of wild type and two synthetic stability mutants of T4 phage lysozyme at silanized silica surfaces. J Colloid. Interface Sci. *182*:292–296 (1996).
18. A. Kato and K. Yutani, Correlation of surface properties with conformational stabilities of wild-type and six mutant tryptophan synthase α-subunits substituted at the same position. Protein Eng. *2*:153–156 (1988).

19. G. Aronsson, L.-G. Mårtensson, U. Carlsson, and B.-H. Jonsson, Folding stability of the N-terminus of human carbonic anhydrase II. Biochemistry 34:2153–2162 (1995).

20. P. Billsten, P.-O. Freskgård, U. Carlsson, B.-H. Jonsson, and H. Elwing, Adsorption to silica nanoparticles of human carbonic anhydrase II and truncated forms induce a molten-globule–like structure. FEBS Lett. 402:67–72 (1997).

21. K. Wannerberger and T. Arnebrant, Comparison of the adsorption and activity of lipases from *Humicola lanuginosa* and *Candida antarctica* on solid surfaces. Langmuir 13: 3488–3493 (1997).

22. K. Wannerberger and T. Arnebrant, Lipases from *Humicola lanuginosa* adsorbed to hydrophobic surfaces—desorption and activity after addition of surfactants. Colloids Surf. B 7:153–164 (1996).

23. T. J. Su, J. R. Lu, R. K. Thomas, Z. F. Cui, and J. Penfold, The effect of solution pH on the structure of lysozyme layers adsorbed at the silica–water interface studied by neutron reflection. Langmuir 14:438–445 (1998).

24. B. Lassen and M. Malmsten, Competitive protein adsorption at plasma polymer surfaces. J. Colloid. Interface Sci. 186:9–16 (1997).

25. M. Malmsten, Ellipsometry studies of the effects of surface hydrophobicity on protein adsorption. Colloids Surf. B 3:297–308 (1995).

26. M. Malmsten, P. Claesson, and G. Siegel, Forces between proteoheparan sulfate layers adsorbed at hydrophobic surfaces. Langmuir 10:1274–1280 (1994).

27. J. N. Israelachvili, *Intermolecular and Surface Forces*. 2nd ed. Academic Press, London, 1992.

28. M. Malmsten, G. Siegel, E. Buddecke, and A Schmidt, Cation-promoted adsorption of proteoheparan sulfate. Colloids Surf. B 1:43–50 (1993).

29. M. Malmsten and A. Veide, Effects of amino acid composition on protein adsorption. J. Colloid. Interface Sci. 178:160–167 (1996).

30. C. Hassinen, K. Köhler, and A. Veide, Polyethylene glycol–potassium phosphate aqueous two-phase systems: insertions of short peptide units into a protein and its effects on partitioning. J. Chromatogr. A 668:121–128 (1994).

31. K. Köhler, C. Ljungquist, A. Kondo, A. Veide, and B. Nilsson, Engineering proteins to enhance their partition coefficients in aqueous two-phase systems. Bio/Technol. 9:642–645 (1991)

32. K. Gustafsson, A. Veide, S.-O. Enfors, and S. Yang, Correlation between DNAk complex formation and proteolysis of recombinant proteins. Bio/Technol. (submitted).

33. K. Berggren, A. Veide, P.-Å. Nygren, and F. Tjerneld, Genetic engineering of protein-peptide fusions for control of protein partitioning in thermoseparating aqueous two-phase systems. Bio/Technol. (submitted).

34. P. Linse and M. Björling, Lattice theory for multicomponent mixtures of copolymers with internal degrees of freedom in heterogeneous systems. Macromolecules 24:6700–6711 (1991).

35. M. Malmsten, N. Burns, and A. Veide, Electrostatic and hydrophobic effects of oligopeptide insertions on protein adsorption. J. Colloid. Interface Sci. (in press).

36. J. J. Ramsden, D. J. Roush, D. S. Gill, R. Kurrat, and R. C. Willson, Protein adsorption kinetics drastically altered by repositioning a single charge. J. Am. Chem. Soc. 117: 8511–8516 (1995).

37. M. Okubo, I. Azume, and Y. Yamamoto, Preferential adsorption of bovine serum albumin dimer onto polymer microspheres having a heterogeneous surface consisting of hydrophobic and hydrophilic parts. Colloid Polym. Sci. 268:598–603 (1990).

38. H. G. W. Lensen, D. Bargeman, P. Bergveld, C. A. Smolders, and J. Feijen, High-performance liquid chromatography as a technique to measure the competitive adsorption of plasma proteins onto latices. J. Colloid. Interface Sci. 99:1–8 (1984).

39. M. C. Wahlgren and T. Arnebrant, Removal of T4 lysozyme from silicon oxide surfaces by sodium dodecyl sulfate: a comparison between the wild type protein and a mutant with lower thermal stability. Langmuir 13:8–13 (1997).

6

Orientation and Activity of Immobilized Antibodies

JAMES N. HERRON, HSU-KUN WANG, VĚRA JANATOVÁ, *
JACOB D. DURTSCHI, and DOUGLAS A. CHRISTENSEN
University of Utah, Salt Lake City, Utah, U.S.A.

KARIN CALDWELL   Uppsala University, Uppsala, Sweden

I-NAN CHANG   Development Center for Biotechnology,
Taipei, Taiwan, Republic of China

SHAO-CHIE HUANG   Diagnostic Products Corporation,
Los Angeles, California, U.S.A.

I. INTRODUCTION

Antibodies have found many applications in biotechnology and clinical medicine, including diagnostics assays, environmental testing, food testing, process monitoring and separations. In all cases the antibody is employed as a molecular recognition element that binds specifically to its antigen with high affinity. Actually the term *antigen* is sort of a misnomer because the mentioned applications are invariably performed in vitro, far removed from the animal which produced the antibodies. Instead, we shall refer to the antigen as the *analyte* in the case of immunoassays and as the *ligand* in the case of chromatography and other separations. In most instances the antibody is immobilized to a solid support—usually a bead or a planar surface —which provides the means for separating the antibody–analyte (or antibody–ligand) complex from unbound antigen and impurities in the sample. Interestingly, this partitioning step is required by both immunoassays and separations—for purposes of quantification in the former and as a means of purification in the latter.

Without question, polysaccharides, polystyrene, and silica are the three most common substrates for antibody immobilization. Polysaccharides are employed in a number of different types of chromatographic supports [1] and more recently as immobilization matrices in biosensors such as BIACORE [2–4]. Polystyrene was the traditional substrate of agglutination assays (latex beads) and the enzyme-linked immunosorbent assay (microtiter plates) [5], but more recently is being employed in biosensors as well [6]. Silica is a very versatile substrate and can be fabricated into

---

Present location: Prague, Czech Republic.

a number of different forms including particles, wafers (e.g., fused quartz microscope slides), and thin films (≤ 1 μm) deposited using plasma chemical vapor deposition. Silica particles are employed in both chromatographic matrices (e.g., HPLC) and immunoassays [1, 7–9], while silica wafers and thin films are finding an increasing number of applications in biosensors [10–12].

Much has been written about the use of immobilized antibodies in affinity chromatography, and several reviews exist [1,7,9,13,14], so this topic will not be given further consideration. Rather, we will focus on applications of immobilized antibodies in immunoassays, with emphasis on the activity, packing, and orientation of these molecules. These topics will be examined from our own perspective of immobilized antibodies employed in biosensors. Many of the experiments are our own, resulting in a definite bias for silica and polystyrene as immobilization substrates because these materials are most often employed in our laboratory. For information about immobilizing antibodies to the carboxymethyl dextran matrix used in the BIAcore biosensor, the reader is referred to articles by Johnsson et al. [2,4] and O'Shannessy et al. [3].

From an operational perspective, the same issues arise in both traditional immunoassays and biosensors, namely, the surface activity of the immobilized antibodies and nonspecific binding (NSB) of various components in the sample to the immunologically active surface. Our definition of surface activity is the percent of immobilized antibody active sites that are capable of binding analyte. Antibodies immobilized by physical adsorption—probably the most common immobilization technique—are often only marginally active (<20%). This can have negative impact on both assay sensitivity and NSB—the former because relatively few active capture molecules are available to bind the analyte, and the latter because sample components are inclined to interact nonspecifically with the preponderance of inactive antibodies which can be partially denatured.

From this discussion it is clear that there are ample reasons to improve upon physical adsorption as an immobilization method. To do so, however, we must first understand why physisorbed antibodies exhibit such low surface activities. The classical viewpoint from the protein adsorption literature is that proteins interact strongly with solid surfaces (especially hydrophobic ones) and tend to unfold or "spread" along the surface in order to maximize the number of interfacial contacts [15]. Thus, an effective immobilization strategy should minimize interfacial contacts between the antibody and the surface in order to maintain the antibody in its native conformation. This can be achieved by making the surface as hydrophilic as possible, a strategy which has the additional benefit of minimizing nonspecific binding with sample components. Steric considerations are important as well and, if possible, the antibody should be oriented so that its active sites are exposed to bulk solution.

Based on these considerations, our laboratory has formulated an immobilization strategy (Fig. 1) in which the surface is first coated with a thin film of hydrophilic material (either polymer or protein) that acts as a matrix for site-specific immobilization of antibodies [6,16–18]. We refer to the thin film as a "passivating layer" because it limits the interaction of proteins and other sample components with the surface. Several materials have been investigated for use in passivating layers, including polyethylene glycol, polymethacrylate hydrogels, and avidin. These studies are presented in this chapter, along with an assessment of the relative merits of each material.

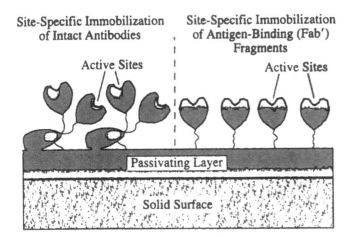

Site-Specific Immobilization of Intact Antibodies

Active Sites

Site-Specific Immobilization of Antigen-Binding (Fab') Fragments

Active Sites

Passivating Layer

Solid Surface

FIG. 1 Strategy for preserving the binding activity of immobilized antibodies. The surface is first coated with a thin film or "passivating layer" of hydrophilic material (either polymer or protein) that acts as immobilization matrix for antibodies and also limits nonspecific interactions with specimen components such as proteins or cells. Either intact antibodies or antigen-binding (Fab') fragments can be attached to the passivating layer using "site-specific" immobilization techniques which orient these molecules so that their active sites are exposed to solvent.

Site-specific conjugation of antibodies to the passivating layer—the other critical element in our immobilization strategy—requires the generation of unique functional groups within a specific region of the antibody molecule, preferably well removed from the active site. Two common approaches have been reported in the literature for achieving this end. The first is based on reproduction of reactive aldehyde groups within the antibody's carbohydrate moieties by mild oxidation with sodium periodate [17–23]. These aldehyde groups are located in the antibody's Fc region about 7 nm away from the active site (Fig. 2) and can react with hydrazido groups incorporated in the passivating layer. In the other approach, the intact antibody is digested with a protease such as pepsin or ficin to produce $F(ab')_2$ fragments (defined in Fig. 2), which in turn are reduced to Fab' fragments using dithiothreitol [24–29]. The reduction produces one or more reactive thiol groups at the C-terminal end of the Fab' fragment (again about 7 nm away from the active site; see Fig. 2) that can react with either maleimide or pyridyl disulfide groups incorporated in the passivating layer [6,18,30–33]. Choosing between the two approaches depends upon several factors, the most important of which is whether Fab' fragments can easily be produced for the particular antibody that is being immobilized. Of the mouse immunoglobulin subclasses, IgG1 is most readily disposed to the production of Fab' fragments, although such fragments can be produced from the IgG2a and IgG3 subclasses as well but with lower yield [24–26].

We have also examined a unique surface chemistry in which the IgG molecule serves as its own passivating layer [16,17,34,35]. This method involves a brief exposure of the antibody to acidic conditions (pH 2.8) for approximately 20 min before immobilization to a hydrophobic surface. Our investigations have shown that such brief acid exposure selectively denatures the Fc region of the antibody and makes it

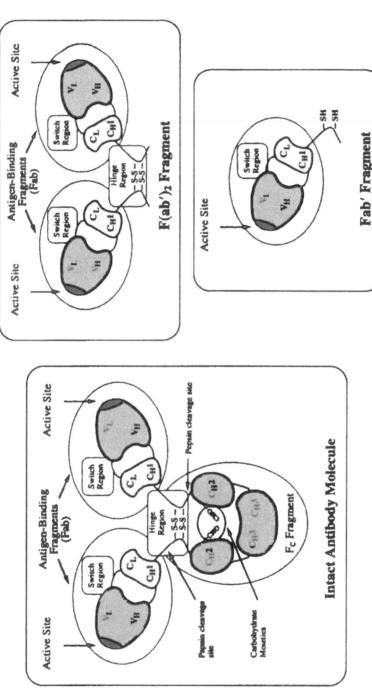

FIG. 2 Immunoglobulin G (IgG) and its antigen-binding fragments. The intact antibody (left panel) consists of (two antigen-binding fragments (Fab) and a constant fragment (Fc). Each of the former comprises four domains (V_L, V_H, C_L, C_{H1}), while the latter comprises two C_{H2} domains and two C_{H3} domains. Each Fab fragment is attached to the Fc fragment by a flexible region of polypeptide referred to as the "hinge region." The molecule has two carbohydrate moieties—one located in each of the two C_{H2} domains—that can be used for site-specific immobilization of the intact antibody to a surface as described in the text. Cleavage sites for two common proteases (papain and pepsin) are also shown. Pepsin cleaves the intact antibody below the disulfide bonds in the C-terminal portion of the hinge region to form a bivalent antigen-binding fragment referred to as a "$F(ab')_2$" (upper right panel). The $F(ab')_2$ can be reduced with dithiolthreitol to form Fab' fragments (lower right panel) with reactive thiol groups that can be used for site-specific immobilization.

more inclined to adsorb to the surface than the Fab' regions [34,35]. These studies will be described in this chapter, and acid pretreatment will be compared and contrasted to other immobilization strategies.

Finally, our approach to investigating the surface activity of the immobilized antibodies and nonspecific binding of analyte to the surface was to immobilize an antihapten antibody (such as antifluorescein) to the surface and then quantify the binding of a hapten–protein conjugate (e.g., fluorescein-labeled bovine serum albumin) to the immobilized antibody. In such case, the unmodified protein (e.g., bovine serum albumin) serves as a control for nonspecific binding because it does not contain the antigenic determinant recognized by the immobilized antibody. This is perhaps the least biased means of evaluating NSB because the immobilized antibody layer is identical for both specific and nonspecific binding assays. Even the same protein analyte is used in both cases, the only difference being the presence of haptenic groups in the specific binding case. Bovine serum albumin (BSA) was chosen as the carrier protein because its high interfacial activity is thought to accentuate nonspecific binding [36,37].

II. EXPERIMENTAL PROCEDURES

A. Preparation of Antibodies and Their Fragments

1. Antibodies

Antibodies of several different specificities were employed in these studies, including antibiotin, antifluorescein, antihuman serum albumin (anti-HSA), and antiovalbumin (anti-OVA). Except for antifluorescein (discussed below), all of these were polyclonal antibodies derived from animal sera—antibiotin, anti-HSA, and anti-OVA (tracer antibody) from goat, anti-OVA (capture antibody) from rabbit. The immunoglobulin G (IgG) fraction of antibiotin and anti-HSA antibodies were obtained from commercial sources (antibiotin from Sigma, St. Louis, MO and anti-HSA from Organon Teknika, West Chester, PA) and used without further purification. Anti-OVA antibodies (both capture and tracer) purified by protein G chromatography were obtained from the U.S. Naval Medical Research Institute (NMRI, Bethesda, MD) and used without further purification.

A murine monoclonal antifluorescein antibody (MAb 9-40) was used to evaluate and compare the different immobilization techniques. It belongs to the IgG_1 (κ) subclass and exhibits an affinity of about $1 \times 10^7 M^{-1}$ at 25°C [38]. The hybridoma cell line that secretes this antibody was obtained from Professor Edward W. Voss, Jr. (University of Illinois at Urbana-Champaign). Hybridoma cells were grown in tissue culture (Delbecco's minimal essential medium with 20% fetal calf serum) for approximately 1 week and then injected I.P. (10^6 cells/mouse) into BALB/c mice, which had been previously injected with pristane. Ascites fluid was collected 14–21 days after inoculation, and the IgG fraction was prepared by precipitation in 50% saturated ammonium sulfate followed by anion exchange chromatography (DEAE-cellulose, Pierce Chemical). Protein A (Pharmacia) chromatography [39,40] was used as the final purification step. The monoclonal antibody eluted in 0.1 M glycine, pH 3.02, was ≥95% pure.

2. Antigen-Binding Fragments (Fab')

Purified MAb 9-40 was digested with pepsin to produce (Fab')₂ fragments [24–27], which were then reduced to Fab' fragments using dithiothreitol (DTT). Specifically, 33 mg of purified MAb 9-40 and 1 mg pepsin (Sigma) were dissolved in 0.1 M sodium acetate buffer (pH 4.2) and the digestion was carried out at 37°C for 16 h. The digestion was terminated adjusting the pH of the reaction mixture to 8.0 with 2 M Tris base. The (Fab')₂ fraction was separated by gel permeation chromatography (Superdex HiLoad, Pharmacia) using phosphate buffer saline (PBS), pH 7.7, as eluent. Fab' fragments were prepared by reducing the (Fab')₂ fragments (1 mg/mL) with 1.75 mM DTT and 3.5 mM ethylenediamine tetraacetate (EDTA) in 0.17 M Tris buffer (pH 7.4) for 45 min at room temperature. After reduction, excess DTT was removed by gel permeation chromatography using a Sephadex G-25 column (Pharmacia) equilibrated in 0.1 M sodium phosphate buffer (pH 6.0) containing 5 mM EDTA.

B. Antigens

Bovine serum albumin was obtained from Sigma. It was labeled with radioiodine and/or fluorescein isothiocyanate as described below. Purified ovalbumin was obtained from NMRI.

C. Chemical Modification of Antibodies and Their Fragments

1. Acid Pretreatment

Acid pretreatment is a method for enhancing the surface concentration of antibody immobilized by physical adsorption [16,17,34,35]. Approximately 8 mg of antibodies were dissolved in 0.5 mL of 0.1 M citric acid/sodium phosphate buffer solution (pH 2.7) and incubated for various times (20 min, 1 h, 5 h). Solutions were then neutralized by passage through a PD-10 column (Pharmacia) which had been equilibrated in phosphate-buffered saline, pH 7.4.

2. Oxidation of Immunoglobulin Carbohydrate Groups

As mentioned previously, the carbohydrate groups in the Fc region can be oxidized and used for site-specific immobilization of antibodies. Several different applications of this strategy are described in this chapter. In some cases, oxidized antibodies were conjugated directly to hydrazido groups present in the passivating layer, while in another case oxidized antibodies were derivatized with biotin and coupled to immobilized avidin.

MAb 9-40 was oxidized using metaperiodate using a procedure described by O'Shannessy et al. [19,20]. Specifically, MAb 9-40 (1–2 mg/mL) was oxidized using 10 mM sodium periodate (in 0.1 M sodium acetate buffer, pH 5.5) for 20 min at 0°C. The reaction was quenched by addition of glycerol to a final concentration of 15 mM. The quenching reaction was allowed to proceed for 5 min at 0°C. Oxidized MAb 9-40 was purified by gel filtration chromatography using Sephadex G-25 equilibrated in 0.1 M sodium acetate buffer, pH 5.5. About 25% of the antibody molecules were inactivated by the oxidation procedure.

3. Biotinylation of Intact IgG and Fab′ Fragments

One of the immobilization strategies examined in this chapter is self-assembled monolayer using immobilized avidin and biotinylated antibodies. Different strategies were used for biotinylating intact MAb 9-40 and its Fab′ fragments. In case of the intact antibody, biotin–LC–hydrazide (Pierce, Rockford, IL) was coupled to oxidized carbohydrate groups in the Fc region using a procedure described by O'Shannessy et al. [19]. In brief, the carbohydrate groups were oxidized as described in the previous section and then biotin–LC–hydrazide (5 mM) was added and allowed to react with agitation for 2 h at room temperature. Unreacted biotin–LC–hydrazide was removed using a Sephadex G-25 column equilibrated in PBS.

In the case of Fab′ fragments, biotin-HPDP [N-(6-[biotinamido])hexyl)-3′-(2′-pyridyldithiol)propionamide, Pierce] was coupled to free thiol groups in the C-terminal region of the protein. A 4 mM stock solution of biotin-HPDP was prepared in dimethylformamide (DMF, Aldrich, Milwaukee, WI). This solution was added to Fab′ fragments in PBS (0.5–1 mg/mL) to give a 20-fold molar excess of biotin-HPDP. The biotinylation reaction was carried for 90 min at room temperature. Biotinylated Fab′ fragments were purified by gel permeation chromatography using Sephadex G-25 equilibrated in PBS.

4. Radiolabeling of Antibodies and Antigens

Antibodies, antigen-binding fragments, and antigens were radiolabeled with ^{125}I using chloramine T as described by Chuang et al. [41]. One-half milliliter of protein solution (1 mg/mL) was mixed with 3 μL of carrier-free ^{125}I (100 mCi/mL, Amersham) and 50 μL of chloramine T solution (4 mg/mL) and then allowed to react at room temperature for 1 min. The reaction was quenched by adding 50 μL sodium metabisulfite solution (4.8 mg/mL) for 2–3 min. Unreacted iodide was removed by gel permeation chromatography (Sephadex G-25, Pharmacia). Radiolabeling efficiency (RE) was determined by precipitating the labeled protein with 20% trichloroacetic acid (TCA, Sigma) in presence of BSA as a carrier, and was calculated using the following equation:

$$RE = \frac{CPM_{solution} - CPM_{super}}{CPM_{solution}} \tag{1}$$

where $CPM_{solution}$ is the number of counts in 5 μL of the labeled protein solution before TCA precipitation, and CPM_{super} is the number of counts in 5 μL supernatant. Samples were counted using a Beckman 170M radioisotope detector. The efficiency was always greater than 95%. The specific activity (SA) of the iodinated protein was determined using a UV-visible spectrophotometer (Beckman, Model 35) at 278 nm. Extinction coefficients ($\varepsilon_{278\,nm}^{0.1\%}$) of 1.35 and 0.67 were used for MAb 9-40 and BSA, respectively. Typical SA values were 0.4–2.0 × 10^{15} counts min^{-1} mole^{-1}.

Radioiodination with chloramine T occasionally results in a loss or decrease in the antigen-binding activity of the antibody (or its fragments). In such cases, antibodies were labeled using Enzymobeads (Bio-Rad Laboratories, Hercules, CA), which produce milder iodination conditions than chloramine T [42–44]. Before labeling, sodium azide (preservative) was removed from the antibody solution, using Sephadex G-25 (Pharmacia), because sodium azide is a potent inhibitor of lactoperoxidase. The following reagents were added to an Enzymobead reaction vial: (1)

50 μL of 0.2 M phosphate buffer (pH 7.2); (2) 500 μL F(ab')$_2$ solution; (3) 1 mCi Na^{125}I; (4) 25 μL of 2% glucose. This mixture was allowed to react for 40 min at room temperature and the unreacted Na^{125}I was removed by gel filtration (Sephadex G-25). The radiolabeling efficiency (with respect to percent of ^{125}I incorporated) was much lower for Enzymobeads than for chloramine-T (3.5–5% vs. 81%); however, the two methods were comparable with respect to the fraction of protein that was labeled with ^{125}I ($\geq$95%).

5. Conjugation of Fluorophores to Analytes and Antibodies

Bovine serum albumin was labeled with fluorescein using the following procedure. Ten milligrams of BSA was dissolved in 1 mL of 0.1 M carbonate–bicarbonate buffer (CBB), pH 9.3. A 10-fold molar excess of fluorescein isothiocyanate (FITC, Molecular Probes, Eugene, OR) was dissolved in 50 μL of DMF and then added to the protein solution. The reaction was allowed to proceed for 2 h at room temperature. The fluorescein-BSA conjugate was purified using a PD-10 column, equilibrated in phosphate buffer, pH 7.5. Both the BSA concentration and the degree of labeling were determined from the conjugate's absorption spectra. Fluorescein absorbs maximally near 490 nm, but also exhibits significant absorption at 280 nm, a wavelength typically used for determining protein concentration. For this reason, the absorbance at 280 nm was corrected for the absorption of fluorescein at this wavelength:

$$[BSA] = \frac{A_{280\,nm} - 0.15A_{490\,nm}}{\varepsilon_{BSA}} \tag{2}$$

$$[Fluorescein] = \frac{A_{490\,nm}}{\varepsilon_{fluorescein}} \tag{3}$$

where the molar extinction coefficient for fluorescein ($\varepsilon_{fluorescein}$) was 73,000 $M^{-1}\,cm^{-1}$ and that for BSA (ε_{Ab}) was 44,890 $M^{-1}\,cm^{-1}$.

A similar procedure was used for labeling antibodies with Cy5 (Amersham), a red-emitting (675 nm, emission maximum) cyanine dye which contains two amine-reactive NHS-ester groups. Following directions given in the package insert, 1 mg protein in 0.5 mL of CBB was added to a vial containing the dry dye. These reagents were mixed thoroughly and allowed to react for 30 min. The labeled antibody was separated from unreacted dye using a PD-10 column, equilibrated in pH 7.5 phosphate buffer. Both the antibody concentration and the degree of labeling were determined from the conjugate's absorption spectra. Cy5 absorbs maximally near 650 nm, but also exhibits some absorption at 280 nm. Thus, in calculating concentrations the absorbance at 280 nm was corrected for the absorpton of fluorescein at this wavelength:

$$[Ab] = \frac{A_{280\,nm} - 0.05A_{650\,nm}}{\varepsilon_{IgG}} \tag{4}$$

$$[Cy5] = \frac{A_{660\,nm}}{\varepsilon_{Cy5}} \tag{5}$$

where the molar extension coefficient for Cy5 (ε_{Cy5}) was 250,000 $M^{-1}\,cm^{-1}$ and that for the antibody (ε_{Ab}) was 202,500 $M^{-1}\,cm^{-1}$.

6. Photoaffinity Labeling of Antibodies

In order to probe the orientation of immobilized antibodies a fluorescent label was attached to antibiotin and antifluorescein antibodies in the vicinity of the active site as described by Chang et al. [45]. Antibodies were then immobilized to a solid surface, and the solvent accessibility of the label was probed using iodide ion, a fluorescence quencher [35]. A rather specialized reagent was used for attachment of the fluorophore near the active site, namely, a photoaffinity label that consists of a hapten antigen (either biotin or fluorescein) at one end, a photoactivatible azide group at the other end, and an internal disulfide bond. The haptenic portion is allowed to bind to the active site and the azide group is activated by exposure to UV light. The activated azide group forms a covalent bond with the antibody and the disulfide bond is reduced using dithiothreitol to form a free thiol group in the vicinity of the active site. Finally, a fluorescent probe (either fluroescein or tetramethylrhodamine) was conjugated to the thiol group. Specific details of this labeling procedure are given in the next few paragraphs.

(a) Preparation of Photoafffinity Cross-Linker.   A functional derivative of the hapten antigen with a primary amine group—either *N*-(2-aminoethyl)biotinamide or 5-[(2-aminoethyl)thioureidyl] fluorescein—was purchased from Molecular Probes. Ten milligrams of the derivatized hapten was dissolved in 2 mL of DMF (predried by molecular sieve and $CaCl_2$) and mixed with 4 μL of triethyleneamine (Aldrich) and 20 mg of sulfosuccinimidyl-2(*m*-azido-*o*-nitrobenzamido)-1,3'-dithiopropionate (SAND, Pierce), a heterobifunctional cross-linker with an internal disulfide bond. This mixture was allowed to react at 20°C for 24 h in the dark, followed by separation on a silica gel column (200–400 mesh, Aldrich) in methanol/acetone (1:1 volume ratio). The product, either biotin-AND or fluorescein-AND (hereafter referred to as "hapten-AND"), was characterized by thin-layer chromatography (TLC), IR, and UV spectroscopy.

(b) Photoaffinity Labeling.   A fourfold molar excess of hapten-AND was added to a 6 mg/mL solution of antibodies in PBS. The concentration of hapten-AND was determined by UV absorption at 320 nm, using an extinction coefficient of 9×10^3 M^{-1} cm^{-1}. The mixture was incubated at room temperature for 2 h in the dark to allow sufficient time for the antibody/hapten-AND complex to form. Irradiation was carried out for 6 min using a mercury lamp (750 W, Schoeffel Instrument Co., NJ) with a 320-nm band-pass filter (Oriel Co.), with the reaction vessel placed 20 cm from the light source. Unreacted hapten-AND was removed using a PD-10 column equilibrated in PBS.

(c) Conjugation of Fluorescent Probes.   The pH of the photoaffinity-labeled antibody preparation was adjusted to 8.0 using 0.2 *M* carbonate–bicarbonate buffer (pH 9.2), and then DTT and EDTA were added to final concentrations of 50 m*M* and 2 m*M*, respectively. The mixture was allowed to react at 37°C for 12 h. An excess of underivatized hapten was then added in order to displace the derivatized hapten (now with a free thiol group) from the active sites. Excess hapten was removed using a PD-10 column. Fluorescein-5-maleimide (FM, Molecular Probes) was conjugated to antibiotin antibodies and tetramethylrhodamine-5-(and-6)-maleimide (TRM, Molecular Probes) was conjugated antifluorescein antibodies. Conjugation reactions were performed in 0.2 *M* acetate buffer (pH 6.0) at 4°C for 3 h. The FM and TRM were

added in six- and 10-fold molar excess, respectively. Unbound fluorophore was removed by a PD-10 column, equilibrated in PBS.

(d) Degree of Labeling. The concentration of the fluorophore was determined by measuring its UV-visible absorption at its absorption maximum (λ_{max}) and then the apparent Ab concentration was determined by measuring its absorption at 280 nm. Because most fluorophores have some absorption at 280 nm, the real antibody concentration was obtained by correcting for the contribution of the fluorophore to the measured absorption at 280 nm as described by Chang et al. [45].

(e) Control Antibodies. As mentioned, DDT treatment produces a sulfhydryl group in the vicinity of the active site by reducing the internal disulfide bond of the photoaffinity label. However, disulfide bonds located in the hinge region of the antibody are also susceptible to the DDT treatment. Thus, fluorescent probes were attached in both the vicinity of the active site and the hinge region of the photoaffinity-labeled antibodies making it necessary to prepare a group of control antibodies which were only fluorescently labeled in the hinge region. Control antibodies (both antibiotin and antifluorescein) were prepared using the same procedures as described above, except that they were not photoaffinity labeled prior to the DTT treatment.

D. Immobilization of Antibodies

1. Silica Surfaces

(a) Physical Adsorption. Because unmodified silica is a rather poor substrate for antibody adsorption [46–48], silica surfaces are usually derivatized with dichloro-dimethylsilane (DDS) as a prelude to antibody immobilization. Such treatment renders the surface more hydrophobic, thus promoting adsorption. Silica samples were silanized as described by Lin et al. [16]. In brief, sample chips (1 cm $\times$ 1 cm $\times$ 1 mm) were cut from fused silica slides (CO grade, ESCO) and the edges were finely polished. Chips were then cleaned in chromic acid at 80°C, rinsed extensively in deionized water and allowed to react with 10% (v/v) DDS (Petrarch) in dry toluene for 30 min at room temperature. Next, they were rinsed with absolute ethanol and cured in a vacuum oven (which had been flushed with nitrogen gas three times) at 120–130°C for 1 h. The quality and uniformity of the DDS layer was characterized by contact angle measurements using the Wilhelmy plate apparatus and ESCA analysis. Typically, DDS-silica surfaces exhibited advancing and receding contact angles of 89° and 65°, respectively. Antibody immobilization was accomplished by immersing the DDS-coated chips in solutions of either native or acid pretreated antibodies for 3 h at room temperature (20°C). Excess antibody was removed by rinsing the chips five times with PBS buffer.

(b) Covalent Attachment.

    Immobilization to a hydrogel film. In this immobilization strategy antibodies were conjugated to a thin layer of hydrogel which was grafted to a silica surface [18]. The hydrogel was composed of polymethacryloyl hydrazide. The pendant hydrazido groups were used for both grafting the polymer to the surface and conjugation of antibody molecules. Prior to grafting of polymer, silica surfaces were modified in a two-step reaction (amination followed by derivatization with glutaraldehyde) to produce a monolayer of reactive aldehyde groups [16,49].

1. Polymer preparation. Polymethacryloylchloride (PMaCl) was prepared by the radical polymerization of methacryloylchloride in dioxane under an inert atmosphere. The reaction mixture contained 21.1 mol% methacryloylchloride (MaCl, Aldrich), 78.1 mol% dioxane (Aldrich), and 0.8 mol% azobisisobutyronitrile (AIBN, Polyscience, Warlington, PA) and was allowed to react for 24 h at 60°C while stirring. Afterward, the reaction mixture was diluted with more dioxane (twice the amount used for the polymerization) and slowly added to an excess of hydrazine hydrate (Sigma). A volumetric ratio of 2:5 was used for polymethacryloylchloride and hydrazine hydrate. The addition was carried out in an ice bath under a nitrogen atmosphere. The mixture was then stirred for 1 h at room temperature. The product [polymethacryloylhydrazine (PMahy)] was purified by evaporation of dioxane and the remaining hydrazine hydrate, followed by washing in distilled water and dialysis (SpectraPor dialysis membrane, 3500 MW cutoff). It was stored in aqueous solution at 4°C under a nitrogen atmosphere. The molecular weight of the polymer was determined by gel permeation chromatography. Specifically, the hydrochloride form of PMahy was prepared by reaction with dry HCl in methanol and applied to a Superose 12 column (Pharmacia) equilibrated in 0.05 M Tris buffer with 0.5 M NaCl, pH 8.0.

2. Chemical activation of silica substrate. A reactive aldehyde layer was formed on the surface of the silica substrate as described by Lin et al. [16,49]. In brief, silica chips (1.0 × 0.9 × 0.1 cm, CO grade, ESCO) were cleaned in chromic acid and treated with 5% (v/v) 3-aminopropyltriethoxy silane (APS, Aldrich) for 20 min at room temperature. Chips were then washed with distilled water and ethanol, and reacted with 2.5% (w/v) glutaraldehyde (Polyscience) for 2 h at room temperature and washed again with distilled water. Chips derivatized in such fashion will be referred to as CHO-silica.

3. Grafting of hydrogel film. CHO-silica chips were immersed in 5% (w/v) polymethacryloyl hydrazide for 2 h at room temperature. The pH of the reaction mixture was adjusted to 5.2 by addition of 10% acetic acid. For immobilization of Fab' fragments, the pendant hydrozine groups were derivatized with 0.1% (w/v) succinimidyl 4-(N-maleimidomethyl)cyclohexane-1-carboxylate (SMCC, Pierce) in DMF for 1 h at room temperature.

4. Antibody immobilization. Oxidized antibodies were coupled for 20 h at 4°C in acetate buffer (pH 5.2). Fab' fragments were coupled for 24 h at 4°C in phosphate buffer with 5 mM EDTA (pH 6.0). A concentration of 1 × 10^{-6} M was used in coupling reactions for both oxidized antibodies and Fab' fragments.

Immobilization to a polyethylene glycol film. In this immobilization strategy, antibodies were immobilized to silica surfaces via linear polyethylene glycol (PEG) chains as described by Huang et al. [49]. The key to this chemistry is a homobifunctional PEG molecule in which both ends are derivatized with hydrazido groups [bis(hydrazido)polyethylene glycol]. The use of the hydrazido function allowed PEG chains to be attached to the same CHO-silica surfaces described above for the hydrogel chemistry.

1. Derivatization of polyethylene glycol. Five grams of either polyethylene glycol (3400 MW, Aldrich) or methoxy polyethylene glycol (5000 MW, Aldrich) were reacted with 1.77 g of p-nitrophenylchloroformate (Aldrich) in 5 mL of benzene at room temperature for 24 h with agitation. Dry ethyl ether was added to precipitate the bis(nitrophenyl)PEG [or methoxy (nitrophenyl)PEG] from the benzene solution.

The precipitate was then redissolved in benzene and precipitated with ethyl ether. This procedure was repeated and the sample was dried overnight under vacuum. The extent of derivatization was determined as described by Huang et al. [49]. Nitrophenyl derivatives were converted to hydrazide derivatives [bis(hydrazido)PEG or methoxy(hydrazido)PEG] by dissolving 5 g of the appropriate nitrophenyl derivative in 2 mL methanol and adding this solution to 2 mL of hydrazine (Aldrich). The mixture was allowed to react for 3 h at room temperature with agitation, after which the polymer was precipitated by addition of 500 mL dry ethyl ether. The polymer was redissolved in methanol and precipitated in ethyl ether. This procedure was repeated until all yellow color had disappeared. After the final precipitation the wet polymer powder was dried overnight under a vacuum.

2.   Attachment of bis(hydrazido)PEG to silica surfaces.   Silica chips with a reactive aldehyde layer (CHO-silica) were prepared as described above. Twenty-four milligrams of bis(hydrazido)PEG powder were dissolved in 1.2 mL of 11% (w/v) K_2SO_4 in pH 5.2 sodium acetate buffer. The high sulfate concentration is thought to salt out the polymer onto the surface, thereby increasing its local concentration and promoting coupling [50]. CHO-silica chips were then immersed in the polymer solution and incubated at 60°C for 24 h. After thorough rinsing in DI water, the level of coupling was examined by both chemical analysis and electron spectroscopy for chemical analysis (ESCA) as described by Huang et al. [49]. Prior to the coupling of antibodies to these surfaces, free aldehyde groups still remaining after grafting of the PEG were blocked by reaction with 0.2 M aqueous ethanolamine. In this process, each chip was immersed in 2 mL of the amine solution and left on a shaker for 30 min at room temperature. The derivatized surface will be referred to as Hz-PEG-silica.

3.   Antibody immobilization.   Two types of conjugation reactions were examined for the intact antibody: (1) random attachment of antibodies (1 mg/mL) to CHO-silica surfaces (the conjugation reaction was carried out in pH 9.2 CBB for 6 h at room temperature); and (2) site-specific immobilization of oxidized antibodies (1 mg/mL) to Hz-PEG-silica surfaces (conjugation reaction was performed in pH 5.2 acetate buffer for three days at 4°C). For conjugation of Fab' fragments, the hydrazido end groups of the immobilized PEG chains were first converted to maleimido groups using SMCC (conversion reaction was carried out for 1 h at room temperature in PBS, pH 7.2). The attachment of Fab' was carried out for 20 h in 5 mM EDTA, pH 6.0, at 4°C.

(c)   Self-Assembled Monolayers.   In this immobilization strategy, avidin was first immobilized to silica surfaces by physical adsorption and then biotinylated antibodies (or Fab' fragments) were allowed to self-assemble on top of the avidin-coated surface. Clean silica examples were immersed in an avidin solution (3×10^{-6} M) for 3 h at room temperature. Silica samples were then washed several times in PBS to remove unadsorbed avidin. Avidin-coated surfaces were then immersed in a 1.5×10^{-7} M solution of either biotin-IgG or biotin-Fab' for 1 h at room temperature. Unbound protein was removed by washing in PBS. In some cases, biotinylated polyethylene glycol (biotin-PEG) was coupled to surfaces following the immobilization of biotin-Fab' fragments. Biotin-PEG was prepared by reaction of N-hydroxysuccinimido-biotin (Sigma) with NH_2-PEG-OCH$_3$ (5000 MW, Sigma) for 2 h in 0.1 M sodium carbonate buffer (pH 9.0), and purified by dialysis using SpectraPor dialysis

membranes (1000 mW cutoff). Silica surfaces (coated with avidin and Fab' fragments) were immersed in a $1 \times 10^{-7}\ M$ solution of biotin-PEG for 1 h at room temperature. Unbound biotin-PEG was removed by washing in PBS.

2. Polystyrene Surfaces

Physical adsorption was the primary method used for immobilizing antibodies to polystyrene (PS) surfaces. The method was particularly convenient because most proteins adsorb strongly to hydrophobic surfaces. This immobilization strategy is analogous to the aforementioned adsorption of antibodies to DDS-silica surfaces.

Prior to immobilization, polystyrene surfaces were immersed in 30 mL of ethanol and washed for 1 min with agitation to remove any residues. After drying, surfaces were rinsed in PBS and then immersed in a $5 \times 10^{-8}\ M$ solution of antibodies in PBS buffer for 2 h at room temperature. Antibody-coated surfaces were immersed in a postcoating solution of 5 mg/mL BSA in PBS for 1 h at room temperature to block any surface sites available for nonspecific binding. Unbound BSA was removed by rinsing the surfaces several times with PBS buffer. Coated surfaces were dried by vacuum desiccation for 1–2 h at room temperature and stored desiccated at 4°C until use.

E. Immunoassays

Both radioimmunoassays and fluoroimmunoassays were employed in studies described in this chapter. The former were used in cases where exact quantification of the surface concentration of antibody or analyte was required, while the latter were used for rapid determination of analyte concentration (e.g., clinical diagnostics assays) and also for probing the orientation of immobilized antibodies.

1. Radioimmunoassays

Radioimmunoassays (RIA) were used to compare the surface activities of antibodies immobilized using the different immobilization techniques and also to assess the nonspecific binding of a given immobilization technique. Surface activity is defined as the fraction of immobilized active sites which are capable of binding analyte. It was calculated by taking the ratio of the surface concentration of analyte (in the presence of a saturating concentration of analyte in bulk solution) to the surface concentration of immobilized antibody active sites. Nonspecific binding was evaluated by measuring the surface concentration of radiolabeled bovine serum albumin.

(a) Surface Concentration of Immobilized Antibodies. Radiolabeled antibodies were immobilized to 1 cm $\times$ 1 cm $\times$ 1 mm silica chips using immobilization procedures described above. Chips were rinsed five times in PBS and then counted for 1 min with a Beckman Model 170M radioisotope detector. Surface antibody concentration (Γ_{Ab}) values were calculated using the following formula:

$$\Gamma_{Ab} = \frac{CPM/SA}{A_{surf}} \qquad (6)$$

where *CPM* is counts per minute, *SA* is the specific activity of the radiolabeled antibody, and A_{surf} is the surface area of the silica chip (typically 2.4 cm^2). Each chip was counted in triplicate in order to determine the mean and standard error of the surface concentration.

(b) Surface Analyte Concentration. Unlabeled antibodies were immobilized to 1 cm × 1 cm × 1 mm silica chips using the immobilization procedures described above. Chips were rinsed five times in PBS and then immersed in a solution of radioiodinated analyte (either FI-BSA for quantification of specific binding or BSA for quantification of NSB) for 1 h at room temperature. Following this incubation, chips were rinsed again five times in PBS and counted for 1 min with a Beckman Model 170M radioisotope detector. Surface analyte concentration (Γ_{Anal}) values were calculated using the following formula:

$$\Gamma_{Anal} = \frac{CPM/SA}{A_{surf}} \tag{7}$$

where *CPM* is counts per minute, *SA* is the specific activity of the radiolabeled analyte, and A_{surf} is the surface area of the silica chip. Each chip was counted in triplicate in order to determine the mean and standard error of the surface analyte concentration.

2. Fluoroimmunoassays

Over the past decade, our group has refined a subclass of fluoroimmunoassays (FIA) based on total internal reflection fluorescence (TIRF), an optical technique that is especially well suited for measuring the concentrations of fluorescent molecules at solid–liquid interfaces. Because optical substrates (e.g., glass, quartz, and some thermoplastics) usually exhibit higher indices of refraction than either water or air, a beam of light traveling through the substrate and striking its edge or face will either be refracted into the aqueous phase or reflected totally back into the substrate, depending on the angle of incidence. In the latter case, the incident and reflected beams interfere to produce a standing wave within the substrate (Fig. 3). This wave has a finite electric field amplitude right at the edge of the substrate, but decays exponentially over a distance of ca. 120 nm as one moves into the aqueous phase. This decaying electric field is referred to as the evanescent field.

In our assay format the optical substrate is a planar waveguide composed of a material with high index of refraction and good optical quality, such as quartz or polystyrene. Planar waveguides are analogous to fiber optics in the sense that light coupled into one end of the waveguide at the correct angle, or "mode," will be confined inside and propagate down the entire length by total internal reflection with little or no loss of radiation (a process called *waveguiding*). One side of the waveguide is exposed to ambient air (and hence kept clean and dry) while the other side is coated with antibody and exposed to the assay solution (Fig. 3). The evanescent field encompasses the immobilized antibody layer and can excite fluorescently labeled "tracer" molecules bound to the antibodies. Fluorescence emission is collected perpendicular to the surface of the waveguide using either a photodetector (e.g., photomultiplier tube) or a CCD camera. The latter enables spatial resolution of different regions on the waveguide's surface. Different antibodies can be immobilized to the regions, allowing several different immunoassays to be performed in parallel on a single specimen.

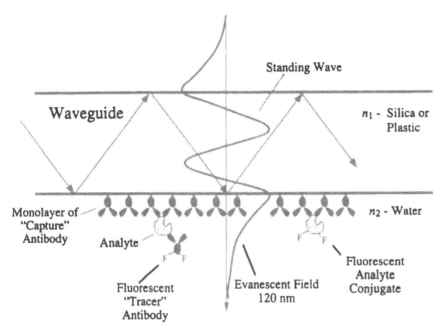

FIG. 3 Concept of total internal reflection and its application to fluoroimmunoassays. Total internal reflection occurs when a beam of light traveling through an optical substance such as a waveguide (with index of refraction n_1) strikes an interface with a less dense medium such as water or air (with lower index of refraction n_2). If the angle of incidence is less than the critical angle, the beam is reflected totally back into the waveguide. Interference between the incident and reflected beams sets up a standing wave in the waveguide, which decays into the aqueous phase producing an "evanescent" electric field. This field is capable of stimulating the emission of fluorophores located within ca. 100 nm of the waveguide. Immunoassays are performed by immobilizing a capture antibody on the surface of the waveguide. In a competitive assay (lower right) a fluorescent analyte conjugate is immobilized along with the antibody. In the absence of unlabeled analyte, the conjugate binds to the antibody that is excited by the evanescent field. Upon addition of analyte, conjugate molecules are displaced by unlabeled analyte molecules and diffuse out of the evanescent field. In a sandwich assay (lower left) a fluorescent tracer antibody is added to the bulk solution. In the absence of analyte, the tracer antibody remains in solution and very little fluorescence is observed. Upon addition of analyte, a sandwich is formed on the waveguide, bringing the tracer antibody within the evanescent field, where it fluoresces.

Two different generations of waveguide immunosensors are shown in Fig. 4. The first generation sensor consisted of a 2.5 cm $\times$ 7.5 cm $\times$ 1 mm fused silica slide that was mounted in a two-channel flow cell [10,11,18]. Light was coupled into the end of the waveguide using external lenses, and fluorescence emission was collected from the side of the waveguide using an imaging system (lens, filter, and CCD camera). One channel of the flow cell (sample channel) was used to perform assays, while the other was used as a reference channel to control for background fluorescence and NSB. The second generation sensor is an injection-molded polystyrene waveguide with integral coupling lens and three sample wells [6,18]. This device has greater alignment tolerance than the first-generation sensor, making it suitable

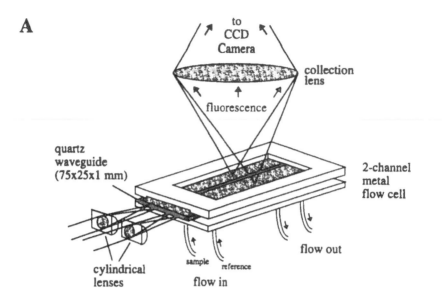

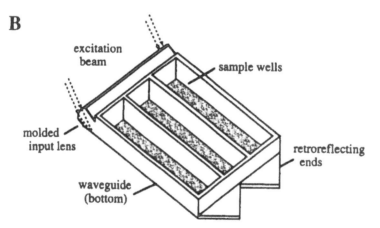

FIG. 4 Two different types of evanescent wave immunosensors. (A) This sensor employs a waveguide and metal flow cell. Light is coupled into the end of the waveguide using two cylindrical lenses and fluorescence is collected from the side of the waveguide using an imaging lens, filter, and CCD camera. The flow cell is divided into two channels by means of a low-index Teflon gasket next to the waveguide which separated the flowing solution into two compartments. Different solutions were injected into each channel—a "sample" solution containing both analyte and fluorescently labeled tracer antibody and a "reference" solution containing only labeled tracer antibody. The latter is used as a control for both nonspecific bonding (NSB) and optical fluctuations within the instrument. (B) This sensor is an injection-molded polystyrene unit consisting of a 0.5-mm-thick waveguide on the bottom, an integral coupling lens, and three sample wells on top. The cylindrical coupling lens is molded to the front end of the waveguide to allow greater tolerance in positioning the sensor in the instrument. Two 45° retroreflectors are also molded into the far end of the waveguide to reflect the waveguided light back on itself; this theoretically can increase the effective excitation intensity by a factor of up to 2. The aforementioned sample and reference solutions are added to two of the wells. The third well can be used for an additional sample or a calibration standard.

for use in in vitro diagnostics assays. One of the sample wells is used as a reference channel, while the other two can be used to perform assays. Alternatively, the second well can be used for assays and the third for calibration purposes.

Either competitive immunoassays or sandwich immunoassays can be performed, depending on the choice of tracer molecule. A comparison of the two indicated that sandwich immunoassays were more sensitive [10]. This was due in part to the lower background fluorescence that can be attained in sandwich assays and also to the slow dissociation rate of analyte complexed to immobilized antibodies [16,17,49]. The latter factor can significantly impair the response time of competitive assays in which the labeled analyte is complexed to the immobilized antibody before the analyte is added (e.g., displacement assays). For these reasons, only the sandwich fluoroimmunoassay will be considered further.

(a) Sandwich Fluoroimmunoassay. Antibodies fulfill two different functions in a sandwich assay—an immobilized antibody is used to capture the analyte, while a soluble, fluorescently labeled antibody is used to detect, or "trace," analyte binding. For sandwich formation, capture and tracer antibodies must bind to different regions of the analyte molecule, far enough removed to prevent competitive binding. For large analytes with repetitive epitopes (e.g., viruses and bacteria), the same antibody can usually be employed in both capture and tracer roles, provided that it is specific for the repetitive epitope. However, such an approach does not work for smaller analytes (e.g., proteins and polysaccharides) which contain only one copy of a given epitope. In such cases, two different antibodies must be used which bind noncompetitively to different epitopes (referred to as a "two-site assay").

One of the studies described in this chapter is a comparison of binding isotherms obtained at different assay times. Ovalbumin (OVA) was employed as a model analyte in these studies. A two-site immunoassay was formulated for OVA in which rabbit anti-OVA was used as the capture antibody, and goat anti-OVA was used as the tracer antibody. Both antibodies were polyclonal preparations purified using protein G. The capture antibody was immobilized to polystyrene waveguide sensors by physical adsorption and the tracer antibody labeled with Cy5 as described in Sections II.D.2 and II.C.5, respectively. A series of OVA solutions was prepared in PBS that varied over a concentration range of 10^{-12} to 10^{-7} M in half-$\log_{10}$ increments. Tracer antibody was added to each solution to give a final concentration of 10^{-9} M. Each analyte solution (referred to as "sample") was introduced into the well of a coated waveguide sensor and assayed every minute over a period of 60 min using the evanescent wave fluorometer described by Christensen and Herron [6]. A reference solution containing only tracer antibody (10^{-9} M in PBS) was introduced at the same time into another well of the waveguide and used as a control. The ratio of sample to reference fluorescence (R) was calculated for each analyte concentration (C) and assay time. Binding isotherms were constructed by plotting R versus C for each assay time and analyzed using the following equation:

$$R = \frac{\Delta R_{max} K_a C}{1 + K_a C} + R_0 \tag{8}$$

where ΔR_{max} is the full-scale increase in fluorescence ratio, R_0 is the fluorescence ratio in absence of analyte, and K_a is the apparent affinity of the binding reaction.

F. Orientation of Immobilized Antibodies

1. Adsorption Isotherms

One way to examine the orientation of immobilized antibodies is to determine their packing density under different immobilization conditions. This can be accomplished by performing quantitative adsorption isotherms in which the exact amount of antibody immobilized is quantified using radiolabels. This approach was used to examine the packing density of acid-pretreated anti-HSA antibodies adsorbed to DDS-silica surfaces. Adsorption isotherms were obtained by varying bulk anti-HSA concentration over a range of 10^{-8} to 10^{-5} M and determining the surface antibody concentration by radioimmunoassay as described by Chang et al. [34]. Isotherms were obtained for native anti-HSA antibodies and those pretreated with acid for 1, 20, 60, and 300 min immediately prior to immobilization.

The classical Langmuir theory of gas adsorption can be applied to proteins in solution provided that (1) the solution is sufficiently dilute, (2) only one class of adsorption site is present, (3) there are no lateral interactions between adsorbed proteins, and (4) adsorption is reversible [15]. In such cases the surface concentration of adsorbed protein (Γ) is given by

$$\Gamma = \frac{\Gamma_{max} K C}{1 + KC} \tag{9}$$

where Γ_{max} is the surface concentration of protein at full monolayer coverage, K is the Langmuir constant, and C is the protein concentration in bulk solution. Previously, our group examined the adsorption of antibodies to DDS-silica surfaces and found that at least two classes of adsorption sites were present [34]. The following two-component Langmuir equation was used to fit the adsorption data:

$$\Gamma = \frac{\Gamma_1 K_1 C}{1 + K_1 C} + \frac{\Gamma_2 K_2 C}{1 + K_2 C} \tag{10}$$

where K_1 and Γ_1 are the adsorption constant and maximum surface concentration, respectively, of a high-affinity class of adsorption sites, and K_2 and Γ_2 are the adsorption constant and maximum surface concentration, respectively, of a low-affinity class of adsorption sites. In the case of adsorption data obtained for acid-pretreated antibodies, an additional parameter (orientational parameter α) was added to Eq. (10) to describe the packing density of the immobilized antibodies [see Eq. (15) in Section III.B.1]. Adsorption data sets obtained for several different acid pretreatment times were fit to Eq. (15) by multivariate analysis using the Scientist by MicroMath (Salt Lake City, UT).

2. Iodide Quenching Studies

Iodide quenching was used to probe the orientation of immobilized acid-pretreated antibodies. In particular, antibiotin and antifluorescein antibodies were labeled in the vicinity of the active site using the photoaffinity labeling technique described in Section II.C.6. Antibodies were then acid pretreated for various times (0, 20, 60, 120, 300 min) and immobilized to DDS-silica waveguides. Fluorescence-quenching experiments were performed using TIRF as described by Chang et al. [35]. In brief, quenching solutions were prepared by adding KI or NaCl to PBS buffer so that the sum of [NaCl] + [KI] was held constant at 0.6 M. Six solutions were prepared with iodide concentrations of 0.0, 0.1, 0.2, 0.3, 0.4, and 0.5 M, respectively. Antibody-coated waveguides were mounted in the two-channel flow cell shown in Fig. 4 and flushed with a 0.6 M NaCl in PBS. The above quenching solutions were then injected into the TIRF cell one at a time in order of increasing concentration. Each solution was allowed to equilibrate with the immobilized antibody for 30 min before the fluorescence intensity was measured. Fluoroescence-quenching data were analyzed using the modified Stern–Volmer equation [Eq. (11)]. Iodide-quenching studies were also performed in bulk solution as a control. Specifically, photoaffinity-labeled antibodies were added to the KI/NaCl solutions described above. A final antibody concentration of $5.3 \times 10^{-6}\ M$ was used in each case.

Iodide-quenching data were analyzed with a modified form of the Stern–Volmer equation [51]:

$$\frac{F_0}{\Delta F} = \frac{1}{f_a K_Q [Q]} + \frac{1}{f_a} \tag{11}$$

where $\Delta F = F_0 - F$, f_a is the fraction of accessible fluorophores, and K_Q is the Stern–Volmer quenching constant. This equation is typically used when one class of fluorophores is inaccessible. Its application to quenching data obtained by TIRF spectroscopy was previously described by Horsley et al. [52] and Chang et al. [35].

III. RESULTS AND DISCUSSION

A. Activity of Immobilized Antibodies

An antibody's activity is its main asset in applications such as affinity chromatography and immunoassays, so it is not surprising that the activity of immobilized antibodies is a topic of some interest to workers in these fields. Unfortunately, the literature is not replete with quantitative examples of the surface activity of immobilized antibodies, which is probably due to difficulties in measuring the exact surface concentrations of the immobilized antibody and bound analyte. This paucity of information motivated us to undertake a comparison study of the surface activity of antibodies immobilized using several different surface chemistries. As mentioned, this necessitated the use of radiolabels in order to quantify the surface concentrations of both the immobilized antibody and also the bound analyte in the presence of a saturating concentration of bulk analyte. Results are presented here for both silica and polystyrene surfaces.

1. Silica Surfaces

Four different immobilization methods were investigated. These included (1) physical adsorption to silica surfaces coated with dichlorodimethyl silane (DDS); (2) covalent attachment to silica surfaces coated with a polymethacrylate hydrogel; (3) covalent attachment to silica surfaces through polyethylene glycol chains; and (4) self-assembled monolayers comprised of immobilized avidin and biotinylated antibodies (or Fab' fragments). Each of these will be discussed in turn.

(a) *Physical Adsorption.* Because physical adsorption is the most common immobilization method, it is a good reference point for a comparison study of such

methods. Monoclonal antifluroescein antibody 9-40 (MAb 9-40) and its Fab' fragments were immobilized to DDS-coated silica surfaces as depicted in Fig. 5. Intact
antibodies received one of two pretreatment procedures prior to adsorption: mild
temperature denaturation (65°C, 1 h) or exposure to acidic pH (pH 3, 1 h). These
pretreatments are often used in solid-phase immunoassays to increase the amount of
IgG which is adsorbed to the substrate [16]. They are thought to expose hydrophobic
pockets in the Fc region of the antibody and make it more inclined to adsorb on
hydrophobic surfaces [34].

Surface concentrations of immobilized antibodies, Fab' fragments and bound
analyte are presented in Table 1. The surface concentrations of immobilized IgG
were 3×10^{-12} mol/cm^2 and 2.2×10^{-12} mol/cm^2 for the thermal and acid pretreatments, respectively. These values are similar and correspond to nearly monolayer
coverage of antibody [34,49]. Although thermal pretreatment resulted in a slightly
higher surface concentration for this particular antibody, our experience with other
antibodies suggests that pretreatment techniques need to be evaluated on a case-by-
case basis. Interestingly, nearly identical surface concentrations of bound analyte
("total binding" in Table 1) were observed for the two pretreatment procedures (6.5
$\times 10^{-13}$ mol/cm^2 for thermal pretreatment versus 6.7×10^{-13} mol/cm^2 for acid
pretreatment). However, because fewer moles of the acid-pretreated antibody were
immobilized, the surface activity of this antibody was slightly higher than its thermal
pretreated counterpart (15% activity for acid pretreatment versus 11% for thermal
pretreatment). Such low surface activity figures are typical for antibodies immobilized by physical adsorption [16,17,48], and, as we will see, they can be significantly
improved through use of site-specific immobilization techniques. Nonspecific binding
was 6.5% of total binding for thermal-pretreated antibodies and 9.9% for acid-pretreated antibodies. Although these values would probably be acceptable for commercial assay development, they can be improved through use of a passivating layer.

A different picture emerged for Fab' fragments immobilized by physical adsorption. In particular, very high levels of nonspecific binding (70% of total binding)
were observed, thus making physical adsorption an unacceptable means of immobilizing Fab' fragments. In addition, the surface concentration of Fab' fragments is
ca. 50% lower than that of the intact IgG, which corresponds to a fourfold lower

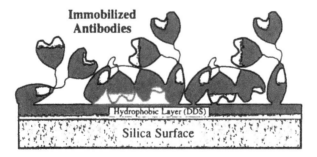

FIG. 5 Immobilization of antibodies by physical adsorption. Antibodies were adsorbed to
silica surfaces that had been "hydrophobicized" by silanization with dichlorodimethylsilane
(DDS). This immobilization technique results in random orientation of the antibodies and low
levels (<15%) of surface activity.

TABLE 1   Immobilization of Antibodies by Physical Adsorption

Surface	Antibody[a] (orientation)	Total binding[b] (moles/cm²) ×10^{-13}	Nonspecific binding[c] (moles/cm²) ×10^{-13}	Relative nonspecific binding (%)	Immobilized antibody (moles/cm²) ×10^{-12}	Surface activity[d] (%)
DDS-silica[e]	IgG (random) (65°C, 1 h)[f]	6.5 ± 0.4	0.42 ± 0.03	6.5 ± 0.90	3.0 ± 0.1	11.0 ± 1.0
DDS-silica	IgG (random) (pH3, 1 h)[f]	6.7 ± 0.2	0.66 ± 0.27	9.9 ± 4.3	2.2 ± 0.1	15.0 ± 1.1
DDS-silica	Fab' (random)	3.7 ± 0.2	2.6 ± 0.5	70.3 ± 17.3	1.3 ± 0.1	28.5 ± 3.7

[a] Antibody: monoclonal antifluorescein antibody 9-40 (MAb 9-40)—either intact immunoglobulin G (IgG) or Fab' fragments (Fab')—was immobilized to the surface. Orientation: antibodies were immobilized either with random orientation (random) or using a site-specific conjugation method (specific).

[b] Total binding: surface concentration of ¹²⁵I-(fluorescein)bovine serum albumin (FL-BSA) bound to immobilized MAb 9-40 or its Fab' fragments after 1-h exposure to a 1.5 × 10⁻⁷ M solution of FL-BSA.

[c] Nonspecific binding: surface concentrations of ¹²⁵I-BSA bound nonspecifically to immobilized MAb 9-40 or its Fab' fragments after 1-h exposure to a 1.5 × 10⁻⁷ M solution of ¹²⁵I-BSA.

[d] Surface activity: percent of immobilized active sites which bound antigen. (IgG has two active sites per molecule, while Fab' fragments have one active site per molecule).

[e] DDS-silica: silica surfaces derivatized with dichlorodimethylsilane.

[f] Intact antibodies were either incubated at an elevated temperature (65°C) or exposed to acidic conditions (pH 3) for 1 h before adsorption. (Such pretreatments are thought to increase the surface concentration of immobilized antibody.)

number of active sites (each IgG contains two active sites). Interestingly, the surface activity of Fab' fragment (28.5%) appears to be higher than that of the intact IgG (11–15%), but a large fraction of this "activity" is probably due to NSB.

(b) Covalent Attachment.

Immobilization to a hydrogel film.   In this immobilization strategy, antibodies are coupled to silica surfaces through an intervening layer of hydrogel, which is comprised of polymethacryloylhydrazide (26,000 MW). The pendant hydrazido groups are used both for grafting the polymer to the surface and for conjugating oxidized antibodies to the hydrogel. In both cases, the hydrazido groups react with aldehyde groups—located either on the silica surface (CHO-silica) or within the antibody's carbohydrate moieties—to form a stable hydrazone linkage which, unlike a Schiff base, does not require further reduction. This reaction can be carried out under slightly acidic conditions (pH 5–6) because the hydrazido group has a pK_a of 2.6 [53,54]. Such conditions help to reduce intramolecular Schiff base formation between oxidized carbohydrate groups and lysine residues within the antibody, a competing reaction which can reduce coupling efficiency [55]. The thickness of the hydrogel layer is controlled by the length and temperature of the grafting reaction. A layer thickness of ca. 5 nm (determined by ellipsometry) was produced using the conditions described in Section II.D.12(b)(1). For conjugation of Fab' fragments instead of oxidized antibodies, the polymer's pendant hydrazido groups are first modified to a maleimido function using SMCC. This enables Fab' fragments to be coupled to the maleimido groups via free thiol groups near their C-termini. The immobilization of antibodies and Fab' fragments to the hydrogel layer is depicted in Fig. 6.

Surface concentrations of immobilized MAb 9-40 and its Fab' fragments are presented in Table 2. Interestingly, significantly higher amounts of the intact antibody could be immobilized with the hydrogel chemistry than with any other immobilization chemistry. Specifically, a surface concentration of 8.0×10^{-12} mol/cm^2 was observed for oxidized antibodies immobilized to the hydrogel surface, while values of 3×10^{-12} mol/cm^2, 2.1×10^{-12} mol/cm^2 and 1.0×10^{-12} mol/cm^2 were observed for physical adsorption (heat-pretreated antibodies), PEG spacers, and self-assembled monolayers, respectively (Tables 1, 3, and 4). The same trend was also observed for immobilized Fab' fragments, but to a lesser degree: a surface concentration of 2.8×10^{-12} mol/cm^2 was observed for Fab' fragments immobilized to hydrogels, while values of 1.3×10^{-12} mol/cm^2 and 1.1×10^{-12} mol/cm^2 were observed for physical adsorption and self-assembled monolayers, respectively (surface concentrations were not determined for Fab' fragments immobilized to PEG spacers). However, the increased surface concentration of antibodies and Fab' fragments on hydrogel surfaces was not an advantage because few if any of the extra sites were available for binding (see below).

Binding assays were performed both with and without a preswelling step (surfaces were soaked in PBS for 30 min prior to addition to analyte). Results are presented in Table 2 and showed that immobilized Fab' fragments were far more effective than immobilized IgG at binding analyte. In fact, the binding capacity of the immobilized Fab' (14×10^{-13} mol/cm^2) was 10-fold higher than that observed for the immobilized IgG (1.2×10^{-13} mol/cm^2). This difference was even more pronounced when the surface activities of the two species were compared. In this

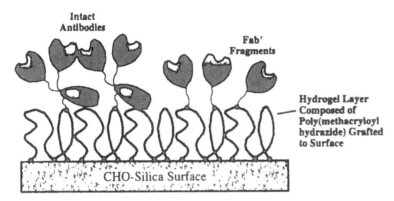

FIG. 6 Immobilization of antibodies and Fab′ fragments to hydrogel-coated silica surfaces. Silica surfaces were first silanized with 3-aminopropyltriethoxysilane and chemically modified with glutaraldehyde to form a monolayer of aldehyde groups (CHO-silica). Poly(methacryloyl hydrazide) (26,000 MW) was then grafted to CHO-silica surfaces through the polymer's pendant hydrazide groups. These groups were also used for site-specific conjugation of oxidized antibodies. For site-specific immobilization of Fab′ fragments, the pendant hydrazide groups were converted to maleimido groups using succinimidyl-4-(N maleimidomethyl)-cyclohexane-1-carboxylate (SMCC). This immobilization technique worked well for Fab′ fragments, resulting in surface activities of ca. 50% and low levels of nonspecific binding. Results for intact antibodies were disappointing, however, because only marginal (<2%) surface activities were observed.

case, less than 1% of the IgG's active sites were available for binding, while ca. 52% of the Fab′ sites were functional. The low surface activity observed for IgG was probably due to its high surface concentration (ca. 8×10^{-12} mol/cm^2). This figure is substantially higher than expected for a fully packed monolayer of IgG ($1-2 \times 10^{-12}$ mol/cm^2 [34,49]). At such high surface concentrations, the immobilized IgG may form a densely packed three-dimensional matrix which may be impenetrable to protein antigens. Conversely, the surface density of the Fab′ fragments (2.8×10^{-12} mol/cm^2) is only about one-half of the expected monolayer coverage (ca. 5×10^{-12} mol/cm^2). This probably results in the antigen-combining sites being more assessable to macromolecular antigens.

Nonspecific binding data are also presented in Table 2. Interestingly, immobilized IgG exhibited somewhat lower levels of nonspecific binding than immobilized Fab′ fragments ($1-3 \times 10^{14}$ versus $3-4 \times 10^{-14}$ mol/cm^2). Also, the preswelling step reduced NSB in both cases. Nonspecific binding values obtained for the preswollen surfaces were lower than those observed for antibodies immobilized by either physical adsorption or PEG tethers (Tables 1 and 3) and are comparable to NSB values observed for self-assembled monolayers.

Immobilization to a polyethylene glycol film. In this immobilization strategy, antibodies are attached to a silica surface via polyethylene glycol chains. A special homobifunctional PEG molecule [bis(hydrazido)polyethylene glycol, 3400 MW] was used that contains reactive hydrazido groups at both ends, one of which was attached to the CHO-silica surface, while the other was conjugated to the oxidized antibody. For conjugation of Fab′ fragments, the hydrazido end groups of the immobilized

TABLE 2 Immobilization of Antibodies by Poly(Methacryloyl Hydrazide) Hydrogels

Surface	Antibody[a] (orientation)	Total binding[b] (moles/cm²) $\times 10^{-13}$	Nonspecific binding[c] (moles/cm²) $\times 10^{-13}$	Relative nonspecific binding (%)	Immobilized antibody (moles/cm²) $\times 10^{-12}$	Surface activity[d] (%)
CHO-silica[e]/hydrogel[f]	IgG (specific)	1.2 ± 0.1	0.3 ± 0.02	25.0 ± 0.4	8.0 ± 0.3	0.8 ± 0.1
CHO-silica/hydrogel (preswollen[g])	IgG (specific)	1.7 ± 0.1	0.1 ± 0.1	5.9 ± 0.2	8.0 ± 0.3	1.1 ± 0.1
CHO-silica/hydrogel	Fab' (specific)	14 ± 0.5	0.4 ± 0.02	2.9 ± 0.05	2.8 ± 0.2	50.8 ± 1.6
CHO-silica/hydrogel (preswollen)	Fab' (specific)	15 ± 0.4	0.3 ± 0.02	2.0 ± 0.2	2.8 ± 0.2	54.5 ± 2.3

[a] See footnote a in Table 1.
[b] See footnote b in Table 1.
[c] See footnote c in Table 1.
[d] See footnote d in Table 1.
[e] CHO-silica: silica surfaces were silanized with 3-aminopropyl triethoxysilane and then chemically modified with glutaraldehyde to form a monolayer of aldehyde groups.
[f] Hydrogel: poly(methacryloyl hydrazide) (26,000 MW) was grafted to CHO-silica surfaces through the polymer's pendant hydrazide groups. These groups were also used for conjugation of oxidized antibodies. For immobilization Fab' fragments, the pendant hydrazide groups were converted to maleimido groups using succinimidyl-4-(N-maleimidomethyl)cyclohexane-1-carboxylate (SMCC).
[g] Hydrogel-coated surfaces were soaked for 30 min in PBS prior to addition of analyte.

TABLE 3 Immobilization of Antibodies by Polyethylene Glycol Spacers

Surface	Antibody[a] (orientation)	Total binding[b] (moles/cm²) ×10^{-13}	Nonspecific binding[c] (moles/cm²) ×10^{-13}	Relative nonspecific binding (%)	Immobilized antibody (moles/cm²) ×10^{-12}	Surface activity[d] (%)
CHO-silica[e]	IgG (random)	5.5	2.1	38.2	3.75	7.3
CHO-silica/PEG[f]	IgG (specific)	5.6	1.1	19.6	2.1	13.6
CHO-silica/PEG (25% Hz₂)[g]	Fab' (specific)	8.7	0.37	4.2	ND[h]	ND
CHO-silica/PEG (100% Hz₂)	Fab' (specific)	7.7	0.58	7.6	ND	ND

[a] See footnote a in Table 1.
[b] See footnote b in Table 1.
[c] See footnote c in Table 1.
[d] See footnote d in Table 1.
[e] See footnote e in Table 2.
[f] PEG: bis(hydrazido)polyethylene glycol (3400 MW) was grafted to CHO-silica surfaces through the polymer's pendant hydrazide groups. These groups were also used for conjugation of oxidized antibodies. For immobilization of Fab' fragments, the pendant hydrazide groups were converted to maleimido groups using succinimidyl-4-(N-maleimidomethyl)cyclohexane-1-carboxylate (SMCC).
[g] 25% Hz₂: bis(hydrazido)polyethylene glycol (3400 MW) and methoxy(hydrazido) polyethylene glycol (5000 MW) were mixed in a 1:3 molar ratio prior to grafting, which effectively reduced the number of potential Fab' attachment sites by 75%.
[h] ND = not determined.

TABLE 4 Immobilization of Antibodies Using Self-Assembled Monolayers

Surface	Antibody[a] (orientation)	Total binding[b] (moles/cm²) ×10⁻¹³	Nonspecific binding[c] (moles/cm²) ×10⁻¹³	Relative nonspecific binding (%)	Immobilized antibody (moles/cm²) ×10⁻¹²	Surface activity[d] (%)
Silica/avidin[e]	biotin-IgG[f] (specific)	7.1 ± 0.1	0.17 ± 0.01	2.4 ± 0.2	1.0 ± 0.1	37.7 ± 1.4
Silica/avidin	biotin-Fab'[g] (specific)	8.5 ± 0.1	0.22 ± 0.01	2.6 ± 0.2	1.2 ± 0.1	70.8 ± 12.6
Silica/avidin (biotin-PEG postcoat)	biotin-Fab' (specific)	7.6 ± 0.2	0.13 ± 0.05	1.7 ± 0.7	1.2 ± 0.1	63.3 ± 8.7

[a] See footnote a in Table 1.
[b] See footnote b in Table 1.
[c] See footnote c in Table 1.
[d] See footnote d in Table 1.
[e] Avidin was immobilized to underivatized silica surfaces by physical adsorption.
[f] Biotin-LC-hydrazide was coupled to oxidized carbohydrate groups in the Fc region of the antibody.
[g] Biotin-HPDP [N-(6-[biotinamido]hexyl)-3'-(2'-pyridyldithiol)propionamide] was coupled to free thiol groups in the C-terminal region of the Fab' fragment.

PEG chains were first converted to maleimido groups using SMCC. The immobilization of antibodies and Fab' fragments via PEG chains is depicted in Fig. 7.

As mentioned in Section II.D.1(b)(2), conjugation of PEG chains to CHO-silica was carried out at pH 5.2 in the presence of 11% K_2SO_4. These conditions produced a PEG surface concentration of 3×10^{-11} mol/cm^2 [49], which gives an average spacing of about 2.4 nm between chain attachment points. Interestingly, this number is comparable to the radius of gyration (R_g) of a 4000 MW PEG chain in solution, suggesting that the tethered PEG chains assume a fairly extended conformation on the surface forming sort of a thick "brush" coating. This is because a more coiled chain conformation would require significantly greater spacing ($\geq 2R_g$) between attachment points [49].

Surface antibody concentration was quantified by conjugating radiolabeled, oxidized MAb 9-40 to Hz-PEG-silica surfaces. In addition radiolabeled, unoxidized MAb 9-40 was conjugated directly to CHO-silica surfaces (without the PEG spaces) as a control. The latter reaction was carried out at pH 9.2 to promote Schiff base formation between the immobilized aldehyde groups and the antibody's lysine residues. It does not result in site-specific coupling because lysine residues are distributed throughout the antibody molecule. The results of these experiments are presented in Table 3. A surface concentration of 2.1×10^{-12} mol/cm^2 was observed for antibodies immobilized via PEG spacers, while a value of 3.75×10^{-12} mol/cm^2 was observed for antibodies immobilized directly to CHO-silica surfaces. The 45% lower surface concentration observed in the former case is probably due to two factors: the steric

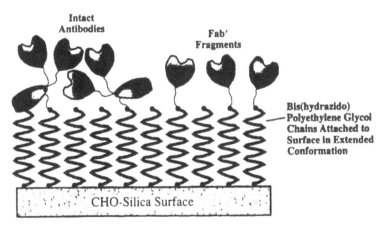

FIG. 7 Immobilization of antibodies and Fab' fragments to silica surfaces coated with polyethylene glycol (PEG). A bifunctional PEG derivative [bis(hydrazido)polyethylene glycol, 3400 MW] was grafted to the CHO-silica surfaces described in Fig. 6 through the polymer's pendant hydrazide groups. These groups were also used for site-specific conjugation of oxidized antibodies. For site-specific immobilization of Fab' fragments, the pendant hydrazide groups were converted to maleimido groups using SMCC. In some cases the bifunctional PEG derivative was mixed with a monofunctional PEG derivative [methoxy(hydrazido)polyethylene glycol, 5000 MW] in a 1:3 molar ratio prior to grafting in order to reduce the number of potential Fab' attachment sites. This procedure actually resulted in higher levels of analyte binding and lower levels of nonspecific binding than were obtained when 100% bifunctional PEG was used.

barrier produced by immobilized PEG chains (which may significantly reduce the rate of the conjugation reaction) and the lower concentrations of reactive groups (the surface concentration of hydrazide in Hz-PEG-silica is lower than that of aldehyde in CHO-silica [49], and the antibody possesses fewer carbohydrate moieties than lysine residues).

The analyte-binding capacity and nonspecific binding of immobilized antibodies were determined using radiolabeled analytes (either FL-BSA or BSA). These data are presented in Table 3 and showed that nearly identical amounts of FL-BSA (5.6 $\times$ 10^{-13} mol/cm^2 for the site-specific/PEG coupling chemistry and 5.5 $\times$ 10^{-13} mol/cm^2 for random attachment without PEG) were bound by the immobilized antibodies, regardless of whether or not PEG was employed as a spacer arm. However, the surface activity of the immobilized antibodies was almost twofold higher when PEG spacers were used. In addition, absolute nonspecific binding was significantly lower when PEG spacers were employed (1.1 $\times$ 10^{-13} mol/cm^2 and 2.1 $\times$ 10^{-13} mol/cm^2 with and without PEG spacers, respectively). Although random attachment resulted in a higher surface antibody concentration than site-specific attachment via PEG spacers, few if any of these additional antibody molecules were able to bind analyte and appeared to increase NSB as well.

For immobilization of intact antibodies, the PEG spacer strategy was somewhat disappointing when compared to physical adsorption and self-assembled monolayers. In fact, the surface concentrations of immobilized antibody and bound analyte obtained when using PEG spacers were nearly identical to those obtained for physical adsorption of acid-pretreated antibodies, although the NSB figures obtained in the case of PEG spacers were substantially worse. From the low surface activity (13.6% for PEG spacers), one must conclude that antibody molecules conjugated to the PEG chains are so closely packed that they sequester a large fraction of the binding sites from interacting with the analyte. Immunoglobulin G is a disk-shaped molecule with a diameter of around 150 Å. A monolayer coverage of IgG in a flat arrangement can therefore be assumed to contain 1.1 $\times$ 10^{-12} mol/cm^2, or about half of the surface antibody concentration observed in the case of PEG spacer and less than one-third of that observed for random coupling. Such crowded surfaces are likely to cause steric hindrance, particularly in the binding of large analytes. Thus, it might prove advantageous to dilute the antibody on the surface for most efficient analyte binding.

In order to examine the question of steric hindrance and the possibility of obtaining higher analyte binding activity, attachment of intact antibodies was abandoned in favor of the much smaller Fab' fragment. In addition to reducing the crowding at the surface by reducing the size of capture molecule, the surface concentration of Fab' fragments was controlled by varying the concentration of coupling sites on the surface. This was accomplished by mixing monofunctional PEG [methoxy-(hydrazido)polyethylene glycol, 5000 MW] with the homobifunctional analog [bis(hydrazido)polyethylene glycol, 3400 MW] at two different stoichiometric ratios (25% and 100% homobifunctional PEG) prior to surface derivatization. Following derivatization, hydrazido groups at the chain ends were modified with SMCC to enable conjugation of Fab' fragments.

As before, analyte binding and NSB were assessed using radiolabeled FL-BSA and BSA, respectively. The results are presented in Table 3 and indicated that Fab' fragments were ca. 40% more effective at binding analyte than were intact antibodies. Moreover, NSB was reduced by nearly twofold in the case of Fab' fragments. Interestingly, further improvements were obtained when the surface concentration of reactive hydrazido groups was reduced to 25% of its maximum value (Table 3). Specifically, surface analyte concentration increased to a value of 8.7×10^{-13} mol/cm^2, while NSB decreased to a value of 3.7×10^{-14} mol/cm^2. These were the best values obtained for this particular immobilization chemistry and rivaled those obtained for self-assembled monolayers (Table 4). Clearly, the often-voiced protein immobilization goal of maximizing the ligand density to achieve the best specific binding characteristics, should be reexamined in light of the present results.

(c) Self-Assembled Monolayers. Self-assembled monolayers (SAMs) have received a lot of attention in recent years as an elegant means of immobilizing proteins to either chromatographic supports or biosensors. Three general classes of SAMs have been described in the literature—lipids, alkane thiols, and alkane silanes and siloxanes. Each of these will be discussed briefly in the next paragraph. For additional information, the reader is referred to recent reviews by Rudolph [56], Mrksich and Whitesides [57], Wink et al. [58], and Wirth et al. [59].

Lipid monolayers form spontaneously at the air–water interface with the polar head groups in the aqueous phase and the acyl chains extending into the air. Such monolayers can be used as a template for self-assembly of protein layers by conjugating ligands such as biotin to a small fraction (<10%) of the polar head groups. Such experiments are usually performed in a Langmuir trough and the Langmuir-Blodgett film, complete with pendant protein layers, can be transferred to a solid surface by dipping the surface in the trough. Working together with Professor Helmut Ringsdorf's group at the University of Mainz in Germany, our laboratory has used this technique to immobilize antibodies to several different substrates including silica, mica, and gold-coated silica [60–64]. Alkane thiols, the second class of SAMs, exhibit high affinity for gold and related metals and thus are often used to immobilize antibodies (or analytes) to gold-coated surfaces or colloidal gold particles [65–68]. Also, a couple of groups have used them as an alternative to carboxymethyldextran for immobilizing antibodies (or analytes) to gold-coated surface plasmon resonance sensors [65,68]. The third class (alkane silanes and siloxanes) can be used to form SAMs on many different types of silica substrates including glass, quartz, and siloxane rubber [69–73]. Often, the strategy is to "hydrophobicize" surface in order to promote adsorption of avidin (or streptavidin), which in turn can act as template for the self-assembly of virtually any biotinylated macromolecule. Alternately, biotinylated BSA can be adsorbed to the hyphobicized surface and act as template for the self-assembly of (strept)avidin–protein conjugates.

In addition to the aforementioned self-assembled lipid systems, our group has investigated a variation of the third class of SAMs in which avidin is allowed to adsorb directly to unmodified silica (instead of to alkylsilane-modified silica). Since this protein has a high isoelectric point, it has a net positive charge at pH 7.4 and adsorbs strongly to underivatized silica, which is negatively charged. Furthermore, our studies showed that the immobilized avidin is properly oriented for binding biotinylated proteins [18]. Once the avidin layer has formed on the surface, a second layer of either biotinylated antibodies or biotinylated Fab' fragments is allowed to self-assemble. Biotinylated polyethylene glycol (5000 MW) can also be added as a postcoat in cases where control of nonspecific binding is important. This immobilization strategy is illustrated in Fig. 8. Radiolabeling studies were used to quantify

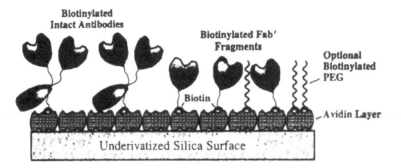

FIG. 8 Immobilization of antibodies and Fab' fragments to silica surfaces using self-assembled monolayers. Avidin was immobilized to underivatized silica surfaces by physical adsorption. The adsorption mechanism is thought to be primarily electrostatic in nature due to avidin's high isoelectric point. Once the avidin layer was formed, either biotinylated antibodies or biotinylated Fab' fragments were allowed to self-assemble on the surface. In some cases, surfaces were postcoated with biotinylated PEG (5000 MW) to reduce nonspeciifc binding. This immobilization technique gave excellent results for both intact antibodies and Fab' fragments.

the surface concentrations of the different layers in the self-assembled system. These studies are described below.

Avidin formed a relatively stable monolayer both on native silica surfaces and on silica surfaces derivatized with DDS. Surface concentrations of 7.8×10^{-12} and 8.2×10^{-12} mol/cm^2 were observed for avidin immobilized on these two surfaces, respectively. Because similar avidin surface concentrations were observed for both native silica and DDS-silica surfaces, we decided to use native silica surfaces in subsequent studies in order to eliminate the silanization step. Based on the molecular dimensions of avidin (60 × 55 × 40 Å [74,75]), full monolayer coverage would be about 1×10^{-11} mol/cm^2. Thus, a surface density of ca. 8×10^{-12} mol/cm^2 represents about 80% of full monolayer coverage. Previous studies have suggested that use of avidin in immunoassays produces unacceptable levels of nonspecific binding [76]; however, this was not observed in our studies. Avidin-coated silica surfaces exhibited relatively low levels of nonspecific binding with BSA (5.3×10^{-14} mol/cm^2). The same was true for antibodies and Fab' fragments immobilized to avidin coated surfaces (see below), where nonspecific binding (measured using BSA) was consistently less than 2.5×10^{-14} mol/cm^2 (Table 4).

Once a stable avidin layer had assembled on the silica surface, either biotinylated antibodies (MAb 9-40) or biotinylated 9-40 Fab' fragments were allowed to self-assemble. This process was quantified using radiolabeled proteins and the results are presented in Table 4. Surface concentrations of 1.0×10^{-12} and 1.2×10^{-12} mol/cm^2 were obtained for the biotinylated IgG and Fab' layers, respectively. Although these values were somewhat lower than observed for antibodies immobilized by the other methods (Tables 1–3), the surface coverage was high enough to give satisfactory performance in immunoassays (see below).

The next step was to determine the analyte-binding capacity and nonspecific binding of immobilized antibodies and Fab' fragments using radiolabeled analytes (either FL-BSA or BSA). Surface analyte concentrations of 7.1×10^{-13} and $8.5 \times$

10^{-13} mol/cm^2 were obtained for FL-BSA bound to immobilized antibodies and Fab′ fragments, respectively (Table 4). When these values were converted to surface activities (ratio of surface analyte concentration to surface active site concentration), a very interesting story emerged—antibodies and Fab′ fragments immobilized by self-assembled monolayers exhibited significantly higher surface activities than those immobilized by any other method. Specifically, intact antibodies immobilized by SAMs exhibited ca. 38% activity, while less than 15% activity was observed for the other immobilization methods; and Fab′ fragments immobilized by SAMs exhibited ca. 70% activity, while other methods resulted in 55% activity or less. Moreover, self-assembled monolayers exhibited lower levels of nonspecific binding (on an absolute scale) than any other immobilization chemistry: 1.7×10^{-14} and 2.2×10^{-14} mol/cm^2 for intact antibodies and Fab′ fragments, respectively (Table 4). We concluded from these experiments that immobilization of Fab′ fragments utilizing SAMs was superor to other immobilization methods examined in these studies.

Masking agents such as BSA, casein, and surfactants are often used in conventional solid-phase immunoassays to block nonspecific binding sites. This often improves both the dynamic range and sensitivity of an immunoassay. Although Fab′ fragments immobilized via SAMs produced low levels of nonspecific binding (2.2×10^{-14} mol/cm^2), we decided to evaluate a masking strategy based on polyethylene glycol to reduce nonspecific binding even further. This polymer is well known for its protein-resistant properties and has been used to coat solid surfaces in order to reduce protein adsorption [77–86]. Biotinylated Fab′ fragments were allowed to self-assemble on avidin-coated silica surfaces, followed by postcoating with biotinylated PEG (5000 MW). Immunoassays were then performed with radiolabeled FL-BSA and BSA to quantify specific and nonspecific binding. Data are presented in Table 4 and showed that masking the surface with PEG reduced absolute nonspecific binding to 1.3×10^{-14} mol/cm^2, a decrease of more than 40%. Considering that the concomitant decrease in total binding was only 11%, the masking step produced a real improvement in nonspecific binding.

Some mention should be made to the length of the PEG chains relative to the dimensions of the Fab′ fragments. X-ray diffraction studies have shown that the dimensions of the Fab′ fragments are ca. $40 \times 40 \times 70$ Å [87]. In order to ascertain the thickness of the PEG layer, we coupled biotin-PEG to avidin-coated silica surfaces (without the Fab′ layer) and measured the thickness of the avidin and PEG layers using ellipsometry. These studies showed that the avidin layer was about 40 Å thick and that the PEG layer was about 20 Å; thus, in a matrix of PEG and Fab′ molecules, the Fab′ fragments will extend 20–50 Å beyond the ends of the PEG chains. In this sort of matrix, we expect that the active sites will be readily available for binding, while the PEG chains will mask nonspecific binding sites between Fab′ molecules.

2. Polystyrene Surfaces

No material is more widely used in traditional solid-phase immunoassays than polystyrene—from the latex beads employed in agglutination assays to the microtiter plates used in the ubiquitous ELISA, polystyrene has become the substrate of choice for immobilizing antibodies. That is not to say, however, that polystyrene is particularly well suited for this role. In fact, studies by Professor John Butler's group at the University of Iowa suggest otherwise. Over the past decade, Butler's group has performed the same sort of investigations with antibodies immobilized to polystyrene as our group has with antibodies immobilized to silica surfaces; and interestingly their results were very similar to ours. Specifically, they found that antibodies immobilized to polystyrene by physical adsorption retained only a small fraction ($\leq 10\%$) of their binding activity [88–92]. Furthermore, surface activity could be significantly enhanced by inserting a passivating layer (typically streptavidin or an antiglobulin antibody) between the capture antibodies and the surface. Considering the commercial success of polystyrene-based immunoassays, one wonders if surface activity is a matter of any practical consequence. It may not be, as long as one only considers traditional solid-phase immunoassays where enzyme amplification or radiolabels can effectively be used to compensate for marginal surface activity. However, in the emerging generation of immunosensors—where assay speed is a matter of paramount importance—the use of enzyme amplification or radiolabels is precluded by the desire to perform assays in 5 min or less; thus placing much greater importance on surface activity as a means of enhancing signal-to-noise ratio. As mentioned, surface activity can be significantly enhanced by elevating capture antibodies off the polystyrene surface. Butler and coworkers employed a system called protein avid biotin capture (PABC) in which a protein such as rabbit IgG is first biotinylated and adsorbed to the surface and then acts as a template for immobilization of biotinylated antibodies via a streptavidin bridge [88,89,91,93]. Actually, this immobilization strategy is a variant of the self-assembled monolayer system described above, so it is not surprising that Butler's group obtained excellent results.

(a) Kinetic Considerations in Assays.   As mentioned, assay speed is of paramount importance in the emerging generation of immunosensors. Therefore, a thorough study of assay kinetics is essential in order to optimize assay parameters and signal levels for very short assay times. Let us consider the first few minutes of a sandwich assay in the absence of sample agitation. As the assay proceeds, the analyte in the region very near the waveguide surface will be depleted due to binding to the immobilized capture antibody. However, the depleted analyte will not be replaced very rapidly from bulk solution because the diffusion of analyte in the unstirred solution is often slow compared to binding kinetics (especially the association or "on" rate). Thus, the amount of analyte bound in the first few minutes will reflect the concentration of depleted analyte near the waveguide surface rather than its true bulk concentration, and assay readings at early times will indicate less bound analyte than if the assay were allowed to proceed to equilibrium.

   If one repeats the above rapid assay for several different bulk analyte concentrations and plots the data on a binding isotherm, it will appear that the curve has shifted toward higher analyte concentrations (and hence lower assay affinity) than if the assays had been allowed to reach equilibrium. These shifts can be a couple orders of magnitude or more for very rapid (<5 min) assays and the shape of the curve will be somewhat distorted. Because of these effects, the affinity constant (K_a) determined from the midpoint of the curve represents an "apparent" affinity rather than the true equilibrium constant. Similar effects would have been observed if binding kinetics were rate limiting rather than diffusion. Thus, an in-depth examination of assay kinetics can reveal either diffusion or binding kinetics limitations and possible ways to compensate for them.

We have developed a mathematical model of our evanescent wave immuno-assay system that incorporates both binding kinetics and mass transport. This model is based on a number of assumptions which are specific to our assay geometry and protocol. Perhaps the most important of these is that the system is unstirred; in particular, the assay solution is injected quickly into the flow cell chamber and then left stagnant with no external agitation or flow. In addition, the temperatures of the assay solution, waveguide, and flow cell are kept constant (within a few degrees Celsius of room temperature) before the start of the assay, suggesting that no appreciable convection is taking place during the assay. Temperature gradients can also be generated if thermal energy is added to the flow cell or waveguide during the course of the assay. The only possible source of thermal energy is conversion of the waveguided excitation laser light (generated by a 10-mW laser diode). However, only a very small fraction of this light is actually adsorbed by the waveguide, flow cell, and assay solution. Based on these considerations, we believe that diffusion is the most probable mass transport mechanism for transferring analyte and tracer antibody to and from the waveguide surface. With this assumption, mass transport of analyte and antibody in the assay solution can be described as Fick's law:

$$\frac{\partial C}{\partial t} = D \frac{\partial^2 C}{\partial x^2} \tag{12}$$

where C is analyte concentration, x is distance along an axis normal to the waveguide, D is the diffusion constant of the analyte in complex with the tracer antibody, and t is time.

We have also made the simplification of one-dimensional mass transport, which is probably valid for our assay geometry where the flow cell chamber is very small (<1 mm) in the dimension (along x) from the waveguide surface to the top of the chamber. Because the width and length of the chamber are much larger (>1 cm), there should be negligible edge effects within the flow cell chamber that would cause deviation from diffusion in only the one dimension. We have also assumed that lateral interactions between binding sites on the waveguide surface are negligible. Unfortunately, the one-dimensional partial differential equation that describes this system does not yield an analytical solution for any but the most simplified models. The complications arise from heterogeneous boundary conditions. In our case, we have chosen a zero-diffusion boundary condition at the upper surface of the flow cell chamber (opposite the waveguide):

$$\frac{\partial C}{\partial x} = 0 \tag{13}$$

Boundary conditions at the waveguide surface were also simplified in our preliminary simulations. Specifically, we assumed that the binding kinetics were very fast compared to diffusion. This assumption allows us to use an equilibrium binding equation rather than one for binding kinetics:

$$K_a = \frac{\Gamma}{C(\Gamma_{max} - \Gamma)} \tag{14}$$

where Γ is the surface concentration of bound analyte, Γ_{max} is the surface concentration of antibody binding sites, and K_a is the equilibrium binding constant. This assumption was removed in later work, and standard binding kinetic expressions for association and dissociation rates (based on k_{on} and k_{off}, respectively) were used in the boundary condition.

The use of the equilibrium equation and concentration term (C) being the same as the boundary concentration in Fick's second law equation yields our model. Unfortunately, this partial differential boundary value problem cannot be solved analytically. A finite difference method known as Crank–Nicholson scheme was used to solve the problem. Simulations were run on a Silicon Graphics Indigo XS24 workstation. The only source of instability in the solution is caused by the boundary condition at the binding surface. Care must be taken that the time step size is not too large for a given spatial step size.

In a comparison between the mathematical model and real assay data, a series of numerical solutions were generated using the following parameter values: (1) equilibrium binding constant (K_a) of $10^9 M^{-1}$; (2) antigen diffusion coefficient (D) of 4×10^7 cm^2/s; (3) surface concentration of immobilized binding sites (Γ_{max}) of 10^{-13} mol/cm^2; and (4) analyte concentration was varied over a range of 10^{-10} to $10^{-7} M$. Figure 9 shows simulated binding isotherms in which the apparent shift and distortion of the curves becomes more pronounced with decreasing assay time. This can be compared with similar curves plotted for actual assay data (Fig. 10) generated from a series of ovalbumin sandwich assays performed for various ovalbumin concentrations and assay times. One sees the similarities between the curves from real

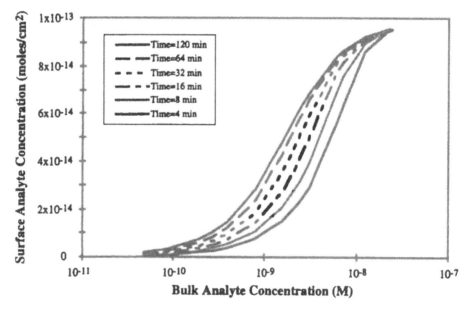

FIG. 9 Computer simulation of the effects of assay time on binding isotherms for a sandwich immunoassay. The computer model contained terms for both mass transport and binding kinetics (see text). The following parameter values were used in the simulation—equilibrium constant (K_a) for analyte binding to antibody: $10^9 M^{-1}$; maximum surface concentration of bound analyte (Γ_{max}): 10^{-13} mol/cm^2; diffusion coefficient (D) for the analyte/tracer antibody complex: 4×10^{-7} cm^2/s; and thickness of the standing liquid layer: 1 mm.

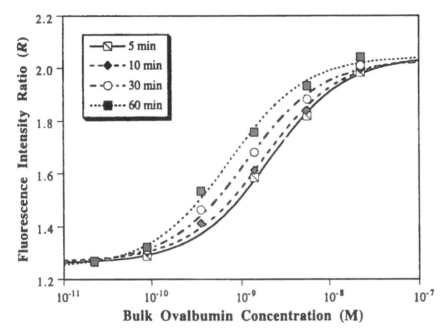

FIG. 10 Impact of assay time on binding isotherms for an ovalbumin sandwich immunoassay. A series of ovalbumin assays was performed using the evanescent wave immunosensor shown in Fig. 4B. Relative fluorescence intensity (R) was plotted versus bulk ovalbumin concentration for four assay times (5, 10, 30, and 60 min). Data obtained at each assay time were fitted using Eq. (8). Apparent association constant values of $4.7 \pm 0.4 \times 10^8 \, M^{-1}$, $5.6 \pm 0.6 \times 10^8 \, M^{-1}$, $8.8 \pm 1.2 \times 10^8 \, M^{-1}$, and $1.4 \pm 0.2 \times 10^9 \, M^{-1}$ were obtained for the 5-, 10-, 30-, and 60-min assay times, respectively.

and model data. Thus, the mathematical model has been shown to follow the general trends of our assays. Additional work fitting assay data to our model should give insight into limitations and possible improvements.

B. Packing and Orientation of Acid-Pretreated Antibodies Immobilized by Physical Adsorption

From the foregoing discussion, we have learned that surface chemistry can have profound impact on the surface activity and nonspecific binding of immobilized antibodies. In particular, we saw that elevation of the capture antibody off the surface with an intervening "passivating" layer could improve both surface activity and NSB. As mentioned in Section I, our group has also investigated a unique surface chemistry in which the IgG molecule serves as its own passivating layer [16,17, 34,35]. Specifically, the antibody is exposed to acidic conditions (pH 2.7) for approximately 20 min before immobilization to a hydrophobic surface, which selectively denatures the Fc region of the antibody and makes it more inclined to adsorb to the surface than the Fab regions [34,35].

These investigations have shown that acid pretreatment can increase the surface concentration of physisorbed antibodies by threefold or more, which produces a concomitant increase in their analyte-binding capacity [16,17]. Even at such high surface concentrations, however, the antibodies appear to adsorb in a single, highly packed monolayer, suggesting that they must be highly ordered on the surface [34]. As mentioned in Section III.A.1(b)(2), high packing density can impair the surface activity of the immobilized antibodies due to steric hindrance (i.e., antibody active sites are too closely packed to allow a macromolecular analyte to bind to each and every one). This observation certainly holds for acid-pretreated antibodies, which exhibited a surface activity of ca. 15% (Table 1).

Still, the packing density and orientation of immobilized antibodies are topics of keen interest, especially when trying to ascertain the mechanism of the acid pre-treatment effect. For this reason, we have developed two different analytical tech-niques for probing the packing density and orientation of immobilized antibodies. The first is a modified two-component Langmuir equation that is used to analyze quantitative adsorption data [34]. This equation contains a special term called the "orientation parameter" that describes the orientation of antibodies on the surface. The second is a special chemical modification technique for placing fluorescence probes in the vicinity of the antibody's active site [45]. Following immobilization, chemical quenchers such as iodide can be used to ascertain whether or not the probe's fluorescence can be quenched. If so, it indicates that the active site is exposed to bulk solvent [35]. Although developed for probing the packing density and orienta-tion of acid-pretreated antibodies, these techniques are generally applicable to other immobilization techniques as well. They will be described in turn in subsequent sections.

1. Two-Component Langmuir Adsorption Analysis with Orientational Parameter

A series of adsorption isotherms was obtained for anti-HSA antibodies which had been acid pretreated for various times prior to adsorption. These data are presented in Fig. 11 and clearly show that surface concentration is maximal for 20 min pre-treatment time. Individual curves were analyzed using the standard two-component Langmuir equation [Eq. (10)]. Although excellent fits were obtained, we were some-what perplexed because the curve-fitting parameters (Γ_1, K_1, Γ_2, K_2) obtained at different pretreatment times did not vary in a systematic fashion. In such cases, multivariate analysis can be useful for identifying common factors within a large number of data sets.

As a prelude to setting up a multivariate analysis for a particular system, it is often useful to review what is known about the model or hypothesis being tested. Previously, we showed that acid pretreatment results in greater exposure of hydro-phobic regions within the antibody to bulk solvent [17] and that the Fc fragment is more susceptible to acid pretreatment than the Fab fragments [17,35]. Thus, for short pretreatment times (e.g., 20 min) the exposed hydrophobic regions are expected to reside principally in the Fc fragment. Our hypothesis is that antibodies pretreated for 20 min will be immobilized primarily through the Fc region, leaving the Fab regions free to bind to analyte. Such an orientation (disklike IgG molecules lined up side by side as dishes in a dish rack) leads to the highest possible packing density on the surface. At longer acid-pretreatment times, hydrophobic regions arise in the Fab regions as well, leading to loss of the preferential orientation and a reduction in surface concentration.

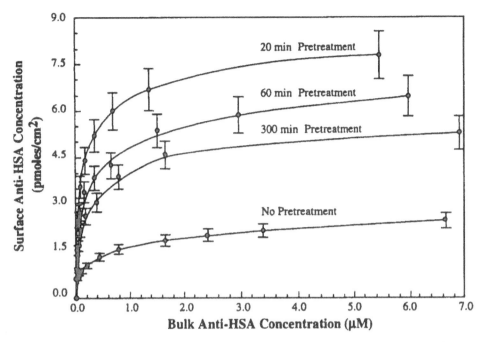

FIG. 11 Langmuir isotherms for the adsorption of acid-pretreated antibodies. Antihuman serum albumin antibodies (anti-HSA) were labeled with ^{125}I and then exposed to acidic conditions (pH 2.7) for various lengths of time (0, 20, 60, and 300 min) before being immobilized to DDS-silica surfaces by physical adsorption. Surface concentration of adsorbed antibody was plotted as a function of bulk antibody concentration. Experimental points and error bars represent the mean and standard error of either three or four independent measurements (n = 3 for native antibody and n = 4 for acid-pretreated antibody). Solid lines are the result of a multivariate analysis which was used to model adsorption data (Table 5).

Based on the above hypothesis, we added an additional parameter (α) to the two-component Langmuir equation [Eq. (10)] that accounts for the orientation of the antibody on the surface:

$$\Gamma = \alpha \left(\frac{\Gamma_1 K_1 C}{1 + K_1 C} + \frac{\Gamma_2 K_2 C}{1 + K_2 C} \right) \tag{15}$$

This parameter is related to the packing density of the immobilized antibodies. A value close to unity represents the highest possible packing density (IgG disks stacked side by side as in a dish rack), while lower values reflect more random orientations which occupy greater surface area per molecule (IgG disks laid out flat on a table). The affinities (K_1, K_2) of the two types of adsorption sites were considered to be independent of acid-pretreatment time, while the maximum concentrations of such sites (Γ_1, Γ_2) and the orientation parameter (α) were allowed to vary with pretreatment time.

Adsorption data sets obtained for four different acid-pretreatment times (0, 20, 60, 300 min) were fit to Eq. (15) using multivariate analysis. The curve fits are shown in Fig. 11 and the corresponding curve-fitting parameters in Table 5. Excellent fits ($r^2 > 0.98$) were obtained for all four data sets. As mentioned, K_1 and K_2 were

TABLE 5 Multivariate Adsorption Analysis of Native and Acid-Pretreated Antibodies on DDS-Silica Surfaces[a]

Linked parameters (held constant in all data sets)
  Adsorption constant, high-affinity sites (K_1): $2.0\ (\pm 0.6) \times 10^7\ M^{-1}$
  Adsorption constant, low-affinity sites (K_2): $1.6\ (\pm 0.2) \times 10^6\ M^{-1}$
  Max. surface concentration, low-affinity sites (Γ_2): $3.7\ (\pm 0.8) \times 10^{-12}$ moles/cm^2

Variable parameters (allowed to vary in data sets obtained for different pretreatment times)

Pretreatment time (min)	Orientational parameter (α)	Max. surface concentration, high-affinity sites (Γ_1) (moles/cm^2)	Goodness of fit (r^2)
0	0.38	$2.4\ (\pm 0.2) \times 10^{-12}$	0.98
20	0.93	$5.0\ (\pm 0.9) \times 10^{-12}$	0.99
60	0.74	$5.0\ (\pm 0.9) \times 10^{-12}$	0.99
300	0.60	$5.0\ (\pm 0.9) \times 10^{-12}$	0.99

[a]Antihuman serum albumin antibodies were labeled with ^{125}I and then exposed to acidic conditions (pH 2.7) for various lengths of time (0, 20, 60, and 300 min.) before being immobilized to DDS-silica surfaces by physical adsorption. A complete adsorption isotherm was obtained for each pretreatment time (Fig. 11). Data collected for all four pretreatment times were fit simultaneously to Eq. (15) using multivariate analysis. This equation was derived from Langmuir adsorption with two classes of adsorption sites (high and low affinity) per antibody molecule. The equation also contains an additional parameter (α) that describes the orientation of antibodies on the surface.

considered to be independent of pretreatment time and therefore were linked across all data sets. Although Γ_1 and Γ_2 were initially thought to depend on pretreatment time, nearly identical Γ_2 values were observed for all data sets, so this parameter was linked as well. Even Γ_1 varied less than expected, with the only distinguishing factor being whether the antibody had been acid pretreated. Thus, Γ_1 was linked in the final analysis across the 20-, 60-, and 300-min data sets. Only orientational parameter α varied significantly across data sets, reaching its maximum value of 0.93 at 20 min pretreatment time and decreasing at longer times.

Our interpretation of these findings is that the native antibody possesses two classes of adsorption sites—high and low affinity—with adsorption constants K_1 and K_2, respectively. Upon acid pretreatment the number of high-affinity sites is increased, resulting in an increase in Γ_1. Apparently, acid pretreatment has little or no effect on the number (or affinity) of the low-affinity sites because K_2 and Γ_2 are invariant. Interestingly, the model predicts that the increased surface concentration at 20 min pretreatment time is due to a combination of orientational effects and the number of high-affinity binding sites in the antibody population. Finally, the minimum value for α (0.38) was observed for native antibodies, suggesting that the difference in adsorbed surface area between the most densely packed case and random adsorption is at most about threefold.

2. Iodide Quenching Studies

Although the above adsorption studies showed that the packing density of immobilized antibodies was highest for the 20-min acid-pretreatment time, these studies were not sufficient to prove that antibodies adsorbed through their Fc regions, leaving the Fab regions exposed to solvent. In order to prove this hypothesis, we employed the photoaffinity labeling technique described in Section II.C.6 to attach covalently a fluorescent probe to the antibody in the vicinity of the active site. Both antibiotin and antifluorescein (MAb 9-40) antibodies were labeled in this fashion, the former with fluorescein and the latter with tetramethylrhodamine. The photoaffinity labeling technique was not strictly specific and placed probes in both the active site and hinge regions of the antibody [45], a caveat which necessitated the use of control antibodies labeled only in the hinge region. Antibodies labeled at both sites are denoted by "(HS)," while those only labeled in the hinge are denoted by "(H)."

All four fluorescently labeled antibody preparations [antibiotin (HS), antibiotin (H), antifluorescein (HS), and antifluorescein (H)] were acid pretreated for various lengths of time (0, 20, 60, and 300 min) and adsorbed to DDS-silica surfaces. In each case the solvent accessibility of the fluorescence probes was examined in fluorescence-quenching studies using iodide as the quencher. Each data set was plotted as $F_0/\Delta F$ versus 1/[iodide] and analyzed using the modified Stern–Volmer equation [Eq. (11)] to determine the fraction of accessible fluorophores (f_a) and the Stern–Volmer quenching constant (K_Q). Typical data sets for native (no pretreatment) antibiotin (HS) antibodies, both in bulk solution and adsorbed to DDS-silica surfaces, are shown in Fig. 12A. The mean and standard error of at least four independent measurements are plotted for each data point. The straight line is the best fit of the data obtained by using weighted linear regression, and the statistical errors of f_a and K_Q were determined from the analysis. Similar analyses were performed for all four antibody preparations at four different pretreatment times, 16 analyses in all.

An in-depth examination of these data is presented by Chang and Herron [35], so the present discussion will be limited to those results most germane to the topic at hand, namely, the orientation of immobilized antibodies. In particular we will focus on a parameter called the "Fab orientation preference factor" (FOPF), which is defined as the ratio of the fraction of accessible fluorophores for an antibody labeled in both active site and hinge regions to that of an antibody labeled in only the hinge region [35]:

$$FOPF = \frac{f_a(\text{HS})}{f_a(\text{H})} \tag{16}$$

This parameter can also be viewed as the solvent accessibility of the active site relative to that of the hinge region and should be unity for antibodies in bulk solution where both locations will be equally exposed to solvent. A value of one is also expected for antibodies immobilized randomly to the surface. Higher values are expected when antibodies are immobilized in such a way that their Fab regions are exposed to solvent.

Fab orientation preference factor was plotted versus acid-pretreatment time in Fig. 12B for antibiotin and antifluorescein antibodies, both in bulk solution (PBS) and physisorbed to DDS-silica surfaces. As expected, both types of antibodies exhibited FOPF values close to unity in bulk solution regardless of pretreatment time. Unity values were also observed for native antibodies adsorbed to DDS-silica surface (where immobilization is thought to be random). In contrast, higher FOPF values were observed for acid-pretreated antibodies, reaching a maximum at 20 min pretreatment time and then decreasing back toward unity at longer times. Maximum

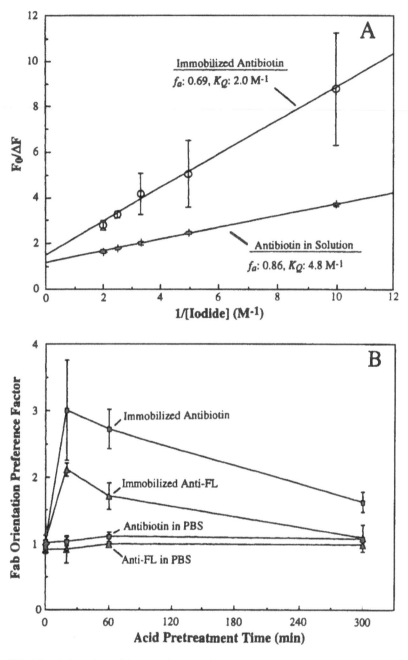

FIG. 12 Orientation of immobilized acid-pretreated antibodies. Antibiotin and antifluorescein (anti-FL) antibodies were photoaffinity labeled with a fluorescent probe in the vicinity of the active site. Antibodies were then acid pretreated for various lengths of time (0, 20, 60, 300 min) and adsorbed to DDS-silica surfaces. A fluorescence quencher (iodide) was used to examine the solvent accessibility of the fluorescent probe. Iodide quenching studies were also performed with photoaffinity-labeled antibodies in bulk solution as a control. (A) Typical modified Stern–Volmer plots obtained for native (0 min pretreatment time) photoaffinity-labeled antibiotin antibodies both in PBS buffer solution and adsorbed to DDS-silica surfaces.

FOPF values of 3.0 $\pm$ 0.8 and 2.1 $\pm$ 0.2 were obtained for antibiotin and antifluorescein, respectively. Considering the larger errors in the former measurement, it is unclear if a higher degree of Fab orientation was achieved in the case of antibiotin. In any case, the same trend was observed for both antibiotin and antifluorescein antibodies. Furthermore, the profile of FPOF versus pretreatment time is almost identical to that of orientation parameter α (Table 5 and Fig. 11), which strongly supports our hypothesis that antibodies acid pretreated for 20 min prior to adsorption are stacked side by side on the surface like dishes in a dish rack with their Fab regions exposed to solvent. Furthermore, we believe that such orientational effects are responsible for both the increased surface concentration and analyte-binding capacity observed when acid-pretreated antibodies are immobilized to DDS-silica surfaces.

C. Applications of Immobilized Antibodies

Immobilized antibodies have achieved widespread use in applications, such as affinity chromatography and immunoassays. The second of these is more significant from an economic perspective, which perhaps explains the immense effort that has been invested in immunoassay research and development over the past two decades (close to 24,000 publications concerning immunoassays exist in the *Medline* database). Although impossible to cover all of these, we will at least enumerate the more common applications and cite pertinent review articles for each. Traditional immunoassays such as the RIA and the ELISA continue to be important (and often essential) research tools in disciplines such as biochemistry, cell biology, immunology, pharmacology, and physiology [94–97]. Moreover, the advent of BIAcore and other sensors has led to a renaissance of new applications in biomedical research over the past few years [98–108]. Clinical immunoassays are omnipresent in modern medical practice and often are the only convenient method for quantifying levels of peptide and protein analytes [109–114]. They can also be used for detecting serum levels of drugs and other low-molecular-weight biomolecules [114,115]. In fact, immunodiagnostics has experienced steady growth over the last two decades, becoming a multibillion dollar worldwide market in the 1990s. This growth is expected to continue into the next decade as time-critical assays are moved out of the reference laboratory and into "point-of-care" settings such as the emergency room, intensive care units, and physicians' offices. Not surprisingly, the use of immunoassays is also widespread in veterinary medicine, both for diagnostic purposes and screening [116–118]. Finally, a number of immunoassays have been reported for environmental monitoring and food testing [118–123], and the demand for such assays is expected to grow rapidly over the next few years with the increasing public awareness of microbial contamination in meat and antibiotics and hormones in milk.

Values for fraction of accessible fluorophores (f_a) and the Stern–Volmer quenching constant (K_Q) are indicated. Similar plots were obtained at other acid pretreatment times as well (data not shown). (B) Fab orientation preference factor (FOPF) of antibiotin and anti-FL antibodies as a function of increasing acid pretreatment time both in bulk solution and adsorbed to DDS-silica surfaces. Preferential exposure to the Fab regions to bulk solution was observed at short (20 min) pretreatment times. (Adapted with permission from Ref. 35. Copyright 1995 American Chemical Society.)

IV. CONCLUSIONS

Although immobilized antibodies play a constitutive role in immunoassays, the immobilized process itself is often poorly understood by workers in the field and rarely examined from a mechanistic standpoint. This situation provided the motivation for our studies of surface activity and nonspecific binding of antibodies immobilized to silica surfaces using several different immobilization techniques. These studies showed that physical adsorption resulted in only marginal (<15%) surface activity, but much higher levels (75% or better) could be obtained by inserting a "passivating" layer between the antibody and the surface. Several different materials were investigated for use in passivating layers, including hydrogels, polyethylene glycol, and self-assembled monolayers comprised of avidin and biotinylated antibodies. Of these, self-assembled monolayers were optimal with respect to both surface activity and nonspecific binding, although acceptable results were also obtained for Fab' fragments (but not the intact IgG) immobilized to either hydrogels or polyethylene glycol spacers.

Although conventional wisdom dictates that densely packed antibody layers should have higher analyte-binding capacities than less crowded antibody layers, our studies showed that this was not necessarily the case. Examination of the total binding figures for the three immobilization techniques which produced monolayer coverage of antibodies (physical adsorption, PEG spacers, and self-assembled monolayers) revealed that surface concentrations of FL-BSA were 6×10^{-13} to 8×10^{-13} mol/cm^2 almost regardless of surface antibody concentration (Tables 1, 3, and 4). This finding was underscored by results from PEG studies in which the number of Fab' conjugation sites on the surface was varied by changing the ratio of mono- to bifunctional PEG molecules. In this case, slightly higher levels of analyte binding were observed for the less crowded antibody layer (Table 3).

Taken together, the above findings suggest that two factors can limit a surface's analyte-binding capacity: the packing density of immobilized antibodies and the size of the analyte molecule. In cases where analyte size exceeds the spacing between immobilized antibodies, steric hindrance between analyte molecules may prevent all of the antibody active sites from being utilized, resulting in an apparent reduction in surface activity. If true, this may also account for the reduced surface activity observed in some cases, especially those in which significantly higher surface antibody concentrations were observed (e.g., physical adsorption and intact antibodies attached to PEG spacers). The more densely packed surfaces also exhibited higher levels of nonspecific binding. As a whole, these observations make a compelling argument for controlling antibody packing density as part of an effective immobilization strategy.

Polystyrene is the most widely used immobilization substrate for antibodies in immunoassays and bears more than a superficial resemblance to hydrophobicized silica with respect to immobilization parameters such as surface activity and nonspecific binding. In fact, Butler and coworkers at the University of Iowa found that antibodies immobilized to polystyrene surfaces by physical adsorption retained less than 10% of their solution binding activity. Their solution to the problem was to employ an immobilization scheme—analogous to our self-assembled monolayers approach—in which a biotinylated protein is first adsorbed to the surface and serves as a passivating layer for the subsequent immobilization of biotinylated capture antibodies through a streptavidin bridge. As with our self-assembled monolayer approach, very high levels (close to 100%) of surface activity were observed for this immobilization strategy.

Mindful of the ever-growing demand for faster assays in immunodiagnostics, we investigated the impact of assay time on assay performance. Our strategy involved developing a computer model of a solid-phase immunoassay that incorporated both mass transport of analyte from bulk solution to the surface and binding of analyte to the immobilized antibodies. The predictions of these simulations were tested using an evanescent wave immunosensor system with fast enough response time to allow binding kinetics to be monitored. A sandwich immunoassay for ovalbumin was used as a model system to mimic immobilization conditions commonly used in the ELISA (intact antibodies immobilized to polystyrene surfaces by physical adsorption). Similar results were obtained from both simulations and experiments, namely, the apparent affinity (K_a, in units of M^{-1}) of an immunoassay decreases with assay time due primarily to mass transport limitations. The effect can be quite large (10- to 100-fold decrease in K_a) at short assay times (2–5 min) compared to assays performed under equilibrium conditions. This finding suggests that high sensitivity and fast assay time can be competing goals in an assay development strategy.

Acid pretreatment provides an attractive means of increasing the surface concentration of immobilized antibody molecules. Using this method, very high surface antibody concentrations could be achieved (ca. 7.5×10^{-12} mol/cm^2 for 20-min pretreatment time; see Fig. 11); yet only a small fraction (ca. 15%) of these molecules were able to bind analyte (which is probably understandable in light of the preceding discussion). Still, we were intrigued by how such high surface concentrations could arise within a single monolayer of antibody molecules, and we devised two new methods to investigate this issue: analysis of adsorption data with a modified Langmuir equation and fluorescence-quenching studies of immobilized, photoaffinity-labeled antibodies. These studies indicated that acid-pretreated antibodies are packed very closely on the surface with the disk-shaped IgG molecules stacked side by side (like dishes in a dish rack) so that the active sites are exposed to solvent and thus able to bind analyte. This orientation resulted in at least a threefold higher packing density than when antibodies were more randomly oriented.

REFERENCES

1. J. Carlsson, J.-C. Janson, and M. Sparrman, Affinity chromatography. In *Protein Purification: Principles, High Resolution Methods, and Applications* (J.-C. Janson and L. Rydén, eds.). VCH Publishers, New York, 1989, pp. 275–329.
2. B. Johnsson, S. Lofas, and G. Lindquist, Immobilization of proteins to a carboxymethyldextran-modified gold surface for biospecific interaction analysis in surface plasmon resonance sensors. Anal. Biochem. *198*:268–277 (1991).
3. D. J. O'Shannessy, M. Brigham-Burke, and K. Peck, Immobilization chemistries suitable for use in the BIAcore surface plasmon resonance detector. Anal. Biochem. *205*: 132–136 (1992).
4. B. Johnsson, S. Lofas, G. Lindquist, A. Edstrom, R. M. Muller Hillgren, and A. Hansson, Comparison of methods for immobilization to carboxymethyl dextran sensor surfaces by analysis of the specific activity of monoclonal antibodies. J. Mol. Recognit. *8*:125–131 (1995).

5. P. Tijssen and A. Adam, Enzyme-linked immunosorbent assays and developments in techniques using latex beads. Curr. Opin. Immunol. *3*:233–237 (1991).

6. D. A. Christensen and J. N. Herron, Optical immunoassay systems based upon evanescent wave interactions. SPIE *2680*:58–67 (1996).

7. S. R. Narayanan and L. J. Crane, Affinity chromatography supports: a look at performance requirements. Trends Biotechnol. *8*:12–16 (1990).

8. M. J. Berry, J. Davis, C. G. Smith, and I. Smith, Immobilization of Fv antibody fragments on porous silica and their utility in affinity chromatography. J. Chromatogr. *587*: 161–169 (1991).

9. P. Cutler, Affinity chromatography. Methods Mol. Biol. *59*:157–168 (1996).

10. J. N. Herron, K. D. Caldwell, D. A. Christensen, S. Dyer, V. Hlady, P. Huang, V. Janatova, H. K. Wang, and A. P. Wei, Fluorescent immunosensors using planar waveguides. SPIE *1885*:28–39 (1993).

11. D. Christensen, S. Dyer, D. Fowers, and J. Herron, Analysis of excitation and collection geometries for planar waveguide immunosensors. SPIE *1886*:2–8 (1993).

12. T. E. Plowman, W. M. Reichert, C. R. Peters, H. K. Wang, D. A. Christensen, and J. N. Herron, Femtomolar sensitivity using a channel-etched thin film waveguide fluoroimmunosensor. Biosensors Bioelectron. *11*:149–160 (1996).

13. S. Ostrove, Affinity chromatography: general methods. Methods Enzymol. *182*:357–371 (1990).

14. R. Hall, P. D. Hunt, and R. G. Ridley, Monoclonal antibody affinity chromatography. Methods Mol. Biol. *21*:389–395 (1993).

15. J. D. Andrade, Principles of protein adsorption. In *Surface and Interfacial Aspects of Biomedical Polymers* (J. D. Andrade, ed.). Plenum Press, New York, 1985, pp. 1–80.

16. J. N. Lin, J. D. Andrade, and I. N. Chang, The influence of adsorption of native and modified antibodies on their activity. J. Immunol. Methods *125*:67–77 (1989).

17. J. N. Lin, I. N. Chang, J. D. Andrade, J. N. Herron, and D. A. Christensen, Comparison of site-specific coupling chemistry for antibody immobilization on different solid supports. J. Chromatogr. *542*:41–54 (1991).

18. J. N. Herron, D. A. Christensen, K. D. Caldwell, V. Janatová, S. C. Huang, and H. K. Wang, Waveguide immunosensor with coating chemistry providing enhanced sensitivity. U.S. Patent No. 5512492. April 30, 1996.

19. D. J. O'Shannessy, M. J. Dobersen, and R. H. Quarles, A novel procedure for labeling immunoglobulins by conjugation to oligosaccharide moieties. Immunol. Lett. *8*:273–277 (1984).

20. D. J. O'Shannessy and R. H. Quarles, Specific conjugation reactions of the oligosaccharide moieties of immunoglobulins. J. Appl. Biocehm. *7*:347–355 (1985).

21. W. L. Hoffman and D. J. O'Shannessy, Site-specific immobilization of antibodies by their oligosaccharide moieties to new hydrazide derivatized solid supports. J. Immunol. Methods *112*:113–120 (1988).

22. R. Abraham, D. Moller, D. Gabel, P. Senter, I. Hellstrom, and K. E. Hellstrom, The influence of periodate oxidation on monoclonal antibody avidity and immunoreactivity. J. Immunol. Methods *144*:77–86 (1991).

23. C. A. Wolfe and D. S. Hage. Studies on the rate and control of antibody oxidation by periodate. Anal. Biochem. *231*:123–130 (1995).

24. E. Lamoyi and A. Nisonoff, Preparation of F(ab')$_2$ fragments from mouse IgG of various subclasses. J. Immunol. Methods *56*:235–243 (1983).

25. P. Parham, On the fragmentation of monoclonal IgG1, IgG2a, and IgG2b from BALB/c mice. J. Immunol. *131*:2895–2902 (1983).

26. E. Lamoyi, Preparation of F(ab')$_2$ fragments from mouse IgG of various subclasses. Methods Enzymol. *121*:652–663 (1986).

27. R. Kurkela, L. Vuolas, and P. Vihko, Preparation of F(ab')$_2$ fragments from monoclonal mouse IgG1 suitable for use in radioimaging. J. Immunol. Methods *110*:229–236 (1988).

28. M. Mariani, M. Camagna, L. Tarditi, and E. Seccamani, A new enzymatic method to obtain high-yield F(ab')₂ suitable for clinical use from mouse IgG1. Mol. Immunol. 28:69–77 (1991).

29. Y. Zou, M. Bian, Z. Yiang, L. Lian, W. Liu, and X. Xu, Comparison of four methods to generate immunoreactive fragments of a murine monoclonal antibody OC859 against human ovarian epithelial cancer antigen. Chin. Med. Sci. J. 10:78–81 (1995).

30. C. Jayabaskaran, P. F. Davison, and H. Paulus, Facile preparation and some applications of an affinity matrix with a cleavable connector arm containing a disulfide bond. Prep. Biochem. 17:121–141 (1987).

31. N. D. Heindel, H. R. Zhao, R. A. Egolf, C. H. Chang, K. J. Schray, J. G. Emrich, J. P. McLaughlin, and D. V. Woo, A novel heterobifunctional linker for formyl to thiol coupling. Bioconjug. Chem. 2:427–430 (1991).

32. T. M. Spitznagel, J. W. Jacobs, and D. S. Clark, Random and site-specific immobilization of catalytic antibodies. Enzyme Microb. Technol. 15:916–921 (1993).

33. B. Catimel, M. Nerrie, F. T. Lee, A. M. Scott, G. Ritter, S. Welt, L. J. Old, A. W. Burgess, and E. C. Nice, Kinetic analysis of the interaction between the monoclonal antibody A33 and its colonic epithelial antigen by the use of an optical biosensor. A comparison of immobilisation strategies. J. Chromatogr. A 776:15–30 (1997).

34. I. N. Chang, J. N. Lin, J. D. Andrade, and J. N. Herron, Adsorption mechanism of acid pretreated antibodies on dichlorodimethylsilane-treated silica surfaces. J. Colloid Interface Sci. 174:10–23 (1995).

35. I. N. Chang and J. N. Herron, Orientation of acid-pretreated antibodies on hydrophobic dichlorodimethylsilane-treated silica surfaces. Langmuir 11:2083–2089 (1995).

36. A. N. Asanov, L. J. Delucas, P. B. Oldham, and W. W. Wilson, Heteroenergetics of bovine serum albumin adsorption from good solvents related to crystallization conditions. J. Colloid Interface Sci. 191:222–235 (1997).

37. D. Cho, G. Narsimhan, and E. I. Frances, Adsorption dynamics of native and pentylated bovine serum albumin at air–water interfaces: surface concentration/surface pressure measurements. J. Colloid Interface Sci. 191:312–325 (1997).

38. R. M. Bates, D. W. Ballard, and E. W. Voss, Jr., Comparative properties of monoclonal antibodies comprising a high-affinity antifluorescyl idiotype family. Mol. Immunol. 22:871–877 (1985).

39. J. W. Goding, Use of staphylococcal protein A as an immunological reagent. J. Immunol. Methods 20:241–253 (1978).

40. J. J. Langone, Applications to immobilized protein A in immunochemical techniques. J. Immunol. Methods 55:277–296 (1982).

41. H. Y. Chuang, W. F. King, and R. G. Mason, Interaction of plasma proteins with artificial surfaces: protein adsorption isotherms. J. Lab. Clin. Med. 92:483–496 (1978).

42. S. Matzku and M. Zoller, Iodination of immunoadsorbed antibodies: chloramine T vs lactoperoxidase. Immunochemsitry 14:367–371 (1977).

43. M. Morrison, Lactoperoxidase-catalyzed iodination as a tool for investigation of proteins. Methods Enzymol. 70:214–220 (1980).

44. L. Marotta, M. Shero, J. M. Carter, W. Klohs, and M. A. Apicella, Radio frequency glow discharge and solid-phase lactoperoxidase-glucose oxidase beads as methods for etching ultra-thin plastic sections for immunoelectron microscopy. J. Immunol. Methods 71:69–82 (1984).

45. I. N. Chang, J. N. Lin, J. D. Andrade, and J. N. Herron, Photoaffinity labeling of antibodies for applications in homogeneous fluoroimmunoassays. Anal. Chem. 67:959–966 (1995).

46. U. Jonsson, M. Malmqvist, and I. Ronnberg, Immobilization of immunoglobulins on silica surfaces. Stability. Biochem. J. 227:363–371 (1985).

47. U. Jonsson, M. Malmqvist, and I. Ronnberg, Immobilization of immunoglobulins on silica surfaces. Kinetcis of immobilization and influence of ionic strength. Biochem. J. 227:373–378 (1985).

48. J. N. Lin, J. Herron, J. D. Andrade, and M. Brizgys, Characterization of immobilized antibodies on silica surfaces. IEEE Trans. Biomed. Eng. *35*:466–471 (1988).

49. S.-C. Huang, K. D. Caldwell, J.-N. Lin, H.-K. Wang, and J. N. Herron, Site-specific immobilization of monoclonal antibodies using spacer-mediated antibody attachment. Langmuir *12*:4292–4298 (1996).

50. E. Kiss, C.-G. Gölander, and J. C. Eriksson, Surface grafting of polyethyleneoxide optimized by means of ESCA. Prog. Colloid. Polym. Sci. *74*:113–119 (1987).

51. S. S. Lehrer, Solute perturbation of protein fluorescence. The quenching of the tryptophan fluorescence of model compounds and lysozyme by iodide ion. Biochemistry *10*:3254–3263 (1971).

52. D. Horsley, J. Herron, V. Hlady, and J. D. Andrade, Fluorescence quenching of absorbed hen and human lysozymes. Langmuir *7*:218–222 (1991).

53. J. K. Inman and H. M. Dintzis, The derivatization of cross-linked polyacrylamide beads. Controlled introduction of functional groups for the preparation of special-purpose biochemical adsorbents. Biochemistry *8*:4074–4082 (1969).

54. P. A. S. Smith, *Derivatives of Hydrazine and Other Hydronitrogens Having N—N Bonds*. Benjamin/Cummings, Reading, MA, 1983.

55. D. J. O'Shannessy, Hydrazido-derivatized supports in affinity chromatography. J. Chromatogr. *510*:13–21 (1990).

56. A. S. Rudolph, Biomaterial biotechnology using self-assembled lipid microstructures. J. Cell. Biochem. *56*:183–187 (1994).

57. M. Mrksich and G. M. Whitesides, Using self-assembled monolayers to understand the interactions of man-made surfaces with proteins and cells. Annu. Rev. Biophys. Biomol. Struct. *25*:55–78 (1996).

58. T. Wink, S. J. van Zuilen, A. Bult, and W. P. van Bennkom, Self-assembled monolayers for biosensors. Analyst *122*:43R-50R (1997).

59. M. J. Wirth, R. W. Fairbank, and H. O. Fatunmbi, Mixed self-assembled monolayers in chemical separations. Science *275*:44–47 (1997).

60. M. Ahlers, D. W. Grainger, J. N. Herron, K. Lim, H. Ringsdorf, and C. Salesse, Quenching of fluorescein-conjugated lipids by antibodies. Quantitative recognition and binding of lipid-bound haptens in biomembrane models, formation of two-dimensional protein domains and molecular dynamics simulations. Biophys. J. *63*:823–838 (1992).

61. H. Ebato, J. N. Herron, W. Müller, Y. Okahata, H. Ringsdorf, and P. Suci, Docking of a second functional protein layer to a streptavidin matrix on a solid support: studies with a quartz crystal microbalance. Angew. Chem. *31*:1087–1090 (1992).

62. J. N. Herron, W. Müller, M. Paudler, H. Riegler, H. Ringsdorf, and P. A. Suci, Specific recognition-induced self-assembly of a biotin lipid/streptavidin/Fab fragment triple layer at the air/water interface: ellipsometric and fluorescence microscopy investigations. Langmuir *8*:1413–1416 (1992).

63. H. Ebato, C. A. Gentry, J. N. Herron, W. Muller, Y. Okahata, H. Ringsdorf, and P. A. Suci, Investigation of specific binding of antifluorescyl antibody and Fab to fluorescein lipids in Langmuir–Blodgett deposited films using quartz crystal microbalance methodology. Anal. Chem. *66*:1683–1689 (1994).

64. D. E. Leckband, T. Kuhl, H. K. Wang, J. Herron, W. Muller, and H. Ringsdorf, 4-4-20 anti-fluorescyl IgG Fab' recognition of membrane bound hapten: direct evidence for the role of protein and interfacial structure. Biochemistry *34*:11467–11478 (1995).

65. D. J. van den Heuvel, R. P. Kooyman, J. W. Drijfhout, and G. W. Welling, Synthetic peptides as receptors in affinity sensors: a feasibility study. Anal. Biochem. *215*:223–230 (1993).

66. C. Duan and M. E. Meyerhoff, Separation-free sandwich enzyme immunoassays using microporous gold electrodes and self-assembled monolayer/immobilized capture antibodies. Anal. Chem. *66*:1369–1377 (1994).

67. F. Pariente C. La Rosa, F. Galan, L. Hernandez, and E. Lorenzo, Enzyme support systems for biosensor applications based on gold-coated nylon meshes. Biosens. Bioelectron. *11*:1115–1128 (1996).

68. G. B. Sigal, C. Bamdad, A. Barberis, J. Strominger, and G. M. Whitesides, A self-assembled monolayer for the binding and study of histidine-tagged proteins by surface plasmon resonance. Anal. Chem. *68*:490–497 (1996).

69. S. Margel, E. A. Vogler, L. Firment, T. Watt, S. Haynie, and D. Y. Sogah, Peptide, protein, and cellular interactions with self-assembled monolayer model surfaces. J. Biomed. Mater. Res. *27*:1463–1467 (1993).

70. H. Morgan, D. J. Pritchard, and J. M. Cooper. Photo-patterning of sensor surfaces with biomolecular structures: characterization using AFM and fluorescence microscopy. Biosens. Bioelectron. *10*:841–846 (1995).

71. J. H. Silver, R. W. Hergenrother, J. C. Lin, F. Lim, H. B. Lin, T. Okada, M. K. Chaudhury, and S. L. Cooper, Surface and blood-contacting properties of alkylsiloxane monolayers supported on silicon rubber. J. Biomed. Mater. Res. *29*:535–548 (1995).

72. J. F. Mooney, A. J. Hunt, J. R. McIntosh, C. A. Liberko, D. M. Walba, and C. T. Rogers, Patterning of functional antibodies and other proteins by photolithography of silane monolayers. Proc. Natl. Acad. Sci. USA *93*:12287–12291 (1996).

73. M. Volker and H. U. Siegmund, Forster energy transfer in ultrathin polymer layers as a basis for biosensors. Exs. *80*:175–191 (1997).

74. L. Pugliese, A. Coda, M. Malcovati, and M. Bolognesi, Three-dimensional structure of the tetragonal crystal form of egg-white avidin in its functional complex with biotin at 2.7 A resolution. J. Mol. Biol. *231*:698–710 (1993).

75. L. Pugliese, M. Malcovati, A. Coda, and M. Bolognesi, Crystal structure of apo-avidin from hen egg-white. J. Mol. Biol. *235*:42–46 (1994).

76. N. M. Green, Avidin. Adv. Protein Chem. *29*:85–133 (1975).

77. K. Bergtrom, E. Osterberg, K. Holmberg, A. S. Hoffman, T. P. Schuman, A. Kozlowksi, and J. H. Harris, Effects of branching and molecular weight of surface-bound poly(ethylene oxide) on protein rejection. J. Biomater. Sci. Polym. Ed. *6*:123–132 (1994).

78. B. E. Rabinow, Y. S. Ding, C. Qin, M. L. McHalsky, J. H. Schneider, K. A. Ashline, T. L. Shelbourn, and R. M. Albrecht, Biomaterials with permanent hydrophilic surfaces and low protein adsorption properties. J. Biomater. Sci. Polym. Ed. *6*:91–109 (1994).

79. B. Wesslan, M. Kober, C. Freij-Larsson, A. Ljungh, and M. Paulsson, Protein adsorption of poly(ether urethane) surfaces modified by amphiphilic and hydrophilic polymers. Biomaterials *15*:278–284 (1994).

80. E. Osterberg, K. Bergstrom, K. Holmberg, T. P. Schuman, J. A. Riggs, N. L. Burns, J. M. Van Alstine, and J. M. Harris, Protein-rejecting ability of surface-bound dextran in end-on and side-on configurations: comparison to PEG. J. Biomed. Mater. Res. *29*: 741–747 (1995).

81. C. J. van Delden, J. M. Bezemer, G. H. Engbers, and J. Feijen, Poly(ethylene oxide)-modified carboxylated polystyrene latices—immobilization chemistry and protein adsorption. J. Biomater. Sci. Polym. Ed. *8*:251–268 (1996).

82. C. Freij-Larsson, T. Nylander, P. Jannasch, and B. Wesslen, Adsorption behavior of amphiphilic polymers at hydrophobic surfaces: effects on protein adsorption. Biomaterials *17*:2199–2207 (1996).

83. D. K. Han, K. D. Park, G. H. Ryu, U. Y. Kim, B. G. Min, and Y. H. Kim, Plasma protein adsorption to sulfonated poly(ethylene oxide)-grafted polyurethane surface. J. Biomed. Mater. Res. *30*:23–30 (1996).

84. D. Beyer, W. Knoll, H. Ringsdorf, J. H. Wang, R. B. Timmons, and P. Sluka, Reduced protein adsorption on plastics via direct plasma deposition of triethylene glycol monoallyl ether. J. Biomed. Mater. Res. *36*:181–189 (1997).

85. J. H. Lee, B. J. Jeong, and H. B. Lee, Plasma protein adsorption and platelet adhesion onto comb-like PEO gradient surfaces. J. Biomed. Mater. Res. *34*:105–114 (1997).

86. L. Litauszki, L. Howard, L. Salvati, and P. J. Tarcha, Surfaces modified with PEO by the Williamson reaction and their affinity for proteins. J. Biomed. Mater. Res. *35*:1–8 (1997).

87. J. N. Herron, A. H. Terry, S. Johnston, X. M. He, L. W. Guddat, E. W. Voss, Jr., and A. B. Edmundson, High resolution structures of the 4-4-20 Fab-fluorescein complex in two solvent systems: effects of solvent on structure and antigen-binding affinity. Biophys. J. *67*:2167–2183 (1994).

88. M. Suter, J. E. Butler, and J. H. Peterman. The immunochemistry of sandwich ELISAs. III. The stoichiometry and efficacy of the protein-avidin-biotin capture (PABC) system. Mol. Immunol. *26*:221–230 (1989).

89. K. S. Joshi, L. G. Hoffman, and J. E. Butler, The immunochemistry of sandwich ELISAs. V. The capture antibody performance of polyclonal antibody-enriched fractions prepared by various methods. Mol. Immunol. *29*:971–981 (1992).

90. J. E. Butler, L. Ni, R. Nessler, K. S. Joshi, M. Suter, B. Rosenberg, J. Chang, W. R. Brown, and L. A. Cantarero, The physical and functional behavior of capture antibodies adsorbed on polystyrene. J. Immunol. Methods *150*:77–90 (1992).

91. J. E. Butler, L. Ni, W. R. Brown, K. S. Joshi, J. Chang, B. Rosenberg, and E. W. Voss, Jr., The immunochemistry of sandwich ELISAs. VI. Greater than 90% of monoclonal and 75% of polyclonal anti-fluorescyl capture antibodies (CAbs) are denatured by passive adsorption. Mol. Immunol. *30*:1165–1175 (1993).

92. J. E. Butler, P. Navarro, and J. Sun, Adsorption-induced antigenic changes and their significance in ELISA and immunological disorders. Immunol. Invest. *26*:39–54 (1997).

93. M. Suter and J. E. Butler, The immunochemistry of sandwich ELISAs. II. A novel system prevents the denaturation of capture antibodies. Immunol. Lett. *13*:313–316 (1986).

94. A. J. Pesce and J. G. Michael, Artifacts and limitations of enzyme immunoassay. J. Immunol. Methods *150*:111–119 (1992).

95. T. Portsmann and S. T. Kiessig, Enzyme immunoassay techniques. An overview. J. Immunol. Methods *150*:5–21 (1992).

96. J. D. Sedgwick and C. Czerkinsky, Detection of cell-surface molecules, secreted products of single cells and cellular proliferation by enzyme immunoassay. J. Immunol. Methods *150*:159–175 (1992).

97. R. Elkins, Immunoassay: recent developments and future directions. Nucl. Med. Biol. *21*:495–521 (1994).

98. L. J. Kricka, Selected strategies for improving sensitivity and reliability of immunoassays. Clin. Chem. *40*:347–357 (1994).

99. D. J. O'Shannessy, Determination of kinetic rate and equilibrium binding constants for macromolecular interactions: a critique of the surface plasmon resonance literature. Curr. Opin. Biotechnol. *5*:65–71 (1994).

100. S. Y. Rabbany, B. L. Donner, and F. S. Ligler, Optical immunosensors. Crit. Rev. Biomed. Eng. *22*:307–346 (1994).

101. A. C. Malmborg and C. A. Borrebaeck, BIAcore as a tool in antibody engineering. J. Immunol. Methods *183*:7–13 (1995).

102. A. Szabo, L. Stolz, and R. Granzow, Surface plasmon resonance and its use in biomolecular interaction analysis (BIA). Curr. Opin. Struct. Biol. *5*:699–705 (1995).

103. D. J. O'Shannessy and D. J. Winzor, Interpretation of deviations from pseudo-first-order kinetic behavior in the characterization of ligand binding by biosensor technology. Anal. Biochem. *236*:275–283 (1996).

104. L. L. Christensen, Theoretical analysis of protein concentration determination using biosensor technology under conditions of partial mass transport limitation. Anal. Biochem. *249*:153–164 (1997).

105. M. M. Davis, D. S. Lyons, J. D. Altman, M. McHeyzer-Williams, J. Hampl, J. J. Boniface, and Y. Chien, T cell receptor biochemistry, repertoire selection and general features of TCR and Ig structure. Ciba Found. Symp. *204*:94–100; discussion 100–104 (1997).

106. P. Schuck, Reliable determination of binding affinity and kinetics using surface plasmon resonance biosensors. Curr. Opin. Biotechnol. *8*:498–502 (1997).

107. P. Schuck, Use of surface plasmon resonance to probe the equilibrium and dynamic aspects of interactions between biological macromolecules. Annu. Rev. Biophys. Biomol. Struct. *26*:541–566 (1997).

108. M. H. Van Regenmortel, D. Altschuh, J. Chatellier, N. Rauffer-Bruyere, P. Richalet-Secordel, and H. Saunal, Uses of biosensors in the study of viral antigens. Immunol. Invest. *26*:67–82 (1997).

109. R. P. Ekins and F. W. Chu, Multianalyte microspot immunoassay—microanalytical "compact disk" of the future. Clin. Chem. *37*:1955–1967 (1991).

110. J. H. Howanitz, Review of the influence of polypeptide hormone forms on immunoassay results. Arch. Pathol. Lab. Med. *117*:369–372 (1993).

111. E. Ishikawa, S. Hashida, T. Kohno, K. Hirota, K. Hashinaka, and S. Ishikawa, Principle and applications of ultrasensitive enzyme immunoassay (immune complex transfer enzyme immunoassay) for antibodies in body fluids. J. Clin. Lab. Anal. *7*:376–393 (1993).

112. L. J. Kricka, Ultrasensitive immunoassay techniques. Clin. Biochem. *26*:325–331 (1993).

113. D. W. Chan, Clinical instrumentation (immunoassay analyzers). Anal. Chem. *67*:519R–524R (1995).

114. C. H. Self and D. B. Cook, Advances in immunoassay technology. Curr. Opin. Biotechnol. *7*:60–65 (1996).

115. A. C. Mehta, A critical appraisal of chromatographic and immunoassay techniques for clinical drug analysis. J. Clin. Pharm. Ther. *71*:325–331 (1992).

116. K. H. Nielsen, P. F. Wright, W. A. Kelly, and J. H. Cherwonogrodzky, A review of enzyme immunoassay for detection of antibody to *Brucella abortus* in cattle. Vet. Immunol. Immunopathol. *18*:331–347 (1988).

117. C. A. Golden, Overview of the state of the art of immunoassay screening tests. J. Am. Vet. Med. Assoc. *198*:827–830 (1991).

118. M. J. Sauer and R. Jackman, Immunoassay and related procedures for analysis of hormones and veterinary products in milk of farm animals. Endocr. Regul. *25*:14–24 (1991).

119. B. M. Kaufman and M. Clower, Jr., Immunoassay of pesticides. J. Assoc. Off. Anal. Chem. *74*:239–247 (1991).

120. B. M. Kaufman and M. Clower, Jr., Immunoassay of pesticides: an update. J. AOAC Int. *78*:1079–1090 (1995).

121. T. Yasumoto, M. Fukui, K. Sasaki, and K. Sugiyama, Determinations of marine toxins in foods. J. AOAC Int. *78*:574–582 (1995).

122. B. Hock, Advances in immunochemical detection of microorganisms. Ann. Biol. Clin. (Paris) *54*:243–252 (1996).

123. J. Sherry, Environmental immunoassays and other bioanalytical methods: overview and update. Chemosphere *34*:1011–1025 (1997).

7

Interactions Between Surfaces Coated with Carbohydrates, Glycolipids, and Glycoproteins

PER M. CLAESSON   Royal Institute of Technology and Institute for Surface Chemistry, Stockholm, Sweden

I. INTRODUCTION

Carbohydrates are one of the main constituents in biological systems, where they have a multitude of functions. The polysaccharides act as major structural units in plants (e.g., cellulose) and insects and crabs (e.g., chitin), as energy storage in plants (e.g., starches) and animals (e.g., glycogen), as important units in glycolipid and glycoprotein receptors, and in protective coatings formed by, e.g., mucins.

The simple monosaccharides have the chemical structure $C_nH_{2n}O_n$ with $n = 5$ for pentoses and $n = 6$ for hexoses. The hexoses adopt conformations that are six-membered (pyranose) rings or five-membered (furanose) rings. The monosaccharide has a large number of chiral sites and this results in the presence of a large number of stereoisomers having the same chemical formula. For instance, glucose, mannose, and galactose all have the chemical formula $C_6H_{12}O_6$, but they differ in the relative orientation of the hydroxide groups (Fig. 1). Each of these monosaccharides also exists in the anomeric α and β form, which differ with respect to the orientation of the hydroxide group connected to the C1 atom (Fig. 1). The α and β anomers are in thermodynamic equilibrium (mutarotation equilibrium) with each other, and for glucose the ratio of the $\alpha{:}\beta$ form is 9:16 [1].

The stereochemistry of the monosaccharides has an important impact on their interactions with each other and with water. For instance, myoinositol is rather soluble in water, whereas scylloinositol is insoluble. As seen in Fig. 2, these compounds only differ in the orientation of one hydroxide group. Further, the surfactant octyl-β-glucoside is very soluble in water and forms micelles above a concentration of about 20 mM, whereas the anomer octyl-α-glucoside has a limited solubility at room temperature. These differences can be explained by differences in crystal energy (differences in possibilities for sugar–sugar hydrogen bonds) as well as by differences in interactions with water. Here it has been argued [2] that the important parameter is how well the monosaccharide fits into the preferential dynamic water structure. The pyranose structure, even if it can exist in chair and boat conformations, is rather inflexible.

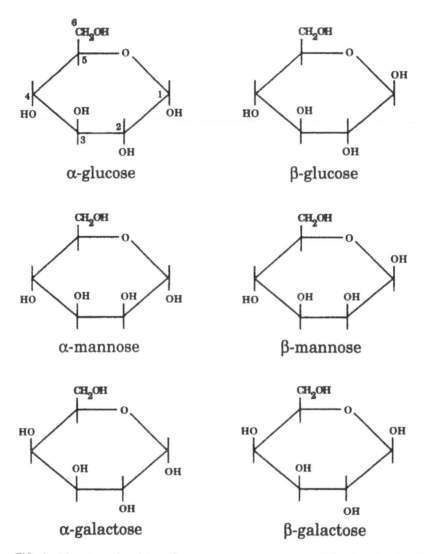

FIG. 1 The stereochemistry of some common pyranoses all having the chemical formula $C_6H_{12}O_6$. The α and β anomers are in thermodynamic equilibrium with each other.

Recently the interactions between water and sugar have been investigated in some detail using quantum mechanical and molecular simulation methods [3–5]. Even though the details of the results are sensitive to the model used to describe the water molecules [3] several conclusions can be drawn from these studies. Specific hydrogen bonds are formed between water and sugar hydroxyl groups, and the formation of these hydrogen bonds significantly influences the conformation and flexibility of the sugar moieties as well as the anomer equilibrium. Most water molecules in the first hydration shell are found to hydrogen bond to more than one sugar hydroxyl group and most hydroxyl groups are hydrogen bonded to more than one water molecule. The typical strength of a hydrogen bond between a sugar hydroxyl group and water is about 6–11 kT at room temperature, and an additional energy of

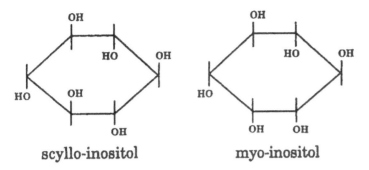

FIG. 2 The stereochemistry of scyllo-inositol and myo-inositol.

about 2 kT per hydrogen bond may be gained by the cooperativity effect of multiple hydrogen bonds [5].

In polysaccharides the monosaccharides are bound together by flexible glycoside linkages joining the C1 atom of one of the sugars via a hemiacetal oxygen to either C1, C2, C3, C4, or C6 of the neighboring monosaccharide. The nature of these bonds is of prime importance for the structure of the polysaccharides and their interactions [6]. In many biologically important polysaccharides, glycolipids, and glycoproteins, the monosaccharides themselves are chemically modified with, e.g., carboxyl, amine, or sulfate groups. Clearly, the number of different polysaccharides that can be formed is enormous, and it is no surprise that this class of compounds has found a wide use in nature.

In this chapter we are concerned with interactions between surfaces coated with simple carbohydrates and complex polysaccharide-containing compounds. It is outlined as follows: First the main types of interactions that may occur between surfaces are introduced, and then the two types of surface force techniques that have been utilized to collect the data presented in this chapter are briefly discussed. In the results section we begin by discussing the forces acting between simple sugar-based surfactants. These data give information about the hydration of the sugar units and their importance for the interaction forces. In the next section, the glycolipids that have a more complex chemical structure are considered, and here we discuss ion binding to charged sugar units as well as how the increased flexibility of the head group may influence the short-range interaction. After this we review some results concerned with interactions between common polysaccharides, and it will be seen how steric forces become important and how the nature of the glycoside linkage influences the adsorbed layer structure. Finally, some data for glycoproteins are discussed and interpreted in terms of what has been learned from simpler systems.

II. SURFACE INTERACTIONS

A. van der Waals Forces

Forces between molecules caused by permanent and induced dipoles and other multipoles are collectively known as van der Waals forces. Hamaker [7] calculated the van der Waals force between particles by integrating the van der Waals forces acting between the individual atoms of the particles and arrived at an expression for the van der Waals force that is a product of a factor which depends on the geometry of the particles and another factor, the Hamaker constant, which depends on the chemical composition of the particles. This treatment of the van der Waals force is rather coarse, but the concept of a Hamaker constant is still being used.

A more accurate theory was developed by Lifshitz et al., who described the van der Waals force as originating from spontaneous electromagnetic fluctuations at interfaces [8]. This theory, in contrast to the Hamaker approach, takes into account many-body effects, temperature dependence, effects due to a surrounding medium, and the finite speed of light. When two interfaces are close to each other, the fluctuating fields associated with them will interact with each. The energy of interaction per unit area (W_{vdW}) between two flat surfaces a distance D apart is given by

$$W_{vdW} = -\frac{A}{12\pi D^2} \tag{1}$$

where A is the nonretarded Hamaker constant, which depends on the dielectric properties of the two interacting particles and the intervening medium. When these properties are known, one can calculate the Hamaker constant. An approximate equation for two particles (subscripts 1 and 3) interacting across a salt-free medium (subscript 2) is

$$A \approx \frac{3kT}{4}\left(\frac{\varepsilon_1 - \varepsilon_2}{\varepsilon_1 + \varepsilon_2}\right)\left(\frac{\varepsilon_3 - \varepsilon_2}{\varepsilon_3 + \varepsilon_2}\right)$$
$$+ \frac{3h\upsilon}{8\sqrt{2}}\frac{(n_1^2 - n_2^2)(n_3^2 - n_2^2)}{\sqrt{n_1^2 + n_2^2}\sqrt{n_3^2 + n_2^2}\,(\sqrt{n_1^2 + n_2^2} + \sqrt{n_3^2 + n_2^2})} \tag{2}$$

where

    k = Boltzmann's constant
    T = absolute temperature
    υ = main absorption frequency in the UV region, often about 3×10^{15} Hz
    h = Planck's constant
    ε = static dielectric constant
    n = refractive index in visible light

We note that a positive Hamaker constant results in a negative (attractive) energy of interaction. Here we have neglected retardation effects, which are due to the finite speed of the propagation of electromagnetic waves. Such effects will lower the van der Waals force at large separations compared to predictions based on Eq. (1). When salt is added to the intervening medium, it results in a decrease of the first term in Eq. (2), whereas the second term is unaffected [9]. For further discussions of the approximations made in order to arrive at Eq. (2) the reader is referred to Ref. 10.

An inspection of Eq. (2) allows several conclusions to be drawn:

1. The van der Waals interaction between two particles of the same kind is always attractive, whereas it can be repulsive between different types of particles.
2. The magnitude of the van der Waals attraction increases with the difference in dielectric properties between the medium and the particles.
3. For particles interacting with each other across an aqueous solution the van der Waals interaction will be less important if the surface is coated

with a layer that contains a large quantity of water (e.g., water-swollen polysaccharide layers).

B. Electrostatic Double-Layer Forces

Electrostatic double-layer forces are always present between charged surfaces in electrolyte solutions. Counterions to the surface (ions with opposite charge to that of the surface) are attracted to it, whereas co-ions are repelled. Hence, outside the surface, in the so-called diffuse layer, the concentration of ions will differ from that in bulk solution, and the charge in the diffuse layer balances the surface charge. An electrostatic double-layer interaction arises when two charged surfaces are so close together that their diffuse layers overlap. When the electrostatic surface potential is small, the free energy of interaction per unit area between flat surfaces (W_{dl}) is [11]

$$W_{dl} = \frac{\kappa^{-1}}{\varepsilon_0 \varepsilon} \left(\frac{(\sigma_1^2 + \sigma_2^2)e^{-\kappa D} + 2\sigma_1 \sigma_2}{e^{\kappa D} - e^{-\kappa D}} \right) \tag{3}$$

which for large distances reduces to

$$W_{dl} = \frac{2\kappa^{-1}\sigma_1 \sigma_2}{\varepsilon_0 \varepsilon} e^{-\kappa D} \tag{4}$$

where

  ε_0 = permittivity of vacuum
  ε = static dielectric constant of the medium
  σ = surface charge density
  κ^{-1} = Debye screening length, given by

$$\kappa^{-1} = \sqrt{\frac{\varepsilon_0 \varepsilon kT}{1000 N_A e^2 \sum_i C_i z_i^2}} \tag{5}$$

where

  e = charge of the proton
  N_A = Avogadro's number
  C_i = concentration of ion i expressed in mol/dm^3
  z_i = valency of ion i

Provided the surface charge is independent of surface separation, one can draw the following conclusions from Eqs. (3)–(5):

1. At large separations the double-layer interaction decays exponentially with surface separation, and the decay length equals the Debye length.
2. The Debye length and consequently the range of the double-layer force decreases with increasing salt concentration and the valency of the ions present.
3. At large separations, surfaces having the same sign on their charge repel each other, whereas surfaces having opposite sign on their charge attract each other.
4. Surfaces having opposite sign on their charge repel each other at sufficiently small separations provided the magnitudes of the surface charges

are not the same. This is due to the loss of entropy of the counterions that are confined to the gap between the surfaces.

A further complication is that if the surface charge originates from weak acidic or basic groups, like carboxylic acids, then the surface charge density for both surfaces will vary with the surface separation, which may result in a complex distance dependence of the double-layer force [12]. The famous *DLVO theory* for colloidal stability [13] assumes that the total force is given by the sum of the double-layer force and the van der Waals force. This additivity approach is not strictly correct [14], but in many cases is a good first approximation.

The Poisson–Boltzmann (PB) approximation is often evoked when describing electrostatic double-layer forces theoretically. In this model the surfaces and the separating medium are treated as continuous media (effects due to the molecular structure of the materials are neglected) characterized by their static dielectric constant. The ions are viewed as point charges (neglecting ion size effects) which distribute themselves in the mean potential created by a uniform surface charge density (disregarding that charges are localized) and a static mean ion distribution. (Hence, effects due to correlations of the location of individual ions are not taken into account.)

Measurements of the forces acting between charged solid surfaces appear to be consistent with calculations based on the nonlinear PB equation. However, this might be a coincidence, since in order to calculate the forces according to the PB model a value of the surface charge density has to be assumed because the true surface charge density is seldom known independently. Hence, the only strict conclusion is that the PB model predicts the functional form of the force versus distance curve, whereas it is unproved that the model predicts the right magnitude of the force for a given surface charge density. More advanced calculations indicate that, for a given surface charge density, the real double-layer force is often smaller than predicted by the PB theory due to the neglect of ion–ion correlation effects. This discrepancy becomes increasingly important at high surface charge densities and in electrolyte solutions containing divalent or multivalent ions [15].

C. Repulsive Short-Range Forces

Rand, Parsegian, and coworkers [16,17] have studied the interactions in a range of different phospholipid systems by measuring the osmotic pressure relative to pure water as a function of bilayer separation. They observed a short-range repulsive force that decays roughly exponentially with the water layer thickness. The decay constant varies somewhat between different systems, but it is typically only a few Angstroms. This short-range repulsion is due to several effects, such as dehydration of the polar groups, giving rise to a hydration force, and steric hindrance of the molecular motion in the normal direction to the interface, producing a steric-type repulsive force contribution. This latter repulsion has an entropic origin and is known as an undulation force when the collective motion of the molecules is considered [18], and as a protrusion force when one considers the molecular motion of individual molecules [19]. There is an ongoing debate on whether the protrusion or the dehydration mechanism is the most important reason for the observed repulsive force.

D. Hydrophobic Interactions

Water molecules tend to associate in clathrate-like structures around hydrophobic molecules (e.g., hydrocarbon) in aqueous solutions. The formation of such dynamic structures minimizes the free energy of the system, and thus makes the unfavorable water/hydrocarbon contact less unfavorable than in the hypothetical case without such structures. When two hydrophobic molecules are brought together, the total contact area toward the aqueous solution decreases and this gives rise to an attractive force known as the hydrophobic interaction. This interaction is the driving force for removal of hydrophobic groups from water and, hence, the reason why amphiphilic compounds associate into micelles, vesicles, and bilayers. The hydrophobic inter-action between molecules is short-range, extending only one or two water molecules out from the hydrophobic moiety [20].

The contact interaction between hydrophobic solid surfaces is also strongly attractive, and this is consistent with the high interfacial tension against water and, thus, related to the hydrophobic interaction. More surprisingly, long-range attractive forces have been measured between many, but not all, types of macroscopic hydro-phobic surfaces immersed in aqueous solutions. There is no agreement about the molecular origin of the long-range part of the observed attraction between macro-scopic water-repellent surfaces, and this is a controversial subject in colloid science. It would lead us too far astray to discuss all ideas that have been proposed as explanations for the long-range attraction between macroscopic hydrophobic sur-faces, but the interested reader is referred to the recent paper by Hato [21] and references therein.

E. Hydrogen Bonds

Hydrogen bonds can be formed between a proton covalently bonded to an electro-negative atom (e.g., oxygen) and another electronegative atom nearby. The hydrogen bond distance is larger than the covalent bond distance, and the energy of the hy-drogen bond formed depends in a rather complex way on the distance between the participating atoms and the angle between the atoms. The hydrogen bond energy is to a large degree due to electrostatic interactions. However, when a hydrogen bond is formed it gives rise to a change in the electron density around the atoms, and thus the hydrogen bond has also a covalent character.

Water molecules can form very strong hydrogen bonds with each other, which explains the preferential tetrahedral arrangement of water molecules in the dynamic structure of liquid water. Water can also form strong hydrogen bonds with polar solutes, like simple sugars. It has been shown that the spatial arrangement of the hydroxyl groups of the sugar unit strongly influences the interaction with water, and most favorable interactions are observed for sugars that fit into the dynamic structure of liquid water [2].

When two sugar units dispersed in water come into contact, it results in the disruption of hydrogen bonds between sugar and water and formation of hydrogen bonds between the sugar units and between the released water. It has been claimed that the sugar–water hydrogen bond is stronger than the water–water hydrogen bond [22] and that glucose is a weak structure promoter [2]. Thus, for glucose one may expect that the process above gives rise to an increase in enthalpy and entropy. The enthalpic part of the interaction is for this case repulsive and the entropic part attractive. Hence, for cases where the entropy part dominates, an attractive interaction originating from changes in hydrogen bonding will occur. This situation arises most easily for polysaccharides that adopt regular structures with internal sugar–sugar hydrogen bonds, an example being the water-insoluble cellulose consisting of glucose units joined by 1,4-β-glucoside linkages.

F. Interactions Due to Adsorbed Uncharged Polymer Layers

Adsorbed polymer layers strongly affect the interaction between surfaces. Factors which determine the sign, range, and magnitude of polymer-induced forces are, e.g., solvent/polymer interaction (the χ parameter), solvent/surface interaction (the surface chi parameter, χ_s), surface coverage, and polymer concentration [23].

Models describing the interaction between particles carrying adsorbed flexible polymers have been developed by de Gennes [24] and by Scheutjens and Fleer [23]. These theories, when used under the assumption of a constant adsorbed amount, are applicable when the polymer adsorption/desorption rate is slow compared with how fast the polymer-coated surfaces approach each other. Under such circumstances the total amount of polymers on the surfaces is independent of the surface separation and the system is not in true equilibrium with bulk solution. However, often the speed of approach is sufficiently slow for the irreversibly adsorbed polymers to adopt the most favorable conformation for each surface separation. Hence, there is an equilibrium within the layer. This situation is referred to as a quasiequilibrium, or restricted equilibrium state. Adsorbed polymers may generate interparticle forces by bridging or by steric interaction.

A polymer gives rise to an attractive bridging interaction when segments belonging to the same molecule are bound to two surfaces. This occurs most easily when the surface coverage is low. The polymer gains entropy when the surfaces come closer together, and this is the molecular mechanism behind the attraction. The bridging attraction may be the dominant force at large separations. However, steric effects (see below) generate repulsive forces at smaller separations. The range of the attraction is determined by the length of the polymer tails extending from the surface.

As the surface coverage increases, the effect of bridging polymers becomes less important, whereas force contributions arising from interactions between polymers adsorbed onto different surfaces become more important. Such contributions are

1. An osmotic interaction that contains contributions from a change in ideal entropy of mixing and changes in solvation of the polymer segments
2. An elastic interaction that arises due to a loss of conformational entropy due to the fact that the number of possible conformations of the polymer chains decrease as the surface separation is reduced

Under quasiequilibrium conditions, the second contribution always becomes increasingly repulsive as the surface separation is decreased, and it will always dominate at small enough separations. The osmotic contribution may be either repulsive or attractive depending on the solvent quality. It is unfavorable for polymer chains to interpenetrate in good solvents, whereas such a process is favorable in poor solvents. The solvent quality, which is a measure of the strength of the interaction between individual segments compared to those between segment and solvent, is usually expressed by the χ-parameter. The χ-parameter is smaller than 0.5 in good solvents and larger than 0.5 in poor ones. Under poor solvency conditions the outer part of the interaction between polymer coated surfaces may thus be attractive.

G. Interactions Due to Adsorbed Polyelectrolyte Layers

The forces acting between surfaces covered with polyelectrolytes have been studied extensively in recent years [25–27]. It is found that under low-ionic-strength conditions highly charged polyelectrolytes adsorb in a very flat conformation on oppositely highly charged surfaces [28]. This is due to the strong electrostatic attraction between the polyelectrolyte and the surface. As the electrostatic forces are decreased by increasing the ionic strength [29] or by reducing the charge density of the polymer [30], the adsorbed layer adopts a more extended structure, and it becomes more like that formed by nonionic polymers. The forces acting between surfaces coated with polyelectrolytes are of the same type as those acting between surfaces coated with nonionic polymers, with the addition that electrostatic double-layer forces are also important for polyelectrolyte coated surfaces.

The bridging mechanism is also slightly different for polyelectrolytes and for nonionic polymers. For nonionic polymers bridging is caused by polymers bound to two surfaces. For polyelectrolytes, however, the polymers do not need to be bound to two surfaces to cause bridging. This is due to the long-range nature of electrostatic forces. Hence, a bridging attraction arises when the polyelectrolyte has some segments sufficiently close to one surface and others close to the other surface to interact with the electrostatic potential emanating from them. Thus, a part of the polyelectrolyte is attracted to one surface and another part to a second surface. As for nonionic polymers this bridging attraction has mainly an entropic origin [28].

III. METHODS

It is possible to measure forces between macroscopic interfaces or colloidal particles with various methods [16,31–36]. In this chapter we present data obtained with two of these techniques, which will be briefly described below. For the measurements of interactions between solid surfaces we utilized the interferometric surface force apparatus, which exists in many versions [31,37–39]. The principle behind the different designs is similar, and depicted schematically in Fig. 3. The preferred substrate in the surface force apparatus (SFA) is muscovite mica, an aluminosilicate mineral that can easily be cleaved into large molecularly smooth sheets. However, other substrates have also been used [40,41].

The main advantage of the interferometric surface force apparatus over noninterferometric techniques is that the use of optical interferometry allows the determination of absolute surface separations, whereas the other techniques report distances relative to a "hard wall." The possibility of measuring absolute distances with the SFA also allows determination of adsorbed layer thicknesses and orientation of asymmetric molecules (e.g., proteins) on surfaces. The optical interference technique also allows studies of a range of phenomena such as surface deformation, local radius of curvature of the surfaces, phase separation, and measurements of refractive index.

Claesson

A)

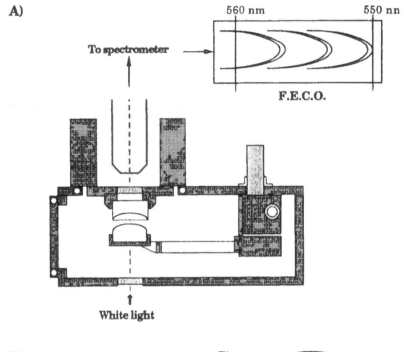

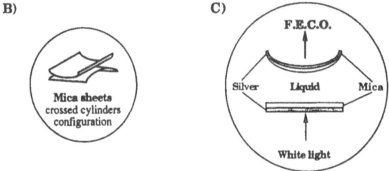

FIG. 3 (A) Schematic of the main components of the interferometric surface force apparatus. The stainless steel measuring chamber contains the two interacting surfaces. One mica surface is glued to a silica disk attached to a piezoelectric crystal (topmost part). The other surface, also glued to a silica disk, is mounted on a double-cantilever force-measuring spring. The surfaces are oriented in a crossed-cylinder configuration (B). White light enters through the window in the bottom of the chamber. It is multiply reflected between the silver layers, and a standing wave pattern, fringes of equal chromatic order (F.E.C.O.), is generated (C). The standing waves exit through the top window, and the wavelength and fringe shape are analyzed in a spectrometer.

For these reasons one may state that this technique still provides the most detailed information of all surface force techniques. The main drawbacks with the SFA are that it is a slow technique and that relatively few types of substrate surfaces fulfill the requirements of being sheetlike, transparent, and molecularly smooth.

Forces between two air–liquid interfaces can be measured in a thin-film balance [36,42]. The porous frit version of this instrument, which is the one used in this investigation, is shown in Fig. 4. It consists of a gas-tight measuring cell. The solution under investigation is placed in the bottom of the cell and it is also contained in a porous glass frit with a capillary tube welded to one of its sides. The foam film is formed in a small hole drilled in the frit. In the flat portion of the film the disjoining pressure (Π) equals the capillary pressure, which in turn equals the difference in the pressure in the gas phase (P_g) and in the liquid phase (P_l):

$$\Pi = P_g - P_l = P_g - P_r + \frac{2\sigma\cos\theta}{r} - \Delta\rho g h \qquad (6)$$

where P_r is the pressure above the capillary, σ is the surface tension of the surfactant solution, θ is the contact angle of the liquid against the capillary wall, r is the radius of the capillary tube, $\Delta\rho$ is the density difference between the solution and the gas, h is the height of solution in the capillary tube above the film, and g is the gravitational constant. Varying the gas pressure in the measuring cell allows the disjoining pressure to be changed, and this, of course, changes the equilibrium film thickness. In this way the repulsive branch of the disjoining pressure isotherm can be determined.

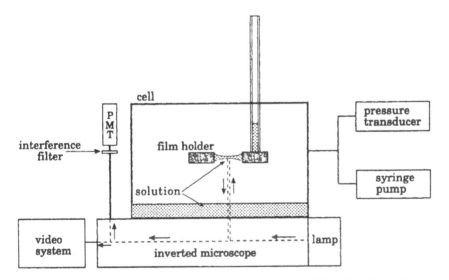

FIG. 4 Schematic figure showing the main components of the thin-film balance. A macroscopic foam film is formed in a hole drilled in a porous glass frit. The surfactant solution is contained in the frit, in the glass capillary, and at the bottom of the closed cell. The film thickness is determined by interferometry. The reflected light is viewed by a videocamera, and the intensity of a selected wavelength is measured with a photomultiplier tube (PMT). The pressure in the measuring cell is varied by means of a syringe pump and measured by a pressure transducer.

IV. RESULTS

A. Sugar-Based Surfactants

Surfactants with sugar units as polar group are becoming of increasing importance as alternatives to commercial surfactants because they are less harmful for the skin and for the environment. In the context of this chapter they are also of interest since the study of interactions between layers of sugar-based surfactants provides some important insight into the forces that are of importance between the building blocks of polysaccharides and the carbohydrate chains of glycoproteins. They can also be regarded as the simplest type of glycolipid.

The interactions between hydrophobic solid surfaces and between air–water interfaces coated with simple sugar-based surfactants have been investigated in some detail [43–46]. The forces acting across a macroscopic foam film stabilized by octyl-β-glucoside, C8β, were deduced using two types of thin-film balances: the Scheludko cell [43] and the porous frit cell [44]; the latter is shown in Fig. 4, and the structure of the surfactant is provided in Fig. 5. Some data obtained with the porous frit technique are illustrated in Fig. 6. Note that the long-range interaction between two air–water interfaces with adsorbed C8β is dominated by an electrostatic double-layer force. This may at first seem surprising since the sugar surfactant is nonionic. However, the charge does not originate from the surfactant itself but is intrinsic to the air–water interface. It has been shown in several studies using electrophoretic [47,48] and thin-film balance techniques [49,50] that the air–water interface is negatively charged. The presence of the negative charge at the air–water interface is commonly assumed to be due to a positive adsorption of hydroxide ions. This hypothesis finds some support by measurements of the pH dependence of the interfacial charge, but in my opinion the origin of the charge at the air–water interface is not yet proved.

The magnitude of the electrostatic repulsion decreases as the surfactant concentration is increased. Once the surface separation is below about 18 nm, an attractive van der Waals force becomes dominant and at this point the film ruptures when the surfactant concentration is well below the critical micelle concentration (cmc) ($C \leq 1/2$ cmc). However, at higher surfactant concentrations, close to the cmc, the action of the van der Waals force instead induces a phase transition from a water-rich common black film to a water-poor Newton black film. The Newton black film is stabilized by short-range steric and hydration forces. The results shown in Fig. 6 demonstrate that the water layer thickness in the Newton black film is less than 1 nm and that the force increases very steeply at shorter separations. This observation provides evidence for the short range of the forces acting between uncharged sugar moieties in aqueous solutions, involving no more than one or two layers of water molecules outside each interface.

The forces acting between two hydrophobized mica surfaces carrying an adsorbed layer of C8β are illustrated in Fig. 7. On this type of surface the sugar groups of the surfactants are directed toward the solution. Unlike the situation at the air–water interface no electrostatic double-layer force is observed between hydrophobic solid surfaces coated with this sugar-based surfactant. A short-range force is observed in the distance regime 2–3 nm. A comparison with data obtained for other nonionic and zwitterionic surfactants shows that the interactions between the sugar groups of C8β are of a shorter range and steeper than those observed for, say, phospholipids [17] and ethylene oxide–based surfactants [51]. An implication of this finding is that,

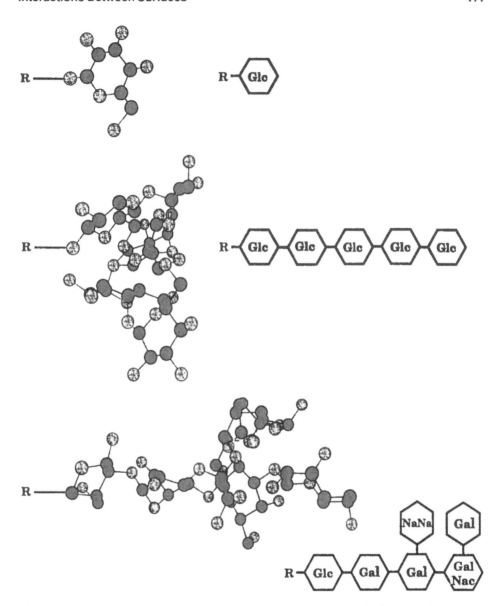

FIG. 5 The chemical structure of octyl-β-glucoside ($C_8\beta$), 1,3-di-O-octyl-2-(β-glyco-syl)pentamaltoside (diC$_8$Mal$_5$), where the glucose units are connected via α-1,4-glucoside linkages, and the glycolipid G$_{M1}$. The negative charge of the G$_{M1}$ is due to the carboxylate group of the N-acetylneuramic acid. Black spheres represent carbon atoms, gray spheres oxygen atoms, and white spheres nitrogen atoms. Hydrogen atoms are not shown. Glc = glucose, Gal = galactose, GalNac = N-acetylgalactoseamine, NaNa = N-acetylneuramic acid.

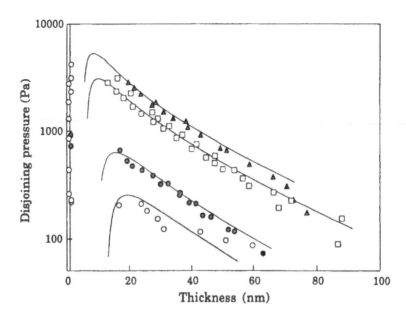

FIG. 6 Disjoining pressure isotherms for $C_8\beta$ in 0.1 mM KBr solution. ▲ = 3 mM surfactant concentration, □ = 10 mM surfactant concentration, ● = 21 mM surfactant concentration, and ○ = 25 mM surfactant concentration. The solid lines are calculated according to DLVO theory using a nonretarded Hamaker constant of 3.7×10^{-20} J. The Debye length used in the calculations is 30.4 nm, in accordance with theoretical predictions. The boundary condition of constant surface charge density was used with the following values for the area per charge: 3 mM $C_8\beta$ 98 nm^2, 10 mM $C_8\beta$ 138 nm^2, 21 mM $C_8\beta$ 318 nm^2, 25 mM $C_8\beta$ 452 nm^2.

in agreement with the data obtained from thin-film balance measurements, the sugar units are not heavily hydrated. The adsorbed layer thickness, including any water of hydration, is 1.5 nm. This value is close to the extended length of the molecule, which further emphasizes the limited range of the hydration force between uncharged sugar units. A similar conclusion can be drawn from the forces measured between adsorbed layers of sucrose or cyclodextrin [52]. The compression of the C8β layer under a high force may have two origins: dehydration and a pressure-induced change of conformation due to a depletion of surfactants in the contact zone.

The forces acting between a sugar-based surfactant with a larger sugar unit, 1,3-di-O-octyl-2-(β-glycosyl)pentamaltoside, abbreviated diC$_8$Mal$_5$, adsorbed on hydrophobic surfaces have also been investigated [53]. The surfactant was synthesized by the group of Professor Hato, Tsukuba, Japan [54], and the structure of the compound is illustrated in Fig. 5. The forces measured across a 1 mM solution of diC$_8$Mal$_5$ (Fig. 7) are, in many respects, similar to the ones obtained for the smaller surfactant C8β. No long-range electrostatic force is present, but at a distance of just below 20 nm an attractive force component gives rise to an inward jump. A short-range steric/hydration repulsion dominates at distances below about 7 nm and the layer can under a high force be compressed to about 5 nm. The adhesion force measured on separation is very similar in strength to that observed for C8β, about

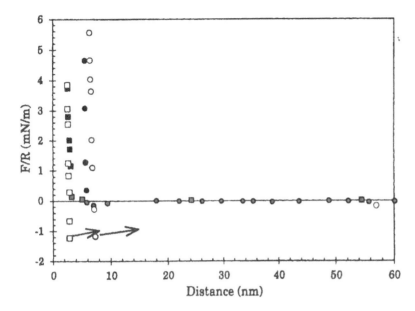

FIG. 7 Force normalized by radius between hydrophobized mica surfaces across surfactant solutions. The forces were determined across a 25 mM solution (■) on approach and (□) on separation and across a 1 mM diC$_8$Mal$_3$ solution (●) on approach and (○) on separation. Arrows indicate outward jumps.

1 mN/m. Hence, an increase in the length of the glucose chain does not remove the attractive interaction between the adsorbed layers of the sugar surfactants.

With these results in mind, let us consider the origin of the attractive interaction. First, a van der Waals attraction is always present between any two surfaces that do not have the same dielectric properties as the solvent. Glucose has a rather high refractive index (about 1.51) compared to water (1.33). Since the structure of the head group of diC$_8$Mal$_3$ is rather compact with no branching points, it is reasonable to assume that the water content of the layer is rather small and one can thus expect a significant van der Waals attraction between the sugar groups of the two layers. In fact, the long-range nature of the attractive force observed strongly points to the importance of the van der Waals force.

Let us also consider the importance of hydrogen bonds for the magnitude of the contact attraction. When two sugar groups come into contact, the hydrogen bond between sugar and water is broken and a hydrogen bond between the sugar units and the released water is formed. As argued in Section II.E, it is not obvious that this gives rise to a decrease in free energy. One should also bear in mind that the energy of hydrogen bonds crucially depends on the mutual orientation of the hydrogen-bonding species. When the molecules are anchored onto surfaces their orientational and conformational freedom are strongly restricted, which limits the possibility of formation of favorable hydrogen bonds between two such surface layers in contact. Hence, in my opinion it is unlikely that a change in hydrogen bonding gives rise to any attractive contribution to the force between octyl-β-glucoside layers adsorbed to hydrophobic surfaces.

B. Glycolipids

The forces acting between two types of glycolipids have been investigated. The galactolipids studied by Marra [55,56], which are abundant in plant thylakoid membranes, have two long hydrocarbon chains as the hydrophobic moiety and uncharged sugar units as polar groups. The gangliosides studied by Parker [57] and Luckham et al. [58] have a more complex chemical structure, with a hydrophobic moiety consisting of a sphingosine residue with an attached C16–C18 carboxylic acid and a branched and charged sugar unit as polar group. The charge originates from the sugar N-acetylneuramic acid, which is a sialic acid.

1. Galactolipids

The forces measured between uncharged digalactosyl diglyceride (DGDG) and monogalactosyl diglyceride (MGDG) layers deposited on hydrophobic mica surfaces reported by Marra [55] are similar to those measured between the sugar-based surfactants described above. No long-range electrostatic force is reported, but the most long-range part of the interaction is dominated by a van der Waals force with a Hamaker constant of 5.3–7.5×10^{-21} J. The attractive minimum in the force curve with a depth of $F/R \approx 2$ mN/m is located 1.3 nm from contact between the anhydrous DGDG layers. At shorter separations a strong steric/hydration force dominates the interaction. For MGDG, which has a smaller polar group, the range of the steric/hydration force is less and the attractive minimum is deeper. Free bilayers of DGDG have also been studied with the osmotic stress technique [17]. In these measurements the water layer thickness of the fully hydrated bilayer was reported to be 1.46 nm, in close agreement with the range of the short-range force reported by Marra. Again, these data show that the hydration of the sugar units are limited and that the short-range force is considerably smaller than for, e.g., dipalmitoyl phosphatidylcholine (DPPC), where the range of the steric/hydration force is about 20 Å below the chain-melting temperature and 33 Å above the melting point as determined by the osmotic stress technique [17]. The measured bilayer adhesion obtained by the pipet aspiration technique using lipid vesicles and estimated from the balance of van der Waals and steric/hydration forces at the maximum swelling of the bilayers in water [17] is also close to those reported by Marra for DGDG.

2. Gangliosides

The polar moiety of gangliosides differ from those of the galactosides in that they contain more sugar units, are branched, and contain negatively charged sialic acid groups. The structure of the G_{M1} head group is shown in Fig. 5. The structural differences between the gangliosides and the galactosides have profound effects on all aspects of the interaction forces. The long-range forces between hydrophobic surfaces coated with the gangliosides G_{M1} and G_{T1b} are in dilute electrolyte solutions ($C < 0.1$ M) dominated by an electrostatic double-layer force originating from the sialic acid residues [57,58]. The interaction profile between deposited G_{M1} layers determined by Luckham, Wood, and Swart [58] is schematically shown in Fig. 8.

Here the double-layer force dominates the interaction down to a separation of about 20 nm relative to the contact between the hydrophobized mica surfaces in air. (Note that in the original paper the separation was reported relative to the contact between bare mica surfaces and thus 4.8 nm larger than those in Fig. 8.) By comparing the magnitude of the measured double-layer force with forces calculated in the Poisson–Boltzmann approximation, one can obtain a value of the apparent charge per ganglioside. The results from such calculations are illustrated in Fig. 9. Clearly, the apparent charge is significantly less than 1 per deposited ganglioside, which indicates that most sialic acid residues are uncharged. This is a rather surprising result considering that the pK_a value for sialic acids is around 2.6–2.7 [59,60] and the pH of unbuffered water is typically about 5.6–5.8. Even though the apparent surface charge deduced from the force measurements may be somewhat too low due to uncertainties in the location of the plane of charge and due to the neglect of ion–ion correlation effects in the PB model [15], it is clear that the charge density of the ganglioside layer is lower than expected from the chemical structure of G_{M1}. There are two possibilities that may rationalize this observation. First, the pK_a value of the sialic acid residue in the layer ought to be higher than for sialic acids in bulk water due to the lower dielectric constant in the layer. Second, counterions may to a large extent be incorporated within the ganglioside layer. From Fig. 9 it is clear that the apparent charge per ganglioside increases with increasing monovalent salt concentration. This is qualitatively as expected for ionizable surface groups and can be explained as an increased screening of the electrostatic repulsion within the layer at higher ionic strengths.

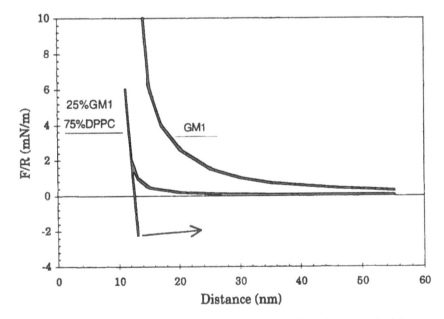

FIG. 8 Force normalized by radius between mica surfaces first coated with a monolayer of DPPC and then with a second layer of G_{M1} (upper line), or a second layer consisting of 25% G_{M1} and 75% DPPC (lower line). The arrow indicates an outward jump. (Redrawn from Ref. 58.)

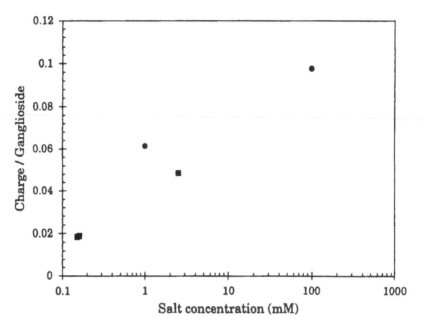

FIG. 9 Charge per deposited ganglioside as a function of salt concentration. (■, Data from Ref. 58; ●, data from Ref. 57.)

Luckham and coworkers [58] found that by increasing the compressive load a steric/hydration repulsion is encountered at distances below about 20 nm. The final separation between the G_{M1} layers was 13.8 nm from contact between the hydrophobized mica surfaces, which corresponds to a thickness of one G_{M1} layer of 6.9 nm. This layer thickness is larger than the 5 nm reported from x-ray scattering. The range of the steric/hydration repulsion, about 6 nm, observed for the gangliosides is significantly larger than that observed for galactosides and for the sugar-based surfactants discussed previously. This can be rationalized by the larger and branched head group that extends further from the interface and that also can take up a considerable amount of water (about 60 water molecules per head group has been claimed for gangliosides [57,58]). Parker, using the surface force apparatus, measured a very small layer thickness for the G_{M1} layer, about 3 nm [57]. It is possible that his finding is due to the larger area per deposited G_{M1} molecule that would allow the head group to adopt a flat conformation or to the G_{M1} molecules tendency to diffuse away from the contact zone under a high pressure. The latter interpretation gains some support from the observation that the measured layer thickness for mixed DPPC/G_{M1} layers, which are in the frozen state, is significantly larger than that of fluid mixed DPPO/G_{M1} layers [57].

The forces acting between layers consisting of mixtures of gangliosides and phospholipids have also been measured [57,58]. An example where a 25:75 mixture of G_{M1}/DPPC was used is shown in Fig. 8 (redrawn from Ref. 58). The repulsive double-layer force and the steric/hydration force are significantly reduced in the mixed system. It is also observed that on separation the two layers adhere. This observation was interpreted as evidence for an attractive interaction between the head groups of the phospholipid and the ganglioside. An alternative interpretation may be that the larger conformational freedom of the ganglioside head group in the mixed layer makes it easier to adopt orientations that give rise to favorable hydrogen bonds between the carbohydrate moieties of the opposing surfaces.

C. Polysaccharides

The properties of polysaccharides depend on the type of sugar units they contain and on the type of linkages formed between the sugar units. The importance of the type of sugar–sugar bond formed is easily visualized by considering the polymers formed by glucose. When the glucose units are linked by β-1,4-glucoside bonds, a straight chain with each sugar unit rotated 180° relative to its neighbors in the chain is formed (Fig. 10). This is the structure of cellulose, which is further stabilized by hydrogen bonds between neighboring glucose units in the chain and between neighboring chains in the fibrils. These structural features make it impossible to dissolve cellulose in water and explain its usefulness as a structural entity in plants. When the glucose units instead are linked by α-1,4-glucoside bonds, a more soluble hollow helix structure is formed (Fig. 10), which is the structure of important polysaccharides used for energy storage, such as the amylose component of starch. The solubility is increased by branching through α-1,6-glucoside linkages. In nature this is used in the amylopectin component of starch (about one branch point for 30 glucose units) and glycogen (about one branch point for 10 glucose units). Still another polysaccharide consisting of only glucose is dextran, where all sugars in the main chain are linked by α-1,6-glucoside bonds, which gives this polymer a large solubility due to its flexible coil structure (Fig. 10).

1.   Short-Range Forces Between Polysaccharide Helices

Rau and Parsegian, using the osmotic stress technique [16], investigated the short-range forces acting between ordered arrays of two stiff, helix-forming polysaccharides: uncharged schizophyllan and charged xanthan [61]. Schizophyllan consists of a backbone of glucose units linked by β-1,3-glucoside bonds with each third of the glucose units having an additional glucose linked to it by a β-1,6-glucoside bond. In aqueous solution a triple helix structure with a diameter of about 2 nm is formed. The forces between the ordered helices are measurable at interaxial distances below 3 nm, i.e., approximately 1 nm from direct contact [61]. The repulsion decays exponentially with decreasing interaxial spacing, with a decay length of 0.34 nm until the interaxial spacing is decreased to 2.2 nm. Hence, a short-range hydration force extending about 1 nm can be inferred from these measurements. The range of this force is thus similar to that observed between simple uncharged sugar surfactants. At even smaller separation a stronger repulsion due to steric interaction between the helices is seen.

The xanthan has a cellulose backbone with linked trisaccharides. One of the trisaccharides is negatively charged glucuronate, pK_a 3.2 [62], and some of the trisaccharides (about 50%) also contain a terminal negatively charged pyruvate group. In aqueous solution single- or double-stranded helices with diameters of about 2.1 nm are formed. The forces between such xanthan helices are stronger than those observed between uncharged schizophyllan, with a range of about 2 nm [61]. Clearly,

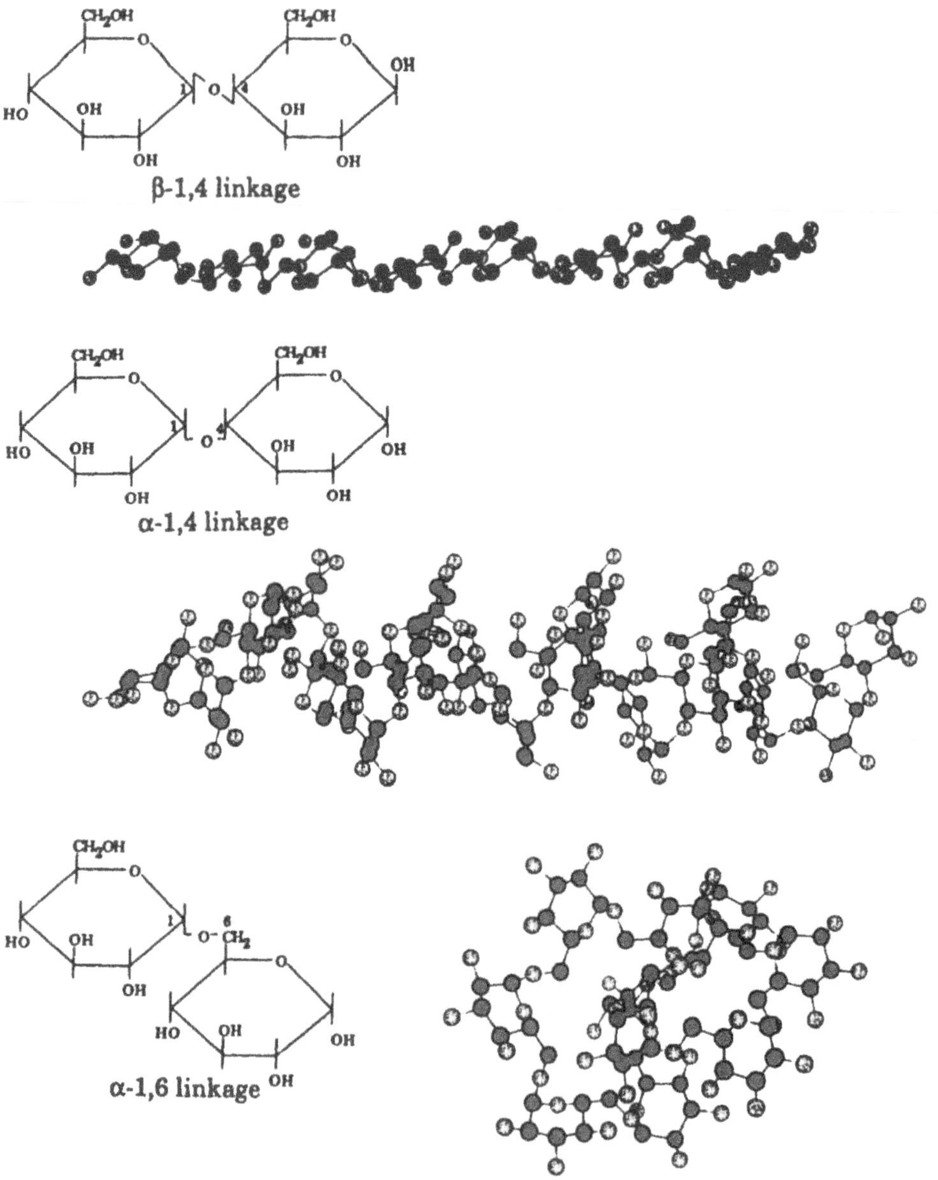

FIG. 10 Typical conformations for polymers of glucose joined by β-1,4-glucoside linkages (top), α-1,4-glucoside linkages (middle), α-1,6-glucoside linkages (bottom).

the charge of the xanthan has profound influence on the short-range forces. The amount of NaCl in the solution (0.1–1 M) and addition of $MgCl_2$ and $CaCl_2$ do not affect these interactions to any large extent, even though the forces in the presence of $CaCl_2$ may be slightly less. This is convincing evidence that the short-range interaction is not due to an electrostatic double-layer force. However, the removal of the charged pyruvate group significantly decreases the repulsive force, demonstrating that the presence of the charged groups influences the hydration of the polysaccharide and, thus, the range of the hydration force. The insensitivity of the short-range force to the type and concentration of simple salts is somewhat surprising since the binding of these ions ought to affect the hydration of the polysaccharide. Rau and Parsegian interpret this as evidence that these ions either do not bind to xanthan or are buried within the helical polysaccharide structure [61]. This observation, together with the apparent low charge of ganglioside layers reported by Luckham et al. [58] and Parker [57], emphasizes the importance of understanding the ion-binding properties of charged polysaccharides. This, in fact, is at the heart of the biological function of some glycoproteins, such as proteoheparan sulfate, as will be discussed.

2. Cellulose

Cellulose, the most abundant polysaccharide on earth, is, as mentioned above, not soluble in water due to the fact that the long straight conformation of the polysaccharide is ideal for formation of intrachain and interchain hydrogen bonds, which promotes crystalline structures. In order to investigate the interactions between model cellulose surfaces, we have applied the Langmuir–Blodgett method [63] to prepare well-defined and smooth surfaces [64]. In the studies presented, 10 layers of tri-methylsilyl cellulose were deposited on hydrophobized mica. After deposition the cellulose was regenerated in a humid HCl atmosphere. It was found that the thickness of each cellulose layer in air was 0.4 nm, increasing to 0.65 nm when the humidity was increased to close to 100%. It is well known that crystalline cellulose takes up a very limited amount of water [5] and hence the rather large uptake of water seen here indicates that our sample is mainly composed of amorphous cellulose. The swelling in liquid water is even larger, and it is seen that under high compression ($F/R \approx 10$ mN/m), each layer is about 0.8 nm. The interaction between cellulose surfaces across water, pH 5.5–6.0, is shown in Fig. 11. No long-range double-layer force is observed, which is expected for unoxidized cellulose, considering that the pK_a of glucose is around 12.4 [65]. However, at higher pH values of 8–10 a small double-layer force is present between the layers, which may indicate a slight oxidation of the cellulose sample or preferential ion binding.

A repulsive force extends to a separation of slightly above 30 nm from the underlying hydrophobized mica surface; i.e., the range relative to that of the compressed layer thickness is about 14 nm. In light of the results discussed, it is clear that this force is much too long range to be a hydration force and that some dangling tails extend from the surfaces, and the force observed is a steric force arising from a compression of these tails. On separation, a weak attractive force is observed. The origin of the attraction is most likely a van der Waals force between the hydrated cellulose layers. The Hamaker constant, from Lifshitz theory, was estimated to be 0.7–0.9 $\times$ 10^{-20} J in water and 8.4 $\times$ 10^{-20} J in dry air [64].

3. Chitosan

Chitin (N-acetylglucoseamine) is the next most abundant polysaccharide on earth, and it is the main structural unit of the exoskeletons of insects and crustacea. It is, like cellulose, built up by β-1,4-glucoside linkages and can thus form long extended structures. When chitin is deacetylated one obtains chitosan [poly(glucoseamine)], which can be dissolved in weak acetic acid solutions. Chitosan is a weak cationic

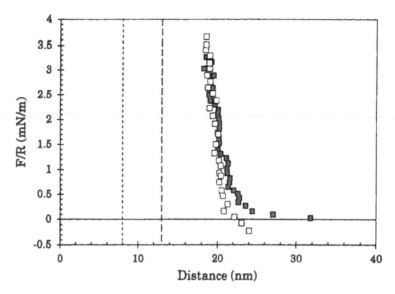

FIG. 11 Force normalized by radius as a function of separation between two hydrophobic surfaces coated by cellulose. Forces measured on approach and on separation are represented by ■ and □, respectively. Dotted and dashed vertical lines represent the thickness of the cellulose layers in dry air and humid air, respectively.

polyelectrolyte with a pK_a value of about 6.3–7 [66]. It adsorbs readily on negatively charged interfaces, and for this reason it can be employed as wet- or dry-strength additive in papermaking [67] or to stabilize negatively charged emulsions under acidic conditions [68].

The pH dependence of the force acting between negatively charged mica surfaces precoated with a layer of chitosan has been investigated in some detail [69], and some data are illustrated in Fig. 12. At low pH the adsorbed chitosan overcompensates the negative charge of the underlying surface, and the interaction at distances above 5 nm is dominated by a weak electrostatic double-layer force. At smaller distances a steric force component is observed. The small range of the steric force component demonstrates that due to the high electrostatic affinity for the surface the large chitosan molecules (MW $\approx$ 7 × 10^5 g/mol) adsorb in a very flat conformation. This is consistent with theoretical predictions [70]. The polyelectrolyte-coated surfaces become uncharged at pH 6.2, as evidenced by the absence of any double-layer force. Instead a weak attractive force appears at a separation between 3 and 2 nm. This force can be explained by a van der Waals attraction, even though it is possible that other attractive forces are present between the segments. At smaller separations a very steep steric force is observed. At higher pH values the affinity between chitosan and the surface decreases and the layer becomes more extended. The steric force now extends to a separation of 5 nm. The adsorbed chitosan molecules can no longer neutralize the surface charge, and for this reason a repulsive double-layer force reappears and dominates the long-range interaction.

When comparing the results obtained for chitosan and cellulose—note that both are linear polysaccharides which adopt straight conformations due to the β-1,4-

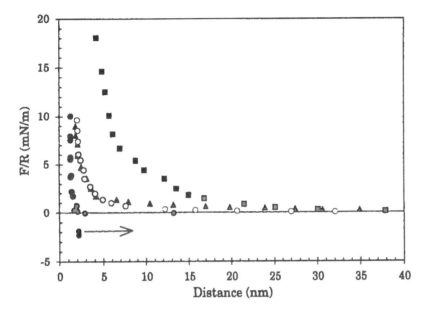

FIG. 12 Force normalized by radius as a function of surface separation between mica surfaces coated with chitosan across a 0.01 wt% acetic acid solution. Symbols represent the forces at various pH values: (▲) pH 3.8, (○) pH 4.9, (●) pH 6.2, (■) pH 9.1.

glucoside linkages—we notice some differences and similarities. In both cases the range of the steric force is rather small, particularly when compared to the results presented below, obtained for polysaccharides that adopt more random conformations in solution. Clearly, the possibilities for forming hydrogen bonds between large regions of the polysaccharides within the layer promotes a compact layer structure on the surface. The surface affinity between cationic chitosan and negatively charged mica surfaces is very large, and this makes the chitosan surface layer even flatter than that of cellulose. On separation, a stronger attractive force is present between the flatter chitosan surfaces under conditions when it is uncharged compared to that for the more swelled cellulose surfaces. This can be rationalized in terms of a stronger van der Waals force in the former case. Note that it is sufficient with the presence of a rather small electrostatic double-layer force to completely remove the adhesive force.

4. Xylan

Xylan is a slightly branched hemicellulose. In plants it is associated with cellulose, where it contributes to the structural units. Xylans have a heterogeneous chemical composition, containing glucose, mannose, galactose, xylose, arabinose, and uronic acids, with the exact chemical composition varying between different sources. Uronic acids are sugar groups containing a carboxylic acid. The pK_a value is 3.2–3.7 for the most common types of uronic acids in hemicelluloses [62,71,72]. The xylan used in the reported experiment was obtained from pine kraft pulp bleached with oxygen and hydrogen peroxide. It has a rather strong negative charge because 9.9% of the sugar units are of the uronic acid type. Despite the net negative charge it has been found that xylan adsorbs to negatively charged mica surfaces, presumably through the uncharged sugar units [52].

The forces acting between the mica surfaces in the presence of xylan are long range and repulsive (Fig. 13). The force does not decay exponentially with surface separation, indicating that the force is mainly due to steric interactions between the extended layers. The reason for the extension of the layer is the electrostatic repulsion between the mica surface and the negative residues of the xylan and the mutual repulsion between the charged segments. This is supported by the fact that less strongly charged xylans give rise to less long-range repulsive forces [73]. As the ionic strength is increased, the electrostatic repulsion between the charged segments and these segments and the surface is reduced. As a consequence, the layer becomes more compact and the range of the steric force decreases (Fig. 13). The presence of a small amount of divalent calcium ions results in a dramatic decrease of the repulsive force observed on approach and the adsorbed layer becomes very compact. On separation, a small adhesive force is observed in the presence of calcium ions.

We note that calcium ions have a large effect on the forces acting between negatively charged xylan layers, whereas hardly any effects on the interaction between negatively charged xanthan helices were reported [61]. The reason for this difference is that the polysaccharides have different structures. Xylan on mica adopts an extended structure and the forces are due to a steric interaction. The layer structure is easily compacted by screening of electrostatic forces and ion binding. On the other hand, the xanthan helices that are compact to start with interact via short-range hydration forces. The compact structure makes it less easy to incorporate calcium ions, and if they are taken up by the compact polysaccharide they have only a minor effect on the hydration forces.

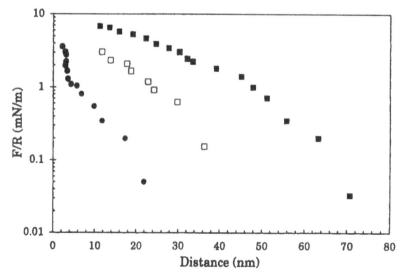

FIG. 13   Force normalized by radius as a function of separation between two mica surfaces across a solution containing 0.9 mg/mL xylan bleached pine kraft. The salts present in the solution were (■) 1 mM KBr, (□) mM KBr, and (●) 10 mM KBr + 0.1 mM $CaCl_2$.

D. Glycoproteins

Glycoproteins consist of at least one protein part and at least one carbohydrate part. One important class is the negatively charged proteoglycans that bind water and cations and form the extracellular medium of connective tissues. Among other things they have functions as viscoelastic modifiers in joints. Some of the glycoproteins have a clear amphiphilic character where the protein moiety is anchored in the cell membrane. The proteoheparan sulfate discussed below is such a glycoprotein. Other glycoproteins, the mucins, are important constituents of the mucus gel that covers many of the internal surfaces in animals.

1. Proteoheparan Sulfate

Proteoheparan sulfate is a type of proteoglycan, i.e., an anionic glycoprotein consisting of a protein core carrying highly carboxylated and sulfonated carbohydrate side chains. The peptide region of proteoheparan sulfate is rather hydrophobic, whereas the strongly negatively charged glycosaminoglycan side chains are hydrophilic. In nature proteoheparan sulfate is found in endothelial cell membranes with the carbohydrate chains protruding into the extracellular solution. It has been suggested that, among other functions, it acts as a flow sensor [74], where with increasing flow rate the conformation of the side chains changes from a random coil to an extended conformation. This in turn is suggested to trigger a release of bound cations that may act as messengers [74].

Ellipsometric measurements have shown that the peptide chain adsorbs strongly to hydrophobic surfaces, whereas the polysaccharide chains do not adsorb in measurable quantities [75]. The interfacial behavior of the proteoheparan sulfate is also strongly influenced by the ionic strength of the solution and by specific cationic effects [76].

In order to study the forces acting between proteoheparan sulfate layers, the protein was allowed to adsorb on hydrophobized mica surfaces. This results in an orientation of the nonadsorbing carbohydrate chains directed into the solution. The adsorption took place from a 0.1 mM NaCl solution containing 0.2 mg/mL proteoheparan sulfate. The forces measured between the proteoheparan sulfate–coated surfaces after replacing the protein solution with a protein-free 0.1 mM NaCl solution [77] are shown in Fig. 14. A repulsive double-layer force dominates the long-range interaction, whereas a steric repulsion is most important at separations below 9 nm. The magnitude of the repulsive double-layer force is low, considering the charge of the proteoheparan sulfate molecule and the adsorbed amount. This gives a strong indication that a large amount of cations are contained within the carbohydrate layer, consistent with the suggested model of proteoheparan sulfate as a flow sensor, where a large number of cations are suggested to be incorporated in the layer under no-flow conditions. As the surfaces are separated, a small attractive minimum is noted.

Addition of CaCl$_2$ to a concentration of 1.25 mM results in a significant decrease in the long-range repulsive force (Fig. 14), and the decay length of the force is no longer consistent with an electrostatic double-layer force. Instead the force, measurable to a separation of about 40 nm, is of steric origin and due to interactions between the carbohydrate chains. We note that the thickness of the layer under a high compressive force also decreases somewhat upon addition of CaCl$_2$, indicating

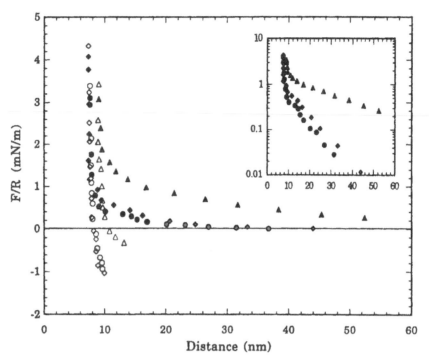

FIG. 14 Forces normalized by radius between proteoheparan sulfate layers adsorbed on hydrophobized mica surfaces as a function of surface separation at different excess electrolyte concentrations. Forces were measured in 0.1 mM NaCl before addition of CaCl$_2$ (▲, △) and after addition of 1.25 mM CaCl$_2$ (◆, ◇) and 2.5 mM CaCl$_2$ (●, ○). Filled and unfilled symbols represent the force measured on approach and separation, respectively. The inset shows the forces measured on compression on a logarithmic force scale.

a small structural change within the layer due to incorporation of calcium ions. A comparatively strong attraction is observed at a separation of 9 nm. A further addition of CaCl$_2$ to 2.5 mM results in a reduction of the long-range steric force but no further decrease in the compressed layer thickness. Note that the double-layer repulsion observed before addition of CaCl$_2$ precludes any determination of the range of the long-range component of the steric force in this case.

Since both xylan and proteoheparan sulfate have negatively charged polysaccharide chains extended into the solution when adsorbed to solid surfaces, it is of some interest to compare the results obtained for these two systems. In both cases it is observed that addition of calcium ions decreases the range of the steric repulsion. This can be rationalized in terms of the decreased electrostatic repulsion between charged segments within the carbohydrate chains, resulting in an entropically driven chain contraction. An addition of calcium ions also increases the adhesive force between both types of polysaccharides. It is not completely clear if this is solely due to a decrease in intralayer repulsion or if specific calcium ion–mediated forces are of importance for the magnitude of this adhesion.

2. Mucin

Mucins are very large linear and flexible glycoproteins with molecular weights of $5-25 \times 10^6$ g/mol. They consist of subunits with a molecular weight of about $2-3 \times 10^6$ g/mol joined end to end. About 80% of the weight is due to oligosaccharides that are clustered in regions flanked by stretches composed predominantly of amino acids. The structure is schematically shown in Fig. 15. It is an important type of glycoprotein since it covers many internal surfaces in the body. Like the xylans discussed previously, the exact chemical composition and charge density vary between different sources, and when comparing the data in the literature one finds that these differences are important for the interfacial behavior [78–80].

The forces between hydrophobic surfaces precoated with a layer of rat gastric mucin (RGM), a weakly charged mucin with a radius of gyration of 190 nm, are shown in Fig. 16. It is likely that it is the carbohydrate-poor regions that adsorb to the hydrophobic surface, with the carbohydrate-rich regions extending into the solution, and the force curves thus reflect the interactions between the carbohydrate-rich moieties. The forces observed are purely repulsive and measurable at distances below about 150 nm. At large distances the force decayed exponentially with a decay constant of 30 nm; at shorter distances a more rapid decay of the force was observed. The force was only weakly dependent on the salt concentration in the range 0.1–150 mM, demonstrating that it is a steric force that dominates the interaction. The

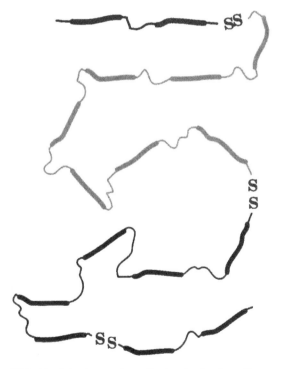

FIG. 15 Mucin consists of subunits having oligosaccharide clusters (thick lines) flanked by bare protein chains (thin lines). The subunits, which have a molecular weight of $2-3 \times 10^6$ g/mol, are joined end-to-end by disulfide bonds.

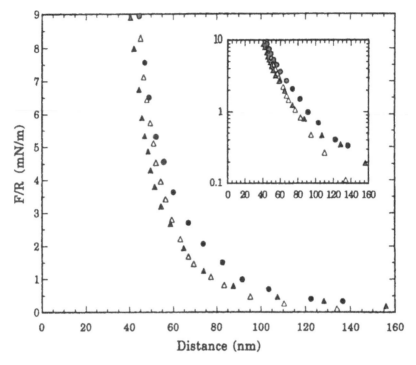

FIG. 16 The force between hydrophobized mica surfaces coated with rat gastric mucin. The forces were measured on approach in (△) 0.1 mM, (▲) 10 mM, and (●) 0.15 M NaCl.

force was the same on approach and on separation, indicating quasiequilibrium conditions; i.e., conformational changes occurred rapidly, whereas the adsorption/desorption kinetics were slow, compared to the time scale of the measurements. For this situation theoretical models describing the forces operating between polymer-coated surfaces under restricted equilibrium conditions should be applicable [23,24].

In contrast to what was observed for RGM, it was very difficult to measure nonhysteretic forces in the presence of the more highly charged pig gastric mucin (PGM). Instead it was found that it took a very long time for the system to reach equilibrium after each change in surface separation. This indicates that the extrusion of material from the contact zone (water and possibly some of the polymer) is a slow process. It may be that such nonequilibrium effects are of importance for the protective function of mucins. Clearly, relaxation effects in mucin are in some cases very important. To increase the understanding of such nonequilibrium phenomena, further development of experimental techniques, e.g., for studies of the relaxation of a polymer layer when kept under a constant force, and theoretical models for nonequilibrium forces is required.

V. SUMMARY

The short-range hydration/steric force acting between uncharged sugar-based surfactants, galactolipids, and ordered polysaccharides has a range of approximately 1–

1.5 nm. Hence, at most three layers of water molecules outside each sugar unit give a significant contribution to the hydration. The presence of charged groups on the carbohydrate gives rise to an electrostatic double-layer force and increases the hydration of the sugar unit and, thus, the range of the hydration force. Likewise, branching of the sugar moiety increases the range of the short-range force, which is due to a larger incorporation of water and a larger loss of conformational entropy for the sugar moiety when confined to the space between two surfaces.

The attractive forces measured between carbohydrate-containing molecules can, to a large extent, be explained as originating from van der Waals forces. Hence, any structural change that results in an increase in the water content of the adsorbed layer reduces this attractive force. It is clear that the possibility for formation of hydrogen bonds between large stretches of polymer chains is important for the structure of polysaccharides present at interfaces, and thus the range of the steric force. However, it is not obvious that hydrogen bonds between two such layers forced into contact are of any significance.

An increase in ionic strength influences the interaction between charged polysaccharides. In addition to the well-understood reduction in the electrostatic double-layer force, there is also a change in conformation of the polysaccharide chains. At higher salt concentrations more compact conformations are favored and, consequently, the range of steric forces is reduced. On the other hand, the addition of salt has only a limited effect on the range of the hydration force. For both gangliosides and negatively charged polysaccharides a lower apparent charge is measured than could be expected from the molecular structure. This indicates that a significant amount of counterions is located within the carbohydrate layer.

ACKNOWLEDGMENT

Andra Dedinaite is gratefully thanked for making the drawings of the molecules. This work was supported by the Swedish Natural Science Research Council (NFR).

REFERENCES

1. A. Sugget, In *Water: A Comprehensive Treatise*, Vol. 4 (F. Franks, ed.). Plenum Press, New York, 519–567, 1975.
2. F. Franks, In *Water: A Comprehensive Treatise*, Vol. 2 (F. Franks, ed.). Plenum Press, New York, 1–54, 1973.
3. J. W. Brady and R. K. Schmidt, The role of hydrogen bonding in carbohydrates: molecular dynamics simulations of maltose in aqueous solutions. J. Phys. Chem. 97:958–966, 1993.
4. F. Zuccarello and G. A. Buemi, Theoretical study of D-glucose, D-galactose, and parent molecules: solvent effect on conformational stabilities and rotational motions of exocyclic groups. Carbohydrate Res. 273:129–145, 1995.
5. C. Fringant, I. Tvaroska, K. Mazeau, M. Rinaudo, and J. Desbrieres, Hydration of α-maltose and amylose: molecular modelling and thermodynamics study. Carbohydrate Res. 278:27–41, 1995.
6. D. A. Rees, *Polysaccharide Shapes*, Chapman and Hall, London, 1977.

7. H. C. Hamaker, The London–van der Waals attraction between spherical particles. Physica *4*:1058–1072, 1937.

8. I. E. Dzyaloshinskii, E. M. Lifshitz, and L. P. Pitaevskii, The general theory of van der Waals forces. Adv. Phys. *10*:165–209, 1961.

9. B. Davies and B. W. Ninham, van der Waals forces in electrolytes. J. Chem. Phys. *56*: 5797–5801, 1972.

10. J. N. Israelachvili, *Intermolecular and Surface Forces*. Academic Press, London, 1991.

11. V. A. Parsegian and D. Gingell, On the electrostatic interaction across a salt solution between two bodies bearing unequal charges. Biophys. J. *12*:1192–1204, 1972.

12. D. Chan, T. W. Healy, and L. R. White, Electrical double-layer interactions under regulation by surface ionization equilibrium—dissimilar amphoteric surfaces. J. Chem. Soc. Faraday Trans. I. *72*:2844–2865, 1976.

13. E. J. N. Verwey and J. Th. G. Overbeek, *Theory of the Stability of Lyophobic Colloids*. Elsevier, Amsterdam, 1948.

14. B. W. Ninham and V. Yaminsky, Ion binding and ion specificity: the Hoffmeister effect, Onsager and Lifshitz theories. Langmir *13*:2097–2108, 1997.

15. P. Attard, J. Mitchell, and B. W. Ninham, Beyond Poisson–Boltzmann: images and correlations in the electric double layer. II. Symmetric electrolyte. J. Chem. Phys. *89*: 4358–4367, 1988.

16. V. A. Parsegian, N. Fuller, and R. P. Rand, Measured work of deformation and repulsion of lecithin bilayers. Proc. Natl. Acad. Sci. USA *76*:2750–2754, 1979.

17. R. P. Rand and V. A. Parsegian, Hydration forces between phospholipid bilayers. Biochim. Biophys. Acta *988*:351–376, 1989.

18. W. Z. Helfrich, Steric interaction of fluid membranes in multilayer systems. Naturforsch. *33a*:305–315, 1978.

19. J. N. Israelachvili and H. Wennerström, Entropic forces between amphiphilic surfaces in liquids. J. Phys. Chem. *96*:520–531, 1992.

20. C. Tanford, *The Hydrophobic Effect*. Wiley, New York, 1973.

21. M. Hato, Attractive forces between surfaces of controlled "hydrophobicity" across water: a possible range of "hydrophobic interactions" between macroscopic hydrophobic surfaces across water. J. Phys. Chem. *100*:18530–18538, 1996.

22. J. B. Taylor and J. S. Rowlinson, The thermodynamic properties of aqueous solutions of glucose. Trans. Faraday Soc. *51*:1183–1192, 1955.

23. G. J. Fleer, M. A. Cohen Stuart, J. M. H. M. Scheutjens, T. Cosgrove, and B. Vincent, *Polymers at Interfaces*. Chapman & Hall, London, 1993.

24. P. G. de Gennes, Polymers at interfaces: a simplified view. Adv. Colloid Interface Sci. *27*:189–209, 1987.

25. P. F. Luckham and J. Klein, Forces between mica surfaces bearing adsorbed polyelectrolyte, poly-*l*-lysine, in aqueous media. J. Chem Soc. Faraday Trans. 1. *80*:865–878, 1984.

26. T. Afshar-Rad, A. I. Bailey, P. F. Luckham, W. MacNaughtan, and D. Chapman, Forces between poly-*l*-lysine of molecular weight range 4000–75000 adsorbed on mica surfaces. Colloids Surf. *25*:263–277, 1987.

27. M. A. G. Dahlgren, How Polyelectrolytes Behave at Solid Surfaces. Ph.D. thesis, Royal Institute of Technology, Stockholm, 1995.

28. M. A. G. Dahlgren, Å. Waltermo, E. Blomberg, P. M. Claesson, L. Sjöström, T. Åkesson, and B. Jönsson, Salt effects on the interaction between adsorbed cationic polyelectrolyte layers—theory and experiment. J. Phys. Chem. *97*:11769–11775, 1993.

29. M. A. G. Dahlgren, H. C. M. Hollenberg, and P. M. Claesson, The order of adding polyelectrolyte and salt affects surface forces and layers structures. Langmuir *11*:4480–4485, 1995.

30. P. M. Claesson, M. A. G. Dahlgren, and L. Eriksson, Forces between polyelectrolyte coated surfaces: relation between surface interaction and floc properties. Colloids Surf. *93*:293–303, 1994.

31. J. N. Israelachvili and G. E. Adams, Measurements of forces between two mica surfaces in aqueous electrolyte solutions in the range 0–100 nm. J. Chem. Soc. Faraday Trans. 1. *74*:975–1001, 1978.

32. W. A. Ducker, T. Senden, and R. M. Pashley, Measurements of forces in liquids using a force microscope. Langmuir *8*:1831–1836, 1992.

33. J. L. Parker, Surface force measurements in surfactant systems. Prog. Surf. Sci. *47*:205–271, 1994.

34. V. Bergeron and C. J. Radke, Equilibrium measurements of oscillatory disjoining pressure in aqueous foam films. Langmuir *8*:3020–3026, 1992.

35. D. L. Sober and J. Y. Walz, Measurement of long range depletion energies between a colloidal particle and a flat surface in micellar solutions. Langmuir *11*:2352–2356, 1995.

36. D. Exerowa, T. Kolarov, and K. Khristov, Direct measurement of disjoining pressure in black foam films. I. Films from an ionic surfactant. Colloids Surf. *22*:171–185, 1987.

37. J. N. Israelachvili and P. M. McGuiggan, Adhesion and short-range forces between surfaces I: new apparatus for surface force measurements. J. Mater. Res. *5*:2223–2231, 1990.

38. J. L. Parker, H. K. Christenson, and B. W. Ninham, Device for measuring the force and separation between two surfaces down to molecular separations. Rev. Sci. Instrum. *60*: 3135–3138, 1989.

39. J. Klein, Forces between mica surfaces bearing adsorbed macromolecules in liquid media. J. Chem. Soc. Faraday Trans. I. *79*:99–118, 1983.

40. R. G. Horn, D. T. Smith, and W. Haller, Surface forces and viscosity of water measured between silica sheets. Chem. Phys. Lett. *162*:404–408, 1989.

41. R. G. Horn, D. R. Clarke, and M. T. Clarkson, Direct measurement of surface forces between sapphire crystals in aqueous solutions. J. Mater. Res. *3*:413–416, 1988.

42. A. Scheludko, Thin liquid films. Adv. Colloid Interface Sci. *1*:391–464, 1967.

43. Å. Waltermo, E. Manev, R. Pugh, and P. M. Claesson, Foam films and surface force studies of aqueous solutions of octyl-β-glucoside. J. Disp. Sci. Techn. *15*:273–296, 1994.

44. V. Bergeron, Å. Waltermo, and P. M. Claesson, Disjoining pressure measurements for foam films stabilized by nonionic sugar-based surfactants. Langmuir *12*:1336–1342, 1996.

45. Å. Waltermo, P. M. Claesson, and I. Johansson, Alkyl glucosides on hydrophobic surfaces studied by surface force and wetting measurements. J. Colloid Interface Sci. *183*: 506–514, 1996.

46. Å. Waltermo, P. M. Claesson, E. Manev, S. Simonsson, I. Johansson, and V. Bergeron, Foam and thin liquid–film studies of alkyl glucoside systems. Langmuir *12*:5271–5278, 1996.

47. C. Li and P. Somasundaran, Reversal of bubble charge in multivalent inorganic salt solutions—effect of magnesium. J. Colloid Interface Sci. *146*:215–218, 1991.

48. C. Li and P. Somasundaran, Reversal of bubble charge in multivalent inorganic salt solutions—effect of lanthanum. J. Colloid Interface Sci. *81*:13–15, 1993.

49. E. D. Manev and R. Pugh, Diffuse layer electrostatic potential and stability of thin aqueous films containing a nonionic surfactant. Langmuir *7*:2253–2260, 1992.

50. R. Cohen and D. Exerowa, Electrosurface properties of lysophosphatidylcholine foam films: effect of pH and Ca^{2+}. Colloids Surf. A *85*:271–278, 1994.

51. P. M. Claesson, R. Kjellander, P. Stenius, and H. K. Christenson, Direct measurements of temperature-dependent interactions between non-ionic surfactant layers. J. Chem. Soc. Faraday Trans. 1. *82*:2735–2746, 1986.

52. P. M. Claesson, H. K. Christenson, J. M. Berg, and R. D. Neuman, Interactions between mica surfaces in the presence of carbohydrates. J. Colloid Interface Sci. *172*:415–424, 1995.

53. M. Hato, S. Ohnishi, and P. M. Claesson, Unpublished data.

54. H. Minamikawa, T. Murakami, and M. Hato, Chemical systhesis of 1,3-di-*O*-alkyl-2-*O*-(β-glycosyl)glycerols bearing oligosaccharides as hydrophilic groups. Phys. Lipids *72*: 111–118, 1994.

55. J. Marra, Controlled deposition of lipid monolayers and bilayers onto mica and direct force measurements between galactolipids bilayers in aqueous solution. J. Colloid Interface Sci. *107*:446–458, 1985.

56. J. Marra, Direct measurement of attractive van der Waals and adhesion forces between uncharged lipid bilayers in aqueous solution. J. Colloid Interface Sci. *109*:11–20, 1986.

57. J. L. Parker, Forces between bilayers containing charged glycolipids. J. Colloid Interface Sci. *137*:571–576, 1990.

58. P. Luckham, J. Wood, and R. Swart, The surface properties of gangliosides. II. Direct measurement of the interaction between bilayers deposited on mica surfaces. J. Colloid Interface Sci. *156*:173–183, 1993.

59. J. K. Dzandu, Role of sialic acid in selective silver staining of red cell glycophorins. Appl. Theor. Electrophor. *1*:137–144, 1989.

60. C. D. Hurd, Acidities of ascorbic and sialic acids. J. Chem. Educ. *47*:481–482, 1970.

61. D. C. Rau and V. A. Parsegian, Direct measurement of forces between linear polysaccharides xanthan and schizophyllan. Science *249*:1278–1281, 1990.

62. A. Haug and B. Larsen, Separation of uronic acids by paper electrophoresis. Acta Chem. Scand. *6*:1395–1396, 1961.

63. M. Schaub, G. Wenz, G. Wegner, A. Stein, and D. Klemm, Ultrathin films of cellulose on silicon wafers. Adv. Mater. *5*:919–922, 1993.

64. M. Holmberg, J. Berg, S. Stemme, L. Ödberg, J. Rasmusson, and P. M. Claesson, Surface force studies of Langmuir–Blodgett cellulose films. J. Colloid Interface Sci. *186*:369–381, 1997.

65. G. G. Fasman, ed., *Handbook of Biochemistry and Molecular Biology. Physical and Chemical Data*, vol. 1. CRC Press, Cleveland, 1976.

66. M. Rinaudo and A. Domard, In *Chitin and Chitosan* (G. Skläk-Bræk, T. Anthonsen, P. Sandford, eds.). Elsevier, London, 1989, pp. 71–86.

67. M. Laleg and I. I. Pikulik, Strengthening of mechanical pulp webs by chitosan. Nordic Pulp Paper Res. J. *6*:174–199, 1992.

68. P. Fält, B. Bergenståhl, and P. M. Claesson, Stabilization by chitosan of soybean oil emulsions coated with phospholipid and glycocholic acid. Colloids Surf. A *71*:187–195, 1993.

69. P. M. Claesson and B. W. Ninham, pH-dependent interaction between irreversibly adsorbed chitosan layers. Langmuir *8*:1406–1412, 1992.

70. M. R. Böhmer, O. A. Evers, and J. M. H. M. Scheutjens, Weak polyelectrolytes between two surfaces: adsorption and stabilization. Macromolecules *23*:2288–2301, 1990.

71. R. Kohn and P. Kovac, Dissociation constants of D-galacturonic and D-glucuronic acids and their *O*-methyl derivatives. Chem. Zvesti. *32*:478–485, 1978.

72. P. Hirsch, The acid strength of glucoronic acid in comparison with that of oxycellulose. Recl. Trav. Chim. Pays-Bas. *71*:999–1006, 1952.

73. J. Laine, P. Stenius, A. Suurnäkki, and P. M. Claesson, The effect of wood components on the adhesive properties of cellulosic materials. Proc. Int. Symp. on Wood and Pulping Chemistry, Montreal, 1997.

74. S. Siegel, M. Malmsten, D. Klüssendorf, A. Walter, F. Schnalke, and A. Kauschmann, Blood-flow sensing by anionic biopolymers. J. Autonomic Nervous Syst. *57*:207–213, 1996.

75. M. Malmsten, G. Siegel, E. Buddecke, and A. Schmidt, Cation-promoted adsorption of proteoheparan sulphate. Colloids Surf. B *1*:43–50, 1993.
76. M. Malmsten and G. Siegel, Electrostatic and ion-binding effects on the adsorption of proteoglycans. J. Colloid Interface Sci. *170*:120–127, 1995.
77. M. Malmsten, P. M. Claesson, and G. Siegel, Forces between proteoheparan sulfate layers adsorbed at hydrophobic surfaces. Langmuir *10*:1274–1280, 1994.
78. E. Perez, J. E. Proust, A. Baszkin, and M. M. Boissonnade, In situ adsorption of bovine submaxillary mucin at the mica/aqueous solution interface. Colloids Surf. *9*:297–307, 1984.
79. E. Perez and J. E. Proust, Forces between mica surfaces covered with adsorbed mucin across aqueous solution. J. Colloid Interface Sci. *118*:182–191, 1987.
80. M. Malmsten, E. Blomberg, P. M. Claesson, I. Carlstedt, and I. Ljusegren, Mucin layers on hydrophobic surfaces studied with ellipsometry and surface force measurements. J. Colloid Interface Sci. *151*:579–590, 1992.

8

Protein Adsorption Kinetics

JEREMY J. RAMSDEN   Institute of Experimental Medicine,
Hungarian Academy of Sciences, Budapest, Hungary

I. INTRODUCTION—THE KINETIC IMPERATIVE

The ubiquity of protein adsorption in natural processes pushes it to a very prominent position among biological phenomena, and its understanding ranks as one of the most essential challenges in the search for the fundamental mechanisms of living processes. Even at the very origins of life, the ordered arrangement of small components via adsorption onto minerals is a serious contender for the emergence of the complex macromolecular assemblies considered to be the precursors of living systems. As far as contemporary cells are concerned, it suffices to recall the classic experiments of Kempner and Miller [1,2], in which, *provided the cells are not homogenized according to standard biochemical protocols*, centrifugation results in a cytosol practically devoid of proteins and other macromolecules, thus convincingly demonstrating that practically all enzymes are located at or in the lipid membranes pervading (eukaryotic) cells. In living matter, protein adsorption to lipid membranes is the most widely considered process, but if additionally adsorption to DNA is considered, then protein adsorption encompasses most of the regulatory processes controlling gene expression. Blood clotting is the result of an exquisitely orchestrated sequence of protein adsorption events [3]. Furthermore, the adsorption of proteins is usually a prerequisite to the adsorption of cells and multicellular organisms to different kinds of surfaces; this aspect encompasses phenomena as diverse as the assembly of primitive multicellular organisms, the immune response, the growth of neurones and neurites, the attachment of mussels and limpets to rocks, the formation of biofilms by bacteria, etc.

The practical applications of protein adsorption are no less numerous, ranging from the chromatographies, blotting techniques, and immunosorbent assays, etc., so indispensable to the isolation, purification, and identification of individual molecules, and hence to most experimental molecular biology; to medical devices, including various kinds of implants (stents, prostheses), dialysis machines, sensors for clinical diagnostics, etc.; and beyond to the storage of pharmaceuticals, design of food processing equipment, and marine antifouling paint.

Two very important features distinguish protein adsorption from the adsorption of other materials at solid–liquid interfaces. The first is the remarkable specificity of adsorption, i.e., one protein will strongly adsorb to a particular surface, yet another one, which may only be very slightly different, will not (we shall discuss more precise definitions of *specificity* in Section IV.F); see Ref. 4 for an overview and some specific results, Ref. 5 for a striking example of a single-point mutation drastically changing the adsorption strength, and Ref. 6 for an example of the specificity of adsorption controlling multicellular assembly. In this last example, specificity appears to be conferred by glyco moieties rather than amino acids; of course under "proteins" we include polyamino acids modified by grafting oligosaccharides, fatty acids, etc., onto them. This specificity is really unique to proteins because, in contrast to other polymers, they have a very specific, template-directed composition.

The second feature distinguishing protein adsorption from many other adsorption phenomena comes from the fact that proteins are complex molecules with multiple stable (hence rememberable) states, and during their interactions with surfaces may undergo transitions between these states. This is in addition to the phenomenon of adsorption-induced denaturation, in which, roughly speaking, bonds between the macromolecule and the surface replace intramolecular bonds without any enthalpic penalty, whereas there is a gain in entropy to drive the process [7]. As will be discussed in more detail in Section IV, this feature plays a considerable influence in protein *desorption*, as well as adsorption. We consider the term *adsorption* to encompass desorption, which conceptually can be considered to be the inverse process.

Finally, why the kinetic imperative? At a deep level, it is enforced by the apparent mortality of all known life forms: there is insufficient time to await the attainment of equilibrium (it is still important to understand the nature of the equilibrium state, even if it is but rarely observed). It follows that it is of great heuristic value in the study of bioprocesses to understand the way a process develops kinetically; it often turns out that the state attained very shortly after a certain process is initiated determines the outcome of a cascade of further steps.

II. FUNDAMENTAL CHARACTERISTICS OF AN ADSORPTION PROCESS

An absorption process can be completely characterized by

1. The radial distribution function of the adsorbed particles, i.e., $g(s)$, the probability that the center of another particle will be found at a distance s from the center of a target particle
2. The fractional surface coverage θ at saturation, denoted here by θ_∞
3. The fluctuations in the numbers of particles ν adsorbed in contiguous or otherwise randomly chosen nonoverlapping areas
4. The kinetics of particle accumulation, i.e., $\nu(t)$

Of these four, only the last will be of particular concern in this chapter. Different types of particles and different mechanisms of adsorption each have more or less distinct *kinetic signatures*, and much of the practical work of investigating protein adsorption involves identifying and quantifying these signatures.

The adsorption process can be broken down into the following successive steps (see Fig. 1):

1. Far from the surface, the protein is in its native conformation and surrounded by solvent having its bulk constitution. If there is any movement

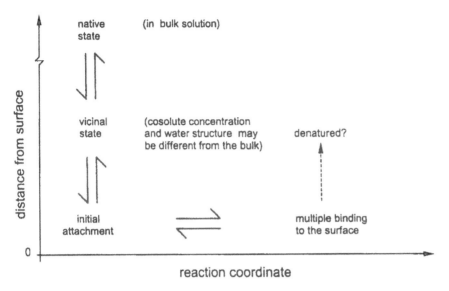

FIG. 1 Sketch of the processes important as a protein (adsorbate) approaches a surface (adsorbent). The reaction coordinate refers to conformational change.

at all of the solvent phase relative to the surface, transport toward the surface is via convection. Note that some proteins are considered to be "natively unfolded" [8]—an example is myristoylated alanine-rich C kinase substrate, MARCKS [9]—meaning that they have no definite conformation under normal physiological conditions.

2. At some distance from the surface convection becomes slower than diffusion. This is the *diffusion boundary* (δ_d) [10], and for proteins, with diffusion coefficients $D \sim 10 \ \mu m^2/s$, moving under low Reynolds number conditions (Re ~ 1–100), the distance of this diffusion boundary from the surface may be tens of micrometers, i.e., well beyond the range of even the farthest reaching interfacial forces. Hence at this boundary protein and solvent still have their bulk constitution. Note that any ions present in solution may significantly affect diffusion of the protein molecules: the motions of the protein's own small, mobile counterions are correlated with those of the protein, which is therefore speeded up while its counterions are retarded, but if additional electrolyte is present, the correlations are decoupled, since the small ions of the same sign as the protein's net charge will tend to replace the protein in the coupled flow, thereby decelerating the protein [11].

3. At a distance of ~ 10 nm the protein begins to sense the effect of interfacial forces (Fig. 2). The longest range one is the so-called solvation force, hydrophilic force, or hydrophobic force [12,13], which in aqueous systems is actually a manifestation of the highly ramified hydrogen-bonding network in liquid water [14,15]. This force decays exponentially with a characteristic length χ, whose value depends on whichever is the largest of the correlation length of the solvent polarization, the width (roughness) of the

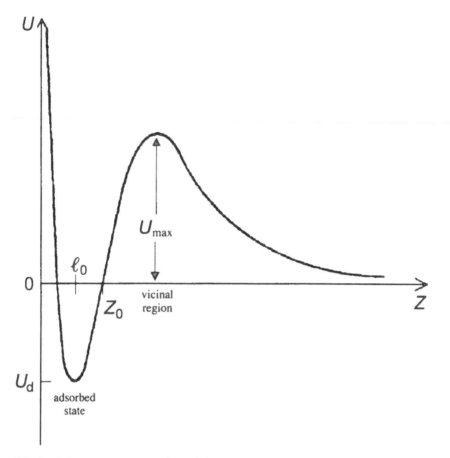

FIG. 2 Schematic representation of the features of adsorbate–adsorbent interaction energy curves.

interface, and the correlation length of polarization fluctuations within the plane of the interface [15]; a typical value of χ is 1 nm. The force is usually repulsive [13].

Next comes the electrostatic force whose length is given by the ionic strength–dependent Debye length κ^{-1}:

$$\kappa^{-1} = \frac{(\varepsilon\varepsilon_0 kT/c_{\text{ion}})^{1/2}}{e} \tag{1}$$

Under typical ionic strengths encountered in a cell the Debye length is less than 1 nm. Since the absolute magnitude of the electrostatic force is typically only around one-tenth of the hydrophilic force [13–16] it may often be neglected in biological adsorption problems.

The Lifshitz–van der Waals force is of short range (subnanometer) and attractive. It is important for the adsorption kinetics insofar as it determines the height U_{max} of the energy barrier between the protein and the surface. U_{max} is often a good approximation for the adsorption probability per unit time [17]; more exact considerations are given in Section IV.B.

Finally, the Born repulsion, of extremely short range, prevents the atoms of adsorbent and adsorbate from fusing with one another and determines the distance of closest approach ℓ_0.

The net repulsion engendered by these forces have the direct kinetic consequence of retarding arrival at the surface proper: the energy barrier characterized by U_{max} has to be surmounted. This in turn results in an accumulation of still dissolved protein molecules in the vicinity of the surface (see Section IV). An indirect consequence is that the ionic composition of the solution may be drastically changed, especially if the surface is electrostatically charged. For example, a negatively charged surface will create an ionic atmosphere enriched in positive ions, including protons; the consequent lowering of pH may be sufficient to cause charge reversal of a negatively charged protein, which is then electrostatically attracted to the surface! This effect has been experimentally demonstrated [5]. More subtle indirect effects include a changed diffusion coefficient due to the changed ionic composition as discussed above and changed protein conformation (which may, however, be too slow to be important; see Table 1 and the accompanying discussion). Protein conformation and diffusivity may also be changed by the altered water structure in the vicinity of the surface [15], which may promote protein aggregation.

4. Once the protein has surmounted the energy barrier and is initially attached to the surface, two fates are possible. It may simply reverse its path and resurmount the energy barrier (usually starting from a deeper position than zero with respect to the bulk solution), or it may undergo other changes of conformation, leading to multiple bonding, which is favored by a gain of entropy [7], and to solvent expulsion, which usually strongly enhances bonding. The multiply bonded "state" is not usually a single one, but a series of increasingly strongly bonded intermediates. The protein may be so strongly altered it no longer can carry out an enzymatic function, and its antigenicity may be completely changed (one reason for the harmfulness of asbestos is the very strong immune reaction invoked by normally immunogenically invisible serum proteins becoming denatured upon adsorption on the asbestos surface [18]). Direct desorption into the bulk solution is then extremely unlikely, and even if it does occur the redissolved protein may not be in its native state.

5. No mention has yet been made of a further difficulty, namely, that of finding space on the surface at which to adsorb. Monolayer adsorption of

TABLE 1 Characteristic Length Scales of Protein Adsorption

Description	Equation	Typical value
Protein diameter	$2r$	3 nm
Reaction length	δ_a [Eq. (6)]	100 μm
Diffusion boundary layer	δ_d [Eq. (2)]	100 μm

a single protein component has dominated experimental and theoretical investigations in the field for at least the last 50 years, even though it corresponds to a situation seldom encountered in nature. When considering the deposition of blood proteins, for example, hundreds of components are present and active, and the detailed investigation of these processes is still at the beginning [19,20]. Blood clotting is an example of an exquisitely regulated process involving dozens of different proteins assembling at a surface which is, in fact, understood in considerable detail [3]; and the formation of so-called amyloid plaques in the brain by prions, leading to debilitating spongiform encephalopathies—through prion proteins altering their conformational state upon adsorption and therefore facilitating the accrual of additional proteins which in turn become altered—has strikingly revived Ramsden's early concept of massed adsorpta [21].

In the case of a single soluble molecule adhering to the surface and not undergoing any conformational change, it is sufficient to compute the probability that space, at a randomly chosen spot, is available (see Section IV.D). Even this "simple" calculation cannot be carried out analytically in two dimensions; at best an analytical approximation can be obtained, which is nevertheless accurate enough for interpreting experimental data [22]. There is presently no comparably universal theory for handling the more complicated, but more realistic, cases of multicomponent, multilayer adsorption. That is a challenge for the future.

The spatial scales (summarized in Table 1) of these different processes (Sections II.B.1–5) are not wholly distinct. Allusion has already been made to the way in which energy barriers lead to an accumulation of molecules in the vicinal layer and hence retard transport from the bulk; binding of proteins to the surface may release previously bound small ions or osmolytes, changing their concentration in the vicinal layer, with further consequences for the protein, etc. (see the more detailed discussion in Section IV). Hence, these spatial scales cannot be completely separated from one another. One the other hand, the characteristic time scales of the adsorption process (Table 2) are more widely separated, and can be more readily varied by the experimenter, as will be discussed in Section VI.

III.   EXPERIMENTAL PREREQUISITES

It cannot be sufficiently emphasized how dramatically the development of a new generation of measuring techniques has changed the state of knowledge of protein adsorption kinetics (see Ref. 23 for a comprehensive review of experimental methods). The new generation is based on perturbation of an evanescent electromagnetic field decaying from the adsorbent surface into the bulk solution. The only requirement is that the polarizability, i.e., the refractive index, of the adsorbate is different from that of the liquid in which it is dissolved. The evanescent field can be generated by total internal reflection at the interface [23,26], by a guided wave, which essentially amounts to the same thing, guiding taking place via repeated total internal reflections (a technique called optical waveguide lightmode spectrometry, OWLS, in the literature) [23,27,28], or by surface plasmons (which require a metal surface,

TABLE 2 Characteristic Time Scales of Protein Adsorption

Description	Equation	Typical value
Time to diffuse its diameter	$4r^2/D$	1 μs
Time to surmount energy barrier	$D/k_a^2 = \delta_a^2/D$	10^3 s
Adsorption time	$\delta_d/(Dc_b\phi a)$	10–10^4 s
Surface residence time	$1/k_d{}^a$	10^3 s
Conformational relaxation time	$1/k_s{}^a$	10^3 s

[a]If the process is Poissonian.
Note: see the list of symbols at the end of the chapter.

typically gold, platinum, or aluminum) [23,29]. The measurement requires no labeling of the adsorbate and is able to yield an accurate determination of the absolute number of particles adsorbed per unit area [30]. This, together with the precision of the measurement (currently typically ± 1 particle per 0.1 μm^2) and the achievable time resolution (≤ 1 ms), has enabled adsorption kinetics to be probed in unprecedented detail, and allowed a far more comprehensive comparison between theory and experiment than was ever possible before.

Formerly, the only way of achieving high sensitivity was to use fine dispersions of particles as the adsorbent in order to achieve an immense surface area in a small volume which could then be probed with conventional spectroscopic and other techniques. The main weaknesses of this approach are or have been the possibility that the small particles have surface properties different from those of the surface of bulk material; the difficulty of making fast kinetic measurements, although this has recently been solved (e.g., [24]); and the difficulty of rigorously analyzing transport in a strongly mixed suspension of particles in a protein solution.

Apart from actual measurement sensitivity and temporal resolution, for many biological and medical questions the ability to measure under physiological conditions is of vital importance. This includes the use of blood and other physiological fluids as the liquid phase and the preparation of surfaces mimicking cell membranes, DNA etc. In this regard silica and titania, which have rather similar surface parameters [15], are very favorable substrates, since not only can nativelike phospholipid bilayer membranes [31] or native DNA [32] be assembled on them, but it is well known that the oxide layer on titanium prostheses plays an essential rôle in assuring the effective integration of the metal into the body, and hence may be considered as nondenaturing surfaces for (some) proteins.

A. Measuring the Rate of Accumulation

The primary goal is to determine the number of adsorbed particles per unit area during successive epochs following exposure of a vacant surface to a biopolymer solution. Ideally the surface should be atomically smooth and planar, and the fluid flowing over it fast enough to ensure that the rate of accumulation is not transport limited, yet slow enough to ensure that flow is laminar, not turbulent. Under such conditions of convective-diffusive transport, and for certain favorable geometries, the thickness δ_d of the diffusion boundary layer can be given exactly [10], e.g., for a tube of cylindrical cross-section A and radius R,

$$\delta_d = \left(\frac{3}{2}\right)^{2/3} \left(\frac{DxRA}{F}\right)^{1/3} \tag{2}$$

where x is the distance from the inlet to the measuring spot and F the volumetric flow rate. Such an expression is essential for computing the number of adsorption attempts per unit area per unit time (see Section IV.C for the further use of this expression), which sets the proper time of any adsorption process (one adsorption attempt corresponding to one unit of dimensionless time).

In recent years much work has been done in which the binding of a soluble protein to a receptor protein covalently attached to a dextran scaffold is measured. The motivation for such an experimental arrangement, which has been especially favored when using surface plasmon resonance (SPR) to detect the binding, appears to have been threefold: dextran renders the gold or other metal required for SPR more biocompatible, i.e., less susceptible to denature proteins; the adsorption zone is converted from a plane to a volume, thereby enhancing the numbers of molecules within the evanescent field decaying exponentially away from the metal surface and hence the sensitivity of detection; and well-established chemical methods for immobilizing proteins to dextran exist. Mass transport is much more complicated, and an analytical solution to the appropriate transport equations cannot be given. Useful approximate methods have however been developed [33]. It should also be pointed out that dextran strongly repels proteins via hydrophilic repulsion, and hence strongly perturbs the binding process [34]. Furthermore, it changes the chemistry of the aqueous medium in which the reactions are taking place (the ratio of free hydroxyls to free lone pairs in water), which may exert additional influence on the binding reaction.

B. Obtaining Complementary Information

Although the accurate measurement of the number of adsorbed particles per unit area is sufficient to characterize the adsorption kinetics (and comparison between theory and experiment allows further inferences regarding the shape, size, and arrangement of particles and postadsorption changes in shape or affinity to be made; see Section VI), the fractional surface coverage at saturation, and fluctuations in the numbers of particles adsorbed in contiguous nonoverlapping areas, complete characterization of the adsorption process may require further information, either to directly confirm inferences made from the kinetics, or to obtain information unobtainable from kinetic measurements. This information can be classified as concerning

1. Spatial arrangement, obtainable by microscopy (e.g., atomic force microscopy for protein-sized particles [35])
2. Chemical bonding, obtainable by vibration spectroscopy (infrared absorption, Raman, sum-frequency, etc.), the surface forces apparatus [36,37], or the atomic force microscope [38,39]
3. Biological function, which may be the most important attribute of an adsorbed protein. Immunogenicity (i.e., conformation and accessibility of an epitope) can be determined by measuring the binding of an appropriate monoclonal antibody [20] (using the same setup as for the initial adsorption process), as well as other biomolecular recognition phenomena. Adsorbed

enzyme activity can also be measured in situ (M. G. Cacace and J. J. Ramsden, to be published).

These techniques generally require fairly long data collection intervals and hence are unsuitable for kinetic measurements, and are mainly useful for characterizing the final state of an adsorption process or slow changes in an adsorbed layer.

In order to study the kinetics of adsorption of a complex mixture, some kind of labeling is required, since the evanescent wave-based techniques simply measure the total polarizability change in the vicinity of the interface. If different proteins are labeled with different fluorescent markers, their deposition may be followed simultaneously by total internal reflection fluorescence microscopy, which suppresses fluorescence from the bulk solution. Naturally, the possible influence of the labels (which may be large, polycyclic aromatic polyvalent ions) on the behavior of the proteins must be carefully scrutinized, e.g., by comparing their adsorption to a membrane with the unlabeled form using one of the evanescent wave-based techniques.

IV. THE KINETIC EQUATION

In this section the minimal mathematical formalism appropriate for the descriptive modelling of protein adsorption is discussed.

A. The Memory Function

During a small (infinitesimal) interval of time Δt_1 at epoch t_1, a number $\Delta \nu = k_a(t_1)$ $c_v(t_1)\, \phi(t_1)\, \Delta t_1$ of molecules will be adsorbed. ν is the number of particles adsorbed per unit area, k_a the adsorption rate coefficient (to be discussed in Section IV.B), c_v the concentration of dissolved protein in the vicinity of the surface (to be discussed in Section IV.C; also see Fig. 1; for the moment it can be considered to be approximately equal to the bulk dissolved concentration c_b), and ϕ is the available area function (to be discussed in Section IV.D). A compact way of expressing ν then is

$$\nu(t) = \int_0^t k_a(t_1)c_v(t_1)\phi(t_1)Q(t, t_1)\, dt_1 \tag{3}$$

where the memory kernel $Q(t, t_1)$ denotes the fraction of molecules adsorbed at t_1 that remain adsorbed at the later epoch t [40]; clearly $Q(t_1, t_1) = 1$. If the desorption probability does not depend on the surface density (cf. Section V), then the time difference $t - t_1$ is suffcient to characterize Q. Under certain circumstances, k_a and c_v may be constant, or close enough, allowing them to be taken out of the integral. For an irreversible process, $Q = 1$; for a simple Poissonian desorption (no memory), $Q(t) = \exp(k_d t)$, where k_d is the desorption rate coefficient. This simple behavior is rarely encountered with proteins. Typically, additional bonding of an initially attached molecule can take place, or a protein can adsorb with different affinities in different orientations (although ultimately all must be present in the orientation with the highest affinity if all orientations occupy the same area at the interface). These situations have been experimentally observed (e.g., for serum albumins adsorbing on oxide surfaces [41,42]). Sometimes the processes have been descriptively modelled as reversible and irreversible adsorption in parallel, or as initial adsorption followed by a surface reaction enhancing the affinity [41,42]. Memory functions can be found for these and other adsorption schemes [40,43]. It is a necessary condition for the system to reach equilibrium that

$$\lim_{t \to \infty} Q(t) = 0 \tag{4}$$

B. The Adsorption Rate Coefficient

An appropriate relation between k_a and the particle-adsorbent interaction potential $U(z)$, where z is the perpendicular distance between the particle (measured from its closest point) and the adsorbent surface is [44]

$$k_a = \frac{D}{\delta_a} \tag{5}$$

where δ_a is the reaction distance, defined as

$$\delta_a = \int_{z_0}^{\infty} \left[\exp\left(\frac{U(z)}{kT}\right) - 1 \right] dz \tag{6}$$

The three forces contributing additively to $U(z)$ have already been introduced in Section II.B.3: Lifshitz–van der Waals (LW), solvation, hydrophilic or hydrophobic (solv), and electrostatic (el). Approximate expressions for them are, respectively (for a sphere of radius r approaching an infinite plane) [14,15,45]:

$$U^{(LW)} = \frac{2\pi\ell_0^2 U^{(LW)\|} r}{z} \tag{7}$$

(ignoring retardation, i.e., accurate up to $z \approx 10$ nm, at which distance the LW force becomes almost negligible),

$$U^{(solv)}(z) = 2\pi\chi U^{(solv)\|} \exp\left(\frac{\ell_0 - z}{\chi}\right) r \tag{8}$$

and

$$U^{(el)} = 4\pi\epsilon_0\epsilon\psi_1\psi_3 \ln[1 + \exp(-\kappa z)]r \tag{9}$$

(assuming constant—and moderate—electrostatic surface potentials ψ, valid for $\kappa r \geq 10$, and still reasonable for $\kappa r > 5$) where $1/\kappa$ is the well-known Debye length:

$$\kappa = e\left(\frac{c_{ion}}{\epsilon\epsilon_0 k_B T}\right)^{1/2} \tag{10}$$

(for a 1:1 electrolyte). The subscripts 1, 2, and 3 refer respectively to the adsorbent, the liquid, and the adsorbate. The interaction energies for infinite parallel sheets (U^ℓ) needed for Eqs. (7) and (8) are related to the interfacial tensions by the Dupré equation [14]:

$$U_{1,2,3}^{(LW \text{ or } solv)\|} = \gamma_{1,3}^{(LW \text{ or } solv)} - \gamma_{1,2}^{(LW \text{ or } solv)} - \gamma_{2,3}^{(LW \text{ or } solv)} \tag{11}$$

Fowkes established that the interfacial tensions are additive [46]:

$$\gamma_{1,2}^{(tot)} = \gamma_{1,2}^{(LW)} + \gamma_{1,2}^{(solv)} \tag{12}$$

They can be obtained from single-substance surface tensions (tabulated, e.g., in [14]) using the Girifalco–Good equation [14,47]

$$\gamma_{12}^{(LW)} = (\sqrt{\gamma_1^{(LW)}} - \sqrt{\gamma_2^{(LW)}})^2 \tag{13}$$

and [25]

$$\gamma_{12}^{(solv)} = 2(\sqrt{\gamma_1^{\oplus}} - \sqrt{\gamma_2^{\oplus}})(\sqrt{\gamma_1^{\ominus}} - \sqrt{\gamma_2^{\ominus}}) \tag{14}$$

Details of the surface potential calculations (using the Healy–White ionizable surface group model) can be found, e.g., in Refs. 15 and 45.

1. The Reliability of Equations (7)–(9)

Although few real proteins are spheres, most soluble globular proteins are ellipsoidal with modest aspect ratios, implying that only small corrections to the interaction potentials should be made on account of their shape. Similarly, even though some protein molecules (e.g., lysozyme and cytochrome c) have strongly asymmetrical spatial distributions of ionized residues on their surfaces, the dominant hydrophilic force $U^{(solv)}$ tends to be rather uniformly distributed [48]. Hence Eqs. (7)–(9) should be reasonably accurate for what may be called "macroscopic" interactions. *Microscopically* (meaning at the nanometer scale!), however, it is well known that proteins are quite irregular, both in shape and surface chemistry, and even close to atomically smooth surfaces of metal oxides may not only contain impurity centers of different polarizability from that of the matrix, but since the oxide is itself at least a binary compound, it is chemically heterogeneous at the nanometer scale. van Oss [49] has pointed out that the repulsive potential between serum albumin and glass calculated using Eqs. (7)–(9) is sufficiently strong to preclude adsorption, whereas experimentally it is well known that adsorption takes place. Ramsden [9] has similarly commented on the adhesion paradox of MARCKS adsorbing on lipid membranes. To resolve this paradox, van Oss has proposed that the overall interaction energy of a protein like serum albumin be considered as the sum of a "macroscopic" repulsion, and a "microscopic" attraction. The latter uses the same Eqs. (7)–(9), but applies them to specific small regions of the protein, e.g., sharp protrusions rich in hydrogen-bond donors, i.e., of markedly different shape and composition from the overall average, and also takes into account "impurities" (e.g., Ca^{2+} in glass or Ti^{4+} in TiO_2), somewhat in analogy to Smoluchowski's solution of the problem of coagulation of colloidal particles [50], in which he refers to the balance between the improbability of adsorption (due to "macroscopic" repulsion) and the probability of adsorption (due to "microscopic" attraction) [51]. It appears that detailed modelling of the adsorption process, e.g., with Brownian dynamics, may be rather useful for elucidating features of the microscopic interaction [75].

C. Transport Phenomena

Good approximations for analyzing the flow regimes can be derived from the equations of Fick and Smoluchowski. If the surface (at $z = 0$, where z is the coordinate normal to the surface) is a perfect sink for the adsorbate, the bulk solution concentration c_b is zero at $z = 0$, the concentration gradient from bulk (i.e., from the diffusion boundary, at which the concentration is maintained at its bulk value) to surface will be approximately linear, and the rate of accumulation is

$$\frac{d\nu}{dt} = \frac{c_b D}{\delta_d} \tag{15}$$

The effect of any energy barrier is to retard adsorption. Hence in the immediate vicinity of the surface, the local (vicinal) bulk concentration c_v will be much higher than zero (although still less than c_b). The rate of accumulation at the surface can be written as the product of c_v and the chemical rate coefficient k_a, i.e., $d\nu/dt = k_a c_v$ in analogy to a first-order reaction in the bulk, and the Fick/Smoluchowski regime (linear concentration gradient) then applies to the zone above this vicinal region. Hence

$$V \frac{dc_v}{dt} = S \left[D \frac{(c_b - c_v)}{\delta_d} - k_a c_v \right] \tag{16}$$

where V and S are unit volume and surface respectively. Strictly speaking the distance of the vicinal layer from the surface should be subtracted from δ_d in the denominator, but since this distance is of the order of molecular dimensions, i.e., only a few nanometers, whereas δ_d is of the order of a few or a few tens of micrometers, this correction can be neglected. A more complete and exact treatment is given in Ref. 43. Letting the left-hand side of Eq. (16) go to zero yields an explicit expression for c_v:

$$c_v = \frac{c_b}{1 + \delta_d k_a / D)} \tag{17}$$

In any real experiment, starting from $\nu(t = 0) = 0$ implies $c_b(t = 0) = 0$, and even if c_b is changed instantaneously to some other (finite, positive!) value, c_v will change only after some delay, and hence $d\nu/dt$ increases from zero up to a maximum before decreasing. Recent experiments fully exploiting the high time resolution possible with evanescent wave-based techniques have been able to probe the short time regime during which c_v is changing rapidly [52].

D. Exclusion Zones

The concept of the exclusion zone is central to understanding monolayer adsorption. In any dimension d, an object deposited on a continuum defines an exclusion zone around it, within which the *center* of another adsorbing object may not be placed. For monosized objects of characteristic size r, the exclusion zone has an area $\sim (2r)^d$. As a result, the available area ϕ (expressed as a fraction of the total area) falls off with the fraction of total area occupied (θ) more quickly than exponential (Langmuirian) filling, in which objects are placed onto receptor sites exactly as big or bigger than the object. As the surface gets filled up beyond $\theta \approx 0.1$, these exclusion zones begin to overlap, introducing polynomial corrections resembling a virial expansion:

$$\phi = 1 - b_1 \theta - b_2 \theta^2 - b_3 \theta^3 - \cdots \tag{18}$$

The problem has been solved exactly in one dimension [53], in which it is called the "random parking" model, referring to cars parked along a street without predefined spaces. The solution also yields the jamming limit θ_J, at which no further particles can adsorb. Approximate expressions have been found for $d = 2$ [54], called random sequential adsorption (RSA) (reviewed in Ref. 55); a selection of coefficients for Eq. (18) is given in Table 3. The RSA model has been extended to include reversible adsorption, or adsorption of particles which are mobile on the surface [56], which slightly changes the higher-order coefficients.

In ideal RSA, any particle unable to find vacant space to adsorb is annihilated; in reality, a protein will remain in solution above the surface and make repeated correlated attempts to adsorb. Numerical simulations of this process have shown that as the surface becomes covered with adsorbed particles, some particles become trapped above growing zones of adsorbed particles which grow faster than the particles can diffuse sideways [57]. This suggests a further generalization to RSA in which particles are allowed to adsorb after having reached a space large enough to accommodate it by a path of correlated lateral diffusion in the immediate vicinity of previously adsorbed proteins. This so-called generalized ballistic deposition (GBD) [60,61] is characterized by the ratio of probabilities $j = p/p'$, p' being the probability of a protein molecule adsorbing on a hitherto unoccupied patch of adsorbent at which it arrives directly from the bulk (i.e., a pure RSA process), and p the probability that it adsorbs after arriving via correlated lateral diffusion. $j = 0$, the lowest possible value, would correspond to pure random sequential adsorption (i.e., binding takes place solely independently of preadsorbed protein at empty patches of surface), and higher values correspond to increasingly favored protein *polymerization* (since the process competing with RSA results in proteins becoming attached to the edges of clusters).

The expansion [Eq. (18)] breaks down at high surface coverages ($\theta > 0.3$), when the available surface is no longer a continuum but consists of discrete small triangular sites. θ then approaches the jamming limit according to a power law,

$$\theta_J - \theta \sim t^{1/2} \tag{19}$$

(For spheres; $\theta > 0.4$ [63].) Interpolating between Eqs. (18) and (19) leads to extremely accurate expressions for the whole range of θ from 0 to θ_J, such as [54]

TABLE 3 Coefficients for the Available Area Function [Eq. (18)]

Adsorption mode	b_1	b_2	b_3	θ_J	Application
Langmuirian	1	0	0	1	Clustering, big receptors
RSA (spheres)[a]	4	3.808	1.407	0.547[b]	Irreversible adsorption
RSA (spheres)	4	3.808	2.424	0.816[a]	Equilibrium adsorption
GBD	$4(1-j)$	$3.808 - 0.180j - 3.128j^2$	$1.407 + 4.679j - 25.58j^2 + 8.550j^3$	0.691[c]	Nucleation and growth

[a]See Refs. 58 and 59 for results for nonspherical particles.
[b]Palásti [62] conjectured that $\theta_{J,d=2} = \theta^2_{J,d=1}$; although not proved, it is rather accurate and hence of heuristic value.
[c]For $j = \infty$. cf. $\theta_J = 0.9069$ for close triangular packing.

$$\phi = \frac{(1 - \theta^*)^3}{1 - 0.8120\theta^* + 0.2336(\theta^*)^2 + 0.0845(\theta^*)^3} \tag{20}$$

where $\theta^* = \theta/\theta_J$.

The introduction of the available area function ϕ implies the modification of some of the kinetic equations introduced earlier—keeping the analogy to homogeneous reactions, they become second order, and, e.g., Eq. (16) becomes

$$V\frac{dc_v}{dt} = S\left[D\frac{(c_b - c_v)}{\delta_d} - k_a c_v \phi\right] \tag{21}$$

E. Postadsorption Changes and Desorption

"Proteins molecules are kinetically unequilibrated mechanical-statistical systems with rigid memory on various organizational levels" (Lifshitz, quoted by Blumenfeld [64]). It is well-established that protein conformational flexibility is essential for biological activity [65]. Hence, unlike rigid colloidal particles, most proteins have rather limited conformational stability and a perturbation such as adsorption to a surface may involve energies comparable to the cohesive energy of the compact form of the molecule (whose conformation is due to the "soft" LW, electrostatic, and solvation forces discussed in Sections II.A.3 and IV.B) and hence may engender conformational changes. Moreover, many surfaces engender changes in the solvent structure, especially of extensively hydrogen-bonded solvents such as water, and these changes may perturb the protein [15,85]. Multiple adsorbate conformations can considerably affect adsorption equilibria and kinetics [66], and phenomenologically are manifested by a marked history dependence of the adsorption kinetics [67].

To a first level of approximation the conformational change can be considered as a first-order reaction $\alpha \rightarrow \beta$, where α and β are the native adsorbed and conformationally altered forms, but whether the change can occur depends on sufficient space being available if the change involves an increase in the area occupied by the molecule at the surface, and hence [68]:

$$\frac{d\nu_\alpha}{dt} = k_a c_v \phi - k_s \nu_\alpha \Psi \tag{22}$$

and

$$\frac{d\nu_B}{dt} = k_s \nu_\alpha \Psi \tag{23}$$

where k_s is the rate coefficient for expansion and Ψ is a function analogous to ϕ giving the probability that space to rearrange is available.

If desorption of protein has to be taken into account, a term with a chemical desorption coefficient k_d should be added to, e.g., Eq. (21), which then becomes

$$V\frac{dc_v}{dt} = S\left[D\frac{(c_b - c_v)}{\delta_d} + k_d(t)\nu - k_a c_v \phi\right] \tag{24}$$

Non-Poissonian desorption is equivalent to a time-dependent desorption rate coefficient, which can be written in terms of the memory kernel $Q(t)$ (Section IV.A) and its derivative $Q'(t) = dQ(t)/dt$ as [40]

$$k_d(t) = \frac{\int_0^t \phi(t_1)Q'(t, t_1)dt_1}{\int_0^t \phi(t_1)Q(t, t_1)dt_1} \tag{25}$$

taking k_a and c_v to be time invariant [cf. Eq. (3)].

V. COLLECTIVE EFFECTS

Until now it has been tacitly assumed that the proteins interact with each other purely via hard-body exclusion, and that the bulk volume concentration is low enough for the effects of such exclusion to be wholly neglected in the liquid phase. Hence in all other respects each molecule behaves as an individual entity. In biological (living) systems, however, there is growing evidence for signal transduction being dependent on the collective action of congeries of molecules gathered closely together. A review of this evidence properly belongs elsewhere; here a few examples of such collective effects in model experimental systems will be described.

The simultaneous occurrence of lateral mobility of the adsorbed protein at a surface and the adhesion of adsorbed proteins to one another allows clustering or crystallization to occur. This has been nicely illustrated for cytochrome P450, the enzyme responsible for broadly specific detoxification of xenobiotics in the liver. The crystallization could be strikingly blocked by increasing the flux of molecules to the surface; this is equivalent to lowering the temperature so that the system gets quenched in a glassy disordered state [69]. In the liver, the P450 molecule is working together with other enzymes such as cytochrome P450 reductase, and presumably mixed crystals are formed. These supramolecular assemblies are a highly effective way of increasing the polyfunctionality and density of active sites of an enzymatic zone.

A similar phenomenon of clustering has been observed for the MARCKS-related protein [70], but only when complexed with calmodulin. Calmodulin appears to be involved in many diverse regulatory functions in cells: triggering the formation of membrane rafts has hitherto not been observed. Aggregation (clustering) has a strong effect on adsorption equilibria [71,72].

The desorption of serum albumin from silica was found to depend inversely on the free area ($\sim\phi^{1/2}$) [41], implying that neighbors mutually promote their desorption. This dependence could form the basis of control elements with interesting properties.

Elongated proteins that are found extensively outside cells, and form so-called basement membranes supporting cells, neurites, etc., usually show much more complex adsorption behavior than relatively simple globular proteins. Fibronectin is one of the most common of these proteins, and is found in a high concentration in the blood, in which it is almost as abundant as albumin. Careful scrutiny of the buildup of a monolayer on silica revealed that constant restructuring of the molecules occurs as additions take place, resulting in a layer of approximately constant density, but steadily increasing mean thickness [73]. At high ionic strength (1 M NaCl), however, the protein adsorbs in an almost ideally random sequential fashion [74]. If deposition takes place at "high" bulk concentration (40 nM), a 3 nmol/m^2 layer irreversible with respect to dilution is formed, but at a bulk concentration ten times less (4 nM), no deposition at all can be measured! This is itself evidence for strong cooperative interactions between the adsorbed molecules.

The environment in cells is crowded and highly organized, which profoundly effects the association of dissolved proteins [34,76]. The dextran matrices popular in binding studies (Section III.A) mimic the cytomatrix with which much of the eucaryotic cell is filled. Hence, for investigating reactions taking place within the cell, these artificial matrices are quite appropriate.

A. The Interrelation of the Kinetic Equation Elements and Their Possible Time Dependences

The principal reason why k_a may be time dependent is because the accumulation of proteins at the surface may change the characteristic energetic parameters of the surface: if the protein does not undergo any conformational change and is adsorbed in a random orientation, one expects that $\gamma(t) = \theta\gamma_3 + (1 - \theta)\gamma_1$, from which the relevant, time-dependent U and hence k_a can be calculated.

The time dependence of c_v follows immediately from Eq. (24) when either or both k_a and ϕ vary with time.

It is a fortunate and fortuitous feature of the random sequential adsorption model that its astonishingly good applicability to protein adsorption in the absence of significant postadsorption changes is due to the fact that two corrections which should be introduced to increase its accuracy, the hydrodynamical interactions and the nonuniformity of adsorption attempts due to diffusion, are practically equal and opposite and hence cancel out [77].

Adsorption of a protein may result in the release (displacement) of small molecules from the surface into the vicinal layer, which may affect the protein conformation.

VI. EVALUATING EXPERIMENTAL DATA

In this section are gathered a few remarks on practical issues of acquiring and interpreting adsorption kinetics data.

A. Adsorption to a Continuum

Plotting the rate of adsorption against the amount adsorbed in direct accordance with

$$\frac{d\nu}{dt} = k_a c_v \phi(\nu) - k_d \nu \tag{26}$$

[cf Eq. (15)] is of great practical value, for it is then usually possible to qualitatively identify the adsorption mechanism immediately by inspection. Four principle types of behavior are observed:

1. $d\nu/dt$ constant, implying that $\phi = 1$, which is characteristic of the formation of isolated aligned chains (e.g., lysozyme dissolved at low ionic strength [43,78]);

2. $d\nu/dt$ concave (i.e., progressively slower), implying that $\phi(\theta)$ is a characteristic polynomial function of θ resulting from the fact that the proteins

interact via excluded volume only, i.e., according to random sequential adsorption (e.g., common blood proteins such as serum albumin, IgG, transferrin [22,41,42,73])

3. dv/dt linear, implying that $\phi = 1 - \theta$ (Langmuirian adsorption), which results from excluded volume with annihilation of exclusion zones, implying two-dimensional clustering or crystallization (e.g., cytochrome P450 at a membrane [69,70])

4. dv/dt concave/convex, implying that $\phi(\theta)$ is a characteristic polynomial function of θ, in turn implying generalized ballistic deposition (e.g., phospholipase A_2 [61])

Equation (26) with an appropriate expression for ϕ can be fitted to numerically differentiated data, or the numerically integrated version of Eq. (26) fitted to the (v, t) data, by a least-squares optimization procedure, with free parameters a, k_a (j for a GBD process), and k_d if adsorption is reversible, in which case it is more robust to globally fit both adsorption and desorption to Eq. (26) with an appropriate change of boundary conditions at the moment of flooding. If the particle-free solution (PFS) is introduced rapidly, then the entire diffusion boundary region will be swept clear at that moment, i.e., $c(z) = 0$ for all $z > 0$. On the other hand, if the PFS is introduced slowly, then $c(z) = 0$ for $z \geq \delta_d$.

It is a good idea to compare the observed rate with the maximum upper limit calculated from Eq. (15) (which might only be exceeded if there is a long-range attractive force between particle and surface).

Evidence for conformational rearrangement, including simple reorientation, can be obtained from a series of adsorption measurements carried out at different bulk concentrations. If the usual method of analysis yields decreasing a with increasing c_b, rearrangement can be inferred, and its characteristic time $\tau_s = 1/k_s$ can be obtained by equating it to the characteristic time for adsorption, $\tau_a = 1/(Jc_b\phi a)$ at the bulk concentration corresponding to the midpoint between the limiting lower and higher areas, where J is the protein flux to the empty surface normalized to unit adsorbent area and unit bulk concentration [22]. Further evidence for slow conformational rearrangements can be obtained by varying the time the system is maintained at its adsorption plateau in the presence of dissolved proteins, before flooding with particle-free solution.

B. Adsorption to Discrete Receptor Sites

This situation is often encountered when investigating the operation of biosensors. Consider a monolayer of receptors (antigens) of area a_R and surface density (i.e., number per unit area) v_R. The two-dimensional projected area of each ligand (i.e., the antibody) is a_L, and its dissolved bulk concentration is c_b as before.

When the receptor layer is exposed to flowing ligand solution, the ligand is transported to the layer via convective diffusion and the binding rate is

$$\frac{dv_L}{dt} = c_v\phi \tag{27}$$

where ϕ gives the probability that an arriving ligand will find a vacant receptor. Unsuccessful molecules (i.e., those not finding a vacant receptor) will remain some time in the vicinity of the surface, which of course reduces the flux from the bulk. If $a_L < a_R$, then ϕ is simply $1 - \theta$, where θ is the fraction of the surface occupied by ligands, i.e.,

$$\theta = a_L \nu_L \tag{28}$$

and has a maximum possible value of $\theta_x = \sigma$, where σ is a dimensionless receptor site density defined by

$$\sigma = a_L \nu_R \tag{29}$$

Under this condition $\nu_L(\infty) = \nu_R$. The same results apply if $a_L > a_R$ but ν_R is very small, and hence the receptors are mostly isolated from each other. Neither of these cases reflects biosensors intended for practical clinical use, however; the antibody (ligand) is usually much larger than the antigen, and in order for the response to be measurable, ν_R should be large.

In this case, both $\nu_L(t)$ and $\nu_L(\infty)$ will depend on both ν_R and a_L [viz. on σ, see Eq. (7)] and on ν_L, i.e., the fraction of occupiable sites occupied. In principle the formalism of Eq. (26) can still be used; instead of $\phi(\nu_L)$ we now have $\phi(\nu_L, \sigma)$. Jin et al. [79,80] have shown that there exists an unexpectedly simple mapping between this random site process (random addition of ligands to randomly distributed receptors) and random sequential addition, for which relevant expressions are already available. Defining

$$\theta^* = \theta/\theta_x \tag{30}$$

where [79,80]

$$\theta_x(\sigma) = \theta_J\left(1 - \frac{1 + 0.3136\sigma^2 + 0.45\sigma^3}{1 + 1.8285\sigma + 0.5075\sigma^3 + \sigma^{7/2}}\right) \tag{31}$$

one has [79,80]

$$\phi(\theta^*, \sigma) = (1 - \theta^*)(1 - B_1\theta^* - B_2\theta^{*2}) \tag{32}$$

where the constants are

$$B_1 = \frac{0.7126 + 1.404\sigma^{1/2}}{1/\sigma + 3.4363 + 2.4653\sigma^{1/2}} \tag{33}$$

and

$$B_2 = \frac{0.07362 + 0.1204\sigma^{1/2}}{1/\sigma + 0.5443 + 0.2725\sigma^{1/2}} . \tag{34}$$

The experimentally available parameters are ν_R and ν_L.

VII. PERSPECTIVES

It has been often pointed out that knowledge of the complete gene sequence (genome) of an organism is only one of many elements necessary for understanding that organism. The set of expressed proteins (the proteome)—characteristic of a given state of an organism at a particular epoch in its life history [81,82]—is derived from the genome but cannot necessarily be predicted from it, and even knowledge of the proteome is, by itself, inadequate for understanding the collection of chemical reactions within a cell (its metabolism); for this, knowledge of all the *interactions* between the proteins (and other cellular components, including DNA) is necessary [83,84]. The word *interactome* has been coined to describe the set of protein–protein interactions. As was pointed out at the beginning of this chapter, the vast majority of proteins are associated with membranes, and hence their place in the interactome is determined inter alia by their membrane affinity, involving the *lipidome*, a word which has been coined to reflect the astonishing variety of lipid molecules in many natural cell membranes. This determines both the likelihood of reacting with a partner and adsorption-induced changes in binding affinity for another protein or reaction specificity. Many important questions concerning human health will doubtless turn out to depend on the *dynamics* of protein–membrane, protein–protein, and protein–DNA interactions, and hence the kind of investigations described in this chapter will assume a central role in molecular medicine.

It is clear that significant progress toward understanding the vastness of the interactome (proteins can interact with several others [84], and all of the reactions should be characterized for the range of conditions under which they are likely to occur, including changes in cosolute concentration, membrane composition, etc.) will require highly creative new approaches, comparable to those which have enabled entire large genomes to be sequenced, including increasing the throughput of data gathering, and improving the effectiveness of extracting information from the data.

ACKNOWLEDGMENTS

The author thanks Dr. K. Tiefenthaler of Artificial Sensing Instruments, Zurich, Switzerland, and Dr. I. Szendrö of MicroVacuum, Budapest, Hungary, for many interesting and stimulating discussions about the use of optical waveguides and grating couplers for protein adsorption measurements. He also thanks the Institute of Experimental Medicine (KOKI) for hospitality during the preparation of this chapter.

LIST OF PRINCIPAL SYMBOLS

a area occupied per adsorbed protein molecule

c concentration

D diffusivity

d (fractal) dimension

e elementary (proton) charge

g pair correlation function

j ratio of probabilities for a particle to adsorb next to a previously adsorbed particle and in isolation

J specific flux toward an adsorbent surface

k Boltzmann's constant, kinetic coefficient

K equilibrium constant

ℓ_0 equilibrium contact distance (distance of closest approach)

Q memory kernel

r protein (hydrodynamic) radius
R gas constant
t time, duration, epoch
T temperature
U protein–surface interaction potential
z perpendicular distance between two surfaces

α dimensionless (receptor) site density
γ surface tension
δ distance
ϵ permittivity
θ fraction of surface covered by adsorbate
κ inverse Debye length
ν number of adsorbed molecules per unit area
σ dimensionless receptor site density
ϕ fraction of area available for adsorption
χ characteristic decay length for donor–acceptor interaction
ψ electrostatic potential
Ψ fraction of area available for spreading

Subscripts and Superscripts

a   adsorbing surface (adsorbent)
a adsorption, adsorbed
b bulk
d desorption, diffusion
el   electrostatic
J jamming
LW   Lifshitz–van der Waals
solv solvation
s spread, spreading
v vicinal
$\oplus$ electron accepting
$\ominus$ electron donating
1   solid
2   liquid
3   protein

REFERENCES

1. E. S. Kempner and J. H. Miller. Exp. Cell Res. 51:141–149, 1968.
2. E. S. Kempner and J. H. Miller. Exp. Cell Res. 51:150–156, 1968.
3. E. W. Davie, K. Fujikawa, W. Kisiel. Biochemistry 30:10363–10370, 1991.
4. J. J. Ramsden. Colloids Surf. A 173:237–249, 2000.
5. J. J. Ramsden, D. J. Roush, D. S. Gill, R. Kurrat, R. C. Willson. J. Amer. Chem. Soc. 117:8511–8516, 1995.
6. G. N. Misevic. Microsc. Res. Tech. 44:304–309, 1999.
7. A. Fernández, J. J. Ramsden. J. Biol. Phys. Chem. 1:70–73, 2001.

8. P. H. Weinreb, W. Zhen, A. W. Poon, K. A. Conway, P. T. Lansbury. Biochemistry *35*: 13709–13715, 1996.

9. J. J. Ramsden. In *Bioelectronic Applications of Photochromic Pigments* (A. Dér and L. Keszthelyi, eds.). IOS Press, Amsterdam, 2001, pp. 244–269.

10. V. G. Levich. *Physicochemical Hydrodynamics*. Prentice Hall, Englewood Cliffs, NJ, 1962.

11. D. G. Leaist. J. Phys. Chem. *93*:474–479, 1989.

12. J. N. Israelachvili, R. Pashley. Nature (Lond.) *300*:341–342, 1982.

13. C. J. van Oss. In *Protein Interactions* (H. Visser, ed.). V. C. H. Verlag, Weinheim, 1992.

14. C. J. van Oss. *Interfacial Forces in Aqueous Media*. Marcel Dekker, New York, 1994.

15. M. G. Cacace, E. M. Landau, J. J. Ramsden. Q. Rev. Biophys. *30*:241–278, 1997.

16. J. J. Ramsden. Colloids Surf. B *14*:77–81, 1999.

17. G. R. Wiese, T. W. Healy. Trans. Faraday Soc. *66*:490–499, 1970.

18. C. J. van Oss, J. O. Naim, P. M. Costanze, R. F. Giese, W. Wu, A. F. Sorling. Clays Clay Minerals *47*:697–707, 1999.

19. L. Vroman, A. L. Adams. Surf. Sci. *16*:438–446, 1969.

20. R. Kurrat, B. Wälivaara, A. Marti, M. Textor, P. Tengvall, J. J. Ramsden, N. D. Spencer. Colloids Surfaces B *11*:187–201, 1998.

21. W. Ramsden. Trans. Faraday Soc. *22*:484–485, 1926.

22. J. J. Ramsden. Phys. Rev. Lett. *71*:295–298, 1993.

23. J. J. Ramsden. Q. Rev. Biophys. *27*:41–105, 1994.

24. A. Docoslis, W. Wu, R. F. Giese, C. J. van Oss. Colloids Surf. B *13*:83–104, 1999.

25. C. J. van Oss, M. K. Chaudhury, R. J. Good. Chem. Rev. *88*:927–941, 1988.

26. P. Schaaf, Ph. Dejardin, A. Schmitt. Revue Phys. Appl. *20*:631–640, 1985.

27. P. K. Tien. Rev. Mod. Phys. *49*:361–420, 1977.

28. K. Tiefenthaler, W. Lukosz. J. Opt. Soc. Amer. B *6*:209–220, 1989.

29. T. Turbadar. Proc. Phys. Soc. *73*:40–44, 1959.

30. J. J. Ramsden. J. Statist. Phys. *73*:853–877, 1993.

31. J. J. Ramsden. Phil. Mag. B *79*:381–386, 1999.

32. M. L. Chiu, P. H. Viollier, T. Katoh, J. J. Ramsden, C. J. Thompson. Biochemistry *40*: 12950–12958, 2001.

33. P. Schuck, A. P. Minton. Anal. Biochem. *240*:262–272, 1996.

34. C. J. van Oss. IJBC *4*:139–152, 1999.

35. P. Lavalle, A. L. DeVries, C. H. C. Cheng, S. Scheuring, J. J. Ramsden. Langmuir *16*: 5785–5789, 2000.

36. J. P. Gallinet, B. Gauthier-Manuel. Colloids Surf. *68*:189–193 (1992).

37. D. E. Leckband. Adv. Biophys. *34*:173–190, 1997.

38. R. Bruinsma. Proc. Natl. Acad. Sci. USA *94*:375–376, 1997.

39. G. N. Misevic. Mol. Biotechnol. *18*:149–153, 2001.

40. J. Talbot. Adsorption *2*:89–94, 1996.

41. R. Kurrat, J. J. Ramsden, J. E. Prenosil. J. Chem. Soc. Faraday Trans. *90*:587–590, 1994.

42. R. Kurrat, J. E. Prenosil, J. J. Ramsden. J. Colloid Interface Sci. *185*:1–8, 1997.

43. J. J. Ramsden. In *Biopolymers at Interfaces*, 1st ed. (M. Malmsten, ed.). Marcel Dekker, New York, 1998, pp. 321–361.

44. L. A. Spielman, S. K. Friedlander. J. Colloid Interface Sci. *46*:22–31, 1974.

45. J. J. Ramsden, In *Encyclopedia of Surface and Colloid Science* (A. Hubbard, ed.). Marcel Dekker, New York, 2001.

46. F. M. Fowkes, J. Phys. Chem. *67*:2538–2541, 1963.

47. L. A. Girifalco, R. J. Good. J. Phys. Chem. *61*:904–909, 1957.

48. C. Calonder, J. J. Ramsden. Phys. Chem. B *105*:725–729, 2001.

49. A. Docoslis, W. Wu, R. F. Giese, C. J. van Oss. Colloids Surf. B *22*:205–217, 2001.

50. M. von Smoluchowski. Z. Phys. Chem. *92*:129–168, 1917.
51. C. J. van Oss, A. Docoslis, W. Wu, R. F. Giese. Colloids Surf. B *14*:99–104, 1999.
52. C. Calonder, P. R. Van Tassel. Langmuir *17*:4392–4395, 2001.
53. A. Rényi. Magy. Tud. Akad. Mat. Kut. Int. Közl. *3*:109–125, 1958.
54. P. Schaaf, J. Talbot. J. Chem. Phys. *91*:4401–4409, 1989.
55. P. Schaaf, J. C. Voegel, B. Senger. Ann. Physique *23*:1–89, 1998.
56. G. Tarjus, P. Schaaf, J. Talbot. J. Chem. Phys. *93*:8352–8360, 1990.
57. P. O. Luthi, J. J. Ramsden, B. Chopard. Phys. Rev. E *55*:3111–3115, 1997.
58. P. Viot, G. Tarjus, S. M. Ricci, J. Talbot. J. Chem. Phys. *97*:5212–5218, 1992.
59. S. M. Ricci, J. Talbot, G. Tarjus, P. Viot. J. Chem. Phys. *7*:5219–5228, 1992.
60. G. Tarjus, P. Viot, H. S. Choi, J. Talbot. Phys. Rev. E *49*:3239, 1994.
61. G. Csúcs, J. J. Ramsden. J. Chem. Phys. *109*:779–781, 1998.
62. I. Palásti. Magy. Tud. Akad. Mat. Kut. Int. Közl. *5*:353, 1960.
63. Y. Pomeau. J. Phys. A *13*:L193–L196, 1980.
64. L. Blumenfeld. J. Theor. Biol. *58*:269–284, 1976.
65. A. Dér, J. J. Ramsden. Naturwissenschaften *85*:353–355, 1998.
66. A. P. Minton. Biophys. J. *76*:176–187, 1999.
67. C. Calonder, P. R. Van Tassel. Proc. Natl. Acad. Sci. USA *98*:10664–10669, 2001.
68. P. R. Van Tassel, L. Guemouri, J. J. Ramsden, G. Tarjus, P. Viot, J. Talbot. J. Colloid Interface Sci. *207*:317–323, 1998.
69. J. J. Ramsden, G. I. Bachmanova, A. I. Archakov. Phys. Rev. E *50*:5072–5076, 1994.
70. G. Vergères, J. J. Ramsden. Arch. Biochem. Biophys. *378*:45–50, 2000.
71. R. C. Chatelier, A. P. Minton. Biophys. J. *71*:2367–2374, 1996.
72. A. P. Minton. Biophys. Chem. *86*:239–247, 2000.
73. L. Guemouri, J. Ogier, J. J. Ramsden. J. Chem. Phys. *109*:3265–3268, 1998.
74. L. Guemouri, J. Ogier, Z. Zekhnini, J. J. Ramsden. J. Chem. Phys. *113*:8183–8186, 2000.
75. S. Ravichandran, J. D. Madura, J. Talbot. J. Phys. Chem. B *105*:3610–3613, 2001.
76. H. Knull, A. P. Minton. Cell Biochem. Function *14*:237–248, 1996.
77. J. Bafaluy, B. Senger, J. C. Voegel, P. Schaaf. Phys. Rev. Lett. *70*:623–626, 1993.
78. V. Ball, J. J. Ramsden. J. Phys. Chem. B *101*:5465–5469, 1997.
79. X. Jin, N. H. L. Wang, G. Tarjus, J. Talbot. J. Phys. Chem. *97*:4256–4258, 1993.
80. X. Jin, J. Talbot, N. H. L. Wang. AIChE J *40*:1685–1696, 1994.
81. N. G. Anderson, L. Anderson. Clin. Chem. *28*:739–748, 1982.
82. J. Vohradský, J. J. Ramsden. FASEB J. *15*:2054–2056, 2001. (full online version; published July 24, 2001: 10.1096/fj.00-0889fje.)
83. E. Conkey. Proc. Natl. Acad. Sci. USA *79*:3236–3240, 1982.
84. D. J. Raine, V. Norris. J. Biol. Phys. Chem. *1*:80–86, 2001.
85. S. N. Timasheff. Rev. Biophys. Biomol. Struct. *22*:67–97, 1993.

9
Mobility of Biomolecules at Interfaces

ROBERT D. TILTON   Carnegie Mellon University, Pittsburgh, Pennsylvania, U.S.A.

I. INTRODUCTION

Can adsorbed proteins move on a surface, or are they strictly immobilized? The observation that adsorption is frequently irreversible might imply the latter. Proteins adsorb tenaciously because of the large number of segmental contacts established between the protein and the surface. While each physisorbed segment contributes approximately 1 kT to the overall adsorption energy, the large number of these "bonds" produces a large total adsorption energy and a large activation barrier to desorption. For a protein to migrate on a surface requires the disruption and reformation of a considerable number of segmental contacts. Considering the tenacity of adsorption, it would be reasonable to expect that proteins should be effectively immobilized on most surfaces. Nevertheless, several systems have been examined where adsorbed proteins are indeed capable of diffusing over distances that are orders of magnitude greater than the length of the protein molecule. Long-range surface diffusion is possible.

Lateral mobility represents one of many dynamic phenomena at work in adsorbed protein layers. Many questions need to be asked about adsorbed protein dynamics in general. It is not difficult to envision a close relationship between protein dynamics and biochemical activity. As challenging as it may be to quantify any of the fundamental, time-averaged structural characteristics of an adsorbed protein layer, i.e., molecular orientation, conformation, or organization, it is equally desirable to quantify the dynamics by which the layer structure evolves as well as the dynamics by which molecules sample the distribution of allowable states within a layer at "equilibrium" (perhaps better described as steady state). This chapter focuses exclusively on the lateral mobility of biological macromolecules, mainly proteins but also oligonucleotides, adsorbed at solid–liquid interfaces. Although certain issues are common to the lateral mobility of adsorbed proteins and of proteins associated with lipid bilayers, the latter has received far more attention in the literature (see, for example, Refs. 1–7) and will not be considered here.

Protein lateral diffusion on solid surfaces may be readily distinguished from the lateral diffusion of intrinsic membrane proteins or even of water-soluble proteins bound to lipid bilayers. Intrinsic membrane protein diffusion can be described adequately in terms of the hydrodynamics of a protein immersed in a finite fluid of high viscosity, bounded on both sides by semi-infinite media of lower viscosity. In real cells, there may be external constraints imposed, for example, by impermeable lipid domains or by cytoskeletal contacts, and lateral interactions between diffusing proteins are important in crowded cell membranes, but the basic diffusion mechanism is governed mainly by the protein's size and the lipid viscosity [8,9]. The physico-chemical properties unique to any particular protein are relatively unimportant. In fact, it is the external control of membrane protein diffusion that is of the greatest interest to cell biologists. Likewise, the lateral diffusion of soluble proteins bound to the outer surface of a lipid bilayer, for example, immunoglobulins specifically bound to hapten-conjugated lipids, is not terribly sensitive to the identity of the particular protein in question. Specifically bound immunoglobulins have been shown to diffuse with the same diffusion coefficient as the lipids themselves [10]. The dominant contributing factor to protein mobility in these cases is the fluidity of the lipid bilayer itself. As we will see, the unique properties of the protein cannot be ignored in the mechanism of surface diffusion on solid surfaces, and as yet no predictive theory exists for absolute values of protein surface diffusion coefficients.

In contrast to proteins associated with lipid assemblies, diffusion of proteins tightly bound to solid surfaces requires the motion of many physisorbed segments relative to the fixed surface. Since the energy barrier to dislocating a single physisorbed segment from the surface is of order kT, the probability that at any given instant some small number of segments are detached is far greater than the probability of all segments detaching simultaneously. Brownian motion of the entire protein can then result from random thermal fluctuations of different parts of the poly-peptide chain, not necessarily requiring the entire molecule to desorb. A simplistic analogy may be drawn to a crawling inchworm. At any instant during its crawl, some part of its body remains fixed on the surface while the remainder is detached and in motion, searching for a new place to land on the surface. Just as a rigid inchworm could not crawl, some flexibility of protein conformation may be required for surface diffusion. Surface diffusion might be linked to the breathing modes of the protein structure.

II. RELEVANCE OF SURFACE DIFFUSION

A. Two-Dimensional Order

If adsorbed proteins are indeed mobile, what is the significance of that observation? First, lateral mobility can profoundly affect the most basic property of an adsorbed layer, the surface excess concentration. It may do so by violating one of the tenets of random sequential adsorption (RSA). The RSA model describes the irreversible adsorption of nonoverlapping particles that are immobile on the surface [11,12]. Once adsorbed, particles cannot desorb, and they must remain exactly where they landed. Since there is no mechanism for rearranging adsorbed particles, they are locked into a random arrangement on the surface. According to the RSA model, there is a maximum surface coverage beyond which further adsorption is impossible. Known as the jamming limit or the random parking limit, this maximum coverage is approximately 55% by area for disks, decreasing somewhat for anisotropic shapes (e.g., to

51% for rectangles of aspect ratio 3.5) [13]. Laterally mobile particles can rearrange themselves after adsorption and thereby achieve more efficient, orderly packing arrangements. Greater monolayer packing efficiency means high surface coverages are attainable, provided the intermolecular forces between neighboring proteins are favorable. So, lateral mobility can enable two-dimensional order and increased surface concentrations.

Particularly elegant examples of order facilitated by adsorbed protein lateral mobility are the two-dimensional crystals formed by certain proteins, such as streptavidin [14] or ferritin [15], after they adsorb to insoluble lipid monolayers spread at the air–water interface. The crystals are large enough to use for crystallographic elucidation of protein structure. The fluidity of the monolayer to which the proteins adsorb is critical for this self-assembly process. While this example falls outside the realm of proteins on solid surfaces, it beautifully illustrates the close relationship between lateral mobility and two-dimensional order.

B. Molecular Recognition

Postadsorption lateral mobility can also affect the kinetics of molecular recognition processes on surfaces, such as enzyme catalysis or receptor–ligand binding. The theory of "reduction of dimensionality" and related receptor binding theories [16–18] describe the potential for surface diffusion to accelerate these processes. In some cases, three-dimensional diffusion is remarkably effective, and surface diffusion (although it never slows a process) may be unnecessary for efficient binding. Whether or not surface diffusion significantly accelerates binding depends on the relative values of bulk and surface diffusion coefficients and on the characteristic length scales over which diffusion occurs, as determined by the density of receptors on the surface. Since surface diffusion coefficients are far more difficult to estimate than are bulk diffusion coefficients, reliable a priori predictions of whether surface diffusion will be significant in the kinetics of catalytic or receptor–ligand binding processes on surfaces are difficult if not impossible. Measurements are required.

Receptor–ligand binding is essential for cell adhesion and spreading on surfaces. Thus, lateral mobility of proteins (receptors or ligands) is a requirement for adhesion in that receptors must randomly fall into registry with ligands. The first instinct might be to consider the binding dynamics to be controlled by the lateral diffusion of intrinsic membrane proteins (receptors) into contact with immobilized proteins (ligands) on the solid surface. It is noteworthy that surface diffusion coefficients available in the literature for proteins on solids are quite comparable to the diffusion coefficients of intrinsic membrane proteins. Adsorbed ligand mobility may therefore accelerate binding kinetics.

C. Relevance to Experimentation: Effect of Protein Mobility on Scanning Probe Microscopy

Two-dimensional ordering, molecular recognition, and catalysis are examples of fundamental biophysical processes that may be affected by the lateral mobility of proteins at interfaces. Lateral mobility can also influence the use of certain tools, particularly scanning probe microscopes, to characterize adsorbed protein layers. The force between the tip and the protein can dramatically alter the two-dimensional arrangement of the layer by dragging the adsorbed molecules [19,20]. This may be overcome by covalently immobilizing proteins prior to imaging or by using noncontact or "tapping-mode" atomic force microscope methods. Even when the tip is not a source of artifacts, the mobility of proteins may cause the microscopic resolution to suffer.

D. Basic Insight into Adsorbed Layer Heterogeneity

Finally, surface diffusion can potentially offer other insights into the fundamentals of adsorption. As described later, surface diffusion measurements have provided additional evidence for the increasingly accepted idea that adsorbed protein layers are heterogeneous. Adsorbed layers may contain distributions of surface diffusion coefficients, appearing as a coexistence of mobile and immobile molecules. Although the cause of heterogeneous mobility is as yet difficult to prove, it may reflect a distribution of conformations, orientations, or states of aggregation.

The complexities of protein adsorption make it essential to discuss surface diffusion in the context of all other molecular level characteristics of adsorbed layers. Consider the factors that may reasonably be expected to influence surface diffusion. Protein flexibility: as noted in the inchworm analogy, flexibility may be important. Momentarily detached segments must be free to flex and randomly reattach elsewhere on the surface. The more flexible the protein conformation is, the larger will be the random steps taken by detached segments. On the other hand, highly flexible proteins tend to experience more significant conformational changes on surfaces [21–24] and may therefore have a larger fraction of segments bound to the surface. This pinning effect could then hinder diffusion. Conformation: since it controls the protein's flexibility and the number of polypeptide segments in contact with the surface, the conformation adopted by an adsorbed protein is critical. In a similar manner, proteins in different orientations may have different surface diffusion coefficients. Organization: the two-dimensional organization of the layer, by controlling the steric hindrance to diffusion, will also be critical. This is discussed in more detail later.

Cause-and-effect relationships between surface diffusion and the various other molecular level characteristics of adsorbed layers are difficult to discern. Apparent cause or effect can go both ways. For example, lateral interactions (crowding effects) between neighboring proteins are often believed to decrease the extent of conformational changes and also the ability of proteins to assume orientations with large projected areas. Both are therefore more favorable at low coverages. Consider the following: surface diffusion can facilitate layer reorganization into close-packed arrangements. This may in turn limit the extent of conformational change, but it will also increase the degree of steric hindrance to diffusion. Will reorganization enhance or diminish lateral mobility? Will surface diffusion enable two-dimensional aggregation only to have that aggregation shut down subsequent surface diffusion? The only thing that is clear is that there is no obvious cause-and-effect relationship. Furthermore, as the aggregation scenario suggests, one can imagine that lateral mobility may be strongly time dependent. Perhaps adsorbed molecules are mobile for only a finite time while aggregation or conformational changes progress. As will become evident, such time dependencies have not received much attention in the literature.

III. INDIRECT EVIDENCE FOR ADSORBED PROTEIN MOBILITY

As discussed above, surface diffusion may allow nonrandom adsorbed layer arrangements. Thus, any evidence for two-dimensional ordering suggests—but does not prove—the occurrence of surface diffusion. The following is a far from thorough discussion of ordering in adsorbed layers and serves merely to establish that there is indeed a body of indirect evidence for adsorbed protein mobility. Let us first consider measurements of surface coverage in excess of the jamming limit as evidence for ordering and, by extension, for surface diffusion. It is quite common for proteins to attain surface concentrations consistent with a close-packed monolayer (see, for example, Refs. 25–27], i.e., well in excess of the RSA jamming limit. To illustrate, a hexagonal close-packed (HCP) array of disks occupies approximately 91% of the available surface area. Ribonuclease A (RNase A) adsorbs to 2.2 mg/m^2 on polystyrene at pH 8 [28]. Given the dimensions ($2.5 \times 2.5 \times 4$ nm) and molecular weight (13,700) of this ellipsoidal molecule, 2.2 mg/m^2 is consistent with hexagonal close packing of molecules oriented vertically. This is the largest coverage possible for a monolayer.

Unfortunately, surface concentrations are not definitive gauges for the degree of order in an adsorbed protein layer. This is because key assumptions, often untested, must be made about the orientation of the proteins relative to the surface. For example, lysozyme adsorbs to 2.2 mg/m^2 on silica from 5 mM, pH 7 triethanolamine buffers [29]. The lysozyme structure is approximately a $3 \times 3 \times 4.5$ nm prolate ellipsoid of molecular weight 14,300. Since a HCP monolayer of *horizontal* lysozyme molecules would give a surface concentration of 2.1 mg/m^2, this may be evidence for complete HCP ordering on silica. On the other hand, a monolayer of *vertical* lysozyme molecules would give 1.9 mg/m^2 at the *RSA jamming limit*. So a 2.2 mg/m^2 lysozyme surface concentration could correspond to a complete HCP monolayer of horizontal molecules, or it could correspond to a mostly random vertical monolayer containing but a small degree of order. Clearly, independent measures of protein orientation are a minimal requirement for reliably interpreting surface concentrations in terms of two-dimensional order. In this particular lysozyme study, independent fluorescence spectroscopic measurements indicated that the adsorbed molecules, preferentially assumed a horizontal orientation and therefore probably did develop an ordered layer on silica.

Of course, molecules that are capable of desorbing from the surface violate another premise of the RSA model. Desorption/readsorption events also can lead to more orderly, denser layers, independent of any protein surface diffusion. Parenthetically, this raises an important point regarding the definition of surface diffusion. The sequential adsorption–desorption–bulk diffusion–readsorption process may be referred to as "hopping diffusion" in that it still results in a random relocation of molecules on the surface. In this chapter, surface diffusion by definition will refer to the random transport of molecules that do not escape the adsorption energy well confining them to the surface; i.e., they diffuse in direct contact with the surface. Experimentally it may be difficult to distinguish between the two mechanisms, although hopping diffusion may lead to non-Fickian or anomalous diffusion, a topic to be discussed later.

Besides consideration of surface concentrations, adsorption kinetic analyses may reveal the growth of ordered layers. RSA kinetic models are clearly distinguishable from models that allow postadsorption mobility and clustering. Ramsden et al. [30] used an integrated optics technique to measure the adsorption of proteins belonging to the cytochrome P450 family and were indeed able to discern clustering kinetics under certain conditions.

There is kinetic evidence for clustering in other systems as well. The initial kinetics of ferritin adsorption silica [31,32] were found to be consistent with a computer simulation [33] based on both bulk and surface diffusion. That simulation allowed for a surface diffusion coefficient three orders of magnitude smaller than the bulk value. As discussed later, two- to three-order-of-magnitude differences between bulk and surface diffusion coefficients have indeed been measured experimentally. Furthermore, using transmission electron microscopy and dried surfaces, Nygren and coworkers [31,32] imaged adsorbed ferritin clusters on hydrophilic and hydrophobized silica surfaces. These clusters changed their fractal morphology over the course of the adsorption process.

Park et al. [34] used bifunctional cross-linking agents to link fibrinogen molecules after adsorption to glass surfaces. After cross-linking, they eluted the fibrinogen and analyzed the size distribution by gel electrophoresis, finding mostly large fibrinogen aggregates. Assuming that only molecules in close proximity may be cross-linked, the prevalence of multimers in the eluant pointed to patchwise adsorption. In the spirit of this chapter, an alternative explanation could be that surface diffusion enabled randomly arranged fibrinogen molecules to be cross-linked during transient, but sufficiently long-lived, collisions. Either way, this cross-linking is circumstantial evidence for protein lateral mobility.

Ordered arrays of adsorbed proteins have been imaged directly using scanning tunneling microscopy. Lenhoff and Haggerty [35] found that lysozyme develops highly ordered arrays when adsorbed to graphite. They noted that the long-range spatial arrangement of adsorbed lysozyme was inconsistent with tip-induced artifacts. Rather than long straight lines that are sometimes induced by tip dragging, large regions of the arrays appeared to be curved, displaying a "ring motif."

The preceding observations all hint at the lateral mobility of proteins adsorbed to solid surfaces. There is evidence for rotational mobility as well. For example, Lee and Belfort [36] used the surface force apparatus and enzyme assays to demonstrate that ribonuclease A changed its average orientation as it adsorbed slowly to mica. Blomberg et al. [37] made similar observations for lysozyme on mica. Employing a total internal reflection fluorescence method based on potential-sensitive emission, Robeson and Tilton [29] observed that lysozyme spontaneously reoriented when it reached a critical coverage while adsorbing to silica. It does not appear that the adsorbed proteins in these studies were free to continuously rotate about an axis parallel to the surface. Rather, their rotation appears to be induced suddenly by lateral interactions among neighboring proteins. Nevertheless, for some time at least, they have a finite rotational mobility.

There is far more than circumstantial evidence for protein mobility on solid surfaces. The remainder of this chapter focuses on experimental methods for the direct measurement of lateral mobility and on the results of such studies.

IV. EXPERIMENTAL METHODS TO STUDY LATERAL MOBILITY ON SOLID SURFACES

A. Fluorescence Recovery After Photobleaching

The technique most often employed to measure lateral diffusion is fluorescence recovery (or "redistribution") after photobleaching (FRAP) [28,38–47]. Many of the cited papers contain detailed instrument descriptions for the interested reader. While this is not a chapter about FRAP, some detailed discussion of the principles underlying this technique is essential, simply because all quantitative studies of protein surface diffusion on solid surfaces published to date are based on FRAP. Thus the technique and the phenomenon are intimately linked. The reader should gain some appreciation of the strengths and limitations of FRAP, so as to ultimately formulate his or her own critical evaluation of the data. Particular attention will be paid here to precautions that must be exercised in order to obtain meaningful results from FRAP data.

The small magnitude of surface diffusion coefficients (10^{-10} to 10^{-7} cm²/s) makes it essential to probe diffusion over rather small length scales. Because the diffusion time depends on the square of the experimental length scale, surface diffusion measurements over even millimeter length scales are impractical. FRAP is designed for measuring transport over more appropriate length scales on the order of micrometers.

FRAP requires the use of proteins that have been covalently labeled with extrinsic fluorophores. Since their excitation spectra closely match the high-power lines of argon ion lasers, fluorescein, eosin, and rhodamine are commonly used for this purpose. Several variants of FRAP are described in the literature, but all are based on the irreversible photobleaching of fluorophores by a pulse of high-intensity laser excitation directed toward a small region of the surface. The photobleaching pulse superimposes a gradient of functional fluorophores upon an otherwise uniform distribution of labeled molecules. The subsequent relaxation (in the case of purely diffusional transport) or drift (in the case of directional flow) of this superimposed fluorescence gradient is detected by low-intensity fluorescence excitation. Put simply, photobleaching destroys some of the fluorophores in a small region. That decreases the observed fluorescence intensity emitted from that region. As unbleached molecules migrate into the bleached region from somewhere else on the surface, the fluorescence emitted from that region recovers.

The transport parameters are obtained from the kinetics and extent of this fluorescence redistribution. One must assume a transport mechanism (i.e., diffusion and/or flow) and test it for consistency with the mathematical form of the fluorescence redistribution. In most cases of interest here, the transport parameters are the diffusion coefficient, D, and the mobile fraction, f. (Some adsorbed proteins in a layer may be immobile.) Minor instrumental details aside, FRAP methods generally differ in two ways, namely, the geometry of the photobleaching illumination and the means for monitoring the subsequent redistribution of fluorescent species.

When periodic patterns are used for photobleaching, such as fringe patterns generated by interfering two coherent laser beams or square-wave patterns formed by imaging a grating onto a surface, the technique is referred to as fluorescence recovery after pattern photobleaching (FRAPP). Otherwise, it is common to employ a circular or an elliptical spot, where the intensity distribution across the spot displays the Gaussian form provided by the TEM_{00} laser mode.

Two methods are most commonly employed to monitor the redistribution after photobleaching. Either the photobleached region is illuminated uniformly and imaged by photographic or video fluorescence microscopy, or the photobleached region is illuminated by the same beam or pattern that was used for photobleaching, although at much lower intensity, while the total integrated fluorescence intensity is measured by a photomultiplier tube (no imaging).

1. Analysis of FRAP Data Under Ideal Conditions

To accurately measure transport parameters it is essential to accurately relate the instantaneous fluorescence intensity to the instantaneous fluorophore concentration profile on the surface. Ideally, the fluorescence intensity is simply a linear reporter of fluorophore concentration. It is therefore usually assumed that the fluorescence emission intensity $F(t)$ is directly proportional to the unbleached fluorophore concentration C, so that

$$F(t) \propto \int_A I(r, \theta) C(r, \theta, t)\, dA \tag{1}$$

where the integral is performed over the entire illuminated area when a photomultiplier tube is used to measure total intensity. $I(r, \theta)$ is the intensity profile of the low intensity excitation illumination, for example, circular Gaussian for spot FRAP or sinusoidal for fringe pattern photobleaching. The fluorophore concentration profile is determined by the solution to the diffusion equation

$$\frac{\partial C}{\partial t} = D\nabla^2 C \tag{2}$$

with the initial conditions dictated by the first-order photobleaching reaction and the geometry, intensity, and duration of the photobleaching pulse. The initial condition is thus

$$C(r, \theta, t = 0) = \bar{C} \exp[-k I_b(r, \theta)] \tag{3}$$

where $\bar{C}$ is the uniform prebleach concentration of fluorescent species, and k lumps the photobleaching rate constant and pulse duration.

For fringe pattern photobleaching, the intensity varies only in the direction perpendicular to the fringes. The intensity profile of the photobleaching fringe pattern in Cartesian coordinates is

$$I_b(x) = I_{b0} \left[1 + A \sin\left(\frac{\pi x}{w}\right) \right] \tag{4}$$

where w, the half-period of the fringe pattern, determines the characteristic length scale for the experiment. It is easily varied by changing the angle of laser beam intersection. Neglecting the Gaussian laser profile, which in reality is superimposed on the fringe pattern, leads to insignificant error as long as w is much less than the diameter of the entire beam [48]. The amplitude, A, describes the fringe pattern contrast. Although often assumed to be unity, A should be measured to accurately analyze FRAPP data.

Solving the diffusion equation with the initial condition dictated by Eqs. (3) and (4), the concentration profile of unbleached fluorophores bound to mobile biomolecules (a fraction f of the total biomolecules) evolves with time after the photobleaching pulse as

$$\frac{C_m(x,\ t)}{\bar{C}} = f\left\{\frac{a_0}{2} + \sum_{n=1}^{\infty} \exp\left(\frac{-n^2\pi^2 Dt}{w^2}\right)\left[a_n \cos\left(\frac{n\pi x}{w}\right)\right.\right.$$
$$\left.\left. + b_n \sin\left(\frac{n\pi x}{w}\right)\right]\right\} \tag{5}$$

where the Fourier coefficients are

$$a_n = \frac{1}{w}\int_{-w}^{w} \exp[-kI_b(x)] \cos\left(\frac{n\pi x}{w}\right) dx \tag{6}$$

$$b_n = \frac{1}{w}\int_{-w}^{w} \exp[-kI_b(x)] \sin\left(\frac{n\pi x}{w}\right) dx \tag{7}$$

Those fluorophores bound to immobile species remain forever in their initial distribution, so

$$\frac{C_{im}(x)}{\bar{C}} = (1 - f)\exp[-kI_b(x)] \tag{8}$$

The evolution of the concentration profile after photobleaching is illustrated in Fig. 1 for $A = 1$ and $f = 1$. It can be seen that higher-order terms in the Fourier

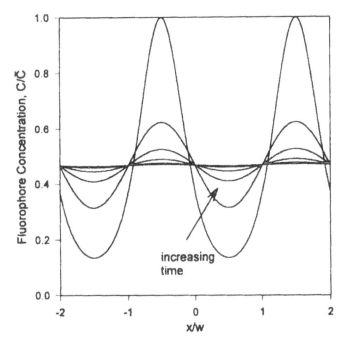

FIG. 1 Evolution of the unbleached fluorophore concentration distribution after fringe pattern photobleaching, for $A = 1$ and $k = 1$.

expansion decay rapidly, so that a simple sinusoidal concentration profile [46] may be safely assumed without introducing significant error into the data analysis. Note also that the concentration does not return to its original, prebleach level. For this reason, the fluorescence recovery will never be "complete" after pattern photo-bleaching. The significance of this fact will be discussed in Section IV.A.2.

If the fluorescence redistribution is monitored by illuminating the surface with the same (but highly attenuated) periodic pattern used for photobleaching [simply Eq. (4) multiplied by a constant $\ll 1$] the total intensity detected by the photomul-tiplier tube is described exactly by

$$\frac{\bar{F} - F(t)}{\bar{F} - F_0} = 1 + \frac{Ab_1 f}{2 - a_0 - Ab_1}\left[1 - \exp\left(\frac{-\pi^2 Dt}{w^2}\right)\right] \tag{9}$$

where $\bar{F}$ and F_0 are the fluorescence intensities before and immediately after pho-tobleaching. Equation (9) results when Eqs. (4)–(8) are inserted into Eq. (1). All unknown proportionality constants cancel in writing Eq. (9).

D and f are experimentally determined by nonlinear least-squares regression of the fluorescence recovery data according to Eq. (9). Sample recoveries are plotted for different mobile fractions in Fig. 2, and data for BSA adsorbed to polystyrene are presented along with the regression result in Fig. 3.

To evaluate a_0 and b_1, the bleaching depth parameter, k, must be determined from the ratio of fluorescence intensities immediately after photobleaching and im-

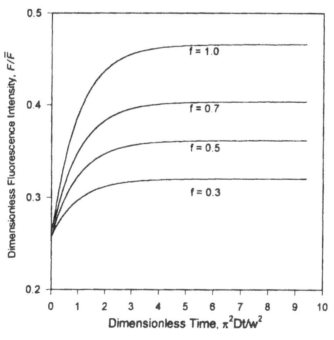

FIG. 2 Theoretical fluorescence recoveries after fringe pattern photobleaching, for $A = 1$ and $k = 1$. Different mobile fractions are noted. Since $F/\bar{F} = 1$ before photobleaching, it is evident that less than 33% of the fluorescence is recovered even for a mobile fraction of unity.

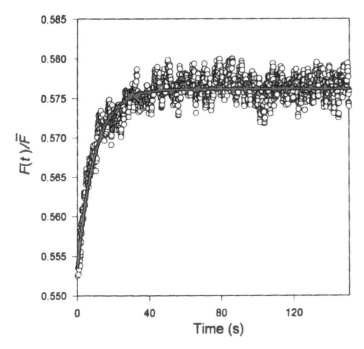

FIG. 3 FRAPP data for BSA adsorbed to PMMA. The curve is the regressed recovery for $D = 6.7 \times 10^{-8}$ cm^2/s and $f = 0.40$.

mediately before photobleaching, $F_0/\bar{F}$. This is most easily done graphically, using Eq. (1) to generate a theoretical plot of k versus $F_0/\bar{F}$.

If the fluorescence redistribution is monitored by direct fluorescence microscopic imaging, one can either illuminate the field uniformly, in which case the local fluorescence intensity is directly proportional to $C_{im}(x) + C_m(x, t)$, or use Gaussian laser illumination, in which case the local fluorescence intensity is directly proportional to the product of a Gaussian times the same quantity. The diffusion coefficient and mobile fraction may then be determined directly from plots of the time-dependent fluorescence profiles.

For spot photobleaching with a circular Gaussian beam,

$$I_b(r) = \frac{2P}{\pi R^2} \exp\left(\frac{-2r^2}{R^2}\right) \tag{10}$$

and the time-dependent fluorescence signal during the recovery is [49,50]

$$\frac{F(t)}{\bar{F}} = f\left(\sum_{n=0}^{\infty} \frac{(-k)^n}{n!} \frac{1}{1 + n(1 + 2t/\tau_s)}\right) + (1 - f)\frac{F_0}{\bar{F}} \tag{11}$$

where τ_s, the spot photobleaching diffusion time, is equal to $R^2/4D$.

2. Relative Merits of Spot FRAP and FRAPP

Spot and pattern photobleaching each have advantages and disadvantages. Aside from matters of convenience, such as ease of varying-length scales for testing transport mechanisms (where FRAPP holds the advantage), there are more fundamental differences between the two methods. If signal-to-noise ratio is a problem, spot FRAP has the advantage over FRAPP. This is because a layer composed of 100% mobile species will produce a complete fluorescence recovery with spot FRAP. In other words, the fluorescence signal will eventually return to its prebleach value. Even with 100% mobility, FRAPP produces *at most* a 33% recovery of the fluorescence destroyed by photobleaching (Fig. 2). Even less recovery results if $A < 1$. Deeper bleaches, i.e., larger k, also produce less recovery. Because of this incomplete recovery, FRAPP data are more likely to be obscured by noise.

Examining Eqs. (9) and (11), one can see that the fluorescence redistribution in spot FRAP is asymptotic, whereas the redistribution in FRAPP is exponential in time. This affords the pattern photobleaching method a distinct advantage for measuring mobile fractions. Furthermore, the value of the diffusion coefficient obtained from spot FRAP is also quite sensitive to the range of times considered in the data analysis. It is very important to accurately measure F_0 and to analyze spot FRAP data for long times. Because of the exponential behavior, the diffusion coefficient does not suffer as much if relatively short time ranges are sampled with FRAPP data. Still, it is recommended that data be sampled for at least three times the value of the exponential time constant, $\tau = w^2/\pi^2 D$, regressed via Eq. (9).

3. Potential FRAP Artifacts

If the fringe contrast parameter A is assumed to be unity without actually measuring it, the mobile fraction can be underestimated. As shown in Fig. 4, the full extent of the fluorescence recovery, plotted as the "fluorescence relaxation" $[\bar{F} - F(t)]/(\bar{F} - F_0)$, described by Eq. (9) is compressed for any value of A less than unity. Since the extent of the recovery controls the regressed value of the mobile fraction, any unaccounted decrease in A incorrectly decreases the calculated mobile fraction. The contrast parameter should be measured either by digital image analysis or by the oscillating fringe method [28].

A less obvious potential source of artifacts is fluorescence concentration quenching. All of the analysis presented in Section IV.A.1 assumes that the fluorescence intensity is directly proportional to the fluorophore concentration, i.e., that the fluorescence quantum yield is independent of fluorophore concentration. If this proportionality does not hold, fluorescence is no longer a simple reporter of local concentration, and the preceding analysis fails. This problem can occur in practice due to fluorescence concentration quenching. A manifestation of nonradian fluorescence energy transfer, concentration quenching causes the quantum yield to decrease with increasing fluorophore concentrations. It is significant only at very large fluorophore concentrations, but these may be attained easily in adsorbed layers, where intermolecular separation distances are on the order of nanometers. The following discussion of the effect that concentration quenching has on FRAPP measurements, and how to correct for it, is based on Ref. 28. Quenching artifacts are not merely a theoretical possibility. Significant concentration quenching has been demonstrated for adsorbed ribonuclease A labeled with fluorescein.

Analysis of FRAPP data in this undesirable situation requires knowledge of the relative quantum yield, q_r, as a function of fluorophore concentration. The relative quantum yield simply indicates the emission intensity per fluorophore at a given

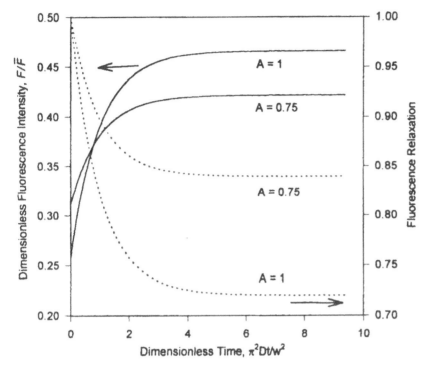

FIG. 4 The fringe contrast parameter has a large effect on the extent of the fluorescence recovery. Solid curves show $F/\bar{F}$, while the dotted curves show the fluorescence relaxation $(\bar{F} - F)/(\bar{F} - F_0)$.

concentration relative to that at infinite dilution and is easily measurable. This quantity is plotted as a function of the two-dimensional concentration of fluorescein labels in the adsorbed ribonuclease A layer in Fig. 5. The key to the data analysis is to incorporate the functional dependence of the relative quantum yield on the local fluorophore concentration, $q_r(C)$, into the integral in Eq. (1), giving

$$F(t) \propto \int_A I(r,\ \theta) C(r,\ \theta,\ t) q_r(C)\ dA \tag{12}$$

Equation (12) can be integrated numerically for the data regression.

Clearly, concentration quenching dictates the (nonlinear) relationship between fluorescence intensity and fluorophore concentration at all times during the fluorescence recovery. If not properly taken into account, it will also alter the determination of the bleaching parameter k from the ratio $F_0/\bar{F}$. This produces a cascade of errors starting from the incorrect initial condition for the diffusion problem. Fortunately, Eq. (12) with the measured $q_r(C)$ dependence is all that is needed to account for quenching at all stages of the analysis.

An examination of the effects of variable relative quantum yields (as measured for fluorescein-labeled ribonuclease A) has indicated that the main consequence of concentration quenching is to invalidate the mobile fraction determination. For pattern photobleaching, the diffusion coefficients extracted from the data are not af-

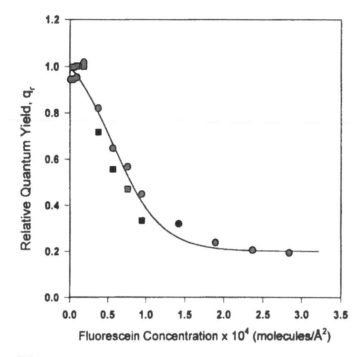

FIG. 5 The relative quantum yield for fluorescein-labeled ribonuclease A (with varying labeling ratios) decays in a sigmoidal fashion with increasing fluorophore concentration on the surface. Circles are for ribonuclease A adsorbed to polystyrene surfaces at 2.2 mg/m²; squares are for adsorption to polydimethylsiloxane surfaces at the same surface concentration.

fected. Figure 6 shows how the mobile fraction can be overestimated if concentration quenching occurs but is not taken into account. This is a plot of the incorrect mobile fraction that would be obtained by the standard data analysis for any given value of the actual mobile fraction. As seen in the figure, higher labeling ratios, i.e., higher average numbers of fluorophores per protein molecule, amplify the errors. Very large mobile fraction overestimates are possible. Since the mobile fraction provides critical information about the adsorbed layer heterogeneity, major qualitative errors can arise from concentration quenching artifacts. At least the diffusion coefficients would not be affected.

Figure 6 is not a "master plot" since relative quantum yields will vary from system to system. Corresponding plots are readily generated via Eq. (12) for any measured $q_r(C)$ behavior. Figure 6 is specific to fluorescein-labeled RNase A adsorption at 2.2 mg/m².

In the case of spot FRAP, concentration quenching is even more troubling. Besides overestimating the mobile fraction, unaccounted concentration quenching can lead to large overestimates of the diffusion coefficient and in certain cases can lead to fluorescence recoveries that disastrously mimic the effect of directed flow superimposed on diffusion.

Concentration quenching will distort fluorescence recovery results regardless of whether a PMT is used to measure the total integrated intensity or whether the changing spatial distribution of fluorescence is imaged microscopically. The tendency

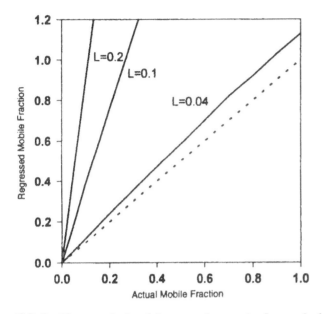

FIG. 6 The magnitude of the error that results from naively regressing FRAPP data under concentration quenching conditions increases with increasing labeling ratio, L.

of concentration quenching to distort the fluorescence microscopy image is illustrated in Fig. 7. The dotted curves show the normal behavior expected in the absence of concentration quenching. The rather distorted curves shown by the solid traces were produced by the $q_r(C)$ relationship for fluorescein-labeled RNase A at 2.2 mg/m^2.

Concentration quenching can cause other surprises in FRAP experiments. Qualitatively, the effect of quenching on fluorescence recovery is as follows. Before photobleaching, the local concentration of active fluorophores is relatively high, thereby depressing the emission per fluorophore (quantum yield). After photobleaching, the local active fluorophore concentration is decreased, relieving some of the quenching and allowing greater emission intensity per fluorophore. Because of this, it is sometimes observed that photobleaching actually increases the fluorescence intensity. This is a sure warning sign of significant concentration quenching. These effects of concentration quenching, originally developed by Robeson and Tilton [28], have been supported by the experiments of Chan et al. [51].

How else can concentration quenching be detected, besides the anomalous fluorescence enhancement by photobleaching? The best way is to measure the fluorescence intensity as a function of the fluorophore/protein labeling ratio at fixed protein surface concentrations. This plot should be linear; negative curvature indicates concentration quenching. Analysis of this plot provides the $q_r(C)$ dependence. Also, if total internal reflection fluorescence (TIRF) is used to monitor adsorption kinetics prior to photobleaching, the fluorescence signal may pass through a gradual maximum in time as the surface concentration reaches sufficiently large values to bring about quenching. This serves as a useful warning to either measure $q_r(C)$ or take steps to eliminate concentration quenching [28]. Smaller proteins are more susceptible to concentration quenching, since they allow smaller distances of closest fluorophore–fluorophore approach, thereby favoring energy transfer.

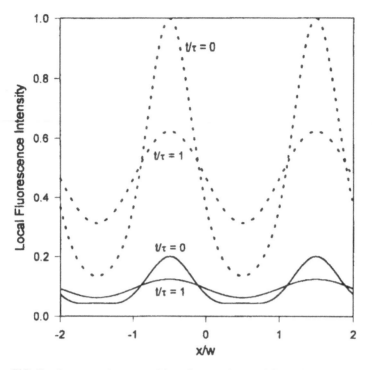

FIG. 7 Concentration quenching distorts the spatial distribution of fluorescence that would be imaged microscopically in a FRAPP experiment. $\tau = w^2/\pi^2 D$.

The best way to prevent the effects of concentration quenching is to only sparingly label the adsorbing biomolecules, so that on average there is significantly less than one label per biomolecule. In this way, even at fixed biomolecule surface concentration, the concentration of fluorophores is comparatively low. The goal is to achieve average interfluorophore distances significantly greater than the Förster energy transfer radius. This is on the order of nanometers but does depend on the particular system in question. Unfortunately, decreasing the labeling ratio may produce unacceptable signal-to-noise ratio problems.

Some uncertainty remains concerning the possibility of concentration quenching artifacts in older FRAP studies because the degree of biomolecule labeling was often not quantified, and of course no measures were taken to inspect for concentration quenching. Unfortunately concentration quenching does not produce an obvious, signature effect on the shape of the fluorescence recovery curves that would alert the reader to potential trouble. While quantitative results of studies that were unfortunately affected by concentration quenching *could* be in question, the qualitative conclusion that at least some fraction of the labeled species was mobile remains valid. Also, diffusion coefficients obtained from pattern photobleaching measurements would remain valid, regardless.

B. Other Techniques

There are other techniques for measuring lateral mobility besides FRAP. For example, autoradiography can be employed to observe the migration of adsorbed proteins from a region of high surface concentration across a bare surface [52]. In addition, the electrophoretic mobility of adsorbed proteins may be measured to estimate the magnitude of the forces holding proteins in place on a surface [53].

Advances in near-field microscopy have recently been made that allow single molecules to be imaged at an interface, either by fluorescence labeling or by colloidal gold labeling [54–56]. Thus single molecules can be tracked over time. Although not necessary for tracking measurements, near-field microscopes have even been used to photobleach single molecules. Using these state-of-the-art single-molecule tracking techniques, one would record the mean-square displacement of single molecules as a function of time and determine the diffusion coefficient from

$$\langle r^2 \rangle_{2-D} = 4Dt \tag{13}$$

Of course, many molecules, or else many steps taken by a single molecule, must be averaged in order to obtain good statistics. (In contrast, FRAP provides an ensemble averaged diffusion coefficient.) At this point, single molecule tracking has not yet been applied to proteins adsorbed on solid surfaces.

V.   WHICH DIFFUSION COEFFICIENT IS MEASURED?

These techniques do not all measure the same type of diffusion coefficient and should not be expected to provide identical results. What kind of diffusion coefficient do they measure? FRAP (or FRAPP) measures the long-time self-diffusion coefficient, or the "tracer" diffusion coefficient. Molecular motions are observed over distances that are large compared to the intermolecular separation, and thus lateral interactions between neighboring molecules are important. Here it is important to keep in mind that the concentration gradient we discussed is only a gradient of labels. The protein concentration is uniform, so there is no relaxation of a protein gradient. Postelectrophoretic relaxation, or any technique that involves the relaxation of an actual protein concentration gradient, provides a mutual diffusion coefficient. Again, migration distances are large and lateral interactions are important, but mutual diffusion coefficients and self-diffusion coefficients respond dramatically differently to intermolecular forces [57–59]. The response to intermolecular forces determines how the diffusion coefficient depends on concentration.

Self-diffusion coefficients decrease for all types of intermolecular interactions, whether attractive or repulsive, and therefore must decrease with increasing concentration. In contrast, mutual diffusion coefficients increase for purely repulsive interactions and decrease for purely attractive interactions. Thus, the mutual diffusion coefficient can either increase or decrease with concentration, depending on the sign of the intermolecular forces. More complex force profiles that display attractive and repulsive interactions over different ranges lead to more complex concentration dependencies. Extrema are possible.

Single-molecule tracking techniques can provide both the short-time and the long-time self-diffusion coefficient. The short-time diffusion coefficient describes the motion of molecules in a region smaller than the intermolecular separation. Such a measurement has perhaps the greatest potential relevance to adsorbed protein conformational dynamics, since protein migration is observed without the influence of lateral interactions.

Perhaps the most important result so far to come from single-molecule tracking (applied to cell membrane components) is that two-dimensional biomolecular transport can be far more complex than is usually assumed when interpreting FRAP data [54,60]. As discussed in Section IV.A.1, one typically assumes a coexistence of immobile species and mobile species that all have the same diffusion coefficient. In contrast to this simple picture, simultaneous occurrence of three distinct transport modes has been observed by single-molecule tracking in cell membranes. Unrestrained Brownian motion (simple self-diffusion), restrained Brownian motion (as can occur when a species is corraled in a segregated membrane microdomain or by cytoskeletal "cages"), and directed flow may all occur, while some fraction may indeed be immobile. If there is a finite probability that a restrained molecule can be released from its cage, this leads to so-called long-tail kinetics or anomalous diffusion [60,61]. The "cage" effect may also refer to momentary binding or immobilization, and anomalous diffusion can result if there is a broad distribution of bound lifetimes. In anomalous diffusion, the mean-square displacement evolves as if the diffusion coefficient were time dependent:

$$\langle r^2 \rangle_{2-D} = 4D(t)t \tag{14}$$

where

$$D(t) = D_0 t^{\alpha-1} \tag{15}$$

The constant α relates to the lifetime of a molecule in a trap. It is unity for simple diffusion. Values less than unity correspond to anomalous diffusion, where the expected second-order dependence of the diffusion time on the length scale fails. Just as the diffusion coefficient is a phenomenological constant that can be traced to the random walk model, α can be traced to the statistics of a random walk with transient constraints. Nagle [61] and Feder et al. [60] have shown that anomalous diffusion provides FRAP data that are difficult to distinguish from the fluorescence recovery produced by normal diffusion with coexisting mobile and immobile species.

Nagle generated simulated spot FRAP data by anomalous diffusion (with 100% mobile species) and then analyzed that data according to the standard spot FRAP analysis. The standard analysis yielded widely varying diffusion coefficients and mobile fractions, depending on the experimental time scale examined. Functional differences between FRAP recoveries produced by anomalous and simple diffusion only became apparent when data were sampled over several decades in time relative to the characteristic recovery time τ_t. Since this may be unfeasible in many FRAP studies, Nagle recommends that care should be taken to examine similar length and time scales when using FRAP in comparative studies of different systems. Certainly, the qualitative conclusions about the biomolecular transport dynamics are much different for the two types of diffusion. An immobile fraction implies traps with infinitely deep potential energy wells. This is considerably different from the case of restrained Brownian motion if "immobile" molecules are in fact highly mobile over small distances (nanometers as probed in single-particle tracking as opposed to micrometers probed in FRAP).

Saxton [62] has used Monte Carlo simulations to examine the influence of immobile obstacles on the type of diffusion. For obstacle concentrations below the percolation threshold, FRAP detects normal Fickian diffusion, but if the obstacle concentration is near the percolation threshold, FRAP may detect anomalous diffusion. According to percolation theory, when more than some critical area fraction of the surface is covered by obstacles, there are no continuous paths available for diffusing molecules to sample all regions of the surface. Above this percolation threshold, the long-time diffusion coefficient becomes zero. The shape of the obstacles is also important; fractal obstacles such as diffusion-limited aggregates produce anomalous diffusion at lower obstacle concentrations. Other factors that might lead to anomalous diffusion, such as transient immobilization, were not considered. Also noteworthy was the observation that if obstacles are the only factor at work, some of them must be absolutely immobile to see anomalous diffusion.

VI. DIRECT MEASUREMENTS OF SURFACE DIFFUSION

It would be nice if the data available in the literature were sufficiently thorough to spawn a predictive theory for surface diffusion coefficients. It is considerably more difficult at this stage to compare protein surface diffusion coefficients than it is to compare bulk diffusion coefficients, owing to the complexity of protein–surface interactions and strong lateral interactions. Add to this the possibility that the layer may contain a distribution of protein mobilities, and it becomes clear that direct comparisons between different systems are not straightforward. At a minimum, if a comparative analysis or predictive theory is the goal, surface diffusion coefficients must be measured at equal surface coverages (area fraction, not mass concentration), or better at a variety of coverages and extrapolated to infinite dilution. Currently, there is just not enough data in the literature to deduce relationships between protein structure, surface chemistry, and lateral mobility. What do exist at this point are data to confirm that surface diffusion of proteins on solid surfaces can be a significant phenomenon in a number of experimental systems.

The first published study of adsorbed protein surface diffusion was carried out by Michaeli et al. in 1980 [52]. In their experiments, half of a glass cover slip was exposed to a pH 7.2, phosphate-buffered saline (PBS) solution of ^{125}I-labeled bovine serum albumin (BSA). After adsorbing BSA for 30 min, they rinsed the cover slip with fresh PBS and then entirely submerged it in protein-free PBS for times ranging from 1 to 16 h. After the prescribed submersion time, the distribution of adsorbed BSA on the surface was imaged by autoradiography. They observed that the protein front migrated in proportion to the square root of time, as expected for diffusion. There was no evidence of protein desorption, suggesting that the migration was due to surface diffusion rather than desorption and subsequent readsorption elsewhere on the surface. Although the BSA surface concentration profile was not quantified, the front migration allowed an order of magnitude estimate for the surface diffusion coefficient. The remarkable finding was that the front migration suggested a diffusion coefficient two orders of magnitude *larger* than that measured for BSA at infinite dilution in solution. The resolution to this puzzle may be that this gradient relaxation experiment examines mutual diffusion. Given the accelerating effect of repulsive interactions on mutual diffusion noted above, the surprisingly rapid front migration probably indicates that adsorbed BSA molecules strongly repel their neighbors.

Shortly afterward, Burghardt and Axelrod [40] published the first quantitative study of protein surface diffusion, in their case BSA on quartz. They used an elliptical spot FRAP method in a total internal reflection configuration to measure simultaneous surface diffusion and exchange between adsorbed and dissolved proteins. The procedure was to photobleach the adsorbed layer while labeled BSA was still present in solution. Thus, two fundamental processes contributed to the fluorescence recovery: surface diffusion of unbleached species from the extremities into the photobleached region and displacement of adsorbed, photobleached species by fluorescent species arriving from the bulk solution. Parameters extracted from the data were the surface diffusion coefficient, the exchange rate constant, and the fraction of proteins existing in strongly versus weakly bound states. The two processes were distinguished by different time constants in the fluorescence recovery. Furthermore, the apparent surface diffusion time constant was sensitive to the characteristic length scale (i.e., the width of the photobleached ellipse), whereas the exchange time constant was length-scale invariant, lending support to the conclusion that the former time constant was indeed influenced by surface diffusion.

Burghardt and Axelrod found that adsorbed BSA existed in at least three distinct states, denoted as irreversible, slowly reversible, and rapidly reversible. The proteins in the rapidly reversible state were the ones capable of surface diffusion, having a best estimate diffusion coefficient of $(5 \pm 1) \times 10^{-9}$ cm^2/s. The distribution of proteins among the three classes was a function of the bulk BSA concentration, with the two reversible classes becoming increasingly important at higher concentrations. In the limit of large concentrations, the protein fractions in the slowly reversible and rapidly reversible states were 0.30 ± 0.02 and 0.23 ± 0.01, respectively. Burghardt and Axelrod concluded from the relative values of the exchange and diffusion time scales that adsorbed BSA was capable of diffusing over distances of at least 1 μm before desorbing.

Chan et al. [51] later measured similar diffusion coefficients for BSA adsorbed to glass surfaces. Using pattern FRAPP, they measured diffusion coefficients of approximately 3×10^{-9} cm^2/s and mobile fractions of 0.20 to 0.28, depending on the bulk BSA concentration.

The pioneering studies of Michaeli et al. and Burghardt and Axelrod clearly demonstrated that adsorbed proteins can indeed be mobile. Since irreversible adsorption is quite common, an important objective remained, namely, to quantify the lateral mobility of proteins irreversibly adsorbed on solid surfaces. Were the mobile proteins only loosely associated with the adsorbed layer (as perhaps implied by the correlation of surface diffusion with rapid reversibility in the case of BSA on quartz) or were even tightly bound proteins capable of surface diffusion? The mobility of tightly bound proteins has more far-reaching consequences for the dynamic nature of the protein–surface interaction.

To answer this question, Tilton et al. [46] employed a FRAPP method based on interference of two coherent beams in total internal reflection at the solid–liquid interface. In this study, BSA was adsorbed to spin-case polymethylmethacrylate (PMMA) films and to cross-linked spin-cast films of polydimethylsiloxane (PDMS). The proteins adsorbed from flowing PBS solutions for 40 min, whereupon the flow cell was flushed with protein-free buffer until the BSA desorption rate was negligible. Finally, the irreversibly adsorbed layer was photobleached. Generally, greater than 90% of the proteins remained adsorbed for many hours, much longer than the duration of any FRAPP measurement.

FRAPP data were interpreted according to the standard model of coexisting mobile and immobile species. Since the potential influence of concentration quenching had not yet been identified, concentration quenching was not examined. It should be noted, though, that the TIRF adsorption data did not display the overshoot that serves as a warning sign for significant concentration quenching. Thus, to the extent that the coexistence of mobile and immobile species is correct, the analysis as conducted would be satisfactory. The key results were the following: When adsorbed to PMMA at a surface concentration of 1.6 mg/m^2, independently measured by a tritium-labeling procedure, the BSA surface diffusion coefficient and mobile fraction were $D = (1.2 \pm 0.3) \times 10^{-9}$ cm^2/s and $f = 0.37 \pm 0.05$. On PDMS at 1.4 mg/m^2, the surface diffusion coefficient and mobile fraction were $D = (2.6 \pm 0.1) \times 10^{-9}$ cm^2/s and $f = 0.28 \pm 0.02$.

To verify that the recovery was due to Fickian surface diffusion, the length-scale dependence of the characteristic fluorescence recovery time was examined. For Fickian diffusion, the characteristic recovery time τ scales as w^2 for FRAPP. Recovery data for BSA on PMMA showed a linear relationship between τ and w^2 for w^2 ranging from 10 to 60 μm^2. Furthermore, the average diffusion coefficient obtained from the slope ($1/\pi^2D$) was 1.1×10^{-9} cm^2/s. This was the same as the average of all diffusion coefficients extracted from each individual experiment.

The second-order dependence of τ on w also ruled out the possibility that rotational diffusion, not translational diffusion, produced the fluorescence recovery. If the photobleaching pulse is short relative to the rotational diffusion time, a polarized laser beam can selectively photobleach fluorophores whose transition dipole moment is aligned with the electric field vector. Then, rotation of unbleached fluorophores into alignment with the electric field vector of the attenuated monitoring laser beam can produce a fluorescence recovery [63]. Rotational diffusion would produce length-scale-invariant fluorescence recovery times, in conflict with the data. This study provided the first evidence that irreversibly adsorbed proteins were mobile and that the transport mechanism was Fickian diffusion.

A. Lateral Interactions: Effect of Coverage

Diffusion coefficients are highly concentration dependent, as has been well documented for diffusion in solution [64–67]. When it comes to surface diffusion, there have been but a few studies of concentration effects on either fluid or solid surfaces. Theoretical efforts to describe concentration-dependent lateral diffusion continue to outpace experimental efforts. Hindered self-diffusion has been demonstrated for cell membrane proteins [9,68] and for antibodies bound to lipid haptens in supported bilayers [3,69,70]. In the case of solid–liquid interfaces, the only data that span a wide concentration range [71] are shown in Fig. 8. When adsorbed irreversibly to PMMA, the BSA long-time self-diffusion coefficient decreased by approximately one order of magnitude as the area fraction coverage (calculated from measured surface concentrations by assuming horizontal orientation) increased from 0.1 to 0.7. Diffusion coefficients in Fig. 8 are normalized by the infinite dilution surface diffusion coefficient $D_0 = (5.6 \pm 0.5) \times 10^{-8}$ cm^2/s, as obtained by extrapolation to zero coverage. Note that the infinite dilution surface diffusion coefficient is an order of magnitude smaller than the infinite dilution bulk diffusion coefficient, 6.7×10^{-7} cm^2/s. While the surface diffusion coefficient depended strongly on the surface concentration, the mobile fraction did not. It remained constant at $f = 0.4 \pm 0.14$ over the full range of surface concentrations.

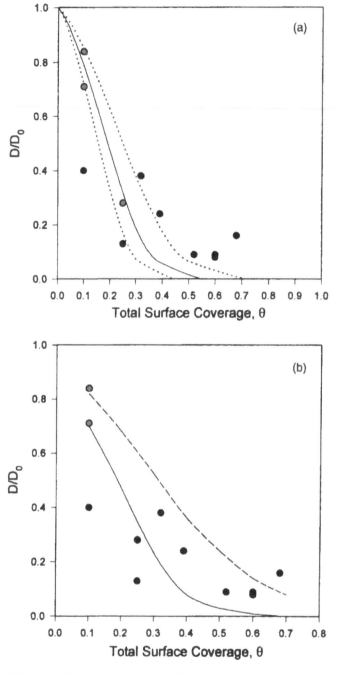

FIG. 8 (a) Data are the surface diffusion coefficients measured for BSA adsorbed to PMMA at 37°C, normalized by the infinite dilution surface diffusion coefficient, D_0. The solid curve is the Saxton continuum model. Dotted curves show the range of uncertainty in the model prediction due to the experimental uncertainty in the mobile fraction. (b) The same data are plotted with Saxton's lattice simulation predictions. The dashed curve is for point particles; the solid curve is for finite-sized particles.

Chan et al. [51] examined the surface diffusion of DNA oligonucleotides adsorbed to methylated glass surfaces under conditions that would not cause concentration-quenching artifacts. They found that as the surface coverage increased from approximately 10 to 30% of a monolayer the diffusion coefficient decreased from approximately 2.2×10^{-9} to 0.5×10^{-9} cm^2/s. The adsorbed oligonucleotide mobile fraction did not depend on the coverage. It remained constant at approximately 0.28. Gaspers et al. [72] also found that the mobile fraction of collagenase on immobilized peptide surfaces was constant while the diffusion coefficient decreased for two surface coverages. These trends are similar to the behavior of BSA or PMMA.

Biophysical models for hindered diffusion often target interacting membrane components, but none of the model assumptions would make them inappropriate for interpreting surface diffusion at solid–liquid interfaces. Most current models do not consider hydrodynamic or specific interactions, and even hard-disk interactions are sufficient to produce strongly concentration-dependent diffusion coefficients. Besides lateral interactions, the coexistence of mobile and immobile species, or other distributions of mobilities, are often taken into account in these models.

The strong concentration dependence of the BSA surface diffusion coefficient can be fairly well described by a modification of Saxton's [73] model for two-dimensional tracer diffusion in an archipelago of immobile obstacles. This model is based in part on continuum percolation theory, which predicts a zero long-time self-diffusion coefficient above the percolation threshold. Two-dimensional continuum percolation theory provides convenient scaling relationships between the long-time tracer diffusion coefficient and the area fraction occupied by obstacles. These relationships fail at low obstacle concentrations, where effective medium approaches are more successful. To model long-time diffusion coefficients over a broad range of obstacle concentrations, Saxton thus employed percolation theory for obstacle concentrations within a critical range of the percolation threshold and effective medium theory near zero obstacle concentration. A cubic polynomial interpolation connected these regimes. This model describes diffusion of noninteracting tracer particles, so the only coverage relevant to the model calculations is that due to immobile obstacles rather than the total coverage of obstacles plus tracers.

To compare the BSA/PMMA data to this model, the immobile proteins may be considered as obstacles. Thus, the area fraction of obstacles, θ_{obs}, is $(1 - f)\theta$, where θ is the total BSA surface coverage. Wherever Saxton's model equations call for the obstacle area fraction, θ_{obs} is inserted to generate the model predictions, but the normalized diffusion coefficients in Fig. 8 are plotted against θ, the total BSA area fraction. In Fig. 8a, the model and data are fairly consistent. The order of magnitude decrease in D is well captured. The most pronounced discrepancy is the absence of a percolation threshold in the data. This model only considered diffusion of noninteracting tracer particles. Only immobile obstacles hinder their diffusion. Since 40% of the adsorbed BSA molecules were mobile, interactions among the diffusing species might be expected to be important as well. Allowing the extreme possibility that all adsorbed proteins, whether mobile or immobile, are equally effective as obstacles, then the diffusion coefficient should drop to zero at a total area fraction of only 0.332, the two-dimensional percolation threshold for obstacles.

To better examine the effects of both mobile and immobile obstacles, Saxton performed a series of Monte Carlo simulations of random walks on a lattice [74], where interactions among the diffusing particles were allowed. In the simulation, area fractions correspond to the fraction of occupied lattice sites. The simulation results (read from Fig. 1 in Ref. 74) are compared to the BSA/PMMA data in Fig. 8b. To use the simulation results, one must know both the total area fraction of all species and the area fraction of immobile obstacles. As above, the latter is taken to be $\theta_{obs} = (1 - f)\theta$. The figure is a plot of the normalized diffusion coefficient versus the total area fraction, θ.

There are two curves in Fig. 8b. The dashed curve is the model prediction based on interacting point particles on the lattice. The solid curve is for finite-size particles (i.e., particles larger than a single lattice site). Although the percolation threshold appears to shift to higher coverages in this figure, this is an artifact arising from the difference between lattice and continuum percolation thresholds. Saxton's models successfully capture the magnitude of the concentration dependence of the BSA surface diffusion coefficient. The apparent absence of a percolation threshold in the data could possibly be evidence for anomalous diffusion rather than true coexistence of mobile and immobile proteins, although the Fickian character of the fluorescence recovery would argue otherwise.

B.   Origin of the Immobile Fraction

The origin of the coexistence of mobile and immobile proteins (or perhaps the co-existence of freely diffusing proteins and some undergoing constrained or anomalous diffusion) is uncertain. It may reflect a distribution of conformational states in the adsorbed layer. Another possibility is that protein aggregates coexist with mono-merically dispersed proteins on the surface. Those proteins in aggregates would display greatly hindered diffusion, perhaps appearing immobile over the time scale of the typical FRAP experiment. To be consistent with the data in Ref. 71, the immobile proteins would necessarily have a diffusion coefficient at least two orders of magnitude smaller than the mobile proteins. The surface aggregation model has been put forward by Wright et al. [3] to explain the lateral mobility of monoclonal antibodies bound to hapten-conjugated lipids in a supported bilayer. The FRAP data were consistent with two populations. One population had a diffusion coefficient of 3.5×10^{-9} cm^2/s, independent of concentration, while the diffusion coefficient of the less mobile population was concentration dependent, decreasing from 1.5×10^{-9} to 0.25×10^{-9} cm^2/s as the lateral density of antibodies increased to approximately the density associated with the crystalline state. The fraction of antibodies in the slowly mobile population increased from 0 to approximately 0.7 as the concentration increased, consistent with concentration-dependent aggregation. This type of aggregation is presumably caused by the antigen-mediated cross-linking of bifunctional antibodies.

What can be learned from the observation that the apparent mobile fraction for adsorbed BSA is independent of surface concentration? First note that this is not a spurious observation. Recall that Chan et al. [51] and Gaspers et al. [72] made similar observations for adsorbed oligonucleotides and collagenase, respectively. Tamm [69] has also observed this phenomenon for monoclonal antibodies bound to lipid haptens in supported phospholipid bilayers. The antibody mobile fraction was constant over a wide range of surface concentrations, provided the antibody–hapten conjugates were uniformly distributed over the bilayer surface. At lower temperatures where phase separation occurred, antibodies were visibly aggregated and immobile.

While aggregation can cause immobilization, it appears not to be responsible for the adsorbed BSA behavior. Simple aggregation models, such as the isodesmic aggregation model [75], indicate that the fraction of molecules in aggregates must increase with increasing concentration. For this model [76], the fraction of molecules bound in aggregates of size i is

$$\phi(i) = \frac{iK^{i-1}}{P_t} \left(\frac{2KP_t + 1 - \sqrt{4KP_t + 1}}{2K^2P_t} \right)^i \tag{16}$$

where K is the equilibrium association constant and P_t is the total concentration of molecules. While the choice of equilibrium constant is arbitrary in Fig. 9, it shows that the fraction of monomeric, i.e., mobile, proteins is a monotonically decreasing function of surface concentration. Aggregation would therefore produce a mobile fraction that decreased with increasing surface coverage. The tentative conclusion for BSA is that the coexistence of mobile and immobile proteins results from a distribution of conformations or orientations on the surface rather than aggregation.

What if the mobile fraction has nothing to do with the protein itself but is the result of a heterogeneity in some surface property? If 40% of the adsorbed proteins are mobile, could 60% of the adsorbed proteins have wandered into irreversible traps on the surface? This is not likely. If 60% of the surface were covered by traps approximately the same size as a protein, the distance between them would be so small that the proteins would diffuse into a trap within seconds of adsorbing. At low

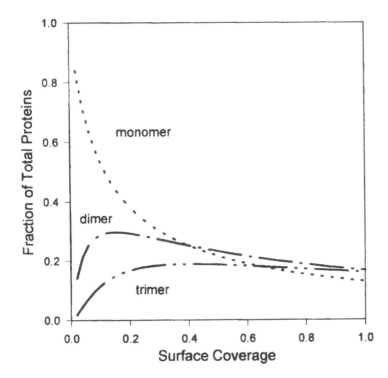

FIG. 9 Illustration of the isodesmic aggregation model. The fraction of proteins existing as monomers must decrease as the total surface concentration increases.

protein surface coverages, where the density of traps exceeds the density of proteins, all proteins should become immobilized. Even at fairly high coverages, such as the 51% jamming limit, all of the proteins should have found a trap. The explanation that remains most consistent with the data, or at least that has not yet faced a counterargument, is that the mobile fraction reflects a heterogeneity of protein conformation or orientation.

C. Effect of Adsorption Time

The time dependence of adsorbed protein lateral mobility has received little attention. The only observation made to date is that the surface diffusion coefficient and mobile fraction of BSA adsorbed at constant surface coverage on PMMA were both independent of adsorption time between 15 min and 7 h [76]. Whatever heterogeneity existed in the adsorbed BSA layer evidently evolved in less than 15 min.

VII. SURFACE DIFFUSION MECHANISM

How do proteins diffuse on a surface? The first step toward answering this question is to attempt to distinguish between true surface diffusion and hopping diffusion (desorption/bulk diffusion/readsorption). The best evidence for true surface diffusion is that irreversibly adsorbed proteins are mobile. In the case of BSA for example, 40% of the adsorbed molecules can diffuse over length scales on the order of 10 μm in tens of seconds, whereas hours to days are required for 40% of the layer to desorb. Nevertheless, one might argue that standard methods for measuring desorption kinetics, e.g., TIRF, reflectometry, or ellipsometry, only detect desorption when desorbed molecules diffuse away from the interface into the bulk. Molecules that manage to break all segmental attachments to the surface but diffuse only a short distance away before readsorbing will appear to have been adsorbed the whole time.

This type of hopping, where diffusion actually occurs in solution with the corresponding bulk diffusion coefficient, may also lead to anomalous diffusion [77]. Since BSA/PMMA fluorescence recoveries obeyed Fickian expectations, hopping diffusion would appear not to have occurred. Further insight into the question of hopping versus true surface diffusion is obtained by comparing surface diffusion coefficients measured in the presence or absence of unlabeled proteins in solution. Without unlabeled protein in solution, desorption leads to "vacancies" which will be filled by either a photobleached molecule or a fluorescent molecule that arrives (by any mechanism) from elsewhere on the surface. With unlabeled proteins in solution, vacancies created by proteins beginning an excursion away from the surface have a greater chance of being filled by a nonfluorescent (either photobleached or unlabeled) protein than in the former case, since the bulk solution is now an additional source for nonfluorescent molecules. In the BSA/PMMA system, FRAPP results for these two cases were indistinguishable, again suggesting the occurrence of true surface diffusion.

Effects of fluid shear stresses on adsorbed protein transport have also been examined for the BSA/PMMA system. Here the use of periodic pattern photobleaching to render the diffusion problem one dimensional is helpful. By alternately orienting the fringes parallel or perpendicular to the direction of shear in a slit flow cell, it is possible to probe anisotropic transport rates. This was done for the BSA/ PMMA system at a wall shear rate $\dot{\gamma}$ of 100 s^{-1}, the highest shear rate attainable with the particular flow system used. There was no anisotropic transport. Diffusion coefficients extracted from the FRAPP data were indistinguishable for either shear orientation. Can this offer any help in distinguishing between hopping and true surface diffusion? In general, perhaps, but in this case the answer is, unfortunately, "no." If we assume that a protein must diffuse by its own length L away from a surface before bulk diffusing to a new location, we can estimate the relative time scales for convection and diffusion over the characteristic length provided by the fringe period:

$$\frac{\tau_{\text{diff}}}{\tau_{\text{conv}}} = \frac{\dot{\gamma}Lw}{2D_{\text{bulk}}} \qquad (17)$$

The length of BSA is of order 10 nm, while the fringes were of order 10 to 20 μm. For the BSA bulk diffusion coefficient of 6.7 $\times$ 10^{-7} cm^2/s, the diffusion time is approximately 5 to 10% of the convection time. Although convection may therefore be expected to have a small but possibly significant effect, a 5 to 10% difference in FRAPP recovery times for the alternating directions would unfortunately be within the experimental error bars on D.

Rabe and Tilton [78] measured the activation energy for BSA surface diffusion on polyhexylmethacrylate (PHMA) at approximately 50% surface coverage over the temperature range of 10 to 30°C. The activation energy was 1.2 $\times$ 10^{-19} J, or approximately 30 kT per molecule. Although this activation energy lumps the effect of temperature on viscosity and cannot separate the effects of lateral interactions from protein–surface interactions, it suggests the involvement of many segment–surface interactions, if each were to contribute approximately 1 kT adsorption energy.

There is an interesting interplay between protein surface diffusion and adsorbed enzyme activity. Gaspers et al. [72] measured the surface diffusion coefficient, mobile fraction (Table 1), and catalysis kinetics for collagenase, a nonautolytic protease, irreversibly adsorbed to modified glass surfaces covered by covalently bound, uncharged peptides known to be cleaved by collagenase. The kinetics of the collagenase-catalyzed peptide cleavage were described by a two-dimensional Smoluchowski-type reactive collision model [79]. The reaction kinetics were found to be under mixed kinetic and transport (surface diffusion) control.

What is the nature of the diffusion process in this reactive system? One possible surface diffusion mechanism could entail collagenase binding to the peptide surface via its active site, followed by cleavage of the bound peptide and the enzyme's hopping to a new site. Since this mechanism should produce anomalous diffusion, it cannot be entirely correct. Collagenase transport was again consistent with Fickian diffusion, as the FRAPP characteristic time scaled with the square of the characteristic length. Furthermore, when inactivated by the presence of the calcium chelator EDTA in the buffer, collagenase had a larger diffusion coefficient than it had in its active state. This was true at two different surface concentrations. Mobile fractions were 80 to 90% in all cases. Thus, processes other than reaction must contribute to the dynamics. Evidently, collagenase surface diffusion involves the dynamics of multiple segmental interactions with the surface, not just those of the active site. The decreased diffusion coefficient measured for the active enzyme may reflect the high binding energy for peptides in the active site. This may make a large contribution to the overall activation energy for diffusion. The effects of the divalent Ca^{2+} cation on surface and intermolecular forces might also be considered, especially if the underlying glass surface created residual negative surface charge. Then the role of EDTA might be quite subtle.

Interestingly, collagenase was entirely immobile in the presence of cysteine, a noncompetitive inhibitor. Cysteine inhibits proteolysis by blocking the active site. The cause of the complete immobilization is unknown but may involve a conformational change brought about by the strong binding of cysteine to the active site.

A. Effect of Surface Motions

Surface diffusion requires the dislocation of some or all of the protein's segmental surface contacts. Although a rolling sort of motion is possible, it is likely that the thermal fluctuations of the polypeptide chain allow the protein to randomly crawl on the surface. In a similar manner, the mobility of the underlying surface can play an important role in the mechanism of surface diffusion. The obvious example where surface mobility is very important for diffusion is the lateral mobility of proteins bound to lipid bilayers. There, the lipid surface with which the protein interacts is fluid, and the individual lipids are free to diffuse over large distances during the time of a typical FRAP measurement. Polymeric materials above their glass transition temperature T_g are somewhat fluid, although the constraints on individual chains severely hinder their diffusion over length and time scales of relevance to FRAP experiments. Nevertheless, thermal motions of polymeric surfaces have been implicated in the mechanism for adsorbed protein surface diffusion.

The evidence for this is the pronounced increase in the BSA surface diffusion coefficient on polybutylmethacrylate (PBMA) films as the temperature passes through the PBMA glass transition at 20°C [78], as shown in Fig. 10. In contrast to PBMA, the surface diffusion coefficient for BSA on PHMA shows only an Arrhenius temperature dependence, with nothing remarkable occurring at 20°C. In terms of surface energetic properties, PBMA and PHMA are indistinguishable by standard capillarity methods. They are only distinguished by their glass transition (T_g for PHMA is −5°C), so PHMA serves as a transition-free baseline for comparison. The difference between PBMA and PHMA at 20°C implicates PBMA chain motions in the BSA surface diffusion mechanism, although the data clearly show that polymer chain motions are by no means entirely responsible for the protein diffusion.

B. Comparison of Adsorbed Proteins

It is extremely difficult to compare results from different surface diffusion studies, partly because of the strong effect of surface concentration on long-time self-diffusion coefficients. Another difficulty is that there simply are not enough independent studies to adequately distill from the existing data the effects of protein conformational flexibility, molecular weight, and hydrophobicity; the effects of surface flexibility, charge, and hydrophobicity; or the effects of lateral interactions. Available data for proteins adsorbed on solid surfaces are presented in Table 1. (This table contains some of the diffusion data from Ref. 80 but excludes data for adsorbed proteins in alcohol/water mixtures.) Buffer compositions in Table 1 may vary, but all are be-

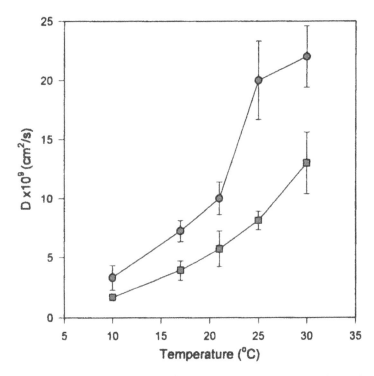

FIG. 10   The diffusion coefficient of BSA increases discontinuously on PBMA (circles) as the temperature exceeds its glass transition temperature. On PHMA (squares), the diffusion coefficient increases smoothly at all temperatures. Error bars are 98% confidence limits based on 8 to 10 repeated measurements at each temperature.

tween pH 7 and 8 and have ionic strengths of at least 50 mM. Some physical properties of the proteins are presented in Table 2.

Surface diffusion coefficients of mobile proteins are on the order of 10^{-10} to 10^{-7} cm^2/s, consistently smaller than the bulk infinite dilution diffusion coefficient. The largest protein surface diffusion coefficient is 1.7×10^{-7} cm^2/s, measured for ribonuclease A at low coverage on polystyrene. RNase A is a small, rigid protein [81] of moderate hydrophobicity [82], and polystyrene is rigid. The smallest values are reported for the larger proteins. Ferritin and collagenase surface diffusion coefficients are on the order of 10^{-10} cm^2/s at low or moderate coverages, although collagenase has the largest measured mobile fractions. Perhaps this relates to the mobility of the underlying peptide surface. Ferritin is an extremely large protein. Fibrinogen, also an extremely large protein, is completely immobile on silica and hydrophobized silica. Ferritin diffusion may be facilitated by its rigid spherical structure. Compared to the flexible, trinodular fibrinogen structure, the rigid sphere would have fewer segmental contacts with the surface. Because of the many factors controlling surface diffusion, a more thorough examination of the specific molecular properties that favor surface diffusion is currently unwarranted. It is encouraging that reproducibility is fairly good between labs that have examined similar systems.

Only comparisons of infinite dilution surface diffusion coefficients would provide viable insight into the specifics of individual protein–surface interactions, but

TABLE 1 Summary of Available Protein Surface Diffusion Data on Solid Surfaces

Protein (label)	Surface	T (°C)	Surface concentration (mg/m^2)	D (cm^2/s)	f	Ref.
BSA (rhodamine)	quartz	NA	5.6 to 22[a]	$(5 \pm 1) \times 10^{-9}$	0.23 ± 0.01	40
BSA (eosin)	PMMA	37	infinite dilution	$(5.6 \pm 0.5) \times 10^{-8}$	0.4 ± 0.14	71
BSA (eosin)	PMMA	37	1.6	$(1.2 \pm 0.3) \times 10^{-9}$	0.37 ± 0.05	71
BSA (fluorescein)	polystyrene	25	1.5	$(8 \pm 2) \times 10^{-9}$	0.28 ± 0.03	28
BSA (fluorescein)	glass	NA	3.3	$(3.3 \pm 0.2) \times 10^{-9}$	0.28 ± 0.02	51
BSA (fluorescein)	APTES glass	NA	NA	$(1.8 \pm 0.4) \times 10^{-9}$	0.25 ± 0.03	51
BSA (fluorescein)	MAPTES glass	NA	NA	$(4.9 \pm 0.6) \times 10^{-9}$	0.19 ± 0.04	51
BSA (fluroescein)	PBMA	21	1.5	$(1 \pm 0.14) \times 10^{-8}$	$> 0.09^{b}$	78
BSA (fluroescein)	PHMA	21	1.5	$(5.7 \pm 1.5) \times 10^{-9}$	> 0.08	78
Ribonuclease A (eosin)	polystyrene	30	0.5	$(1.7 \pm 0.6) \times 10^{-7}$	0.44 ± 0.04 $(\sim 0.15^{c})$	80
Ribonuclease A (fluorescein)	polystyrene	25	2.2	$(4.9 \pm 0.7) \times 10^{-8}$	0.14 ± 0.04	28
Ferritin (fluorescein)	silica	NA	NA	6×10^{-10}	NA	32

Ferritin (fluorescein)	methylated silica	NA	NA	3×10^{-10}	NA	32
Fibrinogen (fluorescein)	silica and methylated silica	NA	NA	immobile	(0)	32
Collagenase (eosin) inactive	FALGPA-glass[d]	NA	0.9	$(9.8 \pm 0.6) \times 10^{-10}$	0.80 ± 0.05	72
Collagenase (eosin) active	FALGPA-glass	NA	0.9	$(4.1 \pm 0.8) \times 10^{-10}$	0.91 ± 0.08	72
Collagenase (eosin) inactive	FALGPA-glass	NA	2.0	$(3.2 \pm 0.5) \times 10^{-10}$	0.91 ± 0.08	72
Collagenase (eosin) active	FALGPA-glass	NA	2.0	$(2.1 \pm 0.8) \times 10^{-10}$	0.82 ± 0.08	72
Collagenase (eosin) cysteine-inactivated	FALGPA-glass	NA	0.9	immobile	(0)	72

[a] This large surface concentration may be in error due to a fluorescence calibration procedure that assumed equal quantum yields for adsorbed tetramethylrhodamine and for adsorbed tetramethylrhodamine-labeled BSA. Later studies have shown that such calibration procedures may be invalid. Nevertheless, calibration errors cancel in the analysis of FRAP data (provided of course that concentration quenching is not in effect), so the conclusions regarding surface diffusion and exchange would not be affected.

[b] Only lower limits could be reported due to experimental difficulty recording F_0 on the particular spot FRAP instrument used in that study. Relative value of diffusion coefficients thus are more reliable than absolute values. Note that a range of temperatures was examined in this study.

[c] The methods developed in Ref. 28 indicate that concentration quenching affected the mobile fraction in this study. The corrected value in parentheses assumes that eosin and fluorescein have similar Förster radii for fluorescence energy transfer.

[d] FALGPA, a peptide cleaved by collagenase, was covalently immobilized on glass.

TABLE 2 Protein Properties

Protein	Molecular weight	$D^0_{2,1} \times 10^7$ (cm²/s)	Hydrophobicity $H\phi$ (cal)	Adiabatic compressibility $10^{12}\ \tilde{\beta}_s$ (cm²/dyn)	Comments
BSA	67,000	6.7	1120	10.5	flexible
RNase A (bovine pancreas)	13,700	12.6	870	1.12	rigid
Ferritin (horse spleen)	747,000 (20 wt % iron)	6.1 (Stokes-Einstein estimate)			heat-stable, 35 Å radius sphere
Fibrinogen (human)	340,000	2.0			trinodular structure
Collagenase (functionally defined zinc proteases from *C. histolyticum*)	SDS-PAGE bands *major*: 106,000 *minor*: 79,000 71,000				

only one such value exists in the literature. The BSA infinite dilution surface diffusion coefficient is one order of magnitude smaller than the corresponding bulk diffusion coefficient.

VIII. SUMMARY AND RECOMMENDATIONS FOR FUTURE RESEARCH

A significant fraction of proteins that are irreversibly adsorbed to solid surfaces may be mobile over large distances within the plane of adsorption. When examined, the lateral transport mechanism has been consistent with Fickian diffusion with a long-time self-diffusion coefficient approximately one or more orders of magnitude less than the bulk diffusion coefficient. To date, no protein/surface combination has been observed that displays 100% mobility. In all cases where adsorbed proteins were mobile, mobile and apparently immobile proteins shared the surface. While 100% mobility has yet to be observed, some systems display zero apparent mobility (translated as having a diffusion coefficient less than 10^{-11} to 10^{-12} cm²/s). The reason for the coexistence of mobile and immobile proteins is not yet established, although it likely reflects a distribution of orientations or altered conformations on the surface.

The surface diffusion mechanism probably involves the thermal fluctuations of the polypeptide backbone that allow the molecule to crawl on a surface without desorbing. In at least one case, the diffusion mechanism involved a coupling between the thermal motions of the protein and the surface to which it was adsorbed. Because intermolecular distances in adsorbed layers of any appreciable surface concentration are rather small, lateral interactions have a great influence on the value of the surface diffusion coefficient, causing as much as an order of magnitude decrease as the coverage increases from the infinite dilution limit to monolayer coverage. This is consistent with current models of hindered diffusion.

Whereas basic consideration of hydrodynamics often can provide a good prediction of bulk diffusion coefficients at infinite dilution, we are not even close to such a predictive capability for protein surface diffusion coefficients. There is still a need for more measurements under carefully controlled and well-defined conditions. For adequate comparisons, data should be acquired at more than one surface coverage and extrapolated to infinite dilution. This isolates the effects of lateral interactions from those of protein–surface interactions. Furthermore, detailed comparisons of self- and mutual diffusion coefficients would highlight the effects of lateral interactions. Whereas both should extrapolate to the same infinite dilution diffusion coefficient, the sensitivity to lateral interactions is quantitatively and qualitatively different. This has been attempted using an imaging reflectometry technique, but the lateral resolution in mapping concentration gradients was insufficient for reliable diffusion analysis [83].

Finally, careful attention must be paid to avoid potential pitfalls in the interpretation of FRAP data, particularly the errors that arise from undetected concentration quenching. In that light, a comparison between FRAP measurements and single-particle tracking measurements on adsorbed proteins would be quite useful. Besides helping to uncover any as yet undiscovered FRAP artifacts, single-particle tracking has the capability to directly observe anomalous diffusion, should it occur.

While this chapter has expressed many concerns and uncertainties regarding the interpretation of FRAP data and the adsorbed biomolecule transport mechanism, it is clear that irreversibly adsorbed proteins and oligonucleotides can be mobile. The consequences of lateral mobility are numerous. Accordingly, the possibility of lateral mobility should be considered as a central molecular level feature of adsorbed layers, no less important than the molecular conformation, orientation, or organization.

REFERENCES

1. R. J. Cherry, Rotational and lateral diffusion of membrane proteins. Biochim. Biophys. Acta 559:289–327, 1979.
2. D. E. Golan and W. Veatch, Lateral mobility of band 3 in the human erythrocyte membrane studied by fluorescence photobleaching recovery: evidence for control by cytoskeletal interactions. Proc. Natl. Acad. Sci. USA 77:2537–2541, 1980.
3. L. L. Wright, A. G. Palmer III, and N. L. Thompson, Inhomogeneous translational diffusion of monoclonal antibodies on phospholipid Langmuir–Blodgett films. Biophys. J. 54:463–470, 1988.
4. W. L. C. Vaz, K. Jacobson, E.-S. Wu, and Z. Derzko, Lateral mobility of an amphipathic apolipoprotein, ApocC-III, bound to phosphatidylcholine bilayers with and without cholesterol. Proc. Natl. Acad. Sci. USA 76:5645–5649, 1979.
5. C. A. Eldridge, E. L. Elson, and W. W. Webb, Fluorescence photobleaching recovery measurements of surface lateral mobilities on normal and SV40-transformed mouse fibroblasts. Biochemistry 19:2075–2079, 1980.

6.   J. H. Hochman, M. Schindler, J. G. Lee, and S. Ferguson-Miller, Lateral mobility of cytochrome c on intact mitochondrial membranes as determined by fluorescence redistribution after photobleaching. Proc. Natl. Acad. Sci. USA *79*:6866–6870, 1982.

7.   J. M. Pachence, S. Amador, G. Maniara, J. Vanderkooi, P. L. Dutton, and J. K. Blasie, Orientation and lateral mobility of cytochrome c on the surface of ultrathin lipid multilayer films. Biophys. J. *58*:379–389, 1990.

8.   P. G. Saffman and M. Delbrück, Brownian motion in biological membranes. Proc. Natl. Acad. Sci. USA *72*:3111–3113, 1975.

9.   R. Peters and R. J. Cherry, Lateral and rotational diffusion of bacteriorhodopsin in lipid bilayers: experimental test of the Saffman–Delbrück equations. Proc. Natl. Acad. Sci. USA *79*:4317–4321, 1982.

10. L. M. Smith, J. W. Parce, B. A. Smith, and H. M. McConnell, Antibodies bound to lipid haptens in model membranes diffuse as rapidly as the lipids themselves. Proc. Natl. Acad. Sci. USA *76*:4177–4179, 1979.

11. J. Feder, Random sequential adsorption. J. Theor. Biol. *87*:237–254, 1980.

12. P. Schaaf and J. Talbot, Surface exclusion effects in adsorption processes. J. Chem. Phys. *91*:4401–4409, 1989.

13. R. D. Vigil and R. M. Ziff, Random sequential adsorption of unoriented rectangles onto a plane. J. Chem. Phys. *91*:2599–2602, 1989.

14. S. A. Darst, M. Ahlers, P. H. Meller, E. W. Kubalek, R. Blankenburg, H. O. Ribi, H. Ringsdorf, and R. D. Kornberg, Two-dimensional crystals of streptavidin on biotinylated lipid layers and their interactions with biotinylated macromolecules. Biophys. J. *59*:387–396, 1991.

15. K. Yase and T. Udaka, Two-dimensional crystallization of ferritin molecules at the air–water interface. J. Cryst. Growth *166*:946–951, 1996.

16. G. Adam and M. Delbrück, In *Structural Chemistry and Molecular Biology* (A. Rich, N. Davidson, eds.), Freeman, San Francisco, 1968, pp. 198–215.

17. H. C. Berg and E. M. Purcell, Physics of chemoreception. Biophys. J. *20*:193–219, 1977.

18. D. Axelrod and M. D. Wang, Reduction of dimensionality kinetics at reaction limited cell surface receptors. Biophys. J. *66*:588–600, 1994.

19. E. Droz, M. Taborelli, P. Descouts, and T. N. C. Wells, Influence of surface and protein modification on immunoglobulin G adsorption observed by scanning force microscopy. Biophys. J. *67*:1316–1323, 1994.

20. G. J. Leggett, C. J. Roberts, P. M. Williams, M. C. Davies, D. E. Jackson, and S. J. B. Tendler, Approaches to the immobilization of proteins at surfaces for analysis by scanning tunneling microscopy. Langmuir *9*:2356–2362, 1993.

21. T. Arai and W. Norde, The behavior of some model proteins at solid–liquid interfaces. 1. Adsorption from single protein solutions. Colloids Surf. *51*:1–15, 1990.

22. A. Kondo and K. Higashitani, Adsorption of model proteins with wide variation in molecular properties on colloidal particles. J. Colloid Interface Sci. *150*:344–351, 1992.

23. P. M. Claesson, E. Blomberg, J. C. Fröberg, T. Nylander, and T. Arnebrant, Protein interactions at solid surfaces. Adv. Colloid Interface Sci. *57*:161–228, 1995.

24. E. Blomberg, P. M. Claesson, and R. D. Tilton, Short-range interaction between adsorbed layers of human serum albumin. J. Colloid Interface Sci. *166*:427–436, 1994.

25. W. Norde, Adsorption of proteins from solution at the solid–liquid interface. Adv. Colloid Interface Sci. *25*:267–340, 1986.

26. W. Norde and J. Lyklema, The adsorption of human plasma albumin and bovine pancreas ribonuclease at negatively charged polystyrene surfaces. 1. Adsorption isotherms. Effects of charge, ionic strength and temperature. J. Colloid Interface Sci. *66*:257–265, 1978.

27. J. L. Brash and D. J. Lyman, Adsorption of plasma proteins in solution to uncharged hydrophobic polymer surfaces. J. Biomed Mater Res. *3*:175–189, 1969.

28. J. L. Robeson and R. D. Tilton, Effect of concentration quenching on fluorescence recovery after photobleaching measurements. Biophys. J. *68*:2145–2155, 1995.

29. J. L. Robeson and R. D. Tilton, Spontaneous reconfiguration of adsorbed lysozyme layers observed by total internal reflection fluorescence with a pH-sensitive fluorophore. Langmuir *12*:6104–6113, 1996.

30. J. J. Ramsden, G. I. Bachmanova, and A. I. Archakov, Kinetic evidence for protein clustering at a surface. Phys. Rev. E *50*:5072–5076, 1994.

31. H. Nygren, Nonlinear kinetics of ferritin adsorption. Biophys. J. *65*:1508–1512, 1993.

32. H. Nygren, S. Alaeddin, I. Lundström, and K.-E. Magnusson, Effect of surface wettability on protein adsorption and lateral diffusion. Analysis of data and a statistical model. Biophys. Chem. *49*:263–272, 1994.

33. M. Stenberg and H. Nygren, Computer simulation of surface-induced aggregation of ferritin. Biophys. Chem. *41*:131–141, 1991.

34. K. Park, S. J. Gerndt, and H. Park, Patchwise adsorption of fibrinogen on glass surfaces and its implication in platelet adhesion. J. Colloid Interface Sci. *125*:702–711, 1988.

35. L. Haggerty and A. M. Lenhoff, Analysis of ordered arrays of adsorbed lysozyme by scanning tunneling microscopy. Biophys. J. *64*:886–895, 1993.

36. C.-S. Lee and G. Belfort, Changing activity of ribonuclease A during adsorption: a molecular explanation. Proc. Natl. Acad. Sci. USA *86*:8392–8396, 1989.

37. E. Blomberg, P. M. Claesson, J. C. Fröberg, and R. D. Tilton, Interaction between adsorbed layers of lysozyme studied with the surface force technique. Langmuir *10*: 2325–2334, 1994.

38. E. L. Elson, Fluorescence correlation spectroscopy and photobleaching recovery. Ann. Rev. Phys. Chem. *36*:379–406, 1985.

39. N. L. Thompson, T. P. Burghardt, and D. Axelrod, Measuring surface dynamics of biomolecules by total internal reflection fluorescence with photobleaching recovery or correlation spectroscopy. Biophys. J. *33*:435–454, 1981.

40. T. P. Burghardt and D. Axelrod, Total internal reflection/fluorescence photobleaching recovery study of serum albumin adsorption dynamics. Biophys. J. *33*:455–467, 1981.

41. R. M. Weis, K. Balakrishnan, B. A. Smith, and H. M. McConnell, Stimulation of fluorescence in a small contact region between rat basophil leukemia cells and planar lipid membrane targets by coherent evanescent radiation. J. Biol. Chem. *257*:6440–6445, 1982.

42. B. A. Smith, Measurement of diffusion in polymer films by fluorescence redistribution after pattern photobleaching. Macromolecules *15*:468–472, 1982.

43. F. Lanni and B. R. Ware, Modulation detection of fluorescence photobleaching recovery. Rev. Sci. Instrum. *53*:905–908, 1982.

44. J. R. Abney, B. A. Scalettar, and N. L. Thompson, Evanescent interference patterns for fluorescence microscopy. Biophys. J. *61*:542–552, 1992.

45. D. Axelrod, D. E. Koppel, J. Schlessinger, E. Elson, and W. W. Webb, Mobility measurement by analysis of fluorescence photobleaching recovery kinetics. Biophys. J. *16*: 1055–1069.

46. R. D. Tilton, C. R. Robertson, and A. P. Gast, Lateral diffusion of bovine serum albumin adsorbed at the solid–liquid interface. J. Colloid Interface Sci. *137*:192–203, 1990.

47. G. W. Gordon, B. Chazotte, X. F. Wang, and B. Herman, Analysis of simulated and experimental fluorescence recovery after photobleaching. Data for two diffusing components. Biophys. J. *68*:766–778, 1995.

48. J. Davoust, P. F. Devaux, and L. Leger, Fringe pattern photobleaching, a new method for the measurement of transport coefficients of biological macromolecules. EMBO J. *1*:1233–1238, 1982.

49. J. Tong and J. L. Anderson, Partitioning and diffusion of proteins and linear polymers in polyacrylamide gels. Biophys. J. *70*:1505–1513, 1996.

50. J. Tong, Partitioning and diffusion of macromolecules in polyacrylamide gels. PhD dissertation, Carnegie Mellon University, Pittsburgh, Pennsylvania, 1995.

51. V. Chan, D. J. Graves, P. Fortina, and S. E. McKenzie, Adsorption and surface diffusion of DNA oligonucleotides at liquid/solid interfaces. Langmuir *13*:320–329, 1997.

52. I. Michaeli, D. R. Absolom, and C. J. van Oss, Diffusion of adsorbed protein within the plane of adsorption. J. Colloid Interface Sci. *77*:586–587, 1980.

53. D. R. Absolom, I. Michaeli, and C. J. van Oss, Electrophoresis of adsorbed albumin. Electrophoresis 2:273–278, 1981.

54. A. Kusumi, Y. Sako, and M. Yamamoto, Confined lateral diffusion of membrane receptors as studied by single particle tracking (nanovid microscopy). Effects of calcium-induced differentiation in cultured epithelial cells. Biophys. J. *65*:2021–2040, 1993.

55. W. P. Ambrose, P. M. Goodwin, J. C. Martin, and R. A. Keller, Single molecule detection and photochemistry on a surface using near-field optical excitation. Phys. Rev. Lett. *72*:160–163, 1994.

56. A. J. Meixner, D. Zeisel, M. A. Bopp, and G. Tarrach, Super-resolution imaging and detection of fluorescence from single molecules by scanning near-field optical microscopy. Optical Eng. *34*:2324–2332, 1995.

57. B. A. Scalettar, J. R. Abney, and J. C. Owicki, Theoretical comparison of the self diffusion and mutual diffusion of interacting membrane proteins. Proc. Natl. Acad. Sci. USA *85*:6726–6730, 1988.

58. J. R. Abney, B. A. Scalettar, and J. C. Owicki, Self-diffusion of interacting membrane proteins. Biophys. J. *55*:817–833, 1989.

59. J. R. Abney, B. A. Scalettar, and J. C. Owicki, Mutual diffusion of interacting membrane proteins. Biophys. J. *56*:315–326, 1989.

60. T. J. Feder, I. Brust-Mascher, J. P. Slattery, B. Baird, and W. W. Webb, Constrained diffusion or immobile fraction on cell surfaces: a new interpretation. Biophys. J. *70*:2767–2773, 1996.

61. J. F. Nagle, Long tail kinetics in biophysics? Biophys. J. *63*:366–370, 1992.

62. M. J. Saxton, Anomalous diffusion due to obstacles: a Monte Carlo study. Biophys. J. *66*:394–401, 1994.

63. M. Velez and D. Axelrod, Polarized fluorescence photobleaching recovery for measuring rotational diffusion in solutions and membranes. Biophys. J. *53*:575–591, 1988.

64. J. L. Anderson, F. Rauh, and A. Morales, Particle diffusion as a function of concentration and ionic strength. J. Phys. Chem. *82*:608–616, 1978.

65. T. J. O'Leary, Concentration dependence of protein diffusion. Biophys. J. *52*:137–139, 1987.

66. N. Muramatsu and A. P. Minton, Tracer diffusion of globular proteins in concentrated protein solutions. Proc. Natl. Acad. Sci. USA *85*:2984–2988, 1988.

67. B. A. Scalettar, J. E. Hearst, and M. P. Klein, FRAP and FCS studies of self-diffusion and mutual diffusion in entangled DNA solutions. Macromolecules 22:4550–4559, 1989.

68. D. W. Tank, E. S. Wu, P. R. Meers, and W. W. Webb, Lateral diffusion of gramicidin C in phospholipid multibilayers: effects of cholesterol and high gramicidin concentration. Biophys. J. *40*:129–135.

69. L. K. Tamm, Lateral diffusion and fluorescence microscope studies on a monoclonal antibody specifically bound to supported phospholipid bilayers. Biochemistry 27:1450–1457, 1988.

70. S. Subramaniam, M. Seul, and H. M. McConnell, Lateral diffusion of specific antibodies bound to lipid monolayers on alkylated substrates. Proc. Natl. Acad. Sci. USA *83*:1169–1173, 1986.

71. R. D. Tilton, A. P. Gast, and C. R. Robertson, Surface diffusion of interacting proteins. Effect of concentration on the lateral mobility of adsorbed bovine serum albumin. Biophys. J. *58*:1321–1326, 1990.

72. P. B. Gaspers, C. R. Robertson, and A. P. Gast, Enzymes on immobilized substrate surfaces: diffusion. Langmuir *10*:2699–2704, 1994.

73. M. J. Saxton, Lateral diffusion in an archipelago. Effects of impermeable patches on diffusion in a cell membrane. Biophys. J. *39*:165–173, 1982.

74. M. J. Saxton, Lateral diffusion in a mixture of mobile and immobile particles: a Monte Carlo study. Biophys. J. *58*:1303–1306, 1990.

75. L. W. Nichol, In *Protein–Protein Interactions* (C. Frieden, L. W. Nichol, eds.), Wiley, New York, 1981, pp. 1–30.

76. R. D. Tilton, Surface diffusion and hydrophobic interactions in protein adsorption. PhD dissertation, Stanford University, Palo Alto, California, 1991.

77. O. V. Bychuk and B. O'Shaughnessy, Anomalous diffusion at liquid surfaces. Phys. Rev. Lett. *74*:1795–1798, 1995.

78. T. E. Rabe and R. D. Tilton, Surface diffusion of adsorbed proteins in the vicinity of the substrate glass transition temperature. J. Colloid Interface Sci. *159*:243–245, 1993.

79. P. B. Gaspers, A. P. Gast, and C. R. Robertson, Enzymes on immobilized substrate surfaces: reaction. J. Colloid Interface Sci. *172*:518–529, 1995.

80. R. D. Tilton, C. R. Robertson, and A. P. Gast, Manipulation of hydrophobic interactions in protein adsorption. Langmuir *7*:2710–2718, 1991.

81. K. Gekko and Y. Hasegawa, Compressibility–structure relationship of globular proteins. Biochemistry *25*:6563–6571, 1986.

82. C. C. Bigelow, On the average hydrophobicity of proteins and the relation between it and protein structure. J. Theoret. Biol. *16*:187–211, 1967.

83. N. S. B. Prasad, Estimation of the intermolecular forces between adsorbed proteins via optical reflectometry and Monte Carlo simulations. MS thesis, Carnegie Mellon University, Pittsburgh, Pennsylvania, 1994.

10

Protein Adsorption in Relation to Solution Association and Aggregation

TOMMY NYLANDER   University of Lund, Lund, Sweden

I. INTRODUCTION

The adsorption of proteins is like other amphiphilic molecules, strongly influenced by their stability in solution. Changes in the media surrounding the protein, such as pH, ionic strength, heat, and the presence of an interface, will affect the conformational stability—that is, how stable the secondary and tertiary structures are toward these changes. The consequence can be aggregation/precipitation of the protein, although it is important to bear in mind that aggregation/association of proteins can occur without or with only minor conformational changes (cf. Ref. 1). An example of this is the self-association of proteins such as insulin and β-lactoglobulin.

The self-association of insulin has important consequences physiologically and in formulations used for diabetes therapy. Insulin is stored in the pancreas β-cells as crystalline arrays of zinc-stabilized hexamers [2]. After being released in the bloodstream the concentration is so low (10^{-8}–10^{-11} M) that the monomeric form prevails. The monomers ($M_w \approx 5800$ Da) are the biologically active species, which interact with the insulin receptor. The formulations used for diabetes therapy often contain colloidal dispersions of amorphous or crystalline insulin in a solution of insulin. The colloidal particles give a slower release of the active monomer and thus a more long-term physiological response, while the dissolved hormone gives a more immediate response [3]. A monomeric insulin mutant (B9Ser $\rightarrow$ Asp and B27Thr $\rightarrow$ Glu) can be obtained by protein engineering [4]. The mutation increases the charge repulsion in the monomer–monomer contact domain, which prevents self-association. The monomeric insulin analog B9Asp+B27Glu has an activity similar to human insulin, but since it does not self-associate this preparation gives an even faster biological response [4].

It has long been known that insulin has a tendency to polymerize noncovalently to form aggregates or fibrils [5]. Fibrillation can be induced by contact with certain (mainly hydrophobic) materials [6,7], by exposure to the air–aqueous interface when the solution is shaken [3,8], by high shear rates [9], and by heat [5]. The formation of fibrils occurs via insulin monomers and requires changes in the monomer conformation [10]. Hydrophobic interactions are involved in the process [5], and the monomers in the fibrils seem to be joined by parallel β-sheet structures [11], resulting in distorted higher oligomers which can grow infinitely. It is clear that the risk for fibril formation can be reduced if the formation for higher insulin oligomers is promoted, e.g., by adding zinc [12] or calcium [13]. Fibril formation can also be reduced by changing the solvent, e.g., by adding glycerol and phenol [6], by using autologous serum [14], or by adding moderate amounts of urea [9]. The addition of urea was also found to inhibit insulin adsorption on polymer surfaces. Polypropylene glycol/ polyethylene glycol block copolymers had a similar effect on insulin adsorption at hydrophobic surfaces since they adsorb to the surface, modifying its properties and in addition inhibiting aggregation of insulin [7]. Thus, the properties of the surfaces interacting with solution are of great importance, and it has been found that hydrophobic materials such as silicone rubber and polypropylene can induce fibrillation [6]. Glass (hydrophilic) surfaces, on the other hand, do not have this effect [6], and insulin was still active after being desorbed from a glass surface where the protein had been adsorbed for months [15].

The other main oligomeric protein to be discussed in this chapter is β-lactoglobulin. β-Lactoglobulin is one of the major whey proteins and shows an intriguing association behavior involving monomers, dimers, and octamers [16–18]. The degree of association is strongly dependent on pH, ionic strength, concentration, and genetic variant of β-lactoglobulin. The biological role of β-lactoglobulin still remains uncertain [19]; hence, the biological implication of its associated behavior has not yet been resolved. The protein has been suggested to take part in the transport of retinol [20–22] and has been shown to bind many other hydrophobic molecules such as phospholipids, fatty acids, and triglycerides [23–27]. Because β-lactoglobulin is the major protein constituent in whey protein preparations, the self-association of β-lactoglobulin is bound to affect the functional properties in terms of aggregation, gelation and surface properties of importance in a range of food products.

Similar to insulin [2,28], the monomer–monomer contact in the β-lactoglobulin dimer contains a shared β-sheet structure [19] where hydrogen bonding plays an important role. These intriguing and very beautiful arrangements of the polypeptide chains give a very tight and specific organization at the monomer–monomer interfaces. However, proteins with less ordered structure can still exhibit self-association. One of the most studied examples, which most of us experience in everyday life, is the caseins in milk. The α_{s1}-, α_{s2}-, β-, and κ-caseins form aggregates, casein micelles, where the individual casein monomers are linked by a complex pattern of interactions, such as hydrophobic interactions, disulfide bridges, and interactions involving colloidal calcium phosphate (cf. Refs. 18 and 29). Thus, these "micelles" are different from what is usually meant by micelles in surfactant solutions, both in terms of the interaction between the monomers and the structure of the micelles. The casein micelles have a very hydrated structure; for instance, a micelle consisting of pure κ-casein has a specific volume of about 6.7 mL per gram of κ-casein [30], compared to 0.87 mL/g [31] for sodium dodecylsulfate in the micellar state. The structure of the casein micelle is still a matter of controversy, where structures with [18,32] and without [29,33] submicelles have been put forward. However, κ-casein is in both models located at the surface of the casein micelle with its hydrophilic part protruding from the surface, giving the micelle a hairy and hydrated structure [34]. Such a structure will, by mainly steric repulsive forces, prevent aggregation of the casein micelles [35,36]. Thanks to their affinity to interfaces as well as the structure of the adsorbed layers, caseins have been used as emulsifiers and foaming agents in applications ranging from food to building materials such as paint, glue, and putty [18]. β-Casein has an amino acid sequence which is divided into one hydrophilic and one hydrophobic domain [17]. This makes it the most amphiphilic casein, and β-casein will therefore be the focus of our discussion. The biological function of β-casein is to stabilize the colloidal form of calcium phosphate in the casein micelle [37,38]. From a colloid and surface chemical point of view the protein resembles block copolymers (or to some extent surfactants), which form monodisperse aggregates or micelles. At room temperature this assembly process starts above 0.1 mg/mL [39,40]. Parallel to the aggregation of ionic surfactants, where the attraction between hydrophobic domains is balanced by the electrostatic repulsion between the hydrophilic domains, the aggregation is promoted by increase of ionic strength [39,41]. The association decreases with temperature, and β-casein exists as monomers at 4°C [39,41].

The third and last part of this chapter will address the relation between heat-induced protein aggregation in solution and the adsorption of the proteins on a solid surface. Food products, e.g., milk, but also pharmaceuticals in liquid form are usually heat treated, mainly to enhance their storage qualities. The physical and chemical properties, in particular for liquids containing heat-sensitive compounds such as proteins, are, in most cases, also affected. Furthermore, the processing causes deposits —fouling—on exposed surfaces (e.g., heat exchanger surfaces, ultrafiltration membranes, tubing, and pipes) of the process equipment. Fouling results in less effective heat transfer in heat exchangers, but it can increase the risk for bacterial infections and seriously damage the processed product (e.g., giving it a burnt taste). Thus, the equipment has to be cleaned frequently, which adds to the processing cost. It is therefore not surprising that a lot of research has been focused on "solving" the fouling problem, particularly in the dairy industry. The deposits formed during heat treatment of milk (at 85–115°C) consist mainly of proteins [42]. β-Lactoglobulin constitutes only about 10% of the total amount of the proteins in milk [18], but still it was found that about half of the protein in milk deposits formed at 70–80°C was made up of β-lactoglobulin [43]. One reason can be the presence of the free sulfhydryl group in β-lactoglobulin [17], which is crucial in the heat-induced aggregation of the protein [44]. The importance of sulfhydryl groups in milk fouling was demonstrated by a decrease in the deposit when these groups were blocked with iodate [45]. It has been proposed, based on model studies, that the fouling of milk should be related to the aggregation rather than the denaturation of β-lactoglobulin [46]. Thus, we will mainly discuss the relation between the aggregation of β-lactoglobulin in solution and the interfacial behavior.

II. ADSORPTION OF OLIGOMERIC GLOBULAR PROTEINS

A. Insulin

Insulin exhibits a complex self-association behavior in solution, where the monomers associate to form dimers, tetramers, and hexamers [2]. High-resolution x-ray diffraction data shown that the hexamer has the shape of a flattened spheroid with a diameter of about 49 Å and a height of about 34 Å; the dimer has an oblong shape of approximately $20 \times 25 \times 40$ Å, and the size of the wedge-shaped monomer is about $20 \times 25 \times 20$ Å [28]. The monomer surface has two domains where hydrophobic amino acids dominate. These domains are involved in the formation of oligomers [47], in which hydrogen bonding also contributes [48]. The monomer–monomer contact involves the shared β-structure [2], so the interaction is fairly strong and only oligomers with an even number of monomers are formed. The dimer–dimer contact in the tetramer is determined by similar forces, but the contact region is less densely packed [2], indicating a weaker interaction.

In contrast to the monomer, the entire surface of the hexamer is more or less hydrophilic. It should thus be expected that the monomers and the different oligomers should behave differently at interfaces. Furthermore, surface properties, such as hydrophobicity, should strongly influence adsorption. A change in equilibrium between oligomers might therefore change the adsorption behavior completely. In zinc-free solution the monomers, dimers, and tetramers dominate. The equilibrium is shifted toward tetramers with increasing concentration, decreasing pH (decreasing charge), and increasing ionic strength (screening of electrostatic repulsion) [49,50]. Human insulin has a net charge of about -2 per monomer at physiological pH 7.4 and an apparent isoelectric point at about pH 6.4 [51]. The addition of divalent ions, like Zn^{2+}, shifts the equilibrium toward the hexamer [2,52,53]. The hexamer has two high-affinity binding sites for zinc located along the symmetry axis in the center of the hexamer [54,55]. Additional zinc can bind with less affinity on the surface of the hexamer [56,57], and it has been shown that the addition of two more zinc ions per hexamer increases the physical stability of the insulin solution [3].

1. Adsorption

The influence of the total insulin concentration on adsorption to a methylated silica (hydrophobic) surface and a bare (hydrophilic) silica surface is shown in Table 1. The data, which include human and modified monomeric insulin, are taken from Nilsson et al. and were obtained by in situ ellipsometry [58]. The adsorption is quite different for each type of surface, reflecting the different properties of the monomers compared to the other oligomeric forms. The amount of human insulin adsorbed to the hydrophobic surface (about 1.4 mg/m^2) is independent of the total insulin concentration within the range studied (3–300 mM). This means that the adsorbing form (monomers, oligomers) must be present at sufficiently high concentration even at the

TABLE 1 Amount of Insulin Adsorbed after 3600 s Versus Concentration at pH 7.4 as Determined by In Situ Ellipsometry

Concentration (μM)	Amount adsorbed (mg/m^2)	
	Hydrophobic surface (methylated silica)	Hydrophilic surface (clean silica)
3	1.2	—
6	1.4	0.0
30	—	0.0
200	—	0.0
300	1.3	0.5
Monomeric insulin, 300	1.3	0.0

Source: Ref. 58.

lowest total insulin concentration. The concentrations of monomers, dimers, tetramers, and hexamers at a total insulin concentration (calculated as monomers) of 6 μM are 1.68, 2.12, 2.24 $\times$ 10^{-2}, and 2.13 $\times$ 10^{-6} μM, respectively [59], calculated according to Pocker and Biswas [60]. In addition, nearly the same adsorption of the monomeric insulin analog was observed. The amount corresponds to a close-packed monolayer of monomers, with one of the hydrophobic domains on the monomer facing the surface [61]. The monomer is also more hydrophobic than the oligomeric forms. Thus, it is almost certain that human insulin sticks to the hydrophobic surface in its monomeric form.

On the hydrophilic silica surface adsorption is determined by the total insulin concentration, and a detectable amount was observed only at the highest concentration used (300 μM). No adsorption of monomeric insulin was observed. The adsorption rate and the amount of human insulin are low (Fig. 1) compared to the hydrophobic surface (Fig. 2). This indicates that the concentration of the adsorbing oligomeric form is low. At a total insulin concentration (calculated as monomers) of 150 μM, the concentrations of monomers, dimers, tetramers, and hexamers are 8.00, 47.94, 11.49, and 2.48 $\times$ 10^{-2} μM, respectively [59], calculated according to Pocker and Biswas [60]. Both the hydrophilic surface, with an isoelectric point of about 2 [62], and the insulin molecule, with an isoelectric point at pH 6.4 [51], carry a negative net charge at pH 7.4. Thus, electrostatic repulsion between the adsorbing protein and the surface is expected to occur. In fact, the amount of human insulin adsorbed on a hydrophilic chromium surface was found to be considerably higher [61]. The chromium surface is expected to be neutral or to carry a slight positive net charge at pH 7.4 since it has an isoelectric point at about pH 7 [62,63]. It should

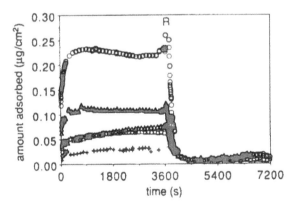

FIG. 1 Amount of human insulin adsorbed on a hydrophilic silica surface versus time as determined by in situ ellipsometry. The amount adsorbed from a solution containing no added salt (+), 0.1 mM Zn^{2+} ($\square$), 0.1 mM Zn^{2+} + 0.3 mM Na^+ ($\triangle$), 0.1 mM Zn^{2+} + 0.1 mM Ca^{2+} ($\blacktriangle$), and 0.2 mM Zn^{2+} ($\circ$) are compared. The mean values after 300 s in mg/m^2 from at least two replicate experiments $\pm$ the maximum deviation are 0.40 $\pm$ 0.10 with no added salt, 0.61 $\pm$ 0.01 with 0.1 mM Zn^{2+}, 0.69 $\pm$ 0.04 with 0.1 mM Zn^{2+} + 0.3 mM Na^+, 1.11 $\pm$ 0.01 with 0.1 mM Zn^{2+} + 0.1 mM Ca^{2+}, and 2.95 $\pm$ 0.60 with 0.2 mM Zn^{2+}. The insulin concentration was 300 μM. The pH was adjusted to 7.4. R indicates rinse with water or corresponding protein-free salt solution. Representative experiments were chosen out of at least two replicates. (From Ref. 58.)

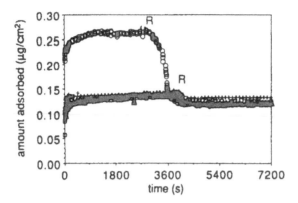

FIG. 2   Amount of human insulin adsorbed on a methylated (hydrophobic) silica surface versus time as determined by in situ ellipsometry. The amount adsorbed from a solution containing no added salt (+), 0.1 mM Zn^{2+} ($\square$), 0.1 mM Zn^{2+} + 0.1 mM Ca^{2+} ($\blacktriangle$), and 0.2 mM Zn^{2+} (o) are compared. The mean values after 300 s in mg/m^2 from at least two replicate experiments $\pm$ the maximum deviation are 1.41 $\pm$ 0.02 with no added salt, 1.33 $\pm$ 0.01 with 0.1 mM Zn^{2+}, 1.38 $\pm$ 0.03 with 0.1 mM Zn^{2+} + 0.1 mM Ca^{2+}, and 2.69 $\pm$ 0.1 with 0.2 mM Zn^{2+}. The insulin concentration was 300 μM. The pH was adjusted to 7.4. R indicates rinse with water or corresponding protein-free salt solution. Representative experiments were chosen out of at least two replicates. (From Ref. 58.)

be borne in mind that even though the net charge of the protein is negative, it still contains four positive charges per monomer unit, of which two are involved in ion pairs with carboxyl groups within the monomer [28,51]. This means that the expected electrostatic repulsion can be minimized by proper orientation of the protein at the interface, so the positively charged groups on the protein can be close enough to the negative sites of the surface. The hexamer is the most hydrophilic oligomeric form. It is also the largest specie at this pH and might therefore have more contact points with the surface. Hence, it is likely that insulin adsorbs to the hydrophilic surface in hexameric form.

As discussed, the addition of divalent ions, e.g., zinc, shifts the equilibrium between the oligomeric forms toward the hexamers [2]. At a total insulin concentration (calculated as monomers) of 150 μM and a Zn^{2+} concentration of 50 μM (i.e., two zinc ions per insulin hexamer), the concentrations of monomers, dimers, tetramers, and hexamers are 0.73, 0.40, 8.15 $\times$ 10^{-4}, and 24.74 μM, respectively [59], calculated according to Pocker and Biswas [60] and using the zinc binding constants reported by Brill and Venable [64]. The influence of ionic strength and the addition of divalent ions on the adsorption to hydrophilic and hydrophobized silica are demonstrated in Figs. 1 and 2, respectively. The adsorption of human insulin to a hydrophilic surface depends heavily on the type and concentration of the added salt. In all cases the adsorbed insulin readily desorbs when flushing the cuvette with pure water or corresponding protein-free salt solution, which demonstrates the weak interaction with this surface. At a zinc concentration of 0.1 mM, corresponding to two zinc ions per insulin hexamer, the plateau value of the amount adsorbed increases considerably relative to the value recorded for the zinc-free insulin solution. A further increase in the zinc concentration to 0.2 mM, corresponding to 4 zinc ions per insulin hexamer, gives a very large increase in the amount adsorbed, about six times the value for the adsorption from zinc-free solution. A similar increase with the amount of added zinc was observed for the insulin adsorption on hydrophilic chromium surfaces [61]. However, no complete desorption was observed in this case, which is well in line with the stronger interaction between the adsorbed molecule and the chromium surface.

To investigate if the large amount of insulin adsorbed to the hydrophilic surface in the presence of high concentrations of zinc was an effect of the specific binding or merely a consequence of the increase in ionic strength, other types of salts of the same ionic strength were added. Also a pure electrostatic screening effect achieved by increasing the ionic strength is expected to increase the adsorption as well as the self-association of insulin. The addition of sodium chloride did not at all change the adsorption from an insulin solution containing 0.1 mM zinc. However, when calcium chloride of the same ionic strength was added instead of sodium chloride, a moderate increase in the adsorbed amount from 0.7 to 1.1 mg/m^2 was observed. In contrast, a dramatic increase in the amount is observed when 0.2 mM zinc is added, giving values of the surface concentration (2.9 mg/m^2) which are a factor of 3 larger than in the presence of 0.1 mM zinc. This clearly demonstrates the specific effect of zinc. It is also noteworthy that no adsorption of monomeric insulin on hydrophilic silica was observed, not even in the presence of 0.2 mM zinc [58]. When considering the constant (versus concentration) amount adsorbed on the hydrophobized silica surface, it is no surprise that almost no effect of added salt was found for the adsorption on this surface (Fig. 2). The only exception is at the highest zinc concentration used (0.2 mM) where the amount is double that of the zinc-free case. The amount is, however, reduced upon rinsing with protein-free salt solution to the same level as in the absence of zinc.

In contrast to the desorption from the hydrophilic surface, no desorption was otherwise observed from the hydrophobic surface. This observation indicates that insulin interacts more strongly with the hydrophobic surface. Since adsorption is mainly independent of the salt concentration, the packing at the hydrophobic surface seems not to be hampered by electrostatic repulsion. The hydrophobic surface domain of the monomer, which is responsible for the monomer–monomer contact in the dimer, is very flat and rich in aromatic residues [28]. It is likely that this domain is directed toward the hydrophobic surface and responsible for the strong interaction between the surface and the protein. Therefore, the monomers will be preferentially adsorbed to the hydrophobic surface, even if few of them are present at the higher zinc concentration. The presence of a loosely bound, desorbable fraction at the highest zinc concentration (0.2 mM) is likely to be adsorbed on the top of the monomer layer. Since the effect occurs only in the presence of zinc and not calcium, the outer layer must be an effect of the specific binding of zinc. One possibility is that the outer layer consists of hexamers, which may bind to the surface when the zinc concentration is high enough.

As mentioned before, the formation of higher oligomers is favored by a decrease of pH. In fact, the formation of oligomers larger than hexamers at high zinc concentration has been reported [52]. Solubility in neutral and slightly alkaline solution also decreases with increasing zinc concentration and decreasing pH, and precipitation starts around pH 7 [3]. The amount of human insulin adsorbed onto hydrophilic chromium from a solution contain 300 μM protein and 0.2 mM zinc

(4Zn^{2+} hexamer) is shown versus pH in Fig. 3. As a consequence of the higher degree of association and in the proximity of precipitation from solution, the amount increases sharply as the pH approaches 7. As on the silica surfaces the desorbable fraction at this high zinc concentration is significant. However, the amount in the remaining irreversibly adsorbed layer seems to be fairly constant. Note that the amount of insulin in this layer roughly corresponds to a monolayer of hexamers, which confirms the conclusion that insulin sticks to hydrophilic surfaces in its hexameric form.

We have so far seen that there are some differences in how the oligomers respond to differences in the properties of the hydrophilic surface. In the following, some complementary examples, using platinum and mica surfaces, employing electrochemical and surface force techniques, respectively, will be discussed. The importance of the surface charge is illustrated in Fig. 4, where the results from ellipsometric measurements of the amount of insulin adsorbed to a platinum surface are shown as a function of the applied potential of the platinum [65]. Practically no protein is adsorbed from a zinc-free solution at potentials between −0.1 and −0.3 V versus SCE (saturated calomel electrode), and the adsorption from insulin solutions containing 0.1 and 0.2 mM is at its minimum in this potential region. The potential of zero charge for polycrystalline platinum is also located in this region, that is, near −0.12 V versus SCE [65]. The amount adsorbed also increases with zinc concentration in this potential region. Furthermore, the minimum is shifted toward more positive potentials as the zinc concentration increases. This is likely to reflect changes in the equilibrium between the oligomeric forms as well as in the charge, resulting from changes in zinc concentration. The amount of insulin adsorbed from both zinc-free and zinc insulin solutions increases on both positive and negative sides. These

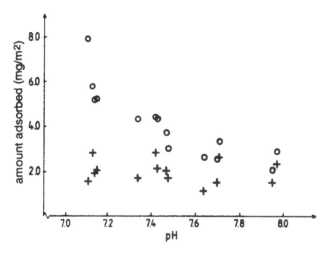

FIG. 3 Amounts adsorbed on bare hydrophilic chromium from 4Zn^{2+}/hexamer human insulin solution as determined by in situ ellipsometry are shown versus pH. The circles are the amounts after 60 min of adsorption and the crosses indicate the values reached after 60 min of rinse with water. The insulin was dissolved in water at a concentration of 300 μM (1.9 mg/mL). (From Ref. 61.)

Γ_{max}, $\mu g/cm^2$

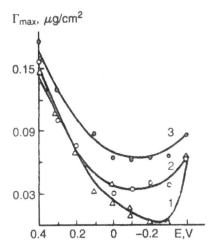

FIG. 4 The plateau value of the adsorption, Γ_{max}, of human insulin on platinum obtained by in situ ellipsometry, shown as a function of the applied potential on the platinum electrode (relative saturated calomel electrode, SCE). The insulin solution contained 300 μM insulin in 0.15 M NaCl of pH 7.4 and 25°C (1), with the addition of $2Zn^{2+}$/insulin hexamer (2), and $4Zn^{2+}$/insulin hexamer (3). (From Ref. 65.)

measurements were performed at a relatively high concentration of background electrolyte, 0.15 M NaCl.

2. Interactions Between Adsorbed Layers

Surface force measurements provide information on adsorption to a surface, but they also give hints about the structure of the adsorbed layer [66]. We will here recapture some of the main features of two studies where surface force measurements between insulin-coated surfaces have been used to reveal the structure of the adsorbed layers [59,67], since they complement the discussion in Section II.A.1.

The buildup of the adsorbed layer on hydrophilic mica from a 6 μM human insulin solution is illustrated in Fig. 5, where the interaction forces between these surfaces are shown versus surface separation [59]. The results show that the steric wall has been shifted out 20 Å 30 min after adding the insulin solution. This corresponds to the presence of one monolayer of monomers between the surfaces. The steric wall shifts toward larger surface separation with increasing incubation time. An equilibrium thickness of about 86 Å is finally reached after about 24 h. This value corresponds roughly to the dimensions of two layers of insulin hexamers, one on each surface. As discussed above, the insulin preferably adsorbs to a hydrophilic surface in its hexameric form. What probably occurs here is an initial adsorption of monomers which are finally replaced by the hexamers. This happens even though the concentration of hexamers is very low, which can explain the time effect, since the transport of these species to the surface then becomes very slow (transport-limited adsorption). These types of exchange reactions are visualized for the competitive adsorption of blood proteins on glass surfaces, known as the Vroman effect [68,69], where low-molecular-weight proteins are replaced with less abundant proteins of

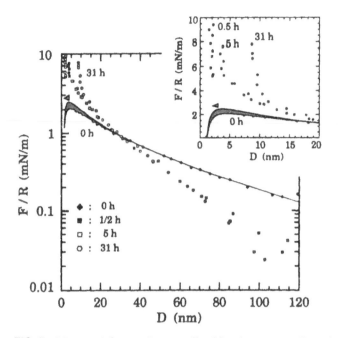

FIG. 5 Measured forces, F, normalized by the mean radius of curvature, R, of the surfaces, as a function of surface separation, D, between crossed hydrophilic mica cylinders immersed in 6 μM human insulin (pH 7.4) at different times after injection: (♦) pure water; (□) half an hour after injection; (■) 5 h; (○) 31 h, where thermodynamic equilibrium has been reached. The inset (linear scale) shows an enlargement of the data at short separations, indicating that the thickness of the adsorbed insulin molecules layers increases with time to reach a plateau of 8–9 nm (formation of one layer of hexamers on each mica surface). On contrast, at large surface separations, the force-distance profiles in presence of insulin keep lying all parallel to each other with the same exponential decay length (κ^{-1} = 23.7 nm; Table 1). Solid lines here and for subsequent figures show DLVO calculation as the sum of the repulsive double-layer force, obtained by numerical solutions of the nonlinear Poisson–Boltzmann equation for both constant surface potential (lower curve) and constant surface charge (upper curve) boundary conditions, and the attractive van der Waals force with a nonretarded Hamaker constant of 2.2 × 10^{-20} J [37,58]. Both diffuse layer boundary and the plane of dispersion forces onset are supposed to be located at the mica–mica contact (D = 0). Parameters for the double-layer force best fit to (○) are Debye length κ^{-1} = 55 nm, and surface potential at infinite separation, Ψ_0^{∞} = −140 mV. The arrowhead indicates the position where the surfaces jump into adhesive minimum. (From Ref. 59.)

higher molecular weight. A similar effect has been observed for human albumin, where the dimers are preferentially adsorbed from a mixture of dimers and monomers [70].

For insulin concentrations of 30 μM or more, the equilibrium in terms of adsorbed layer thickness and force profiles is reached within 12 h. The force curves recorded in the presence of human insulin and the monomeric analog at protein concentrations of 30 and 150 μM are shown in Fig. 6. The steric wall for the interaction in the presence of human insulin is also at this higher insulin concentration located at surface separations of about 90–100 Å, which corresponds to two layers

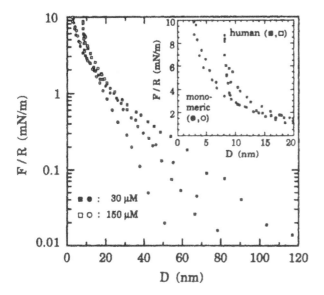

FIG. 6 Normalized force as a function of surface separation for mica surfaces immersed in human insulin (■, □) and monomeric analog B9Asp+B27Glu (●, ○). At large surfaces separations, the force-distance profile is exponential (linear on the semilog scale) for all solutions with a decay length, which depends on the concentration of the different oligomeric forms of insulin as well as their charge. At small surface separations, a steric repulsion is encountered due to loose and adsorbed insulin molecules (inset; linear scale). Filled symbols correspond to a bulk insulin concentration of 30 μM, while open symbols are for solutions more concentrated (150 μM). (From Ref. 59.)

of hexamer. In fact it was observed that this equilibrium thickness as determined by surface force measurements hardly changes when the equilibrium is shifted toward hexamers either by increasing the insulin concentration or by adding zinc [59]. However, with the monomeric analog B9Asp+B27Glu, which exists as a monomer under the same experimental conditions [4,71], the steric wall is located at substantially smaller separations, about 35 Å. This is slightly less than expected from two monolayers of monomers, which indicates that the mica surface is not completely covered with monomeric insulin. It was also found that monomeric insulin, which has a net charge of −4 at pH 7.5, did not adsorb if the ionic strength was increased by replacing water with a 2.5 mM sodium chloride solution [59].

The weak but noticeable adsorption of monomeric insulin on mica is contrary to what happened on the silica surface. Although silica and mica have a strong negative charge at pH 7.5, there are large differences in their interfacial structure. For silica, the charge arises from the ionization of specific hydroxyl groups along the surface, whereas the negative charge of the mica surface comes from dissociation and exchange of potassium ions in the mica lattice with the solution [72,73]. Furthermore, as opposed to the ellipsometry study, the surface force measurements were carried out under nonstirred conditions.

At large separations the interaction between adsorbed insulin layers is dominated by a long-range electrical double-layer repulsion. The extent of the double layer is reduced with increasing protein concentration (Fig. 6). Furthermore, the range of the force at the same protein concentration is substantially shorter for the more charged monomeric analogue (net charge −4) than for human insulin (net charge −2 per monomer). The different oligomers do contribute as polyelectrolytes to the double-layer force [59]. Hence, from the direct measurement of the decay length, the net charge of the species (monomers and oligomers) can be determined if the concentration of the different species can be estimated [59].

As demonstrated in Fig. 7 the buildup of the insulin layer on the hydrophobic surface is very different. In contrast to what was observed on the hydrophilic mica surface, the forces are no longer only repulsive [67]. After 40 min of adsorption an attractive minimum is observed at a surface separation of about 65 Å. After further compression a steric wall at a separation of about 42 Å is found, which roughly corresponds to two monolayers of monomers. After 4 h a second layer is evident from the force curve, where the occurrence of a repulsive force between 85 and 76 Å indicates reorientation and/or compression of the molecules in the outer layer. The inward jump from surface separation of 76 Å to 65 Å on further compression sug-

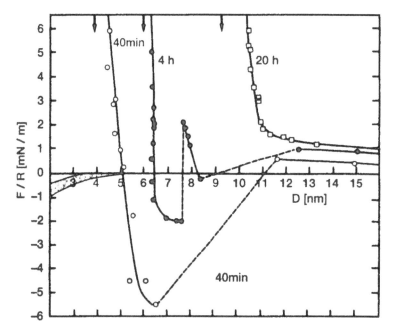

FIG. 7 Measured forces, F, normalized by the mean radius of curvature, R, of the surfaces, as a function of surface separation, D, between adsorbed layers of human insulin at different equilibration times (○, 40 min; ●, 4 h; □, 20 h). The dashed lines do not represent the actual forces but serve only to connect different parts of the force curves. The arrows indicate the thickness of a highly compressed insulin layer after different equilibration times. The shaded region illustrates the van der Waals forces contribution from the DODA-coated mica. The nonretarded van der Waals force has been calculated from $F/R = A/6D^2$ with values of the Hamaker constant (A) between 2.2×10^{-20} J and 1.0×10^{-20} J. The former value corresponds to mica interacting across water (the value for hydrocarbon interacting across water is 0.5– 0.6×10^{-20} J). The insulin solution contained 300 μM (1.9 mg/mL) insulin in pure water at pH 7.3. (From Ref. 67.)

gests that the outer layer is squeezed out. The steric wall has now moved further outward. After 20 h of equilibration the recorded force curve is now entirely repulsive, where the surface separation at maximum compression now has increased to about 90 Å, which is almost the same surface separation as for the interaction between hydrophilic mica in the presence of human insulin. It may indicate that insulin adsorbs in its hexameric form also on hydrophobized mica [67]. However, since the buildup mechanism of the layer is very different, other oligomeric forms adsorbed from solution or formed at the interface may also take part.

B. β-Lactoglobulin

β-Lactoglobulin is present in bovine milk at a concentration of about 3 mg/mL, and at this concentration it mainly exists as dimers with a subunit molecular weight of about 18,350 Da [18]. Each monomer, which has a molecular weight of about 18,350 Da, contains one free cysteine and two disulfide bridges [74]. The dimer resembles two spheres with a diameter of 36 Å, which in contact are impinged by 2.3 Å [75,76]. The structure of β-lactoglobulin, crystallized at pH 6.5, has been determined to 1.8-Å resolution, and the main feature of the structure is that it contains a barrel composed of eight β-strands, which is thought to be the binding side for hydrophobic ligands [19]. The binding of hydrophobic compounds, such as fatty acids, has been shown to affect the interfacial properties of β-lactoglobulin [77]. The self-association of β-lactoglobulin is controlled by factors like pH, temperature, protein concentration, ionic conditions, and genetic variants. The dimers dissociate into monomers below pH 3.5 [78,79]. At about pH 7.5 a conformational transition occurs, the so-called Tanford transition, which features the titration of a carboxylic group with an unusually high pK_a of about 7.3 and a change in the environment of a tyrosine residue [80,81]. The transition also leads to an expansion of the molecule and an increased reactivity of the free sulfhydryl group [82]. As a consequence of the transition, the dimer dissociates above pH 8 [83]. The dissociation in this pH range is found to increase with temperature [84]. The formation of higher oligomers (e.g., octamers) in the pH range 3.7 to 5.2, is favored at low temperatures, with a maximal association at pH 4.6 and close to 0°C [85,86].

The two most common genetic variants of bovine β-lactoglobulin in milk from western cattle are the A and B variants, where only two amino acid residues differ: 64Asp and 118Val in the A variant are replaced by 64Gly and 118Ala in the B variant [17,74]. Consequently, the net charge of the A variant is more negative, −5 per monomer at pH 7, and the pK_a, which is reported to be around 5.2 is slightly lower [16,17,87,88]. The most notable difference between the two genetic variants, is the difference in their association behavior. In solutions where dimers are dominating, the variant A is more dissociated and the dimer dissociation constants for the A and B variants at pH 7 (0.1 M NaCl) have been determined to be 6.3×10^{-5} and 0.8×10^{-5} M, respectively [83]. However, the octamer formation is less favored for the variant B [89,90]. In a series of excellent papers Elofsson et al. present results concerning the influence of the association of β-lactoglobulin on the adsorption at interfaces, making use of the differences in dimerization of the genetic variants A and B [91–94]. We shall discuss the highlights of some of their findings, which demonstrate some common features with the insulin adsorption discussed in the previous sections.

The adsorption isotherms for β-lactoglobulin A and B as well as for the A and B 1:1 mixture on methylated (hydrophobic) silica are shown in Fig. 8 [92]. The isotherms show two distinctive plateaus for both genetic variants. The one occurring at low concentration corresponds to a surface concentration of 1 mg/m^2, the higher plateaus to a concentration of 2.5 mg/m^2. It is tempting to assume that the presence of the two plateaus is related to the concentration of monomers and dimers in solution. If one assumes that β-lactoglobulin adsorbs as monomeric hard spheres (diameters of 36 Å and molecular weight 18,350) packed in hexagonal close-packed monolayers, the surface concentration would be about 2.7 mg/m^2. However, as pointed out by Mackie et al., one also has to take the parking limit (i.e., the fraction of an area that can be occupied by objects of one size randomly placed on a surface) into account [95]. Cooper [96] found a parking limit of about 0.5 area fraction for disks on a plane, which is in agreement with practical experiments with latex spheres on a flat glass surface [97]. Thus, about 1.4 mg/m^2 might be a more realistic value for a monolayer of β-lactoglobulin monomers. Dimers standing up would give twice this surface concentration, provided that no (electrostatic) repulsive forces are present between the adsorbed molecules and no major structural changes occur when the protein adsorbs to the interface, which is far from general [98]. However, as a first assumption it is not unreasonable to believe that the lower plateau observed in Fig. 8 corresponds to a monolayer of monomers and the higher one to a layer of dimers. It is tempting to assume that the presence of the two plateaus is related to the concentration of monomers and dimers in solution, in particular as the isotherm of

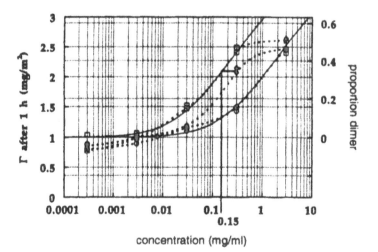

concentration (mg/ml)

FIG. 8 Plateau amount adsorbed, Γ, obtained after 1 h on methylated (hydrophobic) silica, using in situ ellipsometry, versus concentration for β-lactoglobulin A (○) and B (□) and a 1:1 (w/w) mixture of A and B variants (♦) at 25°C in 0.1 M phosphate buffer ($I = 0.017$), pH 7.0. The solid curves show the proportion of dimers in solution, as calculated from dissociation constants of 6.3 × 10^{-5} and 0.8 × 10^{-5}M for the A and B variants, respectively [83]. Superposition of the two isotherms for the 1:1 (w/w) mixture of the A and B variants onto the isotherms of β-lactoglobulin B corresponds to a shift in concentration of a factor of two. This is indicated for the concentration of 0.3 mg/mL of the mixture. (Reprinted with permission from Ref. 92. Copyright © 1997, American Chemical Society.)

the variant A is shifted toward higher protein concentration, which reflects the higher degree of dissociation of the variant A dimers. Furthermore, the increase in the proportion of dimers, as calculated from the dissociation constants for variant A and B dimers [83], almost coincides with the increase of the amount adsorbed (Fig. 8). The isotherm for the mixture of β-lactoglobulin A and B follows more closely the isotherm of variant B, although it is shifted to higher concentration. In fact, the shift corresponds to a reduction of a factor of 2 in concentration, which is the concentration of variant B in the A and B mixture [92]. This is an indication of the preferential adsorption of β-lactoglobulin B dimers, which was further confirmed by the results from a study of the sequential adsorption of variants A and B [92].

Models which feature exchange reactions between adsorbed monomers and dimers from solution were designed by Wahlgren and Elofsson [94], based on the adsorption data for the two β-lactoglobulin variants [92]. They were able successfully to describe the adsorption kinetics for both β-lactoglobulin A and B with the same model for five different concentrations. The obtained rate constant for the two genetic variants differed by a factor of 7.9. This is exactly the same as the ratio between the dissociation constants for β-lactoglobulin A and B [83].

The adsorption isotherms for the different genetic variants of β-lactoglobulin on hydrophilic silica are shown in Fig. 9 [92]. Similar to the adsorption on the hydrophobic surface, the isotherms correlate very well with the monomer-dimer equilibrium for variants A and B. However there is a distinct difference: the lower plateau corresponding to the adsorption of monomers is absent. This means that, as observed for insulin (Section II.A.1), no monomers seems to stick on the hydrophilic silica surface. This is quite interesting and might be a reflection of the similarities in the forces and peptide chain organization (shared β-sheet) involved in the dimerization of insulin and β-lactoglobulin monomers. Furthermore, the amount of β-lactoglob-

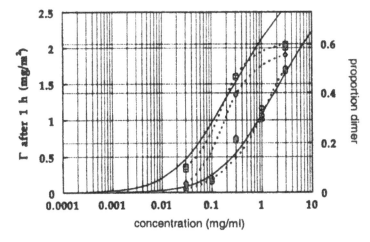

FIG. 9 Plateau amount adsorbed, Γ, obtained after 1 h on bare (hydrophilic) silica, using in situ ellipsometry, versus concentration for β-lactoglobulin A (○) and B (□) and a 1:1 (w/w) mixture of A and B variants (♦) at 25°C in 0.01 M phosphate buffer ($I = 0.017$), pH 7.0. The solid curves show the proportion of dimers in solution, as calculated from dissociation constants of 6.3×10^{-5} and 0.8×10^{-5} M for the A and B variants, respectively [83]. (Reprinted with permission from Ref. 92. Copyright © 1997 American Chemical Society.)

ulin adsorbed on the hydrophilic surface is considerably lower than on the hydrophobic surface. This is likely to be a consequence of charge repulsion between the negatively charged surface and the protein with its negative net charge [92]. One might expect that β-lactoglobulin A, with its slightly more negative net charge, might adsorb to lower extent on the hydrophilic surface than variant B. However, this is not evident from Fig. 9. The isotherm for the β-lactoglobulin A and B mixture closely follows that of variant B. This means that on the hydrophilic surface it is not only variant B that adsorbs from the mixture, since we then would expect a lower amount adsorbed and we do not see the shift of the isotherms by a factor of 2 as observed for the hydrophobic surface. Furthermore, the sequential adsorption of variant B after A gives a higher amount than B alone [92]. Still, variant B seems to be preferentially adsorbed, and now adsorption of A occurs onto a layer of β-lactoglobulin B.

The association of β-lactoglobulin is, as discussed in the introduction of this section, strongly dependent on pH. Elofsson et al. did not observe any strong correlation between the amount adsorbed on a hydrophobic surface and the pH-dependent monomer–dimer equilibrium [92]. The maximum in adsorption was found close to the isoelectric point, which is commonly observed for proteins and could be related to the decrease in repulsion between the adsorbing molecules as well as an increase in structural stability of the protein [98]. A decrease in the amount of β-lactoglobulin adsorbed was observed when increasing the pH from 4.6 to 7.0. This was attributed to the increase in dissociation of β-lactoglobulin oligomers [92], which has been shown to occur in this range [83,84]. β-Lactoglobulin is known to exist as monomers below pH 3.5 [78,79]; however, the amount adsorbed at pH 2.5 was almost twice (1.8 mg/m^2) the value of the lower plateau in the isotherm (Fig. 8) attributed to monomer adsorption. Only a moderate decrease in adsorption was observed for variant B, while β-lactoglobulin A adsorption was almost constant in the pH interval of 2.5 to 4.6. This illustrates the complexity of the interfacial behavior of oligomeric proteins, where not only the equilibrium between the oligomeric forms but a range of other factors contributes, such as changes in the ability to pack at the interface and changes in structure and stability, which all can be induced by pH changes [92]. In addition, rearrangement/decomposition can occur at the interface. This was observed by Elofsson et al. when they studied the adsorption of β-lactoglobulin A under conditions where the octameric form is present (pH 4.6 and 4.6°C) [92]. In contrast to the adsorption at 25°C, an initial maximum in the adsorbed amount versus time curve was recorded within the first 100 s. However, it soon thereafter disappeared, and after less than 1000 s the curve was identical to the one recorded at 25°C. It was concluded that the octameric form of β-lactoglobulin is probably not stable at the surface and rearranges itself into monomers/dimers.

III. ADSORPTION OF SELF-ASSOCIATING PROTEINS WITH UNORDERED STRUCTURE

The self-association of the proteins we discuss here appear in many cases above a certain concentration, corresponding to the critical micellar concentration (cmc) for simple surfactants [99]. In analogy with surfactants and block copolymers, the association pattern should have profound effect on the interfacial behavior as it determines the monomer concentration, which is constant above the cmc. The adsorption isotherms for surfactants usually feature a sharp increase followed by a plateau at concentrations, which can be related to the cmc of the surfactant [63,100–102]. The structure of the surfactant layer at the plateau depends on the surface properties. Monolayers with hydrophobic moiety toward the surface are most likely to occur at hydrophobic surfaces, while bilayers with the hydrophilic head group of the surfactant molecules facing the surface and the aqueous solution are most favorable on hydrophilic surfaces [63,100–102]. A range of modifications of these structures, as incomplete mono- or bilayers as well as bilayers with interpenetrating chains, is plausible. There is also convincing evidence of surfactant micelle adsorption/formation at interfaces [103,104].

The adsorption isotherms for sodium caseinate on hydrophobized chromium and at the air–water interface are compared in Fig. 10 [105]. Although a commercial sample which contains a mixture of caseins was used, the data demonstrate some important features of the interfacial behavior of a group of associating proteins, which is very important in a range of applications. The sharp increase in surface tension reduction followed by a plateau is similar to the behavior of a simple surfactant, and from the data a "cmc" of 0.1 mg/mL can be extrapolated. This is in agreement with the self-associating properties of sodium caseinate, which is considered to form aggregates of 250 kDa [106]. The adsorption isotherm for the hydrophobized chromium surface shows a similar sharp increase, although it occurs at 10 times lower concentration than the surface tension reduction increase. Furthermore, instead of a plateau, a maximum is discernible. Because the caseinate sample contains a mixture of caseins with different affinity to the interface, it may be relevant to correlate these results with those from adsorption experiments using mixed surfactant solutions [63]. The occurrence of a maximum in the adsorption isotherms for

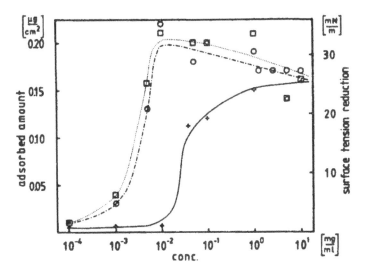

FIG. 10 Amount of sodium caseinate adsorbed on a hydrophobized chromium surface (squares and circles correspond to the two different surfaces used), obtained by in situ ellipsometry, and surface tension reduction (crosses), obtained using the drop-volume technique, versus concentration. The data refer to values attained after 40 min. The sodium caseinate was dissolved in pure water. (From Ref. 105.)

mixtures of surfactants with different affinity for the interface has been explained by
Trogus et al. [107]. They showed that the smaller the surface area per volume of
surfactant solution is, the larger is the maximum in the adsorption isotherm. The
isotherm for the solid–aqueous interface in Fig. 10 was obtained under conditions
where the surface per volume was about 20 times smaller than when recording the
corresponding isotherm for the air–aqueous interface [105]. The model of Trogus et
al. is based on the competition of two processes: adsorption at the surface and solubilization in the micelle [107]. Below the cmc, the monomer concentration increases
with the total surfactant concentration; thus the surfactant with the largest affinity
for the surface will dominate the adsorption process. Once micelles start to form,
this more-surface-active component will also be preferentially included in the micelles. This will lead to a decrease in the amount adsorbed, and a maximum in the
adsorption isotherm appears. Bearing in mind that proteins are far more complex
than surfactants, the model presented above can still be applied qualitatively on a
mixed system of associating proteins.

The concentration dependence of the adsorption of β-casein and the 1-93 segment of β-casein to a methylated (hydrophobic) silica surface is shown in Fig. 11
[108]. A considerably larger amount is adsorbed when increasing the concentration
from 0.1 to 1.0 mg/mL. The curve recorded at the higher concentration features a
plateau which, after about 200 s of adsorption, is followed by an additional increase
in the amount adsorbed. This is not observed at the lower concentration. The additional increase suggests a buildup of additional layer(s) and occurs parallel to the
association of β-casein in solution, which takes place at bulk concentrations larger

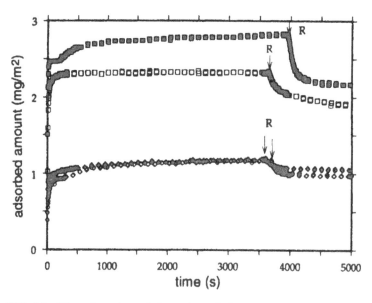

FIG. 11   The adsorption of β-casein and β-casein (1-93) on a hydrophobic silica surface
versus time at 0.1 ($\square$) and 1.0 ($\blacksquare$) mg/mL (42 μM) and 0.046 ($\diamond$) and 0.46 ($\blacklozenge$) mg/mL (42
μM), respectively, from a 0.02 M imidazol-HCl buffer, pH 7.0. The adsorption was recorded
by using in situ ellipsometry and the adsorption was after 1 h followed by flushing the cuvette
with pure buffer as indicated by R. (From Ref. 108.)

than 0.1 mg/mL [39,40]. A sharp increase in the amount of β-casein adsorbed at the air–water interface has previously been observed by Graham and Phillips at bulk concentrations larger than 0.1 mg/mL [109]. The fact that the protein adsorbed in this ad layer is also more easily desorbed (Fig. 11) suggests that the forces interacting with the interface are weaker than those of the molecules adsorbed in the initial stage of the adsorption. The formation of a second layer, as seen with other proteins [61,67,110,111], is therefore most probable. Also note that if the β-casein solutions were allowed to stand for some time before starting the experiment, a large irreproducible increase in the adsorbed amount occurred after a time lag of several minutes ($\approx$30 min) [108]. The initial adsorption up to monolayer coverage was, however, unaffected. In conclusion, the formation of the monolayer seems to take place via monomers and is thus independent of the aggregation state in the solution. This agrees with reports showing that the adsorption behavior is unaffected by a reduction of the temperature to 4°C [109] as well as by the presence of urea [112]. The N-terminal 1-93 peptide of β-casein lacks the end of the C-terminal hydrophobic part of β-casein [108]. Berry and Creamer [113] have shown that removal of the C-terminal 20 amino acid residues from β-casein destroys the latter's ability to associate in solution. As shown in Fig. 11, the adsorption of β-casein (1-93) at 0.046 and 0.46 mg/mL, which are the same molar concentrations as used for β-casein, gives the same final plateau value. However, a similar initial plateau as observed for β-casein also appeared at the higher concentration of β-casein (1-93) (Fig. 11). Therefore, it cannot be ruled out that aggregates are formed at the interface, since the peptide has an amphiphilic character like the native β-casein [108]. This could explain the reversible increase in the amount of protein adsorbed as in the case of β-casein.

A number of studies suggest that β-casein adsorbs with its hydrophobic part anchored to the hydrophobic surfaces, while the hydrophilic part of the molecule protrudes extensively into the aqueous phase [114–119]. A proteolytic enzyme endoproteinase Asp-N, which cleaves specifically at the N-terminal part of aspartate amino acid residues, has been used to test whether these residues [two residues (43 and 47) out of four in the whole protein] in the hydrophilic part of β-casein are accessible for the enzyme when the protein is adsorbed on a surface [108,120]. We will discuss these studies together with data from direct measurements of the interaction between adsorbed β-casein layers [121] to give a plausible picture of the adsorbed layer structure of β-casein on hydrophobic as well as hydrophilic surfaces.

The results for the adsorption of β-casein on silica surfaces, which have been made hydrophobic by gas-phase silanization, are shown in Fig. 12a [120]. The initial adsorption is fast and reaches relatively stable values after about 500 to 1000 s. The amount and protein volume fraction are almost unaffected by rinsing with protein-free buffer solution. It is noteworthy that addition of β-lactoglobulin at this stage has no effect on the adsorbed amount, but β-casein adsorbs to a preadsorbed layer of β-lactoglobulin, in line with β-casein's ability to work as a protective colloid [108]. Adding endoproteinase Asp-N causes a significant decrease in the surface excess. The reduction of β-casein on the surface by 24% corresponds nicely with a cleavage of the protruding N-terminal of the protein between residues 42-43 or 46-47 [108,120].

The forces between β-casein layers adsorbed onto mica surfaces, which have been made hydrophobic by Langmuir–Blodgett deposition of dimethyl-dioctadecyl-ammonium bromide (DDOAB), are shown versus surface separation in Fig. 12b

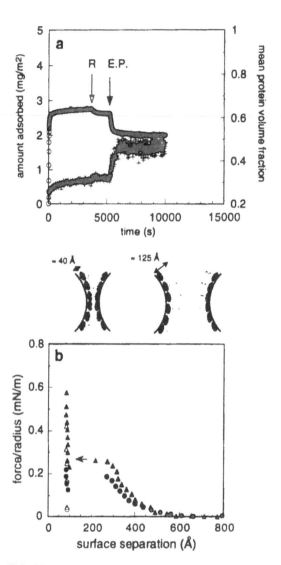

FIG. 12 (a) Time evolution of ellipsometry measurements of the amount adsorbed (○) and the mean protein volume fraction (+) at the hydrophobic silica surface after injection of β-casein (0.1 mg mL^{-1}, 0.02 M imidazol-HCl buffer pH 7.0), during rinsing (R) with protein-free buffer (pH 7), and after injection (E.P.) of endoproteinase Asp-N. The final concentration of endoproteinase Asp-N was 0.04 mg mL^{-1}. β-casein was injected at $t = 0$ and continuous rinsing was performed in the region between 3600 and 4000 s. The mean protein volume fraction was calculated from the mean thickness, amount adsorbed, and a partial specific volume for β-casein of 0.742 mL g^{-1}. The data are adopted from Kull et al. [120]. (b) The normalized force measured between hydrophobized mica surfaces coated with an adsorbed β-casein layer in a solution containing 0.1 mg/mL β-casein in 1 mM NaCl (pH 7.0) (▲, △) and after dilution in 1 mM NaCl (●, ○). Filled and unfilled symbols represent the force on compression and decompression, respectively. Note that the surface separation is given relative the contact between the DDOAB covered surfaces. The thickness of the DDOAB layer was ~20 Å. The arrows indicate jump due to instability of the spring. The solid line represents a DLVO fit, where the plane of charge and origin of the van der Waals forces are placed at the

[121]. A long-range repulsive force, most probably of electrostatic origin, prevails at large surface separation. The presence of double-layer forces is in agreement with the study of Brooksbank et al. [117], who observed that electrostatic repulsive forces contribute in preventing flocculation of β-casein-covered hydrophobic particles. As the surfaces are brought closer into contact, the repulsive force is overcome by an attractive force at a surface separation of about 250 Å (Fig. 12b). This causes the protein-covered surfaces to slide into contact. Further compression was found to not significantly change the surface separation, ≈ 80 Å. The force curve recorded on separation shows that the contact is adhesive, with a measured pull-off force of about 2.4 mN/m. Only a slight decrease in the magnitude of the double-layer force was observed, when the β-casein solution was replaced with pure 1 mM NaCl. Small-angle x-ray scattering data for layers adsorbed at the air–water interface suggest that bulk of the protein is confined in a dense layer within about 20 Å of the surface and that a segment of about 40 amino acid residues, the hydrophilic part, forms a more open and flexible structure and penetrates the solution, reaching about 100 Å out from the interface [115]. This layered structure of the adsorbed β-casein film has been confirmed by neutron reflectivity measurements of β-casein adsorbed to a hydrophobic self-assembled monolayer, formed on a silicon surface from octadecyltrichlorosilane solution [119]. The protein in the inner layer occupied 61% of the volume, while the volume fraction of protein in the outer layer was only about 12%. This means that the hydrophilic segments can well interpenetrate when β-casein-covered surfaces are brought into contact. For the same reason, the hydrophilic part of β-casein is accessible to attack by endoproteinase Asp-N when the protein is adsorbed at a surface.

We can now give a plausible picture of what is likely to happen when two β-casein layers, adsorbed on hydrophobic surfaces, approach each other (Fig. 12b). Initially an electrostatic double-layer force prevails until the surface separation reaches about 250 Å. At this separation we obtain a marked deviation from the double-layer force because the hydrophilic tails come in close contact, indicating that the layer thickness is about 125 Å. The segment densities of the hydrophilic tails are low enough to allow entanglement and bridging, which gives rise to an attractive force. Thus, the surfaces slide into contact. Further compression is not possible, as quite a compact layer is formed. When we subsequently tried to separate the surfaces, an attractive force was apparent, which at least partially depends on entanglement and/or bridging of the polypeptide chains.

Clearly, a β-casein molecule adsorbed to a hydrophobic surface is highly oriented. Thus, the surface properties are bound to affect the structure of the interfacial layer. The adsorption process of β-casein on the hydrophilic surface is much slower

---

onset of the steric wall (surface separation 82 Å), the Debye length is 96 Å and the surface potential is -48 mV. The top schematic figure shows the plausible structures of the adsorbed layers of β-casein on hydrophobized mica. The structure at large separations, where electrostatic repulsive forces dominate, and at small surface separations, where steric interactions dominate are shown. On hydrophobic surfaces a monolayer is assumed, with the hydrophilic chain (shown in dark gray) protruding into the solution and the hydrophobic chain (black) anchored at the surface. The data are adopted from Nylander and Wahlgren [121].

than at hydrophobic silica, and plateau conditions are not fully established within the range of the experiment (ca. 2.5 h) (Fig. 13a) [120]. Furthermore, the adsorbed amount at the end of the adsorption step is 4.3 mg m^{-1}, which is higher than the plateau surface excess of 2.8 mg m^{-2} measured at the hydrophobic surface. Also a substantially larger desorbable fraction is observed on the hydrophilic surface. It is clear that β-casein is much more accessible for endoproteinase Asp-N on the hydrophilic surface, leaving only 1.5 mg m^{-2}, compared to the corresponding reduction of 24% noted for β-casein on the hydrophobic surface. One possible explanation for this observation is the formation of a bilayer structure, where the initial step is the protein–surface interactions. It should be borne in mind that even if β-casein has a net negative charge, it still carries about 19 positive charges per molecule at neutral pH. This means that interaction with the negatively charged silica can occur by proper orientation of the protein at the interface. The protein molecules in the first layer are adsorbed in a flat conformation and are likely to make the surface more hydrophobic. This later step is therefore probably governed by protein–protein interactions, might involve extensive rearrangements within the protein layer, and is likely to be slower than the initial protein–surface interaction.

The interaction between the hydrophilic mica surfaces in the presence of β-casein is entirely repulsive (Fig. 13b), in contrast to corresponding force curves recorded for the interaction between hydrophobized mica (Fig. 12b) [121]. The exponential decay of the force versus distance suggests that double-layer forces contribute to the repulsive force between β-casein layers on hydrophilic surfaces (Fig. 13b). At surface separations close to 300 Å, a slightly lower force than expected for a pure double-layer force is observed, indicating onset of steric forces [121] (not visible in Fig. 13b). This surface separation, which is about 50 Å larger than on the hydrophobic surface, indicates that substantially more β-casein is adsorbed on the pure mica. This agrees with what we observed in the adsorption experiments when comparing the amounts of β-casein adsorbed on hydrophobized silica with those on pure, hydrophilic silica surfaces. When the β-casein solution is replaced with pure buffer, a drastic change in the force versus distance profile is observed for the interaction between β-casein layers on the hydrophilic surface (Fig. 13b) [121]. Not only is the magnitude of the double-layer force reduced, but the fact is that the interaction is no longer entirely repulsive as an attractive force is apparent at surface separations below about 300 Å. The attractive force causes the surfaces to spontaneously move toward each other until they come to the steric force wall, located about 130 Å out from the mica–mica contact (Fig. 13b) [121]. Most remarkable, however, is that the force curve, after rinsing, very much resembles the one recorded between β-casein layers on hydrophobized mica, although the steric wall seems to be located some 47 Å out. The resulting β-casein layers exposed to the aqueous solution apparently look similar on the two types of surfaces. It is therefore tempting to consider formation of a β-casein bilayer on the hydrophilic surface, where the outer layer is similar to the monolayer suggested to occur at the hydrophobic silica surface (Fig. 13b). The steric force observed on compressing a β-casein layer on the hydrophilic surface in the presence of 1 mM NaCl is similar to the corresponding layer on the hydrophobic surface. It seems not to be possible to remove the outer layer by applying a high compressive force. In parallel to the ellipsometry study, the surface force measurements show a larger and more reversible (larger desorption upon rinsing with pure buffer) adsorption on the hydrophilic surface compared to adsorption on the hydrophobized surface. Also, the reduction in surface excess of β-casein upon addition of endoproteinase Asp-N confirmed the difference in structure of the adsorbed layer between the two types of surfaces.

IV.   ADSORPTION IN RELATION TO (HEAT) AGGREGATION OF PROTEINS

The native protein structure is a consequence of a delicate balance of forces, including electrostatic forces, hydrogen bonding, van der Waals forces, conformational entropy, and so-called hydrophobic interactions [122–124]. As a consequence of increased thermal motion of amino acid residues in the polypeptide chain, most proteins undergo major conformational changes (denaturation) at elevated temperatures. The marginal stability of proteins makes the unfolding process very cooperative, and it can often be regarded as a two-stage process from a native to an unfolded state for a protein or a protein domain [122,125,126]. However, the unfolding as well as folding can, under certain conditions, occur via an intermediate state, the molten globule state [127–132]. The protein in this state, which can be achieved by pH changes, heating, denaturation agents, breaking disulfide bonds and removal of ligands, retains its secondary structure, whereas the tertiary is fluctuating. Thus the structure is more expanded, and more hydrophobic domains are exposed. It has also been suggested that the protein can adopt a molten globule state when it interacts with an interface [132]. The unfolding of the molten globule gives rise to little or no heat of absorption, making the process hard to monitor by calorimetric measurements [130]. In many cases thermal unfolding induces aggregation which eventually leads to precipitation of the protein. The thermally induced unfolding of β-lactoglobulin is quite complex due to the dissociation of the dimer at increasing temperature [84], as well as its tendency to aggregate [44,133,134]. In contrast to the other major whey protein, α-lactoglobulin, the unfolding of β-lactoglobulin is considered to be irreversible [135].

Results from several studies indicate that the unfolding of β-lactoglobulin is not a simple two-stage process. For instance, after the transitions at about 70–80°C some structure seems to remain, as an additional transition was observed at 120–140°C [136,137]. The high-temperature transition was related to the rupture of the disulfide bridges [136]. The use of high-resolution calorimeters has enabled the study of unfolding at lower concentration, and it was found that T_{max} of the transition (temperature at which heat capacity peaked) was minimal at concentrations when the β-lactoglobulin is expected to be in its monomeric state [138]. The thermograms recorded for the transition occurring at about 70–80°C were not in agreement with a two-state model [91,138], and the higher transition was suggested to involve the activation of a free-sulfhydryl group [91]. The value of T_{max} also decreases sharply with increasing pH in the neutral pH range [139], which can be correlated with the Tanford transition discussed in the beginning of Section II.B. In fact the rate of aggregation at 63°C was also found to increase with pH and connect to the proportion of unfolded β-lactoglobulin at the same temperature [134]. Several mechanisms have been suggested for the thermally induced aggregation of β-lactoglobulin [16,33,133]. Roefs and de Kruif presented a model analogous to an ordinary polymerization reaction, where the initiation step, involving activation of the sulfhydryl group, is

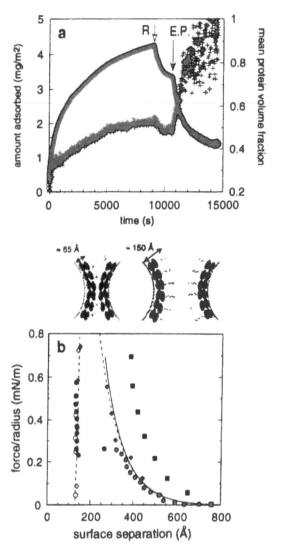

FIG. 13 (a) Time evolution of ellipsometry measurements of the amount adsorbed (○) and the mean protein volume fraction (+) at the hydrophilic (pure) silica surface after injection of β-casein (0.1 mg mL^{-1}, 0.02 M imidazol-HCl buffer pH 7.0), during rinsing (R) with protein-free buffer (pH 7), and after injection (E.P.) of endoproteinase Asp-N. The final concentration of Asp-N was 0.04 mg mL^{-1}. β-Casein was injected at $t = 0$ and continuous rinsing was performed in the region between 9000 and 9400 s. The mean protein volume fraction was calculated from the mean thickness, amount adsorbed, and a partial specific volume for β-casein of 0.742 mL g^{-1}. The data are adopted from Kull et al. [120]. (b) The normalized force measured between pure hydrophilic mica surfaces coated with an adsorbed β-casein layer in a solution containing 0.1 mg/mL β-casein and 1 mM NaCl (pH 7.0) (■) and after dilution in 1 mM NaCl (♦, ◇). The force curve for β-casein on hydrophobized mica (●, ○) after the protein solution has been removed and replaced with pure 1 mM NaCl is also inserted and the curve has been shifted 47 Å outward to facilitate comparison. The unfilled symbols indicate the curve recorded during decompression in 1 mM NaCl. The dashed and solid lines are the calculated DLVO curves for constant surface charge and potential, respectively. The

followed by a propagation step, which involves exchange with only one of the two disulfide bonds to give one new activated sulfhydryl group [44]. The chain of reactions is terminated when two activated sulfhydryl groups react [44]. This will render a linear polymer. However, other studies have shown that aggregation can occur, although to a lesser extent, when the free-sulfhydryl groups are blocked and no sulfhydryl–disulfide bond exchange can occur [140].

The amount of β-lactoglobulin adsorbed on a hydrophilic chromium surface is shown in Fig. 14 as a function of time and temperature [141,142]. The adsorption did not change very much in the temperature range between 25 and 73°C. However, when the temperature of the thermally induced unfolding was approached, a sharp increase is observed after a time lag of several minutes (Fig. 14). As the temperature increases further, the induction period decreases. This is also observed for aggregation of β-lactoglobulin in bulk solution [91,134]. Such interfacial behavior suggests that aggregation at the surface will occur when a sufficient number of molecules have been activated (unfolded) in the bulk solution. It should also be borne in mind that the surface in itself also can activate β-lactoglobulin to aggregate. A time-dependent polymerization of β-lactoglobulin at the oil–water interface of emulsion droplets was observed already at 25°C [143]. Furthermore it was found that no polymerization of the protein took place if the free-sulfhydryl group was blocked by N-ethylmaleimide (NEM) [91]. The activating role of the surface is also suggested by the results from our β-lactoglobulin study, where we found that aggregation at the interface was observed at a lower temperature compared to when visible aggregation was found in bulk solution [142]. Preheating of the solution above the protein-unfolding temperature in the absence of the surface prevented the exponential increase in the amount adsorbed, and the observed plateau values were almost the same as those recorded at room temperature [142]. This observation suggests that activated molecules, maybe with a reactive sulfhydryl, are deactivated during the preheating, as they form aggregates which do not seem to stick to the surface [91]. Similar observations were made by Jeurnink et al., who found that the activated intermediates formed in bulk solution at elevated temperature were necessary for the continuous increase in adsorption of pure β-lactoglobulin as well as of a whey protein mixture [144]. If the temperature was too low, not enough activated molecules were present, and if it was too high the activated molecules aggregated before they reached the interface. The adsorption in both cases was restricted to a protein monolayer [144]. In fact they could correlate the deposition rate with the concentration of activated molecules in solution [144], using the model for the activation and aggregation of β-lactoglobulin presented by Roefs and de Kruif [44]. It is noteworthy that the other main whey protein, α-lactoglobulin, does not show this exponential increase

---

parameters used for the calculations are a surface potential of -64 mV and a Debye length of 96 Å and the onset of the forces is located at the steric wall 130 Å from the mica–mica contact. The top schematic shows the plausible structures of the adsorbed layers of β-casein on pure hydrophilic mica. A bilayer is assumed on the hydrophilic surface, where the inner layer is more dense and makes the surface more hydrophobic, which leads to the adsorption of a second layer with basically the same structure as the protein monolayer on the hydrophobic surface. The data are adopted from Nylander and Wahlgren [121].

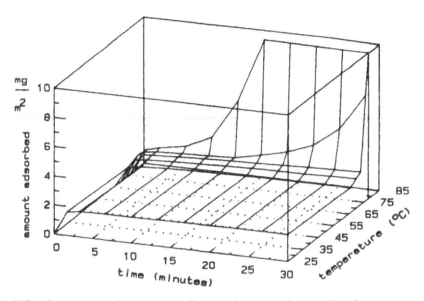

FIG. 14 Amount of β-lactoglobulin adsorbed on a hydrophilic (bare) chromium surface versus time and temperature, using in situ ellipsometry. A protein concentration of 1 mg/mL in phosphate buffered saline (PBS) (pH 6.0, 0.01 M phosphate, 0.15 M sodium chloride, ionic strength 0.17) was used. The data are adopted from Arnebrant et al. [142].

in the adsorbed amount, not even far above temperatures where the thermally induced unfolding of the protein occurs [142]. The interfacial behavior of α-lactalbumin can be related to its low tendency to aggregate when heated, which also causes the weak gelling properties of the protein [139]. Around the unfolding temperature of α-lactalbumin the adsorption instead increased gradually with temperature, and a plateau in the amount adsorbed was always reached [142]. Similar behavior was also observed for insulin in the temperature range close to its unfolding temperature [145].

The strong correlation between aggregation in bulk and at interface has been clearly demonstrated by Elofsson et al., and their results for the adsorption of β-lactoglobulin A and B on a chromium surface as well as the bulk aggregation under similar conditions are presented in Fig. 15. These data were recorded at 68°C and pH 6.9. At this temperature no surface aggregation was observed in the earlier study at pH 6.0 [142], from which the results in Fig. 14 are taken. Different samples of β-lactoglobulin were used in the two studies, but most likely the difference in interfacial behavior can be related to the lower heat stability [139] and larger tendency for the protein to aggregate [134] at higher pH. The onset of aggregation at the interface as well as in bulk solution is delayed for slightly longer time for β-lactoglobulin A compared to variant B (Fig. 15). This can be related to the difference in monomer–dimer equilibrium (Section II.B) [91], which probably is important also at higher temperatures [138]. The A variant is also more negatively charged, which might hamper the initial stage of the aggregation [91]. After the lag period, the

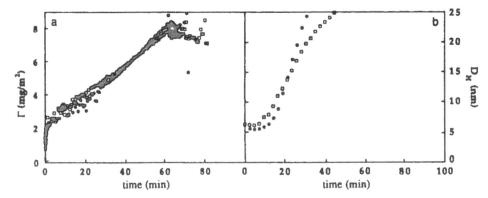

FIG. 15 The correlation between aggregation of β-lactoglobulin A (●) and B (□) on a bare (hydrophilic) chromium surface and the aggregation in bulk solution. The proteins were dissolved in 0.50 M phosphate buffer containing 0.1 M NaCl, pH 6.88 giving a final protein concentration of 8 mg/mL. The temperature was kept at 68°C. (a) The adsorbed amount, recorded by in situ ellipsometry. Rinsing was performed with pure buffer after 60 min for 5 min at a flow rate of 20 mL buffer/min. (b) Hydrodynamic diameter, D_H, in solution as measured by dynamic light scattering. (From Ref. 91.)

aggregation at the interface is faster for the β-lactoglobulin A than the B variant (Fig. 15a), which is also observed for the aggregation in solution (Fig. 15b) as well as the protein unfolding [91].

The importance of the sulfhydryl group for the aggregation of β-lactoglobulin has been discussed previously, and consequently it was found that the presence of NEM, which blocks the free-sulfhydryl group, prevented the buildup of the protein multilayer at the interface [91]. The effect was found to be similar to a decrease in temperature to 65°C, which is below the second thermal transition (onset at about 65°C) assigned to the increased reactivity of the sulfhydryl group [91]. However, if a layer of aggregated β-lactoglobulin was treated with NEM to block available sulfhydryl groups and a second addition of β-lactoglobulin at 68°C was performed, the aggregation at the interface was resumed. The propagation step of the aggregation is thus not controlled entirely by the formation of disulfide bonds. The aggregation of β-lactoglobulin in bulk solution was also found to proceed in the presence of a sulfhydryl blocking agent (NEM), and in this case the aggregation started without any notable time lag [140]. Anyhow, the interaction between the molecules in the surface aggregates is strong and was found to increase with residence time and temperature, as observed from the fraction which was removed by the addition of sodium dodecylsulfate (SDS) [146]. A stronger interaction was observed between β-lactoglobulin and the methylated (hydrophobic) silica surface compared with the hydrophilic chromium surface, both in terms of higher amounts adsorbed and adsorption rate as well as the smaller fraction removed by SDS [147]. However, the main features of the adsorption, in terms of for instance the time lag of the large increase in surface concentration, are similar at the two surfaces and occur at the same temperatures [147].

V. CONCLUDING REMARKS

We can conclude that the protein adsorption at an interface is very much dependent on protein bulk solution behavior. Thus, to fully understand the interfacial behavior of associating proteins, knowledge of their bulk behavior is crucial. From this it follows that it is important to use well-characterized samples, and with proteins one definitely cannot always rely on commercially available samples. The development of theoretical models or general principles is a challenge indeed, bearing in mind the number of variables which have to be considered. However, theoretical models which try to correlate protein aggregation and interfacial behavior, either the equilibrium ones such as those presented by Chatelier and Minton [148] or those which model the adsorption kinetics which were presented by Wahlgren and Elofsson [94], can aid the researcher to get on the right track. The use of genetic variants or mutant proteins with different self-association is, as demonstrated here, a very powerful tool to study interfacial processes. Some general rules, based on this review, can be put forward:

1. Monomers are usually more hydrophobic, as the association into oligomers is usually, to a large extent, determined by hydrophobic interactions. Monolayers of monomers are therefore often formed on hydrophobic surfaces, and, particularly when monomers dominate in solution (e.g., at low total concentration), the adsorption is much stronger to a hydrophobic surface than to a hydrophilic one.
2. The adsorption of large aggregates are usually favored, particularly at hydrophilic surfaces. Even if they are present at small concentration, they usually win the race for the interface, as they, due to their size, can interact with a larger number of binding sites. Of course, the interaction between proteins and the hydrophilic surface is very much controlled by electrostatic forces and here the larger net charge of an oligomer compared to a monomer can be the determining factor.
3. Regarding the effect of heat, it can be concluded that if a protein shows a strong tendency to aggregate in solution when heated, it will certainly form large deposits on a surface under the same conditions. In fact, the aggregation at an interface is often directly correlated with the one occurring in bulk. The surface can also play an activating role on the protein and thus promote aggregation. A solution containing proteins which have been deactivated by aggregate formation during preheating or by adding agents which block reactive groups usually do not give large deposits on the surface.

ACKNOWLEDGMENT

I would like to express my gratitude to Thomas Arnebrant, Hugo Christenson, Per Claesson, Peter Cuypers, Ulla Elofsson, Bengt Ivarsson, Patrick Kékicheff, Kåre Larsson, Barry Ninham, Marie Paulsson, Valdemaras Razumas, Fredrik Tiberg, Magnus Wahlgren, and Marie Wahlgren (have I forgotten somebody?) for very interesting discussions on protein adsorption and many other useful things.

REFERENCES

1. C. Tanford, Protein denaturation. Adv. Protein Chem. *23*:121–275, 1968.
2. T. Blundell, G. Dodson, D. Hodgkin, and D. Mercola, Insulin: the structure in the crystal and its reflection in chemistry and biology. Adv. Protein Chem. *26*:279–402, 1972.
3. J. Brange, B. Skelbaek-Pedersen, L. Langkjaer, U. Damgaard, H. Ege, S. Havelund, L. G. Heding, K. H. Jørgensen, J. Lykkeberg, J. Markussen, M. Pingel, and E. Rasmussen, *Galenics of Insulin. The Physico-Chemical and Pharmaceutical Aspects of Insulin and Insulin Preparations.* Springer-Verlag, Berlin, 1987.
4. J. Brange, U. Ribel, J. F. Hansen, G. Dodson, M. T. Hansen, S. Havelund, S. G. Melberg, F. Norris, K. Norris, L. Snel, A. R. Sørensen, and H. O. Voigt, Monomeric insulins obtained by protein engineering and their medical implications. Nature *333*: 679–682, 1988.
5. D. F. Waugh, D. F. Wilhelmson, S. L. Commerford, and M. L. Sackler, Studies of the nucleation and growth reactions of selected types of insulin fibrils. J. Am. Chem. Soc. *75*:2592–2600, 1953.
6. J. Brange, S. Havelund, P. E. Hansen, L. Langkjaer, E. Sørensen, and P. Hildebrandt, Formulation of physically stable insulin solution for continuous infusion by delivery systems: In *Hormone Drugs* (J. L. Gueriguian, E. D. Bransome, and A. S. Outschorn, eds.). United States Pharmacopeial Convention, Rockville, MD, 1982, pp. 96–105.
7. H. Thurow and K. Geisen, Stabilisation of dissolved proteins against denaturation at hydrophobic surfaces. Diabetologia *27*:212–218, 1984.
8. J. Bringer, A. Heldt, and G. M. Grodsky, Prevention of insulin aggregation by dicarboxylic amino acids during prolonged infusion. Diabetes *30*:83–85, 1981.
9. S. Sato, C. D. Ebert, and S. W. Kim, Prevention of insulin self-association and surface adsorption. J. Pharm. Sci. *72*:228–232, 1983.
10. J. Brange, J. F. Hansen, S. Havelund, and S. G. Melberg, Studies of the insulin fibrillation process. In *Advanced Models for the Therapy of Insulin-Dependent Diabetes* (P. Brunetti, W. Waldhäusl, eds.). Raven Press, New York, 1987, pp. 85–90.
11. M. J. Burke and M. A. Rougvie, Cross-β protein structures. I. Insulin fibrils. Biochemistry *11*:2435–2439, 1972.
12. J. Brange, S. Havelund, E. Hommel, E. Sørensen, and C. Kühl, Neutral insulin solutions physically stabilised by addition of Zn^{2+}. Diabetic Med. *3*:532–536, 1986.
13. J. Brange and S. Havelund, Properties of insulin in solution. In *Artificial Systems for Insulin Delivery* (P. Brunetti, K. G. M. M. Alberti, A. M. Albisser, K. D. Hepp, M. M. Benedetti, eds.). Raven Press, New York, 1983, pp. 83–88.
14. A. M. Albisser, W. Lougheed, K. Perlman, and A. Bahoric, Nonaggregating insulin solutions form long-term glucose control in experimental and human diabetes. Diabetes *29*:241–243, 1980.
15. T. Mizutani, Decreased activity of proteins adsorbed onto glass surfaces with porous glass as a reference. J. Pharm. Sci. *69*:279–282, 1980.
16. H. A. McKenzie, β-Lactoglobulins. In *Milk proteins. 2. Chemistry and Molecular Biology* (C. B. Anfinsen, M. L. Anson, and J. T. Edsall, eds.). Academic Press, New York, pp. 257–330, 1971.
17. W. N. Eigel, J. E. Butler, C. A. Ernstrom, H. M. Farrel Jr., V. R. Harwalkar, and R. M. Whitney, Nomenclature of proteins of cow's milk: fifth revision. J. Dairy Sci. *67*: 1599–1631, 1984.
18. P. Walstra and R. Jenness, *Dairy Chemistry and Physics.* Wiley-Interscience, New York, 1984.
19. S. Brownlow, J. H. Morais Cabral, R. Cooper, D. R. Flowers, S. J. Yewdall, I. Polikarpov, A. C. T. North, and L. Sawyer, Bovine β-lactoglobulin at 1.8 Å resolution—still an enigmatic lipocalin. Structure *5*:481–495, 1997.

20. M. J. Papiz, L. Sawyer, E. E. Eliopoulos, A. C. T. North, B. C. Findlay, R. Sivapra-
    sadarao, T. A. Jones, M. E. Newcomer, and P. J. Kraulis, The structure of β-lactoglob-
    ulin and its similarity to plasma retinol-binding protein. Nature 324:383–385, 1986.
21. L. Sawyer, One fold among many. Nature 327:659, 1987.
22. A. C. T. North, Three-dimensional arrangement of conserved amino acid residues in a
    superfamily of specific ligand binding proteins. Int. J. Biol. Macromol. 81:56–58, 1989.
23. M. C. Diaz de Villegas, R. Oria, F. J. Salva, and M. Calvo, Lipid binding by β-
    lactoglobulin of cow milk. Milchwissenschaft 42:357–358, 1987.
24. P. Puyol, M. D. Perez, J. M. Peiro, and M. Calvo, Effect of binding of retinol and
    palmitic acid to bovine β-lactoglobulin on its resistance to thermal denaturation. J.
    Dairy Sci. 77:1494–1502, 1994.
25. D. K. Sarker, P. J. Wilde, and D. C. Clark, Competitive adsorption of L-α-lysophos-
    phatidylcholine/β-lactoglobulin mixtures at the interfaces of foams and foam lamellae.
    Colloids Surfaces B: Biointerfaces 3:349–356, 1995.
26. L. K. Creamer, Effect of sodium dodecyl sulfate and palmitic acid on the equilibrium
    unfolding of β-lactoglobulin. Biochemistry 34:7170–7176, 1995.
27. A. Kristensen, T. Nylander, M. Paulsson, and A. Carlsson, Calorimetric studies of
    interactions between β-lactoglobulin and phospholipids in solutions. Int. Dairy J. 7:
    87–92, 1997.
28. E. N. Baker, T. L. Blundell, J. F. Cutfield, S. M. Cutfield, D. M. Crowfoot Hodgkin,
    R. E. Hubbard, N. W. Isaccs, C. D. Reynolds, K. Sakabe, N. Sakabe, and N. M. Vijayan,
    The structure of 2Zn pig insulin crystals at 1.5 Å resolution. Phil. Trans. R. Soc. Lond.
    B 319:369–456, 1988.
29. C. Holt, Structure and stability of bovine casein micelles. Adv. Protein Chem. 43:63–
    151, 1992.
30. H. J. Vreeman, J. A. Brinkhuis, and C. A. van der Spek, Some association properties
    of bovine SH-κ-casein. Biophys. Chem. 14:185–193, 1981.
31. T. S. Brun, H. Høiland, and E. Vikingstad, Partial molal volumes and isentropic partial
    molal compressibilities of surface-active agents in aqueous solution. J. Colloid Interface
    Sci. 63:89–96, 1978.
32. D. G. Schmidt, Association of caseins and casein micelle structure. In *Developments
    in Dairy Chemistry, 1* (P. F. Fox, ed.). Applied Science, London, 1982, pp. 61–86.
33. I. Heertje, J. Visser, and P. Smits, Structure formation in acid milk gels. Food Micro-
    structure 4:267–277, 1985.
34. P. Walstra, The voluminosity of bovine casein micelles and some of its implications.
    J. Dairy Res. 46:317–323, 1979.
35. D. S. Horne, Steric stabilization and casein micelle stability. J. Colloid Interface Sci.
    111:250–260, 1986.
36. C. G. de Kruif, Skim milk acidification. J. Colloid Interface Sci. 185:19–25, 1997.
37. C. Holt and L. Sawyer, Primary and secondary structure of the caseins in relation to
    their biological role. Protein Eng. 2:251–259, 1988.
38. C. Holt, M. N. Wahlgren, and T. Drakenberg, Ability of β-casein phosphopeptide to
    modulate the precipitation of calcium phosphate by forming amorphous dicalcium
    phosphate nanoclusters. Biochem. J. 314:1035–1039, 1996.
39. D. G. Schmidt and T. A. J. Payens, The evaluation of positive and negative contribu-
    tions to the second virial coefficient of some milk proteins. J. Colloid Interface Sci.
    39:655–662, 1972.
40. A. L. Andrews, D. Atkinson, M. T. A. Evans, E. G. Finer, J. P. Green, M. C. Phillips,
    and R. N. Robertson, The conformation and aggregation of bovine β-casein A. I.
    Molecular aspects of thermal aggregation. Biopolymers 18:1105–1121, 1979.
41. T. A. Payens and B. W. Markwijk, Some features of the association of β-casein.
    Biochem. Biophys. Acta 71:517–530, 1963.

42. H. Burton, Reviews of the progress of dairy science. Section G. Deposits from whole milk in heat treatment plant—a review and discussion. J. Dairy Res. 35:317–330, 1968.

43. M. Lalande, J-P. Tissier, and G. Corrieu, Fouling of heat transfer surfaces related to β-lactoglobulin denaturation during heat processing of milk. Biotechnol. Prog. 1:131–139, 1985.

44. S. P. F. M. Roefs and K. G. de Kruif, A model for the denaturation and aggregation of β-lactoglobulin. Eur. J. Biochem. 226:883–889, 1994.

45. P. J. Skudder, E. L. Thomas, J. A. Pavey, and A. G. Perkin, Effects of adding potassium iodate to milk before UHT treatment. I. Reduction in the amount of deposit on the heated surfaces. J. Dairy Res. 48:99–113, 1981.

46. S. M. Gotham, P. J. Fryer, and A. M. Pritchard, Model studies of food fouling. In *Fouling and Cleaning in Food Processing* (H. G. Kessler, D. B. Lund, eds.). University of Munich, Munich, 1989.

47. C. Chothia and J. Janin, Principles of protein–protein recognition. Nature 256:705–708, 1975.

48. S. Strazza, R. Hunter, E. Walker, and D. W. Darnall, The thermodynamics of bovine and porcine insulin and proinsulin association determined by concentration difference spectroscopy. Arch. Biochem. Biophys. 238:30–42, 1985.

49. W. Kadima, L. Øgendal, R. Bauer, N. Kaarsholm, K. Brodersen, J. F. Hansen, and P. Porting, The influence of ionic strength and pH on the aggregation properties of zinc-free insulin studied by static and dynamic light scattering. Biopolymers 33:1643–1657, 1993.

50. J. S. Pedersen, S. Hansen, and R. Bauer, Aggregation behaviour of zinc-free insulin studied by small-angle neutron scattering: analysis by use of a thermodynamic equilibrium model. Progr. Colloid Polym. Sci. 98:215–218, 1995.

51. N. C. Kaarsholm, S. Havelund, and P. Hougaard, Ionization behaviour of native and mutant insulins: pK perturbation of B13-Glu in aggregated species. Arch. Biochem. Biophys. 283:496–502, 1990.

52. E. Fredericq, The association of insulin molecular units in aqueous solutions. Arch. Biochem. Biophys. 65:218–228, 1956.

53. V. Ramesh and J. H. Bradbury, ¹H NMR studies of insulin. Reversible transformation of 2-zinc to 4-zinc insulin hexamer. Int. J. Pept. Protein Res. 28:146–153, 1986.

54. M. J. Adams, T. L. Blundell, E. J. Dodson, G. G. Dodson, M. Vijayan, E. N. Baker, M. M. Harding, D. C. Hodgkin, B. Rimmer, and S. Sheat, Structure of rhombohedral 2 zinc insulin crystals. Nature 224:491–495, 1969.

55. M. F. Dunn, S. E. Pattison, M. C. Storm, and E. Quiel, Comparison of the zinc binding domains in the 7S nerve growth factor and the zinc insulin hexamer. Biochemistry 19:718–725, 1980.

56. P. T. Grant, T. L. Coombs, and B. H. Frank, Differences in the nature of the interaction of insulin and proinsulin with zinc. Biochem. J. 126:433–440, 1972.

57. S. O. Emdin, G. G. Dodson, J. M. Cutfield, and S. M. Cutfield, Role of zinc in insulin biosynthesis. Some possible zinc–insulin interactions in the pancreatic B-cell. Diabetologia 19:174–182, 1980.

58. P. Nilsson, T. Nylander, and S. Havelund, Adsorption of insulin on solid surfaces in relation to the surface properties of the monomeric and oligomeric forms. J. Colloid Interface Sci. 144:145–152, 1991.

59. T. Nylander, P. Kékicheff, and B. W. Ninham, The effect of solution behavior of insulin on the interactions between adsorbed layers of insulin. J. Colloid Interface Sci. 164:136–150, 1994.

60. Y. Pocker and S. B. Biswas, Self-association of insulin and the role of hydrophobic bonding: a thermodynamic model of insulin dimerization. Biochemistry 20:4354–4361, 1981.

61. T. Arnebrant and T. Nylander, Adsorption of insulin on metal surfaces in relation to association behavior. J. Colloid Interface Sci. *122*:557–566, 1988.

62. G. A. Parks, The isoelectric point of solid oxides, solid hydroperoxides and aqueous complex systems. Chem. Rev. *65*:177–198, 1965.

63. T. Arnebrant, K. Bäckström, B. Jönsson, and T. Nylander, An ellipsometry study of ionic surfactant adsorption on chromium surfaces. J. Colloid Interface Sci. *128*:303–312, 1989.

64. A. S. Brill and J. H. Venable, The binding of transition metal ions in insulin crystals. J. Mol. Biol. *36*:343–353, 1968.

65. V. Razumas, J. Kulys, T. Arnebrant, T. Nylander, and K. Larsson, Insulin adsorption of platinum. Électrokhimiya *24*:1518–1521, 1988.

66. P. M. Claesson, E. Blomberg, J. C. Fröberg, T. Nylander, and T. Arnebrant, Protein interactions at solid surfaces. Adv. Colloid Interface Sci. *57*:161–227, 1995.

67. P. M. Claesson, T. Arnebrant, B. Bergenståhl, and T. Nylander, Direct measurements of the interaction between layers of insulin adsorbed on hydrophobic surfaces. J. Colloid Interface Sci. *130*:457–466, 1989.

68. L. Vroman and A. L. Adams, Adsorption of proteins out of plasma and solution in narrow spaces. J. Colloid Interface Sci. *111*:391–402, 1986.

69. P. Wojciechowski, P. ten Hove, and J. L. Brash, Phenomology and mechanism of the transient adsorption of fibrinogen from plasma (Vroman effect). J. Colloid Interface Sci. *111*:455–465, 1986.

70. H. G. W. Lensen, D. Bargeman, P. Bergveld, C. A. Smolders, and J. Feijen, High-performance liquid chromatography as a technique to measure competitive absorption of plasma proteins onto lattices. J. Colloid Interface Sci. *99*:1–8, 1984.

71. M. Roy, R. W.-K. Lee, N. C. Kaarsholm, H. Thøgersen, J. Brange, and M. F. Dunn, Sequence-specific ^{1}H-NMR assignments for the aromatic region of several biologically active, monomeric insulins including native human insulin. Biochim. Biophys. Acta *1053*:63–73, 1990.

72. G. L. Gaines and D. Tabor, Surface adhesion and elastic properties of mica. Nature (London) *178*:1304–1305, 1956.

73. G. L. Gaines, The ion-exchange properties of muscovite mica. J. Phys. Chem. *61*:1408–1413, 1957.

74. S. G. Hambling, A. S. McAlpine, and L. Sawyer, β-Lactoglobulin. In *Advanced Dairy Chemistry, Vol. 1: Proteins* (P. Fox, ed.) Elsevier Applied Science, London, 1992, pp. 141–190.

75. D. W. Green and R. Aschaffenburg, Twofold symmetry of β-lactoglobulin molecule in crystals. J. Mol. Biol. *1*:54–64, 1959.

76. J. Witz, S. N. Timasheff, and V. Luzzati, Molecular interactions in β-lactoglobulin. VIII. Small-angle x-ray scattering investigation of the geometry of β-lactoglobulin A tetramerization. J. Am. Chem. Soc. *86*:168–173, 1964.

77. D. C. Clark, F. Husband, P. J. Wilde, M. Cornec, R. Miller, J. Krägel, and R. Wüstneck, Evidence of extraneous surfactant adsorption altering adsorbed layer properties of β-lactoglobulin. J. Chem. Soc. Faraday Trans. *91*:1991–1996, 1995.

78. R. Townend, L. Weinberger, and S. N. Timasheff, Molecular interactions in β-lactoglobulin. IV. The dissociation of β-lactoglobulin below pH 3.5. J. Am. Chem. Soc. *82*:3175–3179, 1960.

79. O. E. Mills and L. K. Creamer, Conformational change in bovine β-lactoglobulin at low pH. Biochim. Biophys. Acta *379*:618–626, 1975.

80. C. Tanford, L. G. Bunville, and Y. Nozaki, The reversible transformation of β-lactoglobulin at pH 7.5. J. Am. Chem. Soc. *81*:4032–4036, 1959.

81. S. N. Timasheff, L. Mescanti, J. J. Basch, and R. Townend, Conformational transitions of bovine β-lactoglobulin A, B and C. J. Biol. Chem. *241*:2496–2501, 1966.

82. J. K. Zimmerman, G. H. Barlow, and I. M. Klotz, Dissociation of β-lactoglobulin near neutral pH. Arch. Biochem. Biophys. *138*:101–109, 1970.

83. H. A. McKenzie and W. H. Sawyer, Effect of pH on β-lactoglobulins. Nature *214*: 1101–1104, 1967.

84. C. George, S. Cuinand, and T. Tonnelat, Etude thermodynamique de la dissociation réversible de la beta-lactoglobuline B pour des pH supérieurs á 5.5. Biochim. Biophys. Acta *59*:737–739, 1962.

85. R. Townend, R. J. Winterbottom, and S. N. Timasheff, Molecular interactions in β-lactoglobulin. II. Ultracentrifugal and electrophoretic studies of association of β-lactoglobulin below its isoelectric point. J. Am. Chem. Soc. *82*:3161–3168, 1960.

86. R. Townend and S. N. Timasheff, Molecular interactions in β-lactoglobulin. III. Light scattering investigation of the stoichiometry of association between pH 3.7 and 5.2. J. Am. Chem. Soc. *82*:3168–3174, 1960.

87. J. J. Basch and S. N. Timasheff, Hydrogen ion equilibria of the genetic variants of bovine β-lactoglobulin. Arch. Biochem. Biophys. *118*:37–47, 1967.

88. H. Swaisgood, Chemistry of milk protein. In: *Developments in Dairy Chemistry, I* (P. Fox, ed.). Applied Science, London, pp. 1–59, 1982.

89. T. F. Kumosinski and S. N. Timasheff, Molecular interactions in β-lactoglobulin. X. Stoichiometry of the β-lactoglobulin mixed tetramerization. J. Am. Chem. Soc. *88*: 5635–5642, 1966.

90. H. A. McKenzie, W. H. Sawyer, and M. B. Smith, Optical rotary dispersion and sedimentation in the study of association-dissociation: Bovine β-lactoglobulins near pH 5. Biochim. Biophys. Acta *147*:73–92, 1967.

91. U. M. Elofsson, M. A. Paulsson, P. Sellers, and T. Arnebrant, Adsorption during heat treatment related to the thermal unfolding/aggregation of β-lactoglobulins A and B. J. Colloid Interface Sci. *183*:408–415, 1996.

92. U. M. Elofsson, M. A. Paulsson, and T. Arnebrant, Adsorption of β-lactoglobulin A and B in relation to self-association: effect of concentration and pH. Langmuir *13*: 1695–1700, 1997.

93. U. M. Elofsson, M. A. Paulsson, and T. Arnebrant, Adsorption of β-lactoglobulin A and B in relation to self-association: effects of ionic strength and phosphate ions. Colloids Surfaces B: Biointerfaces *8*:163–169, 1997.

94. M. Wahlgren and U. Elofsson, Simple models for the adsorption kinetics and their correlation to the adsorption of β-lactoglobulin A and B. J. Colloid Interface Sci. *188*: 121–129, 1997.

95. A. R. Mackie, J. Mingins, R. Dunn, and A. N. North, Preliminary studies of β-lactoglobulin adsorbed on polystyrene latex. In *Food, Polymers, Gels, and Colloids*, 82 (E. Dickinson, ed.). Royal Society of Chemistry, Cambridge, 1991, pp. 96–112.

96. D. W. Cooper, Parking problem (sequential packing) simulations in two and three dimensions. J. Colloid Interface Sci. *119*:442–450, 1987.

97. G. Y. Onoda and E. G. Liniger, Experimental determination of the random-parking limit in two dimensions. Phys. Rev. A *33*:715–716, 1986.

98. W. Norde, Adsorption of proteins from solution at the solid–liquid interface. Adv. Colloid Interface Sci. *25*:267–340, 1986.

99. B. Lindman and H. Wennerström, Micelles. Amphiphile aggregation in aqueous solution. Topics Current Chem. *87*:1–83, 1980.

100. R. M. Pashley and J. N. Israelachvili, A comparison of surface forces and interfacial tension of mica in purified surfactant solutions. Colloids Surfaces *2*:169–187, 1981.

101. J. F. Scamehorn, R. S. Schechter, and W. H. Wade, Adsorption of surfactants on mineral oxide surfaces from aqueous solution. J. Colloid Interface Sci. *85*:463–478, 1982.

102. P. Somasundran and J. T. Kunjappu, In-situ investigations of adsorbed surfactants and polymers on solids in solution. Colloids Surfaces *2*:245–268, 1989.

103. F. Tiberg, B. Jönsson, J. Tang, and B. Lindman, Ellipsometry studies of the self-assembly of nonionic surfactants at the silica–water interface: Equilibrium aspects. Langmuir 10:2294–2300, 1994.

104. F. Tiberg, B. Jönsson, and B. Lindman, Ellipsometry studies of the self-assembly of nonionic surfactants at the silica-water interface: kinetic aspects. Langmuir 10:3714–3722, 1994.

105. T. Arnebrant, T. Nylander, P. A. Cuypers, P.-O. Hegg, and K. Larsson, Relation between adsorption on a metal surface and monolayer formation at the air/water interface from amphiphilic solutions. In Surfactants in Solution, II (K. L. Mittal and B. Lindman, eds.). Plenum, New York, pp. 1291–1299, 1984.

106. L. K. Creamer and G. D. Berry, A study of the dissociated bovine micelles. J. Dairy Res. 42:169–183, 1975.

107. F. J. Trogus, R. S. Schechter, and W. H. Wade, A new interpretation of adsorption maxima and minima. J. Colloid Interface Sci. 70:293–305, 1979.

108. T. Nylander and N. M. Wahlgren, Competitive and sequential adsorption of β-casein and β-lactoglobulin on hydrophobic surfaces and the interfacial structure of β-casein. J. Colloid Interface Sci. 162:151–162, 1994.

109. D. E. Graham and M. C. Phillips, Proteins at liquid interfaces. II. Adsorption isotherms. J. Colloid Interface Sci. 70:415–426, 1979.

110. T. Arnebrant, B. Ivarsson, K. Larsson, I. Lundström, and T. Nylander, Bilayer formation at adsorption of proteins from aqueous solutions on metal surfaces. Prog. Colloid Polymer Sci. 70:62–66, 1985.

111. P. Kékicheff, W. A. Ducker, B. W. Ninham, and M. P. Pileni, Multilayer adsorption of cytochrome c on mica around isoelectric pH. Langmuir 6:1704–1708, 1990.

112. D. G. Dalgleish, The conformations of proteins on solid/water interfaces—caseins and phosvitin on polystyrene lattices. Colloids Surfaces 46:141–155, 1990.

113. G. P. Berry and L. K. Creamer, The association of bovine β-casein. The importance of the C-terminal region. Biochemistry 14:3542–3545, 1975.

114. D. G. Dalgleish and J. Leaver, The possible conformations of milk proteins adsorbed on oil/water interfaces. J. Colloid Interface Sci. 141:288–294, 1991.

115. A. R. Mackie, J. Mingins, and A. N. North. Characterisation of adsorbed layers of a disordered coil protein on polystyrene latex. J. Chem. Soc. Faraday Trans. 87:3043–3049, 1991.

116. E. Dickinson, D. S. Horne, J. S. Phipps, and R. M. Richardson, A neutron reflectivity study of the adsorption of β-casein at fluid interfaces. Langmuir 9:242–248, 1993.

117. D. V. Brooksbank, C. M. Davidson, D. S. Horne, and J. Leaver, Influence of electrostatic interactions on β-casein layers adsorbed on polystyrene lattices. J. Chem. Soc. Faraday Trans. 89:3419–3425, 1993.

118. P. J. Atkinson, E. Dickinson, D. S. Horne, and R. M. Richardson, Neutron reflectivity of adsorbed β-casein and β-lactoglobulin at the air–water interface. J. Chem Soc. Faraday Trans. 91:2847–2854, 1995.

119. G. Fragneto, R. K. Thomas, A. R. Rennie, and J. Penfold, Neutron reflectivity of bovine β-casein adsorbed on OTS self-assembled monolayers. Science 267:657–660, 1995.

120. T. Kull, T. Nylander, F. Tiberg, and M. Wahlgren, Effect of surface properties and added electrolyte on the structure of β-casein layers adsorbed at the solid/aqueous interface. Langmuir 13:5141–5147, 1997.

121. T. Nylander and N. M. Wahlgren, Forces between adsorbed layers of β-casein. Langmuir 13:6219–6225, 1997.

122. P. L. Privalov and S. J. Gill, Stability of protein structure and hydrophobic interaction. Adv. Protein Chem. 39:191–234, 1988.

123. K. A. Dill, Dominant forces in protein folding. Biochemistry 29:7133–7155, 1990.

124. P. K. Ponnuswamy, Hydrophobic characteristics of folded proteins. Prog. Biophys. Molec. Biol. *59*:57–103, 1993.

125. P. L. Privalov, Stability of proteins—small globular proteins. Adv. Protein Chem. *33*: 167–241, 1979.

126. P. L. Privalov, Stability of proteins. Proteins which do not present a single cooperative system. Adv. Protein Chem. *35*:1–104, 1982.

127. D. A. Dolgikh, R. I. Gilmanshin, E. V. Brazhnikov, V. E. Bychkova, G. V. Semisotnov, S. Y. Venyaminov, and O. B. Ptitsyn, α-Lactalbumin: compact state with fluctuating tertiary structure? FEBS Lett. *136*:311–315, 1981.

128. M. Ohgushi and A. Wada, "Molten-globule state": a compact form of globular proteins with mobile side-chains. FEBS Lett. *164*:21–24, 1983.

129. D. A. Dolgikh, L. V. Abaturov, I. A. Bolotina, E. V. Brazhnikov, V. E. Bychkova, R. I. Gilmanshin, Y. O. Lebedev, G. V. Simisotnov, E. I. Tiktopulo, and O. B. Ptitsyn, Compact state of a protein molecule with pronounced small-scale mobility: bovine α-lactalbumin. Eur. Biophys. J. *13*:109–121, 1985.

130. K. Kuwajima, The molten globule state as a clue for understanding the folding and cooperativity of globular-protein structure. Proteins Struct. Funct. Genet. *6*:87–103, 1089.

131. O. B. Ptitsyn, R. H. Pain, G. V. Semisotnov, E. Zerovnik, and O. I. Razgulyaev, Evidence for a molten globule state as a general intermediate in protein folding. FEBS Lett. *262*:20–24, 1990.

132. E. Dickinson and Y. Matsumura, Proteins at liquid interfaces: role of the molten globule state. Colloids Surfaces B: Biointerfaces *3*:1–17, 1994.

133. W. G. Griffin, M. C. A. Griffin, S. R. Martin, and J. Price, Molecular basis of thermal aggregation of bovine β-lactoglobulin A. J. Chem. Soc. Faraday Trans. *89*:3395–3405, 1993.

134. U. Elofsson, P. Dejmek, and M. Paulsson, Heat induced aggregation of β-lactoglobulin studied by dynamic light scattering. Int. Dairy J. *6*:343–357, 1996.

135. M. Rüegg, U. Moor, and B. Blanc, A calorimetric study of the thermal denaturation of whey proteins in simulated milk ultrafiltrate. J. Dairy Res. *44*:509–520, 1977.

136. J. N. de Wit and G. Klarenbeek, A differential scanning calorimetric study of bovine β-lactoglobulin at temperatures up to 160°C. J. Dairy Res. *48*:293–302, 1981.

137. M. Paulsson, P.-O. Hegg, and H. B. Castberg, Thermal stability of whey proteins studied by differential scanning calorimetry. Thermochim. Acta *95*:435–440, 1985.

138. X. L. Qi, S. Brownlow, C. Holt, and P. Sellers, Thermal denaturation of β-lactoglobulin: effect of protein concentration at pH 6.75 and 8.05. Biochim. Biophys. Acta *1248*:43–49, 1995.

139. M. Paulsson and P. Dejmek, Thermal denaturation of whey proteins in mixtures with caseins studied by differential scanning calorimetry. J. Dairy Sci. *73*:1–10, 1990.

140. U. M. Elofsson, Protein adsorption in relation to bulk phase properties: β-lactoglobulins in solution and at the solid/liquid interface. PhD thesis, Lund University, Lund, Sweden, 1996.

141. T. Nylander, Proteins at the metal/water interface—adsorption and solution behaviour. PhD thesis. University of Lund, Lund, 1987.

142. T. Arnebrant, K. Barton, and T. Nylander, Adsorption of α-lactalbumin and β-lactoglobulin on metal surfaces versus temperature. J. Colloid Interface Sci. *119*:383–390, 1987.

143. E. Dickinson and Y. Matsumura, Time-dependent polymerization of β-lactoglobulin through disulphide bonds at the oil–water interface in emulsions. Int. J. Biol. Macromol. *13*:26–30, 1991.

144. T. Jeurnink, M. Verheul, M. Cohen Stuart, and C. G. de Kruif, Deposition of heated whey proteins on a chromium oxide surface. Colloid Surfaces B: Biointerfaces *6*:291–307, 1996.

145. S. M. MacDonald and S. Roscoe, Electrochemical studies of the interfacial behavior of insulin. J. Colloid Interface Sci. *184*:449–455, 1996.

146. C. A.-C. Karlsson, M. C. Wahlgren, and A. C. Trägårdh, Time and temperature aspects of β-lactoglobulin removal from methylated silica surfaces by sodium dodecyl sulphate. Colloid Surfaces B: Biointerfaces *6*:317–328, 1996.

147. C. A.-C. Karlsson, M. C. Wahlgren, and A. C. Trägårdh, β-Lactoglobulin fouling and its removal upon rinsing by SDS as influenced by surface characteristics, temperature and adsorption time. J. Food Eng. *30*:43–60, 1996.

148. R. C. Chatelier and A. P. Minton, Adsorption of globular proteins on locally planar surfaces: models for the effect of excluded surface area and aggregation of adsorbed protein on adsorption equilibria. Biophys. J. *71*:2367–2374, 1996.

11

Mechanism of Interfacial Exchange Phenomena for Proteins Adsorbed at Solid–Liquid Interfaces

VINCENT BALL* University of Basel, Basel, Switzerland

PIERRE SCHAAF Institut Charles Sadron (CNRS–ULP), Strasbourg, France

JEAN-CLAUDE VOEGEL Faculté de Chirurgie Dentaire, INSERM—Unité 424, Strasbourg, France

I. INTRODUCTION

The adsorption or macromolecules on solid surfaces presents some specific aspects, in particular its irreversibility toward dilution, which has been analyzed theoretically [1]. For biopolymers, and in particular for proteins, clear theoretical explanations are not yet available. However, it is known that they interact with surfaces by forming a large number of physical interactions (electrostatic interactions, dispersion forces, hydrophobic interactions that can be described by means of Lewis acid–base forces [2], hydrogen bonds). Moreover, proteins display the ability to change their conformation [3–5] during the adsorption process, thus allowing a variation of the enthalpy and entropy of the system. These two characteristic features are believed to induce specific properties for protein adsorption processes.

    1. Their adsorption on solid surfaces is often partially, if not totally, irreversible, particularly on hydrophobic surfaces [6,7]. Hence, when a layer of adsorbed macromolecules is brought in contact with a pure solvent or buffer, no (or only small) desorption is observed over experimental time scales extending over hours or days. This is a direct consequence of the number of interaction points that a protein can establish with the chemical groups of a given surface. Even if the energy associated with each link is only of the order of kT, the overall interaction energy between the protein and the surface can become very important [8], particularly if the mean residence time of the protein increases [9], allowing for the increase in the number of interaction points by means of structural modifications in the adsorbed protein. Thus, the desorption probability should be rigorously described by means of memory functions [10]. All the links between the adsorbed protein and the surface have to be broken for allowing desorption. The activation energy for such a process may be

---

Current affiliation: Institute of Chemistry, LIMBO, Strasbourg, France.

very high [11], rendering the desorption rate constant very low, typically of the order of 10^{-4} to 10^{-6} s^{-1}, depending strongly on the protein, the surface, and the buffer used [9,12–14]. Moreover, lateral interactions between the adsorbed proteins may render the desorption process non-Poissonian [12]; i.e., it cannot be fitted by an exponential or a sum of exponential functions.

2. Even if the adsorption process is irreversible, adsorbed macromolecules, denoted by A, can usually still be removed from the surface by bringing the adsorbed layer in contact with a solution of macromolecules of type B. Gradually molecules of type A are replaced by molecules of type B by a process called the *exchange process*. This kind of cooperative replacement, which needs macromolecules dissolved in the bulk, has been observed very early in the case of polymers [15,16]. For proteins its occurrence has been suspected after the experiments performed by Vroman's group [17], where serum was put in contact with solid surfaces and, subsequently, antibodies directed against one specific protein in the adsorbed state. The adsorbed antibodies were detected by means of the variation of optical layer thickness. After different adsorption times, it was found that the amount of adsorbed antibodies progressively decreased: this observation was originally attributed to a denaturation of the target antigens. Only later has it been recognized that the antigens initially adsorbed can be displaced or exchanged by proteins presenting greater affinity for the surface. This transient adsorption of one component from a complex protein mixture, later called the *Vroman effect* [18], has been observed not only for fibrinogin but also for other serum proteins. Related effects can also be observed at constant adsorption time by changing the dilution of the plasma and observing the adsorption of one labeled component in the mixture or in narrow spaces (in the lens on surface configuration [19,20]. Many studies have been performed using plasma or serum [21–25] or simplified mixtures [26–33] in order to find the proteins responsible for the displacement of the protein of interest (mainly fibrinogen, which plays a big role in blood clotting and thrombosis [34–36]. Situations in which only one kind of protein is present in a mixture of different-sized aggregates [37–39] belong to such competitive adsorption experiments. However, only a qualitative picture has emerged from this complicated dynamical process, and a full description of the Vroman effect can be found elsewhere [40]. Some attempts have been made recently to adjust the transient adsorption curves by kinetic models [41–44]. In all these models, the exchange reaction was considered to be of *first order with respect to both the adsorbed protein and the molecules in the bulk.* The experimental curves could be relatively well fitted; in particular, the occurrence of a maximum in the adsorbed amount of one given protein could be described. However, these qualitative models have their limitations since the existence of a bimolecular rate-limiting step for the exchange reaction was not rigorously demonstrated. Indeed, some examples of the contrary can be found in the literature: the desorption rate of bovine serum albumin (BSA) preadsorbed onto hydrophobic poly dimethyl siloxane is not increased in the presence of BSA in the bulk even at very high concentrations [45]. Moreover, the whole protein pool of the plasma has sometimes been described as acting as a homogeneous solution of concentration C_B [43]. This assumption totally overlooks the fact that some proteins, for instance, high-molecular-weight kininogen (HMWK), even present at trace concentrations, can easily displace preadsorbed fibrinogen [21]. The surface chemistry seems also to play an important role as illustrated on low-temperature isotropic carbon surfaces: once a protein has reached this

surface it passivates it against further adsorption by other plasma proteins and no Vroman effect is detected [46,47].

Hence, a better understanding and model of the exchange mechanism reaction for adsorbed proteins needs further studies in which all experimental parameters are well controlled. Among these is the composition of the solution put in contact with a well-characterized surface. It is the aim of this chapter to describe the experiments performed for this purpose and to summarize the main results obtained.

The occurrence of exchange processes in relatively simple systems has been clearly demonstrated in experiments reported by Brash et al. [48–50], principally with serum albumin onto polyethylene surfaces [48] and fibrinogen on glass [49] or polyelectrolytes differing in their surface charge and hydrophilic/hydrophobic balance [50]. In these early studies, some important observations for understanding the exchange reaction at a molecular level have been obtained, as described later.

For synthetic polymers, Pefferkorn et al. [51] observed also identical replacement processes. In the case of ^{3}H-labeled polyacrylamide adsorbed onto aluminosilicated glass beads, these authors found that the desorption rate of adsorbed polymers increases linearly with the bulk concentration of the same polymers dissolved in the bulk.

According to the first qualitative model of Jennissen [8], one can imagine the exchange process as represented in Fig. 1: desorption of adsorbed proteins can only occur by a simultaneous disruption of all the physical interactions between the surface and the adsorbed protein. However, these adsorbed molecules can break the bonds with the surface in a gradual way by a progressive replacement of the preadsorbed molecule by a molecule from the bulk in a reptation-like motion. It is assumed that this replacement process would need a much lower activation energy than required by a "pure" desorption—that is, desorption in the presence of a pure buffer.

II. EXPERIMENTAL TECHNIQUES FOR STUDIES OF COMPETITIVE AND SEQUENTIAL ADSORPTION REACTIONS

During an exchange process, adsorbed proteins are replaced on the surface by proteins from the bulk. At saturation of the surface, when maximal adsorption is reached, the exchange process constitutes yet another way for proteins to reach the surface at

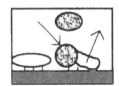

Desorption in pure buffer   Cooperative desorption in presence of proteins in the bulk

FIG. 1 Comparison between desorption in pure buffer (left-hand panel) and desorption in presence of proteins in the bulk, represented by black circles (right-hand panel). In the transition state for the exchange situation (cooperative desorption) the incoming protein reduces the activation energy for desorption by displacing progressively the bounds established between the preadsorbed protein and the surface (according to ideas developed in Ref. 8).

constant surface coverage, i.e., in steady-state conditions. Hence, the total amount of adsorbed material and the refractive index of the adsorbed layer is thus not expected to change widely during the exchange process. The first experimental techniques that were used to demonstrate the existence of exchange (occurrence was detected by immunological detection experiments) [17] were based on the labeling of the proteins, either the initially adsorbed ones or the molecules from the solution in contact with the surface or both types of molecules. In this last case, the adsorbed and dissolved proteins carried different isotopes [48,49]. The molecules can be radiolabeled using ^{131}I, or more often ^{125}I, or can carry fluorescent labels as in the total internal reflectance fluorescence (TIRF) technique [45]. For example, Brash and Samak used the differences in the emission spectra of ^{131}I and ^{125}I to measure simultaneously the on and off rates in release experiments of albumin at polyethylene surfaces [48]. In general, the proteins are then labeled by means of aromatic substitution on tryptophan or tyrosine groups with the method of McFarlane [52,53], using ICl as iodinating agent or softer oxidizing agents like chloramine-T [54] or lactoperoxidase [55]. These kinds of experiments allow demonstration of the presence of true exchange with equal adsorption and desorption rates in a rather straightforward manner. However, they are difficult to perform particularly when ^{131}I labels are used since the radioactive lifetime is only eight days. Moreover, one has to use a detection apparatus operating in larger spectral domains or use two different detectors to simultaneously measure both signals.

In general, an important difficulty when extrinsic labels are used is that great care has to be taken since the added atom or molecule is able to modify both the *hydrodynamic properties* and the *affinity* of the proteins for the surface. Fluorescent labels, consisting mainly of interacting aromatic rings [56], are expected to introduce a rather important hydrodynamic perturbation (the effect is important if the studied protein is small). However, this method does not introduce additional heterogeneity in the protein population since all molecules can easily be labeled [45,57]. In fluorescence experiments, the labeling ratio should not be too important in order to avoid effects like excimer formation (interaction between fluorophores in an excited state in the same molecule) which is able to modify the fluorescence quantum yield (i.e., the intensity of emission at constant incident irradiance) and complicate the quantification of the adsorbed amounts of proteins. In any case, when the fluorophores are sensitive to pH or polarity modification close to surfaces, both the quantum yield and the wavelength of maximum emission will again change [58]. This effect is expected not to be too important for TIRF because one measures only the rate of fluorescence decrease when adsorbed proteins undergo desorption. Such phenomena are not encountered when radiolabeling reactions are employed, using ^{125}I or ^{131}I and methods derived from that of MacFarlane [52,53] or others [54,55]. Only a small fraction of the proteins are substituted (mainly disubstituted on the same tyrosine group, with the iodine labeling method in accordance with Holeman's rule for electrophilic aromatic substitution [59]). In such an approach great care has to be taken and one has to verify that labeled molecules do not adsorb preferentially over nonlabeled ones. For metallic surfaces where electron donor–acceptor interactions between iodine atoms or tyrosine or tryptophan groups and metallic electron acceptor sites on the surface are possible, such behavior has been observed [60]. These effects are sometimes difficult to detect if one uses the radioactive dilution method [61,62], and ideally one should measure depletion in the particle-containing fluid by another analytical method (UV absorption or methods like those described by Bradford [63] or Lowry et al. [64]). The availability of commercial kits makes these test measurements relatively easy. The calibration curve used for these measurements should be made with the protein of interest. Indeed, the calibration curve for the Lowry method is strongly protein dependent [65].

Note that standard methods that allow each protein to be labeled with a ^{14}C isotope by means of reductive methylation are also available and well characterized [66]. However, artifacts in protein adsorption studies cannot be excluded since the methylation may increase local hydrophobicity and, thus, affinity for the adsorbent [67].

When radiolabeled proteins are used one should always check for the stability of the bond between the labeling group or isotope and the chemical function to which it is bound in order to be sure that the measured decrease in surface fluorescence or radioactivity corresponds to the removal of the whole protein and not just to the removal of the radioactive atom [48].

In polymer physics preferential adsorption behavior of some labeled groups has also been observed. ^{2}H-labeled polymers have been used to measure the exchange rate of polycarboxylates by using the different absorption frequencies of $-COO^{1}H$ ($2950 \ cm^{-1}$) and $-COO^{2}H$ ($2100-2300 \ cm^{-1}$) groups [68]. In this case, the deuterated groups reveals a higher affinity for the surface [69], which induces a strong driving force for exchange in the case of long and strongly labeled polymers.

Optical techniques like ellipsometry [70], scanning angle reflectometry [71] or optical waveguide mode spectroscopy [72] are only sensitive to variations in either the thickness of the adsorbed layer or its refractive index. These physical quantities remain unchanged in a 1:1 stoichiometrical process, when the molecules in the adsorbed and unbound state are identical. However, in the case of synthetic polymers or polyelectrolytes having different refractive index increment values, such studies are possible [73,74]. For different-sized proteins (like fibrinogen, albumin, or IgG molecules) ellipsometry has been used in combination with TIRF [75,76]. These optical techniques are well suited for the measurement of surface concentrations with great accuracy when the flat surface is put in contact with a solution containing one type of protein. However, qualitative information relative to the adsorption dynamics in two- or three-component systems [77,78] has already been obtained by these techniques. TIRF has also been used in combination with FRAP (fluorescence recovery after photobleaching) to investigate the exchange behavior of adsorbed proteins as a function of temperature to obtain thermodynamic parameters of the exchange process [57].

The use of labeled proteins is not an absolute necessity if one only wants to determine the composition of the adsorbed layer after a given reaction time. The surface composition can be evaluated by the reactivity of antibodies specific for one of the species present in the adsorbed state. Optical techniques are very useful in this context. Such experiments have allowed the measurement of the transient adsorption behavior of plasma proteins [17], the antibody directed against a certain adsorbed protein being detected by variation of ellipsometric thickness. This approach has been used more recently with surfaces presenting a wettability gradient [79] in order to determine the effect of the surface energetics on the exchange process. Even if the antibody recognition method avoids the use of labels, either the directly adsorbed species or the molecules taken up by an exchange mechanism can lose a part of their biological activity. In his pioneering studies, Vroman explained the reduction in antibody binding over time by a progressive denaturation of the adsorbed target antigens [17]. Only later, it became obvious that part of this biological activity loss should be attributed to a displacement of the preadsorbed proteins. However, these incoming antibodies can even replace an adsorbed antigen through an exchange process, even if the probability of such a process is expected to be a few orders of magnitude lower than direct antibody–antigen reactions. Moreover, the antigenic epitopes can be directed toward the surface and thus not recognized by the antibodies of the bulk. Some studies have addressed this question [80–84], and no clear answer has emerged. The behavior may be strongly dependent on the surface chemistry of either the protein or the sorbent or even buffer dependent.

With proteins having different isoelectric points (pI), like α-lactalbumin (pI = 4.3) and lysozyme (pI = 11.1), the measurement of the streaming potential combined with reflectometry has highlighted the role of electrostatic interactions in sequential adsorption (and thus exchange processes) [85].

The surface force apparatus has been used to monitor the displacement of high-molecular-weight polystyrene chains adsorbed onto crossed mica cylinders by shorter chains. The displacement process resulted in a modification of the force profile between the two surfaces [86]. To our knowledge such a technique has not yet been employed for dynamic studies with adsorbed proteins. Such a method could yield interesting information on the structural evolution of the adsorbed layer in the case of sequential adsorption of two different-sized proteins. Indeed, this technique has proven to be very efficient in the study of the structure of adsorbed layers (like the lectin Concanavalin A [87] or BSA [88] or other proteins [89,90]).

The use of paramagnetic labels could also be of great help in all these studies: it has been shown that ESR (electron spin resonance) can allow both estimation of the adsorbed amounts [91] and structural modifications (however, only in the vicinity of the spin label). One could also plan to do such experiments by adsorbing spin-labeled proteins and to induce exchange by introducing nonlabeled molecules.

Exchange processes can be studied not only by using different experimental techniques but also by using different experimental strategies, and one must distinguish between *competitive adsorption* studies and *sequential adsorption* experiments. In the first case a mixture of proteins is directly brought in contact with the surface (this is also the case when a given protein displays an association behavior in the bulk). The adsorption process is then the result of a complex interplay between the transport of the molecules to the surface and the interfacial reaction. Studies where the protein solution is the plasma, which contains more than 200 different proteins, constitutes an extreme example of this kind of experiment. This type of experiment is usually intended to reproduce situations in which a biological fluid comes in contact with a biomaterial [92].

In sequential adsorption studies different proteins are added in sequential steps separated by buffer rinsing. These kinds of experiments are well adapted for investigating the influence of factors such as the residence time on the exchange process or for determining exchange kinetic laws in relatively simple situations.

We detected an important difficulty that arises in exchange studies of the sequential adsorption type, namely, that an additional adsorption process can occur even when a saturated layer of adsorbed proteins is brought in contact with a protein solution after removal of the first solution and rapid rinsing with buffer [93,94]. This process was not only observed with colloidal particles, where it could be partially attributed to deflocculation of bridged particles [93] during sedimentation of the particles–rinsing–readdition of buffer or a new protein solution, but also during adsorption of fibrinogen on the surface of a cylindrical silica tube [94]. Moreover, when using sulfated latex particles great care was taken to avoid significant particle aggregation during the adsorption of IgG molecules onto their surface [93]. Such an additional adsorption has also been observed by means of optical waveguide light-mode spectroscopy for the adsorption of hen egg-white lysozyme onto an $Si(Ti)O_2$ layer [95]. In this set of experiments, after saturation of the surface with lysozyme, the adsorption cell was rinsed with buffer (in laminar conditions, with a wall shear rate close to 17 s^{-1}), and subsequently a protein solution of the same concentration was injected and more than 20% additional adsorption was found relative to the first plateau when the 10 mM Hepes buffer (pH 7.4) contained 100 mM NaCl. There seems to be no clear explanation for such a phenomenon. It could be related to fundamental aspects of the protein adsorption behavior like heterogeneity [96] of the adsorbed layers, for instance rearrangements in the adsorbed layer during the de-sorption process before the layer is again brought in contact with a new protein solution [97]. This additional adsorption process contradicts the observation that when adsorption is performed stepwise, the amount of adsorbed protein per unit area is usually smaller than what is obtained when the surface is directly brought in contact with the final protein solution. This is particularly the case for protein ad-sorption on hydrophobic surfaces [98].

III. STUDIES OF THE REACTION MECHANISM OF EXCHANGE: GENERAL RESULTS

The existence of exchange processes has been proven by a large number of exper-iments [48–51] and appears clearly in the Vroman effect [18–21]. On the other hand, very few studies were devoted to establish kinetic laws governing these pro-cesses. In which situations can the replacement process be described by a bimolecular rate equation with partial orders of one for both the adsorbed and the bulk proteins? We will try to summarize the conclusions from these studies [29,30,57,93,94,99–102] and establish relations with other studies in which the exchange kinetics were not directly investigated. We will also try to summarize the studies aimed to find the relevant physicochemical parameters implied in this kind of surface processes.

Most exchange experiments reported up to now concerned homogeneous re-placement (A displaced by A). When a solution of B proteins (either of the same nature as the preadsorbed ones or different but differently labeled as these adsorbed molecules) is brought in contact with a layer of already adsorbed molecules of type A, one often first observes a rapid "additional" adsorption, as mentioned in the previous section. In addition to this rapid adsorption process, an amount $\Delta\Gamma_B^r$ of proteins of type B reaches the surface through a slower process. It has been shown that $\Delta\Gamma_B^r$ is often, within experimental precision, equal to the amount of proteins of type A released from the surface [48,49,99,101]. For instance, for adsorbed IgG molecules in contact with a fibrinogen solution, IgGs are displaced approximately in a one-to-one ratio with incoming fibrinogen [99]. All these experimental observations constitute strong indications for the occurrence of an exchange process.

Let us first discuss the *kinetic aspects* of the exchange reaction: if a layer of adsorbed proteins of type A is brought in contact with a solution of proteins of type B, the amount of proteins of type A released from the surface after a time t typically follows a curve as in Fig. 2. It can be seen that the release kinetics becomes faster when the concentration of proteins of type B in solution is increased. This is a typical signature of an exchange process. From these kinetic curves one can also conclude that *several time scales* are involved in the release mechanism. In order to analyze the kinetics of the release process, the experimental functions $\Gamma(t)$ corresponding to the amount of proteins remaining on the surface after a time t were approximated by the sum of different exponentials.

$$\Gamma(t) = \sum_i \Gamma_i \exp(-k_i t) + \Gamma_\infty \tag{1}$$

where Γ_i corresponds to the concentration of type i proteins reversibly adsorbed at time $t = 0$, and k_i is the corresponding rate constant. This sum should be extended over all time scales involved in the release mechanism. The Γ_∞ term corresponds to the amount of proteins which are neither exchangeable nor desorbable: they are thus called irreversibly adsorbed proteins. Now, Eq. (1) constitutes nothing else than a fitting function and its validity has not been theoretically proven, even if it represents a typical response function.

For all systems investigated and reported in the literature, the release kinetics of adsorbed proteins in the presence of a protein solution could be correctly described by one or two characteristic time scales and three adsorbed populations [57,101] so Eq. (1) reduces to:

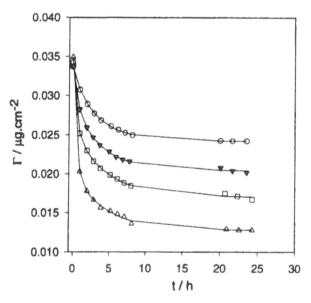

FIG. 2 Typical exchange kinetics of labeled adsorbed lysozyme on TiO_2 surfaces (data taken from Ref. 101) with identical nonlabeled proteins in PBS buffer at various bulk concentrations: $\bigcirc$, in the presence of pure buffer; $\blacktriangledown$, $(1.37 \pm 0.10) \times 10^{-2}\%$ (w/w); $\square$, $(2.56 \pm 0.10) \times 10^{-2}\%$ (w/w); $\triangle$, $(3.13 \pm 0.10) \times 10^{-2}\%$ (w/w). Continuous lines correspond to the fit of Eq. (2) to these data.

$$\Gamma(t) = \Gamma_1 \exp(-k_1 t) + \Gamma_2 \exp(-k_2 t) + \Gamma_\infty \qquad (2)$$

We denote by Γ_1 (resp. Γ_2) the population which is rapidly (resp. slowly) desorbable or exchangeable. In some cases the amount of B proteins (carrying a label) which adsorb on the surface after a time t could also be followed.

It has also been found experimentally that, for all investigated systems, the constant k_1 varies, within experimental errors, linearly with the concentration C_B of the proteins in the solution in contact with the adsorbed layer [93,94,99–103]. One exception has been found for apo-α-lactalbumin adsorbed in the first plateau of the sigmoidal adsorption isotherm onto a TiO_2 surface [102].

A typical evolution of k_1 with C_B is represented in Fig. 3. This figure also shows that a desorption process takes place even in the absence of dissolved proteins and that its rate constant corresponds to the nonzero value of k_1 at $C_B = 0$. On the other hand, the kinetic constant k_2 is often found independent of C_B. These results prove that in this case the rapid exchange process is of order 1 with respect to the proteins in solution, whereas the slow-release process is of order 0 with respect to C_B.

The slow process thus seems not to be a true exchange process but rather a desorption/adsorption process, the proteins of type B filling the vacancies created by the molecules of type A desorbed from the surface. On the other hand, the rapid process seems to be a true exchange process in which the release mechanism strongly depends upon the presence of proteins in the solution. The fact that the amount of proteins of type A adsorbed on the surface at time t can be fitted by a function of

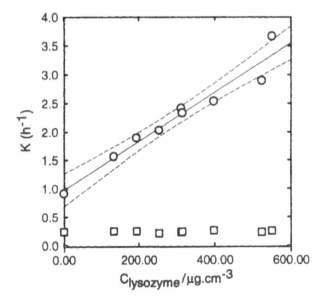

FIG. 3 Variation of the rate constants K for the homogeneous exchange process of adsorbed ^{125}I-labeled lysozyme molecules by nonlabeled lysozyme [101] of concentration $C_{lysozyme}$. The adsorption was performed from PBS buffer (pH = 7.5) onto colloidal TiO_2 particles. (O) and ($\square$) are rate constants K_1 and K_2 for the fast and slow kinetic regimes, respectively, obtained from the experimental data issued from Fig. 2 and fitted by Eq. (3).

type (2) seems also to indicate that the rapid and slow release processes can be assumed to be of first order with respect to the different populations of adsorbed molecules. Similar results have been observed for the exchange process of synthetic polymers [51]. It can be noticed that the exchange process seems not of first order with respect to all the adsorbed proteins. This result corroborates an observation of Elgersma et al. [103], who found that for small amounts of IgGs adsorbed on polystyrene lattices no displacement by BSA molecules could be observed, whereas such a displacement became apparent for higher amounts of adsorbed IgGs. This indicates that if only a small amount of proteins is adsorbed these latter are more likely to be found in an irreversible adsorbed state and can thus no longer be displaced from the surface. This could well be in relation to spreading effects whose modelization can in some circumstances fit the experimental data very well [104]. This also agrees with the observation that steric hindrance during an adsorption process can prevent the molecules from denaturation and also undergoing the transition from a reversible to an irreversible state on the surface. One can thus state that in a certain number of experimental systems [93,94,99–102] the kinetic constant of the fact release process is given by

$$k_1 = k_{e1}C_B + k_{d1} \tag{3}$$

where k_{d1} corresponds to the fast desorption constant and k_{e1} to the exchange rate constant. This result has been used in all the empirical models that have been proposed to describe adsorption processes of proteins on solid surfaces, but it is only recently that their validity has been tested systematically for a large variety of systems, mostly on rather hydrophilic surfaces [94,99–102]. Table 1 lists the values of the constants k_{e1}, k_{d1}, and k_2 for various systems investigated in our lab and by other groups (Table 2) [48,57]. It must be noticed that, for the systems that exhibit three adsorbed populations, the constant k, was independent of C_B. Since this has only been reported for two cases [94,101], one cannot conclude that it represents a general

TABLE 1 Values of Bimolecular Rate Constants for Cooperative Exchange Processes (Exchange) and First-Order Desorption Rate Constants

Proteins	Surfaces	k_{e1} (10^4 L·mol^{-1}·s^{-1})	k_{d1} (10^{-4} s^{-1})	k_{d2} (10^{-4} s^{-1})	Ref.
IgG$_{ads}$/IgG	Latex-OSO$_3^-$	1.61 ± 0.27	nonmeasured	nonmeasured	93
IgG$_{ads}$/IgG	Ti-TiO$_2$	2.64	0.25	—	100
IgG$_{ads}$/Fib	glass tube	106.7	1.11 ± 0.02	6.26	99
Fib$_{ads}$/Fib	glass tube	8.25	0.58	—	94
Lys$_{ads}$/lys	TiO$_2$ rutile	17.4 ± 0.3	2.70	0.68	101
apoα$_{ads}$/apoα, first plateau	TiO$_2$ rutile	—	1.67	—	102
apoα$_{ads}$/apoα, 2nd plateau	TiO$_2$ rutile	0.92	0.96	—	102

Note: k_{d1} and k_{d2} are the rate constants for the fast and slow desorption processes, respectively. In most cases, only one kinetic regime was necessary to fit the data (see text). In this case the rate constants are k_{d1} and k_{e1} for the desorption and cooperative release, respectively. In column 1 the subscript "ads" refers to the protein adsorbed in the first step in the sequential adsorption experiment.

TABLE 2 Apparent First-Order Rate Constants for Exchange Processes Taken from the Literature

Protein (homogeneous system)	Sorbent	k_1 (s^{-1})	k_2 (s^{-1})	Ref.
Human serum albumin	Polyethylene	$(4.90 \pm 0.01) \times 10^{-6}$ at $\gamma^a = 250$ s^{-1}	—	48
Hen egg white lysozyme	OTS treated quartz	0.1 s^{-1}	6.7×10^{-4} s^{-1}	57

$^a\gamma$ denotes the interfacial shear rate in the polyethylene tubing.

law, but it is still an indication that at long adsorption times only desorption is observed; exchange is no longer seen. There is no explanation for this rather unexpected observation.

The validity of a kinetic function of type (2) is based on the assumption of the existence of three (or fewer) *independent* adsorbed populations on the surface. This classification of the adsorbed molecules in different reservoirs is now a classical idea in the field of protein adsorption [44,104,105], but it can only be a very crude approximation, as we will see. This, however, implies also that the amount Γ_i, with $i = 1$, 2, and ∞, is independent of the adsorption time or of C_B but depends only on the amount of adsorbed proteins at time $t = 0$ (beginning of the release experiment). However, this is not confirmed experimentally. It is typically found that Γ_∞ decreases when the mean residence time of the adsorbed molecules decreases [101] or when C_B increases [99,101,106]. For example, the system of adsorbed lysozyme on titanium oxide surfaces in the presence of a lysozyme solution [101] was found to follow a kinetic function of type (2). In this case, it was observed that Γ_1 increases with C_B while Γ_∞ decreases, Γ_2 remaining constant with C_B (Fig. 4). This suggests, for this case, that proteins from population 1 evolve with time into irreversibly fixed proteins. The experiments of Jennissen with phosphorylase B adsorbed at butyl sepharose also displayed such a behavior [106]. The same observation also appears in the exchange reaction of polyacrylamide adsorbed on aluminosilicated glass beads [51a]. Thus, proteins evolve with the adsorption time into more strongly bound states. This effect has also been observed in competitive adsorption experiments of fibrinogen with plasma where the rate of displacement of adsorbed fibrinogen decreased with residence time [23].

This aging process has been proposed for a long time [44] and seems to be partly related to a denaturation of the macromolecule on the surface. Many different models were proposed in order to account for this transition from a reversibly to an irreversibly fixed state of the adsorbed proteins [44,104]. The time at which proteins reach the naked sorbent surface seems indeed to play a big role. Schmidt et al. [57], by studying the exchange process of lysozyme on methylated quartz surfaces, found that the irreversibly adsorbed population seems to correspond to the proteins which first reached the surface and could change their structure in such a way as to optimize their interactions with the sorbent. This was supported by the observation of a loss of enzymatic activity (against *Micrococcus lysodeikticus*) corresponding to the fraction of these lysozyme molecules adsorbed at the very beginning (the first seconds)

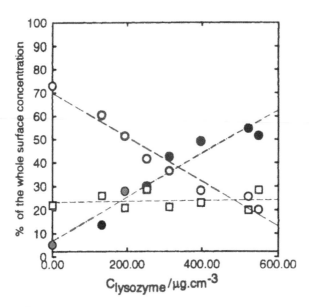

FIG. 4 Evolution with C_{lysozyme} of the relative population of 1, 2, and 3 (in %): ●, population 1; □, population 2; ○, population 3. (Data taken from Ref. 101.)

of the adsorption kinetics. As in the work performed on hydrophilic TiO_2 [101], Schmidt et al. [57] found two populations of proteins which could be released from the sorbent. The characteristic desorption time scales are respectively about 10 s and 10^3 s (Table 2) for the rapidly and slowly releasable populations. With our radiotracer technique [99], one is not able to measure very fast characteristic times: typically one can only discriminate processes with time scales of the order of 10^3 s. The very fast desorption measured by means of TIRF [57], in the presence of lysozyme in solution, was attributed to fast "exchange" of molecules adsorbed in a second layer. However, we prefer to use the term "releasable" even if these authors used the term "exchangeable." Indeed, Schmidt et al. never proved that the release process they observed is indeed an exchange process because they did not report measurements as a function of the concentration of the protein solution in contact with the adsorbed layer. Thus one cannot know if the observed kinetics correspond to a real exchange process, as we have previously defined it, or if it corresponds to a desorption/adsorption process as we have demonstrated it for the slow release process of hen egg white lysozyme on TiO_2 particles [101]. This explains why we tabulated the rate constants obtained by Schmidt et al. [57] and those obtained by Brash and Samak [48], who found only one characteristic time scale for albumin adsorbed on polyethylene, with units corresponding to a pseudo–first order process (i.e., units of time^{-1}). Indeed, in these studies it has not been demonstrated (but rather suggested) that the process is of first order against the proteins in solution. To compare the rate constants obtained in our studies with those of Schmidt et al. and Brash and Samak, one should subtract their values from the desorption rate constant in the presence of pure buffer and divide the result by the bulk concentration, according to Eq. (3).

Another clear demonstration of the role of structural modifications in the description of the exchange process has been provided by the study of the competitive adsorption of a mixture of proteins—in particular fibrinogen (Fb) and high-molec-ular-weight kininogen (HMWK) on a glass surface by Déjardin et al. [29,30]. They have shown that the ratio of the adsorbed amount of HMWK to the adsorbed amount of fibrinogen after an infinite adsorption time varies as

$$\frac{\Gamma_{HMWK}}{\Gamma_{Fb}} = AC_{HMWK} \tag{4}$$

where A is a constant. From a theoretical analysis they could relate the constant A to the exchange constant k_1 of adsorbed fibrinogen by HWMK and to the kinetic constant k_{irr} describing the transition from the reversible to the irreversible state of adsorbed fibrinogen:

$$A = \frac{k_1}{k_{irr}} \tag{5}$$

Let us now discuss the influence of various physicochemical parameters on the exchange process. We will start with the *influence of temperature*. Schmidt et al. [57] investigated the temperature effect on the release process for adsorbed lysozyme and observed that the process speeded up as temperature increased. This conclusion applies for the rapid- and the slow-release processes. It indicated that these processes are thermally activated. Assuming a two-state process with an activation barrier of height ΔE, they determined the value of the activation energy from an Arrhenius plot. They found for the slowly releasable population an activation energy ΔE of the order of 92 kJ·mol^{-1}. Furthermore, by assuming that the system was in equilibrium, they could determine the binding enthalpy of the slowly releasable class of adsorbed lysozyme and found it to be of the order of 16 kJ·mol^{-1}. Le et al. [25] found also that the times at which maximal adsorption of fibrinogen from plasma onto glass was obtained are temperature dependent.

In their pioneering work, Brash et al. found "no obvious correlation between exchange behavior and the varying characteristics of the adsorbing surface, namely the charge and the hydrophobicity." Wojciechowski et al. came to the same conclu-sion after studying competitive adsorption processes [23]. It must, however, be no-ticed that it is very difficult from competitive adsorption experiments to extract pre-cise information on a single aspect of the exchange process because the whole process is driven by the interplay of many processes, such as the diffusion of proteins to the surface, the residence of different proteins on the surface, and the denaturation of these latter on the surface. Since these experiments were performed, it has been recognized that the exchange process takes place more readily on a hydrophilic surface than on a hydrophobic one [79]. This result also seems to emerge from the study of Slack and Horbett [43], even if these authors do not discuss this point and do not give the values of the contact angles of the various materials on which they have performed their displacement experiments. On hydrophobic surfaces the release of adsorbed proteins either by desorption or by exchange can be totally absent [46,47]. In particular, it has been shown that, although albumin does not generally dominate in competitive adsorption over, e.g., fibrinogen or IgG proteins, when it is preadsorbed on a hydrophobic surface from a pure albumin solution, its displacement from the surface by either fibrinogen or IgG is not observed to a detectable extent [75,76]. The experiments performed on a wettability gradient [79] showed that the exchange reaction was indeed dependent on the interfacial surface tension: it was more important on the hydrophilic than on the hydrophobic part of the gradient. The effect of sorbent wettability seems to vary from one protein to another since fibrinogen adsorbed on the hydrophobic side could not be displaced by IgGs from the bulk, whereas IgGs adsorbed on the same side of the gradient could be exchanged by fibrinogen from the bulk solution. Note that human serum albumin adsorbed onto polyethylene, a hydrophobic surface, can be exchanged by the same protein in the bulk [48]. However, bovine serum albumin, a relatively close to human serum protein, seems only to undergo a desorption/adsorption process when adsorbed on polydimethylsiloxane [45] because the concentration of dissolved albumin does not change the rate constants for desorption. Hence the effect of the surface hydrophobicity on the exchange process or desorption in pure buffer is not clear. The presence of hydrophobic patches on the protein surface as well as its structural stability can certainly influence its behavior at the sorbent surface.

The effect of the *surface charges* of both the adsorption surface and the proteins on the displacement process seems to exist even if it is not an essential parameter: it does not induce by itself the displacement process. However, if the displacement process exists, its extent depends to some degree on the electrostatic interaction between the respective proteins and the surface and between the proteins present in the system [103]. In some solution conditions and for certain surfaces it is even possible to build up layers of proteins alternating in the sign of their global charge by means of strong electrostatic attractive forces [106]. On the other hand, in competition or in sequential adsorption experiments, charge effects can play an important role because they influence the adsorption kinetics and thus the sequence of the adsorption of the proteins as well as the extent of replacement of a preadsorbed protein [85].

Chan and Brash have analyzed the influence of the *salt concentration* on the homogeneous exchange process of human fibrinogen preadsorbed on borosilicate glass [49]. With this system it was found that the fraction of exchangeable proteins increases by increasing either the concentration of TRIS buffer at constant NaCl or the concentration of NaCl at constant buffer content. These authors, however, did not measure the effect of salt concentration on the rate of the exchange process. The result relative to the fraction of exchangeable molecules is surprising because both the protein and the surface are negatively charged at pH 7.35 (isoelectric point of fibrinogen, 5.8 [108]); the point of zero charge of silica is around pH 2.5–3.0 [109], and then one would expect that the layer becomes more and more strongly anchored when the ionic strength increases, owing to shielding of the electrostatic interactions, rendering the desorption or exchange probability lower. Keeping this argument in mind, one may thus assume that the increase in salt concentration changes the structure of the protein in a manner to reduce its interactions with the surface. Note that in this study the fibrinogen was preadsorbed in 0.05 $mol \cdot L^{-1}$ of TRIS, the ionic strength being modified only in the second step of the sequential adsorption experiment. However, an increase in ionic strength was also found to reduce steady surface concentration of fibrinogen, which is also unexpected considering only the electrostatic interactions.

For β-casein adsorbed at an air–water interface from phosphate buffer containing either 0.1 $mol \cdot L^{-1}$ or 1.0 $mol \cdot L^{-1}$ NaCl, it was observed that the kinetics of the exchange process is slowed down with increasing the salt concentration [110].

But in this study it is difficult to rationalize the effect of the relative charge densities of the protein and the pure air–water interface because the measurements of the surface potential of an air–water interface gave either positive or negative values [111].

To our knowledge no competitive or sequential adsorption experiments have been done on a given surface for a given proteinous system at a given ionic strength except by changing either the chemical nature of the cation or the anion. Such experiments would be very important because it is known that certain ions can stabilize or destabilize the native structure of a protein [112], and, hence, its ability to unfold on the sorbent surface.

The influence of the *molecular weight* of the displacing proteins on the exchange process has not been systematically studied. It has been found that fibrinogen usually exchanges albumin or IgG molecules more easily than the reverse situation, fibrinogen being a protein of larger molecular mass than albumin or fibrinogen. On the other hand, fibrinogen is displaced on most surfaces more readily than HMWK [21] even if the molecular weight (1.10×10^5 g·mol^{-1}) for HMWK is smaller than that of fibrinogen (3.40×10^5 g·mol^{-1}). The only experiment in which the mass of the exchanging protein has been varied without changing its chemical nature is due to the group of Norde [103], who investigated the displacement of IgG proteins on polystyrene lattices by monomeric and dimeric BSA. They found that dimeric BSA displaced larger amounts of adsorbed IgG molecules than monomeric BSA even though the molar concentration of dimeric BSA was half that of the monomer. The effects of the molecular weight on adsorption and exchange are discussed in other chapters of this volume.

One other important parameter which often governs interfacial reactions is the *shear rate* at the interface. This parameter, which depends on the flow rate and the geometry of the experimental cell, influences the value of the Nernst layer thickness into which the diffusion rate of the adsorbing particles is faster than the convection due to the macroscopic motion of the bulk. From the studies performed up to now the influence of this parameter is also not clear. On the other hand, the homogeneous exchange process of fibrinogen on a borosilicate glass tube was found to be independent of the shear rate [49], but, on the other hand, the homogeneous exchange rate and the fraction of exchangeable albumin adsorbed on polyethylene increased significantly with the interfacial shear rate [48]. The exchange rate constant of human IgGs adsorbed on a silica tube and displaced by human fibrinogen in solution has been found to be several orders of magnitude smaller than the Lévêque constant [99] describing a reaction kinetics limited by the transport to the surface. This again suggests that the rate-limiting step of the exchange reaction could be the formation of a *transient complex* between the preadsorbed and the incoming molecules.

The exchange process is typically due to the fact that macromolecules are not rigid objects but can change their conformation. In particular, it has been proposed that the adsorbed protein A* forms a complex A*B with protein B, the asterisk denoting the molecule in direct contact with the surface. This complex then turns and thus becomes B*A, the protein A being subsequently removed from the complex and liberated into the solution [30]. Such a process has never been directly observed for protein exchange at an interface, but suggested in the case where it was found that transport was not the rate-limiting process [99].

A somewhat similar mechanism has been reported recently [83]. A layer of human IgG proteins (we will denote them by A) adsorbed on a silica surface was brought in contact with a solution of rabbit IgGs, denoted by B, which were antibodies against the human IgG proteins. The layer structure was followed by scanning angle reflectometry [71], allowing one to determine its structural evolution. The incoming proteins B appeared first to embed themselves in the preadsorbed layer A*. The structure of the layer relaxed with time, becoming both thicker and less dense. The final protein layer was surprisingly thick—as much as three times the characteristic size of an IgG molecule. Moreover, the ratio of the number of proteins B that reacted with one molecule A* was as high as 4. Finally, it could be shown that the refractive index profile goes through a maximum at a nonzero distance from the silica interface. All these optical results led to the hypothesis of the formation, on the surface, of an A*B complex that subsequently turns, as postulated for the exchange process. This exposes the protein A to the solution, and an additional protein B can thus interact with A, allowing the formation of a triple layer B*AB. Such a mechanism was only possible due to the polyclonal nature of the antigen A, and it demonstrates the dynamic behavior of an adsorbed protein layer.

IV. PERSPECTIVES AND CONCLUDING REMARKS

In view of the previous section, it seems necessary to obtain experimental data in order to have better insight about the important physicochemical parameters in exchange reactions. It would then be interesting to study the exchange reaction in a given protein system and, for instance, change the surface charge density [polystyrene latexes with different densities of charged groups or oxides having different zero-point charges (PZC)] or the hydrophilicity. Note that surface charge density and hydrophilicity cannot be varied in an independent manner: by increasing the surface charge density of latex particles one expects its hydrophilicity to increase. In the case of oxides, one can change the PZC by modifying the nature of the metal cation [109], but one then also changes the Lewis acid–base interaction contributing to the surface tension of the oxide [2a]. The influence of the interfacial shear rate, of the ionic strength and of the nature of the anion or the cation at constant ionic strength should also be investigated on a given surface for a given protein.

Moreover, experimental data comprising more points, i.e., better time resolution, in one kinetic experiment and characterized by better signal-to-noise ratio are needed in order to test for the validity of more complicated kinetic models. Thus, the use of fluorescently labeled proteins should be recommended since they can statistically all be labeled [45,57,58] in contrast to most of the radiolabeling methods. But in this case one should remember that the quantum yield of the fluorescence emission may strongly be modified in the vicinity of an interface. Thus, *only relative variations* of emission intensity can be measured, as stated in the experimental section.

By means of optical waveguide light mode spectroscopy it is possible to achieve real-time monitoring of desorption in pure buffer with an excellent precision (a few $ng \cdot cm^{-2}$) and time resolution to the order of a few seconds [72]. This allows one to take memory effects into account for the description of desorption in pure buffer. Then one could better understand the progressive transition from apparently "reversible" to "irreversible" populations. Indeed, taking into account effects such as spreading, Van Tassel et al. [113] were able to describe experimental results concerning the adsorption and desorption kinetics [114] for lysozyme adsorbing on a SiO_2 surface.

From the reported studies [93,94,99–102], it turns out that the observed exchange processes have mainly been investigated on a macroscopic level and that a realistic picture at a molecular level is missing. We are still thinking about the scheme proposed by Jennissen [8] (Fig. 1), and indeed the arguments in favor of the formation of a transient complex [30,83] at the surface seem to confirm that. But this little piece of evidence is absolutely not a proof.

An accurate way to observe transition states would be to perform fluorescence energy transfer experiments in a TIRF (total internal reflectance fluorescence) configuration where the adsorbed molecules carry a donor group D that can only excite the fluorescence of an acceptor group A when A is sufficiently close to D (the interaction scales like r^{-6}, where r is the distance between A and D) and when both groups are well oriented [115]. Rapid progress in atomic force microscopy, particularly in the reduction of tip size, could also allow us to observe interfacial molecular complexes. Other fluorescence techniques, such as fluorescence recovery after photobleaching (FRAP), allow one to measure diffusion constants in the adsorbed state [116]. In such experiments, two kinds of population are usually found on solid surfaces [117] or supported bilayers [118]. It would be of great interest to investigate if there is a correlation between the surface mobility of adsorbed proteins and their ability to be exchanged or desorbed.

A better understanding of the protein exchange mechanism would require study of the time evolution of the radial distribution function of adsorbed molecules during their exchange with bulk molecules. Indeed, one may ask whether the exchange process is a signature of the protein distribution in the adsorbed layer. For instance, the amounts of adsorbed ferritin molecules that progressively leave the surface through exchange in presence of apoferritine molecules (which do not carry a ferrihydrite core and cannot be detected directly without negative staining) can be observed by means of electron microscopy [119,120]. Since it is possible that ferritin forms clusters in the adsorbed state (observed by means of transmission electron microscopy [120] at solid–liquid interfaces), one should expect that exchange processes *occur mainly at the border of such islands* where bulk proteins would have a greater probability of interacting with available surface sites. This would be of very general signification since it has been demonstrated that proteins cluster on artificially created surface defects during their adsorption [121].

It seems also necessary to study more deeply the dynamic behavior of proteins that have very long average residence times: does the whole adsorbed amount progressively enter in the irreversible state? In order to reduce the effect of structural modifications in the study of the exchange process, it would also be of interest to study the behavior of proteins strongly resistant to denaturing conditions like ferredoxin or the $(\alpha\beta)$ barrel phosphoribosyl anthranilate isomerase of thermophilic microorganisms [122a,122b] which can be produced in rather important amounts by means of polymerase chain reaction.

From a practical point of view it would be interesting to have more insight into exchange processes at fluid–fluid interfaces. To our knowledge very few data, particularly from a kinetic point of view, are available [110,123–127 and references therein] at such interfaces. The values of the free energies for incorporation of amino acids modified with fatty acids of different lengths into bilayers of different composition, as determined by Peitzsch and McLaughlin [128], suggest that a rather fast replacement of myristoylated (alkyl chains with 14 carbon atoms) peptides or proteins by the palmitoylated (alkyl chain with 16 carbon atoms) counterparts should occur. The enthalpy variations of these kinds of processes can be directly measured by high-sensitivity titration calorimetry [129] by injecting palmitoylated proteins in a solution containing liposomes already covered with myristoylated proteins. Of particular interest would be to adsorb proteins modified with a lipid anchor (for instance, the myristoylated MARCKS signal transduction protein [130] or the respiratory chain component cytochrome c [131,132]) onto bilayers and to study the exchange by the same protein carrying either lipid chains of the same length or longer chains. In this case complicating effects like spontaneous desorption or strong structural modifications should be strongly reduced, allowing again more accurate modelization of the process. Moreover, the understanding of exchange processes onto the surfaces of vesicles is of particular interest for the improvement of drug transport toward cells (for example, cancerous cells) to reduce the "opsonization" of drug-carrying liposomes [133,134]. In such situations unwanted exchange processes of the lipid-imbedded targeting proteins by plasma proteins can drastically reduce the efficiency of the therapy. The increasing use of phospholipids as biomaterials [135,136] and as protein-resistant materials should constitute a starting point for additional studies concerning the dynamics of protein mixtures at such surfaces.

The eventual observation of exchange between an adsorbed protein and one of its mutants introduced subsequently in the bulk would be of highest interest for the whole field of protein dynamics at interfaces, since recent studies have demonstrated that apparently minor mutations can induce dramatic modifications in the interfacial behavior [13,137,138].

Finally, polymer–protein exchange studies seem necessary since some proteins appear able to displace polymers [139] aimed to act as steric agents against protein adsorption. In such studies the molecular weight of the adsorbed chain should be varied in order to see if there exist power laws like those observed in polymer–polymer exchange experiments [140 and references therein]. These experiments would, moreover, be useful in the improvement of protein-repelling surfaces [141] when the polymers are not covalently attached to the sorbent surface.

Another interesting set of experiments would be to detect the occurrence of exchange reactions for surfaces covered with polysaccharides or polynucleotides [142]. Note that exchange processes have to be taken into account for the description of polymerization-depolymerization of actin filaments [143]. In these conditions one has to consider the adsorption of monomers at growing fibrils as a one-dimensional process.

Last, but not least, practically no work has been done in the theoretical description of surface exchange processes [144,145]. The experiments performed in the group of Pefferkorn for polyacryl amide [51] were at the origin of the work of de Gennes [144]. This author found a bimolecular rate equation (with partial reaction orders of one with respect to both the adsorbed and dissolved polymers) by assuming that the dissolved polymer has to tunnel across the preadsorbed film to reach the sorbent. But, to our knowledge, no theoretical work has been done for the description of the exchange mechanism for adsorbed proteins.

Clearly, in the exciting field of protein–protein or protein–polymer exchange processes a lot remains to be done (both from experimental and theoretical points of view) for a better understanding of interfacial events, which play a big role in biochemical regulation (blood clotting, cell adhesion, etc.) and technical applications.

ACKNOWLEDGMENTS

V. Ball is supported by a Lavoisier fellowship (Paris, France) and is indebted to Dr. M. Winterhalter for his critical reading of the manuscript.

REFERENCES

1. M. A. Cohen Stuart, J. M. H. M. Scheutjens, and G. J. Fleer, Polydispersity effects and the interpretation of polymer adsorption isotherms. J. Polym. Sci. Polym. Phys. 18:559–573, 1980.
2. a. C. J. van Oss, M. K. Chaudhury, and R. J. Good, Interfacial Lifshitz—van der Waals and polar interactions in macroscopic systems. Chem. Rev. 88:927–941, 1988. b. C. J. van Oss, W. Wu, and R. F. Giese, Macroscopic and microscopic interactions between albumin and hydrophilic surfaces. ACS Symp. Ser. 602:80–91, 1995.
3. a. W. Norde, Adsorption of proteins at solid–liquid interfaces. Cells Mater. 5:97–112, 1995. b. M. Kleijn and W. Norde, The adsorption of proteins from aqueous solution on solid surfaces. Heter. Chem. Rev. 2:157–172, 1995.
4. J. Buijs, W. Norde, and J. W. Th. Lichtenbelt, Changes in the secondary structure of adsorbed IgG and F(ab')$_2$ studied by FTIR. Langmuir 12:1605–1613, 1996.
5. A. Ball and R. A. L. Jones, Confirmational changes in adsorbed proteins. Langmuir 11:3542–3548, 1995.
6. F. McRitchie, The adsorption of proteins at solid–liquid interface. J. Colloid Interface Sci. 38:484–488, 1972.
7. G. Penners, Z. Priel, and A. Silberberg, Irreversible adsorption of triple-helical collagen monomers from solution to glass and other surfaces. J. Colloid Interface Sci. 80:437–444, 1981.
8. H. P. Jennissen, The binding and regulation of biologically active proteins on cellular interfaces: model studies on enzyme adsorption on hydrophobic binding site lattices and biomembranes. Adv. Enzyme Reg. 19:377–406, 1981.
9. M. E. Soderquist and A. G. Walton, Structural changes in proteins adsorbed on polymer surfaces. Colloid Interface Sci 75:386–397, 1980.
10. J. Talbot, Time dependent desorption: a memory function approach. Adsorption 2:89–94, 1996.
11. D. Sarkar and D. K. Chattoraj, Activation parameters for kinetics of protein adsorption at silica–water interface. J. Colloid Interface Sci. 157:219–226, 1993.
12. R. Kurrat, J. J. Ramsden, and J. E. Prenosil, Kinetic model for serum albumin adsorption: experimental verification. J. Chem. Soc. Faraday Trans. 90:587–590, 1994.
13. J. J. Ramsden, D. J. Roush, D. S. Gill, R. Kurrat, and R. C. Willson, Protein adsorption kinetics drastically altered by repositioning a single charge. J. Am. Chem. Soc. 117:8511–8517, 1995.
14. R. Kurrat, J. E. Prenosil, and J. J. Ramsden, Kinetics of human and bovine serum albumin adsorption at silica-titania surfaces. J. Colloid Interface Sci. 185:1–8, 1997.
15. C. Thies, Competitive adsorption of polymers on solid surfaces. J. Phys. Chem. 70:3783–3789, 1966.

16. K. Furusawa and K. Yamamoto, Measurement of surface adsorbed polymer displacements. Bull. Chem. Soc. Jpn. *56*:1958–1962, 1983.

17. L. Vroman and A. L. Adams, Identification of rapid changes at plasma–solid interfaces. J. Biomed. Mater. Res. *3*:43–67, 1969.

18. T. A. Horbett, Mass action effects on competitive adsorption of fibrinogen from hemoglobin solutions and from plasma. Thromb. Haemostas *51*:174–181, 1984.

19. H. B. Elwing, L. Li, A. R. Askendal, G. S. Nimeri, and J. L. Brash, Protein displacement phenomena in blood plasma and serum studied by the wettability gradient method and the lens on surface methods. ACS Symp. Ser. *602*:138–149, 1995.

20. L. Vroman and A. L. Adams, Adsorption of proteins out of plasma and solutions in narrow spaces. J. Colloid Interface Sci. *111*:391–402, 1986.

21. L. Vroman, A. L. Adams, G. C. Fischer, and P. C. Munoz, Interaction of high molecular weight kininogen, factor XII, and fibrinogen in plasma at interfaces. Blood *55*:156–159, 1980.

22. S. Slack and T. A. Horbett, Physicochemical and biochemical aspects of fibrinogen adsorption from plasma and binary protein solutions onto polyethylene and glass. J. Colloid Interface Sci. *124*:535–551, 1988.

23. P. Wojciekowsky, P. Ten Hove, and J. L. Brash, Phenomenology and mechanism of the transient adsorption of fibrinogen from plasma (Vroman effect). J. Colloid Interface Sci. *111*:455–465, 1986.

24. R. M. Cornelius, P. W. Wojciekowski, and J. L. Brash, Measurement of protein adsorption kinetics by an in situ, "real-time," solution depletion technique. J. Colloid Interface Sci. *150*:121–133, 1992.

25. M. T. Le, J. N. Mulvihill, J.-P. Cazenave, and Ph. Dejardin, Transient adsorption of fibrinogen from plasma solutions flowing in silica capillaries. ACS Symp. Ser. *602*:129–139, 1995.

26. G. E. Stoner, S. Srinivasan, and E. Gileadi, Adsorption inhibition as a mechanism for the antithrombogenic activity of some drugs. I. Competitive adsorption of fibrinogen and heparin on mica. J. Phys. Chem. *75*:2107–2111, 1971.

27. T. Arnebrant and T. Nylander, Sequential and competitive adsorption of β-lactoglobulin and κ-casein on metal surfaces. J. Colloid Interface Sci. *111*:529–533, 1986.

28. A. V. Elgersma, R. L. J. Zsom, J. Lyklema, and W. Norde, Adsorption competition between albumin and monoclonal immuno gamma globulins on polystyrene lattices. J. Colloid Interface Sci. *152*:410–428, 1992.

29. Ph. Dejardin, P. Ten Hove, X. J. Yu, and J. L. Brash, Competitive adsorption of high molecular weight kininogen and fibrinogen from binary mixtures to glass surface. Langmuir *11*:4001–4007, 1995.

30. Ph. Dejardin and M. T. Le, Ratio of final interfacial concentrations in exchange processes. Langmuir *11*:4008–4012, 1995.

31. M. Deyme, A. Baskin, J. E. Proust, E. Perez, G. Albrecht, and M. M. Boissonnade, Collagen at interfaces. II. Competitive adsorption of collagen against albumin and fibrinogen. J. Biomed. Mater. Res. *21*:321–328, 1987.

32. A. Baszkin and M. M. Boissonnade, Competitive adsorption of albumin and fibrinogen at solution-air and solution-polyethylene interfaces. In situ measurements. ACS Symp. Ser. *602*:209–227.

33. B. Wälivaara, A. Askendal, I. Lundström, and P. Tengvall, Imaging of the early events of classical complement activation using antibodies and atomic force microscopy. J. Colloid Interface Sci. *187*:121–127, 1997.

34. R. F. Doolittle, Fibrinogen and Fibrin. Ann. Rev. Biochem. *53*:195–229, 1984.

35. W. G. Pitt, K. Park, and S. L. Cooper, Sequential protein adsorption and thrombus deposition on polymeric biomaterials. J. Colloid Interface Sci. *111*:343–362, 1986.

36. E. W. Davie, K. Fujikawa, and W. Kisiel, The coagulation cascade. Initiation, maintenance, and regulation. Biochemistry *30*:10363–10370, 1991.

37. R. L. J. Zsom, Dependence of preferential bovine serum albumin oligomer adsorption on the surface properties of monodisperse polystyrene lattices. J. Colloid Interface Sci. *111*:434–445, 1986.

38. W. F. Weinbrenner and M. R. Etzel, Competitive adsorption of α-lactalbumin and bovine serum albumin to a sulfopropyl ion-exchange membrane. J. Chromat A *662*: 414–419, 1994.

39. C. E. Giacomelli, M. J. Avena, and C. P. de Pauli, Adsorption of bovine serum albumin onto TiO_2 particles. J. Colloid Interface Sci. *188*:387–395, 1997.

40. S. M. Slack and T. A. Horbett, The Vroman effect. A critical overview. ACS Symp. Ser. *602*:112–128, 1995.

41. A. Nadarajah, C. F. Lu, and K. K. Chittur, Modeling the dynamics of protein adsorption to surfaces. ACS Symp. Ser. *602*:181–194, 1996.

42. C. F. Lu, A. Nadarajah, and K. K. Chittur, A comprehensive model of multiprotein adsorption on surfaces. J. Colloid Interface Sci. *168*:152–161, 1994.

43. S. M. Slack and T. A. Horbett, Changes in the strength of fibrinogen attachment to solid surfaces: an explanation of the influence of surface chemistry on the Vroman effect. J. Colloid Interface Sci. *133*:148–165, 1989.

44. I. Lundström and H. Elwing, Simple kinetic models for protein exchange reactions on solid surfaces. J. Colloid Interface Sci. *136*:68–84, 1990.

45. Y. L. Cheng, S. A. Darst, and C. R. Robertson, Bovine serum albumin adsorption and desorption rates on solid surfaces with varying surface properties. J. Colloid Interface Sci. *118*:212–223, 1987.

46. L. Feng and J. D. Andrade, Protein adsorption at low temperature isotropic carbon. III. Isotherms, competitivity, desorption and exchange of human albumin and fibrinogen. Biomaterials *15*:323–333, 1994.

47. J. A. Chinn, R. E. Phillips, K. R. Lew, and T. A. Horbett, Tenacious binding of fibrinogen and albumin to pyrolite carbon and biomer. J. Colloid Interface Sci. *184*:11–19, 1996.

48. J. L. Brash and Q. M. Samak, Dynamics of interactions between human albumin and polyethylene surface. J. Colloid Interface Sci. *65*:495–504, 1978.

49. B. M. C. Chan and J. L. Brash, Adsorption of fibrinogen on glass: reversibility aspects. J. Colloid Interface Sci. *82*:217–225, 1981.

50. J. L. Brash, S. Uniyal, C. Pusineri, and A. Schmitt, Interaction of fibrinogen with solid surfaces of varying charge and hydrophobic–hydrophilic balance. II. Dynamic exchange between surface and solution molecules. J. Colloid Interface Sci. *95*:28–36, 1983.

51. a. E. Pefferkorn, A. Carroy, and R. Varoqui, Dynamic behavior of flexible polymers at a solid/liquid interface. J. Polym. Sci. Polymer Phys. Ed. *23*:1997–2008, 1985. b. T. Cosgrove and J. W. Fergie-Woods, On the kinetics and reversibility of polymer adsorption. Colloids Surf. *25*:91–99, 1987.

52. A. S. McFarlane, Labeling of plasma proteins with radioactive iodine. Biochem. J. *62*: 135–143, 1956.

53. A. S. McFarlane, Efficient trace labelling of proteins with iodine. Nature *182*:53, 1958.

54. A. A. Hussain, J. A. Jona, A. Yamada, and L. W. Dittert, Chloramine-T in radiolabeling techniques. Anal. Biochem. *224*:221–226, 1995.

55. a. M. Morrison, G. S. Bayse, and R. G. Webster, Labeling of proteins by using the catalitic activity of lactoperoxydase. Immunochemistry *8*:289–293, 1971. b. M. Morrison, Lactoperoxidase-catalysed iodination as a tool for investigation of proteins. Methods in Enzymol. *70*:214–220, 1980.

56. R. S. Davison, Application of fluorescence microscopy to a study of chemical problems. Chem. Soc. Rev. 25:241–252, 1996.

57. C. F. Schmidt, R. M. Zimmermann, and H. E. Gaub, Multilayer adsorption of lysozyme on a hydrophobic substrate. Biophys. J. 57:577–588, 1990.

58. V. Hlady and J. D. Andrade, Fluorescence emission from adsorbed bovine serum albumin and albumin-bound 1-anilinonaphthalene-8-sulfonate studied by TIRF. Colloids Surf. 32:359–369, 1988.

59. C. H. Li, Iodination of tyrosine groups in serum albumin and pepsin. J. Am. Chem. Soc. 67:1065–1069, 1945.

60. W. H. Grant, L. E. Smith, and R. R. Stromberg, Radiotracer techniques for protein adsorption measurements. J. Biomed. Mater. Res. Symp. 8:33–38, 1977.

61. A. T. Van der Scheer, J. Feijen, J. Klein Elhorst, P. G. L. C. Krugers Dagneaux, and C. A. Smolders, J. Colloid Interface Sci. 66:136–145, 1978.

62. B. S. Murray, Adsorption kinetics of nonradiolabeled lysozyme via surface pressure-area isotherms. Langmuir 13:1850–1852, 1997.

63. M. M. Bradford, A rapid and sensitive method for the quantitation of microgram quantities of protein utilizing the principle of protein dye binding. Anal. Biochem. 248–254, 1976.

64. O. H. Lowry, N. J. Rosebrough, A. L. Farr, and R. J. Randall, Protein measurement with the Folin phenol reagent. J. Biol. Chem. 193:265–275, 1951.

65. S. Duinhoven, PhD Thesis, Agricultural University of Wageningen, The Netherlands, 1992.

66. N. Jentoft and D. G. Dearborn, Labeling of proteins by reductive methylation using sodium cyanoborohydride. J. Biol. Chem. 254:4359–4365, 1979.

67. S. Staunton and H. Quiquampoix, Adsorption and conformation of bovine serum albumin on montmorillonite: modification of the balance between hydrophobic and electrostatic interactions by protein methylation and pH variation. J. Colloid Interface Sci. 166:89–94, 1994.

68. D. J. Kuzmenka and S. Granick, Kinetics of polymer adsorption measured in situ at the solid–liquid interface: utility of the infrared total internal reflection method. Colloids Surf. 31:105–116, 1988.

69. D. C. Leonhardt, H. E. Johnson, and S. Granick, Adsorption isotope effect for protio and deuteropolystyrene at a single solid surface. Macromolecules 23:685–690, 1990.

70. J. A. de Feijter, J. Benjamins, and F. A. Veer, Ellipsometry as a tool to study the adsorption behavior of synthetic and biopolymers at the air–water interface. Biopolymers 17:1759–1772, 1978.

71. P. Schaaf, Ph. Dejardin, and A. Schmitt, Reflectometry as a technique to study the adsorption of human fibrinogen at the silica–solution interface. Langmuir 3:1131–1135, 1987.

72. K. Tiefenthaler, Integrated optical couplers as chemical waveguide sensors. Adv. Biosensors 2:261–289, 1992.

73. J. C. Dijt, M. A. Cohen Stuart, and G. J. Fleer, Surface exchange kinetics of chemically different polymers. Macromolecules 27:3229–3237, 1994.

74. N. G. Hoogeveen, M. A. Cohen Stuart, and G. J. Fleer, Polyelectrolyte adsorption on oxides. II. Reversibility and exchange. J. Colloid Interface Sci. 182:146–157, 1996.

75. M. Malmsten and B. Lassen, Competitive protein adsorption at phospholipid surfaces. Colloids Surf. B Biointerfaces 4:173–184, 1995.

76. M. Malmsten and B. Lassen, Competitive protein adsorption studied with TIRF and ellipsometry. J. Colloid Interface Sci. 179:470–477, 1996.

77. B. Lassen and M. Malmsten, Competitive protein adsorption at plasma polymer surfaces. J. Colloid Interface Sci. 186:9–16, 1997.

78. B. Lassen and M. Malmsten, Structure of protein layers during competitive adsorption. J. Colloid Interface Sci. *180*:339–349, 1996.
79. H. Elwing, A. Askendal, and I. Lundström, Protein exchange reactions on solid surfaces studied with a wettability gradient method. Prog. Colloid Polym. Sci. *74*:103–107, 1987.
80. R. K. Sandwick and K. J. Schray, Conformational states of enzymes bound to surfaces. J. Colloid Interface Sci. *121*:1–12, 1988.
81. E. Lutanie, J.-C. Voegel, P. Schaaf, M. Freund, J.-P. Cazenave, and A. Schmitt, Competitive adsorption of human immunoglobulin G and albumin: consequences for structure and reactivity of the adsorbed layer. Proc. Natl. Acad. Sci. USA *89*:9890–9894, 1992.
82. S. A. Darst, C. R. Robertson, and J. A. Berzofsky, Adsorption of the protein antigen myoglobin affects the binding of conformation specific monoclonal antibodies. Biophys. J. *53*:533–539, 1988.
83. L. Henrich, E. K. Mann, J. C. Voegel, G. J. M. Koper, and P. Schaaf, Scanning angle reflectometry study of the structure of antigen-antibody layers adsorbed on silica surfaces. Langmuir *12*:4857–4867, 1996.
84. S. Duinhoven, R. Poort, G. Van der Voet, W. G. M. Agterof, W. Norde, and J. Lyklema, Driving forces for enzyme adsorption at solic liquid interfaces. I. The serine protease savinase. J. Colloid Interface Sci. *170*:340–350, 1995.
85. H. Shirahama, J. Lyklema, and W. Norde, Comparative protein adsorption in model systems. J. Colloid Interface Sci. *139*:177–187, 1990.
86. J. Klein, Long range surface forces: the structure and dynamics of polymers at interfaces. Pure Appl. Chem. *64*:1577–1584, 1992.
87. B. Gauthier-Manuel and J.-P. Gallinet, Refractive index measurements to study the structural behavior of anisotropic macromolecules confined between two mica surfaces. J. Colloid Interface Sci. *175*:476–483, 1995.
88. J.-P. Gallinet and B. Gauthier-Manuel, Adsorption–desorption of serumalbumin on bare mica surfaces. Colloids Surf *68*:189–193, 1992.
89. P. M. Claesson, E. Blomberg, J. C. Fröberg, T. Nylander, and T. Arnebrant, Protein interactions at solid surfaces. Adv. Colloid Interface Sci. *57*:161–227, 1995.
90. P. M. Claesson, Interactions between surfaces coated with carbohydrates, glycolipids, and glycoproteins. In *Biopolymers at Interfaces* (M. Malmsten, ed.). Marcel Dekker, New York, 1998, pp. 281–320.
91. R. Nicholov, R. P. Veregin, A. W. Neumann, and F. DiCosmo, Protein adsorption at the solid–liquid interface monitored by electron spin resonance spectroscopy. Colloids Surf. A Physicochem. Eng. Aspects *7*:159–166, 1993.
92. J. D. Andrade and V. Hlady, Plasma protein adsorption: the big twelve. Ann. NY Acad. Sci. 158–172, 1987.
93. V. Ball, Ph. Huetz, A. Elaissari, J.-P. Cazenave, J.-C. Voegel, and P. Schaaf, Kinetics of homomolecular exchange processes in the adsorption of proteins on solid surfaces. Proc. Natl. Acad. Sci. USA *91*:7330–7334, 1994.
94. V. Ball, PhD dissertation, chapter III. Université Louis Pasteur, Strasbourg, France, 1996.
95. V. Ball and J. J. Ramsden, unpublished observations.
96. A. Sadana, Protein adsorption and inactivation on surfaces. Influence of heterogeneities. Chem. Rev. *92*:1799–1818, 1992.
97. J. Talbot, P. R. Van Tassel, P. Viot, G. Tarjus, and J. J. Ramsden, Enhanced saturation coverages in adsorption-desorption processes, submitted.
98. E. Brynda, M. Houska, and F. Lednicky, Adsorption of human fibrinogen onto hydrophobic surfaces: the effect of concentration in solution. J. Colloid Interface Sci. *113*:164–171, 1986.

99. Ph. Huetz, V. Ball, J.-C. Voegel, and P. Schaaf, Exchange kinetics for a heterogeneous protein system on a solid surface. Langmuir: *11*:3145–3152, 1995.

100. V. Ball, A. Bentaleb, J. Hemmerlé, J.-C. Voegel, P. Schaaf, Dynamic aspects of protein adsorption onto titanium surfaces: mechanism of desorption into buffer and release in the presence of proteins in the bulk. Langmuir *12*:1614–1621, 1996.

101. A. Bentaleb, V. Ball, Y. Haïkel, J.-C. Voegel, and P. Schaaf, Kinetics of the homogeneous exchange of lysozyme adsorbed on titanium oxide surfaces. Langmuir *13*:729–735, 1997.

102. A. Bentaleb, Y. Haïkal, J.-C. Voegel, and P. Schaaf, Kinetics for the homogeneous exchange in the adsorption of proteins on a titanium oxide surface. J. Biomed. Mater. Res., in press.

103. A. V. Elgersma, R. L. J. Zsom, J. Lyklema, and W. Norde, Adsorption competition between albumin and monoclonal immuno-gamma-globulins on polystyrene lattices. J. Colloid Interface Sci. *152*:410–428, 1992.

104. P. R. Van Tassel, P. Viot, and G. Tarjus, A kinetic model of partially reversible protein adsorption. J. Chem. Phys. *106*:761–770, 1997.

105. R. L. Beissinger and E. F. Leonard, Immunoglobulin sorption and desorption rates on quartz: evidence for multiple sorbed states. ASAIO J. *3*:160–175, 1980.

106. H. P. Jennissen, Protein binding to two dimensional binding site lattices: sorption kinetics of phosphorylase b on immobilized butyl residues. J. Colloid Interface Sci. *111*:570–586, 1986.

107. Y. Lvov, K. Ariga, I. Ichinose, and T. Kunitake, Assembly of multicomponent protein films by means of electrostatic layer by layer adsorption. J. Am. Chem. Soc. *117*:6117–6123, 1995.

108. L. Feng and J. D. Andrade, Structure and adsorption properties of fibrinogen. ACS Symp. Ser. *602*:66–79, 1995.

109. J.-P. Jolivet, De la solution à l'oxyde. Condensation des cations en solution aqueuse. In *Chimie de surface des oxydes*. Interéditions/CNRS Editions, Paris, 1994.

110. J. R. Hunter, R. G. Carbonell, and P. K. Kilpatrick, Coadsorption and exchange of lysozyme/β-casein mixtures at the air/water interface. J. Colloid Interface Sci. *143*:37–53, 1991.

111. H. Brockman, Dipole potential of lipid membranes. Chem. Phys. Lipids *73*:57–79, 1994.

112. P. H. Von Hippel and K. Y. Wong, On the conformational stability of globular proteins. J. Biol. Chem. *240*:3909–3923, 1965.

113. P. R. Van Tassel, P. Viot, and G. Tarjus, A kinetic model of partially reversible protein adsorption. J. Chem. Phys. *106*:761–770, 1997.

114. M. C. Wahlgren, T. Arnebrant, and I. Lundström, The adsorption of lysozyme to hydrophilic silicon oxide surfaces: comparison between experimental data and models for adsorption kinetics. J. Colloid Interface Sci. *175*:506–514, 1995.

115. T. P. Burghardt and D. Axelrod, Total internal reflection fluorescence study of energy transfer in surface-adsorbed and dissolved bovine serum albumin. Biochemistry *22*:979–985, 1983.

116. L. K. Tamm and E. Kalb, Molecular luminescence spectroscopy (S. G. Schulman ed.). Chemical Analysis Series, Vol. 77, Wiley, New York, 1993, pp. 253–305.

117. R. D. Tilton, C. R. Robertson, and A. P. Gast, Lateral diffusion of bovine serum albumin adsorbed at the solid–liquid interface. J. Colloid Interface Sci. *137*:192–203.

118. E. Kalb and L. K. Tamm, Incorporation of cytochrome b_5 into supported phospholipid bilayers by vesicle fusion to supported monolayers. Thin Solid Films *210/211*:763–765, 1992.

119. V. Ball, PhD dissertation, Chapter V, Université Louis Pasteur, Strasbourg, France, 1996.

120. H. Hygren, Nonlinear kinetics of ferritin adsorption. Biophys. J. 65:1508–1512, 1993.
121. A. P. Quist, A. Petersson, C. T. Reimann, A. A. Bergman, D. D. N. Barlo Daya, A. Hallen, J. Carlsson, S. O. Oscarsson, and B. U. R. Sundqvist, Site selective molecular adsorption at nanometer scale MeV atomic ion induced surface defects. J. Colloid Interface Sci. 189:184–187, 1997.
122. a. W. Pfeil, U. Gesierich, G. R. Kleemann, and R. Sterner, Ferredoxin from the hyperthermophile *Thermotoga maritima* is stable beyond the boiling point of water. J. Mol. Biol. 272:591–596, 1997. b. R. Sterner, G. R. Kleemann, H. Szadkowski, A. Lustig, M. Hennig, and K. Kirschner, Prot. Sci. 5:2000–2008, 1996.
123. M. F. Lecompte, I. R. Miller, Reversibility of prothrombin adsorption on lipid monolayers. J. Colloid Interface Sci. 123:259–266, 1988.
124. P. Suttiprasit, V. Krisdashima, and J. McGuire, The surface activity of α-lactalbumin, β-lactoglobulin, and bovine serum albumin. I. Surface tension measurements with single-component and mixed solutions. J. Colloid Interface Sci. 154:316–326, 1992.
125. J.-L. Courthaudon, E. Dickinson, and W. W. Christie, Competitive adsorption of lecithin and β-casein in oil in water emulsions. J. Agricult. Food Chem. 39:1365–1368, 1991.
126. F. MacRitchie and A. E. Alexander, Kinetics of adsorption of proteins at interfaces. III. The role of electrical barriers in adsorption. J. Colloid Sci. 18:464–469, 1963.
127. E. Dickinson, Adsorption of sticky hard spheres: relevance to protein competitive adsorption. J. Chem. Soc. Faraday Trans. 88:3561–3565, 1992.
128. R. M. Peitzsch and S. McLaughlin, Binding of acylated peptides and fatty acids to phospholipid vesicles: pertinence to myristolyated proteins. Biochemistry 32:10436–10443, 1993.
129. J. Seelig, Titration calorimetry of lipide–peptide interactions. Biophys. Biochim. Acta 1331:103–116, 1997.
130. G. Vergères, S. Manenti, T. Weber, and C. Stürzinger, The myristoyl moiety of myristoylated alanine-rich C kinase substrate (MARCKS) and MARCKS-related protein is embedded in the membrane. J. Biol. Chem. 270:19879–19887, 1995.
131. M. Rytömaa and P. K. J. Kinnunen, Reversibility of the binding of cytochrome c to liposomes. J. Biol. Chem. 270:3197–3202, 1995.
132. K. Kakinoki, Y. Maeda, K. Hasegawa, and H. Kitano, Kinetics of binding processes of cytochrome c onto liposome surfaces. J. Colloid Interface Sci. 170:18–24, 1995.
133. M. Malmsten, B. Bergenstahl, M. Masquelier, M. Palsson, and C. Peterson, Adsorption of apolipoprotein B at phospholipid model surfaces. J. Colloid Interface Sci. 172:485–493, 1995.
134. M. Malmsten, Protein Adsorption in Intravenous Drug delivery. In *Biopolymers at Interfaces* (M. Malmsten, ed.). Marcel Dekker, New York, 1998, pp. 561–596.
135. D. Chapman, Biomembranes and new hemocompatible materials. Langmuir 9:39–45, 1993.
136. D. Chapman and P. I. Harris, Biomembranes, ion channels and new biomaterials. Biochem. Soc. Trans. 24:329–340, 1996.
137. B. Singla, V. Krisdhasima, and J. McGuire, Adsorption kinetics of wild type and two synthetic stability mutants of T4 phase lysozyme at silanized silica surfaces. J. Colloid Interface Sci. 182:292–296, 1996.
138. M. Malmsten, T. Arnebrant, and P. Billsten, Interfacial behavior of protein mutants and variants. In *Biopolymers at Interfaces* (M. Malmsten, ed.). Marcel Dekker, New York, 1998, pp. 119–142.
139. J.-C. Voegel, E. Pefferkorn, and A. Schmitt, Steric protection of a modified glass surface with adsorbed polyacrylamide: interactions with blood proteins. J. Chromat. Biomed. Appl. 428:17–24, 1988.
140. J. F. Douglas, H. E. Johnson, and S. Granick, A single kinetic model of polymer adsorption and desorption. Science 262:2010–2012, 1993.

141. K. L. Prime and G. M. Whitesides, Self-assembled organic monolayers: model systems for studying adsorption of proteins at surfaces. Science 252:1164–1167, 1991.

142. V. Chan, D. J. Graves, P. Fortina, and S. E. McKenzie, Adsorption and surface diffusion of DNA oligonucleotides at liquid/solid interfaces. Langmuir 13:320–329, 1997.

143. A. Wegner, Head to tail polymerization of actin. J. Mol. Biol. 108:139–150, 1976.

144. P. G. de Gennes, Pénétration d'une chaîne dans une couche adsorbée: échanges solution/adsorbat et pontage de grains colloïdaux. C. R. Acad. Sci. 301[série II]:1399–1403, 1985.

145. Y. Wang, R. J. Diermeier, and R. Rajogopalan, Exchange kinetics in spherical geometry. Langmuir 13:2348–2353, 1997.

12

Interactions Between Proteins and Surfactants at Solid Interfaces

MARIE WAHLGREN and CAMILLA A.-C. KARLSSON   University of Lund, Lund, Sweden

STEFAN WELIN-KLINTSTRÖM   Linköping University, Linköping, Sweden

I. INTRODUCTION

The interactions between proteins and surfactants at solid surfaces are of notable importance in our daily life. The first thing that comes to mind is probably the use of detergents in removal of protein-rich soil, both in industry and at home. Surfactants are used for cleaning dishes, clothes, process equipment, teeth, contact lenses, etc. Other important applications of surfactants where knowledge of interactions between surfactants and proteins is of importance are in bioanalytical methods and in bioseparation. Furthermore, the removal of proteins by surfactants has been used for investigations into the nature of protein adsorption itself [1–7]. Still our knowledge about what governs the interaction between adsorbed proteins and surfactants or the competitive adsorption of these to solid interfaces is far from complete. This chapter aims to describe what is known so far on a basic level about the interaction of proteins and surfactants at solid interfaces. It also gives some examples of practical applications where these interactions are of importance. In the latter section there is no attempt to completely cover all the different fields but rather to give a few examples of each subject. We do not give any details on the interaction between proteins and surfactants in solution [8] or the adsorption of surfactants to interfaces. Furthermore, work in related areas such as protein and surfactant adsorption at air–water interfaces are only referred to when directly relevant to the topic of solid interfaces [9–11].

II. GENERAL FEATURES OF THE INTERACTIONS BETWEEN PROTEINS AND SURFACTANTS AT SOLID INTERFACES

The factors determining the events when proteins and surfactants are present at interfaces are of course numerous. In the next section we discuss the effects of protein, surfactant, and surface properties in more detail. Here the attempt is to give a general description of the effect of surfactant on adsorbed proteins and the competitive adsorption of surfactant and protein. One factor of great importance for the outcome is the order of addition of proteins and surfactants. In Fig. 1, the result of adding sodium dodecyl sulfate (SDS) after the adsorption of β-lactoglobulin onto a chromium oxide surface is compared with the adsorption from a mixture of both species [12]. As can be seen, SDS only removes a small amount of the preadsorbed protein, but in the mixture it is able to hinder adsorption completely. The difference observed could be due to irreversible changes in the adsorbed protein or that adsorption to the interface hinders surfactant–protein interactions. Due to these differences we choose to divide this section into one part dealing with removal of preadsorbed proteins and one dealing with competitive adsorption between proteins and surfactants.

A. Removal of Proteins from Solid Interfaces by Surfactants

As pointed out, when a protein adsorbs to an interface it is possible for it to adapt to the interface, rendering it more difficult to remove. Still surfactants are often capable of removing large amounts of preadsorbed proteins. The following can occur when surfactants are added to a preadsorbed layer of proteins:

1.   Surfactant adsorbs or binds to the protein molecules at the interface. This is common for most ionic surfactants [8,13,14].
2.   Surfactants adsorb to the interface. Surfactants will absorb to the interface providing there is a favorable interaction between them and the interface. Thus, they will absorb to most hydrophobic interfaces. On hydrophilic interfaces the character of the hydrophilic group will decide whether or

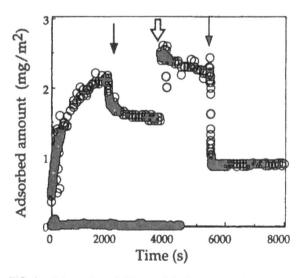

FIG. 1   Adsorption of β-lactoglobulin onto a chromium surface in the presence of SDS (filled circles) and followed by addition of SDS (open circles). The concentration of SDS was twice the cmc in the water and of β-lactoglobulin 1 mg/mL. Filled arrows indicate rinsing with buffer; the open arrow indicates the addition of surfactant. (From Ref. 12.)

not the surfactant will absorb. Thus, ionic surfactants will adsorb to surfaces of opposite charge [15,16–18]. Another example is that the ethylene oxide type of surfactants can absorb to silanol groups [19–23].

3. Surfactants do not interact with either component of the interface. This could be the case for some nonionic surfactants at hydrophilic interfaces.

Provided that the surfactant concentration is not extremely low, these processes are generally very fast, within seconds of the surfactant addition. Furthermore, the surfactant is generally reversibly adsorbed or bound to the surface layer.

Points 1 and 2 can lead to removal of the protein from the interface, although even in these cases the removal of protein might be incomplete or nonexistent. The reason for incomplete removal will be discussed in Section III.B. Surfactant-induced removal of proteins follows two patterns, illustrated in Figs. 2 and 3. In Fig. 2 two ellipsometry experiments are used to illustrate these patterns [12]. The protein and surface are the same in both experiments, β-lactoglobulin and silicon oxide, respectively. The variations are in the surfactant used. In the left panel the surfactant is SDS, and in the right cetyl trimethylammonium bromide (CTAB). As can be seen in the figure, SDS gives a clean surface immediately after surfactant addition. The addition of CTAB leads to an increase in the adsorbed amount which is due to the adsorption of the surfactant to the interface. After the system is rinsed, the surfactant desorbs, leaving a clean surface. The two patterns observed made Wahlgren and Arnebrant suggest two mechanisms for removal [12], one where surfactant binds to the protein and solubilizes it, and one where the protein is removed by surfactant replacing it at the surface (Fig. 3).

It is obvious from the results in Fig. 2a that surfactants can remove proteins from solid interfaces without adsorbing to the interface, thus validating the solubilization model. This mechanism demands that the surfactant interact with the protein, which is the case for most ionic surfactants but not generally for nonionic ones [8].

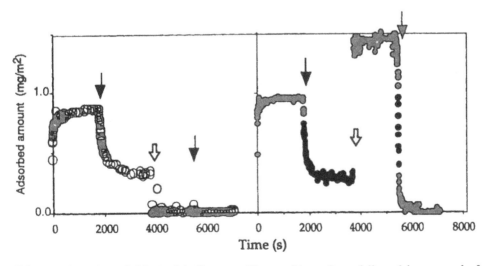

FIG. 2 Adsorption of β-lactoglobulin onto silicon oxide surfaces followed by removal of SDS (left panel) and CTAB (right panel). Filled arrows indicate rinsing with buffer; open arrows indicate the addition of surfactant.

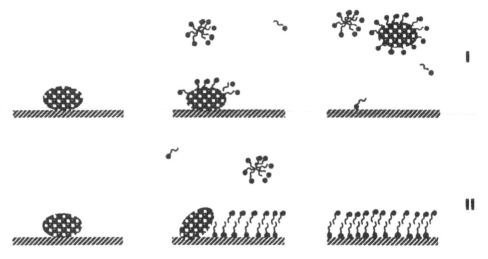

FIG. 3 Schematic illustration of the two suggested models for protein removal by surfactants.

The binding isotherm of surfactants to protein usually contains all or some of the following steps [24]:

1. Binding of surfactant by electrostatic or hydrophobic interactions to specific sites in the protein, such as for β-lactoglobulin [24–26] and serum albumin [24,27,28]
2. Cooperative adsorption of surfactant to the protein without gross conformational changes
3. Cooperative binding to the protein followed by conformational changes [13,28–31]

It is not obvious at which of these steps the addition of surfactant causes protein removal. However, an increase in ionic strength will have different effects on the binding of surfactant, depending on whether the binding occurs as in step 1 or step 3. It has been observed for T4 lysozyme that the removal starts at a lower surfactant concentration when the ionic strength is increased. This behavior is in line with the effect of ionic strength on the associative behavior of surfactants [32]. This indicates that at least for this case when surface and protein had opposite charges, the removal starts when the surfactant associates cooperatively to the adsorbed protein.

To verify the removal of proteins by the displacement mechanisms, one needs to study a system where it is likely that the surfactant only interacts with the surface, not the protein. This is, for example, true for most nonionic surfactants. The only types of interactions observed between these types of surfactants and proteins have been the binding to specific sites, such as hydrophobic domains in serum albumin. These surfactants will allow good removal of most proteins from hydrophobic interfaces [33–35], indicating that the removal mechanism is plausible.

B. Competitive Adsorption of Proteins and Surfactants

Experimentally it is observed that the presence of surfactants in protein solutions may influence the amount of proteins adsorbed to solid surfaces in three ways:

1. Complete hindrance of protein adsorption
2. Reduced amounts of protein adsorbed compared to adsorption from pure protein solution
3. Increased amounts adsorbed due to cooperative adsorption of protein and surfactant

Protein adsorption will not occur for protein/surfactant mixtures if the surfactant has a higher affinity than the protein for the surface or a protein/surfactant complex without affinity for the surface is formed.

Reduced amounts of protein might be found when a protein/surfactant complex is formed that still has an affinity for the surface. The presence of surfactant could, for example, due to sterical reasons, lead to a reduction in the amount of protein adsorbed.

An increase in the amount adsorbed has been observed at surfactant concentrations below [36,37] and above [38–40] critical micelle concentration (cmc). The presence of bound surfactant might change the charge or the structure of the adsorbing protein. This might lead to higher affinity for the surface or changes in the packing of the protein at the surface. There has also been the suggestion that low amounts of surfactant can make the surface more hydrophobic, thus promoting adsorption [41]. This mechanism is probably an exception and it is equally possible that the surfactants interact directly with the protein, thus influencing adsorption.

Dickinson and coworkers have used models from polymer–surfactant interaction to describe the competitive adsorption of proteins and low-molecular-weight surfactants [42,43]. Their main experimental work was done at the air–water or oil–water interfaces, but the models can also be used for solid–water interfaces. They used Monte Carlo models to simulate the competitive adsorption of molecules. Systems studied included flexible chains together with displacer molecules (relevant, for example, for caseins) [43] and competitive adsorption between large spheres and small ones (relevant for globular proteins) [42]. The models tested are one continuum model, where each adsorbed molecule is represented by spheres of different size, and one lattice model, where the individual polymer segments (amino acids) or a single low-molecular-weight species occupies one square in a lattice [44]. These models are gross oversimplifications but still give some basic understanding into the effect of concentration and strength of interaction to the surface and between protein and surfactant. For the flexible chain model information on the effect of chain length is also obtained. The models can describe many of the experimentally observed behaviors of mixed systems.

III. INFLUENCES OF SURFACTANT, PROTEIN, AND SURFACE PROPERTIES

A. Surfactant

1. Type of Surfactant

The character of the surfactant head group is, of course, of great importance, especially for hydrophilic surfaces, where it determines if the surfactant will interact with the interface and the strength of this interaction. As described above, surfactant can remove proteins both by solubilization and displacement. Thus, adsorption onto the surface is not mandatory for protein removal. In fact the removal of proteins from negative surfaces (e.g., silicon oxide) is often larger for SDS and other negatively charged surfactants than for trimethylammonium bromides (TABs) [12,35,39]. This is most likely due to the effect of the head group on the interaction between the surfactant and the protein. As a rule of thumb the interaction between protein and surfactant decreases in the order SDS > cationic surfactants >> nonionic surfactants. Normally, nonionic surfactants do not interact with proteins, which could explain why they seldom have any large cleaning effect on hydrophilic surfaces [35,45].

In addition, on hydrophobic interfaces it has been reported that the head group influences the amount removed. Rapoza and Horbett [4] have observed that the elutability (the relative amount of protein removed) was higher for surfactants with small head groups (SDS, DAHCl, $C_{12}E_4$) than for Tween 20, which has a large sorbitol ring. They also found that the positively charged DAHCl was the most efficient of these four surfactants in removing fibrinogen from polymeric surfaces. The explanation could be that DAHCl has a smaller head group than the other surfactants and thus might adsorb in denser layers to the interface. Similar effects have been seen for BRIJ type of surfactants used as a coating in capillary electrophoresis [46]. It has also been observed that approaching the cloud point of $C_{12}E_5$ increases its competitive power [33]. This might also be related to the packing efficiency of the surfactant.

There are indications that at least on hydrophobic surfaces the chain length of the surfactant is of less importance than the head group [4,39,46]. For example, Rapoza and Horbett [4] found no difference in elutability of fibrinogen when investigating alkyl sulfates with varying chain lengths. However, the hydrophobic part of the surfactant might not always be composed of a single straight hydrocarbon chain. Welin-Klintström [35] has observed that surfactants with complex and bulky hydrophobic parts, such as CHAPS and deoxycholic acid, will be less efficient on hydrophobic and hydrophilic surfaces.

Chattoraj and coworkers have studied the influence of surfactant type on the adsorption from mixtures of proteins and surfactants onto interfaces [38,47]. They found that the degree of cooperative adsorption of gelatin and TABs to alumina passed through a maximum with increasing chain length of the surfactant [38]. Similar results have also been observed for fibrinogen and TABs at silicon oxide surfaces [39]. Samanta and Chattoraj studied systems where the presence of surfactant reduced the amount of protein adsorbed [47,48]. For alumina they found that CTAB is more efficient in hindering adsorption of a globular protein (BSA) than SDS [47], while others have found that SDS is better in blocking protein adsorption at silica surfaces [39,49]. The material is not large enough for reaching any general conclusion, but one could speculate that the surfactant is most efficient in hindering protein adsorption when it has the same charge as the surface. Thus, hindering adsorption through solubilization of the protein instead of competition for the interface might be more efficient at hydrophilic interfaces. Similar trends have been observed in the surfactant-induced removal of proteins from solid surfaces [33,35,49].

2. Concentrations Below and Above Critical Micelle Concentration

There are few investigations into how the surfactant concentration affects the amount removed [4,32,34,50–52], and from these it is obvious that removal does not occur until a threshold value has been reached. When the threshold has been reached, the removal increases rapidly with increasing surfactant concentration and soon reaches a plateau value. In Fig. 4 the dependence of protein removal on surfactant concentration is illustrated [34]. The threshold value is usually well below the cmc for the surfactant but is probably linked to the cooperative behavior of the surfactant.

Factors that would influence the cooperativity, such as surfactant chain length [4] or ionic strength of the solution [4,32], move the onset of desorption in the same direction as they would move the cmc for the surfactant. Furthermore, it is known that the presence of a surface or a polymer usually decreases the critical association concentration with several magnitudes. Samanta and Chattoraj [48], for example, noted that massive binding of SDS to BSA occurred at the alumina interface at a concentration that was nearly 100 times lower than the one expected for cmc.

Claesson and coworkers have used surface force measurements to study the interaction between lysozyme adsorbed to mica and anionic surfactants [51,52]. They found that the introduction of the surfactants first leads to an increase in the apparent interfacial charge, and close to cmc they observe a removal of the protein. They did not detect any major conformational changes in the adsorbed proteins due to the adsorption of surfactants [below the critical association constant (cac)]. They also noted that the layer thickness did not increase upon adsorption of surfactants to the protein, indicating that the surfactants have penetrated the protein layer.

In the case of removal by displacement Wahlgren and Arnebrant have observed, for lysozyme adsorbed to methylated silica, that the removal followed the adsorption of surfactant to a clean surface [34] (Fig. 4).

Note that the start of removal is not only dependent on the surfactant but also on the protein layer. It has been found that the removal started at different concentrations for mutant forms of T4 lysozyme [32] and lipase [50]. The T4 lysozyme investigated had a mutation that gave larger conformational changes of the protein compared to the wild-type protein upon adsorption to interface. This was thought to favor binding of surfactant at low concentrations, leading to solubilization and removal at lower surfactant concentrations than for the wild-type protein [32]. The lipases differed in their hydrophobicities. It was found that removal from methylated

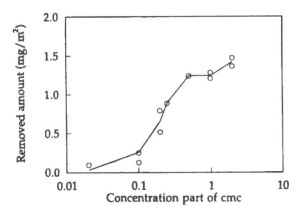

FIG. 4 Dependence of protein removal on surfactant concentration. Removal of lysozyme from methylated silica by $C_{12}E_5$. (Data from Ref. 34.)

silica started at higher surfactant concentrations for the more hydrophobic protein, probability due to stronger interaction between protein and surface [50]. Other investigators have observed that surfactant-induced removal can differ between proteins [4]. It might even differ for the same protein, depending on the adsorption conditions. Horbett and Rapoza have observed a difference in the critical surfactant concentration for removal between fibrinogen layers formed at different concentrations [4]. This was probably due to difference in conformation or orientation of the adsorbed proteins. These issues are discussed in more detail elsewhere in this book.

The same concentration dependency is of course also observed for the competitive adsorption of protein and surfactant [37,41,47,48]. The two main differences between competitive adsorption and removal of preadsorbed proteins are that

1. Proteins and surfactants are present in the solution at the same time, and if the surfactant binds to the protein this might influence the amount of free surfactant molecules, resulting in an apparent shift in cmc and cac of the surfactant.
2. The protein has not adsorbed to the interface before it binds surfactants. This means that protein–surfactant interactions will not be influenced by the presence of the surface.

The adsorption isotherms for proteins and surfactants often differ substantially (Fig. 5). Protein usually adsorbs in a broader concentration interval than surfactants and at lower total concentrations. However, the surfactants are often more surface active than the protein when the concentration is above the critical association concentration. Thus, at least for hydrophobic interfaces the surfactants will start to compete with the protein above this critical concentration, but the proteins will adsorb below it. A practical consequence of this is that if the mixture is diluted the protein might start to adsorb even though the ratio between protein and surfactant is constant [37].

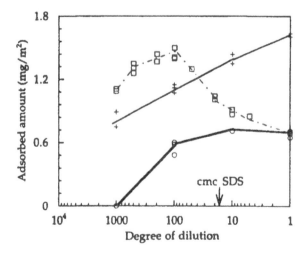

FIG. 5 The competitive adsorption of β-lactoglobulin and SDS (squares) and the adsorption of the pure components, SDS (open circles) and β-lactoglobulin (crosses), to a methylated silica surface. A degree of dilution of 1 represents a surfactant concentration of 2 cmc in water and a protein concentration of 1 mg/mL. (From Ref. 37.)

B. Protein

1. Protein Properties

The ability of a surfactant to remove proteins from a surface varies considerably, depending on the protein in question. Due to the complexity of the issue and the large variations between the different proteins, it has been difficult to pinpoint the crucial protein characteristics. Wahlgren et al. [53] made an attempt to study the effect of protein properties on the removal of proteins by dodecyltrimethylammonium bromide (DTAB). The correlations were not very good, but it seemed that at hydrophobic methylated silica surfaces a decrease in the degree or removal might be observed when the size and shell hydrophobicity of the proteins increased. At hydrophilic silicon oxide surfaces some influence of the protein's size, isoelectric charge, and structural stability was evident. Later studies using mutants of T4 lysozyme have verified the importance of conformational stability both on hydrophilic and hydrophobic silica [54,55]. The effect of charge is, however, far more complicated than what was indicated from earlier studies, and factors such as location of the charge, etc., might play a considerable role [56]. Wannerberger and Arnebrant studied the lipase *Humicola lannuginosa* and a mutant that had increased hydrophobicity. They found in agreement with the earlier study that increased hydrophobicity decreased the capability of surfactants SDS and $C_{12}E_5$ to remove protein from methylated silica [50]. A more thorough discussion of mutant protein adsorption is given elsewhere in this book.

2. Structure of the Protein Layer

The structure of the protein layer might also influence the degree of removal by surfactants. It is often observed that only a fraction of the adsorbed proteins can be removed by surfactants. This has been taken as an evidence for a heterogeneous surface layer. The heterogeneity could be due to different orientation of the proteins, different degree of conformational changes of the adsorbed proteins or heterogeneity of the surface [57]. The surface coverage is one of the factors that are thought to influence orientation and conformational changes of the protein molecules. Thus, if the layer formed is below a monolayer there is usually a higher possibility for the protein to adapt to the surface by changing its conformation, and this might lead to lower elutability [4]. However, it has also been observed, in the case of collagen, that high protein density makes the protein more difficult to remove. This could be due to interactions between the adsorbed proteins [58].

A few proteins may form bi- or multilayers when they adsorb to an interface when the bulk concentration is high. This could also influence the removal by surfactants. Figure 6 demonstrates this for the removal of lysozyme by nonionic surfactant $C_{12}E_5$ and SDS. Furthermore, identical experiments have been performed for bovine serum albumin (BSA) (Wahlgren, unpublished data). Lysozyme is known to form bilayers at higher concentrations, while BSA only forms monolayers. When the bulk concentration of the protein is increased, $C_{12}E_5$ seems to become less efficient in removing the lysozyme, but this does not happen to BSA. The decrease seen for lysozyme could thus be due to the formation of a second layer of protein. The formation of bilayers did not seem to affect the cleaning efficiency of SDS. This surfactant binds, in contrast to $C_{12}E_5$, directly to the protein and could remove proteins both by solubization and displacement.

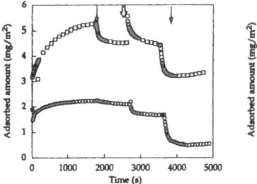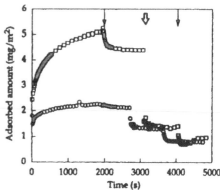

FIG. 6 Effect of multiple layers of adsorbed protein. Adsorption of lysozyme 1 mg/mL (circles) and 10 mg/mL (squares) followed by desorption by $C_{12}E_5$ (left panel) and SDS (right panel). Adsorption was measured at 25°C in a 0.01 M phosphate buffer pH 7. Filled arrows indicate rinsing with buffer; open arrows indicate the addition of surfactant.

There are few studies into the structure of adsorbed protein layers both prior to and after surfactant-induced protein removal. Feng et al. [59] has used atomic force microscopy to study the adsorption and SDS-induced removal of high-density lipoproteins (HDL). They found that HDL adsorbed onto mica surfaces existed both as single proteins and as clusters of two or three particles. After SDS exposure, the number of adsorbed HDL particles had decreased drastically and no clusters were observed. Furthermore the proteins remaining after removal seem to be scattered randomly over the surface.

3. Adsorption Time and Temperature

As pointed out, structural changes of the adsorbed protein tend to reduce the amount of protein removed by surfactant. The same is probably true also for restructuring the adsorbed layer. These changes will of course be time [2,5] and temperature dependent [59–62]. Rapoza and Horbett have measured a decrease in surfactant-induced removal of fibrinogen for up to 5 days [5]. The time effects have also been observed on a much faster time scale, i.e., within hours or less from the initial adsorption [3,7]. The time-dependent change in elutability has been correlated to structural changes of the fibrinogen molecule as measured by infrared spectroscopy [7]. It has been observed that the rate of conversion of fibrinogen to a nonelutable state is dependent on the character of the surface [3,7,63,64]. Other proteins, such as lysozyme, BSA, IgG [2], β-lactoglobulin [61], and high-density lipoproteins [59], have also shown a time dependence in their elutability. Thus, time-dependent changes are probably a rather general phenomenon for many proteins.

McGuire and coworkers used elutability data in combination with simple models for protein adsorption to estimate the conversion of protein from elutable to nonelutable form [54,56,65,66]. They refined the model to contain not only the conversion of the protein to a nonelutable state but also a rate constant for the removal of protein by surfactant [65]. The latter can in some cases be the rate-determining step, and then elutability data will not give any good information on the strength of protein adhesion to the interface [65]. Thus, elutability data cannot always be used as an indirect measurement of the strength in the interaction between surface and protein.

Increased temperature can lead to a higher rate in the time-dependent structural changes of the protein. It can also induce denaturation of the protein and aggregation at the interface and in solution. This will lead to high amounts of protein adsorbed, and the surfactant-induced removal will decrease drastically [59–62] (Fig. 7). The figure illustrates that these effects occur in a narrow temperature interval, as would be expected for temperature-induced protein denaturation. Karlsson et al. [60] have shown that the effect of temperature varies considerably for different surfaces, such as chromium, steel, and methylated silica. For a methylated silica surface the amount removed decreases with increasing temperature between 25 and 80°C, although the largest decrease is between 60 and 73°C. The metal surface first has an increase in the amount removed by a combination of rinsing and surfactants, when the temperature is increased to 73°C. After this the removal decreases when the temperature is further increased. Thus, there might be an optimal temperature for protein removal that differs among different types of surfaces.

The temperature has also been shown to effect the competitive adsorption of surfactant and proteins, especially at surfactant concentrations below cmc. Wahlgren et al. [33] observed for fibrinogen and $C_{12}E_5$ that an increase in the temperature from 22 to 34°C increased the competitive power of the surfactant. $C_{12}E_5$ has a cloud point around 30°C, and it is plausible that the change in the surfactant explains this temperature dependence.

C. Surface

Attempts have been made to correlate surface properties with the degree of surfactant-mediated protein removal. One parameter of interest is, of course, the chemical

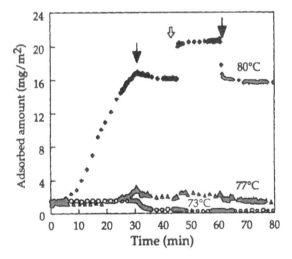

FIG. 7 Temperature dependence on the adsorption of 1 mg/mL of β-lactoglobulin onto methylated silica followed by removal by twice the cmc of SDS. Filled arrows indicate rinsing with buffer; the open arrow indicates the addition of surfactant.

composition of the material, and Ertel et al. [2] have found for radiofrequency plasma-deposited polystyrene that the change is elutability with time increased with increasing surface oxygen. Other parameters of interest would be surface roughness, surface heterogeneity and the molecular rigidity of, in particular, polymer surfaces. However, it has often been difficult to give simple explanations for the observed results [3,12]. This is not surprising considering that the mechanism of removal, the interaction between surface and surfactant and protein all vary among different surfaces. Still, Horbett and coworkers have successfully correlated the degree of removal induced by SDS (elutability) to the biocompatibility of surfaces [63]. A high degree of fibrinogen removal has been seen for surfaces with good biocompatibility.

Some work that best illustrates the complexity of the surface influences has been done by Elwing and coworkers, using surfaces with a hydrophobicity gradient [33,35,64,67]. These gradients are based on silicon oxide surfaces that were made hydrophobic in different degrees by the use of dimethylochlorosilane, rendering one end of the surface strongly hydrophobic (contact angle >90°) and one end, where no silanization occurred, strongly hydrophilic (contact angle <10°).

The degree of surfactant-induced removal over the gradient surface varies considerably with the type of surfactant [35]. We will discuss two of these surfactants. In both cases, fibrinogen is adsorbed to the gradient surface, and, as can be observed, fibrinogen adsorption is strongly dependent on surface hydrophobicity.

Figure 8 shows the effects of SDS added to the preadsorbed fibrinogen. Sodium dodecyl sulfate in itself only adsorbs on the very hydrophobic end of the gradient (Fig. 9) but removes nearly all fibrinogen from the hydrophilic and hydrophobic ends. In the middle of the gradient the surfactant is less efficient, and some fibrinogen

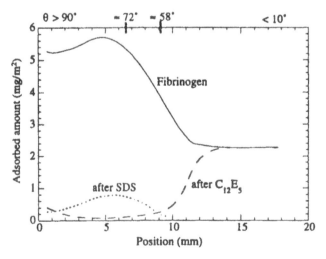

FIG. 8 Adsorption and surfactant-induced removal of fibrinogen at a wettability gradient surface. The contact angle to the water, θ, at different positions along the gradient is indicated at the top. Solid line indicates fibrinogen adsorption. Dashed lines indicate amount of fibrinogen left after incubation with surfactant, SDS, and $C_{12}E_5$, as indicated. (Data from Ref. 35.)

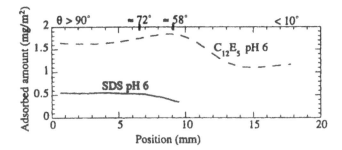

FIG. 9 Adsorption of SDS and $C_{12}E_5$ onto a wettability gradient surface. The contact angle to the water, θ, at different positions along the gradient is indicated at the top.

remains adsorbed to the surface after cleaning. It could be that the removal mechanism shifts over the gradient from displacement at the hydrophobic end to solubilization at the hydrophilic one. The removal is efficient in the extreme hydrophobic end because the surfactant adsorbs strongly to the interface, a factor that is decreasing rapidly when the surface becomes more hydrophilic (Fig. 9). On the hydrophilic end of the gradient the surface is negatively charged, making repulsion between the strongly charged SDS/protein complex and the surface high, thus facilitating removal by solubilization. The more hydrophobic the surface becomes, the less charged it will be. Thus the reason for the lower cleaning efficiency in the middle of the gradient could be that neither removal mechanism is functioning optimally.

Figure 8 also shows the effect of $C_{12}E_5$ on the preadsorbed fibrinogen. As can be seen, the surfactant removes all the fibrinogen in the middle of the gradient, while the amount removed decreases at the hydrophilic and hydrophobic ends. For $C_{12}E_5$ there is no interaction between the protein and the surfactant, so the removal mechanism is displacement. However, the surfactant adsorbs over the whole gradient (Fig. 9). It is likely that the failure of the surfactant to remove protein at the hydrophilic end is due to the rather weak interaction between the surface and the surfactant. This cannot be the case at the hydrophobic end. However, fibrinogen is known to strongly interact with hydrophobic surfaces, which leads to time-dependent structural changes in the adsorbed protein, making it more difficult to remove from the surface. Thus, the reason for failure at the hydrophobic end probably reflects the strong interaction between surface and protein.

The hydrophobicity of the surface not only affects the interaction between the adsorbed species, but it has also been observed that the rate of removal might vary with position at a gradient surface. Welin-Klintström et al. found that, for $C_{12}E_5$-induced removal of fibrinogen, the rate of removal was at its lowest in the intermediate region [68]. Another factor observed to vary over the gradient surface is the critical concentration at which surfactant is able to hinder protein adsorption [33]. The concentration needed is lower at the more hydrophobic end of the gradient.

The gradient surfaces only give information on how chemical difference in a surface affects removal. Another factor that might be of importance is surface morphology; e.g., for polymeric surfaces the rigidity of the polymer has been found to be important [3]. The less rigid polymers were more difficult to clean. The reason might be that these can interact more strongly with the protein.

IV. PRACTICAL USE OF SURFACTANT–PROTEIN INTERACTIONS

A. Cleaning Food Processing Equipment

Fouling, the deposition or accumulation of unwanted material on a surface, is significant in industrial processing. In the food industry, severe cases are often encountered. Proteins are frequently to blame for heavy fouling at elevated temperatures [69,70], as high temperatures cause denaturation and aggregation of proteins. Common consequences of fouling include decreased operating efficiency and unsatisfactory product quality. This is, of course, most important for the thick layers formed at high temperatures, but for specific processes such as micro- and ultrafiltration fouling at low temperatures might also have a detrimental effect. (The latter applications are discussed more extensively elsewhere in this book.) Furthermore, the microbiological safety of the food can be endangered. For these reasons, frequent and thorough cleaning of food processing equipment is required.

Cleaning is a complicated multistage process, involving mass transfer and chemical reactions in the bulk and the deposit. The cleaning process is strongly influenced by the nature of the deposit, surface, and cleaning agent and temperature and flow conditions [71,72]. Unheated surfaces are often cleaned with formulated alkaline detergents, which, in addition, generally contain mainly sequestering agents and surfactants. Traditionally, deposits on heated surfaces are removed by two-stage cleaning procedures involving alkali and acid. Usually, the acid is nitric or phosphoric acid, whereas the alkali may be pure sodium hydroxide, although suitable additives are often used.

Alkalis remove protein and fat, whereas acids are efficient in dissolving mineral deposits. Sequestering agents can keep hard-water ions in solution and remove mineral deposits. Among surfactants, anionics are the most widely used. They can often remove various soil types by solubilization or displacement. Nonionic surfactants may be efficient in fat removal, but usually do not interact with protein. Cationic surfactants have poor cleaning characteristics but are often effective components in disinfectants. Formulated acid detergents are generally inferior to alkaline types. This was demonstrated when comparing the abilities of alkaline anionic, acid anionic, and acid cationic detergents to remove skim milk deposits from stainless steel [73]. Single-stage cleaning of heat exchangers using a formulated alkaline detergent may prove more efficient than a traditional two-stage procedure. The superiority of the former was demonstrated when cleaning pasteurizers which had been fouled by milk heated to 72°C [74] and sterilization equipment where milk had been heated to 138°C [75].

Grasshoff [76,77] has investigated the efficiency of various additives in alkaline detergents. The addition of surfactants resulted in only minor cleaning efficiency improvements [77]. In contrast, Karlsson et al. [78] observed enhanced removal of the whey protein β-lactoglobulin from stainless steel when surfactant was added to alkali. The anionic surfactant SDS was rather inefficient on its own, but the combination of SDS and sodium hydroxide resulted in significantly higher cleaning efficiency than with pure alkali alone. While Grasshoff studied the removal of approximately 0.3-mm-thick deposits, the investigation by Karlsson et al. [78] concerned protein layers about 10 nm thick.

Enzymes such as proteases can also be used in formulated detergents, and Munoz-Aguado et al. [79] have shown that pretreating fouled membranes with chymotrypsine (a protease) before cleaning with CTAB improves the surfactant cleaning capacity. They also showed that Terg-A-Zyme, a commercial detergent containing proteases, had good cleaning efficiency. The enzyme probably decreases the molecular weight of the adsorbed protein species, thus rendering them more soluble or easier to remove by competition. In creating enzyme-based detergents it is probably important to consider the effect of the surfactant molecules on the enzyme. High concentrations of surfactant might hinder enzyme substrate interactions [80] or denature the enzyme. (For a thorough review on enzymes in detergency see Ref. 81.)

The removal of food soils is, in general, improved at higher cleaning temperatures. As an example, when cleaning steel surfaces fouled by skim milk, all formulated detergents tested showed improved performance at 75°C compared to at 20°C [73]. However, heating the milk films resulted in higher resistance to subsequent desorption. In most investigations the fouling temperature has been fixed, whereas the temperature during subsequent cleaning has been varied. Karlsson et al. [60–62,78] have studied β-lactoglobulin removal by SDS and sodium hydroxide, employing the same temperature during fouling as during cleaning. The cleanability of methylated silica and, above 73°C, of metal oxides was observed to decrease with increasing temperature. The effects were attributed to temperature-dependent changes in the deposit [60,61,78]. Results indicate that the outcome of a cleaning procedure can be appreciably affected by relatively small temperature deviations during fouling.

B. Dental Applications

The formation of plaque on teeth and restorative materials is known to be the causative factor in the development of caries and periodontal disease. It is known that the saliva proteins adsorb to surfaces in the oral cavity and form a conditioning film. This film will strongly influence the following buildup of, for example, plaque. Thus it is of great interest to study this film. In addition, surfactant-induced removal of the film is interesting to investigate, as removal of the salivary proteins most likely will lead to removal of secondary adsorbed species such as bacteria.

Arnebrant and coworkers have used surfactants to study adsorbed salivary protein. They investigated both standard types of surfactants, such as SDS and CTAB [82–84], and the surface-active dental drug Delmopinol [85,86]. Depending on the surfactant used the removal was found to follow solubilization- or replacement-type mechanisms [82]. The degree of removal was strongly dependent on the concentration of saliva in the experiment [83] and from which glands the saliva was collected [86]. The saliva contains a large variety of proteins, and the variance between different experimental conditions could be due to difference in the protein composition of the salivary film.

Vassilakos et al. also showed that the surfactant-induced removal of proteins adsorbed from six different saliva fractionations varied considerably [84]. They found that for removal by CTAB, the most difficult fraction to remove was the one containing the high-molecular-weight saliva proteins. Furthermore, the fraction that contained the most hydrophobic proteins had the largest difference in elutability between hydrophilic and hydrophobic surfaces. While these proteins could be completely removed by CTAB from hydrophilic surfaces, only 50% or so we were removed from a hydrophobic silica surface.

C. Biochemical Applications

Surfactants are utilized in various biochemical methods, such as the purification and analysis of proteins, in analytical methods based on enzymes or immunological techniques, and in cleaning and regenerating chromatographic columns, biosensors, etc. Their ability to hinder protein adsorption is used both to reduce the depletion of the substance that should be analyzed due to adsorption to the walls of test tubes, etc. [87], and to hinder nonspecific adsorption of protein in, for example, immunological methods and chromatography [88]. The latter issues are addressed in greater detail in Chapter 27.

There are also examples in which low amounts of surfactants have been used to increase the adsorption of proteins to surfaces. One such example is the adsorption of hemoglobin to silver electrodes [36]. In this case the surfactant might induce a changed structure of the protein. The combined effects gave an increase in voltammetric response and thus a higher sensitivity in the analysis [36].

1. Immobilization of Proteins

Various biosensor technologies use surface-bound protein, i.e., enzymes or immunoglobulins for analytical purposes. Immobilized enzymes are also used in other biotechnological processes, such as enzyme reactors, affinity chromatography, and immunoassays. There are several benefits in immobilization of proteins. For example, the enzymes can be reused in a process, and the stability of the enzymes might be improved. Surfactants can be used to improve the immobilization of these proteins and to test the strength of the attachment of the protein to the surface [89]. (*See also* Chapter 27.) Covalent attachment is one method to immobilize proteins to biosensors. This might increase the stability of an electrode, and spontaneously adsorbed proteins are an unwanted contamination of these electrodes. Surfactants can be used to reduce noncovalently bound proteins. Willams and Blanch [90] showed that the reduction of noncovalent attached protein was more efficient if the surfactant was present during the covalent attachment than if it was used to remove the protein after the linkage.

2. Immunoassays

Antibodies are used in several analytical methods, such as radioimmunoassay (RIA), ELISA, etc. These methods usually involve binding the antigen or antibody to a solid interface by spontaneous adsorption or by immobilization techniques. The strength of the protein attachment is important, and surfactants have been used to probe it [91,92]. Surfactants can also be used to restore the sensitivity of an assay when performed in a complex solution. This was, for example, observed for zwitterionic ions in an ELISA analysis of staphylococcal enterotoxin B in cheddar cheese homogenate [93]. Immunoassays are described in greater detail elsewhere in this book.

3. Chromatography

Surfactants have been used in various ways in chromatographic processes. They can be divided into systems in which surfactants are used to improve the chromatographic material and those in which they are used during the purification steps.

In the first case, surfactants improve affinity chromatographic columns by removing nonspecifically adsorbed proteins after immobilization to the solid supports. This reduces protein leakage from the support material [94]. Furthermore, surfactants coat chromatography material, changing the physical properties of the columns, e.g., polyoxyethylene-type nonionic surfactants have been used to coat reversed-phase supports. This was found to hinder protein adsorption at moderate or low ionic strength but still allow it at high ionic strength, making it possible to use the columns for hydrophobic interaction chromatography [95].

In a chromatographic column surfactants can hinder nonspecific adsorption. For example, 3-[(3-cholamidoprophyl)dimethylammonio]-1-propanesulfonate (CHAPS) and SDS have been used to reduce the nonspecific adsorption of proteoglycans to chromatographic media, such as Sepharose CL-2B, cellulose, and controlled pore glass (CPG) [88]. They can, of course, also be part of the elution buffer, thus utilizing their ability to dissolve proteins. For example, Warren et al., used CHAPS to elute and stabilize glucocorticoid receptors [87].

4. Capillary Zone Electrophoresis

Capillary zone electrophoresis is a rapidly growing method for separating biological molecules, such as DNA, peptides, and proteins. One problem in the method is the adsorption of these molecules to the surface of the capillary walls (usually fused silica tubing). This adsorption leads to peak broadening. Several methods have been employed to avoid this problem, among them the use of surfactants [46,96–100], including dynamic coating with a cationic surfactant [98,99] or zwitterionic surfactant [101] in the separation buffer; hydrophobic modification of the silica followed with coating by nonionic surfactants [46,97,102]; and using SDS to solubilize the proteins [96,103]. Furthermore, SDS and CTAC have been used to improve the separation in a polyacrylamide-coated silica capillary [104]. Nonionic coating surfactants have also been applied to other types of capillary material, e.g., hollow propylene fibers [100]. As can be expected, methods where the surfactant is present during separation are more reproducible than those in which the capillary has been precoated with surfactant [46,97]. This is due to desorption of the surfactant from the capillary during use [97].

IV. CONCLUDING REMARKS

Clearly, numerous experimental results in this field have been obtained. However, fundamental physical models are still needed, but obtaining these models is difficult because of the complexity of the issue. These types of models must take into account the protein–surfactant interaction in solution as well as the interaction of the two components and the surface. We believe that there is a need driven by the increasing use of surfactants and proteins for such a deeper understanding of the subject. We have pointed out several areas in which, especially in biotechnology, surfactants are used in protein-containing systems. These will probably increase in size and number. One further issue that will put large demands on the surfactant industry is the request for environmentally safe detergents.

ACRONYMS

BSA	Bovine serum albumin
cac	Critical association constant
CHAPS	3-[(3-cholamidopropyl)dimethylammonio]-1-propanesulfonate
cmc	Critical micelle concentration
CTAB	Cetyl trimethylammonium bromide
$C_{12}E_4$	Tetraethyleneglycol mono n-dodecyl ether
$C_{12}E_5$	Pentaethyleneglycol mono n-dodecyl ether
DAHCl	Dodecylammonium chloride
DTAB	Dodecyltrimethylammonium bromide
ELISA	Enzyme ligand immunosorbent assay
HDL	High-density lipoproteins
IgG	Immunoglobulin G
RIA	Radioimmunoassay
SDS	Sodium dodecyl sulfate
TABs	Trimethylammonium bromides

REFERENCES

1. J. L. Bohnert and T. A. Horbett, Changes in adsorbed fibrinogen and albumin interaction with polymers indicated by decreases in detergent elutability. J. Colloid Interface Sci. *111*:363, 1986.
2. S. I. Ertel, B. D. Ratner, and T. A. Horbett, The adsorption and elutability of albumin, IgG, and fibronectin on radiofrequency plasma deposited polystyrene. J. Colloid Interface Sci. *147*:433, 1991.
3. R. J. Rapoza and T. A. Horbett, Post-adsorptive transitions in fibrinogen influence of polymer properties. J. Biomed. Mat. Res. *24*:1263, 1990.
4. R. J. Rapoza and T. A. Horbett, The effects of concentration and adsorption time on the elutability of adsorbed proteins in surfactant solutions of varying structures and concentrations. J. Colloid Interface Sci. *136*:480, 1990.
5. R. J. Rapoza and T. A. Horbett, Changes in the SDS elutability of fibrinogen adsorbed from plasma to polymers. J. Biomater. Sci. Polymer Edn. *1*:99, 1989.
6. R. J. Rapoza and T. A. Horbett, Mechanisms of protein interactions with biomaterials: effect of surfactant structure on elutability. Polym. Mater. Sci. Eng. *59*:249, 1988.
7. T. J. Lenk, T. A. Horbett, B. D. Ratner, and K. K. Chittur, Infrared spectroscopic studies of time-dependent changes in fibrinogen adsorbed to polyurethanes. Langmuir *7*:1755, 1991.
8. K. P. Ananthapadmanabhan, Protein–surfactant interactions. In *Interactions of Surfactants with Polymers and Proteins* (K. P. Ananthapadmanabhan and E. D. Goddard, eds.). CRC Press, Boca Raton, FL, 1993, p. 319.
9. T. Nylander and B. Ericsson, Interactions between proteins and polar lipids. In *Food Emulsions* (S. Friberg and K. Larsson, eds.). Marcel Dekker, New York, 1997, p. 189.
10. E. Dickinson and C. M. Woskett, Competitive adsorption between proteins and small molecule surfactants in food emulsion. In *Food and Colloids* (R. D. Bee, ed.). Royal Society of Chemistry, London, 1989, p. 74.
11. E. Dickinson, B. S. Murray, and G. Stainsby, Protein adsorption at the air–water and oil–water interfaces. In *Advances in Food Emulsions and Foams*. Elsevier Applied Science, London, 1988, p. 123.

12. M. C. Wahlgren and T. Arnebrant, Interaction of cetyltrimethylammonium bromide and sodium dodecylsulphate with β-lactoglobulin and lysozyme at solid surfaces. J. Colloid Interface Sci. *142*:503, 1991.

13. C. A. Nelson, The binding of detergent to proteins. J. Biol. Chem. *246*:3895, 1971.

14. J. A. Reynolds and C. Tanford, The gross conformation of protein–sodium dodecyl sulfate complexes. J. Biol. Chem. *245*:5161, 1970.

15. T. Arnebrant, T. Nylander, P. A. Cuypers, P.-O. Hegg, and K. Larsson, Relation between adsorption on a metal surface and monolayer formation at the air/water interface from amphiphilic solutions. In *Surfactants in Solution* (K. L. Mittal and B. Lindman, eds.). Plenum, New York, 1984, p. 1291.

16. P. J. Wängnerud and B. Jönssou, Ionic Surfactants at the charged solid/water interface: significance of premicellar aggregation. Langmuir *10*:3542, 1994.

17. P. J. Wängnerud and B. Jönssou, The adsorption of ionic amphiphiles as bilayers on charged surfaces. Langmuir *10*:3268, 1994.

18. P. J. Wängnerud and G. Olofssou, Adsorption isotherms for cationic surfactants on silica determinated by in situ ellipsometry. J. Colloid Interface Sci. *153*:392, 1992.

19. F. Tiberg, Self-assembly of nonionic amphiphiles at solid surfaces. Thesis, University of Lund, Sweden, 1994.

20. F. J. Tiberg, B. Jönssou, J. Tang, and B. Lindman, Ellipsometry studies of the self-assembly of nonionic surfactants at the silica–water interface: equilibrium aspects. Langmuir *10*:2294, 1994.

21. F. J. Tiberg, B. Jönssou, and B. Lindman, Ellipsometry studies of the self-assembly of nonionic surfactants at the silica–water interface: kinetic aspects. Langmuir *10*:3714, 1994.

22. J. T. Brinck and F. Tiberg, Adsorption behaviour of two binary nonionic surfactant systems at the silica–water interface. Langmuir *12*:5042, 1996.

23. F. Tiberg and M. Landgren, Characterization of thin non-ionic surfactant films at the silica/water interface by means of ellipsometry. Langmuir *9*:927, 1993.

24. C. Tanford, *The Hydrophobic Effect: Formation of Micelles and Biological Membranes*. Wiley, New York, 1980.

25. M. N. Jones and A. Wilkinson, The interaction between β-lactoglobulin and sodium *n*-dodecyl sulphate. Biochem. J. *153*:713, 1976.

26. G. C. Kresheck, W. A. Hargraves, and D. C. Mann, Thermometric titration studies of ligand binding to macromolecules. Sodium dodecyl sulfate to β-lactoglobulin. J. Phys. Chem. *81*:532, 1977.

27. B. Ericsson and P.-O. Hegg, Surface behaviour of adsorbed films from protein–amphiphile mixtures. Prog. Colloid Polym. Sci. *70*:92, 1985.

28. Y. Nozaki, J. A. Reynolds, and C. Tanford, The interaction of a cationic detergent with serum albumin and other proteins. J. Biol. Chem. *249*:4452, 1974.

29. Y.-Y. T. Su and B. Jirgensons, Further studies of detergent-induced conformational transitions in proteins. Arch. Biochem. Biophys. *181*:137, 1977.

30. M. Subramanian, B. S. Sheshadri, and M. P. Venkatappa, Interaction of proteins with detergents: binding of cationic detergents with lysozyme. J. Biosci. *10*:359, 1986.

31. A. V. Few, R. H. Ottewill, and H. C. Parreira, The interaction between bovine plasma albumin and dodecyltrimethylammonium bromide. Biochem. Biophys. Acta *18*:136, 1955.

32. M. Wahlgren and T. Arnebrant, Removal of T4 lysozyme from silicon oxide surfaces by sodium dodecyl sulphate SDS: a comparison between wild type protein and a mutant with lower thermal stability. Langmuir *13*:8, 1997.

33. M. Wahlgren, S. Welin-Klintström, T. Arnebrant, A. Askendal, and H. Elwing, Competition between fibrinogen and nonionic surfactant at adsorption to a wettability gradient surface. Colloid Surfaces B *4*:23, 1995.

34. M. Wahlgren and T. Arnebrant, Removal of lysozyme from methylated silica surfaces by a nonionic surfactant pentaethyleneglycol mono n-dodecyl ether ($C_{12}E_5$). Colloid Surfaces B 6:63, 1996.

35. S. Welin-Klintström, A. Askendal, and E. Elwing, Surfactant and protein interactions on wettability gradient surfaces. J. Colloid Interface Sci. 158:188, 1993.

36. G. Li, H. Chen, and D. Zhu, A new method for the voltammetric response of hemoglobin. J. Inorganic Biochem. 63:207, 1996.

37. M. Wahlgren and T. Arnebrant, The concentration dependence of adsorption from a mixture of β-lactoglobulin and sodium dodecyl sulfate onto methylated silica surfaces. J. Colloid Interface Sci. 148:201, 1992.

38. A. Samanta and D. K. Chattoraj, Simultaneous adsorption of gelatin and long-chain amphiphiles at solid–water interface. J. Colloid Interface Sci. 116:168, 1987.

39. M. Wahlgren, T. Arnebrant, A. Askendal, and S. Welin-Klinström, The elutability of fibrinogen by sodium dodecyl sulphate and alkyltrimethylammonium bromides. Colloids Surf. A 70:151, 1993.

40. B. Folmer, K. Holmberg, and M. Svensson, Interaction of rhizomucor miehei lipase with an amphoteric surfactant at different pH values. Submitted, 1997.

41. J. W. T. Lichtenbelt, W. J. M. Heuvelsland, M. E. Oldenzeel, and R. L. J. Zsom, Adsorption and immunoreactivity of proteins on polystyrene and on silica. Competition with surfactants. Colloids Surf. B 1:75, 1993.

42. E. Dickinson and S. R. Euston, Computer simulation of competitive adsorption between polymers and small displacer molecules. Mol. Phys. 68:407, 1989.

43. E. Dickinson, Monte Carlo model of competitive adsorption between interacting macromolecules and surfactants. Mol. Phys. 65:895, 1988.

44. E. Dickinson, *Proteins in Solution and Interfaces: Interactions of Surfactants with Polymers and Proteins*. CRC Press, Boca Raton, FL, 1993, p. 295.

45. D. Sarkar and D. K. Chattoraj, Kinetics of desorption from the surface of protein-coated aluminum by various desorbing reagents. J. Colloid Interface Sci. 178:606, 1996.

46. J. K. Towns and F. E. Regnier, Capillary electrophoretic separations of proteins using nonionic surfactant coatings. Anal. Chem. 63:1126, 1991.

47. D. Sarkar and D. K. Chattoraj, Effect of denaturants and stabilisers on protein adsorption at solid–liquid interfaces. Ind. J. Biochem. Biophys. 31:100, 1994.

48. A. Samanta and D. K. Chattoraj, Mutual adsorption of protein and detergent at the aluminia–water interface. Progr. Colloid Polym. Sci. 68:144, 1983.

49. M. Wahlgren and T. Arnebrant, Adsorption of β-lactoglobulin onto silica, methylated silica and polysulphone. J. Colloid Interface Sci. 136:259, 1990.

50. K. Wannerberger and T. Arnebrant, Lipase from *Humicola lannuginosa* adsorbed to hydrophobic surfaces—desorption and activity after addition of surfactants. Colloids Surf. B 7:153, 1996.

51. R. D. Tilton, E. Blomberg, and P. M. Claesson, Effect of anionic surfactant on interactions between lysozyme layers adsorbed on mica. Langmuir 9:2102, 1993.

52. E. Blomberg, Surface forces studies of adsorbed proteins. Thesis, Kungliga tekniska Högskolan, Stockholm, Sweden, 1993.

53. M. C. Wahlgren, M. A. Paulsson, and T. Arnebrant, Adsorption of globular model proteins to silica and methylated silica surfaces and their elutability by dodecyltrimethylammonium bromide. Colloids Surf. A 70:139, 1993.

54. J. McGuire, M. Wahlgren, and T. Arnebrant, Structural stability effects on adsorption and dodecyltrimethylammonium bromide–mediated elutability of bacteriophage T4 lysozyme at silica surfaces. J. Colloid Interface Sci. 170:182, 1995.

55. J. McGuire, V. Krisdhasima, M. Wahlgren, and T. Arnebrant, Comparative adsorption studies with synthetic, structural stability and charge mutants of bacteriophage T4 ly-

sozyme. In *Proteins at Interfaces* (J. Brash and T. Horbett, eds.). ACS Symposium Series 602, Washington, DC, 1995, p. 52.

56. J. McGuire, M. Wahlgren, and T. Arnebrant, The influence of net charge and charge location on adsorption and dodecyltrimethylammonium bromide–mediated elutability of bacteriophage T4 lysozyme at silica surfaces. J. Colloid Interface Sci. *170*:193, 1995.

57. T. A. Horbett and J. L. Brash, Proteins at interfaces: current issues and future prospects. In *Proteins at Interfaces—Physicochemical and Biochemical Studies* (J. L. Brash and T. A. Horbett, eds.). American Chemical Society, Washington, DC, 1987, p. 1.

58. P. A. Dimilla, S. M. Albelda, and J. A. Quinn, Adsorption and elution of extracellular matrix proteins on non-tissue culture polystyrene petri dishes. J. Colloid Interface Sci. *153*:212, 1992.

59. M. Feng, A. B. Morales, T. Beugeling, A. Bantjes, K. van der Werf, G. Gosselink, B. de Grooth, and J. Greve, Adsorption of high density lipoproteins (HDL) on solid surfaces. J. Colloid Interface Sci. *177*:364, 1996.

60. C. A.-C. Karlsson, M. C. Wahlgren, and A. C. Trägårdh, β-Lactoglobulin fouling and its removal upon rinsing and by SDS as influenced by surface characteristics, temperature and adsorption time. J. Food Eng. *30*:43, 1996.

61. C. A.-C. Karlsson, M. C. Wahlgren, and A. C. Trägårdh, Time and temperature aspects of β-lactoglobulin removal from methylated silica surfaces by sodium dodecyl sulphate. Colloids Surf. B *6*:317, 1996.

62. C. A.-C. Karlsson, Surfactant cleaning of solid surfaces fouled by protein. Thesis, University of Lund, Sweden, 1996.

63. J. A. Chinn, S. E. Posso, T. A. Horbett, and B. D. Ratner, Post-adsorptive transitions in fibrinogen: influence of polymer properties. J. Biomed. Mater. *26*:757, 1992.

64. H. Elwing, A. Askendal, and I. Lundström, Desorption of fibrinogen and γ-globulin from solid surfaces induced by nonionic detergent. J. Colloid Interface Sci. *128*:296, 1989.

65. P. Vinaraphong, V. Krisdhasima, and J. McGuire, Elution of proteins from silanized silica surfaces by sodium dodecylsulfate and dodecyltrimethylammonium bromide. J. Colloid Interface Sci. *174*:351, 1995.

66. V. Krisdhasima, P. Vinaraphong, and J. McGuire, Adsorption kinetics and elutability of α-lactalbumin, β-casein, β-lactoglobulin, and bovine serum albumin at hydrophobic and hydrophilic interfaces. J. Colloid Interface Sci. *161*:325, 1993.

67. H. Elwing and C.-G. Gölander, Protein and detergent interaction phenomena on solid surfaces with gradients in chemical composition. Adv. Colloid Interface Sci. *32*:317, 1990.

68. S. Welin-Klintström, R. Jansson, and H. Elwing, An off-null ellipsometer with lateral scanning capability for kinetic studies at solid–liquid interfaces. J. Colloid Interface Sci. *157*:498, 1993.

69. M. Lalande, J.-P. Tissier, and G. Corrieu, Fouling of heat transfer surfaces related to β-lactoglobulin denaturation during heat processing of milk. Biotech. Prog. *1*:131, 1985.

70. P. J. Skudder, E. L. Thomas, J. A. Pavey, and A. G. Perkin, Effects of adding potassium iodate to milk before UHT treatment. I. Reduction in the amount of deposit on the heated surfaces. J. Dairy Res. *48*:99, 1981.

71. W. G. Jennings, A. A. McKillop, and J. R. G. Luick, Circulation cleaning. J. Dairy Sci. *40*:1471, 1957.

72. W. G. Jennings, Circulation cleaning. III. The kinetics of a simple detergent system. J. Dairy Sci. *42*:1763, 1959.

73. T. J. Nisbet and A. G. Langdon, Milk protein interactions at stainless steel/aqueous interfaces. N. Z. Dairy Sci. Technol. *12*:83, 1977.

74. D. A. Timperley and C. N. M. Smeulders, Cleaning of dairy HTST plate heat exchangers: comparison of single- and two-stage procedures. J. Soc. Dairy Technol. *40*:4, 1987.

75. D. A. Timperley, A. P. M. Hasting, and G. De Goederen, Developments in the cleaning of dairy sterilization plant. J. Soc. Dairy Technol. *47*:44, 1994.

76. A. Grasshoff, Environmental aspects on the use of alkaline cleaning solutions. In *Fouling and Cleaning in Food Processing* (H. G. Kessler and D. B. Lund, eds.). Prien Chiemsee, Federal Republic of Germany, 1989, p. 107.

77. A. Grasshoff, Reinigung von Plattenwärmeubertragern in der Molkereiindustrie. Tenside Surf. Det. *27*:130, 1990.

78. C. A.-C. Karlsson, M. C. Wahlgren, and A. C. Trägårdh, The removal of β-lactoglobulin from stainless steel surfaces at high and low temperature as influenced by the type and concentration of cleaning agent. J. Food Proc. Eng., in press, 1998.

79. M. J. Munoz-Aguado, D. E. Wiley, and A. G. Fane, Enzymatic and detergent cleaning of a polysulphone ultrafiltration membrane fouled with BSA and whey. J. Membrane Sci. *117*:175, 1997.

80. K. Wannerberger, M. Wahlgren, and T. Arnebrant, Adsorption from lipase–surfactant solutions onto methylated silica surfaces. Colloids Surf. B *6*:27, 1996.

81. J. van Ee, H. O. Misset, and E. J. Baas, *Enzymes in Detergency*. Marcel Dekker, New York, 1997.

82. T. Arnebrant and T. Simonsson, The effect of ionic surfactants on salivary proteins adsorbed on silica surfaces. Acta Odontol. Scand. *49*:281, 1991.

83. N. Vassilakos, T. Arnebrant, and P.-O. Glantz, Interaction of anionic and cationic surfactants with salivary pellicles formed at solid surfaces *in vitro*. Biofouling *5*:277, 1992.

84. N. Vassilakos, T. Arnebrant, J. Rundegren, and P.-O. Glantz, *In vitro* interactions of anionic and cationic surfactants with salivary fractions on well defined solid surfaces. Acta Odontol. Scand. *50*:179, 1992.

85. T. Simonsson, T. Arnebrant and L. Petersson. The effect of delmopinol on salivary pellicles, the wettability of tooth surfaces *in vivo* and bacterial cell surfaces *in vitro*. Biofouling *3*:251, 1991.

86. N. Vassilakos, T. Arnebrant, and J. Rundegren, *In vitro* interactions of delmopinol hydrochloride with salivary films adsorbed at solid/liquid interfaces. Caries Res. *27*:176, 1993.

87. B. Warren, S. P. Kusk, and R. G. Wolford, Purification and stabilization of transcriptional active glucocorticoid receptor. J. Biol. Chem. *271*:11434, 1996.

88. L. J. J. Hronowski and T. P. Anastassiades, Nonspecific interaction of proteoglycans with chromatography media and surfaces: effect of this interaction on the isolation efficiencies. Anal. Biochem. *191*:50, 1990.

89. J. P. Chen, D. Kiaei, and H. A. S., Activity of horseradish peroxidase adsorbed on radiofrequency glow discharged-treated polymers. J. Biomater. Sci. Polym. Edn. *5*:167, 1993.

90. R. A. Williams and H. W. Blanch, Covalent immobilization of protein monolayers for biosensor applications. Biosens. Bioelectron. *9*:159, 1994.

91. A. Safranj, D. Kiaei, and A. S. Hoffman, Antibody immobilization onto glow discharge treated polymers. Biotechnol. Prog. *7*:173, 1991.

92. E. Delamarche, G. Sundarababu, H. Biebuyck, B. Michel, C. Gerber, H. Sigrist, H. Wolf, H. Ringsdorf, N. Xanthopoulos, and H. J. Mathieu, Immobilization of antibodies on a photoactive self-assembled monolayer on gold. Langmuir *12*:1997, 1996.

93. C. Morissette, J. Goulet, and G. Lamoureux, Rapid and sensitive sandwich enzyme-linked immunosorbent assay for detection of staphylococcal enterotoxin B in cheese. Appl. Environ. Microbiol *57*:836, 1991.

94. D. Leckband and R. Langer, An approach for stable immobilization of proteins. Biotechnol. Bioeng. *37*:227, 1991.

95. Y. L. K. Sing, Y. Kroviarski, S. Cochet, D. Dhermy, and O. Bertrand, High-performance hydrophobic interaction chromatography of proteins on reversed-phase supports coated with non-ionic surfactants of polyoxyethylene type—purification of a fungal aspartic proteinase. J. Chromatogr. *598*:181, 1992.

96. J. C. Olivier, M. Taverna, C. Vauthier, P. Couvreur, and D. Baylocq-Ferrier, Capillary electrophoresis monitoring of the competitive adsorption of albumin onto the orosomucoid-coated polyisobutylcyanoacrylate nanoparticles. Electrophoresis *15*:234, 1994.

97. B. P. Salmanowicz, Capillary electrophoresis of seed albumins from Vici species using uncoated and surface-modified fused silica capillaries. Chromatographia, *41*:99, 1995.

98. A. Emmer, M. Jansson, and J. Roeraade, Improved capillary zone electrophoretic separation of basic proteins, using a fluorosurfactant buffer additive. J. Chromatogr. *547*: 544, 1991.

99. W. G. H. M. Muijselaar, C. H. M. M. De Bruijn, and F. M. Everaerts, Capillary zone electrophoresis of proteins with a dynamic surfactant coating: influence of a voltage gradient on the separation efficiency. J. Chromatogr. *605*:115, 1992.

100. M. W. F. Nielen, Capillary zone electrophoresis using a hollow polypropylene fiber. J. High Resolution Chromatogr. *16*:62, 1993.

101. H. K. Kristensen and S. H. Hansen, Micellar electrokinetic chromatographic separation of basic polypeptides with equal mass to charge ratio using dynamically modified silica capillaries. J. Liq. Chromatogr. *16*:2961, 1993.

102. X. W. Yao and F. E. Regnier, Polymer-coated and surfactant-coated capillaries for isoelectric focusing. J. Chromatogr. *632*:185, 1993.

103. Q. Wu, H. A. Claessens, and C. A. Cramers, The influence of surface treatments on the electroosmotic flow in micellar electrokinetic capillary chromatography. Chromatographia *33*:303, 1992.

104. M. A. Strege and A. L. Lagu, Capillary electrophoretic protein separations in polyacrylamide-coated silica capillaries and buffers containing ionic surfactants. J. Chromatogr. *630*:337, 1993.

13

Toward Functionalized Polyelectrolyte Biofilms

PIERRE SCHAAF Institut Charles Sadron (CNRS–ULP),
Strasbourg, France

JEAN-CLAUDE VOEGEL Faculté de Chirurgie Dentaire,
INSERM—Unité 424, Strasbourg, France

I. INTRODUCTION

Ten years ago, Decher and coworkers [1,2] proposed a simple and original method
to realize supramolecular architectures on almost any charged solid substrate. These
auto-assembled structures are obtained by dipping the substrate alternately in a polycation and a polyanion solution, the different layer depositions being separated by
rinsing steps. The motor of the buildup of the multilayer films is the charge excess
that appears after each new polyelectrolyte deposition [3–5]. This then allows a
subsequent adsorption of a polyelectrolyte of opposite charge. Using this property,
one can build multilayer films constituted of up to several hundreds of bilayers. After
the determination of the buildup mechanism, the structures of such films were investigated by means of x-ray and neutron reflectivity experiments. For the experimental conditions used, Kiesig fringes [2,6–8] and neutron reflectivity fringes [9,10]
were observed. From these results it was concluded that the investigated films had
an organized structure, each individual polyelectrolyte layer penetrating into neighboring ones [11].

Numerous studies were then conducted in order to analyze the influence on the
film structure of the various parameters that can be changed during the buildup
process, and several general rules seemed to emerge. For films constituted of strong
polyelectrolytes, the pH of the polyelectrolyte solutions has almost no effect on the
film structure. This is, on the other hand, affected by the ionic strength of the polyelectrolyte solutions during the buildup process: increasing the salt concentration
leads to thicker films [5,12]. For weak polyelectrolytes, the thickness and the deposited amount of polyelectrolytes per bilayer depend strongly on the pH of the
solution during construction [13]. The buildup structure is mainly governed by the
most weakly charged polyelectrolyte of the system: the smaller the charge of one of
the two polyelectrolytes, the larger the thickness and deposited amount per bilayer
and the larger the roughness of the film [3,14]. In the absence of salt, when both
polyelectrolytes are fully charged, one can even reach for some systems pH values
where it becomes impossible to construct a multilayer film. The molecular picture
behind these rules is quite simple: in the absence of salt, fully charged polyelectro-

lytes (strong or weak) have a large persistence length. Such polyelectrolytes adopt quite flat conformations on the surface and can form strong polyelectrolyte/polyelectrolyte complexes where most of the monomers of each chain are in direct interaction with monomers of the chain of opposite charge. The charge overcompensation must thus be small after each buildup step. On the contrary, when the ionic strength of the polyelectrolyte solutions is increased or when the pH is such that one of the polyelectrolytes is only partially charged, the polyelectrolytes adopt more loopy configurations on the surface, and the interactions between the monomers of the two chains of opposite charge are much weaker. This leads to more extended bilayers with larger charge overcompensation. Larger amounts of polyelectrolytes are then needed to overcompensate these charges.

The structure of multilayer films can also be modified after their construction by abrupt pH or ionic strength changes during the rinsing steps. It was observed that when a film is formed by deposition of polyelectrolytes from aqueous salt containing solutions, the surface roughness of the film is increased substantially by rinsing it with pure water, and one can even form discrete nanometer-sized pores [15]. The optical thickness of a film also increases if the final rinsing step is performed with a solution of lower ionic strength than during the film buildup. The reverse effect (film shrinkage) is observed when the rinsing is performed with a solution of higher ionic strength [5]. This clearly indicates that polyelectrolyte multilayers are usually not in an equilibrium but rather in a metastable state. Recent studies performed by the groups of Rubner [16] on the poly(acrylic acid)/poly(allylamine) system also showed the possibility to induce microporous morphologies in the corresponding films. The porosities can be obtained by constructing the films in conditions leading to a large roughness followed by a rinsing step with a strong acidic solution (pH < 2.5). One then observes an important increase of the film thickness and the formation of pores having characteristic sizes of the order of 100 to 500 nm. The structural changes in the system seem to result from a protonation of a large fraction of carboxylate groups inducing a rupture of an important fraction of $-COO^-/-NH_3^+$ bonds, which leads to local structural film reorganizations and to a microphase separation.

These results illustrate the large flexibility of polyelectrolyte multilayers to construct films with different properties and constitutes one of the reasons for their numerous potential applications, in particular for the design of targeted biofilms. The aim of this chapter is to review some of the results that have been obtained in this field, and we will largely focus on results obtained by our groups. Our main goals consist in the development of new implant coatings with targeted properties such as cellular adhesion control, antimicrobial, or biomineralization triggering properties. Such goals can be attained in different ways such as (1) proper selections of the polyelectrolytes used for the film buildup, (2) specific protein insertion in the film architectures, and (3) polyelectrolyte functionalization with appropriate peptides, like adhesion peptides (to favor cellular adhesion), antibacterial peptides (to inhibit bacterial adhesion), or calcium chelating peptides (to trigger apatite nucleation). This raises however numerous fundamental problems concerning (1) the behavior of proteins adsorbed or embedded in the multilayer films, (2) the ability of cells to interact with embedded proteins or peptides bound to a polyelectrolyte and embedded in the architecture, (3) the ability of the same architectures to induce calcium phosphate salts nucleation, and (4) the biocompatibility properties of the polyelectrolyte films. We will present several answers dealing with these questions.

II. BEHAVIOR OF PROTEINS ADSORBED ON OR EMBEDDED IN POLYELECTROLYTE MULTILAYERS

A. Interactions of Proteins and Polyelectrolytes in Solution

Before analyzing the interactions of proteins with polyelectrolyte multilayers it is interesting to briefly summarize some results relative to their interactions in solution. Proteins are known to interact with polyelectrolytes in solution and to form poly-electrolyte/protein complexes [17,18]. To understand the nature of these interactions one must take into account the fact that a polyelectrolyte behaves usually as a charged and flexible chain and that a protein can in general be viewed as a compact globular entity with an outer surface bearing several positive and negative charge patches. A polyelectrolyte can thus create several links with a given protein molecule leading to strong interactions. Dubin and coworkers undertook a systematic study of protein/polyelectrolyte systems in solution [19–21]. They found, by gradually changing the pH, the existence of two transition points. There is first a well-defined pH at which binding between the polyelectrolyte chains and the proteins starts. Usually the protein and the polyelectrolyte have a similar charge at this point. This pH is defined as pH_{crit}. At this pH small aggregates must form and the binding between proteins and polyelectrolytes is reversible [22]. The fact that pH_{crit} is often on the wrong side of the isoelectric point must originate from the existence of local protein domains with effective charges opposite in sign to the net protein charge. While the pH is further changed in the same direction, a second transition point denoted as pH_ϕ is observed. It is characterized by the formation of large droplets, usually described as coacervates, which must originate from strong interactions between the polyelectrolytes and the proteins leading to the formation of stable bridging. It was also shown that an increase of the ionic strength I of the solution leads to a weakening of the protein/polyelectrolyte interaction and thus influences pH_{crit} and pH_ϕ accordingly. By using isothermal titration microcalorimetry, Ball et al. [23] investigated thermodynamic aspects of these complexation processes. Figure 1 represents a typical example of the amount of energy that has to be introduced in the form of heat into a system during the addition of bovine serum albumin (BSA) to a poly(allylamine) (PAH) solution at pH 7.4 in order to maintain the temperature within the reactor equal to the temperature in a reference reactor. Each new BSA addition in the reactor is accompanied by a positive heat peak which clearly indicates that the polyelectrolyte/protein complexation process is endothermic. The enthalpy associated with the binding of BSA molecules with PAH chains present in large excess in the solution could be estimated to be of the order of 400 kJ/mol of BSA molecules for solutions containing 0.1 M and 0.01 M of NaCl; it becomes much smaller when the NaCl concentration is 1 M. The formation of PAH/BSA complexes at pH 7.4 being a spontaneous process, its associated Gibbs free energy must be negative. The complexation process must thus be driven by entropy. Similar results were found for interactions between DNA and proteins. For these systems it was demonstrated that the driving force of the complexation process is the increase of entropy associated with the release of counterions from interacting proteins and polyelectrolytes [24].

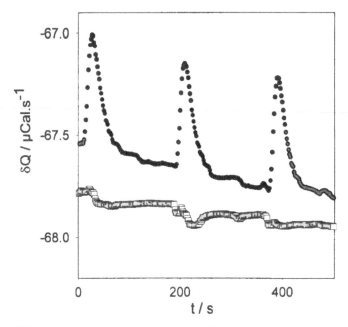

FIG. 1 Evolution of the heat consumed during successive BSA injections. (●) (1.3 mg/cm³) in a 5.05 mg/mL PAH solution in Tris-NaCl (0.1 *M*) buffer. The lower curve (□) represents the dilution heat of BSA in the buffer in the absence of PAH. For clarity, this curve is shifted with respect to the curve corresponding to the injection of the BSA solution in the PAH solution (from Ref. 23). Notice the positive heat peaks which indicate that the BSA/PAH binding process is endothermic.

A similar effect must take place for the BSA/PAH complexation and more generally for most of the protein/polyelectrolyte complex formation.

B. Adsorption of Proteins on Polyelectrolyte Multilayers

From the above results one can expect that proteins also interact with polyelectrolyte multilayers and this is indeed the case. Complex polyelectrolyte/protein multilayers such as enzyme/polyelectrolyte microreactor films have, for example, been reported [25] and are reviewed in this book. Caruso and coworkers studied the adsorption of anti-IgG proteins on PAH/PSS multilayers [26] and even the formation of anti-IgG/ PSS multilayers [27]. They found that the amount of anti-IgG molecules adsorbed on (PAH/PSS)$_n$ multilayers was identical on films constituted by 2 and 5 bilayers. They thus concluded that the proteins that adsorb on the outermost layer are not able to penetrate deep into the film. Ladam et al. [28] investigated the interaction of a series of globular proteins (human serum albumin, HSA; α-lactalbumin, αLA; myoglobin, MGB; α$_1$-acid glycoprotein, αAgly; ribonuclease A, Rnase; and lysozyme, LSZ) with PSS/PAH multilayer films. The adsorption experiments were performed at pH 7.35 with a NaCl concentration of 0.15 *M*. Figure 2 represents (Γ_{PAH} − Γ_{PSS})/max(Γ_{PAH},Γ_{PSS}) as a function of the isoelectric point (pH$_i$) of the proteins, Γ_{PAH} (resp. Γ_{PSS}) corresponding to the protein amount adsorbed on a PAH (resp. PSS) ending multilayer whereas max(Γ_{PAH},Γ_{PSS}) is the maximum value of the two amounts.

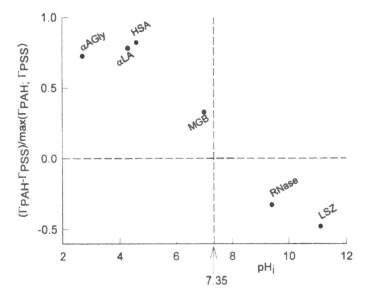

FIG. 2 Evolution of $(\Gamma_{PAH} - \Gamma_{PSS})/\max(\Gamma_{PAH}, \Gamma_{PSS})$ as a function of the isoelectric point (pH_i) for various proteins. Γ_{PAH} (resp. Γ_{PSS}) corresponds to the protein amount adsorbed on a PEI-$(PSS\text{-}PAH)_3$ [resp. PEI-$(PSS\text{-}PAH)_3$-PSS] multilayer film, and $\max(\Gamma_{PAH}, \Gamma_{PSS})$ is the maximum value of the two amounts. (From Ref. 28.)

As expected, the adsorbed protein amount is higher when the protein bears a net charge of opposite sign to the excess charge of the multilayer. Müller et al. [29,30] came to a similar conclusion for a series of proteins adsorbed on various multilayers. These results clearly demonstrate the important role of electrostatic interactions as the driving force for protein adsorption on polyelectrolyte multilayers. However, even a highly negatively charged protein such as αAgly or a highly positively charged protein like LSZ can respectively adsorb on PSS (polyanion) or PAH (polycation) ending multilayer films [28], in accordance with complexation results found in solution. Figure 3 represents the ratio L_{opt}/L_{dim}, where L_{opt} represents the optical thickness of these adsorbed protein layers and L_{dim} the largest dimension of the protein molecules as a function of the isoelectric point of the protein. It appears that proteins usually adsorb in monolayers on polyelectrolyte films. When the outer layer of the film and the protein bear charges that are opposite in sign one can nevertheless find protein layers which present an optical thickness that largely exceeds the largest size of the protein. This suggests the formation of protein/polyelectrolyte complexes on the surface as it is found in the bulk when proteins are mixed with polyelectrolytes. Bridging of polyelectrolytes with proteins can explain the formation of such thick protein layers on the multilayer film.

A more extensive adsorption study has been performed for HSA adsorbed on PSS/PAH multilayers [31]. In order to analyze the effect of the ionic strength on the binding of this protein on multilayers, the HSA adsorption process was followed by a first rinse with a buffer solution at a similar NaCl concentration (0.15 M) to that during adsorption. Only a very small fraction, on the order of 5% of the adsorbed

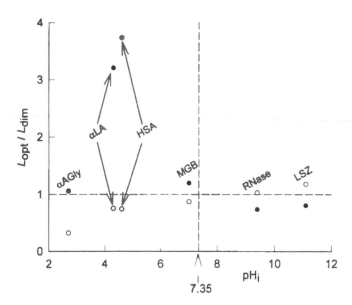

FIG. 3 L_{opt}/L_{dim} versus protein isoelectric point pH_i for different protein layers adsorbed on a PEI-(PSS-PAH)$_3$ (O) or PEI-(PSS-PAH)$_3$-PSS (●) multilayer (from Ref. 28). L_{opt} represents the optical thickness of the adsorbed protein layers determined by scanning angle reflectometry, and L_{dim} corresponds to the largest dimension of the protein.

proteins, was released on both PSS and PAH terminating multilayers. This first rinse was then followed by a second rinse. When this second rinse was conducted with a 10^{-2} M NaCl buffer solution no change in the adsorbed protein amount was detected. Only when a 2 M NaCl buffer solution was used for this second rinse, about 55% (resp. 40%) of the albumin molecules adsorbed on a PAH (resp. PSS) terminating multilayer film were released. This indicates that large protein releases can only be induced with buffer solutions of high ionic strength, higher than for the adsorption step. The protein adsorption process on a multilayer thus takes place in such a way that only the protein/polyelectrolyte interactions that are strong enough to appear irreversible over the experimental time scales, for the used ionic strength, are efficient. By increasing the ionic strength of the solution once the adsorption took place, one decreases enough of these interactions to render them reversible over the experimental time scales.

The buildup of the adsorbed protein layer is mainly driven by electrostatic interactions and should thus also be influenced during the adsorption process by the ionic strength of the protein solution. This is indeed the case, as can be seen in Fig. 4 where the evolution of the albumin amount adsorbed on PSS/PAH multilayer films is represented as a function of the logarithm of the NaCl concentration of the solution, the pH of the solution being adjusted to ±7.35. On a negatively charged PSS ending multilayer the adsorbed amount increases with the NaCl concentration. A similar result was obtained for anti-IgG molecules immobilized on PSS ending multilayers by increasing the MnCl$_2$ concentration in the solution [26]. This effect is not unexpected for negatively charged proteins adsorbing on a negatively charged sur-

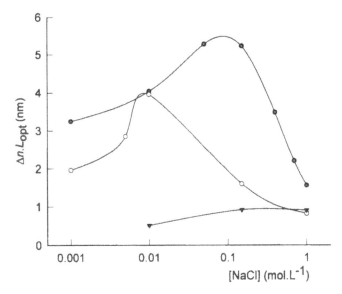

FIG. 4 Effect of the salt concentration on the adsorption of HSA on polyelectrolyte multi-layers. The amount of adsorbed HSA is represented as a function of the NaCl concentration (in logarithm scale) in solution. The adsorption process was carried out at pH 7.35 in the following conditions: (●) HSA concentration 22 mg/100 mL on PEI-(PSS-PAH)$_3$; (○) HSA concentration 4.5 mg/100 mL on PEI-(PSS-PAH)$_3$; (▼) HSA concentration 4.5 mg/100 mL on PEI-(PSS-PAH)$_3$-PSS. The lines are added for clarity.

face. A different behavior takes place when HSA interacts with a PAH ending film. The amount of adsorbed albumin first strongly increases with the NaCl concentration, passes through a maximum, and then decreases continuously until salt concentration reaches 1 M. It was found that the salt concentration corresponding to the maximum amount of adsorbed HSA is also dependent upon protein concentration. The maximum is shifted from 10^{-1} M to 10^{-2} M NaCl when the protein concentration is reduced from 22 mg/100 mL to 4.5 mg/100 mL. It was found that the optical thickness of the adsorbed protein layer varies similarly to the adsorbed amount going through a maximum when the adsorbed amount is maximal. Such a maximum with salt concentration for both the adsorbed amount and the layer thickness on the PAH ending multilayer films seems to indicate that the forces responsible for the formation of the HSA layers must be counterbalanced by interprotein repulsive interactions probably also of electrostatic origin. It would be of great interest to verify if the results obtained for albumin can be extended to other proteins and are thus more general.

C. Structure of Proteins Adsorbed on or Embedded in Polyelectrolyte Multilayers

Adsorption and insertion of proteins in polyelectrolyte multilayer films present quite larger interest in terms of biomedical and biotechnological applications if the adsorbed or inserted proteins keep their biological activity. This requires that the structure of the proteins remains preserved in the presence of the polyelectrolytes. Several results strongly suggest that this is indeed the case. Proteins seem to keep their secondary structure, which can even be stabilized, when brought in contact with polyelectrolytes in the bulk [32,33]. Anderson and Hatti-Kaul [34] showed, for example, by using circular dichroism in the far UV range, that polyethyleneimine offers protection to lactate dehydrogenase against complete loss of secondary structure during protein storage. Caruso et al. [26] built PAH/PSS multilayer films on which they adsorbed or in which they embedded anti-IgG molecules. They demonstrated that the anti-IgG molecules adsorbed on or embedded in the films could, under some restrictions, interact with IgG molecules from the solution and thus retain their biological activity. Moreover the absence of infrared (IR) bands at 1625 and 1520 cm^{-1} in Fourier transform infrared reflection absorption spectra (FTIR-RAS), which are characteristic for denaturated proteins, led them to conclude that no significant denaturation of anti-IgG occurs in the multilayer films. Müller et al. [30] came to the same conclusion by a simple analysis of attenuated total reflection Fourier transform infrared (ATR-FTIR) protein spectra. A more precise study of protein structures, fibrinogen in this case, adsorbed on or embedded in PSS/PAH multilayers was recently performed by Schwinté et al. [35], who investigated in detail the evolutions of ATR-FTIR spectra. Attention was paid to the amide I band originating predominantly from the C=O stretching vibrations of the peptide bond groups. It was shown that at least five components were necessary to decompose the fibrinogen amide I band (i.e., 1616 cm^{-1}, intermolecular β-sheets; 1663 cm^{-1}, intramolecular β-sheets; 1650 cm^{-1}, α-helix; 1668 cm^{-1}, turn structures; 1683 cm^{-1}, antiparallel intermolecular β-sheets). The results are summarized in Table 1. Only slight variations in the amide I band between the spectra of fibrinogen dissolved in solutions at 2 $mg \cdot mL^{-1}$ and fibrinogen adsorbed on or embedded in polyelectrolyte multilayers were detected. This indicates that the proteins keep their secondary structure in the presence of the polyelectrolyte multilayers despite the fact that in the adsorbed state, the local fibrinogen concentration can be of the order of 160 $mg \cdot mL^{-1}$. For such high protein concentrations, the amide I spectrum of the fibrinogen solution contains large contributions of intermolecular β-sheet structures. Such contributions were not found when fibrinogen was adsorbed on polyelectrolyte films. This indicates that polyelectrolyte multilayer films seem to prevent the formation of such intermolecular structures and thus also hinder direct protein/protein interactions.

The influence of the polyelectrolytes on the thermal stability of fibrinogen adsorbed on or embedded in multilayers has been analyzed following the evolution of the amide I spectra during heating over temperature domains ranging from 28 to 85°C. These spectra were compared to their counterparts in solution. Whereas two transitions are observed in solution, one occurring between 40–45°C and the other between 70–75°C only, the lower temperature transition remains when fibrinogen molecules were in contact with PSS/PAH multilayers. Moreover, while the onset temperatures of the structural changes corresponding to the lower temperature transition are very similar for fibrinogen in solution, for fibrinogen adsorbed on multilayers or embedded in the polyelectrolyte architectures, the structural changes appear to take place over a broader temperature range for the embedded proteins compared to the proteins adsorbed onto multilayers or in solution. One also observes a lower intermolecular β-sheet content after the temperature has been increased for fibrinogen molecules embedded in multilayer films as compared to fibrinogen in solution. These

TABLE 1 Secondary Structure Content of Fibrinogen in Solution (raws 1–3), Adsorbed on PAH or PSS Ending Multilayers (raw 4 and 5) and Embedded in PSS/PAH Multilayers (raw 6–9) as Determined from the Decomposition of the Amide I Band of the IR Spectra

	Intermolecular β-sheet	Intramolecular β-sheet	α-Helix	Turn	Intermolecular β-sheet
Fibrinogen in solution at 2 mg·mL^{-1}	1616–1618 4–6	1633–1635 42–50	1650–1653 35–37	1668–1670 8–11	1683–1688 3–5
Fibrinogen in solution at 5 mg·mL^{-1}	1621 4	1635 39	1652 35	1668 19	1683 4
Fibrinogen in solution at 50 mg·mL^{-1}	1617 9	1633 25	1652 32	1669 21	1684 12
—PAH-fib	1615–1616 5–7	1635 42–43	1652–1653 36	1668–1669 9–10	1680–1681 6–7
—PSS-fib	1611–1613 2	1635 44–46	1652–1653 34–37	1668–1669 10–12	1681 6–7
—PAH-fib-PAH—	1618–1622 0–3	1636 46–53	1652–1653 25–39	1668–1669 12–17	1682–1683 1–3
—PSS-fib-PSS—	1621–1624 1	1636 39–40	1652 35–36	1669 20	1681–1683 3–4
—PAH-fib-PSS—	1618–1623 0–2	1636 40–44	1652 33–37	1669–1670 19	1682–1683 1–4
—PSS-fib-PAH—	1623 2–3	1636 39–43	1652 34–37	1668–1669 18	1680–1682 3–4

Notes: Fib represents fibrinogen. Each column corresponds to the specific structure defined in the head of the columns. For each band, the range of the central frequency (cm^{-1}) (upper numbers) and the relative intensity (%) to the total amide I intensity (lower numbers) are systematically given.
Source: Ref. 35.

results strongly suggest that polyelectrolytes not only protect the proteins from aggregating at room temperature, but also thermally stabilize them.

D. Diffusion of Proteins on and in Polyelectrolyte Multilayers

Due to the multiple links that proteins can establish with polyelectrolytes, one would expect them to remain immobile once adsorbed on or embedded in polyelectrolyte multilayer films. Fluorescence recovery after photobleaching (FRAP) was used to obtain first results relative to the diffusion process of HSA molecules adsorbed on PSS/PAH polyelectrolyte multilayers. Some of the results are summarized in Table 2. It was found that at least two populations of adsorbed proteins exist on the top or within the multilayers. One population, which represents typically 50–70% of the adsorbed proteins, corresponds to proteins that are able to diffuse laterally along the surface on or in the multilayers, while the second population appears almost immobile over the experimental time scale. The diffusion behavior of the negatively charged HSA appears different under the experimental conditions when the proteins are adsorbed on PAH (polycation) or PSS (polyanion) ending films. On PAH ending films one observes that the diffusion coefficient strongly decreases with the protein surface concentration ranging from 1.5×10^{-9} cm^2·s^{-1} for low surface concentration

TABLE 2 Diffusion Coefficients of HSA Adsorbed on or Embedded in Different Film Architectures

Multilayer architecture	Γ_{HSA} (μg/cm^2)	D (cm^2/s)	f
PEI(PSS/PAH)$_3$-HSA	0.42	9.7×10^{-11}	0.7
PEI(PSS/PAH)$_3$-HSA	0.03	1.5×10^{-9}	0.9
PEI(PSS/PAH)$_2$PSS-HSA	0.04	6.4×10^{-11}	0.9
PEI(PSS/PAH)$_3$-HSA-(PAH/PSS)$_3$	0.08	9.0×10^{-10}	0.3
PEI(PSS/PAH)$_2$PSS-HSA-(PSS/PAH)$_3$	0.17	1.1×10^{-10}	0.6
PEI(PSS/PAH)$_2$PSS-HSA-(PSS/PAH)$_3$	0.03	1.0×10^{-10}	0.7
PEI(PSS/PAH)$_2$PSS-HSA-(PAH/PSS)$_3$	0.18	1.0×10^{-10}	0.5

Notes: Γ_{HSA} represents the amount of HSA adsorbed on or embedded in the film; D represents the HSA diffusion coefficient; and f corresponds to the mobile fraction of HSA molecules on the surface.
Source: Ref. 37.

up to 10^{-10} cm$^2 \cdot$s^{-1} for high surface concentrations, whereas on a PSS terminating film the lateral diffusion coefficient of HSA is equal to 0.6×10^{-10} cm$^2 \cdot$s^{-1} independent of the protein surface concentration. It is thus very surprising to find that negatively charged proteins diffuse more rapidly on a positively charged polyelectrolyte layer than on a negatively charged one. Moreover, the intrinsic diffusion coefficient (low concentration) of HSA on PAH terminating films is of the same order of magnitude than on a hydrophilic solid surface such as glass, whereas on PSS it becomes similar to what is found on hydrophobic polymeric surfaces like polymethylmethacrylate (PMMA) [38]. When the proteins are embedded in multilayer structures the fraction of immobile proteins usually increases. However, the diffusion coefficients of the diffusing proteins remain of the same order of magnitude as when the proteins are adsorbed on the multilayers. In the case of proteins adsorbed on PSS ending films it is even observed that the diffusion coefficient increases after embedding. The reasons for this behavior are not fully clear, but a temptative explanation based on the wrapping of the HSA molecules by PAH chains was proposed.

III. TOWARD BIOFILMS

A. Construction of Polyelectrolyte Multilayers with Natural Polyelectrolytes

Multilayer films can also be constructed by using polyelectrolytes, at least one being a natural polyelectrolyte. These films are usually developed in order to render surfaces bioinert. Natural polyelectrolytes present the advantage of high biocompatibility and are also degradable allowing to confer specific biological properties to the surface for a limited period of time.

Picart et al. [39] studied the formation of multilayer films based on the Poly-(L-lysine)/hyaluronic acid (PLL/HA) system. Using streaming potential measurements they could show that, similarly to "conventional" polyelectrolyte multilayer systems, the driving force of the buildup process of such films is also the alternate overcompensation of the surface charge after each new PLL or HA deposition.

Whereas polyelectrolyte systems such as PSS/PAH usually form stratified multilayers, each polyelectrolyte being deposited on top of the previous layer, it was found that the buildup of the $(PLL/HA)_n$ films by the alternate deposition of PLL and HA takes place over two regimes. The first regime extends, for an ionic strength of about 0.15 M, roughly up to about 8 bilayers. The initial PLL layer deposited on a clean glass slide appeared homogeneously distributed and exhibited a low surface roughness of about of 0.35 nm. After the first HA deposition leading to a PLL/HA layer, two kinds of structures became visible by atomic force microscopy (AFM) imaging: large ones, called "islands" having typical sizes of the order of few micrometers and a height of the order of 1 μm, and smaller "islets" whose characteristic size was of the order of 1 μm or even smaller. These islets and islands most probably result from the interaction of PLL with HA which may well be able to form complex coacervates on the surface, as it was suggested for a similar PLL/alginate system [40]. As the number of deposited bilayers increased, the islands became larger whereas the smaller droplets became more seldom. A typical AFM image of a $(PLL/HA)_4$ film is represented in Fig. 5A. The structures generated during the first PLL/HA multilayer buildup regime resembled qualitatively to structures formed by liquid droplets on a solid surface during a continuous condensation from a supersaturated vapor and which are called "breath figures" [41]. After the deposition of the eighth PLL/HA bilayer neither individual islands nor islets were visible anymore. All these structures seemed to have coalesced, leading to the formation of an almost uniform film which became very flat with a roughness root-mean-square value of the order of 10 nm (see Fig. 5B). It is thus reasonable to assume that these droplets were formed of a homogeneous PLL/HA liquidlike material. After the formation of the homogeneous and flat surface, additional depositions led to the presence of very small dots for the layer ending with PLL, and of larger dots for the layer ending with HA. AFM imaging of films extending the $(PLL/HA)_{10}$ multilayer could not be performed because the tip was locally strongly attracted or repulsed from the surface, indicating the presence of large density fluctuations on the surface. However, quartz microbalance measurements revealed that the deposition of further polyelectrolyte layers goes on and the deposited mass seems even to increase exponentially, as it was found by Elbert et al. [40] for the PLL/alginate system. Optical waveguide lightmode spectroscopy seemed to indicate that in this second buildup regime, when a $(PLL/HA)_n$ film was brought in contact with a PLL solution, PLL chains can diffuse into the multilayer similarly to a diffusion in a porous material. The diffusion inside the film appeared to take place up to the first initially deposited PLL layer. The interior of the film being neutral, it is expected that these chains do not form new strong complexes with the HA chains, but rather interact weakly with them and in particular with the interfacial hydration layer of HA. The PLL chains do however not only diffuse into the interior of the film during this adsorption step, but they also form new PLL/HA complexes with the HA chains constituting the outer layer of the multilayer. During the rinsing step after PLL deposition, a fraction of the PLL chains that diffused in the interior of the film diffused out of it. Once this film was brought in contact with a HA solution, HA chains interacted with the PLL chains lying at the outer part of the film. Moreover, OWLS signals indicated that all the free PLL chains remained in the film at the end of the rinsing step also diffused toward the outer film surface that now acted as a perfect sink. As soon as these chains reached the surface, they interacted with HA chains and thus also contributed to the buildup

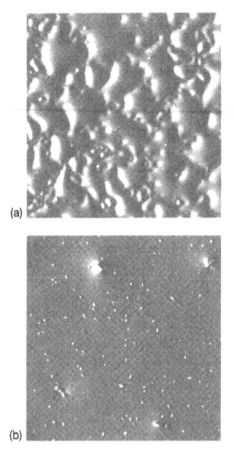

FIG. 5 50 × 50 μm^2 AFM images obtained for two characteristic stages of the buildup process of a (PLL/HA)$_n$ multilayer on a silica surface. (a) Image corresponding to a (PLL/HA)$_4$ film. One can observe that the surface is covered by large droplets whose vertical and horizontal size is typically of the order of 1 to several micrometers and also of droplets of much smaller size. The buildup mechanism corresponds to an increase in size and the coalescence of these droplets until a continuous film is reached. This is achieved for our system after the deposition of eight bilayers. (b) A typical film corresponding to (PLL/HA)$_8$-PLL.

of the layer. One can assume that the amount of free PLL chains remaining in the film at the end of the rinsing step of the PLL solution by pure buffer was proportional to the film thickness. This should then lead, as observed, to an exponential increase of the total mass of the film in the second buildup regime of the film. We believe that the observations found for the PLL/HA system should also be valid for systems like PLL/alginate.

B. Interaction of Polyelectrolyte Multilayers with Cells

Elbert et al. [40] used polyelectrolyte multilayers to build polymeric barriers on tissue surfaces to improve postsurgical healing or on the surfaces of tissue-engineered implants. To create such a barrier two fundamental obstacles have to be overpassed.

First, the buildup of self-assembled films onto biological surfaces is hindered by the heterogeneity of the chemical groups on such surfaces. Second, the self-assembled architecture must itself be biologically inert. This second aspect is particularly important since most of the polymeric materials induce a foreign-body response once implanted. The PLL/alginate system was used [40] to build multilayers on proteinaceous surfaces. These surfaces were designed to be extremely adhesive for proteins and cells, such as the extracellular matrix produced by fibroblast cells. Moreover, they exhibited the typical heterogeneities of biological surfaces. The PLL/alginate system is known to form complex coacervates in solution at physiological conditions and very thick gel complexes at surfaces [42]. It was shown that this system is suitable for alternative depositions of the polycation (PLL) and of the polyanion (alginate). The structure of the film was however strongly dependent on the way in which the rinsing step was performed. Under strong rinsing conditions the buildup of a multilayer film whose thickness increases exponentially with the number of bilayers—the thickness of a film constituted by 15 PLL/alginate bilayers being of the order of 150 nm—was observed. This exponential growth in thickness was explained by the formation of complex gels on the surface. According to Elbert et al. [40] the formation of such a complex gel is likely due to the ability of PLL and/or alginate to diffuse through a solution of the other polyelectrolyte without precipitating while slowly interacting. Then, following the deposition of PLL or alginate on the surface, the rinsing step does not remove all of the nonadsorbed polyelectrolyte in the vicinity of the surface. The viscous layer of the polyelectrolyte solution near the film interacts then with the oppositely charged polyelectrolyte and forms a complex gel which leads to a thick layer. Such a process, which can describe the formation of thick bilayers can however, in our opinion, not explain the exponential growth of the film thickness during the different steps of the buildup process. This explanation should merely account for the fact that, under gentle rinsing conditions, structures with thicknesses per bilayer of the order of tenths of a nanometer were obtained. In our opinion the formation of PLL/alginate films in the exponential regime is similar to what was observed by Picart et al. [39] for the HA/PLL system and is due to PLL diffusion in and out of the film during the adsorption and rinsing steps.

These PLL/alginate multilayers were then used to evaluate fibroblast spreading on the surface. It was found that, under strong rinsing conditions, fibroblast cell spreading decreased monotonically with the number of deposited bilayers and spreading was inhibited after deposition of the fifth bilayer, alginate constituting the outer layer of the multilayer film. Moreover, after eight bilayers, cell spreading was prevented to the same extent whatever the nature of the outer polyelectrolyte layer (PLL or alginate). Furthermore, by testing the viability of the cells on contact with the multilayers, Elbert et al. demonstrated that the inhibition of cell spreading could not result from toxic cellular effects which could eventually take place. Soluble polycations are indeed known to be toxic to cells at low concentrations in solution. The inhibition of cell spreading should thus be due to the fact that proteins establish only limited interactions with the hydrogel-like PLL/alginate films.

Serizawa et al. [43] investigated the possibility of conferring to multilayer films anti- or procoagulant activities. To this aim they investigated the alternating anti- versus procoagulant activity of dextran sulfate (Dex) and chitosan multilayer films. Dextran sulfate was selected for its anticoagulant properties that come from its

charged sulfate groups, whereas chitosan, a polycation, is known to exhibit procoagulant activity. The alternate deposition of dextran and chitosan in the absence of
NaCl leads to ultrathin films having a thickness of 3.7 nm for 10 Dex/chitosan
bilayers. Such a small film thickness indicates that each layer does not fully cover
the entire surface. On the other hand, as it is generally observed, the deposition of
such multilayers from 1 M NaCl polyelectrolyte solutions gave very thick films (227
nm for 10 Dex/chitosan bilayers) with a film thickness that seems to increase exponentially with the number of deposited bilayers. It was found that the assemblies
prepared in the presence of 1 M NaCl showed clear alternating anti- and procoagulant
activity. On the other hand, surface assemblies prepared in the absence of NaCl did
not seem to retain any of the specific bioactivity of each layer, neither totally anticoagulant nor fully procoagulant activity were found. Serizawa et al. also investigated
the activity of [(chitosan–Dex)$_2$–chitosan]–Dex–chitosan and (chitosan–Dex)$_3$–chi-
tosan–Dex assemblies in which [(chitosan–Dex)$_2$–chitosan] and (chitosan–Dex)$_3$
were prepared in the presence of 1 M NaCl solutions, the subsequent polyelectrolyte
layers being deposited in the absence of NaCl. In this case blood coagulation was
observed on both chitosan and Dex terminating films. This shows that for this system,
the outermost surface is the key factor in the anticoagulant activity observed rather
than the total film thickness. This result may be due to the fact that at high salt
concentrations, the polyelectrolyte layer becomes so thick that the underlying layer
does not emerge out of the outer layer, whereas in the absence of salt the layers are
so thin and the interactions between the two polyelectrolytes are so strong that chemical groups of the two polyelectrolytes from at least the two outer layers emerge
toward the solution and are thus able to interact with the cells.

Brynda and Houska [44] developed new kinds of biofilms based on polyelec-
trolyte/protein multilayers in which the proteins were crosslinked with glutatalde-
hyde. In such way they could, for example, build albumin/heparin multilayer films.
Both being negatively charged at the physiological pH the film construction was
performed at a pH lying below the isoelectric point of the protein so that the protein
was positively charged while heparin remains negatively charged. Once the film was
built up, it was brought in contact with a glutaraldehyde solution. Glutaraldehyde
forms covalent bonds between the amine groups of the basic amino acid residues of
albumin and some of the unsubstituted amino groups of heparin. Once the multilayer
was fixed it could be brought into contact with a solution at physiological conditions.
This led to an expulsion of all the unbound macromolecules due to strong electrostatic repulsion. A similar procedure was also used in order to build purely proteinaceous assemblies such as, for example, pure immunoglobulin (IgG) layers. Thus,
first multilayer films were prepared by using a polyelectrolyte making the joint between the protein layers, but it was not incorporated into the assembly during the
fixation step. Dextran sulfate constitutes such a polyelectrolyte. Brynda and Houska
[44] selected the albumin/heparin system in order to achieve biocompatible and anticoagulant coatings. Albumin is often adsorbed or covalently fixed on surfaces in
order to improve their blood compatibility by reduction of surface protein adsorption.
Heparin is known for its anticoagulant properties and it acts by increasing catalyti-
cally the rate at which antithrombin inhibits proteases coagulation. Beside its anticoagulation properties, heparin possesses also antibacterial and antiviral activities. In
vivo tests on (albumin)$_n$ and (albumin/heparin)$_n$ coatings indicated improved blood-
compatible properties expressed by decreased platelet adhesion and fibrinogen ad-

sorption. The adhesion was reduced on prostheses coated with the neat albumin assembly, heparin being known as platelet adhesion promotor. Moreover, negligible amounts of IgG molecules are found on (albumin/heparin)$_n$ structures. An active role of immobilized heparin in the inhibition of blood coagulation on these later coatings was indicated by the binding of antithrombin. These coatings were also shown to be stable over periods of at least two months. By coating the heart ventricle implant of a goat, no thrombus was observed even after 1 month.

Recently Tryoen-Toth et al. [45] investigated the viability, adhesion, stability of osteoblast phenotype and inflammatory response of two types of cells, SaOs-2 osteoblast-like human carcinoma cells and human periodontal ligament (PDL) cells in contact with different multilayer architectures. All the architectures were first constituted of an initial PEI-(PSS/PAH)$_2$ film, PEI representing polyethyleneimine. On this precursor film they added the following architectures: PGA, (PGA/PLL)$_2$, PSS, or (PSS/PEI). The results are summarized in Table 3, where + (resp. −) indicates a favorable (resp. unfavorable) cell response. From this table it comes out that the cell response is very dependent on the film architecture. A PEI terminating film presents, for all the investigated responses, very poor properties. This is consistent with the fact that PEI is known to be cytotoxic. The PGA ending multilayer appeared to be a film architecture with a large positive spectrum. In particular, the fact that the cell phenotypes are maintained for PGA and PLL ending films seems very promising for future biological applications. The conclusions summarized in Table 3 are simple observations, but we do not possess any explanation for the different behaviors.

Targeted polyelectrolyte multilayers can also be obtained by embedding polyelectrolytes, onto which active peptides have been covalently bound, into the multilayer structure. One of the central questions is whether cells can communicate with the peptides that are deeply embedded in the multilayer films. Chluba et al. [46] selected PLL and PGA as the polyelectrolyte system and murine melanoma cells as a biological test model system to address a first answer to this question. These cells respond specifically to a small peptide hormone, α-melanocortin, which constitutes

TABLE 3 Summary of the Responses of SaOS-2 Osteoblast-Like Cells and Human Periodontal Ligament (PDL) Cells on Different Polyelectrolyte Multilayer Films

	—PEI	—PAH	—PGA	—PLL	—PSS
Cell viability	−	+	+	+	+
ALP expression of SaOS-2 cells	−	−	+	+	+
ALP expression of PDL cells	−	−	+	+/−	+
OC expression of SaOS-2 cells	−	+	+	+	+
OC expression of PDL cells	−	−	+	+/−	+
IL-8 production in SaOS-2 cells	−	−	+	+	+
IL-8 production in PDL cells	−	−	+/−	+/−	+

Notes: —PEI represents PEI-(PSS/PAH)$_2$-(PSS/PEI)$_2$; —PAH, PEI-(PSS/PAH)$_2$; —PGA, PEI-(PSS/ PAH)$_2$-PGA; —PLL, PEI-(PSS/PAH)$_2$-(PGA/PLL)$_2$; and —PSS, PEI-(PSS/PAH)$_2$-PSS. + (resp. −) indicates a favorable (resp. unfavorable) cellular response. Alkaline phosphatase (ALP) and osteocalcin (OC) are two osteoblast phenotypic markers. IL-8 is a cytokine secreted in the proinflammatory process (+ indicates that no IL-8 secretion could be observed).
Source: Ref. 45.

a potent stimulator of melanogenesis in mammalian melanocytes and in melanoma cells. It binds to a specific receptor on the cell surface, melanocortin-1 receptor (MCl-R) and induces activation of tyrosinase, the key enzyme for melanin formation. In the chain of reactions leading to the formation of melanin, one observes after 1–2 h the formation of cyclic AMP. The short time response of the cells to α-melanocortin can thus be quantified by determining the amount of cyclic AMP produced after a few hours of contact between the cells and α-melanocortin. The long time response (more than 2 days of contact) can be analyzed by quantifying the amount of melanin produced by the cells. A synthetic α-melanocortin derivative has been covalently coupled to PLL. We will denote this modified polyelectrolyte by PLL-M. Using optical waveguide lightmode spectroscopy, it has first been shown that PLL-M can be embedded in a $(PLL/PGA)_n$ multilayer. When PLL-M forms the outer layer of a multilayer film α-melanocortin remains as biologically active as the free hormone. When PLL-M is embedded in $(PLL/PGA)_n$ architectures, the long-term activity, measured by melanin production is maintained even when it lies under 20 PLL/PGA bilayers (see Fig. 6). Its short time activity depends on integration depth (see Fig. 7). This seems to indicate that the embedded PLL-M chains do not merge to the outer layer of the multilayer film but that the cells must find a way to communicate through the film with the embedded hormone. Several hypotheses can be advanced:

1. The PLL-M bonds break with time so that the film acts as a controlled release architecture. It is however unexpected that this can explain the observed cell response because these films were active for several weeks, and a rapid breaking of the PLL-M bonds should then also lead to a rapid loss of the film activity.

2. Cells in contact with the film could also produce enzymes that are able to degrade locally the film and thus come in direct contact with the embedded hormone. In order to test this hypothesis Chluba et al. [46] designed experiments excluding any possibility of film degradation. They integrated nonbiodegradable polyelectrolytes into the multilayers. The insertion of four PSS/PAH bilayers did not lead to a significant reduction of melanin production. In addition, the insertion of eight bilayers with no biodegrad-

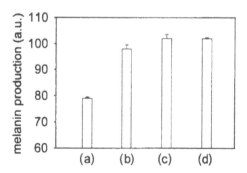

FIG. 6 Melanin production in B16-F1 cells in contact for 4 days with different functionalized multilayers. (a) $PEI(PSS/PAH)_2(PGA/PLL-M)$; (b) $PEI(PSS/PAH)_2(PGA/PLL-M)(PGA/PLL)_5$; (c) $PEI(PSS/PAH)_2(PGA/PLL-M)(PGA/PLL)_{10}$; (d) $PEI(PSS/PAH)_2(PGA/PLL-M)(PGA/PLL)_{20}$. (From Ref. 46, where further details can be found.)

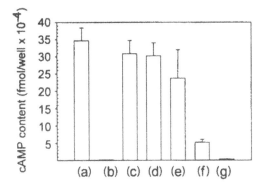

FIG. 7 Cyclic AMP production in B16-F1 cells in contact for 3 h with different functionalized multilayers. (a) Plastic + free M (α-melanocortin analog), which serves as a positive control; (b) plastic (negative control); (c) PEI(PSS/PAH)$_2$(PGA/PLL-M); (d) PEI(PSS/PAH)$_2$ (PGA/PLL-M)(PGA/PLL)$_5$; (e) PEI(PSS/PAH)$_2$(PGA/PLL-M)(PGA/PLL)$_{10}$; (f) PEI(PSS/ PAH)$_2$(PGA/PLL-M)(PGA/PLL)$_{20}$; (g) PEI(PSS/PAH)$_2$(PGA/PLL-M)(PGA/PLL)$_{25}$. (From Ref. 46, where further details can be found.)

able PSS/PAH polyelectrolytes between PLL-M and the outer layer lead to a 25% reduction of the melanin production of the melanoma cells in contact with these surfaces. This demonstrates the ability of the cells to find a way to communicate through the polyelectrolyte multilayers without degrading the films. It further shows that PSS/PAH layers might form a barrier depending on layer thickness as described by others [40].

3. Porosity of the (PLL/PGA)$_n$ films could also explain access of cells to embedded layers in the film architecture. Further investigations are necessary in order to get a better understanding of the recognition mechanism of embedded hormones by cells. Whatever the nature of the recognition mechanism, these results constitute a key that opens the route for the design of complex cell modulation systems on biomaterial surfaces. One could, for example, incorporate specific adhesion sequences in order to target certain cell types to induce desired signal transduction pathways and simultaneously to prevent adhesion of undesired organisms.

IV. BIOMINERALIZATION INDUCTION BY POLYELECTROLYTE MULTILAYER FILMS

The deposition of weakly soluble salts on polymers is of major importance not only for fundamental research dealing with biomineralization, but also for practical applications such as the design and development of materials suitable for prosthetic uses in bone and tooth diseases. The structural macromolecules contained in the biological matrix, mainly carboxylates, are believed to control nucleation, growth, polymorphism, shape, chemical composition, orientation, and crystal texture [47]. Acidic macromolecules present in solution usually inhibit the nucleation and growth of calcium phosphate salts, whereas they promote calcification once adsorbed on a substrate. The synthesis of inorganic/organic hybrids by mimicking biomineralization has attracted large attention, but only discrete mineral crystals have generally been obtained on organic matrixes [48]. Several factors could influence mineral nucleation and crystal growth on a polymer film. Particularly important seems to be the degree of supersaturation of the solution and the surface charges on the polymer film. The success of a biomimetic surface comes from its ability to promote strongly heterogeneous nucleation on the film layer so that homogeneous nucleation in solution is fully avoided. Strong interactions between mineral surfaces and films promote heterogeneous nucleation. High degrees of supersaturation of the mother solution favors on the other hand direct homogeneous nucleation from solution [49]. Other parameters like polar acid–base or strong electrostatic interactions and stereochemical recognition between inorganic crystals and film architecture promote heterogeneous nucleation. One key factor for a successful mimic of biomineralization is to increase the charge density or polarity on the film surface.

Thus, Zhang and Gonsalves [49] investigated calcium carbonate formation on chitosan or chitosan/poly(acrylic acid) (PAA) films. Poly(acrylic acid) proved to be very efficient in promoting heterogeneous nucleation of calcium carbonates and in avoiding totally its homogeneous nucleation. At low PAA surface concentrations heterogeneous nucleation and further crystal growth occurred leading to large spherical deposits made up by numerous smaller crystallites. In the presence of high PAA amounts, even though nucleation did occur, further growth was inhibited by the mobile carboxylic anions of PAA.

Apatites (in particular hydroxyapatite) are considered as the model compounds for the inorganic constituents of bones and teeth. Apatite formation in contact of biofilms have also recently been investigated. Bigi et al. [50] studied the role of stretching and of sodium polyacrylate onto nucleation and growth of hydroxyapatite in the film architectures. These authors found amorphous calcium phosphate aggregates in the absence of stretching, whereas ordered deposition of crystalline hydroxyapatite was observed in stretched polyacrylate/gelatin films. Layered structures with thicknesses up to 2 μm were found. Also Ngankam et al. [51] investigated calcium phosphate nucleation on PSS/PAH films and compared these data with those obtained on a naked silica surface [52]. The process was followed by means of an optical "in situ" method (scanning angle reflectometry) which indicated that both the positive (PAH) and negative (PSS) ending multilayer architectures induced nucleation at lower supersaturation than naked silica. Also, formed crystals as shown by infrared spectroscopy were different on a PAH and PSS ending film. Hydroxyapatite or octacalcium phosphate was found in presence of multilayer films, whereas dicalcium phosphate dihydrate precipitated on the naked silica surface. The authors explained their observation by assuming a strong calcium (or a weak proton) accumulation on the negatively charged ending film, and an important phosphate or hydroxyl ions accumulation close to the positively ending film. These studies furnish useful information toward biomaterial coatings favoring osseointegration of implants in contact with hard biological tissues through heterogeneous nucleation. Including in the film architectures mineralization-inducting proteins, covalently bound peptides that are active in induction of the nucleation, or even small hydroxyapatite nuclei could even further improve the ability of polyelectrolyte multilayers to trigger after deposition on implant surfaces heterogeneous apatite nucleation.

V. CONCLUSIONS

Polyelectrolyte multilayer films appear to be a very versatile tool to construct surface coatings with targeted properties. This seems to be particularly the case in the biological field. We only mentioned in this chapter film buildups on macroscopic flat surfaces. However, another very active research area connected to this field concerns the preparation of microcapsules that can be used for example for drug delivery purposes. A further development of this field will require a better understanding of the relation between the polyelectrolyte systems, the construction conditions, and the film structure. Indeed, films can be smooth or rough, compact or microporous, and it will be of great interest to make use of these different structures to confer to the film specific properties. Also, a more fundamental understanding of the interactions of cells with such films would be needed. How do cells establish interactions with the polyelectrolytes constituting the multilayer? How do the mechanical properties of the film play a role in cellular adhesion? Answers to such questions would be of greatest importance in the design of highly specific biomaterials.

ACKNOWLEDGMENTS

The authors thank the program Adhésion Cellules-Matériaux, the program CNRS Physique et Chimie du Vivant, and the ACI Ingénérie Tissulaire no. 4G005F for financial support. Moreover, we are indebted to V. Ball, F. Cuisinier, G. Decher, P. Lavalle, G. Ogier, C. Picart, and B. Senger for their stimulating discussions.

REFERENCES

1. G. Decher and J.-D. Hong, Buildup of ultrathin multilayer films by a self-assembly process. 1. Consecutive adsorption of anionic and cationic bipolar amphiphiles on charged surfaces. Makromol. Chem. Macromol. Symp. 46:321–327, 1991.
2. G. Decher, J.-D. Hong, and J. Schmitt, Thin solid films. 210/211:831, 1992.
3. N. G. Hoogeveen, M. A. Cohen-Stuart, G. J. Fleer, and M. R. Böhmer. Formation and stability of multilayers of polyelectrolytes. Langmuir 12:3675–3681, 1996.
4. F. Caruso and H. Möhwald, Protein multilayer formation on colloids through a stepwise self-assembly technique. J. Am. Chem. Soc. 121:6039–6046, 1999.
5. G. Ladam, P. Schaad, J. C. Voegel, P. Schaaf, G. Decher, and F. Cuisinier. In situ determination of the structural properties of initially deposited polyelectrolyte multilayers. Langmuir 16:1249–1255, 2000.
6. G. Decher and J. Schmitt, Fine-tuning of the film thickness of ultrathin multilayer films composed of consecutively alternating layers of anionic and cationic polyelectrolytes. Progr. Colloid Polymer Sci. 89:160–164, 1992.
7. Y. Lvov, G. Decher, and H. Möhwald, Assembly, structural characterization, and thermal behavior of layer-by-layer deposited ultrathin films of poly(vinyl sulfate) and poly(allylamine). Langmuir 9:481–486, 1993.
8. H. Hong, D. Davidov, Y. Avny, H. Chayet, E. Z. Faraggi, and R. Neumann. Electroluminescence and x-ray reflectivity studies of self-assembled ultra-thin films. Adv. Mater. 7:846–849, 1995.

9. J. Schmitt, T. Grünewald, G. Decher, P. S. Pershan, K. Kjaer, and M. Lösche. Internal structure of layer-by-layer adsorbed polyelectrolyte films: a neutron and x-ray reflectivity study. Macromolecules 26:7058–7063, 1993.

10. D. Korneev, Y. Lvov, G. Decher, J. Schmitt, and S. Yaradaikin. Neutron reflectivity analysis of self-assembled film superlattices with alternate layers of deuterated and hydrogenated polystyrenesulfonate and polyallylamine. Physica B 213/214:954, 1995.

11. G. Decher, Fuzzy nanoassemblies: toward layered polymeric multicomposites. Science 277:1232–1237, 1997.

12. J. Ruths, F. Essler, G. Decher, and H. Riegler. Polyelectrolytes. I: Polanion/polycation multilayers at the air/monolayer/water interface as elements for quantitative polymer adsorption studies and preparation of hetero-superlattices on solid surfaces. Langmuir 16:8871–8878, 2000.

13. D. Yoo, S. S. Shiratori, and M. F. Rubner, Controlling bilayer composition and surface wettability of sequentially adsorbed multilayers of weak polyelectrolytes. Macromolecules 31:4309–4318, 1998.

14. S. S. Shiratori and M. F. Rubner. pH-dependent thickness behavior of sequentially adsorbed layers of weak polyelectrolytes. Macromolecules 33:4213–4219, 2000.

15. A. Fery, B. Schöler, T. Cassagneau, and F. Caruso. Nanoporous thin films formed by salt-induced structural changes in multilayers of poly(acrylic acid) and poly(allylamine). Langmuir 17:3779–3783, 2001.

16. J. D. Mendelson, C. J. Barret, V. V. Chan, A. J. Pal, A. M. Mayes, and M. F. Rubner. Fabrication of microporous thin films from polyelectrolyte multilayers. Langmuir 16: 5017–5023, 2000.

17. C. Tribet, I. Porcar, P. A. Bonnefont, and R. Audebert. Association between hydrophobically modified polyanions and negatively charged bovine serum albumin. J. Phys. Chem. B 102:1327–1333, 1998.

18. M. M. Andersson, R. Hatti-Kaul, and W. Brown. Dynamic and static light scattering and fluorescence studies of the interactions between lactase dehydrogenase and poly-(ethyleneimine). J. Phys. Chem. B 104:3660–3667, 2000.

19. J. Xia, P. L. Dubin, Y. Kim, B. B. Muhoberac, and V. J. Klimkowski, Electrophoretic and quasi-elastic light scattering of soluble protein–polyelectrolyte complexes. J. Phys. Chem. 97:4528–4534, 1993.

20. K. W. Mattison, P. L. Dubin, and I. J. Brittain, Complex formation between bovine serum albumin and strong polyelectrolytes: effect of polymer charge density. J. Phys. Chem. B 102:3830–3836, 1998.

21. K. R. Grymonpré, B. A. Staggemeier, P. L. Dubin, and K. W. Mattison. Identification by integrated computer modeling and light scattering studies of an electrostatic serum albumin–Hyaluronic acid binding site. Biomacromolecules 2:422–429, 2001.

22. K. Kaibara, T. Okazaki, H. B. Bohidar, and P. L. Dubin. pH-induced coacervation in complexes of bovine serum albumin and cationic polyelectrolytes. Biomacromolecules 1:100–107, 2000.

23. V. Ball, M. Winterhalter, P. Schwinté, Ph. Lavalle, J.-C. Voegel, and P. Schaaf, Complexation mechanism of bovine serum albumin and poly(allylamine hydropchloride). J. Phys. Chem. B 106:2357–2364, 2002.

24. M. T. Record, Jr., C. F. Anderson, and T. Lohman. Thermodynamic analysis of ion effects on the binding and conformational equilibria of proteins and nucleic acids: the roles of ion association or release, screening, and ion effects on water activity. Quaterly Rev. Biophys. 11:103–178, 1978.

25. K. Ariga and T. Kunitake, Sequential catalysis in organized multienzyme films. In Protein Architecture (Y. Lvov and H. Möhwald, eds.). Marcel Dekker, New York, 2000, pp. 169–191.

26. F. Caruso, K. Niikura, D. N. Furlong, and Y. Okahata, Assembly of alternating poly-electrolyte and protein multilayer films for immunosensing. Langmuir *13*:3427–3433, 1997.

27. F. Caruso, D. N. Furlong, K. Ariga, I. Ichinose, and T. Kunitake, Characterization of polyelectrolyte–protein multilayer films by atomic force microscopy, scanning electron microscopy, and Fourier transform infrared reflection-absorption spectroscopy. Langmuir *14*:4559–4565, 1998.

28. G. Ladam, P. Schaaf, F. J. G. Cuisinier, G. Decher, and J. C. Voegel, Protein adsorption onto auto-assembled polyelectrolyte films. Langmuir *17*:878–882, 2000.

29. M. Müller, T. Rieser, K. Lunkwitz, and J. Meier-Haack, Polyelectrolyte complex layers: a promising concept for anti-fouling coatings verified by in-situ ATR-FTIR spectroscopy. Macromol. Rapid Commun. *20*:607–611, 1999.

30. M. Müller, T. Rieser, P. L. Dubin, and K. Lunkwitz. Selective interaction between pro-teins and the outermost surface of polyelectrolyte multilayers: influence of the polyanion type, pH, and salt. Macromol. Rapid Commun. *22*:390–395, 2001.

31. G. Ladam, C. Gergely, B. Senger, G. Decher, J. C. Voegel, P. Schaaf, and F. J. G. Cuisinier. Protein interactions with polyelectrolyte multilayers: interactions between hu-man serum albumin and polystyrene sulfonate/polyallylamine multilayers. Biomacro-molecules *1*:674–687, 2000.

32. T. D. Gibson, J. N. Hulbert, B. Pierce, and J. I. Webster, The stabilisation of analytical enzymes using polyelectrolytes and sugar derivatives. In *Stability and Stabilisation of Enzymes* (W. J. J. van den Tweel, A. Harder, and R. M. Buitelaar, eds.). Elsevier, Am-sterdam, 1993, pp. 337–346.

33. J. L. Xia, P. L. Dubin, E. Kokufuta H. Havel, and B. B. Muhoberac, Light scattering, CD and ligand binding studies of ferrihemoglobin complexes. Biopolymers *50*:153–161, 1999.

34. M. M. Andersson and R. Hatti-Kaul, Protein stabilizing effects of polyethyleneimine. J. Biotechnology *72*:21–31, 1999.

35. P. Schwinté, J. C. Voegel, C. Picart, Y. Haikel, P. Schaaf, and B. Szalontai, Stabilizing effects of various polyelectrolyte multilayer films on the structure of adsorbed/embedded fibrinogen molecules. An ATR-FTIR study. J. Phys. Chem. B *105*:11906–11916, 2001.

36. L. Szyk, P. Schaaf, C. Gergely, J. C. Voegel, and B. Tinland. Lateral mobility of proteins adsorbed on polyelectrolyte multilayers studied by fluorescence recovery after photo-bleaching. Langmuir *17*:6248–6253, 2001.

37. L. Szyk, P. Schwinté, J. C. Voegel, P. Schaaf, and B. Tinland, Dynamical behavior of human serum albumin adsorbed on or embedded in polyelectrolyte multilayers. J. Phys. Chem. B *106*:6049–6055, 2002.

38. R. D. Tilton, Mobility of biomolecules at interfaces. In *Biopolymers at Interfaces* (M. Malmstem, ed.). Marcel Dekker, New York, 1998, pp. 363–407.

39. C. Picart, Ph. Lavalle, P. Hubert, F. J. G. Cuisinier, G. Decher, P. Schaaf, and J.-C. Voegel, Build-up mechanism for hyaluronic acid/polylysine films onto a solid interface. Langmuir *17*:7414–7424, 2001.

40. D. L. Elbert, C. B. Herbert, and J. A. Hubbell. Thin polymer layers formed by poly-electrolyte multilayer techniques on biological surfaces. Langmuir *15*:5355–5362, 1999.

41. D. Fritter, D. Beysens, and C. M. Knobler, Growth of breath figures. Phys. Rev. A *43*: 2858–2869, 1991.

42. G. M. O'Shea and A. M. Sun, Encapsulation of rat islets of Langerhans prolongs xe-nograft survival in diabetic mice. Diabetes *35*:943–946, 1986.

43. T. Serizawa, M. Yamaguchi, T. Matsuyama, and M. Akashi, Alternating bioactivity of polymeric layer-by-layer assemblies: anti- vs procoagulation of human blood on chitosan and dextran sulfate layers. Biomacromolecules *1*:306–309, 2000.

44. E. Brynda, and M. Houska, Ordered multilayer assemblies: albumin/heparin for biocompatible coatings and monoclonal antibodies for optical immunosensors. In *Protein Architecture* (Y. Lvov and H. Möhwald, eds.). Marcel Dekker, New York, 2000, pp. 251–285.

45. P. Tryoen-toth, D. Vautier, Y. Haikel, J.-C. Voegel, P. Schaaf, J. Chluba, and J. Ogier. Viability, adhesion, and bone phenotype of osteoblast-like cells on polyelectrolyte multilayer films. J. Biomed. Mater. Res. *60*:657–667, 2002.

46. J. Chluba, J.-C. Voegel, G. Decher, P. Erbacher, P. Schaaf, and J. Ogier. Peptide hormone covalently bound to polyelectrolytes and embedded into multilayer architectures conserves full biological activity. Biomacromolecules *2*:800–805, 2001.

47. S. Weiner and W. Traub, Bone structure: from angstroms to microns. Faseb J. *6*:879–885, 1992.

48. P. Calvert and S. Mann, The negative side of crystal growth. Nature *386*:127–128, 1997.

49. S. Zhang and K. E. Gonsalves, Influence of the chitosan surface profile on the nucleation and growth of calcium carbonate films. Langmuir *14*:6761–6766, 1991.

50. A. Bigi, E. Boanini, S. Panzavolta, and N. Roveri, Biomimetic growth of hydroxyapatite on gelatin films doped with sodium polyacrylate. Biomacromolecules *1*:752–756, 2000.

51. P. A. Ngankam, Ph. Lavalle, J. C. Voegel, L. Szyk, G. Decher, P. Schaaf, and F. J. G. Cuisinier. Influence of polyelectrolyte multilayer films on calcium phosphate nucleation. J. Am. Chem. Soc. *122*:8998–9005, 2000.

52. P. A. Ngankam, P. Schaaf, J. C. Voegel, and F. J. G. Cuisinier, Heterogeneous nucleation of calcium phosphate salts at a solid/liquid interface examined by scanning angle reflectometry. J. Crystal Growth *197*:927–938, 1999.

14

Self-Assembly of Functional Protein Multilayers: From Planar Films to Microtemplate Encapsulation

KATSUHIKO ARIGA ERATO Nanospace Project, Japan Science and Technology Corporation, Tokyo, Japan

YURI M. LVOV Institute for Micromanufacturing, Louisiana Tech University, Ruston, Louisiana, U.S.A.

I. INTRODUCTION: ALTERNATE LAYER-BY-LAYER ADSORPTION AS A NOVEL METHODOLOGY FOR PROTEIN ORGANIZATION

Currently, molecular devices and nanodevices are being energetically researched as one of the most important key technologies in the 21st century [1–4]. However, nature perfected much more sophisticated devices more than three billion years ago [5,6]. Ultrafine information processors, energy converters, and machines are all constructed by functional molecules through self-assembling processes. Nature relegates the important role to proteins that can operate with extremely high specificity and efficiency. Highly sophisticated functions such as energy conversion and electron transport result from specified spatial organization of enzymes. Multiple numbers and kinds of proteins often work sequentially and/or cooperatively, for example, a receptor for signal recognition and an effector for signal amplification effectively coupled on a plasma membrane [7–9]. Protein functional relay also plays important roles in photosynthesis [10,11]. Learning and mimicking biological protein arrays would lead to fruitful approaches in nanotechnology.

Protein immobilization has been investigated using various support materials. In particular, lipid membranes such as lipid bilayer [12–14] and Langmuir–Blodgett (LB) films [15–18] are useful thin-layer media for protein immobilization because they are straightforward mimics of plasma membranes. However, the lipid membranes sometimes form a well-packed structure which causes undesirable difficulties such as limited substrate diffusion [19]. In order to allow small molecules to easily access the immobilized proteins, a less ordered structure of polymer thin films would be advantageous. As a good example, polyion films might be excellent media for immobilizing proteins because proteins are effectively fixed through strong electrostatic interaction.

An alternate layer-by-layer adsorption method recently has attracted much attention as a novel preparation method for molecular films of polyions [20,21]. The concept of the alternate layer-by-layer adsorption was first proposed for charged colloidal particles in 1966 by Iler [22]. Decher and coworkers developed this concept and demonstrated various assemblies, mainly using linear polyions or bolaamphiphiles [23–29]. The alternate layer-by-layer adsorption is basically conducted through electrostatic interaction, although several modified methods such as stereocomplex assembly [30], charge-transfer assembly [31], and biospecific assembly [32,33] have been recently proposed. The basic concept of this method is illustrated in Fig. 1a. This technique is driven by two fundamental phenomena, charge neutralization and resaturation, upon adsorption of charged materials on oppositely charged surfaces. These processes result in an alternative change in the surface charge

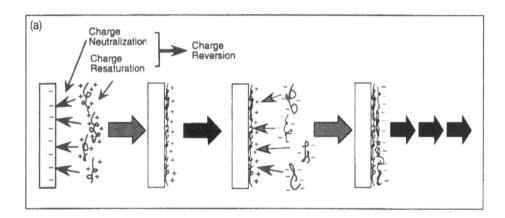

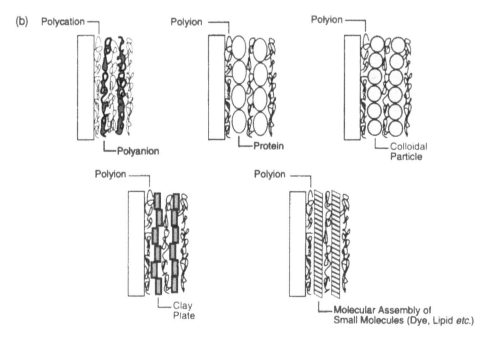

FIG. 1 (a) Basic mechanism of the alternate layer-by-layer adsorption. Effective charge reversion based on charge neutralization and charge resaturation leads to continuous assembly of a polycation and a polyanion. (b) Various examples of the alternately assembled films.

and lead to a continuous assembly between positively and negatively charged materials. Kunitake et al. proved the charge neutralization and resaturation during the assembly process by a surface force measurement [34].

Recent progress with this technique revealed that various materials such as polyions [35–40], inorganic materials [41–54], and aggregated small molecules [55–61] can be assembled (Fig. 1b). The alternate layer-by-layer adsorption is also available for various biorelated materials. We demonstrated the general applicability of this method to a wide range of proteins [62,63]. Successful assemblies of other biopolymers such as DNA and charged polysaccharides were also reported by us and others [64–76]. We also prepared nano-sized reactors composed of multiple kinds of enzymes that could be the first attempt at a protein-related nanodevice obtained by alternate layer-by-layer adsorption [77–80]. In this chapter, we summarize alternate assemblies between polyions and proteins and the preparation of multiprotein reactors. Recent progress in protein assembly on spherical microtemplates and enzyme encapsulation is also described.

II. PRACTICAL PROCEDURE AND ADVANTAGES

Figure 2 illustrates practical procedures of the alternate layer-by-layer adsorption and examples of polyions generally used in this technique. As one can recognize in this illustration, the simplicity of the assembling procedure is one of the most prominent advantages of this method. Instruments required in this procedure are only beakers and tweezers. The assembling process is quite simple: A solid support with a charged surface (negatively charged in this case) is first immersed in a solution containing oppositely charged polyions (polycations in this case). Because the concentrations of the polyions are relatively high, the excess polyions are adsorbed, resulting in effective reversal of the surface charge. The solid support is rinsed in pure water, and then the support is immersed in a solution of an oppositely charged protein (anionic protein in this case). The surface charge is again reversed. The adsorption amount is spontaneously regulated by the charge repulsion in the excess adsorbed layer. Repeating both the adsorption processes leads to alternation of the surface charge and multilayer formation of polyions and proteins in alternate sequence. This simple procedure can be conducted by automated machines. Clark and Hammond [81] and Shiratori and Yamada [82] reported an automatic dipping machine for the alternate layer-by-layer adsorption.

Application of the alternate layer-by-layer adsorption to protein immobilization provides various advantages in protein organization. Typical advantageous points are

1. Wide applicability of proteins. Because this method is driven by quite general electrostatic interaction, many kinds of proteins can be assembled. Most proteins, especially water-soluble proteins, have charged sites on their surface, and thus they are basically assembled by the electrostatic layer-by-layer adsorption.

2. Rich variety of layered structure. The film thickness (the number of layers) can be controlled simply by the number of dipping cycles. A film composed of different protein layers can be obtained if we use a suitable combination of polyions and proteins. Layering sequence is also easily controlled.

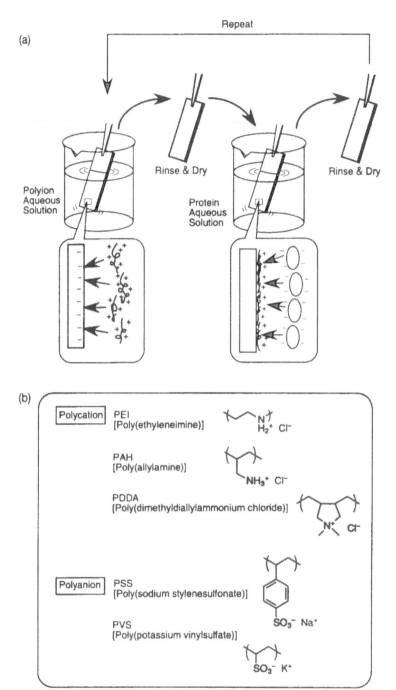

FIG. 2 (a) Practical procedure for the alternate layer-by-layer adsorption is illustrated. Assembly between a cationic polyion and an anionic protein on a negatively charged solid support is shown as an example. (b) Examples of polyions generally used in this technique.

Therefore, we can prepare protein–polyion films with desirable thickness and layering sequences.

3. Minimization of protein denaturation. The assembling procedure does not require chemical reactions and severe physical conditions. The proteins are simply adsorbed from their aqueous solution. Mildness of the adsorption conditions avoids unnecessary disturbance of the protein structure upon immobilization.

III. EVALUATION OF ASSEMBLING PROCESS

Evaluation methods of the assembling process and the prepared films are summarized in Table 1 [83–87]. The adsorption amount can be detected by ultraviolet and visible spectroscopy (UV-vis) and Fourier transform infrared (FTIR) spectroscopies. The increase in the adsorption bands specific for proteins was measured as the assembling process proceeded. A linear relationship between the band strength and the number of the protein layers was confirmed. The thickness of the adsorbed layers can be evaluated by x-ray reflectivity, surface plasmon resonance (SPR), and scanning angle reflectometry (SAR). Using a quartz crystal microbalance (QCM) is a unique technique to evaluate the assembling process. It provides the weight increase upon film preparation. Because all kinds of adsorbents have weight, this technique is theoretically applicable to all of the assembled materials.

Analytical methods that are applicable to other thin films can be generally used for evaluation of the alternately assembled films. Cross-sectional observation of the protein–polyion film by scanning electron microscopy (SEM) revealed the constant thickness of the film [62]. Surface observation of the corresponding films by atomic force microscopy (AFM) showed that the surface roughness of the films is comparable only to the protein dimensions [86]. These observations indicate that a quite uniform texture can be obtained even in thicker films.

As described above, the QCM technique is one of the most powerful and convenient methods for evaluation of the assembling process. Examples of QCM analyses of the alternate layer-by-layer adsorption are described here. The adsorption of proteins and polyions can be quantitatively evaluated using the QCM technique, because the resonant frequency of the QCM sensitively changes due to the mass adsorption on its electrodes [88,89]. The frequency shift $(-\Delta F)$ is given below for the 9 MHz AT-cut QCM:

$$-\Delta F = \frac{18.3M}{A} \qquad (1)$$

where M and A respectively represent the mass increase (ng) and the area (m^2) of the electrode on the QCM plate. Because the error in the frequency measurement is within a few hertz, we can evaluate the weight of every layer with precision at a nanogram level. On-line measurement of the frequency during the adsorption process is possible, and the kinetics of the adsorption steps can also be analyzed [62].

In the examples shown in Figure 3, alternate assemblies of GOD-PEI [glucose oxidase–(poly(ethyleneimine))] and PEI-GA (glucoamylase) films were investigated in two kinds of assembling sequences. The QCM plate was used as a solid support for the film assembly, and its frequency shift was monitored in air after the adsorption

TABLE 1 Analytical Methods for Alternate Layer-by-Layer Adsorption

Method[a]	Information and parameters	Example[b]	Ref.
UV-vis	Amount of adsorbent	Linear increase of soret band in Mb-PSS assembly	62
FTIR	Amount of adsorbent	Linear increase of amide bands in anti-IgG-PSS assembly	71
QCM	Adsorption weight	Linear increase of adsorbent weight in GOD-PEI assembly	62
X-Ray Reflectivity	Film thickness	Carnation mottle virus adsorption on PAH surface	83
SPR	Film thickness	IgG adsorption on PSS surface	70
SAR	Film thickness	αLA adsorption on PAH surface	84
SEM	Film morphology	Cross-section of GOD-PEI film	62
TEM	Film morphology	Image of urease-embedded PSS-PAH microcapsule	85
AFM	Surface morphology	Surface image of GOD-PEI film	96
XPS	Film composition	Elemental analysis of GA, GOD-PEI film	80
ζ-Potential	Surface charge	Alternative changes of surface charge in BSA-PDDA assembly on latex particle	87
CV	Electrochemical activity	Redox behavior of Mb in Film assembled with PSS	10
SPLS	Film thickness on particle	Linear increase of film thickness in BSA-PDDA assembly on latex particle	87

[a]UV-vis, UV/vis spectroscopy; FTIR, Fourier transform infrared spectroscopy; QCM, quartz crystal microbalance; SPR, surface plasmon resonance; SAR, scanning angle reflectometry; SEM, scanning electron microscopy; TEM, transmission electron microscopy; AFM, atomic force microscopy; XPS, x-ray photoelectron spectroscopy; CV, cyclic voltammetry; SPLS, single particle light scattering.

[b]Mb, Myoglobin. IgG: Immunoglobulin G; αLA, α-lactalbumin; GOD, glucose oxidase; BSA, bovine serum albumin; GA, glucoamylase.

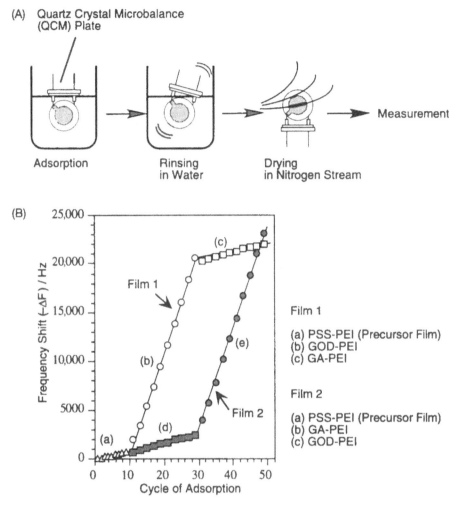

FIG. 3 (A) Evaluation process of adsorption amount by a QCM method. (B) Frequency shifts during the adsorption process of GOD and GA with polycation PEI—*film 1*: (a) precursor PEI-PSS layers, (b) GOD-PEI layers, and (c) GA-PEI layers; *film 2*: (a) precursor PEI-PSS layers, (d) GA-PEI layers, and (e) GOD-PEI layers. [PEI], 1.5 mg mL^{-1}; [PSS], 3 mg mL^{-1}; [enzyme], 1-2 mg mL^{-1}.

and rinsing process (see Fig. 3A). On a bare QCM electrode, the precursor film (four layers of PEI-PSS [poly(sodium styrenesulfonate) film] was assembled (Fig. 3B, process a). The precursor layer is indispensable because it provides a sufficient amount of surface charges and facilitates further protein assembly. In film 1, GOD-PEI layers were assembled on the precursor film (process b), and constant film growth was observed. The average frequency change for adsorption of a single GOD-PEI layer was ca. 2000 Hz. Simple calculation using the molecular mass of the protein revealed that GOD probably formed aggregates and adsorbed in three to four layers in a single adsorption under this assembly condition. This assembly process was repeated and stopped at the PEI adsorption step of the 10th cycle. A GA-PEI assembly was next conducted and constant film growth was repeatedly observed with an average frequency change of ca. 145 Hz per cycle (process c). The latter value corresponds to a single monolayer adsorption of GA.

Assembly with reversed protein sequence was also examined as film 2. Initially, the GA-PEI film was assembled on the precursor film (process d), and then the GOD-PEI film was assembled (process e). The observed frequency changes for each assembly were essentially identical to the corresponding process for film 1. Therefore, the assembly processes of both the protein–polyion layers proceed independently and do not interfere with each other. These results encourage us to prepare variously assembled films without interference from previously adsorbed layers.

Heterogeneous assembly between protein and materials other than conventional polyions is also possible. However, direct assembly between relatively rigid components is difficult [62] except for some instances [90]. For example, alternate assembly between inorganic materials and proteins was usually unsuccessful. The aid of an intermediate adsorption of a flexible polyion sometimes plays the role of a glue to adhere the rigid components together. In the example shown in Fig. 4, GOD and montmorillonite microplates were assembled in one film with a sandwiched layer of PEI [62].

Because the QCM analysis was so simple and convenient, the assembling behavior of various proteins was demonstrated. Alternate assemblies of 18 proteins with oppositely charged polyions are summarized in Table 2 where the analyzed

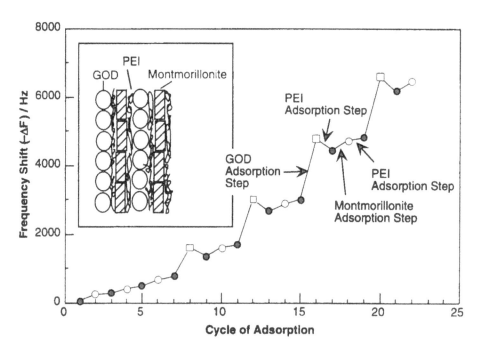

FIG. 4 Quartz crystal microbalance frequency changes in the three-component alternate layer-by-layer adsorption between GOD, PEI, and montmorillonite at 22°C: [montmorillonite], 0.3 mg mL^{-1}; [PEI], 1.5 mg mL^{-1}; [GOD], 2 mg mL^{-1}; adsorption time of 20 min. Open squares, solid circles, and open circles represent GOD, PEI, and montmorillonite adsorption processes, respectively.

TABLE 2 Alternate Assembly Between Protein and Polyion

Protein	Molecular weight	Isoelectric point	pH used	Charge	Alternate with	Protein monolayer mass coverage (mg m^{-2})	Thickness of protein-polyion (nm)	Protein globule dimensions (nm)
Cytochrome c	12400	10.1	4.5	+	PSS$^-$	3.6	2.4 + 1.6	2.5 × 2.5 × 3.7
Lysozyme	14000	11	4	+	PSS$^-$	3.5	2.3 + 1.9	3.0 × 3.0 × 4.5
Histone f3	15300	11	7	+	PSS$^-$	3.3	2.2 + 2.0	Diameter 3.4
Myoglobin	17800	7.0	4.5	+	DNA, PSS$^-$	6	4.0 + 2.0	2.5 × 3.5 × 4.5
Bacteriorhodopsin	26000	6	9.4	−	PDDA$^+$	7.5	5.0 + 1.0	Diameter 5.0
Carbonic anhydrase	29000	5.5	8.3	−	PEI$^+$	2.8	Total 2.2	1.5 × 2.2 × 2.5
Pepsin	35000	1	6	−	PDDA$^+$	4.5	3.0 + 0.6	Diameter 3.0
Peroxidase (POD)	42000	8.0	4.2	+	PSS$^-$	5.3	Total 3.5	Diameter 3.5
Hemoglobin	64000	6.8	4.5	+	PSS$^-$	26	17.5 + 3.0	5.0 × 5.5 × 6.5
			9.2	−	PEI$^+$	27	Total 18.2	
Albumin	68000	4.9	8	−	PDDA$^+$	23	16.0 + 1.0	11.6 × 2.7 × 2.7
			3.9	+	heparin	30	20.0 + 1.0	
Glucoamylase (GA)	95000	4.2	6.8	−	PDDA, PEI$^+$	4	2.6 + 0.5	Diameter 6.3
Photosynthetic RC	100000	5.5	8	−	PDDA$^+$	13	9.0 + 1.0	13 × 7.0 × 4.0
Concanavalin	104000	5	7	−	PEI$^+$	8.6	5.7 + 0.8	3.9 × 4.0 × 4.2
Alcohol dehydrogenase	141000	5.4	8.5	−	PDDA$^+$	12.2	8.5 + 1.0	9.0 × 4.0 × 4.0
IgG	150000	6.8	7.5	−	PSS$^-$	15	Total 10.0	14 × 10 × 5
Glucose oxidase (GOD)	186000	4.1	6.8	−	PDDA$^+$	12	Total 8.0	Diameter 8
			6.5	−	PEI$^+$	51	34.4 + 0.8	
Catalase	240000	5.5	9.2	−	PEI$^+$	9.6	6.4 + 0.8	Diameter 9.0
Diaphorase	600000	5	8	−	PEI$^+$	31	Total 21.0	Diameter 11.5

Source: Refs. 62–74.

thicknesses of the adsorbed layer are listed and compared with protein dimensions [62–74]. In all of the combinations, the assembling process proceeds for an unlimited number of cycles. The mass increase evaluated from the QCM frequency was highly reproducible. It is noteworthy that the estimated thickness of the protein layer is approximately comparable to the protein dimensions except for some examples (GOD-PEI, etc.). These data suggest the formation of a relatively uniform monolayer for a wide range of proteins, although the orientation, packing density, and hydration of adsorbed proteins are not considered in these estimations.

Control of solution pH is important in these assemblies. In order to achieve successful assembly, the pH of the protein solution should be set apart from the isoelectric point of the protein. Interestingly, the same protein can be assembled in both positively charged form and negatively charged form. Hemoglobin is positively charged at pH 4.5 and was assembled with anionic PSS. Negatively charged hemoglobin at pH 9.2 was assembled alternately with cationic PDDA [poly-(diallyldimethylammonium chloride)]. Similarly albumin was assembled with anionic heparin and cationic PDDA at pH 3.9 and 8, respectively.

IV. PLANAR FILM–TYPE ENZYME NANOREACTOR

Prior to constructing a multienzyme reactor, the activity of the enzyme assembled in a thin film was investigated, because the stability of the enzyme is crucial for reactor design. The experimental setup and reaction scheme for the simplest reactor are illustrated in Fig. 5. The PEI-GOD films were assembled on a precursor film on a quartz plate, and the activity of the reactor was examined by the coupled enzymatic reaction of GOD and peroxidase (POD). In this reaction, GOD in the film converts D-glucose and O_2 to D-glucono-δ-lactone and hydrogen peroxide (H_2O_2), respectively, and POD in solution oxidizes DA67 (indicator) using H_2O_2 as the oxidant. The reaction can be easily followed by monitoring the strong absorbance of the oxidized DA67 at 665 nm.

Because the long-term stability of the reactor is an important requirement for practical usage, the storage stability was first examined (Fig. 6) [78]. Experiments were carried out under three different conditions: (1) stored in water at 25°C; (2) stored in 0.1 M PIPES buffer (pH 7) at 4°C; and (3) allowed to stand in air at 4°C. The enzymatic activity drastically decreased in water at 25°C, and approximately 70% of the activity was lost after 4 weeks. Bacterial growth may be the cause of the deterioration of the film. In contrast, the film retained in the buffer at 4°C kept its activity over 14 weeks. The films kept in air at 4°C showed a 10% decrease in the activity in the first week, probably due to drying of the film. However, the activity was preserved during the following 13 weeks. These results indicate that activity of the enzyme in the alternately assembled films can be maintained under suitable storage conditions.

We also examined the thermostability of the enzyme immobilized in alternate assembled films (Fig. 7) [78]. A one-layer GOD film on a quartz plate was incubated in water at a given temperature for 10 min. The enzymatic activity was measured at 25°C immediately after the incubation period and after storing the film in air at room temperature for 30 min. Activity was similarly measured for GOD dissolved in water. The dissolved GOD lost its activity even at 30–40°C, and it became almost inactive

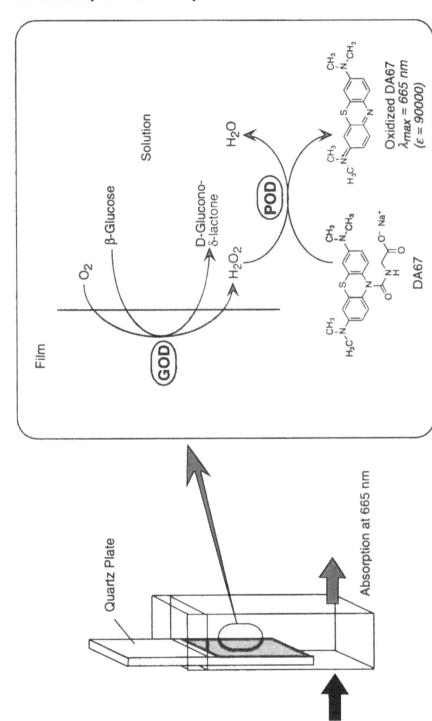

FIG. 5 Evaluation method for enzymatic activity of GOD assembled in a film on a quartz plate. GOD-catalyzed reaction in the film is coupled with POD-catalyzed reaction in solution. GOD activity can be indirectly evaluated from the absorbance increase upon formation of the DA67 oxidized form.

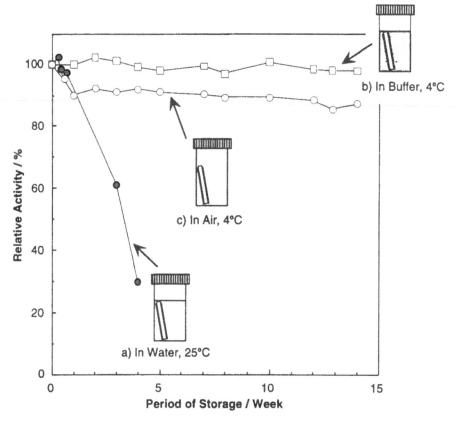

FIG. 6 Storage stability of GOD activity in GOD-PEI films: (a, ●) storage in pure water at 25°C. (b, □) storage in PIPES buffer (pH 7) at 4°C. (c, ○) storage in air at 4°C.

at 50°C (curve a). The observed loss in activity was partially recovered by returning the solution to room temperature (curve b), probably due to a reversible denaturation mechanism. Curve c demonstrates a remarkable improvement in thermostability of GOD in the film. A significant decrease in the activity was not detected even after incubation at 50°C. Recovery of the reduced activity was not observed in the film sample at any incubation temperature (curve d). Suppression of the conformational mobility in the film would cause improvement in activity and absence of the recovery.

With the knowledge of the advantages in film construction and improvement in enzymatic activity, we prepared nanoreactors composed of multiple kinds of enzymes [77,80]. Figure 8 illustrates the structure of the multienzyme reactors containing GOD and GA prepared on an ultrafilter and by the enzymatic sequential reaction. Starch is the substrate and is hydrolyzed into glucose by GA. The produced glucose is converted to gluconolactone by GOD. A coproduct (H_2O_2) can be detected by the POD-catalyzed reaction in solution. An aqueous solution of water-soluble starch in 0.1 M PIPES buffer (pH 7.0) was placed on the enzyme-immobilized ultrafilters, and filtration was started by applying pressure to the upper cups with a syringe.

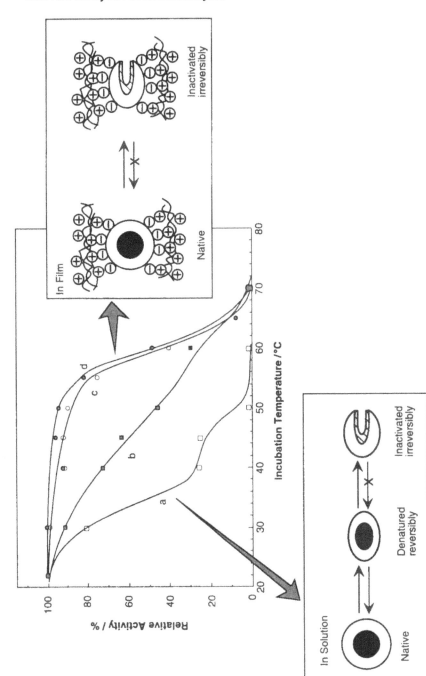

FIG. 7 Thermostability of GOD in solution and in the film: (a) the relative activity of aqueous GOD immediately after incubation. (b) The relative activity of aqueous GOD after incubation and storage in air for 30 min at room temperature. (c) The relative activity of GOD in the film immediately after incubation. (d) The relative activity of GOD in the film after incubation and storage in air for 30 min at room temperature. The activity relative to that at 22°C is plotted as a function of incubation temperature. Plausible mechanism of thermal denaturation is also illustrated.

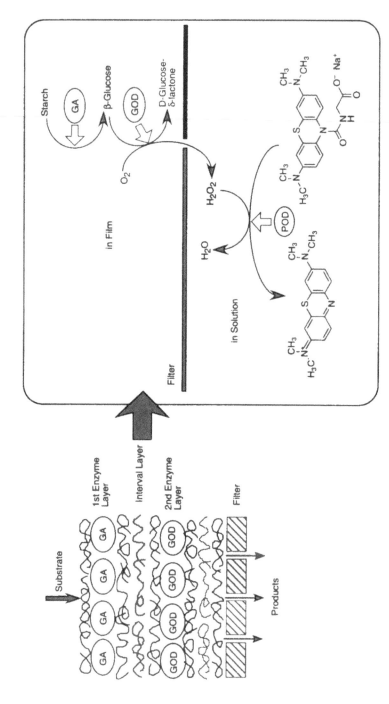

FIG. 8 Structure of a multienzyme reactor composed of GA and GOD is illustrated on the left. Sequential reaction catalyzed by the reactor is shown on the right. In order to evaluate the reaction efficiency, DA67 oxidation by POD in the filtrate is coupled with the sequential reaction in the film.

In order to investigate the effect of film organization on the reaction efficiency, the following enzyme films were prepared,

Film I: ultrafilter + (PEI/PSS)$_4$ + (PEI/GOD)$_2$ + (PEI/PSS)$_{10}$ + (PEI/GA)$_2$
            + PEI
Film II: ultrafilter + (PEI/PSS)$_4$ + (PEI/GOD)$_2$ + (PEI/PSS)$_2$ + (PEI/GA)$_2$
            + PEI
Film III: ultrafilter + (PEI/PSS)$_4$ + (PEI/GA)$_2$ + (PEI/PSS)$_2$ + (PEI/GOD)$_2$
            + PEI
Film IV: ultrafilter + (PEI/PSS)$_4$ + (PEI/MIX)$_2$ + PEI

where MIX represents adsorbed layers from an aqueous solution of an equimolar mixture of GOD and GA. The reaction efficiencies of these reactors are summarized in Fig. 9a.

The highest yield of the products (H$_2$O$_2$) was obtained with film I, and film II showed the second highest yield. They are apparently superior to film III. The substrate of the coupled catalysis (starch) has difficulty in diffusing through the film because of its high molecular weight. Therefore, the film construction seriously affects the reaction efficiency. Film III has a protein layering sequence opposite to the reaction sequence. This unfavorable structure resulted in low activity. The importance of the separation between the two enzyme layers is suggested from the difference in activity observed between film I and film II. This cannot be explained by the ease of substrate diffusion. Inhibition of GA activity by gluconolactone was reported [91,92], and larger separation between the GOD and GA layers would reduce this inhibition. Film IV showed the lowest yield. Coexistence of GA and GOD might result in the lowest activity due to the inhibition of GA by gluconolactone.

Characteristics of the enzyme nanoreactors prepared by the alternate layer-by-layer adsorption are schematically summarized in Fig. 9b. Reactor ability can be optimized by adjusting the number of layers, the layering sequence, and layer separation. A solid support for the film preparation can be freely selected if some extent of charges can be induced onto the support surface. The latter characteristic is advantageous for combining the enzyme film and fabricated device tips. A flexible nature of the film construction by the alternate layer-by-layer adsorption would open the way to developing novel types of protein-based nanodevices.

V. PROTEIN ASSEMBLY ON MICROTEMPLATE

The assembly process elaborated for planar solid supports can be adapted for nano- and microtemplates (colloid particles with sizes of 0.1 to 5 μm, e.g., latex spheres, lipid microtubules, microcrystals, biological cells, and other colloids) [93–99]. In this process, a polycation solution is added to the suspension of colloid particles, and after adsorption saturation, the particles are separated from free polycations in solution. A polyanion layer is then deposited. In the same manner, one can deposit any number of polyion or nanoparticle layers on the shell (Fig. 10).

Urease multilayers were assembled onto submicrometer-sized polystyrene spheres by the sequential adsorption of urease and polyion, in a predetermined order, utilizing electrostatic interactions for layer growth. The catalytic activity of the bio-colloids increased proportionally with the number of urease layers deposited on the

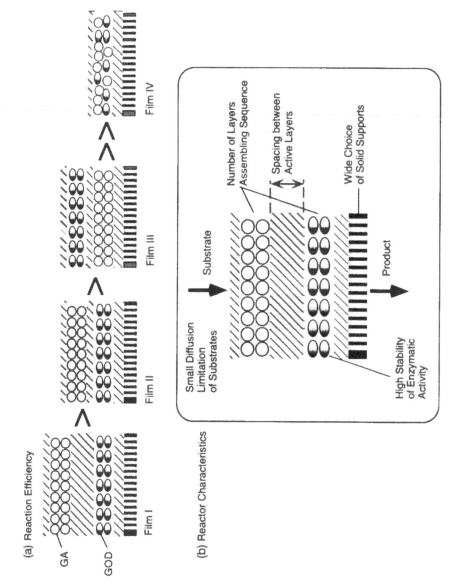

FIG. 9 (a) Reaction efficiency of films with various structures. (b) Characteristics of the multienzyme reactor produced by the alternate layer-by-layer adsorption.

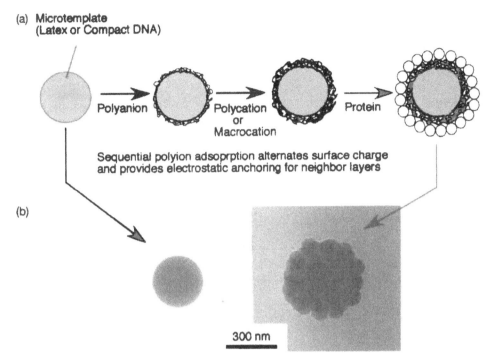

(a) Microtemplate
(Latex or Compact DNA)

Polyanion

Polycation
or
Macrocation

Protein

Sequential polyion adsoprption alternates surface charge
and provides electrostatic anchoring for neighbor layers

(b)

300 nm

FIG. 10 (a) Scheme of polyion/nanoparticle shell on latex. (b) Corresponding TEM images of a 300-nm diameter core with a 75-nm silica shell.

particles, demonstrating that biocolloid particles with tailored enzymatic activities can be produced. It was further found that precoating the latex spheres with nano-particles (40-nm silica or 12-nm magnetite) enhanced both the stability (with respect to adsorption) and enzymatic activity of the urease multilayers. The presence of the magnetite nanoparticle coating also provided a magnetic function that allowed the biocolloids to be easily and rapidly separated with a permanent magnet [100]. The fabrication of such colloids opens new avenues for the application of bioparticles and represents a promising route for the creation of complex catalytic particles.

The growth of the urease multilayers on the polystyrene (PS) particles was followed by microelectrophoresis. Figure 11a gives the zeta (ζ)-potential of the latex particles coated with polyions, nanoparticles, and urease [PDDA/PSS/PDDA/40-nm silica or 12-nm magnetite/PDDA/(urease/PDDA)$_{1-4}$]. The ζ-potential alternates be-tween negative and positive values, corresponding to the sequential adsorption of cationic and anionic species, respectively. This indicates a reversal in surface charge of the PS particles. At step four, after silica or magnetite nanoparticle adsorption, a slightly less negative ζ-potential was observed compared to the PSS-outermost par-ticles, but the subsequent adsorption of PDDA restored the higher positive ζ-potential of ca. +45 mV. At the next (sixth) step, urease was adsorbed and the ζ-potential decreased to −8 mV. This value is close to what would be expected because the measurements were performed at ca. pH 5.6 and the isoelectric point of urease is 5. The seventh PDDA adsorption step again recharged the particle surface to ca. +40 mV. Repeatable and alternating ζ-potentials were observed with further deposition

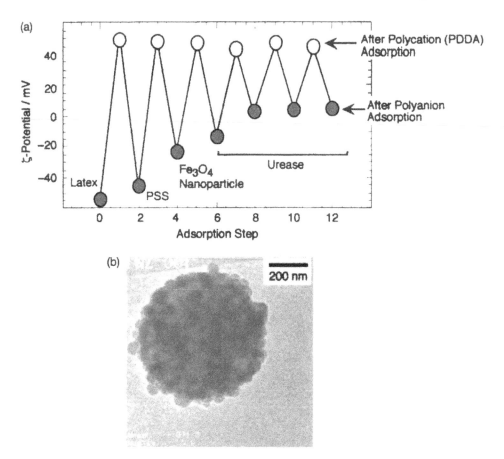

FIG. 11 (a) Monitoring of the particle surface potential during the polyion/magnetite/urease shell assembly. (b) TEM image of this biocolloid.

of urease and PDDA. Overall, Fig. 11 qualitatively indicates the alternate adsorption of the deposited species with an associated change in the surface charge. This is a key aspect of the alternate layer-by-layer adsorption technique.

Figure 11b shows a transmission electron microscope (TEM) image of a PS sphere coated with [PDDA/PSS/PDDA/40-nm silica/PDDA/(urease/PDDA)₃]. A rather uniform distribution of silica nanoparticles is obtained on the surface of the PS spheres. The diameter of the coated particles is approximately 575 nm, which corresponds to the initial PS particle diameter of 470 nm with approximately two sets of the polyion/silica nanoparticle/urease shells (50 nm × 2). Polystyrene particles prepared in the same way with 12-nm magnetic particles showed a less uniform coating of nanoparticles.

Nevertheless, the absolute enzymatic activity of the urease in the magnetite-containing particles was similar to that of the corresponding particles with a layer of 40-nm silica and ca. 30% of urease. A 3 kG permanent magnet approaching the solution resulted in the collection of all of the modified latex spheres (on a wall region closest to the magnet) in about 30 s. Immersing the permanent magnet into the solution containing the magnetite/urease-coated particles induced collection of the particles on the magnet. As demonstrated in these examples, the additional magnetic function is particularly useful in applications where separation and reuse of the particles is required. In addition, multifunctional biocolloid nanoreactors prepared by the alternate layer-by-layer adsorption have the benefits of high surface area and enzymatic stability. These facts make them attractive for use in various applications.

VI. ENZYME-CONTAINING POLYION CAPSULE

A novel approach to encapsulate various materials invented in the last couple of years is also based on the layer-by-layer adsorption of oppositely charged macromolecules onto colloidal particles [85,93–95,99–103]. Different templates with sizes ranging from 50 nm to tens of millimeters, such as organic and inorganic colloidal particles, protein aggregates, and drug nanocrystals can be coated with nano-organized multilayer films. The colloidal templates can then be decomposed under conditions where the polymer shell is stable, which leads to the formation of hollow capsules with a defined size, shape, and shell thickness (Fig. 12). The method has the capacity to employ a great variety of substances as shell components, as well as core materials [85,99–102]. Typically, the diameter of the shells used was between 0.5 and 5 μm, and the skin (four to eight polyion bilayer) thickness was 20–40 nm. Figure 13 shows an AFM image of a dried collapsed shell; folds in the shell are quite visible in the image.

The permeability of the capsule wall and release of the encapsulated materials depend on the wall thickness and composition, and they may be regulated by pH and the ionic strength of the solution. As shown in Ref. 102, polyion multilayer shells with eight layers of the PSS/PAH pair assembled around a low-molecular-weight dye core provide barrier properties for release under conditions where the core is dissolved (assembly at pH 4 and release at pH 7). This is a promising approach for fabrication of systems with prolonged and controlled release properties. Usually, a polyion shell is nonpenetrable by macromolecules with molecular weights above 4000. Therefore, most enzymes encapsulated in the polyion shells could be preserved, and the shell protects these enzymes from microbes or other high-molecular-weight inhibitors.

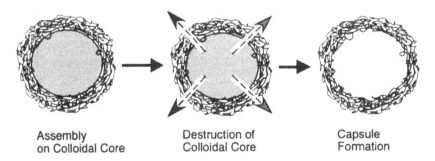

Assembly   Destruction of   Capsule
on Colloidal Core   Colloidal Core   Formation

FIG. 12 Decomposition of the colloidal core results in suspension of polyion capsules.

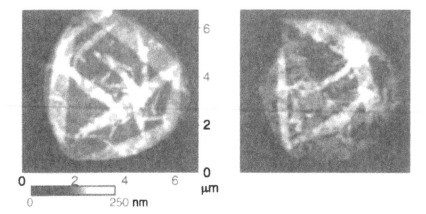

FIG. 13 Atomic force microscopy image of two 5-μm-diameter polyion shells with 20-nm wall thickness (dried on mica).

Three recent approaches on macromolecule encapsulation into the polyion shells are

1. The pH-controlled encapsulation and release of macromolecules from polyion capsules of a few millimeters diameter. In Ref. 101, capsules were prepared via alternate adsorption of oppositely charged polymers, PAH and PSS, onto decomposable melamine formaldehyde cores. The capsule pores were opened for macromolecule penetration at below pH 6 and closed at above pH 8. An operation with opening/closing capsule walls composed of the polyion multilayers is based on Rubner's recent finding that varying the solution pH can induce a charge balance between polycations and polyanions in the multilayer, resulting in opening of 50-nm pores [104]. The pore opening/closing depends on the pH at which the primary assembly was performed and on the ionization properties of the assembled polyions. The detailed structural changes, control of the pore sizes, and the threshold are still not understood and deserve in-depth studies.

2. Pore opening/closing based on varying the solution dielectric constant is demonstrated. In the system drawn in Fig. 14, the pores in 5-μm diameter capsules with 20-nm thick walls of a (PSS/PAH)$_4$ composition were controlled to open and close by changing the surrounding solvents (open in water/ethanol mixture and closed in water), resulting in encapsulation of urease [85]. The enzymes were captured in the capsule based on the molecular weight–selective shell permeability, while small substrates and products of enzymatic reactions can freely penetrate the capsule wall. An encapsulation of α-chymotrypsin into PSS/PAH microcapsules was also demonstrated [101]. There is a possibility of entrapping macromolecules not only inside the polyion capsules but in the capsule walls. The situation with the wall attachment of enzymes and drug molecules is unfavorable and has to be avoided.

3. Use of an active material itself (e.g., a drug or an enzyme) as a microtemplate was also demonstrated. The approach was tested using encapsulation

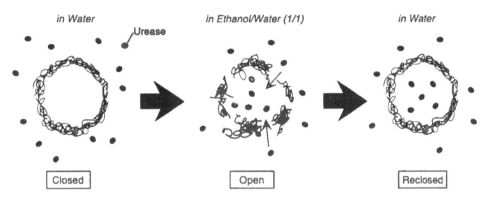

FIG. 14 Scheme of urease loading in a 5-μm-diameter (PSS/PAH)$_4$ shell.

of protein nanocrystals [96]. This approach looks most promising for drug/ enzyme encapsulation. It is possible to assemble a polyion shell on drug microcores with the wall composition controlled at nanometer precision. On the outermost layer of such a capsule, a layer of antibodies targeted to a certain antigen can be assembled to provide targeted delivery of the encapsulated drug.

VII. FUTURE PERSPECTIVE

Construction of a nano-sized or a similarly miniaturized system is a big research target in current and future technology. Approaches based on self-assembling processes have the potential to achieve molecularly precise fabrication. Because proteins are well-designed molecules with excellent functions such as catalysis and recognition, their artificial organization is one of the most interesting targets in this field. However, random assembly of the proteins would be less fruitful in development of useful nanodevices. Organization with regularity or controllability will provide direction-specified signaling and conversion systems.

Preparation of planar multienzyme films prepared by the alternate layer-by-layer adsorption opened a way to construction of a protein functional relay in one dimension (film z-direction). An ultrathin multienzyme reactor might be one of the most elegant demonstrations of functional expression of the alternately assembled structure. However, these films have macroscopic dimension in the film x-y plane, and no structural regulation in the two-dimensional plane is done in this approach. Therefore, it can still be regarded as a primitive example from the viewpoint of nanofabrication.

Assembly on a particle template is more advanced nanofabrication than conventional film assemblies, because the dimension of the assembly is controlled in all directions. The particles and capsules with assembled films can be regarded as a highly functionalized dot (zero-dimensional system) from a microscopic view. The dot has the function of controlled drug release and enzymatic molecular conversion as described above. Relaying these functions between several dots would lead to development of highly functionalized ultrasmall devices. Therefore, controlled positioning of these dots is quite an attractive research target. In current technology, arranging, assembling, and patterning of small particles are already available techniques. Combining our capsules with particle-organizing techniques would provide a hierarchical structure of protein functions. Preparation of a hierarchical organization is also important for extending molecular-level functions to macroscopically available signals in a sophisticated way. Assembling structures will be a key technique in the next generation of the alternate layer-by-layer adsorption methodology.

REFERENCES

1. Special Issue on Nanotech. Sci. Am. *September*, 2001.
2. H. S. Nalwa, ed. *Handbook of Nanostructured Materials and Nanotechnology*, Vols. 1–5. Academic Press, San Diego, 1999.
3. R. Dagami, Chem. Eng. News 2000 Feb. 28:36–38, 2000.
4. http://www.nano.gov/
5. D. Goodsell, American Scientist *88*:230–237, 2000.
6. D. G. Goodsell, *Our Molecular Nature: The Body's Motors, Machines and Messages.* Springer-Verlag, New York, 1996.
7. M. Rodbell, Biosci. Rep. *15*:117–133, 1995.
8. A. G. Gilman, Biosci. Rep. *15*:67–97, 1995.
9. C. S. Zuker and R. Ranganathan, Science *283*:650–651, 1999.
10. B. Albert, D. Bray, J. Lewis, M. Raff, K. Roberts, and J. D. Watson, *Molecular Biology of the Cell*, 2nd ed. Garland Publishing, New York, 1989, pp. 377–379.
11. D. Gust, T. A. Moore, and A. L. Moore, Acc. Chem. Res. *34*:40–48, 2001.
12. E. Sackmann, Science *271*:43–48, 1996.
13. T. H. Watts, H. E. Gaub, and H. M. McConnell, Nature *320*:179–181, 1986.
14. J. Kikuchi, K. Ariga, and K. Ikeda, Chem. Commun. 547–548, 1999.
15. P. Fromherz, In *Electron Microscopy at Molecular Dimension* (W. Baumeister, W. Vogell, eds.). Springer-Verlag, Berlin, 1980, p. 338.
16. C. A. Nicolae, S. Cantin-Riviere, A. El Abed, P. Peretti, Langmuir *13*:5507–5510, 1997.
17. L. Marron-Brignone, R. M. Morelis, J.-P. Chauvet, P. R. Coulet, Langmuir *16*:498–503, 2000.
18. N. Damrongchai, E. Kobatake, T. Haruyama, Y. Ikariyama, and M. Aizawa, Bioconjugate Chem. *6*:264–268, 1995.
19. Y. Okahata, T. Tsuruta, K. Ijiro, and K. Ariga. Thin Solid Films *180*:65–72, 1989.
20. G. Decher, Science *277*:1232–1237, 1997.
21. W. Knoll, Curr. Opin. Colloids Interfaces 1:137–143, 1996.
22. R. K. Iler, J. Colloid Interface Sci. *21*:569–594, 1996.
23. M. L. Sche, J. Schmitt, G. Decher, W. G. Bouwman, and K. Kjaer. Macromolecules *31*:8893–8906, 1998.
24. D. Korneev, Y. Lvov, G. Decher, J. Scmitt, and S. Yaradaikin, Physica B *213/214*:954–956 (1995).
25. G. Decher, J. D. Hong, and J. Schmitt, Thin Solid Films *210/211*:831–835, 1992.
26. G. Decher, Y. Lvov, and J. Schmitt, Thin Solid Films *244*:772–777, 1994.
27. Y. Lvov, G. Decher, and H. Möhwald, Langmuir *9*:481–486, 1993.
28. G. Decher and J.-D. Hong, Ber. Bunsenges Phys. Chem. *95*:1430–1434, 1991.
29. G. Decher and J.-D. Hong, Makromol. Chem. Macromol. Symp. *46*:321–327, 1991.
30. T. Serizawa, K. Hamada, T. Kitayama, N. Fujimoto, K. Hatada, and M. Akashi, J. Am. Chem. Soc. *122*:1891–1899, 2000.
31. Y. Shimazaki, R. Nakamura, S. Ito, and M. Yamamoto, Langmuir *17*:953–986, 2001.

32. Y. Lvov, K. Ariga, I. Ichinose, and T. Kunitake, J. Chem. Soc. Chem. Commun. 2313–2314, 1995.
33. S. V. Rao, K. W. Anderson, and L. G. Bachas. Biotech. Bioeng. 65:389–396, 1999.
34. P. Bernt, K. Kurihara, and T. Kunitake, Langmuir 8:2486–2490, 1992.
35. Y. Lvov, K. Ariga, M. Onda, I. Ichinose, and T. Kunitake, Colloid Surf. A 146:337–346, 1999.
36. I. Ichinose, S. Mizuki, S. Ohno, H. Shiraishi, and T. Kunitake, Polym. J. 31:1065–1070, 1999.
37. Y. Lvov, S. Yamada, and T. Kunitake, Thin Solid Films 300:107–112, 1997.
38. A. C. Fou, O. Onitsuka, M. Ferreira, M. F. Rubner, and B. R. Hsieh. J. Appl. Phys. 79:7501–7509, 1996.
39. L. Krasemann and B. Tieke, Langmuir 16:287–290, 2000.
40. M. Onoda and K. Yoshino, Jpn. J. Appl. Phys. 34:L260–L263, 1995.
41. H. Lee, L. J. Kepley, H.-G. Hong, S. Akhter, and T. E. Mallouk, J. Phys. Chem. 92:2597–2601, 1988.
42. K. Ariga, Y. Lvov, I. Ichinose, and T. Kunitake, Appl. Clay Sci. 15:137–152, 1999.
43. I. Ichinose, H. Tagawa, S. Mizuki, Y. Lvov, and T. Kunitake, Langmuir 14:187–192, 1998.
44. Y. Lvov, K. Ariga, M. Onda, I. Ichinose, and T. Kunitake, Langmuir 23:6195–6203, 1997.
45. K. Ariga, Y. Lvov, M. Onda, I. Ichinose, and T. Kunitake, Chem. Lett. 125–126, 1997.
46. Y. Lvov, K. Ariga, I. Ichinose, and T. Kunitake, Langmuir 12:3038–3044, 1996.
47. T. Yonezawa, H. Matsune, and T. Kunitake, Chem. Mater. 11:33–35, 1999.
48. Y. M. Lvov, J. F. Rusling, D. L. Thomsen, F. Papadimitrakopoulos, T. Kawakami, and T. Kunitake, Chem. Commun. 1229–1230, 1998.
49. F. Caruso, R. A. Caruso, and H. Möhwald, Science 282:1111–1114, 1998.
50. J. W. Ostrander, A. A. Mamedov, and N. Kotov, J. Am. Chem. Soc. 123:1101–1110, 2001.
51. J. H. Fendler, Chem. Mater. 8:1616–1624, 1996.
52. J.-A. He, R. Valluzzi, K. Yang, T. Dolukhanyan, C. Sung, J. Kumar, S. K. Tripathy, L. Samuelson, L. Balogh, and D. A. Tomalia, Chem. Mater. 11:3268–3279, 1999.
53. M. Gao, X. Zhang, B. Yang, F. Li, and J. Shen, Thin Solid Films 284/285:242–245, 1996.
54. G. S. Lee, Y.-J. Lee, and K. B. Yoon, J. Am. Chem. Soc. 123:9769–779, 2001.
55. A. Toutianoush, F. Saremi, and B. Tieke, Mater. Sci. Eng. C8-9:343–352, 1999.
56. I. Ichinose, K. Fujiyoshi, S. Mizuki, Y. Lvov, and T. Kunitake, Chem. Lett. 257–258, 1996.
57. Y. Lvov, F. Essler, and G. Decher, J. Phys. Chem. 97:13773–13777, 1993.
58. K. Ariga, Y. Lvov, and T. Kunitake, J. Am. Chem. Soc. 119:2224–2231, 1997.
59. S. Yamada, A. Harada, T. Matsuo, S. Ohno, I. Ichinose, and T. Kunitake, Jpn. J. Appl. Phys. 36:L1110–L1112, 1997.
60. K. Araki, M. J. Wagner, and M. S. Wrigton, Langmuir 12:5393–5398, 1996.
61. T. Cooper, A. Campbell, and R. Crane, Langmuir 11:2713–2718, 1995.
62. Y. Lvov, K. Ariga, I. Ichinose, and T. Kunitake, J. Am. Chem. Soc. 117:6117–6123, 1995.
63. Y. Lvov, In *Protein Architecture: Interfacing Molecular Assemblies and Immobilization Biotechnology* (Y. Lvov, and H. Möhwald, eds.). Marcel Dekker, New York, 2000, pp. 125–167.
64. F. Caruso, In *Protein Architecture: Interfacing Molecular Assemblies and Immobilization Biotechnology* (Y. Lvov and H. Möhwald, eds.). Marcel Dekker, New York, 2000, pp. 193–227.

65. X. Zwang, J. Shen, In *Protein Architecture: Interfacing Molecular Assemblies and Immobilization Biotechnology* (Y. Lvov and H. Möhwald, eds.). Marcel Dekker, New York, 2000, pp. 229–250.
66. E. Brynda, M. Houska, In *Protein Architecture: Interfacing Molecular Assemblies and Immobilization Biotechnology* (Y. Lvov, H. Möhwald, eds.). Marcel Dekker, New York, 2000, pp. 251–285.
67. J. Kong, W. Sun, X. Wu, J. Deng, Y. Lvov, R. Desamero, H. Frank, and J. Rusling, Bioelectrochem. Bioenerg. *48*:101–107, 1999.
68. S. Keller, H. Kim, and T. Mallouk, J. Am. Chem. Soc. *116*:8817–8818, 1994.
69. J. Hodak, R. Etchenique, E. Calvo, K. Singhal, and P. Bartlett, Langmuir *13*:2708–2716, 1997.
70. F. Caruso, K. Niikura, N. Furlong, and Y. Okahata, Langmuir *13*:3427–3433, 1997.
71. F. Caruso, N. Furlong, K. Ariga, I. Ichinose, and T. Kunitake, Langmuir *14*:4559–4565, 1998.
72. Y. M. Lvov, Z. Lu, X. Zu, J. Schenkman, and J. Rusling, J. Am. Chem. Soc. *120*: 4073–4080, 1998.
73. J.-A. He, L. Samuelson, L. Li, J. Kumar, and S. Tripathy, Langmuir *14*:1674–1679, 1998.
74. J. Kong, Z. Lu, Y. Lvov, R. Desamero, H. Frank, and J. Rusling, J. Am. Chem. Soc. *120*:7371–7372, 1998.
75. Y. Lvov, M. Onda, K. Ariga, and T. Kunitake, J. Biomater. Sci. Polym. Ed. *9*:345–355, 1998.
76. K. Ariga, M. Onda, Y. Lvov, and T. Kunitake, Chem. Lett. 25–26, 1997.
77. K. Ariga and T. Kunitake. In *Protein Architecture: Interfacing Molecular Assemblies and Immobilization Biotechnology* (Y. Lvov and H. Möhwald, eds.). Marcel Dekker, New York, 2000, pp. 169–191.
78. M. Onda, K. Ariga, and T. Kunitake, J. Biosci. Bioeng. *87*:69–75, 1999.
79. M. Onda, Y. Lvov, K. Ariga, and T. Kunitake, Biotech. Bioeng. *51*:163–167, 1996.
80. M. Onda, Y. Lvov, K. Ariga, and T. Kunitake, J. Ferment. Bioeng. *82*:502–506, 1996.
81. S. L. Clark and P. T. Hammond, Adv. Mater. *10*:1515–1519, 1998.
82. S. S. Shiratori and M. Yamada, Polym. Adv. Technol. *11*:810–814, 2000.
83. Y. Lvov, H. Haas, G. Decher, H. Möhwald, A. Mikhailov, B. Mtchedlishvily, E. Morgunova, and B. Vainstein, Langmuir *10*:4232–4236, 1994.
84. G. Ladam, P. Schaaf, F. J. G. Cuisinier, G. Decher, and J.-C. Voegel, Langmuir *17*: 878–882, 2001.
85. Y. Lvov, A. A. Antipov, A. Mamendov, H. Möhwald, and G. B. Sukhoruv. Nano Letters *1*:125–128, 2001.
86. M. Onda, Y. Lvov, K. Ariga, and T. Kunitake, Jpn. J. Appl. Phys. *36*:L1608–L1611, 1997.
87. F. Caruso and H. Möhwald, J. Am. Chem. Soc. *121*:6039–6046, 1999.
88. G. Sauerbrey, Z. Physik. *155*:206–222, 1959.
89. K. Ariga, K. Endo, Y. Aoyama, and Y. Okahata, Colloids Surf. A *169*:177–186, 2000.
90. Y. Lvov, B. Munge, O. Giraldo, I. Ichinose, S. L. Suib, and J. F. Rusling, Langmuir *16*:8850–8857, 2000.
91. M. Dixon and E.-C. Webb, *Enzymes*, 2nd ed. Academic Press, New York, 1964, p. 315.
92. M. Ohnish, T. Yamashita, and K. Hiromi, J. Biochem. *79*:1007–1012, 1976.
93. E. Donath, G. Sukhorukov, F. Caruso, S. Davis, and H. Möhwald, Angew Chem. Int. Ed. *37*:2202–2205, 1998.
94. F. Caruso, R. Caruso, and H. Möhwald, Science *282*:1111–1114, 1998.
95. G. Sukhorukov, M. Brumen, E. Donath, and H. Möhwald, J. Phys. Chem. B, *103*: 6434–6440, 1999.

96. F. Caruso, D. Trau, H. Möhwald, and R. Renneberg. Langmuir *16*:1485–1488, 2000.
97. Y. Lvov, R. Price, A. Singh, J. Selinger, and J. Schnur, Langmuir *16*:5932–5935, 2000.
98. Y. Lvov and R. Price, Colloids Surfaces Biointerfaces *165*:674–679, 2001.
99. S. Leporatti, A. Voigt, G. Sukhorukov, E. Donath, and H. Möhwald, Langmuir *16*: 4059–4063, 2000.
100. Y. Lvov and F. Caruso, Anal. Chem. *73*:4212–4217, 2001.
101. O. Tiourina, G. Sukhorukov, Y. Lvov, and H. Möhwald, Macromol. Biosci. *1*:209–214, 2001.
102. A. Antipov, G. Sukhorukov, E. Donath, and H. Möhwald, J. Phys. Chem. B *105*:2281–2284, 2001.
103. G. Sukhorukov, A. Antipov, A. Voigt, D. Donath, and H. Möhwald, Macromol. Rapid Commun. 22:44–46, 2001.
104. J. Mendelson, C. Barrett, V. Chan, A. Pal, A. Mayes, and M. Rubner, Langmuir *16*: 5017–5023, 2000.

15

Biological Activity of Adsorbed Proteins

THOMAS A. HORBETT   University of Washington,
Seattle, Washington, U.S.A.

I. INTRODUCTION

This chapter focuses on the biological activity of adsorbed proteins as related to the bodily responses to biomedical devices used in clinical medicine. In that context, the currently accepted approach is that further understanding of how adsorbed proteins affect cellular interactions with foreign materials will eventually result in design criteria for improved biomaterials that are more accepted in the body. Thus the chapter provides mechanistic insights into how the body responds to synthetic biomaterials as derived from an understanding of the effects of adsorbed proteins on cell interactions. Physicochemical aspects of protein adsorption are reviewed elsewhere [1–6].

To address the biological activity of adsorbed proteins, a brief description of the fundamental biochemical mechanisms of cellular interactions with foreign materials is provided by describing adhesion proteins and cell adhesion receptors. Then the major concepts used to understand the biological activity of adsorbed adhesion proteins are presented, namely, the presence or absence of the adhesion protein on the surface as determined by its affinity for the surface and variations in the molecular potency of the adhesion protein. The term *molecular potency*, introduced by Steele et al. [7], is used to indicate that the fraction of adsorbed adhesion protein molecules that express cell adhesive activity is typically less than 1, and depends on the adsorbing surface and adsorption conditions, as described in more detail later. When presenting the concepts of affinity and molecular potency, examples of each concept are also briefly described. Following presentation of these concepts, a detailed review of more recent studies illustrating each concept is given. The earlier literature in this area has been discussed in more detail in previous reviews by the author [8–10].

II. BIOCHEMICAL BASIS OF THE BIOLOGICAL ACTIVITY OF ADSORBED PROTEINS

The biochemical basis of the biological activity of adsorbed proteins in regard to biomedical implants is that most of the cells in our body have adhesion receptors.

As a result, the adsorption of their cognate ligands, the adhesion proteins, to the surface of an implant often results in the adhesion and activation of cells on the implant. These adhesive and activation events are thought to affect the way the implant is or is not accepted into the body. In this regard, the three major types of reaction of the body to implants, namely, blood clotting, the foreign body reaction, and healing of orthopedic implants into the bone, all are thought to be strongly affected by adsorbed adhesion proteins. Thus understanding of the adhesion receptors and their ligands is essential to understanding how adsorbed proteins affect the success of implants. However, this area of biochemistry has grown in recent years to a rather large industry in its own right, so here only a brief general description and a few very recent advances are given, along with references to several recent reviews on the adhesion receptors and their proteins [11,12].

The integrins are the major cell adhesion receptors involved in adhesion of cells to extracellular matrices and to the adsorbed proteins present on the surfaces of biomaterials. The other cell adhesion receptors (cadherins, selectins, and IgG superfamily) are primarily involved in cell–cell adhesive events. The calcium-dependent α, β-heterodimeric integrins come in 24 different varieties, and their ligands include the four major adhesive proteins in the plasma, namely, fibrinogen, fibronectin, vitronectin, and von Willebrand's factor, as well as other proteins such as osteopontin. Osteopontin has not usually been considered an important protein in biomaterials research, but it does interact with integrins and may be involved in bone bonding to metal implants [13] and in the foreign body reaction [14]. The role of the plasma proteins in mediating cell adhesion to biomaterials was originally suggested by the fact that their preadsorption as purified proteins to a surface mediates cell attachment, whereas adsorption of all other proteins prevents cell attachment. In recent years, more direct evidence of a physiologic role for the these proteins was obtained by selective depletion and repletion studies of the complex mixtures of proteins that implants actually encounter. In the typical study of this type, removal of one of these proteins from plasma or serum used to preadsorb a surface caused cell attachment to greatly decrease, and readdition of the depleted protein restores adhesion (reviewed by the author in Ref. 15). Thus, for example, my lab has used afibrinogenemic plasma or serum for selective removal of fibrinogen, and gelatin or immunoadsorbent columns for selective removal of fibronectin or vitronectin, to show that only fibrinogen removal from plasma affects platelet adhesion to a fairly wide variety of surfaces [16–18]. The immunoadsorbent studies in my lab were done in collaboration with McFarland and Steele, who developed the immunoadsorbent technology for fibronectin and vitronectin. Steele and his collaborators originally used the immunoadsorbent columns to show that vitronectin rather than fibronectin is the key protein in mediating adhesion of several kinds of cells to some surfaces [3]. Tang et al. have shown a key role for fibrinogen in the acute phase of the foreign body reaction in vivo using the enzyme ancrod to selectively deplete fibrinogen in mice [19].

While the integrins and their adhesive partners were originally described as playing primary roles in adhesive events in both the developing and adult animals, it is now well known that both "outside-in" and "inside-out" cell signaling processes are mediated by the integrins and their ligands, i.e., a wide variety of physiological changes in the adherent cell can be induced. Among the most important of the inside-out processes are the control of an integrin's ability to bind its ligand, i.e., some integrins like the GP IIb/IIIa in platelets are ordinarily unable to bind their ligands unless the cell has been stimulated with an agonist that turns the integrin on via still mysterious intracellular events. However, GP IIb/IIIa is able to bind to adsorbed fibrinogen (but not to soluble fibrinogen) without prior agonist activation, i.e., un-activated platelets bind to fibrinogen adsorbed to surfaces. Recent progress on the integrins includes the first crystal structure, that of $\alpha_v\beta_3$ [20]. This structure reveals many interesting aspects of how these receptors may work. Thus, the structure observed includes a severely bent conformation (the "genu" site, where the molecule resembles the bending of a leg during genuflection). The bent conformation is seen in only a minority of rotary shadowed molecules, suggesting the existence of a highly flexible site in the integrin that would allow the molecule to assume drastically different conformations that may be involved in integrin activation. In addition, the occurrence of novel and unpredicted structural domains established that this integrin possesses a mosaic of domains that could function in controlling integrin clustering and their interactions with other membrane-bound receptors.

III. MAJOR MECHANISMS AFFECTING BIOLOGICAL ACTIVITY OF ADSORBED PROTEINS

The major mechanisms affecting biological activity of adsorbed adhesion proteins are the affinity of the adhesion protein for a surface, affecting how much of the proteins is present on a particular surface, and the degree to which the adsorbed adhesion protein expresses its biological activity, referred to here as its molecular potency. Affinity is affected by molecular properties of the protein as well as the surface it is adsorbing to. Affinity of a protein affects adsorption under competitive conditions, i.e., it determines how successfully the adhesion protein occupies the surface in the face of competition for the limited binding sites from the typically large excesses of other proteins in the surrounding medium, as discussed in Section III.A. Variations in molecular potency of the absorbed adhesion proteins are probably caused by a number of factors, including the orientation and conformation of the adhesion protein. The molecular potency appears to be affected by the chemistry of the adsorbing surface, the presence of coadsorbed albumin, and how long the protein has been on the surface ("residence time"), as discussed further in Section III.B.

A. Affinity of the Adhesion Proteins

What do we know about the affinity of the adhesion proteins for surfaces? Unfor-tunately, the answer to this question is not very complete nor is it easily quantitated. Thus careful studies of this basic question have been relatively few in number. Sec-ond, the affinity of adhesion proteins for surfaces cannot be directly measured with the equilibrium constant typically used for other molecular binding process because protein adsorption is essentially an irreversible process on many surfaces. However, characterization of the relative affinity of the adhesion proteins under competitive conditions in mixtures with other proteins provides some relevant insight into this question. *Relative affinity* is a term introduced by the author some years ago to denote how well a given protein competes for a surface against another protein. In binary mixtures of two proteins, equal weight concentrations of an adhesion protein and a competing protein would be expected to result in equal surface concentrations of the two proteins unless one has a higher affinity relative to the other, and so binary mixture studies in which these are varied is one way to estimate competitive or relative affinity. When protein adsorption from more complex mixtures such as blood plasma or serum are measured, it is still possible to estimate the affinity of the protein by comparing its surface excess (or deficit) in comparison to what is expected if it adsorbed in proportion to its weight fraction in the bulk phase. Both types of studies have been done with adhesion proteins, showing that the affinity of the major adhesion proteins often is in the following order: fibrinogen > vitronectin > fibronectin >> albumin. Thus, Steele et al. [3,7,21–24] and others [25,26] have shown that on many (but not all) surfaces, vitronectin adsorption is in excess over fibronectin. Consequently, vitronectin plays a much more important role than fibronectin in mediating adhesion to many surfaces under serum adsorption conditions typical in many cell culture studies. Studies from the author's lab [27,28] and others [26] with binary mixtures of fibrinogen have shown it typically strongly outcompetes other proteins including albumin and IgG, while fibronectin is only mildly more surface active than these two major plasma proteins, and vitronectin appears to have a surface affinity in between those of fibrinogen and fibronectin.

B. Molecular Potency

Molecular potency is a term coined by J. Steele and used here to mean that the ability of an adhesion protein to mediate adhesive events varies a great deal with how it is adsorbed and is not simply related to the number of adsorbed adhesion molecules per unit area. Thus, for example, fibronectin adsorbed to polystyrene supports cell adhesion, but only at higher adsorbed amounts than when it is adsorbed to the wettable version of polystyrene that has been surface modified for tissue culture [7,29–31]. In addition, coadsorbed albumin enhances the potency of the adsorbed fibronectin [29–32]. Similar variations in the molecular potency of adsorbed fibrinogen have been reported on different surfaces [33], when albumin is coadsorbed [34], or as a function of how long the protein has been on the surface (residence time) [35,36]. These variations in potency are thought to be related to variations in molecular spreading (degree of unfolding) of the adsorbed adhesion proteins, but may also (or instead) be related to variations in the orientation or accessibility of the cell binding domains on the molecules or to how tightly the protein is held to the surface [8,37]. The author has reviewed the relevant literature on molecular potency in greater detail in a prior review [10], and recent examples are discussed in Section V.

IV. AFFINITY VARIATIONS IN ADSORPTION OF ADHESION PROTEINS: NONFOULING MATERIALS

A. Types of Nonfouling Materials

The most important recent findings about variations in the affinity of the adhesion proteins have been made by the search for so-called nonfouling materials that would reduce the affinity for most proteins, including the adhesion proteins, to very low levels. Considerable research on materials which would resist protein adsorption and so be nonfouling has been done in recent years [39–42]. Polysaccharides, lipidlike surfaces including surfaces with phosphorylcholine [43]; self-assembled monolayers (SAMs) on gold substrates made with ethylene glycol–terminated thiol compounds [44–46], hydroxyl-terminated silanes, or addition of polyethylene glycol to the silanated surfaces [47]; and polyethylene glycol are some of the most prominent nonfouling coatings attempted. The resulting coatings sometimes show excellent resistance to protein uptake ("zero" protein uptake on some ethylene glycol–terminated self-assembled monolayers on gold [44]), but in other cases still show considerable uptake (e.g., 100 ng/cm^2 fibrinogen uptake from plasma to BBA-mPEG coatings on SR in studies by Defife et al. [48] and ca. 160 protein spots detected in SDS washes of PEGylated polycyanoacrylate particles by two-dimensional gel electrophoresis compared to ca. 250 spots in the untreated particles [49]). Lipidlike surfaces also exhibit variable protein repellency. For example, polyurethanes with incorporated phosphorylcholine reduce fibrinogen uptake from buffered fibrinogen solutions to about 30 ng/cm^2 [50], while methacryloyloxethyl phosphorylcholine polymers grafted to cellulose hemodialysis membranes reduced total protein adsorption from blood plasma from 1300 to about 300 ng/cm^2 [51]. In the following sections, only the poly(ethylene oxide) (PEO)-like materials will be considered further.

B.   Poly(Ethylene Oxide)–Like Nonfouling Materials

Protein adsorption to PEO-like materials has been studied extensively in recent years. Many attempts to understand the mechanisms of its low protein adsorption have been made, and several reviews have been written [38–42]. A brief summary of the author's view of the state of the art is given here. Four topics will be covered:

1.   The mechanisms of nonfouling of PEO
2.   The amounts of fibrinogen adsorbed to PEO materials
3.   Blood interactions with PEO materials
4.   The possible role of PEO in enhancing the biological activity of adhesion proteins adsorbed to PEO surfaces

1.   Mechanisms of PEO Protein Repellency

Steric exclusion and coverage of the surface appear to be the prime factors necessary to prevent protein uptake by PEO-coated surfaces. Steric exclusion arises from the unfavorable entropy that occurs when longer PEO chains become more crowded as protein molecules attempt to fill spaces between adjacent chains. Many early studies supported this idea because longer-chain PEOs often gave lower adsorptivity than lower-molecular-weight versions. However, more recent studies done with tightly packed, self-assembled monolayers have shown that even very short chains terminated in ethylene glycol groups are capable of greatly reducing protein uptake. The SAM studies thus suggest that when the surface coverage is very complete and the layer is consequently very dense, the surface cannot be "seen" by approaching protein molecules. Malmsten and Muller proposed that both steric and shielding effects contribute to protein repellency: "Once a sufficiently thick adsorbed layer is obtained, attractive interactions between a protein and the underlying surface is essentially fully screened. At this stage, the protein adsorption is typically quite low and only weakly dependent on the PEO molecular weight" [52].

Some workers have proposed that PEO mixtures which are polydisperse with respect to molecular weight will have the greatest efficacy in steric stabilization of colloidal particles [53] and in protein repellency [54]. Thus mixtures of long- and short-chain PEO/PPO copolymers adsorbed to gold substrate surface plasmon resonance slides reduced albumin adsorption more than when copolymers of one molecular weight were used [54]. Plasma deposited tetraglyme surfaces probably contain mixtures of PEO chains of varying length, and this may be one of the reasons these materials have shown very good protein repellency.

Recent theoretical studies of the mechanisms of protein adsorption suggests which molecular properties of the PEO groups are important to protein repellency [55,56]. In this model, PEO polymers attracted to the surface compete with protein for surface sites and are effective in preventing protein adsorption at equilibrium. In contrast, if the polymer chains are not attracted to the surface, they are expected to be effective in a kinetic sense because the extended chains prevent the close approach of the protein molecules to the substrate. However, at long times the proteins can still sneak through random openings in the extended chains to find a surface site and so these are not good in preventing equilibrium adsorption. This theory also implies the potential utility of surfaces with mixtures of PEO-containing polymeric chains. Morra has provided an overview of theories of fouling resistance in which he stresses the shortcomings of the steric repulsion models [38]. Morra reviews several recent studies showing the likely importance of specific chemical (hydrogen bonding) and conformational (helix versus random coil) features of PEO that the "physical" or steric exclusion theories do not consider. Thus, there is recent evidence that the specific conformation of the PEO chain strongly influences its fibrinogen binding and that some conformers of PEO are actually highly adsorptive to fibrinogen [58].

Implicit in many of the theories of PEO protein repellency is the role of "structured water" created by the strong binding of the water to PEO. Thus, the adsorption of proteins to the PEO chain would require the displacement of a great deal of tightly bound water, effectively causing dehydration of the PEO chains, an event that would be energetically unfavorable as the destruction of all the structured water would create positive entropy.

2. Fibrinogen Adsorption and Nonfouling Surfaces

Fibrinogen's relatively high affinity for surfaces and high concentration in blood plasma cause it to adsorb in substantial quantities on most surfaces. Furthermore, we [59] and others [60–62] have observed that even small amounts of adsorbed fibrinogen, much below a monolayer or the levels that normally adsorb, are still quite sufficient to support platelet [59] and monocyte [63] adhesion. Thus, while efforts to prevent cell attachment by using so-called nonfouling surfaces that reduce protein uptake appear to be a fundamentally sound approach to improvement of biocompatibility, it appears it will require the development of extremely good nonfouling technology, i.e., ultralow protein adsorptive surfaces. In reviewing success to date with this approach, it appears that the technology used in many labs does not come very close to this goal. How well do the various types of PEO-based protein resistant materials achieve the design criteria of extremely low fibrinogen uptake? A review of the literature shows that many of the PEO materials have not been evaluated in regard to fibrinogen adsorption, but rather have been evaluated against other proteins.

For example, Irvine et al. used the adsorption of albumin and cytochrome c to demonstrate the protein repellency of their linear and star PEO grafts [64] and many others have similarly used albumin [46,54] or other proteins such as fibronectin [65,66] to characterize repellency. While it is likely that these materials would also exhibit fibrinogen repellency had they been tested, the lack of direct data makes it extremely difficult to assess whether they would achieve the ultralow fibrinogen adsorptivity criteria. Thus, the state of the art is that only a few PEO-like materials have been shown to meet the ultralow fibrinogen uptake criteria, as is now shown.

Studies from K. Park's lab have shown their PEO materials sometimes meet this criteria. Park's group reported extensive adsorption data from 0.1 mg/mL fibrinogen solutions in buffer to a series of radiation-grafted PEO surfaces [67]. Triblock copolymers (Pluronic™) of the general formula (ethylene oxide)$_n$–(propylene oxide)$_m$–(ethylene oxide)$_n$ with segments of various lengths were used. On most of the PEO-treated glass surfaces, regardless of the PEO chain length, fibrinogen adsorption was below 0.02 μg/cm^2, compared to 0.47 μg/cm^2 on the control glass surface. Grafting of Nitinol wire with Pluronic PF127 reduced adsorption to 0.06 μg/cm^2 compared to 0.5 μg/cm^2 on Nitinol. Grafting of PF127 to pyrolytic carbon (PC) reduced the adsorption to 0.57 μg/cm^2 compared to 0.86 μg/cm^2 on the PC itself. Increasing concentrations of surfactant used during grafting also decreased fibrinogen adsorption, but the lowest adsorption achieved in the concentration studies were approximately 0.05 μg/cm^2 on glass surfaces treated with 15 mg/mL PP1053.

Park's group later reported fibrinogen adsorption to expanded polytetrafluoroethylene (ePTFE), Silastic, and silanized glass after grafting with a PEO–polybutadiene (PB)–PEO triblock copolymer (COP5000, containing a 5000 MW PEO and 750 MW PB block) that they synthesized [68]. Fibrinogen adsorption to silanized glass grafted with COP5000 was very low (ca. 0.01 μg/cm^2) but was still substantial on Silastic (ca. 0.07 μg/cm^2) and ePTFE (ca. 0.25 μg/cm^2) that had been grafted with COP5000.

Malmsten and Muller reported that fibrinogen adsorption to PEO/polylactide copolymers adsorbed to methylated silica was reduced to nearly zero, as detected ellipsometrically, although the protein concentration in the adsorbing solution was not specified [52]. Liu et al. reported bovine plasma fibrinogen adsorption from 4 mg/mL solutions varied from 5.8 to 0.4 μg/cm^2 (using amide I adsorption obtained from Fourier transform infrared spectroscopy), decreasing as the amount of PEO grafted onto polyurethanes increased [69]. In the latter study, many of the adsorption values are far above a theoretical monolayer and thus very far above the ultralow fibrinogen uptake criteria. Thus the absolute values are somewhat suspect, although the tenfold reduction in fibrinogen uptake is consistent with other studies of PEO surfaces and is probably correct. Kim et al. also reported bovine fibrinogen uptake from 30 mg/mL solutions in buffer onto PEO-grafted polyurethane/polystyrene interpenetrating networks, calculating adsorption from the depletion of the bulk phase by the film samples [70]. Adsorption to the PEO surfaces was reported to be about sevenfold less than the control, but the absolute values reported (ca. 2000 μg/cm^2) make little sense, as does the use of a depletion method, which is not capable of detecting changes in bulk concentration unless particulate samples are used.

Plasma deposited glyme-based materials exhibit ultralow fibrinogen adsorption and platelet adhesion [71]. In studies in our group of fibrinogen adsorption to plasma deposited tetraglyme samples, glow discharge deposition of tetraglyme monomer onto glass, PTFE, and polyethylene was found to cause resistance to fibrinogen uptake and platelet attachment [71]. The resistance was very good, although it did depend on the substrate that was coated. Thus fibrinogen adsorption from 0.1% plasma to tetraglyme-coated PTFE was about 2 ng/cm^2, but was about 15 ng/cm^2 on tetraglyme-coated glass surfaces. In recent studies in our lab, the earlier work has been verified: tetraglyme-treated fluorinated ethylene-propylene (FEP) showed a 95 to 100% reduction of fibrinogen adsorption, depending on the particular lot of samples used [63]. In these studies, the fibrinogen concentration in the original plasma was quite high (4.39 mg/mL), and the adsorption was done from 1% plasma because the Vroman effect causes maximal adsorption if more diluted plasma is used. Adsorption to FEP controls averaged 116 ng/cm^2, which is quite typical for fibrinogen adsorption to hydrophobic polymers. In contrast, fibrinogen adsorption to the five different lots of tetraglyme-treated samples was quite low, varying from 0 to 6.4 ng/cm^2. In these studies, the five lots were made under identical conditions, but there was still some variation in fibrinogen adsorption. In preparing coatings that will exhibit ultralow fibrinogen adsorption, we conclude that complete coverage of the substrate with the PEO-like polymerized tetraglyme is important. Second, production of chains of at least three or four ethylene glycol repeat units in a row before being cross-linked or having other nonether carbons is needed for fibrinogen repellency. In addition, retention of a high degree of ether carbon content by minimizing cross-linking is also a critical factor unique to plasma polymerized surfaces.

3. Blood Interactions with PEO Surfaces

Although most PEO polymers have exhibited reduced interactions with blood, especially with regard to platelets in in vitro studies, the results have varied widely, and none of these types of materials have yet been shown to achieve high blood compatibility when studied in vivo for longer times. Here a few examples that support these summary statements are given. Detailed reviews of this topic are available [39–42].

Kinam Park's group reported that PEO-containing triblock copolymers grafted to glass, ePTFE, or Silastic did not exhibit greatly improved blood compatibility in an acute phase canine ex vivo series shunt model when compared to albumin preadsorbed Tygothane control segment, even though in vitro fibrinogen adsorption and in vitro platelet deposition were lowered [68]. Thus they found that platelet deposition to the PEO grafts was about 35% less than the control, whereas much larger reductions in platelet adhesion to these materials had been observed in their in vitro studies. These authors proposed that these differences were likely ascribable to the use of anticoagulation in vitro, as well as temperature and shear effects. Because of the well-known effect of anticoagulants on platelet reactivity, we have avoided their use in our past studies preferring instead to use platelet/red cell suspensions containing physiological calcium and magnesium levels at 37°C. Interestingly, Park's group did report that the platelet deposition in the in vivo study was approximately proportional to the residual fibrinogen adsorption. The latter finding supports the hypothesis that ultralow fibrinogen adsorption will improve blood compatibility. Sung Wan Kim's group also found that segmented polyurethanes with PEO reduced platelet adhesion in vitro but did not prolong occlusion times when used as arteriovenous shunts in rabbits [72]. In contrast, studies of platelet interactions with poly(ethylene oxide) networks by Merrill and his coworkers showed that platelet adhesion could be as low as 1 platelet per 1000 μm^2 after 1 h of blood contact in an ex vivo baboon shunt model, depending on the molecular weight of the PEO and the PEO content of the network [73]. Many other studies have shown reduced platelet adhesion and blood reactivity to PEO surfaces. For example, comblike PEO gradient surfaces made by graft copolymerization of poly(ethylene glycol) monomethacrylate macromers onto corona discharge–treated polyethylene showed decreasing platelet adhesion with increasing graft [74]; PEO attached to polypropylene oxide (PPO) by UV initiated cross-linking reduced platelet adhesion by about 50% compared to PPO [75]; PEO grafted onto Biomer vascular grafts had increased patency compared to Biomer control [76]; and PEO coating of various polymers by a mutual solvent polymeric entanglement method reduced adhesion by up to 20-fold [77]. Wagner's group used PEO in a novel way by coupling it to fibrinogen already adsorbed to a surface, which resulted in 94–96% reduction of platelet adhesion [78]. In our group, plasma deposited tetraglyme samples exhibiting ultralow fibrinogen uptake have also shown excellent resistance to platelet adhesion in vitro [71].

Some literature suggests that PEO polymers may be thrombogenic in vivo but not thromboadhesive, i.e., the clots they cause are rapidly embolized. In Sefton's lab, PEO addition to poly(vinyl alcohol) (PVA) grafts reduced protein adsorption, but were "incapable of reducing the platelet consumptive properties of PVA hydrogel . . ." [39,79]. These observations are consistent with older data for a series of acrylic hydrogels varying in water content, where Hanson et al. observed very little adherent clot but increasing rates of platelet consumption on higher-water-content hydrogels [80]. However, neither Sefton's materials nor the acrylic hydrogels were shown to have ultralow fibrinogen adsorption. The studies do suggest the necessity of studying thromboembolization and platelet consumption in addition to platelet adhesion or clot deposition.

4. Does PEO Enhance the Biological Activity of Adsorbed Adhesion Proteins?

The biological activity of adsorbed adhesion proteins can be enhanced by the coadsorption of other proteins and also by "substrate activation" (reviewed by the author in Refs. 9 and 10). It is thought that coadsorbed proteins fill in sites on the surface near the adhesion protein, preventing them from completely unfolding as more and more contacts with the surface are made. Substrate activation refers to the enhancement of the biological activity of the adhesion protein when it is adsorbed. Thus, for example, platelets do not bind to fluid phase fibrinogen unless the platelets have first been exposed to an agonist such as ADP or thrombin, while unstimulated platelets bind readily to adsorbed fibrinogen and other adsorbed proteins [81–83].

Two recent studies suggest activation phenomena may occur with PEO and adsorbed adhesion proteins. Thus fibronectin adsorbed to surfaces with moderate amounts of PEO on them cause greater adhesion of fibroblasts than occurs for fibronectin adsorbed to the same surface in the absence of any PEO at all [66]. In addition, cells interacting with fibronectin adsorbed to surfaces with intermediate degrees of PEO exhibited enhanced fibrillar fibronectin deposition by the cells [65]. Liu et al. reported large differences in the percent of denaturation of BSA adsorbed

to surfaces carrying various amounts of PEO and that thrombus deposition was minimal at an intermediate degree of PEO content [69]. These observations are potentially of great importance in the design of nonfouling surfaces because they indicate
that PEO may actually enhance the ability of adsorbed proteins to interact with cells.
Thus PEO surfaces that are incompletely repellant due to incomplete coverage or
insufficient steric repulsion may in fact be more reactive with the partner cell than
the starting substrate.

V. VARIATIONS IN MOLECULAR POTENCY OF ADSORBED ADHESION PROTEINS

In this section, studies of the molecular potency of adsorbed adhesion proteins are
reviewed, with an emphasis on the more recent literature. Since the main methods
to study variations in molecular potency have been the use of antibodies or cells,
these topics are reviewed separately.

A. Antibody Binding Studies

Self-assembled monolayers of alkylthiolates on gold have been used in several recent
studies to examine the effect of changes in surface chemistry on the biological activity of adsorbed fibronectin. Fibronectin adsorption to methyl or carboxyl terminated monolayers measured with ^{125}I radiolabeled fibronectin was found to vary and
was higher on the carboxyl terminated SAMs than on methyl terminated SAMs [84].
The biological activity was studied with a monoclonal antibody to the RGD cell
binding domain. As expected from the ^{125}I results, the antibody binding to fibronectin
on the methyl terminated SAM was less than that to fibronectin on the carboxyl
terminated SAM, but the binding was even lower than expected. Calculation of the
fraction of the adsorbed fibronectin able to bind the antibody showed that the ratio
of antibody bound to fibronectin adsorbed to the COOH terminated SAM was 25%
(mol/mol) compared to 10% on the CH_3 terminated SAM surface. When albumin
was coadsorbed, the ratios increased to 39 and 21% on the COOH and CH_3 surfaces,
respectively. These studies, from Grainger's lab, are summarized in the first entry in
Table 1.
    Another study of antibody binding to fibronectin adsorbed to SAMs terminated
in CH_3, OH, COOH, or NH_2 is summarized in the second entry in Table 1 [57]. In
these studies, from Garcia's lab, it was found that the antibodies employed bound
much better to fibronectin adsorbed in low amounts to OH terminated SAM than to
the other SAMs tested, although antibody binding to fibronectin on all surfaces
reached high levels when high amounts of fibronectin were present on the surfaces.
Thus these studies indicate that the epitope to which the antibody binds is present
when fibronectin is adsorbed to any of the surfaces, but the antibodies' binding
affinity is much higher for fibronectin adsorbed on the OH SAM than on the other
surfaces. The differences in antibody binding affinity were attributed to differences
in conformation of the adsorbed fibronectin and were well correlated with differences
in cellular responses (reviewed below). As in the first study in Table 1 from Grainger's lab, fibronectin adsorption was measured with ^{125}I radiolabeled form of the
protein. In this study, the binding of one of the antibodies studied (3E3) was also

TABLE 1 Changes in Biological Activity of Adsorbed Adhesion Proteins Detected by Antibody Binding

Protein	Surface	Antibody	Antibody binding behavior	Ref.
FN	CH₃ SAM	III-10	10% of RGD sites bind the Ab	84
	COOH SAM	III-10	25% of RGD sites bind the Ab	
FN	CH₃ SAM	3E3	Both antibodies ca. 20-fold	57
	OH SAM	HFN7.1	better to FN on OH SAM	
	COOH SAM		than to other SAMs	
	NH₂ SAM			
FN	TCPS, PS	HFN7.1	Bound ca. tenfold better to FN on TCPS than on PS	31
		pAb	Bound ca. tenfold better to FN on TCPS than on PS	
		3E1	Bound about same to FN on both TCPS and PS	
		4B2	Bound about same to FN on both TCPS and PS	
FN	Immulon	III-10	1.2 (soln.) 2.2 (adsbd.)	87
		III-9	2.2 (soln.) 0.6 (adsbd.)	
		III-4	1.8 (soln.) 1.8 (adsbd.)	
FN	TCPS, PS	pAb	pAb binding to FN is much higher for FN adsorbed in low amounts to TCPS than on PS	97
FBGN	PS	2G-5 (γ 373-385; RIBS-1)	mAb binding to adsorbed FBGN minimally inhibited by solution FBGN	98
FBGN	PS	9F9 (α 87-100; RIBS-2)	mAb binding to adsorbed FBGN moderately inhibited by solution FBGN	86
FBGN	PS	DSB2	mAb binding to adsorbed FBGN not inhibited by solution FBGN	85

Note: SAM, self-assembled monoloayers; TCPS, tissue culture polystyrene; PS, polystyrene; FN, fibronectin; FBGN, fibronogen.

greater on the COOH SAM than on the CH₃ SAM, as reported in the study from Grainger's lab, but the differences were much less than observed between the OH and other surfaces. In contrast, the other antibody (HFN7.1) used in the study from Garcia's lab did not show much difference in binding to fibronectin adsorbed to the COOH and CH₃, although it was reported to differ in the study of these surfaces from Grainger's lab.

The third study listed in Table 1 was also done by Garcia and collaborators [31]. It was done with four types of antibodies to fibronectin, three monoclonals, and one polyclonal. Fibronectin was adsorbed to either untreated or tissue culture grade polystyrene. As summarized in the table, one of the monoclonal antibodies and the polyclonal antibody bound much better to fibronectin adsorbed to tissue culture grade polystyrene than to fibronectin adsorbed on plain polystyrene, while the other two monoclonal antibodies bound about the same to fibronectin on either surface. However, as in the studies with SAMs, the differences were evident only at lower amounts of adsorbed fibronectin, i.e., the antibodies bound equally well to fibronectin on all surfaces when fibronectin adsorption was high enough. Thus the differences are not due the complete absence of the epitope for the antibody, but thought by the authors to be due to differences in the affinity of the antibody for the adsorbed fibronectin. The fact that differences in binding to adsorbed fibronectin are noted with some but not all antibodies suggests that the adsorption process affects different regions of the molecule in different ways, e.g., some regions of the adsorbed molecule are more denatured than others. This idea is strongly supported by studies in our laboratory on adsorbed fibrinogen with a panel of ten monoclonal antibodies [36]. In the latter studies, it was found that antibody binding to the adsorbed fibrinogen changed as time elapsed after the protein had been adsorbed, but the changes varied greatly with antibody type. Thus, the binding of some antibodies increased, others decreased, and others remained unchanged with residence time on the surface [36].

Although changes in the affinity of the antibody for adsorbed fibronectin was the explanation given by the Garcia and his collaborators for the fact that antibody binding differences depended on the amount of fibronectin adsorbed to the surface, an alternative explanation is that the differences in antibody binding are due to the presence of two forms of fibronectin on the surface, one that binds the antibody and one that does not. Given the exquisite sensitivity of antibody binding to structural motifs in their antigens, it may be easier to imagine that slight structural alterations completely eliminate the affinity for some of the adsorbed proteins rather than alter the affinity for all of the adsorbed molecules to the same degree. In the all-or-none model, binding of the antibody to the adsorbed fibronectin only occurs at high loading of fibronectin on certain surfaces because most of the molecules loaded at lower densities are indeed unable to bind the antibody, but at higher loadings the adsorbed fibronectin is in a different form due to crowding by previously adsorbed fibronectin molecules. In light of the strong effects of coadsorbed albumin in increasing cell and antibody binding to adsorbed fibronectin, it seems reasonably plausible that at higher fibronectin loadings, some of the adsorbed fibronectin has a similar effect as coadsorbed albumin, preventing the loss of the antibody binding site by preventing complete spreading of the later adsorbed molecules onto the surface.

Several other less recent studies with monoclonal antibodies that show the formation of hidden sites after adsorption of fibrinogen [85,86] and fibronectin [87] to surfaces are also summarized in Table 1. In the fourth and fifth studies listed, differences in the amount of binding of antibodies to fibronectin in the surface phase were detected, while in the last three studies it was shown that excess fibrinogen addition to the antibody solution did not prevent the binding of the antibody to adsorbed fibrinogen. In the fourth study of fibronectin listed in Table 1, three different monoclonal antibodies to different domains in fibronectin were used; one bound better to adsorbed fibronectin, one bound less well, and one was the same for adsorbed and dissolved fibronectin. In studies of this type, a preliminary screen of the many different monoclonal antibodies is done to find those that seem to discriminate between surface and dissolved proteins. Only a few of the many antibodies tested show this property, presumably reflecting the fact that only a small subset of the many epitopes possible in proteins is affected by the adsorption process. Clearly, changes in the adhesion protein occur upon adsorption, but they are not necessarily extensive throughout the entirety of the molecule. Instead, the changes in the adhesion protein are more subtle and limited.

B. Cell Interactions

Variations in the molecular potency of adsorbed adhesion proteins affecting cells have been observed in two distinct experimental settings, and so the author has previously used the terms *substrate activation* and *modulation* in connection with these different kinds of studies. The adsorption of adhesion proteins to surfaces often greatly increase their cell adhesive properties, a phenomenon referred to as substrate activation. For example, unstimulated platelets will adhere to adsorbed fibrinogen but platelets do not bind soluble fibrinogen unless the platelets have first been activated with an agonist such as ADP or thrombin [83]. Other types of cells have been shown to bind more avidly to adsorbed fibronectin than to soluble fibronectin (reviewed in Ref. 10). Substrate activation was thought originally to be due to the higher local concentration of adhesion proteins in the adsorbed layer and the consequent increase in binding provided by multiple interactions with the cell. However, recent work suggests that it is due to the exposure of novel binding sites in the adsorbed protein that are normally hidden in the soluble protein, as reviewed above. Related to substrate activation is modulation of the biologic activity of adhesion proteins induced by adsorption on different surfaces. Modulation refers to the fact that chemically different surfaces with similar amounts of an adsorbed adhesion protein exhibit differences in cell attachment or spreading. For example, fibrinogen adsorbed to CF_3 rich fluorocarbon gas plasma treated surfaces does not support platelet adhesion as well as fibrinogen adsorbed to tetrafluoroethylene (TFE), even though the surfaces adsorb similar amounts of fibrinogen [33]. Modulation of the degree of interaction of adsorbed fibrinogen with platelets [88,89] and of adsorbed fibronectin with several types of cells [29,32] show that molecular potency of the adhesion protein is affected by surface chemical differences. A more detailed review of modulation effects is available [10]. In the remainder of this section, more recent studies of variations in molecular potency of adhesion proteins adsorbed to surfaces are reviewed in greater detail.

Tang's group recently reported a correlation between phagocyte adhesion after 1-day implantation in the intraperitoneal cavity of mice and the amount of MAC-1 binding sites on fibrinogen adsorbed to five different polymers [90]. MAC-1 is an integrin receptor (also designated CD11b/CD18) on phagocytes that mediates their adhesion to fibrinogen. In these studies, the polymers used were polyethylene terephthalate (PET), polyvinyl chloride (PVC), polyethylene (PE), polydimethylsiloxane (PDMS), and a polyetherurethane (PEU). After 16 h of implantation, large differences in phagocyte accumulation on the surfaces were noted, e.g., the surface bound enzyme activity (used to measure phagocyte accumulation) was 6 on PET, 4 on PVC, 2 on PE, and around 0.5 on PDMS and PEU. In previous studies, Tang and his collaborators had shown that phagocyte accumulation was mediated by adsorbed fibrinogen [19] and that the peptide sequence 190–202 in fibrinogen's gamma chain, designated P1, is involved in binding of the phagocytes through their MAC-1 integrin [91]. Another peptide sequence, 377–395 of the gamma chain (P2), is also involved in binding to MAC-1. Since these epitopes are hidden in soluble fibrinogen, but expressed in fibrin, it was proposed that exposure of these sites makes the implant, with its adsorbed fibrinogen, appear to the phagocytes to be the fibrin normally present at wound site, prompting them to adhere. Using monoclonal antibodies that bind specifically to the P1 or P2 epitopes, Tang and his collaborators studied the state of the adsorbed fibrinogen in several ways. First, they showed that fibrinogen adsorbed to PET bound both the P1 and P2 antibodies, and that the addition of soluble fibrinogen to the antibody solution did not block binding of the antibodies to the adsorbed fibrinogen, although addition of either the P1 peptide or the P2 peptide did block the binding. The inability of soluble fibrinogen to block anti-P1 or anti-P2 binding to the adsorbed fibrinogen confirm that P1 and P2 are neoepitopes that are only exposed after adsorption, another example of substrate activation. The binding of these antibodies to fibrinogen adsorbed to the five polymers also varied a great deal, e.g., anti-P1 binding was high for fibrinogen adsorbed to PET (ca. 0.09) and much lower for fibrinogen adsorbed to PDMS (ca. 0.01). When phagocyte accumulation to the five surfaces was plotted against anti-P1 or anti-P2 binding to fibrinogen adsorbed to these surfaces, there was a roughly linear correlation with reasonably good correlation coefficients (0.9 for P1; 0.70 for P2). These studies provide one of the best and clearest examples of variations in molecular potency of adsorbed adhesion proteins, in the form of both substrate activation (exposure of the P1/P2 sites and phagocyte binding only after fibrinogen is bound) and modulation of the biological activity of the adsorbed protein by differences in surface chemistry (variation in phagocyte and anti-P1/P2 binding to fibrinogen adsorbed to the series of polymers).

Garcia and his colleagues have studied the modulation of cell proliferation and differentiation by substrate-dependent changes in fibronectin conformation on both polystyrene based surfaces [31] and on alkylthiolate SAMs [57]. As reviewed above, these authors were able to show differences in the binding of antibodies to fibronectin adsorbed on these surfaces. In the cell studies, they showed myoblast proliferation and differentiation was affected by the state of the adsorbed fibronectin. Although initial cell adhesion and morphology of the myoblast was similar on both plain polystyrene and tissue culture grade polystyrene, after 3 days in culture, the cells on polystyrene had grown to confluence but very few of the cells exhibited the bipolar morphology characteristic of myotube formation. On tissue culture polystyrene, proliferation had reached the subconfluent stage, and more of the cells appeared bipolar. Using an immunofluorescent stain for sarcomeric myosin, a muscle specific marker, only 6% of the cells on polystyrene were musclelike, while 21% of the cells on tissue culture polystyrene were musclelike. Further studies of cellular interactions with these surfaces were done using a novel method that cross-links the integrins of the cell with the adsorbed fibronectin if the integrin and the fibronectin are in fact binding. The cross-linked integrins are collected by SDS extraction and analyzed by gel electrophoresis and antibody staining. The cross-linking studies showed distinct differences in types of cellular integrins engaged by the fibronectin adsorbed to the various surfaces, namely, there was increased engagement of the $\alpha_5\beta_1$ integrin in comparison to the $\alpha_v\beta_3$ integrin by cells interacting with fibronectin on the tissue culture polystyrene in comparison to polystyrene. [The cross-linking studies were done with a fibroblast cell line (IMR-90) rather than myoblasts because integrin expression is constant in fibroblasts, while it varies greatly in the myoblast line as they begin to differentiate. The differentiation-induced changes in integrin expression would have made the cross-linking studies uninterpretable had myoblasts been used.] The differences in types of cellular integrin engaged by the fibronectin adsorbed on the various surfaces correlated with the changes in differentiation. Since the authors had also shown that function blocking antibodies to α_5-blocked differentiation of the cells into muscle cells, while an antibody to α_v did not reduce differentiation, the differences in which type of integrin in the cell was engaged by the fibronectin adsorbed to the two kinds of surfaces and the corresponding differences in degree of differentiation are consistent with a model in which the conformation of the adsorbed fibronectin affects differentiation because it affects the type of integrin engaged. In separate studies, Garcia and Boettiger also showed that the enhanced engagement of the $\alpha_5\beta_1$ integrin by fibronectin results in a proportional increase in the phosphorylation of focal adhesion kinase (FAK), a tyrosine kinase involved in early integrin-mediated signaling [92]. The results thus extend previous studies of modulation effects on cell attachment or spreading to a fairly deep level of differential integrin engagement and consequent changes in cell signaling.

In the work from Garcia's group using SAMs, the integrin cross-linking method was used to show differences in the α_5 engaged by fibronectin adsorbed by the various SAMs used, and there were also differences in focal adhesion components inside the cells as studied with an immunofluorescent staining method [57]. Finally, this group also showed that cell adhesion strength, as studied with a spinning disk device, was greater when fibronectin was adsorbed to bioactive glasses than on control glasses, despite similar amounts of adsorbed fibronectin, again supportive of the idea that substrate chemistry modulates the molecular potency of the adsorbed fibronectin [93].

Grainger and his colleagues have also done extensive work on the modulation of fibroblast adhesion, spreading, and proliferation on SAMs [84] and on other surfaces [94,95]. As reviewed above, antibody binding to fibronectin adsorbed on COOH and CH_3 terminated SAMs showed that the molecular potency of fibronectin was greater when it was adsorbed on the COOH terminated surface. They also found that Swiss 3T3 fibroblast attachment and spreading were greater on the COOH SAM than on the CH_3 terminated SAM, provided the fibronectin adsorption was done from serum or in the presence of excess albumin, i.e., surfaces adsorbed with purified fibronectin did not display differences in attachment or spreading. Fibroblast proliferation, done in the presence of serum, was greater on the COOH terminated SAM. Immunofluorescent staining of the cells for filamentous actin, paxillin, and phophotyrosine, markers for focal adhesion formation, was also done. As observed in the attachment and spreading studies, there was a difference in actin and paxillin and phosphorylation only when the surfaces were preadsorbed with serum or mixtures of fibronectin with albumin, and as in the other studies cells on the COOH surface exhibited greater staining for these markers of focal adhesion. The authors concluded that for the COOH SAM, high levels of fibronectin cell binding domain accessibility (detected with an antibody) correlate with high degrees of cell attachment, spreading, and growth, while for the CH_3 SAM, lower amounts of fibronectin and cell binding domain availability correlates with lower cell interactions. The authors also found that although coadsorbed albumin enhanced the ability of fibronectin adsorbed to either surface to bind an antibody specific to the cell binding domain of fibronectin, in the case of the CH_3 SAM this enhancement was not enough to overcome the accompanying decrease in total amount of adsorbed fibronectin. In other studies of cell interactions with these same SAMs, the same group studied intracellular signaling events by measuring the levels of integrin-regulated GTPase RhoA [96]. Consistent with the effects on cell spreading and growth, the RhoA became approximately twice as activated and membrane localized when the fibroblasts were cultured on the COOH terminated SAM in comparison to the CH_3 SAM.

VI. SUMMARY AND CONCLUSIONS

The biological activity of adsorbed proteins is most clearly expressed by the large and specific effects that adsorption of adhesion proteins have on the interaction of cells with solid surfaces. It is fair to say that the expression of this activity is the major recognition system that allows the body to react to foreign materials that themselves lack any intrinsic recognition motif. Thus the more-or-less accidental and incidental adsorption of the adhesion proteins to the surfaces of implanted biomaterials is nonetheless fundamental to how the body responds to biomaterials. The integrins and their cognate ligands, the adhesion proteins, provide the biochemical underpinnings of cell responses to implanted biomaterials, and form a sound theoretical basis for the development of ultralow protein adsorption materials that would be biocompatible by virtue of completely resisting the uptake of adhesion proteins. However, for most materials, models which attempt to predict or influence the cellular response by considering only the absolute amount of an adhesion protein on the surface are insufficient. Instead, there is abundant evidence to show that the biological activity of the adsorbed adhesion proteins strongly depends on the type of surface to which it adsorbed, so that both the amount and the molecular potency of the adsorbed adhesion protein have to be considered.

REFERENCES

1. V. Hlady and J. Buijs, Local and global optical spectroscopic probes of absorbed proteins, In *Biopolymers at Interfaces* (M. Malmsten, ed.). Marcel Dekker, New York, 1998, pp. 181–220.
2. J. L. Brash, Role of plasma protein adsorption in the response of blood to foreign surfaces. In *Blood Compatible Materials and Devices* (C. P. Sharma and M. Szycher, eds.). Technomic Publishing, Lancaster, 1991, pp. 3–24.
3. J. G. Steele, T. R. Gengenbach, G. Johnson, C. McFarland, B. A. Dalton, P. A. Underwood, R. C. Chateiler, and H. J. Griesser, Mechanism of the initial attachment of human vein endothelial cells onto polystyrene-based culture surfaces and surfaces prepared by radiofrequency plasmas. In *Proteins at Interfaces*: II. *Fundamentals and Applications*, Vol. 602 (T. A. Horbett and J. L. Brash, eds.). American Chemical Society, Washington, D.C., 1995, pp. 436–449.
4. C. A. Haynes and W. Norde, Globular proteins at solid/liquid interfaces. Coll. Surf. B: Biointerfaces 2:517–566, 1994.
5. W. Norde, Adsorption of proteins at solid–liquid interfaces. Cells Materials 5:97–112, 1995.
6. M. Mrksich and G. M. Whitesides, Using self-assembled monolayers to understand the interactions of manmade surfaces with proteins and cells. Annu. Rev. Biophys. Biomol. Struct. 25:55–78, 1996.

7. J. G. Steele, B. A. Dalton, G. Johnson, and P. A. Underwood, Polystyrene chemistry affects vitronectin activity: an explanation for cell attachment to tissue culture polystyrene but not to unmodified polystyrene. J. Biomed. Mater. Res. 27:927–940, 1993.
8. T. A. Horbett, The role of adsorbed adhesion proteins in cellular recognition of biomaterials. BMES Bull. 23:5–9, 1999.
9. T. A. Horbett, Principles underlying the role of adsorbed plasma proteins in blood interactions with foreign materials. Cardiovasc. Pathol. 2:137S–148S, 1993.
10. T. A. Horbett, The role of adsorbed proteins in animal cell adhesion. Coll. Surf. B: Biointerfaces 2:225–240, 1994.
11. D. R. Phillips, L. Nannizzi-Alaimo, and K. S. Prasad, β_3 tyrosine phosphorylation in $\alpha_{Ib}\beta_3$ (platelet membrane GP IIb-IIIa) outside-in integrin signaling. Thromb. Haemostas. 86:246–258, 2001.
12. D. G. Woodside, S. Liu, and M. H. Ginsberg, Integrin activation. Thromb. Haemostas. 86:316–323, 2001.
13. M. D. McKee and A. Nanci, Osteopontin at mineralized tissue interfaces in bone, teeth, and osseointegrated implants: ultrastructural distribution and implications for mineralized tissue formation, turnover, and repair. Microsc. Res. Tech. 33:141–164, 1996.
14. S. Ashkar, G. F. Weber, V. Panoutsakopoulou, M. E. Sanchirico, M. Jansson, S. Zawaideh, S. R. Rittling, D. T. Denhardt, M. J. Glimcher, and H. Cantor, Eta-1 (osteopontin): an early component of type-1 (cell mediated) immunity. Science 287:860–864, 2000.
15. T. A. Horbett, The role of adsorbed adhesion proteins in cellular recognition of biomaterials. BMES Bull. 23:5–9, 1999.
16. W.-B. Tsai and T. A. Horbett, Human plasma fibrinogen adsorption and platelet adhesion to polystyrene. J. Biomed. Mater. Res. 1997.
17. W.-B. Tsai and T. A. Horbett, The role of fibronectin in platelet adhesion to plasma preadsorbed polystyrene. J. Biomater. Sci. Polym. Ed. 10:163–181, 1999.
18. W.-B. Tsai, J. M. Grunkemeier, C. D. McFarland, and T. A. Horbett, Platelet adhesion to polystyrene-based surfaces preadsorbed with plasmas selectivity depleted in fibrinogen, fibronectin, vitronectin or von Willebrand's Factor. J. Biomed. Mater. Res. 60:348–359, 2002.
19. L. Tang and J. W. Eaton, Fibrin(ogen) mediates acute inflammatory responses to biomaterials. J. Exp. Med. 178:2147–2156, 1993.
20. J.-P. Xiong, T. Stehle, B. Diefenbach, R. Zhang, R. Dunker, D. L. Scott, A. Joachimiak, S. L. Goodman, and M. A. Arnaout, Crystal structure of the extracellular segment of integrin aVB3. Sciencexpress 1–13, 2001.
21. J. G. Steele, G. Johnson, W. D. Norris, and P. A. Underwood, Adhesion and growth of cultured human endothelial cells on perfluorosulphonate: role of vitronectin and fibronectin in cell attachment. Biomaterials 12:531–539, 1991.
22. J. G. Steele, G. Johnson, and P. A. Underwood, Role of serum vitronectin and fibronectin in adhesion of fibroblasts following seeding onto tissue culture polystyrene. J. Biomed. Mater. Res. 26:861–884, 1992.
23. J. G. Steele, G. Johnson, and P. A. Underwood, Role of serum vitronectin and fibronectin in adhesion of fibroblasts following seeding onto tissue culture polystyrene. J. Biomed. Mater. Res. 26:861–884, 1992.
24. J. G. Steele, G. Johnson, C. Mcfarland, B. A. Dalton, T. R. Gengenbach, R. C. Chatelier, P. A. Underwood, and H. J. Griesser, Roles of serum vitronectin and vitronectin in initial attachment of human vein endothelial cells and dermal fibroblasts on oxygen- and nitrogen-containing surfaces made by radiofrequency plasmas. J. Biomater. Sci. Polym. Ed. 6:511–532, 1994.
25. M. D. Bale, L. A. Wohlfahrt, D. F. Mosher, B. Tomasini, and R. C. Sutton, Identification of vitronectin as a major plasma protein adsorbed on polymer surfaces of different copolymer composition. Blood 74:2698–2706, 1989.

26. D. J. Fabrizius-Homan and S. L. Cooper, A comparison of the adsorption of three adhesive proteins to biomaterial surfaces. J. Biomater. Sci. Polym. Ed. *3*:27–47, 1991.

27. T. A. Horbett, P. K. Weathersby, and A. S. Hoffman, The preferential adsorption of hemoglobin to polyethylene. J. Bioeng. *1*:61–78, 1977.

28. T. A. Horbett, Mass action effects on competitive adsorption of fibrinogen from hemoglobin solutions and from plasma. Thromb. Haemostas. *51*:174–181, 1984.

29. F. Grinnell and M. K. Feld, Adsorption characteristics of plasma fibronectin in relationship to biological activity. J. Biomed. Mater. Res. *15*:363–381, 1981.

30. D. K. Pettit, T. A. Horbett, and A. S. Hoffman, Influence of the substrate binding characteristics of fibronectin on corneal epithelial cell outgrowth. J. Biomed. Mater. Res. *26*: 1259–1275, 1992.

31. A. J. Garcia, M. D. Vega, and D. Boettiger, Modulations of cell proliferation and differentiation through substrate-dependent changes in fibronectin conformation. Mol. Biol. Cell *10*:785–798, 1999.

32. K. Lewandowska, N. Balachander, C. N. Sukenik, and L. A. Culp, Modulation of fibronectin adhesive functions for fibroblasts and neural cells by chemically derivatized substrata. J. Cell. Physiol. *141*:334–345, 1989.

33. D. Kiaei, A. S. Hoffman, T. A. Horbett, and K. R. Lew, Platelet and monoclonal antibody binding to fibrinogen adsorbed on glow-discharge-deposited polymers. J. Biomed. Mater. Res. *29*:729–739, 1995.

34. J. A. Chinn, S. E. Posso, T. A. Horbett, and B. D. Ratner, Postadsorptive transitions in fibrinogen adsorbed to polyurethanes: changes in antibody binding and sodium dodecyl sulfate elutability. J. Biomed. Mater. Res. *26*:757–778, 1992.

35. J. A. Chinn, S. E. Posso, T. A. Horbett, and B. D. Ratner, Post-adsorptive transitions in fibrinogen adsorbed to biomer: changes in baboon platelet adhesion, antibody binding, and sodium dodecyl sulfate elutability. J. Biomed. Mater. Res. *25*:535–555, 1991.

36. T. A Horbett and K. R. Lew, Residence time effects on monoclonal antibody binding to adsorbed fibrinogen. J. Biomater. Sci. Polym. Ed. *6*:15–33, 1994.

37. J. A. Chinn, R. E. Phillips, K. R. Lew, and T. A. Horbett, Tenacious binding of fibrinogen and albumin to pyrolite carbon. J. Coll. Interf. Sci. *184*:11–19, 1996.

38. M. Morra, On the molecular basis of fouling resistance. J. Biomater. Sci. Polym. Edn. *11*:547–570, 2000.

39. G. R. Llanos and M. V. Sefton, Review: does polyethylene oxide possess a low thrombogenicity? J. Biomater. Sci. Polym. Ed. *4*:381–400, 1993.

40. D. Leckband, S. Sheth, and A. Halperin, Grafted poly(ethylene oxide) brushes as nonfouling surface coatings. J. Biomater. Sci. Polym. Ed. *10*:1125–1147, 1999.

41. J. H. Lee, H. B. Lee, and J. D. Andrade, Blood compatibility of polyethylene oxide surfaces. Prog. Polym. Sci. *20*:1043–1079, 1995.

42. A. S. Hoffman, Non-fouling surface technologies. J. Biomater. Sci. Polym. Ed. *10*:1011–1014, 1999.

43. K. Ishihara and Y. Iwasaki, Reduced protein adsorption on novel phospholipid polymers. J. Biomater. Applications *13*:111–127, 1998.

44. K. L. Prime and G. M. Whitesides, Self-assembled organic monolayers: model systems for studying adsorption of proteins at surfaces. Science *252*:1164–1167, 1991.

45. R. L. C. Wang, H. J. Kreuzer, and M. Grunze, Molecular conformation and solvation of oligo(ethylene glycol)-terminated self assembled monolayers and their resistance to protein adsorption. J. Phys. Chem. B *101*:9767–9773, 1997.

46. H. B. Lu, C. T. Campbell, and D. G. Castner, Attachment of functionalized poly(ethylene glycol) films to gold surfaces. Langmuir *16*:1711–1718, 2000.

47. G. Mao, D. G. Castner, and D. W. Grainger, Polymer immobilization to alkylchlorosilane organic monolayer films using sequential derivatization reactions. Chem. Mater. *9*:1741–1750, 1997.

48. K. M. DeFife, M. S. Shive, K. M. Hagen, D. L. Clapper, and J. M. Anderson, Effects of photochemically immobilized polymer coatings on protein adsorption, cell adhesion and the foreign body reaction to silicone rubber. J. Biomed. Mater. Res. *44*:298–307, 1999.

49. M. T. Paracchia, S. Harnisch, H. Pinto-Alphandary, A. Gulik, J. C. Dedieu, D. Desmaele, J. d'Angelo, R. H. Muller, and P. Couvreur, Visualization of in vitro protein-rejecting properties of PEGylated stealth polycyanoacrylate nanoparticles. Biomaterials *20*:1269–1275, 1999.

50. L. Ruiz, E. Fine, J. Voros, S. A. Makohliso, D. Leonard, D. S. Johnston, M. Textor, and H. J. Mathieu, Phosphorylchlorine-containing polyurethanes for the control of protein adsorption. J. Biomater. Sci. Polym. Ed. *10*:931–955, 1999.

51. K. Ishihara, T. Shinozuka, Y. Hanazaki, Y. Iwasaki, and N. Nakabayashi, Improvement of blood compatibility on cellulose hemodialysis membrane: IV. Phospholipid polymer bonded to the membrane surface. J. Biomater. Sci. Polym. Ed. *10*:271–282, 1999.

52. M. Malmsten and D. Muller, Interfacial behavior of 'new' poly(ethylene oxide)-containing copolymers. J. Biomater. Sci. Polym. Ed. *10*:1075–1087, 1999.

53. V. Stenkamp and J. Berg, The role of long tails in steric stabilization and hydrodynamic layer thickness. Langmuir *13*:3827–3832, 1997.

54. K. D. Pavey and C. J. Olliff, SPR analysis of the total reduction of protein adsorption to surfaces coated with mixtures of long- and short-chain polyethylene oxide block copolymers. Biomaterials *20*:885–890, 1999.

55. T. B. McPherson, A. Kidane, I. Szleifer, and K. Park, Prevention of protein adsorption by tethered poly(ethylene oxide) layers: experiments and single-chain mean-field analysis. Langmuir *14*:176–186, 1998.

56. J. Satulovsky, M. A. Carignano, and I. Szleifer, Kinetic and thermodynamic control of protein adsorption. Proc. Natl. Acad. Sci. USA *97*:9037–9041, 2000.

57. B. G. Keselowsky, D. M. Collard, and A. J. Garcia, Surface chemistry modulates fibronectin conformation and directs integrin binding and specificity to control cell adhesion, submitted (2002).

58. P. Harder, M. Grunze, R. Dahint, G. M. Whitesides, and P. E. Laibinis, Molecular conformation in oligo(ethylene glycol)-terminated self-assembled monolayers on gold and silver surfaces determines their ability to resist protein adsorption. J. Phys. Chem. B *102*:426–436, 1998.

59. W.-B. Tsai, J. M. Grunkemeier, and T. A. Horbett, Human plasma fibrinogen adsorption and platelet adhesion to polystyrene. J. Biomed. Mater. Res. *44*:130–139, 1999.

60. K. Park, F. W. Mao, and H. Park, Morphological characterization of surface-induced platelet activation. Biomaterials *11*:24–31, 1990.

61. K. Park, F. W. Mao, and H. Park, The minimum surface fibrinogen concentration necessary for platelet activation on dimethyldichlorosilane-coated glass. J. Biomed. Mater. Res. *25*:407–420, 1991.

62. G. A. Adams and I. A. Feuerstein, How much fibrinogen or fibronectin is enough for platelet adhesion? Trans. Am. Soc. Artif. Int. Organs *27*:219–224, 1981.

63. M. Shen, Y. V. Pan, M. S. Wagner, K. D. Hauch, D. G. Castner, B. D. Ratner, and T. A. Horbett, Inhibition of monocyte adhesion and fibrinogen adsorption on glow discharge plasma deposited tetraethylene glycol dimethyl ether. J. Biomater. Sci. Polym. Ed. *12*:961–978, 2001.

64. D. J. Irvine, A. M. Mayes, S. K. Satija, J. G. Barker, S. J. Sofia-Allgor, and L. G. Griffith, Comparison of tethered star and linear poly(ethylene oxide) for control of biomaterials surface properties. J. Biomed. Mater. Res. *40*:498–509, 1998.

65. G. Altankov, V. Thom, T. Groth, K. Jankova, G. Jonsson, and M. Ulbright, Modulating the biocompatibility of polymer surfaces with poly(ethylene glycol): effect of fibronectin. J. Biomed. Mater. Res. *52*:219–230, 2000.

66. V. Thom, G. Altankov, T. Groth, K. Jankova, G. Jonsson, and M. Ulbricht, Optimizing cell–surface interactions by photografting of poly(ethylene glycol). Langmuir *16*:2756–2765, 2000.

67. T. B. McPherson, H. S. Shim, and K. Park, Grafting of PEO to glass, Nitinol, and pyrolytic carbon surfaces by gamma irradiation. J. Biomed. Mater. Res. *38*:289–302, 1997.

68. A. Kidane, G. C. Lantz, S. Jo, and K. Park, Surface modification with PEO-containing triblock copolymer for improved biocompatibility: in vitro and ex vivo studies. J. Biomater. Sci. Polym. Ed. *10*:1089–1105, 1999.

69. S. Q. Liu, Y. Ito, and Y. Imanishi, Synthesis and non-thrombogenicity of polyurethanes with poly(oxyethylene) side chains in soft segment regions. J. Biomater. Sci. Polym. Ed. *1*:111–122, 1989.

70. J. H. Kim, M. Song, H. W. Roh, Y. C. Shin, and S. C. Kim, The in vitro blood compatibility of poly(ethylene oxide)-grafted polyurethane/polystyrene interpenetrating polymer networks. J. Biomater. Sci. Polym. Ed. *11*:197–216, 2000.

71. G. P. Lopez, B. D. Ratner, C. D. Tidwell, C. L. Haycox, R. J. Rapoza, and T. A. Horbett, Glow discharge plasma deposition of tetraethylene glycol dimethyl ether for fouling-resistant biomaterial surfaces. J. Biomed. Mater. Res. *26*:415–439, 1992.

72. K. D. Park, W. G. Kim, H. Jacobs, T. Okano, and S. W. Kim, Blood compatibility of SPUU-PEO-heparin graft copolymers. J. Biomed. Mater. Res. *26*:739–756, 1992.

73. E. L. Chaikof, E. W. Merrill, J. E. Coleman, K. Ramberg, R. J. Connolly, and A. D. Callow, Platelet interaction with poly(ethylene oxide) networks. AIChE J. *36*:994–1002, 1990.

74. J. H. Lee, B. J. Jeong, and H. B. Lee, Plasma protein adsorption and platelet adhesion onto comb-like PEO gradient surfaces. J. Biomed. Mater. Res. *34*:105–114, 1997.

75. J. G. F. Bots, A. Bantjes, and L. van der Does, Small diameter blood vessel prostheses from blends of polyethylene oxide and polypropylene oxide. Biomaterials *7*:393–399, 1986.

76. C. Nojiri, T. Okano, H. A. Jacobs, K. D. Park, S. F. Mohammed, D. B. Olsen, and S. W. Kim, Blood compatibility of PEO grafted polyurethane and HEMA/styrene block copolymer surfaces. J. Biomed. Mater. Res. *24*:1151–1171, 1990.

77. N. P. Desai and J. A. Hubbell, Solution technique to incorporate polyethylene oxide and other water-soluble polymers into surfaces of polymeric biomaterials. Biomaterials *12*:144–153, 1991.

78. C. R. Deible, P. Petrosko, P. C. Johnson, E. J. Beckman, A. J. Russell, and W. R. Wagner, Molecular barriers to biomaterial thrombosis by modification of surface proteins with polyethylene glycol. Biomaterials *20*:101–109, 1999.

79. G. R. Llanos and M. V. Sefton, Immobilization of poly(ethylene glycol) onto a poly(vinyl alcohol) hydrogel: 2. Evaluation of thrombogenicity. J. Biomed. Mater. Res. *27*:1383–1391, 1993.

80. S. R. Hanson, L. A. Harker, B. D. Ratner, and A. S. Hoffman, In vivo evaluation of artificial surfaces with a nonhuman primate model of arterial thrombosis. J. Lab. Clin. Med. *95*:289–304, 1980.

81. B. S. Coller, Interaction of normal, thrombasthenic, and bernard-soulier platelets with immobilized fibrinogen: defective platelet-fibrinogen interaction in thrombasthenia. Blood *55*:169–178, 1980.

82. J. A. Chinn, T. A. Horbett, and B. D. Ratner, Baboon fibrinogen adsorption and platelet adhesion to polymeric materials. Thromb. Haemostas. *65*:608–617, 1991.

83. B. Savage and Z. M. Ruggeri, "Selective recognition of adhesive sites in surface-bound fibrinogen by glycoprotein IIb-IIIa on nonactivated platelets. J. Biol. Chem. *266*:11227–11233, 1991.

84. K. B. McClary, T. P. Ugarova, and D. W. Grainger, Modulating fibroblast adhesion, spreading, and proliferation using self-assembled monolayer films of alkylthiolates on gold. J. Biomed. Mater. Res. *50*:428–439, 2000.

85. J. Soria, C. Soria, M. Mirshahi, C. Boucheix, A. Aurengo, J. Y. Perrot, A. Bernadou, M. Samama, and C. Rosenfeld, Conformational change in fibrinogen induced by adsorption to a surface. J. Coll. Interf. Sci. *107*:204–208, 1985.

86. T. P. Ugarova, A. Z. Budzynski, S. J. Shattil, Z. M. Ruggeri, M. H. Ginsberg, and E. F. Plow, Conformational changes in fibrinogen elicited by its interaction with platelet membrane glycoprotein GPIIb-IIa. J. Biol. Chem. *268*:21080–21087, 1993.

87. T. P. Ugarova, C. Zamarron, Y. Veklich, R. D. Bowditch, M. H. Ginsberg, J. W. Weisel, and E. F. Plow, Conformational transitions in the cell binding domain of fibronectin. Biochemistry *34*:4457–4466, 1995.

88. J. N. Lindon, G. McManama, L. Kushner, E. W. Merrill, and E. W. Salzman, Does the conformation of adsorbed fibrinogen dictate platelet interactions with artificial surfaces? Blood *68*:355–362, 1986.

89. J. H. Silver, H.-B. B. Lin, and S. L. Cooper, Effect of protein adsorption on the blood-contacting response of sulphonated polyurethanes. Biomaterials *14*:834–844, 1993.

90. W. J. Hu, J. W. Eaton, and L. Tang, Molecular basis of biomaterial-mediated foreign body reactions. Blood *98*:1231–1238, 2001.

91. L. Tang, T. P. Ugarova, E. F. Plow, and J. W. Eaton, Molecular determinants of acute inflammatory responses to biomaterials. J. Clin. Invest. *97*:1329–1334, 1996.

92. A. J. Garcia and D. Boettiger, Integrin–fibronectin interactions at the cell–material interface: initial integrin binding and signaling. Biomaterials *20*:2427–2433, 1999.

93. E. Heilmann, P. Friese, S. Anderson, J. N. George, S. R. Hanson, S. A. Burstein, and G. L. Dale, Biotinylated platelets: a new approach to the measurement of platelet life span. Br. J. Haematol. *85*:729–735, 1993.

94. D. W. Grainger, G. Pavon-Djavid, V. Migonney, and M. Josefowicz, Assessment of fibronectin conformation adsorbed to polytetrafluoroethylene surfaces from serum protein mixtures and correlation to support of cell attachment in culture. J. Biomater. Sci. Polym. Ed. (in press), 2001.

95. A. L. Koenig, V. Gambillara, and D. W. Grainger, Correlating fibronectin adsorption with endothelial cells adhesion and signaling on polymer substrates. J. Biomed. Mater. Res. (in press), 2001.

96. K. B. McClary and D. W. Grainger, RhoA-induced changes in fibroblasts cultured on organic monolayers. Biomaterials *20*:2435–2446, 1999.

97. F. Grinnell and M. K. Feld, Fibronectin adsorption on hydrophobic and hydrophilic surfaces detected by antibody binding and analyzed during cell adhesion in serum containing medium. J. Biol. Chem. *257*:4888–4893, 1982.

98. C. Zamarron, M. H. Ginsberg, and E. F. Plow, Monoclonal antibodies specific for a conformationally altered state of fibrinogen. Thromb. Haemostas. *64*:41–46, 1990.

16

Protein Interactions with Monolayers at the Air–Water Interface

DAVID W. BRITT   Utah State University, Logan, Utah, U.S.A.

G. JOGIKALMATH and VLADIMIR HLADY   University of Utah, Salt Lake City, Utah, U.S.A.

I. INTRODUCTION: LANGMUIR FILMS AS MEMBRANE MIMICS

A. Motivation

Proteins vary widely in size, shape, and function, yet they all have the common feature of being surface-active macromolecules. This high interfacial activity arises from the amphipathic nature of proteins, which are composed of polar, hydrophobic, and charged residues. When left unchecked, they accumulate at interfaces, reaching surface concentrations far greater than the corresponding bulk concentration. In most applications this is an undesirable property leading to fouling of surfaces and possible denaturation, aggregation, and loss of activity of the proteins.

Inside living organisms, protein surface activity is primarily controlled by the properties of the surfaces (mainly lipid membranes) which they encounter. Understanding how protein adsorption is regulated on biological membranes and attempting to mimic these systems in the laboratory are topics relevant to all medical and biotechnology fields and are major focuses of current research [1–3].

This chapter reviews protein adsorption to Langmuir monolayers, emphasizing the advantages of using insoluble amphiphile films as model membranes and illustrating the benefits offered by mixed monolayers. Methods of tailoring film properties to promote or discourage protein adsorption, as depicted in Fig. 1, are introduced. The first half of the chapter reviews Langmuir monolayers, protein–monolayer interactions, and relevant instrumentation. This is followed by the presentation of a novel means of enhancing the electrostatic binding of proteins to mixed charged/neutral amphiphile monolayers. The remainder of the chapter is dedicated to the so-called protein-repellent surfactant films, focusing on the influence of monolayer packing density, surfactant conformation, and osmotic (steric) barriers to protein adsorption.

B. Model Membranes

The term *Langmuir monolayer* describes a floating surfactant film situated at an air–water or oil–water interface. These films were first characterized in the late 1800s

415

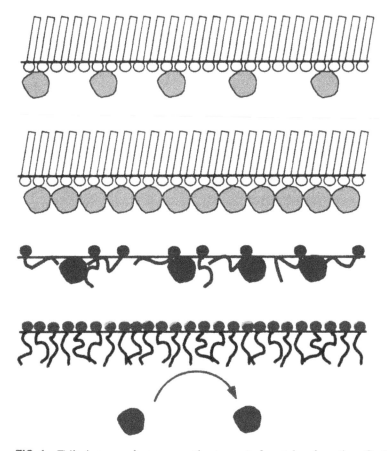

FIG. 1 Tailoring monolayer properties to control protein adsorption. Optimizing the electro-static binding of a charged protein to an oppositely charged monolayer by diluting the charged amphiphile in a neutral amphiphile matrix. The key is to select the right ratio of charged and neutral amphiphiles (upper panels). A PS-PEO monolayer can be changed from a protein-adsorbing to protein-repellent form by increasing the surface density of PS-PEO and thus forcing PEO chains into a brush configuration (lower panels).

in the kitchen of Agnes Pockles, who studied soap films using a water-filled pan with a compression barrier and primitive surface balance—the predecessor of the modern Langmuir trough. This technology rapidly came to maturity in the early to mid-1990s through work by Katherin Blodgett and Irving Langmuir, who developed methods of transferring films to solid supports and performed the preliminary inves-tigations of protein films at the air–water interface [4].

      In addition to Langmuir monolayers, a variety of membrane mimics, such as black lipid membranes, supported bilayers, and single- and multilamellar vesicles, have been employed in protein–membrane investigations. While vesicles are perhaps the most popular systems, they have some fundamental limitations with regard to controlling amphiphile packing density, membrane composition, phase behavior, and curvature [5]. The planar geometry of black lipid membranes [6,7] and supported bilayers [8] eliminates the curvature problem, and these films are superior for integral membrane–protein studies and measuring membrane permeability, yet they do not offer the simplicity, stability, or control over film properties attainable with Langmuir monolayers.

The main deficit of Langmuir films as membrane mimics is their asymmetry, representing only one-half the bilayer leaflet. But with this limitation comes certain advantages. For instance, many amphiphiles (phospholipids, fatty acids, polymers, proteins, etc.) that cannot be prepared as vesicles can be spread as Langmuir films. Moreover, in contrast to bilayers, monolayer membranes lend themselves to the direct potentiometric measurement of the surface potential [5,9,10], while the Langmuir trough barrier system allows the packing density and phase state of the films to be precisely controlled. The continued popularity of Langmuir monolayers as membrane mimics is a testament to their versatility and unique properties [5,11,12].

II. LANGMUIR FILMS AND INSTRUMENTATION

A. Monolayer Characterization: Surface Pressure and Surface Potential

The collective response of insoluble amphiphiles at the air–water interface as they are compressed from a disordered, two-dimensional "gaslike" state through any phase transitions to a close-packed condensed state provides a unique signature in the form of the well-known surface pressure (π) versus molecular area (A) isotherm. The π-A isotherm, usually measured with a surface balance (Wilhelmy plate), provides an indication of interamphiphile as well as amphiphile subphase interactions. For mixed amphiphile films miscibility diagrams can be constructed from a series of isotherms collected for various ratios of the two amphiphiles [13]. Such diagrams highlight any deviation of mixed film properties from "ideal" additive mixing behavior. For molecular area, deviation from ideality upon mixing (ΔA_{mix}) is calculated at a given surface pressure according to

$$\Delta A_{mix}(\pi) = A_{12}(\pi) - [\chi_1 A_1(\pi) + \chi_2 A_2(\pi)] \tag{1}$$

where A_{12} is the average amphiphile molecular area measured for the mixture at a given surface pressure, and A_1 and A_2 are the molecular areas of the pure components at the same surface pressure. Multiplying A_1 and A_2 by their respective mole fractions in the mixture, χ_1 and χ_2, yields their expected contributions to the net area based on ideal mixing. Additional parameters such as surface potential (ΔV) and effective dipole moment (μ) can also be evaluated in the same manner to yield $\Delta\Delta V_{mix}$ and $\Delta\mu_{mix}$.

The surface potential provides an indication of amphiphile dipole orientation (alkyl-tail tilt angle), lateral density, and charge state. Moreover, ΔV contains contributions from oriented water dipoles, ions in the electric double layer or proteins accumulated beneath the film. The vibrating plate condenser (Kelvin method) is the most common means to measure ΔV. In this method, a capacitance developed across the interface separating a platinum electrode in the subphase and an oscillating electrode held 1–2 mm above the water surface is compensated by an applied DC voltage [9,10]. ΔV can be normalized in terms of molecular area to remove the influence of amphiphile packing density and expressed as the effective dipole moment according to the Helmholtz equation:

$$\mu = \varepsilon_o \Delta V A \tag{2}$$

where ε_O is the permittivity of the vacuum, and A is the average area available per molecule (determined from the π-A isotherm).

It must be stressed that μ is not an absolute value, rather it is an effective dipole moment, resulting from a change of the mean dipole moment density *normal* to the interface related to one amphiphile with respect to the clean water surface. Spreading a monolayer on a clean air–water interface creates two new interfaces, namely, the alkyl-tail–air interface and the head group–water interface. Due to the low dielectric surrounding the alkyl-tail terminal methyl groups, their contribution to the effective dipole moment is quite large: for close-packed hydrocarbon tails, Vogel and Möbius calculated this contribution to be 0.35 D (corresponding to 660 mV of the measured surface potential) [9]. Consequently, any changes in the alkyl-tail tilt angle can lead to significant changes in ΔV and μ, as discussed in Section III.D.

When considering the interactions of proteins or other molecules with amphiphile head groups, the contribution of the alkyl-tail–air interface (or for transferred films, the alkyl-tail–substrate interface) is often ignored. However, as demonstrated by Cordroch et al., the chemistry and local environment of the terminal groups of the alkyl-tails can dramatically shift the protonation equilibria of head groups by 1–2 units [14,15], which could certainly influence protein interactions at the head group region.

B. Protein Interfacial Activity and Surface Tension

Before discussing protein–monolayer interactions, we first consider protein adsorption to a monolayer-free interface. In this case, a decrease in the initial surface tension from that of the pure liquid, γ_o (~72 mN/m for water), to that of liquid plus protein, γ_p, provides an indication of protein surface activity. This difference in surface tension is the surface pressure, π.

$$\pi = \gamma_o - \gamma_p \tag{3}$$

Surface pressure arises from the lowered chemical potential of the surface layer of water as it mixes with proteins (or other surfactants). Bulk water molecules diffuse to the region of lower chemical potential, bringing about the increase in surface pressure. If a monolayer of amphiphiles is already present at the interface, then the protein must work against the surface pressure exerted by this film. In this case the net surface pressure is given by

$$\pi = \gamma_o - \gamma_m - \gamma_p \tag{4}$$

where γ_m is the surface tension of the interface plus monolayer. Note that the term $\gamma_o - \gamma_m$ is the initial surface pressure (π_o) due to the monolayer, which determines the extent (γ_p) to which protein can penetrate the film.

For close-packed films (high π_o), the surface balance has the major disadvantage of being insensitive to proteins that adsorb, but cannot penetrate or otherwise induce a change in the film. It has been calculated that a protein needs some 10 to 15 Å^2 (6–10 amino acid residues) cleared at the air–water interface in order to penetrate the film [16]. For this reason measurements are often restricted to less densely packed films (low γ_m), which permit some extent of protein insertion. When working with close-packed films, techniques such as surface plasmon resonance or light reflection must be used, as discussed next.

C. Light Reflection at the Air–Water Interface

Detecting proteins that adsorb to a Langmuir monolayer but do not induce a change in surface pressure requires more advanced methods (usually optical or gravimetric) than the surface balance. Here we discuss two optical techniques used in our laboratory, light reflection and surface plasmon resonance, for monitoring protein interfacial activity.

Light reflection has traditionally been used for measuring the spectra of dye molecules at the air–water interface as a function of packing density and subphase conditions [17]. Fixed wavelength adsorption kinetics of soluble light-absorbing molecules, such as proteins, to the interface can also be attained with this technique [18]. The enhancement of reflection at normal incidence from the interface due to the presence of dye (or protein) is given by [17]

$$\Delta R = A(R_i)^{0.5} + \rho^2 \qquad (5)$$

where R_i is the reflectivity of the monolayer free interface, ρ^2 the reflectivity of the monolayer plus protein interface, and A the absorption of the protein. It is often the case that ρ^2 is negligible compared to $A(R_i)^{0.5}$, in which event reflection is equivalent to absorption. Since the contribution of protein in the subphase is accounted for in the reference section of the trough, sensitivity is restricted to protein adsorbed at the interface, as depicted in Fig. 2.

For such a differential measurement, it is important to have protein evenly distributed in the subphase. Injecting the protein into a stirred subphase does not

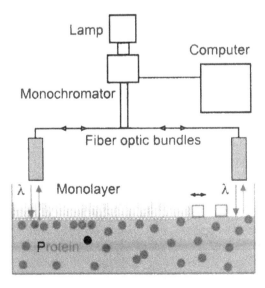

FIG. 2 Measuring protein–monolayer interactions as a difference in light reflection between the monolayer-containing interface and a monolayer-free reference.

lead to a rapid and uniform distribution of protein under both the reference and measuring sections of the trough. Thus, for the reflectivity data presented in this chapter, a 0.01 mg/mL protein (horse spleen ferritin) solution was gently poured into a 18×56 cm^2 Teflon trough, the surface aspirated clean, and a baseline measurement made at 360 nm. Next, the monolayer was spread from chloroform and rapidly compressed to the desired packing density while measuring ΔR [18].

D. Surface Plasmon Resonance at the Air–Water Interface

Since another chapter in this monograph is devoted to the surface plasmon resonance (SPR) technique (Chapter 22), we limit discussion to our adaption of this method to monitoring protein interaction with insoluble poly(styrene)–poly(ethylene oxide) (PS-PEO) block copolymer Langmuir films. Commercial Spreeta SPR sensors (Texas Instruments, Inc.) were used for SPR measurements. These devices combine the sensor surface with all the optic and electronic components required for SPR experiments in a compact assembly with polarized near-infrared light (840 nm) as the light source. The gold surface of the SPR sensor was first cleaned in an oxygen plasma and then modified by octadecanethiol (2 mM in absolute ethanol for 10 min), followed by rinsing with absolute ethanol and then double-distilled water.

Prior to the protein adsorption experiment, the hydrophobic SPR sensor was preinitialized in air, calibrated in double-distilled water, and an initial reading of refractive index (RIU) obtained by measuring the PBS buffer solution used as the subphase. The PS-PEO monolayer was then prepared on this subphase and the sensor lowered into full contact with the monolayer. The SPR signal was first recorded as a baseline for at least 10 min to ensure the integrity of the monolayer as well as the stability of the signal (in control experiments a stable signal from the PS-PEO monolayer was obtained over 5 h).

Once the baseline signal was recorded, 5 mL of protein (HSA, 1 mg/mL) was injected into the PBS subphase where a small Teflon stir bar mixed the subphase and helped HSA toward the surface, as depicted in Fig. 3. The final HSA concentration in the PBS subphase was 0.008 mg/mL. The SPR signal acquisition was initiated simultaneously with the protein injection and, in parallel, a change of surface pressure was measured using a Wilhelmy plate positioned in close proximity to the SPR probe. Both signals were monitored until a plateau in the SPR signal was observed (typically within 2 h).

III. MONOLAYERS AND PROTEIN ADSORPTION

A. Electrostatic Binding of Proteins to Monolayers

Proteins adsorb to floating monolayers and other interfaces to satisfy their amphipathic nature. Adsorption reduces surface energy, allowing physical bonds between protein and amphiphile to form while entropy is increased through the release of counterions [19,20]. The binding of charged molecules to oppositely charged films is a common model system since the monolayer charge density and subphase ionic strength can be tuned to increase or decrease adsorption [21–25]. However, caution must be exercised when interpreting the data since changing the ionic strength, am-

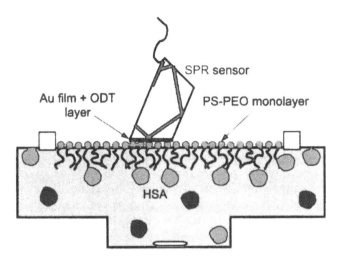

FIG. 3 Surface plasmon resonance technique adapted to the air–water interface utilizes an SPR sensor in contact with the monolayer.

phiphile ratios, or the amphiphile packing density also influences monolayer phase behavior and domain structure.

B. Amphiphile Miscibility, Phases, and Protein Adsorption

The electrostatic binding of charged proteins to oppositely charged monolayers is generally governed by the lateral charge density in the monolayer, with more charged amphiphiles leading to greater protein adsorption. However, other parameters, such as amphiphile miscibility and monolayer phase behavior, play critical roles. This was highlighted in recent studies of cytochrome c (cyt-c) adsorption to mixed monolayers of zwitterionic and anionic phospholipids by Maierhofer and Bucknall and by Käsbauer and Bayler [25,26]. As expected, cyt-c adsorption increased proportionally to the fraction of anionic amphiphile in the mixed monolayer. However, when the miscibility of the two amphiphile components was systematically varied, some interesting cyt-c adsorption behavior emerged. The authors observed enhanced cyt-c binding to immiscible mixtures of the two amphiphiles (mismatched chain lengths) compared to miscible mixtures (identical chain lengths) of the same amphiphiles, but only when the films were held in the liquid condensed (LC) state. When the miscible and immiscible films were held in the liquid expanded (LE) state, they bound similar amounts of cyt-c. Clearly, local charge density and amphiphile phase differences (on the size scale of LC domains) governed protein binding, while the net amphiphile (charge) density played a lesser role.

This example suggests phase behavior–dependent binding (and likely lateral distribution) of cyt-c. The axial distribution (i.e., penetration depth into the lipid) of cyt-c also depends on monolayer phase behavior and charge density. A study by Teisse demonstrated that when the monolayer packing density was decreased or the ratio of anionic to zwitterionic lipid increased, cyt-c penetrated more deeply into the films [27]. Likewise, an increase in cyt-c insertion with decreasing ionic strength was observed. These results were explained based on enhanced electrostatic interactions between amphiphile and protein; however, as the author pointed out, the smaller anionic phospholipid head group compared to the zwitterionic head group could also affect the interpretation of depth of insertion. Furthermore, increasing the ionic strength in order to reduce protein–lipid electrostatic binding also screens the like-charged lipids from each other, resulting in a more condensed monolayer as well as altering domain morphology [28–31]. A typical example is the calcium-induced segregation of DPPS into domains from a DPPS/DPPC mixed monolayer [29].

A final consideration regarding monolayer phase behavior and protein adsorption is protein-induced amphiphile demixing [19] and phase transitions [32]. Penetration of the protein into the monolayer reduces the area available to the amphiphiles and effectively compresses the amphiphiles. A manifestation of this process is given in Section IV, where π is observed to change discontinuously upon albumin insertion into PS-PEO films. A recent review by Vollhardt and Fainerman details this remarkable event [32]. These examples should remind the reader of the subtleties associated with even simple model systems of just one or two amphiphile types and a single type of subphase protein.

C. Enhanced Protein Adsorption to Mixed Monolayers: Ferritin Binding to Mixed Cationic/Neutral Monolayers

The previous examples of the electrostatic binding of cytochrome c to oppositely charged monolayers followed the generally observed trend of increased protein adsorption with increasing fraction of charged amphiphile. In this section we show that just the opposite can happen, where decreasing the monolayer charge density by mixing a cationic amphiphile with neutral amphiphiles increases the amount of anionic protein adsorbed. The reason for this behavior will become apparent from miscibility diagrams, which reveal more positive surface potentials and effective dipole moments in the mixtures despite decreased charge densities.

We have recently investigated the adsorption behavior of negatively charged horse spleen ferritin (isoelectric point 4.5) to cationic monolayers of varying charge density held at the air–water interface of the reflection trough [18]. In these experiments, the charge density was varied by two different means. In the first method, the cationic amphiphile, dioctadecyldimethylammonium bromide [$(C_{18})_2N(CH_3)^+Br^-$] (DOMA) was mixed with the neutral amphiphile, stearic acid methyl ester ($C_{18}COOCH_3$) (SME). These amphiphiles were prepared at various molar ratios, spread from chloroform, then rapidly compressed to 30 mN/m on preequilibrated subphases of 0.01 mg/mL ferritin in water. A second means of controlling charge density was to keep the amphiphile composition fixed but vary the packing density (surface pressure) of the amphiphiles. In this technique, single-component films of either DOMA or eicosylamine ($C_{20}NH_2$) were spread on ferritin subphases then compressed to surface pressures of 10, 20, or 30 mN/m. Ferritin adsorption was recorded continuously after film spreading as a change in light reflection at the interface.

The ferritin binding kinetics on $C_{20}NH_2$ films prepared at different surface pressures are given in Fig. 4. It is apparent that the reflectivity signal increased during compression, a result of ferritin adsorption/insertion to the film before the feedback surface pressure (indicated by the arrows) was reached. As might be anticipated for an electrostatically driven adsorption, films prepared a higher surface pressures (higher charge densities) bound more protein than "loosely" packed films (compare

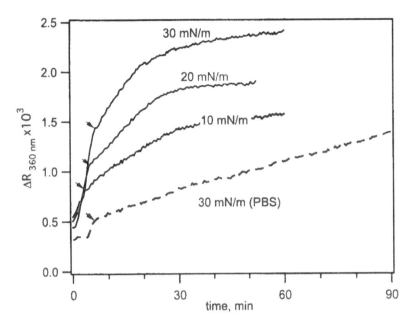

FIG. 4 Ferritin adsorption kinetics to eicosylamine films as a function of monolayer charge density (surface pressure). The monolayers were prepared on 0.01 mg/mL ferritin containing subphases in either water (solid lines) or 150 mM PBS (dashed line). The arrows mark when the films reached the indicated surface pressures.

the plateau ΔR values). Protein-induced changes in surface pressure were not observed above 20 mN/m, suggesting the large ferritin particles (diameter 12.5 nm) were unable to penetrate compressed films. From the ΔR data it is noted that the rate of adsorption also increased with increasing packing density (compare the slopes preceding the plateaus). A similar trend was seen for ferritin binding to DOMA films (data not shown).

These results indicate a direct correlation between monolayer charge density (controlled by surface pressure) and ferritin adsorption. However, when monolayer charge density was varied by diluting the cationic amphiphile in a neutral amphiphile matrix, an opposite trend emerged. This is seen in Fig. 5, where ferritin adsorption kinetics to a series of SME:DOMA mixtures prepared at 30 mN/m are presented. Comparing the plateau ΔR values at 90 min a trend of increased binding with increasing SME fraction is apparent. A maximum enhancement in binding for the SME:DOMA 6:1 ratio is noted, yet even the SME:DOMA 10:1 ratio binds more ferritin than pure DOMA. This unique behavior is highlighted in the inset of the graph, where ferritin adsorption at 90 min is plotted versus mole fraction SME.

A reduced rate of ferritin binding from PBS (dashed lines in Figs. 4 and 5) is indicative of the electrostatic nature of the protein–monolayer interaction. However, we should caution that a plateau was never reached from the PBS subphase, and at longer times the amount of ferritin adsorbed may very well surpass the coverages reported for adsorption from pure water [23,24]. To minimize salt-induced aggregation of ferritin (in bulk and the adsorbed state) as well as to avoid salt-driven amphiphile phase separation [28,29], salt-free subphases were favored.

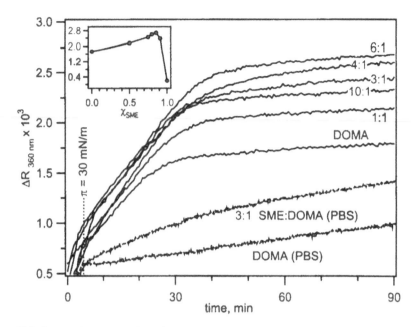

FIG. 5 Ferritin adsorption kinetics to the indicated SME:DOMA mixtures held at 30 mN/m (same subphase conditions as in Fig. 4). The inset in this graph compares the amount of ferritin adsorbed at 90 min versus the mole fraction of SME in the monolayer. (Reproduced by permission of The Royal Society of Chemistry on behalf of the PCCP Owner Societies.)

From a packing density standpoint, a pure DOMA film held at 30 mN/m corresponds to a charge density of about one charge per 60 $Å^2$. For DOMA mixed with SME in a 6:1 ratio, the charge density is reduced to one charge per 180 $Å^2$. Thus, a threefold reduction of charged amphiphile density resulted in a ~1.5-fold increase in bound ferritin after 90 min of adsorption (see Fig. 4). Furthermore, ferritin adsorption to SME films from water subphases was negligible, suggesting SME alone did not contribute to ferritin binding under these conditions [33]. Obviously SME acts as more than a charge diluent when mixed with DOMA. In fact, as is shown in the following miscibility analysis section, SME actually amplifies the surface potential in spite of reducing the monolayer charge density.

D. Amphiphile Miscibility and Mixing Diagrams

The unique ferritin adsorption behavior on SME:DOMA monolayers suggests that these mixtures must exhibit properties that are not a simple average of the two constituent amphiphiles. In this section we demonstrate that this is indeed the case by evaluating SME:DOMA miscibility in terms of mixing induced changes in monolayer area, surface potential, and effective dipole moment.

1. Excess Surface Potential and Ferritin Adsorption Trends

A first indication of the unique SME:DOMA mixing behavior is found in the π-A and ΔV-A isotherms, presented in Fig. 6. From the isotherms, it is noted that the

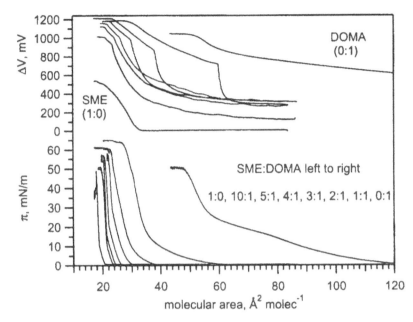

FIG. 6 Surface potential– and surface pressure–area isotherms for the indicated SME:DOMA mixtures on water. (Reproduced by permission of The Royal Society of Chemistry on behalf of the PCCP Owner Societies.)

mixtures collapse at higher surface pressures (π_c) and surface potentials (ΔV_c) than the pure components. The increase in π_c may arise from a more efficient packing structure in the mixture, where smaller SMEs ($\sim$20 Å^2/molec) fill in the interstitial spaces between the larger DOMAs ($\sim$60 Å^2/molec). The large increases in ΔV_c values for the mixtures, however, are quite unusual, since they indicate the surface potential becomes more positive (by hundreds of millivolts!) as cationic DOMA is diluted in neutral SME.

In Fig. 7 miscibility diagrams constructed from the isotherms clearly illustrate the increase in ΔV (top diagram) and improved packing density (middle diagram) with SME content. Moreover, μ (bottom diagram) follows the same trend as ΔV, indicating that an increase in amphiphile dipole density ($\Delta A_{mix} < 0$) cannot explain the excess positive character of the mixtures since μ normalizes ΔV in terms of area [recall Eq. (2)]. Other factors, such as a more orthogonal alkyl-tail dipole orientation in the mixture, changes in the electric double layer, and/or changes in the interfacial water structure must act to increase the surface potential, which in turn leads to enhanced protein adsorption.

2. Origins of Excess Properties of Mixed Monolayers

If the tilt angle of the alkyl-tail dipoles in the mixture is reduced compared to the pure film, then increases in ΔV and μ are expected due to the increase in the normal component of the dipoles (refer to Section II.A). In transferred films, DOMA has a sizeable tilt angle of 45°, which is a result of the size mismatch between the large

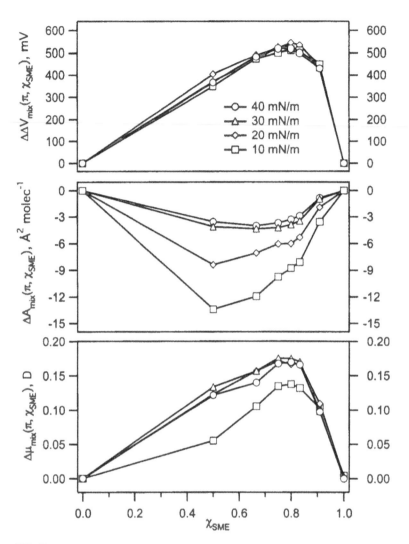

FIG. 7 SME:DOMA miscibility diagrams constructed from the isotherms in Fig. 6.

DOMA head group and the alkyl-tails [34]. Introducing SME into a DOMA mono-layer may effectively compensate for the free space among DOMA tails, thus or-dering the tails and reducing the DOMA tilt angle. Such an event has been observed for other mixed monolayers [35,36], and we have recently verified that this is also the case for SME:DOMA [37]. However, it turns out that ordering of the tails and reduction of the DOMA tilt angle in mixtures can only increase the surface potential by ~100 mV [18], indicating that the remaining excess potential of up to 400 mV (see Fig. 7) must arise from mixing induced changes at the head group–water in-terface.

Interfacial water molecules spontaneously orient with the oxygen atoms di-rected toward the air, thus polarizing the interface and creating a potential jump on the order of -100 to -200 mV [9]. An amphiphile monolayer will depolarize this water in a manner dependent on the amphiphile head group chemistry [38,39]. Vogel and Möbius proposed that amphiphiles with somewhat hydrophobic (methylated) head groups like SME disrupt the interfacial water structure and, when mixed with charged amphiphiles, increase the charge-hydration shell distance [40]. This would effectively amplify the surface potential by reducing the ability of water dipoles (and counterions) to collectively screen amphiphile dipoles and charges. Although the carbonyl dipole [41] could also contribute more strongly to the surface potential in the mixture, infrared reflectivity analysis did not reveal any changes in the SME carbonyl band upon mixing [37].

The influence of head group hydrophobicity was tested by investigating the properties of DOMA mixed with octadecanol (ODOH) as well as DOMA mixed with steric acid ethyl ester (SEE). The mixing diagrams for these films are shown in Fig. 8. Interestingly, ODOH does not lead to as compact a film as either SME or SEE (middle diagram). However, it is clearly seen that for the ODOH:DOMA mixtures $\Delta\mu_{mix}$ remains near zero, while the SEE:DOMA mixtures, like the SME:DOMA mixtures, have excess positive $\Delta\mu_{mix}$ values. Although ODOH:DOMA mixtures have $\Delta\Delta V_{mix}$ values on the order of 250 mV, recall that this can arise from increased dipole density upon mixing ($\Delta A_{mix} < 0$) as well as mixing induced changes in amphiphile tilt angle and alkyl-tail order. Thus, we would not anticipate enhanced ferritin binding to ODOH:DOMA monolayers as was observed for DOMA:SME monolayers.

In summary, the trend of increased ferritin binding as DOMA was diluted in SME is primarily attributed to amplification of DOMA charge by SME-induced disruption of oriented water dipoles. Reduced amphiphile tilt angles and ordering of alkyl-tails also contribute to the enhanced positive character of the mixtures, albeit to a lesser extent. Combined, these effects can significantly enhance protein adsorption in a manner not expected based on a simple monolayer charge dilution model. Drawing from these insights we propose in the next section that using a mixture of PEO chain lengths in a PS-PEO monolayer can improve the protein-repellent character of the film.

IV. PROTEIN REPELLENT MONOLAYERS

A. PEO Surface Density Versus PEO Brush Length

Biomaterials surfaces, when modified with attached poly(ethylene oxide) (PEO) chains, display low levels of protein adsorption, especially when the surface density of PEO is sufficiently high [42–44]. This resistance to protein adsorption has been attributed to (1) steric (or osmotic) repulsion due to the compression of the PEO chains by an approaching protein molecule and/or (2) a unique property of ethylene oxide (EO) segments capable of repelling proteins. Recently, a "two-state" polymer model was also used to explain PEO interactions with water [45] and with dissolved proteins [46,47]. How effective the end-attached PEO chains are in preventing proteins from adsorbing to the underlying surface depends on both the length of the PEO chains and their surface density [48–51]. These two parameters are interrelated as the increase of surface density of PEO also stretches the individual chains and extends their length. For small globular proteins the PEO surface density plays a

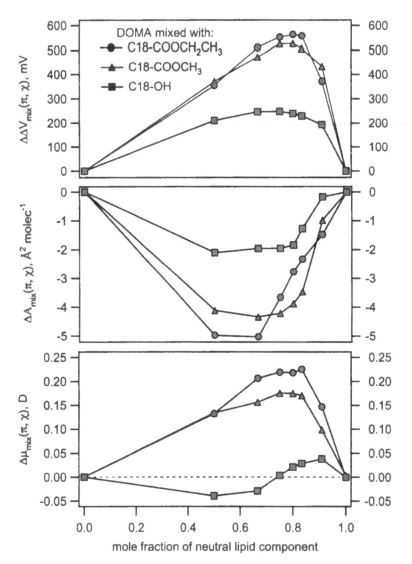

FIG. 8 Miscibility diagrams (at 30 mN/m) of DOMA mixed with ODOH (squares), SME (triangles), and SEE (circles).

more important role than the length of PEO chains, while for large proteins or particles the length of PEO chains is the key element to control. In reality, however, it is rather difficult to fully control *both* the PEO surface density and the brush length as the majority of PEO surface immobilization processes are self-limited by the same effect they aim to achieve: steric self-repulsion. Because of that, there is a considerable interest to design experiments in which both the control of a wider range of densities of surface-attached PEO chains and the measurement of protein adsorption in situ without intermediary wash steps—that could remove weakly adsorbed molecules—are available.

B. In Situ SPR Measurements of PS-PEO Monolayer Protein Interactions

We have recently conducted a set of in situ protein adsorption experiments on surface-attached PEO chains using insoluble poly(styrene)–poly(ethylene oxide) (PS-PEO) block copolymer monolayers at the air–buffer solution interface [52]. Monolayers were designed to cover a wide range of PEO surface densities, from isolated PEO chains to the packed PEO chains in the so-called brush configuration. Protein adsorption onto such monolayers was measured using an in situ surface plasmon resonance (SPR) method adapted to work at the air–solution interface, as outlined in Section II.D. By depositing varying amounts of the PS-PEO block copolymer (MW_{PS} 12.2 kDa, 117 monomers; MW_{PEO} 23.9 kDa, 543 monomers, Polymer Source Inc., Canada) at the interface and by keeping the predetermined barrier position constant, it was possible to set the surface pressure in the monolayer to a desired magnitude and thus achieve different surface density of PS-PEO molecules. Protein adsorption experiments were conducted at several surface pressures, $\pi_{PS\text{-}PEO}$, that resulted in different PEO surface density, σ_{PEO}, and different PEO chain configurations (Table 1) [53,54].

Figure 9 shows the results of human serum albumin (HSA) adsorption kinetics measured with the SPR probe (RIU change versus time, upper panel) and parallel

TABLE 1 PS-PEO Monolayer Surface Pressure, the Resulting PEO Surface Density, and a Description of PEO Chain Configuration

Surface pressure, $\pi_{PS\text{-}PEO}$ (mN/m)	PEO surface density, σ_{PEO} (nm^{-2})	PEO chain configuration (after Refs.)
1.6	0.00645	*Very low surface density regime.* PEO chains are individually adsorbed at air–buffer interface.
5.3	0.01287	*Low surface density regime.* PEO chains adsorbed at the air–buffer interface start to force each other to submerge into the buffer subphase.
9.7	0.01934	*Intermediate surface density regime.* Pseudo–first order phase transition conditions where the PEO chains are confined to a smaller and smaller area, thereby forcing other neighboring chains to leave the air–water interface and extend into the buffer subphase.
10	0.03245	*Onset of the high surface density regime.* PEO chains are desorbed from the air–buffer interface and all become extended in the subphase; the onset of a "brush" configuration.
10.5	0.05155	
11	0.09747	
12.7	0.11947	*High surface density regime.* Brush configuration.
27	0.14286	*High surface density.* PEO chains are even more compressed and stretched in the brush configuration.

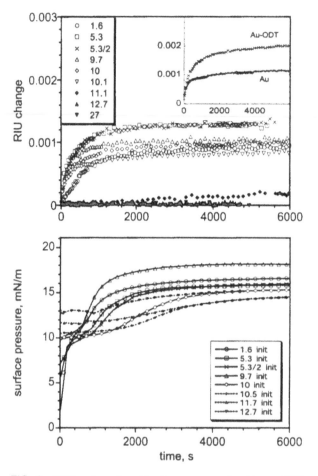

FIG. 9 HSA adsorption kinetics measured with an SPR probe (upper panel) and parallel surface pressure changes measured using the Wilhelmy plate (lower panel). The inset in the upper panel shows the control HSA adsorption kinetics onto the unmodified and hydrophobically modified SPR probe gold surfaces.

surface pressure changes measured using the Wilhelmy plate (surface pressure versus time, lower panel). The SPR method is a surface-sensitive technique with a sensing depth of a fraction of a micrometer [55]; hence, the SPR signal is proportional to the mass of protein accumulated below (i.e., *adsorbed*) or inserted into the PS-PEO monolayer. One finds the HSA adsorption kinetics curves distributed between two distinct groups: one group below and other above $\pi \approx 11$ mN/m in the PS-PEO monolayer (Fig. 9, upper panel). Knowing that the surface pressure in the PS-PEO monolayer is also determining the surface density of PEO chains, one concludes that there is a threshold PEO surface density above which HSA will neither adsorb onto nor insert into the monolayer. Indeed, for $\sigma_{PEO} > 0.0975$ PEO chains nm^{-2} (i.e., one chain per 10.26 nm^2), there is only negligible adsorption/insertion of HSA. By comparing this area per PEO chain with the radius of gyration of an isolated PEO chain in water (R_g = 7.67 nm) [56], one finds that at the threshold surface density the PEO chains are compressed against each other and extended in solution.

C. Protein–PS-PEO Monolayer Interactions Sensed by the Surface Pressure Changes

It is interesting that in the regime of lower PEO surface density the steady-state adsorption kinetics of HSA detected by SPR are grouped at a similar adsorption level; i.e., there is only a weak dependence of HSA adsorption on PEO surface density in the low surface density regime. At these conditions, HSA molecules are able to insert themselves into the monolayer and/or adsorb onto it. The surface pressure changes give more insight into what happens to HSA molecules. In the low surface density regime all surface pressure kinetics display a kink after the initial rise of surface pressure (solid line kinetics, lower panel in Fig. 9). The interpretation of such a kink in not straightforward as the surface pressure change is not proportional to the amount of HSA adsorbed or inserted into the monolayer. It is speculated here that the initial rise of the surface pressure up to the kink corresponds to the insertion of the HSA molecules into voids of the PS-PEO monolayer, while the changes of the surface pressure beyond each kink correspond to protein molecules unfolding at the interface. This speculation is supported by two findings: (1) each kink in surface pressure kinetics occurs at a very similar surface pressure, roughly around $\pi \approx 10$ mN/m, and (2) the total change of the surface pressure from initial surface pressure until the step (kink) becomes smaller with the increase of the initial surface pressure. As the HSA molecules adsorb to the interface, they increase the surface pressure in the monolayer and thus force the PEO to stretch into the subphase. At the adsorption time corresponding to the kink, the SPR kinetics level off, approaching the steady state (Fig. 9, upper panel). The SPR signal will not increase past that stage, yet the surface pressure continues to rise due to the protein. When the process of HSA insertion starts at higher initial surface pressures there is less area available for the insertion of HSA into the monolayer. Finally, at high initial surface pressures, $\pi > 10.5$ mN/m, the kink disappears altogether and is replaced by an induction period (dashed line kinetics, Fig. 9, lower panel). The absence of any surface pressure change in the first 30 min can be taken as another sign that the PS-PEO layer is resisting protein insertion. However, at longer times a smaller increase of surface pressure typically occurs, suggesting that, although there is only very small adsorption measured by SPR, even such a small amount of adsorbed protein can lead to a surface pressure increase.

Figure 10 shows the average HSA layer thickness, calculated from the SPR results, plotted as a function of PEO surface density. The layer thickness is calculated from the magnitude of the RIU changes at the steady state in each adsorption kinetics [55]. This film thickness should not be interpreted as a physical thickness of a continuous-protein thin film attached to the PS-PEO monolayer, but as an optically measured increase of the HSA mass at the interface. One important piece of information from Fig. 10 is that there is a threshold PEO surface density above which protein does not adsorb to the interface. One expects that the threshold PEO surface density is protein size dependent. That is, small proteins may penetrate the grafted PEO chains more easily than very large protein molecules; hence, a higher threshold PEO surface density may be required for preventing small size protein adsorption.

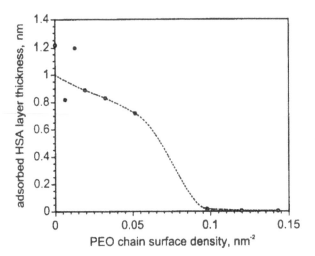

FIG. 10 Average HSA layer thickness calculated from the SPR results and plotted as a function of the PEO surface density. The layer thickness is calculated from the magnitude of the RIU changes at the steady-state level of adsorption kinetics.

Experiments with a large spherical protein, ferritin (*MW* 680 kDa, data not shown), indicated that the PEO surface density threshold between adsorption and no adsorption shifts to a lower value (G. Jogikalmath and V. Hlady, in preparation).

In order to make a surface resistant to adsorption of different proteins from a multiprotein mixture such as blood plasma, one may explore the so-called bimodal PEO brush. In a bimodal PEO brush, the shorter PEO chains are interdispersed between longer PEO chains and expected to force the long PEO chains to extend even further from the interface [57,58]. In such a case, a dual protein repulsion action is expected: (1) large proteins are intercepted by extended long PEO chains, and (2) smaller proteins, which may penetrate in between the long PEO molecules, are intercepted by the interdispersed shorter PEO chains. Preliminary results in our laboratory indicate that the bimodal PEO brush layers are indeed more effective in preventing protein adsorption than either of the two monomodal parent PEO brush layers at the same surface density.

ACKNOWLEDGMENTS

The authors wish to acknowledge financial support from NIH R01-HL44538 (for the PS-PEO protein binding study) and thank Dr. X. Du for confirming the reported ferritin adsorption behavior using the SPR technique.

REFERENCES

1. D. Möbius and R. Miller, eds. *Proteins at Liquid Interfaces*. Elsevier, New York, 1998, pp. 1–498.
2. A. Watts, ed. *Protein–Lipid Interactions*. Elsevier, New York, 1993, pp. 1–379.

3. Interactions of biopolymers with model membranes. International Bunsen Discussion Meeting, Halle, Germany, 2000.
4. C. H. Giles, S. D. Forrester, and G. G. Roberts. In *Langmuir–Blodgett Films* (G. G. Roberts, ed.). Plenum Press, New York, 1990, pp. 1–15.
5. H. Brockman. Current Opin. Struct. Biol. *9*:438–443, 1999.
6. H. T. Tien and A. L. Ottova. J. Membrane Sci. *4886*:1–35, 2001.
7. H. T. Tien. *Bilayer Lipid Membranes (BLM): Theory and Practice.* Marcel Dekker, New York, 1974.
8. E. Sackmann. Science *271*:43–48, 1996.
9. V. Vogel and D. Möbius. Thin Solid Films *159*:73–81, 1988.
10. H. Brockman. Chem. Phys. Lipids *73*:57–79, 1994.
11. R. Maget-Dana. Biochim. Biophys. Acta *1462*:109–140, 1999.
12. J. H. Fendler. *Membrane-Mimetic Approach to Advanced Materials.* Springer-Verlag, Berlin, 1994, pp. 1–203.
13. G. L., Jr., Gaines. *Insoluble Monolayers at Liquid–Gas Interfaces.* Wiley-Interscience, New York, 1966, pp. 1–198.
14. W. Cordroch. *Polarisation unpolarer Moleküle in Monofilmen und Beeinflussung chemischer Gleichgewichte durch geplante Dipolschichten*, in Mathematisch-Naturwissenschaft. Ph.D. dissertation, Georg-August-Universität, Göttingen, 1991.
15. D. Möbius, W. Cordroch, R. Loschek, L. F. Chi, D. Dhathathreyan, and V. Vogel. Thin Solid Films *178*:53–60, 1989.
16. F. MacRitchie. J. Colloid Interface Sci. *57*:393–397, 1976.
17. D. Möbius. In *Langmuir–Blodgett Films* (G. Roberts, ed.). Plenum Press, New York, 1990, pp. 223–272.
18. D. W. Britt, D. Möbius, and V. Hlady. Phys. Chem. Chem. Phys. *2*:4594–4599, 2000.
19. S. May, D. Harries, and A. Ben-Shaul. Biophys. J. *79*:1747–1760, 2000.
20. K. Wagner, D. Harries, S. May, V. Kahl, J. O. Rädler, and A. Ben-Shaul. Langmuir *16*: 303–306, 2000.
21. N Ben-Tal, B. Honig, C. Miller, and S. McLaughlin. Biophys. J. *73*:1717–1727, 1997.
22. K. S. Mayya and M. Sastry. Langmuir *14*:74–78, 1998.
23. F. Höök, M. Rodahl, P. Brzezinski, and B. Kasemo. J. Colloid Interface Sci. *208*:63–67, 1998.
24. C. A. Johnson, Y. Yuan, and A. M. Lenhoff. J. Colloid Interface Sci. *223*:261–272, 2000.
25. M. Käsbauer and T. M. Bayerl. Biochemistry *38*:15258–15263, 1999.
26. A. P. Maierhofer, D. G. Bucknall, and T. M. Bayerl. Biophys. J. *79*:1428–1437, 2000.
27. J. Teissie. Biochemistry *20*:1554–1560, 1981.
28. Y. Yuan and A. M. Lenhoff. Langmuir *15*:3021–3025, 1999.
29. M. Ross, C. Steinem, H.-J. Galla, and A. Janshoff. Langmuir *17*:2437–2445, 2001.
30. W. Knoll, G. Schmidt, H. Rötzer, T. Henkel, W. Pfeiffer, E. Sackmann, S. Mittler-Neher, and J. Spinke. Chem. Phys. Lipids *57*:363–374, 1991.
31. J. Marra. J. Phys. Chem. *90*:2145–2150, 1986.
32. D. Vollhardt and V. B. Fainerman. Adv. Colloid Interface Sci. *86*:103–151, 2000.
33. X. Du, D. W. Britt, and V. Hlady. Unpublished results.
34. K. Okuyama, Y. Soboi, N. Iijima, K. Hirabayashi, T. Kunitake, and T. Kajiyama. Bull. Chem. Soc. Jpn. *61*:1485–1490, 1988.
35. B. Casson and C. D. Bain. J. Phys. Chem. B. *103*:4678–4686, 1999.
36. G. Brezesinski, M. Thoma, B. Struth, and H. Möhwald. J. Phys. Chem. *100*:3126–3130, 1996.
37. D. W. Britt, C. Selle, D. Möbius, and M. Lösche. (Manuscript in preparation.)
38. J. Kim, G. Kim, and P. Cremer. Langmuir *17*:7255–7260, 2001.
39. M. Colic and J. D. Miller. In *Interfacial Dynamics* (N. Kallay, ed.). Marcel Dekker, New York, 2000, pp. 35–82.

40. V. Vogel and D. Möbius. J. Colloid Interface Sci. *126*:408–420, 1988.
41. S. Diaz, F. Amalfa, A. C. Biondi de Lopez, and E. A. Disalvo. Langmuir *15*:5179–5182, 1999.
42. J. M. Harris, ed. *Poly(Ethylene Glycol) Chemistry Biotechnical and Biomedical Applications*. Plenum Press, New York, 1992.
43. D. L. Elbert and J. A. Hubbell. Annu. Rev. Mater. Sci. *26*:365–394, 1996.
44. J. D. Andrade, V. Hlady, and S. I. Jeon. Adv. Chem. Ser. *248*:51–59, 1996.
45. R. L. C. Wang, H. J. Kreuzer, and M. Grunze. J. Phys. Chem. B *101*:9767–9773, 1997.
46. A. Halperin and D. E. Leckband. C. R. Acad. Sci. Paris t1 *IV*:1171–1178, 2000.
47. D. E. Leckband, S. Sheth, and A. Halperin. J. Biomater. Sci. Polym. Ed. *10*:1125–1147, 1999.
48. S. I. Jeon and J. D. Andrade. J. Colloid Interface Sci. *142*:159–166, 1991.
49. S. I. Jeon, J. H. Lee, J. D. Andrade, and P. G. de Gennes. J. Colloid Interface Sci. *142*:149–158, 1991.
50. S. J. Sofia, V. Premnath, and E. W. Merrill. Macromolecules *31*:5059–5070, 1998.
51. A. Halperin. Langmuir *15*:2525–2533, 1999.
52. G. Jogikalmath. Ph.D. dissertation, University of Utah, Salt Lake City, UT, 2002.
53. H. D. Bijsterbosch, V. O. de Haan, A. W. de Graaf, M. Mellema, F. A. M. Leermakers, M. A. Cohen Stuart, and A. A. van Well. Langmuir *11*:4467–4473, 1995.
54. M. C. Faure, P. Bassereau, M. A. Carignano, I. Szleifer, Y. Gallot, and D. Andelman. Eur. Phys. J. *B(3)*:365–375, 1998.
55. L. S. Jung, C. T. Campbell, T. M. Chinowsky, M. N. Mar, and S. S. Yee. Langmuir *14*:5636–5648, 1998.
56. K. Devanand and J. C. Selser. Macromolecules *24*:5943–5947, 1991.
57. A. M. Skvortsov, A. A. Gorbunov, F. A. M. Leermakers, and G. J. Fleer. Macromolecules *32*:2004–2015, 1999.
58. E. P. K. Currie, M. Wagemaker, M. A. Cohen Stuart, and A. A. van Well. Macromolecules *32*:9041–9050, 1999.

17

Local and Global Optical Spectroscopic Probes of Adsorbed Proteins

VLADIMIR HLADY   University of Utah, Salt Lake City, Utah, U.S.A.

JOS BUIJS   Uppsala University, Uppsala, Sweden

I. INTRODUCTION

The conformation of adsorbed proteins has a significant influence on the bioactivity of the proteins or subsequent processes like biofouling, implant rejection, and others. The design of protein-contacting materials requires a detailed knowledge of protein conformation in the adsorbed state as a function of the physical and chemical properties of the sorbent surface. In addition to the protein biological function, it has been shown that conformational changes are a crucial factor in describing the adsorption process. Nowadays, there are many experimental methods available to directly access information on the conformation of adsorbed proteins. Among them, one of the most prominent techniques is optical spectroscopy. In this chapter we review the principles of optical spectroscopy as they apply to a population of protein molecules adsorbed at an interface. We start by reviewing the methods for enhancing the signal from the interfacial molecules and follow with a short description of three spectroscopic techniques most often used in protein adsorption studies: fluorescence, infrared absorbance, and circular dichroism spectroscopy.*

A. Interaction of Protein Molecules with Light

The interaction between light and molecules represents one of the fundamental problems in quantum optics [4]. It is impossible to visualize a single light quantum, a photon, propagating at the speed of light through space because of the inherent uncertainty about the photon's position. Nevertheless, one can imagine a stream of photons moving through space: a light beam. The electromagnetic field associated

---

*This choice is based solely on the authors' experience and preferences rather than on the applicability of spectroscopic techniques to protein adsorption studies or on the quality of spectroscopic information about adsorbed proteins. The reader is referred to recent monographs [1–3] which describe other optical spectroscopic techniques such as surface enhanced Raman scattering, second harmonics generation, and others.

with each proton will induce a time-dependent charge redistribution of a finite dimension in its surrounding. If the space through which the stream of photons moves is occupied by a protein molecule, the entities interacting with the photons' electromagnetic field may be a covalent bond between two atoms, a molecular orbital, some delocalized states, or some vibrational or rotational modes involving a part or whole protein molecule. The complexity of these interactions is simplified by modeling the molecule as a collection of dipoles which interact with the propagating electromagnetic field [5]. It will take the electromagnetic field traveling at the speed of light a mere 3×10^{-18} s to zip through a protein of average size. And yet the effect of the traveling electromagnetic field on the surrounding dipoles in the molecule produces experimentally measurable quantities: changes in transmitted light intensity, polarization, and rotation and differences of energy and angle of the scattered light. Measuring and interpreting these changes falls in the realm of optical spectroscopy.

B. Surface Versus Bulk Sensitivity

The ultimate success of every experimental technique depends on how well the population of molecules of interest is represented in the experimental signal. Translated to the language of protein adsorption: how easy is it to distinguish the signal related to the adsorbed protein population from the signal originating from everything else in the system? Here, it is illustrative to compare the protein number density per unit volume with the number density per unit area. As an example, let's take a 0.1-cm-thick flow cell interfaced at two opposing sides by two optically transparent, flat 1-cm^2 area surfaces. If a 1-mg/mL protein solution flows through such a cell, adsorption to these surfaces may develop, leading to a fraction of μg/cm^2 of adsorbed proteins. We assume that the adsorbed amount is a hefty 0.4 μg/cm and find the protein bulk-to-interface ratio to be above 100. When such a cell is placed into a spectrometer and analyzed in transmission mode, less than 1% of the signal will originate from adsorbed molecules: the signal will be dominated by the bulk solution protein molecules. Clearly, a method is needed for suppressing the signal from the bulk dissolved protein molecules, while enhancing the signal from the interfacial protein population. The suppression of the bulk signal can be achieved by making the cell much thinner than 0.1 cm, which, at some point, becomes quite impractical. The second approach, namely, to enhance the signal from adsorbed molecules, can be achieved either by enhancing the electromagnetic field right at the interface and/or by confining it to the interfacial region only. Nearly all *surface-sensitive* spectroscopic techniques use some form of electromagnetic field enhancement; some by using an evanescent surface wave, others by using highly reflective and metal surfaces.

1. Optics of Total Internal Reflection and Evanescent Surface Wave

A light beam propagating through a solid and encountering a solid/liquid interface will totally reflect from it when the angle of incidence becomes larger than the "critical angle," θ_c:

$$\theta_c = \sin^{-1}\left(\frac{n_1}{n_3}\right) \tag{1}$$

where n_1 and n_3 are the refractive indices of the liquid and the solid, respectively.

Even at the total internal reflection conditions, some of the electromagnetic field penetrates the interface and delivers the electromagnetic energy to the interfacial region in the liquid by propagating parallel to the surface by a Goos–Hansen shift. The theoretical basis of evanescent wave optics has been well described [6–8] and only a simple description of the evanescent wave will be given here. The electric field amplitude of the evanescent wave, E, decays exponentially in solution with distance normal to the interface (z-direction):

$$E = E_0 e^{-z/d} \tag{2}$$

where E_0 is the electric field amplitude right at the interface and the characteristic penetration depth, d, is defined by the angle of incidence, θ, wavelength of light in free space, λ_0, and refractive indices of the solid and the liquid (Fig. 1a):

$$d = \lambda_0 (2\pi)^{-1} (n_3^2 \sin^2\theta - n_1^2)^{-1/2} \tag{3}$$

The magnitude of d is on the order of λ_0 or smaller except at the critical angle conditions where $d \to \infty$. Hence, the molecules in the region outside of the evanescent surface wave (at $z \gg d$) will not feel the presence of the electromagnetic field. Moreover, the intensity of the evanescent surface wave, I_e, is proportional to E^2, and increases to nearly twice the incident light intensity as one approaches the surface. This enhancement of the intensity combined with the exponentially decaying field improves the surface sensitivity of spectroscopic methods which utilize the evanescent surface wave. At low surface coverage the presence of protein molecules at the interface will not affect the optics of total internal reflection (TIR). However, when a dense protein layer is formed, it will present itself to the electromagnetic field as another optically distinct layer with a refractive index, n_2, affecting the local electric field amplitude. The details on the TIR optics of three- and multilayered interfaces can be found elsewhere [8–10].

2. Surface Selection Rules and Metal Surface Enhancement of the Electromagnetic Field

Smooth metal surfaces reflect light very efficiently. In the external reflection from metal surfaces at the near-grazing incidence angle, light experiences a 90° phase shift upon the reflection. A shift in phase by 90° for the light polarized perpendicularly to the metal surface will nearly double the electric field amplitude at the site of reflection, while the electric field amplitude of the light polarized parallel to the metal surface will cancel itself upon reflection. The process results in an electric field enhancement for only one light polarization: the reflection at the metal surface produces a *surface selection rule*. Only those molecules whose transition dipoles are oriented perpendicularly to the surface will effectively interact with the incoming light (Fig. 1b) [5]. The advantages of external reflection spectroscopy are based on this surface selection rule; the main disadvantage is the presence of the electromagnetic field in the bulk phase outside of the metal, which precludes the use of this method in situations where dissolved protein molecules are present in the bulk phase.

Some metal surfaces, like metallic silver, will, when roughened on the scale of tens of nanometers, enhance the electromagnetic field in their very close proximity. The enhancement is quite large (up to a factor of 10^6!) and decays rapidly at very small distances away from the metal surface (Fig. 1c). This electromagnetic field

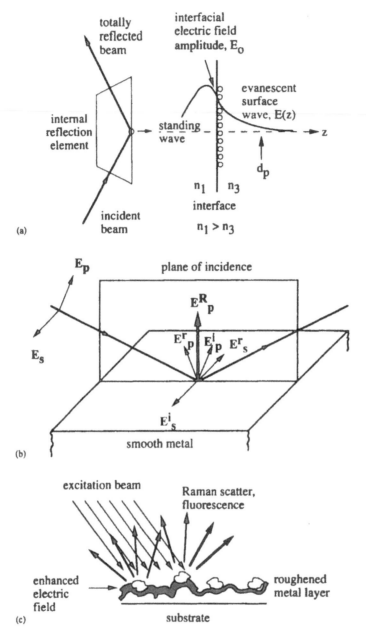

FIG. 1 Three mechanisms of enhancement of the electromagnetic field at interfaces. (a) The optics of total internal reflection creates an evanescent surface wave which decays on the optically rare side of the interface with a characteristics length of d_p. (b) The phase shift upon the reflection of light at a smooth metal surface cancels the electric field components (E_s) oriented perpendicular to the plane of incidence while nearly doubling the components oriented parallel to the plane (E_p). The resulting electric field at the site of reflection (E_p^R) will preferentially interact with the molecular transition dipoles that are oriented in the same way (i.e., perpendicular to the surface). (c) Roughened metal surfaces or small metal islands will locally enhance the strength of the incident electromagnetic field. The enhanced field will, in turn, interact with the molecules that are in contact with the metal. The enhancement effect is quite large and is often used in Raman scattering and fluorescence spectroscopy.

enhancement is frequently utilized in surface enhanced Raman scattering (SERS) experiments [11]. The advantage of such a large electromagnetic field enhancement is found in compensation for the high number/density ratio between the bulk and adsorbed proteins. Because of the use of a metal surface, SERS is often combined with some electrochemical means of studying protein adsorption [12,13]. The main disadvantage is the need for a roughened metal surface, although a thin overlayer of a dielectric has been shown to leave enough of the enhanced electromagnetic field at the dielectric/air (or liquid) interfaces [5].

The choice is now for the researcher to decide which enhancement mechanism to use (Fig. 1a–c): (1) evanescent surface wave of given energy (λ), (2) external reflection at a smooth metal surface and its surface selection rule, and/or (3) a roughened metal surface enhancement. These three types of enhancements will not all be applicable to every spectroscopic method. For example, fluorescence emission is often quenched near metal surfaces, and a dielectric overlayer is required to displace the protein layer away from the metal to avoid quenching. A roughened metal surface will not support specular reflection and the reflectance/absorbance measurements can be difficult. Furthermore, the optical requirements for the sorbent surfaces will limit the variety of sorbent materials suitable to study protein adsorption. Similar to the optical spectroscopy of proteins in the bulk phases, the use of a single spectroscopic technique is less likely to succeed in probing different responses of adsorbed protein molecules. A more comprehensive, multitechnique approach is always preferred and recommended.

C. Proteins' Private and Public Lives

Which information about the protein structure and dynamics is experimentally accessible to optical spectroscopy measurements? Optical spectroscopy may seem to be a poor choice for studying protein structure when compared with x-ray diffraction and NMR spectroscopy. Yet, these two techniques require milligrams of protein, either in an ordered, crystalline state (x-ray diffraction) or in solution (NMR) and will both provide only time-average structural information [14]. Although optical spectroscopic techniques do not give comprehensive structural information, they do require much less protein and provide structure-related information in a dynamic, time-resolved fashion that depends on the duration of a particular spectroscopic event.

In terms of a molecule's dynamics we differentiate between a protein's "private" and "public" life [15]. In a protein's private life we include the events that are occurring on picosecond time scales; in other words, the events that are short enough to involve the protein itself or allow only a very limited coupling with the protein's surrounding. Absorption of a photon by a protein chromophore or excitation of various vibrational and rotational modes of the molecule are examples of a protein's private life. Other spectroscopic events may occur on time scales of a nanosecond and longer, which will allow enough time for the protein's public life, such as ligand binding, diffusion to/from surfaces, and other even slower events, to "show" through. For example, the lifetime of a fluorophore in a protein is on the order of a few nanoseconds, a sufficient amount of time for fluorescence spectroscopy to sample changes in the dielectric environment around the fluorophore or monitor how the local (fluorophore) and global (protein) rotation influence the polarization of the emitted light.

D. Local Versus Global Spectroscopic Probes of Protein Structure and Dynamics

The interaction of light with a protein molecule may occur locally at the level of few covalent bonds, as well as more globally when a larger set of covalent bonds interacts with the light's electromagnetic field in a collective fashion. Each observed spectroscopic phenomenon thus contains information about its locus of origin. In order to simplify the analysis of spectroscopic observations, we divide them into two large groups: local and global spectroscopic probes.* The techniques which primarily provide local spectroscopic information about a protein molecule are fluorescence and SERS spectroscopy. Absorption of light, especially in infrared and circular dichroism spectroscopy will be considered here as global spectroscopic probes because they sample and sum multiple local interactions.

Figure 2 shows a protein structure of human growth hormone (hGH) with the various loci of global and local spectroscopic information. For example, a single tryptophan residue in the protein molecule (Fig. 2a) will be the locus of fluorescence emission in UV. Alternatively, an externally added fluorophore, like dansyl or some other, may also be used to obtain local spectroscopic information. Global techniques, like circular dichroism (CD) operated in the far UV, will selectively interact with optically active moieties in the protein molecule, such as the α-carbon atoms in the protein's polypeptide backbone. The secondary structures in which these atoms participate, such as α-helix, β-sheet, or random coil, will produce differences in the CD spectra (Fig. 2b). CD sampling of other asymmetric moieties in the protein molecule will occur when the light wavelength coincides with the moiety absorption band. The SERS spectroscopy may locally sample the polypeptide backbone and amino acid side chains in one of the α-helices (Fig. 2b). Infrared (IR) spectroscopy will sample IR-active vibrational modes of a protein's covalent bonds. In some cases these modes are affected by the protein secondary structure. For example, the $C{=}O$ and $N{-}H$ groups of the peptide bond form hydrogen bonds whose arrangement depends on the particular type of protein secondary structure (Fig. 2c). The information about the protein secondary structure is typically sought in the absorption bands of a protein's amide groups [16].

Both global and local spectroscopic probe techniques will provide valuable information about protein structure and dynamics. In applying these techniques to the adsorbed protein molecules, one is interested in the changes of protein structure and dynamics caused by the process of adsorption to a particular interface. Hence, one needs a reference state. In most cases, the reference protein state is its native conformation in bulk solution, although an adsorbed protein at some "standard" interface can also be designated as a reference state.

---

*We keep in mind that such a sharp distinction between the two types of protein probes may not be entirely justified in all cases.

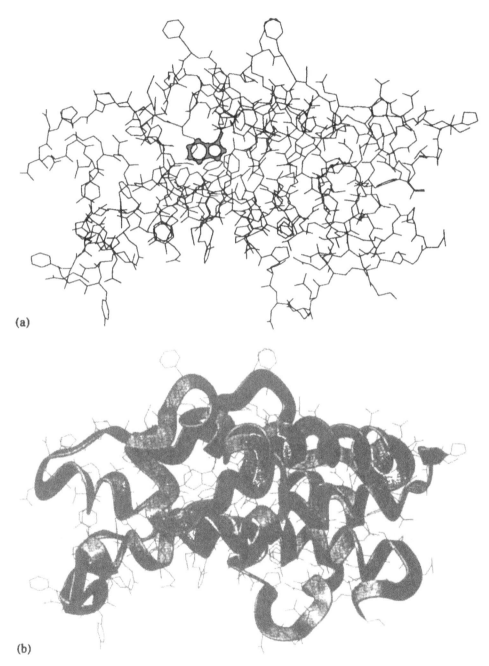

(a)

(b)

FIG. 2a, b Examples of local and global optical spectroscopic probes in a protein molecule (human growth hormone, hGH). (a) Single Trp residue (Trp86) in hGH molecule acts as a local fluorescent probe. (b) Ribbon drawn through the hGH polypeptide backbone indicates various secondary structural elements which affect the hGH CD spectrum in the far UV.

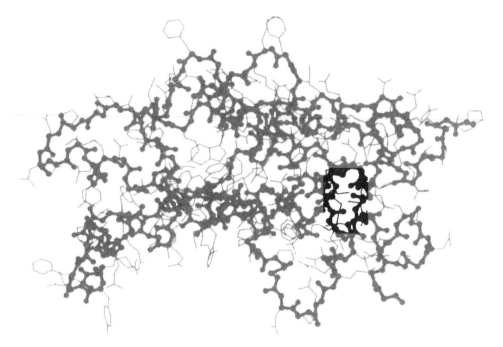

FIG. 2c Ball-and-stick model of the hGH polypeptide backbone containing the C=O and N—H groups used as probes in IR spectroscopy.

II. FLUORESCENCE AS A LOCAL SPECTROSCOPIC PROBE OF PROTEIN MOLECULES

The success of fluorescence spectroscopy in studying dynamics of biological macromolecules lies in its ability to access local fluorophores in biological macromolecules and use the lifetime of their excited states as a time-window for the observation of molecule dynamics [17,18]. In particular, the nanosecond and longer fluorescence time-window enables one to view the evidence of a protein's "public" life, i.e., its interactions with the surrounding solvent molecules, ligands, substrates, and interfaces. In the following section we will review the essential features of various fluorescence probes as they apply to the adsorbed protein structure and dynamics.

A. Protein Fluorescence

Nearly all proteins fluoresce naturally. The intrinsic protein fluorescence has been the subject of several monographs [17–19]. The presence of tryptophan (Trp), tyrosine (Tyr), and phenylalanine (Phe) residues in protein structure is responsible for intrinsic protein fluorescence in UV. In some proteins, prosthetic groups, like pyridoxyl phosphate in phosphorylase B or the heme group in cytochrome c, are also fluorescent. In UV, the fluorescence from Phe is so weak that its contribution can be safely neglected. The protein absorption band at 280 nm is due to Trp and Tyr residues. Between the two, the Trp residue has a larger extinction coefficient; however, there are usually more Tyr than Trp residues in proteins. The two also participate in an energy transfer process which may complicate the interpretation of observed fluorescence. In order to avoid problems of Tyr → Trp energy transfer, one often excites protein UV fluorescence at 295 nm, which will predominantly excite Trp residues. The most commonly used feature of protein fluorescence originates from the Trp residues, whose emission is a sensitive probe of the polarity of each residue's environment and its dipolar relaxation. Several spectral forms of protein Trp residue fluorescence have been categorized and experimentally observed in proteins (Table 1) [19].

The spectral form A has been found to exist only in two proteins: azurin and bacteriorhodopsin. The spectral form S originates from Trp forming a 1:1 excited state complex (exciplex) with a polar group inside a hydrophobic protein core. The form I is from a Trp residue which forms a 2:1 exciplex with some polar group. The larger Stokes shifts of the spectral forms II and III originate from Trp residues in contact with water molecules. The difference between the two forms comes from the mobility of water molecules in contact with Trp: in the form II the Trp indol moiety is in contact with water molecules whose mobility is much smaller than the mobility of the bulk water. In the spectral form III the Trp residue is in contact with the bulk water molecules. Protein unfolding usually leads to a full exposure of Trp residues and the emergence of the Trp spectral form III.

The changes of fluorescence emission maximum have to be carefully considered when interpreting the intrinsic fluorescence from adsorbed proteins. For example, a red shift in adsorbed protein fluorescence emission (i.e., a shift toward longer wavelengths) can be incorrectly interpreted as a (partial) protein unfolding occurring upon adsorption. As stated above, it has been shown that the Trp residue reports on the local changes in its environment as sampled during the lifetime of its excited state [20]. The absorption of a photon by the indole ring in Trp residue instantaneously changes its transition dipole, which, in turn, forces all other dipoles in the vicinity of the residue to respond. If the induced dipolar relaxation is fast, on the time scale of the excited state, the transition dipole–induced dipole relaxation will take some energy away from the excited state and thus shift the energy of emitted photons toward the longer wavelengths. On the contrary, if the induced dipolar relaxation is slow on the time scale of the excited state, the transition dipole will cease to exist before the dipolar relaxation will be able to take energy from the excited state, and thus no wavelength shift will occur. In these two and other similar

TABLE 1 Emission Maxima and Bandwidths in Different
Spectral Forms of Proteins' Trp Fluorescence

Spectral form	λ_{em} maximum (nm)	Bandwidth (nm)
A	307 ($_{sp}$300)[a]	—
S	316–317	—
	($_{sh}$305–307, $_{sh}$320–330)[a]	
I	330–332	48–50
II	340–342	53–55
III	350–353	59–61

[a]sp: secondary peak; sh: shoulder.

scenarios,* the protein's global structure does not have to change in order for the spectral changes to be observed: the only change needed is in the very local environment of the excited state. The rate of local dipolar relaxation may be affected by the presence of an interface and various surface forces. For example, the interface to which protein molecules adsorb may be charged. The electrostatic field of a charged interface will affect both the orientation of the protein's dipoles and the rates by which they will respond to the newly created transition dipole moment of the excited state.

The excited-state lifetime is sometimes taken as a measure of protein unfolding in solution [17]. However, in the interfacial protein studies it is important to recognize that the lifetime of any excited state is also affected by the proximity of a dielectric interface [21–23]. Accordingly, one expects the process of proteins' adsorption to result in a changed fluorescence quantum yield, as noted in several studies [24–27].

1. Extrinsic Fluorophores

In the experimental protein adsorption studies it is frequently convenient or necessary to add an external fluorophore to the protein molecule. Such a fluorophore may have some experimentally advantageous spectral property, like an adsorption band which coincides with the wavelength of laser lines, a large Stokes shift, a high fluorescence quantum yield, or a high sensitivity to its polarity. Typically, two kinds of extrinsic fluorescent probes for protein studies are distinguished: noncovalent probes which bind to protein molecules as ligands in a reversible fashion, and covalent probes that can be bound to protein molecules in an essentially irreversible fashion through a covalent bond. The various properties of the external fluorescent probes have been summarized in the literature [28–30].

Covalently attached fluorescent probes are nowadays commonly used in fluorescence microscopy for the visualization of protein spatial distribution within the sample [31]. Chemically patterned adsorption substrates will often display differences in adsorbed protein density that can be visualized by fluorescence microscopy [32]. Many concerns about the fluorescence emitted by species associated with interfaces already made above apply to extrinsic protein fluorophores as well. Although different fluorescent labels have been frequently used in protein adsorption studies, there is always a concern about the differences in the interfacial activity between labeled and nonlabeled proteins. In addition, a noncovalently bound probe may have a different affinity for the adsorbed protein than for the same protein in solution [33]. On the other hand, the sensitivity of some extrinsic probes to the polarity of their local

---

*Several scenarios are possible. Assume that in the reference state (i.e., protein in solution) the Trp residue excited state experiences a partial loss of the excited state energy into its dipolar environment which results in the emission of a fluorescence photon at the wavelength, λ_{sol} (the "partially relaxed excited state"). Consider the first case: upon protein adsorption the Trp fluorescence lifetime becomes shorter and/or the dipolar relaxation around the excited residue becomes slower. As a result of the shorter lifetime less energy can be taken from the excited state, resulting in a blue-shifted fluorescence ($\lambda_{ads} < \lambda_{sol}$). Here is the second case: upon protein adsorption the fluorescence lifetime becomes longer and/or the dipolar relaxation around it becomes faster. The resulting energy of the emitted photon becomes lower as seen in a red-shifted fluorescence ($\lambda_{ads} > \lambda_{sol}$). We leave to the reader to explore other scenarios starting from (1) initially fully relaxed Trp excited state and (2) initially fully unrelaxed Trp excited state.

environments [34] and to the state of protein aggregation in the adsorbed layer [35] makes them an attractive choice for protein interfacial studies.

2. Fluorescence Energy Transfer and Polarization

The fluorescence emission from an excited state is rather sensitive to the presence of other nearby molecules, as manifested by ready quenching of fluorescence by external quenchers. One of the quenching mechanisms occurs via singlet–singlet energy transfer [36] that takes place when an excited fluorophore (a "donor") interacts via dipole–dipole interactions with another chromophore (an "acceptor") and transfers its excited-state energy before it can emit a fluorescence photon and return to its ground state. Because the efficiency of energy transfer is very sensitive to the distance from the donor to the acceptor, the fluorescence energy transfer has been called a spectroscopic "ruler" [37] and used to experimentally measure the distances between the residues involved in the transfer. When the energy transfer pair is in an isotropic dilute medium, the efficiency of the energy transfer is related to the inverse sixth power of the separation distance between the donor and the acceptor, spectral property of each chromophore, their orientation, the refractive index of the medium between them, and the donor fluorescence quantum yield [17].

The ability to attach extrinsic fluorophores to proteins' selected sites has resulted in a number of experimental studies in which the energy transfer has been used for measuring distances in dissolved and membrane proteins [38]. The application of the "spectroscopic ruler" to a population of adsorbed protein molecules is, however, complicated by the presence of a dielectric interface [21]. The process of adsorption concentrates proteins at the interface and thus brings more than one acceptor in the close proximity of a potential energy donor. The two-dimensional arrangements of donors and acceptors may result in a different energy transfer versus average distance law than observed in dilute solutions [39]. The fluorescence of the donor and the absorption by the acceptor may both be affected by the adsorption and need to be independently determined for the adsorbed molecule [40].

The polarization of fluorescence is a powerful tool in protein adsorption studies since it is affected by the flexibility, mobility, and the state of aggregation of adsorbed molecules [41]. A careful choice of extrinsic fluorophore in terms of its limiting polarization and lifetime enables one to fine tune the time window during which these dynamic properties of adsorbed molecules can be evaluated. In general, upon a polarized excitation the emitted fluorescence will be depolarized if the fluorophores gain an excess local mobility or if the adsorption results in a high local concentration of fluorophores leading to a homotransfer of the excitation energy.

B. Total Internal Reflection Fluorescence Spectroscopy

A technique which has been successfully applied to study proteins at interfaces is total internal reflection fluorescence (TIRF) spectroscopy [42,43]. The main reasons for this success are the high sensitivity of fluorescence spectroscopy in combination with the selectivity of the evanescent surface wave for excitation of proteins, which are at or close to the adsorbent surface allowing thus for an in situ monitoring of small amounts of adsorbed proteins. Current developments in both the method of excitation and the detection of the fluorescence signals made it possible to monitor the time-resolved fluorescence of a single molecule at an interface [44]. The popularity of TIRF spectroscopy also lies in its ability to monitor more than one important aspect of the protein adsorption process. The fluorescence emitted by the adsorbing proteins can be quantified and followed *in real time* to study adsorption and desorption kinetics. If fluorescence is emitted by an intrinsic protein fluorophore, like Trp, one can also monitor changes in the polarity of the local environment of the fluorophore and relate them to changes in protein dynamics and/or conformation. Furthermore, fluorescence polarization measurements can be performed to study the orientation and rotational mobility of adsorbed proteins, while photobleaching experiments allow one to study protein exchange processes at the interface and lateral mobility. Lateral mobility studies are discussed in Chapter 9.

The basic principle of TIRF spectroscopy is the excitation of fluorophores by an exponentially decaying evanescent wave formed at an interface (Section I.B.1) and the subsequent detection of emitted fluorescence. To perform a TIRF experiment one needs the basic spectroscopic instrumentation, such as an excitation source, a photon detection system, lenses to control and focus the excitation and emission light, and a total internal reflection element. Addition of the wavelength-selective components, such as monochromators or cutoff filters, which enable accurate selection of the wavelength of the excitation and emission, are required for the spectroscopic analysis of adsorbed protein fluorescence. Higher intensities of monochromatic light for excitation can be obtained using a laser as an excitation source. The disadvantage is the limited choice of laser line wavelengths. A continuous light source such as a high-pressure Xe arc lamp in combination with a monochromator allows for a selection of the excitation wavelength and enables experiments in UV and visible spectral ranges. The charge-coupled devices' (CCD) photon detection systems extend the TIRF capabilities to both spectral and spatial analyses of the fluorescence. Additional optical components such as polarizers can be added in the optical pathway of the excitation and emission source for polarization measurements. In designing the TIRF instrument one faces the concerns about the instrumental sensitivity which are similar to standard solution fluorescence spectroscopy [17].

At given excitation and emission wavelengths, the fluorescence emitted by molecules in the evanescent wave region is proportional to the probability of light absorption characterized by the molecular extinction coefficient, the concentration of molecules, the intensity of the evanescent wave, and the probability of emission characterized by the fluorescence quantum yield [42]. The concentration and intensity are both functions of the distance from the interface into the cell. Another parameter which determines the detected fluorescence intensity is the instrumental factor, which includes the efficiency of fluorescence collection and the instrumental sensitivity. In TIRF experiments this factor can be obtained by a calibration procedure using fluorescence standards of relatively high concentrations. Once the instrumental factor is known, the observed fluorescence signal from adsorbed proteins can be recalculated into a number proportional to the surface density of adsorbed protein molecules by the ratio between the fluorescence quantum yields of adsorbed proteins and fluorescence standards used [24]. It is, however, also possible to quantify the amount of adsorbed protein by using independent calibration methods [24,45,46]. Many studies of protein adsorption kinetics are performed using extrinsic covalent probes, like fluorescein isothiocyanate (FITC), which absorbs and emits light in the visible wavelength region and has a relatively high quantum yield.

TIRF spectroscopy has been used in numerous protein adsorption studies, and basic principles for the study of adsorption kinetics and description of the earlier work using TIRF for both intrinsic and extrinsic fluorophores are reviewed in the literature [8,42,43]. More recently, TIRF spectroscopy has been employed to study more complex protein adsorption systems. The versatility of a CCD camera coupled to the output of a monochromator was employed to study spatially resolved protein adsorption kinetics [47] on heterogeneous surfaces, like chemical gradients [48,49] and patterned surfaces [32]. Another TIRF example is the use of energy transfer probes attached covalently to the same protein molecule. The labeled protein was allowed to adsorb onto a dichlorodimethylsilylsilica (DDS) gradient surface, and TIRF spectroscopy was used to sample the energy transfer efficiency as a function of time in a spatially resolved way (Brynda and Hlady, unpublished). Figure 3a shows the comparison between the fluorescence spectra of IgG molecules which were double labeled with fluorescein (used as a fluorescence donor, approx. 3.6 fluoresceins/IgG) and tetramethylrhodamine (used as a fluorescence acceptor, approx. 2.7 rhodamins/IgG) recorded for the IgG in dilute solution and when it was adsorbed to a DDS surface. The fluorescence spectra recorded during the first 15 s of the double-labeled IgG adsorption onto the DDS surface are shown in Fig. 3b. It is evident that the initially adsorbed IgG molecules undergo rapid changes that lead to a more efficient energy transfer process than the transfer in dissolved protein.

In addition to the adsorption kinetics studies, other aspects of the protein adsorption process can be inferred from the fluorescence emitted by protein's Trp residues [50]. As stated above this residue acts as a local probe of the protein's structure with its fluorescence emission being very sensitive to the polarity and dipolar relaxation of its local surroundings [51]. Another approach is the addition of a fluorescence quencher to a population of adsorbed proteins. The quencher will adsorb the energy of the excited state of the fluorophore, thereby inhibiting the fluorescence signal. This type of energy transfer is only possible if the quencher comes in close contact with the fluorophore; hence, the quenching reveals the accessibility of a given fluorophore to the quencher. One should take into account the possibility that the concentration of the quencher near the interface can be affected by the surface, especially if both the sorbent surface and the quencher are charged [52]. Yet another way to analyze fluorescence from adsorbed proteins is to follow the decay of the excited state in adsorbed proteins [35,53,54]. Fluorescence lifetime studies reveal information on the deexcitation processes in much more detail than can be observed using the steady-state fluorescence. The time-dependent measurements, however, do require very sensitive fluorescence instrumentation and are not as easily employed to study adsorbed proteins as the steady-state fluorescence methods.

1. Orientation and Mobility of Adsorbed Proteins Studied with TIRF

TIRF spectroscopy can give information on the rotational mobility of fluorescent moieties by monitoring the polarization of emission (anisotropy) as a function of the polarization of the excitation light. In principle, this method measures the average motion of the transition dipole during the excited-state lifetime of a fluorophore. In the TIRF geometry the measured anisotropy reflects all rotational motions relative to the planar interface, which can originate from wobbling of the fluorophore inside the protein as well as from rotation of the whole protein molecule. The analysis of

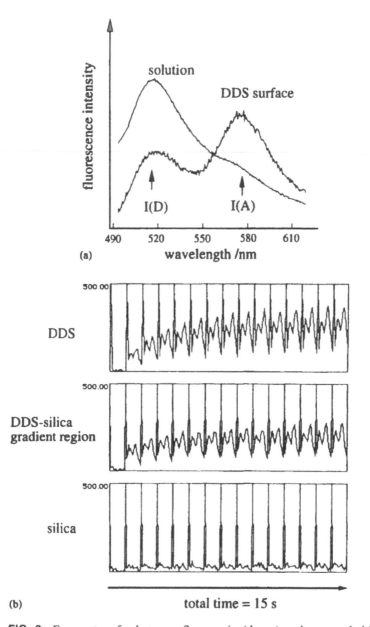

FIG. 3 Energy transfer between fluorescein (donor) and tetramethylrhodamine (acceptor) attached to a single IgG molecule (F-T-IgG). The fluorophores were attached to the IgG using the isothiocyanate chemistry (approx. 3.6 fluoresceins and 2.7 rhodamins per IgG). (a) Comparison between the fluorescence spectra of F-T-IgG in dilute solute and adsorbed to a methylated silica (DDS) surface. The ratio between the fluorescence intensity of the acceptor and the donor, $I(A)/I(D)$, was 0.4 and 1.6 for the solution and adsorbed protein, respectively. The unfolding of F-T-IgG in urea solution did not change the solution spectrum. (b) Fluorescence spectra recorded in 1-s intervals during the first 15 s of the F-T-IgG adsorption onto the DDS–silica gradient surface. The spectra were simultaneously recorded on the three distinct surfaces: on the DDS (upper panel) and silica ends (lower panel) of the gradient surface and on the silica surface, which was approximately half covered with the methyl groups (middle panel). The $I(A)/I(D)$ ratio was found to be different for the three surfaces. It increased as a function of time in the case of the DDS surfaces indicating that rapid changes in the structure of the adsorbed IgG layer are initially taking place (Brynda and Hlady, unpublished data).

rotational mobility from the anisotropy measurements in a two-dimensional system differs from those in a three-dimensional system as described by Morrison and Weber [41]. It is often found that the protein mobility in the adsorbed state is reduced relative to the mobility in dilute solution and can result in a negligible rotational motion during the fluorescence lifetime [50].

Orientation of adsorbed proteins can also be determined using TIRF spectroscopy. The orientation measurements are performed by varying the polarization of the excitation source, which results in a change in the direction of the electric field components of the evanescent wave. The variation in the direction of the electric field results in a selective excitation of those transition dipole moments of fluorophores which are aligned with the field. Consequently, the observed fluorescence intensity depends on the orientation of the fluorophores. The method, first used to study the orientation of dye molecules in phospholipid monolayers [55], was later employed to study the average orientation of adsorbed proteins [56]. By measuring a second variable, such as the polarization of the emitted fluorescence, an orientation distribution of fluorophores can be obtained. Polarized excitation studies of the orientation of adsorbed proteins has been carried out in several laboratories using the porphyrin group of cytochrome c as a model fluorophore [57–60]. The porphyrin group in cytochrome c has a well-defined and relatively large transition dipole moment which is more or less in a fixed position relative to the protein structure.

Full information on the orientation of adsorbed proteins also requires knowledge of the distance between the protein and the surface. As the fluorescence intensity is proportional to the intensity of the evanescent wave and the fluorophore concentration, a density profile of the fluorophore can in principle be found by monitoring the fluorescence as function of the evanescent wave intensity profile. Variation of the angle of incidence [see Eq. (3)] changes the intensity profile of the evanescent surface wave in solution, thereby offering the possibility for determining the fluorophore density profile. For monolayers of adsorbed proteins, however, this method has only a limited accuracy [61] because the evanescent wave penetration depth is two orders of magnitude larger than the size of most proteins. Consequently, changing the angle of incidence results in a rather small variation in overall fluorescence intensity from an adsorbed protein layer and the interpretation of experimental data requires an extremely high accuracy of the measurement. A more recent approach is the introduction of an additional experimental variable, namely, the electrical double layer which develops at a charged interface. The electrical double layer has a sharper gradient in the z-direction than the evanescent surface wave and can be controlled by the solution's ionic strength. It has been demonstrated that this sharper profile can be employed in a TIRF study to determine the protein orientation using fluorescein-labeled protein, where fluorescein acts as a pH-sensitive probe [62].

2. Adsorption-Induced Conformational Changes in hGH Monitored by Fluorescence Spectroscopy

In this section we present an example of how the fluorescence signal, emitted by a single intrinsic Trp residue in a model protein, hGH, is used to study adsorption-induced conformational changes. For this purpose, hGH was adsorbed onto a hydrophobic methylated silica surface from a solution containing 0.1 mg/mL protein and buffered with 0.15 M PBS at pH 7. More experimental details and a similar study of the hGH adsorption onto a variety of surfaces, including self-assembled mono-layers and Langmuir–Blodgett, layers can be found elsewhere [50,63]. TIRF spec-troscopy was employed to selectively excite this single Trp residue in adsorbed hGH molecules and subsequently measure its fluorescence emission. The Trp residue is located in the hydrophobic interior of the hGH molecule and is H-bonded to another α-helix strand [64]. The location of the residue makes it a very sensitive local probe for the protein's conformational changes. The effect of adsorption on the Trp fluo-rescence emission is demonstrated in Fig. 4a in which both the fluorescence spectrum of hGH in solution and adsorbed onto a methylated silica surface are shown. The fluorescence emission from the adsorbed hGH is blue shifted, indicating that either

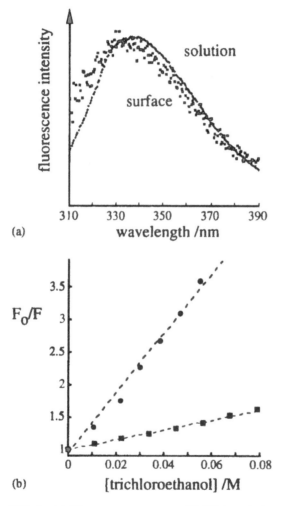

FIG. 4 (a) Fluorescence spectra of hGH in solution (+) and after 0.5 h adsorption onto a methylated silica surface (x). hGH was dissolved in 0.15 M PBS at pH 7. (b) Stern–Volmer plot of quenching by trichloroethanol of hGH fluorescence in solution (squares) and adsorbed onto a methylated silica surface (circles). hGH was dissolved in 0.15 M PBS at pH 7. (From Ref. 50.)

the H-bonding between the Trp residue and the nearby α-helix has been disrupted or the process of dipolar relaxation around the residue has extracted less energy from the excited state (see Section II.A). One should note the importance of the position of this single Trp residue in hGH; similar conformational changes in other proteins would be more difficult to follow if their Trp residue were located on the protein exterior or if more than one Trp residues were present in the protein molecule.

The use of fluorescence quenchers provides additional information about the conformation of adsorbed hGH. Recall that the quencher has to come in close contact with the fluorescence probe in order to absorb the energy of the excited state. In the case of hGH, the quencher has to penetrate into the protein structure to reach the single Trp residue. Hence, the changes in fluorescence quenching reveal the changes in the permeability of a particular hGH conformation. In the present example, quenching experiments are performed by using increasing concentrations of trichloroethanol dissolved in buffer solution. Collisional quenching of hGH fluorescence is described by the Stern–Volmer equation [18], $F_0/F = 1 + K[Q]$, in which F_0 and F are the fluorescence intensities in the absence and presence of the quencher, respectively, $[Q]$ is the concentration of the quencher, and K is the Stern–Volmer constant. The hGH experimental quenching data are shown as a Stern–Volmer plot in Fig. 4b for the protein molecule in solution and adsorbed onto the methylated surface. The difference between slopes in Fig. 4b clearly indicates that the quenching of the adsorbed hGH is characterized by a larger Stern–Volmer constant; i.e., the quenching is more efficient at any given concentration of the quencher for adsorbed than for dissolved protein. The combination of the emission (Fig. 4a) and quenching (Fig. 4b) results indicates that the conformation around the tryptophan residue has been changed and the permeability of hGH conformation for trichloroethanol has been considerably increased upon adsorption on a hydrophobic surface.

III. INFRARED SPECTROSCOPY AND CIRCULAR DICHROISM AS GLOBAL PROBES OF PROTEIN MOLECULES

Although local protein probe spectroscopy techniques such as fluorescence can be used to monitor protein conformational changes, the information obtained is restricted to the microenvironment of the fluorescent probe(s). In order to obtain more global information on the adsorbed protein secondary structure, the following two spectroscopic techniques are frequently used: infrared (IR) spectroscopy and circular dichroism (CD). Both techniques are based on the absorbance of light by the protein molecule. Different variants of infrared spectroscopy are nowadays used in surface adsorption studies. In external reflection–adsorption IR spectroscopy, for example, one uses the geometry of external reflection to measure the reflectance changes by the surface films. The internal reflection IR spectroscopy relies on various types of internal reflection elements to conduct light to the interface where the surface film attenuates its intensity by absorption.*

---

*The field is rich in acronyms. By combining the words infrared, reflection, adsorption, external, spectroscopy, glancing, attenuated, total, and Fourier transform, one can explain all acronyms found in the literature: RAIS, IRAS, FT-IRAS, IRRS, IRRAS, RAIRS, FTIR, ATR, ATR-FT-IR, GIR, IR-ERS, RAS, RAIR, and perhaps a few others [5].

A. Infrared Spectroscopy

In IR spectroscopy, one utilizes the vibrational states of chemical bonds, especially those in the protein backbone, to study the protein structure. The mid-infrared region between 700 and 4000 cm^{-1} is the spectral region most frequently studied. The single vibrational energy is usually accompanied by a number of rotational energy transitions which broaden the absorption band. In the case of macromolecules the contributions from many atoms to a particular vibration complicates the analysis of the IR spectrum. The analysis of the spectral region between 1100 and 1700 cm^{-1} has proven to be a valuable tool to gain information on global properties of the polypeptide conformation and is actually one of the earliest methods employed to study the secondary structure of proteins [65]. The amide IR bands of polypeptides consist of nine characteristic vibrational modes or group frequencies [66]. One of the most useful is the amide I band, located in the wavenumber region approximately between 1600 and 1700 cm^{-1} (i.e., at wavelengths around 6 μm). The amide I band primarily represents the C=O stretching vibrations of this chemical group in the protein backbone (coupled to the in-plane N—H bending and C—N stretching modes). The frequency of this vibration depends on the nature of the hydrogen bonding in which the C=O group participates. This, in turn, is highly sensitive to the secondary structure adopted by the polypeptide chain, e.g., α-helices, β-sheets, turns, and random coil structures and thus provides fingerprints for the protein secondary structure elements [16]. The characteristic frequencies of the different types of secondary structures have been established as a result of systematic studies on a number of protein samples [66–68] and by theoretical calculations [69–71].

In principle, it is also possible to study the protein structure using the amide II band, which represents the N—H bending, and the amide III band, which primarily represents absorption from chemical groups in the side chains. The sensitivity of the amide II region for the polypeptide conformation is relatively small. The amide II region can be useful, however, for quantifying the amount adsorbed [72–74]. The use of the amide III region (1100–1500 cm^{-1}), although sensitive to the secondary nature, is troublesome because these bands are rather weak. Still, the amide II and III regions can be employed to obtain additional information on the results generated by analysis of the amide I region [75,76]. The amide I and II bands in an IR spectrum if IgG molecules adsorbed on a silicon cylindrical internal reflectance crystal are shown in Fig. 5a.

Although IR spectroscopy is one of the earliest methods used for estimating the protein structure, its practical use was limited until the development of the interferometer-based, Fourier-transform infrared (FTIR) instrumentation. FTIR has two

---

FIG. 5 (a) Infrared absorbance spectrum in the amide I and amide II spectral regions of IgG molecules adsorbed on a silica surface. The spectrum consists of an average of 300 scans and is smoothed by a Gaussian filtering over a frequency range of 35 cm^{-1} [118]. The IgG molecules are adsorbed for 1 h from an aqueous solution containing 50 μg/mL IgG and 5 mM phosphate buffer at pH 6. (b) Second-derivative spectrum of the amide I spectral region of adsorbed IgG molecules shown in (a). (c) Curve fitting of individual absorption bands (dashed lines) to the original amide I spectral region (solid line) shown in (a). The curve-fit procedure assumed a Lorentzian shape of the line broadening and is described in Ref. 72.

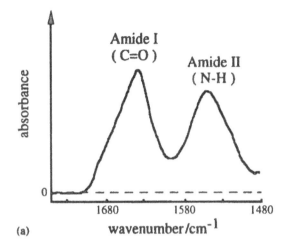

(a)

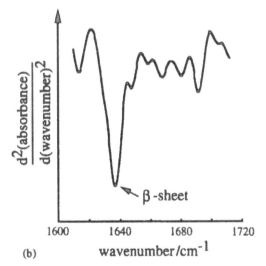

(b)

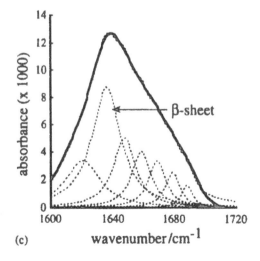

(c)

distinct advantages over the classical IR spectroscopy instruments which were based on the grating monochromators: (1) multiplexing advantage because of the simultaneous detection of all frequencies and (2) luminosity advantage because of the higher light throughput of interferometers over the grating monochromators. Consequently, the IR spectra are acquired more rapidly and with a better signal-to-noise ratio with FTIR instruments. The high sensitivity of FTIR in combination with the totally internally reflected IR beam, commonly referred to as attenuated total reflectance (ATR), offer a valuable method for studying proteins at interfaces [77,78]. The ATR-FTIR technique has proven its value in the analysis of secondary structure elements of proteins in close proximity of the sorbent surface. Examples include studies in which the secondary structure of adsorbed proteins were subjected to alterations due to different adsorbents [72,73,79,80], protein aggregation [81,82], various solvents [83,84], degree of solvation [85], and temperature variations [86]. Similar to the experiments performed with TIRF, ATR-FTIR is also used to access several different characteristics of the protein adsorption process. The infrared absorption intensity can be calibrated [87,88] and transformed in the adsorbed mass/ area amounts, which allows for the study of adsorption kinetics [72,79,89,90]. Since each protein has its specific structure and thus a specific infrared adsorption spectrum, an additional advantage of FTIR over TIRF is the possibility of studying adsorption from protein mixtures without any labeling requirements [91–93]. Furthermore, it is possible to use polarized excitation in an ATR-FTIR geometry and study the orientation and order of adsorbed proteins [94].

Even though ATR-FTIR spectroscopy has proven its value in the study of proteins at interfaces, it is important to realize that the technique also has some practical difficulties. For most research purposes, protein adsorption is preferentially studied in aqueous solutions and involves the problem of interference of the very strong and broad water absorption band around 1640 cm^{-1} with the amide I region. To overcome this problem many infrared studies on protein structures have been performed in D_2O. Using modern sensitive instrumentation, however, it is possible to separate the intense water absorption band from the protein bands. Most of the infrared absorption from solution can be accounted for by subtracting the background spectrum of the solution from the protein absorption spectra. Still a more precise subtraction of solution signal is required as adsorbed proteins displace water molecules from the surface. Fluctuations in the physical properties of the system, such as temperature fluctuations, can alter the optical path length of the IR evanescent wave, thereby changing the intensity of absorption by solution molecules. The simplest and most often used method to overcome this problem is to use a scaling factor in the subtraction of the absorption by the solution utilizing a horizontal baseline in the spectral region where proteins do not absorb (1730–1800 cm^{-1}). Somewhat more advanced algorithms, all based on the subtraction of the solution signal in the spectral region between 1720 and 2700 cm^{-1}, have also been developed [95,96]. Other features which might complicate the analysis of the IR absorption bands sources are the traces of water vapor in the instrument's cell compartment and the overlap between the amide I and amide II bands. The overlap can result in an overestimate of the amount of the structural component contribution commonly determined in the lower wavenumber region of the amide I band. In this same low wavenumber region it is quite likely that some of the IR energy is absorbed by side chains of the proteins [97]. The vibrational motions of the chemical groups can also be affected by the types of interactions these groups can have with the sorbent surface [98], although this is more likely to happen to the side chain groups than to the chemical groups in the protein backbone.

B. Analysis of Protein Conformation from IR Spectra

1. Resolution Enhancement

Although the IR spectrum of an adsorbed protein contains all of the information on the protein's secondary structure, the analysis of an FTIR spectrum for a qualitative estimation, and even moreso for a quantitative estimation, of the protein secondary structure is complicated because of the overlap in absorption bands originating from different structural components. The bands of a protein's structural components are generally blended in one broad-shaped band in the experimentally measured spectrum, as can be seen in Fig. 5a. Improvement of the instrumental resolution is not helpful as the resolution is limited by the natural bandwidth of the absorption bands. However, mathematical procedures are available which will improve the discrimination between the individual bands. Two commonly used procedures for computational resolution enhancement are the second derivative and Fourier self-deconvolution calculations [99]. Both methods have successfully been applied in the analyses of the secondary structure of several proteins [67,68]. It should be emphasized that the resolution of the spectra is enhanced at the expense of a decreased signal-to-noise ratio, and great care has to be taken when analyzing the resolution-enhanced spectra. Besides mathematical resolution enhancement it is also possible to separate overlapping bands by adding experimental variables [75].

2. Peak Assignment to Structural Elements

After establishing the individual absorption bands, the next step is to relate those bands unambiguously to the secondary structure components. This is generally not a problem for a few major structural components, but it might be more troublesome for some smaller peaks located at both extremes of the amide I region. The amide I components associated with β-strands are located in the low wavenumber range between 1620 and 1640 cm^{-1}, and a peak around 1635 cm^{-1} is generally ascribed to antiparallel β-sheets. Components in the lower region may originate from subtle differences in hydrogen bonding patterns in the antiparallel β-strands or from parallel β-sheets [66,67]. In addition, the side chain absorption could contribute to absorption in the low wavenumber part of the amide I region [97]. In most FTIR studies, resolution-enhanced protein spectra show a peak close to 1644 cm^{-1}, which is generally assigned to random coil segments. A component around 1654 cm^{-1} is frequently observed and associated with α-helices. Absorption bands of turns are located in the high wavenumber domain [71], but components associated with β-sheets can also be found in this region. An exact distinction between the location of turns and β-sheets in this high wavenumber domain is not yet unambiguously established. Note that the above-mentioned peak assignments are based on proteins in an aqueous environment and that the frequency position of the absorption bands can shift in D_2O or other solutions. A second-derivative spectrum of IgG adsorbed from aqueous solution onto a silica surface is shown in Fig. 5b. IgG is known to have a high content of β-sheet structure whose absorption peak is clearly visible as a minimum in the second-derivative spectrum. A somewhat different approach has been used in the case of dried protein films on metal surfaces which were studied by an external IR reflection technique (FT-IRAS) [100,101]. The peak assignments have been aided by an independent IRAS study of amino acid adsorbed on metal surfaces [102,103]. However, the inferred orientation of particular protein groups, such as carboxylate, could also have been created by the process of drying and preparing the film for external reflection spectroscopy.

3. Quantitative Analysis

Resolution-enhanced infrared spectra allow a qualitative and semiquantitative identification of the various secondary structures present in the adsorbed proteins. A more quantitative analysis can be performed using curve-fitting methods. The general procedure is to decompose the spectrum into its constituent bands, specifying their shapes, positions, and amplitudes by use of some optimization technique. Then the experimental spectrum is expressed as a superposition of a number of individual absorption bands. For the vibrational band it is generally assumed that they have either a Gaussian or Lorentzian shape or that the spectrum is a convolution of the two. Each Gaussian (and/or Lorentzian) band is initially positioned at the frequency position of the band as observed in the second derivative or Fourier self-deconvolved IR spectrum, followed by an iterative band-fitting procedure. The fitted individual bands are then assigned to the different secondary structure elements and subsequently quantified by using the integrated areas of the individual peaks. The last step of the analysis implies that the extinction coefficients of the structural components are assumed to be equal. This assumption has been validated in an extensive study with 21 proteins in which a good correlation was found between the secondary structures as obtained from FTIR spectroscopy and x-ray crystallographic data, respectively [68].

Resolution-enhanced spectra are mainly used to determine the frequency position of the individual structural components. However, spectral deconvolution can also be used to estimate the bandwidth by varying the input parameters of the deconvolution process. Information on the bandwidth and absorption intensity of the individual bands can also be derived from the second-derivative spectra [72]. Bandwidths of the individual structural components are $15-28$ cm^{-1}. Second-derivative or deconvolved spectra of proteins often reveal that the amide I band consists of six or seven components [68,72,76,104]. Nevertheless, many interpretations are focused on two or three major components. Fitting an absorption band to as many individual bands as can be found might result in a more exact outcome of the quantitative amount of the various structural components. However, it also increases the chance of obtaining physically incorrect results in the fitting procedure. Fluctuations in the spectra which are not correlated to specific chemical structures are also enhanced in the deconvolved and second-derivative spectra and could accidentally be ascribed to an individual absorption band. Furthermore, one needs to take into account that more fitting parameters generally yield a better fit to the original spectrum but do not necessarily lead to a better physical picture of adsorbed components. As quantitative analysis only enhances the possibility of erroneous results, careful spectral interpretation is required. Hence, FTIR spectroscopy is more safely employed to follow protein conformational changes than to quantify protein structural elements in absolute amounts.

In Fig. 5c, the result of a fitting procedure to the IR spectrum of adsorbed IgG (Fig. 5a) is shown [72]. For this fitting procedure, Lorentzian band shapes were assumed and the initial frequency position, bandwidth, and intensity were obtained from the second-derivative spectrum as shown in Fig. 5b. Assignment of the individual peaks to the protein's structural components and quantification of the integrated areas of the individual absorption bands can then be used to study the secondary structure of adsorbed proteins.

4. Adsorption-Induced Conformational Changes in IgG Studied by FTIR Spectroscopy

As an example of an FTIR study of global, secondary structure of adsorbed protein we cite the study performed on the monoclonal IgG molecules and their $F(ab')_2$ fragments whose adsorption was monitored as a function of solution pH [72]. The FTIR results were used to obtain the equilibrium amount adsorbed and the relative amount of random coil structure as concluded from the integrated area of the resolved peak at 1645 cm^{-1}. The IgG molecules were adsorbed onto a circular silicon crystal from a flowing solution containing 50 μg/mL protein and 5 mM buffer. The equilibrium adsorbed amounts are shown in Fig. 6a, and the percentage of random coil structure is shown in Fig. 6b. A maximum in the amount adsorbed occurred near the isoelectric point (iep) of the proteins (which is at pH 5.8 for the monoclonal IgG and at pH 5.9 for its $F(ab')_2$ fragment). The maximum adsorption is often found around the protein's iep, and it is generally explained by the effect of a low net charge density on the protein which results in a relatively high structural stability and hence a more compact protein structure in the adsorbed state. In addition, a decreased electrostatic repulsion between protein molecules allows for a tight protein packing arrangement on the surface. A higher protein net charge density is expected to result in a lower structural stability caused by internal electrostatic repulsion. This hypothesis is supported by the analysis of the adsorbed IgG IR spectra, which show that the amount of random coil structure increases when the pH is further away from the iep. Note that the net charge–related increase in the random coil structure is much larger for the whole IgG than for its $F(ab')_2$ fragment, indicating that the Fc part of the IgG molecule loses more of its structural conformation than the $F(ab')_2$ part. It is well known that under conditions of electrostatic repulsion between protein and a hydrophilic surface, the protein primarily adsorbs due to an entropy gain resulting from an increase in its random coil structure. Thus, the lower amounts of adsorbed $F(ab')_2$ fragments at the pH values above the iep might be related to the effect that the adsorbed $F(ab')_2$ molecules experience a smaller adsorption-induced increase in their random coil structure.

C. Circular Dichroism

As described in the introduction, spectroscopic techniques employed to study the adsorption and conformation of proteins at interfaces are often based on the concept of an electromagnetic field enhancement to selectively excite proteins in close proximity of surfaces. It is, however, also possible to obtain information on the adsorbed state of proteins using techniques and instrumentation originally developed to study proteins in solution, like circular dichroism or spectrofluorometry. Circular dichroism

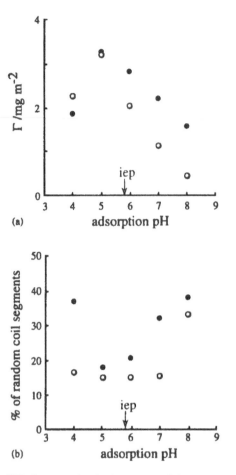

FIG. 6 (a) Adsorbed amount of (b) relative amount of random coil structure of IgG (open circles) and their $F(ab')_2$ fragments (closed circles) adsorbed at a silica surface at different pH values. The equilibrium adsorbed amounts were obtained after 6 h of adsorption, while the random coil structures were obtained for proteins which were adsorbed for an average time of 1 h. The isoelectric point (iep) is indicated on the x-axis. (From Ref. 72.)

(CD) is frequently used to address problems concerning global properties of the protein conformation and is especially sensitive to protein molecule α-helix content [105]. Because of the large number of successful CD studies on protein conformation in solution, the question arises how to most efficiently employ CD spectroscopy to study the conformation of adsorbed proteins. Indirect information on adsorption-induced conformational changes has been obtained by applying CD spectroscopy to proteins desorbed from a surface [106,107]. To study proteins in situ in the adsorbed state a sorbent surface has to be present in the sample volume interrogated by the CD light beam. This sorbent surface should not interfere with the CD measurement (i.e., it should be transparent) and should adsorb enough proteins to obtain a reasonable signal-to-noise ratio. To meet these requirements, experiments have been initially performed using the geometry of stacked quartz plates [108].

Another way to record CD spectra from adsorbed proteins is to employ small colloidal particles, often referred to as ultrafine or nanoparticles, which, when suspended in solution, can serve as a sorbent material for proteins. Because of their small size (<30 nm) nanoparticles provide a large surface area per sample volume for protein adsorption while displaying low to moderate light scattering. The first CD spectra of different proteins adsorbed to small silica particles were reported in the early 1990s [109,110]. These studies clearly confirmed the well-known hypotheses that proteins lose some of their secondary structure upon adsorption and that the extent of conformational changes is correlated with the structural stability of the proteins.

Although the use of these small colloidal particles seems to offer a large expansion in the accessibility of information on the adsorbed protein conformations, there are some limitations which primarily arise from the optical and size requirements for the dispersed particles. It is not possible to produce particles with the required small dimensions from all desired materials. In these experiments the colloidal stability of the dispersion should be rather high to avoid aggregation, which greatly increases the intensity of scattered light. The latter requirement is difficult to achieve for more hydrophobic particles and for particles which are covered with proteins. Finally, to obtain a high sensitivity the particles should consist of a material which is largely transparent in the wavelength region of the applied spectroscopic method. To fulfill these requirements, most CD and fluorescence studies have been performed using small silica particles [109–112]. Recently, the CD studies have been performed on adsorbed proteins using silica coated with aluminum [113] or with particles having a more hydrophobic nature like polystyrene [114] and polytetrafluoroethylene (PTFE) [54]. Another advantage of the small colloidal particles is that their use is not restricted only to CD spectroscopy but can be extended to other spectroscopic methods such as transmission FTIR [115], employed to study the global conformation of adsorbed proteins and fluorescence spectroscopy [116,117] for local information on the protein conformation.

IV.   REMAINING PROBLEMS AND FUTURE OUTLOOK

The large number of papers published during the last few years testifies that the use of the three optical spectroscopy techniques reviewed in this chapter in studying adsorbed proteins is nowadays a very common practice. However, some of the problems still remain and need to be addressed. We discuss two of these below.

1.   Protein Conformation.   A discrepancy between the information provided about adsorbed proteins by the local and global probes can be quite large. Smaller changes in protein structure can produce dramatic effects in fluorescence and no effects in IR or CD spectra. On the other hand, a fractional increase (or decrease) in one of the protein's structural components, as seen by IR spectroscopy (or CD), is difficult to localize to a particular part of the protein. The gap between the local and global information needs to be closed by a careful site-directed labeling and/or mutagenesis which will selectively introduce local spectroscopic probes to designated locations in the protein molecule. Resonance techniques can be used to interrogate these local probes and yield information that will close the existing gap.

2.   Adsorbed Protein Heterogeneity.   Any protein population is characterized by some degree of the conformational heterogeneity. It is likely that once the proteins are adsorbed to an interface, the degree of their conformational heterogeneity will change. In addition, a new orientational heterogeneity appears together with a different supramolecular organization of proteins in the adsorbed layer. In other cases, a balance between (weak) adsorption forces and lateral protein interactions may result in an increase in the order of the adsorbed layer. The supramolecular organization of adsorbed protein layers is presently experimentally accessible in situ, using the scanning probe techniques such as atomic force microscopy and its many variants. Similar local probe methods and techniques are needed for the analysis of the degree of adsorbed conformational heterogeneity. The heterogeneity problem becomes multidimensional if two or more proteins coadsorb to an interface in a dynamic fashion.

What will the future bring? The experiments using photons as probes of molecule structure and dynamics are undergoing a new vigorous development phase. The photon has traditionally been used as an energy probe within its time window of interactions with a molecule. Light, however, has lacked spatial resolution: the limit was set by its long wavelength. Recent advances in scanning near-field optical microscopy (SNOM) and the capability of single-molecule detection have brought to use a new technology that can be used to monitor spectroscopic properties of a single adsorbed molecule [44]. By directing light through a submicrometer-sized aperture of a scanning spectroscopic probe positioned above the sample molecule(s), one combines the advantages of photon and local scanning probe techniques. One can expect that the future evaluation of a single adsorbed molecule with such scanning spectroscopic probes will provide new, in situ information about the dynamic properties of adsorbed proteins, like the rate and magnitude of structural fluctuations, the height of energy barriers, the pathways of interface-induced conformational changes, and perhaps many others. Photons will continue to provide a high resolution in energy in a time-resolved fashion, while the scanning spectroscopic probe will localize the photon–molecule interactions to a very small area, providing thus for a very high spatial resolution. The future really looks bright!

ACKNOWLEDGMENTS

The authors gratefully acknowledge the fellowship from the Netherlands Organization for Scientific Research (NWO) (J.B.), NIH Grant RO1 HL44538 (V.H.) and the support from the Center for Biopolymers at Interfaces.

REFERENCES

1.   J. J. Laserna, ed., *Modern Techniques in Raman Spectroscopy.* Wiley, Chichester, 1996, p. 427.
2.   W. Demtröder, *Laser Spectroscopy.* 2nd ed. Springer, Berlin, 1996, p. 924.
3.   J. M. Hollas, *Modern Spectroscopy.* 3rd ed. Wiley, Chichester, 1996, p. 931.
4.   L. Mandel and E. Wolf, *Optical Coherence and Quantum Optics.* Cambridge University Press, Cambridge, 1995, p. 1166.
5.   M. K. Debe, Optical probes of organic thin films: photons-in, photons-out. Prog. Surf. Sci. *24*:1–282, 1987.

6. N. J. Harrick, *Internal Reflection Spectroscopy.* Wiley-Interscience, New York, 1967, p. 327.
7. D. Axelrod, T. P. Burghardt, and N. L. Thompson, Total internal reflection fluorescence. Ann. Rev. Biophys. Bioeng. *13*:247–268, 1984.
8. D. Axelrod, E. H. Hellen, and R. M. Fulbright, Total internal reflection fluorescence. In *Topics in Fluorescence Spectroscopy*, Vol. 3. *Biochemical Applications* (J. R. Lakowicz, ed.). Plenum Press, New York, 1992, pp. 289–343.
9. W. N. Hansen, Electric fields produced by the propagation of plane coherent electromagnetic radiation in stratified medium. J. Opt. Soc. Am. *58*:380–392, 1968.
10. W. M. Reichert, Evanescent detection of adsorbed films: assessment of optical considerations for absorbance and fluorescence spectroscopy at crystal/solution and polymer/solution interfaces. CRC Crit. Rev. Biocomp. *5*:173–205, 1989.
11. R. K. Chang and T. E. Furtak, eds., *Surface Enhanced Raman Scattering.* Plenum Press, New York, 1982, p. 423.
12. T. M. Cotton, Surface enhanced Raman spectroscopy of biological macromolecules. In *Surface and Interfacial Aspects of Biomedical Polymers*, Vol. 2. *Protein Adsorption* (J. D. Andrade, ed.). Plenum Press, New York, 1985, pp. 161–187.
13. D. Hobara, K. Niki, C. Zhou, G. Chumanov, and T. M. Cotton, Characterization of cytochrome c immobilized on modified gold and silver electrodes by surface-enhanced Raman spectroscopy. Colloids Surf. A *93*:241–250, 1994.
14. J. N. Herron, W. Jiskoot, and D. J. A. Crommelin, eds., *Physical Methods to Characterize Pharmaceutical Proteins.* Plenum Press, New York, 1995, p. 362.
15. J. Jiskoot, V. Hlady, J. J. Naleway, and J. N. Herron, Application of fluorescence spectroscopy for determining the structure and function of proteins. In *Physical Methods to Characterize Pharmaceutical Proteins* (J. N. Herron, W. Jiskoot, and D. J. A. Crommelin, eds.). Plenum Press, New York, 1995, pp. 1–64.
16. M. Levitt and J. Greer, Automatic identification of secondary structure in globular proteins. J. Mol. Biol. *114*:181–293, 1977.
17. J. R. Lakowicz, *Principles of Fluorescence Spectroscopy.* Plenum Press, New York, 1983, p. 496.
18. A. P. Demchenko, *Ultraviolet Spectroscopy of Proteins.* Springer-Verlag, Berlin, 1986, p. 311.
19. E. A. Permyakov, *Luminescent Spectroscopy of Proteins.* CRC Press, Boca Raton, 1993, p. 164.
20. A. P. Demchenko, Fluorescence and dynamics in proteins. In *Topics in Fluorescence Spectroscopy*, Vol. 3. *Biochemical Applications* (J. R. Lakowicz, ed.). Plenum Press, New York, 1992, pp. 65–111.
21. R. R. Chance, A. Prock, and R. Sibley, Molecular fluorescence and energy transfer near interfaces. Adv. Chem. Phys. *37*:1–65, 1978.
22. W. Lukosz and R. E. Kunz, Fluorescence lifetime of magnetic and electric dipoles near a dielectric surface. Opt. Comm. *20*:195–199, 1977.
23. E. H. Hellen and D. Axelrod, Fluorescence emission at dielectric and metal film interfaces. J. Opt. Soc. Am. B *4*:337–350, 1987.
24. V. Hlady, D. R. Reinecke, and J. D. Andrade, Fluorescence of adsorbed protein layers I. Quantitation of total internal reflection fluorescence. J. Colloid Interface Sci. *111*: 555–569, 1986.
25. C.-G. Gölander, V. Hlady, K. Caldwell, and J. D. Andrade, Adsorption of human lysozyme and adsorbate enzyme activity as quantified by means of TIRF, 125-I labelling and ESCA. Colloids Surf. *50*:113–130, 1990.
26. A. Sanders, Kinetics studies of the adsorption mechanism of bovine fibrinogen on native and modified quartz (in German). Ph.D. dissertation, University of Essen, Essen, 1996.

27. A. Martin-Rodriguez, J. Liu, and V. Hlady, Fibrinogen adsorption to modified silica surfaces, in preparation 1998.

28. G. S. Beddard and M. A. West, eds., *Fluorescent Probes*. Academic Press, London, 1981, p. 235.

29. R. P. Haugland, Covalent fluorescent probes. In *Excited States of Biopolymers* (R. F. Steiner, ed.). Plenum Press, New York, 1983, pp. 29–58.

30. R. P. Haugland, M. T. Z. Spence, and I. D. Johnson, *Handbook of Fluorescent Probes and Research Chemicals*, 6th ed. Molecular Probes, Inc., Eugene, 1996, p. 679.

31. W. T. Mason, ed., *Fluorescent and Luminescent Probes for Biological Activity*. Academic Press, London, 1993, p. 433.

32. J. Liu and V. Hlady, Chemical pattern on silica surface prepared by UV irradiation of 3-mercaptopropyltriethoxy silane layer: surface characterization and fibrinogen adsorption. Colloids Surf. B *8*:25–37, 1996.

33. V. Hlady and J. D. Andrade, A TIRF titration study of 1-anilinonaphthalene-8-sulfonate binding to silica-adsorbed bovine serum albumin. Colloids Surf. *42*:85–96, 1989.

34. V. Hlady and J. D. Andrade, Fluorescence emission from adsorbed bovine serum albumin and albumin-bound 1-anilinonaphthalene-8-sulfonate studied by TIRF. Colloids Surf. *32*:359–369, 1988.

35. P. Suci and V. Hlady, Fluorescence lifetime components of Texas Red-labelled bovine serum albumin: comparison of bulk and adsorbed states. Colloids Surf. *51*:89–104, 1990.

36. T. Föster, Intermolecular energy migration and fluorescence. Ann. Phys. (Leipzig) (Translated by R. S. Knox) 2:55–75, 1948.

37. L. Stryer, Fluorescence energy transfer as a spectroscopic ruler. Ann. Rev. Biochem. *47*:819–846, 1978.

38. B. W. van der Meer, G. Coker III, and S.-Y. S. Chen, *Resonance Energy Transfer: Theory and Data*. VCH, New York, 1994, p. 177.

39. D. E. Koppel, P. J. Fleming, and P. Strittmatter, Intramembrane positions of membrane-bound chromophores determined by excitation energy transfer. Biochemistry *18*: 5450–5457, 1979.

40. T. P. Burghardt and D. Axelrod, Total internal reflection fluorescence study of energy transfer in surface-adsorbed and dissolved bovine serum albumin. Biochemistry *22*: 979–985, 1983.

41. L. E. Morrison and G. Weber, Biological membrane modeling with a liquid/liquid interface. Probing mobility and environment with total internal reflection excited fluorescence. Biophys. J. *52*:367–379, 1987.

42. V. Hlady, R. A. Van Wagenan, and J. D. Andrade, Total internal reflection intrinsic fluorescence (TIRIF) spectroscopy applied to protein adsorption. In *Surface and Interfacial Aspects of Biomedical Polymers*, Vol. 2. *Protein Adsorption* (J. D. Andrade, ed.). Plenum Press, New York, 1985, pp. 81–119.

43. Y.-L. Cheng, B. K. Lok, and C. R. Robertson, Interactions of macromolecules with surfaces in shear fields using visible wavelength total internal reflection fluorescence. In *Surface and Interfacial Aspects of Biomedical Polymers*, Vol. 2. *Protein Adsorption* (J. D. Andrade, ed.). Plenum Press, New York, 1985, pp. 121–160.

44. J. J. Macklin, J. K. Trautman, T. D. Harris, and L. E. Brus, Imaging and time-resolved spectroscopy of single molecules at an interface. Science *272*:255–258, 1996.

45. B. K. Lok, Y.-L. Cheng, and C. R. Robertson, Total internal reflection fluorescence: a technique for examining interactions of macromolecules with solid surfaces. J. Colloid Interface Sci. *91*:87–103, 1983.

46. T. P. Burghardt and D. Axelrod, Total internal reflection/fluorescence photobleaching recovery study of serum albumin adsorption dynamics. Biophys. J. *33*:455–467, 1981.

47. V. Hlady, Spatially resolved adsorption kinetics of immunoglobulin G onto the wettability gradient surface. Appl. Spectrosc. *45*:246–252, 1991.

48. Y. S. Lin and V. Hlady, The desorption of ribonuclease A from charge density gradient surfaces studied by spatially resolved total internal reflection fluorescence. Colloids Surf. B *4*:65–75, 1995.

49. Y. S. Lin and V. Hlady, Human serum albumin adsorption onto octadecyldimethylsilylsilica gradient surface. Colloids Surf. B *2*:481–491, 1994.

50. J. Buijs and V. Hlady, Adsorption kinetics, conformation, and mobility of the growth hormone and lysozyme on solid surfaces, studied with TIRF. J. Colloid Interface Sci. *190*:171–181, 1997.

51. M. R. Eftink and C. A. Ghiron, Exposure of tryptophanyl residues in proteins. Quantitative determination of fluorescence quenching studies. Biochemistry *15*:672–680, 1976.

52. D. Horsley, J. Herron, V. Hlady, and J. D. Andrade, Fluorescence quenching of adsorbed hen and human lysozymes. Langmuir *7*:218–222, 1991.

53. B. Crystall, G. Rumbles, T. A. Smith, and D. Phillips, Time-resolved evanescent wave induced fluorescence measurements of surface-adsorbed bovine serum albumin. J. Colloid Interface Sci. *155*:247–250, 1993.

54. M. C. L. Maste, E. H. W. Pap, A. v Hoek, W. Norde, and A. J. W. G. Visser, Spectroscopic investigation of the structure of a protein adsorbed on a hydrophobic latex. J. Colloid Interface Sci. *180*:632–633, 1996.

55. N. L. Thompson, H. M. McConnell, and T. P. Burghardt, Order in supported phospholipid monolayers detected by the dichroism of fluorescence exited with polarized evanescent illumination. Biophys. J. *46*:739–747, 1984.

56. J. G. E. M. Fraaije, J. M. Kleijn, M. van der Graaf, and J. C. Dijt, Orientation of adsorbed cytochrome c as a function of the electrical potential of the interface studied by total internal reflection fluorescence. Biophys. J. *57*:965–975, 1990.

57. M. A. Bos and J. M. Kleijn, Determination of the orientation distribution of adsorbed fluorophores using TIRF. 1. Theory. Biophys. J. *68*:2566–2572, 1995.

58. M. A. Bos and J. M. Kleijn, Determination of the orientation distribution of adsorbed fluorophores using TIRF. 2. Measurements on porphyrin and cytochrome c. Biophys. J. *68*:2573–2579, 1995.

59. P. L. Edminston, J. E. Lee, S.-S. Cheng, and S. S. Saavedra, Molecular orientation distributions in protein films. 1. Cytochrome c adsorbed to substrates of variable surface chemistry. J. Am. Chem. Soc. *119*:560–570, 1997.

60. L. L. Wood, S.-S. Cheng, P. L. Edminston, and S. S. Saavedra, Molecular orientation distributions in protein films. 2. Site-directed immobilization of yeast cytochrome c on thiol-capped, self-assembled monolayers. J. Am. Chem. Soc. *119*:571–576, 1997.

61. W. M. Reichert, P. A. Suci, J. T. Ives, and J. D. Andrade, Evanescent detection of adsorbed protein concentration-distance profiles: Fit of simple models to variable-angle total internal reflection fluorescence data. Appl. Spectrosc. *41*:503–508, 1987.

62. J. L. Robeson and R. D. Tilton, Spontaneous reconfiguration of adsorbed lysozyme layers observed by total internal reflection fluorescence with a pH-sensitive fluorophore. Langmuir *12*:6104, 1996.

63. J. Buijs, D. W. Britt, and V. Hlady, Human growth hormone adsorption kinetics and conformation on self-assembled monolayers. Langmuir *14*:335–341, 1998.

64. A. M. de Vos, M. Ultsh, and A. A. Kosiakoff, Human growth hormone and extracellular domain of its receptor: Crystal structure of the complex. Science *255*:306–312, 1992.

65. A. Elliott and E. J. Ambrose, Structure of synthetic polypeptides. Nature *165*:921–922, 1950.

66. W. K. Surewicz and H. H. Mantsch, New insight into protein secondary structure from resolution-enhances infrared spectra. Biochem. Biophys. Acta *952*:115–130, 1988.

67. H. Susi and D. M. Byler, Protein structure by Fourier transform infrared spectroscopy: second derivative spectra. Biochem. Biophys. Res. Commun. *115*:391–397, 1983.

68. D. M. Byler and H. Susi, Examination of the secondary structure of proteins by deconvolved FTIR spectra. Biopolymers *25*:469–487, 1986.

69. J. F. Rabolt, W. H. Moore, and S. Krimm, Vibrational analysis of peptides, polypeptides, and proteins. 3. alpha-poly(L-alanine). Macromolecules *10*:1065–1074, 1977.

70. A. M. Dwivedi and S. Krimm, Vibrational analysis of peptides, polypeptides, and proteins. 19. Force fields for α-helix and β-sheet structures in a side-chain point-mass approximation. J. Phys. Chem. *88*:620–627, 1984.

71. S. Krimm and J. Bandekar, Vibrational analysis of peptides, polypeptides, and proteins. V. Normal vibrations of β-turns. Biopolymers *19*:1–29, 1980.

72. J. Buijs, W. Norde, and J. W. Th. Lichtenbelt, Changes in the secondary structure of adsorbed IgG and $F(ab')_2$ fragments studied by FTIR spectroscopy. Langmuir *12*: 1605–1613, 1996.

73. J. S. Leon, R. P. Sperline, and S. Raghavan, Quantitative analysis of adsorbed serum albumin on segmented polyurethane using FT-IR/ATR spectroscopy. Appl. Spectrosc. *46*:1644–1648, 1992.

74. D. J. Fink, T. B. Hutson, K. K. Chittur, and R. M. Gendreau, Quantitative surface studies of protein adsorption by infrared spectroscopy. II. Quantification of adsorbed and bulk proteins. Anal. Biochem. *165*:147–154, 1987.

75. M. Müller, R. Buchet, and U. P. Fringeli, 2D-FTIR ATR spectroscopy of thermo-induced periodic secondary structural changes of poly-(L)-lysine: A cross-correlation analysis of phase-resolved temperature modulation spectra. J. Phys. Chem. *100*:10810–10825, 1996.

76. B. R. Singh, M. P. Fuller, and G. Schiavo, Molecular structure of tetanus neurotoxin as revealed by Fourier transform infrared and circular dichroic spectroscopy. Biophys. Chem. *46*:155–166, 1990.

77. R. J. Jakobsen and S. W. Strand, Biological applications of attenuated total reflection (ATR) spectroscopy. In *Practical Spectroscopy Series*, Vol. 15, *Internal Reflection Spectroscopy. Theory and Applications* (F. M. Mirabella Jr., ed.). Marcel Dekker, New York, 1992, pp. 107–140.

78. B. R. Singh and M. P. Fuller, FT-IR in combination with the attenuated total reflectance technique: a very sensitive method for the structural analysis of polypeptides. Appl. Spectrosc. *45*:1017–1021, 1991.

79. S.-S. Cheng, K. K. Chittur, C. N. Sukenik, L. A. Culp, and K. Lewandowska, The conformation of fibronectin of self-assembled monolayers with different surface composition: an FTIR/ATR study. J. Colloid Interface Sci. *162*:135–143, 1994.

80. M. Müller, C. Werner, K. Grundke, K. J. Eichhorn, and H. J. Jacobash, ATR-FT-IR spectroscopy of proteins adsorbed on biocompatible cellulose films. Mikrochim. Acta *14*:671–674, 1997.

81. A. Ball and R. A. L. Jones, Conformational changes in adsorbed proteins. Langmuir *11*:3542–3548, 1995.

82. H. H. Bauer, M. Müller, J. Goette, H. P. Merkle, and U. P. Fringeli, Interfacial adsorption and aggregation associated changes in secondary structure of human calcitonin monitored by ATR-FTIR spectroscopy. Biochemistry *33*:12276–12282, 1994.

83. F. M. Wasacz, J. M. Olinger, and R. J. Jakobsen, Fourier transform infrared studies of proteins using nonaqueous solvents. Effects of methanol and ethylene glycol on albumin and immunoglobulin G. Biochemistry *26*:1464–1470, 1987.

84. E. Dufour, P. Robert, D. Bertrand, and T. Haertlé, Conformation changes of β-lactoglobulin: an ATR infrared spectroscopic study of the effect of pH and ethanol. J. Prot. Chem. *13*:143–149, 1994.

85. R. J. Jakobsen, F. M. Wasacz, J. W. Brasch, and K. B. Smith, The relationship of bound water to the IR Amide I bandwidth of albumin. Biopolymers 25:639–654, 1986.

86. J. L. Kirsch and J. L. Koenig, The variable-temperature FT-IR study of the secondary structure of gamma-globulin, chymotrypsin, serum albumin, and β-lactoglobulin in aqueous solution. Appl. Spectrosc. 43:445–451, 1989.

87. R. P. Sperline, S. Muralidharan, and H. Freiser, New quantitative technique for attenuated total reflection (ATR) spectrophotometry; calibration of the "CIRCLE" ATR device in the infrared. Appl. Spectrosc. 40:1019–1022, 1986.

88. R. P. Sperline, S. Muralidharan, and H. Freiser, In situ determination of species adsorbed at a solid-liquid interface by quantitative infrared attenuated total reflectance spectrophotometry. Langmuir 3:198–202, 1987.

89. F. N. Fu, M. P. Fuller, and B. R. Singh, Use of Fourier transform infrared/attenuated total reflectance spectroscopy for the study of surface adsorption of proteins. Appl. Spectrosc. 47:98–102, 1993.

90. J. Van Straaten and N. A. Peppas, ATR-FTIR analysis of protein adsorption on polymeric surfaces. J. Biomater. Sci. Polymer Edn. 2:113–121, 1991.

91. K. K. Chittur, D. J. Fink, R. I. Leininger, and T. B. Hutson, Fourier transform infrared spectroscopy/attenuated total reflection studies of protein adsorption in flowing systems: approaches for bulk correction and compositional analysis of adsorbed and bulk proteins in mixtures. J. Colloid Interface Sci. 111:419–433, 1986.

92. M. R. Nyden, G. P. Forney, and K. Chittur, Spectroscopic quantitative analysis of strongly interacting systems: human plasma protein mixtures. Appl. Spectrosc. 42:588–594, 1988.

93. R. I. Leininger, D. J. Fink, R. M. Gendreau, T. M. Hutson, and R. J. Jakobsen, Spectroscopic probes of blood–surface interaction. Trans. Am. Artif. Intern. Organs 29:152–157, 1983.

94. C. F. C. Ludlan, I. T. Arkin, X.-M. Liu, M. S. Rothman, P. Rath, S. Aimoto, S. O. Smith, D. M. Engelman, and K. J. Rothschild, Fourier transform infrared spectroscopy and site-directed isotope labeling as a probe of local secondary structure in the transmembrane domain of phospholamban. Biophys. J. 70:1728–1736, 1996.

95. F. Dousseau, M. Therrien, and M. Pézolet, On the spectral subtraction of water from the FT-IR spectra of aqueous solutions of proteins. Appl. Spectrosc. 43:538–542, 1989.

96. J. R. Powell, F. M. Wasacz, and R. J. Jakobsen, An algorithm for the reproducible spectral subtraction of water from the FT-IR spectra of proteins in dilute solutions and adsorbed monolayers. Appl. Spectrosc. 40:339–344, 1986.

97. Y. N. Chirgadze, O. V. Fedorov, and N. P. Trushina, Estimation of amino acid residue side-chain adsorption in the infrared spectra of protein solutions in heavy water. Biopolymers 14:679–694, 1975.

98. B. W. Morrissey and R. R. Stromberg, The conformation of adsorbed blood proteins by infrared bound fraction measurements. J. Colloid Interface Sci. 46:152–164, 1974.

99. J. K. Kauppinen, D. J. Moffat, H. H. Mantsch, and D. G. Cameron, Fourier transform in the computation of self-deconvoluted and first-order derivatives spectra of overlapped bands contours. Anal. Chem. 53:1454–1457, 1981.

100. B. Liedberg, B. Ivarsson, and I. Lundström, Fourier transform infrared reflection absorption spectroscopy (FT-IRAS) of fibrinogen adsorbed on metal and metal oxide surfaces. J. Biochem. Biophys. Meth. 9:233–243, 1984.

101. B. Liedberg, B. Iversson, P.-O. Hegg, and I. Lundström, On the adsorption of beta-lactoglobulin on hydrophilic gold surfaces. J. Colloid Interface Sci. 114:386–395, 1986.

102. B. Liedberg, I. Lundström, C. R. Wu, and W. R. Salaneck, Adsorption of glycine on hydrophilic gold. J. Colloid Interface Sci. 108:123–132, 1985.

103. B. Liedberg, C. Carlsson, and I. Lundström, An infrared reflection-adsorption study of amino acids adsorbed on smooth metal surfaces: L-histidine and L-phenylalanine on gold and copper. J. Colloid Interface Sci. 120:64–75, 1987.

104. P. W. Holloway and H. H. Mantsch, Structure of cytochrome b5 in solution by Fourier-transform infrared spectroscopy. Biochemistry 28:931–935, 1989.

105. M. Bloemendal and W. C. J. Johnson, Structural information on proteins from CD spectroscopy: possibilities and limitations. In *Physical Methods to Characterize Pharmaceuticals Proteins* (J. N. Herron, W. Jiskoot, and D. J. A. Crommelin, eds.). Plenum Press, New York, 1995, pp. 65–100.

106. M. E. Soderquist and A. G. Walton, Structural changes in proteins adsorbed on polymer surfaces. J. Colloid Interface Sci. 75:386–397, 1980.

107. A. G. Walton and M. E. Soderquist, Behavior of proteins at interfaces. Croat. Chem. Acta 53:363–372, 1980.

108. C. R. McMillin and A. G. Walton, A circular dichroism technique for the study of adsorbed protein structure. J. Colloid Interface Sci. 48:345–349, 1974.

109. A. Kondo, S. Oku, and K. Higashitani, Structural changes in protein molecules adsorbed on ultrafine silica particles. J. Colloid Interface Sci. 143:214–221, 1991.

110. W. Norde and J. P. Favier, Structure of adsorbed and desorbed proteins. Colloids Surf, 64:87–93, 1992.

111. A. Kondo, S. Oku, F. Murakami, and K. Higashitani, Conformational changes in protein molecules upon adsorption on ultrafine particles. Colloids Surf. B 1:197–201, 1993.

112. P. Billsten, M. Wahlgren, T. Arnebrandt, J. McGuire, and H. Elwing, Structural changes of T4 lysozyme upon adsorption to silica nanoparticles measured by circular dichroism. J. Colloid Interface Sci. 175:77–82, 1995.

113. A. Kondo and T. Urabe, Temperature dependence of activity and conformational changes in alpha-amylases with different thermostability upon adsorption on ultrafine silica particles. J. Colloid Interface Sci. 174:191–198, 1995.

114. A. Kondo, F. Murakami, and K. Higashitani, Circular dichroism studies on conformational changes in protein molecules upon adsorption on ultrafine polystyrene particles. Biotech. Bioeng. 40:889–894, 1992.

115. B. L. Steadman, K. C. Thompson, C. R. Middaugh, K. Matsuno, S. Vrona, E. Q. Lawson, and R. V. Lewis, The effects of surface adsorption in the thermal stability of proteins. Biotech. Bioeng. 40:8–15, 1992.

116. S. R. Clark, P. Billsten, and H. Elwing, A fluorescence technique for investigating protein adsorption phenomena at a colloidal silica surface. Colloids Surf. B 2:457–461, 1994.

117. D. Kowalczyk, S. Slomkowski, and F. W. Wang, Changes in conformation of human serum albumin (HSA) and gamma globulins (gG) upon adsorption to polystyrene and poly(styrene/acrolein) latexes: studies by fluorescence spectroscopy. J. Bioact. Comp. Polym. 9:282–309, 1994.

118. P. Marchand and L. Marmet, Binomial smoothing filter: a way to avoid some pitfalls of least-squares polynominal smoothing. Rev. Sci. Instrum. 54:1034–1041, 1983.

18

Proteins on Surfaces: Methodologies for Surface Preparation and Engineering Protein Function

KRISHNAN K. CHITTUR   University of Alabama in Huntsville, Huntsville, Alabama, U.S.A.

I. INTRODUCTION

There has been a call to reexamine many of our current approaches to controlling biological interactions at surfaces [1,2]. Instead of preparing a surface, having proteins adsorb onto such surfaces, and examining the consequence, the new paradigm requires that we try to mimic what nature does all the time—i.e., design surfaces that will elicit precisely the biological reactions we want and no more. It is not easy to engineer surfaces that will elicit only specific reactions from the biological milieu they are in contact with. A truly biocompatible material is one that will be integrated with the host or one which allows only those pathways that will allow wound healing [1,2]. An obvious conclusion from such requirements is that engineered surfaces must inhibit all forms of nonspecific interactions, including the rejection of proteins that will not be involved in the wound-healing process.

It is clear that we will need surface analytical tools that can help us verify the design of the surface and help evaluate the structure and organization of proteins on such surfaces. In this chapter we discuss how Fourier transform infrared spectroscopy (FTIR) in a number of different sampling modes and atomic force microscopy (AFM) are contributing to such studies.

The introduction of an interferometer for data collection along with rapid computational techniques for the calculation of the Fourier transform changed infrared spectroscopy in a dramatic fashion. Compact lasers, high-sensitivity detectors, such as those made from mercury-cadmium-telluride (MCT), and the ongoing revolution in solid-state electronics and materials manufacturing (availability of germanium-attenuated total internal reflection crystals) now provide the biomaterials scientist with a technique that provides a unique look at the interface between the biomaterial surface and the protein solution it is interacting with. A discussion of the different sampling modes used with FTIR will be first given. This is followed by an extended discussion of attenuated total internal reflection (ATR) techniques and how they can be used to characterize the surface and the proteins adsorbed to the surface. We will discuss the problems related to the use of H_2O and how D_2O can help with the interpretation of the spectral changes. FTIR with ATR remains a powerful technique

467

with which we can get information about adsorbed protein secondary structure, but we inject some important cautions about this approach.

While FTIR/ATR can provide us with a look at the forest when it comes to adsorbed proteins on surfaces, atomic force microscopy can provide an unparalleled look at some of the trees in this forest. The AFM technique is rapidly evolving as a surface analytical tool, but investigators need to be aware of its many limitations. Newer sampling modes like the noncontact tapping mode in aqueous environments are allowing us to image soft surfaces such as proteins on surfaces. A brief discussion of how AFM is helping biomaterial scientists with both engineering surfaces and examining those surfaces in the biological milieu is presented.

II. FTIR TECHNIQUES AND SURFACES

A. Sampling Modes

There are three basic sampling modes used by researchers using FTIR techniques for designing and studying surfaces.

1. Transmission: The sample analyzed is placed between two infrared trans-missing windows (with or without a spacer) or directly on an infrared transmitting window, like calcium fluoride (CaF_2) or sodium chloride. The infrared spectra of soluble proteins (10 mg/mL or higher) can be obtained for aqueous (H_2O) solutions with CaF_2 transmission windows and spacers no thicker than about 6 μ.

2. Internal reflection: By appropriate choice of internal reflecting elements (IRE) (materials such as germanium and zinc selenide with high refractive indices and high transmittivities) infrared light can be made to travel down the element as a waveguide. At each point where the infrared light reflects totally internally, it creates an evanescent field in the rare medium (lower refractive index). It is the interaction of this evanescent field with infrared-active materials that results in the spectrum. The surface sensitivity of internal reflection techniques comes from the multiple internal reflection of the radiation as it passes through the waveguide. Spectra obtained using linearly polarized light can be used to obtain molecular order information for molecules adsorbed to the surface of the IRE.

3. External reflection: Typically this is a single-reflection technique. Infrared light is reflected off a surface at angles approximately 65 to 70° from the surface normal. The component of light polarized perpendicular to the plane of incidence undergoes a 180° phase shift at the surface. Hence, infrared-active molecules that lie on the surface will be invisible to the beam. This is the so-called surface selection rule of external reflection infrared. Since the magnitude of the infrared band is proportional to the product of the dipole moment and the electric field intensity at the surface, this technique can be used to obtain information about molecular orientation.

Of the three sampling modes mentioned, only the ATR technique allows for the study of proteins on surfaces in contact with aqueous medium.

B. Basics of Attenuated Total Internal Reflection

Consider a trapezoidal ATR crystal (also called an IRE) shown in Fig. 1. With the help of an appropriate optical setup, infrared light is focused onto one of the faces of the ATR crystal. If the angle θ at which the infrared light impinges upon the interface between the ATR crystal (the dense medium) and the air (or water, buffer with or without proteins—the rare medium) is greater than the critical angle, then the light will totally internally reflect within the IRE. By appropriately choosing the thickness W and the length L of the crystal, we can control the total number of reflections N the light will undergo as it propagates down the crystal before emerging at the other end:

$$N = \frac{L}{W} \cot(\theta) \tag{1}$$

At each reflection, an evanescent electric field E is generated in the rare medium whose intensity decays exponentially with distance z into the rare medium, i.e.,

$$E = E_o e^{-z/d_p} \tag{2}$$

where E_o is the intensity of the incident radiation, and d_p, the depth of penetration, represents the distance at which the evanescent wave drops to $1/e$ times the intensity at the surface. The magnitude of d_p gives us an idea of how deep we can see into the rare medium. This is true for both the electric and the magnetic components of the light. Using a device called a polarizer we can selectively absorb one of these components. Perpendicularly polarized light (or the transverse electric wave perpendicular to the plane of incidence on the germanium IRE, the xz-plane) when traveling through the ATR crystal will result in an electric field on the surface of the ATR crystal with a component only on the surface of the crystal, i.e.,

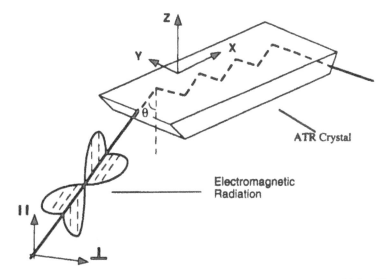

FIG. 1 Linearly polarized light as it travels through a trapezoidal ATR crystal. The light enters the face normally and hits the interface between the crystal and the top of an angle θ.

$$E_o = E_y \tag{3}$$

When the light is polarized parallel to the incident plane, then

$$E_o^2 = E_x^2 + E_z^2 \tag{4}$$

Considering the two components of the traveling light beam is important for understanding and interpreting calculations of molecular orientations on the ATR surface.

The intensity of the reflected light at each reflection point will be reduced by the presence of infrared-absorbing material in the rare medium. The surface sensitivity of ATR results from the addition of a number of small absorbances from each reflection.

The depth of penetration can be calculated using the following formula [3]:

$$d_p = \frac{\lambda}{2\pi n_1 \sqrt{(\sin^2\theta - n_{21}^2)}} \tag{5}$$

where λ is the wavelength of the light (note that the depth of penetration is a function of wavelength), n_1 is the refractive index of the ATR crystal (for germanium it is 4.0 and independent of wavelength), n_2 is the refractive index of the rare medium (this value is a function of wavelength), θ is the angle of incidence, and n_{21} is n_2/n_1. Using typical values for the refractive index of proteins (1.5 at 1550 cm^{-1}) we can calculate d_p at 1550 cm^{-1} to be 0.428 μm for germanium crystals. From a knowledge of the electric field intensities on the surface of the IRE [3], an effective thickness equivalent to that a material obtained by transmission spectroscopy that will give the same absorbance can also be calculated (this value is different for the two polarizations).

C. Quantification of Adsorbed Proteins

The density of absorbed proteins can be calculated in terms of experimentally observable quantities such as the integrated absorbance per reflection, a molar absorptivity for the protein (from transmission spectroscopy), and depth of penetration and the effective thickness. Application of these techniques to Langmuir–Blodgett monolayers [4] and for albumin adsorption to polyurethanes have been presented [5].

Harrick [3] has defined the effective thickness d_e as the thickness of a material which will give the same absorbance in transmission spectra at normal incidence as that found from ATR:

$$d_e = \frac{n_{21}E_o^2 d_p}{2\cos\theta} \tag{6}$$

Using the expressions for calculating the electric field components along x, y, and z directions, it is a straightforward matter to determine the effective thicknesses for the perpendicular and parallel polarizations. Consider now a thin film (like an adsorbed protein) adjacent to the ATR crystal in contact with a protein-containing or protein-free solution. The absorbance per reflection can be written as

$$\frac{A}{N} = \frac{n_{21}E_o^2\varepsilon}{\cos\theta} \int_0^\infty C(z) \exp\left(\frac{-2z}{d_p}\right) dz \tag{7}$$

where A is the integrated absorbance due to internal reflections, N is the total number of internal reflections, ε is the integrated molar absorptivity (L/cm^2 mol, from transmission spectroscopy and using Beer's law), z is the distance from the IRE surface (cm), and $C(z)$ is the concentration as a function of distance from the IRE (mol/L). Let the thickness of the adsorbed protein layer be t. From a distance of 0 to t, $C(z) = C_i$, where C_i is the adsorbed protein concentration, and from t to ∞ we can write $C(z) = C_b$, where C_b is the concentration of protein in solution. We are ignoring transients and are considering only the steady-state case. Because the thickness of the adsorbed layer (50 to 100 Å) is usually much smaller than d_p, we can approximate the calculation of the integral, given by Eq. (7), and arrive at an expression composed entirely of experimentally measurable quantities for the amount of protein adsorbed to the surface (Γ, mol/cm^2):

$$\Gamma = \frac{A/N - \varepsilon C_b d_e}{1000\varepsilon(2d_e/d_p)} \tag{8}$$

A similar expression can be derived for the case of protein adsorption onto polymer-coated germanium crystals.

Investigators have used other techniques to relate infrared absorbances to the amount of adsorbed protein. For example Pitt [6] showed that there is a linear correlation between a defined amount of protein dried on a germanium crystal and the area under the amide II band. These correlations are then used to predict the surface adsorbed protein concentrations when the protein is in contact with aqueous medium. The major assumption involved with this approach is that the drying and rehydration of protein films does not result in irreversible alterations in structure and changes in the extinction coefficient. Experiments have shown that the degree of hydration of proteins in solution, as dried thin films and after rehydration could result in nontrivial differences in the spectra [7].

Our laboratory has been using a different technique for relating infrared absorbance measurements to the actual amounts of protein on the surface. The technique relies on a calibration curve generated using radiolabeled proteins [8]. The intensities of the amide II band for radiolabeled albumin, IgG, and fibrinogen adsorbed to germanium ATR crystals were determined at different surface coverings. These amide II intensities were normalized to account for day-to-day variations in the flow cell assembly that resulted in changes in the *effective path length*. We showed that the relationship between the actual amounts of protein adsorbed on the surface and the amide II band is a straight line.

A detailed discussion of the water band normalization procedure has been described [8]. Briefly, the intensity of the H—O—H bending vibration at 1640 cm^{-1} is taken to reflect the effective path length. We have arbitrarily chosen to normalize all amide II intensities to an effective path length of 700 milliabsorbance units of the H—O—H absorbance in the buffer spectrum. Actual amide II intensities were scaled to a water band intensity of 700 milliabsorbance units to correct for experiment-to-experiment variations in alignment and flow cell assembly. For example, if the intensity at 1640 cm^{-1} for a given experiment was 0.65 absorbance units and the amide II intensity was 0.015 absorbance units, the normalized value was obtained as (0.7 × 0.015)/0.65, or equal to 0.0162 absorbance unit. This normalized intensity of the amide II band can then be converted to actual surface concentration using appropriate correlation curves as discussed below.

Experiments with albumin, IgG, and fibrinogen at various surface coverages allowed us to relate normalized infrared amide II absorbances to the actual amounts of protein on the surface as determined by radiolabeled techniques. Data for the three proteins albumin, IgG, and fibrinogen have been presented in Ref. 8. These correlation curves are shown in Fig. 2. For albumin and IgG, normalized amide II intensity correlates linearly with actual amounts of adsorbed proteins, from submonolayer to a monolayer covered surface. The correlation curve for fibrinogen was poor and had fewer data points. We examined the possibility of predicting the slope of an absorbance surface concentration correlation for one protein from another by using the ratio of extinction coefficients obtained from transmission experiments. As the dashed lines in Fig. 2 indicate, the slope of the albumin correlation curve could be accurately predicted from the IgG curve; the prediction was very poor for fibrinogen.

The technique of normalizing to a water band intensity of 700 milliabsorbance units is a convenient method to lump a number of optical parameters in an ATR experiment (number of reflections, angle of internal reflection, refractive index of protein) into something that is easy to obtain experimentally. Quantitative comparisons of protein amounts on ATR surfaces cannot be made without correcting for differences in path length.

Water is an excellent infrared absorber and has a strong, broad band (due to the H—O—H bending vibration) at 1640 cm^{-1}. This band is approximately one

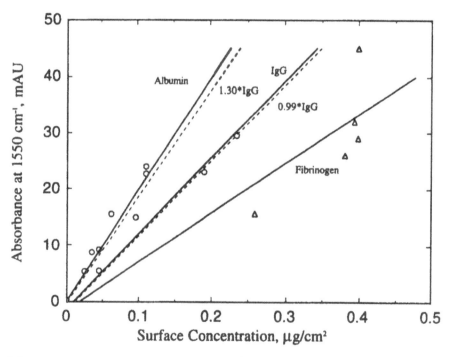

FIG. 2 Correlation of the amide II (1550 cm^{-1}) band intensities with estimated surface concentrations for albumin, IgG, and fibrinogen. The dashed line shown as 1.30*IgG is the predicted line for albumin and the line shown as 0.99*IgG is the predicted curve for fibrinogen. (Reproduced with permission from Academic Press, from *Analytical Biochemistry*, Vol. 165, pp. 147–154, 1987.)

order of magnitude more intense than and overlaps the smaller protein amide I band between 1680 to 1620 cm^{-1}. A discussion of the problems with water subtraction is given later. We will be able to subtract the spectrum of the water band from that of protein in water only if the water absorbance is below approximately 0.8 absorbance unit. This can be achieved by using a high refractive index material like germanium and many reflections or a lower refractive index IRE like zinc selenide and fewer reflections.

D.   Surfaces that Can Be Studied by FTIR/ATR

Protein adsorption can also be studied onto polymer [9,10], metal, or bioceramic surfaces [11] if such materials can be coated onto the IRE surface. These coatings must be thin (100 to 200 Å) so that the per reflection absorbance remains small and the signal leaving the crystal can be detected easily. Studying protein adsorption on bioceramic and metal coatings does pose some unique problems. We have demonstrated the feasibility of studying protein adsorption to bioceramic coatings deposited onto germanium ATR crystals [11]. Argon ions were used to bombard an hydroxyapatite (HA) target and the stream of HA particles was then allowed to deposit onto germanium ATR crystal for 10 min. Ellipsometry measurements indicated a coating thickness of about 250 Å. FTIR/ATR experiments showed that these coatings allowed sufficient energy to be transmitted through the crystal and spectra with reasonable signal to noise could be obtained. The thickness of the coating that can be used will depend greatly on the number of reflections and the absorbance per reflection. Theoretical calculations and experimental verifications of evanescent field attenuation by thin copper films deposited on germanium indicate that the deposition of biopolymers can be followed at the solution–metal interface [12]. There were, however, some unusual observations made in those experiments. For example, the authors noticed a significant enhancement of infrared absorption bands that correlated with the transformation of a continuous metallic film to islands on the germanium substrate. In addition, such surface-enhanced infrared absorptions led to anomalous shifts in absorbance bands of water and the appearance of dispersive band shapes. We did not observe any unusual band shapes or intensity enhancements in our study of protein adsorption to bioceramic substrates [11]. A clearer understanding of the surface enhancement phenomena due to the presence of a metal film on germanium (or other internal reflection elements) will be required before we can expand the use of FTIR/ATR techniques to understand protein adsorption to metallic substrates.

III.   CAN H$_2$O BE SUBTRACTED FROM PROTEIN SPECTRA?

Infrared spectroscopy and aqueous solutions are no longer incompatible with each other. Liquid water is a very strong infrared absorber with three prominent bands centered at 3400 cm^{-1} (H—O stretching band), 2125 cm^{-1} (water association band), and one at 1640 cm^{-1} (the H—O—H bending vibration). The amide I band (due to the protein backbone) for proteins absorbs between 1620 and about 1680 cm^{-1}, overlapping directly with the H$_2$O bending vibration band at 1640 cm^{-1}. The intensity of the water band at 1640 cm^{-1} is about an order of magnitude higher than the amide I bands of proteins and yet precise subtractions of the H$_2$O band are possible because of the frequency precision in FTIR data. The subtraction of a very large H_2O band from a very large absorbance spectrum of protein in H_2O to get a small spectrum of protein was either very difficult or close to impossible with older dispersive infrared spectrometers. The ability to subtract the spectrum of water does not mean, however, that water subtraction is routine or without any problems. We have illustrated this in Figs. 3 and 4.

Figure 3 illustrates the infrared absorbance spectra of phosphate buffered saline (PBS), bovine serum albumin (BSA) (30 mg/mL) dissolved in PBS, and the spectrum of BSA after subtraction of the PBS absorbance spectrum. These absorbance spectra were obtained by placing a drop (about 100 μL) of the solution on a calcium fluoride infrared transmitting window, placing another calcium fluoride window on top and tightening the assembly in a transmission cell. Infrared spectra can be obtained for proteins in aqueous solutions only if the assembled cell path length is about 6 μm and the solution protein concentration is at least 10 mg/mL. At higher path lengths and lower protein concentrations, it will be difficult to obtain the absorbance spectrum of the protein by subtraction.

The similarity of the spectra of PBS and BSA in PBS with the exception of a small *bump* near 1550 cm^{-1} in the BSA in PBS spectrum is evident. Figure 3 also shows clearly how small the BSA absorbance bands are relative to the large water band at 1640 cm^{-1}. It should be obvious from Fig. 3 that the scale factor for PBS subtraction will affect the intensity of the largest of the BSA band around 1655 cm^{-1}. This is more clearly illustrated in Fig. 4.

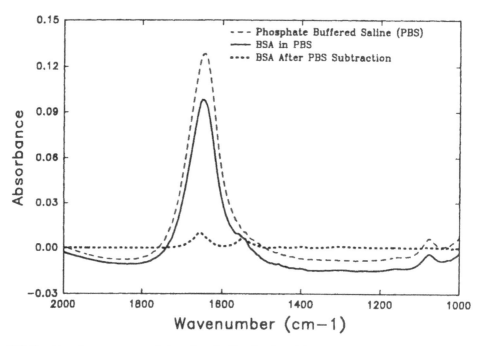

FIG. 3 Absorbance spectra of phosphate buffered saline (PBS), 30 mg/mL solution of bovine serum albumin (BSA) in PBS, and a PBS subtracted spectrum of BSA obtained using transmission spectroscopy.

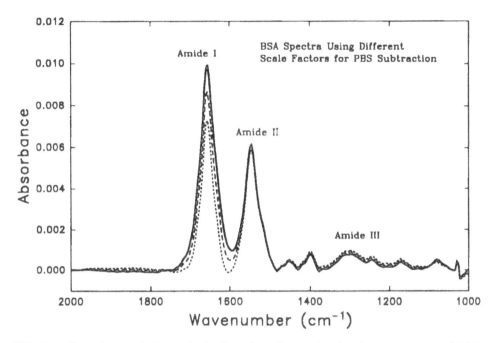

FIG. 4 Effect of the scale factor for buffer subtraction on the absorbance spectrum of BSA.

The absorbance spectra of BSA shown in Fig. 4 were obtained by subtracting different amounts of PBS from the absorbance spectrum of BSA in PBS. The subtraction scale factor has the greatest influence on the shape and the intensity of the amide I band. The protein amide I band is a composite created by bands due to different secondary structures such as α-helix, β-sheet, turns, and random coils. A number of techniques are available to calculate the relative contributions of such structures under the amide I band. We expect, therefore, that the scale factor for water subtraction will have an effect on such calculations. A discussion of how quantitative information about protein secondary structure can be obtained by analysis of the composite amide I band is given later.

Researchers have developed techniques to automatically subtract the spectrum of water to obtain a spectrum of the protein [13,14]. The objective of these methods was to eliminate possible individual bias and uncertainty in the selection of the scale factor due to baseline variations. The method used by Pezolet [14] and his group is based on the assumption that the 2125 cm^{-1} water association band is not affected by the presence of protein. This assumption appears to be a good one for most proteins except when they are highly charged [14]. Powell et al. [13] approached the water subtraction problem through analysis of spectral changes in the region from 1990 to 1790 cm^{-1} as the scale factor is changed. A comparison of these two methods showed, however, that neither had any clear advantage over the other [15]. In fact, in one study we found that experienced spectroscopists were able to subtract water from protein spectra with much better precision than the program described by Powell (unpublished). In a study with cytochrome b$_5$, the authors report that the positions and the relative intensities of component bands (under the amide I and amide II envelope) are maintained even with deliberate over- or undersubtraction of the H_2O peak [16]. This may not be the case with all proteins, and caution is advised. Minor differences in the scale factor for water subtraction could have significant effects on the contribution of different component bands under the amide I band. The question remains then: What is the correct way to subtract the spectrum of H_2O from the spectrum of a protein in H_2O? We recommend that users subtract the water band until the region from 1900 to 1740 cm^{-1} is flat. It has been our experience that this rather subjective technique if applied consistently works well with most protein and surface combinations.

IV. FTIR AND PROTEIN SECONDARY STRUCTURE

An important problem in biomaterials research is the determination of protein structure and orientation at the material–blood or material–tissue interface. For example, studies have clearly shown that often information on the amounts of a particular protein at the interface are insufficient to explain cellular interactions, so information about structure is essential. For example, a number of studies had suggested that surfaces with different chemical compositions which had the same amounts of adsorbed fibronectin (a cell adhesion protein) had different effects on the differentiation responses of fibroblasts and neuronal cells [17,18]. The researchers concluded that these differences were due to differences in fibronectin conformation due to different surface chemistries. Using FTIR/ATR we could detect distinct differences in the conformation of adsorbed fibronectin on these surfaces [19].

A. Relationship to X-Ray Crystallography

It is well accepted that x-ray crystallography provides the most detailed structure for proteins. In some cases, the structure of the protein as determined in the crystalline state may not be exactly the same as when the protein is in solution or on surfaces at much smaller concentrations. NMR spectroscopy has been used in the past by researchers to obtain the solution structure of proteins with known x-ray structure. Such studies indicate that for most proteins, there is very little difference between the proteins in solution and in the crystalline state [20,21].

The database of proteins which have been studied with x-ray crystallography and NMR is unlikely to increase at any rapid rate because of the complexity of doing NMR spectroscopy with most large proteins. In addition, some membrane proteins are likely to adopt their functional form only in the presence of a hydrophobic lipid environment, and knowledge of their x-ray structure in the absence of lipid may not be useful. Infrared spectroscopy, on the other hand, can be used to study the spectra of proteins in both solution and the crystalline state and in the presence of lipids. Spectra can be recorded for small proteins, peptides, and very high-molecular-weight proteins [22].

The practical difficulties involved with the development of high-resolution structural information about proteins has stimulated the development of a number of low-resolution techniques such as circular dicroism and vibrational spectroscopies like Raman and FTIR, which can provide global insights into the overall secondary structure of proteins without precise location of the different structural elements [22].

FTIR spectra can be easily obtained for proteins in solution using transmission techniques by placing a solution of the protein between two infrared transmitting windows such as CaF_2. One particular advantage of the FTIR technique is the ease and the rapidity of acquiring high-quality spectra from very small amounts of protein. What is unique about the FTIR technique (with a microscope) is that spectra can be obtained of both protein crystals and the solution from which the crystals grow [21].

B. Infrared Spectra Structural Correlations

Infrared spectroscopy is now well established as the method for the analysis of protein secondary structure in solution and when adsorbed to surfaces. FTIR spectra can be obtained for proteins in a wide range of environments, in solution and on surfaces including polymers, metals, and bioceramics that are of interest in biomaterials.

There is a wealth of information that can be used to derive structural information by analyzing the shape and position of bands in the amide I region of the spectrum. The presence of a number of amide I band frequencies has been correlated with the presence of α-helical, antiparallel and parallel β-sheets, and random coil structures [23–28]. It is now well accepted that absorbance in the range from 1650 to 1658 cm^{-1} is generally associated with the presence of α-helices in aqueous (H_2O) environments. Precise interpretation of bands in this region are difficult because there is significant overlap of the α-helical structures with random structures. One way to resolve this issue is to exchange the hydrogen from the peptide N—H with deuterium. If the protein contains a significant amount of random structure, the H—D exchange will result in a large shift in the position of the random structure (will now absorb at around 1646 cm^{-1}) and only a minor change in the position of the helical band. The β-sheet vibrations have been shown to absorb between 1640 and 1620 cm^{-1} (though there is considerable disagreement on assignments for parallel and antiparallel β-sheets). Absorbances centered at 1670 cm^{-1} have been assigned to β-turns, or simply turns by some investigators.

Infrared absorbances are greatly affected by the local microenvironment of the different structural groups like α-helix and β-sheet, however, and hence unique assignments are often difficult [29,30]. When helical proteins are associated with membranes, the band appears from 1656 to 1658 cm^{-1}, whereas in soluble proteins this band occurs at lower wave numbers, approximately 1650 to 1655 cm^{-1}. With highly solvent-exposed helices in D_2O, the amide band can shift to a frequency as low as 1644 cm^{-1} [29,30]. It is therefore very important to remember that simple infrared spectra secondary structure correlations do not exist. FTIR can be a powerful tool if it is used in combination with other techniques, such as circular dichroism (CD) [31–33].

C. Quantitation of Secondary Structure

Protein secondary structural elements like the α-helix and β-sheet have broad absorbance bands in the amide I region and their overlap is significant. The valley between two adjacent peaks of equal intensity must be 20% lower than the peak tops to be considered just resolved [34,35]. The widths of protein bands are inherently broad, between 16 to 20 cm^{-1}, and lie close to each other in the amide I region;

thus, obtaining spectra at a very high resolution cannot be used to resolve the amide I into individual components limiting the resolution we can achieve. Deconvolution or second-derivative techniques can be used to mathematically enhance the resolution of the spectrum; i.e., the process separates bands that are intrinsically broader than the instrument resolution [35–37]. Iterative least-squares curve-fitting techniques can then be used for quantitative estimates of the different secondary structures [23–25].

A completely different approach to the problem of quantifying protein secondary structure is based on the assumption that any protein can be considered as a linear sum of a few fundamental secondary structural elements. There is a growing collection of proteins whose structures have been determined by x-ray crystallography. The fundamental structural elements identified by x-ray crystallography include α-helix, β-sheet (parallel and antiparallel), β-turns, and others simply termed random coils. FTIR spectra for such proteins have also been obtained under aqueous conditions. We can then use techniques such as factor analysis to extract a minimal basis set that can represent the entire set of protein spectra. Not surprisingly such analyses indicate that a set of five fundamental spectra can be identified to represent different proteins [38,39]. These spectra can then be used to predict the structure of an unknown protein from its infrared spectrum. Application of such techniques to infrared spectra of complex proteins has been described by a number of investigators [38,39,40]. Principal component regression and partial least-squares techniques have also proven to be versatile for working with protein FTIR spectra.

D. Frequency Precision in FTIR

With an FTIR spectrometer, the reproducibility of the frequency scale (wave number) is very high and independent of the resolution at which the data are collected. This allows us to observe the subtle changes in the infrared spectra of biomolecules in response to external stimuli. We have described, for example, the time-dependent transitions in an adsorbed layer of fibrinogen on two different polyurethanes [9]. There was a clear correlation between the shift in the center-of-gravity frequency of the amide II band with the amount of fibrinogen retained on the surface after exposure to a denaturing environment. Over a period of about 48 h, the maximum shift in the center of gravity was about 2 cm^{-1} in spectra that were obtained at a nominal resolution of 4 cm^{-1}. A discussion of how current FTIR instrumentation can allow us to draw significance of a 2 cm^{-1} shift in 4 cm^{-1} spectra is available [41].

E. FTIR and Blood Proteins

The proteins of interest to biomaterials researchers include albumin, IgG, and fibrinogen, for which only low-resolution x-ray crystallographic data are available, if any. Using a database of infrared spectra of proteins with known x-ray structure to predict the structure of blood proteins like fibrinogen assumes that the database can accurately reflect different structural elements and their environments. This assumption can often lead to wrong conclusions. Quantitation of protein secondary structures using factor analysis methods should be approached cautiously. A combination of full-spectrum techniques such as PLS and PCA with careful analysis of the number of overlapping bands in the amide I region followed by curve fitting is recommended.

F. FTIR and Orientation

The interaction of proteins or cells with proteins occurs in most cases through specific interactions. Biomaterial scientists have known, for example, that a tripeptide sequence composed of arginine (R)–glycine (G)–aspartate (D) is found on proteins like fibronectin, vitronectin, fibrinogen, and collagen. These proteins can mediate the attachment of cells to surfaces through the RGD sequence [42]. Studies have also shown that the deposition of peptides with the RGD cell adhesion sequences on a variety of surfaces can promote cell attachment [43–49]. While the availability of the RGD binding region in cell adhesion proteins like fibronectin is in many cases sufficient for cell adhesion, the rest of the protein provides a number of necessary secondary interactions which stabilize binding to cells. It is obvious, therefore, that the orientation of a protein like fibronectin and its conformation when adsorbed to surfaces will have a great influence on cellular interactions. Studies with a battery of antibodies that recognize a variety of cell binding sites on a protein like fibronectin may be required before we can get a good idea of how protein conformation and orientation can affect cell interaction [42,50,51].

FTIR techniques with polarizers can help us obtain some information on molecular order and molecular orientation. The calculation of molecular orientation of adsorbed proteins with FTIR/ATR is quite complicated, however. Infrared linear dichroism of the appropriate amide I bands can be used to obtain information on how specific portions of the protein molecules are oriented on the surface. If we know the three-dimensional structure of the protein from x-ray diffraction and if the protein retains its structure on adsorption, we can then use linear dichroism to obtain information about how the protein is oriented on the surface.

Consider a protein that is 100% helix and oriented along the z-axis. The angle made by the C=O vibration (which is primarily responsible for the amide I band of proteins) in the amide linkages along the protein backbone has been reported to be approximately 27° to the center line of the α-helix [52–55]. If we assume that all the helices are oriented along the z direction (Fig. 1), we will have a number of amide C=O vibrations aligned approximately 27° with the z-axis. The intensity of the amide I band obtained with perpendicular polarized light will be very small under these conditions because the component of the dipole moment along the y direction will be a small fraction of that along the z-axis. If now the helix is aligned along the y direction, the amide I band with parallel polarized light will be small because now the dipole moment along the z and x directions will be small. The ratio of absorbances obtained with parallel polarized light with perpendicular polarized light (called the dichroic ratio) can be used to estimate the angle made by the C=O dipole (and therefore the α-helix) with the z-axis. These calculations assume a uniaxial orientation for the adsorbed protein molecules (all proteins oriented in one way) which may or may not be justifiable. More rigorous criteria that can be used to discriminate between uniaxial and biaxial distributions using FTIR/ATR have been described [56].

Equations for calculating electric field intensities and absorption equation variables for analyzing the interaction of plane polarized light with thin oriented protein layers on ATR elements have been published in several forms and can be used to calculate molecular orientation [54,55,57]. For monolayers and very thin films, the equations published by Cornell et al. [54] or those by Axelsen et al. [58] can be used. These calculations require refractive indices for the protein layer and the aqueous medium the protein is in contact with. Analysis of molecular orientation by calculating the dichroic ratio of other structural elements like β-sheet can also be done, but the interpretation is even more difficult. Investigators have assumed that the β-sheet vibration is perpendicular to the α-helix.

Calculations of molecular order from polarized FTIR/ATR are critically dependent on the values of the evanescent electric field amplitudes in a protein thin film [59]. Several expressions used for calculating evanescent electric field amplitudes in supported lipid membranes were recently examined [59]. The authors tested the validity of these expressions by measuring the infrared dichroism of poly-γ-benzyl-L-glutamate and poly-β-benzyl-aspartate under conditions where their molecular order was known. Their analysis suggests that the thin-film approximation used by a number of investigators in the literature for calculation of molecular orientation on ATR crystals is not valid. The electric field amplitudes within a thin film on ATR substrates appears to be determined more by the dielectric properties of the bulk phase than by the properties of the thin film. Thus, blind usage of published techniques to calculate molecular order from linear dichroism measurements even for proteins that are known to be completely helical is to be avoided. Orientation analysis of large blood proteins like albumin, IgG, and fibrinogen, which contain a significant amount of nonhelical structure, is thus likely to be more problematical. Infrared linear dichroism can be useful if independent measurements on molecular orientation can be obtained by using monoclonal antibody binding measurements to specific sites on proteins or other techniques.

V. FTIR AND MULTIPLE PROTEIN SOLUTIONS

A major limitation of the FTIR technique for protein adsorption studies is the difficulty of distinguishing different proteins from each other with the help of their infrared spectra alone. In some of our earlier publications, we have provided some data suggesting that this may be possible [60–62]. A close examination of these papers will reveal the major assumptions we had to make to arrive at those conclusions. For example, we showed that the kinetics of adsorption of a binary mixture of albumin and IgG can be followed on germanium ATR crystals [60]. Infrared spectra obtained as a function of time were analyzed using a simple matrix multiplication technique that relied on a calibration matrix generated from mixtures of the two proteins in solution. Today, a number of sophisticated multivariate techniques such as partial least squares are available but are unlikely to improve upon the analysis we have presented. The results from the application of any quantitative analytical tool will depend entirely on the set of spectra used during the calibration phase of the analysis. Analysis of protein mixtures on surfaces using their infrared spectra alone will require a calibration set that models mixture spectral changes with concentration and with conformation. Analysis of changes in single protein conformation from their infrared spectra are challenging to begin with and hence analysis with mixtures will be very difficult. The ability to tag the protein of interest in a mixture with an infrared label by minimally altering its secondary structure and its adsorption behavior will go a long way to solving this problem. The label chosen should adsorb in the wave number region from approximately 1800 to 2800 cm^{-1}, away from interference from both other proteins and the large absorbance bands of water.

A nitrile (—CN) group which absorbs around 2256 cm^{-1} can be introduced into proteins through a simple addition reaction. Acrylonitrile or tetracyanoethylene when added to proteins in solution will couple to the primary amines in proteins, and this reaction rate is known to increase with pH [63]. Unreacted label can then be removed from the reaction mixture by dialysis against buffer. The coupling reaction of acrylonitrile with protein is

$$\text{Protein} - \text{NH}_2 + \text{CH}_2 = \text{CHCN} \rightarrow \text{Protein} - \text{N} - \text{CH}_2\text{CH}_2\text{CN} \tag{9}$$

The disappearance of primary amines can be monitored by a fluorescence assay that uses fluorescamine [64]. The fluorescamine assay requires small amounts of protein, is easily reproducible, and does not require the removal of unreacted acrylonitrile groups. Our experiments with albumin showed that approximately 70% of the lysine groups in albumin could be linked with the —CH$_2$CH$_2$CN label. FTIR/ATR experiments with the labeled albumin showed that the label (2256 cm^{-1}) and the protein (1550 cm^{-1}) approached the surface at a similar rate (indicating covalent coupling) (data not shown).

We are convinced that this approach of using a specific marker on a protein of interest can be used to develop accurate estimates of the concentration of a specific protein adsorbed to surfaces from mixtures. The ability to extract the spectrum of the tagged protein from the mixture [61,62] could then allow analysis of changes in secondary structure of that particular protein if a comprehensive calibration set of spectra are prepared.

VI. ATOMIC FORCE MICROSCOPY AND SURFACES

The AFM measures the forces of interaction between a probe and features on a surface. In a typical configuration, a gold-coated silicon cantilever with a known force constant supports a silicon nitride tip. A piezoelectric scanner is used to raster the sample with this tip. There are three independently controlled piezos, one each for the x-, y-, and z-axes (with z being the distance between the tip and the surface). A laser beam is reflected off the top of the cantilever to a four-quadrant photodetector. As the scanner rasters the sample, the cantilever will be deflected up or down depending on the surface features encountered by the tip. This motion of the cantilever is accurately tracked by the laser beam. The AFM can be operated in a constant-height or a constant-force mode. In the constant-force mode, the computer in the AFM will adjust the z piezo such that the distance between the silicon nitride tip and the surface is maintained constant based on the information obtained from the photodetector. When the cantilever is deflected up because the tip encounters a raised surface feature, this will change the intensities of the laser beam detected by the two halves of the photodetector. The AFM computer will immediately adjust the voltage to the z piezo so that these two signal intensities are again back to what they should be. The voltage required to adjust the z piezo provides the information for the calculation of height of the surface features. Since the cantilever obeys Hooke's law, from a knowledge of the spring constant and known deflections, forces between the tip and the surface can be measured quite accurately. The operation of the AFM in what is called the lateral force mode can be explained similarly. The twisting of the cantilever due to changes in the way the tip interacts with the surface will result in differences in the intensities of the laser beam as detected by the left and the right halves of the four-quadrant photodetector. This difference can be translated into the actual lateral forces experienced by the cantilever if the physical properties of the cantilever are accurately known. Illustrations describing the operating principles of the AFM (and related techniques) have been published [65–68].

The necessity of maintaining a close contact between a hard probe tip such as silicon nitride and a surface for imaging is a force of continuing problems for the scientist imaging *soft* materials such as proteins [69]. This can be overcome to some extent by using tapping mode (TM) AFM. In TM-AFM the tip is vibrated at a specific frequency and brought into close contact with the sample surface. If the tip interacts with a surface feature, the amplitude of vibration is changed. The piezoelectric crystals are automatically adjusted by the AFM computer to recover the original amplitudes. The tip can thus be rastered over a surface to generate topographic information. While AFM operated in the tapping mode does not destroy proteins and other soft molecules, the image resolutions are not as good as that obtained while using contact AFM.

Images obtained by the AFM in the contact mode reflect features of the surface and the tip itself. Since the tip has finite dimensions, imaging small molecules or surface features of dimensions similar to that of the tip are *contaminated* by the tip itself [70]. Techniques to account for the shape of the tip and improve the resolution of the images so obtained have been described in the literature. The idea is to image an object of known dimensions with the tip, create the shape of the tip, and use that to improve the resolution of the final image. The procedure for calculating and using tip shapes have been described as deconvolution [71,72] and by using the language of mathematical morphology [73–75].

Atomic force microscopy and related techniques offer an unparalleled view of individual protein molecules or small collections of such molecules adsorbed to surfaces [66–68,76–78]. AFM has also been used recently to follow the adsorption of proteins directly at the solid–liquid interface [79].

A closely related technique is called chemical force microscopy [80]. The procedure is similar to the regular AFM except that the tips are chemically derivatized. In this paper, the authors show that the spatial arrangement of functional groups on a surface prepared using self-assembled monolayer techniques can be determined. Silicon nitride tips that were covalently modified with self-assembled monolayers terminated in either a hydrophobic (methyl) or a hydrophilic (carboxylic acid) group were used. When this modified tip was scanned across a surface patterned with methyl and carboxylic acid groups, the interactions between the tip and the surface were revealed as differences in frictional force.

The ability to precisely engineer a gold surface with alkanethilates by a simple technique using a stamp and self-assembled monolayers has been described by Whitesides et al. [81]. An elastomer stamp made from poly(dimethyl siloxane) was first prepared using techniques perfected by the electronics industry. This stamp was then used to prepare surfaces patterned with well-defined regions that had distinct chemical and physical properties. Atomic force microscopy was then used to image those surfaces patterned by what is termed microcontact printing [65]. Patterns having dimensions of 0.2 to 100 μm could be easily prepared on surfaces using this process. Chemical force microscopy revealed the surface chemical features with a detail that cannot be matched by any other technique.

VII. SURFACE MODIFICATION STRATEGIES

Nature has given us a number of examples from which we can draw to engineer surfaces [1,2]. These can be broadly divided into two groups: (1) order through recognition, enzyne–substrate, antibody–antigen, DNA–RNA–protein, and (2) order through self-assembly, the formation of the cell wall. Scientists have used this idea of recognition and specificity and designed these into biomaterials. Some of these include

- The preparation of surfaces incorporating peptide sequences like RGD on surfaces for improving cell adhesion [43,46,47,82,83]
- Modified polystyrene surfaces that exhibit heparin-like behavior [84–86]
- The nanopatterning of surfaces for control of cell behavior at interfaces [45,87,88]
- The design of intelligent polymers that can respond to changes in the environment [89]
- A designed protein monolayer by chemically linking site-directed mutant proteins with reactive cysteine to thiol specific functional groups [90]

There are situations where the best strategy appears to be the preparation of surfaces that will prevent all protein adsorption (intraocular lenses). Quite often, however, we need a very strong, specific, and engineered reaction to implants to accelerate the wound-healing process.

We still do not have a clear molecular biological definition of biocompatibility. The rational way to approach this problem appears to be to understand the mechanisms by which the body heals itself and consequently design surfaces that can participate in the wound-healing process.

Understanding the complex processes that occur when cells interact with material surfaces in the presence of complex, multicomponent solutions requires the following [91]:

1. Biomaterial surfaces well characterized with respect to chemical composition, morphology and structure.
2. Understanding of the macromolecular conditioning film structure, conformation, and molecular orientation of proteins that will adsorb.
3. Quantitation of cell processes such as retention, adhesion, differentiation, and growth.
4. The measurement of short-term and long-term effects. It is difficult, however, to predict the long-term performance of an implant from early cell adhesion data [92].

A number of techniques have been used in the literature to examine the effect of surface modification on protein adsorption and cell interactions. These techniques include the use of self-assembled monolayers [17,18,48,87,93,94], radiofrequency plasma techniques for surface modification and polymerization [51,95], the coupling of immobilization of protein-rejecting molecules like poly(ethylene glycol) [96–

100], and, following the work of Pierschbacher and Ruoslahti [82], the deposition of peptides with cell adhesion sequences on a variety of surfaces [43–49]. Radio-frequency plasma modification techniques offer an excellent means to modify surface properties while leaving bulk properties intact, but surface properties are difficult to predict. The surface that results after plasma modification could contain a number of functional groups, the surface could also change with time. Self-assembled mono-layers, however, provide a method by which we can have molecular level control of surface properties.

A. Self-Assembled Monolayers for Surface Modification

Self-assembled monolayers (SAMs) are molecular assemblies that are formed spon-taneously by immersion of a substrate in a solvent. Currently there are two major routes used by scientists to get SAMs for study. These include the use of alkane-thiol adsorption onto gold [93,101–103] and organosilicon agent adsorption onto hydroxyl-containing surfaces [104,105]. In each case, the head group (thiols, chlo-rosilanes, for example) chemisorbs with surface groups (gold, hydroxyls) in an exo-thermic reaction. Hydrophobic interaction between the acyl chains then helps to bring molecular order to this monolayer.

SAMs formed using octadecyltrichlorosilane (OTS) are very stable in the pres-ence of acids and can also be washed with detergents. These SAMs are, of course, not stable in basic environments because of the hydrolysis of the Si—O bond.

Sagiv first described this approach to the preparation of self-assembled layered structures using surfactants with $SiCl_3$ at one end and a chemically convertible apolar terminal function at the other [104]. These self-assembled monolayers are easy to deposit and are stable enough to be used in ex vivo blood contacting situations [104–106]. A range of functionalities can be tolerated in the deposition process and a wide range of chemical transformations can be achieved once the monolayer has been deposited (see references in Ref. 105). These functionalities include olefins, esters, ethers, thioethers, and thioesters. With these built-in functional groups, in situ trans-formations have included the conversion of terminal olefin in alcohol, dibromide or acid, ester to acid or alcohol, and a host of others. They have deposited siloxane-anchored monolayered films with Br, CN, SCN, and $SCOCH_3$ funtionalities. The packing and integrity of the monolayers was confirmed by relatively small water contact angle hysteresis. The surface chemical functionality was confirmed by XPS and IR techniques. While the deposition of a stable hydrophobic monolayer on sur-faces with long-chain alkyltrichlorisilanes is simple in principle, we still do not understand the mechanism [107]. The authors show that the preparation of an ordered film on the surface requires the presence of surface moisture, which in turn allows condensation between neighboring —Si—OH groups of the trihydroxy head groups of the alkyltrichlorisilanes. The authors show that there is insignificant bonding be-tween —Si—OH and SiO_2. SAMs formed with chlorosilanes are proving to be more difficult to prepare. Instead of the chlorosilane functionalities forming covalent links with the surface hydroxyls, they could cross-link with other similar acyl chlorosilanes forming a polymer on the surface and a disordered film. Unless the experimenter is very, very careful, monolayers of alkylsiloxanes tend to result in a film that is in-herently more disordered than thiolated monolayers on gold.

FTIR using both external reflection (ER) and attenuated total internal reflection (ATR) can be very useful in obtaining information about the order in chlorosilane monolayers on surfaces. Distinct shifts can be seen in the position of the CH_2 stretching bands as a function of molecular order (see the table). Since the CH_3 vibration is parallel to the acyl chain, in a perfectly ordered film on the surface, the CH_3 intensity will be stronger in ER than in ATR.

Wave number range	Description of vibration	Notes
2918	Crystalline, antisymmetric CH_2 stretch	Intense in ATR
2924	Liquid, antisymmetric CH_2 stretch	Intense in ATR
2851	Crystalline, symmetric CH_2 stretch	Intense in ATR
2855	Liquid, symmetric CH_2 stretch	Intense in ATR
2960 ± 2	Crystalline, antisymmetric CH_3 stretch	Intense in ER Weak in ATR
2960 ± 2	Liquid, antisymmetric CH_3 stretch	Weak in ER Weak in ATR

Infrared linear dichroism techniques have been used to study the orientational ordering of n-octadecyltrichlorisilane on silicon ATR crystal [108]. While measurements of the orientation of alkyl chains are difficult, changes in the orientation could be detected as the surface coverage increased. The ordering of molecules on the surface increased significantly as monolayer coverage was reached. The authors detected changes in the dichroic ratio and a shift toward lower wave number as the orientational order increased.

Monolayers containing more than one component can also be prepared on hydroxyl-containing surfaces [109]. Conditions in the solution (solvent type) and the surface can lead to large fluctuations in the composition of the mixed monolayer, however. While the initial ionic binding between monolayers and a surface like germanium occur within a few seconds, several minutes are required for the covalent binding of OTS to the surface [110].

Groups as different as OH and CH_3 have been shown to behave independently at the monolayer–air interface in mixed monolayer systems [101]. The composition of SAMs formed on the surface parallels that in solution but are not the same [102,111]. The differences between the compositions of mixed SAMs appear due to factors important during the formation of the monolayers rather than subsequent equilibrium [102]. Multicomponent monolayers did not appear to phase-segregate into single-component domains that were large enough to be detectable by contact angle [111]. Techniques designed for preparing mixed monolayer systems with precise control of surface functionality on hydroxyl containing surfaces have also been described [112].

B. SAMs in Biomaterials Research

Self-assembled monolayer technique provides biomaterials scientists with the ability to precisely control surface chemistry through surface compositions. Surfaces prepared through self-assembly techniques have been shown to be appropriate for studying both in vitro [19,113,114] and a blood-contacting situation [106]. The propensity to activate the intrinsic pathway of the plasma coagulation cascade changes exponentially with surface coverage of polar functional groups, with some interesting exceptions [113,114]. In this study, the propensity to coagulation was studied using glass disk procoagulants bearing close-packed SAMs terminated in 12 different functional groups. SAMs terminated in $-CO_2H$ and NH_3^+ showed a notable deviation from the trend, suggesting chemically specific interaction with proteins of the contact activated system.

We have shown that protein adsorption experiments with FTIR/ATR techniques can be done on functionalized SAMs [19]. Previous studies had suggested that surfaces with different compositions affect the differentiation responses of fibroblasts and neuronal cells by affecting the conformation of the bound fibronectin [17,18]. This was the motivation for our study with FTIR/ATR to help obtain direct information on how changing the surface affects the secondary structure of adsorbed fibronectin [19]. SAMs functionalized with carboxylic acid, thiols, bromide, methyl, and nitrile groups were prepared on germanium ATR crystals; details of these and other techniques have been given [19,115]. We examined the FTIR/ATR data for surface-specific effects on the structure of adsorbed fibronectin. Hydrophobic surfaces (SAMs terminated by -bromide and -methyl) did not support the formation of neurites, while on surfaces terminated by carboxylic acid significantly greater neurites were seen [17]. We found an interesting correlation in the FTIR data. In Fig. 5, the ratio of β-sheet to β-structures is shown for fibronectin adsorbing to the different SAMs. There is a rapid change in the ratio of the two types of β-structure during the first 20 min and then a plateau to a steady-state value. The ratio of β-sheet to β-turn structure was much larger on the hydrophilic (acidic) than on the hydrophobic surface we examined. The presence of β-sheet structure appears to support cell spreading and neurite formation. It is interesting that the total amount of fibronectin adsorbed on all the surfaces was not significantly different from each other [19].

Our FTIR/ATR study clearly demonstrated a measure of surface-dependent conformational change for fibronectin that was correlated with independent measurements on cell behavior on these surfaces.

C. RGD Peptides on Surfaces

In order to control cell adhesion and growth, we must control the binding, attachment, and structure of cell binding proteins such as fibronectin and vitronectin. In a large number of cases, it turns out that instead of the entire protein, we need to immobilize only a part of the protein involved in cell binding. The tripeptide consisting of arginine–glycine–aspartic acid (RGD), for example, has been known to mimic the functions of adsorbed fibronectin [116]. The RGD sequence has been found in vitronectin, fibronectin, von Willebrand factor, fibrinogen, and collagen. This peptide has been shown to play a crucial role in mediating cell attachment and subsequent spreading. When RGD peptides are immobilized to surfaces, they appear to behave like adsorbed fibronectin. The ability of RGD antibodies to bind to adsorbed fibrinogen could account for many, but not all, differences in the growth of corneal epithelial cells [50]. Thus, while the availability of the RGD binding region is in many cases sufficient for cell adhesion, the rest of the fibronectin appears to

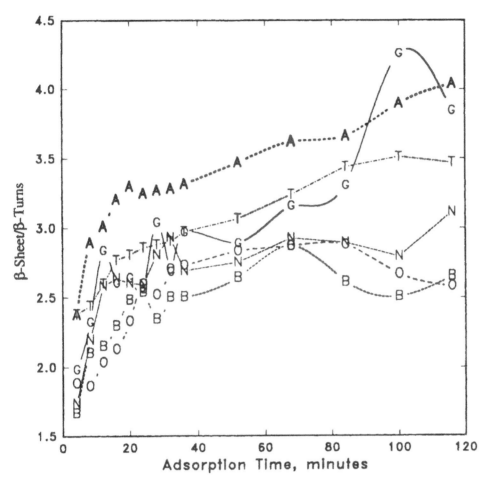

FIG. 5 A plot showing how the ratio of β-sheet to β-turn bands changes with time for a number of surfaces. Bare germanium is shown as G; the other surfaces had a self-assembled monolayer coated onto germanium substrate with different functional groups facing fibronectin: —COOH (A), —CH₃ (O), bromide (B), and nitrile (N). (Reproduced with permission from Academic Press, from *Journal of Colloid and Interface Science*, Vol. 162, pp. 135–143, 1994.)

provide important secondary actions which stabilize binding to cells. The relationship between protein conformation and cell interaction is thus a complex one. Studies with a battery of antibodies that recognize a variety of cell binding sites on a protein may be required [42,50,51].

The density of RGD peptides adsorbed onto a surface turns out to be very important [43]. For example, at a density of 0.1 fmol/cm^2, fibroblasts are shown to adhere but did not spread. At a surface density of 1 fmol/cm^2, the fibroblasts could spread. At much higher densities of 10 fmol/cm^2, the fibroblasts showed evidence for the clustering of an integrin receptor and assembly of cytoskeletal elements. The density of cell adhesion molecules on the surface should be such that the cells that adhere are allowed to move about on the surface.

Why use peptides such as RGD and others when we can use entire proteins? Selectivity is the main reason. Take the example of fibronectin. The protein can adsorb in a number of different conformations and orientations, and only a small fraction of those would be useful for cell adhesion. With the help of peptides, we get a degree of control that is impossible with proteins. For example, the peptide sequence REDV (arginine–glutamic acid–aspartic acid–valine) can be immobilized onto a poorly adhesive substrate and can bind to endothelial cells. It is interesting that this peptide does not encourage the binding of smooth muscle cells or fibroblasts but only endothelial cells [83].

A photochemical technique that can place RGD peptides onto a surface in a controlled manner has been developed [117]. The photoreactive octapeptide (Gly-Gly-Gly-Arg-Gly-Asp-Ser-Pro) (containing the RGD sequence) was derivatized with 4-azidobenzoylsuccinimide at the N-terminal. UV irradiation was used to attach the peptide to a poly(vinyl alcohol) film.

D. Preventing Protein Adsorption

Poly(ethylene glycol) (PEG) is perhaps the best-known hydrophilic polymer that can be used to prevent protein adsorption to surfaces [94,96,99,118,119]. PEG surfaces can be prepared by many techniques, including physical adsorption (high molecular weights), covalent coupling, and graft copolymerization [99,100,119]. Perhaps the definitive study on the factors that affect the protein-rejecting abilities of PEG was done by Prime and Whitesides [94]. They showed that very low protein adsorbing surfaces can be obtained even with very short oligo(ethylene oxide)s on gold. It is not the molecular weight (and so the size) of the PEG that is important but the surface coverage. Surfaces covered completely with low-molecular-weight PEGs were almost as effective as preventing protein adsorption as those covered with higher-molecular-weight PEGs. The ability of PEGs to self-exclude often prevents researchers from preparing surfaces which are completely covered with this hydrophilic polymer.

PEG can be used to prepare surfaces containing RGD peptides that will reject all nonspecific protein adsorption but allow the adhesion and the growth of endothelial cells [44], giving researchers a measure of control over surface properties.

VIII. CONCLUSIONS

Manipulating the nature of the host–implant interaction will require the fabrication of surfaces with a well-defined layer of proteins or with specific protein adsorption properties. This will require the development of technology that will allow the biomaterial scientist to precisely design and build surfaces and evaluate such surfaces. In this chapter we have described how Fourier transform infrared spectroscopy (FTIR) can be a valuable tool for this purpose. We have also given a brief description of atomic force microscopy and how it can provide a level of detail that cannot be obtained using FTIR techniques. A discussion of how self-assembled monolayers on gold- or hydroxyl-containing surfaces can provide the biomaterial scientist with a very high degree of control on surface properties was made.

The measurement of protein adsorption at solid–liquid interfaces under aqueous conditions continues to offer significant challenges to biomaterials researchers. Radiolabeled proteins are still widely used for measuring protein adsorption to surfaces [120–124]. This remains a very sensitive and a reliable technique for obtaining quantitative information about proteins adsorbed to surfaces but offers no information about protein conformation. In addition, waste disposal has become quite complicated and expensive.

FTIR/ATR is a powerful technique with which we can get relevant information for understanding protein adsorption to surfaces. Currently available FTIR spectrometers can provide reproducible IR spectra of proteins, in solution and on surfaces and under aqueous (H_2O) conditions. The speed and sensitivity of commercial FTIR systems with commercially available ATR optics and software for data collection allows the study of protein adsorption kinetics at high concentration and at different wall shear rates.

We have discussed how changes in protein structure can be found by careful analysis of changes in protein spectra. We have refrained from providing tables that contain infrared spectra structure correlations because we feel that information like that could be misleading. Analysis of changes in secondary structure of adsorbed proteins by analysis of the FTIR spectra must be approached with caution. Conclusions about the presence or absence of α-helical, β-sheet, β-turn, and random coil bands could be drawn from spectra obtained in H_2O and D_2O, but these bands are sensitive to the microenvironment. Wherever possible, infrared correlations must be supported using other techniques.

A number of tools developed by the electronics and the biotechnology industry are now being used by scientists designing surfaces. This, coupled with the explosion in surface analytical tools like atomic force microscopy and related techniques, will accelerate the process toward a rational design of biomaterials.

ACKNOWLEDGMENTS

Support from NIH Grant HL-38936, an NSF-EPSCoR grant (OSR-9550480), and the Department of Chemical and Materials Engineering from the University of Alabama in Huntsville is gratefully acknowledged.

REFERENCES

1. B. D. Ratner, New ideas in biomaterials science—a path to engineered biomaterials. J. Biomed. Mater. Res. 27(7):837–850, 1993.
2. B. D. Ratner, The quandary in contemporary biomaterials. In *Surface Control of Biology: A Path to the Next Generation of Biomaterials—Workshop*. 5th World Biomaterials Congress, 1996.
3. N. J. Harrick, *Internal Reflection Spectroscopy*. Interscience, 1967.
4. W. H. Yang and J. D. Miller, Verification of the internal reflection spectroscopy adsorption density equation by Fourier transformation infrared spectroscopy analysis of transferred Langmuir–Blodgett films. Langmuir 9:3159–3165, 1993.

5. J. S. Jeon, R. S. Sperline, and S. Raghavan, Quantitative analysis of adsorbed serum albumin on segmented polyurethane using FTIR/ATIR spectroscopy. Appl. Spectrosc. 46:1644–1648, 1992.

6. W. G. Pitt and S. L. Cooper, In *Proteins at Interfaces: Physicochemical and Biochemical Studies* (J. L. Brash and T. A. Horbett, eds.), Volume 343. American Chemical Society, Washington, D.C., 1987.

7. M. Jackson and H. H. Mantsch, Artifacts associated with the determination of protein secondary structure by ATR-IR spectroscopy. Appl. Spectrosc. 46(4):699–701, 1992.

8. D. J. Fink, T. B. Hutson, K. K. Chittur, R. I. Leininger, and R. M. Gendreau, Quantitative surface studies of protein adsorption by infrared spectroscopy. II. Quantitation of adsorbed and bulk proteins. Anal. Biochem. 167:147, 1987.

9. T. J. Fink, B. D. Ratner, K. K. Chittur, and R. M. Gendreau, Infrared spectroscopic studies of time-dependent changes in fibrinogen adsorbed to polyurethanes. Langmuir 7:1755, 1991.

10. T. J. Lenk, B. D. Ratner, K. K. Chittur, and R. M. Gendreau, IR spectral changes of bovine serum albumin upon surface adsorption. J. Biomed. Mater. Res. 23:549, 1989.

11. J. L. Ong, K. K. Chittur, and L. C. Lucas, Dissolution/reprecipitation and protein adsorption studies of calcium phosphate coatings by FT-IR/ATR techniques. J. Biomed. Mater. Res. 28:1337–1346, 1994.

12. K. P. Ishida and P. R. Griffiths, Theoretical and experimental investigation of internal reflection at thin copper films exposed to aqueous solutions. Anal. Chem. 66:522–530, 1994.

13. J. R. Powell, F. M. Wasacz, and R. J. Jakobsen, An algorithm for the reproducible spectral subtraction of water from the FTIR spectra of proteins in dilute solutions and adsorbed monolayers. Appl. Spectrosc. 40:339–344, 1986.

14. F. Dousseau, M. Therrien, and M. Pezolet, On the spectral subtraction of water from the FTIR spectra of aqueous solutions of proteins. Appl. Spectrosc. 43:538–542, 1989.

15. T. J. Lenk, *Infrared studies of protein adsorption and transitions on polyurethanes*. Ph.D. dissertation, University of Washington, 1994.

16. P. W. Holloway and H. H. Mantsch, Structure of cytochrome b_5 in solution by Fourier transform infrared spectroscopy. Biochemistry 28:931–935, 1989.

17. C. N. Sukenik, K. Lewandowska, N. Balachander, and L. A. Culp, Modulation of fibronectin adhesive functions for fibroblasts and neural cells by chemically derivatized substrate. J. Cell. Physiol. 141:334, 1989.

18. C. N. Sukenik, N. Balachander, L. A. Culp, K. Lewandowska, and K. Merritt, Modulation of cell adhesion by modification of titanium surfaces with covalently attached self-assembled monolayers. J. Biomed. Mater. Res. 24:1307, 1990.

19. S.-S. Cheng, K. K. Chittur, C. N. Sukenik, L. A. Culp, and K. Lewandowska, The conformation of fibronectin on self-assembled monolayers with different surface composition: an FT-IR/ATR study. J. Colloid Interface Sci. 162:135, 1994.

20. H. H. Mantsch and D. Chapman, *Infrared Spectroscopy of Biomolecules*. Wiley-Liss, New York, 1996.

21. J. M. Hadden, D. Chapman, and D. C. Lee, A comparison of infrared spectra of proteins in solution and crystalline forms. Biochim. Biophys. Acta 1248:115–122, 1995.

22. W. K. Surewicz, H. H. Mantsch, and D. Chapman, Determination of protein secondary structure by Fourier transform infrared spectroscopy: a critical assessment. Biochem. 32:389–394, 1983.

23. W.-J. Yang, P. R. Griffiths, D. M. Byler, and H. Susi, Protein conformation by infrared spectroscopy: resolution enhancement by Fourier self deconvolution. Appl. Spectrosc. 39(2):282–287, 1985.

24. D. M. Byler and H. Susi, Examination of the secondary structure of proteins by deconvolved FTIR spectra. Biopolymers 25:469–487, 1986.

25. H. Susi and D. M. Byler, Fourier deconvolution of the amide I Raman of proteins as related to conformation. Appl. Spectrosc. *42(5)*:819–825, 1988.

26. K. Kato, T. Matsui, and S. Tanaka, Quantitative estimation of alpha-helix coil content in bovine serum albumin by Fourier transform-infrared spectroscopy. Appl. Spectrosc. *41(5)*:861–865, 1987.

27. R. J. Jakobsen, F. M. Wasacz, J. M. Brash, and K. B. Smith, The relationship of bound water to the IR bandwidth of albumin. Biopolymers *25*:639–654, 1986.

28. F. M. Wasacz, J. M. Olinger, and R. J. Jakobsen, Fourier transform infrared studies of proteins using non-aqueous solvents. Effects of methanol and ethylene glycol on albumin and immunoglobulin-G. Biochem. *26*:1464–1470, 1987.

29. D. Chapman, M. Jackson, and P. I. Haris, Investigation of membrane protein structure using Fourier transform infrared spectroscopy. Biochem. Soc. Trans. *17*:617–619, 1989.

30. M. Jackson, P. I. Haris, and D. Chapman, Fourier transform infrared spectroscopic studies of Ca^{2+} binding proteins. Biochem. *30*:9681–9686, 1991.

31. S. M. Kelly and N. C. Price, The application of circular dichroism to studies of protein folding and unfolding. Biochim. Biophys. Acta *1338*:161–185, 1997.

32. N. J. Greenfield, Methods to estimate the conformation of proteins and polypeptides from circular dichroism data. Anal. Biochem. *235*:1–10, 1996.

33. M. Bloemendal and W. C. Johnson, Jr., Structural information on proteins from circular dichroism spectroscopy possibilities and limitations. Pharm. Biotechnol *7*:65–100, 1995.

34. P. R. Griffiths and J. A. deHaseth, *Fourier Transform Infrared Spectrometry*. Wiley-Interscience, New York, 1986.

35. B. C. Smith, *Fundamentals of Fourier Transform Infrared Spectroscopy*. CRC Press, Boca Raton, FL, 1996.

36. J. K. Kauppinen, D. J. Moffatt, H. H. Mantsch, and D. G. Cameron, Fourier self-deconvolution: a method for resolving intrinsically overlapped bands. Appl. Spectrosc. *35*:271–276, 1986.

37. D. G. Cameron and D. J. Moffatt, Deconvolution, derivation and smoothing of spectra using Fourier transforms. J. Test. Eval. *12*:78–85, 1984.

38. F. Dousseau and M. Pezolet, Determination of the secondary structure content of proteins in aqueous solutions from their amide I and amide II infrared bands. Comparison between classical and partial least-squares methods. Biochemistry *29*:8771–8779, 1990.

39. R. W. Sarver and W. C. Krueger, Protein secondary structure from Fourier transform infrared spectroscopy: a data base analysis. Anal. Biochem. *194*:89–100, 1991.

40. D. C. Lee, P. I. Haris, D. Chapman, and R. C. Mitchell, Determination of protein secondary structure using factor analysis of infrared spectra. Biochemistry *29*:9185–9193, 1990.

41. D. G. Cameron, J. K. Kauppinen, D. J. Moffatt, and H. H. Mantsch, Precision in condensed phase vibrational spectroscopy. Appl. Spectrosc. *36*:245–250, 1982.

42. T. A. Horbett, The role of adsorbed proteins in animal cell adhesion. Coll. Surf. B Biointerfaces 2:225–240, 1994.

43. S. P. Massia and J. A. Hubbell, Covalent surface immobilization of Arg-Gly-Asp and Tyr-Ile-Gly-Ser-Arg containing peptides to obtain well-defined cell-adhesive substrates. Anal. Biochem. *187*:292, 1990.

44. S. P. Massia and J. A. Hubbell, Human endothelial cell interactions with surface-coupled adhesion peptides on a non-adhesive glass substrate and two polymeric materials. J. Biomed. Mater. Res. *25*:223–242, 1991.

45. T. Matsuda and T. Sugawara, Development of surface photochemical modification method for micropatterning of cultured cells. J. Biomed. Mater. Res. *29*:749–756, 1995.

46. H. B. Lin, Z. C. Zhao, C. Garcia-Echeverria, D. H. Rich, and S. L. Cooper, Synthesis of a novel polyurethane copolymer containing covalently attached RGD peptide. J. Biomater. Sci. Polym. Ed. *3(3)*:217–227, 1992.

47. H. B. Lin, C. Garcia-Echeverria, S. Asakura, W. Sun, D. F. Mosher, and S. L. Cooper, Endothelial-cell adhesion of polyurethanes containing covalently attached RGD-peptides. Biomaterials *13(13)*:905–914, 1992.

48. S. Margel, E. A. Vogler, L. Firment, T. Watt, S. Haynie, and D. Y. Sogah, Peptide protein, and cellular interactions with self-assembled monolayer model surfaces. J. Biomed. Mater. Res. *27(12)*:1463–1476, 1993.

49. D. A. Puleo, Activity of enzyme immobilized on silanized Co-Cr-Mo. J. Biomed. Mater. Res. *29*:951–957, 1995.

50. T. A. Horbett and K. R. Lew, Residence time effects on monoclonal-antibody binding to adsorbed fibrinogen. J. Biomater. Sci. Polym. Ed. *6(1)*:15–33, 1994.

51. D. Kiaei, A. S. Hoffman, T. A. Horbett, and K. R. Lew, Platelet and monoclonal-antibody binding to fibrinogen adsorbed on glow-discharge-deposited polymers. J. Biomed. Mater. Res. *29(6)*:729–739, 1995.

52. E. Nabedryk, M. P. Gingold, and J. Breton, Orientation of gramicidin A transmembrane channel. Infrared dichroism study of gramicidin in vesicles. Biophys. J. *38*:243–249, 1982.

53. K. J. Rothschild and N. A. Clark, Polarized infrared spectroscopy of oriented purple membrane. Biophys. J. *25*:473–488, 1979.

54. D. G. Cornell, R. A. Dluhy, M. S. Briggs, J. McKnight, and L. M. Gierasch, Conformations and orientations of a signal peptide interacting with phospholipid monolayers. Biochemistry *28*:2789–2797, 1989.

55. R. Ishiguro, N. Kimura, and S. Takahashi, Orientation of fusion-active synthetic peptides in phospholipid bilayers: determination by Fourier transform infrared spectroscopy. Biochemistry *32*:9792–9797, 1993.

56. D. J. Ahn and E. I. Franses, Orientations of chain axes and transition moments in Langmuir–Blodgett monolayers determined by polarized FTIR/ATR spectroscopy. J. Phys. Chem. *96*:9952–9959, 1992.

57. S. Frey and L. K. Tamm, Orientation of mellitin in phospholipid bilayers. A polarized attenuated total reflection infrared study. Biophys. J. *60*:922–930, 1991.

58. P. H. Axelsen, W. D. Braddock, H. L. Brockman, C. M. Jones, R. A. Dluhy, B. K. Kaufman, and F. J. Puga, Use of internal reflectance infrared spectroscopy for the in situ study of supported lipid monolayers. Appl. Spectrosc. *49*:526–531, 1995.

59. M. J. Citra and P. H. Axelsen, Determination of molecular order in supported lipid membranes by internal reflection Fourier transform infrared spectroscopy. Biophys. J. *71*:1796–1805, 1996.

60. K. K. Chittur, D. J. Fink, R. I. Leininger, and T. B. Hutson, FTIR/ATR studies of protein adsorption in flowing systems. Approaches for bulk correction and compositional analysis in mixtures. J. Colloids Interface Sci. *111*:419, 1986.

61. M. Nyden, G. Forney, and K. K. Chittur, Spectroscopic quantitative analysis of strongly interacting systems: human plasma protein mixtures. Appl. Spectrosc. *42*:588–594, 1988.

62. M. Nyden and K. K. Chittur, Component spectrum reconstruction from partially characterized mixtures. Appl. Spectrosc. *43*:123–128, 1988.

63. H. R. Bosshard, K. H. Jorgensen, and R. E. Humbel, Preparation and properties of cyanoethylated insulin. An insulin derivative with blocked amino and imidazole groups. Eur. J. Biochem. *9*:353–362, 1969.

64. G. E. Means and R. E. Feeney, In *Chemical Modification of Proteins*. Holden-Day, San Francisco, 1971, pp. 114–117.

65. J. L. Wilbur, H. A. Biebuyck, J. C. MacDonald, and G. M. Whitesides, Scanning force microscopies can image patterned self-assembled monolayers. Langmuir *11*:825–831, 1995.

66. D. Sarid and V. Ellings, Review of scanning force microscopy. J. Vac. Sci. Technol. B. *9*:431–437, 1991.

67. O. Marti and M. Amrein, Academic Press, San Diego, CA, 1993.

68. L. Haggerty, B. A. Watson, M. A. Barteau, and A. M. Lenhoff, Ordered arrays of proteins on graphite observed by scanning tunneling microscopy. J. Vac. Sci. Technol. B *9*(2):1219–1222, 1991.

69. J. D. Andrade, V. Hlady, A. P. Wei, C. H. Ho, A. S. Lea, S. I. Jeon, Y. S. Lin, and E. Stroup, Proteins at interfaces: principles, multivariate aspects, protein resistant surfaces, and direct imaging and manipulation of adsorbed proteins. Clin. Mater. *11*:67, 1992.

70. S. J. Eppell, F. R. Zypman, and R. E. Marchant, Probing the resolution limits and tip interactions of atomic force microscopy in the study of globular proteins. Langmuir *9*: 2281–2288, 1993.

71. P. Markiewicz and M. C. Goh, Atomic force microscopy probe tip visualization and improvement of images using a simple deconvolution procedure. Langmuir *10*:5–7, 1994.

72. P. Markiewicz and M. C. Goh, Atomic force microscope tip deconvolution using calibration arrays. Rev. Sci. Instrum. *66*(*5*):3186–3190, 1995.

73. D. L. Wilson, K. S. Kump, S. J. Eppell, and R. E. Marchant, Morphological restoration of atomic-force microscopy images. Langmuir *11*(*1*):265–272, 1995.

74. J. S. Villarrubia, Scanned probe microscope tip characterization without calibrated tip characterizers. J. Vac. Sci. Technol. B *14*:1518–1521, 1996.

75. J. S. Villarrubia, Morphological estimation of tip geometry for scanned probe microscopy. Surf. Sci. *321*:287–300, 1994.

76. C. A. Siedlecki, S. J. Eppell, and R. E. Marchant, Interactions of human, von Willebrand factor with a hydrophobic self-assembled monolayer studied by atomic-force microscopy. J. Biomed. Mater. Res. *28*(*9*):971–980, 1994.

77. S. J. Eppell, S. R. Simmons, R. M. Albrecht, and R. E. Marchant, Cell-surface receptors and proteins on platelet membranes imaged by scanning force microscopy using immunogold contrast enhancement. Biophys. J. *68*(*2*):671–680, 1995.

78. L. Feng and J. D. Andrade, Surface atomic and domain-structures of biomedical carbons observed by scanning tunneling microscopy (STM). J. Biomed. Mater. Res. *27*: 177–182, 1993.

79. S. L. S. Stipp, In situ, real-time observations of the adsorption and self-assembly of macromolecules from aqueous solution onto an untreated, natural surface. Langmuir *12*:1884–1891, 1996.

80. C. D. Frisbie, L. F. Rozsnyai, A. Noy, M. S. Wrighton, and C. M. Lieber, Functional-group imaging by chemical force microscopy. Science *265*(*5181*):2071–2074, 1994.

81. A. Kumar, H. A. Biebuyck, and G. M. Whitesides, Patterning self-assembled monolayers: applications in material science. Langmuir *10*:1498–1511, 1994.

82. M. D. Pierschbacher and E. Ruoslahti, Cell attachment activity of fibronectin can be duplicated by small synthetic fragments of the molecule. Nature *309*:30–33, 1984.

83. J. A. Hubbel, S. P. Massia, N. P. Desai, and P. D. Drumheller, Endothelial cell-selective materials for tissue engineering in the vascular graft via a new receptor. Biotechnology *9*:568–572, 1991.

84. L. Stanislawski, H. Serne, M. Stanislawski, and M. Jozefowicz, Conformational changes of fibronectin induced by polystyrene derivatives with a heparin-like function. J. Biomed. Mater. Res. *27*:619–626, 1993.

85. Y. B. J. Aldenhoff and L. H. Koole, Studies on a new strategy for surface modification of polymeric biomaterials. J. Biomed. Mater. Res. *29*:917–928, 1995.

86. T. H. Zhang and R. E. Marchant, Novel polysaccharide surfactants—synthesis of model compounds and dextran-based surfactants. Macromolecules *27*(25):7302–7308, 1994.

87. D. A. Stenger, J. H. Georger, C. S. Dulcey, J. J. Hickman, A. S. Rudolph, T. B. Nielsen, S. M. McCort, and J. M. Calvert, Coplanar molecular assemblies of amino- and perfluorinated alkylsilanes: characterization and geometric definition of mammalian cell adhesion and growth. J. Am. Chem. Soc. *114*:8435–8442, 1992.

88. S. K. Bhatia, J. L. Teixeira, M. Anderson, L. C. Shriverlake, J. M. Calvert, J. H. Georger, J. J. Hickman, C. S. Dulcey, P. E. Schoen, and F. S. Ligler, Fabrication of surfaces resistant to protein adsorption and application to 2-dimensional protein patterning. Anal. Biochem. *208(1)*:197–205, 1993.

89. P. S. Stayton, T. Shimoboji, C. Long, A. Chilkoti, G. Chen, J. M. Harris, and A. S. Hoffman, Control of protein-ligand recognition using a stimuli-responsive polymer. Nature *378*:472–474, 1995.

90. M. Jiang, B. Nolting, P. S. Stayton, and S. G. Sligar, Surface-linked molecular monolayers of an engineered myoglobin: structure, stability and function. Langmuir *12*: 1278–1283, 1996.

91. P. C. Schamberger and J. A. Gardella, Jr., Surface chemical modification of materials which influence animal cell adhesion—a review. Coll. Surf. B: Biointerfaces 2:209–223, 1994.

92. C. J. Kirkpatrick, T. Otterbach, D. Anderheiden, J. Schiefer, H. Richter, H. Hocker, C. Mittermayer, and A. Dekker, Quantitative scanning electron-microscopy (SEM) to study the adhesion and spreading of human endothelial cells to surface-modified poly(carbonate urethane)s. Cell Mater. 2(2):169–177, 1992.

93. K. L. Prime and G. M. Whitesides, Self-assembled organic monolayers: model systems for studying adsorption of proteins at surfaces. Science *252*:1164–1167, 1991.

94. K. L. Prime and G. M. Whitesides, Adsorption of proteins onto surfaces containing end-attached oligo(ethylene oxide): a model system using self-assembled monolayers. J. Am. Chem. Soc. *115*:10714–10721, 1993.

95. G. P. Lopez, B. D. Ratner, R. J. Rapoza, and T. A. Horbett, Plasma deposition of ultrathin films of poly(2-hydroxyethyl methacrylate)—surface-analysis and protein adsorption measurements. Macromolecules *26(13)*:3247–3253, 1993.

96. M. S. Sheu, A. S. Hoffman, B. D. Ratner, J. Feijen, and J. M. Harris, Immobilization of polyethylene oxide surfactants for non-fouling biomaterial surfaces using an Argon glow-discharge treatment. J. Adhes. Sci. Technol. *7(10)*:1065–1076, 1993.

97. D. W. Grainger, T. Okano, and S. W. Kim, Protein adsorption from buffer and plasma onto hydrophilic–hydrophobic poly(ethylene oxide)–polystyrene multiblock copolymers. J. Coll. Interf. Sci. *132*:161–175, 1989.

98. M. Amiji and K. Park, Surface modification of polymeric biomaterials with poly(ethylene oxide)—a steric repulsion approach. ACS Symp. Ser. *540*:135–146, 1994.

99. K. Holmberg, K. Bergstrom, C. Brink, E. Osterberg, F. Tiberg, and J. M. Harris, Effects on protein adsorption, bacterial adhesion and contact angle of grafting peg chains to polystyrene. J. Adhes. Sci. Technol. *7(6)*:503–517, 1993.

100. Adsorption of poly(ethylene glycol) amphiphiles to form coatings which inhibit protein adsorption. J. Coll. Interf. Sci. *172*:502–512, 1996.

101. P. E. Laibinis and G. M. Whitesides, ω-Terminated alkanethiolate monolayers on surfaces of copper, silver and gold have similar wettabilities. J. Am. Chem. Soc. *114*: 1990–1995, 1992.

102. P. E. Laibinis, M. A. Fox, J. P. Folkers, and G. M. Whitesides, Comparisons of self-assembled monolayers on silver and gold: mixed monolayers derived from $HS(CH_2)_{21}X$ and $HS(CH_2)_{10}Y$ (X, Y = CH_3, CH_2OH) have similar properties. Langmuir 7:3167–3173, 1991.

103. L. Bertilsson and B. Liedberg, Infrared study of thiol monolayer assemblies on gold: preparation, characterization and functionalization of mixed monolayers. Langmuir 9: 141–149, 1993.

104. L. Netzer and J. Sagiv, A new approach to construction of artificial monolayer assemblies. J. Am. Chem. Soc. *105*:674–676, 1983.

105. N. Balachander and C. N. Sukenik, Monolayer transformation by nucleophilic substitution applications to the creation of new monolayer assemblies. Langmuir *6*:1621–1627, 1990.

106. J. H. Silver, R. W. Hergenrother, J. C. Lin, F. Lim, H. B. Lin, T. Okada, M. K. Chaudhury, and S. L. Cooper, Surface and blood-contacting properties of alkylsiloxane monolayers supported on silicone-rubber. J. Biomed. Mater. Res. *29(4)*:535–548, 1995.

107. D. L. Allara, A. N. Parikh, and F. Rondelez, Evidence for a unique chain organization in long chain silane monolayers deposited on two widely different solid substrates. Langmuir *11*:2357–2360, 1995.

108. R. Banga, J. Yarwood, and A. M. Morgan, Determination of the ordering of alkyl-halogenosilanes on a silicon substrate using FTIR-ATR spectroscopy. Langmuir *11*: 618–622, 1995.

109. J. Sagiv, Organized monolayers by adsorption. I. Formation and structure of oleophobic mixed monolayers and solid surfaces. J. Am. Chem. Soc. *102*:92, 1980.

110. J. Gun and J. Sagiv, On the formation and structure of self-assembling monolayers. III. Time of formation, solvent retention and release. J. Colloid. Interf. Sci. *112(2)*:457, 1986.

111. C. D. Bain, J. Evall, and G. M. Whitesides, Formation of monolayers by the coadsorption of thiols on gold: variation in the head group, tail group and solvent. J. Am. Chem. Soc. *111*:7155–7162, 1989.

112. S. S. Cheng, Ph.D. project, Case Western Reserve University, 1994.

113. E. A. Vogler, J. C. Graper, G. R. Harper, H. W. Sugg, L. L. Lander, and W. J. Brittain, Contact activation of the plasma coagulation cascade. I. Procoagulant surface chemistry and energy. J. Biomed. Mater. Res. *29*:1005–1016, 1995.

114. E. A. Vogler, J. C. Graper, H. W. Sugg, L. M. Lander, and W. J. Brittain, Contact activation of the plasma coagulation cascade. II. Protein adsorption to procoagulant surfaces. J. Biomed. Mater. Res. *29*:1017–1028, 1995.

115. N. Balachander and C. N. Sukenik, Monolayer transformation by nucleophilic substitution: applications to the creation of new monolayer assemblies. Langmuir *6*:1621, 1990.

116. T. G. Vargo, E. J. Bekos, Y. S. Kim, J. P. Ranieri, R. Bellamkonda, P. Aebischer, D. E. Margevich, P. M. Thompson, F. V. Bright, and J. A. Gardella, Jr., Synthesis and characterization of fluoropolymeric substrata with immobilized minimal peptide sequences for cell adhesion studies: I. J. Biomed. Mater. Res. *29*:767–778, 1995.

117. T. Sugawara and T. Matsuda, Photochemical surface derivatization of a peptide containing Arg-Gly-Asp (RGD). J. Biomed. Mater. Res. *29*:1047–1052, 1995.

118. J. H. Lee, B. J. Jeong, and H. B. Lee, Plasma protein adsorption and platelet adhesion onto comb-like PEO gradient surfaces. J. Biomed. Mater. Res. *34*:105–114, 1997.

119. E. Osterberg, K. Bergstrom, K. Holmberg, T. P. Schulman, J. A. Riggs, N. L. Burns, J. M. VanAlstine, and J. M. Harris, Protein-rejecting ability of surface-bound dextran in end-on and side-on configurations: comparison to PEG. J. Biomed. Mater. Res. *29*: 741–747, 1995.

120. S. M. Slack and T. A. Horbett, Changes in the strength of fibrinogen attachment to solid surfaces: an explanation of the influence on surface chemistry on the vroman effect. J. Coll. Interf. Sci. *133*:148, 1989.

121. S. M. Slack and T. A. Horbett, J. Coll. Interf. Sci. *124*:533, 1988.

122. M. D. Bale, D. F. Mosher, L. Wolfarht, and R. C. Sutton, Competitive adsorption of fibronectin, fibrinogen, immunoglobulin, albumin and bulk plasma proteins on polystyrene latex. J. Coll. Interf. Sci. *125*:516, 1988.

123. P. Wojciechowski, P. ten Hove, and J. L. Brash, Phenomenology and mechanisms of the transient adsorption of fibrinogen from plasma (Vroman effect). J. Coll. Interf. Sci. *111*:455, 1986.

124. R. M. Cornelius, P. W. Wojciechowski, and J. L. Brash, Measurement of protein adsorption-kinetics by an in situ, real-time, solution depletion technique. J. Coll. Interf. Sci. *150(1)*:121–133, 1992.

19

Studies on the Conformation of Adsorbed Proteins with the Use of Nanoparticle Technology

PETER BILLSTEN* and UNO CARLSSON   Linköping University, Linköping, Sweden

HANS ELWING   Göteborg University, Göteborg, Sweden

I.   CONFORMATIONAL CHANGES OF PROTEINS AT SOLID SURFACES

Conformational changes of proteins in solution are well-known phenomena and are, in many cases, important for substrate binding and regulation of enzyme activity. Important biological phenomena such as blood coagulation and immune complement activation depend upon protein activation resulting from conformational changes at solid surfaces. Examples of such proteins are fibrinogen, which exposes binding sites for platelets when adsorbed to solid surfaces [1], and complement factor 3, which exposes binding sites for cofactors upon adsorption [2].

Conformation changes of adsorbed proteins have been studied by immunological methods, infrared spectroscopy [3], TIRF [4,5], fluorescence [6], differential scanning calorimetry (DSC) [7,8] and circular dichroism (CD) [9,10]. Although CD is very sensitive to protein conformations, use of this technique for studying conformational changes at solid surfaces is problematic. One suggested method utilizes stacked quartz slides with adsorbed proteins [10]. This places constraints on the experimental setup and limits resolution. A breakthrough came when Kondo et al. studied the conformation of proteins adsorbed onto colloidal silica nanoparticles using regular CD spectroscopy [11]. The silica nanoparticles were small enough (average diameter 20 nm) to not cause significant scattering of the wavelengths of interest, thereby permitting the use of CD in a regular way.

Some interesting results with nanoparticle methodology have been published. The different degree of conformational change upon adsorption to silica nanoparticles for proteins with different structural stability has been studied. The "hard," structurally more stable, protein undergoes less conformational change than the "soft," structurally less stable, protein [12]. In another study, Kondo et al. found a correlation

---

*_Current affiliation_: Analysis and Formulations, Astra Draco AB, Lund, Sweden.

between the enzymatic activity of some enzymes adsorbed to silica nanoparticles and the extent of structural change, measured by CD. This demonstrates that conformational changes upon adsorption are one of the important factors that reduce the enzyme activity [13]. In addition to silica nanoparticles, particles that are partially hydrophobic have been used [14–16]. Norde et al. made an interesting observation using negatively charged Teflon latex nanoparticles. They found, unexpectedly, that adsorption of subtilisin to these particles increased the content of α-helix at low surface coverage [16].

II. PROTEIN FOLDING

Studies of conformational changes of proteins at solid surfaces are intimately related to protein-folding research. The usual way to study protein folding is to place the protein under conditions that favor the denatured state and then to study the regain of structure when the protein is placed under conditions that favor the native state. Folding of small globular proteins is a highly cooperative process [17,18] that usually is completed within seconds or minutes from the time that the proteins have been placed under native conditions. An initial event in folding is the aggregation of hydrophobic residues (hydrophobic collapse) to a more compact structure [19,20]. The size of this structure is reported to be similar to that of the native state [21]. The secondary structure has been suggested to occur as a consequence of the hydrophobic collapse [22,23]. The hypothesis that the hydrophobic collapse initiates the folding reaction is in agreement with the long-range nature of this type of interaction. However, it has been argued that the formation of secondary structure elements is the initial event in folding and that these elements guide the collapse of the polypeptide chain; this because it has been experimentally shown that the secondary structure is capable of being formed before the long-range tertiary contacts are made [24].

The next step in the folding process is packing the tertiary structure [25], which occurs on a slower time scale than the initial phase. The formation of the tertiary structure is visualized by the regain of a native near-UV CD spectrum.

III. THE MOLTEN GLOBULE STATE

The molten globule state is a kinetic and equilibrium intermediate in the folding of proteins [26,27]. The equilibrium molten globule state has been found for several proteins at low pH, high temperature, or in moderate concentrations of denaturing agents. For some proteins the removal of complexed ions is sufficient for the formation of the molten globule state. The molten globule was first predicted to be a kinetic intermediate in the folding process [26], and the term "molten globule" was coined later by Ohgushi and Wada [28]. The molten globule state of α-lactalbumin has been studied at the oil/water interface. It was found that the molten globule state adsorbs to a greater extent than the native state at these surfaces [29]. The molten globule state can be considered an energetically favorable state when the native structure is ruptured.

The molten globule state can be characterized according to the following criteria:

Size: The molten globule state is less compact than the native state but more compact than the unfolded state. The volume of the molten globule is approximately 50% larger than the native state, whereas the unfolded state has about 300% volume increase compared to the native state. The 50% increase in the volume of the protein, when it obtains a molten globule state, suggests that water penetrates the inside of the protein [30].

Secondary structure: The molten globule state is reported to have a high content of secondary structure as measured by far-UV CD [31,32] and IR spectroscopy [33].

Tertiary structure: The near-UV CD spectrum of proteins in the molten globule state is absent or greatly reduced [31,32]. This is due to the fluctuating tertiary structure of the molten globule state.

Hydrophobicity: The hydrophobic groups are exposed in patches to which the hydrophobic dye 8-anilino-1-naphthalenesulfonic acid (ANS) binds [34–36]. The binding of ANS has been interpreted as an indication of the existence of a molten globule state, since the ANS binding is abolished when the molten globule is further unfolded [35,36]. Kinetic experiments have, however, been criticized when ANS is used since it has been shown that the ANS molecules may perturb the refolding process [37].

IV. SITE-DIRECTED MUTAGENESIS

An interesting approach to produce proteins with only small deviations in their structure for adsorption studies is the use of site-directed mutagenesis. This methodology can be used to determine the importance of the intrinsic properties of the protein for the adsorption process. The use of proteins with site-specific mutations was an important experimental approach for material in this work.

The aim of this review is to describe conformational changes of some proteins adsorbed to silica nanoparticles by using CD, DSC, and fluorescence methods. Special attempts were made to relate conformational changes at silica nanoparticles with conformational changes that occur during the folding process of proteins. The review is essentially a concentrate of four papers [6,38–40] that were included in a Ph.D. thesis, "Studies on the conformation of adsorbed proteins," by Peter Billsten, Göteborg University, Sweden, 1997.

V. PROTEINS USED IN ADSORPTION STUDIES

A. Bovine Serum Albumin

Bovine serum albumin (BSA) is a three-domain protein held together by 17 disulfide bonds [41]. The protein consists of 583 amino acid residues and has a molecular weight of 66,430 Da [42]. BSA is an ellipsoidal protein with the dimensions 140 × 40 × 40 Å [43]. Adsorption studies using CD have revealed that BSA has a great tendency to change its conformation on adsorption to solid surfaces [11,12,44] and has therefore been classified as a "soft" protein.

B. Hen Egg-White Lysozyme

Hen egg-white lysozyme (HEL) is a 14,600-Da large monomeric protein consisting of 129 amino acid residues [45]. The crystal structure of HEL is known, and it has a roughly ellipsoidal shape with dimensions 4.6 × 3.0 × 30 Å [46]. The adsorption of HEL to solid surfaces does not involve any major conformational changes [12], and it has therefore been classified as a "hard" protein.

C. Bacteriophage T4 Lysozyme

T4 lysozyme, produced by the virus bacteriophage T4, consists of 164 amino acid residues and has a molecular weight of 18,700 Da [47,48]. Its crystal structure is known to a resolution of 1.7 Å [49]. The enzyme is devoid of disulfide bridges, and the only occurring Cys residue is located at position (97 48). T4 lysozyme consists of two domains, a C- and an N-terminal domains, and the active site is located in the region between these domains [50]. The enzyme molecule has a roughly ellipsoidal shape and is 54 Å long. The diameters of the C- and N-terminal lobes are 24 and 28 Å, respectively [49].

1. Site-Directed Mutants

T4 lysozyme is one of the most thoroughly characterized enzymes using site-specific mutagenesis, and more than 2000 single–amino acid substitutions have been performed [51]. In this work the wild-type enzyme and two site-directed mutants have been used. The site of mutation was position 3, which, in the wild-type enzyme, is occupied by an isoleucine (Ile3). This residue contributes to the major hydrophobic core of the C-terminal domain and helps to link the N- and C-terminal domains [52]. One of the mutants has increased thermal stability compared to the wild-type enzyme. This mutant had the Ile residue in position 3 replaced by a Cys. This leads to a formation of a disulfide bond with Cys in position 97. Thermal unfolding of the Cys mutant revealed that the formation of the disulfide bond stabilizes the mutant by 1.2 kcal/mol relative to the wild type. In the other mutant, Ile in position 3 was replaced by the larger amino acid Trp. This decreases the thermal stability of this mutant with 2.8 kcal/mol relative to the wild-type enzyme [53]. This mutant is one of the least stable T4 lysozyme mutants characterized.

D. Human Carbonic Anhydrase II

Human carbonic anhydrase II (HCAII) consists of 259 amino acid residues and has a molecular weight of 29,300 Da [54]. The crystal structure of the enzyme has been solved to a resolution of 1.54 Å [55]. The enzyme has a roughly ellipsoidal shape with dimensions 55 × 42 × 39 Å. Due to the small amount of α-helix in HCAII it has rather weak bands in the far-UV CD region, and these bands are partially obscured by contributions from aromatic amino acid residues, which also contribute in this region [56]. HCAII is devoid of any disulfide bridges and has only one Cys residue in position 206. The amino acid Trp is usually uncommon in proteins, but in HCAII there are seven Trp residues. This complicates measurements of intrinsic Trp fluorescence, since the energy may be transferred between different Trp residues. The fluorescence energy transfer between the different Trp of HCAII has been characterized [57]. In near-UV CD the many Trp residues give rise to strong bands which are characteristic for the native structure of the enzyme [56].

Folding and unfolding of HCAII has been extensively studied [58–63]. Moderate concentrations of GuHCl induce a molten globule state [58,62,64].

1. Site-Directed Mutants

The pseudo wild type (HCAII$_{pwt}$) was used as a template for all mutants. This variant has the naturally occurring Cys residue (position 206) replaced by serine. Cys residues can now be introduced in other places in the structure without the formation of non-native disulfide bonds with Cys 206. HCAII$_{pwt}$ is not significantly different from the wild-type enzyme regarding activity and stability [59].

The truncated variants used, trunc 5 and 17, have had the 4 and 16 most N-terminal amino acid residues removed using site-directed mutagenesis [63]. The amino acid residues are oddly numbered because the numbering originates from HCAI. Thus, HCAII starts with amino acid number 2. The truncations destabilize these variants, but, interestingly, the native state is destabilized relative to the molten globule state, whereas the molten globule state is not significantly destabilized relative to the unfolded state, as measured by GuHCl unfolding [63].

2. Colloidal Silica Nanoparticles

The nanoparticles had an average diameter of 9 nm, which is approximately twice the diameter of the proteins used except for BSA (Fig. 1). In all experiments involving nanoparticles, a 1:1 molar ratio of proteins to particles was used to ensure that there was enough surface area for all protein molecules to be adsorbed.

The nanoparticle stock solution had a concentration of 5.1×10^{17} particles/mL, which gives a surface area of 130 m^2/mL. The nanoparticles are made from pure amorphous silicon oxide and have a net negative charge (Fig. 2).

VI. METHODS FOR STUDYING CONFORMATION OF ADSORBED PROTEINS

A. Circular Dichroism

CD is a method for determining changes in the protein structure. Furthermore, structural changes in proteins can be measured with millisecond time resolution. However, it is not possible to determine the three-dimensional structure of a protein by using CD. Instead the CD spectrum gives information of the fractional content of different elements of secondary structure in the protein. A CD spectrometer is similar to an ordinary spectrophotometer in that it measures the ability of the sample to absorb light of different wavelengths. However, the CD spectrometer also measures the uneven absorption of left- and right-hand circularly polarized light, which is typical for chiral molecules. Circular dichroism measurements of proteins are usually performed in two wavelength regions: far UV and near UV.

Far-UV CD is measured at wavelengths below 250 nm. The content of secondary structure may be empirically calculated by comparing the CD spectra of proteins with known secondary structure from x-ray diffraction studies of protein

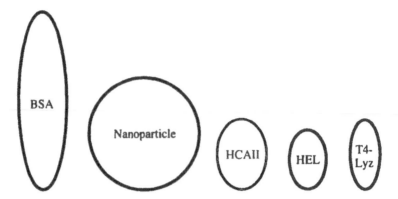

FIG. 1   The relative size of the proteins and nanoparticle (≈9 nm) used in the study.

crystals [65,66]. In addition, theoretical calculations have been performed to assign the CD spectrum of proteins to different secondary structure elements [67].

Near-UV CD is usually measured at wavelengths between 250 and 320 nm and reflects the tertiary structure of the protein. The near-UV CD spectrum reflects the contributions from aromatic amino acid residues and disulfide bonds, although the main contributors to this spectrum are the Trp residues. The near-UV CD spectra are widely used as a fingerprint of the native conformations of proteins, since these spectra are unique for the native conformation of proteins.

B. Fluorescence

8-Anilino-1-naphthalenesulfonic acid (ANS) is a small amphiphilic molecule that fluoresces when excited with light of 360 nm [34]. The wavelength of the emitted light of the fluorophore is dependent on the local environment. ANS is widely used to identify the molten globule state in protein-folding studies [35,36]. This state exposes hydrophobic patches, which ANS binds to, that are not present in the native or the fully unfolded state. The binding of ANS to these patches causes fluorescence to increase and the wavelength of maximum intensity to become blue-shifted.

5-Dimethylaminonaphthalene-1-sulfonamide (DNSA) is a naphthalene dye which fluoresces when excited with light of 320 nm. Sulfonamides bind to the active site of HCAII and function as inhibitors. The binding of DNSA to the active site of HCAII causes the emission to increase and become blue-shifted. DNSA is very

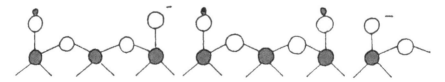

FIG. 2   The surface of amorphous silica consists of polar silanol groups (Si-OH) and charged silica acid groups (Si-O). Large filled circles: Si atoms; small filled circles: H atoms; large open circles: O atoms. (Redrawn from Ref. 77.)

sensitive for the native conformation of the active site, and the binding of DNSA is, therefore, indicative of a native conformation of the active site [61,68].

Tryptophan (Trp) is one of the three naturally occurring aromatic amino acid residues that fluoresce when excited with UV light. When a Trp residue is located in a hydrophobic environment, the fluorescence is blue-shifted relative to the emission when it is located in a hydrophilic environment [69]. When a protein changes its conformation, the exposure of Trp residues to the solvent may change and thereby affect the fluorescence. Trp fluorescence can therefore be used to monitor conformational changes of proteins [5,63].

C. Differential Scanning Calorimetry

The basic principle of calorimetry is that the energy is added to the sample and the energy required to raise the temperature of the sample be measured. This parameter is called the heat capacity (C_p). In a calorimetric measurement C_p is measured as a function of temperature. The enthalpy change (ΔH) can be calculated, from a thermogram, by measuring the area under the peak that a difference in the heat capacity will render [18,70].

DSC is performed on instruments equipped with two identical cells, one filled with sample solution and the other filled with a reference solution, usually water. The temperature of the reference and the sample cell is increased (scanned), and the energy required to maintain equal temperature between sample and reference is monitored. This difference in energy defines C_p.

VII. TRYPTOPHAN FLUORESCENCE OF BSA AND HEL ADSORBED ONTO SILICA NANOPARTICLES

The innovation of Kondo et al. [11] to measure the CD of proteins adsorbed onto colloidal silica nanoparticles was successfully used to measure the intrinsic Trp fluorescence of adsorbed proteins. The "soft" protein BSA together with the "hard" protein HEL were adsorbed on silica nanoparticles with an average diameter of 9 nm. The results demonstrate that BSA produces the largest fluorescent change with both a large fluorescence intensity decrease and a blue shift of 10 nm. Being more structurally stable, HEL shows no shift in the wavelength of maximum intensity upon adsorption (Fig. 3). This is in agreement with previously reported CD data [12].

Interestingly, the Trp fluorescence of BSA adsorbed onto silica nanoparticles correlated well with TIRF measurements of adsorbed BSA at flat surfaces [5]. This illustrates the validity of the use of silica nanoparticles for adsorption studies.

VIII. CONFORMATIONAL CHANGES OF STABILITY MUTANTS OF BACTERIOPHAGE T4 LYSOZYME ADSORBED ONTO COLLOIDAL SILICA PARTICLES

Adsorption of site-directed mutants of T4 lysozyme to silica nanoparticles was investigated. Wild-type protein was used together with two mutants that had the naturally occurring Ile residue in position 3 replaced by a Cys or Trp. The Cys mutant has increased stability compared to the wild-type enzyme, since the Cys residue

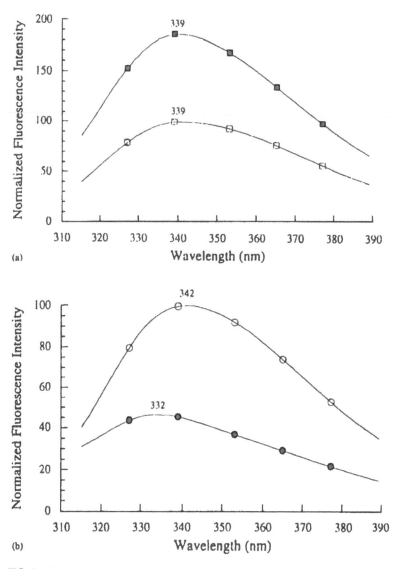

FIG. 3 Fluorescence emission spectra of protein before adsorption (open symbols) and after adsorption (filled symbols) onto silica nanoparticles. Equal volumes of the particles and protein stock solutions were mixed to yield a final concentration of 4.4×10^{15} particles/mL and 0.1 mg/mL protein. Incubation was performed for 2 h at 20°C before measurements. (a) 0.2 mg/mL HEL in 0.01 M ammonia buffer, pH 11. (b) 0.2 mg/mL BSA in 0.01 M acetate buffer, pH 5.0.

forms a non-native disulfide bridge. In contrast, the replacement of Ile with the large amino acid Trp renders the variant less stable than the wild-type enzyme.

The CD spectra of the different variants were measured in the far-UV region. Representative CD spectra for the proteins adsorbed onto the silica nanoparticles are shown in Fig. 4.

The degree of conformational change was estimated by the method of Pro-

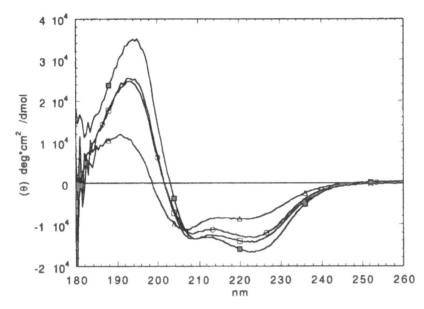

FIG. 4 CD spectra of T4 lysozyme variants. The tryptophan mutant in the absence of particles (■) and the tryptophan mutant (△), cystein mutant (□), and wild-type (○) adsorbed onto 9-nm silica nanoparticles, a 1:1 protein-to-particle ratio with approximately 0.1 mg/mL T4 lysozyme in 10 mM phosphate buffer, pH 7.0. The T4-lysozyme variants were allowed to interact with the particles for 90 min at 20°C before the spectra were recorded.

vencher and Glöckner, by calculating the α-helix content with the CONTIN program [66]. To verify that the proteins had adsorbed to the silica nanoparticles, the protein in solution and the protein particle mixtures were separated by using size exclusion chromatography. No protein appeared in the fraction corresponding to unadsorbed protein, which was taken as an indication that most protein had adsorbed to the particles.

In the absence of particles, all variants of T4 lysozyme had a similar α-helix content. The loss of α-helix upon adsorption correlated with the thermal stability of the different variants in solution (Table 1).

Thus, the Trp mutant lost most α-helix upon adsorption, whereas the Cys mutant kept most of its α-helix. This suggests that protein stability in solution is of great importance for the degree of conformational change it undergoes at a solid surface. This is in agreement with studies performed on the same set of mutants [71]. However, those studies were made by a method involving ellipsometry and elution with detergents. This set of methods only gives indirect information on the conformation of the adsorbed proteins. In addition, Claesson et al. investigated the Trp mutant and the wild-type enzyme of T4 lysozyme adsorbed onto mica surfaces, using the surface force apparatus. It was found that the layer thickness of the Trp mutant was approximately one-third compared to that of the wild type [72]. These findings indicate that the Trp mutant significantly changed conformation and occupied more space at the surface than the wild-type enzyme.

McGuire et al. studied adsorption kinetics for these mutants to flat silica sur-

TABLE 1 Calculation of the α-Helix Content in Adsorbed and Nonadsorbed T4 Lysozyme from CD Spectra Using the CONTIN Program

	Wild type	Cysteine mutant	Tryptophan mutant
% α-helix in absence of particles	57 ± 3	58 ± 2	59 ± 3
% α-helix in presence of particles	45 ± 3	49 ± 2	30 ± 3
% loss in α-helix upon adsorption	12	9	29
$\Delta\Delta G$ (kcal/mol)	0	1.2	-2.8

The amount of α-helix lost upon adsorption is presented as well as the difference in $\Delta\Delta G$ of thermal unfolding for the different proteins.

faces with the use of ellipsometry. The Trp mutant displayed the slowest adsorption kinetics. The wild-type adsorption kinetics were faster than the kinetics for the Cys mutant [71]. The fast adsorption kinetics, observed by McGuire et al., and slow kinetics of the conformational change for the wild type, observed by us, suggest that the conformational change occurred after the protein adsorbs to the particle. However, the fast initial conformational change of the Trp mutant cannot exclude the possibility that part of the conformational change occurred instantaneously.

In contrast, Norde has suggested that the instantaneous conformational change of proteins is an important driving force for protein adsorption [73,74]. However, Norde's suggestion is based on theoretical assumptions that are not necessarily applicable to this case.

In conclusion, the results presented suggest that the use of site-directed mutants with small and controlled variation is important for understanding the impact of intrinsic properties of the protein for the conformation of the adsorbed protein.

IX. TRUNCATED VARIANTS OF HCAII ADSORBED TO NANOPARTICLES OBTAIN A MOLTEN GLOBULE-LIKE STATE

Equilibrium unfolding studies of HCAII in solution in the presence of GuHCl have revealed the existence of an equilibrium folding intermediate, a so-called molten globule state [62–64]. We investigated if such a molten globule state is also induced when HCAII is adsorbed to silica nanoparticles.

HCAII$_{pwt}$ and two truncated variants (with the 4 and 16 most N-terminal amino acids removed) were adsorbed onto silica nanoparticles. Those truncations are reported to destabilize the native state of the enzyme relative to the molten globule state [63]. CD, intrinsic Trp fluorescence, the fluorescent active sites probe DNSA, and the hydrophobic dye ANS were used to study the conformation of the adsorbed variants.

Both truncated variants obtained a molten globule–like state after adsorption to the nanoparticles. This is indicated by the following results:

1. The binding and subsequent fluorescence of ANS indicate the exposure of

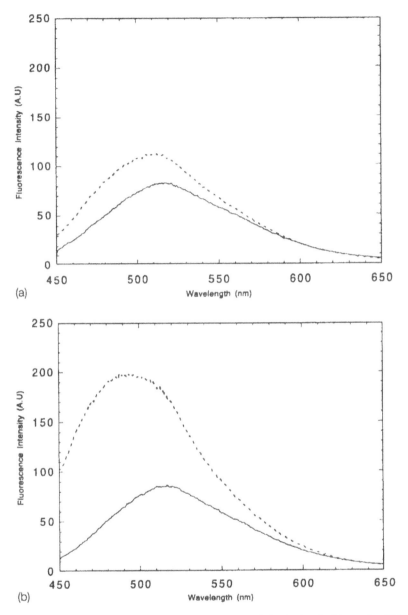

FIG. 5 Binding and subsequent fluorescence of the hydrophobic probe ANS to HCA$_{pwt}$ and to mutants thereof: (a) HCAII$_{pwt}$; (b) Trunc 5; (c) Trunc 17. Protein without particles (solid lines) and after 24 h equilibrium with particles (broken lines). The protein solutions contained 3.4 μM protein in 10 mM phosphate buffer, pH 7.5, and had a 200-fold molar excess of ANS.

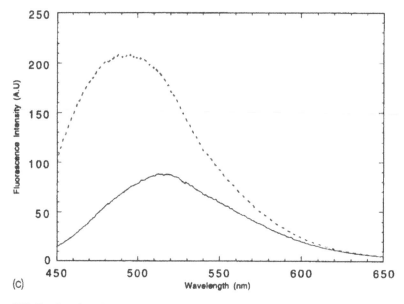

(c)

FIG. 5 Continued

hydrophobic patches. These patches are only present in the native or the denatured state in a low concentration. An increase in the binding of ANS is commonly used as an indication of the molten globule state (Fig. 5) [36].

2. The active sites probe DNSA did not bind to the enzyme. This is in agreement with the DNSA measurements on the GuHCl-induced molten globule state of HCAII (Table 2) [61].

TABLE 2 Fluorescence Emission Maxima (nm) Obtained in the HCAII Experiments

	Without particles	With particles
Intrinsic tryptophan fluorescence		
HCAII$_{pwt}$	334	338
Trunc 5	336	343
Trunc 17	333	342
ANS fluoresence		
HCAII$_{pwt}$	516	510
Trunc 5	515	493
Trunc 17	516	492
DNSA fluoresence		
HCAII$_{pwt}$	461	461
Trunc 5	470	522
Trunc 17	471	521

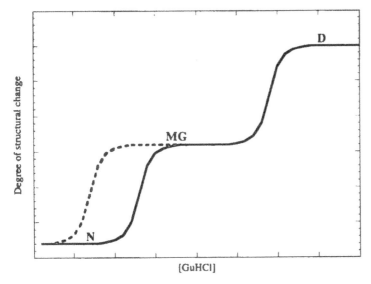

FIG. 6 The relationship between the degree of structural change of a protein and the concentration of a denaturating agent. The unfolding of HCAII is known to be a three-state process: N = native form, MG = molten globule state, and D = denatured state. Truncations destabilize the native state relative to the molten globule state. However, the transition from MG to D is not significantly affected by the truncation. (Redrawn from Ref. 63.)

3. The near-UV CD spectra indicate a fluctuating tertiary structure since the asymmetry around the aromatic residues was lost (data not shown).
4. There were considerable changes in the far-UV CD spectrum. Since the aromatic amino acid residues in the native state contribute to the far-UV as well as to the near-UV spectra, the loss of the ordered tertiary structure observed with near-UV CD also affected the far-UV CD spectrum. In addition, comparison with model spectra suggests the existence of a secondary structure [75]. Furthermore, the negative ellipticity around 200 nm of HCAI is reported to be much larger for the molten globule state, induced by low pH, than for the native state [76] (data not shown).
5. The intrinsic Trp fluorescence was similar to that of the molten globule state of the truncated variants induced with GuHCl (Table 2).

TABLE 3 Temperature at Which the Heat Capacity Reaches a Maximum (T_m) and the Area of the Transition Peak, ΔH, Denotes the Stability of the Protein[a]

HCAII variant	T_m (°C) without nanoparticles	T_m (°C) adsorbed to nanoparticles	ΔH (kcal/mol) without nanoparticles	ΔH (kcal/mol) adsorbed to nanoparticles
HCAII$_{pwt}$	59	52	180	80
Trunc 5	52	no transition	130	0
Trunc 17	53	no transition	130	0

[a]68 μM protein in 10 mM phosphate buffer, pH 7.5, was used.

The conformation of HCAII$_{pwt}$ adsorbed to the silica nanoparticles was found to be intermediate to that reported for the molten globule and the native state, as indicated by

1. A small increase of the ANS binding ability, which indicated a small increase in the exposure of hydrophobic patches (Table 2).

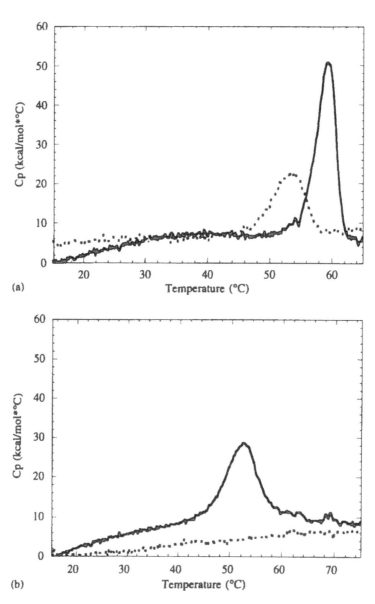

(a)

(b)

FIG. 7 Differential scanning calorimetry (DSC) thermograms of HCAII$_{pwt}$ variants in the absence of nanoparticles (solid lines) and after adsorption to nanoparticles (broken lines): (a) HCAII$_{pwt}$; (b) Trunc 5. In the experiments, 68 μM protein in 10 mM phosphate buffer, pH 7.5, was used.

2. The protein still has a native conformation of the active site, indicated by the binding of DNSA (Table 2).
3. The near-UV CD spectrum suggests a small change in the content of tertiary structure (data not shown).
4. Small changes in the far-UV CD spectrum that are probably due to the change in tertiary structure (data not shown).
5. The intrinsic Trp fluorescence is intermediate to that of the native state and that of the molten globule (Table 2).

It is important to note that truncation of HCAII resulted in a native state that is more sensitive to denaturation compared to that of the wild type. The energy needed for the unfolding of the molten globule state is, however, similar to that of the wild type (Fig. 6).

X. CONFORMATION OF HCAII VARIANTS ADSORBED TO NANOPARTICLES MEASURED BY DSC

It was unclear whether the changes observed for HCAII$_{pwt}$ above were due to the fact that all protein molecules obtained a conformation similar to that of the native state or that only a fraction of the protein obtained a molten globule–like state, whereas the remainder still was native.

To gain additional information on the conformation of HCAII$_{pwt}$ and variants thereof, DSC was used to complement the spectroscopic measurements. HCAII$_{pwt}$, trunc 5 and trunc 17 were adsorbed to colloidal silica nanoparticles. DSC was used to characterize the different variants before and after adsorption to the nanoparticles. Without particles, all variants of HCAII displayed significant transitions upon heating, as shown in Table 3. HCAII$_{pwt}$ was more stable than the other variants, since its T_m value and ΔH were significantly larger.

After the adsorption trunc 5 and 17 displayed no significant thermal transition. However, HCAII$_{pwt}$ displayed a thermal transition with a significantly lower T_m and ΔH (Fig. 7). The lower T_m and ΔH indicate that HCAII$_{pwt}$ absorbed on nanoparticles obtained a distinct and collective state with a thermal stability different from that of the native and molten globule states. This observation excluded the possibility that only a fraction of the protein molecules changed conformation upon interaction with the nanoparticles. Thus, binding of HCAII$_{pwt}$ lowered the stability of the protein, creating a state that has the tertiary structure roughly maintained. This conformation may only exist when the protein is adsorbed to a solid surface.

ACKNOWLEDGMENTS

This work was supported by the Swedish Research Council for Engineering Science (TFR), Swedish Natural Science Council (NFR), Magnus Bergvalls Stiftelse, and The Swedish Foundation for Strategic Research (MASTEC).

REFERENCES

1. H. Nagai, M. Handa, Y. Kawai, K. Watanabe, and Y. Ikeda, Evidence that plasma fibrinogen and platelet membrane GPIIb-IIa are involved in the adhesion of platelets to an artificial surface exposed to plasma. Thromb. Res. *71*, 1993.
2. U. R. Nilsson, K. E. Storm, H. Elwing, and B. Nilsson, Conformational epitopes of C3 reflecting its mode of binding to an artificial polymer surface. Mol. Immunol. *30*:211–219, 1993.
3. B. Liedberg, B. Ivarsson, I. Lundström, and W. R. Salaneck, Fourier transform infrared reflection absorption spectroscopy (FT-IRAS) of some biologically important molecules adsorbed on metal surfaces. Progr. Colloid Polym. Sci. *70*:67–75, 1985.
4. D. R. Lu and K. Park, Effect of surface hydrophobicity on the conformational changes of adsorbed fibrinogen. J. Colloid Interface Sci. *144*:271–281, 1991.
5. V. Hlady and J. D. Andrade, Fluorescence emission from adsorbed bovine serum albumin and albumin-bound 1-anilinonaphthalene-8-sulfone studied by TIRF. Colloids Surf. *32*:359–369, 1988.
6. S. Clark, P. Billsten, and H. Elwing, A fluorescence technique for investigating protein adsorption phenomena at a colloidal silica surface. Colloids Surf. B *2*:457–461, 1994.
7. T. Zoungrana, G. H. Findenegg, and W. Norde, Structure, stability and activity of adsorbed enzymes. J. Colloid Interface Sci. *190*:437–448, 1997.
8. B. L. Steadman, K. C. Thompson, R. C. Middaugh, K. Matsuno, S. Vrona, E. Q. Lawson, and R. V. Lewis, The effects of surface adsorption on the thermal stability of proteins. Biotech. Bioeng. *40*:8–15, 1992.
9. A. Kondo, S. Oku, F. Murakami, and K. Higashitani, Conformational changes in protein molecules upon adsorption on ultrafine particles. Colloids Surf. B *1*:197–201, 1993.
10. C. R. McMillin and A. G. Walton, A circular dichroism technique for the study of adsorbed protein structure. J. Colloid Interface Sci. *48*:345–349, 1974.
11. A. Kondo, S. Oku, and K. Higashitani, Structural changes in protein molecules adsorbed on ultrafine silica particles. J Colloid Interface Sci. *143*:214–221, 1991.
12. W. Norde and J. P. Favier, Structure of adsorbed and desorbed proteins. Colloid Surf. *64*:87–93, 1992.
13. A. Kondo, F. Murakami, M. Kawagoe, and K. Higashitani, Kinetic and circular dichroism studies of enzymes adsorbed on ultrafine silica particles. Appl. Microbiol. Biotechnol. *39*:726–731, 1993.
14. A. Kondo, T. Urabe, and K. Yoshinaga, Adsorption activity and conformation of α-amylase on various ultrafine silica particles modified with polymer silane coupling agents. Colloids Surf. A *109*:129–136, 1996.
15. A. Kondo, F. Murakami, and K. Higashitani, Circular dichroism studies on conformational changes in protein molecules upon adsorption on ultrafine polystyrene particles. Biotechnol. Bioeng. *40*:889–894, 1992.
16. M. C. L. Maste, E. H. W. Pap, A. van Hoek, W. Norde, and A. J. W. G. Visser, Spectroscopic investigation of the structure of a protein adsorbed on a hydrophobic latex. J. Colloid Interface Sci. *180*:632–633, 1996.
17. C. Tanford, Protein denaturation. Adv. Protein Chem. *23*:121–282, 1968.
18. P. L. Privalov, Stability of proteins: small globular proteins. Adv. Protein Chem. *33*:167–241, 1979.
19. T. R. Sosnick, S. Jackson, R. R. Wilk, S. W. Englander, and W. F. DeGrado, The role of helix formation in the folding of a fully alpha-helical coiled coil. Proteins: Struct. Funct. Genet. *24*:427–432, 1996.
20. T. R. Sosnick, L. Mayne, and S. W. Englander, Molecular collapse: the rate limiting step in two-state cytochrome c folding. Proteins: Struct. Funct. Genet. *24*:413–426, 1996.

21. A. D. Miranker and C. M. Dobson, Collapse and cooperativity in protein folding. Curr. Opin. Struct. Biol. *6*:31–42, 1996.

22. P. S. Kim and R. L. Baldwin, Specific intermediates in the folding reactions of small proteins and the mechanism of protein folding. Annu. Rev. Biochem. *51*:459–489, 1982.

23. H. S. Chan and K. A. Dill, Origin of structure in globular proteins. Proc. Natl. Acad. Sci. USA *87*:6388–6392, 1990.

24. S. Williams, T. P. Causgrove, R. Gilmanshin, K. S. Fang, R. H. Callender, W. H. Woodruff, and R. B. Dyer, Fast events in protein folding: helix melting and formation in a small peptide. Biochemistry *35*:691–697, 1996.

25. M. Levitt, M. Gerstein, E. Huang, S. Subbiah, and J. Tsai, Protein folding: the end game. Annu. Rev. Biochem. *66*:549–579, 1997.

26. O. B. Ptitsyn, Molten globule and protein folding. Adv. Protein Chem. *47*:83–229, 1995.

27. K. Kuwajima, The molten globule state as a clue for understanding the folding and cooperativity of globular-protein structure. Proteins: Struct. Funct. Genet. *6*:87–103, 1989.

28. M. Ohgushi and A. Wada, Molten-globule state: a compact form of globular proteins with mobile side-chains. FEBS Lett. *164*:21–24, 1983.

29. E. Dickinson and Y. Matsumura, Proteins at liquid interfaces: role of the molten globule. Colloids Surf. B *3*:1–17, 1994.

30. O. B. Ptitsyn, *Protein Folding*, W. H. Freeman, New York, 1992, pp. 243–300.

31. K.-P. Wong and C. Tanford, Denaturation of bovine carbonic anhydrase B by guanidine hydrochloride. A process involving separable sequential conformational transitions. J. Biol. Chem. *248*:8518–8523, 1973.

32. K. Kuwajima, K. Nitta, M. Yoneyama, and S. Sugai, Three-state denaturation of alpha-lactalbumin by guanidine hydrochloride. J. Mol. Biol. *106*:359–373, 1976.

33. D. A. Dolgikh, L. V. Abaturov, I. A. Bolotina, E. V. Brazhnikov, V. E. Bychkova, V. N. Bushuev, R. I. Gilmanshin, Y. O. Lebedev, G. V. Semisotnov, E. I. Tiktopulo, and O. B. Ptitsyn, Compact state of a protein molecule with pronounced small-scale mobility: bovine alpha lactalbumin. Eur. Biophys. J. *13*:109–121, 1985.

34. L. Stryer, The interaction of a naphthalene dye with apomyoglobin and apohemoglobin. J. Mol. Biol. *13*:482–495, 1965.

35. O. B. Ptitsyn, R. H. Pain, G. V. Semisotnov, E. Zerovnik, and O. I. Razgulyaev, Evidence for a molten globule as a general intermediate in protein folding. FEBS Lett. *262*:20–24, 1990.

36. G. V. Semisotnov, N. A. Rodinova, O. I. Razgulyaev, V. N. Uversky, and R. I. Gilmanshin, Study of the "molten globule" intermediate state in protein folding by a hydrophobic fluorescent probe. Biopolymers *31*:119–128, 1991.

37. M. Engelhard and P. A. Evans, Kinetics of interaction of partially folded proteins with a hydrophobic dye: evidence that molten globule character is maximal in early folding intermediates. Protein Sci. *4*:1553–1562, 1995.

38. P. Billsten, P.-O. Freskgård, U. Carlsson, B.-H. Jonsson, and H. Elwing, Adsorption to silica nanoparticles of human carbonic anhydrase II and truncated forms induce a molten-globule like state. FEBS Lett. *402*:67–72, 1997.

39. P. Billsten, M. Wahlgren, T. Arnebrant, J. McGuire, and H. Elwing, Structural changes of T4 lysozyme upon adsorption to silica nanoparticles measured by circular dichroism. J. Colloid Interface Sci. *175*:77–82, 1995.

40. P. Billsten, U. Carlsson, B.-H. Jonsson, G. Olofsson, and H. Elwing, Conformation of human carbonic anhydrase II variants adsorbed to silica nanoparticles (submitted).

41. J. R. Brown and P. Shockley, *Lipid–Protein Interactions*, Vol. 1, Wiley, New York, 1982, pp. 25–68.

42. K. Hirayama, S. Akashi, M. Furuya, and K.-I. Fukuhara, Rapid conformation and re-

vision of the primary structure of bovine serum albumin by ESIMS and frit-FAB LC/MS. Biochem. Biophys. Res. Commun. *173*:639–646, 1990.

43. T. J. Peters, Serum albumin. Adv. Protein Chem. *37*:161–245, 1985.

44. M. E. Soderquist and A. G. Walton, Structural changes in proteins adsorbed on polymer surfaces. J. Colloid Interface Sci. *75*:386–397, 1980.

45. R. E. Canfield, Amino acid sequence of hens egg white lysozyme. J. Biol. Chem. *238*: 2698–2707, 1963.

46. K. P. Wilson, B. A. Malcolm, and B. W. Matthews, Structural and thermodynamic analysis of compensating mutations within the core of chicken egg white lysozyme. J. Biol. Chem. *267*:10842–10849, 1992.

47. B. W. Matthews, F. W. Dahlquist, and A. Y. Maynard, Crystallographic data of lysozyme from bacteriophage T4. J. Mol. Biol. *78*:575–576, 1973.

48. A. Tsugita, M. Inouye, E. Terzaghi, and G. Streisinger, Purification of bacteriophage T4 lysozyme. J. Biol. Chem. *234*:391–397, 1968.

49. L. H. Weaver and B. W. Matthews, Structure of bacteriophage T4 lysozyme refined at 1.7 Å resolution. J. Mol. Biol. *193*:189–199, 1987.

50. H. R. Faber and B. W. Matthews, A mutant T4 lysozyme displays five different crystal conformations. Nature *348*:263–266, 1990.

51. B. W. Matthews, Studies on protein stability using T4 lysozyme. Adv. Protein Chem. *46*:249–278, 1995.

52. S. J. Remington, W. F. Anderson, J. Owen, L. F. Ten Eyck, C. T. Grainger, and B. W. Matthews, Structure of the lysozyme from bacteriophage T4: an electron density map at 2.4 Å resolution. J. Mol. Biol. *118*:81–98, 1978.

53. M. Matsumura, W. J. Becktel, and B. W. Matthews, Hydrophobic stabilization in T4 lysozyme determined directly by multiple substitution of Ile 3. Nature *334*:406–410, 1988.

54. L. E. Henderson, D. Henriksson, and P. O. Nyman, Primary structure of human carbonic anhydrase. J. Biol. Chem. *251*:5457–5463, 1976.

55. K. Håkansson, M. Carlsson, L. A. Svensson, and A. Liljas, Structure of native and apo carbonic anhydrase II and structure of some of its anion-ligand complexes. J. Mol. Biol. *227*:1192–1204, 1992.

56. P.-O. Freskgård, L.-G. Mårtensson, P. Jonasson, B.-H. Jonsson, and U. Carlsson, Assignment of the contribution of the tryptophan residue to the circular dichroism spectrum of human carbonic anhydrase II. Biochemistry *33*:1011–1021, 1994.

57. L.-G. Mårtensson, P. Jonasson, P.-O. Freskgård, M. Svensson, U. Carlsson, and B.-H. Jonsson, Contribution of individual tryptophan residues to the fluorescence spectrum of native and denatured form of human carbonic anhydrase II. Biochemistry *34*:1011–1021, 1995.

58. L.-G. Mårtensson, B.-H. Jonsson, P.-O. Freskgård, A. Kihlgren, M. Svensson, and U. Carlsson, Characterisation of folding intermediates of human carbonic anhydrase II: probing substructure by chemical labeling of SH groups introduced by site-directed mutagenesis. Biochemistry *32*:224–231, 1993.

59. P.-O. Freskgård, U. Carlsson, L.-G. Mårtensson, and B.-H. Jonsson, Folding around the C-terminus of human carbonic anhydrase II. Kinetic characterization by use of a chemically reactive SH-group introduced by protein engineering. FEBS Lett. *289*:117–122, 1991.

60. U. Carlsson and B.-H. Jonsson, Folding of β-sheet proteins. Curr. Opin. Struct. Biol. *5*: 482–487, 1995.

61. D. Andersson, P.-O. Freskgård, B.-H. Jonsson, and U. Carlsson, Formation of local native-like tertiary structure in the slow refolding reaction of human carbonic anhydrase II as monitored by circular dichroism of tryptophan mutants. Biochemistry *36*:4623–4630, 1997.

62. M. Svensson, P. Jonasson, P.-O. Freskgård, B.-H. Jonsson, M. Lindgren, L.-G. Mårtensson, M. Gentile, K. Boren, and U. Carlsson, Mapping the folding intermediate of
human carbonic anhydrase II. Probing substructure by chemical reactivity and spin and
fluorescence labeling of engineered cystein residues. Biochemistry 34:8606–8620, 1995.

63. G. Aronsson, L.-G. Mårtensson, U. Carlsson, and B.-H. Jonsson, Folding and stability
of the N-terminus of human carbonic anhydrase II. Biochemistry 34:2153–2162, 1995.

64. M. Lindgren, M. Svensson, P.-O. Freskgård, U. Carlsson, B.-H. Jonsson, L.-G. Mårtensson, and P. Jonasson, Probing local mobility in carbonic anhydrase: EPR of spin-
labelled SH groups introduced by site-directed mutagenesis. J. Chem. Soc. Perkin Trans.
2:2003–2007, 1993.

65. C. W. J. Johnson, Protein secondary structure and circular dichroism: a practical guide.
Proteins: Struct. Funct. Genet. 7:205–214, 1990.

66. S. W. Provencher and J. Glöckner, Estimation of globular protein secondary structure
from circular dichroism. Biochemistry 20:33–37, 1981.

67. R. W. Woody, Circular Dichroism and Conformational Analysis of Biomolecules, Plenum Press, New York, 1996, pp. 25–68.

68. R. W. Henkens, B. B. Kitchell, S. C. Lottich, and T. J. Williams, Detection and characterization using circular dichroism and fluorescence spectroscopy of a stable intermediate conformation formed in the denaturation of bovine carbonic anhydrase with
guanidinium chloride. Biochemistry 21:5918–5923, 1982.

69. F. X. Schmid, Spectral method of characterizing protein conformation and conformational changes. In Protein Structure: A Practical Approach (T. E. Creighton, ed.). Oxford
University Press, Oxford, 1989, pp. 251–285.

70. J. M. Sturtevant, Biochemical applications of differential scanning caliometry. Ann. Rev.
Phys. Chem. 38:463–488, 1987.

71. J. McGuire, M. Wahlgren, and T. Arnebrant, Structural stability effects on adsorption
and dodecyltrimethylammonium bromide–mediated elutability of bacteriophage T4 lysozyme at silica surfaces. J Colloid Interface Sci. 170:182–192, 1995.

72. P. M. Claesson, E. Blomberg, J. C. Fröberg, T. Nylander, and T. Arnebrant, Protein
interactions at solid surfaces. Adv. Colloid Interface Sci. 57:161–227, 1995.

73. W. Norde, Adsorption of proteins from solutions at the solid–liquid interface. Adv.
Colloid Interface Sci. 25:267–340, 1986.

74. C. A. Haynes and W. Norde, Globular proteins at solid/liquid interfaces. Colloids Surf.
B 2:517–566, 1994.

75. N. Greenfield and G. D. Fasman, Computed circular dichroism spectra for the evaluation
of protein conformation. Biochemistry 8:4008–4116, 1969.

76. M. V. Jagannadham and D. Balasubramanian, The molten globular intermediate form in
the folding pathway of human carbonic anhydrase B. FEBS Lett. 188:326–330, 1985.

77. C. Eriksson, E. Blomberg, P. Claesson, and H. Nygren, Reactions of two hydrophilic
surfaces with detergents, proteins and whole human blood. Colloids Surf. B 9:67–97,
1997.

20

Scanning Calorimetry in Probing the Structural Stability of Proteins at Interfaces

KARIN FROMELL and KARIN CALDWELL   Uppsala University, Uppsala, Sweden

SHAO-CHIE HUANG   Diagnostic Products Corporation, Los Angeles, California, U.S.A.

I. INTRODUCTION—PROTEINS AT INTERFACES

Whether proteins are intentionally immobilized to solid surfaces to serve, e.g., an analytical or preparative purpose or whether they have just by happenstance come in contact with a foreign surface, the steep gradient in chemical composition that reigns at the interface can have a more or less debilitating influence on their structure. Because of the enormous practical importance of biologically active surfaces in areas such as immunodiagnosis, biosensor development, and the design of selective adsorbents for protein mapping, recent decades have witnessed a large research effort devoted to broadening our understanding of how to maintain structural integrity and functionality of interfacial proteins. In addition to the maintenance of active antibodies, enzymes, affinity anchors, and cell adhesion molecules on solid surfaces, much attention has also been devoted to the suppression of unwanted adsorption of inconsequential or destructively interfering proteins.

In order to understand and control interfacial protein behavior several physical and chemical analysis techniques have been developed, and it is often through a combination of methods with different information content that the greatest insight is gained. Among physical methods to probe protein structure the spectroscopic techniques often give detailed information regarding adsorption-induced shifts in some region of the molecule that is specifically interrogated by light of the selected frequency. Thus, IR analysis provides information regarding changes in the peptide backbone [1], while fluorescence studies can indicate polarity shifts in the environment of a fluor, whether intrinsic or introduced, that occur as a result of the protein's interaction with the surface [2]. Circular dichroism is singularly informative regarding changes in protein secondary structure, as it reports the relative contents of a α-helix, β-sheet, β-turn, and random structure in the protein under examination [3]. However, one practical complication with the spectroscopic techniques is their very specific requirements regarding the properties of the surface itself, as it must not scatter the informative radiation. This puts strict limitations on the types of substrate that can be studied in conjunction with protein adsorption, and spectroscopic techniques are therefore often supplemented with other means of observation.

II. TRANSITION ENTHALPIES AS STABILITY CHARACTERISTICS

An alternative to spectroscopic visualizations of the structural status of a protein is to perform a calorimetric assessment of its energy content. Calorimetry, in its various forms, is a fundamental approach to characterizing the equilibrium thermodynamics of bulk chemical processes. However, it was not until recent decades that instruments became available whose sensitivities allow the studies of protein energetics, with the protein present in submilligram amounts [4–7]. Calorimetric observations are performed following either of two experimental approaches. The first type includes the measurement of heats of reaction at constant temperature (isothermal titration calorimetry, ITC). The ITC analysis is typically built around the repetitive addition of aliquots of a reactant, say an affinity ligand, to a given amount of protein, and a recording of the heat removed, or supplied, to maintain the temperature at a constant level [8–11]. From knowledge of the exact amounts of protein and ligand that have reacted under evolution of a given amount of heat one determines the molar heat of reaction, ΔH. If one continues to add aliquots of ligand until there is no further heat change, one determines the ligand concentration at saturation and thereby the binding constant and the Gibbs free energy ΔG for the binding reaction. The difference between ΔG and ΔH gives the entropy change for the reaction, which is thereby thermodynamically fully characterized. This type of analysis is readily performed with soluble reactants whose interactions rapidly reach equilibrium. However, it is less straightforward for slow reactions in inhomogenous systems, where heat evolution becomes weak and mixing processes interfere with the reaction. This is one reason why ITC is rather infrequently used to analyze the adsorption of protein to solid matter.

The second type of analysis is based on differential scanning calorimetry (DSC), or the measurement of shifts in heat capacity of samples undergoing a controlled temperature increase. The DSC approach is particularly well suited for determinations of the thermal stability of proteins, as it provides a direct measure of the transition enthalpy, ΔH_{tr}, associated with the heat-induced shift from its native to its denatured state. The value for ΔH_{tr} is a sensitive reflection of environmental effects on the protein structure. Thus, structural effects of pH shifts [12,13], shifts in ionic strength and composition [14–16], as well as effects due to the presence of structure forming/breaking compounds [17–19] have been quantified by DSC. In the present context we will examine the ability of DSC to measure the effect of surface composition and attachment mode on the stability of a variety of immobilized proteins. The technique is far less substrate sensitive than the spectroscopic characterization methods discussed above, the only requirement being that the substrate itself is free from thermal transitions in the temperature interval of protein relevance (typically 20–100°C) and that enough protein–substrate complex for a measurable signal can be accommodated in the measuring cell. In order to carry out the analysis under reproducible conditions, one preferably chooses as the substrate a particulate material to which the protein has been immobilized with a well-characterized surface concentration.

III. THE DSC MEASUREMENT OF PROTEINS IN SOLUTION

Solutions of single proteins in their native state represent collections of molecules with narrow distributions of conformational energies. The gradual heating of such a solution from room temperature, or some temperature where the molecules are biologically fully active, at first does little other than add to their kinetic energy. At some point, however, the added heat begins to weaken interactions and break the noncovalent bonds that hold the protein structure together. This thermal unfolding of the molecules is therefore associated with an excess heat capacity, which is the subject of measurement in the DSC analysis [12]. Once the transition is completed, additional heating merely adds kinetic energy to the system, as before. For single-domain proteins this thermal transition is highly cooperative, with the entire structural change being accomplished in a narrow temperature interval, as illustrated schematically in Fig. 1. Proteins consisting of multiple domains tend to display more complicated heat denaturation curves, as each domain undergoes its own transition, more or less influenced by events in other parts of the molecule [20].

The scanning calorimeter is constructed to ramp up the temperature of the measuring cell at a constant, preselected rate and to record the heat flux required to maintain this constant temperature increase. Under conditions where the heat capacity in the cell is changing, the heat flux is modified accordingly. The relationship between heat flux and temperature constitutes the raw data set, which by subtracting the corresponding heat flux to a reference cell containing pure buffer or, as in the situation of interest here, a suspension of substrate particles free of protein, is converted into the relevant *thermogram*. From such plots of heat capacity versus temperature one finds the transition enthalpy ΔH_{tr} by integrating the excess (baseline-adjusted) heat capacity $C_p'(T)$ over the temperature interval of the transition (see Fig. 1a):

$$\Delta H_{tr} = \int_{-\infty}^{+\infty} C_p'(T) \, dT \tag{1}$$

While this procedure assumes that the heat-induced transition is reversible, it makes no assumption regarding, e.g., domain structure in the protein. The ΔH_{tr} value determined this way is therefore referred to as a "calorimetric ΔH." In graphic terms, this value equals the area under the heat capacity curve A adjusted by division with the amount of protein under study so as to express a molar quantity. The enthalpy change $\Delta H(T)$ associated with bringing the sample from temperatures below the transition to any temperature T within the transition zone is similarly found by integration of the recorded heat capacity curve (shaded area in Fig. 1a), with T replacing positive infinity as the upper limit of integration in Eq. (1).

The DSC-based analysis of protein structure has in large measure been developed by Privalov and colleagues [12,21–24]. Indeed, the existence of structural domains in proteins of high molecular weight was demonstrated through their insightful DSC work. In addition, their careful studies of the thermal transitions of a series of

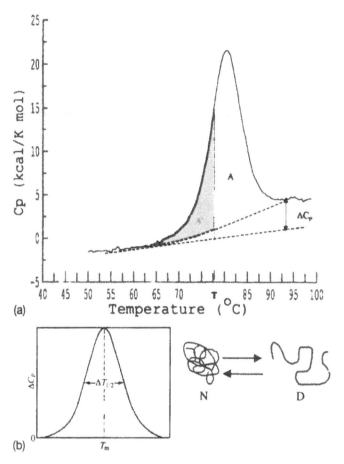

FIG. 1 (a) Thermal transition of a single domain protein. The area A under the baseline-corrected excess heat capacity curve reflects unfolding of the entire population of molecules, while the shaded area A' represents the fraction of the population that has unfolded at temperature T. The difference in heat capacity between the native and denatured states is indicated as ΔC_p in the figure. (b) Schematic illustration of a hypothetical two-state transition, from the native form N to the denatured form D. In an idealized, two-state transition with negligible ΔC_p the transition curve is symmetrical around the average transition temperature, T_m. The width of the transition, identified as the peak width at half height, is labeled as $\Delta T_{1/2}$.

small, single-domain proteins has led to the development of models that allow the extraction of much useful thermodynamic data from a thermogram of the type shown in Fig. 1. The following is a brief summary of the analytical approach that can be taken to reach this information.

As mentioned above, Fig. 1 represents a schematic illustration of a hypothetical single-step transition from the native form, **N**, of a single domain protein to its denatured form **D**. The transition is assumed to be reversible, as seen in the diagram. The temperature at the transition midpoint is referred to as the melting temperature T_m. The higher this temperature, the more stable is the protein.

With the thermal transition completed, area A under the apparent heat capacity curve in Fig. 1a is then proportional to the total number of molecules in the sample, while the shaded area A' in the figure represents the number of molecules having undergone transition at temperature T. The equilibrium constant K for the transition can therefore be found at any temperature as

$$K(T) = \frac{D(T)}{N(T)} = \frac{\alpha}{1 - \alpha} \tag{2}$$

where α replaces the ratio A'/A, i.e., the fraction of molecules being denatured at temperature T.

From the relationships between the equilibrium constant K and molecular change in Gibbs free energy ΔG°, on the one hand and the Gibbs–Helmholtz relationship between the temperature derivative of $(\Delta G^\circ/T)$ and $\Delta H^\circ/T^2$ on the other stems the van't Hoff equation [Eq. (3)], which relates the temperature variation of the equilibrium constant to the standard enthalpy of a reaction. It is given here in its differential (3a) as well as its integrated (3b) form:

$$\frac{d \ln K}{dT} = \frac{\Delta H^\circ}{RT^2} = \frac{1}{\alpha(1 - \alpha)} \frac{d\alpha}{dT} \tag{3a}$$

$$\ln \left(\frac{K(T_1)}{K(T_2)} \right) = \frac{\Delta H^\circ}{R} \left(\frac{1}{T_2} - \frac{1}{T_1} \right) \tag{3b}$$

At the midpoint of the transition $T = T_m$ and $\alpha = 0.5$, so that $K(T_m)$ takes the value of unity. At this temperature the dimensionless differential $(d\alpha/dT)_{T_m}$ in Eq. (3a) can be replaced by the product of the observed (baseline-adjusted) specific heat at this temperature $C_p'(T_m)$ and the protein's molecular weight M, normalized by division with ΔH° (see Fig. 1) [12]. With this substitution, Eq. (3a) takes the form

$$\frac{4MC_p'(T_m)}{\Delta H^\circ} = \frac{\Delta H^\circ}{RT_m^2}$$

or, equivalently,

$$\Delta H_{vH} = 2[RMC_p'(T_m)]^{0.5}T_m \tag{4}$$

Here, R symbolizes the gas constant. The standard enthalpy obtained in this manner is usually given the subscript vH to indicate its origin.

Equation (3b) has been used to cast ΔH°_{vH} in a different form [25] which illustrates the relationship between the transition enthalpy on the one hand and the protein's stability, i.e., melting temperature T_m, as well as the cooperativity, measured as the width of the transition, on the other.

$$\Delta H^\circ_{vH} = 6.9 \left(\frac{T_m^2}{\Delta T_{1/2}} \right) \tag{5}$$

Here, $\Delta T_{1/2}$ is the width of the baseline-corrected heat capacity peak measured at half height. For a highly cooperative transition the peak is narrow, but it increases in width as the heterogeneity of unfolding structures increases. Figure 2 is an illustration of how a chemical destabilization of the protein lysozyme affects both its melting temperature and the width of the transition.

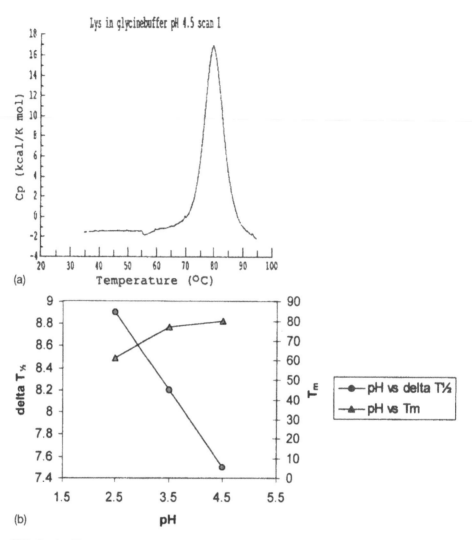

FIG. 2 (a) Thermogram for hen egg white lysozyme (3 mg) in 0.2 M glycine buffer at pH 4.5; scan rate 0.5°C/min. (b) Observed pH dependence of the melting temperature T_m (triangles) and the width of the transition, $\Delta T_{1/2}$ (circles) of lysozyme in glycine buffer obtained from DSC data.

IV. HEAT CAPACITY INCREASES DUE TO THE INCREASE IN HYDROPHOBICITY DURING DENATURATION

The change in heat capacity ΔC_p between a protein in its native form, i.e., at temperatures below the transition region, and in the denatured form found at temperatures above this region has been studied in detail by Privalov and coworkers [26] and found to be quite protein specific. Indeed, as seen in Fig. 3 [27], it appears to correlate linearly with the cumulative surface area of those hydrophobic residues that are buried in the core of the native structures but become exposed to the aqueous environment upon unfolding [28].

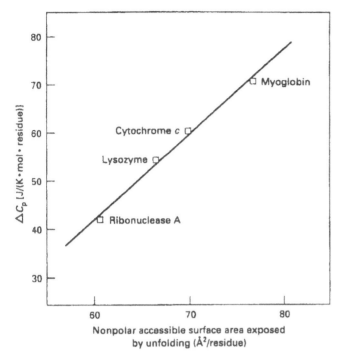

FIG. 3 Relationship between the measured change in heat capacity upon unfolding of several proteins and the nonpolar surface area that is buried in the interior of these proteins assumed to be exposed to the aqueous surroundings upon denaturation. (Reproduced from Ref. 27 with permission.)

In principle, values for ΔC_p can be determined from a single thermogram, as illustrated in Fig. 1, i.e., by identification of the baselines in the C_p versus T curve before and after the melting transition, and computation of the difference between the two at some fixed temperature, as indicated in the figure. In practice, this procedure is error prone due to difficulties in identifying stable baselines. Instead, one can rely on the definition of ΔC_p as the temperature differential of the enthalpy function at constant pressure, or

$$\Delta C_p = \left(\frac{\partial \Delta H}{\partial T} \right)_p \tag{6}$$

and base the determination on observed shifts in ΔH with shifts in T_m, developed, e.g., through shifts in the chemical environment of the protein (which must remain structurally intact under all conditions tested).

The exposure of hydrophobic surfaces during unfolding often leads to aggregation, an exothermic reaction, and a serious distortion of the transition curve, as seen in Fig. 4a. Such aggregation is more severe at higher protein concentrations. It can, in principle, be circumvented if the thermal analysis can be performed on the protein while adsorbed to some nondenaturing surface, where it is unable to come

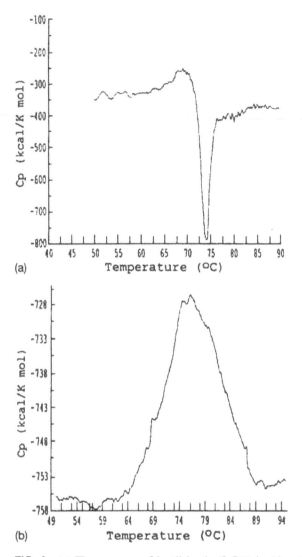

FIG. 4 (a) Thermogram of lentil lectin (LCA) in 10 mM phosphate buffer. The exothermic reaction is due to aggregation of the protein molecules when the hydrophobic surfaces become exposed during unfolding. (b) Thermogram of LCA in Sepharose DVS mannose gel. Aggregation is prevented due to the proteins inability to contact neighboring molecules.

in contact with neighboring molecules (see Fig. 4b). Although the surface is likely to exert some effect on the protein (see below), the melting curve gives at least a qualitative picture of the stability of the biopolymer.

V. PROTEINS AT INTERFACES AS SEEN BY DSC

Whether by accident or design, proteins are often forced to come in contact with solid surfaces where they adsorb with more or less significant impact on their structure. The effect exerted by the surface can conveniently be analyzed by microcalorimetry, especially if the surface is that of a particulate substrate and hence associated with a small volume. Particulate substrates are used in a wide variety of fields ranging from cell biology, immunodiagnostics, and solid phase enzymology, where the protein is to be used in immobilized form, to separation science which aims at designing surfaces for a fleeting, yet selective, contact with the protein. In the latter case an important consideration is the ability to accomplish selectivity without permanently altering the structure of the separated proteins.

A. Substrate Effects on Conformation

Over a long period of time it has become well established that proteins tend to adsorb less tenaciously to hydrophilic than to hydrophobic surfaces [29–31]. Early on, Norde and Lyklema [32] postulated that adsorption, at least to the hydrophobic surface, is entropy driven and based on a combination of effects due to substrate dehydration and structural rearrangements within the protein molecules. In a series of recent studies Norde and coworkers have utilized the DSC technique to demonstrate the clear differences in protein structure that stem from adsorption to hydrophobic and hydrophilic substrates, respectively [33–36]. The two similarly sized single-domain proteins hen egg lysozyme (LSZ, MW 14,400 Da) and α-lactalbumin (αLA, MW 14,200 Da) provide excellent examples of the different behaviors caused by hydrophobic [poly(styrene), PS] and hydrophilic (hematite, αFe_2O_3) substrates. Figure 5 compares the two proteins in terms of their behavior in solution as well as on the two different surfaces. In all cases, transition enthalpies (ΔH_{N-D}) were determined and plotted versus the temperatures at which the transitions were taking place. The figure demonstrates three important facts: First, the magnitudes of the ΔH_{N-D} values differ significantly for the two proteins in solution, with values for αLA being significantly lower than those for LSZ, indicative of a less stable structure. Second, and related to this observation, the values for ΔC_p determined from the slopes of the linear plots in accord with Eq. (6) were lower for αLA than for LSZ. This lower value indicates that the native αLA structure has a comparatively smaller amount of buried, structure-stabilizing hydrophobic residues, as discussed above. Third, the ΔC_p values were reduced upon adsorption for both proteins on both surfaces, although the reductions on the hydrophobic surface were significantly larger than those on the hydrophilic. This suggests that the adsorption had involved hydrophobic amino acid residues from the core of each protein, such that once adsorption had occurred these residues were no longer available for added contact with the surrounding water.

In a different study [37] the same two proteins were analyzed, in terms specifically of their electrostatic interactions with silica particles ($d_p = 11$ nm), under conditions of high and low ionic strengths as well as in the presence or absence of divalent cations (Ca^{2+}). In the case of lysozyme the transition enthalpy at low ionic strength was much smaller than that measured at the high value, which remained remarkably similar to the value measured for the protein free in solution. The low value presumably reflected a stronger Coulombic interaction between the positively charged protein and the negatively charged silica surface at the low charge screening.

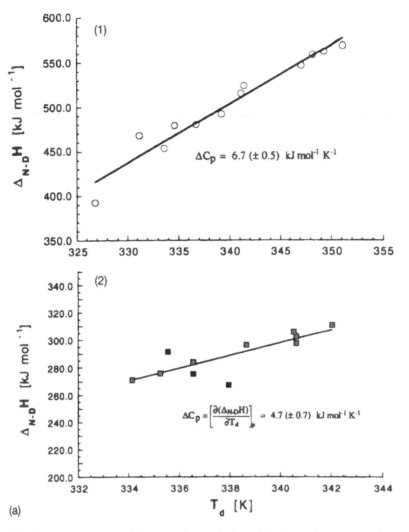

FIG. 5a Dependence of denaturation enthalpy ($\Delta H_{N \cdot D}$) on the denaturation temperature T_d for (1) native lysozyme and (2) native α-lactalbumin dissolved in 50 mM PBS. The slope of the line gives the differential heat capacity increase ΔC_p for the denaturation process.

Following adsorption to silica at high ionic strength, in the presence of divalent cations, the thermal transition occurred at a lower T_m value and over a wider temperature interval than for the free enzyme.

The αLA behaved very differently in that it showed no transition peak when adsorbed at low ionic strength, while the transition became visible at the higher I value. Unlike in the case of lysozyme, the presence of Ca^{2+} apparently led to a stabilization of the adsorbed αLA, demonstrated in the form of a higher value for T_m than that measured for the free protein. It is clear that the buffer composition plays a significant role in regulating the structural integrity of proteins adsorbing to hydrophilic surfaces.

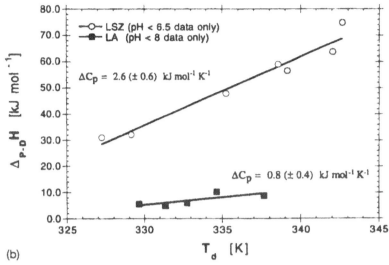

FIG. 5b   Perturbed-state denaturation enthalpies ($\Delta H_{\text{P-D}}$) for lysozyme (LSZ) and α-lactal-bumin (LA) irreversibly adsorbed to negatively charged latices of polystyrene in 50 mM KCl solution.

B.   Proteins on Polystyrene Surfaces

Despite the documented structural destabilization by hydrophobic PS surfaces, the proteins in Fig. 5b both retained some measurable degree of structure on this surface. This is not always the case, as seen in Fig. 6. Here a mannose-binding lectin (LOA) from sweet pea (*Lathyrus odoratus*) had been adsorbed to PS nanoparticles (240 nm in diameter). The presence of the protein on the particles was demonstrated in the fractograms from sedimentation field-flow fractionation shown in Fig. 6a, where the documented shift in elution volume that took place upon adsorption was indicative of a surface concentration of 0.997 mg/m^2 [37]. Although LOA in solution showed a clearly measurable transition at 74.4°C with an unfolding enthalpy of 1020 kcal/mol, the thermogram for the coated particles analyzed in Fig. 6b showed no transition in the temperature interval of 30–90°C. Structural damage of similar type was reported earlier from this laboratory for several other proteins [39].

C.   Interaction with Separation Matrices

It is clearly no coincidence that stationary phases for protein chromatography are largely hydrophilic in nature. An early DSC demonstration of the relationship between phase hydrophobicity and the resulting changes in protein conformation upon contact was given by Steadman et al. (40). In this work, a series of silica phases, both underivatized as well as derivatized with hydrophobic (methyl-, phenyl-, and butyl- groups) or hydrophilic (carboxysulfone) groups, was incubated with seven well-characterized proteins, including lysozyme. The destructive effect of the former, particularly the methyl and phenyl phases, was evidenced by shifts in the protein transition temperatures to lower values than those found in solution. The transition

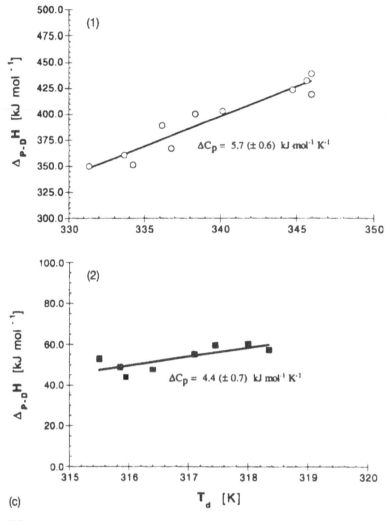

FIG. 5c Perturbed-state denaturation enthalpies ($\Delta H_{P \cdot D}$) for (1) lysozyme and (2) α-lactalbumin irreversibly adsorbed to a colloidal dispersion of hematite (αFe_2O_3) in 50 mM KCl solution. (Adapted from Ref. 33, with permission.)

curves were also significantly broadened and the areas under these curves were much reduced.

The fate of lysozyme on silica phases of different hydrophobicity was similarly probed by Hanson et al. [8] using a combination of frontal analysis chromatography and isothermal calorimetry. Here, the chromatographic studies gave apparent binding constants, and hence ΔG values for the interactions between lysozyme and three surfaces coated with p-2-hydroxyethylmethacrylate, p-ethylmethacrylate, and p-octylmethacrylate, respectively. The calorimetric (ITC) measurements, in turn, provided values for the interaction enthalpies, so that the two techniques together gave a complete thermodynamic characterization of the interactions. Although the ΔG values were comparable for adsorption to all three media, the study clearly demonstrated large differences in values for both ΔH and ΔS, differences that increased with phase

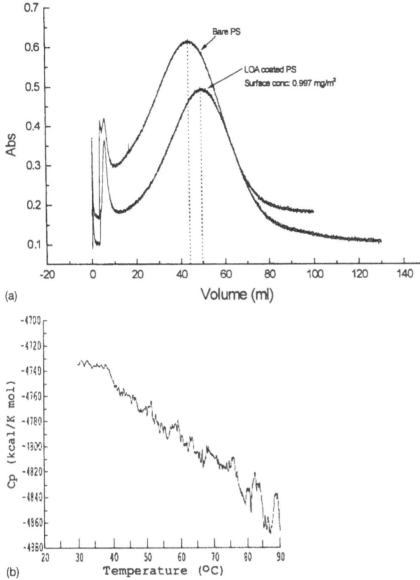

FIG. 6 (a) Sedimentation field-flow fractionation fractogram of bare and LOA (lectin from *Lathyrus odoratus*) coated PS latices 240 nm in diameter. The difference in elution volume between the curves indicates a surface concentration of 0.997 mg/m² (see Ref. 38). (b) Thermogram for LOA adsorbed to 240 nm PS latex particles showing no transition. The PS is highly hydrophobic and frequently leads to a denaturation of the adsorbed protein, as in this case.

hydrophobicity and accordingly suggested a progressively higher structural impact with higher hydrophobicity.

D. Stabilization Through Formation of Solid Phase Affinity Complexes

The formation of an affinity complex between a protein and its complementary ligand is often associated with a stabilization of the protein structure [41]. Such stabilizations can be substantial, as discussed below for the streptavidin–biotin complex [42,43], or they can be slight, as in the case of the mannose-specific lectin LOA in Fig. 7. This lectin shows an increase in T_m of only 1.9°C upon binding to a mannose-derivatized affinity matrix used for its separation.

The polysaccharide gel matrices that are the basic constituent of so many protein separation media are obviously selected for their low impact on protein structure, exemplified by the mannose–sepharose: LOA adsorption complex in Fig. 7. However, for applications other than separation, affinity anchors are often attached to nonpolar surfaces, such as poly(styrene), because of the many favorable physical properties of the polymer, including transparency, a high refractive index, easy molding, and hardness at ambient conditions due to a high glass transition temperature. The PS matrix, as discussed above, is highly hydrophobic and frequently leads to structural changes in the adsorbed protein, as demonstrated pictorially in Fig. 6b. Fortuitously, the protein streptavidin (SA) constitutes a significant exception to this structural fragility [39,42,43]. The bacterial protein SA (tetrameric molecule from *Staphylococcus aureus*, with a MW of 60,000 Da) has an extraordinary affinity (solution binding constant $K_A = 10^{15}$ M^{-1}) for the small (MW 244 Da) vitamin H referred to as d-biotin. Due to this strong affinity, the pair has found widespread use as a coupling link for any component slated for surface immobilization. In such coupling processes the SA is typically first adsorbed to the polymer surface. In a separate reaction, the molecule to be immobilized is provided with a biotin handle and then simply allowed to come in contact with the SA to which it rapidly binds with a practically irreversible bond. Due to the small size of the biotin handle, this linker is easily attached causing very little steric hindrance. Clearly, this immobilization process is crucially dependent on the SA maintaining its structural integrity and ability to bind the biotin linker.

Contrary to most proteins surveyed by us and others, SA actually appears somewhat stabilized by the adsorption to PS. This can be seen in Fig. 8, where the thermogram of SA in solution displays a broader transition peak than that of the adsorbed form. The stabilization is further indicated by Table 1, which lists transition enthalpies for SA in solution and adsorbed, with different surface concentrations, to PS particles of different diameters. Here, the soluble form displays the lowest value for ΔH, with the adsorbed forms showing slightly larger values, irrespective of surface concentration and substrate curvature. Also included in the table is a value for the transition enthalpy of avidin, a glycoprotein from hen egg white with the same affinity for biotin as that of SA and with comparable characteristics of its thermal transition [42].

Formation of the affinity complex between SA, or avidin, and biotin leads to an extraordinary stabilization of both proteins. Indeed, in the case of SA-biotin the transition temperature increases from 84°C for SA alone to above the maximum

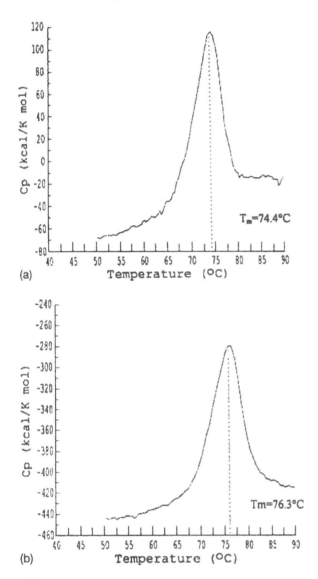

FIG. 7 (a) Thermogram for *Lathyrus odoratus* lectin in solution (10 m*M* phosphate buffer) and (b) adsorbed to its biospecific affinity matrix mannose–Sepharose.

observation temperature (110°C) of the DSC instrument (Calorimetry Sciences Corporation) that was used to generate the data in Table 1, and the SA-biotin thermogram therefore appears featureless. However, an indication of the magnitude of the stabilization can be inferred from the quoted values for avidin listed in the table. Apparently, the avidin molecule experiences a 47°C increase in T_m and more than a threefold increase in ΔH upon binding the biotin affinity ligand. As for the adsorption complex between PS latex spheres and SA-biotin there is no visible transition in the temperature interval 40–110°C, and the affinity pair therefore shows no measurable sign of being destabilized by the adsorption. This desirable condition has been uti-

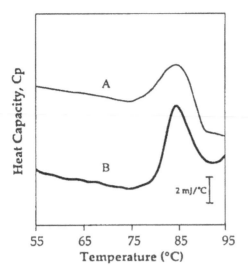

FIG. 8 Thermogram for Streptavidin in solution (A) and adsorbed to PS nanoparticles (B). (Adapted from Ref. 39, with permission.)

lized extensively for surface derivatization mediated by the SA-biotin pair, as noted above.

E. Antibodies as Affinity Anchors

A large variety of bioanalytical procedures, ranging from purification protocols to methods for clinical diagnostics, are based on the capture of an antigen analyte by a surface bound antibody. Very often, as in common immunoassay tests, the antibody is simply adsorbed to PS latex particles or microtiter plates of polystyrene. What then is the impact of the substrate on the immobilized protein? The adsorption of a polyclonal goat anti-human IgG to PS particles is seen in Fig. 9 to have a substantial effect on the structure of the antibody. Both traces in the figure represent identical

TABLE 1 Transition Temperatures and Enthalpies for Streptavidin in Solution and Adsorbed to Latex Particles of Different Size—Surface Concentration, Γ, Determined by Amino Acid Analysis

	Γ (mg/m^2)	T_m (°C)	ΔH (kcal/mol)
Streptavidin in solution	NA	84 ± 2	269 ± 20
PS 165-Streptavidin	9.6 ± 0.3	83 ± 1	277 ± 17
PS 214-Streptavidin	6.0 ± 0.2	83 ± 2	317 ± 14
PS 261-Streptavidin	4.8 ± 0.1	82 ± 1	291 ± 16
PS 314-Streptavidin	4.8 ± 0.1	79 ± 1	316 ± 13
Avidin in solution*	NA	85	300
Avidin-biotin in solution*	NA	132	1050

*Data adapted from Ref. 42.

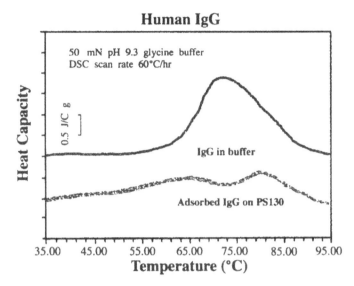

FIG. 9 Thermograms for a polyclonal goat anti-human IgG in solution (upper trace) and adsorbed to PS particles (lower trace).

amounts of protein, with the upper one describing the behavior in solution and the lower one protein adsorbed to latex particles with a diameter of 133 nm. The adsorption complexes had been carefully washed prior to loading the measuring cell, and all protein in this cell was therefore in the adsorbed form. It is clear that a significant amount of structure (ca. 60%) has been lost in the adsorption, specifically that contributed by higher melting elements. A population of affinity-purified antibodies of the type used in Fig. 9 represents a mixture of molecules belonging to different subclasses with different stability characteristics. Norde and coworkers [44,45] have performed thermal stability studies on solutions of monoclonal IgG samples of both subclasses IgG1 and IgG2b, demonstrating significant differences in their response to heating. While a thermogram for IgG1 shows but a single transition at 74°C (ΔH 12.7 J/g), one determined for IgG2b shows two distinct transitions: at 61°C ($\Delta H = 12.5$ J/g) and at 71°C ($\Delta H = 4.5$ J/g). The transition at the lower temperature was suggested to stem from denaturation of the F_{ab} fragments of the molecule, with the upper one deriving from the F_c fragment. Since the adsorption-induced losses of structure in Fig. 9 appear to affect primarily the high-melting region, i.e., the F_c fragment, it stands to reason that a certain antigen-binding effect is still retained following adsorption. However, it might be suspected that different monoclonal antibodies will show different behaviors in the adsorbed form. A broader cataloging of these differences is desirable as it would facilitate the design of highly responsive bioanalytical surfaces.

F. Direct Adsorption versus Tethered Attachment

The attachment of fragile macromolecular ligands to hydrophobic substrates, whether by adsorption or covalent linkage, is obviously fraught with problems related to structural changes. These changes can be significantly reduced, or even eliminated,

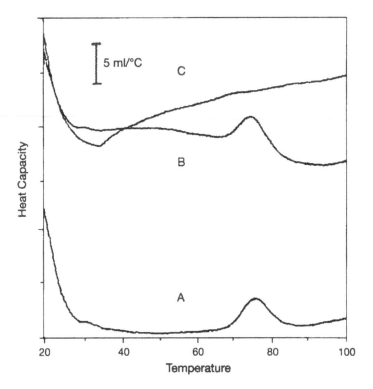

FIG. 10 Thermogram for lysozyme in solution (A), lysozyme covalently attached to PS particles precoated with EGAP (see text) (B), and PS adsorbed to the same particle (C).

TABLE 2 Calorimetric Evaluation of Protein Stability in Solution and After Attachment to PS Surfaces

	Γ (mg/m^2)	T_m (°C)	ΔH (kcal/mol)
FN in PBS	NA	55–80	570
PS-FN	1.79	NA	NA
PS-EGAP-FN	1.36	55–80	440
HSA in PBS	NA	60–75	580
PS-HSA	1.31	NA	NA
PS-EGAP-HSA	1.16	60–75	460
Lyz in water	NA	76	320
PS-Lyz	1.71	NA	NA
PS-EGAP-Lyz	1.52	76	310

Note: The attachment involves either adsorption or coupling via the EGAP (see text) tether. FN, human fibronectin, HSA, human serum albumin, Lyz, hen egg white lysozyme.

if the attachment can proceed via a hydrophilic tether, which prevents the macromolecule from direct contact with the surface. In our laboratory this linkage issue has been solved by allowing the immobilization to proceed in two steps, namely, first a covering of the surface through adsorption of a polymeric surfactant with chemically activated end groups, and second a replacement of the activator with the desired ligand [46]. Our choice of surfactant has been a block copolymer of the poloxamer type, i.e., a molecule consisting of a hydrophobic center block, consisting of poly(propylene oxide) (PPO), surrounded by two equally sized blocks of the highly water-miscible poly(ethylene oxide) (PEO). This compound, sold under the trade name of Pluronic F108, can been activated, e.g., by the introduction of a pyridyl disulfide group into one of its PEO chain ends [47]. The resulting "end group–activated Pluronic," or EGAP, forms monolayers on hydrophobic substrates by adsorption of its PPO block to the surface while it exposes its PEO flanking blocks to the aqueous surroundings. The resulting coating is hydrolysis stable, but releases the pyridyl group, in the form of a thiopyridone, upon contact with a thiol-containing molecule. Ligands immobilized in this manner are structurally quite well protected, as seen in Fig. 10, which shows a DSC comparison of lysozyme in solution (bottom trace) with lysozyme adsorbed to PS latex particles (upper trace) and covalently attached to the same type of particles that had been precoated with EGAP (middle trace). From the figure it is immediately obvious that direct adsorption under the chosen conditions is highly destructive, as no transition is detectable around the 76°C melting temperature that characterizes the soluble form of the enzyme. By contrast, the tethered form behaves in a manner that is very similar to that of the free form, both in terms of T_m and ΔH. This form of ligand attachment has been used with good results in a variety of applications [46,48]. Some details regarding the measured structural stabilities of lysozyme and the plasma proteins human serum albumin (HSA) and fibronectin (FN), all in free and tether-linked forms, are given in Table 2. The tethered FN has been used to support the culturing of anchor-dependent cells of different types [49].

VI. CONCLUSION

Proteins attached to surfaces are being used in a wide variety of contexts. Often losses of biological activity are observed, which arise suspicions that the structure might have been negatively impacted by the attachment. While optical methods may give very specific information related to the location and nature of a structural transition in a surface-bound protein, DSC, while less specific, gives a universal measure of the structural energetics of the system. With modern calorimeters being sensitive enough to allow the analysis of fractions of a milligram of protein in a wide variety of environments, the DSC technique should continue to grow in importance as a means of understanding the structural fate of proteins at interfaces.

ACKNOWLEDGMENTS

Much of the work reported here was performed with support from the Center for Biopolymers at Interfaces (University of Utah, Salt Lake City) and from the Center

for Surface Biotechnology (University of Uppsala, Sweden). This support is gratefully acknowledged.

REFERENCES

1. J. Buijs, W. Norde, and J. W. T. Litchenbelt, Changes in the secondary structure of
    adsorbed IgG and $(F'_{ab})_2$ studied by FTIR spectroscopy. Langmuir *12*:1605–1613, 1996.
2. V. Hlady and J. Buijs, Local and global optical spectroscopic probes of adsorbed pro-
    teins. In *Biopolymers at Interfaces* (M. Malmsten, ed.). Marcel Dekker, New York, 1998,
    Ch. 7.
3. G. D. Fasman, *Circular Dichroism: The Conformational Analysis of Biomolecules*. Ple-
    num Press, New York, 1996.
4. P. L. Privalov and S. A. Potekhin, Scanning microcalorimetry in studying temperature-
    induced changes in proteins. Methods Enzymol. *131*:4–51, 1986.
5. E. Freire, Differential scanning calorimetry. Methods Mol. Biol. *40*:191–218, 1995.
6. V. V. Plotnikov, J. M. Brandts, L.-N. Lin, and J. F. Brandts, A new ultrasensitive scanning
    calorimeter. Anal. Biochem. *250*:237–244, 1997.
7. J. M. Sturtevant and G. Velicelebi, Calorimetric study of the interaction of lysozyme
    with aqueous 1-propanol. Biochemistry *20*:3091–3096, 1981.
8. M. Hanson, K. K. Unger, R. Denoyel, and J. Roquerol, Interactions of lysozyme with
    hydrophilic and hydrophobic polymethacrylate stationary phases in reversed phase chro-
    matography (RPC). J. Biochem. Biophys. Methods *29*:283–294, 1994.
9. F.-Y. Lin, W.-Y. Chen, R.-C. Ruaan, and H.-M. Huang, Microcalorimetric studies of
    interactions between proteins and hydrophobic ligands in hydrophobic interaction chro-
    matography: effects of ligand chain length, density and the amount of blood protein. J.
    Chromatogr. A *872*:37–47, 2000.
10. F.-Y. Lin, W.-Y. Chen, and M. T. W. Hearn, Microcalorimetric studies on the interaction
    mechanism between proteins and hydrophobic solid surfaces in hydrophobic interaction
    chromatography: effects of salts, hydrophobicity of the sorbent, and structure of the
    protein. Anal. Chem. *73*:3875–3883, 2001.
11. C. E. Giacomelli, A. W. P. Vermeer, and W. Norde, Adsorption of immunoglobulin G
    on core-shell latex particles precoated with chaps. J. Colloid Interface Sci. *231*:283–
    288, 2000.
12. P. L. Privalov and N. N. Khechinashvili, A thermodynamic approach to the problem of
    stabilization of globular protein structure: a calorimetric study. J. Mol. Biol. *86*:665–
    684, 1974.
13. M. J. Marcos, R. Chehin, J. L. Arondo, G. G. Ghadan, E. Villar, and V. L. Schnyrov,
    pH-dependent thermal transitions of lentil lectin. FEBS Lett. *443*:192–196, 1999.
14. Y. V. Griko and D. P. Remeta, Energetics of solvent and ligand-induced conformational
    changes in α-lactalbumin. Protein Science *8*:554–561, 1999.
15. C. Giancola, C. De Sena, D. Fessas, G. Graziano, and G. Barone, DSC studies on bovine
    serum albumin denaturation: effects of ionic strength and SDS concentration. Int. J.
    Biol. Macromol. *20*:193–204, 1997.
16. F. Conjero-Lara, P. L. Mateo, F. X. Aviles, and J. M. Sanchez-Ruiz, Effect of Zn^{2+} on
    the thermal denaturation of carboxypeptidase B. Biochemistry *30*:2067–2072, 1991.
17. Y. Liu and J. M. Sturtevant, The observed change in heat capacity accompanying the
    thermal unfolding of proteins depends on the composition of the solution and on the
    method employed to change the temperature of unfolding. Biochemistry *35*:3059–3062,
    1996.
18. L. D. Creveld, W. Meijberg, H. J. C. Berndsen, and H. A. M. Pepermans, DSC studies

of *Fusarium solani pisi* cutinase: consequences for stability in the presence of surfactants. Biophys. Chem. *92*:65–75, 2001.

19. T. Kamiyama, Y. Sadahide, Y. Nogusa, and K. Gekko, Polyol-induced molten globule of cytochrome c: an evidence for stabilization of hydrophobic interaction. Biochim. Biophys. Acta *1434*:44–57, 1999.

20. W. Pfeil and P. I. Ohlsson, Lactoperoxidase consists of domains: a scanning calorimetric study. Biochim. Biophys. Acta *872*:72–75, 1986.

21. P. L. Privalov and S. J. Gill, Stability of protein structure and hydrophobic interaction. Adv. Prot. Chem. *39*:191–234, 1988.

22. E. I. Tiktopulo, P. L. Privalov, S. N. Borisenko, and G. V. Troitskii, Microcalorimetric study of domain organization of serum albumin. Mol. Biol. *19*:884–889, 1985.

23. P. L. Privalov and G. I. Makhatadze, Contribution of hydration and noncovalent interactions to the heat capacity effect of protein unfolding. J. Mol. Biol. *224*:715–723, 1992.

24. V. V. Novokhatny, S. A. Kudinov, and P. L. Privalov, Domains in human plasminogen. J. Mol. Biol. *179*:215–232, 1984.

25. J. T. Edsall and H. Gutfreund, *Biothermodynamics*. J. Wiley and Sons, Chichester, NY, 1983, pp. 217–224.

26. P. L. Privalov, Stability of proteins: small globular proteins. Adv. Prot. Chem. *33*:167–241, 1979.

27. P. L. Privalov and G. I. Makhatadze, Heat Capacity of Proteins. II. Partial Molar Heat Capacity of the Unfolded Polypeptide Chain of Proteins: Protein Unfolding Effects. J. Mol. Biol. *213*:385–391, 1990.

28. F. M. Richards, Areas, volumes, packing, and protein structure. Ann. Rev. Biophys. Bioeng. *6*:151–176, 1977.

29. M. E. Soderquist and A. G. Walton, Structural changes in proteins adsorbed on polymer surfaces. J. Colloid Interface Sci. *75*:386–397, 1980.

30. D. Horsley, J. Herron, V. Hlady, and J. D. Andrade, Fluorescence quenching of adsorbed hen and human lysozyme. Langmuir *7*:218–222, 1991.

31. B. Lassen and M. Malmsten, Competitive protein adsorption at plasma polymer surfaces. J. Colloid Interface Sci. *186*:9–16, 1995.

32. W. Norde and J. Lyklema, Thermodynamics of protein adsorption. J. Colloid Interface Sci. *71*:350–366, 1979.

33. C. A. Haynes and W. Norde, Structures and stabilities of adsorbed proteins. J. Colloid Interface Sci. *169*:313–328, 1995.

34. T. Zoungrana, G. H. Findenegg, and W. Norde, Structure, stability, and activity of adsorbed enzymes. J. Colloid Interface Sci. *190*:437–448, 1997.

35. W. Norde and T. Zoungrana, Surface-induced changes in the structure and activity of enzymes physically immobilized at solid/liquid interfaces. Biotechnol. Appl. Biochem. *28*:133–143, 1998.

36. W. Norde, Proteins at solid surfaces. In *Physical Chemistry of Biological Interfaces* (A. Baszkin and W. Norde, eds.). Marcel Dekker, New York, 2000, pp. 115–135.

37. H. Larseriksdotter, S. Oscarsson, and J. Buijs, Thermodynamic analysis of proteins adsorbed on silica particles: electrostatic effects. J. Colloid Interface Sci. *237*:98–103, 2001.

38. K. D. Caldwell, J.-T. Li, and D. G. Dalgleish, Adsorption behavior of milk proteins on polystyrene latex—a study based on sedimentation field-flow fractionation and dynamic light scattering. J. Chromatogr. *604*:63–71, 1992.

39. G.-Y. Yan, J.-T. Li, S.-C. Huang, and K. D. Caldwell, Calorimetric observations of protein conformation at solid/liquid interfaces. In *Advances in Chemistry 602: Proteins at Interfaces II* (T. Horbett and J. Brash, eds.). ACS Press, Washington, 1995, pp. 256–268.

40. B. L. Steadman, K. C. Thompson, C. R. Middaugh, K. Matsuno, S. Vrona, E. O. Lawson, and R. V. Lewis, The effects of surface adsorption on the thermal stability of proteins. Biotechnol. Bioeng. *40*:8–15, 1992.

41. C. Borrebaeck and B. Mattiasson, A study of structurally related binding properties of Concanavalin A using differential scanning calorimetry. Eur. J. Biochem. *107*:67–71, 1980.

42. J. W. Donowan and K. D. Ross, Biochemistry *12*:512–517, 1973.

43. M. González, C. E. Argaraña, and G. D. Fidelio, Extremely high thermal stability of streptavidin and avidin upon biotin binding. Biomol. Eng. *16*:67–72, 1999.

44. A. W. P. Vermeer and W. Norde, The thermal stability of immunoglobulin: unfolding and aggregation of a multi-domain protein. Biophys. J. *78*:394–404, 2000.

45. A. W. P. Vermeer, C. E. Giacomelli, and W. Norde, Adsorption of IgG onto hydrophobic teflon. Differences between the F_{ab} and F_c domains. Biochim. Biophys. Acta *1526*:61–69, 2001.

46. J. A. Neff, P. A. Tresco, and K. D. Caldwell, Surface modification for controlled studies of cell–ligand interactions. Biomaterials *20*:2377–2393, 1999.

47. J.-T. Li, J. Carlsson, J. N. Lin, and K. D. Caldwell, Chemical modification of surface active poly(ethylene oxide)–poly(propylene oxide) triblock copolymers. Bioconjugate Chemistry *7*:592–599, 1996.

48. C.-H. Ho, L. Limberis, K. D. Caldwell, and R. J. Stewart, A metal-chelating pluronic for immobilization of histidine-tagged proteins at interfaces: immobilization of firefly luciferase on polystyrene beads. Langmuir *14*:3889–3894, 1998.

49. K. Webb, K. D. Caldwell, and P. A. Tresco, A novel surfactant-based immobilization method for varying substrate-bound fibronectin. J. Biomed. Mat. Res. *52*:509–518, 2000.

21

Ellipsometry and Reflectometry for Studying Protein Adsorption

MARTIN MALMSTEN   Institute for Surface Chemistry and Royal
Institute of Technology, Stockholm, Sweden

I. INTRODUCTION

Ever since Drude summarized his theory for the reflection of polarized light, thereby
forming the foundation for ellipsometry and reflectometry [1,2], polarized light has
held promise as a useful tool for characterizing surfaces and for probing interfacial
structures and processes. It was only about 50 years ago, however, that ellipsometry
was initially developed as an experimental technique [3]. This was mainly due to
the requirement for a numerical procedure for solving the equations and determining
the adsorbed layer thickness and refractive index. During the last few decades, however, both ellipsometry and reflectometry, but also other techniques based on reflection such as optical waveguide spectroscopy, total internal reflection microscopy, and
Brewster angle microscopy, have found versatile use in studies of interfacial structures and events [3–7]. Perhaps of particular importance in relation to this has been
the use of these techniques for studying protein and biopolymer adsorption. In the
following chapter, the basis for ellipsometry and reflectometry is briefly outlined and
some experimental considerations discussed. Following this, a few selected examples
are provided where ellipsometry and reflectometry have proven useful for describing
and analyzing protein and biopolymer adsorption. Note, however, that a complete
coverage of the field is not intended. Instead, examples are chosen to illustrate more
general points.

II. METHODOLOGY

The physical basis of both ellipsometry and reflectometry is the change in the state
of polarization on reflection at an interface. These two techniques are strongly related
to each other, and comparable results are obtained. Also, the instrument designs are
somewhat similar, and the requirements regarding surfaces and chemicals used are
essentially identical. In the following sections the basis for the two techniques will
be discussed together with a number of experimental aspects regarding substrate
surfaces and surface modifications, as well as the purity of chemicals used.

A. Ellipsometry

Ellipsometry used for studying interfacial processes and structures involves a complete characterization of the change of the state of polarization of light reflected at the interface. From an experimental perspective, there are several different ways to do this, as previously described in some detail by Azzam and Bashara [3]. In protein adsorption studies, the most frequently employed ellisometry approach is *null ellipsometry*. This approach can be employed in several ways, but a frequently used instrumental setup is schematically illustrated in Fig. 1. Light from a lamp or a laser first passes a polarizer which renders the light planarly polarized. After this, the light is phase shifted through the use of a so-called compensator. After passing the polarizer and the compensator, the light is elliptically polarized. At reflection at the surface under investigation, the state of polarization changes once more, and at a given ellipticity of the light prior to reflection, it becomes linearly polarized after reflection. The state of polarization of this linearly polarized light can then easily be determined by a second polarizer (frequently referred to as the analyzer) followed by light intensity detection, e.g., with a photomultiplier tube. During one measurement of the change in the state of polarization of light on reflection at the surface, the polarizer and the analyzer are moved (keeping the compensator fixed) until the light intensity reaching the photomultiplier is minimized (hence the term null ellipsometry). Note that for a given system there are four sets of positions for the polarizer, the analyzer, the compensator which correspond to minimum intensity. These are usually referred to as the four zones, and by averaging the raw data over the zones most optical component imperfections can be eliminated [3].

Since polarized light can be regarded as the superposition of waves oscillating in perpendicular directions, the light reflection at an interface may be divided into the reflections of its two base components. Generally, the latter are chosen as parallel (p) and perpendicular (s) to the plane of incidence. Doing this, the changes in the state of polarization on reflection can be described by the (complex) amplitude reflection and transmission coefficients for the p- and s-components. These are expressed as the amplitude ratio between the reflected and the incident waves and the transmitted and incident waves, respectively. The overall ellipsometric response from measurements on a bare substrate surface can be written as

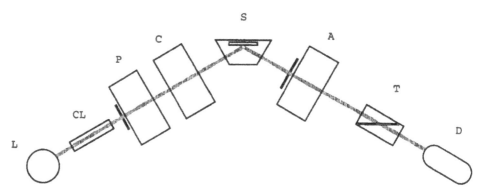

FIG. 1 Typical experimental setup for null ellipsometry: L, light source; CL, collimator; P, polarizer; C, compensator; S, surface; A, analyzer; T, alignment telescope; D, detector.

$$\tan \Psi e^{i\Delta} = \rho(N_0, N_2, \phi_0) \qquad (1)$$

where Δ and $\tan \Psi$ describe the relative phase difference and the (real) amplitude ratio of the p- and s-components on reflection, respectively, and where N_0, N_2, and ϕ_0 refer to the refractive index of the ambient, the (complex) refractive index of the surface, and the angle of incidence, respectively. Δ and Ψ are obtained from the positions of the polarizer, the compensator, and the analyzer. By applying Snells's law, the complex refractive index of the surface may then be determined.

When a thin film is present in the surface, the situation changes somewhat. When light is reflected at such a system, reflection and refraction will occur at each interface (Fig. 2). This multiple reflection is taken care of by introducing a set of reflection ($r_{ij,p}$ and $r_{ij,s}$) and transmission ($t_{ij,p}$ and $t_{ij,s}$) coefficients, where subindices ij refer to the interface between i and j.

Summing up the geometric series of reflections yields overall reflection coefficients according to

$$R_i = \frac{r_{01i} + r_{12i}e^{-i\delta 1}}{1 + r_{01i}r_{12i}e^{-i\delta 1}} \qquad (2)$$

where i = s or p, and subindices 01 and 12 refer to the interface between medium 0 and 1, and between medium 1 and 2, respectively. As before, Ψ and Δ are related to the reflection coefficients according to

$$\frac{R_p}{R_s} = \rho = \tan \Psi e^{i\Delta} \qquad (3)$$

Thus,

$$\tan \Psi e^{i\Delta} = \rho(N_0, N_1, N_2, d_1, \phi_0, \lambda) \qquad (4)$$

where λ is the wavelength of the light, and N_1 and d_1 the refractive index and thickness of the film, respectively. If ϕ_0 and λ are known, N_0 determined from measurements of the refractive index of the ambient and N_2 determined from a separate measurement for the bare surface, two parameters, i.e., d_1 and N_1 (or, rather, its real component n_1), sometimes referred to also as n_f and δ_{el}, respectively, are

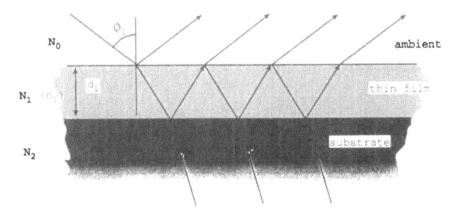

FIG. 2 Light reflection and refraction in a system with film on surface.

obtained from Ψ and Δ. These parameters are obtained numerically by minimizing the imaginary part of the adsorbed layer thickness (since this is a real quantity).

Note that this optical model assumes that the adsorbed layer refractive index is a real quantity (i.e., the absorption is zero). If this is not the case in the experimental system, the latter is underdetermined, and Ψ and Δ must be measured at several wavelengths and/or angles of incidence in order to resolve all unknowns. Note further that the analysis assumes regions of uniform refractive index and completely planar interfaces.

The values of the refractive index and the thickness of the adsorbed layer can be used to calculate the adsorbed amount (Γ). Cuypers et al. [8] derived a method for obtaining the adsorbed amount based on the Lorentz–Lorentz relation, requiring input in terms of the ratio between the molar weight and the molar refractivity (M_d/A_d), as well as the specific volume (v_d) according to

$$\Gamma = \frac{3d_1 f(n_1)}{\dfrac{A_d}{M_d} - v_d \dfrac{n_m^2 - 1}{n_m^2 + 2}} (n_1 - n_m) \tag{5}$$

where

$$f(n_1) = \frac{n_1 + n_m}{(n_1^2 + 2)(n_m^2 + 2)} \tag{6}$$

and n_m is the refractive index of the ambient. According to the model by de Feitjer [9], a somewhat simpler expression relates Γ to n_1 and d_1:

$$\Gamma = \frac{d_1(n_1 - n_m)}{\left(\dfrac{dn}{dc}\right)} \tag{7}$$

where (dn/dc) is the refractive index increment of the adsorbing component (e.g., protein). The two expressions give adsorbed amounts which agree well with each other, as well as with those obtained with other methods (e.g., radioactivity measurements; see further below).

As is described further below, silica is the preferred substrate for ellipsometry and reflectometry measurements for several reasons. At the same time, however, it is important to note that this substrate has a layered structure, and therefore any analysis in terms of adsorbed layer thickness and refractive index must take this into account in the optical model. Frequently, this is not done, but rather the silica surface is described by an effective complex refractive index. This still facilitates extraction of the adsorbed amount (10), but all information on the adsorbed layer thickness and refractive index is lost. If information about the latter two parameters is desired, the initial measurements of the bare substrate surface should be performed in such a way that correct information on both the silicon complex refractive index, as well as on the thickness and refractive index of the oxide film, is obtained. Since these are four unknowns, and since one ellipsometry measurement provides only one set of Ψ and Δ, the system is underdetermined, and in order to resolve all unknowns, measurements should be performed either at different wavelengths or different angles of incidence, or at different ambient refractive index. While the former of these are used in spectroscopic ellipsometry (see below), and the second approach forms the base for scanning angle reflectometry (11), ellipsometry measurements on protein, polymer, and surfactant adsorption, over the last few years in particular, have been increasingly based on the third approach [10,12–16]. Practically, this is performed through measuring Ψ and Δ (with four-zone averaging in order to improve the precision in the measurements) in both air and water prior to adsorption of the protein. For silica with a thick or moderately thick oxide layer, such measurements are possible since there is no further oxidation on changing from air to water, and hence the measurements in the two ambients correspond to the same system.

Particularly during the early days of ellipsometry, measurements were frequently performed in air with dry adsorbed films, mainly due to the simplicity of such measurements. It is important to note, however, that such measurements may provide little information on the structure, or even the adsorbed amount (in the case of Langmuir–Blodgett transfer from the solid–liquid to the air–liquid interface or vice versa), for the adsorbed layer in contact with solution. If the aim of ellipsometry experiments is to obtain such information, it is evident that measurements should be performed with the adsorbed layer in contact with the solution rather than in the dried state.

Due to a very strong covariance in n_1 and d_1, the scattering in the adsorbed amount [essentially the product of n_1 and d_1; see Eq. (7)] is much smaller than those in n_1 and d_1 themselves [13]. Also, the adsorbed amount is much less sensitive than the adsorbed layer thickness and refractive index regarding the choice of optical model. In fact, adsorbed amounts obtained for layered substrates such as silica and hydrophobized silica come out essentially identical whether or not the layered structure of silica is taken into account [10]. Hence, the adsorbed amount is a robust parameter, and therefore investigations also with quite complex systems are possible as long as information on the adsorbed amount is sufficient.

B. Reflectometry

In many ways, reflectometry is closely related to ellipsometry in that many of the same experimental concerns have to be made, and that the information obtained in many ways is comparable. Overall, however, reflectometry is a somewhat simpler methodology, and with a generally robuster (and cheaper) experimental setup. This has without doubt contributed to the strongly increased use of reflectometry over the last decade.

The experimental setup used in a number of single angle of incidence reflectometry studies of protein and (bio)polymer adsorption is schematically illustrated in Fig. 3. Generally, a He/Ne laser is used as the light source. The light from the laser hits a measuring cell at a fixed angle (frequently 45°), after which it is reflected at the interface under study (often silica) and eventually leaves the cell through the prism. For detection, the outgoing light beam is split into its s- and p-polarized components, both of which are detected with photodiodes. These signals are divided to give the measuring signal, usually referred to as S, according to

$$S = \frac{I_p}{I_s} \tag{8}$$

where I_p and I_s denote the reflected intensities of the p- and s-polarized component, respectively. The latter can be expressed as

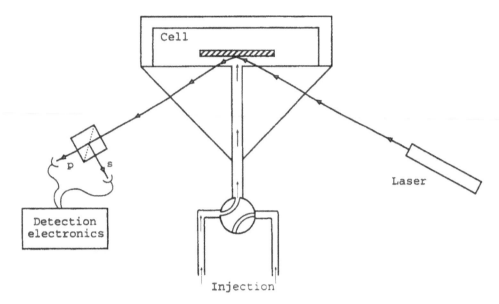

FIG. 3 Typical experimental setup for single angle reflectometry measurement. (Adapted from Ref. 4.)

$$I_i = \phi_i R_i I_i^0 \tag{9}$$

where I_i^0, R_i, and ϕ_i represent the intensity of the incident light of polarization i, the corresponding reflection coefficient, and the corresponding loss factor, respectively, where the latter originates from losses at the surface and at other interfaces in the beam pathway. The detected signal S can therefore be expressed as

$$S = f \frac{R_p}{R_s} \tag{10}$$

where f is a combined factor which is constant at a constant angle of polarization of the laser beam, and which can be determined from the values of R_p/R_s (as given by the optical model) and the measured signal S prior to adsorption. On adsorption, there will be a change in reflectivity, and hence also in the detected signal. From calculations of the reflectivity, it can be shown that the adsorbed amount (Γ) can be obtained from this shift in signal ΔS according to

$$\Gamma = \frac{\Delta S}{S_0} \frac{1}{A_s} \tag{11}$$

where A_s is a sensitivity factor which can be calculated from the optical model [4]. Alternatively, A_s can be found by external calibration of the change in S for a reference system of known adsorption. As an example of the latter, Muller et al. used the adsorption of the nonionic surfactant $C_{12}E_5$ for calibration in reflectometry studies [16].

In scanning angle reflectometry, a slightly different experimental approach is used. When light polarized parallel to the interface hits an ideally sharp (Fresnel) interface at the Brewster angle, the reflectivity is zero. When proteins or other mol-

ecules adsorb at the interface the refractive index profile changes and as a result the
surface is no longer ideally sharp, and a small fraction of the light is reflected.
However, also the entire reflectivity curve (reflected intensity as a function of incidence angle) changes shape and position (Fig. 4). It is these changes in the reflectivity
curve which are analyzed to extract information about the adsorbed layer characteristics. Since the change in the reflectivity close to the Brewster angle is monitored,
these measurements may be very sensitive even to low adsorption. From an instrumental perspective, developments were performed to the scanning angle reflectometry method, e.g., by Schaaf et al. (17), and later on by Leermakers and Gast [18]
to allow for time-resolved adsorption measurements. In the latter methodology, for
example, the reflectivity profile of p-polarized light was obtained by focusing the
incident light beam at the interface and simultaneously monitoring the intensities of
spatially resolved reflected intensities with a photodiode array. Since the whole range
of angles is measured simultaneously, the time resolution of the instrument is limited
only by the integration time of the array.

C. Surfaces

There are several requirements on surfaces or surface modifications to be used for
ellipsometry and reflectometry investigations of protein adsorption. Naturally, the
surface must be smooth enough, with a root mean square roughness of ideally less
than a nanometer or so. Also, since these methods are global in the sense that they
register all changes at the interface, it is essential that the surfaces are chemically
stable and not undergoing chemical oxidation or other chemical transformation, nor

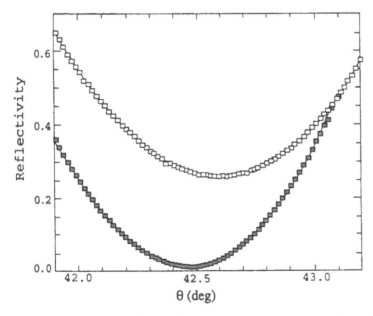

FIG. 4 Typical reflectivity profile obtained from scanning angle reflectometry for a bare
surface in contact with solution (filled squares), and a surface with adsorbed protein (open
squares).

partial dissolution or otherwise reconfigurating during the measurements. Moreover, since different surface modifications constitute interesting ways to control protein adsorption, it is essential that the surface can be further modified in different ways, e.g., through covalent binding of a surface layer to the background surface. Also, a crucial requirement on the substrate surfaces in ellipsometry and reflectometry is that these are highly reflective, and also optically isotropic (although the latter is not an absolute requirement [19]). Together, these considerations have resulted in silica being the substrate of choice for ellipsometry and reflectometry.

Silica surfaces fulfill all requirements for a good substrate for reflectometry and ellipsometry. Some concerns, however, should be addressed prior to the successful use of this substrate in adsorption investigations with these techniques. First, silicon undergoes spontaneous oxidation in both air and aqueous solution to form an oxide layer [3,20]. Therefore, silicon surfaces are not stable in air or aqueous solution, but rather undergo a continuous drift in their optical response. For this reason, silicon surfaces are frequently forcefully oxidized in an oxygen-rich atmosphere at elevated temperature to reach a sufficiently high oxide thickness [10,20]. There is, however, another reason for preparing a thicker oxide layer. For ellipsometry, the response in Ψ is only weakly dependent on the adsorption when the oxide layer is thin, and the adsorption essentially linearly dependent on Δ. Although this does not detrimentally affect the determination of the adsorbed amount, it precludes accurate determination of the thickness and refractive index of the adsorbed layer. For this reason, ellipsometry measurements involving determination of the latter parameters are frequently based on the use of silicon surfaces with an oxide layer thickness of about 300 Å. For reflectometry, the situation is comparable in the sense that the presence of an oxide layer increases the sensitivity of the measurements. In general, oxide layers of a thickness in the range 200–2000 Å are used for this type of measurement [4].

Despite frequent difficulties in using other surfaces than silica for detailed ellipsometry and reflectometry studies of structural aspects of adsorbed layers, other substrates may indeed be successfully used. For example, metal surfaces (e.g., chromium, gold, aluminum, zirconium, and platinum) have been extensively used in ellipsometric investigations of protein adsorption (see, e.g., Refs. 21–26). Just to mention a couple of examples, Ivarsson et al. investigated the adsorption of lysozyme and ovalbumin at platinum, titanium, and zirconium surfaces and found that the nature of the metal significantly affected the structure of the adsorbed protein layer formed [21]. The results obtained were inferred to be related to the electrostatic interaction between protein charges and the characteristic charge of the metal, although indications on the importance of nonelectrostatic forces were also observed. Also, Cuypers et al. used ellipsometry for studying protein adsorption at metallic, and in particular chromium, surfaces [22,25]. Investigations of protein adsorption at metallic surfaces are interesting not least from a practical perspective, since protein adsorption at metal surfaces plays an important role in a number of technical applications and participates in determining, e.g., the physiological response to metallic implants, the buildup of proteinaceous deposits (biofouling) in heat exchangers in the dairy industry, deposition of algae and other marine (micro)organisms on ships, etc. At the same time, however, recent investigations with spectroscopic ellipsometry have shown that at least gold is a quite complex substrate surface for ellipsometry [26,27]. Thus, Mårtensson et al. found that a simple three-layer model was insufficient to correctly describe adsorbed protein layers. Instead, a fourth layer, corresponding to an electron-deficient surface layer, had to be introduced. Whether or not this is the case also for other metallic substrates is still an open question, but may be a possibility.

Mineral surfaces such as hydroxyapatite are of some interest to protein adsorption investigations, not the least in relation to bone-related issues (e.g., guided tissue growth) or in odontological problems, such as the influence of salivary protein adsorption on the deposition of bacteria at the tooth surface and following development of odontological diseases, such as caries and periodontitis. In a number of investigations, Arnebrant et al. have used ellipsometry and surface force measurements to investigate the deposition of saliva and salivary protein fractions at silica and mica surfaces [28–31], and useful information has been gained on the effects of, e.g., saliva concentration, collection method, effects of surface properties, salt concentration, and effects of proteolytic degradation, etc. Later on, Arnebrant and coworkers investigated the saliva adsorption at hydroxyapatite surfaces and found that generally a good agreement was obtained between hydroxyapatite and silica surfaces, but also that the hydroxyapatite surface displayed some peculiar features, including calcium phosphate deposition, which depended on the nature of the conditioning salivary protein film [32]. Also, Ericson et al. polished enamel for investigating salivary protein adsorption by ellipsometry [33]. Although these measurements are very difficult, and the data do not facilitate any detailed analysis, such measurements are possible if a sufficiently smooth surface is used and if information on the adsorbed amount is sufficient. For further discussions on the adsorption properties of salivary proteins, see the chapter by Arnebrant in this volume.

In relation to the use of hydroxyapatite surfaces for ellipsometry measurements of protein adsorption, it is interesting to note that ceramic surfaces also have proven possible for ellipsometry investigations. For example, Bergström et al. investigated the optical properties of a number of ceramic materials, including ZrO_2, β-Si_3N_4, α-Al_2O_3, Y_2O_3, sapphire, MgO, and $MgAl_2O_4$, by spectroscopic ellipsometry [34]. The main focus of interest in these studies was not to investigate protein adsorption, but rather to use the optical information obtained from spectroscopic ellipsometry to calculate material constants for these surfaces. It was found that the Hamaker constants obtained through this approach agree rather well with those obtained with other methods. Given the interest in ceramics in the implant area, the field should now be open for ellipsometric investigations on the protein adsorption in such systems.

Another class of surfaces of interest in a range of biomedical applications is plastics. These offer real challenges in ellipsometry and reflectometry measurements. There are several reasons for this, including surface roughness, sometimes also optical anisotropy, a low or very low absorption and reflection, and overlying signals from front and back reflections. Although it is possible to address these issues— e.g., by only working with very smooth plastic surfaces of optically isotropic materials, by using very thick surfaces or otherwise trying to remove interference from the back reflection, and by using colored plastic substrates—the overall experience seems to be that plastic substrates are not very suitable for ellipsometry and reflectometry investigations. Nevertheless, some studies have been reported where colored polystyrene slides were used as substrate surface for polymer and protein adsorption. For example, Malmsten and Tiberg investigated the temperature-dependent adsorption of ethyl(hydroxyethyl)cellulose and found good agreement with parallel studies with methylated silica regarding the adsorbed amount increase and the contraction of the adsorbed layer with increasing temperature [35]. Furthermore, Tiberg et al. studied the effect of temperature on the thickness of surface-grafted poly(ethylene oxide)–poly(propylene oxide) copolymers, as well as its consequence on both the capacity of such layers to reduce IgG adsorption, and the extent of covalent grafting in parallel immobilization experiments [36]. As can be seen in Fig. 5, increasing temperature yields a solvency-induced collapse of the polymer layer, which results in a cancellation of its protein-rejecting capacity and hence in a higher IgG adsorption.

Interesting also in respect to polymer or plastic surfaces is that polymer, e.g., polystyrene, surfaces suitable for ellipsometry and reflectometry can be obtained by spin-coating silica/glass or methylated silica/glass surfaces. For example, using this approach, Malmsten et al. investigated the adsorption of C3 and its subsequent cleavage at polystyrene and poly(methacrylic acid) surfaces by ellipsometry and TIRF [37]. Furthermore, Elgersma et al. studied the adsorption from both single protein systems and protein mixtures at polystyrene using spin-coated silicon slides and reflectometry [38]. For example, measurements were performed on the initial adsorption of bovine serum albumin (BSA) as a function of pH. As can be seen in Fig. 6, it was found that the initial adsorption rate of BSA displays a maximum at pH $\approx$ 5, i.e., close to the isoelectric point of this protein. At this pH, the saturation adsorption is also the highest. With increasing pH, the net negative charge of BSA increases and as a result the initial adsorption rate (and the saturation adsorption) decreases. Reflectometry for investigating protein adsorption at PS-coated silicon surfaces was also applied by, e.g., Lichtenbelt et al. [39].

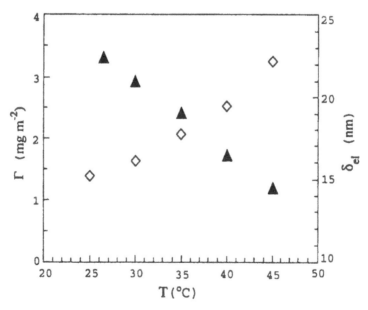

FIG. 5 Adsorbed layer thickness of a tetrabranched EO_{80}/PO_{55} block copolymer layer grafted at a polystyrene surface as a function of temperature (triangles), as well as the amount IgG adsorbed at this surface (diamonds). (Data from Ref. 36.)

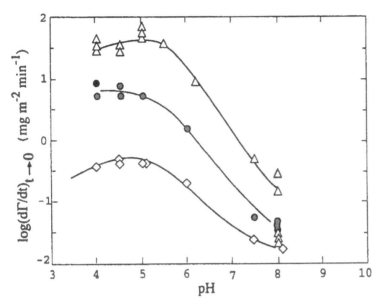

FIG. 6 Initial protein adsorption rate for bovine serum albumin (BSA) at polystyrene obtained with reflectometry. The concentrations used were 1 (diamonds), 10 (circles), and 100 (triangles) mg/m³. (Data from Ref. 38.)

D. Surface Modifications

Due to the limitations in the substrate materials available for ellipsometry and reflectometry measurements, surface modifications play an important role in protein adsorption studies with these techniques. These surface modifications are of different types and complexity, ranging from controlling surface charge, wettability, and chemical functionality through covalent modification to plasma polymer depositions; modification with water-soluble polymers in order to reduce protein adsorption, cellulose layers, layers of phospholipids for mimicking the cell surface; and ultimately layers of cells stratified at the substrate surface.

1. Covalent Modification

Covalent modification of surfaces offers good ways to control more basic surface parameters, such as surface charge, hydrophobicity, and chemical functionality. For silica (and glass) surfaces various silane coupling agents are generally used for this purpose [40]. By the use of such agents, it is possible to control the surface wettability from very hydrophobic ($\theta_a \approx 110°$) to very hydrophilic ($\theta_a < 20°$), and from negatively charged to positively charged. Although these types of surface modifications are quite widely used both for optical investigations and elsewhere, and although silane reactions have been extensively investigated over several decades, some care should still be taken when using them for modifying surfaces to be used in ellipsometry or reflectometry measurements. In particular, this relates to the possibility of water-induced reactions between silane molecules [40]. Thus, if the silane used is di- or trifunctional and residual water is present during the reaction, poly-

merization will occur, resulting in a (frequently thick) physisorbed, rather than chemi-
sorbed, hydrophobic deposit at the surface. Such a layer causes problems, particularly
in ellipsometry, in several different ways. First, due to the presence of the thick
polymerized layer, this must be taken into account in the optical model. This is not
straightforward and also tends to increase the uncertainty in the later-determined
adsorbed layer thickness and refractive index. Second, if the ellipsometry or reflectometry measurements involve the use of surfactants or highly surface-active polymers and proteins, there is a risk that these effectively solubilize the silane polymer
and cause it to desorb, thereby precluding all further analysis of the measurement.
To avoid these problems, the silanization reaction is frequently performed at low
water content, and physisorbed polymers removed, e.g., by solvent rinsing or ultra-
sonication. Another possibility is to use monochlorosilane agents in order to avoid
polymerization altogether. However, frequently such chemicals do not react very fast
with silica surfaces, which may result in poor reaction yield and hence in an insufficient surface modification.

An interesting development of the silanization of silica surfaces are the so-
called gradient surfaces. Thus, through the use of silane coupling agents and a two-
phase procedure, surfaces which are hydrophilic in one end and hydrophobic in the
other may be produced, with a gradient width determined, e.g., by the silane concentration and the reaction time [41]. In fact, not only hydrophobicity gradients, but
also charge and polymer density gradient surfaces may be prepared in this way [42].
Since the length scale of the gradient is macroscopic, generally of the order of
millimeters, it is possible to follow the adsorption in a space-resolved manner over
the gradient simply by connecting the surface holder to a motor and moving the
surface in relation to the light beam. In fact, if a sufficiently fast technique is used,
simultaneously good time and space resolution may be obtained. This, however,
frequently involves techniques other than null ellipsometry, such as off-null ellipsometry [43]. As an example of investigations with the gradient surfaces, one should
mention the work by Elwing and coworkers, who studied, e.g., the adsorption of
fibrinogen at wettability gradient surfaces precoated with human γ-globulin and
found that for this protein the adsorption decreases monotonically with a decreasing
hydrophobicity [41]. Similar investigations have also been performed for, e.g., C3
[44], and also for investigating protein–surfactant systems [45].

For gold and silver surfaces in particular, so-called self-assembling monolayers (SAMs) of (end-capped) alkyl thiols form the basis for the main method for
surface modification. With this approach oriented and highly ordered monolayers
covalently attached to the metal surface may be obtained through simple adsorption
from solution [46–51]. Also, the reproducibility and stability of the films are generally very good. Although alkyl thiols with a methyl end group are frequently used
for the purpose of surface hydrophobization, alkyl thiols with a polar end group can
also be used, and today a number of different polar end groups have been used
[46–51].

Naturally, through the use of SAM mixtures, parameters such as surface wettability, charge, and chemical functionality may be varied in a systematic way. Such
surfaces have also been used in protein adsorption studies, although the main part
of this work has been performed with surface plasmon resonance spectroscopy (SPR)
rather than ellipsometry and reflectometry (see, e.g., Ref. 50).

2. Plasma Polymerization

Plasma polymer deposition offers a powerful way to modify essentially any type of surface in order to obtain the desired surface properties or functionality. First, it is essentially independent of both the chemical nature and the geometry of the underlying substrate surface. Second, as a gas phase technique it avoids the use of solvents, is fast, and is also interesting for larger scale industrial applications. Also, a range of chemical functionalities may be introduced simply by changing the nature or composition of the gas or reaction conditions such as flow, pressure, and electric field strength [52]. Plasma polymer deposition as a tool for surface modification and for controlling protein adsorption has previously been used in conjunction with ellipsometry and TIRF [53–55]. Although successful results were obtained a general observation seems to be that while some plasma polymer depositions are well suited for ellipsometry, with a good stability and a lack of light absorption, some are less stable in water, and also slightly colored. The latter constitutes a major limitation for ellipsometry work where information about the adsorbed layer thickness and refractive index is desired, since the analysis at one angle of incidence and one wavelength only allows analysis if the plasma polymer film is transparent. In the case of an absorbing plasma polymer film, spectroscopic measurements should be used.

3. Modifications with Water-Soluble Polymers

The by far most widely employed method for reducing protein adsorption is through attachment of water-soluble polymers, such as poly(ethylene oxide) or polysaccharide derivatives at surfaces [56–59]. The attachment can take place in different ways, including covalent immobilization through end-capped polymers and physisorption of block or graft copolymers. Largely independent of the method for attachment, however, once the chains are properly immobilized the protein-rejecting capacity depends on the chain length, i.e., the layer thickness, and the grafting density. Both the chain immobilization and the effects of the polymer chains on the protein adsorption are well suited for investigations by ellipsometry and reflectometry. These and other aspects are discussed in more detail in the chapter by Malmsten in this volume.

4. Cellulose Surfaces

For many applications within pulp and paper, cellulose surfaces play an important role. Also, protein and polymer adsorption to such surfaces is important, e.g., in smoothing of cellulose fibers or cellulose surfaces through the use of cellulases, the latter, e.g., for surface modification of paper or in the retention step in papermaking. However, detailed studies on this adsorption have been scarce, largely due to lack of a suitable model substrate. A few years ago, however, Greber and Paschinger [60], as well as Cooper et al. [61], developed a methodology for preparing such surfaces involving Langmuir–Blodgett deposition of silylated cellulose at a hydrophobized surface, followed by desilylation and regeneration of the cellulose. With this procedure, cellulose layers with a roughness of significantly less than 1 nm (in air) can be obtained, making such surface modifications suitable for ellipsometry and reflectometry, but also for other methods, such as surface force measurements and atomic force microscopy (AFM) [62–65]. Spectroscopic ellipsometry has been used for investigating the dielectric behavior of these surfaces and to determine the Hamaker constant of the cellulose–water–cellulose, and cellulose–air–cellulose systems [65]. Using a five-layer model, Bergström et al. measured the optical response at different numbers of cellulose layers. From this, the dielectric response could be determined with good accuracy. Furthermore, ellipsometry has been used for investigating the adsorption of cationic polymers, e.g., APTAC to both these model cellulose surfaces and to the same surfaces made anionic by treatment with succinic anhydride [66]. To the best of the author's knowledge, however, no ellipsometry or reflectometry study has so far been performed concerning the effects of proteins, e.g., cellulases, on these cellulose surfaces. Such measurements should be of significant interest.

5. (Phospho)Lipid Layers

A central aspect in a number of biological processes and biotechnological applications, not the least drug delivery, is the interaction between proteins and cells. Since cell membranes can be expected to play an important role in this, stratified layers of polar lipids, and phospholipids in particular, have received some interest in relation to protein adsorption [67–77]. Such stratified phospholipid layers can be achieved in a number of ways, including Langmuir–Blodgett deposition, spin-coating, adsorption of liposomes, and adsorption from a mixed micellar solution containing the phospholipid and a nonadsorbing surfactant with a high critical micellization concentration (cmc). It has been found that the different deposition techniques yield comparable results in cases where the deposition is successful [76]. However, these methods also have their drawbacks and difficulties from an experimental point of view. For example, Langmuir–Blodgett deposition offers the possibility to deposit lipid mono- or multilayers in a controlled variation. On the other hand, this methodology is very time consuming, and the layers prepared from phospholipids in at least some cases suffer from stability problems [77]. Spin-coating, on the other hand, is a very fast deposition technique, but offers little control over the lipid layer structure. Adsorption from liposomes is probably the most extensively used method for depositing phospholipid layers at surfaces, is simple, and is fast, but again may suffer from reproducibility problems. Adsorption from mixed micellar solutions followed by dilution cycles is slightly more time consuming than liposome adsorption or spin-coating, but results in very well structured mono- or bilayers [78,79].

Stratified phospholipid layers as surface modification for ellipsometry studies have been reported in the literature by a number of different groups. For example, in a series of investigations, Malmsten studied the effect of the polar head group on the adsorption of a range of serum proteins, and found that the nature of the polar head group plays a major role for the adsorption of a range of serum proteins [67–69,76]. Similar findings were found by Cuypers et al. [70] and also by nonellipsometric techniques [80–82]. These aspects are discussed more in detail in the chapter by Malmsten in this volume.

6. Stratified Cells

In many ways, investigations with live cells stratified at a planar interface constitute the ultimate goal for more physiologically oriented investigations of adsorption and/or binding, e.g., of proteins or polymers. That optical methods and polarized light have potential for investigating such cell surfaces was suggested already by Azzam

[83]. As an example of such an investigation we could note the crucial role played by lipoprotein deposition at the surface of the endothelial cells in the blood vessels for the further development of atherosclerosis. In a couple of investigations, Malm-sten et al. have investigated the deposition of a number of key lipoproteins, both directly at endothelial cells and at proteoheparan sulfate, one of the major binding sites of lipoproteins at the endothelium cell walls. They found indications for cal-cium-induced lipoprotein deposition [84–86]. The major finding, in this context was that it was at all possible to use live cells stratified at methylated silica slide as substrate in ellipsometry experiments. At the same time, however, it should be noted that such cell layers may be extremely sensitive, and substantial experimental opti-mization may be required for such measurements with a particular cell line.

E. Sample Purity

An important feature of both ellipsometry and reflectometry is that experiments with these techniques are usually performed at a very low surface area–to–bulk volume ratio. This means that fractionation effects are much more pronounced than in bulk or at adsorption in a colloidal system. This in turn means that the adsorbed layer composition in a multicomponent system may differ dramatically from the average composition. It also means that ellipsometry and reflectometry measurements are quite sensitive to the presence of impurities (see, e.g., Ref. 87). Special care should therefore be taken to obtain sufficiently pure water and to identify efficient cleaning procedures. It also means that the requirements on sample and chemicals purity is much higher than when working with bulk experiments. For example, a 99% pure protein sample may be insufficiently pure for controlled ellipsometry and reflectom-etry measurements, particularly if the impurity is highly surface active, as in the case of surfactants. In such cases, the measurements may be completely dominated by the impurity and provide no information on the adsorption of the pure protein, or even of the unpure protein sample at a larger surface area–to–bulk volume ratio. The use of sufficiently pure samples and chemicals is therefore essential in this type of experiment.

F. Comparison with Other Methods

Particularly for surfactants there are extensive data available for comparison of re-sults obtained with ellipsometry and reflectometry on one hand and those found by methods based on entirely different principles on the other. Such comparisons re-garding the information provided by different methods are also most suitably per-formed for these systems due to their well-defined chemical structure, their avail-ability in high purity, and the absence of second-order effects such as surface-induced conformational changes, competitive spreading, and the like, which could be ex-pected to depend critically on mass transport conditions, surface area–to–bulk vol-ume ratio, etc. In the absence of such effects the information provided by different methods may be straightforwardly compared. Using surfactants, ellipsometry has in a number of investigations been found to provide comparable values of both the adsorbed amount and the adsorbed layer thickness, the latter within a few angstroms, with results obtained by, e.g., neutron and x-ray scattering, neutron reflectivity, and surface force measurements [88–93]. Also for a number of polymer and copolymer systems, a number of studies have shown good agreement between ellipsometry and other methods, such as solution depletion methods, photon correlation spectroscopy, surface force measurements, and atomic force microscopy (see, e.g., Refs. 35, 58, 94–97).

For protein systems the situation is somewhat more complex due to the occurrence in many systems of conformational changes, competitive spreading, etc., but also due to the sensitivity of biopolymer and protein systems for both excess electrolyte concentration and pH. It is therefore not as easy to find data for direct comparison, and also the variations between different reported studies are frequently substantial. However, regarding the adsorbed amount, several studies indicate a good agreement between the adsorbed amounts obtained with ellipsometry on one hand and other methods, such as radioactive labeling [98,99], neutron reflectivity [100], and TIRF [101], on the other. Also, the thicknesses obtained with ellipsometry and reflectometry agree well with each other, and also with those obtained with other methods (see, e.g., Refs. 10, 15, 17, 100, 102, 118).

In many more biologically oriented investigations, different types of immunoassays constitute an alternative to optical investigations of protein adsorption. In these cases it is difficult to quantitatively compare the data obtained with the two approaches, but qualitative comparison is frequently possible. An example of this is shown in Fig. 7, which shows the adsorption of fibrinogen at a surface coated with poly(ethylene oxide) (PEO) as a function of the interfacial chain density of PEO. As can be seen, the decrease in the protein adsorption with increasing PEO chain density of the surface is captured in both methods, and the two methods give the same critical PEO chain density for eliminating fibrinogen adsorption [58]. In the general

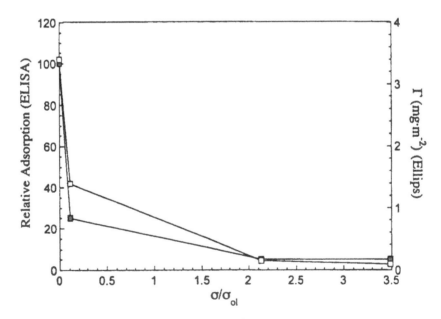

FIG. 7 Amount of fibrinogen adsorbed ($\Delta\Gamma_{tot}$) as a function of the interfacial PEG chain density for monoepoxy-PEG 5000 grafted at amino-functionalized silica determined with ellipsometry (open squares) and ELISA (filled squares). The interfacial overlap concentration (σ_{ol}) corresponds 0.033 chains/nm^2. (Data from Ref. 58.)

case, however, some care should be used when working with immunoassays due to possible complications regarding the linearity between the antigen concentration and the final signal, e.g., generated by an enzyme-labeled second antibody (see also below) [103].

III. INVESTIGATIONS OF ADSORBED LAYER STRUCTURE

One of the advantages with ellipsometry and reflectometry is that information is obtained not only on the adsorbed amount, but also on the thickness and mean refractive index of the adsorbed layer. Hence, simultaneous information is obtained on the surface density and the adsorbed layer structure. Typically, this information is obtained with a time resolution of a few seconds and sometimes even better. To correctly obtain this information, a number of factors should be taken into account, as discussed both above and below.

A. Diffuse and Heterogeneous Layers

A potential problem in the evaluation of ellipsometry or scanning angle reflectometry data is that the models generally used are based on stacked layers of uniform thickness and characterized by a mean refractive index, separated by planar interfaces. In real protein systems, this idealized picture is seldom reached. Instead, adsorbed layers are frequently characterized by heterogeneity both laterally and normal to the interface. The issue of what information ellipsometry provides regarding the thickness of diffuse adsorbed layers, such as those formed by homopolymers, has been addressed by several authors [104,105], and the analyses show that the thickness obtained by ellipsometry corresponds to the first moment of the segment density distribution [105]. This means that ellipsometry probes the mean layer thickness, and therefore ellipsometry gives lower thicknesses for such systems than methods sensitive to the diffuse outer part of the adsorbed layer, such as hydrodynamic or electrokinetic methods [106]. A general concern when working with polymer adsorption is therefore to be aware of what part of the adsorbed layer one is probing. However, for most globular proteins at least, the segment density distributions are steeper, and therefore the differences between the thicknesses obtained with different methods are smaller than for very diffuse layers.

B. Multilayers

Multilayers may occur in biological systems, e.g., as a result of specific interaction or of general electrostatic or other interactions. From an applications point of view, such multilayers hold promise in biotechnology (see the chapter by Ariga and Lvov in this volume) [107]. From the perspective of ellipsometry or reflectometry measurements, the most straightforward parameter for describing multilayers is without any doubt the adsorbed amount. The reason for this is primarily the robustness of this parameter regarding the choice of the optical model. Also, the refractive index increment is very similar for a large number of proteins, and the adsorbed amount only depends weakly on this parameter [Eq. (7)]. For ellipsometry and reflectometry experiments in such systems the effective adsorbed amount is close to the total adsorbed amount. The thickness of the adsorbed layer assembly, on the other hand, is generally much more sensitive to the optical model. Thus, for a more detailed analysis of the layer structure, the layering should be properly accounted for. Practically, this means building up the multilayer successively and determining the thickness and refractive index of the outermost layer by keeping the parameters of all but the outermost layers constant. Even if the layered structure is taken into account by this approach, however, there may still be problems relating, e.g., to diffuse interfaces between the layers.

C.   Examples of Adsorbed Layer Structure in Biopolymer Systems

Particularly over the last few years, the number of investigations focusing on structural aspects of adsorbed biopolymer layers and the formation of adsorbed layers in such systems using ellipsometry and reflectometry has increased. In what follows, only a few examples illustrating a few different aspects of biopolymer adsorption are given. Further examples are provided later in this chapter.

In principle, the simplest types of proteins regarding adsorbed layer structure and formation are those which undergo little surface-induced conformational changes on adsorption. Typical examples of such "hard" proteins are RNase, cytochrome c, subtilisin, and lysozyme. Of these, particularly the latter have been extensively investigated, also with ellipsometry. Indeed, the thickness of the adsorbed layer formed by lysozyme at silica indicates that the adsorbed layer formation in this case is rather straightforward, with the (almost spherical) molecules oriented "side-on" at the surface [108]. Largely analogous conclusions were reached also by Blomberg et al. for mica surfaces using surface force measurements [109], by Su et al. using neutron reflectivity [100], and by Radmacher et al. using atomic force microscopy [110]. Note, however, that in this seemingly simple system there are also numerous complications, since this protein has been reported to display, e.g., a side-on to end-on transition [109], conformational changes on adsorption [111], a pH-dependent interfacial aggregation [110], and a concentration-dependent multilayer adsorption [100, 108,109].

On the other end of the spectrum of protein structure, there are the flexible, or chainlike, proteins, such as mucus glycoproteins [112], proteoglycans [113], and gelatin [114]. Such systems can be expected to behave similarly in many ways to polymers and polyelectrolytes in general. Since structural aspects of the latter systems have been investigated successfully with ellipsometry in particular, one would expect optical methods to be useful also for investigating structural aspects of adsorbed layers formed by biopolymers and flexible chainlike proteins. That this is indeed the case is illustrated below for competitive adsorption in gelatin–surfactant systems. As another example of this, Malmsten et al. used ellipsometry and surface force measurements in order to probe a Ca^{2+}-induced contraction of proteoheparan sulfate adsorbed at a hydrophobic surface [115,116]. On addition of Ca^{2+} the preadsorbed proteoheparan sulfate layer was found with both techniques to contract, presumably as a result of both electrostatic screening and reduction of the linear charge density of the polysaccharide side chains of the glycoprotein due to specific Ca^{2+} binding. Analogous results for a much simpler polyelectrolyte system were obtained with ellipsometry by Poncet et al., who found for hydrophobe-modified poly(acrylate) at methylated silica that increasing the concentration of NaCl resulted in an adsorbed layer contraction, despite a higher adsorbed amount at the higher electrolyte concentration [117].

Ellipsometry and reflectometry can be used also for probing structural effects of complex formation in protein and biopolymer systems. For example, Heinrich et al. studied the adsorbed layers formed at silica in antigen–antibody systems by scanning angle reflectometry [118]. Initial adsorption studies of human IgG at silica revealed that this protein adsorbs in an end-on orientation, in agreement with ellipsometry findings by Malmsten [108]. On subsequent addition of rabbit α-human IgG, there is an increase in the total amount adsorbed as well as in the adsorbed layer thickness (Fig. 8) as a consequence of interfacial complex formation between the antigen and the antibody. Interestingly, both the adsorbed amount and the adsorbed layer thickness indicated that multiple binding of α-IgG to IgG occurs at the surface.

Another type of interfacial complexation which has been investigated with ellipsometry is that between different biopolymers on one hand and surfactants on the other. As one example of this, mixed systems of gelatin and an anionic surfactant are discussed below. Furthermore, surfactant-induced compaction of DNA with cationic surfactants is important, e.g., in gene therapy [119] and has therefore been extensively studied in bulk solution [120,121]. Furthermore, such DNA compaction at interfaces has been investigated with ellipsometry. Thus, Eskilsson et al. studied the interfacial properties of complexes formed between DNA and the cationic surfactant cetyltrimethylammonium bromide (CTAB) [122]. For both silica and methylated silica, indications were found for an interfacial association between DNA and CTAB. For methylated silica, the adsorption of DNA was found to be substantial also in the absence of CTAB, but on addition of the surfactant the interfacial com-

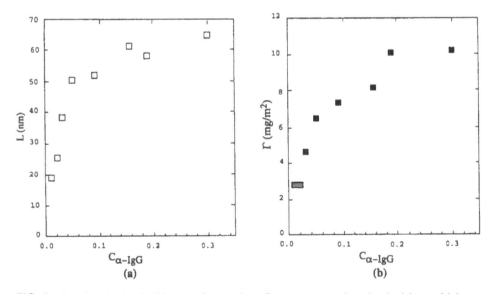

FIG. 8 Results obtained with scanning angle reflectometry on the adsorbed layer thickness (a) and the adsorbed amount (b) on addition of α-IgG after preadsorption of IgG at silica followed by rinsing. (Data from Ref. 118.)

plexation resulted both in a dramatic increase in the adsorbed amount and in a collapse of the adsorbed layer (Fig. 9). From these results it can be concluded that surfactant-induced compaction of DNA occurs also at surfaces, and that these effects may be investigated with ellipsometry.

Yet another type of interfacial association which can be studied with ellipsometry or reflectometry is that in which proteins undergo self-assembly. For such systems, information of the occurrence of surface-bound aggregates can be found from the adsorbed amount. For example, Elofsson et al. studied the adsorption of the two β-lactoglobulin variants A and B, differing primarily in their self-association behavior, and found that the adsorption of the two β-lactoglobulin variants displayed a shift in the concentration-dependent adsorption corresponding to the difference in dissociation constant for the two proteins, suggesting that the dimer adsorbs preferentially over the monomer [123]. More direct information on the occurrence of surface-bound protein assemblies can be obtained from the adsorbed layer thickness. Using this approach, Kull et al. [15] and Nylander et al. [124] used ellipsometry, surface force measurements, and neutron reflectivity to investigate the adsorption of β-casein. It was found that at hydrophobic surfaces, this protein adsorbed in a monolayer, showing no indications of surface-bound micelles. On the hydrophilic surfaces, on the other hand, there were clear indications of adsorbed micelles. This difference between hydrophilic and hydrophobic surfaces regarding surface-bound micelles has previously been observed, e.g., with ellipsometry and reflectometry, for both surfactant [13,93,125–128] and block copolymer [16,94,97,129] systems.

IV. INVESTIGATIONS OF OPTICAL PROPERTIES OF ADSORBED PROTEINS

For proteins with a pronounced absorption, the adsorbed layer is not simply characterized by a layer thickness and a real refractive index, but also by an absorption term ($k \neq 0$), and hence a single measurement of Ψ and Δ is insufficient to resolve all these unknowns. In order to do this, the latter two parameters should be determined at either different wavelengths or different angles of incidence for the same system. In particular, spectroscopic ellipsometry, which measures the optical response as a function of wavelength, has proven to be quite interesting for probing adsorbed protein layers. For example, Mårtensson et al. investigated the adsorption of lactoperoxidase and were able to extract information about the refractive index and absorption coefficient as a function of photon energy [130]. In particular, an absorption peak was detected at 3 eV for the adsorbed protein, which correspond well to the Soret band from the heme group.

Spectroscopic ellipsometry may also prove valuable for characterizing protein adsorption at substrates where the interface plays an important role optically, such as gold. For example, Mårtensson et al. investigated the adsorption of ovalbumin and ferritin and found that a three-phase model is insufficient for describing the layers formed in these systems below a photon energy of 2.5 eV and leads to large errors in the determination of the thickness and refractive index of the protein layers. Instead, a four-layer model was proposed in which an electron-depleted interfacial zone is added [26,27].

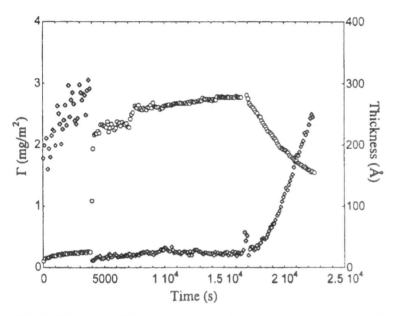

FIG. 9 Effects of CTAB on the adsorbed amount (circles) and adsorbed layer thickness (diamonds) for preadsorbed DNA. At $t = 5000$ s, CTAB was added to a final concentration of 1.5×10^{-6} M to a system where DNA was preadsorbed at methylated silica. At $t = 17,000$ s, rinsing was initiated. (Data from Ref. 122.)

V. INVESTIGATIONS OF ADSORPTION KINETICS

A. Mass Transport

In many biomedical applications, kinetic aspects of the interfacial behavior of proteins is of large interest, and it is the temporal behavior rather than equilibrium aspects which determine the performance of a particular procedure or device, be it in biosensor, diagnostic, biomaterial, or drug delivery applications. It is therefore essential that not only equilibrium or quasistatic aspects of protein adsorption is investigated, but that a significant emphasis is placed also on investigations of adsorption kinetics and of interfacial processes, such as the mechanisms for adsorbed layer formation, transient effects in multicomponent protein systems, degradation and synthesis in interfacial layers, etc. Here, ellipsometry and reflectometry have proven to be quite useful.

As for the investigations of adsorption kinetics, the good time resolution of ellipsometry and reflectometry, but also of other optical techniques, such as optical waveguide spectroscopy [5] and SPR [131], make these methods useful for investigations of adsorption kinetics in both protein and other systems. Such investigations of protein adsorption are described in some detail elsewhere in this volume and will not be further discussed here. In the present section, we merely make a few notes relating to the use of ellipsometry and reflectometry in investigations of adsorption kinetics.

In most ellipsometry experiments reported in the literature today, an open cuvette is used in which the surface is submersed and in which mixing is achieved through a stirrer bar in the bottom of the cuvette. For practical reasons, the dimensions of the cuvette are such that both turbulent flow conditions and convection are present [132]. Although this at first sight might appear as insufficient control of mass transport conditions during the ellipsometry experiment, this is not the case. Instead, such systems are characterized by an unstirred layer closest to the surface. Under typical conditions used in ellipsometry experiments, the thickness of this layer is of the order of 100 μm [133]. Due to the presence of the unstirred layer, mass transport to the surface is determined by diffusion, which in turn is dictated by the concentration gradient over the unstirred layer. Using this setup, the adsorption of nonionic low-molecular-weight surfactants have been analyzed in detail by Tiberg et al. [133] and by Brinck et al. [134–136]. It was found that the adsorption kinetics in such systems can be well described by a model including the concentration of monomers and micelles and their diffusion coefficient in bulk solution. In fact, also the chain length–dependent micelle dissociation time could be determined by this approach, and a good agreement was found with results obtained with other methods [133]. Also kinetic aspects of adsorption in mixed micellar systems could be successfully measured and accounted for. Although this kind of detailed kinetic measurement of mass transport to the surface in protein systems is less frequent, it is sufficient to note here that adequate control of mass transport to the surface is allowed with the experimental setup typically used in ellipsometry measurements.

However, in the setup discussed above mass transport can only be controlled by the concentration gradient, as given by the bulk concentration and the concentration in the surface phase (given by the adsorbed amount). No control of mass transport to the surface through flow is possible. In relation to this, it is interesting to note the use of the impingement jet cell in reflectometry measurements [4,137]. In this the solution is pumped to the cell in an inlet perpendicular to the surface, generating a stagnant point at the surface to which the mass transport is well controlled and, more importantly, can be varied not only via the solution concentration, but also by the flow rate. This in turn allows for detailed kinetic measurements of adsorption in polymer [137] and protein [138] systems (see also discussion below).

B. Affinity Measurements

In the classic biochemical approach, the affinity between two molecules, e.g., antibody–antigen or ligand–receptor pairs, is determined from kinetic experiments in which the "on rate" and the "off rate" are analyzed with adsorption/desorption cycles, where the binding and debinding kinetics are followed. The most widespread technique for following this today is most likely SPR, particularly after the introduction of the commercial Biacore™ instruments [131]. Naturally, such experiments can be performed also with ellipsometry and reflectometry. Irrespective of the method used for this type of investigation, however, it is important to recognize the importance of the interaction between the binding substance and the background surface for the apparent affinity observed. In the case of unspecific binding, clearly the affinity determined by the kinetic approach may be wrong, particularly if the (geometry-weighted) unspecific interaction is stronger than the specific one. Usually this issue is addressed through the use of a surface coating, e.g., based on polysaccharides, poly(ethylene oxide) derivatives, or phospholipids, aimed to reduce the unspecific binding. However, although adsorption/binding is reduced or eliminated, such interactions may affect the affinity determined through this approach, particularly if the interaction is long range, such as in the case of electrostatic interactions in a low screened system.

VI. INVESTIGATIONS OF INTERFACIAL PROCESSES

While the time-resolved adsorption, i.e., $\Gamma(t)$, is extremely useful to follow mass transport from the bulk to the surface, and hence also for coupling kinetic behavior to the absorption isotherm [106,137], and for obtaining indirect information about, e.g., surface-induced conformational changes, ellipsometry and scanning angle reflectometry offer opportunities beyond this, since information is obtained not only about the adsorbed amount, but also on the adsorbed layer thickness and refractive index. This provides a means for probing in larger detail interfacial processes such as adsorbed layer formation and adsorption in multicomponent systems (the latter particularly in combination with a composition-sensitive technique).

A. Adsorbed Layer Formation

Optical techniques such as ellipsometry or scanning angle reflectometry are ideal for investigating the formation of adsorbed protein layers for several reasons. First, simultaneous information is obtained about several parameters describing the adsorbed layers. Second, these methods allow kinetic investigations with a time resolution of the order of 1 s, thereby allowing a large number of data points to be accumulated during a typical adsorption process in a protein system. Third, the sensitivity of the methods allows measurements to follow the adsorbed layer formation essentially from the start (typically from $\Gamma \geq 0.1$ mg/m^2).

When proteins adsorb at interfaces, there are a number of processes occurring, including packing, reorientation, and conformational changes. All of these could be expected to depend on the adsorbed amount, since the space available for interfacial spreading, packing-induced reorientations, and large-scale reorganizations all could be expected to depend on the interfacial protein density. Therefore, studying the area per molecule and the adsorbed layer thickness and protein concentration may be an efficient tool for gaining information about the interplay between these parameters and about the adsorbed layer formation.

Malmsten previously investigated the adsorbed layer formation for a number of proteins at silica and methylated silica surfaces with ellipsometry [10,108,139–141]. Depending on both the surface and the protein, different patterns were observed during the adsorbed layer formation. As one extreme, IgG was found to adsorb with an essentially constant adsorbed layer thickness during the entire adsorption process (Fig. 10a). On the other hand, the mean adsorbed layer refractive index was found to vary essentially linearly with the adsorbed amount. Since the adsorbed layer thickness indicated adsorption in an end-on configuration, the adsorption was inferred to proceed through attachment of IgG molecules in an end-or orientation, and with a higher adsorbed amount being reached simply by packing the end-on oriented IgG molecules more densely at the surface [10].

A similar behavior was found also by Heinrich et al., who investigated the adsorption of IgG at silica through scanning angle reflectometry [118]. Also quan-

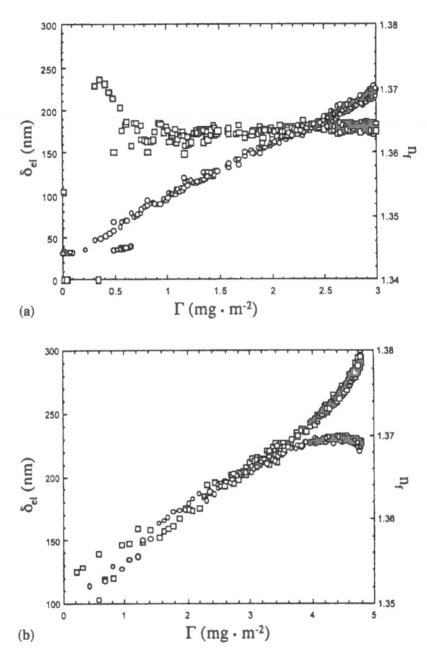

FIG. 10 Adsorbed layer thickness (squares) and mean adsorbed layer refractive index (circles) versus the adsorbed amount for IgG (a) and fibrinogen (b) adsorbing at methylated silica from 0.01 M phosphate buffer, 0.15 M NaCl, pH 7.4. (Data from Ref. 10.)

titatively, these authors found indications for an end-on adsorption of IgG molecules and that an increased adsorbed amount is reached through packing these molecules more closely at the surface. Also, the latter authors described the adsorbed layer formation through so-called optical invariants, the idea of which is to obtain model-independent information about the adsorbed layer characteristics [142].

The rodlike protein fibrinogen, on the other hand, was found to display an entirely different mechanism for the adsorbed layer formation [10]. (Comparable results on the adsorbed layer thickness for fibrinogen at silica were obtained by Schaaf et al. [143].) Thus, as can be seen in Fig. 10b, an increase in the adsorbed amount is facilitated by a simultaneous increase in both the adsorbed layer thickness and the mean refractive index of the adsorbed layer. This means that an increased adsorption is reached by an increasingly dense packing within the adsorbed layer, but also by a reorganization to lower the adsorbed layer protein concentration through increasing the adsorbed layer thickness. The latter can be a consequence of a transition from side-on to a larger degree of end-on orientation, but also of interfacial conformational changes, or both.

In fact, the adsorbed layer formation in the case of fibrinogen is rather similar to that of flexible polymers. Thus, from both kinetic ellipsometric information of the dependence of the adsorbed layer thickness on the adsorbed amount and the equilibrium thickness as a function of the equilibrium adsorbed amount, the thickness of the adsorbed layer has been found to be largely determined by the adsorbed amount [106,144]. Thus, as the adsorbed amount increases, the osmotic pressure builds up, and as a consequence there is a reorganization within the adsorbed polymer layer to take a larger volume into account by increasing the adsorbed layer thickness, thereby effectively reducing the osmotic pressure within the adsorbed layer.

A common feature in time-resolved measurements $[\Gamma(t)]$ of protein adsorption is the occurrence of a maximum of the adsorbed amount versus time curve. In many cases, this is an indication of an insufficiently pure protein sample and the occurrence of competitive and/or sequential adsorption events (see below). Also for sufficiently pure protein samples, however, such an effect in the time-resolved adsorption curve may be observed. The origin of the adsorption maximum at intermediate adsorption time is so-called competitive spreading, by which we mean a transformation of adsorbed protein molecules to make better contact with the surface. As a consequence, the adsorbed protein molecules take up a larger surface area, and hence the adsorbed amount decreases through desorption of some of the more loosely adsorbed protein molecules.

The occurrence of competitive spreading during protein adsorption may be investigated straightforwardly, e.g., by reflectometry or some other optical technique, provided that sufficient control can be exercised regarding the mass transport to the surface. For example, van Eijk and Cohen Stuart investigated the adsorption kinetics of savinase as a function of the interfacial "spreading" time [138]. It was found that the limiting adsorbed amount increases with the flux to the surface (Fig. 11). This is expected, since a higher flux provides less time for initially adsorbing molecules to spread at the surface due to crowding. Thus, at high flux, the area occupied by each adsorbed molecule is smaller, and the adsorbed amount is higher. By a detailed investigation of the effects of the protein concentration and hydrodynamic conditions on the adsorption and the adsorption kinetics and by modelling the data with a "growing disc" model, the authors were able to estimate the spreading rate and

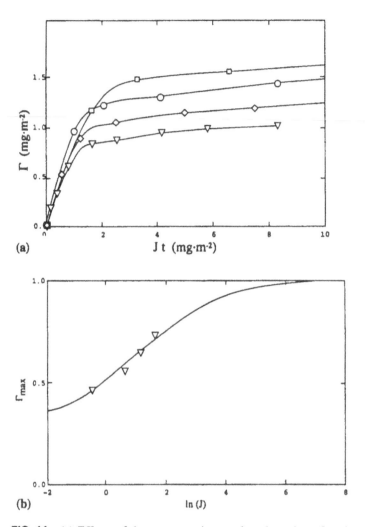

FIG. 11 (a) Effects of the concentration on the adsorption of savinase at silica at pH 8 and $I = 10$ mM. Shown are results obtained for 2.5 (triangles), 7.5 (diamonds), 12.5 (circles), and 20 g/m^3 (squares). (b) Maximum adsorbed amount as a function of the flux obtained experimentally (triangles) and from the "growing disc" model calculations (solid line). (Data from Ref. 138.)

extent of spreading of savinase at silica. Naturally, the growing disc concept is also compatible with frequently observed adsorbed amount maxima in time-resolved adsorption experiments, as discussed previously, e.g., by Cohen Stuart in this volume.

The biological function of adsorbed proteins can be expected to depend on parameters such as the degree of surface-induced conformational changes, crowding-dependent interfacial reorientation, and interfacial packing density, and hence the biological activity should vary also during the adsorbed layer formation. For example, Lee and Belfort investigated the adsorption of the structurally quite rigid RNase A at mica [145]. From surface force measurements it was found that initially the adsorption occurred side-on, which by electropotential considerations was suggested to be due to electrostatic interactions between a positively charged domain of the protein and the negatively charged mica surface. Since the active site resides in the positive potential region, it is not surprising that the enzymatic activity displayed at these conditions was quite low. With an increasing adsorption time, however, the adsorbed amount increases, causing a reorganization of the adsorbed layer from side-on to end-on adsorption. Since this exposes the active site this reorganization results in an increased enzymatic activity. [Although an alternative explanation of these findings could be a crowding-dependent conformational change on adsorption and concomitant activity loss, as described, e.g., by Kondo et al. for a number of protein systems [146,147], this seems less likely considering the structural stability of RNase A.]

Furthermore, Wannerberg and Arnebrant studied the adsorption of lipase as well as its surface-dependent activity with ellipsometry. In particular, the importance of the protein–surface interaction for protein adsorption and interfacial activity was addressed through comparison between the wild type and a mutant of lipase from *Humicola lanuginosa*, where the latter was modified from the former by an Asp96 →Leu mutation [148,149]. Thus, the mutant is more hydrophobic than the wild-type. It was found that the higher hydrophobicity of the mutant resulted in a higher adsorbed amount at methylated silica compared to the wild type. The effect of the surface hydrophobicity on the adsorption of the two proteins was investigated over a range of contact angles ($<10° \leq \theta \leq 90°$). For both proteins, the adsorbed amount was found to decrease with a decreasing contact angle. For silica, with a contact angle of less than $10°$, a very low adsorbed amount (<0.10 mg/m^2) was observed for both proteins. The latter finding indicates that the higher adsorption of the mutant at methylated silica really is due to a larger hydrophobic protein–surface attraction and not due to solvency effects. Considering this, the finding that the nondesorbable fraction after surfactant addition and subsequent rinsing was larger for the mutant than for the wild type [149] is quite expected. In parallel to the absorption measurements, Wannerberger and Arnebrant monitored the interfacial enzymatic activity with respect to the hydrolysis of a water-soluble substrate (*p*-nitrophenylacetate, PNPA), and found that the specific activity of the mutant was lower than that of the wild type. The latter finding may possibly be explained by the stronger hydrophobic interaction between the protein and the surface, causing the mutant to adsorb with its more hydrophobic active site surroundings oriented toward the surface in order to maximize the attractive interaction. This is further supported by the finding of an increasing specific activity with a decreasing surface contact angle, allowing other orientations than those with the active site surroundings directed toward the surface to a higher extent.

B. Multicomponent Systems

Many biotechnological applications involve multicomponent protein systems. Such systems offer a great versatility and display many interesting and important effects relating to both interfacial exchange, multilayer adsorption, and interfacial complex formation. These effects, and particularly interfacial exchange, have been discussed at length in numerous contexts before [38,103,150–154], and are also discussed elsewhere in this volume. Therefore, only some aspects of adsorption in multicom-

ponent systems which are related to the use of ellipsometry and reflectometry are
addressed here.

1. Protein Mixtures

Perhaps the most important and common situation regarding adsorption in multicomponent protein systems is that of competitive adsorption from a complex protein
mixture. This is the case, e.g., for implants in contact with blood, in extracorporeal
therapy and dialysis, and for colloidal drug carriers in parenteral drug delivery, where
the outcome of the competitive adsorption process determines the biological response
and hence also the performance of the implant, the extracorporeal procedure, and
the drug delivery system, respectively. Also in biosensors and solid phase diagnostics,
as well as biotechnological separation methods such as chromatography or ultrafiltration, these events are of crucial importance.

From an analytical perspective, competitive adsorption from multicomponent
systems is challenging, and several approaches have been used to obtain information
on the adsorption kinetics, extent and rates of exchange, etc. Of these, labeling of
one or several proteins in the protein mixture, e.g., through radioactive or fluorescent
groups, is probably the most extensively used approach, and much valuable information has over the years been obtained this way [152,153]. Optical methods such
as ellipsometry and reflectometry also offer some possibilities for investigating protein adsorption from multicomponent systems. A general problem with the latter
methods, however, is that they do not differentiate between different proteins, but
rather provide information on the total adsorbed amount and the overall thickness
and refractive index of the adsorbed layer. Therefore, these methods do not in themselves provide direct information on the composition of the adsorbed layer. Instead,
this must be inferred, either by deliberately varying the protein mixture composition
or by combining these techniques with another method providing composition-related
information.

An approach frequently taken in relation to the use of optical methods for
investigating adsorption from multicomponent protein systems is to use antibodies
for probing the adsorbed layer composition. In such investigations, the protein mixture is generally first adsorbed, whereafter excess protein is removed by rinsing,
followed by exposure of the adsorbed layer to different antibodies. The idea is that
the antibodies bind to their respective protein antigen, which results in an increase
in the total adsorbed amount corresponding to the amount of the probed protein at
the surface. By performing this type of experiment for a number of relevant antibodies, the adsorbed layer composition may in principle be determined.

Using this approach, Vroman et al. employed antibodies for characterizing the
composition of the adsorbed layer formed from plasma, with particular emphasis on
fibrinogen adsorption [150]. Furthermore, Elwing et al. used the same approach in
a number of investigations, e.g., of the mechanisms for complement activation, a
process of great importance and interest in implant, extracorporeal therapy, and parenteral drug delivery applications [44]. Later Tengwall et al. expanded and developed
these investigations further, in particular by placing special emphasis on the temporal
aspects of the deposition of various complement proteins [155]. For example, these
authors investigated the deposition of complement components onto immobilized
human colostrum immunoglobulin (Ig)A and human IgG on methylated silica. In

parallel, soluble complement components such as iC3b, C4d, and Bb were detected by ELISA. By this approach, these authors could show that for the IgA-coated surface, complement activation occurred by the alternative pathway, whereas the IgG-coated surfaces resulted in a fast classic activation.

There are a number of potential difficulties and interferences with the antibody approach, however, and care should be taken to avoid these. For example, in the case of multilayer formation, the inner layers may not be accessible for antibody binding, and therefore the antibody binding only reflects the composition of the outer part of the adsorbed layer. Furthermore, surface-induced conformational changes of the adsorbing protein molecules may render these unrecognizable for the antibodies. Also, the antibody may recognize a certain part of the protein molecules, and if this is oriented toward the surface rather than the bulk solution, again lack of binding may be the result. Moreover, lateral crowding also within the outer part of the adsorbed layer may cause steric restrictions for binding of a large antibody to the surface. Also well known from solid phase diagnostics is the occurrence of false positives, which generally are caused by unselective adsorption of the detection step antibody [103]. Naturally, the same thing may also occur in antibody-based ellipsometry measurements, and hence lead to identification of a protein in the adsorbed layer which is not really there, or overestimation of the adsorbed layer concentration of a given protein, or both. Although there are no foolproof ways to completely avoid these difficulties, one way to try to address this is to use several antibodies for the same protein, screened in a different manner (e.g., with different surface materials or without the use of surfaces) as well as antibodies that recognize different parts of the same protein. Also, the total adsorbed amount may be helpful for determining whether or not there is a risk for multilayer adsorption.

Another approach for addressing the fact that ellipsometry, reflectometry, and most other optical techniques do not provide any information on the adsorbed layer composition is to combine these with a composition-sensitive method not based on antibodies, but rather on labeling. In principle, any labeling method may be used, but the one most commonly used in combination with ellipsometry is fluorescence labeling, particularly in the form of total internal reflection spectroscopy (TIRF). This approach suffers from some difficulties relating to quantum yield effects and concentration-dependent quenching (see Tilton's chapter in this volume) and also to the fact that use of the fluorescent labeling may affect the surface properties of the protein. In order to address the latter of these two, it is important that the labeling density is as low as possible and also that the issue of whether or not the labeling affects the protein surface activity is addressed through separate adsorption or surface tension measurements. Also, the intensity of the excitation light should be kept as low as possible, and measurement points should not be too many and be kept well apart in order to avoid fluorescence bleaching effects [55]. With these precautions, however, TIRF does provide a valuable methodology to complement ellipsometry and reflectometry for investigations of both competitive and sequential adsorption processes in multicomponent protein systems.

For example, Lassen and Malmsten investigated the competitive adsorption of human serum albumin (HSA), IgG, and fibrinogen at different plasma polymer surfaces with a combination method of ellipsometry and TIRF [53–55]. The concentrations used were those in blood diluted 100 times. Since HSA is the smallest of these proteins and also has the highest concentration, the initial adsorption from such a mixture is dominated by HSA adsorption, followed by a larger or smaller exchange of HSA with the later-arriving IgG and fibrinogen molecules. From such investigations it was concluded that while HSA is not extensively replaced by IgG or fibrinogen in the case of a hydrophobic surface prepared through plasma polymerization of hexamethyl disiloxane (HMDSO), the hydrophilic and either positively or negatively charged surfaces prepared through plasma polymerization of acrylic acid and diaminocyclohexane, respectively, were characterized by a larger extent of interfacial exchange. These results are in agreement with those obtained from investigations of sequential adsorption.

Ellipsometry and reflectometry combined with TIRF may also be used successfully in investigations of sequential adsorption by proteins. In fact, studies of sequential adsorption processes are even more suitable for this method combination than simultaneous competitive adsorption. The reason for this is that the intensity in fluorescent measurements is a rather poorly defined parameter due to both concentration-dependent quenching and quantum yield effects, which makes comparison between experiments in which different protein components have been labeled one at a time somewhat difficult. In sequential adsorption experiments, e.g., where a labeled protein is first adsorbed, followed by exposure of the adsorbed layer to a second (unlabeled) protein or a protein mixture, these effects are less of a problem, since the relative change in the fluorescence intensity within the experiment is the main parameter of interest.

The method combination of ellipsometry and TIRF was used for studies of sequential adsorption in multiprotein systems, e.g., by Malmsten et al. [156]. From these investigations, which involved both homogeneous and heterogeneous exchange of preadsorbed HSA, it was concluded that exchange by both these mechanisms is quite limited at hydrophobic surfaces such as methylated silica/glass and HMDSO plasma polymer–modified silica or glass surfaces. On the other hand, exchange of preadsorbed HSA by IgG and fibrinogen at hydrophilic surfaces such as silica/glass and phospholipid-modified silica/glass was found to be more extensive.

Apart from interfacial exchange effects, sequential adsorption of proteins may result in multilayer structures. In fact, such multilayer assemblies are of major technical interest, as described in detail in the chapter by Ariga and Lvov in this volume. Whether or not such multilayers appear and the detailed structure of these are complex issues which we will not address further here. For the present discussion it is sufficient to say that ellipsometry and reflectometry may be used for investigating such multilayer adsorption. For example, Spaeth et al. investigated multilayer adsorption in the biotin–avidin system with spectroscopic ellipsometry [157]. Assuming a simple homogeneous layer, these authors investigated the growth of the layer, and found that each incubation resulted in an increase in the thickness of the layer by 18.75 nm (Fig. 12). The analysis was also found to be limited by high correlation coefficients for less than five incubation steps.

2. Protein–Surfactant Mixtures

Adsorption in mixed protein–surfactant systems bears several similarities with the adsorption from multicomponent protein mixtures. For example, both surfactants and proteins are generally surface active, and hence competitive adsorption will occur in

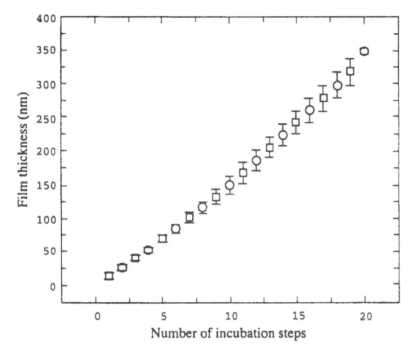

FIG. 12 Thickness of the biotin–avidin film on silica as a function of the number of incubation steps. (Data from Ref. 157.)

such mixed systems. At the same time, however, there are many differences. For example, surfactants are generally significantly more surface active than proteins, which frequently makes the competition more uneven than in multiprotein systems. Also, surfactants are generally much smaller than proteins, which makes both their mass transport to the surface and their adsorption much faster than that of proteins in the general case. Furthermore, the situation is complicated by the fact that proteins and surfactants frequently form aggregates both in solution and in adsorbed layers [158]. These and many other aspects of adsorption from mixed protein–surfactant systems are discussed in the chapter by Wahlgren et al. in this volume. Again, our interest in the present discussion lies mainly in the use of ellipsometry and reflectometry in investigations of the adsorption in such systems.

Analogous to the situation in multicomponent protein systems, ellipsometry and reflectometry investigations of the adsorption in protein–surfactant systems suffer from the methodological problem that these methods do not provide straightforward composition information. However, due to the difference in adsorption and desorption rates between proteins and surfactants, and the fact that most surfactants desorb essentially quantitatively on rinsing [133–136] whereas many proteins display effectively irreversible adsorption [159–160], some information may be obtained simply by rinsing after different steps in the sequential adsorption experiments and by comparing the results with those obtained for the protein(s) and the surfactant(s) alone. Such experiments are described in some detail in the chapter of Wahlgren et al. in this volume.

Also analogous to the multicomponent protein systems, investigations of the competitive adsorption in protein–surfactant systems may be addressed by combining, e.g., ellipsometry and TIRF. As an example of this, Muller et al. studied the adsorption of gelatin at hydrophobic methylated silica surfaces, as well as the effect of exposure of the preadsorbed gelatin layer to an anionic surfactant, sodium dodecylbenzene sulfonate (SDBS) [161]. As can be seen from Fig. 13, initial addition of

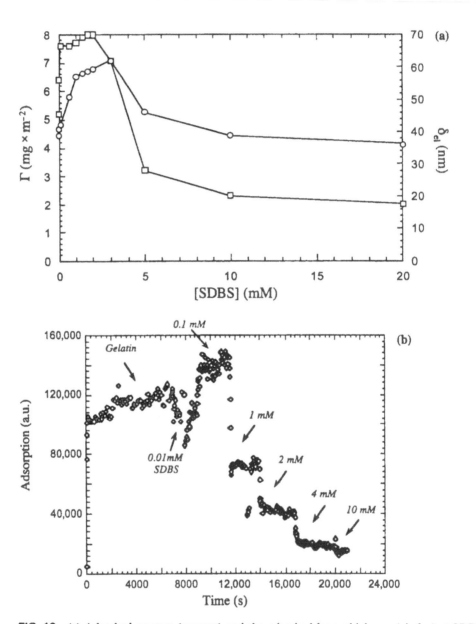

FIG. 13 (a) Adsorbed amount (squares) and the adsorbed layer thickness (circles) at SDBS addition after gelatin adsorption from 0.1 wt%, pH 6.5, at methylated silica for 2 hr. (b) Effect of SDBS addition on gelatin adsorption as determined by TIRF. (Data from Ref. 161.)

SDBS causes an increase in the adsorbed amount, as well as a swelling of the adsorbed layer. This is mainly a result of SDBS binding to the adsorbed gelatin molecules, which results in an effective charging of these, thus causing the layer to swell due to repulsive electrostatic interactions. At these conditions, TIRF results show limited desorption of gelatin. At higher SDBS concentrations, on the other hand, there is a dramatic decrease in both the total adsorbed amount and the amount gelatin adsorbed, as determined from ellipsometry and TIRF, respectively. This decrease could in principle be caused both by an increased binding of SDBS to the gelatin, thereby effectively reducing its adsorption tendency through introduction of more negative charges, and competitive adsorption. In order to establish which of these effects dominates in this particular system, the desorption behavior was investigated as a function of pH. Since the critical desorption concentration was found to decrease with decreasing pH, it was inferred that it is the SDBS binding to the adsorbed gelatin molecules rather than competitive adsorption effects which causes the desorption at higher SDBS concentrations. Interestingly, however, a significant amount of gelatin remains at the surface even at high SDBS concentrations, as inferred from both the total adsorbed amount and the adsorbed layer thickness (ellipsometry) and the residual fluorescence intensity of adsorbed labeled gelatin (TIRF).

C. Interfacial Degradation and Synthesis

The interfacial behavior of enzymes also includes their ability to be active when immobilized at an interface or otherwise exposed to an interface. This is important in a number of technical applications, such as solid phase synthesis, biotechnological and food processing, proteomics, and detergency, to mention just a few. The presence of the surface may induce enzymatic activity, as in the case of, e.g., certain lipases, but more frequently this induces an activity loss due to adsorption-induced conformational changes, accessibility restrictions (particularly important for enzymes with macromolecular substrates such as proteolytic enzymes), orientation-dependent activity losses, etc. Ellipsometry and reflectometry may provide valuable information on these dependencies, however, these techniques also offer possibilities to investigate interfacial degradation or synthesis processes in real time. Such studies were pioneered by Trurnit, who investigated the adsorption/reaction rates in chymotrypsin/ bovine serum albumin systems [162]. As another example of this type of investigation, Hahn et al. investigated the proteolytic degradation of gelatin layers by trypsin and krillase, the latter being an enzyme mixture extracted from Antarctic krill, by ellipsometry and TIRF [163]. Using ellipsometry, the enzymatic degradation of the preadsorbed gelatin layer on addition of krillase could be straightforwardly observed. For example, Fig. 14 shows the effect of adding krillase to gelatin preadsorbed at methylated silica. As can be seen addition of thermally deactivated krillase, which does not affect the surface activity of these proteins, results in only a marginal reduction in the total adsorbed amount. Addition of native enzyme, on the other hand, yielded a strong reduction in both the adsorbed amount and the adsorbed layer thickness. Hence, it is clear that the effect of krillase addition is due to degradation of the gelatin layer rather than interfacial exchange or a combination thereof. In the general case, however, these effects are probably not as straightforwardly decoupled.

The issue of what eventually causes the degradation process to come to a halt is an interesting and complex one. Krillase is known to undergo little autolysis, and

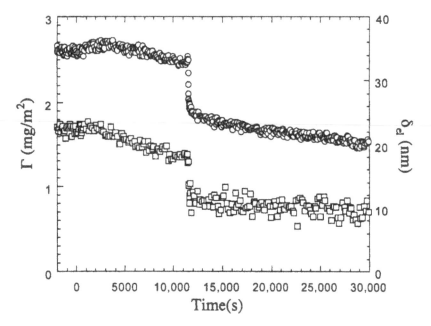

FIG. 14 The total adsorbed amount (circles) and mean adsorbed layer thickness (squares) versus time on methylated silica at 34°C after preadsorption of gelatin and addition of inactivated krillase to a final concentration of 0.01 mg/mL at $t = 0$. At $t = 11,400$ s native krillase was added to a final concentration of 0.01 mg/mL. Gelatin was preadsorbed from a 0.1 wt% solution during 3 hr followed by 1 hr of bulk rinsing, prior to the addition of krillase. (Data from Ref. 163.)

therefore the enzyme solution displays activity also at times long after the interfacial degradation has essentially come to a complete stop. Instead, a clue to the mechanism behind the saturation in the interfacial degradation process can be obtained from information on the adsorbed layer thickness and mean refractive index. As can be seen in Fig. 14, the decrease in the adsorbed amount is smaller than the corresponding decrease in the adsorbed layer thickness, which means that the adsorbed layer after (partial) degradation is denser than before. This is also what is found from the refractive index, which increases after exposure to krillase. Therefore, a possible mechanism behind the saturation behavior could be that the adsorbed gelatin layer is denser closer to the surface, and that a limited substrate accessibility slows down the degradation drastically. This mechanism is also supported by results from small-angle neutron scattering studies, showing adsorbed gelatin layers to be denser closer to the surface than in the outer part of the layer [164,165].

On the other hand, trypsin is an enzyme undergoing fast autolysis [166,167]. The effect of the autolysis, as well as the interplay between the autolysis and the protein–surface interaction is illustrated in Fig. 15. As can be seen, trypsin adsorption at silica displays a dramatic maximum, followed by an adsorbed layer decrease continuing to very low trypsin adsorbed amounts. Addition of a second injection of fresh trypsin at this point results in a new maximum identical to the first [163]. Furthermore, increasing the bulk trypsin concentration or lowering the temperature results in a longer time being required before the adsorbed amount decreases after

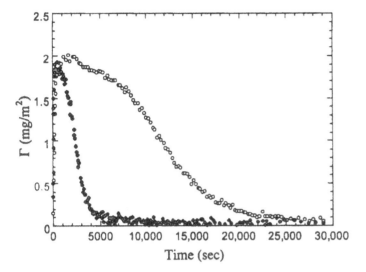

FIG. 15 Adsorption of trypsin at silica at 22°C (circles) and 34°C (diamonds). At $t = 0$ s trypsin was added to the cuvette to a final concentration of 0.01 mg/mL. (Data from Ref. 163.)

reaching the maximum. These findings all indicate that trypsin autolysis in solution is the main origin of the adsorbed amount decrease at longer times. At methylated silica, on the other hand, no such effect is observed. Instead, the adsorption follows the monotonous increase with time that is generally observed for proteins in the absence of competitive spreading or impurity effects. This indicates that the trypsin molecules adsorbed at the hydrophobic methylated silica surface do not respond to the decrease in the chemical potential of trypsin as a consequence of autolytic degradation. This agrees also with previous findings that proteins are more straightforwardly exchanged through both homogeneous and heterogeneous processes at hydrophilic surfaces such as silica or glass than at hydrophobic ones such as methylated silica or glass (see also above).

The effect of proteolytic enzymes on preadsorbed protein layers may also be used in order to obtain information on the structure of the adsorbed layer. For example, Kull et al. probed the adsorbed layer structures formed by β-casein at silica and methylated silica, through investigating how addition of endopeptidase Asp-N, which cleaves the N-terminal of the protein between residues [42,43,46,47], affects the adsorbed amount and adsorbed layer thickness of β-casein at these surfaces [15]. Based on the ellipsometric data obtained, the authors were able to suggest adsorbed layer structures for β-casein at silica and methylated silica. Analogous results were obtained from neutron reflectivity measurements [124]. Note, however, that although the "enzymatic sensor" approach seems to have worked well in this case, care should be taken when using multicomponent protein systems since these frequently display complex adsorption patterns, including competitive adsorption, associative adsorption, and multilayer adsorption. If any of these processes should occur in the type of experiment described above, interpretation of the experimental results in terms of structural parameters is precluded.

In principle, ellipsometry and reflectometry should also be well suited for following interfacial synthesis processes and in particular synthesis of macromolecules, e.g., polysaccharides or DNA. To the best of the author's knowledge, no such studies have been reported in the literature. However, given the importance of DNA and DNA polymerases in biotechnical applications one hopes this situation will soon be remedied.

ACKNOWLEDGMENT

This work was financed by the Institute for Research and Competence Holding AB (IRECO), Sweden, and KK-stiftelsen, Sweden.

REFERENCES

1. P. Drude, Ueber oberflächeschichten. I. Theil. Ann. Phys. Chem. 272:532–560, 1889.
2. P. Drude, Bestimmung der optischen constanten der Metalle. Ann. Phys. Chem. 275: 481–554, 1890.
3. R. M. A. Azzam and N. M. Bashara, *Ellipsometry and Polarized Light*. North-Holland, Amsterdam, 1989.
4. J. C. Dijt, M. A. Cohen Stuart, and G. J. Fleer, Reflectometry as a tool for adsorption studies. Adv. Colloid Interface Sci. *50*:79–101, 1994.
5. W. Lukosz, D. Clerc, Ph. M. Nellen, Ch. Stamm, and P. Weiss, Output grating couplers on planar optical waveguides as direct immunosensors. Biosens. Bioelectron. *6*:227–232, 1991.
6. D. C. Prieve and N. A. Frej, Total internal reflection microscopy: a quantitative tool for measurement of colloidal forces. Langmuir *6*:396–403, 1990.
7. S. Hénon and J. Meunier, Brewster angle microscopy and ellipsometry. In *Modern Characterization Methods of Surfactant Systems*. (B. P. Binks, ed.). Marcel Dekker, New York, 1999.
8. P. A. Cuypers, J. W. Corsel, M. P. Janssen, J. M. M. Kop, W. T. Hermens, and H. C. Hemker, The adsorption of prothrombin to phosphatidylserine multilayers quantified by ellipsometry. J. Biol. Chem. *258*:2426–2431, 1983.
9. J. A. de Feijter, J. Benjamins, and F. A. Veer, Ellipsometry as a tool to study the adsorption behavior of synthetic and biopolymers at the air–water interface. Biopolymers *17*:1759–1772, 1978.
10. M. Malmsten, Ellipsometry studies of protein layers adsorbed at hydrophobic surfaces. J. Colloid Interface Sci. *166*:333–342, 1994.
11. R. D. Tilton, Scanning angle reflectometry and its application to polymer adsorption and coadsorption with surfactants. In *Colloid–Polymer Interactions: From Fundamentals to Practice* (R. S. Farinato and P. L. Dubin, eds.). Wiley, New York, 1999.
12. M. Landgren and B. Jönsson, Determination of the optical properties of Si/SiO_2 surfaces by means of ellipsometry, using different ambient media. J. Phys. Chem. *97*: 1656–1660, 1993.
13. F. Tiberg and M. Landgren, Characterization of thin nonionic surfactant films at the silica/water interface by means of ellipsometry. Langmuir *9*:927–932, 1993.
14. T. Nylander, F. Tiberg, and N. M. Wahlgren, Evaluation of the structure of adsorbed layers of β-casein from ellipsometry and surface force measurements. Int. Diary J. *9*: 313–317, 1999.
15. T. Kull, T. Nylander, F. Tiberg, and N. M. Wahlgren, Effect of surface properties and

added electrolyte on the structure of β-casein layers adsorbed at the solid/liquid interface. Langmuir *13*:5141–5147, 1997.

16. D. Muller, M. Malmsten, S. Tanodekaew, and C. Booth, Adsorption of diblock copolymers of poly(ethylene oxide) and poly(lactide) at hydrophilic silica from aqueous solution. J. Colloid Interface Sci. *228*:317–325, 2000.

17. P. Schaaf, P. Déjardin, and A. Schmitt, Reflectometry as a technique to study the adsorption of human fibrinogen at the silica/solution interface. Langmuir *3*:1131–1135, 1987.

18. F. A. M. Leermakers and A. P. Gast, Block copolymer adsorption studied by dynamic scanning angle reflectometry. Macromolecules *24*:718–730, 1991.

19. D. Beaglehole, E. Z. Radlinska, B. W. Ninham, and H. K. Christenson, Molecular layering in a liquid adsorbed film at room temperature. Langmuir *7*:1843–1845, 1991.

20. C. R. Helms and B. E. Deal, eds., *The Physics and Chemistry of SiO₂ and the Si–SiO₂ Interface* 2. Plenum Press, New York, 1993.

21. B. A. Ivarsson, P.-O. Hegg, I. Lundström, and U. Jönsson, Adsorption of proteins on metal surfaces studied by ellipsometric and capacitance measurements. Colloids Surf. *13*:169–192, 1985.

22. P. A. Cuypers, J. W. Corsel, M. P. Janssen, J. M. M. Kop, H. C. Hemker, and W. T. Hermens, Quantitative ellipsometry of protein adsorption at solid–liquid interfaces. In *Surfactants in Solution*, Vol. 2 (K. L. Mittal and B. Lindman, eds.). Plenum Press, New York, 1984.

23. T. Arnebrant and T. Nylander, Adsorption of insulin on metal surfaces in relation to association behaviour. J. Colloid Interface Sci. *122*:557–566, 1988.

24. B. W. Morrissey, L. E. Smith, R. R. Stromberg, and C. A. Fenstermaker, Ellipsometric investigation of the effect of potential on blood protein conformation and adsorbance. J. Colloid Interface Sci. *56*:557–563, 1976.

25. P. A. Cuypers, W. T. Hermens, and H. C. Hemker, Ellipsometry as a tool to study protein films at liquid–solids interfaces. Anal. Biochem. *84*:56–67, 1978.

26. J. Mårtensson and H. Arwin, Interpretation of spectroscopic ellipsometry data on protein layers on gold including substrate–layer interactions. Langmuir *11*:963–968, 1995.

27. J. Mårtensson, H. Arwin, H. Nygren, and I. Lundström, Adsorption and optical properties of ferritin layers on gold studied with spectroscopic ellipsometry. J. Colloid Interface Sci. *174*:79–85, 1995.

28. T. Nylander, T. Arnebrant, and P.-O. Glantz, Interactions between salivary films adsorbed on mica surfaces. Colloids Surf. A *129–130*:339–344, 1997.

29. T. Nylander, T. Arnebrant, R. E. Baier, and P.-O. Glantz, Interactions between layers of salivary acidic proline rich protein 1 (PRP 1) adsorbed on mica surfaces. Prog. Colloid Polymer Sci. *108*:34–39, 1998.

30. N. Vassilakos, T. Arnebrant, and P.-O. Glantz, Adsorption of whole saliva onto hydrophilic and hydrophobic solid surfaces. The influence of concentration, ionic strength and pH. Scand. J. Dent. Res. *100*:346–353, 1992.

31. L. Lindh, T. Arnebrant, P.-E. Isberg, and P.-O. Glantz, Concentration dependence of adsorption from human whole resting saliva at solid/liquid interfaces: an ellipsometric study. Biofouling *14*:189–196, 1999.

32. I. C. Hahn, U. M. Elofsson, A. Joiner, J. Malmsten, and T. Arnebrant, Salivary protein adsorption onto hydroxyapatite and SDS-mediated elution studied in situ by ellipsometry. Biofouling *17*:173–187, 2001.

33. T. Ericson, K. M. Pruitt, H. Arwin, and I. Lundström, Ellipsometric studies of film formation on tooth enamel and hydrophilic silicon surfaces. Acta Odontol. Scand. *40*: 197–201, 1982.

34. L. Bergström, A. Meurk, H. Arwin, and D. J. Rowcliffe, Estimation of Hamaker con-

stants of ceramic materials from optical data using Lifshitz theory. J. Am. Chem. Soc. 79:339–348, 1996.

35. M. Malmsten and F. Tiberg, Adsorption of ethyl(hydroxyethyl)cellulose at polystyrene. Langmuir 9:1098–1103, 1993.

36. F. Tiberg, C. Brink, M. Hellsten, and K. Holmberg, Immobilization of protein to surface-grafted PEO/PPO block copolymers. Colloid Polym. Sci. 270:1188–1193, 1992.

37. M. Malmsten, B. Lassen, J. Westin, C.-G. Gölander, R. Larsson, and U. R. Nilsson, Adsorption of complement protein C3 at polymer surfaces. J. Colloid Interface Sci. 179:163–172, 1996.

38. A. V. Elgersma, R. L. J. Zsom, J. Lyklema, and W. Norde, Kinetics of single and competitive protein adsorption studied by reflectometry and streaming potential measurements. Colloids Surf. 65:17–28, 1992.

39. J. W. Th. Lichtenbelt, W. J. M. Heuvesland, M. E. Oldenzeel, and R. L. J. Zsom, Adsorption and immunoreactivity of proteins on polystyrene and on silica. Competition with surfactants. Colloids Surf. B 1:75–82, 1993.

40. K. L. Mittal, ed., Silanes and Other Coupling Agents. VSP, Zeist, 1992.

41. H. Elwing, A. Askendal, and I. Lundström, Protein exchange reactions on solid surfaces studied with a wettability gradient method. Prog. Colloid Polym. Sci. 74:103–107, 1987.

42. Y. S. Lin, V. Hlady, and C.-G. Gölander, The surface density gradient of grafted poly(ethylene glycol): preparation, characterization, and protein adsorption. Colloids Surf. B 3:49–62, 1994.

43. S. Welin-Klintström, R. Jansson, and H. Elwing, An off-null ellipsometer with lateral scanning capability for kinetic studies at liquid–solid interfaces. J. Colloid Interface Sci. 157:498–503, 1993.

44. L. Lui and H. Elwing, Complement activation on solid surfaces as determined by C3 deposition and hemolytic consumption. J. Biomed. Mater. Res. 28:767–773, 1994.

45. M. C. Wahlgren, T. Arnebrant, A. Askendal, and S. Welin-Klintström, The elutability of fibrinogen by sodium dodecyl sulphate and alkyltrimethylammonium bromides. Colloids Surf. A 70:151–158, 1993.

46. C. D. Bain and G. M. Whitesides, Modeling organic surfaces with self-assembled monolayers. Angew Chem. Int. Ed. Engl. 28:506–512, 1989.

47. R. G. Chapman, E. Ostuni, L. Yan, and G. M. Whitesides, Preparation of mixed self-assembled monolayers (SAMs) that resist adsorption of proteins using the reaction of amines with a SAM that presents interchain carboxylic anhydride groups. Langmuir 16:6927–6936, 2000.

48. N. Horan, L. Yan, H. Isobe, G. M. Whitesides, and D. Kahne, Nonstatistical binding of a protein to a clustered carbohydrates. PNAS 96:11782–11786, 1999.

49. E. Ostuni, L. Yan, and G. M. Whitesides, The interaction of proteins and cells with self-assembled monolayers of alkanethiolates on gold and silver. Colloids Surf. B 15:3–30, 1999.

50. K. L. Prime and G. M. Whitesides, Adsorption of proteins onto surfaces containing end-attached oligo(ethylene oxide): a model system using self-assembled monolayers. J. Am. Chem. Soc. 115:10714–10721, 1993.

51. C. Roberts, C. S. Chen, M. Mrksich, V. Martichonok, D. E. Ingber, and G. M. Whitesides, Using mixed self-assembled monolayers presenting RGD and (EG)$_3$OH groups to characterize long-term attachment of bovine capillary endothelial cells to surfaces. J. Am. Chem. Soc. 120:6548–6555, 1998.

52. H. Yasuda, Plasma Polymerization. Academic Press, Orlando, 1985.

53. B. Lassen and M. Malmsten, Competitive protein adsorption at plasma polymer surfaces. J. Colloid Interface Sci. 186:9–16, 1997.

54. B. Lassen and M. Malmsten, Structure of protein layers during competitive adsorption. J. Colloid Interface Sci. *180*:339–349, 1996.

55. B. Lassen and M. Malmsten, Competitive protein adsorption studied with TIRF and ellipsometry. J. Colloid Interface Sci. *179*:470–477, 1996.

56. S. I. Jeon, J. H. Lee, J. D. Andrade, and P. G. de Gennes, Protein–surface interactions in the presence of polyethylene oxide. I. Simplified theory. J. Colloid Interface Sci. *142*:149–158, 1991.

57. S. I. Jeon and J. D. Andrade, Protein–surface interactions in the presence of polyethylene oxide. II. Effect of protein size. J. Colloid Interface Sci. *142*:159–166, 1991.

58. M. Malmsten, K. Emoto, and J. M. Van Alstine, Effect of chain density on inhibition of protein adsorption by poly(ethylene glycol) based coatings. J. Colloid Interface Sci. *202*:507–517, 1998.

59. A. Halperin. Polymer brushes that resist adsorption of model proteins: design parameters. Langmuir *15*:2525–2533, 1999.

60. G. Greber and O. Pashinger, Silylderivate der cellulose. Das Papier *35*:547–554, 1981.

61. G. K. Cooper, K. R. Sandberg, and J. F. Hinck, Trimethylsilyl cellulose as precursor to regenerated cellulose fiber. J. Appl. Polymer Sci. *26*:3827–3836, 1981.

62. M. Holmberg, J. Berg, S. Stemme, L. Ödberg, J. Rasmusson, and P. Claesson, Surface force studies of Langmuir-Blodgett cellulose films. J. Colloid Interface Sci. *186*:369–381, 1997.

63. M. Österberg and P. M. Claesson, Interactions between cellulose surfaces: effect of solution pH. J. Adhesion Sci. Technol. *14*:603–618, 2000.

64. M. Österberg, R. Wigren, R. Erlandsson, and P. M. Claesson, Interactions between cellulose and colloidal silica in the presence of polyelectrolytes. Colloids Surf. A *129–130*:175–183, 1997.

65. L. Bergström, S. Stemme, T. Dahlfors, H. Arwin, and L. Ödberg, Spectroscopic ellipsometry characterization and estimation of the Hamaker constant of cellulose. Cellulose *6*:1–13, 1999.

66. V. Buchholz, G. Wegner, S. Stemme, and L. Ödberg, Regeneration, derivatization and utilization of cellulose in ultrathin films. Adv. Mater. *8*:399–402, 1996.

67. M. Malmsten, Ellipsometry studies of protein adsorption at lipid surfaces. J. Colloid Interface Sci. *168*:247–254, 1994.

68. M. Malmsten, Protein adsorption at phospholipid surfaces. J. Colloid Interface Sci. *172*:106–115, 1995.

69. M. Malmsten and B. Lassen, Competitive protein adsorption at phospholipid surfaces. Colloids Surf. B *4*:173–184, 1995.

70. P. A. Cuypers, J. W. Corsel, M. P. Janssen, J. M. M. Kop, W. T. Hermens, and H. C. Hemker, The adsorption of prothrombin to phosphatidylserine multilayers quantified by ellipsometry. J. Biol. Chem. *258*:2426–2431, 1983.

71. H. A. M. Andree, W. T. Hermens, and G. M. Willems, Testing protein adsorption models by off-null ellipsometry: determination of binding constants from a single adsorption curve. Colloids Surf. A *78*:133–141, 1993.

72. P. L. A. Giesen, G. M. Willems, H. C. Hemker, and W. T. Hermens, Membrane-mediated assembly of the prothrombinase complex. J. Biol. Chem. *266*:18720–18725, 1991.

73. P. L. A. Giesen, G. M. Willems, H. C. Hemker, M. C. A. Stuart, and W. T. Hermens, Monitoring of unbound protein in vesicle suspension with off-null ellipsometry. Biochim. Biophys. Acta *1147*:125–131, 1993.

74. J. W. Corsel, G. M. Willems, J. M. M. Kop, P. A. Cuypers, and W. T. Hermens, The role of intrinsic binding rate and transport rate in the adsorption of prothrombin, albumin, and fibrinogen at phospholipid bilayers. J. Colloid Interface Sci. *111*:544–554, 1986.

75. J. M. M. Kop, P. A. Cuypers, T. Lindhout, H. C. Hemker, and W. T. Hermens, The adsorption of prothrombin to phospholipid monolayers quantitated by ellipsometry. J. Biol. Chem. *259*:13993–13998, 1984.

76. M. Malmsten, Serum protein adsorption at phospholipid surfaces in relation to intravenous drug deliver. Colloids Surf. A *159*:77–87, 1999.

77. R. M. Swart, Monolayers and multilayers of biomolecules. In *Langmuir–Blodgett Films* (G. Roberts, ed.). Plenum Press, New York, 1990, pp. 273–316.

78. F. Tiberg, I. Harwigsson, and M. Malmsten, Formation of model lipid bilayers at the silica–water interface by co-adsorption with non-ionic dodecyl maltoside surfactant. Eur. Biophys. J. *29*:196–203, 2000.

79. J. Fagefors, K. Wannerberger, T. Nylander, and O. Söderman, Adsorption of soybean phosphatidylcholine/N-dodecyl β-D-maltoside dispersions at liquid/solid and liquid/air interfaces. Prog. Colloid Polym. Sci. (in press).

80. A. Chonn and P. R. Cullis, Ganglioside GM1 and hydrophilic polymers increase liposome circulation times by inhibiting the association of blood proteins. J. Liposomes Res. *2*:397–410, 1992.

81. A. Chonn, S. C. Semple, and P. R. Cullis, Association of blood proteins with large unilamellar liposomes in vivo. J. Biol. Chem. *267*:18759–18765, 1992.

82. A. Chonn, S. C. Semple, and P. R. Cullis, Separation of large unilamellar liposomes from blood components by a spin column procedure: Towards identifying plasma proteins which mediate liposome clearance in vivo. Biochim. Biophys. Acta *1070*:215–222, 1991.

83. R. M. A. Azzam, Use of a light beam to probe the cell surface in vitro. Surf. Sci. *56*:126–133, 1976.

84. G. Siegel, M. Malmsten, D. Klüssendorf, and W. Leonhardt, Physicochemical binding properties of the proteoglycan receptor for serum lipoproteins. Atherosclerosis *144*:59–67, 1999.

85. M. Malmsten, G. Siegel, and W. G. Wood, Ellipsometry studies of lipoprotein adsorption. J. Colloid Interface Sci. *224*:338–346, 2000.

86. M. Malmsten and G. Siegel, A model substrate for ellipsometry and microscopy studies of lipoprotein deposition at the endothelium. J. Colloid Interface Sci. *240*:372–374, 2001.

87. K. Schillén, P. M. Claesson, M. Malmsten, P. Linse, and C. Booth, Properties of poly(ethylene oxide)–poly(butylene oxide) diblock copolymers at the interface between hydrophobic surfaces and water. J. Phys. Chem. B *101*:4238–4252, 1997.

88. M. R. Böhmer, L. K. Koopal, R. Janssen, E. M. Lee, R. K. Thomas, and A. R. Rennie, Adsorption of nonionic surfactants on hydrophilic surfaces. An experimental and theoretical study on association in the adsorbed layer. Langmuir *8*:2228–2239, 1992.

89. A. Gellan and C. H. Rochester, Adsorption of *n*-alkylpolyethylene glycol non-ionic surfactants from aqueous solution to silica. J. Chem. Soc. Faraday Trans. *181*:2235–2245, 1985.

90. E. M. Lee, R. K. Thomas, P. G. Cummings, E. J. Staples, J. Penfold, and A. R. Rennie, Determination of the structure of a surfactant layer adsorbed at the silica/water interface by neutron reflection. Chem. Phys. Lett. *162*:196–202, 1989.

91. D. C. McDermott, J. R. Lu, E. M. Lee, R. K. Thomas, and A. R. Rennie, Study of the adsorption from aqueous solution of hexaethylene glycol monododecyl ether on silica substrates using the technique of neutron reflection. Langmuir *8*:1204–1210, 1992.

92. M. Rutland and T. J. Senden, Adsorption of the poly(oxyethylene) nonionic surfactant $C_{12}E_5$ to silica: a study using atomic force microscopy. Langmuir *9*:412–418, 1993.

93. F. Tiberg. Physical characterization of nonionic surfactant layers adsorbed at hydrophilic and hydrophobic surfaces by time-resolved ellipsometry. J. Chem. Soc. Faraday Trans. *92*:531–538, 1996.

94. M. Malmsten, P. Linse, and T. Cosgrove, Adsorption of PEO-PPO-PEO block copolymers at silica. Macromolecules 25:2474–2481, 1992.

95. T. J. Barnes and C. A. Prestidge, PEO-PPO-PEO block copolymers at the emulsion droplet–water interface. Langmuir 16:4116–4121, 2000.

96. C. Wu, T. Liu, H. White, and B. Chu, Atomic force microscopy study of $E_{99}P_{69}E_{99}$ triblock copolymer chains on silicon surface. Langmuir 16:656–661, 2000.

97. D. Muller, Modification of surface properties through polymer adsorption. Ph.D. Thesis, University of Lund, 2000.

98. P. W. Wojciechowski and J. L. Brash, Fibrinogen and albumin adsorption from human blood plasma onto chemically functionalized silica substrates. Colloids Surf. B 1:107–117, 1993.

99. H. G. W. Lensen, D. Bargeman, P. Bergveld, C. A. Smolders, and J. Feijen, High-performance liquid chromatography as a technique to measure the competitive adsorption of plasma proteins onto latices. J. Colloid Interface Sci. 99:1–8, 1984.

100. T. J. Su, J. R. Lu, R. K. Thomas, Z. F. Cui, and J. Penfold, The effect of solution pH on the structure of lysozyme layers adsorbed at the silica–water interface studied by neutron reflection. Langmuir 14:438–445, 1998.

101. V. Hlady, J. Rickel, and J. D. Andrade, Fluorescence of adsorbed layers. II. Adsorption of human lipoproteins studied with total internal reflection intrinsic fluorescence. Colloids Surf. 34:171–183, 1988/89.

102. B. W. Morrissey and C. Han, The conformation of γ-globulin adsorbed on polystyrene latices determined by quasielastic light scattering. J. Colloid Interface Sci. 65:423–431, 1978.

103. M. Malmsten, B. Lassen, K. Holmberg, V. Thomas, and G. Quash, Effects of hydrophilization and immobilization on the interfacial behaviour of immunoglobulins. J. Colloid Interface Sci. 177:70–78, 1996.

104. C. M. Marques, J. M. Frigerio, and J. Rivory, Effective ellipsometric thickness of an interfacial layer. J. Opt. Soc. Am. B 8:2523–2528, 1991.

105. J. C. Carmet and P. G. de Gennes, Ellipsometric formulas for an inhomogeneous layer with arbitrary refractive index profile. J. Opt. Soc. Am. 73:1777–1784, 1983.

106. G. J. Fleer, M. A. Cohen Stuart, J. M. H. M. Scheutjens, T. Cosgrove, and B. Vincent, *Polymers at Interfaces*, 1st ed. Chapman & Hall, London, 1993.

107. Y. Lvov and H. Möhwald, eds., *Protein Architecture: Interfacing Molecular Assemblies and Immobilization Biotechnology*. Marcel Dekker, New York, 2000.

108. M. Malmsten, Ellipsometry studies of the effects of surface hydrophobicity on protein adsorption. Colloids Surf. B 3:297–308, 1995.

109. E. Blomberg, P. M. Claesson, J. C. Fröberg, and R. D. Tilton, Interaction between adsorbed layers of lysozyme studied with the surface force technique. Langmuir 10:2325–2334, 1994.

110. M. Radmacher, M. Fritz, J. P. Cleveland, D. A. Walters, and P. K. Hansman, Imaging adhesion forces and elasticity of lysozyme adsorbed on mica with the atomic force microscope. Langmuir 10:3809–3814, 1994.

111. S. F. Oppenheim, J. O. Rich, G. R. Buettner, and V. G. J. Rodgers, Protein structure change on adherence to ultrafiltration membranes: an examination by electron paramagnetic resonance spectroscopy. J. Colloid Interface Sci. 183:274–279, 1996.

112. I. Carlstedt, H. Lindgren, and J. K. Sheehan, The macromolecular structure of human cervical-mucus glycoproteins. Biochem. J. 213:427–435, 1983.

113. G. Siegel, A. Walter, M. Bostanjoglo, A. W. H. Jans, R. Kinne, L. Piculell, and B. Lindman, Ion transport and cation–polyanion interactions in vascular biomembranes. J. Membrane Sci. 41:353–375, 1989.

114. M. Djabourov, Architecture of gelatin gels. Contemp. Phys. 29:273–297, 1988.

115. M. Malmsten and G. Siegel, Electrolyte effects on proteoheparan sulphate adsorption. Uzbek J. Phys. 3:7–15, 2001.
116. M. Malmsten, P. Claesson, and G. Siegel, Forces between proteoheparan sulfate layers adsorbed at hydrophobic surfaces. Langmuir 10:1274–1280, 1994.
117. C. Poncet, F. Tiberg, and R. Audebert, Ellipsometric study of the adsorption of hydrophobically modified polyacrylates at hydrophobic surfaces. Langmuir 14:1697–1704, 1998.
118. L. Heinrich, E. K. Mann, J. C. Voegel, G. J. M. Koper, and P. Schaaf, Scanning angle reflectometry study of the structure of antigen–antibody layers adsorbed on silica surfaces. Langmuir 12:4857–4865, 1996.
119. A. P. Rolland, From genes to gene medicines: recent advances in nonviral gene delivery. Crit. Rev. Ther. Drug Carrier Syst. 15:143–198, 1998.
120. S. M. Melnikov, V. G. Sergeyev, and K. Yoshikawa, Cooperation between salt induced globule–coil transition in single duplex DNA complexed with cationic surfactant and sphere–rod transition of surfactant micelles. Prog. Colloid Polym. Sci. 106:209–214, 1997.
121. S. M. Melnikov, V. G. Sergeyev, and K. Yoshikawa, Transition of double-stranded DNA chains between random coil and compact globule states induced by cooperative binding of cationic surfactant. J. Am. Chem. Soc. 117:9951–9956, 1995.
122. K. Eskilsson, C. Leal, B. Lindman, M. Miguel, and T. Nylander, DNA surfactant complexes at solid surfaces. Langmuir 17:1666–1669, 2001.
123. U. M. Elofsson, M. A. Paulsson, and T. Arnebrant, Adsorption of β-lactoglobulin A and B in relation to self-association: effect of concentration and pH. Langmuir 13:1695–1700, 1997.
124. T. Nylander, F. Tiberg, T.-J. Su, J. R. Lu, and R. K. Thomas, β-Casein adsorption at the hydrophobized silicon oxide–aqueous solution interface and the effect of added electrolyte. Biomacromolecules (in press).
125. P. Levitz, Aggregative adsorption of nonionic surfactants onto hydrophilic solid/water interface. Relation with bulk micellization. Langmuir 7:1595–1608, 1991.
126. P. Levitz, H. van Damme, and D. Keravis, Fluorescence decay of the adsorption of nonionic surfactants at the solid–liquid interface. 1. Structure of the adsorption layer on a hydrophilic solid. J. Phys. Chem. 88:2228–2235, 1984.
127. M. R. Böhmer and L. K. Koopal, Association and adsorption of nonionic flexible chain surfactants. Langmuir 6:1478–1484, 1990.
128. L. M. Grant, T. Edert, and F. Tiberg, Influence of surface hydrophobicity on the layer properties of adsorbed nonionic surfactants. Langmuir 16:2285–2291, 2000.
129. K. Eskilsson, L. M. Grant, P. Hansson, and F. Tiberg, Self-aggregation of triblock copolymers at the solid silica–water interface. Langmuir 15:5150–5157, 1999.
130. J. Mårtensson, H. Arwin, I. Lundström, and T. Ericson, Adsorption of lactoperoxidase on hydrophilic and hydrophobic silicon dioxide surfaces: an ellipsometric study. J. Colloid Interface Sci. 155:43–47, 1993.
131. U. Jönsson and M. Malmqvist, Real time biospecific interaction analysis: the integration of surface plasmon resonance detection, general biospecific interface chemistry and microfluidics into one analytical system. Adv. Biosensors 2:291–336, 1992.
132. H. Elwing, B. Ivarsson, and I. Lundström, Complement deposition from human sera on silicon surfaces studied in situ by ellipsometry: the influence of surface wettability. Eur. J. Biochem. 156:359–365, 1986.
133. F. Tiberg, B. Jönsson, and B. Lindman, Ellipsometry studies of the self-assembly of nonionic surfactants at the silica–water interface: kinetic aspects. Langmuir 10:3714–3722, 1994.
134. J. Brinck, B. Jönsson, and F. Tiberg, Kinetics of nonionic surfactant adsorption and desorption at the silica water interface: one component. Langmuir 14:1058–1071, 1998.

135. J. Brinck, B. Jönsson, and F. Tiberg, Kinetics of nonionic surfactant adsorption and desorption at the silica water interface: binary systems. Langmuir *14*:5863–5876, 1998.

136. J. Brinck and F. Tiberg, Adsorption behaviour of two binary nonionic surfactant systems at the silica–water interface. Langmuir *12*:5042–5047, 1996.

137. J. C. Dijt, M. A. Cohen Stuart, J. E. Hofman, and G. J. Fleer, Kinetics of polymer adsorption in stagnation point flow. Colloids Surf. *51*:141–158, 1990.

138. M. C. P. van Eijk and M. A. Cohen Stuart, Polymer adsorption kinetics: effects of supply rate. Langmuir *13*:5447–5450, 1997.

139. M. Malmsten, Ellipsometry studies of fibronectin adsorption. Colloids Surf. B *3*:371–381, 1995.

140. M. Malmsten, B. Bergenståhl, M. Masquelier, M. Pålsson, and C. Peterson, Adsorption of apolipoprotein B at phospholipid model surfaces. J. Colloid Interface Sci. *172*:485–493, 1995.

141. M. Malmsten, B. Lassen, J. M. Van Alstine, and U. R. Nilsson, Adsorption of complement proteins C3 and C1q. J. Colloid Interface Sci. *178*:123–134, 1996.

142. L. Heinrich, E. K. Mann, J. C. Voegel, and P. Schaaf, Characterization of thin protein films through scanning angle reflectometry. Langmuir *13*:3177–3186, 1997.

143. P. Schaaf, P. Déjardin, and A. Schmitt, Reflectometry as a technique to study the adsorption of human fibrinogen at the silica/solution interface. Langmuir *3*:1131–1135, 1987.

144. S. Stemme, L. Ödberg, and M. Malmsten, Effect of colloidal silica and electrolyte on the structure of an adsorbed cationic polyelectrolyte layer. Colloids Surf. A *155*:145–154, 1999.

145. C.-S. Lee and G. Belfort, Changing activity of ribonuclease A during adsorption: a molecular explanation. Proc. Natl. Acad. Sci. USA *86*:8392–8396, 1989.

146. A. Kondo and J. Mihara, Comparison of adsorption and conformation of hemoglobin and myoglobin on various inorganic ultrafine particles. J. Colloid Interface Sci. *177*: 214–221, 1996.

147. A. Kondo, F. Murakami, M. Kawagoe, and K. Higashitani, Kinetic and circular dichroism studies of enzymes adsorbed on ultrafine silica particles. Appl. Microbiol. Biotechnol. *39*:726–731, 1993.

148. K. Wannerberger and T. Arnebrant, Comparison of the adsorption and activity of lipases from *Humicola languinosa* and *Candida antarctica* on solid surfaces. Langmuir *13*: 3488–3493, 1997.

149. (a) K. Wannerberger and T. Arnebrant, Lipases from *Humicola lanuginosa* adsorbed to hydrophobic surfaces—desorption and activity after addition of surfactants. Colloids Surf. B *7*:153–164, 1996. (b) K. Wannerberger, S. Welin-Klintström, and T. Arnebrant, Activity and adsorption of lipase from *Humicola lanuginosa* on surfaces with different wettabilities. Langmuir *13*:784–790, 1997.

150. L. Vroman and A. L. Adams, Identification of rapid changes at plasma–solid interfaces. J. Biomed. Mater. Res. *3*:43–67, 1969.

151. E. F. Leonard and L. Vroman, Is the Vroman effect of importance in the interaction of blood with artificial materials? J. Biomater. Sci. Polym. Ed. *3*:95–107, 1991.

152. S. M. Slack and T. A. Horbett, Changes in the strength of fibrinogen attachment to solid surfaces: an explanation of the influence of surface chemistry on the Vroman effect. J. Colloid Interface Sci. *133*:148–165, 1989.

153. T. A. Horbett and J. L. Brash, eds., *Proteins at Interfaces*. II. *Fundamentals and Applications*. ACS Symposium Series 602, American Chemical Society, Washington, D.C., 1995.

154. A. V. Elgersma, R. L. J. Zsom, J. Lyklema, and W. Norde, Adsorption competition between albumin and monoclonal immunogammaglobulins on polystyrene latices. J. Colloid Interface Sci. *152*:410–428, 1992.

155. P. Tengwall, A. Askendal, and I. Lundström, Temporal studies on the deposition of complement on human colostum IgA and serum IgG immobilized on methylated silicon. J. Biomed. Mater. Res. *35*:81–92, 1997.

156. (a) M. Malmsten, D. Muller, and B. Lassen, Sequential adsorption of human serum albumin (HSA), immunoglobulin G (IgG), and fibrinogen (Fgn) at HMDSO plasma polymer surfaces. J. Colloid Interface Sci. *193*:88–95, 1997. (b) M. Malmsten and B. Lassen, Competitive adsorption at hydrophobic surfaces from binary protein systems. J. Colloid Interface Sci. *166*:490–498, 1994.

157. K. Spaeth, A. Brecht, and G. Gauglitz, Studies on the biotin–avidin multilayer adsorption by spectroscopic ellipsometry. J. Colloid Interface Sci. *196*:128–135, 1997.

158. E. D. Goddard and K. P. Ananthapadmanabhan, eds., *Interactions of Surfactants with Polymers and Surfactants*. CRC Press, Boca Raton, FL, 1993.

159. W. Norde, Adsorption of proteins from solution at the solid–liquid interface. Adv. Colloid Interface Sci. *25*:267–340, 1986.

160. C. A. Haynes and W. Norde, Globular proteins at solid/liquid interfaces. Colloids Surf. B *2*:517–566, 1994.

161. D. Muller, M. Malmsten, B. Bergenståhl, J. Hessing, J. Olijve, and F. Mori, Competitive adsorption of gelatin and sodium dodecylbenzenesulfonate at hydrophobic surfaces. Langmuir *14*:3107–3114, 1998.

162. H. J. Trurnit, Studies on enzyme systems at a solid–liquid interface. II. The kinetics of adsorption and reaction. Arch. Biochem. Biophys. *51*:176–199, 1954.

163. C. Hahn, D. Muller, T. Arnebrant, and M. Malmsten, Ellipsometry and TIRF studies of enzymatic degradation of interfacial proteinaceous layers. Langmuir *17*:1641–1652, 2001.

164. T. Cosgrove, J. H. E. Hone, A. M. Howe, and R. K. Heenan, A small-angle neutron scattering study of the structure of gelatin at the surface of polystyrene latex particles. Langmuir *14*:5376–5382, 1998.

165. C. N. Likos, K. A. Vaynberg, H. Löwen, and N. J. Wagner, Colloidal stabilization by adsorbed gelatin. Langmuir *16*:4100–4108, 2000.

166. R. L. Smith and E. Shaw, Pseudotrypsin: a modified bovine trypsin produced by limited autodigestion. J. Biol. Chem. *244*:4704–4712, 1969.

167. S. Maroux and P. Desnuelle, On some autolyzed derivatives of bovine trypsin. Biochim. Biophys. Acta *181*:59–72, 1969.

22

Surface Plasmon Resonance Spectroscopies for Protein Binding Studies at Functionalized Surfaces

EVA-KATHRIN SINNER   Max-Planck Institute for Biochemistry, Martinsried, Germany

KAZUTOSHI KOBAYASHI   Hitachi Chemical Company, Ltd., Ibaraki, Japan

THOMAS LEHMANN   Wacker Polymer Systems, Burghausen, Germany

THOMAS NEUMANN, FANG YU, and WOLFGANG KNOLL
Max-Planck Institute for Polymer Research, Mainz, Germany

BIRGIT PREIN   Graz University of Technology, Graz, Austria

JÜRGEN RÜHE   University of Freiburg, Freiburg, Germany

I. INTRODUCTION

Surface plasmon spectroscopy (SPS) has matured into a rather widespread surface optical technique with numerous examples demonstrating its potential for interfacial studies and thin film characterization [1]. This gain in popularity originates, in part, also from the availability of commercial instruments [2] which helped to attract customers from new communities. Foremost, however, it was recognized that surface plasmons constitute an interfacial light source in an evanescent wave format that offers a number of advantages over complementary optical techniques, e.g., ellipsometry. Despite the high sensitivity of SPS for monitoring, on-line and in real time, small changes in the thickness of surface coatings, e.g., by adsorption/desorption processes, it is a relatively simple technique. Essentially, one needs a linearly polarized (laser) light source; a coupling element, typically a glass prism or a grating structure, both coated with a thin (noble) metal layer as an optical resonator and in contact to a measuring cell; and a photodetector module. Thickness changes in the angstrom range can be easily evaluated quantitatively with a time resolution (in extreme cases) down to nanoseconds [3]. For laterally heterogeneous thin film samples a spatial resolution—in a microscopic mode—of a few micrometers can be obtained [4].

The fact that measurements in air (or vacuum) are as easily performed as at a solid–liquid interface gives the method its particular value and attraction in the context of biorecognition and binding studies with functional surfaces in contact to aqueous solutions of biomolecules [5]. In this brief summary, we concentrate on examples of studies with proteins adsorbing and/or binding from solution to either nonspecific surface "sites" or to surface-attached ligands for which they have a particular affinity.

We discuss such association/dissociation studies with SPS for four different sample formats: the first one is an example of a polymer brush of poly(ethyloxazoline) grown by the "grafting-from" technique at a Au surface, rendering this substrate, for sufficiently long end-grafted polymer chains, almost completely inert for nonspecific adsorption of fibrinogen from solution. In the second example, some of the thiol derivatives of a self-assembled monolayer (SAM) carry the heme group of the protein cytochrome b_{562} (cyt b_{562}) as a specific binding site. We will demonstrate the reversible reconstitution of this surface-attached prosthetic group of cyt b_{562} with its apoprotein from the aqueous phase. The next example concerns the interaction of a DNA binding protein with the single and double strands of oligonucleotides coupled to the Au surface via the biotin–streptavidin affinity pair. And, finally, we demonstrate the potential of surface plasmon optics for the study of protein binding to membrane-integral components, in our case the binding of recombinant yeast oxysterol binding proteins to different oxidized sterol molecules incorporated into a lipid bilayer tethered to the Au surface [6].

In addition to the "classical" use of SPS we describe a novel extension of the method surface plasmon fluorescence spectroscopy (SPFR), i.e., the combination of fluorescence detection schemes with the field-enhancement mechanisms operating at resonant surface plasmon excitation [7]. This concept has recently been shown to offer substantial sensitivity enhancements, e.g., for monitoring hybridization reactions at solid–solution interfaces [8]. We demonstrate here that it also provides specific advantages for studies of protein binding.

II. SURFACE PLASMON SPECTROSCOPY AND SURFACE PLASMON FIELD-ENHANCED FLUORESCENCE SPECTROSCOPY

A plasmon surface plariton [or surface plasmon (SP) for short] [9] can be excited at a metal–dielectric interface, typically by a monochromatic light source, e.g., by a helium–neon laser. In the setup given in Fig. 2, it is recorded as a sharp and deep minimum in the intensity of p-polarized light reflected off the base of a metal-coated coupling prism as the angle of incidence θ is scanned (cf. also Fig. 3). One can think of a plasmon as light which is bound to and propagates along a metal surface with the typical characteristics of an electromagnetic mode. The fields associated with such a surface wave extend into the media adjacent to the interface, decaying exponentially away from it (cf. Fig. 1). Consequently, SPs are sensitive to changes in the optical architecture near the interface and hence offer great potential as sensing probes. Surface plasmons already have been shown to be useful for gas sensing [10], biosensing [11], immunosensing [12], and for electrochemical studies [13].

An important feature of surface plasmons is their electromagnetic field enhancement at the interface compared with the incoming light [9]. This enhancement can reach a factor of 100 for a smooth flat Ag surface ($\lambda = 633$ nm), and suggests the use of SPs for surface spectroscopies, like surface plasmon field-enhanced Raman spectroscopy [14] or fluorescence spectroscopy [7,15]. At resonant excitation of a surface plasmon mode the emission from fluorescently labeled molecules—binding to the surface—is strongly enhanced due to the high electromagnetic field associated with the surface plasmon.

A. Surface Plasmons

A surface plasmon is an electromagnetic mode coupled to a charge-density oscillation, propagating along the interface between a metal and a dielectric. The electromagnetic field components associated with the surface plasmon can be found from Maxwell's (Fresnel's) equations describing the dielectric response of the two media by their corresponding complex dielectric functions [9,16].

By their very nature of being surface electromagnetic modes, surface plasmons do not couple directly to free photons, i.e., plane waves of electromagnetic radiation. This situation is similar to optical waveguides. Among the various formats proposed to overcome this problem, the Kretschmann configuration (cf. Fig. 2) is the most convenient setup. Here, a thin metal film is deposited onto a high refractive index glass slide—optically coupled to a prism by a matching fluid. A schematic illustration of this is given in detail in Fig. 1. As one changes the angle of incidence θ of the incoming light of fixed wavelength or frequency, at some angle θ_m, the projection of the lightwave vector along the metal film equals that of the wave vector of a SP at the same frequency. At this matching point the light energy can be effectively transferred to the SP; the laser light "drives" the surface plasmon mode resonantly. Like any other resonance phenomenon the phase of the driven wave changes continuously as one sweeps through the resonance, i.e., as one scans the angle θ.

If the light couples to SPs at the resonance angle, then the reverse must also be true—plasmons can reradiate back via the prism at the same angle. Thus, if we vary the angle of incidence, the phase difference between the reflected light and the light reradiated by SPs will change, and for a particular angle θ_R the two waves are completely out of phase and for a particular thickness of the Au layer of around 50 nm almost equal in amplitude [17]. They will destructively interfere and no light will be detected in reflection. We should point out that the angle of resonance θ_m, i.e., the angle at which the optical field amplitude of the surface mode goes through a maximum is always slightly lower than this angle of minimum reflectivity θ_R, and both are always greater than the critical angle of total internal reflection θ_c of the prism–dielectric interface (cf. Fig. 3a).

B. Monitoring Protein Adsorption/Desorption Processes by SPR

The implementation of this optical concept for protein binding studies is schematically summarized in the experimental setup given in Fig. 2a. The main part is the goniometer that allows for the computer control of the angular scan in SPR spectroscopy. In our setup two laser beams can be coupled alternatively via the prism, and the reflected light is monitored with a photodiode. A typical measuring curve, reflectivity R versus incidence angle θ, is given in Fig. 3a.

The position and the shape of the whole reflectivity curve are very sensitive

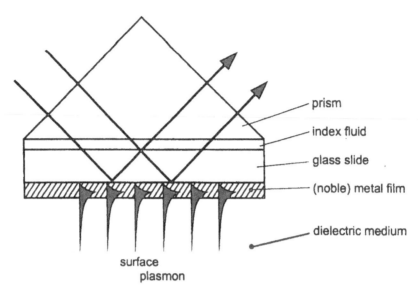

surface
plasmon

FIG. 1 The Kretschmann configuration for the excitation of surface plasmons at a metal–dielectric interface via a prism coupler. Note the evanescent character of the surface plasmon mode extending into the dielectric medium.

to any change of the properties of the surface and the media next to it. It is the basis for the use of surface plasmon resonance techniques for chemical and biological sensing.

For this application of surface plasmon optics, a flow cell can be easily attached to the coupling prism and the metal-coated glass slide, as is schematically given in Fig. 2b. This way, the solution in contact to the sensor surface can be easily exchanged, and protein binding studies can be followed in situ and in real time.

In general, any thin layer of a dielectric medium, e.g., a protein layer, with a refractive index greater than the medium in contact to the metal layer—in the context of this chapter typically an aqueous solution—will shift the dispersion of the surface plasmon [1,9] to higher momentum and, hence, results in a reflectivity curve that is shifted to higher angles. This is shown schematically in Fig. 3a (dashed curve). This shift can be evaluated quantitatively based on a Fresnel algorithm by simulating the shifted resonance curve. The result is an optical thickness of the coating from which the geometrical thickness of the layer is derived provided the refractive index of this layer is known. Typically, this will not be the case; however, a refractive index of $n = 1.45$ for a protein layer is a reasonable assumption. The thickness of the adsorbed layer can then be determined to an accuracy of 0.1–0.2 nm.

If the reflectivity is monitored at a fixed angle of incidence during the adsorption (or desorption) of proteins the corresponding change of the reflected intensity can be used to obtain kinetic information about this process. This is schematically given in Fig. 3b for an adsorption process resulting in a time-dependent increase in reflectivity. For small thickness changes, i.e., changes in R that follow the near-linear slope of the shifting reflectivity curve, these time-dependent changes can be directly compared to simulated curves obtained from an adsorption/desorption model, e.g.,

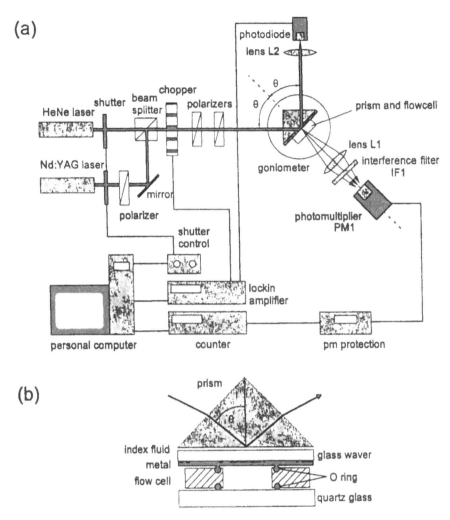

FIG. 2 (a) Surface plasmon resonance (SPR) and surface plasmon fluorescence spectroscopy (SPFS) setup that can be run with two different laser wavelengths. (b) Flow cell used in the set-up of (a) for on-line and in situ measurements of protein binding.

based on Langmuir behavior. But, of course, more complex models can be applied and the resulting time-dependent reflectivity changes evaluated as well.

C. Simultaneous SPR and SPFS Studies

The resonance character of surface plasmon excitation gives rise to a strong enhancement of the optical field at the metal/dielectric interface (cf. Fig. 3a). Owing to the evanescent character of the surface plasmon mode this enhanced field decays exponentially into the dielectric medium, with a $1/e$ penetration depth of ca. 150–200 nm. Any fluorescently labeled biomolecule that reaches this evanescent field will be excited and emits fluorescence light. Labeled molecules that bind to the func-

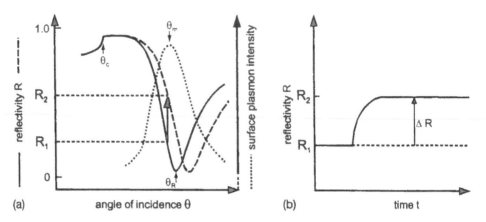

FIG. 3   (a) SPR reflectivity curve as a function of the angle of incidence θ with a minimum in reflectivity at θ_R (—). The corresponding optical field at the metal–dielectric interface strongly peaks at the resonance angle θ_m (•••), which is slightly shifted to angles smaller than θ_R. Any thin dielectric coating, e.g., a protein layer will shift the resonance curves to higher angles (– – – –) provided the refractive index of this layer is larger than the ambient medium. (b) Time-dependent change of the reflectivity monitored at a fixed angle of incidence in order to obtain kinetic data on protein binding or desorption.

tionalized surface are thus excited by a strongly enhanced optical field, which gives rise to the substantial sensitivity enhancement in detecting their binding behavior [8,15].

Care has to be taken, however, that the chromophores are not too close to the metal surface because the well-known Förster energy transfer from the excited chromophore to the (acceptor states of the) metal layer [18] will lead to a dramatic loss of fluorescence intensity. The relevant separation distances are given by the Förster radius, which is in the range of 5–7 nm [18], i.e., much closer to the surface than the extent of the evanescent surface plasmon field. A protein with a conjugated fluorescence label which is kept away from the surface by 1–2 Förster radii can then be monitored with a strongly enhanced sensitivity. This concept is schematically depicted in Fig. 4.

With this extension of surface plasmon spectroscopy one gains sensitivity and exceeds detectability limits whenever the mere label-free recording of a biomolecule is not possible, because the units are too small to add significantly to the thickness of the layer architecture or if their binding to the interface occurs at high lateral dilution (and hence, again, only very small averaged thickness increment—too low to be detected). In these cases, the recording of the fluorescence emitted from labeled proteins at the surface can result in a signal that can be used to evaluate quantitatively their binding behavior.

A particular advantage can be the simultaneous recording of the SPR signal as well as the fluorescence signal. Two examples will be presented. In the first case the binding of a very small unit, a 15-mer target oligonucleotide, too small to modify the surface plasmon resonance, can be followed via the chromophore that is covalently attached, while the competitive binding of a protein to the surface-attached probe strands can be followed directly by its influence on the surface plasmon res-

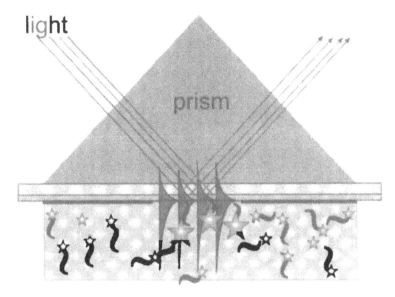

FIG. 4 The concept of using the resonantly enhanced optical field of a surface plasmon for the excitation of chromophore-labeled biomolecules bound to the interface.

onance. In this case then two signals give independent information about the binding behavior of the two systems. In the second case the specific binding of a protein to membrane-integral receptors is screened by the huge SPR signal originating from nonspecific binding of other biopolymers. Here, the combination of fluorescence detection of labeled antibodies with the specificity of antibody–antibody interactions will allow us to differentiate specific versus nonspecific binding.

III. MINIMIZING NONSPECIFIC ADSORPTION TO POLYMER-MODIFIED SURFACES

Before we introduce the various examples of protein recognition and binding to particular sites at specifically designed supramolecular interfacial layers we first discuss the general problem of controlling nonspecific binding (NSB) of proteins to any kind of interface.

Generally speaking, this is still the most demanding challenge for (bio)materials surface science because the fundamentals of NSB are far from being understood— not even in part. However, it has a significant practical importance because eventually it will determine, e.g., the stable integration of medical implants and it is still the major complication in the proper and reliable function of biosensors.

Numerous attempts to control NSB to artificial substrate surfaces have been reported in the literature. Most of the reported work has been performed with poly(ethylene oxide) (PEO). This polymer exhibits a very low surface energy against water (and blood) and is water soluble. PEO coatings can suppress NSB almost completely, which was attributed to the specific structure of PEO in water. A proof of this hypothesis, however, has never been given.

A more general explanation of why (swollen) polymer coatings could reduce protein adsorption was given by Nagaoka [19]: if a polymer brush or mushroom [20] at a solid surface is approached by a protein from solution and then binds to this surface, the resulting compression of the polymer chain leads to a decrease of the configurational entropy and, hence, results in a repulsive force. This model should be largely independent of the details of the chemical structure of these employed polymers. However, a certain length of the individual macromolecular chains is required for that mechanism to operate.

We have performed a few experiments along these lines by coating a solid substrate with a layer of end-grafted polymers prepared by the so-called grafting-from strategy [21]. This concept is given schematically in Fig. 5. In order to be able to use SPR for monitoring the surface reactions, we coated a Au substrate with a monolayer of a tosylate derivative of an alkyl disulfide. The resulting self-assembled monolayer is activated in the presence of n mol of ethyloxazoline and $2n$ mol of piperidine, resulting in the growth of a polymer brush composed of piperidine-capped

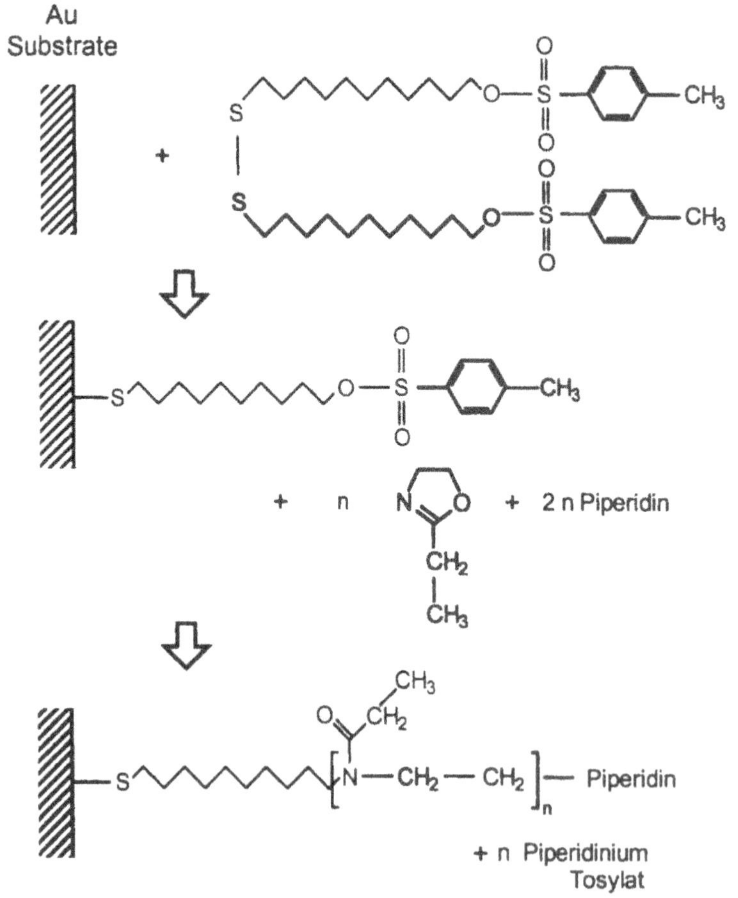

FIG. 5 Growing a polymer brush by the "grafting-from" concept. In this case oligo(ethyloxazoline) chains of different degrees of polymerization P_n are prepared.

oligo-ethyloxazolines of a particular degree of polymerization P_n, covalently attached via the alkyl spacer to the Au surface. The degree of polymerization P_n thus controlled could be varied between $n = 0$ and $n = 150$. These surfaces were then exposed to a fibrinogen solution (0.1 wt% in 0.02 M phosphate buffer at pH 7.5) and the increase in protein thickness monitored as a function of time.

Fibrinogen was used for the adhesion studies as a model protein because it is one of the most abundant plasma proteins (ca. 2–3%) with concentration in the blood of 2–4 g/L. It is considered to be very sticky; it is involved in wound healing and adsorbs strongly to most surfaces. Figure 6 shows the SPR data of the kinetic runs in our adhesion experiments. After injection of the fibrinogen solution into the sample cell ($t = 0$) a rapid rise in the adsorbed layer thickness indicates the high affinity of this protein to solid surfaces. Within minutes (diffusion-controlled) a constant biofilm thickness is obtained which is stable over the next few hours.

If we plot the final layer thickness as a function of the respective degree of polymerization of the functional brush, i.e., the chain length, as is done in Fig. 7, we find that the adsorption to a thin brush is rather pronounced whereas with increasing degree of polymerization P_n, the thickness of the adsorbed fibrinogen layer decreases to virtually zero. This means that a functional coating of such a poly(ethyloxazoline) (PEOx) layer of sufficient chain length, i.e., $n > 100$, can completely block NSB.

Just for comparison we show in Fig. 8 the corresponding contact angle measurements performed with a water droplet on the various PEOx brushes. From the high contact angles at low degrees of polymerization which are dominated by the alkyl chain segments and the piperidine end groups (cf. structure formula given in Fig. 5) one reaches very low values at the higher chain lengths. Obviously, PEOx is able to provide a hydrophilic, presumably soft swollen surface coating that is able

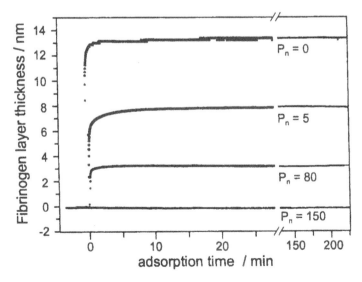

FIG. 6 Adsorption kinetics of fibrinogen to poly(ethyloxazoline) brushes of different degrees of polymerization P_n monitored by SPR. Injection of the protein solution was at $t = 0$.

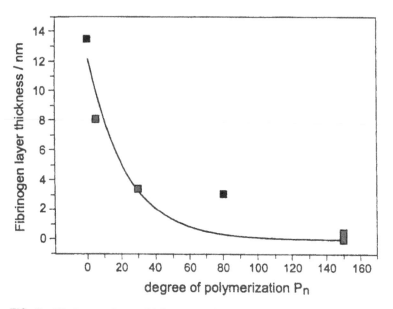

FIG. 7 Fibrinogen layer thickness as determined by SPR as a function of the degree of polymerization of the poly(ethyloxazoline) brush to which it adsorbed.

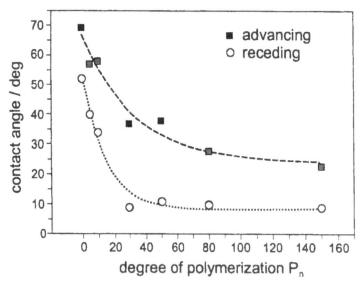

FIG. 8 Advancing and receding water contact angle on poly(ethyloxazoline) brushes of different degrees of polymerization.

—just as PEO—to completely suppress NSB. This strongly supports the generic model of an entropically determined mechanism that prevents protein adsorption.

IV. SURFACE RECONSTITUTION OF CYTOCHROME b_{562}

The manipulation of proteins by physical, (bio)chemical, or genetic methods is an important aspect in protein engineering. For example, future bioelectronics devices or modules needed for emerging nanobiotechnology will require modifications of native biocomponents in order to interface them with technical structures, e.g., electrodes, optical transducers, microchannels, etc., and to adapt them to or compatibilize them with the non-native functional environments, e.g., enhanced operation temperatures, adapting energy levels by tuning of redox potentials, modifying spectral conditions, etc.

Along these lines, we launched a research program on characterizing and manipulating a redox-active protein, cytochrome b_{562}, a member of the family of heme proteins.

These proteins are a particularly interesting class of functional biosystems owing to their unique electronic and optical properties associated with the porphyrin-based prosthetic group. Cyt b_{562} was chosen in this study as a model protein for the fabrication of monolayers and their application for bioelectronic devices. It is a small (MW 12,000 Da) water-soluble protein found in *Escherichia coli*, consisting of 106 amino acid residues. The structure of *E. coli* cyt b_{562} has been revealed by x-ray crystallography to a resolution of 0.25 nm, indicating that it consists of four α-helices (cf. Fig. 9). The prosthetic group of cyt b_{562} (the heme) is noncovalently bound by ligation to the polypeptide chain through methionine 7 on the N-terminal helix and histidine 102 on the C-terminal helix. For this reason, this heme can be easily extracted from the protein, modified by a synthetic-chemical procedure, and then reconstituted with the original apoprotein in vitro. For example, we could synthesize a heme derivative with octadecyl chains, reconstitute holo-cyt b_{562} from apo-cyt b_{562} and its modified heme, and demonstrate the formation of a stable monolayer of the reconstituted cyt b_{562} on a water surface [22].

The concept for the surface reconstitution experiments that we discuss here is shown in Fig. 9. We start with the formation of a so-called mediator layer, which is often used in electrochemical studies of surface-adsorbed proteins and serves as a protective coating preventing the biopolymer from denaturing at the bare Au electrode surface (Fig. 9a). We used 2,2'-dithiodiacetic acid (DTDAA) and followed the self-assembly process by SPS (cf. Fig. 10A). After rinsing (arrow in Fig. 10), a monolayer thickness of ca. 1.0 nm was obtained. Next, a reconstitution product assembled in solution from the heme-free apoprotein and a thiolated derivative of the native heme [obtained by coupling the natural hemin with 2,2'-dithiobis-(ethylamine)hydrochloride—structure formula given in Fig. 11] is injected into the flow cell resulting in a partial replacement of some of the DTDAA molecules and the formation of a mixed SAM, now containing a substantial amount of the surface-attached reconstituted cyt b_{562} (cf. Fig. 9b). This assembly process can be easily followed by SPS and is given in Figure 10 as trace B. After rinsing (arrow) an effective layer thickness (based on Fresnel simulations with a refractive index of n = 1.50 for both the protein and the DTDAA layer) of ca. 4.0 nm is obtained. This

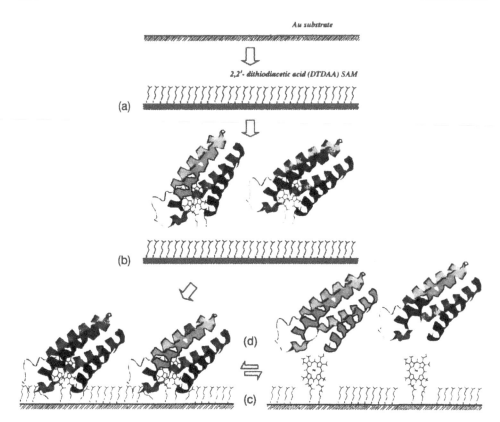

FIG. 9 Schematic drawing of the interfacial architectures prepared for protein resonstruction: Onto a bare Au surface a mediator layer of 2,2'-dithiodiacetic acid (DTDAA) is (self-)assembled (a). Exposure to a solution of cyt b$_{562}$ with a modified dithio-derivatized (cf. Fig. 11) heme results in the binding of some of the proteins by partial replacement of DTDAA molecules (b). Diassembly (c) and reconstitution (d) of the apoprotein to the surface-attached heme can be followed by SPR.

might be compared to the known dimensions of the protein, which can be approximated by an elongated ellipsoid with 5.0 nm along its major axis and 2.5 nm along its minor axis.

As we could show in some detail, the addition of 8 M guanidinium chloride at pH 5.0, which is known to weaken the heme–apoprotein interaction results in the dissociation of the apoprotein from the surface-bound heme derivative (cf. Fig. 9c) [23]. The slow detachment of the apoprotein again can be followed by the kinetic mode of SPS and is shown in Fig. 10c). It is important to note that only the apoprotein and not the whole reconstituted complex dissociates. This has been confirmed by a number of possible tests with native cyt b$_{562}$, pure apoprotein, etc., but most convincingly can be demonstrated by monitoring the changes in the optical and electrochemical properties of the surface heme layer: both the spectral features of this chromophore and its redox potentials are changing in a very characteristic way upon the dissociation or reassociation with the apoprotein [24]. Hence, we could show that the readsorption of the apoprotein that can be followed in Fig. 10, trace

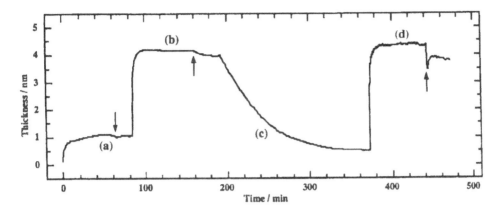

FIG. 10 Kinetic SPR run of the assembly of the DTDAA SAM (a), the binding of reconstituted cyt b_{562} modified with the dithio-heme of Fig. 11 (b), the disassembly of the apoprotein
from the surface-attached heme (c), and the reconstruction of heme and apoprotein at the
surface (d). The arrows indicate rinsing steps.

d, indeed results in the formation of a functional cyt b_{562} complex at the electrode–
electrolyte interface. A schematic of this process is also given in Fig. 9.

The reversible surface reassembly constitutes an important step toward studies
aiming at elucidating the structure–function relationship of proteins at interfaces.
The possibility of reconstitution at a solid surface allows for the application of a

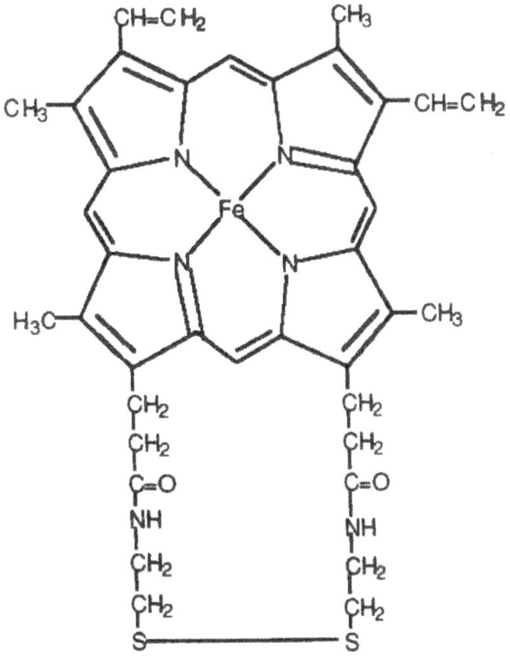

FIG. 11 Structure formula of the dithio-derivatized heme prosthetic group of cytochrome b_{562}.

broad spectrum of analytical tools aiming at characterizing the structural details of surface-bound bio-objects. In addition, functional parameters can be evaluated at the same time and correlated with the molecular and supramolecular properties of these layers.

V. INTERACTION OF SINGLE STRAND BINDING PROTEINS WITH SURFACE-ATTACHED OLIGONUCLEOTIDES

The next example for the application of surface plasmon optics for the detection of binding events at biofunctionalized surfaces concerns the interaction of single strand binding proteins (SSBs) with surface-attached oligonucleotides and their competition with complementary target strands from solution [25].

Single strand binding proteins play an essential role in DNA replication, recombination, and repair. They are present in high concentrations in vivo and are presumed to function by binding preferentially to single-stranded DNA (ssDNA). In the process of replication these proteins stabilize the single-stranded regions which are formed from a double strand by simultaneous action of helicases and Repproteins in the typical replication fork. More than one single strand binding protein are known, but the one from the bacterium *E. coli* is unique in that it forms homotetramers in solution ($4 \times 18{,}843$ Da) [26]. These tetramers are stable over a wide range of solution conditions. Although binding of individual monomers was observed, the tetrameric state seemed to be the predominant species which interacts with the oligonucleotide chain.

In earlier studies we had been working at developing the methodologies and the same formats capable of monitoring hybridization reactions in solution or at surfaces. We focused on the preparation of surface-attached catcher probes making use of different strategies for the interfacial coupling of probe sequences, and on the application of surface plasmon optics for the quantitative detection of hybridization reactions in situ at the solid–solution interface between these probes and complement strands from solution [27].

Figure 12 summarizes the chemical structure formulas of the molecules employed for the buildup of the interfacial architectures (cf. also Fig. 13) used in the SPR and SPFS experiments. First, a binary mixed monolayer of 10 mol% of a biotinylated thiol and 90 mol% of an OH-terminated thiol was prepared on the Au substrate by a self-assembly step. Next, the thiol layer was exposed to a dilute streptavidin solution, which resulted in the formation of a monomolecular layer of streptavidin by specifically binding to the biotin moieties. Thus, a generic binding matrix is obtained that can be used for the coupling of other biotinylated functional molecules like antibody (fragments) [28] or biotinylated oligonucleotides [8], as in our case.

The experiments described here were carried out with a catcher probe of the following molecular structure: to a 15-mer sequence of thymines, T_{15}, used as a spacer sequence in order to decouple the hybridization of a complement to the 15-mer sequence P2 (cf. Fig. 12), a biotin group is attached at the 5′-end, allowing this catcher strand to be bound to the streptavidin matrix. A corresponding probe molecule, labeled at its 5′-end with a fluorophore, Cy5 (structure formula also given in Fig. 12), is called $BioT_{15}Cy5$.

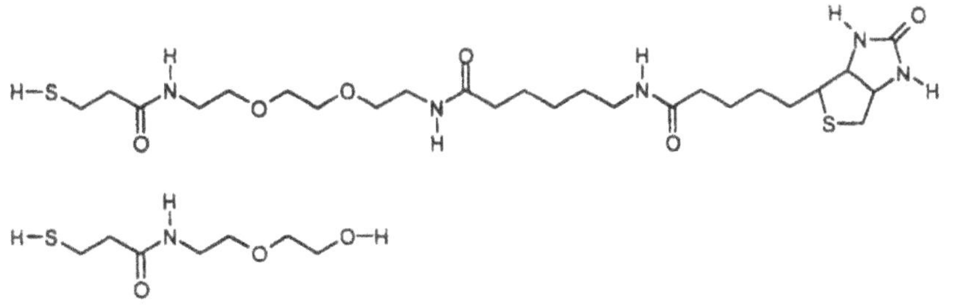

5'-Biotin-T$_{15}$ TGT ACA TCA CAA CTA - 3'   P2

5'-Biotin-T$_{15}$ TGT ACA TCA CAA CTA - Cy5- 3'   BioCy5T15

3'- ACA TGT AGT GTT GAT - Cy5 -5'   T2 (MM0)

3'- ACA TGC AGT GTT GAT - Cy5 -5'   T1 (MM1)

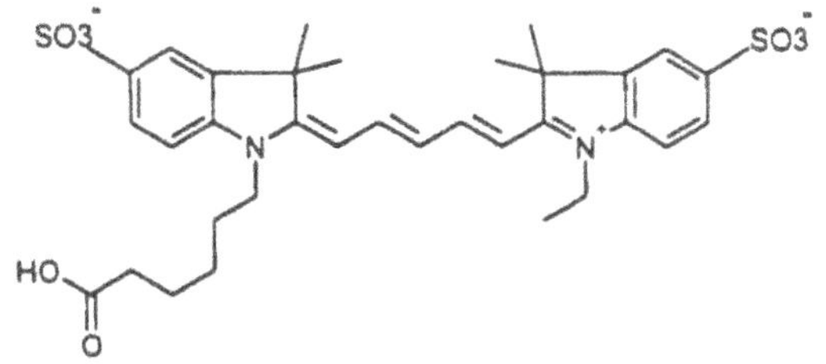

Cy5

FIG. 12 Structural formulas of thiols, oligonucleotides, and fluorescent dye used in the experiments.

Figure 14 displays example of two typical experiments, an angular scan (Fig. 14a), and a kinetic scan (Fig. 14b) (cf. also Fig. 3). The data shown were taken before, during, and after the hybridization of a fluorescently labeled complementary 15-mer strand from solution to a surface-attached probe strand (cf. the structure formulas in Fig. 12). The reflectivity R taken as a function of the angle of incidence θ before hybridization is virtually identical to the curve taken after hybridization because of the little mass (optical thickness) added by the complement strands binding to a very dilute matrix of catcher probes (cf. Fig. 13) with an upper limit of ca. 1 probe strand per 40 nm^2 on the sensor surface. Consequently, the kinetic mode of the SPR does not indicate any change during hybridization (cf. Fig. 14b) and, hence, cannot be used as a sensor signal. However, since all of the complements carry a fluorophore which upon hybridization reaches the high optical fields generated at the interface upon resonant excitation of surface plasmons and hence emits fluorescence light, this fluorescence intensity when taken as a function of time contains the kinetic information of the hybridization reaction and can be analyzed in terms of the corresponding rate constant, k_{on} for the association and k_{off} for the dissociation, respectively. This can be seen also in Fig. 14b. High photon counts, well in excess of 10^6 cps even after rinsing off any nonspecifically adsorbed complementary strands (cf. arrow in Fig. 14b) are observed. The corresponding angular scan of the fluorescence displayed in Fig. 14a shows the very pronounced enhancement of the fluorescence

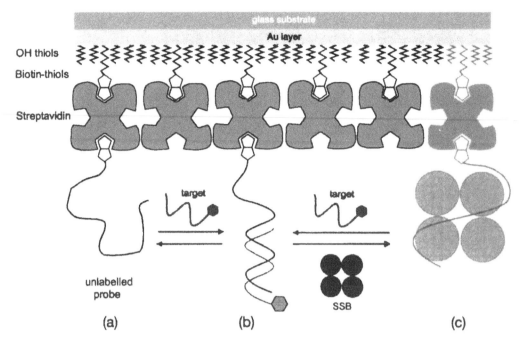

FIG. 13 Surface architecture prepared for studies of the binding of single strand binding proteins (SSBs) to unlabeled probe oligonucleotides (a), resulting in an oligonucleotide–protein complex (b). The target strand is coupled via a biotin to a monolayer of streptavidin. The latter is prepared by self-assembling a binary mixture of 10 mol% biotinylated thiol and 90 mol% OH-terminated thiols (structure formula given in Fig. 12) exposed to a streptavidin solution. Exposing the probe–protein complex to a solution of fluorescently labeled targets (of different mismatches in their base sequence) can result in a competitive replacement of the protein by the formation of oligonucleotide double-stranded hybrids (c). Challenging these hybrids by SSBs can result in the re-formation of the probe–protein complex.

intensity at resonant excitation of surface plasmons with the typical angular displacement between the maximum emission and the minimum in reflectivity (cf. Section II).

These investigations where then extended to studies of the association and dissociation of single strand binding proteins and their binding behavior in competition to complement strands of different mismatches.

With the SSBs being of sufficient molar mass, thus giving rise upon binding to the surface-attached single-stranded oligonucleotide catcher probe to an appreciable change in the surface plasmon resonance angle, and with the target complement oligonucleotides carrying a fluorescence label, we can compare the two signals obtained by the classical SPR and the SPFS directly.

In order to investigate the replacement kinetics of SSGBs by oligonucleotide targets a streptavidin matrix was prepared as schematically depicted in Fig. 13. First, the extent of unspecific binding was determined by applying a 0.1 μM $E.$ $coli$ SSB solution in PBS buffer to the unloaded streptavidin layer. As shown by the corresponding SPR kinetics scan in Fig. 15, a reflectivity jump due to the change in the bulk solution refractive index was observed after injection of the protein solution.

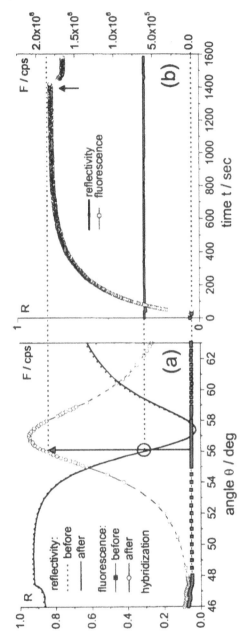

FIG. 14 Angular reflectivity and fluorescence scan curves and the corresponding reflectivity and fluorescence kinetics curves before and after adsorption of fluorescently labeled DNA targets onto the sensor surface. Due to the low molecular weight of the analyte no change in reflectivity is seen at 56°, while the excited fluorescence causes a clear signal increase in both the angular scan and in the kinetic scan.

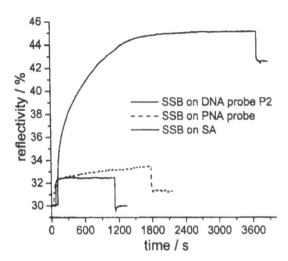

FIG. 15 Adsorption kinetics of SSB onto unmodified streptavidin matrix, a PNA probe, and the DNA probe P2 (cf. Fig. 12).

However, the initial reflectivity was recovered after rinsing the flow cell with protein-free PBS buffer. Thus, nonspecific adsorption of SSBs to the streptavidin surface can be excluded in the further investigation.

Next, biotinylated peptide nucleic acid (PNA) probes were immobilized on the streptavidin from a 1 μM solution in PBS and the adsorption of SSBs investigated by SPR. PNA is a DNA mimic with a backbone that consists of uncharged pseudopeptide linkages. Figure 15 shows the reaction after injection of a 0.1 μM SSB solution in the same buffer. Similar to the unloaded streptavidin surface the initial reflectivity was recovered after rinsing, indicative of a negligible binding of SSB to PNA. This result seems reasonable because the backbone of PNA is uncharged due to the lack of negative phosphate groups. Thus, the essential electrostatic interaction with the protein surface is missing.

In contrast to this, a significant optical thickness increase is obtained during the reaction of SSBs with the immobilized DNA probe P2, the structure of which is given in Fig. 12. The resulting architecture is shown in Fig. 13b: SSBs are bound to the surface-attached oligonucleotide probes. Upon rinsing again a drop in reflectivity was seen, similar to that for the PNA and the unloaded streptavidin case.

Subsequently, MM0 and MM1 labeled targets were tested for their ability to replace the proteins from the surface by binding to the probes (Fig. 13c). The replacement kinetics were monitored by SPR and by SPFS and both the reflectivity and the corresponding fluorescence signal are plotted in Fig. 16.

Upon injection of a 1 μM solution of MM0 target in PBS buffer, a substantial decrease in reflectivity was observed as a function of time. This was attributed to the removal of the SSB tetramers from the surface. Simultaneously, the fluorescence signal increased due to hybridization of the labeled targets to the immobilized probes. However, if a SSB-loaded surface was treated with a 1 μM solution of labeled MM1 target (also shown in Fig. 16), the change in both reflectivity and fluorescence was almost negligible compared to the recorded value for fully complementary sequences.

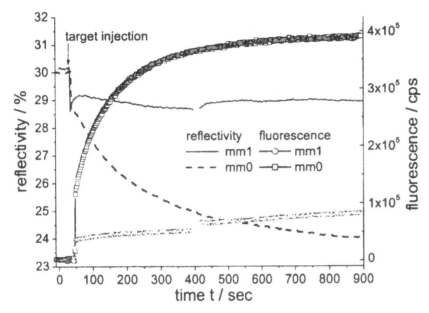

FIG. 16 Competitive replacement of SSBs bound to the probe oligonucleotide strand P2 by injecting fluorescently labeled targets of mismatch 0 (T2, cf. Fig. 12) or mismatch 1 (T1). Simultaneous recording of the time-dependent SPR signal (—, – – –) and the fluorescence intensities (–○–, –□–), respectively.

The gradual loss in reflectivity and the slight increase in fluorescence intensity occur on a time scale similar to the intrinsic desorption rate constant of SSB even in the absence of any competing target strands.

VI. RECOGNITION OF STEROL MOLECULES IN TETHERED MEMBRANES

As the last example for the use of surface plasmon optics for protein binding studies the putative yeast oxysterol binding protein (OSBP) Osh5p [29] is shown to specifically bind to oxidized sterol (OS) molecules [30,31] incorporated into an artificial tethered lipid membrane [6]. We show that the specificity of antibody–antibody interaction and the sensitivity of SPFS allows for the distinct binding properties of a recombinant oxysterol binding protein and of a mutant derivative to be studied.

Artificial phospholipid bilayers mimic a native environment for apolar or amphiphilic molecules such as sterols. If incorporated into an artificial membrane, the highly hydrophobic molecules maintain their structural conformation and, consequently, their biochemical properties [32,33,34]. For on-line studies of protein binding to functional membranes by surface plasmon optical techniques, lipid bilayers were assembled with the proximal monolayer being tethered to the gold substrate and the distal layer being directed toward the bulk aqueous solution [6]. For the construction of the inner leaflet, a hydrophilic peptide (CSRARKQAASIKVAVSADR) was attached to the gold surface through chemisorption of its amino-terminal cysteine thiol moiety.

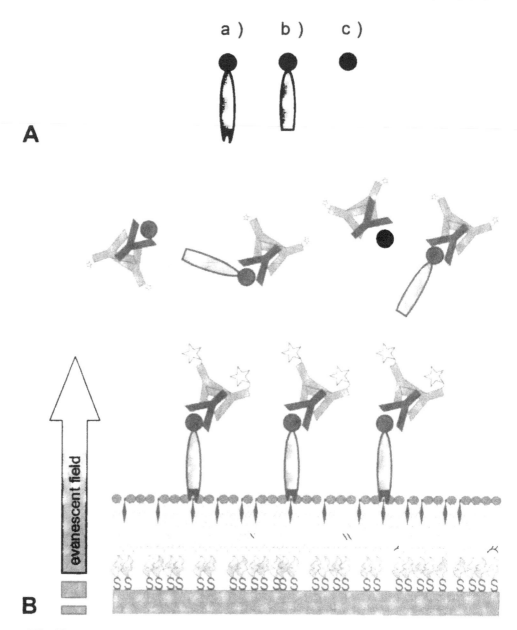

FIG. 17 (A) Schematic drawing of the GST (glutathione S-transferase) fusion of the oxy-sterol-binding protein Osh5p (a), the truncated mutant of the fusion protein (b), and of the mere GST (c). (B) Schematic drawing of the peptide-tethered lipid bilayer with incorporated sterol molecules. To some of them the GST-Osh5p fusion binds specifically. The complex can be identified by SPFS via a GST-specific antibody and a fluorescently labeled anti-GST antibody. The solution can also contain labeled GST or the (nonbinding) truncated mutant of the GST-Osh5p fusion which are, however, rinsed away.

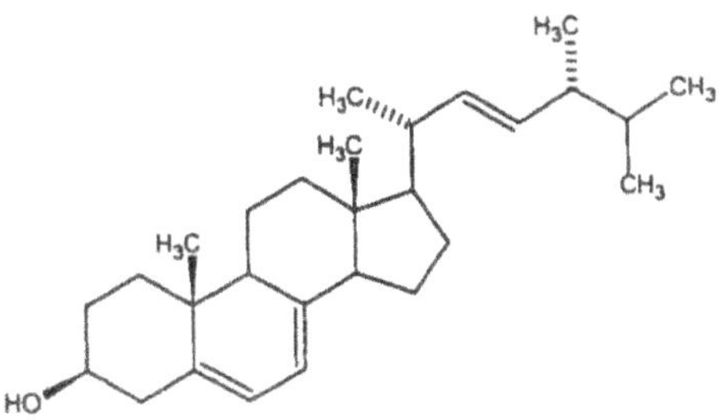

ergosterol (erg)

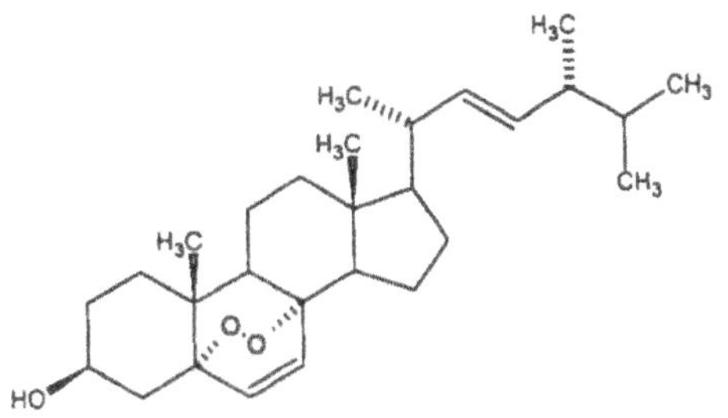

ergosterol endoperoxide (EEP)

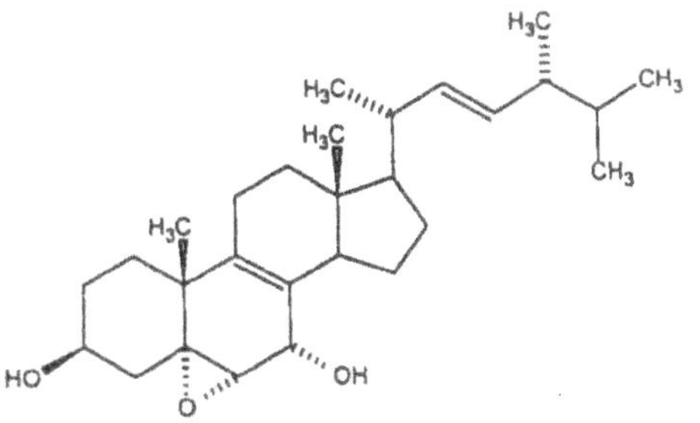

5α,6α-epoxy-(22E)-ergosta-8,22-dien-
3β,7α-diol (8-DED)

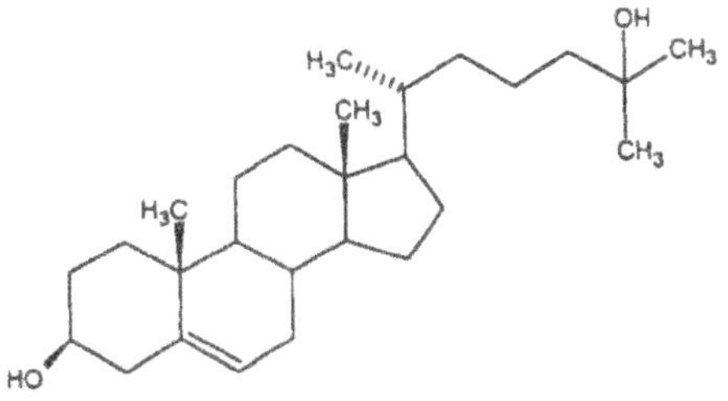

25-hydroxycholesterol (25OH)

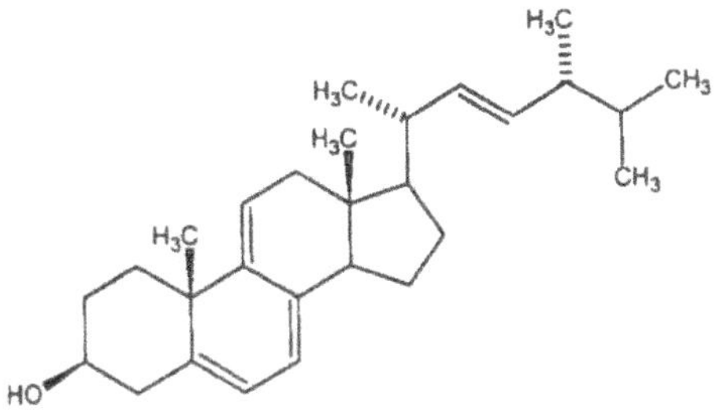

9(11)-dehydroergosterol (DHE)

FIG. 18 Structure formulas of the various (oxy)sterols used in this study.

Subsequent amino-coupling of dimyristoyl phosphatidylethanolamine (DMPE) to the carboxy-terminus of the peptide provided a peptide-tethered monolayer with a highly hydrophobic surface that facilitated vesicle spreading [35]. In order to prepare an "oxysterol-functionalized" lipid membrane, OS molecules were incorporated into vesicles otherwise containing phosphatidylcholine lipid (PC). The OS/PC vesicles were then spread onto the DMPE monolayer in the flow cuvette in order to allow for the formation of the outer leaflet of an artificial membrane. Figure 17B gives a schematic of such a functionalized tethered lipid bilayer membrane. Figure 18 summarizes the structures of the (oxy)sterols used in this study. 25-Hydroxycholesterol is from mammalian cells; all other sterols are typical for yeast cells [30].

It is also assumed that in artificial membranes the incorporated sterol molecules can be recognized specifically by Osh5p. In order to study this binding and to quantify the affinity of the protein for different sterols, soluble GST (glutathione S-trans-

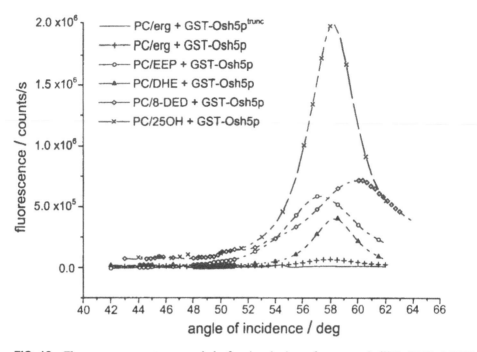

FIG. 19 Fluorescence spectra recorded after incubation of ergosterol, EEP, DHE, 8-DED, and 25OH functionalized membranes with cell extracts containing GST-Osh5p, GST-Osh5p[trunc] proteins.

ferase) fusions of Osh5p or of a truncated Osh5p protein lacking the putative oxysterol binding motif (cf. Fig. 17A) were rinsed across an artificial lipid layer doped with various types of oxysterol molecules. The deletion of the putative oxysterol binding domain in Osh5p was expected to result in a protein that is unable to recognize sterols.

In these studies, whole cell extracts are used which contain the binding proteins of interest but also a large number of other cell components with a certain affinity for nonspecific binding to membranes. Therefore, SPR as a technique for monitoring the specific binding events of Osh5p to sterols is not suited because the specific signal is totally screened by the unspecific adsorption processes. The mere surface plasmon shifts, hence, cannot be evaluated quantitatively. However, the existence of the GST tag fused to the Osh5p provides a scheme for the specific antibody detection of the bound GST fusion protein. A fluorescently labeled antibody directed against the anti-GST antibody then allowed for the correlation of the resulting fluorescence signal with the affinity of the OSBP for oxysterol moieties.

Figure 19 displays the results obtained with the various sterols (cf. Fig. 18). Shown are the angular scans of the fluorescence intensity emitted from the antibody–Osh5p–sterol complex (cf. Fig. 17) as one sweeps through the surface plasmon resonance of the respective sample. Thickness variations in the various tethered membrane preparations as well as the fluctuations in the degree of protein layer formation by nonspecific binding processes of components from the cell extract to the bilayer mentioned above result in a variation of the angular position of the respective fluorescence resonances. However, the peak intensities are a quantitative measure of the degree of specific recognition and binding of Osh5p to the different sterols.

GST-Osh5p displays the highest binding affinity for the mammalian oxysterol 25-hydroxycholesterol. The recognition of a mammalian oxysterol by yeast Osh5p supports the hypothesis that the highly conserved OSBP motif represents the actual oxysterol binding site. The yeast oxysterols 8-DED, EEP, and DHE are bound with decreasing affinity, with the nonoxidized ergosterol showing the lowest affinity. From these results we conclude that binding involves an oxysterol-induced change in the membrane architecture rather than the specific recognition of single oxysterol molecules in membranes. The fluorescence of antibodies bound to the mere GST tag or to the truncated Osh5p mutant lacking the putative binding domain (cf. Fig. 17a) injected into the flow cell as a reference experiment could be easily washed away by even a mild rinsing step.

VII. CONCLUSIONS AND OUTLOOK

The presented results demonstrate the great potential of surface plasmon optics for the monitoring of protein binding to surfaces—either specific or nonspecific—and for the quantitative evaluation of on-line recorded kinetic data and affinity values. In cases where the binding of proteins of relatively low molecular mass or at high lateral dilution result in only a minor angular shift of the surface plasmon resonance, the combination with fluorescence detection schemes offers a substantial increase in sensitivity. The simultaneous observation of the SPR signal of one component and the fluorescence emitted by a second (small) biomolecule turned out to be particularly powerful in the evaluation of competitive binding assays. In another example for the use of SPFS, i.e., in the combination with specific antibody–antibody binding assays the mere SPR signal could not be used to differentiate between nonspecific binding and the highly specific recognition and binding events between a ligand and a protein, whereas the fluorescence signal gave a clear signature for the specificity and selective affinity between the interacting biomolecules. It is to be expected that further studies will give additional evidence for the potential and usefulness of this novel extension of surface plasmon optics for protein (and other biopolymer) binding studies.

Another obvious mode of operation for binding assays in an array format uses surface plasmon microscopy for the multiple and parallel readout of binding processes [36]. The physics of surface plasmon optics allows the individual sensor field to be as small as ca. $20 \times 20 \ \mu m^2$, which is equivalent to a rather high integration density on a SPM chip, and all this without any loss of thickness sensitivity.

Finally, we should mention the various hybrid setups in which surface plasmon optics are combined with the measurement of another physical property of the interface: monitoring the mass load on a quartz crystal microbalance (QCM) simultaneously with the optical thickness has been demonstrated [37] to give additional information about details of the formation of surface layers or of binding processes.

Another most promising combination mode is realized in a SPR setup coupled to an electrochemical cell [13]; the Au substrate that guides the surface plasmon mode is also used at the same time as the working electrode for cyclic voltammetry [38] or impedance spectroscopy [35]. For example, the formation of a functional surface coating by the electropolymerization of a peptide-derivatized monomer unit could be followed and characterized by the CV response of the electrochemical part of the experiment while simultaneously the optical thickness of the growing film could be monitored by SPR. The activity of the peptide chain for antibody binding was then shown directly by replacing the monomer solution in the liquid cell by the corresponding protein solution [38]. With the growing interest in surface plasmon optics we can expect more interesting studies to be performed.

ACKNOWLEDGMENTS

The helpful discussions with many colleagues are gratefully acknowledged. In particular we are grateful to R. Herrmann, D. Kambhampati, S. Kohlwein, T. Liebermann, M. Liley, R. Naumann, P. Nielsen, H. Ringsdorf, P. Sluka, J. Spinke, and M. Zizlsperger.

REFERENCES

1. W. Knoll. Ann. Rev. Phys. Chem. *49*:569–638, 1998.
2. B. Liedberg, C. Nylander, and I. Lundström. Sensors Actuators *4*:299–304, 1983.
3. S. Herminghaus and P. Leiderer. Appl. Phys. Lett. *58*:352–354, 1991.
4. E. Aust, M. Sawodny, S. Ito, and W. Knoll. Scanning *16*:353–361, 1994.
5. W. Knoll, M. Zizlsperger, T. Liebermann, S. Arnold, A. Badia, M. Liley, D. Piscevic, F.-J. Schmitt, and J. Spinke. Colloids Surf. A *161*:115–137, 2000.
6. W. Knoll, C. W. Frank, C. Heibel, R. Naumann, A. Offenhäusser, J. Rühe, E. K. Schmidt, W. W. Shen, and A. Sinner. Rev. Molec. Biotechnol. *74*:137–158, 2000.
7. T. Liebermann, and W. Knoll. Colloids Surf. A *171*:115–130, 2000.
8. T. Liebermann, W. Knoll, P. Sluka, and R. Herrmann. Colloids Surf. A *169*:337–350, 2000.
9. H. Raether. *Surface Plasmons on Smooth and Rough Surfaces and on Grating*, Vol. 111. Springer-Verlag, Berlin, 1988.
10. T. R. E. Simpson, M. J. Cook, M. C. Petty, S. C. Thorpe, and D. A. Russell. Analyst *121*(10):1501–1505, 1996.
11. E. K. Schmidt, T. Liebermann, M. Kreiter, A. Jonczyk, R. Naumann, A. Offenhäusser, E. Neumann, A. Kukol, A. Maelicke, and W. Knoll. Biosens. Bioelectron. *13*(6):585–591, 1998.
12. F. Deckert and F. Legay. J. Pharmaceut. Biomed. Anal. *23*(2–3):403–412, 2000.
13. A. Badia, S. Arnold, V. Scheumann, M. Zizlsperger, J. Mack, G. Jung, and W. Knoll. Sensors Actuators *54*:145–165, 1999.
14. A. Nemetz and W. Knoll. J. Raman Spectrosc. *27*:587–592, 1996.
15. J. P. Attridge, P. B. Daniels, J. K. Deakon, G. A. Robinson, and G. P. Davidson. Biosens. Bioelectron. *6*:201–214, 1991.
16. E. Burstein, W. P. Chem, and Y. J. Hartstein. J. Vac. Sci. Technol. *11*:1004–1019, 1972.
17. S. Herminghaus, M. Klopfleisch, and H. J. Schmidt. Optics Lett. *19*(4):293–295, 1994.
18. H. Kuhn, D. Möbius, and H. Bücher. In *Physical Methods of Chemistry* (A. Weissberger and B. W. Rossiter, eds). Wiley Interscience, New York, 1972, Chapter 7.
19. Nakaoka, siehe T. Lehmann.
20. J. Rühe and W. Knoll. In *Supramolecular Polymers* (A. Ciferri, ed.). Marcel Dekker, New York, 2000, pp. 565–614.

21. O. Prucker and J. Rühe. Macromolecules *31*:592–601, 1998.
22. K. Kobayashi, T. Nagamune, T. Furuno, and H. Sasabe. Bull. Chem. Soc. Jpn. 72:691–696, 1999.
23. E. Itagaki, G. Palmer, and L. P. Hager. J. Biol. Chem. *242*:2272–2277, 1967.
24. K. Kobayashi, M. Shimizu, T. Nagamune, H. Sasabe, Y. Fang, and W. Knoll. (in preparation).
25. T. Neumann and W. Knoll. Isr. J. Chem. *41*(1):69–78, 2001.
26. W. Bujalowski and T. M. Lohman. J. Molec. Biol. *271*:63–74, 1991.
27. D. Kambhampati, P. E. Nielsen, and W. Knoll. Biosens. Bioelectron. *16*:1109–1118, 2001.
28. J. Spinke, M. Liley, H. J. Guder, L. Angermaier, and W. Knoll. Langmuir *9*:1821–1825, 1993.
29. C. T. Beh, L. Cool, J. Phillips, and J. Rine. Genetics *157*(3):1117–1140, 2001.
30. T. Bocking, K. D. Barrow, A. G. Netting, T. C. Chilcott, H. G. L. Coster, and M. Hofer. Eur. J. Biochem. *267*(6):1607–1618, 2000.
31. G. J. Schroepfer Jr. Physiol. Rev. *80*:361–554, 2000.
32. J. Huang and G. W. Feigenson. Biophys. J. *76*(4):2142–2157, 1999.
33. A. M. Smondyrev and M. L. Berkowitz. Biophys. J. *80*(4):1649–1658, 2001.
34. P. L. Yeagle. Biochim. Biophys. Acta *822*(3–4):267–287, 1985.
35. R. Naumann, E. K. Schmidt, A. Jonczyk, K. Fendler, B. Kadenbach, T. Liebermann, and W. Knoll. Biosens. Bioelectron. *14*:651–662, 1999.
36. M. Zizlsperger and W. Knoll. Prog. Colloid Polym. Sci. *109*:244–253, 1998.
37. A. Laschitsch, B. Menges, and D. Johannsmann. Appl. Phys. Lett. 77:2252–2254, 2000.
38. S. Kienle, S. Lingler, W. Kraas, A. Offenhäusser, W. Knoll, and G. Jung. Biosens. Bioelectron. *12*:779–786, 1997.

23
Neutron Reflection Study of Protein Adsorption at the Solid–Solution Interface

J. R. LU   University of Manchester Institute of Science and Technology, Manchester, England

I. INTRODUCTION

The deposition of protein at the solid–aqueous solution interface is of widespread occurrence and is a problem, for example, in the fouling of food processing plant equipment, the deposition of blood proteins onto cardiovascular implants, and in biomedical separations such as the isolation of individual proteins from mixtures using ceramic membranes. A different strand of interest on protein deposition comes from biosensor development and tissue engineering, where protein deposition is desired. In these applications the adsorption kinetics and the biophysical state of the adsorbed protein are likely to dictate the subsequent cellular and tissue responses.

Adsorption of protein often results in a heterogeneous layer that is typically less than 100 Å and that is heavily mixed with water. At the solid–aqueous solution interface, this layer is buried. Although many optical techniques can access it, their sensitivity to the structure and composition of the layer is very limited. Many researchers have undertaken alternative approaches by performing either physical measurements or biochemical assays at the air–solid interface using dried samples. However, sample rinsing, drying, and other treatment such as staining may alter the amount and biophysical state of the adsorbed protein. There is thus little prospect of correlating these measurements to the in situ structure and composition at the solid–solution interface. In addition, a common limitation to most existing physical measurement is the inability to identify individual components across the interface when more than two species are involved, e.g., peptide binding onto protein.

Neutron reflection is a recently developed technique able to probe detailed information about the structure and composition of adsorbed layers [1–4]. Its characteristic feature is that its signal can be altered by isotopic substitution. In a normal chemical system isotopic labeling can be made by deuterium substitution of the sample or the solvent, thus making the neutron measurement sensitive to the labeled species across the interfacial layer. Many recent studies have demonstrated the power of neutron reflection combined with deuterium labeling in studying adsorption of surfactants, synthetic polymers, and their mixtures at planar interfaces [4]. A few recent reviews have outlined the advances in these areas [5–9]. In comparison, there are very limited cases of the use of neutron reflection in studying model biological systems. Some recent progress on the use of neutron reflection and related methods in the study of biomolecules at wet interfaces has been reviewed in Refs. 10 and 11. It has been perceived that neutron reflection provides coarse structural profiling and that little useful information can be gained from such work. In addition, from the experimental point of view most proteins or enzymes cannot be deuterated, and this has been thought to further limit the resolution of the experiment. Over the past 5 years some trial experiments have been made to explore the potential of neutron reflection in revealing molecular features of adsorbed protein layers. This preliminary work has highlighted some interesting structural details that are highly relevant to the understanding of the complex processes of interfacial adsorption. The cases to be described in this chapter demonstrate two important features: the high depth resolution and the simultaneous determination of layer structure and composition. Because biointerfacial systems are complex and difficult to handle, these technical advantages are useful for quantitative measurements of various biointerfacial processes.

II. NEUTRON REFLECTION AND DATA ANALYSIS

Experimentally, neutron reflection is performed more or less the same as light reflection. Essentially, a parallel beam is directed onto a flat surface at the incidence angle θ, and the beam is specularly reflected off (specular reflection). Neutron reflectivity R, defined as the ratio of incoming and exiting beam intensities, is usually plotted as a function of momentum transfer κ ($\kappa = 4\pi \sin \theta / \lambda$, where λ is the incidence beam wavelength). The level and shape of the reflectivity profile is determined by the thickness of the layer covered on the surface τ and the variation in refractive indices across the film ρ. Figure 1 shows the schematic geometry of the beam passage, generic to both air–liquid and solid–liquid interfaces. The characteristic feature is the difference in path length between the beam reflected off the upper part of the film from that from the lower part of the film. The basic principle can

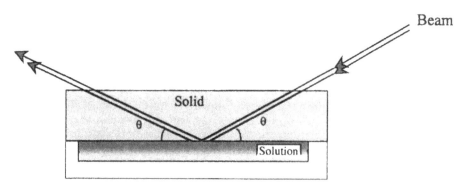

FIG. 1 Schematic diagram to show the geometry of incoming neutron beams impinged onto the solid–solution interface and reflected off specularly.

be better understood from the reflection at the air–solution interface where a thin film is adsorbed. For monochromatic radiation the phase shift varies systematically with the angle of reflection, leading to alternating constructive and destructive interference as the two partially reflected beams move in and out of phase. Measurement of the reflected monochromatic radiation as a function of angle of reflection hence gives a series of interference fringes with constructive peaks at angular spacings [1] determined by

$$n\lambda = 2\tau \sin \theta \tag{1}$$

with n being an integer. The thickness of the film (τ) can be deduced from the separation of the fringes. The intensity fluctuations in the fringes mirror the dependence of reflectivity profile on the thickness of the film and the variation of refractive index profile across the film.

The refractive index in neutron reflection is usually substituted by a more common quantity known as scattering length density ρ, which is related to the interfacial composition [1] by

$$\rho = \sum n_i b_i \tag{2}$$

with b_i being the scattering length of nuclear species i and n_i its number density. Neutron reflectivity is a complex function of the distribution of ρ along the surface normal direction. Change in the distribution of ρ along the surface normal direction will alter neutron reflectivity. Because b_i is of opposite sign between hydrogen and deuterium, deuterium labeling is effective for manipulating neutron reflectivity. A simple example of applying this principle is the preparation of a mixed water containing 8.1 vol% D_2O, giving zero scattering length. If neutron reflection is performed at the air–water interface, no specular reflectivity will be detected from the surface of the mixed water. If, however, a protein is adsorbed, the specular reflectivity will entirely arise from the protein layer. This water is usually called null reflecting water (NRW). For the protein layer adsorbed on the surface of NRW, a simple uniform layer model is usually appropriate for describing the structure of the adsorbed protein layer, and the measured reflectivity profile can be simply modelled using the optical matrix approach [11,13]. The parameters used in the modelling include layer thickness and scattering length density. Because roughness effect is coupled with layer thickness and its influence on layer structure can be incorporated in multilayer modelling through the choice of scattering length density distributions, it is preferable not to introduce it as an additional variable. Thus, for the adsorption of a single uniform layer on NRW, the area per molecule A (Å^2) can be derived from the best fitted τ and ρ through

$$A = \frac{\sum m_p b_p}{\rho \tau} \tag{3}$$

where $\sum m_p b_p$ is the total scattering length for the protein. The surface excess or the adsorbed amount (mg m^{-2}) is then calculated by

$$\Gamma = \frac{MW}{6.02A} \tag{4}$$

where MW is protein's molecular weight.

An important experimental feature of neutron reflection is the measurement of reflectivity profiles under different isotopic contrasts. At the air–water interface, isotopic contrast variations can be easily achieved by varying the ratios of H_2O and D_2O. Whilst the parallel measurements enhance the certainty of the structural information, they also provide information about the extent of immersion of protein layers into water at the air–water interface.

The system that is of more direct relevance to this review is the adsorption of proteins at the solid–water interface. For this type of experiment, a neutron transparent solid substrate is usually used. The neutron beam is directed into the substrate and is specularly reflected back onto the opposite end, as indicated in Fig. 1. As in the case of analyzing the data measured from the air–water interface, data analysis for reflectivity profiles measured at the solid–water interface also proceeds with layer modelling. Because the protein layer in this case is fully immersed in water, Eq. (3) should be modified to include the contribution of water:

$$\rho = \frac{\sum m_p b_p + n_w b_w}{A\tau} \tag{5}$$

where n_w is the number of water associated with the protein, and b_w is its scattering length. The number of water in the layer should satisfy the volume restriction requirement

$$A\tau = V_p + n_w V_w \tag{6}$$

where V_p and V_w are the volumes of protein and water molecules, respectively. The scattering length density of the layer can be calculated from

$$\rho = \phi_w \rho_w + \phi_p \rho_p \tag{7}$$

where ρ_p and ρ_w represent the scattering length densities of protein and water and ϕ_p and ϕ_w are the respective volume fractions ($\phi_p + \phi_w = 1$).

III. EXCHANGE OF LABILE HYDROGENS

Unlike most surfactants or synthetic polymers, proteins contain a substantial amount of labile hydrogens which may or may not fully exchange with bulk water. This uncertainty affects scattering lengths, which would subsequently undermine the accurate determination of the adsorbed amount. Proteins contain two types of labile hydrogens, those on the amide groups in the polypeptide chain and those on the side chains of amino acid groups.

In a more general perspective, monitoring labile H/D exchange is of immense interest to biomolecular research as this process bears indications on kinetic processes of protein folding and unfolding at interfaces. Before the relevance of H/D exchange to interfacial adsorption is discussed it is helpful to review the significance of this topic in bulk solution, as extensive research has already been carried out.

For most fibrous proteins the molecular structure is fairly random and flexible and labile hydrogens are easily exchangeable with bulk water. However, for many globular proteins whose globular structures are retained in aqueous solution, the H/D exchange can be hindered. Much of the obstruction is caused by the steric isolation related to the formation of α-helices and β-sheets within the globular framework [14]. The extent of labile hydrogens with surrounding water has long been used as a useful means to probe the folding and unfolding pathways as any transient change in the structure leads to the access of the blocked labile hydrogens to bulk water. Thus, the ease of exchange can be used as a measure of the degree of masking for the portions of the polypeptide in the folding and unfolding process. The exchange of labile hydrogens in globular proteins with D_2O has been extensively investigated by Hvidt and Nielsen [15] and more recently by Radford et al. [16]. In the case of bovine serum albumin (BSA) there are about 1015 potentially labile hydrogen atoms within each BSA molecule, and some 750 of these will exchange almost instantly at pH 7 and at 0°C [15]. A further 250 exchange over a period of 2 hr. The remaining 30 to 50 appear not to exchange within a further 24 hr. Thus, over 90% of the labile hydrogens are relatively rapidly exchanged. Hvidt and Nielsen have also found that the rate of exchange is also dependent on pH, with all of the labile hydrogens exchanging below pH 3.5 and above pH 9. The exchange rate also rises rapidly with temperature. The hydrogen atoms that are most difficult to exchange are those on the amide groups of the peptide chains because their exchange may be prevented by the formation of the structured hydrophobic environment, the so-called hydrophobic encapsulation and the formation of a hydrogen bonding network. However, most of the labile hydrogens on the backbone are on the outer surface of the globular structure and are therefore readily exchanged, only a relatively small fraction being expected to be difficult to exchange.

For protein molecules adsorbed onto surfaces and interfaces, structural deformation accompanying adsorption may cause sufficient disturbance to the encapsulated peptide chain fragments to allow further exchange of backbone hydrogens. The response of the calculated surface excess to changes in the neutron scattering length density of the layer is approximately linear and any uncertainty in the extent of exchange will convert linearly into an error in the surface adsorbed amount, although the magnitude of the error will be different at each water contrast, as can be shown from Eqs. (3)–(7). Since the difference in scattering length for the labile hydrogens is largest in D_2O, the effect caused by incomplete exchange will be greatest for the D_2O measurements but much less for the other contrasts. Thus, on the basis of the work of others it is unlikely that the extent of exchange is less than about 90%, leading to a maximum error in the surface coverage of 10%. As will be discussed later, the good fit of a given structural model to several reflectivity profiles measured under different contrasts provides some evidence to suggest that exchange must more or less proceed to completion and any uncertainty may not affect the interpretation of the data.

The prospect of experimental monitoring of H/D exchange by neutron reflection can be realized from some simple calculations. We take the adsorption of BSA at the hydrophilic silicon oxide–aqueous solution interface as an example. Out of the total 1015 labile hydrogens, 580 are associated with the polypeptide chain [17]. According to the discussion given above most labile hydrogens on the amino acid side chains will more or less exchange instantly. Some of those on the polypeptide chain may experience difficulty in exchange with D_2O. For simplicity, if 25% of the total 1015 labile hydrogens do not exchange at all, the scattering length of BSA in D_2O is then 0.228 Å. For comparison, if it is assumed that the exchange reaches completion, the scattering length is then 0.257 Å. The total volume of BSA can be estimated from the known amount of amino acids and peptide fragments [18,19], and this value is about 79,100 $\mathring{A}^3$. From the values of scattering length and molecular volume, ρ is calculated to be $2.90 \times 10^{-6} \mathring{A}^{-2}$ for incomplete exchange and $3.25 \times 10^{-6} \mathring{A}^{-2}$ for complete exchange. Obviously, uncertainty in the volume of each individual fragment will affect the reliability of the total volume and thus the value of ρ, but the trend will be the same. If the volume fraction of protein in the layer is taken to be 0.5, the scattering length densities calculated from Eq. (7) are $4.63 \times 10^{-6} \mathring{A}^{-2}$ and $4.80 \times 10^{-6} \mathring{A}^{-2}$, respectively. As the difference in ρ between 75 and 100% exchange is just above the level of experimental error, neutron reflectivity measurement is therefore not so sensitive to the extent of exchange. It should be noted that the possible effect of H/D exchange is greatest in D_2O and the direct effect is on the adsorbed amount. As the difference in ρ is about $2 \times 10^{-7} \mathring{A}^{-2}$ the typical error in surface excess is then about $\pm 5\%$. Under other contrasts, however the effect is expected to be less significant.

The situation described above is largely concerned with the equilibrated systems. It is also of interest to explore if any useful information of the exchange process might be identified at the early stage of dissolution of sample into solvent. The work by Hvidt and Nielsen [15] suggests that a significant fraction of labile hydrogens exchanges within the first 30 min of dissolution of the sample in D_2O. A typical neutron reflection experiment using white beam neutron sources can be done within about 10 min for sufficient statistics to be obtained. Thus neutron reflectivity measurement with time may be a viable option to follow time-dependent exchange. Because lysozyme has very robust structure, it is appropriate to use it as model protein for this test. Again, if it is assumed that half of its labile hydrogens exchange at the point of completion of first reflectivity measurement and the protein volume fraction in the adsorbed layer is 0.5, the scattering length densities for the protein layer are about $4.7 \times 10^{-6} \mathring{A}^{-2}$ for half exchange and $5.0 \times 10^{-6} \mathring{A}^{-2}$ for complete exchange. The reflectivity profiles shown in Fig. 2 were calculated assuming that the layer is adsorbed at the hydrophilic silicon oxide–D_2O interface, with the native oxide layer being 12 $\mathring{A}$ thick. It can be seen from Fig. 2 that there is a gap between the middle part of the reflectivity profiles and the difference is just about measurable. We have tried to follow the actual exchange process of lysozyme by measuring neutron reflectivity with time on the CRISP reflectometer at ISIS Neutron Facility [1,3]. The first reflection profile was obtained 15 min after lysozyme was dissolved in D_2O. Figure 3 compares the first reflectivity profile with that measured after the lysozyme solution was left for about 3 days at room temperature. The two reflectivity profiles are virtually identical, suggesting that the exchange at the solid–D_2O interface must predominantly occur within minutes. This observation is in broad agreement with the results obtained by Radford et al. [16]. From their NMR studies of exchange of labile hydrogens within lysozyme with D_2O, Radford et al. found that of the total 260 labile hydrogens in lysozyme about 180 exchange with D_2O within seconds. Some 30 will exchange within minutes, and some further 20–30 may take hours or days to exchange. The remaining 20–30 may never exchange with D_2O, possibly as a result of hydrophobic encapsulation or restriction caused by hydrogen bonding. As discussed previously, neutron reflectivity is insensitive to the possible presence of the remaining 20–30 unexchanged labile hydrogens within lysozyme.

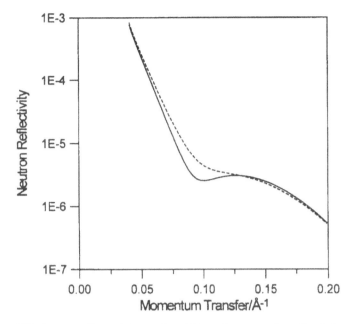

FIG. 2 The effect of incomplete H/D exchange on the reflectivity profiles from the adsorption of a 30-Å lysozyme layer at the hydrophilic silicon oxide–D_2O interface. The volume fraction of the protein in the adsorbed layer is taken to be 0.5 and the reflectivity profiles were calculated assuming 75% exchange (dashed line) and complete exchange (continuous line).

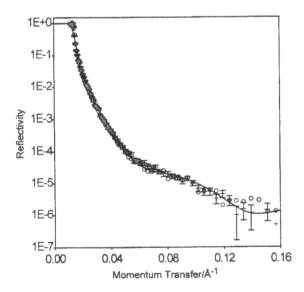

FIG. 3 Reflectivity profiles measured from lysozyme adsorption at the hydrophilic silicon oxide–D_2O interface at 1 gdm^{-3} and pH 7 after the solid sample was dissolved in D_2O for 15 min (O), and 3 days before the reflectivity measured (+). Both reflectivity profiles are identical, suggesting that the H/D exchange must predominantly occur within the first few minutes. The continuous line was calculated with two 30-Å lysozyme layers and a total surface adsorbed amount of 3.6 mgm^{-2}.

IV.   ADSORPTION AT THE HYDROPHILIC SILICON
OXIDE–WATER INTERFACE

Neutron reflection offers a convenient route for determining the in situ structural conformation of biomolecules adsorbed at a given solid–solution interface. These measurements allow monitoring of the effects of surface properties of the solid substrate on the conformational structure changes of proteins during adsorption and the possible correlation of such information to their biological states. In comparison with measurements at the air–water interface, the main difference is the choice of contrasts because the scattering length densities of solids are generally nonzero. Although both silicon and quartz are found to be appropriate solid substrates for neutron reflection experiments (the scattering length densities are 2.1×10^{-6} Å^{-2} for silicon and 4.2×10^{-6} Å^{-2} for crystalline quartz), quartz surfaces are more difficult to polish and are thus rougher. In contrast, the native oxide layer on the surface of silicon can be made very smooth and the thickness is usually controlled between 10–20 Å. The structure and composition of the oxide layer is usually characterized before protein adsorption under different ratios of H_2O and D_2O. The uniformity of the native oxide layer is indicated by the presence of a negligible amount of water inside it. The subsequent protein adsorption is measured also from several water contrasts to enhance interfacial resolution. When the H/D ratio is adjusted to match silicon, the signal is predominantly from the protein layer. However, this signal is weak because the scattering length densities for most proteins are around $2.5-3 \times 10^{-6}$ Å^{-2}. Because most protein layers contain some 50% water, water mixing will further reduce the scattering length density for the layer. Figure 4 shows the neutron reflectivity of a lysozyme layer adsorbed from D_2O (a); in the aqueous solution with its deuterium content adjusted to give a water scattering length density of 4.0×10^{-6} Å^{-2} (CM4) (b); and contrast matched to silicon, with $\rho = 2.1 \times 10^{-6}$ Å^{-2} (CMSi) (c) at 1 gdm^{-3} and pH 7 [20,21]. In each case the adsorbed layer is highlighted differently by the variation of bulk water contrasts. This makes it possible to follow with some accuracy the effects of protein concentration, pH, and ionic strength on both surface coverage and layer thickness. Simultaneous fitting of the three reflectivity profiles gives two-layer interfacial conformations for adsorbed lysozyme with 30 Å each. The inner layer contains more protein than the outer layer and the total adsorbed amount is around 3.6 mgm^{-2}. Since lysozyme has an approximate dimension of 30 × 30 × 45 Å, the formation of the two 30-Å layers at the solid–solution interface suggests the adoption of two sideways-on molecular layers. Other information to be described in the following supports the retaining of globular entity of the protein after adsorption.

Protein adsorption at a hydrophilic solid–water interface is often found to be irreversible. This is usually shown as hysteresis in the adsorption isotherm, i.e., the surface excess with respect to the increase and decrease of bulk solution concentration is not equal [22]. Hysteresis in the adsorbed amount is likely to be related to the number of contacts between protein and surface. Although the fraction of segments in contact with the surface may be small and typically less than a few percent, the adsorption energy can easily be in excess of 100 kJ mol^{-1} because of the large total number of contacts. This is not easily overcome by the weak driving force of a concentration gradient. Thus, irreversible adsorption is not necessarily associated with denaturation of the protein. Alteration of the pH may have a large enough effect

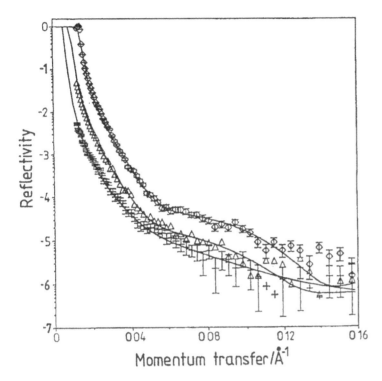

FIG. 4 A two-layer fit to the reflectivity profiles at pH 7 in the presence of 1 gdm^{-3}: ($\Diamond$) D$_2$O; ($\triangle$) CM4 ($\rho = 4 \times 10^{-6}$ Å^{-2}); (+) CMSi ($\rho = 2.1 \times 10^{-6}$ Å^{-2}). The continuous lines were calculated using surface adsorbed amount of 3.6 mgm^{-2}; $\tau_1 = \tau_2 = 30$ Å.

on the interaction of the protein with the surface to provide a driving force for desorption and could therefore be used to provide a more complete test for irreversibility, although it has rarely been used in this manner. This is particularly the case when the protein is stable with respect to pH change in the bulk solution.

Lysozyme has a well-defined equilibrium structure in bulk solution within a significant pH range [23,24]. The effects of pH on lysozyme adsorption at the hydrophilic solid–water interface have been systematically studied by Su et al. [20,21]. The variation of surface excess and layer thickness with solution pH was monitored by following the change of the reflectivity profiles. Figure 5 shows the variation of neutron reflectivity profiles at a fixed protein concentration of 0.03 gdm^{-3}. The measurements were made by varying the solution pH in two cycles. In the first cycle the pH started at 4, increased to 7, then 8, and back to 4; Fig. 5 shows the reflectivity profiles in D$_2$O measured at these four values of the pH. The two reflectivity curves at pH 4 are almost identical, showing that adsorption is reversible with respect to this particular pattern of pH variation. Figure 5 also shows quite clearly that the adsorbed amount increases at pH 7 and reaches its highest value at pH 8. The effects of the reversed cycle, i.e., starting at pH 7, moving down to 4, and returning to 7, are similar, again showing that adsorption is completely reversible and that the adsorbed amount at pH 4 is less than that at pH 7. Neutron reflectivity therefore shows that the adsorption depends only on the actual pH and not on the route to a given

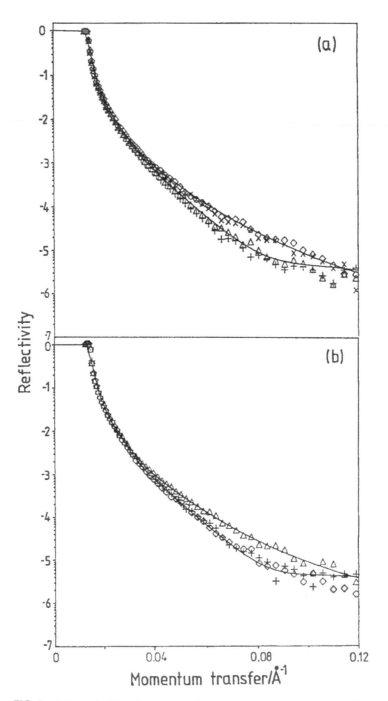

FIG. 5 Effect of pH on lysozyme adsorption measured as the variation of reflectivity in D_2O at 0.03 gdm^{-3}. (a) The solution pH is initially 4 ($\diamond$), then raised to 7 ($\triangle$), followed by 8 (+), before returning to 4 ($\times$). (b) The pH is initially 7 ($\diamond$), lowered to 4 ($\triangle$), and finally returned to 7 (+). The continuous lines are the uniform layer fits to the measured reflectivities at pH 4 and pH 7 and are drawn for guidance. The results suggest completely reversible adsorption.

pH. The reversibility with respect to pH suggests that no denaturation of lysozyme occurs at this interface. This is supported by the fact that over the entire pH range studied the adsorbed layers can be modelled into a single uniform layer and the layer thicknesses are comparable to the axial lengths of the globular structure.

It is interesting to note that the surface excess at pH 8 is 1.9 mgm^{-2}, which is almost twice the value at pH 4, showing that pH has a strong effect on the lysozyme adsorption. However, the decrease of surface excess with pH is opposite to what one might expect just from electrostatic considerations because lysozyme is positively charged and the surface of silicon oxide is negatively charged within the normal pH range [24]. According to Iler [25] the negative charge density on the silica surface is approximately constant between pH 4 and 8, and the net positive charge on lysozyme increases from 8 at pH 8 to 10 at pH 4 [23,24], and hence the electrostatic attraction between lysozyme and the surface should increase with decreasing pH. The observed trend of the variation of surface excess is opposite to the level of electrostatic interaction between the protein and the solid surface, suggesting that the electrostatic attraction between the surface and lysozyme is less important than the lateral electrostatic repulsion between protein molecules within the layer, i.e., the pH dependence of lysozyme adsorption is governed more by protein–protein interaction than by protein–surface interaction.

That adsorption is reversible with respect to pH does not necessarily mean that the structure of the protein is not affected by adsorption. Direct contacts between protein and solid surface may lead to partial breakdown of fragments of α-helices or β-sheets, which may generate further contacts with the surface. However, the native state of lysozyme in aqueous solution is highly ordered with most of the polypeptide backbone having little or no rotational freedom and, although structural rearrangement may occur upon adsorption, the internal coherence of the lysozyme should prevent it from unfolding into loose random structures on the surface [23,26]. Upon adsorption, the thicknesses of adsorbed layers of globular proteins are therefore expected to be comparable with their dimensions in aqueous solution and measurements of the layer thickness can then be used to assess the orientation of the protein molecules on the surface and whether they are distorted to any extent. For example, Claesson et al. [27] have adopted this approach in their determination of the conformation of lysozyme at the solid–water interface using a combination of surface force apparatus and ellipsometry. The advantage of neutron reflection is that it is more sensitive than other techniques to the structural distributions normal to the interface. This has already been supported by the reflectivity measurements shown in Fig. 4 at different water contrasts. The fit of the set of reflectivity profiles at a lower concentration of 0.03 gdm^{-3} to a single uniform layer model of 30 Å for the protein layer, after allowance has been made for the exchange of labile hydrogen atoms, also suggests that the protein molecules are adsorbed sideways-on at this concentration. Similar measurements at different water contrasts at pH 4 also show that the thickness of the layer is about 35 Å at this pH, although the amount of protein in the layer is less. The increase in the thickness of the layer suggests that the layer is slightly tilted. This may not be surprising considering that the charge density within the protein layer increases as the pH is lowered [20,21].

It is interesting to note that we have previously studied the adsorption of lysozyme at the air–water interface. We found that at neutral pH, adsorption changes from sideways-on at very low lysozyme concentration to end-on at higher concentrations. The changeover clearly enables the surface to accommodate more protein molecules. However, an end-on monolayer would give a monolayer thickness of 45 Å, and such a change is not observed at the silica–water interface at the higher coverage found at pH 7. As can be deduced from Fig. 4, neutron reflection measurements are sensitive to the difference between 30 and 45 Å and, even when the surface excess is allowed to float to obtain the best possible match between calculated and observed reflectivities, the fit for a thickness of 45 Å is unacceptable [20,21].

The effects of pH on lysozyme adsorption on the hydrophilic surface are summarized schematically in Fig. 6. The surface coverage decreases with pH as a result of increased repulsion between the molecules inside the monolayer, and this increased level of repulsion is also reflected in the increased percentage of end-on orientations. At higher bulk concentrations, adsorption produces a sideways-on bilayer at pH 7, and the measurements suggest that there is less lysozyme in the outer layer. Only monolayer adsorption occurs at pH 4 because of the strong electrostatic repulsion within the protein layer, but the higher coverage induced by higher bulk lysozyme concentration leads to a higher fraction of molecules adopting the end-on orientation.

The adsorption of BSA and HSA at the hydrophilic silica–water interface has also been investigated using neutron reflection [28,29]. The isoelectric point of BSA and HSA is 4.8 and the effect of shifting solution pH away from the isoelectric point on the adsorption at the hydrophilic solid–water interface can be easily examined. It was found that under a fixed bulk concentration the surface excess decreases when the pH is moved away from the isoelectric point. The effect of pH is more intensified when the bulk concentration is increased. The results from the measurements on HSA are the same in pattern, except that HSA has a slightly lowered surface excess. The difference is consistent with the observation at the air–water interface, confirming that BSA has a higher surface activity. It should be noted that comparison of the possible difference in the surface activity of BSA and HSA has long been of interest. It is worthwhile to mention that Kurrat et al. [30] have recently studied the adsorption of BSA and HSA on the surface of an optical guide comprising SiO_2 (TiO_2) using an optical reflection method. These authors have found that the surface excess of HSA is almost twice as high as that of BSA. This result is clearly opposite to that

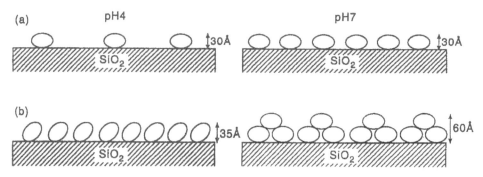

FIG. 6 Schematic diagram to illustrate the variation of surface coverage and structural conformation of lysozyme adsorbed at the silica–water interface with solution pH and bulk concentration. The lysozyme concentrations are (a) 0.03 gdm^{-3} and (b) 1 gdm^{-3}.

obtained by neutron reflection. Furthermore, Kurrat et al. [30] have derived values
of the thicknesses of the adsorbed layers and the dimension of the layers which are
appropriate to a headways-on adsorption. Given that the adsorbed layer is very thin
and that the volume fraction of protein with the adsorbed layer is less than 50%, it
is very unlikely to form a headways-on monolayer.

The general pattern of the adsorption of globular proteins at the hydrophilic
solid–water interface is remarkably similar to what has been described for the adsorption of these globular proteins at the air–water interface. In addition, the surface
excesses at the two interfaces are close and the difference is mostly less than 1
mgm^{-2}. These observations suggest that the structure and composition of the protein
layers is dominated by the extent of electrostatic repulsion within the adsorbed layers
and is less affected by the interaction between the protein and the substrate.

The reversibility of albumin adsorption at the hydrophilic solid–water interface
has also been studied and the adsorption is found to be reversible with respect to
pH at the BSA concentration below 0.15 gdm^{-3}, indicating that there is no denaturation [28]. This is confirmed by the ability to model the reflectivity using a single
uniform layer model. Denaturation would lead to a more fragmented peptide distribution and hence layers of different density. Comparison of the layer thickness with
the dimensions of the ellipsoidal structure of the globular solution structure again
indicates that the molecules adsorb sideways-on. Nevertheless, the layer thickness is
mostly less than the short axis length of 40 Å [17], suggesting that adsorption onto
the hydrophilic surface results in some structural deformation. An increase of layer
thickness with bulk concentration also suggests that the extent of the distortion is
reduced as the lateral repulsion between protein molecules increases. At a concentration of 2 gdm^{-3} the adsorption is found to be irreversible to pH. The irreversibility
is evidenced by the higher surface excess found when pH is increased from 5 to 7
than the surface excess obtained from the direct BSA adsorption onto the bare oxide
surface at pH 7. This difference in surface excess is likely to be caused by the
unavailability of sufficient energy to detach all the contacts simultaneously when the
total surface excess is high.

V. ADSORPTION AT THE HYDROPHOBIC SOLID–WATER INTERFACE

The results described above have shown that adsorption of globular proteins onto
the hydrophilic silicon oxide–water interface does not lead to the breakdown of the
globular framework. It is useful to examine how the adsorbed layer is affected when
the hydrophobicity of the solid surface is increased. Although a great deal of effort
has been made to study the adsorption of globular proteins onto different solid substrates, it is difficult to derive any simple roles from such work because almost all
of these studies were made using solid substrates which are of different origin
[22,23]. For example, change in surface hydrophobicity is usually mixed with a
change in the charge density on the surface. Surface hydrophobicity can be varied
in a more controlled manner by grafting an organic layer onto either quartz or silicon
and confer on the outer part of this layer an appropriate chemical functionality. Using
this approach we have been able to graft monolayers of alkyl chains attached on the
outer surface with functional groups such as OH, NH_2, $PO_4C_2H_4N(CH_3)_3$ (phospho-

rylcholine, PC). While many of these studies are still in progress we show in this section the adsorption of lysozyme onto the hydrophobic solid–water interface to illustrate some of the basic features [31]. The hydrophobic surface was prepared by self-assembly of a monolayer of octadecyltrichlorosilane (OTS) and, for the experiment to be discussed, the hydrophobic layer was coated from $C_6H_{13}C_{12}D_{24}SiCl_3$.

The effect of pH on lysozyme adsorption at the hydrophobic solid–water interface was investigated in a similar way to that described in the previous section at the hydrophilic solid–water interface. Thus, the lysozyme concentration was fixed at 0.03 gdm^{-3} with a constant ionic strength of 2 mM and the layer structure studied as a function of pH. The first part of the experiment established that adsorption of lysozyme on this surface is irreversible with respect to concentration and pH, in contrast to its behavior at the bare silica surface. Thus if the pH were varied in two opposite directions, (1) from 7 to 4 and back to 7 and (2) from 4 to 7, different reflectivity profiles were obtained, corresponding to different final states depending on the pH path. Figure 7a shows the reflectivity profiles in D$_2$O measured for the first three values of the pH with the profile for the partially deuterated OTS/D$_2$O interface included for comparison. The shift in the minimum toward a lower κ value in the reflectivity at the initial pH 7 than for OTS on its own indicates that there is adsorption of lysozyme onto the hydrophobic surface and a further shift in reflectivity upon the first change in pH from 7 to 4 suggests an increase in the adsorbed amount of lysozyme. The change in pH from 4 back to 7 has little further effect on the adsorbed amount, suggesting that the adsorbed protein layer is no longer able to respond to a change in pH. Thus the protein has been irreversibly adsorbed to a saturation limit. Two of the reflectivity profiles for the reverse cycle of pH change are shown in Fig. 7b. Although there is some response of the adsorbed layer upon pH change from 4 to 7, the level of the change is less than for the first pH change in the opposite direction shown in Fig. 7a. The small difference in this case is either caused by adsorption, desorption, or structural rearrangement, and the exact structural change upon pH shift can be obtained by modelling the surface layer distribution.

The reflectivity profiles at the initial pH 7 and at the initial pH 4 are similar, which suggests that the initial adsorption is largely driven by the hydrophobic surface and is indifferent to the solution pH. Although in some cases the uniform layer model does not fit the intensity at the minimum in the reflectivity profile exactly, it does reproduce the position of the minimum well. In general, the uniform layer model usually offers an accurate estimate of the surface excess, even when it does not account too well for the detailed shape of the reflectivity. The conclusion from the monolayer fit is that the thickness of the layer is between 11 and 15 Å for all of the pH conditions studied.

The strong hydrophobic interaction between the peptide fragments and the solid surface may lead to the breakdown of the globular assembly. If the denatured protein completely unfolds, there will be regions of loose random structure on the surface. Depending on the extent of denaturation hydrophobic fragments will be segregated at the OTS surface allowing hydrophilic fragments to extend into the aqueous solution. Information regarding the state of the adsorbed lysozyme can be obtained by examining the structural distribution of the protein layer on the OTS surface. The work on the adsorption of lysozyme at the hydrophilic surface shows that in all circumstances the thickness of the layer could be related to the known dimensions

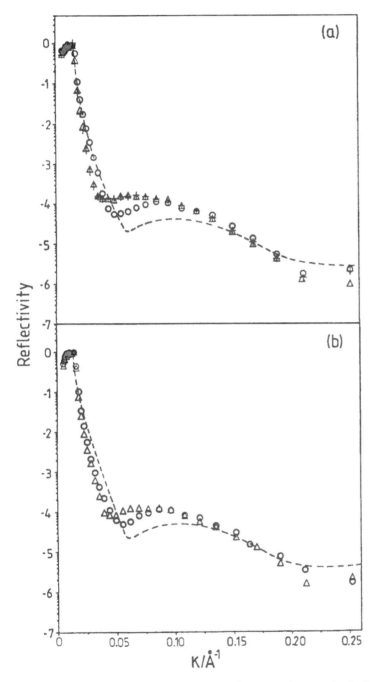

FIG. 7 Effect of pH on the reflectivity for lysozyme layers adsorbed at the partially deuterated OTS–D$_2$O interface with its bulk concentration fixed at 0.03 gdm^{-3}. (a) The solution pH was initially 7 (O), then lowered to 4 (Δ), finally back to 7 (+). (b) The pH was initially 4 (O), then increased to 7 (Δ). The reflectivity from the OTS–D$_2$O interface in the absence of protein is included for comparison (dashed lines).

of the protein in solution and this is taken to support the tertiary structure of the protein remaining largely intact on adsorption. However, for the hydrophobic surface no model of the interfacial layer where the tertiary structure of the lysozyme was retained could be made to fit the observed profiles. This, even without more detailed fits, further confirms that lysozyme is denatured at the hydrophobic OTS surface.

Once the constraint of using the dimension of the unperturbed globular protein is discarded it is found to be relatively straightforward to fit the reflection data to structural profiles characteristic of synthetic polymers. Lu et al. [31] found that if a single uniform layer model was used, the shape of the reflectivities could be approximated by a protein layer with the thickness of 10 Å and a protein volume fraction of 0.85. Note that an added advantage of the labeled hydrophobic layer is that the composite thickness of protein and hydrophobic layer gives interference features in an accessible range of momentum transfer. A layer only 10 Å thick would normally be too thin to measure with any accuracy. The thickness of 10 Å is equivalent to the length of the side chains of two average amino acids. This suggests that the adsorbed protein layer is in the form of peptide chains with the hydrophobic side chains adsorbed on the surface of OTS and the poor fits in the uniform layer model must be caused by the neglect of the hydrophilic side chains extending into aqueous solution. Further evidence for this is the correlation of the variation of reflectivity in the low κ region with pH. This is particularly the case at pH values of 4 and 7 following initial deposition of the protein. Lu et al. [31] have demonstrated that if a two-layer model is used with the outer layer having lower volume fraction to account for the diffuse layer the fitting to the reflectivity profiles is improved. The importance of these small, but measurable, differences between the two models shows how important it is to maximize the contrast between the protein layer and the two bulk phases. The results from the two-layer model show that although the adsorbed amount of lysozyme at pH 4 and 7 is the same, the second layer is thicker at pH 4. This variation of the diffuse layer with solution pH suggests that the tail region contains charged groups. As described previously, the net charge on a lysozyme molecule depends on solution pH, and the number of charges decreases from 10 at pH 4 to 8 at pH 7 [23,24]. Thus at pH 4, the charge density within the protein layer is higher and, once the protein is denatured, the greater freedom of movement will allow the stronger repulsion to cause the layer to be more diffuse. At pH 7, this repulsion is reduced and the protein layer becomes more dense. The structural changes within this lysozyme layer are quite different from those on the hydrophilic surface and provide further strong evidence that the protein is denatured.

Although the two-layer model fits all of the reflectivity profiles in different pH conditions, it is a very coarse grained model and one would expect the actual segment distribution profiles to be smoother, although the intrinsic resolution of the neutron reflection experiment is not great enough to distinguish the two-layer model from a continuous distribution. In previous work on the adsorption of the polyethylene oxide (PEO) at the air–water interface Lu et al. have shown that an equivalent effect can be obtained by further division of the two layers [32]. Figure 8 shows volume fraction profiles where several layers have been used to fit the observed reflectivities and where the distributions are further smoothed by joining the midpoints of each block in the histogram. Not surprisingly, these protein distributions fit all the sets of data at least as well as the simpler structure. The loops or tail fragments in the diffuse

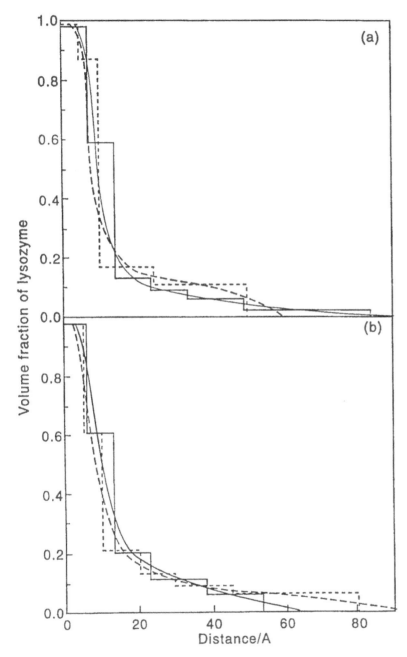

FIG. 8 Volume fraction distributions obtained by smoothing the structural distributions from the two-layer models using six- to seven-layer models. The solution pH was changed from (a) 7 (dotted line) to 4 (continuous line) and (b) from 4 (dotted line) to 7 (continuous line). The histograms were smoothed by drawing a curve through the middle position of each individual slab at a given pH. Each structural profile fits three to four reflectivity profiles from different isotopic contrasts at the same pH.

region are composed of either the hydrophobic fragments, which are forced into the diffuse layer because of steric constraints, or peptide chains which contain more polar or charged groups and which dislike the OTS surface. The response of the diffuse layer distribution to the change in solution pH suggests the latter alternative. If polar and charged groups dominate the composition of the diffuse layer, the scattering length for the protein fragments in this region would be different from the average value. It should be noted that it is the average value that has been used in the fitting described above. In the case of D_2O the scattering length of the protein fragments in the diffuse layer will tend to be higher due to exchange of a greater number of labile hydrogens and the current treatment might then underestimate the amount of protein in the diffuse region. It is difficult to estimate such errors because they depend on the exact content of the components in the diffuse layer, which is not known. However, since the fraction of protein in the diffuse layer is in all cases less than 0.2 of the total surface excess the uncertainty will not lead to serious errors in the values of the total surface composition.

Adsorption of lysozyme onto hydrophobic surfaces has also been studied using ellipsometry [33]. It was generally found that the adsorbed layers could be modelled using a uniform layer model with a thickness of around 100 Å. Such a large discrepancy is likely to be caused by the insensitivity of the elliptical signal to the protein layer that is heavily mixed with water. X-ray reflection has also been used to study protein adsorption, e.g., a recent study by Petrash et al. [34] on the adsorption of HSA on a self-assembled monolayer of hexadecyl trichlorosilane on silica. In comparison with neutron reflection, a typical x-ray reflection can measure reflectivity up to 10^{-8} with κ extending out to about 0.6 Å^{-1}. This higher sensitivity renders x-ray reflection a greater structural resolution, but the main drawback for x-ray reflection at the moment is that such experiments can only be done ex situ and the subtle changes characteristic of proteins in an aqueous environment are lost.

VI. ADSORPTION AT THE C$_{15}$OH−WATER INTERFACE

In a systematic investigation of the effect of surface hydrophobicity on the adsorption of proteins at the solid−solution interface, we have examined the adsorption behavior of model globular proteins at the C$_{15}$OH−water interface [35]. C$_{15}$OH represents pentadecyl trichlorosilane with terminal hydroxyl groups [Cl$_3$Si(CH$_2$)$_{15}$OH, abbreviated to C$_{15}$OH]; its attachment onto planar silicon oxide surface produces a self-assembled monolayer of C$_{15}$OH with hydroxyl groups on the outer surface. The advancing contact angle (θ_a) for C$_{15}$OH was 53° and is approximately in the middle of that of bare silicon oxide (ca. 0°) and that for OTS (ca. 110°). We have shown that whilst the OTS surface deteriorates the globular assembly of proteins, like egg white lysozyme, the globular framework is retained when these protein molecules are adsorbed on the hydrophilic silicon oxide. It is hence of interest to examine the effect of surfaces with intermediate hydrophobicity. It is relevant to mention that short alkyl chains bearing hydroxyl groups are readily incorporated into polymeric hydrogels. These hydroxyl groups are known to facilitate water uptake, which is essential for many biomaterials applications. A typical example of these hydrogel polymers is poly(2-hydroxyethyl methacrylate), which is widely known as HEMA polymer. The water content inside HEMA polymer can vary from 30 to 80 wt%, depending on its exact chemical nature and physical structure. An important function of these polymers is to reduce nonspecific protein fouling. As protein adsorption onto biomaterials is one of the primary events, leading to thrombus formation at the blood–biomaterial interface, it is useful to explore some molecular characteristics related to the interactions between hydroxyl surfaces and model protein molecules. This structural information can be of relevance to the interpretation of the complex processes relating to the exposure of polymeric hydrogels to blood. An added advantage of using the chemically grafted $C_{15}OH$ monolayer is its fixed surface composition. This is in contrast to the swelling and component redistributions associated with polymeric hydrogels.

Because spectroscopic ellipsometry is quick at following the time-dependent process, it has been used to characterize the dynamic adsorption of lysozyme at the $C_{15}OH$–solution interface. The ellipsometric measurements were carried out at an incidence angle around 75° and over a typical wavelength range of 350–650 nm using a Woollam WVASE32 [36]. We show in Fig. 9 the ellipsometric profiles recorded for the adsorption of 1 gdm^{-3} lysozyme at pH 7 onto the surface of silicon

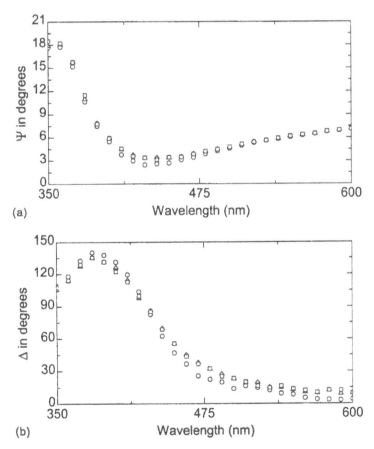

(a)

(b)

FIG. 9 Ellipsometric profiles of ψ (a) and Δ (b) recorded after 2 min (□) and 30 min (△) for the adsorption of 1 gdm^{-3} lysozyme at pH 7. The measurements at the $C_{15}OH$–buffered water interface (○) are also shown for comparison.

oxide chemically coated with a monolayer of $C_{15}OH$. In a spectroscopic measurement, the two ellipsometric angles ψ and Δ are usually recorded as a function of wavelength λ. The angle ψ measures the change in the amplitude of the polarized beam after reflection, whilst Δ measures the change in phases [37]. The first set of ψ and Δ was measured when the surface was in contact with the buffer solution (marked as circles). The second set of ψ and Δ was recorded 2 min after the hydroxyl surface was in contact with the lysozyme solution (squares), and the third set after 30 min (triangles). Clear difference can be seen between the first two measurements, but little difference is observed between the second and third sets of the data, suggesting that the adsorbed amount varies little after just 2 min. These measurements show that the time-dependent adsorption at the $C_{15}OH$–solution interface only occurs within the first few minutes. Similar measurements were also carried out for adsorption of lysozyme and BSA at the solid–solution interfaces where the solid substrates were either bare silicon oxide or OTS. It was found that the time-dependent absorption for these proteins at solid–solution interfaces is much shorter than observed at the air–solution interface.

The variation of adsorbed amounts at different lysozyme concentrations is shown in Fig. 10. The exact amount of adsorption can be reliably extracted from fitting refractive index profiles to ψ and Δ simultaneously using the optical matrix formula, although it is less reliable to derive information about the thickness of the layers from the ellipsometric data. It can be seen from Fig. 10 that adsorption at the higher lysozyme concentration shows a slower process to equilibrium but the adsorbed amount under both concentrations reaches constant values after some 10 min.

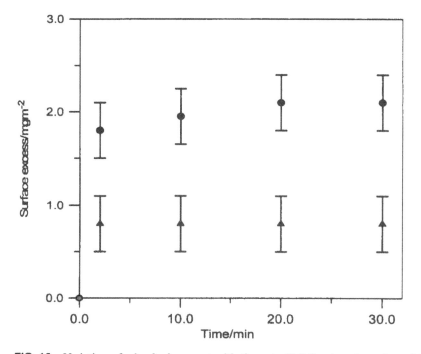

FIG. 10 Variation of adsorbed amount with time at pH 7 for the adsorption of 0.03 (▲) and 4 gdm^{-3} (●) lysozyme at the $C_{15}OH$–buffered aqueous solution interface.

Similar measurements have been carried out under different pH conditions. Again, it was found that it only took minutes to reach a constant amount of adsorption.

Because neutron reflection provides better resolution to the structure of the adsorbed protein layer than ellipsometry, it was used to characterize the structure of bare oxide and subsequently the $C_{15}OH$ layer under different ratios of H_2O and D_2O. The simultaneous fitting of reflectivity profiles showed that the thickness of the oxide layer τ was 20 Å, whilst its scattering length density ρ was fixed at 3.4×10^{-6} Å^{-2}. As the value of ρ is the same as that of amorphous oxide, the result suggests a negligible amount of water in the oxide layer. Chemical grafting through silane reaction produces a hydrogenated $C_{15}OH$ layer with its scattering length close to zero. This organic layer is well highlighted under D_2O. Figure 11 compares the reflectivity profiles at the solid–D_2O interface before and after $C_{15}OH$ grafting. The best fit (continuous line) was produced with a thickness of 16 Å and $\rho = (0.2 \pm 0.2) \times 10^{-6}$ Å^{-2}. The errors quoted here reflect the range of variation beyond which obvious deviations between the measured and calculated reflectivities occur. That the thickness of the layer is shorter than the fully extended length of 20 Å for the hydroxyl layer suggests that the layer is on average tilted. The scattering length density for the pure liquid pentadecanol is -2×10^{-7} Å^{-2}, and the deviation from this reflects the extent of mixing of water into the layer. The small difference thus suggests that within the experimental error the layer is close to liquid alkanol. It should be noted that in comparison with the attachment of organic monolayers on

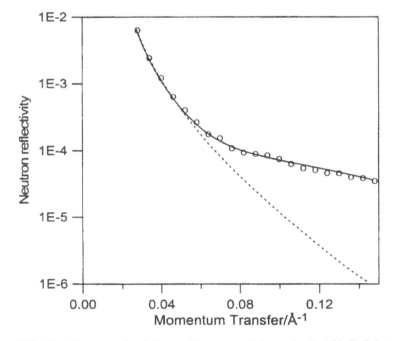

FIG. 11 Neutron reflectivity profile measured from the $C_{15}OH$–D_2O interface. The continuous line was calculated assuming that pentadecyl-1-ol layer is 16 Å thick with liquidlike density. The dashed line is the calculated reflectivity for the bare SiO_2–D_2O interface with the thickness of the oxide layer taken to be 20 Å thick.

gold through thiol reaction (see the work of Prime et al. [38]), the coupling of trichlorosilane groups with the hydroxy groups on the surface of silicon oxide is less selective. The poor binding on silicon oxide has resulted in a less well packed organic layer, evidenced by the high advancing contact angle of 53°, as compared with some 10 to 30° for the hydroxyl layers coated on gold. Neutron reflection offers a reliable measurement of the average volume fraction of the $C_{15}OH$ layer, but is insensitive to the possible presence of defects within the surface region. Other techniques such as IR can offer detailed information about the *trans/gauche* conformations of the underlying carbon chains on gold, but are completely ineffective when such thin layers are grafted on silicon oxide.

Lysozyme adsorption was subsequently carried out in D_2O at pH 7, and the resultant reflectivity profiles are shown in Fig. 12. At the lowest concentration of 0.03 gdm^{-3}, a small difference was observed between the reflectivity profiles in the presence and absence of lysozyme, suggesting little lysozyme adsorption. As lysozyme is increased to 1 gdm^{-3}, noticeable deviation occurs around $\kappa = 0.05$ $Å^{-1}$ where a broad interference fringe is clearly visible. Further increase to a lysozyme concentration of 4 gdm^{-3} shifts the broad peak toward lower κ, indicating the thickening of the adsorbed layer and an increased level of adsorption. The reflectivity profile at 1 gdm^{-3} was best fitted by a uniform layer distribution with the layer thickness of 38 Å. The volume fraction of lysozyme within the layer is below 0.2, showing that the adsorbed layer is very loose. This layer thickness is between the two axial lengths of crystalline structure for lysozyme, indicating the formation of a

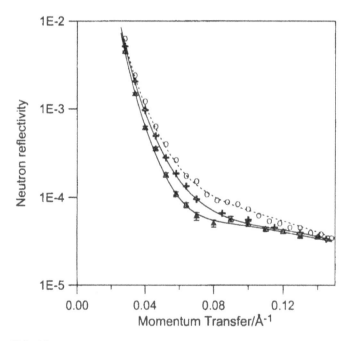

FIG. 12 Neutron reflectivity profiles measured from the $C_{15}OH-D_2O$ interface after the adsorption of lysozyme at 0.03 (o), 1 (+), and 4 gdm^{-3} (▲) at pH 7. The continuous lines are the best fits. The dashed line is the best fit to the measured reflectivity before lysozyme adsorption.

tilted monolayer. That no density gradient is required to fit the data suggests that the adsorbed lysozyme also retains its globular framework. The sensitivity of the measured reflectivity to the thickness of the layer can be tested by varying τ away from the optimal value, while ρ is varied accordingly so that minimum deviation is caused between the calculated and the measured reflectivities. We show in Fig. 13 the comparison of the fits for the adsorption at 1 gdm^{-3} and pH 7. The long dashed line was calculated assuming that the layer is 50 Å thick, and the short dashed line for a layer of 25 Å. The large variation of the shape and level of the reflectivity with layer thickness strongly indicates that the measurement is sensitive to the thickness of the layer within the quoted error of a few angstroms.

The formation of a tilted monolayer of 38 Å at 1 gdm^{-3} and pH 7 is in contrast to the sideways-on adsorption under the same solution condition at the SiO_2–water interface. There are two main factors contributing to the observed difference. First, the presence of $C_{15}OH$ layer changes the level of van der Waals interaction between the surface and the protein layer [39]. Second, the coating of the $C_{15}OH$ provides a spacing between SiO_2 and the protein layer. As a result, the adsorbed lysozyme molecules are no longer in direct contact with the weakly negatively charged surface, and the preferred conformation may result from the reduced electrostatic interaction between lysozyme and the coated surface. As the lysozyme concentration is increased to 4 gdm^{-3}, the density distribution can be modelled into a bilayer. The inner layer is about 42 Å and is thicker than the layer formed at the lower concentration of 1 gdm^{-3}, suggesting that while more lysozyme is adsorbed the layer is further tilted

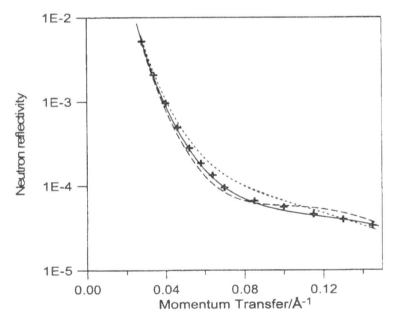

FIG. 13 Neutron reflectivity measured from the $C_{15}OH$–D_2O interface after adsorption of 1 gdm^{-3} lysozyme at pH 7. The continuous line is the best fit obtained by taking $\tau = 38$ Å and $A = 2300$ Å^2. The long dashed line was calculated assuming that the lysozyme layer is 50 Å^2 thick and the short dashed line using the layer thickness of 25 Å.

toward a headways-on conformation. The formation of an almost headways-on inner layer is consistent with the area per molecule of 1150 Å^2, which is smaller than the minimum value required for sideways-on adsorption but above the minimum value required for headways-on adsorption.

The concentration dependence of lysozyme adsorption at pH 4 under the otherwise same solution conditions was also examined. Overall, it was found that the level of adsorption is substantially reduced as the pH is lowered. No adsorption was detected at 0.03 gdm^{-3}. As the concentration was increased to 1 gdm^{-3}, small difference was observed between the reflectivity profiles in the presence and absence of lysozyme, suggesting weak adsorption. At 4 gdm^{-3}, visible difference between reflectivity profiles was seen, suggesting an increased adsorption. The thickness of the adsorbed layer at 4 gdm^{-3} was found to be 35 Å and was indicative of the formation of a slightly tilted monolayer.

Lysozyme adsorption was irreversible with respect to its bulk concentration at both hydrophilic and hydrophobic solid–water interfaces. The same situation was found at the C_{15}OH–water interface, suggesting the relatively high energy barrier for desorption under dilution. The extent of reversibility with respect to pH is however affected by surface hydrophobicity. We found that, for lysozyme adsorbed at the hydrophilic silicon oxide–water interface, the adsorption is reversible with respect to pH over a wide range of concentration. The level of adsorption depends only on the final solution pH, regardless of how it has been reached. At the OTS-coated solid–water interface, however, the adsorption is completely irreversible, possibly as a result of the increased contacts among fragmented peptides and between the peptides and the surface. It is thus of interest in examining the reversibility of adsorption on the surface of C_{15}OH.

The reversibility with respect to pH was examined by fixing lysozyme concentration at 1 gdm^{-3} and the routine of pH variation was the same as used previously, that is, the measurement started at pH 4, followed by pH shifting from 4 to 7, and then back to 4. Each of these reflectivity profiles at the shifted pH was compared with that corresponding to the first exposure of the fresh surface under the same pH. Each reflectivity pair was identical, suggesting that the adsorption is completely reversible. A similar experiment was then performed at a higher lysozyme concentration of 4 gdm^{-3}. The two reflectivity profiles shown in Fig. 14a were obtained at pH 4 and are virtually identical within the experimental error, which would suggest a reversible adsorption. However, there is a large discrepancy between the two reflectivity profiles measured at pH 7 (Fig. 14b), suggesting that the adsorption is irreversible. The position of the broad interference fringe corresponding to the shift of pH from 4 to 7 is moved to lower κ, suggesting more adsorption than that measured at initial pH 7. The dashed line shown in Fig. 14 represents the reflectivity from the C_{15}OH-coated solid–D_2O interface, and the deviation from this curve is indicative of the level of adsorption. It is then clear from Fig. 14 that, in spite of the discrepancy between the two reflectivity profiles at pH 7, the overall level of adsorption at pH 7 is greater than at pH 4, a trend similar to that observed at the lower lysozyme concentration of 1 gdm^{-3}. The irreversible adsorption observed at 4 gdm^{-3} when solution pH was shifted from 4 to 7 is likely to be caused by the preadsorbed lysozyme at pH 4. When the pH was increased to 7, the preadsorbed layer was largely retained, thus presenting a surface that was different from the hydroxyl groups. This preadsorbed lysozyme layer must have induced further lyso-

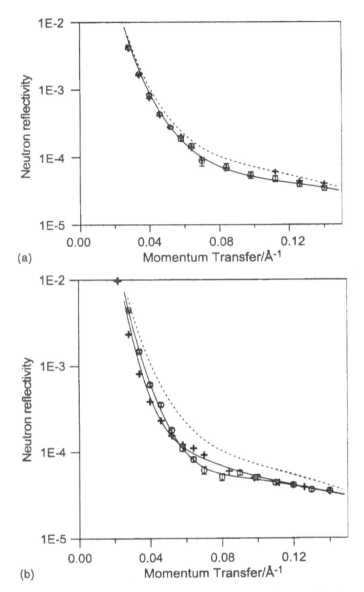

FIG. 14 Neutron reflectivity measured at pH 4, followed by increasing pH to 7 (+), and then back to 4, with lysozyme concentration fixed at 4 gdm^{-3}. (a) The two reflectivity profiles measured at the initial pH 4 (○) and the value shifted from 7 to 4 (+) are compared. (b) The corresponding two measurements at pH 7 are compared. Whilst the two reflectivity profiles at pH 4 are identical, the two at pH 7 are not, hence indicating irreversible adsorption. The continuous lines are the best fits and the dashed line is the best fit to the interface before lysozyme adsorption.

zyme adsorption, resulting in a total amount greater than that obtained at the fresh $C_{15}OH$–solution interface.

The effect of surface hydrophobicity on protein adsorption has been vigorously examined by Prime et al. [37,40]. In their work the functionalized surfaces were formed by self-assembly onto Au and Ag through thiol reaction. Lysozyme adsorption on the surfaces with terminal hydroxyl groups, with ethoxylate units bearing terminal OH and OCH_3, and with the mixtures of the ethoxylates and alkyl chains was investigated. Whitesides et al. [37,40,41] found that θ_a for self-assembled monolayers with terminal hydroxyl groups may vary over a wide range, but mostly below 30°, as compared with some 50° observed on silica, suggesting that θ_a is affected by the packing of the coated monolayer and that the layers are overall better packed through thiol reaction than through silane coating. The effectiveness of the hydroxyl surfaces in prohibiting protein adsorption were found to be comparable to those grafted with ethoxylate units. In addition, these authors found that layers coated on the surface of Ag give higher level of *trans* conformation and hence lower contact angles, indicating better packing on the outer surface by oxygen atoms. The surface excess for lysozyme was, however, found to be much greater on the coated Ag surfaces, suggesting that the surfaces with greater hydrophilicity produce higher surface excesses. A further piece of information that supports the trend of adsorption is our recent study of protein adsorption on poly(methylmethacrylate) (PMMA). The contact angle of PMMA is about 80° and it was found that the surface excesses at the PMMA–solution interface are also lower than the corresponding values on OTS and SiO_2. It should however be noted that the correlation between the level of surface adsorption and surface hydrophobicity through contact angle is very empirical, because contact angle itself offers no information about the structure and chemical nature of the surface. An intermediate value of contact angle can be obtained through different means. For example, a contact angle of 50° can be achieved by coating a loose layer of short chain alkanes by forming a mixed layer of alkyl chains and hydrophilic groups. It is unclear how the adsorption is affected by the nature of these surfaces under different surface conditions.

VII. ADSORPTION AT THE PC MONOLAYER–SOLUTION INTERFACE

Many studies in the literature have shown that coatings of certain polymers can reduce nonspecific protein deposition [42,43]. Amongst many polymer coatings so far studied, those containing phosphorylcholine groups are extremely effective at reducing protein deposition in vitro (e.g., see Refs. 44–48). Similar protein-repelling effects of these PC polymers have also been observed in the work by Campbell et al. [49] and in a number of other literature studies referred to in Ref. 50 when the polymers were exposed to blood. From a biophysical viewpoint, the advantage in using pure proteins instead of blood is that the extent of protein deposition can be related to the structure and composition of the coated layer and that complication relating to the unknown composition of blood is avoided. This approach enables some delicate physical characterization to be made toward understanding the nature of surface biocompatibility. The encouraging results from PC polymers have motivated us to explore whether a similar antifouling effect is rendered by a molecular monolayer of phospholipids chemically grafted onto a solid substrate. It has however been widely perceived that a chemically anchored phospholipid monolayer would not offer a high level of biocompatibility because the attachment of these molecules onto a solid substrate constrains their mobility. Self-assembly of phospholipid bilayer through physical attachment has been reported to show high efficiency in reducing protein adsorption [51–53]. However, the main shortcoming with the strategy of mimicking labile phospholipid bilayer structure is the lack of stability stemming from the physical coating.

In the last few years, we have shown that contrary to the common belief, a chemically anchored organic monolayer with terminal PC groups can reduce protein adsorption very effectively. The PC compounds used for surface coating were assembled in bulk solution and the surface layers were formed via dip coating. A typical monomer synthesized was APTMS-APC that was made by refluxing 3-aminopropyl trimethoxysilane (APTMS) with acryloyloxyethylphosphorylcholine (APC) in isopropanol for 2 hr [54]. Since the PC compound has a labile hydrogen on the secondary amine group, its dimer form can be obtained by coupling two monomers with a bridging spacer such as a diisocyanate. The coupling was again carried out in isopropanol by refluxing the monomer and the spacer for a further 2 hr. The chemical structure of the APTMS-APC dimer is shown in Fig. 15. An obvious advantage of this reaction scheme is the direct control of the surface coverage of the PC groups by the length of the spacer. It should be remembered that Hayward et al. [55] have explored the concept of the chemical grafting of PC monolayer onto silicon oxide some 20 years ago. But in their case, the attached alkyl chain–bearing terminal PC groups were unstable as a result of the presence of an oxygen between carbon and silicon, the C-O-Si connection. The resulting alkyl silicate structure readily hydrolyses, with loss of the alkyl chains and their PC functionality. Our approach is similar to that of Hayward et al., but we have sought direct chemical bonding between the carbon and silicon so that stable chemical grafting is obtained. In comparison with C-O-Si bonding, the formation of Si-O-Si connections at the end of the organic monolayer is much more stable. This part of the layer is identical to the underlying silica layer in chemical composition although their structures might be different.

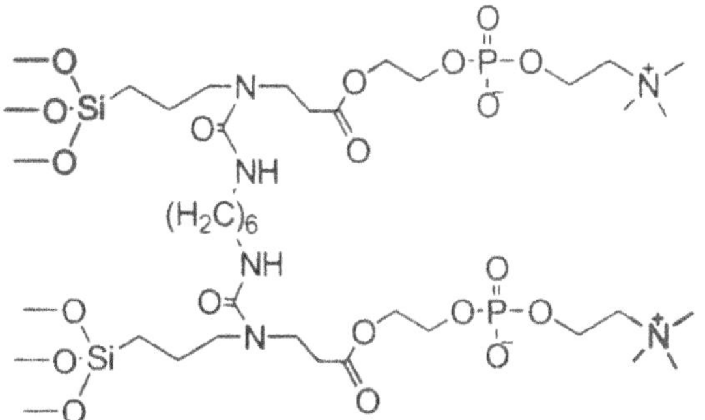

PC Dimer

FIG. 15 Molecular structure of a PC dimer assembled in bulk solution and its formation of monolayer or bilayer coatings is manifested through dip coating.

The procedure used for the characterization of protein adsorption was the same as that developed for OTS and $C_{15}OH$ surfaces. The oxide layer in this case was found to be 16 Å thick without any defects and the PC layer to be 18 Å containing 35% water. The higher water content in the PC layer was likely to be caused by the bulky PC head groups and the loose packing possibly occurred underneath the PC outer layer. Protein adsorption was first characterized using lysozyme solutions with the lysozyme concentrations of 0.03, 1, and 4 gdm^{-3}, all at pH 7 and in D_2O. For clarity, only the measurements at the lowest and highest lysozyme concentrations are shown in Fig. 16. The reflectivity profile from the coated solid–pure D_2O interface is shown (dashed line) for comparison. Because the difference between reflectivity profiles is small, the scale range was expanded so that the change caused by lysozyme adsorption is more easily visible. The main observation from these measurements is that the adsorbed lysozyme layer is thick but very diffuse. At the lowest lysozyme concentration of 0.03 gdm^{-3}, the layer is some 60 Å thick and the volume fraction of protein is 0.08. The corresponding surface excess is 0.8 mgm^{-2}. At the highest concentration of 4 gdm^{-3}, $\tau = 68$ Å, $f_p = 0.17$, and $\Gamma = 1.8$ mgm^{-2}. The feature of the formation of thick and diffuse protein layers is remarkably similar to the observation found on the PC polymer surfaces in our previous work [46,47], but is very different from the structural conformations observed on the bare oxide and OTS surfaces, as will be discussed later.

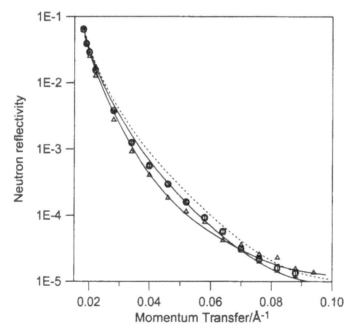

FIG. 16 Neutron reflectivity measured at the PC monolayer coated silicon oxide–D_2O interface with lysozyme concentration at 0 (dashed line), 0.03 gdm^{-3} (○), and 4 gdm^{-3} (Δ). The continuous lines were calculated by taking the surface adsorbed amount to be 0.9 and 1.8 mgm^{-2} and the lysozyme layer thicknesses to be 60 and 68 ± 5 Å at 0.03 and 5 gdm^{-3} lysozyme, respectively. The solution pH was controlled at 7 and the total ionic strength at 0.02 M.

After lysozyme adsorption study, the solution was drained and the surface was cleaned with pure buffered solution to see if any adsorbed protein could be removed. It was found that rinsing with water or buffer solution could result in partial protein removal. The fraction of lysozyme removed was greater than that achieved on other surfaces previously described. Complete removal of adsorbed proteins could be achieved by exposing PC dimer surfaces to a cationic surfactant, dodecyl trimethyl ammonium bromide ($C_{12}TAB$). In comparison, only partial removal was obtained for lysozyme layers adsorbed on OTS and hydrophilic silicon oxide surfaces. These differences may suggest that the proteins adsorbed on PC surfaces were loosely attached.

The adsorption of BSA was measured at 0.05 and 2 gdm^{-3} and at pH 5. The first addition of BSA at 0.05 gdm^{-3} causes an observable shift in the reflectivity, and the subsequent increase in BSA concentration to 2 gdm^{-3} has much smaller effect. The structural parameters obtained from data modelling also showed the formation of a thick but diffuse BSA distribution. The layer thickness at both BSA concentrations was found to be 80 Å with $f_p = 0.08$ and $\Gamma = 1$ mgm^{-2}. Similar adsorption measurement was subsequently made using fibrinogen, at the concentrations of 0.1 and 1 gdm^{-3}, all at pH 5 in D_2O, and it was clear that the reflectivity changed very little after the first addition of fibrinogen, again showing that there was little effect of protein concentration. The result thus suggests a lower level of adsorption of fibrinogen on the PC surface. The data analysis shows that the loosely adsorbed protein could also be approximated to a diffuse layer of some 80 Å with $f_p = 0.08$ and $\Gamma = 1$ mgm^{-2}.

VIII. SUMMARY

The direct characterization of the in situ structural conformation of a protein layer adsorbed at the solid–solution interface has been difficult. Through a number of examples given here we have demonstrated that neutron reflection is effective at determining the structure and composition of an adsorbed protein layer. Isotopic contrast variation helps to improve the resolution of the experiment. Accurate determination of the layer structure allows us to infer not only the structural conformation of protein molecules under different solution conditions, but also the possible role played by protein molecules during different interfacial processes.

The examples given in this chapter have also indicated that because of its molecular range structural resolution, neutron reflection can be used to investigate the effects of substrate surface properties on the structure and composition of adsorbed protein layers. Table 1 lists the key structural parameters obtained from neutron reflectivity analysis [54]. These results show that apart from varying adsorbed amount on different interfaces, the protein layer thickness and volume fraction also vary substantially. This information together with further interfacial characteristics such as inhomogeneous density distribution and pH reversibility enables us to distinguish one interface from the other. These types of structural details should be of immense relevance to the understanding of various roles played by proteins at interfaces.

Finally, it is of relevance to mention that neutron reflection together with limited ellipsometric data suggests that blood protein adsorption at the solid–solution

TABLE 1 Comparison of Adsorption of Three Model Proteins onto Different Substrates Measured at the Solid–Water Interface

Protein	C (gdm^{-3})	pH	Γ_{SiO_2} (mgm^{-2})	Γ_{OTS} (mgm^{-2})	Γ_{PC100B} (mgm^{-2})	Γ_{C15OH} (mgm^{-2})	Γ_{PCm1} (mgm^{-2})	Γ_{PCm2} (mgm^{-2})	Γ_{PCbi} (mgm^{-2})
Lysozyme	0.03	7	1.0	1.9	0.4	0.7	0.2	0.7	0.6
	1	7	3.7	4.2	1.2	1.6	1.2	1.3	0.8
BSA	0.05	5	2.0	2.2	0.4	1.2	0.4	1.0	0.4
	2	5	2.8	3.8	1.0	1.2	1.0	1.1	0.7
Fibrinogen	0.1	7	5.0	5.2	0.3	1.0	0.3	0.7	0.4
	1	7	5.8	6.4	0.7	1.0	0.7	0.7	0.5

Γ_{SiO_2}, Γ_{OTS}, Γ_{PC100B}, and Γ_{PC15OH} denote surface adsorbed amount on the surfaces of bare silicon oxide [20,21,28,29,56]; octadecyl trichlorosilane (OTS) terminally grafted onto silicon oxide [31]; PC polymer with methacrylate backbone bearing dodecyl chains, isopropyl groups, silyl cross-linkers, and PC groups (PC100B) [46,47]; and pentadecyl-1-ol terminally grafted on silicon oxide (C15OH) [35]. The surface excesses adsorbed on small PC molecule–coated surfaces are denoted by Γ_{PCm1} (grafted by stepwise connection [57]). Γ_{PCm2} and Γ_{PCbi} (monolayer and bilayer coatings formed by the one-step coating scheme using APTMS-APC dimer). The experimental errors are about ± 0.3 mgm^{-2}.

interface can reach equilibration within minutes. This is in contrast to the much slower process observed at the air–solution interface. Neutron reflection is expected to contribute to the dynamic changes of protein structure in the future.

REFERENCES

1. J. Penfold and R. K. Thomas, J. Phys. Cond. Matter 2:1369, 1990.
2. J. R. Lu, E. M. Lee, and R. K. Thomas, Acta Cryst. A52:42, 1996.
3. J. Penfold, R. M. Richardson, A. Zarbakhsh, J. W. P. Webster, D. G. Bucknall, A. R. Rennie, R. A. L. Jones, T. Cosgrove, R. K. Thomas, J. S. Higgins, P. D. I. Fletcher, E. J. Dickinson, S. J. Roser, I. A. McLure, A. R. Hillman, R. W. Richards, E. J. Staples, A. Burgess, E. A. Simister, and J. W. White, J. Chem. Soc. Faraday Trans. 93:3899, 1997.
4. J. R. Lu and R. K. Thomas, J. Chem. Soc. Faraday Trans. 94:995, 1998.
5. R. K. Thomas, In Scattering Method in Polymer Science (R. W. Richards, ed.) Ellis Horwood, Chichester, 1995.
6. R. K. Thomas, J. R. Lu, and J. Penfold, Materials Sci. Forum 154:153, 1994.
7. J. R. Lu, R. K. Thomas, and J. Penfold, Adv. Colloid Interface Sci. 84:143, 2000.
8. G. J. Fleer, M. A. Cohen-Stuart, J. M. H. M. Scheutjens, T. Cosgrove, and B. Vincent, Polymers at Interfaces, Chapman & Hall, London, 1993.
9. T. P. Russell, Mat. Sci. Rep. 5:171, 1990.
10. J. R. Lu and R. K. Thomas, In Physical Chemistry of Biological Interfaces (A. Baszkin and W. Norde, eds.). Marcel Dekker, New York, 2000.
11. J. R. Lu, Ann. Rep. Prog. Chem. Sect. C95:3, 1999.
12. O. S. Heavens, Optical Properties of Thin Solid Films. Dover, New York, 1965.
13. M. Born and E. Wolf, Principles of Optics. Pergamon, Oxford, 1970.
14. A. D. Miranker and C. M. Dobson, Curr. Opinion Struct. Biol. 6:31, 1996.
15. A. Hvidt and S. O. Nielsen, Adv. Protein Chem. 21:287, 1966.
16. S. E. Radford, M. Buck, K. D. Topping, C. M. Dobson, and P. A. Evans, Proteins: Struct. Funct. Genet. 14:237, 1992.
17. T. Peters, Adv. Protein Chem. 37:161, 1985.
18. D. W. Van Krevelen, Properties of Polymers, 3rd ed. Elsevier, 1990.
19. T. Chalikian, M. Totrov, R. Abagyan, and K. Breslauer, J. Mol. Biol. 260:588, 1996.
20. T. J. Su, J. R. Lu, R. K. Thomas, Z. F. Cui, and J. Penfold, Langmuir 14:438, 1998.
21. T. J. Su, J. R. Lu, R. K. Thomas, Z. F. Cui, and J. Penfold, J. Colloid Interface Sci. 203:419, 1998.
22. T. A. Horbett and J. L. Brash, Protein at Interfaces II, ACS Symp. Ser. 602, Washington, 1995.
23. C. A. Haynes and W. Norde, Colloid Surf. B 2:517, 1994.
24. C. Tanford and R. Roxby, Biochemistry 11:2192, 1972.
25. R. K. Iler, The Chemistry of Silica. Wiley, New York, 1972.
26. C. A. Haynes, E. Sliwinski, and W. Norde, J. Colloid Interface Sci. 164:394, 1994.
27. P. M. Claesson, E. Blomberg, J. C. Froberg, T. Nylander, and T. Arnebrant, Adv. Colloid Interface Sci. 57:161, 1995.
28. T. J. Su, J. R. Lu, R. K. Thomas, Z. F. Cui, and J. Penfold, J. Phys. Chem. B 102:8100, 1998.
29. T. J. Su, J. R. Lu, R. K. Thomas, Z. F. Cui, and J. Penfold, J. Phys. Chem. B 103:3727, 1999.
30. R. Kurrat, J. E. Prenosil, and J. J. Ramsden, J. Colloid Interface Sci. 185:1, 1997.
31. J. R. Lu, T. J. Su, R. K. Thomas, A. R. Rennie, and R. Cubitt, J. Colloid Interface Sci. 206:212, 1998.

32. J. R. Lu, T. J. Su, R. K. Thomas, J. Penfold, and R. W. Richards, Polymer *37*:109, 1996.
33. C. Malmsten, J. Colloid Interface Sci. *166*:333, 1994.
34. S. Petrash, A. Liebmann-Vinson, M. D. Foster, L. M. Lander, W. J. Brittain, E. A. Vogler, and C. F. Majkrzak, Biotechnol. Prog. *13*:635, 1997.
35. T. J. Su, R. J. Green, Y. Wang, E. F. Murphy, J. R. Lu, R. Ivkov, and S. K. Satija, Langmuir *16*:4999, 2000.
36. D. A. Styrkas, V. Butun, J. R. Lu, J. L. Keddie, and S. P. Armes, Langmuir *16*:5980, 2000.
37. R. M. A. Azzam and N. M. Bashara, *Ellipsometry and Polarised Light*, North-Holland, Amsterdam, 1977.
38. K. Prime and G. M. Whitesides, Science *252*:1164, 1991.
39. C. M. Roth and A. M. Lenhoff, Langmuir *9*:962, 1993.
40. K. Prime and G. M. Whitesides, J. Am. Chem. Soc. *115*:10714, 1993.
41. P. Harder, M. Grunze, R. Dahint, G. M. Whitesides, and P. E. Laibinis, J. Phys. Chem. B *102*:426, 1998.
42. J. D. Andrade, *Surface and Interfacial Aspects of Biomedical Polymers*, Vol. 2. Plenum, New York, 1988.
43. L. Feng and J. D. Andrade, J. Biomed. Mater. Res. *28*:735, 1994.
44. Y. Iwasaki, A. Fujike, K. Kurita, K. Ishihara, and N. J. Nakabayashi, Biomater. Sci. Polymer. Ed. *8*:91, 1996.
45. Y. P. Yianni, In *Structural and Dynamic Properties of Lipids and Membranes* (P. J. Quinn and R. J. Cherry, eds.). Portland Press, London, 1992.
46. E. F. Murphy, J. R. Lu, A. L. Lewis, J. Brewer, J. Russell, and P. Stratford, Macromolecules *33*:4545, 2000.
47. E. F. Murphy, J. R. Lu, J. Brewer, J. Russell, and J. Penfold, Langmuir *15*:1313, 1999.
48. E. F. Murphy, J. L. Keddie, J. R. Lu, J. Brewer, and J. Russell, Biomaterials 20:1501, 1999.
49. E. J. Campbell, V. O'Byrne, P. Stratford, I. Quirk, T. A. Vick, M. C. Miles, and Y. P. Yianni, ASAIO J. *40*:M853, 1994.
50. A. L. Lewis, Colloid Surf. B, Biointerfaces *18*:261, 2000.
51. P. A. Cuypers, J. W. Corsel, M. P. Janssen, J. M. M. Kop, W. T. Hermens, and H. C. Hemker, J. Biol. Chem. *258*:2426, 1983.
52. M. Malmsten. J. Colloid Interface Sci. *172*:106, 1995.
53. M. Malmsten, J. Colloid Interface Sci. *168*:247, 1994.
54. J. R. Lu, E. F. Murphy, T. J. Su, A. L. Lewis, P. Stratford, and S. K. Satija, Langmuir *17*:3382, 2001.
55. (a) J. A. Hayward, A. A. Durrani, C. Shelton, D. C. Lee, and D. Chapman, Biomaterials 7:126, 1987. (b) J. A. Hayward, A. A. Durrani, Y. Lu, C. R. Clayton, and D. Chapman, Biomaterials 7:252, 1986.
56. E. F. Murphy, Ph.D. Thesis, University of Surrey, 1999.
57. Y. Wang, T. J. Su, R. J. Green, Y. Tang, D. Styrkas, T. N. Danks, R. Bolton, and J. R. Lu, Chem. Commun. *587*, 2000.

24

XPS, ToF-SIMS, and MALDI-MS for Characterizing Adsorbed Protein Films

HANS J. GRIESSER   Ian Wark Research Institute, University of South Australia, Mawson Lakes, Australia

SALLY L. McARTHUR, MATTHEW S. WAGNER, and DAVID G. CASTNER   University of Washington, Seattle, Washington, U.S.A.

PETER KINGSHOTT   Risø National Laboratory, Roskilde, Denmark

KEITH M. McLEAN   Commonwealth Scientific and Industrial Research Organisation, Clayton, Australia

I. INTRODUCTION

The rapid, irreversible adsorption of proteins onto solid surfaces from biological media is a well-known phenomenon. Upon implantation of a biomedical device, for example, a layer of adsorbed protein immediately forms on the surface of the biomaterial [1]. This layer of adsorbed proteins directs subsequent biological responses to the material. Nonspecific adsorption of proteins in biomedical applications can result in a variety of undesirable consequences including bacterial adhesion [2–4], calcium-containing deposits on biomaterial surfaces [5], and biomedical device toxicity [6]. The adsorbed protein layer can also influence animal cell adhesion [7–9] and subsequent tissue integration or rejection [10] of the implant. The composition, concentration, conformation, orientation, and spatial distribution of adsorbed proteins all affect and mediate subsequent biological reactions to the surface. Protein adsorption is important also in for example the biofouling of water purification, transport, and storage systems and on marine structures and vessels both static and moving.

The problems caused by indiscriminate, nonspecific protein adsorption can be addressed in principle by the design and fabrication of synthetic materials or surface coatings that either can resist the adsorption of all proteins ("nonfouling surfaces") [11–17] or selectively adsorb a specific protein whose presence at the interface causes predictable, desired consequences [18–22]. In either case, the rational design of such surfaces is predicated upon an improved fundamental understanding of how and why various proteins adsorb at various interfaces. This need has led to an enormous body of literature on protein adsorption, as discussed elsewhere in this book,

641

and concurrently to the development of a range of methods for characterizing adsorbed protein layers.

A wide variety of physical methods has been used to characterize adsorbed protein films. Some of these techniques are based on spectroscopic principles and use the adsorption or emission of specific electromagnetic radiation; others employ mass spectrometry to characterize molecules or molecular fragments. Some of these techniques were developed for ultrahigh vacuum surface analysis in materials science due to their low detection limits for elemental and molecular signals. Widely used for example in the semiconductor industry for characterizing surface layers and verifying stringent requirements for the cleanliness of surfaces, such surface analysis techniques have also found utilization in biomaterials research. In this chapter, the spectroscopic method x-ray photoelectron spectroscopy (XPS) and the mass spectrometric methods static time-of-flight secondary ion mass spectrometry (static ToF-SIMS) and matrix-assisted laser desorption ionization mass spectrometry (MALDI-MS) are discussed for the characterization of adsorbed protein layers. XPS and ToF-SIMS are well described in a number of reviews and books on surface analysis (e.g., Ref. 23), and the reader is referred to such texts for a general introduction to the methods and instrumentations. The application of MALDI-MS to the analysis of adsorbates on materials is very recent, on the other hand; accordingly, a brief outline of the methodology will be provided.

The three techniques possess different advantages and disadvantages and thereby provide complementary information for characterizing protein adlayers. XPS has been employed widely in materials science for the quantitative characterization of the elemental composition of a variety of solid materials. Its capability to detect low levels of surface contaminants also makes it well suited to the detection of adsorbed proteins on synthetic materials. Detecting photoelectrons emitted from all elements except H and He, XPS offers quantitative analysis of the elemental composition of surface layers and adlayers, but provides little detailed molecular information on macromolecules such as proteins. Generally XPS is incapable of identifying particular proteins. Static ToF-SIMS, on the other hand, is advantageous in terms of exceptionally low detection limits for adsorbed proteins. Its detection of molecular fragments can provide information on which proteins are present in the adsorbed layer. MALDI-MS complements ToF-SIMS by providing the mass signal of the entire macromolecule (e.g., protein) rather than low molecular fragments, but does not contain information about the structure of the adsorbed protein layer. The intensities of MALDI-MS signals likewise are only quantifiable under specific circumstances, and at present the method is semiquantitative at best. These and other aspects of these techniques are discussed in their respective sections. Examples of representative studies and results are also provided.

When performing protein adsorption experiments on biomaterials in vitro or implanting a biomedical device in vivo, the surface composition of the starting synthetic material must also be examined. This is because adventitious surface contaminants, processing aids, lubricants, surface-migrating additives, phase separation, polymer surface rearrangement, etc., all can markedly change the surface composition and properties of the material. It is therefore essential to perform adequate surface characterization of the material itself to ensure reliable and valid interpretation of observed interfacial responses. An illustrative example is the ToF-SIMS identification of the lubricant *bis*-ethylene-stearamide on a commercial polyurethane [24]; this additive may play a key role in the good bioresponses of that material.

Since these methods can detect adsorbed proteins on a variety of surfaces, they can also play an important role in the development of covalently immobilized protein layers for bioactive coatings. This area of biomaterials research has received much interest over the last decade, with many papers reporting protein immobilization by various linkage chemistries. Again, XPS and ToF-SIMS are eminently suited to verifying and quantifying (in the case of XPS) surface-immobilized proteins and providing feedback for the optimization of linkage chemistries, while MALDI-MS can, in conjunction with XPS, assess whether the proteins have indeed been covalently immobilized or simply adsorbed, as reviewed below. XPS, static ToF-SIMS, and MALDI-MS provide complementary information that is critical to characterizing the composition, concentration, and structure of adsorbed protein films.

II. X-RAY PHOTOELECTRON SPECTROSCOPY

A. General Considerations

X-ray photoelectron spectroscopy (XPS) is the most widely used ultrahigh vacuum (UHV) surface analysis technique. XPS is also called electron spectroscopy for chemical analysis (ESCA), and the two names are used interchangeably. The popularity of XPS can be attributed to the high level of information it provides, its ability to analyze a wide variety of samples, and its sound theoretical basis. This section focuses on the use of XPS to characterize adsorbed protein films. The theory of XPS, the technique, instrumentation, and its range of applications have been reviewed elsewhere (e.g., Ref. 25).

Briefly, in an XPS experiment the surface to be analyzed is placed in an UHV environment and irradiated with x-rays. The adsorption of x-rays by the atoms in the sample leads to ejection of core and valence electrons (photoelectrons). These photoelectrons have energies that are unique to each element and sensitive to the chemical state of the element. If the sample is homogeneous, the intensity of the photoelectrons is proportional to the concentration of the element it is ejected from. XPS has been used in a wide range of biomedical surface analysis studies and applications. Some of the features and analytical capabilities that make XPS particularly well suited for characterizing the adsorption of proteins to biomaterials are [26,27]:

1. *High surface sensitivity.* XPS has a sampling depth of 10 nm when the sample surface is positioned normal to the detector, making the sampling depth larger than the dimensions of most adsorbed proteins. This allows for signals from the substrate and protein overlayer to be detected simultaneously (see Fig. 1).

2. *Range of elements analyzed.* XPS detects all elements (except H and He) present within the sampling depth. In general, the nitrogen content of the protein is utilized to determine the presence and quantify the amount of protein on a surface. The presence of nitrogen in the substrate complicates quantification but may be overcome using other elements in the protein such as Fe, Zn, or S. In addition, if the substrate contains an element not

SIMS sampling depth   XPS sampling depth
(10-15 Å)   (80-100 Å)

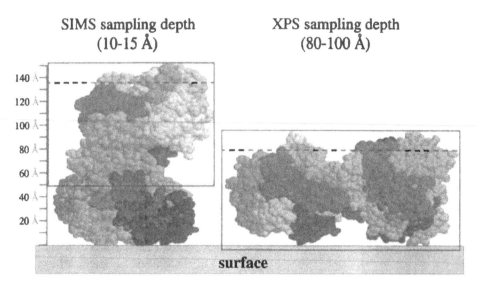

FIG. 1 The sampling depth of XPS and static SIMS are significantly different. Depending on the electron take-off angle, XPS is sensitive to the outermost 20–100 Å, while static SIMS is sensitive to the outermost 10–15 Å. The lower sampling depth of static SIMS enables it to sample the conformation and orientation of the adsorbed proteins more readily that XPS. [Protein structure obtained from www.expasy.ch (SWISS-PROT: P24627 TRFL-BOVIN).]

found in the protein, signal attenuation may be used to quantify the amount of adsorbed protein.

3. *Angle dependent XPS.* Angle dependent XPS (ADXPS) allows the analysis depth to be varied between 2 to 10 nm, enabling detection of very small amounts (~ 10 ng/cm^2) of surface adsorbed protein [28] and enabling the protein film thickness to be investigated. In combination with ^{125}I radiolabeling, the homogeneity and distribution of protein on the surface can also be assessed. If the surface is porous, the distribution of protein *within* the outer 10 nm of the material may also be monitored.

4. *Freeze hydration XPS* [27,29,30]. One of the limitations of XPS is the need to perform analysis on dehydrated surface under UHV, which may alter the protein conformation and thus layer thickness. Analysis can be performed on frozen hydrated films to avoid this, although it is a complex experiment that requires specialized equipment.

B. Characterization of Protein Adsorption

XPS has been used to monitor and characterize protein adsorption on a range of both model and "real" biomaterial surfaces. Early work in this field and much of the underlying theory has been reviewed in detail by Paynter et al. [26]. More recently, model substrates such as mica, glass, silicon wafers, and a variety of simple polymers like polystyrene (PS) have been used to investigate the adsorbed amount, thickness, stability, and conformation of adsorbed protein layers [31–38]. In more application-specific studies, in vitro and in vivo protein adsorption onto a range of experimental and commercially available products has also been studied [35,39–42].

Proteins contain mostly carbon, oxygen, and nitrogen, with substantially lower levels of other elements such as metals, phosphorus, and sulfur incorporated into specific structures. Paynter et al. [26] list both the theoretical atomic composition and the measured XPS atomic composition of a number of proteins, but in many situations the exact composition of a protein film is unknown. However, the overall elemental compositions of most proteins of interest are quite similar; based on data collated from the literature, a reasonable estimate for a "typical" protein composition is approximately 65% C, 20% O, and 15% N. In theory, the presence of minor elements (sulfur, iron, etc.) in the protein other than carbon, oxygen, and nitrogen may enable detection and/or identification. In practice their distribution within the protein and their low concentrations often means that they are at or below the effective detection limits for most XPS instruments even when thick protein films (>10 nm) are being analyzed.

Detection of proteins at interfaces using XPS generally involves the detection of nitrogen from the proteins and/or measurement of the attenuation of a unique substrate elemental signal (e.g., fluorine) that becomes attenuated due to the presence of an adsorbed protein overlayer. In a study investigating the XPS detection limits for adsorbed protein films, Wagner et al. used a ratio of the nitrogen signal to a unique substrate element to determine the level at which a significant difference from the control could be detected [28]. Correlation of these results with ^{125}I data indicated that the detection limits on non-nitrogen containing substrates were ~10 ng/cm^2. In the same study, protein adsorption to two nitrogen-containing radiofrequency glow discharge (RFGD) polymers was also analyzed. As the substrate contained the same atoms as the adsorbing proteins (though in different atomic concentrations), changes in the N/C ratio were used to evaluate the limits of detection. In the first instance, the nitrogen content of the substrate was ~10% and a significant difference between sample and control was only detected at protein surface concentrations of ~150 ng/cm^2. In the second case, a higher substrate nitrogen concentration of ~13% precluded the detection of protein from the elemental ratios below a surface concentration of ~1000 ng/cm^2.

Complementary to providing the elemental composition of a protein film, high-resolution acquisition of elemental signals, particularly the C 1s and the N 1s peaks, can provide additional quantitative information [26]. Carbon atoms in different bonding structures give rise to spectral components located at slightly different binding energies. For the detection of proteins, analysis for a C 1s component indicative of the amide groups of amino acids is particularly useful. Unless the biomaterial itself contains substantial amounts of amide or similar chemical groups (e.g., polyurethanes) or groups such as carboxyls whose signals arise at binding energies very close to amides and thereby cause considerable overlap, the detection of an "amide band" in C 1s spectra is diagnostic and can be quantified by curve-fitting protocols. High-resolution N 1s spectra also can be useful; amide N is expected to produce a component at 400.0 eV, whereas amine groups for instance are located at ~399.2 eV. In Fig. 2 the C 1s spectra recorded with thick films of albumin and fibrinogen clearly show three distinct environments of C atoms: hydrocarbon (285.0 eV), C—N and C—O (285.7 to 286.5 eV), and N—C=O (amide; 288.2 eV), representing the major structural elements present in the peptide backbone of the proteins

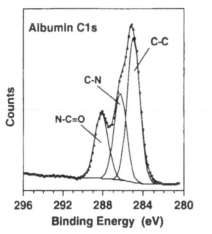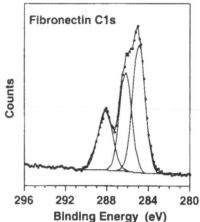

FIG. 2 High-resolution C 1s XPS spectra of thick films of albumin and fibronectin. The hydrocarbon, amine, and amide contributions to the C 1s peak are readily distinguishable using peak fitting algorithms. (From Ref. 43.)

[43]. Even when there is some overlap between the signals from the protein and the underlying substrate, C 1s spectra can be used to assess protein adsorption onto surfaces. The example in Fig. 3 shows C 1s spectra collected from adsorbed fibrinogen on a nitrogen-containing substrate. Using the N/C ratio it was not possible to detect unambiguously the presence of protein until a monolayer of protein was adsorbed, but the introduction of an amide peak (N—C=O, 288.2 eV) in the C 1s spectra was clearly detectable already at a fibrinogen concentration of ~150 ng/cm^2 [28].

Attenuation of a specific element from the substrate or the introduction of a specific element by the adsorption of a protein can be utilized to calculate the protein film thickness or determine the intercalation of protein into a porous substrate. Algorithms utilizing the x-ray emission angle θ, theoretical composition of the protein film or substrate I_∞, and inelastic mean free path of the emitted photoelectron of a specific element λ enable the calculation of protein film thickness d using Eq. (1):

$$I = I_\infty \exp^{-d/\lambda \cos \theta} \tag{1}$$

A number of other algorithms exist, but most assume that the protein film is homogeneous and continuous. If the XPS data do not fit the form of Eq. (1) this indicates the protein film is incomplete or patchy. Paynter et al. have shown that it is possible to incorporate a fractional coverage term into Eq. (1), but the quantity of protein adsorbed to the surface must be established from another technique such as radiolabeling [26].

These algorithms are valid for smooth, flat samples. Often, however, nonideal surface topographies are encountered when analyzing commercial polymers and biomedical devices. For curved or rough substrates, instead of a single value there is a distribution of photoelectron take-off angles present in the sampled area. Chatelier et al. have discussed the incorporation of slope frequency histograms of the surface topography features, obtained by AFM or STM, into the XPS analysis of overlayer

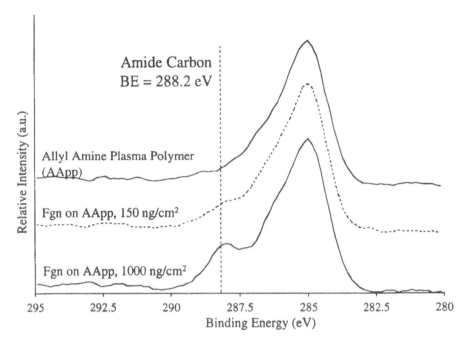

FIG. 3 High-resolution C 1s XPS spectra of fibrinogen adsorbed onto a nitrogen-containing allylamine plasma polymer substrate. The XPS spectrum of the substrate does not show an appreciable amount of the amide C 1s peak at a binding energy of 288.2 eV. Introduction of this peak with the adsorption of fibrinogen enables the fibrinogen adsorption to be tracked, resulting in a detection limit of ∼150 ng/cm². (From Ref. 28.)

thicknesses [44]. The method, which adds summation terms to Eq. (1), is particularly useful for determining the thickness of coatings (<10 nm) on biomedical devices such as contact lenses. Approximate approaches and limitations are also discussed in Ref. 44; for "moderate" surface topographies it is possible to use approximations such as the average slope to determine accurate overlayer thicknesses.

Blomberg et al. used XPS and the surface force apparatus (SFA) to investigate the orientation and structural stability of lysozyme and human serum albumin (HSA) adsorbed onto mica [31]. The N 1s and K 2p signals from the protein film and the substrate were compared and the amount of adsorbed protein calculated using the known nitrogen content of the protein. In these calculations it was assumed that both the protein film and the distribution of nitrogen within the film were homogenous. The adsorbed amount was monitored as a function of solution concentration and then combined with the known dimensions of the protein molecule and the thickness of the adsorbed layer as determined by SFA to assess the relative stability and packing density of the adsorbed protein.

In a similar study, XPS was used to monitor the adsorption of lysozyme to a n-heptylamine radiofrequency glow discharge (RFGD) polymer modified Teflon® surface as a function of protein solution concentration (Fig. 4) [45]. The RFGD film was deliberately kept thin (1–2 nm) so that a fluorine signal from the underlying Teflon substrate was still detectable. Thus, both the F 1s and N 1s signals were

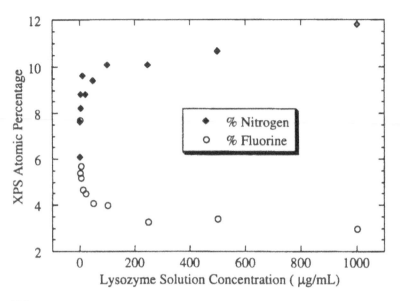

FIG. 4 Fluorine and nitrogen surface atomic percentages as measured by XPS using an electron take-off angle of 0° (measured from the surface normal) of lysozyme adsorbed onto a thin, nitrogen-containing n-heptylamine plasma polymer layer deposited onto a FEP substrate. Despite the presence of nitrogen in the plasma polymer film, the adsorption of protein is readily detectable in a reduction of the fluorine signal from the FEP substrate and an increase in the nitrogen signal due to the adsorption of protein. (From Ref. 45.)

monitored during protein adsorption. The adsorption of lysozyme resulted in an increase in the N 1s signal from lysozyme and a concomitant decrease in the F 1s signal from the substrate.

C. Limitations of the Technique

One limitation in all of these studies is that the XPS data are generated in a dehydrated state, which may induce further denaturation and alter the thickness of the protein film. Freeze hydration XPS, developed by Ratner and coworkers [30], involves the rapid freezing of a wet sample within the XPS entry chamber and can be used to circumvent these problems. Subsequent exposure to UHV at a temperature of approximately −100°C etches the ice from the surface via sublimation. The sample temperature is then lowered below −120°C for XPS analysis [46]. The process has been utilized in a number of protein adsorption studies, specifically to investigate the role of substrate chemistry in the distribution and orientation of adsorbed proteins [27,29,36]. Baty et al. used freeze hydration XPS to study the adsorption of mussel adhesive protein (MAP) onto polystyrene (PS) and polyoctadecyl methacrylate (POMA). AFM images indicated that the surfaces produced different protein packing and analysis of the nitrogen distribution using ADXPS indicated that the protein films varied on each surface. Both surfaces were hydrophobic, but had significantly different surface free energies. On PS, nitrogen was enriched at the outermost level of the protein film, whilst on POMA, nitrogen enrichment occurred at the polymer interface [36]. The differences are thought to be due to variations in the orientation or conformation of the protein at the two interfaces.

In addition to the effects of UHV on film thickness, there are other limitations of XPS for the study of protein adsorption. Similarities in the elemental composition of most proteins mean that XPS cannot be used to differentiate between proteins adsorbed from complex mixtures such as plasma or tears. Hence, multitechnique approaches are generally required to fully characterize the adsorbed protein film and adsorption characteristics of the surface. Notwithstanding these limitations, there are a number of possible advances in XPS that may prove useful for the analysis of surface associated proteins in the future. Recent developments in imaging XPS have opened up opportunities for imaging of proteins patterned onto surfaces. This improved spatial resolution may enable visualization of protein coverage, which, in combination with depth profiling, could give further insight into effects of surface chemistry on protein adsorption.

III. STATIC TIME OF FLIGHT SECONDARY ION MASS SPECTROMETRY

A. General Considerations

Static time of flight secondary ion mass spectrometry (static ToF-SIMS) has been used extensively to address biological problems in both the spectroscopic and imaging modes. Detailed reviews of static ToF-SIMS spectroscopy [47–49] and imaging [50] are available. Briefly, in common with XPS, ToF-SIMS is an UHV technique ($\sim 10^{-9}$ torr), but instead of by x-rays, the surface is bombarded by an energetic (~ 1 keV) primary ion source. The primary ion transfers its energy to the atoms in the sample via a series of collision cascades, some of which return to the surface and result in emission of secondary ions. Only a small proportion (<1% [51]) of the emitted particles are ionized, but the plume contains atomic and molecular ions as well as molecular fragments. Several monatomic primary ion sources are used, including Cs^+, Ga^+, Ar^+, and Xe^+, as well as polyatomic primary ion sources such as SF_5^+ [51]. It should be noted that the type of primary ion source used can influence the observed ToF-SIMS fragmentation patterns.

SIMS is a destructive technique since material is sputtered from the sample. To maintain surface sensitivity, limit sample damage, and ensure that the same region of the surface is not sampled more than once in each scan, an empirical limit on the primary ion dose, termed static limit, has been determined. For polymeric samples this has been determined to be approximately 10^{13} primary ions/cm^2 [52]. Under these conditions, the sampling depth of static SIMS is generally accepted to be approximately 10 Å [53] with a sensitivity of 10^7 to 10^{11} atoms/cm^2 (Fig. 1) [54].

The development of time-of-flight mass analyzers for SIMS [55–57], has greatly enhanced the mass resolution ($5{,}000 \leq m/\Delta m \leq 10{,}000$) of instruments and resulted in a theoretically limitless mass range. Both of these factors are critical for the detection and identification of proteins on surfaces. Enhanced mass resolution enables different species with the same nominal mass to be separated, discriminating nitrogen-containing fragments from hydrocarbon- or oxygen-containing fragments. The high-mass range allows for the detection of large fragments and whole molecules. Other developments such as laser postionization of the sputtered plume of material [58–60] have resulted in improved quantitation via the ionization of more material in the sputtered plume.

Static ToF-SIMS is useful for the characterization of adsorbed protein films due to its chemical specificity and surface sensitivity. Due to its basis in mass spectrometry, ToF-SIMS yields information about the molecular structure of the surface. ToF-SIMS has been used in the characterization of biomaterial surfaces [54,61], biomolecules in tissue [62], and adsorbed protein films [28,43,63–66]. SIMS has also been used to generate images of cells [67] and pharmaceutical distributions in tissue [68,69]. However, gaining useful information (such as the difference between two adsorbed protein films) is difficult due to the large number of peaks in the ToF-SIMS spectrum. In cases where there is an absence of unique peaks for different samples, the relative intensities of the peaks in the SIMS spectra are important for distinguishing the spectra [70].

B. Static ToF-SIMS Analysis of Peptides and Proteins

There are five major objectives for static ToF-SIMS analysis of proteins and peptides:

1. *Identification.* Static ToF-SIMS, with its combination of surface sensitivity and chemical specificity, has the potential to identify which proteins are present in an adsorbed film [64]. However, since most proteins have the same amino acid fragments the relative intensities of these fragments must be analyzed for protein identification [63].

2. *Quantification.* Using static ToF-SIMS spectra from single component protein films as references, it is possible to determine the relative surface composition of simple mixed protein films (e.g., binary films) [66]. However, due to the similarity of protein spectra it is not currently possible to quantify all proteins present in complex films (e.g., films deposited from plasma).

3. *Spatial distribution.* Using the imaging mode of static ToF-SIMS it is possible to characterize patterned protein films at spatial resolutions on the micrometer scale.

4. *Conformation and orientation.* Since the static ToF-SIMS sampling depth is smaller than the dimensions of most proteins, it is possible to examine the conformational and orientational changes for proteins which have a heterogeneous distribution of amino acids in their three-dimensional structure.

On the most fundamental level, static ToF-SIMS has been used to investigate films of adsorbed proteins and their constituent amino acids. In the late 1970s and early 1980s Benninghoven's group described the molecular ions formed during static ToF-SIMS analysis of single amino acids [71,72]. The technique has also been used to detect differences in two peptides with the same composition but differing conformations (α-helix versus β-sheet) [73] and differences in the solid state oxidation of the sulfur in the side chain of a methionine-containing peptide [74].

Some of the most significant work to date on the interpretation of static SIMS spectra from adsorbed protein films was that of Mantus et al. [64], who examined the fragmentation patterns of amino acid homopolymers and described the major secondary ions formed during analysis of each of the amino acids. The work was performed using a quadrupole mass analyzer but was later repeated using a ToF mass analyzer for improved mass resolution [75]. Since that study, static ToF-SIMS has been used in the detection [28,76,77], identification [63], and quantification [32,78] and in studying the conformation [32,43] of adsorbed proteins.

Static ToF-SIMS analysis of adsorbed protein samples has proved to be relatively straightforward in terms of the generation of spectra, but data analysis has proven to be significantly more complex. Mantus et al. showed that a static SIMS spectrum of a protein film is comprised of at least two peaks from each of the 20 amino acids [64], but most authors have selected only a few "representative" peaks from the static ToF-SIMS spectrum for their protein data analysis. Lhoest [63] and Ferrari [78] were the first to apply multivariate analysis techniques (utilizing many of the peaks from the static ToF-SIMS spectra) to characterize the differences between static ToF-SIMS spectra of different adsorbed protein films and the amount of adsorbed protein, respectively.

C. Challenges and Advances in Static ToF-SIMS Analysis of Proteins

Static ToF-SIMS analysis has been applied to many studies involving peptides and proteins, but there remain a number of significant challenges to be resolved. A primary challenge is the requirement for UHV. While adsorption of a protein onto a surface may result in some denaturation, the removal of bound water and subsequent dehydration of the protein under UHV likely causes further denaturation. Cold stage static SIMS, in which the sample is maintained in a frozen-hydrated state during analysis, has been developed to circumvent this issue [79]. While a freeze fracture technique has been applied successfully to the analysis of cell membranes [67,80], it is yet to be applied to the analysis of proteins or peptides. In addition, issues related to the effects of vacuum on adsorbed proteins and peptides are currently being investigated and methods for protecting the protein structures are also being developed [81]. Despite the effects of the vacuum process on an adsorbed protein film, a number of studies have shown that chemically relevant information can still be obtained and utilized in detection and characterization of adsorbed protein films. Recent studies have shown that static ToF-SIMS is capable of detecting adsorbed protein at surface concentrations comparable to standard ESI-MS and matrix-assisted laser description/ionization–mass spectrometry (MALDI-MS) detection limits (attomole) (Table 1) [28].

The complexity of the protein mass spectra presents a considerable challenge to researchers (Fig. 5). Mantus et al. showed that each amino acid has a complex static ToF-SIMS fragmentation pathway comprised of the immonium ion ($^+NH_2$=CHR) and several other fragmented species [64]. Since proteins can be made up from any or all of the 20 basic amino acids, the protein static ToF-SIMS spectrum is a convolution of the individual fragmentation pathways of each of the constituent amino acids. Tidwell et al. [43] showed by univariate analysis that for albumin and fibronectin the intensities of different immonium ions were enough to distinguish one protein from another. But in general the multidimensional aspect of static ToF-SIMS spectra makes univariate analysis difficult due to the need for the selection of peaks from the spectra for data analysis.

The application of multivariate analysis methods for static ToF-SIMS data anal-

TABLE 1   Summary of Detection Limits for Proteins on a Variety of Substrates in ToF-SIMS

Substrate	Positive ion ToF-SIMS (ng fibrinogen/cm²)	Negative ion ToF-SIMS (ng fibrinogen/cm²)
Mica	0.1	—[a]
PTFE	100	100/2[b]
Heptyl amine plasma polymer	2	15
Allylamine plasma polymer	10	1

[a] Negative ion ToF-SIMS was not utilized to analyze fibrinogen on mica.
[b] Results are dependent on whether the F⁻ or F_2^- peak is used in the data analysis.
Source: Ref. 28.

ysis enables high-throughput analysis of large amounts of data while retaining much of the data in the entire mass spectrum. In a number of recent studies it has been shown that static ToF-SIMS in conjunction with the multivariate analysis method principal component analysis (PCA) is able to distinguish between different adsorbed protein films on a variety of surfaces [63,65]. PCA, in particular, reduces large, multidimensional data sets into fewer, more manageable plots that concisely describe the major variations in the static ToF-SIMS data. Using PCA, the static ToF-SIMS

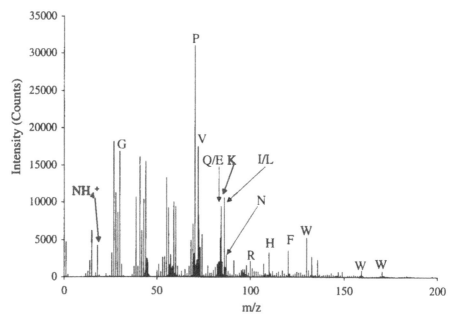

FIG. 5   Static positive ion ToF-SIMS spectrum of immunoglobulin G adsorbed onto mica from a 100 μg/mL solution. Several peaks indicative of individual amino acids are shown in this figure.

spectra of several different proteins can be readily differentiated (Fig. 6) [65]. Furthermore, the same protein from several different species can be distinguished despite a high degree of structural and sequential similarity between the proteins (Fig. 7) [65]. In more complex systems, Lhoest et al. have demonstrated that static ToF-SIMS and PCA were able to track the changing surface composition of binary fibronectin–albumin protein films [63].

A major challenge in the analysis of proteins by static ToF-SIMS is the inability to produce secondary ions larger than a single amino acid. No multiresidue fragments are observed in standard static ToF-SIMS analysis of adsorbed proteins. The lack of poly(amino acid) fragments in the spectra increases the difficulty of identifying individual proteins present in single and multicomponent films. Michel et al. demonstrated the use of a novel sodium cationization technique to assist in the formation of molecular ion from a small peptide, renin substrate (MW = 1759) [82]. In these studies, proteins were adsorbed onto COONa-terminated self-assembled monolayers (SAMs). The resulting spectra indicated significantly improved signal intensity from the protein molecular ions compared to etched silver substrates. In an alternative approach, matrix-assisted laser desorption/ionization mass spectrometry matrices have been mixed with protein solutions and applied as thick films to substrates. This procedure has been demonstrated to enhance molecular ion formation in SIMS for

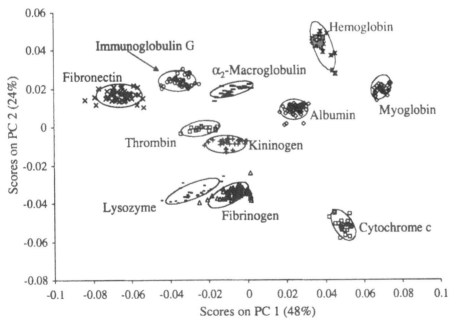

FIG. 6 Scores on the first two principal components (PCs) for the positive ion static ToF-SIMS spectra of different proteins adsorbed onto mica substrates from 100 μg/mL single protein solutions. The first two PCs capture the first and second largest directions of variance in the data (and capture 72% of the total variance in the data set). Each point in this plot is an individual spectrum, with each protein having several replicate spectra obtained. The spectra are well resolved from one another using this method. (From Ref. 81.)

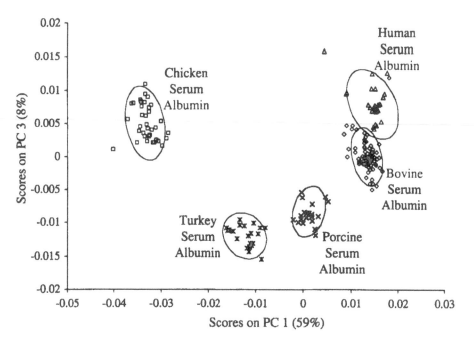

FIG. 7 Scores on the first and third PCs for the positive ion static ToF-SIMS spectra of serum albumin from different species adsorbed onto mica substrates from 100 μg/mL single protein solutions. Although these proteins are highly conserved both structurally and sequentially, they are readily differentiated by their ToF-SIMS spectra. For example, human serum albumin and bovine serum albumin are ~75% conserved in their amino acid sequence and ~90% in their amino acid composition. (From Ref. 81.)

peptides and proteins as large as bovine insulin (MW = 5733) [83,84]. However, this technique has not been successfully extended to adsorbed protein films.

D. Static ToF-SIMS Chemical State Imaging

The patterning of proteins and other biomolecules, such as DNA, on surfaces plays an important role in the development of diagnostic assays and sensors as well as scaffolds for tissue engineering [85]. In addition, there is increasing interest in visualizing the distribution of small molecules and proteins within tissue samples [86]. While there are a number of labeling techniques that exist to enable the visualization of biomolecules within these structures, static ToF-SIMS imaging presents a unique opportunity to detect and spatially map proteins present on the surface in very low concentrations. Static ToF-SIMS also enables the surrounding environment to be characterized for contaminants and continuity.

Producing spatially resolved images of biomolecules on surfaces is not trivial. It is possible to produce images with a full mass spectrum at every pixel, but there is a balance that must be reached between mass resolution and spatial resolution. A number of instrumental factors such as type of primary ion source and ion dose can be manipulated to maintain mass resolution without sacrificing spatial resolution, but under these conditions it may be difficult to acquire an image within the static ion dose limit. At present, using a liquid metal ion source such as Ga^+, it is possible to maintain a mass resolution of >4000 m/z with spatial resolutions of 5 to 10 μm. Below 5 μm, the mass resolution can decay by as much as a factor of 10.

As discussed earlier, most proteins either do not produce unique molecular ions or only produce them at low intensity. As a result, there may be overlap between ion signals from the substrate and the protein or, in the instances where there are a number of proteins present, overlap with amino acid fragments from different proteins. By utilizing a stable isotope label on a specific protein within a multicomponent protein micropattern, spatial resolution was achieved within the image by collecting ion images from $C^{15}N^-$ and CN^- species [85]. This system works well when it is possible to introduce a label to one or multiple components of a system, but this is not always desirable or practical.

Remembering that it is possible to collect static ToF-SIMS images with entire mass spectra at every pixel, it seems practical to apply some of the same data processing principles used in the analysis of individual spectra to image analysis. Individual species in a static ToF-SIMS image typically have both low signal-to-noise ratios and low ion intensities, making it difficult to resolve the pattern or specific molecules from the background. Low pass filters are commonly applied as a first stage of image processing to improve the signal-to-noise ratio, but this generally causes some loss in spatial resolution [87]. Since one of the greatest problems with static ToF-SIMS imaging is the low ion intensity from individual ions in individual pixels, it seems practical to use a multivariate analysis technique to determine the relationships between spectra from each image pixel. By determining the ions responsible for the greatest degrees of variation in the image, it is possible to combine these ions, with optimal weighting, to reconstruct a chemical image of the surface. By combining the signals of multiple ion fragments from a single protein a spatially resolved image of that protein's distribution on a patterned surface can be formed.

E. Emerging Technologies and Applications

The surface sensitivity of static ToF-SIMS and its utility in probing protein conformation and orientation remains the focus of current research. In a study examining static ToF-SIMS spectra from proteins adsorbed at different temperatures, Tidwell et al. have shown significant variations in the intensities of characteristic amino acid peaks [43]. These results suggested that in addition to identifying the protein, analysis of static ToF-SIMS spectra could provide information about the orientation and conformation of the protein. Recent studies by Xia et al. have indicated that it is possible to limit the dehydration effects of UHV on the conformation of proteins by incorporating sugar into the protein film [81]. Studies using surface plasmon resonance (SPR) and static ToF-SIMS showed that antibody films dried in the presence of sugar adsorbed antigen at levels comparable to undried antibody film. In contrast, there was a significant reduction in activity of the adsorbed antibody when it was dehydrated without sugar. Comparison of the static ToF-SIMS spectra from the samples dried with and without sugar indicated that their surface amino acid composition were different. For films dried with sugar a relative enhancement of fragments from hydrophilic amino acids was detected, while for films dried without sugar a relative enhancement of fragments from hydrophobic amino acids was detected [81].

Recent advances in SIMS instrumentation may fundamentally improve the capabilities of static ToF-SIMS for characterizing adsorbed proteins. Key advances lie in the development of polyatomic primary ion sources, which have been demonstrated to improve the yields of molecular ions for small organic compounds [88] and small molecules in protein matrices [89]. In addition they may also enable depth profiling of adsorbed protein layers, finally adapting the dynamic SIMS modality to organic surfaces. Development of polyatomic sources for liquid metal ion guns would also provide enhanced molecular ion signals at high spatial resolutions. These sources have the potential to improve the surface and detection sensitivity for proteins on surfaces, factors that are critical if researchers are to characterize and image increasingly complex biological or biomimetic systems.

IV. MALDI MASS SPECTROMETRY

A. General Considerations

MALDI-MS has only very recently been applied to the characterization of adsorbed protein films [45,90–95] and offers a unique, complementary, and versatile means of characterization. However, much development work is still needed to fully exploit the potential benefits of MALDI-MS and turn it into a truly routine method. MALDI-MS was originally developed to address the mass spectrometry challenges of analyzing high-molecular-weight macromolecules, not as a surface analysis method. MALDI-MS was introduced in the late 1980s for the rapid analysis of macromolecular analytes [96], for the accurate determination of their molecular mass [97], and for the detection of impurities at very low concentrations. In contrast to other mass spectrometry techniques, which aim to identify the chemical structure of analytes from fragmentation patterns, the main aim of MALDI-MS is the detection of parent molecules (as ions).

Sample preparation for "conventional" MALDI-MS is carried out by mixing analyte and "matrix" molecules in an appropriate solvent mixture, followed by deposition of a drop of solution onto a sample stub. As the solvents evaporate, matrix crystals with encapsulated analyte molecules form on the stub [98,99]. Both the analyte and the matrix molecules must be dissolvable under the same conditions to prevent preferential precipitation of one of the components [100]. The sample stub is then introduced into the spectrometer, and the dried crystals are irradiated with a pulsed laser. Pulsed nitrogen UV lasers (337 nm) are commonly used. Absorption of the laser light by the matrix molecules leads to rapid heating and evaporation of the matrix crystals with embedded analyte molecules. Photoactivated reactions lead to ionization of both matrix and analyte molecules [101]. Volatilized analyte ions (and matrix molecule ions) are then detected by a mass analyzer. A time-of-flight mass analyzer is the detection system of choice for macromolecular analytes. Matrix molecules are typically aromatic acids or aromatic carbonyl compounds (e.g., sinapinic acid and α-hydroxycinnamic acid), selected on the basis of a high absorption coefficient for the laser wavelength and an ability to rapidly convert the adsorbed laser energy into heat. The matrix must also be able to ionize the analyte molecules, for instance by excited-state proton transfer [102,103].

In the surface analysis mode, such as in the detection of adsorbed proteins [45,90–95,104–107], the MALDI-MS sample preparation method consists of pipetting a drop of matrix solution onto a protein-covered biomaterial surface. The solvent is allowed to evaporate to allow formation of matrix crystals, the sample is mounted on a holder and introduced into the spectrometer. The analysis then proceeds as in conventional MALDI-MS (Fig. 8).

The key features of interest for surface-mode MALDI-MS characterization of adsorbed protein films are

1. *Detection of macromolecular ions.* In contrast to the extensive fragmentation that occurs in static ToF-SIMS analysis, the matrix crystals achieve rapid and "soft" volatilization and ionization of embedded macromolecular analytes in MALDI-MS. Most analytes do not fragment. Thus, the method allows the detection of the parent molecule (as an ion) and hence the determination of its molecular mass. The molecular weights of proteins up to 550,000 Da have been determined using MALDI-MS [108].

2. *Low detection limits.* Attomolar concentrations and nanograms of analyte are routinely detected in conventional MALDI-MS [109]. New sample preparation methods including microspot deposition [110] and automation [111] allow improved sensitivity of detection; recently useful signals were reported with ~5 analyte molecules per square micrometer or 25,000 molecules of Substance P [112]. The limits of detection for adsorbed proteins on biomaterials are not yet known but initial indications are that for some proteins amounts of 1 ng/cm^2 or less may be readily detectable [45,92,94]. The detection limits are only partially determined by instrumental factors; as discussed below, surface binding strength and sample preparation methodology play a large role in the ability to detect adsorbed proteins by MALDI-MS in the surface analysis mode.

3. *High mass resolution.* Since MALDI-MS and ToF-SIMS both use a ToF detector, the same mass resolution considerations should apply to both techniques. However, the ToF analyzer is typically operated in different modes (linear versus reflectron) for the two methods. The mass resolution ($\Delta m/m$) for MALDI-MS can be of the order of 10^{-3} to 10^{-4} [113], but

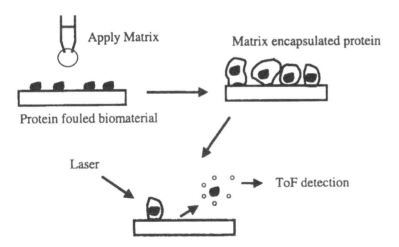

FIG. 8 MALDI-MS in the surface analysis mode as used for the characterization of adsorbed protein films.

sample matrix effects can markedly increase peak widths. Nevertheless, different proteins that coadsorb from multicomponent solutions usually are well separated [94].

4. *Direct detection of adsorbed proteins.* MALDI-MS can directly detect adsorbed proteins on various synthetic material surfaces (metals, ceramics, polymers), including everyday biomedical devices such as contact lenses [45,90–95,104–107] without the need for elution, transfer of proteins, or other steps that might lead to loss of analyte.

5. *Structural information.* Unfortunately soft ionization generally produces few fragments that can be used to determine the structure or sequence of proteins. This can be partly overcome by the use of postsource decay [114] to yield some sequence information [115]. Franzen et al. have recently discussed the information content and challenges in using MALDI-MS for high-throughput sequencing [116].

Thus, MALDI-MS complements XPS and SIMS as a means of investigating protein adsorption processes with its unique ability to acquire molecular mass information, which often can be used to identify which protein(s) is adsorbed onto a sample. MALDI-MS has also been employed to determine the mass of proteins electroblotted from PAGE gels onto porous transfer membranes [117,118], for the analysis of the hybridization of oligonucleotides on biochips [119,120], and for the detection of adsorbed lipids, polymer additives, surface contaminants, and other adsorbed (non-protein) compounds [121].

B. Characterization of Adsorbed Protein Films

The MALDI-MS spectrum from lysozyme adsorbed onto the surface of a contact lens contains multiple peaks (Fig. 9). The molecular ion ($\{M+H\}^+$) signal usually dominates the spectrum, and its mass can be used to identify the protein. The detection of intact molecular ions indicates that the surface-mode MALDI sample prep-

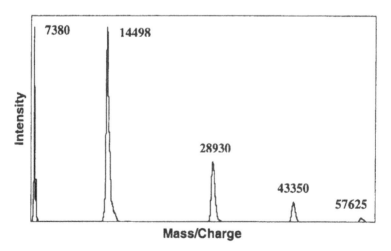

FIG. 9 MALDI-MS spectrum of lysozyme adsorbed onto the surface of a contact lens. The m/z region below 7000 is omitted for clarity; matrix ions are observed at low masses.

aration, volatilization, and ionization processes do not cause cleavage of the proteins. Even for proteins as large as IgG, a significant fraction of the molecules adsorbed onto biomaterials surfaces have been volatilized and detected by MALDI-MS without fragmentation [95]. A second mass signal in the lysozyme spectrum at a m/z value of ~7.2 kDa originates from the doubly charged monomer ion ($\{M+2H\}^{2+}$). Dimer ions ($\{2M+H\}^+$) at an m/z value twice the molecular weight are also commonly observed. Higher oligomers can also occasionally be observed [45]. When several proteins are present, each protein usually gives rise to the multiple peaks described above, so the first task is to group the signals into m/z integers of parent ions. Then by comparison to reference spectra the proteins can be identified.

The peaks from a given protein may not line up as exact integer multiples of the main peak. The ToF detector is calibrated for sample analysis in the standard stub geometry. When a biomaterial sample is placed on the stub, there is an indeterminate shift in the flight geometry due to the finite thickness of the biomaterial and thus a shift in the m/z scale. Adding an internal calibrant (a protein of known mass) to the matrix solution, or shifting the mass scale such that the peaks line up as multiples, compensates for this artifact.

As for conventional MALDI-MS, the detection limits of surface-mode MALDI-MS are extremely low, but factors such as tight binding can suppress the protein signal (see below). Lysozyme has been detected with a high S/N ratio at ~30 ng/cm^2 (~15% of monolayer coverage) on contact lenses [92,104] and on an amine radiofrequency glow discharge polymer (RFGD) coating immersed for only 5 s into a lysozyme solution [45,91]. XPS indicated that for this 5-s time point the lysozyme coverage was <10 ng/cm^2. In both cases the high signal-to-noise ratio of the MALDI-MS lysozyme peak suggested that the ultimate detection limit might be substantially lower. It is likely that lysozyme concentrations of the order of 1 ng/cm^2 and perhaps substantially lower can be readily detected.

With its ability to simultaneously detect signals from several proteins, MALDI-MS is well suited to the analysis of multicomponent adsorbed layers such as those found on biomedical devices retrieved from in vivo experiments. For MALDI-MS the retrieved biomaterial or biomedical device only requires a thorough rinsing to remove loosely bound material and salts prior to the application of the matrix solution. The analysis by MALDI-MS of biofilms on contact lenses retrieved from human wearers [92,95,104,107] illustrates the capabilities of this method. An example is shown in Fig. 10 comparing spectra from pHEMA contact lenses worn for 10 min to a spectrum of tears from the same patient [95,105]. A number of peaks with different m/z values are evident. The signal at m/z = 14,500 Da can be assigned to the molecular ion peak ($\{M+H\}^+$) of human tear lysozyme, a well-known major constituent of fouling layers on contact lenses. The signal at m/z = 7200 is the doubly charged monomer ion of lysozyme. A number of other signals of substantial intensity are also present, indicating that in addition to lysozyme other low-molecular-weight proteins adsorb from human tear fluid onto contact lenses. These proteins had not been detected in earlier studies on contact lenses using other techniques. Conventional MALDI-MS analysis of tear fluid (Fig. 10) yielded all the peaks observed on the lens surfaces plus some additional peaks [95]. Clearly, contact lens fouling is much more complex than had been assumed since a number of different proteins simultaneously adsorb within short time periods. It is interesting to note that larger proteins are not detected at early stages of lens fouling.

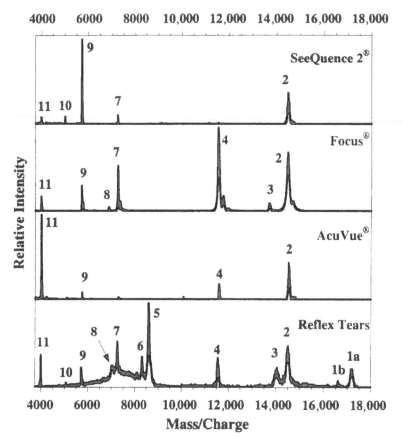

FIG. 10   Comparison of MALDI-MS spectra collected on commercial, pHEMA-based contact lenses after 10 min wear by one patient to the reflex tear MALDI-MS spectrum of the same patient.

It is important to verify assignments and interpretations by comparison to "control" spectra. This includes spectra acquired from a biomaterial exposed to single protein solutions and spectra acquired from proteins deposited onto a standard steel stub by the conventional mode by mixing a protein solution with a matrix solution. Control spectra should contain the $\{M+H\}^+$, $\{M+2H\}^{2+}$, and oligomer peaks with similar relative intensities [45]. It is also important to verify that the absence of a MALDI-MS signal for a particular protein is due to the absence of that protein in the adsorbed film, not the inability to detect a signal from that adsorbed protein. It is advisable to use complementary techniques such as ELISA to probe for specific proteins that might be present on the surface but not detectable by MALDI [95].

Unique advantages of MALDI-MS characterization of complex adsorbed protein layers are its high mass resolution, which separates proteins of similar masses, and the detection of proteins with masses <10 kDa. Moreover, only a small biomaterial sample ($\sim$1 mm^2) is needed for analysis. This avoids having to pool samples and allows study of time effects and patient variability of contact lens fouling [95]. Biomaterials with proteins adsorbed from other biological fluids (blood, urine, saliva, etc.) are also amenable to detailed analysis by MALDI-MS; an example of a spectrum recorded after serum adsorption is shown in Fig. 11 [121].

C. Study of Competitive Protein Adsorption and Displacement

MALDI-MS can be used to probe protein competition events from complex in vitro biological media, which may enable faster development of surface coatings intended to selectively attract target proteins from multicomponent solutions. MALDI-MS analysis of competitive adsorption can be used to find out which proteins adsorb, and therefore one can elucidate correlations or "rules" linking protein properties (e.g., pI) and surface properties (e.g., surface potential) in competitive adsorption.

An example of such research is the exposure of polysaccharide coatings to human tear fluid proteins [45,90]. The coatings were fabricated by covalent immobilization of partially oxidized dextran, hyaluronic acid, chondroitin sulfate, and carboxymethylated dextrans [40,122,123]. MALDI-MS spectra showed remarkable differences in the adsorption patterns [45,90]. On a hyaluronic acid coating, signals from lysozyme, lactoferrin, and an impurity at m/z = 22,788 were observed. The impurity was introduced from the commercially sourced lactoferrin. Signals from albumin and IgG, on the other hand, were not observed. On the dextran coating, signals assignable to lysozyme and lactoferrin were observed whereas the impurity peak was absent. On the chondroitin sulfate coating, only lysozyme adsorbed. When this coating was immersed in a solution containing no lysozyme (but all the other components), the surface MALDI mass spectrum revealed that lactoferrin and the impurity in lactoferrin adsorbed [45]. Thus, the absence of these two proteins on the chondroitin sulfate coating exposed to the full artificial tear fluid is a reflection of their inability to compete in adsorption, not of an inability of the method to detect them. Carboxymethylated dextrans with different degrees of carboxyl substitution produced a series of coatings with varying charge [95] and allowed study of charge

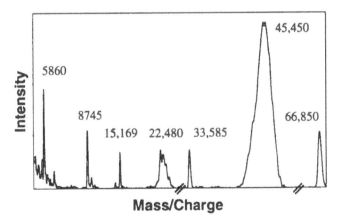

FIG. 11 MALDI-MS spectrum of protein adsorption from 20% fetal bovine serum to an RFGD plasma modified surface. The peaks at 66,850 and 33,585 are the molecular ion ($\{M+H\}^+$) and the doubly charged ion ($\{M+2H\}^{2+}$) of bovine serum albumin. The peaks at 5860 may be the molecular ions ($\{M+H\}^+$) of insulin and APO C-III, respectively. The other peaks are yet to be identified.

and electrosteric effects. A highly charged coating adsorbed only lysozyme and lac-toferrin, while a lower charged coating also adsorbed albumin and IgG. These results were independently confirmed by ELISA [124] and interpreted with the aid of AFM surface force curves. This illustrates that interpreting protein adsorption in terms of separable physicochemical factors such as charge can be quite complex.

MALDI-MS can also be used to study time-dependent changes in protein concentrations by comparing the relative peak intensities of the different proteins. Competitive adsorption of lysozyme and albumin onto RFGD coatings [105] indicated that both proteins adsorbed initially and that albumin did not fully displace lysozyme after long time periods. When one protein was adsorbed first and the other added later, various effects were observed, but generally lysozyme was surprisingly persistent. However, MALDI-MS signal intensities are subject to various, incompletely understood, factors, and quantitation is fraught with challenges (see below). We believe at present it is only warranted to compare relative peak intensities for very similar cases (e.g., adsorption from two-component solutions). Even then care is advised.

D. Present Limitations

One limitation is that for some surface/protein combinations, adsorbed proteins such as human serum albumin could be detected by techniques such as XPS and ELISA but not by MALDI-MS [121]. One factor is the time elapsed between adsorption and analysis. Samples analyzed immediately after removal from protein solutions, in our experience, usually provide much stronger spectral signals than samples that have been stored—dry, in water, or in phosphate-buffered saline—before MALDI-MS analysis. For some proteins such as lysozyme a delay of a few days is acceptable; for other proteins a MALDI-MS signal is no longer detectable after that time. This may be due to surface-induced denaturation of the protein since lysozyme would be expected to denature much less than albumin. The matrix solvent (typically water/acetonitrile/trifluoroacetic acid) needs to dissolve proteins off the biomaterial surface for successful incorporation into matrix crystals and subsequent volatilization. So it is reasonable that a tightly bound, denatured protein may resist incorporation into the matrix. Moreover, sticky proteins such as vitronectin may have sufficiently strong affinity to many surfaces even when not denatured. It is therefore reasonable to interpret failure to detect an adsorbed protein as a consequence of excessively tight binding to the biomaterial surface. The surface-mode MALDI-MS experiment thus represents a situation in which the protein–surface affinity has to be in the "right" range. Different matrices and solvents can be used to adjust this range. McComb et al. reported that wheat glutenin proteins could be detected without loss of spectral intensity up to 14 days after adsorption to polyurethane carrier films [125].

Another limitation arises from the upper mass limit of currently available ToF detectors, which is insufficient for the detection of the parent ions of some of the proteins of biomedical interest, particularly some of those involved in thrombus formation. However, larger proteins may be difficult to desorb anyway due to their surface affinity; to date we have not been able to detect any protein larger than IgG [121].

Another possible cause for not detecting an adsorbed protein is that the protein may not be fully incorporated into the matrix crystal. Proteins expelled from the growing matrix crystals may be located at the very edge of the matrix crystals and beyond the laser impingement area. This was investigated using scanning electron microscopy, laser confocal microscopy, and fluorescently labeled proteins [95,126]. Protein incorporation and thus MALDI signals depend on the location and the matrix [126] as well as the substrate biomaterial [95]. Matrix/analyte crystals formed upon solvent evaporation tend to concentrate at the end of the droplet, with a thinly distributed layer of crystals covering the central region of the sample. Individual crystals varied in size and shape with the matrix. Figure 12 illustrates crystals formed with α-cyano-4-hydroxycinnamic acid and sinapinic acid. While different matrix compounds are available, at present insufficient knowledge exists for selection of the optimal matrix. Thus, empirical optimization of matrices and solvent systems is required and can be quite time consuming.

As in static ToF-SIMS, matrix effects can strongly affect peak intensities. This poses a challenge for quantitative analysis. It has been demonstrated that the influence of surface–peptide binding affinity on MALDI ion signals is extremely resistant to changes in the method of sample preparation [127] and surprisingly sensitive to

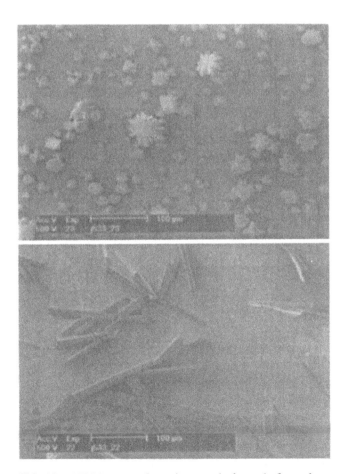

FIG. 12   SEM images of matrix crystals formed after solvent evaporation. Left: α-cyano-4-hydroxycinnamic acid crystals; right: sinapinic acid crystals. (From Ref. 95.)

the nature of the chemical interaction between the surface and the peptide [128]. The reduction in protein ion signal can be directly correlated with the classically defined surface–protein retention affinity (the quantity of surface-bound protein that cannot be removed by vigorous washing of the surface with SDS) [129]. This insight led to the development of a MALDI-based standard additions methodology for quantitation of surface–protein retention affinities. MALDI-MS has also been used for the quantification of lysozyme adsorbed to Ti dental implants [106], with a linear relationship between MALDI signal intensity and the amount of adsorbed lysozyme.

E. Application to Immobilized Protein Layers

The immobilization of protein layers on synthetic bulk materials has received much attention as a means of achieving improved biocompatibility. To prevent desorption/ exchange effects when the device is placed in the biological environment the protein molecules must be covalently immobilized to the surface. It can be difficult to distinguish between covalent binding and adsorption, or to assess what fractions of the surface-located proteins are covalently linked. MALDI-MS analysis provides a unique means of studying this [39], as adsorbed proteins can be embedded into matrix crystals, while covalently linked proteins are not amenable to matrix incorporation. Thus, in principle, adsorbed proteins should be detectable by MALDI-MS (bearing in mind, however, the limitations discussed above) but not covalently immobilized proteins. This was demonstrated by immobilizing various proteins onto polysaccharide layers using carbodiimide chemistry [39]. By leaving out the carbodiimide, protein adsorption instead of protein immobilization occurred. Quantitative analysis by XPS showed that after rinsing, comparable amounts of proteins were present on both types of samples. MALDI-MS showed striking differences; the samples produced without carbodiimide gave protein signals with high S/N ratios, whereas no protein signals could be detected from samples produced with carbodiimide. It was estimated that >97–99% of the protein coverage on the carbodiimide samples was covalently immobilized [39]. However, when analyzing immobilized triple helical assemblies of collagen-like synthetic oligopeptides it was determined by MALDI-MS that not all strands of a triple helix could be pinned covalently [130]. Hence, the MALDI-MS method allows the direct probing of the effectiveness and completeness of covalent immobilization reactions.

V. CONCLUSIONS

XPS, SIMS, and MALDI-MS are versatile and useful techniques for the characterization of adsorbed protein layers. XPS instrumentation and protocols are well developed and the technique is suitable for routine application. Interpretation and quantitation are well secured; coupled with experimental flexibility this makes it eminently suitable as a prime technique in biomaterials research. SIMS provides very low detection limits and information about the molecular structure of adsorbed proteins, but experimentation and data interpretation are more involved than for XPS. It is unique among the three techniques for its ability to probe conformation and orientation of adsorbed proteins. The spatial imaging capability of SIMS instruments is also of much interest for biomaterials research; the spatial resolution of XPS imaging has also advanced with modern instruments. MALDI-MS instrumentation is well developed, detection limits are very low, and experimental protocols simple, but adoption of the method to surface analysis including characterization of adsorbed protein layers is very new and requires further developments. MALDI-MS is particularly attractive for its ability to detect and distinguish coadsorbed proteins from complex biological fluids.

ACKNOWLEDGMENTS

HJG, SLM, PK, and KMcL gratefully acknowledge partial support from the Cooperative Research Centre for Eye Research and Technology and from the Ciba-Vision Corporation (Atlanta, GA) for some of the experimental work described herein. DGC, SLM, and MSW acknowledge generous support from the National ESCA and Surface Analysis Center for Biomedical Problems (NESAC/Bio, funded by NIH grant RR-01296 from the National Center for Research Resources) during the preparation of this manuscript and for some of the experimental work described herein.

REFERENCES

1. R. E. Baier and R. C. Dutton. J. Biomed. Mater. Res. *3*:191–206, 1969.
2. Y. H. An and R. J. Friedman. J. Biomed. Mater. Res. *43*:338–348, 1998.
3. J. D. Bryers. Colloids Surf. B: Biointerfaces 2:9–23, 1994.
4. W. Rudnicka, B. Sadowska, A. Ljungh, and B. Rozalska. FEMS Immunol. Med. Microbiol. *19*:7–14, 1997.
5. S. L. Vasin, I. B. Rosanova, and V. I. Sevastianov. J. Biomed. Mater. Res. *39*:491–497, 1998.
6. S. I. Ertel, B. D. Ratner, A. Kaul, M. B. Schway, and T. A. Horbett. J. Biomed. Mater. Res. *28*:667–675, 1994.
7. T. A. Horbett. Colloids Surf. B: Biointerfaces 2:225–240, 1994.
8. C. H. Thomas, C. D. McFarland, M. L. Jenkins, A. Rezania, J. G. Steele, and K. E. Healy. J. Biomed. Mater. Res. *37*:81–93, 1997.
9. W. J. Kao, J. A. Hubbel, J. M. Anderson. J. Mater. Sci.: Mater. Med. *10*:601–605, 1999.
10. A. G. Gristina. Science *237*:1588–1595, 1987.
11. G. P. Lopez, B. D. Ratner, C. D. Tidwell, C. L. Haycox, R. J. Rapoza, and T. A. Horbett. J. Biomed. Mater. Res. *26*:415–439, 1992.
12. M.-S. Sheu, A. S. Hoffman, J. G. A. Terlingen, and J. Feijen. Clin. Mater. *13*:41–45, 1993.
13. M.-S. Sheu, A. S. Hoffman, B. D. Ratner, J. Feijen, and J. M. Harris. J. Adhes. Sci. Technol. 7:1065–1076, 1993.
14. M. Morra, E. Occhiello, and F. Garbassi. Clin. Mater. *14*:255–265, 1993.
15. L. Deng, M. Mrksich, and G. M. Whitesides. J. Am. Chem. Soc. *118*:5136–5137, 1996.
16. P. Harder, M. Grunze, R. Dahint, G. M. Whitesides, and P. E. Laibinis. J. Phys. Chem. B *102*:426–436, 1998.
17. P. Kingshott, H. Thissen, and H. J. Griesser. Biomaterials *23*:2043–2056, 2002.
18. M. Boeckl, T. Baas, A. Fujita, K.-O. Hwang, A. L. Bramblett, B. D. Ratner, J. W. Rogers, and T. Sasaki. Biopolymers *47*:185–193, 1998.

19. J. L. Bohnert, B. C. Fowler, T. A. Horbett, and A. S. Hoffman. J. Biomater. Sci. Polym. Ed. *1*:279–297, 1990.

20. D. Kiaei, A. S. Hoffman, and T. A. Horbett. J. Biomater. Sci. Polym. Ed. *4*:35–44, 1992.

21. B. D. Ratner, T. Boland, E. E. Johnston, and C. D. Tidwell. In *Thin Films and Surfaces for Bioactivity and Biomedical Applications* (C. D. Tidwell, ed.). Materials Research Society, Boston, MA, 1995.

22. H. Shi, W.-B. Tsai, M. D. Garrison, S. Ferrari, and B. D. Ratner. Nature *398*:593–597, 1999.

23. J. C. Vickerman, Ed. *Surface Analysis—The Principal Techniques*, John Wiley & Sons, Chichester, England, 1997.

24. D. Briggs. In *Comprehensive Polymer Science*, Vol. 1: *Polymer Characterization* (C. Booth and C. Price, eds.). Pergamon Press, Oxford, 1989, pp. 543–599.

25. B. Ratner and D. Castner. In *Surface Analysis—The Principal Techniques* (J. C. Vickerman, ed.). John Wiley & Sons, Chichester, England, 1997, pp. 43–98.

26. R. W. Paynter and B. D. Ratner. In *Surface and Interfacial Aspects of Biomedical Polymers* (J. D. Andrade, ed.). Plenum Press, New York, 1985, pp. 189–216.

27. B. D. Ratner, T. A. Thomas, D. Shuttleworth, and T. A. Horbett. J. Colloid Interface Sci. *83*:630–642, 1981.

28. M. S. Wagner, S. L. McArthur, M. Shen, T. A. Horbett, and D. G. Castner. J. Biomater. Sci.: Polym. Ed. (in press).

29. R. W. Paynter, B. D. Ratner, T. A. Horbett, and H. R. Thomas. J. Colloid Interface Sci. *101*:233–245, 1984.

30. K. Lewis and B. Ratner. J. Colloid Interface Sci. *159*:77–85, 1993.

31. E. Blomberg, P. Claesson, and J. Froberg. Biomaterials *19*:371–386, 1998.

32. J.-B. Lhoest, E. Detrait, P. van den Bosch de Aguilar, and P. Bertrand. J. Biomed. Mater. Res. *41*:95–103, 1998.

33. B. J. Tyler. Ann. NY Acad. Sci. *831*:114–126, 1997.

34. R. Margalit and R. P. Vasquez. J. Protein Chem. *9*:105–108, 1990.

35. P. Kingshott, R. C. Chatelier, H. A. W. St. John, and H. J. Griesser, Polym. Mater. Sci. Eng. *76*:77–78, 1997.

36. A. M. Baty, P. A. Suci, B. J. Tyler, and G. G. Geesey. J. Colloid Interface Sci. *177*: 307–315, 1996.

37. H. Fitzgerald, P. F. Luckham, S. Eriksen, and K. Hammo. J. Colloid Interface Sci. *149*: 1–9, 1992.

38. M. S. Shen, Y. V. Pan, M. S. Wagner, K. D. Hauch, D. G. Castner, B. D. Ratner, and T. A. Horbett. J. Biomater. Sci.: Polym. Ed. *12*:961–978, 2002.

39. K. M. McLean, S. L. McArthur, R. C. Chatelier, P. Kingshott, and H. J. Griesser. Colloids Surf. B: Biointerfaces *17*:23–35, 2000.

40. S. L. McArthur, K. M. McLean, P. Kingshott, H. A. W. St. John, R. C. Chatelier, and H. J. Griesser. Colloids Surf. B: Biointerfaces *17*:37–48, 2000.

41. H. Ichijima, T. Kawai, K. Yamamoto, and H. D. Cavanagh. CLAO J. *26*:18–20, 2000.

42. D. Hart, M. DePaolis, B. D. Ratner, and N. Mateo. CLAO J. *19*:169–173, 1993.

43. C. D. Tidwell, D. G. Castner, S. L. Golledge, B. D. Ratner, K. Meyer, B. Hagenhoff, and A. Benninghoven. Surf. Interface Anal. *31*:724–33, 2001.

44. R. C. Chatelier, H. A. W. St. John, T. R. Gengenbach, P. Kingshott, and H. J. Griesser. Surf. Interface Anal. *25*:741–746, 1997.

45. P. Kingshott. Interfacial aspects of protein and lipid adsorption to contact lens surfaces. Ph.D. Thesis, University of New South Wales, Sydney, Australia, 1998.

46. H. A. W. St. John, T. R. Gengenbach, P. G. Hartley, H. J. Griesser. In *Surface Analysis Methods in Material Science*, 2nd Ed. (D. O'Connor, B. Sexton, and R. Smart, eds.). Springer-Verlag, Berlin, 2002.

47. J. C. Vickerman and A. Swift. In *Surface Analysis—The Principal Techniques* (J. C. Vickerman, ed.). John Wiley & Sons, Chichester, England, 1997, pp. 135–214.

48. A. Adriaens, L. vanVaeck, and F. Adams. Mass Spectrom. Rev. *18*:48–81, 1999.

49. L. vanVaeck, A. Adraens, and R. Gijbels. Mass Spectrom. Rev. *18*:1–47, 1999.

50. M. L. Pacholski and N. Winograd. Chem. Rev. *99*:2977–3005, 1999.

51. L. Hanley, O. Kornienko, E. T. Ada, E. Fuoco, and J. L. Trevor. J. Mass Spectrom. *34*:705–723, 1999.

52. G. Marletta, S. M. Catalano, and S. Pignataro. Surf. Interface Anal. *16*:407–411, 1990.

53. M. J. Hearn, D. Briggs, S. C. Yoon, and B. D. Ratner. Surf. Interface Anal. *10*:384–391, 1987.

54. D. Leonard and H. J. Mathieu. Fresenius J. Anal. Chem. *365*:3–11, 1999.

55. B. T. Chait and K. G. Standing. Int. J. Mass Spectrom. Ion Phys. *40*:185–193, 1981.

56. P. Steffens, E. Niehuis, T. Friese, and A. Benninghoven. In *Ion Formation from Organic Solids* (A. Benninghoven, ed.). Springer-Verlag, Munster, Germany, 1982.

57. A. Benninghoven, B. Hagenhoff, and E. Niehuis. Anal. Chem. *65*:630A–640A, 1993.

58. S. M. Daiser. IEEE Circuits Devices *7*:27–31, 1991.

59. H. J. Mathieu and D. Leonard. High Temp. Mater. Processes *17*:29–44, 1998.

60. M. Wood, Y. Zhou, C. L. Brummel, and N. Winograd. Anal. Chem. *66*:2425–2432, 1994.

61. B. D. Ratner. Surf. Interface Anal. *23*:521–528, 1995.

62. C. M. John, R. W. Odom. Int. J. Mass Spectrom. Ion Processes *161*:47–67, 1997.

63. J.-B. Lhoest, M. S. Wagner, C. D. Tidwell, and D. G. Castner. J. Biomed. Mater. Res. *57*:432–440, 2001.

64. D. S. Mantus, B. D. Ratner, B. A. Carlson, and J. F. Moulder. Anal. Chem. *65*:1431–1438, 1993.

65. M. S. Wagner and D. G. Castner. Langmuir *17*:4649–4660, 2001.

66. M. S. Wagner, M. Shen, T. A. Horbett, and D. G. Castner. J. Biomed. Mater. Res. (submitted).

67. T. L. Colliver, C. L. Brummel, M. L. Pacholski, F. D. Swanek, A. G. Ewing, and N. Winograd. Anal. Chem. *69*:2225–2231, 1997.

68. J. Clerc, C. Fourre, and P. Fragu. Cell Biol. Int. *21*:619–633, 1997.

69. C. Fourre, J. Clerc, and P. Fragu. J. Anal. Atomic Spectrom. *12*:1105–1110, 1997.

70. M. E. Kargacin and B. R. Kowalski. Anal. Chem. *58*:2300–2306, 1986.

71. A. Benninghoven and W. K. Sichtermann. Anal. Chem. *50*:1180–1184, 1978.

72. W. Lange, M. Jirikowsky, and A. Benninghoven. Surf. Sci. *136*:419–436, 1984.

73. C. G. Worley, E. P. Enriquez, E. T. Samulski, and R. W. Linton. Surf. Interface Anal. *24*:59–67, 1996.

74. L. Sun and J. J. A. Gardella. Pharm. Res. *17*:859–862, 2000.

75. S. Bartiaux. Etude de poly(oxides amines) par ToF-SIMS et XPS. Application a l'analyse d'une proteine. In *Faculte des Sciences Agronomiques, Unite de Chimie des Interfaces*. Universite Catholique de Louvain, Louvain-la-Neuve, 1995.

76. J. Davies, C. S. Nunnerley, and A. J. Paul. Colloids Surf. B: Biointerfaces *6*:181–190, 1996.

77. C. M. Pradier, P. Bertrand, M. N. Bellon-Fontaine, C. Compere, D. Costa, P. Marcus, C. Poleunis, B. Rondot, and M. G. Walls. Surf. Interface Anal. *30*:45–59, 2000.

78. S. Ferrari and B. D. Ratner. Surf. Interface Anal. *29*:837–844, 2000.

79. C. Derue, D. Gibouin, F. Lefebvre, B. Rasser, A. Robin, L. LeSceller, M. C. Verdus, M. Demarty, M. Thellier, and C. Ripoll. J. Trace Microprobe Tech. *17*:451–460, 1999.

80. D. M. Cannon, M. L. Pacholski, N. Winograd, and A. G. Ewing. J. Am. Chem. Soc. *122*:603–610, 2000.

81. N. Xia, C. J. May, S. L. McArthur, and D. G. Castner. Langmuir *18*:4090–4097, 2002.

82. R. Michel, R. Luginbuhl, D. J. Graham, and B. D. Ratner. J. Vac. Sci. Technol. A *18*: 1114–1118, 2000.
83. K. J. Wu and R. W. Odom. Anal. Chem. *68*:873–882, 1996.
84. K. Wittmaack, W. Szymczak, G. Hoheisel, and W. Tuszynski. J. Am. Soc. Mass Spectrom. *11*:553–563, 2000.
85. A. M. Belu, Z. Yang, R. Aslami, and A. Chilkoti. Anal. Chem. *73*:143–150, 2001.
86. M. Pacholski, D. Cannon, A. Ewing, and N. Winograd. Rapid Commun. Mass Spectrom. *12*:1232–1235, 1998.
87. B. Tyler. In *ToF SIMS: Surface Analysis by Mass Spectrometry* (J. C. Vickerman and D. Briggs, eds.). SurfaceSpectra Limited, Manchester, 2001, pp. 475–493.
88. A. Appelhans and J. Delmore. Anal. Chem. *61*:1087–1093, 1989.
89. G. Gillen and S. Roberson. Rapid Commun. Mass Spectrom. *12*:1303–1312, 1998.
90. P. Kingshott, H. A. W. St. John, R. C. Chatelier, and H. J. Griesser. Polym. Mater. Sci. Eng. *76*:81–82, 1997.
91. P. Kingshott, H. A. W. St. John, R. C. Chatelier, F. Caruso, and H. J. Griesser. Am. Chem. Soc. Polym. Prep. *38*(1):1008–1009, 1997.
92. H. A. W. St. John, P. Kingshott, and H. J. Griesser. Polym. Mater. Sci. Eng. *76*:83–84, 1997.
93. P. Kingshott, H. A. W. St. John, T. Vaithianathan, K. McLean, and H. J. Griesser. Trans. Am. Soc. Biomater. *21*:253, 1998.
94. P. Kingshott, H. A. W. St. John, and H. J. Griesser. Anal. Biochem. *273*:156–162, 1999.
95. S. L. McArthur. Fouling characteristics of contact lenses: the role of surface modification and chemistry in controlling protein adsorption. Ph.D. Thesis, University of New South Wales, Sydney, Australia, 2000.
96. M. Karas and F. Hillenkamp. Anal. Chem. *60*:2299–2301, 1988.
97. R. A. Zubarev, P. A. Demirev, P. Håkansson, and B. U. R. Sundqvist. Anal. Chem. *67*: 3793–3798, 1995.
98. O. Vorm, P. Roepstorff, and M. Mann. Anal. Chem. *66*:3281–3287, 1994.
99. K. Strupat, M. Karas, and F. Hillenkamp. Int. J. Mass Spectrom. Ion Processes *111*: 89–102, 1991.
100. D. C. Muddiman, A. I. Gusev, and D. M. Hercules. Mass Spectrom. Rev. *14*:383–429, 1995.
101. H. Ehring and B. U. R. Sundqvist. J. Mass Spectrom. *30*:1303–1310, 1995.
102. Y. F. Zhu, K. L. Lee, K. Tang, S. L. Allman, N. I. Taranenko, and C. H. Chen. Rapid Commun. Mass Spectrom. *9*:1315–1320, 1995.
103. M. E. Gimon, L. M. Preston, T. Solouki, M. A. White, and D. H. Russell. Org. Mass Spectrom. *27*:827–830, 1992.
104. P. Kingshott, H. A. W. St. John, R. C. Chatelier, and H. J. Griesser. J. Biomed. Mater. Res. *49*:36–42, 2000.
105. T. C. Vaithianathan. Development of methods for the accurate study of protein adsorption onto biomaterials. Ph.D. Thesis, University of New South Wales, Sydney, Australia, 1999.
106. E. Leize, M. Leize, J.-C. Voege, and A. Dorsselaer. Anal. Biochem. *272*:19–25, 1999.
107. S. L. McArthur, K. M. Mclean, H. A W. St. John, and H. J. Griesser. Biomaterials *22*: 3295–3304, 2001.
108. T.-W. Chan, A. W. Cilburn, and P. J. Derrick. Org. Mass Spectrom. *27*:53–56, 1992.
109. S. Jespersen, W. M. A. Niessen, U. R. Tjaden, J. Vandergreef, E. Litborn, U. Lindberg, and J. Roeraade. Rapid Commun. Mass Spectrom. *8*:581–584, 1994.
110. M. Schuerenberg, C. Luebbert, H. Eickhoff, M. Kalkum, H. Lehrach, and E. Nordhoff. Anal. Chem. *72*:3436–3442, 2000.

111. T. Miliotis, S. Kjellstrom, J. Nilsson, T. Laurell, L. E. Edholm, and G. Marko-Varga. Rapid Commun. Mass Spectrom. *16*:117–126, 2002.
112. B. O. Keller and L. Li. J. Am. Soc. Mass Spectrom. *12*:1055–1063, 2001.
113. A. Overberg, M. Karas, and F. Hillenkamp. Rapid Commun. Mass Spectrom. *5*:128–131, 1991.
114. R. Kaufmann, D. Kirsch, and B. Spengler. Int. J. Mass Spectrom. Ion Processes *131*: 355–385, 1994.
115. R. Kaufmann, B. Spengler, and F. Lutzenkirchen. Rapid Commun. Mass Spectrom. *7*: 902–910, 1993.
116. J. Franzen, R. Frey, A. Holle, and K. O. Krauter. Int. J. Mass Spectrom. *206*:275–286, 2001.
117. M. M. Vestling and C. C. Fenselau. Mass Spectrom. Rev. *14*:169–178, 1995.
118. J. A. Blackledge and A. J. Alexander. Anal. Chem. *67*:843–848, 1995.
119. G. Narayanaswani and R. Levis. J. Am. Chem. Soc. *119*:6888–6890, 1997.
120. M. O'Donnell, K. Tang, H. Koster, C. L. Smith, and C. R. Cantor. Anal. Chem. *69*: 2438–2443, 1997.
121. K. M. McLean, S. L. McArthur, and H. J. Griesser, unpublished results.
122. L. Dai, P. Zientek, H. A. W. St. John, P. Pasic, R. C. Chatelier, and H. J. Griesser. In *Surface Modification of Polymeric Biomaterials* (B. D. Ratner and D. G. Castner, eds.). Plenum Press, New York, 1996, pp. 147–164.
123. H. J. Griesser, R. C. Chatelier, L. Dai, H. A. W. St. John, T. Davis, and R. Austen. Polym. Mater. Sci. Eng. *76*:79, 1997.
124. S. L. McArthur, K. M. McLean, P. Kingshott, H. A. W. St. John, and H. J. Griesser. Trans. Am. Soc. Biomater. *22*:243, 1999.
125. M. E. McComb, R. D. Oleschuk, A. Chow, H. Perreault, R. G. Dworschak, M. Znamirowski, W. Ens, K. G. Standing, and K. R. Preston. Can. J. Chem. *79*:437–447, 2001.
126. Y. Dai, R. M. Whittal, and L. Li. Anal. Chem. *68*:2495–2500, 1996.
127. A. Walker, Y. Wu, R. Timmon, K. Nelson, and G. Kinsel. J. Mass Spectrom. *34*:1205–1207, 1999.
128. A. Walker, H. Qui, Y. Wu, R. Timmons, and G. Kinsel. Anal. Biochem. *271*:123–130, 1999.
129. A. Walker, C. Land, G. Kinsel, and K. Nelson. J. Am. Soc. Mass Spectrom. *11*, 2000.
130. H. J. Griesser, K. M. McLean, G. J. Beumer, X. Gong, P. Kingshott, G. Johnson, and J. G. Steele. Mater. Res. Symp. Proc. *544*:9–20, 1999.

25

Interaction of Proteins with Polymeric Synthetic Membranes

GEORGES BELFORT Rensselaer Polytechnic Institute, Troy, New York, U.S.A.

ANDREW L. ZYDNEY University of Delaware, Newark, Delaware, U.S.A.

I. INTRODUCTION

A. Synthetic Membranes and Biotechnology

Pressure-driven membrane processes are being increasingly integrated into existing reaction, isolation, and recovery schemes for the biological production of valuable molecules. These processes can be distinguished by their increasing volumetric permeation rates and have been called reverse osmosis, nanofiltration, ultrafiltration, and microfiltration. They are commonly used to retain all ions, bivalent from univalent ions, macromolecules, and suspended colloids and particles, respectively. Membrane systems are being used for both upstream (involving the bioreactor) and downstream (involving product recovery) processing, taking advantage of their permselectivity, high surface area per unit volume, and their potential for controlling the level of contact and/or mixing between two phases. These processes are now commonly being used to concentrate proteins, exchange buffers, clarify suspensions for cell harvesting, and sterile filter liquids to remove viruses and bacteria [1].

Advances in materials, module design, and operation have led to better performance. Membrane materials have evolved from hydrophobic polymers such as polypropylene, polyvinylidene fluoride (PVDF), and polysulfone to hydrophilic cellulosics, acrylate modified PVDF, and polysulfone-containing additives in the casting melt, such as polyvinylpyrrolidone (PVP). Using a relatively standard filtration protocol (of buffer, protein solution, and then buffer again), Koehler and Belfort [2,3] compared the performance of a polysulfone ultrafiltration membrane with a hydroxyethyl methacrylate (HEMA) surface-modified polysulfone membrane. The results are shown in Fig. 1, where it can be clearly seen that the drop in volume flux on exposure to the lysozyme solution by the unmodified polysulfone membrane is significantly higher than for the modified one. The solute flux remains below 5 μmol/m^{2}h for the former membrane, while it reaches a value of 150 μmol/m^{2}h for the latter. The interaction of the protein with the membrane and, more importantly, its interference with membrane performance obviously need to be understood and dealt with in order to successfully use this technology in bioprocessing.

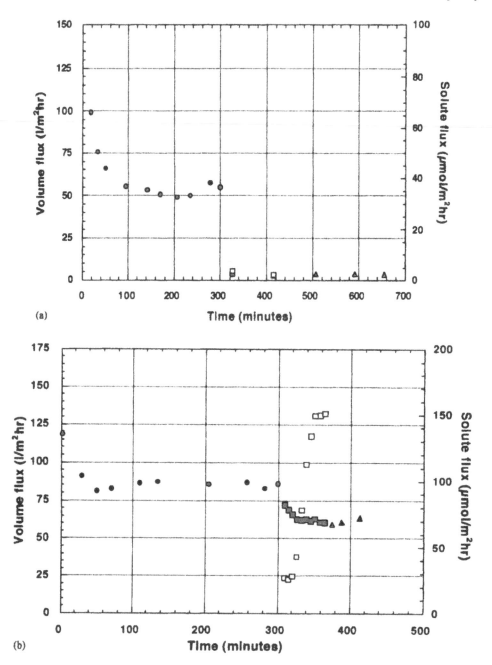

FIG. 1 Membrane filtration with 50 mg/L lysozyme solution under constant transmembrane pressure. Buffer flux of clean membrane (●); volume flux of lysozyme solution (■); solute flux of lysozyme solution (□); and buffer flux of fouled membrane (▲), all at pH 11: (a) GR81PP polysulfone membrane and (b) HEMA modified GR81PP polysulfone membrane. (From Refs. 2 and 3.)

In this chapter, we address the question of how proteins interact with synthetic membrane polymeric materials through a phenomenological understanding and through direct measurements of the intermolecular forces between model proteins and polymeric membranes. It is important to point out that in addition to diffusional processes, convection through pressure-driven flows exists during the interaction of proteins in the feed solutions and the membranes. We conclude the chapter with a brief description of various approaches used to minimize the degradative effect of protein–polymer adhesion on membrane performance (often called fouling).

B. Analyzing the Problem

As discussed above, protein-membrane interactions play a critical role in determining the overall performance of many ultrafiltration and microfiltration systems used for bioprocessing and biomedical applications. These interactions are referred to as "membrane fouling," which is the irreversible alteration in the membrane caused by specific physical and/or chemical interactions between the membrane and various components present in the process stream. Fouling typically results in a decay in filtrate flux and/or an alteration in membrane selectivity. These changes often continue throughout the process and eventually require extensive cleaning or replacement of the membrane.

The effects of membrane fouling on the filtrate flux and solute sieving can often be very similar to those associated with concentration polarization, which is a reversible boundary layer phenomenon associated with the accumulation of retained solutes/particles at the membrane surface [4]. Osmotically active solutes can reduce the effective pressure-driving force across the membrane, while cake- or gel-forming materials can provide an additional hydrodynamic resistance to flow. The extent of concentration polarization is determined by the bulk mass transfer characteristics of the membrane device. Therefore, the effects of this polarization can be reduced, and in some cases eliminated, by sufficiently increasing the rate of solute mass transfer back away from the membrane so that there is effectively no boundary layer accumulation of retained material.

In contrast, membrane fouling occurs because of specific physical and/or chemical interactions between the proteins and the membrane surface. The rate and extent of membrane fouling will be a function of the device fluid mechanics, but fouling cannot, in general, be eliminated simply by increasing the rate of solute mass transfer. It is also important to note that the term "irreversible" with respect to fouling can often be reversed by backflushing, application of very high cross-flow velocities, or chemical cleaning. However, the mechanical and/or chemical forces required to reverse the fouling must be generated in separate "cleaning" cycles.

Zeman and Zydney [4] and Marshall et al. [5] have provided excellent reviews of protein fouling during both ultrafiltration and microfiltration, focusing specifically on the effects of fouling on the permeate flux, protein retention, and selectivity of these membrane processes. These studies clearly demonstrate that all of the factors affecting protein adsorption on nonporous surfaces can also play a role in membrane systems. This includes the specific physicochemical characteristics of the protein [6] and membrane [7], the solution pH and ionic strength [8,9], temperature [10], and the presence of various stabilizing or chaotropic agents in solution [11]. In addition, protein fouling is also a function of the detailed fluid mechanics in the particular membrane device (the transmembrane pressure drop, local shear stress, and secondary flows) as well as the detailed pore size and morphology of the membrane [12,13].

C. Overview of Phenomena

Membrane fouling is composed of pore plugging, pore narrowing, and cake deposition. All of these phenomena share two important factors in the filtration of biological fluids: the positive interaction between dissolved protein and itself (aggregation) and between dissolved protein and the membrane surface (adsorption). The mechanisms that underlie these attractive forces at the molecular level are unknown. This is significant because, in principle, membranes could be produced that exhibit smaller attractive forces between the membrane surface and the protein. This, in turn, should yield membranes that have a longer operational life and exhibit higher performance characteristics (i.e., improved retention and flux).

To date, a great deal of work has been performed on membrane filtration using various macromolecules including proteins (such as bovine serum albumin, BSA), dextrans, polyethylene glycols, and others. Different flux-decline rates occurred for all the various cases showing a solute-dependent process. Thus, each new membrane or macromolecule system must be individually analyzed. Among the many reasons proposed for this result are

1. Protein adsorption (specifically protein–membrane and protein–protein interactions and their dependence on pH and ionic strength) [9,14–19]
2. Reduced driving force due to an osmotic back-pressure from solute buildup at the membrane surface [20–25]
3. Increased resistance due to protein deposition and cake formation along with the increased viscosity of the fluid near the membrane surface [26–30]

Selecting BSA as a model protein, experiments have shown that fluxes are higher when the pH of the solution is not equal to the pI of the protein [9,15,16,27,30], when the ionic strength of the solution is low [27] and when the surface of the membrane is hydrophilic [31]. Zydney and coworkers [6,32] and Tracey and Davis [33] have proposed a two-step mechanism to describe BSA fouling and flux decline in stirred-cell microfiltration. Large BSA aggregates are convectively dragged toward the membrane surface constricting and blocking the pores, with these aggregates serving as nucleation or attachment sites for further BSA deposition. By filtering (1) prefiltered BSA solutions or (2) nonaggregating BSA (BSA in which the free sulfhydryl is capped with a cysteinyl or carboxymethyl group), Kelly and Zydney [6,32] have shown that minimal flux decline occurs. When they used a prefiltered BSA solution, they were able to show that in the absence of aggregates, there was little, if any, fouling. Also, the rate of mixing and hence back-migration had no effect on the results. With mixtures of unfiltered and prefiltered BSA (through a 100-kD molecular-weight cutoff ultrafiltration membrane), they showed that "flux-decline was determined entirely by the deposition of aggregates and was unaffected by the concentration of native (monomeric) BSA" [32].

Kim and Fane [34] have compared the performance of four commercial ultrafiltration (UF) membranes with the same nominal molecular-weight cutoff but with different hydrophilicity using a 0.1 wt% BSA solution in a cross-flow test cell. They obtained "enhanced UF fluxes with slower flux loss and lower solute resistance" for the hydrophilic membranes as compared to the unmodified hydrophobic membranes (as measured by contact angle). They also report that the "hydrophilic membranes were not necessarily easier to clean." In addition, they observed, as others have before, that UF fluxes were greater when the solution pH was away from the pI.

Maa and Hsu [11] provide convincing evidence that the presence of aggregates in solution is not the only potential cause of fouling with recombinant human growth hormone (rhGH). They suggest that "aggregation/adsorption in the filter pores during filtration is a better explanation for membrane fouling." High pH, low salt, and the presence of a nonionic detergent all resulted in improved flux.

These phenomenological studies are extremely useful in suggesting the causes of fouling, and they are discussed in detail in Section II. To understand the mechanisms and to provide direct evidence of fouling, intermolecular force measurements were performed by Koehler and Belfort [2,3] between a model protein (lysozyme) and a thin, hydrophobic polymer film (polysulfone). Polysulfone was chosen as the polymer because it is a commonly used membrane material in industry due to its chemical and structural stability. The forces were compared with three permeation flux criteria from membrane filtration results in an attempt to relate the molecular scale measurements with the macroscopic observations. Their correlation is presented here, and it shows a simple relation between the forces and flux decline. Details of these direct measurements and correlations are presented in Section III. For completeness, we discuss in Section IV various chemical approaches to minimize protein fouling. We conclude in Section V with a summary of our understanding of protein–polymer interactions and membrane filtration performance and some thoughts on future directions for research.

II. PHENOMENOLOGICAL UNDERSTANDING OF PROTEIN–MEMBRANE INTERACTIONS

The rate and extent of protein fouling in any given system will be determined by the strength of the intermolecular interactions between the proteins and the membrane in combination with the effects of the various hydrodynamic (and body) forces acting on the proteins. The hydrodynamic forces not only determine the rate of protein transport toward the membrane (and thus the local protein concentration in the immediate vicinity of the membrane surface), but they can directly contribute to membrane fouling through the physical deposition of proteins (and protein aggregates) on the membrane surface by creating shear conditions within or near the pores for induced aggregation and through shear-induced denaturation. Thus, it is useful to examine two distinct types of protein–membrane interactions: protein adsorption, which refers to the specific intermolecular interactions that occur in the absence of filtration (i.e., in the absence of hydrodynamic forces), and protein deposition that is directly associated with the filtrate flow through the membrane.

A. Protein Adsorption

The effects of protein adsorption on membrane transport depend critically on the relative size of the protein and the membrane pores. The discussion below is divided into three subsections: (1) adsorption to microfiltration membranes with pores that are much larger than the size of the proteins, (2) adsorption to semipermeable ultrafiltration membranes with pores that are comparable in size to the proteins of interest, and (3) adsorption to fully retentive membranes that have pores that are considerably smaller than the proteins.

1. Microfiltration Membranes

Experimental studies of protein adsorption on microfiltration (MF) membranes are generally consistent with the behavior seen with nonporous surfaces (see discussion in Chapter 3 of this volume). Typically, monolayer protein adsorption is found throughout the internal pore surface of the MF membranes [4,12]. This has been shown for a variety of proteins (bovine serum albumin, hemoglobin, yeast alcohol dehydrogenase, and β-lactoglobulin) on a number of different MF membranes (hydrophobic poly(vinylidene fluoride), polycarbonate, polyethersulfone, aluminum oxide, nylon, and cellulose acetate). For example, Bowen and Hughes [35] showed that the adsorption of BSA on aluminum oxide membranes could be effectively described by a Langmuir adsorption isotherm with $m_{max} = 2.5$ mg/m^2 and K_{ads} on the order of 1 mg/L. There was also some evidence for a weakly bound multilayer adsorption at high protein concentrations, similar to the behavior seen by Lee and Ruckenstein for BSA adsorption on nonporous surfaces [36]. Protein adsorption on highly hydrophilic microfiltration membranes can be significantly less than a monolayer [37]. However, this reduced adsorption may be due, at least in part, to desorption of loosely bound protein during the rinsing steps used prior to evaluating the amount of adsorbed protein. Belfort et al. [17], using a pore constriction model to describe the reduction in flux during filtration of 10% BSA through a 0.45-μm hydrophilic polysulfone membrane, suggested that either there were approximately 21 layers of BSA adsorbed in the membrane pores, or protein aggregation and pore blocking occurred.

The effect of monolayer protein adsorption on the solvent flux through a microfiltration membrane will generally be quite small. This can be seen most easily by modeling the flux using the Hagen-Poiseuille equation for flow through a membrane composed of an array of uniform cylindrical pores:

$$J_v = \frac{N \pi r_p^4 \Delta P_{TM}}{8 \mu \delta_m} \tag{1}$$

where N is the membrane pore density, r_p is the pore radius, μ is the viscosity, ΔP_{TM} is the transmembrane pressure, and δ_m is the membrane thickness. Thus, the flux through an idealized MF membrane with uniform 0.2-μm pores would be decreased by only 6% after adsorption of a protein monolayer 30 Å thick.

Although protein adsorption has little effect on the hydraulic permeability of an MF membrane, it can dramatically alter the membrane surface charge and chemical characteristics. Figure 2 shows data for the zeta potential of a 0.2-μm track-etched polycarbonate membrane with essentially uniform cylindrical pores. The streaming potential across the membrane (E) was measured as a function of the transmembrane pressure drop (ΔP_{TM}) using Ag/AgCl electrodes placed in 0.01 M KCl solutions on the two sides of the membrane. The zeta potential (ζ) was then calculated from the Smoluchowski equation:

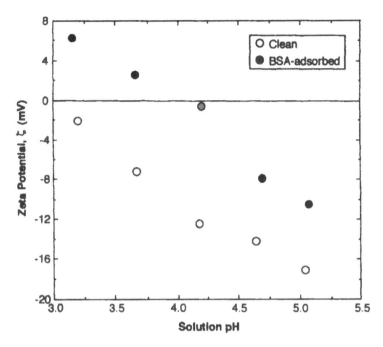

FIG. 2 Effect of BSA adsorption on the zeta potential of a 0.2-μm track-etched polycarbonate membrane. Data obtained using 0.001 M KCl solutions.

$$\zeta = \frac{\mu km}{\varepsilon_0 \varepsilon} \qquad (2)$$

where m is the slope of the E versus ΔP_{TM} data, μ is the solution viscosity, k is the solution conductivity, ε is the permittivity of a vacuum, and ε_0 is the dielectric constant. Data are shown for a single membrane, before and after BSA adsorption, as a function of the solution pH. The zeta potential of the clean polycarbonate membrane is negative over the entire pH range, with the isoelectric point (the pH at which the membrane has no net charge) around pH 3. In contrast, the zeta potential of the same membrane after protein adsorption is positive at low pH, with an isoelectric point of approximately 4.1. This shift in the membrane surface charge is a direct result of the adsorption of BSA, which has an isoelectric point of about 4.7 under these conditions. This change in the membrane surface charge can dramatically alter subsequent electrostatic interactions between proteins and the membrane, affecting the rate of membrane fouling [38] as well as the transmission of charged proteins through the membrane pores [39].

2. Semipermeable Ultrafiltration Membranes

Protein adsorption to ultrafiltration membranes with pores that are only slightly larger than the size of the protein can be affected by "steric" interactions in the narrow pores. These steric effects include a loss of accessible surface area due to simple geometric constraints as well as an increase in the free energy of the adsorbed protein caused by unfavorable intermolecular interactions with the surrounding membrane surface. For example, a protein might adsorb to a nonporous surface in an orientation such that certain negatively charged amino acids are oriented away from specific negatively charged groups on the substrate. This type of free-energy minimization will be much less effective in a narrow pore since the negatively charged amino acids will be in close proximity to the charged groups on the opposite surface of the pore wall.

Robertson and Zydney [18] directly evaluated such complicating effects on the adsorption of BSA to semipermeable polyethersulfone (PES) ultrafiltration membranes in a series of static-soak experiments, with protein uptake evaluated using radioactively labeled (^{125}I) BSA. Equilibrium adsorption in this system was not attained for approximately 12 h due to the slow diffusive transport of BSA through the porous membrane structure. Lee and Belfort [40] have reported that about 24 h were needed for ribonuclease A to reach orientation equilibrium through a rotational diffusion process during adsorption. These slow transients demonstrate that considerable care must be taken in interpreting literature data presented for "equilibrium" adsorption. Representative results are summarized in Table 1 along with the data for a 0.16-μm microfiltration membrane cast from the same base polymer [41]. The amount of protein adsorbed per external area increased with increasing pore size for the smaller pore membranes, attaining its maximum value for the 300,000 molecular-weight cutoff (300 kD) membrane. The increase in protein uptake for the smaller pore membranes was attributed to the increase in accessibility of the pores, while the drop in adsorption for the two largest pore membranes was caused by the decrease in total pore (surface) area associated with the increase in the pore size. The mass of BSA adsorbed per total pore area by the 300-kD, 1000-kD, and 0.16-μm membranes were similar, and the calculated surface coverages were in good agreement with those obtained for monolayer adsorption of albumin on a variety of nonporous surfaces [42]. BSA adsorption to the 100-kD membrane was about 50% less than that for the larger pore membranes due to steric interactions in the narrow membrane pores. There was almost no measurable albumin adsorption in the porous skin of the 50-kD membrane, although there was still considerable adsorption to the porous substructure and matrix [18].

The final column in Table 1 shows the ratio of the filtrate flux (at a constant transmembrane pressure) after albumin adsorption to that determined for the clean membrane. The maximum reduction in the permeability occurred for the 1000-kD membrane, with the change in permeability for the smallest (50-kD) and largest

TABLE 1 Protein Adsorption on Semipermeable Ultrafiltration Membranes

Membrane	Mass adsorbed per external area (mg/m^2)	Mass adsorbed per pore area (mg/m^2)	Hydraulic permeability ratio (L_p/L_{p0})
50,000	20 $\pm$ 50	0.1 $\pm$ 0.2	0.68 $\pm$ 0.14
100,000	230 $\pm$ 90	2.0 $\pm$ 0.8	0.54 $\pm$ 0.11
300,000	440 $\pm$ 150	4.5 $\pm$ 1.5	0.20 $\pm$ 0.30
1,000,000	240 $\pm$ 150	4.0 $\pm$ 2.5	0.06 $\pm$ 0.03
0.16 μm	130 $\pm$ 30	3.0 $\pm$ 0.4	0.79 $\pm$ 0.10

(0.16-μm) pore membranes being fairly small. These data can be used to evaluate the effective reduction in the membrane pore size using the simple cylindrical pore model, Eq. (1). The results for the 300-kD, 1000-kD, and 0.16-μm membranes were again consistent with the adsorption of a protein monolayer, with the calculated thickness of the adsorbed layer being on the order of 50 Å. In contrast, the flux data for the 50-kD and 100-kD membranes would correspond to reductions in pore radius of only 3 and 7 Å, respectively, which is really just an artifact of modeling the complex adsorption behavior in the highly constricted pores of these membranes using a single cylindrical pore size.

The extent of protein adsorption on semipermeable UF membranes is also a function of the solution pH. Experimental data for the adsorption of β-lactoglobulin to polysulfone membranes indicate that the maximum protein uptake and the maximum reduction in filtrate flux both occurred at the isoelectric pH of β-lactoglobulin [43]. Hanemaaijer et al. [43] also found much lower levels of protein adsorption on more hydrophilic cellulosic membranes, in agreement with data reported in a number of other investigations [5]. Koehler and Belfort [2,3] have shown how the intermolecular forces can be correlated with BSA fouling of PES UF membranes as a function of pH (see discussion in Section III).

Not surprisingly, protein adsorption also has a significant effect on the sieving (or rejection) characteristics of semipermeable UF membranes. Figure 3 shows data for the asymptotic dextran sieving coefficients, S_∞, for a 100-kD (bottom panel) and a 50-kD (top panel) molecular-weight cutoff PES membrane obtained before and after equilibrium adsorption of BSA [44]. The asymptotic sieving coefficient is defined as the ratio of the solute flux across the membrane (under conditions dominated by convection) to the filtration velocity divided by the solute concentration at the upstream surface of the membrane [4]. The sieving coefficients after BSA adsorption were uniformly smaller than those for the clean membrane; e.g., the sieving coefficient for a 20K dextran through the 100-kD membrane decreased from about $S_\infty =$ 0.55 to less than 0.22 after adsorption of BSA. BSA adsorption has a much greater effect on the sieving for the larger molecular-weight cutoff membrane, which is consistent with the protein uptake and hydraulic permeability data reported in Table 1. Mochizuki and Zydney [44] analyzed these data for membrane sieving, in combination with results for hindered dextran diffusion and solvent flow. They concluded that protein adsorption occurred preferentially in the larger pores of the membrane pore size distribution and that the radii of these larger pores were effectively reduced by the protein adsorption.

3. Fully Retentive Ultrafiltration Membranes

Protein adsorption to fully retentive UF membranes is primarily a surface phenomenon since the membrane pores are virtually inaccessible to the proteins. Several studies have shown that protein adsorption to these fully retentive membranes does reduce the filtrate flux, with this effect generally attributed to some type of pore blockage [45]. It is, however, difficult to quantitatively verify this hypothesis, given the very tortuous and interconnected pore structure of most asymmetric UF membranes. Even if part of the membrane surface were completely "blocked" by the adsorbed protein, the filtrate would be able to flow through the area around the blocked pores and then percolate down through the membrane. This would include at

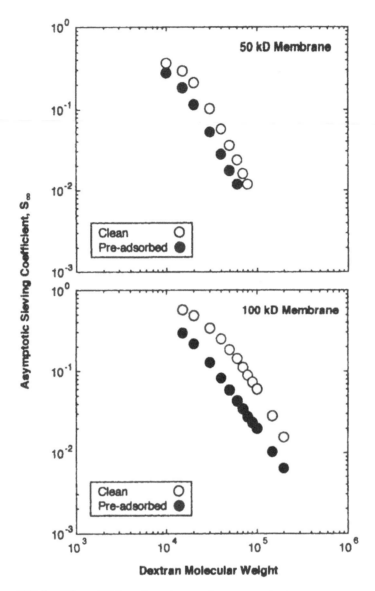

FIG. 3 Effect of BSA adsorption on the asymptotic dextran sieving coefficients for a 50-kD (upper panel) and a 100-kD (lower panel) polyethersulfone membrane. (Data from Ref. 44.)

least some of the pore space under the blocked pores, due to the interconnected nature of the pore network.

More recent work by Boyd et al. [46] has provided a very different physical picture of protein adsorption to fully retentive membranes. Boyd et al. proposed that the adsorbed protein exists in a "porous" layer on the upper surface of the membrane. This protein layer provides an additional resistance to both solute and solvent transport in series with that provided by the membrane. The transport properties of such protein-coated membranes could be accurately described using a two-layer membrane model, in which the flux through each layer is described by standard membrane transport equations with appropriate matching conditions applied at the interface between the two layers.

B. Protein Deposition

The flux decline observed during actual protein filtration can be much more dramatic than that caused by static protein adsorption. Typical experimental data for the filtrate flux during filtration of 10 g/L BSA solutions through a fully retentive (30-kD), a partially retentive (100-kD), and a large-pore (0.16-μm) polyethersulfone membrane at a constant pressure of 35 kPa are shown in Fig. 4 [38]. In each case, the membranes were first preadsorbed with BSA for a minimum of 12 h; thus the flux decline in Fig. 4 is due to protein deposition on (or within) the membrane. The initial flux through the 0.16-μm MF membrane is more than a factor of 10 greater than the flux through the fully retentive 30-kD membrane, due to the much greater pore size (and hydraulic permeability) of the larger-pore MF membrane. The flux through the MF membrane declines quite rapidly, decreasing by over an order of magnitude within the first 500 s of filtration. The net result is that the filtrate fluxes through these membranes after 1 h of filtration differ by less than 20%, with the largest flux actually obtained with the partially permeable 100-kD membrane.

Several studies have shown that the large flux decline observed during protein filtration through large-pore MF membranes is due to the formation of a relatively thick protein deposit on the upper surface of the membrane [4,12]. This protein deposit is able to fully span the large pores present on the membrane surface, pro-

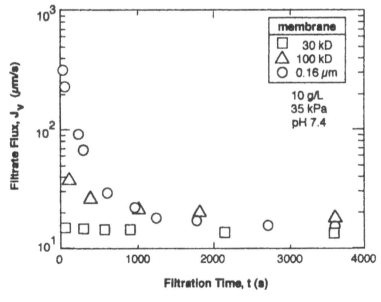

FIG. 4 Flux decline during BSA filtration through fully retentive (30-kD), partially retentive (100-kD), and initially nonretentive (0.16-μm) polyethersulfone membranes. (Data from Ref. 38.)

viding an additional hydraulic resistance to filtrate flow in series with the underlying membrane. The physical properties of this type of protein deposit are discussed in the next section.

There has been considerable controversy in the literature over the mechanism by which the proteins actually form such a deposit. Recent work in this area has demonstrated that the initial fouling is generally caused by the deposition of small amounts of aggregated and/or denatured protein present in the bulk protein solutions [6,12]. Such protein aggregates are clearly visible in SEM images of fouled membranes [4]. These aggregates can be formed during the initial purification or fractionation of the proteins or by the various processing operations used in the particular filtration (particularly when high temperatures or shear stresses are used).

Figure 5 shows the effects of BSA aggregates on fouling of a 0.16-μm PES membrane during BSA microfiltration [6]. The flux decline observed with the two commercial BSA preparations were dramatically different. The much more rapid fouling seen with the heat shock–precipitated BSA was due to the greater amount of aggregated and denatured protein in this sample. The protein aggregates could be removed by prefiltering the BSA solution through a 100,000 molecular-weight cutoff UF membrane. The prefiltered BSA solution exhibited almost no decline in filtrate flux over a 30-min filtration period. Stirring the BSA solution at an elevated temperature (33°C) caused an increase in the number of protein aggregates and significantly increased the rate of protein fouling [6]. Similar results were obtained by Meireles et al. [10], who found that the rate of BSA aggregation and denaturation increased with increasing temperature and cross-flow velocity. BSA aggregation can also occur in the pumps used in most cross-flow systems due to the high interfacial shear stresses [4].

Kelly and Zydney [47] have demonstrated that BSA aggregation in bulk solution, and in turn BSA fouling, occurs through the formation of intermolecular thiol-disulfide linkages between BSA molecules. BSA has 35 cysteine amino acid residues, 34 of which are linked pairwise via 17 intramolecular disulfide bonds. The one remaining cysteine provides a free-sulfhydryl (—SH) group near the NH_2-terminal end of the molecule. This free sulfhydryl can serve as a nucleophile and attack an existing internal disulfide linkage in a separate BSA molecule [47]. The reaction product is a BSA dimer which has two free-sulfhydryl groups that can participate in further thiol-disulfide interchange reactions, leading to the formation of large protein aggregates. Kelly and Zydney [47] have shown that the capping of this free-sulfhydryl group by a cysteinyl group (giving an —S—S— bond) or a carboxymethyl group (giving an —S—CH_2—COOH group) completely eliminates the flux decline seen during BSA microfiltration.

Although much of the literature data on protein fouling has focused on the behavior of BSA, similar results have also been obtained with other proteins. Figure 6 shows data for the filtration of 2 g/L solutions of five different proteins through 0.22-μm pore size poly(vinylidene fluoride) MF membranes [48]. The upper panel shows the normalized filtrate flux, while the lower panel shows the instantaneous rate of flux decline, defined as

$$K = -\frac{1}{J_v}\left(\frac{dJ_v}{dt}\right) \tag{3}$$

The rate of flux decline for pepsin, cys-BSA, and lysozyme decreases throughout

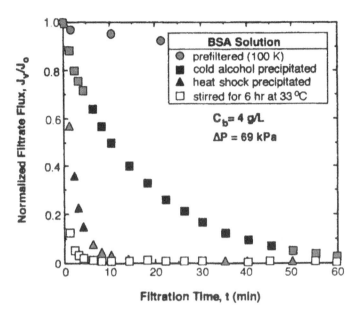

FIG. 5 Normalized filtrate flux as a function of time for BSA filtration through 0.16-μm polyethersulfone MF membranes. Data are shown for several different BSA preparations. (From Ref. 6.)

the filtration. In contrast, the rate of flux decline for BSA and β-lactoglobulin initially increases with time, passing through a maximum value before declining at long times. Kelly and Zydney [47,48] attributed this initial increase in the rate of flux decline to the increase in the rate of chemical addition of native protein as the surface deposit grows. This "chemical" fouling was eliminated by blocking the free-sulfhydryl group on the BSA; i.e., the flux decline for cys-BSA occurred only by aggregate deposition.

The solid curves in Fig. 6 represent model correlations which explicitly account for these two distinct fouling mechanisms: poor blockage associated with the deposition of large protein aggregates and the chemical attachment of native (monomeric) BSA to the growing deposit through the formation of intermolecular disulfide linkages. The final expression for the filtrate flux is

$$\frac{J_v}{J_0} = \frac{\dfrac{\beta N_0}{\alpha} + 1 - \dfrac{J_{ss}}{J_0}\left\{1 - \exp\left[J_0 C_b \dfrac{\alpha A}{N_0}\left(1 + \dfrac{\beta N_0}{\alpha}\right)t\right]\right\}}{\dfrac{\beta N_0}{\alpha} + \exp\left[J_0 C_b \dfrac{\alpha A}{N_0}\left(1 + \dfrac{\beta N_0}{\alpha}\right)t\right]} \qquad (4)$$

where J_{ss} is the steady-state filtrate flux obtained at long times, N_0 is the pore density of the clean membrane, α is related to the rate of aggregate deposition, and β is related to the rate of chemical attachment of native BSA to the growing deposit. This model is in excellent agreement with experimental data for all five proteins. The data for the proteins with no free sulfhydryls (pepsin, cys-BSA, and lysozyme) were all described with $\beta = 0$; i.e., there was no chemical attachment step. In contrast,

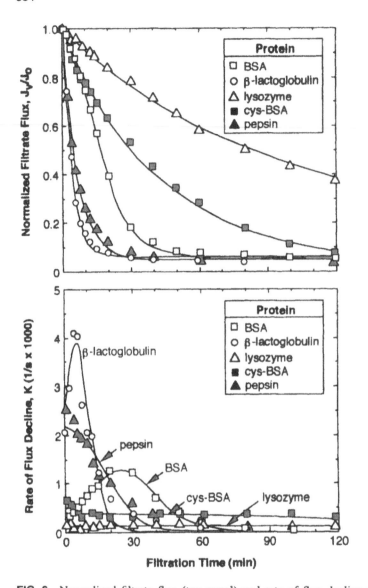

FIG. 6 Normalized filtrate flux (top panel) and rate of flux decline (bottom panel) for the constant-pressure filtration of 2-g/L solutions of several different protein solutions through 0.2-μm PVDF membranes [48].

the results for BSA and β-lactoglobulin, both of which have a single free-sulfhydryl group, show distinct positive values of β (corresponding to a finite rate of chemical addition of native protein to the growing deposit).

Although this model is in excellent agreement with experimental data for both the filtrate flux and the instantaneous rate of flux decline, it is not currently possible to obtain a priori estimates of the two key fouling parameters, α and β, for different proteins, nor is it possible to quantify the effects of membrane pore size, morphology, and surface chemistry on these fouling parameters. In addition, this analysis completely neglects the complex changes in cake resistance and protein polarization that can occur during the course of the protein microfiltration. These phenomena will clearly need to be considered in the development of a more complete model for protein fouling during membrane microfiltration.

The filtrate flux data in Figs. 5 and 6 suggest that the steady-state flux during protein microfiltration approaches a quasi-steady value at long times. Palecek and Zydney [38] hypothesized that a quasi-steady flux was attained when the hydrodynamic drag on the proteins associated with the filtrate flow toward the membrane is no longer sufficient to overcome the repulsive (primarily electrostatic) interactions that exist between the native (bulk) protein and the protein deposit on the membrane surface. As long as the convective drag force is greater than the electrostatic repulsion, proteins continue to add to the deposit and the flux continues to decline. Palecek and Zydney [38] estimated the electrostatic repulsive interaction using an expression of the force between two charged spheres, while the hydrodynamic force was evaluated using the Stokes drag law. The quasi-steady flux was then determined by equating these forces, yielding [38]

$$J_v = J_{pI} + \omega\sigma^2 \exp(-\kappa h) \tag{5}$$

where σ is the protein surface charge density, κ^{-1} is the Debye length, h is the critical distance between the proteins at which deposition occurs, ω is a proportionality constant related to the ion valence and overall ionic strength, and J_{pI} is the flux at the protein isoelectric point, i.e., at the pH where the protein has no net charge.

The quasi-steady fluxes obtained during microfiltration of a number of different proteins in 0.15 M ionic strength solutions are shown in Fig. 7 as a function of the square of the protein surface charge density [38,48]. The net charge for the different proteins was obtained from literature titration data [38,48]. The quasi-steady fluxes

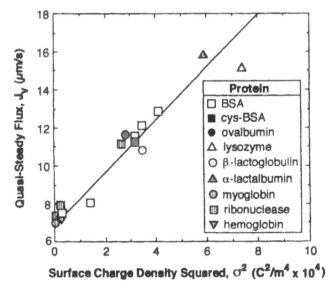

FIG. 7 Steady-state filtrate flux as a function of the square of the protein surface charge density. Solid line is given by Eq. (5). (From Ref. 48.)

were very well correlated by Eq. (5) with $J_{pI} \approx 6 \; \mu m/s$. This includes data obtained with proteins that are positively (lysozyme) and negatively (ovalbumin, α-lactalbumin, β-lactoglobulin, and BSA) charged, as well as results for ribonuclease which include data at pH both above and below the pI. Although Eq. (5) is clearly phenomenological in nature, it provides a very effective means of correlating, and interpreting, quasi-steady flux data in protein microfiltration.

In order to extend these theoretical treatments to protein filtration through partially or fully retentive membranes, it is necessary to account for the effects of protein retention and concentration polarization on both the rate of fouling and the flux. Chudacek and Fane [49] applied a simple cake filtration model to protein fouling during ultrafiltration by simply incorporating a term for the diffusive back protein transport associated with the polarization phenomena. Howell et al. [45] also used a cake filtration model to describe the flux during protein ultrafiltration. However, the rate of cake formation was assumed to be controlled by a chemical polymerization reaction at the membrane surface. Suki et al. [50] also used a cake filtration model for protein ultrafiltration, but the rate of protein deposition was determined by the difference between the instantaneous and steady-state values of the cake mass. There is currently no quantitative data demonstrating the validity of any of these models.

C. Properties of Protein Deposits

Lee and Merson [51] examined the structure of the protein deposits formed on 0.4-μm Nucleopore polycarbonate membranes using scanning electron microscopy. BSA and β-lactoglobulin formed sheetlike deposits on these membranes, while immunoglobulin G formed granules which stacked into layers, creating a porous matrix. The detailed structure of these deposits was also affected by the solution pH and ionic strength [52]. Kelly and Zydney [47] and Tracey and Davis [33] have shown that BSA deposits formed during stirred-cell microfiltration consist of protein aggregates in an amorphous protein matrix that fully covers many of the pores, with similar results reported by Kim et al. [53] for BSA deposits formed during protein ultrafiltration. These protein deposits are generally about 1 μm thick, although Glover and Brooker [54] observed whey protein deposits that were as much as 30 μm thick.

Several studies have demonstrated that the hydraulic resistance provided by such protein deposits increases with increasing applied pressure [41,49]. This behavior is generally described by the simple power-law relationship

$$\alpha_{cake} = \alpha_0 (\Delta P_{TM})^n \tag{6}$$

which has been used extensively in the analysis of flow through compressible porous media. The power-law exponent for BSA deposits is approximately $n = 0.5$ under a variety of conditions [41,49]. The hydraulic resistance determined for these BSA deposits is also in relatively good agreement with that predicted by the Kozeny-Carman equation with a porosity of about 50% at an applied pressure of approximately 10 psi [41].

The hydraulic resistance provided by the protein deposit is also a function of solution pH and ionic strength [8,12,19]. This dependence on the solution environment is due to changes in the protein packing density associated with electrostatic repulsive forces between charged proteins within the deposit [2,3,19], alterations in the conformation of the individual proteins associated with intramolecular electrostatic effects, and/or the development of an electroosmotic counterflow through the charged protein layers [19]. The hydraulic resistance provided by the protein deposit is generally greatest at the protein isoelectric point due to the increase in packing density caused by the reduction in intermolecular electrostatic repulsion as the pH approaches the pI [8]. Similarly, the hydraulic resistance tends to increase with increasing salt concentration at pH ≠ pI due to the increased electrostatic shielding between charged proteins [19].

In addition to being compressible, the protein deposits display a relatively slow (dynamic) response to changes in operating conditions due to the slow rearrangement and/or deformation of the proteins within the closely packed protein layers [19,41,55]. Typical time constants for compression of these protein deposits range from about 660 s for ribonuclease A to as much as 1400 s for BSA [19,55]. The adiabatic compressibility of these proteins are 1.12×10^{-12} cm^2/dyne and 10.5×10^{-12} cm^2/dyne, respectively. The dynamic response of these protein layers to changes in solution pH or salt concentration is more complex, including both a slow rearrangement of the proteins in response to the change in intermolecular electrostatic repulsion and a much more rapid alteration in the magnitude of the electroosmotic counterflow through the protein deposit [19,55].

Mochizuki and Zydney [55] studied the sieving characteristics of deposited layers of bovine serum albumin, using polydisperse dextrans, with the results summarized in Fig. 8. The data were obtained with a BSA deposit formed on the surface of a 0.16-μm PES MF membrane during a constant-pressure stirred-cell filtration. The microfiltration membrane itself had no measurable retention for any of these

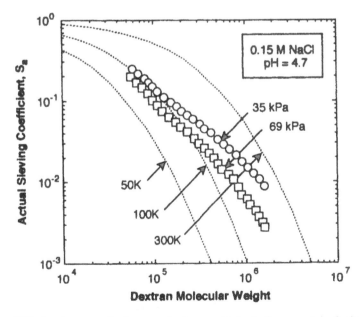

FIG. 8 Asymptotic dextran sieving coefficients for a protein-fouled 0.16-μm polyethersulfone membrane. Solid curves represent data obtained at two applied pressures. Dotted curves represent data obtained for a series of UF membranes with different molecular-weight cutoffs. (From Ref. 55.)

dextrans; thus, the sieving coefficients shown in Fig. 8 are attributable to the presence of the BSA deposit. The asymptotic sieving coefficients, S_∞, for this BSA deposit were similar to those of a 100,000-molecular-weight-cutoff UF membrane. Thus, the protein-fouled MF membrane actually has retention characteristics comparable to that of a UF membrane. The dextran sieving coefficients decrease with increasing pressure due to the tighter packing of the proteins in the deposit at higher pressures. This effect is quite pronounced, with the sieving coefficients at 69 kPa being about a factor of 2–4 smaller than those at 35 kPa. The dextran sieving coefficients were also a function of solution pH and salt concentration [55], analogous to the behavior discussed previously for the hydraulic resistance.

III. DIRECT MEASUREMENTS OF INTERACTIONS BETWEEN PROTEINS AND POLYMER FILMS

A. Direct Measurements of Intermolecular Forces—Background

Atomic force microscopy (AFM), total internal reflection microscopy (TIRM), and surface forces apparatus (SFA) can be used to measure intermolecular forces directly. Of the three, SFA is the only method, to our knowledge, that has been used to study the behavior of intermolecular forces between adsorbed proteins (on mica) and smooth polymeric films [2,3,56] with the goal of understanding membrane fouling. The purpose of conducting these studies is to compare the force measurements, such as the adhesion of protein to polymer, with membrane filtration results in an attempt to relate the molecular scale measurements with macroscopic membrane fouling observations. Before presenting a summary of these results in Section III.D, we review the literature on the measurements of intermolecular forces between proteins (Section III.B) and between several molecules of biological origin such as polysaccharides, carbohydrates, and nucleosides (Section III.C). Finally, we discuss protein–polymer interactions with relevance to membrane fouling and present some of the first correlations that link adhesion forces to filtration performance with protein solutions. For a more comprehensive discussion on surface force measurements in these systems, see Chapter 7.

B. Protein–Protein Interactions

Forces between two adsorbed protein layers have been measured in aqueous solutions to determine the protein–protein interactions both in their native and denatured states [57–73]. Hydrophobic interactions measured between two protein layers were first observed by Afshar-Rad and coworkers [58] for myelin basic protein. An adhesive jump into contact was seen at a distance of approximately 17 nm. All of the other proteins studied followed Derjaguin-Landau-Vervey-Overbeek (DLVO) theory for most of the separation distance [61,63,65,66,69,71]. At small separations, the forces deviated from DLVO due to compressibility of the layers and hydration forces.

Weak long-range attractive interactions between mucin protein layers were observed to start at about 500 nm surface separation, but were not easily explained [59]. Gallinet and Gauthier-Manuel [67] and Fitzpatrick and coworkers [66] studied the structural transition of concanavalin A and bovine serum albumin, respectively, under applied pressure and different solution conditions. At low pH values, a highly unfolded state of the protein can be observed in the system. Similarly, Kékicheff and coworkers [61] have measured multilayer adsorption of cytochrome c on mica around the isoelectric pH of the protein.

Adhesion forces were measured between human serum albumin layers [69]. At low protein surface coverages, adhesion was measured, while at high surface coverages no adhesion was observed, indicating bridging of the protein. Blomberg and coworkers [70] studied the adhesion between two adsorbed lysozyme layers. The adhesion was approximately 1 mN/m at pH 5.6 for when the layers were formed from a solution with an initial concentration of 50 mg/L. Tilton and coworkers [68] found that the adhesion forces between two opposing lysozyme surfaces were eliminated after the addition of a sodium dodecane sulfonate surfactant solution [68].

Lee and Belfort [40] described the orientation of adsorbed ribonuclease A on the surface by activity measurements. They discovered that initially the protein had its major axis parallel to the surface and, after about 24 h, the protein had rotated on the surface (with the square root of time suggesting rotational diffusion) so that the major axis was perpendicular to the surface. Forces between the denatured forms of bovine serum albumin have been studied by Pincet and coworkers [72]. They discovered that the denatured form of the proteins behave like flexible polymers in solution and that they follow DeGennes's scaling theory for polymers.

Direct measurements between receptors and ligands have been obtained to elucidate the interactions between the moieties [74–77]. Specific attractive interactions were seen over relatively short ranges as compared with nonspecific interactions. Streptavidin–biotin interactions were seen to have a much stronger interaction than any of the other receptor–ligand interactions measured. The streptavidin and biotin were both immobilized to the mica through a supported lipid bilayer. Leckband and coworkers deduced the following: (1) long-range protein–polymer interactions with immobilized ligands were controlled by both the protein and the bilayer surface compositions, and (2) short-range specific binding was affected by both the protein structure and the bilayer interfacial properties.

Protein adsorption on surfaces has also been measured using the AFM. The adhesion forces and elasticity of lysozyme on mica were measured [78]. At pH 4, single molecules were able to be imaged, whereas at pH 6.4 only aggregates could be observed. Young's modulus of the lysozyme molecules agreed well with the macroscopic compressibility of wet lysozyme crystals.

Receptor–ligand interactions (streptavidin–biotin) have also been measured with the AFM [79]. Researchers immobilized biotin to the mica surfaces by adsorbing biotinylated bovine serum albumin to mica. Streptavidin was subsequently introduced to the system so that it adsorbed onto one of the coated mica surfaces. No attractive interaction was seen upon bringing a layer of the receptor and a layer of the ligand into contact, but a measurable receptor–ligand interaction was measured, pulling the surfaces apart.

C. Biological Molecule Interactions (Nonproteins)

Several types of biological molecule interactions have been studied with SFA. Claesson and coworkers [73] measured interactions between mica surfaces in the presence of carbohydrates. An increase in the amount of neutral carbohydrate (cyclodextrin and sucrose) yielded a decrease in the charge of the mica surfaces which suppressed the double-layer interaction. A negatively charged carbohydrate (xylan) was able to adsorb in higher quantities with increases in the ionic strength. At the limit of high ionic strength, the multilayers of carbohydrate behaved like a polymer brush in a good solvent. These and similar investigations are discussed in some detail in Chapter 7.

Forces between adsorbed gelatin layers were measured by Kawanishi and coworkers [80]. The interactions are similar to those seen for globular proteins. Below the pI, the gelatin adsorbed in a flat conformation. Above the pI of the gelatin, the adsorbed gelatin had a more extended configuration. The adhesion of the two gelatin layers was the greatest at the pI of the gelatin.

Luckham and coworkers [81,82] examined the interactions of gangliosides with gangliosides and phosphocholines. Ganglioside–ganglioside interactions were repulsive with no adhesion and followed DLVO theory. Ganglioside–phosphocholine interactions were repulsive in nature with adhesion upon separation, indicating some sort of specific interaction.

Pincet and coworkers [56] have measured attractive interactions indicating specific recognition between nucleosides. The interactions were determined to be long-range (about 600 Å) hydrogen bonding. Frank and Belfort [83] measured the intermolecular forces between two layers of adsorbed extracellular polysaccharides (EPS) on silica and between EPS and a smooth RTV film using an atomic force microscope with a modified cantilever. Polymer bridging occurred between negatively charged EPS and was absent with neutral EPS. It was also observed for an anionic model dextran. Light scattering measurements confirmed a contraction of the EPS in free solution from large to small dimensions with an increase in ionic strength. As the ionic strength was increased to marine salt concentrations (about 0.5 M), adsorbed anionic EPS also changed from an extended to a contracted configuration and collapsed onto the solid substrate surface forming a strongly adhered layer. Presumably, this EPS layer is an important component of the conditioning surface onto which bacteria and ultimately higher organisms eventually adhere.

D. Protein–Polymer Interactions

Very little research has been performed on protein–polymer interactions even though they are of tremendous relevance in engineering [57]. Since it is well known that the chemistry of synthetic polymer membranes affects the filtration performance of protein solutions (Figs. 1–3 and 7), it is of interest to obtain quantitative measurements of the interaction between model proteins, such as ribonuclease A, lysozyme, and serum albumin, and typical membrane polymers, such as relatively hydrophilic cellulose acetate and mostly hydrophobic polysulfone.

1. Hydrophilic Membrane Polymers

Pincet and coworkers [84] measured the forces between a hydrophilic polymer (cellulose acetate) and different proteins (ribonuclease A and human serum albumin). The interaction was seen to be repulsive in all cases with no measured adhesion. Comparison of forces between two protein layers with those between a protein layer and a cellulose acetate (CA) film showed that at high pH both proteins retained their native conformation on interacting with the CA film, while at the isoelectric point (pI) or below the tertiary structure of the proteins was disturbed. These measurements provided the first molecular evidence that disruption of protein tertiary structure could be responsible for the reduced permeation flows observed during membrane filtration of protein solutions and suggested that operating at high pH values away from the pI of proteins will reduce such fouling. Phenomenological confirmation of this recommendation can be found in the literature [5,8,38,43].

2. Hydrophobic Membrane Polymers

Koehler and Belfort [2,3] focused on measuring the interactions between (1) a hydrophobic polysulfone film (static captive bubble contact angle, $\theta_s = 91 \pm 1°$) that is known to induce severe fouling and a hydroxyethyl methacrylate (HEMA) low-temperature plasma-modified hydrophilic film ($\theta_s = 53 \pm 3°$), and (2) lysozyme (Lz), in order to explain membrane fouling on a molecular level. Using the SFA, they measured these interactions and correlated these results with filtration data obtained using the same polymers (as membranes) and the same protein (Lz).

The normalized forces between the adsorbed Lz layers and polysulfone and modified polysulfone films were measured below, at, and above the pI of lysozyme and compared with four different permeation fluxes obtained from ultrafiltration experiments. The intermolecular forces between two protein layers were also measured at the different pH values. Adsorption kinetics of Lz onto mica were also obtained. Buffer and Lz solutions at similar pH values and concentrations were filtered with 6-kD polysulfone membranes to obtain flux-decline and, hence, fouling measurements.

Hydrophobic membranes, such as polysulfone, exhibit extremely long-range attractive interactions (on the order of 1500–2000 Å) during compression (when the surfaces approach each other) with proteins such as Lz (Fig. 9b). This long-range attractive interaction is absent when the polymer is surface modified with hydrophilic HEMA, as shown in Fig. 9c. Even in the presence of electrostatic repulsion at pH values above the isoelectric point of lysozyme (when both Lz and polysulfone were negatively charged), a long-range attractive interaction of around 210 μN/m was observed. Such interactions were absent at all pH values with measurements between adsorbed Lz-Lz layers (Fig. 9a).

From these measurements, simple linear correlations were found relating the normalized forces to the volume and solute fluxes from ultrafiltration experiments. For example, from the data in Table 2 and the plot in Fig. 10, it appears that the reduction in flux due to protein fouling (J_p/J_{wi}) could be correlated with the protein–polymer and protein–protein intermolecular forces according to the following expression:

$$\frac{J_p}{J_{wi}} = \frac{A}{\gamma(|E_{ad}^{PSU-Lz}|)} + \frac{B}{\gamma(|E_{ad}^{Lz-Lz}|)} \tag{7}$$

where A and B are weighting factors ($A + B = 100$); A_{ad}^{PSU-Lz} and E_{ad}^{Lz-Lz} are the mean adhesion forces listed in Table 2 for PSU-Lz and Lz-Lz, respectively, and γ is the ratio of the maximum amount adsorbed onto the surface at equilibrium at a given pH value divided by the maximum adsorption at the highest pH value (pH 12). Fitting the force data to the flux ratio data (Fig. 10a), we obtained values for A and B of 47 and 53, respectively, with an excellent fit. The fact that the slope of the

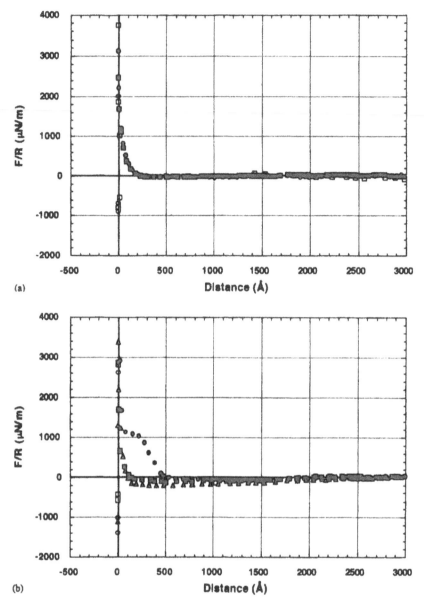

FIG. 9 Normalized force versus separation distance at the pI of lysozyme between (a) two adsorbed lysozyme layers, (b) polysulfone film and adsorbed lysozyme layer, and (c) HEMA-modified polysulfone film and adsorbed lysozyme layer, at $T = 21°C$ and 10^{-2} M KOH/HNO$_3$. Lysozyme concentration was 50 mg/L with 4-h protein adsorption time. Compression: filled symbols. Decompression: open symbols. (a) pH 10.8 [time after end of Lz adsorption, first run: 30 min (●, ○); second run: 1 h 50 min (■, □)]. (b) pH 10.9 [time after end of Lz adsorption, first run: 50 min (●, ○); second run: 2 h (■, □); third run: 2 h 40 min (▲, △)]. (c) pH 11.00 [time after end of lysozyme adsorption, first run: 20 min (●, ○), second run: 2 h 30 min (■, □)]. (From Refs. 2 and 3.)

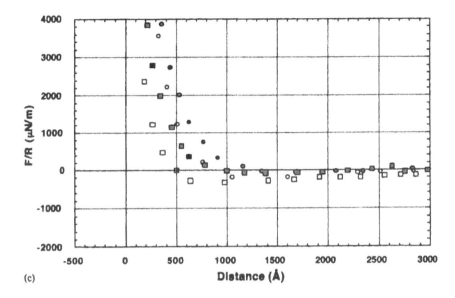

(c)

correlation is close to unity and the y-intercept is almost zero gives credence to the correlation. This suggests that both protein–polymer and protein–protein interactions are important when dealing with membrane fouling and that their contributions are approximately equal.

Similar correlations to that of Eq. (7) for the ratios of the buffer fluxes after and prior to exposure to the protein solution in the presence and absence of transmembrane pressure have also been obtained. For the solute flux, Eq. (7) needs to be adjusted because of units as follows:

$$\frac{J_{solute}\bar{v}_{solute}}{J_{w1}} = \left(\frac{A}{\gamma(|E_{ad}^{PSU\text{-}Lz}|)} + \frac{B}{\gamma(|E_{ad}^{Lz\text{-}Lz}|)}\right) C_0 \bar{v} \tag{8}$$

where $\bar{v}_{solute}$ is the molar volume (mL/mol) (see footnote in Table 2), and C_0 is the initial Lz feed concentration in g/L. This correlation is tested with the data yielding a decent fit with A and B equaling 81 and 19, respectively (Fig. 10b). In this case, however, the slope was approximately 1.6 and the y-intercept was 36. The significance of these values, however, is not clear. The values of the correlation variables, A and B, suggest that the most important interaction for protein transport in the pores is the protein–membrane interaction. It should be stressed, however, that this will most likely depend upon the size of the protein in relation to the pore.

With respect to fouling, both surface chemistry of the membrane and solution conditions could be chosen to minimize fouling for specific protein solutions. Hence, as a result of this study, fouling of polysulfone membranes with lysozyme solutions can be reduced if (1) filtration is conducted at pH values above the pI of lysozyme (approximately 10.8–11.0) and (2) the membranes are modified such that long-range attractive interactions are reduced. These results support those from previous phenomenological studies on membrane filtration of protein solutions (Section II) and

TABLE 2 Summary of Membrane Filtration and SFA Measurements

pH	SFA adhesion (μN/m)		Ratio of permeation fluxes (effect of protein filtration)			No P
			During ultrafiltration			
	PSU-Lz	Lz-Lz	J_p/J_{w1}	J_{w2}/J_{w1}	$\dfrac{J_{solute}\bar{v}_{solute}}{J_{w1}} \times 10^{8a}$	J_{w2}/J_{w1}
12.0	-513 ± 56	-184 ± 7	0.356 ± 0.019	0.38 ± 0.01	480.0 ± 9.6	0.49 ± 0.03
11.6	-371 ± 88	-191 ± 100	0.205 ± 0.009	0.22 ± 0.04	457.7 ± 7.0	0.33 ± 0.01
11.0 (pI)	-1019 ± 56	-841 ± 30	0.058 ± 0.015	0.07 ± 0.02	38.77 ± 15.5	0.13 ± 0.01
6.6	-10300 ± 100	-282 ± 15	0.094 ± 0.007	0.11 ± 0.03	119.4 ± 7.0	0.19 ± 0.01

[a] Creighton [126] reports a value of 0.703 mL/g for the partial specific volume, $\bar{v}$, $\bar{v}_{solute} = \bar{v} \times 14{,}400$, where Lz molecular weight = 14,400 Da.
Source: Ref. 2.

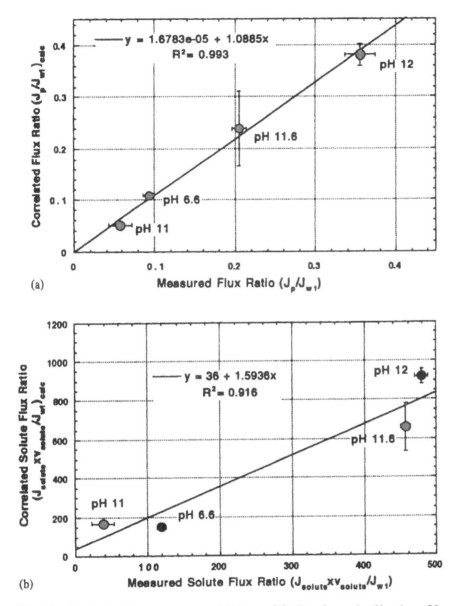

FIG. 10 Correlation between measured indices of fouling from ultrafiltration of lysozyme solutions and calculated flux ratios according to Eqs. (7) and (8) obtained from measured intermolecular forces using SFA and ratios of adsorbed amounts for (a) J_p/J_{w1} and (b) $J_{solute}\bar{v}_{solute}/J_{w1}$. (From Ref. 2.)

are the first evidence relating intermolecular force interactions with macroscopic events in membrane fouling.

IV. CHEMICAL APPROACHES TO MINIMIZING PROTEIN FOULING

As discussed previously, protein fouling during membrane filtration is determined by the physicochemical characteristics of the protein and the membrane surface, the detailed composition of the filtering solution, and the hydrodynamic behavior of the membrane device. The use of different hydrodynamic methods to reduce membrane fouling has been discussed in some detail by Belfort et al. [12] and Winzeler and Belfort [13] and is not covered here; thus, the following discussion is focused on the modification of the filtering solution and/or the membrane in order to minimize the extent of protein fouling.

A. Control of Solution Conditions

The effects of pH and solution ionic strength on protein adsorption and deposition have been discussed previously. In general, adsorption is maximum at the protein isoelectric point due to the minimization of intermolecular electrostatic repulsion between the adsorbed molecules and/or the decrease in the stability of the protein in free solution. This results in maximum adhesion at the pI, especially for hydrophobic polymers. Protein adsorption also increases at high salt concentrations [4,5] due to the increased shielding provided by the bulk electrolyte. The quasi-steady flux also increases with increasing protein charge (Fig. 7) and decreases with increasing salt concentration, in both cases due to the change in the electrostatic repulsion between the proteins in the bulk solution and those already deposited on the membrane surface.

Of even greater interest is the possibility of minimizing fouling by reducing the rate of the sulfhydryl-mediated reactions that are known to cause protein aggregation (and chemical attachment of native protein to the growing deposit). Figure 11 shows data for the effect of adding a metal chelator (either EDTA or citrate) on BSA fouling during constant-pressure filtration through a 0.22-μm PVDF membrane [47]. The flux decline is significantly smaller in the presence of the metal chelators due to the reduction in the concentration of free-metal cations (most probably Fe^{3+} present in trace amounts in the phosphate buffer), which can serve as catalysts for these sulfhydryl-mediated reactions [47]. Similar effects have been seen in whey filtration where the removal of divalent Ca^{2+} yields a significant improvement in the filtrate flux [52]. Similarly, removal of Mg^{2+} from buffer and peptone salts medium for the fermentation of *B. polymyxa* resulted in vastly improved fluxes [85].

Just as significantly, the presence of trace amounts of specific cations can significantly increase the rate of protein fouling. For example, Kelly and Zydney [47] showed that micromolar levels of Cu^{2+} caused a large increase in BSA fouling, with this effect attributed to the known catalytic activity of Cu^{2+} for the sulfhydryl-mediated reactions involved in BSA aggregation. Maa and Hsu [11] found a large increase in fouling during the sterile filtration of recombinant human growth hormone upon addition of small amounts of Zn^{2+}, which is known to induce aggregation of this protein.

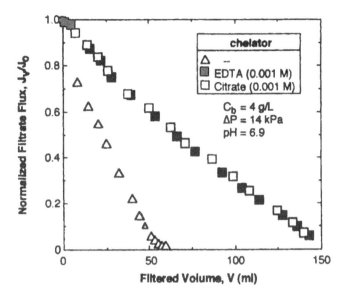

FIG. 11 Effect of metal chelators on the flux decline observed during BSA filtration through 0.22-μm PVDF membranes [47].

It is also possible to dramatically alter the extent of protein fouling by using surfactants and other stabilizing agents. For example, Maa and Hsu [11] found that 0.2% Tween 20 significantly improved the flux during sterile filtration of recombinant human growth hormone. In this case, the Tween 20 apparently reduced the number and size of the protein aggregates (as measured by quasi-elastic light scattering) and, in turn, the extent of aggregate deposition. Howell et al. [45] found that the addition of 30 mM cysteine reduced fouling during the ultrafiltration of ovalbumin solutions, with this effect apparently due to the reversible blocking of the free-sulfhydryl groups in ovalbumin. Further discussion on the effects on protein interfacial behavior of the oligomerization processes is given in Chapter 10.

B. Use of Surface Modification Techniques

The investigation of novel methods for reducing protein adsorption onto polymeric surfaces is an important area of research which has applications in the medical uses of polymers [86,87] and for membrane filtration [50,88,89]. To date, the best approach to reduce fouling of membranes by proteins has been to use hydrophilic polymers bearing hydroxyl, amine, or carboxylic acid groups as membrane materials. Unfortunately, such hydrophilic polymers are not very stable in organic solvents, often exhibit narrow tolerances of pH and temperature, and are difficult to form into ultrafiltration and microfiltration membranes. Exceptions are cellulosic and polyamide polymers. Most commercially available ultrafiltration and microfiltration membranes are made from hydrophobic materials with low surface energies such as poly(propylene) (surface energy of 29.5 mN/m), poly(vinylidene fluoride) (42–47 mN/m), poly(ethylene terephthalate) (24.2 mN/m), and poly(arylsulfone) (approximately 49 mN/m). These materials adsorb proteins strongly and hence show large reductions of the permeation flux with time. It is generally acknowledged that hydrophilic surfaces are less susceptible to fouling [90–92]. Thus, there has been considerable interest in developing general strategies for the introduction of hydrophilic groups onto the surface of commercial ultrafiltration membranes as effective means for reducing the nonselective adsorption of proteins.

Two general approaches have been used for decreasing the unwanted adsorption of proteins on the membrane surface. The first approach involves the a priori choice of low-protein-adsorbing polymers, usually hydrophilic or polar materials such as regenerated cellulose, cellulose acetate, or polyacrylonitrile. A variant of this approach is to modify the polymer prior to casting the membrane [93]. The second approach relies on the post facto modification of the surface chemistry of many apolar commercial membranes. These approaches have significantly reduced the adsorption of proteins onto the membrane surface. Several patents have been issued to membrane manufacturers for modifying well-known polymers or precast porous films in this regard. Examples include the partial hydrolysis of polyacrylonitrile membranes [94], surface coating of hydroxylpropyl cellulose onto polyvinylidene fluoride [95], inclusion of polyvinyl pyrrolidone in the casting solution of polysulfone or polyetherimide membrane materials [96], the attachment of hydroxylpropyl cellulose polymers to the surface of polysulfone [97], and the use of ionizing (high-energy γ-rays or x-rays) radiation for the random grafting of hydroxyalkyl acrylate or methacrylate monomers onto polyamide (nylon 66) [98].

Several surface modification techniques for commercially available ultrafiltration membranes have been developed [99–117]. These methods can be divided into six classifications: (1) coating a hydrophilic thin layer of a protein onto the surface of a membrane followed by curing with heat and/or radiation [100–102], (2) coating a hydrophilic thin layer of a polysaccharide onto the surface of membrane which is held only by physical adsorption [103–106], (3) chemical attachment of selected polymers and/or monomers [107–114], (4) coating with an individually oriented monolayer using Langmuir–Blodgett techniques [115] or self-assembly methods [116], (5) deposition of a hydrophilic layer from a glow discharge plasma onto the membrane [117,118], and (6) initiating an in situ polymerization by radiation and grafting reactions onto a support membrane to impart hydrophilic properties [119,120]. The choice of the specific surface modification technique depends upon the chemical structure of the given support membrane and on the desired characteristics of surface modification. It is of substantial commercial interest to develop an inexpensive and effective site-specific method to modify polymers such as poly(aryl) sulfone membranes so as to render them "low fouling."

Because of their wide application, improving the performance of poly(aryl-sulfone) membranes for filtration application with protein solutions is an important goal for the membrane industry. Three approaches have been taken by Belfort and coworkers to reduce protein fouling through increased wettability of poly-(arylsulfone) membranes. They include the use of photochemical irradiation (UV) methods [121–123], low-temperature plasma treatment and subsequent grafting of hydrophilic monomers [124,125], and premodification of poly(arylsulfone) polymer prior to casting [93]. A summary of the salient results from each approach is given below.

1. Photochemical (UV) Modification of Poly(arylsulfone) Precast Membranes

A novel and general method for modifying hydrophobic poly(arylsulfone) ultrafiltration membranes to produce highly hydrophilic surfaces was developed. This method consists of the direct UV irradiation of poly(arylsulfone) membranes in the presence of water or methanol soluble monomers. It was discovered that the poly(arylsulfone) is intrinsically photoactive and that no photoinitiators are required for this process [121–123]. Membranes were obtained in which polymeric segments derived from the hydrophilic vinyl monomers are directly bound to the poly(arylsulfone) chains by direct chemical bonds. A mechanism involving a photochemically induced free-radical cleavage of the poly(arylsulfone) chains has been proposed. Table 3 shows the receding contact angles for the unmodified (original) and modified PES membranes for different types of monomers obtained by 10 min of UV irradiation in water [122]. The contact angle of the original unmodified membrane was $44 \pm 4°$. It can be seen that the receding contact angle, θ, for each modified membrane is appreciably lower than the original unmodified membrane. This suggests that the modified membranes are more hydrophilic than the original membrane. Table 3 also shows the degree of grafting for each of the monomers grafted onto PES membranes. Up to 9.74% (vinyl imidazole) increase in weight of the membrane can be obtained in a standard grafting period of 10 min. In general, grafting increases as the concentration of the vinyl monomer increases. Some monomers such as GMA and its derivatives and vinyl imidazole are more efficiently grafted onto the membrane than the others shown in the table. It should be noted that those monomers that undergo the most efficient grafting are those whose propagating radical chain ends are the most stabilized. As a consequence, such grafting polymerizations are less prone to termination reactions, which limit the kinetic chain length and lower the molecular weight of the grafts. The effects on filtration of BSA solutions of the irradiation conditions and concentration of two monomers, 2-hydroxyethyl methacrylate (HEMA) and methacrylic acid (MAc), have been examined [122]. Single-solute BSA adsorption isotherms showed that the amount of BSA adsorbed onto the HEMA-modified PES membrane prepared in water is about 43 and 28% less than for the unmodified original membrane for 1 and 2 mg/L of BSA in solution, respectively. This resulted in increased protein solution fluxes (increased by 20–27%) at comparable solute rejections and hence better membrane performance.

2. Low-Temperature Plasma Treatment and Subsequent Grafting on Poly(arylsulfone) Precast Membranes

Low-temperature plasma-induced surface modifications of polyacrylonitrile (PAN) and polysulfone (PSf) ultrafiltration (UF) membranes were studied [124,125]. Treatment with water and with He plasma drastically and almost permanently increased the surface hydrophilicity of polysulfone UF membranes. However, in contrast to the behavior of polyacrylonitrile UF membranes [124], polysulfone surface pore structure was also changed, as indicated by altered water permeabilities and reduced protein retentions [125]. The lower permeability polysulfone membranes (nominal molecular-weight cutoff of 10 kD) showed slower but more extended conversion due to plasma excitation and stronger indications of pore etching effects in comparison with 30-kD cutoff membranes. Polymer peroxides on PAN and polysulfone mem-

TABLE 3 Receding Contact Angle of Modified PES Membranes[a]

Monomer[b]	Monomer conc. (wt%)	Degree of grafting (%)	Receding contact angle, θ^c (deg)
Original PES (control)	—	—	44 ± 4
HEMA	0.5	2.96	35 ± 2
	1.0	2.07	27 ± 4
	2.5	3.34	28 ± 2
AAm	0.5	0.93	19 ± 3
	1.0	0.93	19 ± 6
	2.5	0.93	19 ± 3
MAc	0.5	0.92	21 ± 3
	1.0	2.45	17 ± 2
	2.5	2.66	20 ± 3
GMA	0.5	5.59	27 ± 2
	1.0	4.65	30 ± 2
	2.5	7.27	32 ± 5
GMA-OH	0.5	2.1	29 ± 2
	1.0	3.98	25 ± 3
	2.5	6.26	27 ± 2
GMA-glucosamine	0.5	1.65	27 ± 5
	1.0	1.94	29 ± 6
	2.5	6.15	24 ± 7
GMA-DEAE	0.5	5.59	31 ± 3
	1.0	4.65	28 ± 8
	2.5	7.27	31 ± 4
HPMA	0.5	1.34	30 ± 2
	1.0	1.50	26 ± 2
	2.5	3.30	32 ± 3
Vinyl imidazole	0.5	2.85	28 ± 2
	1.0	2.81	28 ± 4
	2.5	9.74	27 ± 3
Vinyl pyrrolidinone	0.5	3.06	29 ± 2
	1.0	2.9	37 ± 4
	2.5	3.13	31 ± 6

[a] A 10-min UV irradiation time was used for all the experiments, and water was the solvent except for GMA (water/MeOH = 1/1).
[b] HEMA = 2-hydroxyethyl methacrylate, GMA = glycidyl methacrylate, MAc = methacrylic acid, AAm = acrylamide, HPMA = 2-hydroxypropyl methacrylate.
[c] Receding contact angle for an inverted air bubble in water.

branes created by plasma excitation were monitored with the 2,2-diphenyl-1-picryl hydrazyl (DPPH) assay. Graft polymerization of hydrophilic monomers such as 2-hydroxyethyl methacrylate (HEMA) and acrylic or methacrylic acid onto polyacrylonitrile and polysulfone UF membrane surfaces could be initiated via thermal decomposition of peroxides. The degree of modification could be adjusted by polymerization conditions. Graft polymer–modified surfaces were characterized using FTIR-ATR and ESCA spectra. The hydrophilicities were improved as compared with the parent membranes, indicated by contact angles ($\theta_{octane/water}$) of, e.g., 34° for polyacrylonitrile-grafted HEMA (1.1 μmol/cm^2) and 43° for polysulfone-grafted HEMA (1.4 μmol/cm^2). A clear dependency of polyacrylonitrile UF membrane water permeability on the amount of grafted monomer was observed. The monomer type influenced the water permeation flux per mole of grafted acrylate via specific swelling of the graft polymer layer in water. Hydrophilic polyacrylonitrile membranes, modified either by plasma treatment [124] or HEMA graft polymerization, showed significantly reduced fouling due to static protein adsorption, and improved protein UF performance. In particular, for water-plasma-treated polyacrylonitrile membranes with high initial retention, higher fluxes (up to 150%) with the same or even improved retentions were achieved. Hydrophilized polysulfone-grafted HEMA membranes can provide improved performance in protein ultrafiltration over unmodified PSf UF membranes because pore etching effects are compensated for by the grafted layer ($\geq$0.5 μmol/cm^2), yielding both improved filtrate flux (by >30%) and protein retention of bovine serum albumin. An example of these improvements can be seen in Table 4. Notice how the water and protein solution fluxes are increased with HEMA-grafted polysulfone. In some cases the protein retention, R, is also increased!

In summary, plasma-induced graft polymer modification of UF membranes can be used to improve membrane performance by simultaneously controlling the surface hydrophilicity and permeability.

3. Premodification of Poly(arylsulfone) Polymer Prior to Casting

Here we have modified polysulfone prior to casting the membrane [93]. Five different chemically modified versions of polysulfone were prepared via two homogeneous chemical reaction pathways. They, together with the base polymer, were cast as membranes by a phase inversion process. The surface energies of these membranes, as measured by contact angles, were used to characterize the different membranes. Streaming potential measurements were obtained to probe the surface charge of the membranes. The surface roughness of each membrane was also determined by atomic force microscopy. Each membrane was exposed to deionized water, 0.08 g/L bovine serum albumin solution, and then deionized water using a standard filtration procedure to simulate protein fouling and cleaning potential.

Both the *chemistry* and the *size* of the grafted molecules were correlated with respect to volumetric flux during ultrafiltration of protein solutions [93]. Surface roughness seemed to be important for filtering pure water. Hysteresis between advancing and receding contact angles increased with hydrophilicity of the membrane surfaces. One possible explanation could be that surface reorientation was more likely with hydrophilic than with hydrophobic membranes. The membrane modified by direct sulfonation had the lowest surface energy and the shortest grafted chain length and exhibited the highest volumetric flux with BSA solution. It was also the easiest to clean and exhibited the highest initial flux recovery by stirring (91%) and backflush (99%) methods with deionized water. In most cases, backflushing rather than stirring was more effective in recovering the water flux.

TABLE 4 Effect of He Plasma Treatment and Plasma-Induced Graft Polymerization on PSf-UF Membrane Performance

Membrane	J_w (L/m²h); 0.1 MPa[a]	Lz-UF J_f (L/m²h)[b,c] ($RFR_{UF} = 1 - J_f/J_w$)[e]	Lz-UF R (%)	BSA-UF J_f (L/m²h)[b,d] ($RFR_{UF} = 1 - J_f/J_w$)[e]	BSA-UF R (%)
4: original	435	101 (0.77)	96	124 (0.72)	93
5: PSf-pl	380	118 (0.69)	86		
6: PSf-g-HEMA 0.7 μmol/cm²	612	202 (0.67)	92	163 (0.73)	97

PM 30: He plasma, 0.1 torr, 25 W, 30 s; grafting: 10 min in air, then 1 h at 50°C in 5% HEMA solution in water. BSA = bovine serum albumin; Lz = lysozyme.

[a] Measured in stirred UF cells (50 mL).

[b] Measured in cross-flow thin-channel UF unit (200 mL): 0.15 MPa, 1 m/s.

[c] Feed concentration $c_0 = 1$ g/L, pH = 4.5.

[d] Feed concentration $c_0 = 1$ g/L, pH = 5.8.

[e] $RFR_{UF} = 1 - J_{w, \text{adsorbed}}/J_{w, \text{clean membrane}}$, J_f = protein solution flux.

Source: Ref. 125.

V. SUMMARY

The presence of proteins in solution when treated by synthetic polymeric membranes during filtration have a noticeable and often negative effect on performance, including permeation flux and solute retention. In addition to describing the commercial relevance of this problem, we have attempted to describe the various aspects associated with its understanding, including a definition of the generic problem of protein fouling and a description of the underlying phenomena.

Most of the understanding of this problem has come from phenomenological studies of protein–membrane interactions, while only recently have direct measurements between proteins and polymer membrane films been obtained. Thus, protein adsorption and deposition onto membranes and the properties of protein deposits have been reviewed in considerable detail. With respect to direct measurements, some background is provided on protein–protein and other biological molecule interactions. In addition, recent results relating protein adhesion with filtration performance are described. For completeness, we summarize current chemical approaches (omitting hydrodynamic approaches) to minimizing fouling. This involves controlling the solution conditions and the surface chemistry of the membranes. Exciting new progress is being made in all the above-mentioned fields, and understanding the interaction of proteins with polymeric synthetic membranes remains a rich and fertile field of research.

ACKNOWLEDGMENTS

We thank our families for their patience, understanding, and realization that love and passion for research is what drives us. We also thank our collaborators, past and present graduate students, and academic mentors for their assistance, advice, and direction. Finally, we appreciate the financial support from the U.S. and state governments, private foundations, and companies for helping make all this possible.

REFERENCES

1. G. Belfort and C. A. Heath, New developments in membrane bioreactors. In: *Membrane Processes in Separation and Purification* (J. G. Crespo and K. W. Boddeker, eds.), Kluwer Academic Publishers, The Netherlands, 1994, pp. 9–26.
2. J. A. Koehler, M. Ulbricht, and G. Belfort, Intermolecular forces between proteins and polymer films with relevance to filtration. Langmuir, *13*:4162–4171, 1997.
3. J. A. Koehler and G. Belfort, Intermolecular forces between proteins and plasma-modified polymer films with relevance to filtration, 1998.
4. L. J. Zeman and A. L. Zydney, *Microfiltration and ultrafiltration: principles and applications.* Marcel Dekker, New York, 1996.
5. A. D. Marshall, P. A. Munro, and G. Tragardh, The effect of protein fouling in microfiltration and ultrafiltration on permeate flux, protein retention and selectivity: a literature review. Desalination *91*:65–108, 1993.
6. S. T. Kelly, W. S. Opong, and A. L. Zydney, The influence of protein aggregates on the fouling of microfiltration membranes during stirred cell filtration. J. Membr. Sci. *80*:175–187, 1993.

7. K. M. Persson, G. Capannelli, A. Bottino, and G. Tragardh, Porosity and protein ad-sorption of four polymeric microfiltration membranes. J. Membr. Sci. 76:61–71, 1993.
8. S. P. Palecek, S. Mochizuki, and A. L. Zydney, Effect of ionic environment on BSA filtration and the properties of BSA deposits. Desalination 90:147–159, 1993.
9. A. G. Fane, C. J. D. Fell, and A. Suki, The effect of pH and ionic environment on the ultrafiltration of protein solutions with retentive membranes. J. Membr. Sci. 16:195–210, 1983.
10. M. Meireles, P. Aimar, and V. Sanchez, Albumin denaturation during ultrafiltration: effects of operating conditions and consequences on membrane fouling. Biotech. Bioeng. 38:528–534, 1991.
11. Y.-F. Maa and C. C. Hsu, Membrane fouling in sterile filtration of recombinant human growth hormone. Biotech. Bioeng. 50:319–328, 1996.
12. G. Belfort, R. H. Davis, and A. L. Zydney, The behavior of suspensions and macro-molecular solutions in crossflow microfiltration. J. Membr. Sci. 96:1–58, 1994.
13. H. B. Winzeler and G. Belfort, Enhanced performance for pressure-driven membrane processes: the argument for fluid instabilities. J. Membr. Sci. 80:35–47, 1993.
14. E. Matthiasson, The role of macromolecular adsorption in fouling of ultrafiltration membranes. J. Membr. Sci. 16:23, 1983.
15. M. Nystrom, M. Laatikainen, M. Turku, and P. Jarvinen, Resistance to fouling accom-plished by modification of ultrafiltration membranes. Progr. Colloid Polym. Sci. 82:321, 1990.
16. R. McDonogh, H. Bauser, N. Stroh, and H. Chmiel, Concentration polarization and adsorption effects in cross-flow ultrafiltration of proteins. Desalination 79:217, 1990.
17. G. Belfort, J. M. Pimbley, A. Greiner, and K.-Y. Chung, Diagnosis of membrane fouling using rotating annular filter. 1. Cell culture media. J. Membr. Sci. 44:161–181, 1988.
18. B. C. Robertson and A. L. Zydney, Protein adsorption in asymmetric ultrafiltration membranes with highly constricted pores. J. Colloid Interface Sci. 134(2):563, 1990.
19. S. P. Palecek and A. L. Zydney, Hydraulic permeability of protein deposits formed during microfiltration: effect of solution pH and ionic strength. J. Membr. Sci. 95:71–81, 1994.
20. H. Nabetani, M. Nakajima, and A. Watanabe, Effects of osmotic pressure and adsorp-tion on ultrafiltration of ovalbumin. AICHE J. 36(6):907, 1990.
21. R. L. Goldsmith, Macromolecular ultrafiltration with microporous membranes. Ind. Eng. Chem. Fundam. 10:113, 1971.
22. W. F. Leung and R. F. Probstein, Low polarization in laminar ultrafiltration of mac-romolecular solutions. Ind. Eng. Chem. Fundam. 18(3):274, 1979.
23. V. L. Vilker, C. K. Colton, and K. Smith, The osmotic pressure of concentrated protein solutions. Effect of concentration and pH in saline solutions of bovine serum albumin. J. Colloid Interface Sci. 79(2):548, 1981.
24. G. Jonsson, Boundary layer phenomena during the ultrafiltration of dextran and whey protein solutions. Desalination 51:61, 1984.
25. J. G. Wijmans, S. Nakao, and C. A. Smolders, Flux limitation in ultrafiltration: osmotic pressure and gel layer model. J. Membr. Sci. 20:115, 1984.
26. W. F. Blatt, A. Dravid, A. S. Michaels, and L. Nelson, Solute polarization and cake formation in membrane ultrafiltration. Causes, consequences and control techniques. Membrane Science and Technology (J. E. Flinn, ed.), Plenum Press, New York, 1970, pp. 47–97.
27. A. G. Fane, C. J. D. Fell, and A. G. Waters, Ultrafiltration of protein solutions through partially permeable membranes. The effect of adsorption and solution environment. J. Membr. Sci. 16:211, 1983.
28. S.-I. Nakao, T. Nomura, and S. Kimura, Characteristics of macromolecular gel layer formed on ultrafiltration tubular membrane. AICHE J. 25:615, 1979.

29. S.-I. Nakao and S. Kimura, Analysis of solute rejection in ultrafiltration. J. Chem. Eng. Jpn. *14*(1):32, 1981.

30. N. P. Tirmizi, Ph.D. Thesis, Study of ultrafiltration of macromolecular solutions. Rensselaer Polytechnic Institute, Troy, New York, 1990.

31. J. H. Hannemaaijer, T. Robbertsen, T. van den Boomgaard, and J. W. Gunnink, Characterization of clean and fouled ultrafiltration membranes. Desalination *68*:93, 1988.

32. S. T. Kelly and A. L. Zydney, Mechanisms for BSA fouling during microfiltration. J. Membr. Sci. *107*:115, 1995.

33. E. M. Tracey and R. H. Davis, BSA fouling of track-etched polycarbonate microfiltration membranes. J. Colloid Interface Sci. *167*:104, 1994.

34. K.-J. Kim and A. G. Fane, Performance evaluation of surface hydrophilized novel ultrafiltration membranes using aqueous proteins. J. Membr. Sci. *99*:149, 1995.

35. W. R. Bowen and D. T. Hughes, Properties of microfiltration membranes. Adsorption of bovine serum albumin at aluminum oxide membranes. J. Membr. Sci. *51*:189–200, 1990.

36. S. H. Lee and E. Ruckenstein, Adsorption of proteins onto polymeric surfaces of different hydrophilicities—a case study with bovine serum albumin. J. Colloid Interface Sci. *125*:365, 1988.

37. A. Pitt, The nonspecific protein binding of polymeric microporous membranes. J. Parenteral Sci. Technol. *41*:110, 1987.

38. S. P. Palecek and A. L. Zydney, Intermolecular electrostatic interactions and their effect on flux and protein deposition during protein filtration. Biotech. Prog. *10*:207–213, 1994.

39. N. S. Pujar and A. L. Zydney, Electrostatic and electrokinetic interactions during protein transport through narrow pore membranes. Ind. Eng. Chem. Res. *33*:2473–2482, 1994.

40. C. S. Lee and G. Belfort, Changing activity of ribonuclease A during adsorption: a molecular explanation. Proc. Natl. Acad. Sci. USA *86*:8392, 1989.

41. W. S. Opong and A. L. Zydney, Hydraulic permeability of protein layers deposited during ultrafiltration. J. Colloid Interface Sci. *142*:41–60, 1991.

42. W. Norde, F. MacRitchie, F. G. Nowicka, and J. Lyklema, Protein adsorption at solid-liquid interfaces: reversibility and conformation aspects. J. Colloid Interface Sci. *112*: 447, 1986.

43. J. H. Hanemaajer, T. Robbertsen, T. van den Boomgard, and J. W. Gunnink, Fouling of ultrafiltration membranes: the role of protein adsorption and salt precipitation. J. Membr. Sci. *40*:199–217, 1989.

44. S. Mochizuki and A. L. Zydney, Effect of protein adsorption on the transport characteristics of asymmetric ultrafiltration membranes. Biotech. Prog. *8*:553–561, 1992.

45. J. A. Howell, O. Velicangil, M. Lowell, and A. L. Herrera Zeppelin, Ultrafiltration of protein solutions. Ann. NY Acad. Sci. *369*:355, 1981.

46. R. Boyd, L. J. Langsdorf, and A. L. Zydney, A two-layer model for the effects of blood contact on membrane transport in artificial organs. Trans. Am. Soc. Artif. Intern. Organs *40*:M864–M869, 1994.

47. S. T. Kelly and A. L. Zydney, Effect of thiol-disulfide interchange reactions on albumin fouling during membrane microfiltration. Biotech. Bioeng. *44*:972–982, 1994.

48. S. T. Kelly and A. L. Zydney, Protein fouling during microfiltration: comparative behavior of different model proteins. Biotech. Bioeng. *55*:91–100, 1997.

49. M. W. Chudacek and A. G. Fane, The dynamics of polarization in unstirred and stirred ultrafiltration. J. Membr. Sci. *21*:145, 1984.

50. A. Suki, A. G. Fane, and C. J. D. Fell, Flux decline in protein ultrafiltration. J. Membr. Sci. *21*:269, 1984.

51. D. N. Lee and R. L. Merson, Examination of cottage cheese whey by scanning electron microscopy: relationship to membrane fouling during ultrafiltration. J. Dairy Sci. *58*: 1423, 1974.

52. D. N. Lee and R. L. Merson, Chemical treatments of cottage cheese whey to reduce fouling of ultrafiltration membranes. J. Food Sci. *41*:778–786, 1976.

53. K. J. Kim, A. G. Fane, C. J. D. Fell, and D. C. Joy, Fouling mechanisms of membranes during protein ultrafiltration. J. Membr. Sci. *68*:79–91, 1992.

54. F. A. Glover and B. E. Brooker, The structure of the deposits formed during the concentration of milk by reverse osmosis. J. Dairy Res. *41*:89, 1974.

55. S. Mochizuki and A. L. Zydney, Sieving characteristics of albumin deposits formed during microfiltration. J. Colloid Interface Sci. *159*:136–145, 1993.

56. F. Pincet, E. Perez, L. Lebeau, and C. Mioskowski, Long range H-bond specific interactions between nucleosides. J. Chem. Soc. Faraday Trans. *91*:4329–4330, 1995.

57. J. A. Koehler, Ph.D. Thesis, Intermolecular forces between proteins and polymers films with relevance to filtration. Rensselaer Polytechnic Institute, November, 1996.

58. T. Afshar-Rad, A. I. Bailey, P. F. Luckham, W. MacNaughtan, and D. Chapman, Forces between proteins and model polypeptides adsorbed on mica surfaces. Biochem. Biophys. Acta *915*:101–111, 1987.

59. E. Perez and J. E. Proust, Forces between mica surfaces covered with adsorbed mucin across aqueous solution. J. Colloid Interface Sci. *118*:182–191, 1987.

60. P. M. Claesson, T. Arnebrant, B. Bergenståhl, and T. Nylander, Direct measurement of the interaction between layers of insulin adsorbed on hydrophobic surfaces. J. Colloid Interface Sci. *130*:457, 1989.

61. P. Kékicheff, W. A. Ducker, B. W. Ninham, and M. P. Pileni, Multilayer adsorption of cytochrome c on mica around isoelectric point. Langmuir *6*:1704–1708, 1990.

62. P. Kékicheff and B. W. Ninham, The double layer interaction in asymmetric electrolytes. Europhys. Lett. *12*:471–477, 1990.

63. P. F. Luckham and M. A. Ansarifar, Biomedical aspects of the direct measurement of the forces between adsorbed polymers and proteins. Br. Polym. J. 22:233–243, 1990.

64. G. Belfort and C. S. Lee, Attractive and repulsive interactions between and within adsorbed ribonuclease A layers. Proc. Natl. Acad. Sci. USA *88*:9146, 1991.

65. E. Blomberg, P. M. Claesson, and C. G. Gölander, Adsorbed layers of human serum albumin investigated by the surface force technique. J. Disp. Sci. Tech. *12*:179–200, 1991.

66. H. Fitzpatrick, P. F. Luckham, S. Eriksen, and K. Hammond, Bovine serum albumin adsorption to mica surfaces. Colloid Surf. *65*:43–49, 1992.

67. J.-P. Gallinet and B. Gauthier-Manual, Structural transitions of concanavalin A adsorbed onto bare mica plates: surface force measurements. Eur. Biophys. J. 22:195, 1993.

68. R. D. Tilton, E. Blomberg, and P. M. Claesson, Effect of anionic surfactant on interactions between lysozyme layers adsorbed on mica. Langmuir *9*:2102–2108, 1993.

69. E. Blomberg, P. M. Claesson, and R. D. Tilton, Short-range interaction between adsorbed layers of human serum albumin. J. Colloid Interface Sci. *166*:427–436, 1994.

70. E. Blomberg, P. M. Claesson, J. C. Fröberg, and R. D. Tilton, Interaction between adsorbed layers of lysozyme studied with the surface force technique. Langmuir *10*: 2325–2334, 1994.

71. T. Nylander, P. Kékicheff, and B. W. Ninham, The effect of solution behavior of insulin on interactions between adsorbed layers of insulin. J. Colloid Interface Sci. *164*:136, 1994.

72. F. Pincet, E. Perez, and G. Belfort, Do denatured proteins behave like polymers? Macromolecules *27*:3424–3435, 1994.

73. P. M. Claesson, H. K. Christenson, J. M. Berg, and R. D. Neuman, Interactions between mica surfaces in the presence of carbohydrates. J. Colloid Interface Sci. *172*:415–424, 1995.

74. D. E. Leckband, J. N. Israelachvili, F.-J. Schmitt, and W. Knoll, Long-range attraction and molecular rearrangements in receptor–ligand interactions. Science 255:1419–1421, 1992.

75. D. Leckband and J. Israelachvili, Molecular basis of protein function as determined by direct force measurements. Enzyme Microb. Technol. 15:450–459, 1993.

76. D. E. Leckband, F.-J. Schmitt, J. N. Israelachvili, and W. Knoll, Direct force measurements of specific and nonspecific protein interactions. Biochem. 33:4611–4624, 1994.

77. D. E. Leckband, T. Kuhl, H. K. Wang, J. Herron, W. Müller, and H. Ringsdorf, 4-4-20 Anti-Fluorescyl IgG Fab' Recognition of membrane bound hapten: direct evidence for the role of protein and interfacial structure. Biochem. 34:11467–11478, 1995.

78. M. Radmacher, M. Fritz, J. P. Cleveland, D. A. Walters, and P. K. Hansma, Imaging adhesion forces and elasticity of lysozyme adsorbed on mica with the atomic force microscope. Langmuir 10:3809–3814, 1994.

79. G. U. Lee, D. A. Kidwell, and R. J. Colton, Sensing discrete streptavidin-biotin interactions with atomic force microscopy. Langmuir 10:354–357, 1994.

80. N. Kawanishi, H. K. Christenson, and B. W. Ninham, Measurement of the interaction between adsorbed polyelectrolytes: gelatin on mica surfaces. J. Phys. Chem. 94:4611–4617, 1990.

81. P. Luckham, J. Wood, and R. Swart, Direct measurement of the forces between gangliosides deposited on mica surfaces. Thin Solid Films 210/211:696–698, 1992.

82. P. Luckham, J. Wood, and R. Swart, The surface properties of gangliosides. II. Direct measurement of the interaction between bilayers deposited on mica surfaces. J. Colloid Interface Sci. 156:173–183, 1993.

83. B. P. Frank and G. Belfort, Intermolecular forces between extracellular polysaccharides measured using the atomic force microscope. Langmuir 13:6234–6240, 1997.

84. F. Pincet, E. Perez, and G. Belfort, Molecular interactions between proteins and synthetic membrane polymer films. Langmuir 11:1229–1235, 1995.

85. N. Nagata, K. J. Herouvis, D. M. Dziewulski, and G. Belfort, Cross-flow membrane microfiltration of a bacterial fermentation broth. Biotech. Bioeng. 34:447–466, 1989.

86. A. T. Horbett and J. L. Brash, Proteins at interfaces: current issues and future prospects. ACS Symp. Ser. 343:1, 1987.

87. A. S. Hoffman, Blood–biomaterial interactions: an overview, biomaterials: interfacial phenomena and applications. Adv. Chem. Series 199:3, 1982.

88. E. Matthiasson, The role of macromolecular adsorption in fouling of ultrafiltration membranes. J. Membr. Sci. 16:23, 1983.

89. A. G. Fane and C. J. D. Fell, A review of fouling control in filtration. Desalination 62:117, 1987.

90. A. S. Michaels and S. L. Matson, Membranes in biotechnology: state of the art. Desalination 53:231, 1985.

91. W. Norde, Adsorption of proteins from solution at the solid–liquid interface. Adv. Colloid Interface Sci. 25:267, 1986.

92. M. Kim, K. Saito, S. Furusaki, T. Sugo, and J. Okamoto, Water flux and protein adsorption of a hollow fiber modified with hydroxyl groups. J. Membr. Sci. 56:289, 1991.

93. A. Nabe, E. Staude, and G. Belfort, Surface modification of polysulfone ultrafiltration membranes and fouling by BSA solutions. J. Membr. Sci. 133:57–72, 1997.

94. L. T. Hodgins and E. Samuelson, Hydrophilic article and method of producing same. U.S. Patent 4,906,379, 1990.

95. M. J. Steuck and N. Reading, Porous membrane having hydrophilic surface and process. U.S. Patent 4,618,533, 1986.

96. H. D. W. Roesink and C. A. Smolders, Process for the preparation of hydrophilic membranes and such membranes. U.S. Patent 4,798,847, 1989.

97. J. M. Henis, M. K. Tripodi, and D. I. Stimpson, Modified polymeric surfaces and process for preparing same. U.S. Patent 4,794,002, 1988.

98. P. J. Degen, Fluid treatment system having low affinity for proteinaceous materials. U.S. Patent 4,959,150, 1990.

99. M. Ulbricht, H. Matuschewski, A. Oechel, and H.-G. Hicke, Photo-induced polymerization surface modifications for the preparation of hydrophilic and low-protein-adsorbing ultrafiltration membranes. J. Membr. Sci. *115*:31–47, 1996.

100. F. F. Stengaard, Characteristics and performance of new types of ultrafiltration membrane with chemically modified surfaces. Desalination *70*:207, 1988.

101. K. B. Hvid, P. S. Nielsen, and F. F. Stengaard, Preparation and characterization of a new ultrafiltration membrane. J. Membr. Sci. *53*:189, 1990.

102. M. Nyström and P. Jarvinen, Modification of polysulfone ultrafiltration membranes with UV irradiation and hydrophilicity increasing agents. J. Membr. Sci. *60*:275, 1991.

103. K. J. Kim, A. G. Fane, and C. J. D. Fell, The performance of ultrafiltration membranes pretreated by polymers. Desalination *70*:229, 1988.

104. A. Jonsson and B. Jonsson, The influence of nonionic surfactants on hydrophobic and hydrophilic ultrafiltration membranes. J. Membr. Sci. *56*:49, 1991.

105. M. Nyström, Fouling of unmodified and modified polysulfone ultrafiltration membranes by ovalbumin. J. Membr. Sci. *44*:183, 1989.

106. M. Nyström, M. Laatikainen, K. Turku, and P. Jarvinen, Resistance to fouling accompanied by modification of ultrafiltration membranes. Progr. Colloid Polym. Sci. *82*:321, 1990.

107. M. Warhlgen, B. Sivik, and M. Nyström, Dextran modifications of polysulfone UF membranes: streaming potential and BSA fouling characteristics. Acta Polytech. Scan. Chem. Ser. *194*:1, 1990.

108. A. Higuchi, N. Iwata, M. Tsubaki, and T. Nakagawa, Surface-modified polysulfone hollow fibers. J. Appl. Polym. Sci. *36*:1753, 1988.

109. A. Higuchi, N. Iwata, and T. Nakagawa, Surface modified polysulfone hollow fibers. II: Fibers having $CH_2CH_2CH_2SO_3^-$ segments and immersed in HCl solution. J. Appl. Polym. Sci. *40*:709, 1990.

110. A. Higuchi, S. Mishima, and T. Nakagawa, Separation of proteins by modified polysulfone membranes. J. Membr. Sci. *57*:183, 1991.

111. L. Breitbach, E. Hinke, and E. Staude, Heterogeneous functionalizing of polysulfone membranes. Die Angew Makromol. Chem. *184*:183, 1991.

112. M. D. Guiver, J. W. Apsimon, and O. Kutowy, The modification of polysulfone by metalation. J. Polym. Sci. Part C, Polym. Lett. *26*:123, 1988.

113. M. D. Guiver, S. Croeau, J. D. Hazlett, and O. Kutowy, Synthesis and characterization of carboxylated polysulfones. Br. Polym. J. *23*:183, 1990.

114. K. J. Kim, A. G. Fane, and C. J. D. Fell, The effect of Langmuir–Blodgett layer pretreatment on the performance of ultrafiltration membranes. J. Membr. Sci. *43*:187, 1989.

115. F. Vigo, M. Niccia, and C. Uliana, Poly(vinyl chloride) ultrafiltration membranes. J. Membr. Sci. *36*:213, 1988.

116. G. Belfort and P. Boehme, Modification of porous and non porous materials using self-assembled monolayers, U.S. application, July 9, 1996.

117. M. Ulbricht and G. Belfort, Surface modification of ultrafiltration membranes by low temperature plasma. 1. Treatment of polyacrylonitrile. J. Appl. Polym. Sci. *56*:325–343, 1995.

118. B. Keszler, G. Kovacs, A. Toth, I. Bertoti, and M. Hegyi, Modified polyethersulfone membranes. J. Membr. Sci. *62*:201, 1991.

119. M. Kim, K. Saito, S. Furusaki, T. Sugo, and J. Okamoto, Water flux and protein adsorption of a hollow fiber modified with hydroxyl groups. J. Membr. Sci. *56*:289, 1991.

120. B. Keszler, G. Kovacs, A. Toth, I. Bertoti, and M. Hegyi, Modified polyethersulfone membranes. J. Membr. Sci. *62*:201, 1991.

121. H. Yamagishi, J. V. Crivello, and G. Belfort, Development of a novel photochemical technique for modifying poly(arylsulfone) ultrafiltration membranes. J. Membr. Sci. *105*:237–247, 1995.

122. H. Yamagishi, J. V. Crivello, and G. Belfort, Evaluation of photochemically modified poly(arylsulfone) ultrafiltration membranes. J. Membr. Sci. *105*:249–259, 1995.

123. J. V. Crivello and G. Belfort, Low fouling ultrafiltration and microfiltration aryl polysulfone. U.S. Patent 5,468,390, 1995.

124. M. Ulbricht and G. Belfort, Surface modification of ultrafiltration membranes by low temperature plasma. I. Treatment of polyacrylonitrile. J. Appl. Polym. Sci. *56*:325–343, 1995.

125. M. Ulbricht and G. Belfort, Surface modification of ultrafiltration membranes by low temperature plasma. II. Graft polymerization onto polyacrylonitrile and polysulfone. Membr. Sci. *111*:193–215, 1995.

126. T. E. Creighton, *Proteins: Structures and Molecular Properties*, W. H. Freeman, New York, 1993, p. 266.

26

Protein Adsorption in Intravenous Drug Delivery

MARTIN MALMSTEN Institute for Surface Chemistry and Royal Institute of Technology, Stockholm, Sweden

I. INTRODUCTION

Intravenous administration of drugs is the route of choice in several cases. For example, when the patient is unconscious or cannot retain anything administered orally or when an acute treatment is necessary, intravenous administration may be the only possibility. Furthermore, for a range of substances which are poorly absorbed when given orally, e.g., as a consequence of their degradation in the gastrointestinal tract or poor uptake due to their physicochemical properties, parenteral administration is a direct way of administration, despite the somewhat limited patient compliance.

Frequently, there is a need for a colloidal carrier of the drug to be administered intravenously. Reasons for this can include poor drug solubility, drug toxicity, hydrolysis of the drug in aqueous solution, the need for a controlled or sustained release, or a desire to be able to control the drug circulation time and tissue distribution. However, when a colloidal drug carrier is administered intravenously, it is frequently observed that the carrier is cleared from circulation rapidly by the reticuloendothelial system (RES) and accumulated in certain tissues, such as liver, spleen, and marrow [1-3]. The main reason for this is that the particle uptake is performed to a large extent by macrophages, which are located primarily in these tissues. Although the preferential uptake in RES-related tissues may be favorable in certain instances, it is generally a problem for intravenous administration of colloidal drug carriers. First, since the circulation time is often short and the drug solubility low, the uptake of the drug can be quite limited, resulting in a low bioavailability in tissues other than those mentioned above. Furthermore, since large parts of the dosage are accumulated in these tissues, the local concentration of the drug may be quite high, resulting in a dose-limiting local toxicity. Not surprisingly, therefore, it has been found in numerous cases that a long circulation time and an even tissue distribution is advantageous from a therapeutic point of view. This is the case for poly(ethylene glycol)– (PEG) modified colloidal drug carriers (see below).

It has been found that the uptake in RES of colloidal drug carriers depends on various factors, notably the size of the carrier and its surface properties (e.g., wettability, charge, and chemical functionality [1-5]). In particular, the uptake increases with increasing particle size, hydrophobicity, and charge, but is also dependent on, e.g., the specific chemical functionality of the carrier and, for phospholipid-containing drug carriers, the fluidity of the carrier surface. These issues will be discussed more extensively.

A growing body of experimental results indicates that the RES uptake is initiated by adsorption of certain serum proteins, called opsonins, at the carrier surface [1-4,6-11]. Among these, complement proteins such as C3 and C1q can adsorb at the carrier surface and activate the complement cascade. Also, immunoglobulins (e.g., IgG) may facilitate the uptake of colloidal drug carriers, e.g., through interaction with the complement system (C1q) or by a direct interaction between the adsorbed IgG molecules and the Fc receptors present at macrophage surfaces [1-4,12-18]. Yet other proteins which may facilitate uptake of a colloidal drug carrier through adsorption at its interface include "adhesion" proteins such as fibrinogen and fibronectin. In addition, other proteins, e.g., the Hageman factor and CRP, are believed to act as opsonins.

Chonn et al. found in a nice series of investigations that there is an inverse correlation between the total adsorption of serum proteins on a drug carrier, on one hand, and the circulation time, on the other [4,8-11]: thus, the smaller the adsorption, the longer the circulation time (Fig. 1). Furthermore, carriers adsorbing little serum protein, and therefore circulating in the bloodstream for a long time (e.g., PEG-modified carriers), tend to have a more even tissue distribution than carriers cleared rapidly from circulation (see below). However, until quite recently, little has been known about which serum proteins are actually involved in the opsonization. This chapter therefore focuses on some recent developments regarding these issues, based particularly on model systems. Throughout, the adsorption pattern of various surfaces will be compared to the biological behavior of the corresponding drug carrier.

II. DIFFERENT TYPES OF COLLOIDAL CARRIERS

A. Micelles

One of the most interesting types of colloidal drug carriers is that of block copolymer micelles [19-31]. While most other colloidal drug carriers, e.g., liposomes and emulsions, are thermodynamically unstable, block copolymer micelles are typically thermodynamically stable, yielding excellent storage properties and easy preparation. Furthermore, block copolymers are typically extremely long-lived below critical micelle concentration (cmc), which means that they take a very long time to disintegrate even on severe dilution, e.g., in connection with intravenous administration. A particularly well characterized system is PEG-poly(aspartate) conjugates containing, e.g., adriamycin (ADR) [20-27]. In this system small micelles (15-60 nm) are formed in which ADR is solubilized and anchored in the micelle interior. Due to the extremely slow release of unimers from the micelles (~days), the release of ADR from the micelles proceeds largely through the biodegradation of the poly(aspartate) segments. The micelles display a very long circulation time in the bloodstream and a slow uptake in liver, spleen, and marrow. Although ADR possesses a considerable anticancer activity, the survival time is reduced at concentrations higher than about 10 mg/kg due to the occurrence of side effects. For the micellar formulation, on the other hand, the side effects occur at 1-2 orders of magnitude higher ADR concentration. Therefore, the use of these micelles increases the maximum ADR concen-

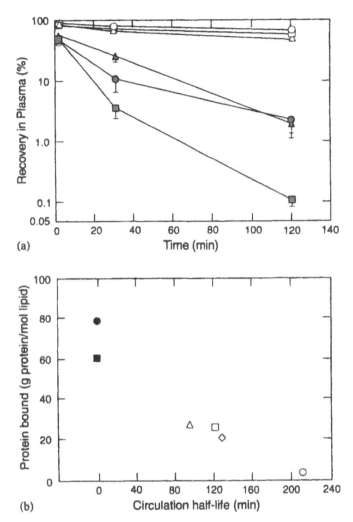

FIG. 1 (a) Plasma clearance of LUVs containing trace mounts of [³H]cholesterylhexadecyl ether administered intravenously in CD1 mice at a dose of about 20 μmol total lipid per 100 g of mouse weight. Results are shown for liposomes containing SM:PC:GM1 (72:18:10) (open circles), PC:CH (55:45) (open squares), PC:CH:plant PI (35:45:20) (open triangles), PC:CH: dioleoylphosphatidylserine (DOPS) (35:45:20) (filled triangles), PC:CH:dioleoylphosphatidic acid (DOPA) (35:45:20) (filled circles), and PC:CH:DPG (35:45:20) (filled squares). (b) Correlation between total amount of protein adsorbed and circulation time. (Data from Ref. 9.)

tration possible to use in therapy without causing toxic side effects. Due to this and to the longer circulation time in the bloodstream, the cytotoxicity of the block copolymer conjugate micelles is superior to that of ADR in itself.

B. Liposomes

Ever since the discovery of liposomes about 30 years ago, they have been thought to be of promise for drug delivery, and not the least for parenteral administration.

Reasons for this have included the capacity of liposomes to solubilize both water-soluble and oil-soluble substances and the possibility of a sustained and controlled release, as well as some more superficial ones, such as the structural similarity between the liposome membrane and those of cells. Over the decades, however, liposomes have been fraught with difficulties related to elaborate preparation, poor storage stability, limited solubilization capacity, limited release control, and rapid clearance on intravenous administration, with concomitant problems related to, e.g., poor drug bioavailability and local toxicity in RES-related tissues. Nevertheless, liposomes have experienced a renaissance during the last few years or so, and particularly through the development of PEG-modified liposomes (so-called stealth liposomes), several of these problems have been circumvented or at least reduced [32–35]. Therefore, several liposomal products have recently appeared on the market, and many more are currently tested and documented. For a compilation of liposome-based drug formulations, see Ref. 34.

C. Emulsion Droplets

Emulsions are one of the industrially most extensively used types of colloidal drug formulation systems, and also within parenteral applications oil-in-water (o/w) emulsions have been employed [36–43]. One of the applications most extensively relying on parenteral administration in commercial systems so far has been that of nutrition emulsions [38]. In these, soybean, cottonseed, or safflower oil is typically emulsified with a phospholipid (mixture) in an aqueous solution containing also, e.g., carbohydrates. A number of these systems exist on the market, with Intralipid, Lipofundin, Travemulsion, and Liposyn as a few examples. Another commercially used type of emulsion for parenteral administration is artificial blood. These are perfluorochemical (PFC) emulsions [39–42] typically consisting of droplets about 100–200 nm in diameter. In many respects these emulsions resemble lipid emulsions used for parenteral nutrition, although other types of emulsifiers are occasionally used. However, a range of drugs has also been formulated in o/w emulsions for parenteral use [43].

D. Others

Although liposomes, emulsion droplets, and block copolymer micelles may be the most extensively studied colloidal drug carriers, they are certainly not the only ones of interest. Numerous other types of colloidal drug carriers are of interest and have been investigated. For example, solid nanoparticles are of interest to a range of drug delivery conditions [44–50]. One of these types of particles is biodegradable polymers, which may be loaded with bioactive substances. On chemical or enzymatic degradation the drug is released, and since the degradation rate may be controlled through the choice, e.g., of the (co)polymer composition, the release rate may also be controlled [49]. Although these carriers have been studied mainly in relation to oral administration, e.g., of vaccines, they are certainly interesting also for parenteral administration.

Furthermore, so-called cubosomes have potential for parenteral administration of both water-soluble and oil-soluble substances [51]. Cubosomes are dispersed particles of cubic phase, which are stabilized, e.g., by a PEG-containing copolymer. It has been found that these particles circulate in the bloodstream for an extensive period of time, at the same time as both water-soluble and oil-soluble substances may be solubilized in the particles, and released in a controlled manner. Yet other types of colloidal drug carriers of interest for parenteral drug delivery are solid lipid nanoparticles, prepared, e.g., from precipitation from emulsion systems [52,53], and solid drug nanoparticles [54].

A completely different type of colloidal drug (carrier) relevant for intravenous drug delivery is that constituted by protein and peptide drugs. Especially for genetically engineered proteins, these are recognized as exogeneous, which leads to side effects related, e.g., to degradation and to immunicity response and a consecutive rapid clearance of the protein from bloodstream circulation [55–57]. Also for this type of colloidal entity, a surface modification (through covalent attachment of PEG chains) may strongly affect the bloodstream circulation time.

III. SERUM PROTEIN ADSORPTION AT LIPID SURFACES

A. Methodology

Despite the importance of protein adsorption at phospholipid surfaces for parenteral administration of colloidal drug carriers such as liposomes, emulsions, lipoprotein mimics, solid lipid nanoparticles, etc., little is known about this process. To a large extent this is probably due to experimental difficulties related to studying protein adsorption at phospholipid surfaces. Traditionally, the adsorption of proteins to (phospho)lipid membranes has been studied by Langmuir balance-based methods [58–61]. However, in the majority of these studies, the adsorbed amount was not determined, and the adsorption instead was inferred from the change in surface pressure on introduction of the protein in the subphase. Although qualitative results may be inferred from these types of investigations, it is difficult to quantify, e.g., the influence of the (phospho)lipid properties on the protein adsorption. Nevertheless, in a range of studies, the adsorbed amount was also monitored, e.g., through radiolabeling of the protein [60,61].

Ideally, one would like to investigate the adsorbed amount of various proteins directly at the colloidal drug carrier of interest. In principle, there are several ways to do this, but most of them are so-called solution-depletion techniques, based in one way or another on measuring the solution concentration of the protein of interest before and after equilibration with the colloidal carriers together with the colloidal particle size (see Ref. 62). However, there are several potential difficulties associated with this type of measurement. While the adsorbed amount may readily be obtained at plateau conditions in the adsorption isotherm for the protein(s) investigated (Table 1), half coverage may be difficult to study due to colloidal instability caused by bridging flocculation. Furthermore, particularly for low-adsorbing phospholipids, care must be taken not to desorb the adsorbed protein during the separation (typically centrifugation) following the adsorption. Moreover, colloidal drug carriers such as emulsion droplets, liposomes, etc., are typically polydisperse, which means that there is a considerable uncertainty in the available surface area. Therefore, the adsorbed amount in terms of mass per unit surface area is subject to this uncertainty. Naturally, this uncertainty remains also if the adsorption is determined by detecting the adsorbed amount directly rather than the depleted amount (see, Refs. 11 and 63). Despite this uncertainty, however, the adsorbed amounts obtained with solution-deple-

TABLE 1 Adsorbed Amount (mg/m²) of Fgn, IgG, and HSA at PC and
PA Obtained with Ellipsometry for Flat Plates and Solution Depletion for
the Corresponding Emulsion System

Phospholipid/method	Γ_{HSA}	Γ_{IgG}	Γ_{Fgn}
PC/ellipsometry	0.30	0.20	0.05
PC/emulsion	0.20	0.25	0.10
PA/ellipsometry	0.90	0.80	5.20
PA/emulsion	1.35	0.70	6.70

Source: Ref. 71.

tion techniques at plateau conditions show fairly good agreement with those obtained
with other techniques (Table 1).

Another approach used by several investigators is to deposit (phospho)lipid
layers at a flat solid substrate, and then follow the adsorption with optical methods
such as ellipsometry, surface plasmon resonance spectroscopy (SPR), or total internal
reflectance fluorescence spectroscopy (TIRF) [6,7,64–71]. This approach has the
advantage of allowing detailed studies of adsorption kinetics, determinations of quite
low adsorbed amounts, studies of exchange kinetics, etc., at well-defined conditions.
The main drawback with this approach is that flat surfaces are used; i.e., there might
be curvature effects changing the adsorption at these surfaces from that at colloidal
carriers. These curvature effects could be expected to be particularly pronounced for
large proteins and/or small carriers, whereas the surface of a carrier much larger than
a protein will appear as effectively flat for the protein, and curvature effects will be
largely absent. In order to illustrate this effect, Table 1 shows the adsorbed amounts
of a few serum proteins at a few phospholipid surfaces, as obtained with ellipsometry
for flat plates and as obtained for the corresponding emulsion system with a solution-
depletion method. As can be seen, quite comparable adsorbed amounts are obtained
with the two approaches, indicating that at least for the proteins investigated, the flat
plate results have bearing also for the adsorption at colloids with a diameter of 300
nm [71].

B. Effects of Wettability

It has been found in numerous biomedical applications that the response to a material
in contact with a biological system depends on the wettability of the material in
contact with the biological system [72,73]. Also, macrophage uptake has been found
to depend strongly on the wettability of colloidal particles. More precisely, for very
hydrophilic particles, a low uptake is observed. As the particles become more hydrophobic, their uptake increases, but for very hydrophobic particles a lower macrophage uptake is observed once more [2].

Motivated by this, but even more so by issues related to the acceptance of
implanted biomaterials as well as other biomedical applications, numerous studies
have been performed over the last few decades regarding the effects of surface
hydrophobicity on protein adsorption. Although the rough general trend seems to be
that most proteins tend to adsorb more extensively at hydrophobic than at hydrophilic

surfaces, there are numerous systems not displaying this behavior [6,7,74–76]. However, for very hydrophilic and uncharged surfaces, a low adsorbed amount is generally observed. Examples of surfaces displaying this behavior are those modified by PEG-containing polymers or cellulose ethers (see below), and zwitterionic phospholipids and polymer mimics of the latter (see below) [6,7,77,78].

An interesting feature displayed by at least several hydrophobic surfaces is that the Vroman effect [74,75,79–81] is, at least to some extent, absent or reduced. Thus, while the adsorption at a hydrophilic surface (e.g., glass) from a complex protein mixture such as blood is characterized by an initial adsorption of small and abundant proteins (e.g., human serum albumin, HSA) and a subsequent replacement of these by larger proteins of a lower serum concentration but with a higher adsorption affinity (e.g., high-molecular-weight kininogen, HMWK) [74,75,79], this exchange seems to be more limited at hydrophobic surfaces. This behavior is particularly clear if HSA is allowed to preadsorb at hydrophobic surfaces, since this dramatically reduces the adsorption of other serum proteins, such as immunoglobulin G (IgG) and fibrinogen (Fgn) [64]. In fact, this effect has been extensively used in practical applications, since serum albumin is frequently used as a blocking agent, the role of which is to reduce or eliminate nonspecific adsorption of subsequently added proteins, such as IgGs [82]. Interfacial exchange is discussed more extensively by Ball, Schaaf, and Voegel in Chapter 11.

C. Effects of Charge

The surface charge of colloidal drug carriers has been found to be important to the macrophage uptake and the resulting bloodstream circulation time and tissue distribution of the carrier. In particular, as the charge increases, so does the uptake [2]. There have been numerous studies of the effects of the surface charge on protein adsorption. The interested reader is referred to previous reviews [74,75] and other chapters of this volume. Regarding phospholipid surfaces, one generally observes a higher serum protein adsorption at (at least some) charged phospholipids (e.g., phosphatidic acid, PA, disphosphatidylglycerol, DPG—also called cardiolipin—and phosphatidylserine, PS) than at, e.g., zwitterionic ones (e.g., phosphatidylcholine, PC, phosphatidylethanolamine, PE, and sphingomyelin, SM) (Table 2, Figs. 2b and 3) [4,6–11,70,83–85]. This effect seems to be strongly related to the more extensive uptake and shorter circulation time observed for PA, DPG, and PS than for PC, PE, and SM (Figs. 1 and 2) [4,5,9,86–90], the latter thus being directly attributable to the extensive adsorption of proteins at the carrier surface. In fact, Chonn et al. have in a series of investigations directly shown the existence of an inverse correlation between the total amount of protein adsorbed at the drug carrier, on one hand, and the bloodstream circulation time of the latter, on the other [4,8–11]. Furthermore, a number of candidate proteins responsible for this process have been identified, including IgG, Fgn, C3, C1q, C-reactive protein, β2-glycoprotein I, and fibronectin (Fnk) [4,6,7,16,83–85]. From these findings it would seem that surface charge favors serum protein adsorption, despite the sign of charge for the surface and the proteins typically being the same.

On the other hand, the serum protein adsorption at surfaces where the negative charge is "shielded" by an uncharged carbohydrate residue, such as phosphatidylinositol (PI) and ganglioside GM1, is, with a few exceptions, quite low (Table 2).

TABLE 2 Adsorbed Amount (mg/m^2) of HSA, IgG, Fgn, Fnk, C3, and C1q at Various Surfaces

	Γ_{HSA}	Γ_{IgG}	Γ_{Fgn}	Γ_{Fnk}	Γ_{C3}	Γ_{C1q}
Silica-CH$_3$	0.80	3.00	4.90	1.90	0.40	2.60
Silica	0.35	1.10	2.90	1.90	3.10	1.90
PC	0.30	0.20	0.05	0.10	0.65	0.10
PE	0.15	0.05	0.15	0.05	—	—
SM	1.80	0.05	0.20	0.30	—	—
PI	0.40	0.25	0.15	0.05	0.05	0.15
PC/GM1[a]	0.15	0.05	0.10	0.05	—	—
PA	0.90	0.80	5.20	2.20	2.30	0.25
DPG	1.20	3.80	4.90	1.10	—	—
PS	0.70	0.10	0.50	0.10	—	—
F127[b]	0.05	0.05	0.20	±0.00	0.05	0.05
PEI-PEG[c]	±0.00	±0.00	±0.00	±0.00	±0.00	±0.00
C$_{18:1}$-EO$_{151}$[d]	−0.20	0.10	0.05	0.05	−0.20	±0.00

[a] PC/GM1 = 7/1.
[b] Obtained from Γ_{tot} − Γ(F127) after rinsing (c_{eq} < 10 ppm).
[c] Obtained from Γ_{tot} − Γ(PEI-PEG) after rinsing (c_{eq} < 10 ppm).
[d] Obtained from Γ_{tot} − Γ(C$_{18:1}$-EO$_{151}$) after rinsing (c_{eq} = cmc).
Source: Refs. 7 and 85.

Analogously to this, carriers consisting of these lipids are taken up much less extensively and cleared less rapidly than, e.g., PA and DPG [4,91–94] (Fig. 1). This seems to indicate that not only the charge and wettability of the surface, but also more specific structural aspects determine the protein adsorption. For extensive protein adsorption to occur, the protein–surface interaction should be attractive, or at least not too repulsive [74,75,96]. Although there may be an attractive electrostatic interaction between the negatively charged surface and positively charged domains of the proteins, the Debye length at physiological conditions is about 7 Å [97]. Only at about this distance of separation between the surface and the protein will there be a significant electrostatic interaction. In the case of PI and ganglioside GM1, the carbohydrate residue protrudes about this distance into the solution, and since the hydration or protrusion force due to these residues could be expected to be steeper than the (weak) electrostatic interaction, the former will dominate the protein–surface interaction at distances smaller than this. (For a more in-depth discussion of the interaction forces in, e.g., ganglioside GM1 systems, the reader is referred to Chapter 7.) Consequently, the adsorbed amount of most serum proteins is low at surfaces consisting of PI and ganglioside GM1. For PA and DPG, on the other hand, the hydration interactions are of less importance, and hence the adsorption may be quite high, especially since the high degree of electrostatic screening favors electrostatic interactions between positively charged domains of the protein and the negatively charged surface. In conclusion, although the surface charge has a clear influence on the interaction with serum proteins, electrostatic effects are partially masked by other effects, and due to this and the quite different adsorption behavior of different proteins at a given surface, this dependence is not trivial.

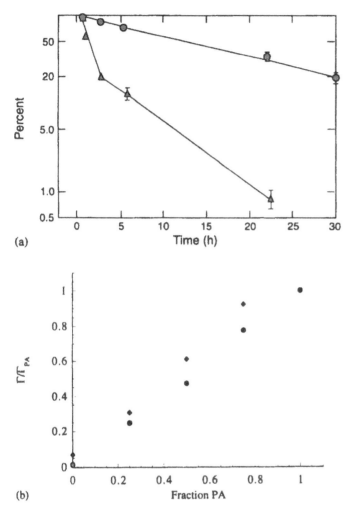

(a)

(b)

FIG. 2   (a) Clearance of SUV liposomes consisting of distearoylphosphatidylcholine (DSPC)/
CH (1:1) (circles) and DSPC/DSPA/CH (0.75:0.25:1) (triangles) from bloodstream circulation
after injection in mice at a dose of about 57 mg phospholipid per kg body weight. (Data from
Ref. 5.) (b) Adsorption of Fgn (circles) and ApoB (diamonds) at PC/PA mixed phospholipid
surface versus the phospholipid mixture composition. (Data from Refs. 71 and 84.)

D. Mixed Phospholipid Systems

In practical formulations, mixed phospholipid systems are frequently used, e.g., due
to impurities in the commercial lecithin samples, due to the desire to increase the
carrier–carrier repulsive interaction, to reduce the leakage rate from liposomes, etc.
It can be expected that the biological response to carriers consisting of such a mixed
phospholipid layer depends on the phospholipid composition [5]. It is therefore interesting to investigate the adsorption at mixed phospholipid surface. (For the corresponding discussion for PEG-modified surfaces, see Section IV.) As an example,
Fig. 2b shows the adsorption of Fgn and apolipoprotein B (ApoB) at mixed PC/PA

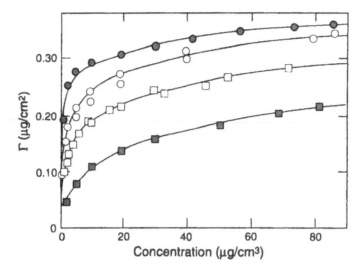

FIG. 3 Adsorption isotherms of prothrombin at mixed phospholipid layers of DOPS and dioleoylphosphatidylcholine (DOPC) as a function of the phospholipid composition. The DOPS:DOPC ratio was 0.20:0.80 (filled squares), 0.40:0.60 (open squares), 0.80:0.20 (open circles), and 1.0:0 (filled circles). (Data from Ref. 70.)

layers as a function of the phospholipid composition. As can be seen, the adsorption increases essentially linearly with an increasing PA content for both proteins. Similarly, Fig. 3 shows the adsorption of prothrombin at mixed layers of dioleoylphosphatidylcholine (DOPC) and dioleoylphosphatidylserine (DOPS) as a function of the layer composition, and indicates an increasing adsorption with an increasing DOPS content [70]. The main point to be made here, however, is that the protein adsorption to mixed phospholipid layers depends on their composition, that these effects may be readily studied with, e.g., ellipsometry, and that there is again a good agreement between the circulation time of colloidal drug carriers on one hand, and the adsorption of, e.g., Fgn, on the other.

E. Effects of Lipid Fluidity

In principle, the protein adsorption could be expected to depend on the fluidity of the (phospho)lipid layer [4], and particularly if the proteins are not only adsorbed but also partly penetrating into the lipid layer this could be expected to be of major importance. Most likely the degree of interpenetration will vary between proteins, but at least for some proteins, e.g., serum albumin, α-lactalbumin, and apolipoprotein A-I, A-II, and A-IV, lipid layer penetration has been observed experimentally [59–61]. This is bound to have effects also on the uptake of colloidal drug carriers, and it is therefore interesting to note that carriers containing sphingomyelin, consisting of long and saturated acyl groups, behave similarly to those consisting of phosphatidylcholine [8] of an identical polar head group but with different lipid layer fluidity [98]. In another study, however, liposomes consisting of saturated acyl groups containing more than 16 carbons have been found to adsorb large quantities of serum proteins and to be cleared more rapidly from circulation [4]. Despite the possible importance of these effects in relation to protein adsorption, very few studies have addressed these issues, which therefore are still largely unclear.

F. Adsorption from Serum and Plasma

From competitive adsorption studies in binary protein systems it was found that, although preadsorbed HSA is not exchanged by Fgn at PA, the adsorption of the latter protein is little affected by the presence of preadsorbed HSA, nor was surface blocking observed for PC [64]. Thus, although a Vroman effect in the classical sense was not observed, these results seem to indicate that opsonins will have relatively free access to phospholipid surfaces, e.g., on intravenous administration of a liposome or emulsion formulation. This is an important finding also from a methodological point of view, since it shows that model experiments with single-protein solutions have some bearing on physiologically more relevant experiments. Note, however, that this seems to depend on the surface properties, and some authors have indeed used albumin adsorption in particular in order to reduce macrophage uptake of colloidal carriers [2,99]. To obtain more information, we have investigated the total adsorption from serum and plasma at different phospholipid and other surfaces [71]. Overall, the adsorption from serum and plasma was found to be analogous to that of a few key serum proteins. For example, PA was previously found to promote Fgn adsorption also for mixed protein systems, at the same time as this surface adsorbs quite high amounts of several proteins, including Fgn, Fnk, and C3 (Table 2). Consequently, the high total adsorption at PA from serum and plasma was expected for this surface (Table 3). The same argument also applies for silica. On methylated silica, on the other hand, it was previously shown that the exchange of HSA in mixed protein systems is quite limited [64,80,81]. Therefore, qualitatively and quantitatively, the serum adsorption experiments seem to agree with an adsorption of essentially nonreplaceable HSA and the outline scenario. PC and different PEG modifications (see below) were found in the single-protein experiments to result in low adsorbed amounts for a range of serum proteins (Table 2), and therefore the low adsorbed amount obtained also on adsorption from plasma and serum is quite expected (Table 3).

TABLE 3 Limiting Adsorption (mg/m^2) at Various Surfaces from Serum/Plasma (1/100) in PBS

Surface	Serum	EDTA plasma	Heparin plasma	Citrate plasma
Silica	5.2	3.2	5.0	4.7
Silica-CH$_3$	1.2	0.7	1.1	0.4
PA	2.7	2.6	2.0	1.8
PC	0.4	0.3	0.1	0.1
PEI-PEG*	<0.05	<0.05	<0.05	<0.05

*Obtained from $\Gamma_{tot} - \Gamma$(PEI-PEG) after rinsing ($c_{eq} < 10$ ppm).
Source: Ref. 71.

G. Complement Activation

As mentioned, the correlation between serum protein adsorption and clearance from circulation of colloidal drug carriers has been shown in studies by Chonn et al., who found an inverse relationship between the total amount of protein adsorbed and the circulation time [8–11]. As indicated, several different mechanisms have been identified for the initiation of the phagocytosis. In particular, complement activation has been found to be important for opsonization [4,12–18]. However, in what way this mechanism depends on the carrier properties is a somewhat complex issue, since the adsorption of complement components may or may not be associated with an activation of the complement cascade. As can be seen in Fig. 4, there is a clear correlation between the C1q adsorption and C1q-related activation in the case of DPG. However, the coupling between adsorption and activation could also be weak or essentially nonexistent, e.g., if C3 adsorbs at a surface without its subsequent cleavage into C3a and C3b [95]. An issue which has been addressed in relation to this is to what extent complement activation depends on the carrier surface charge. In this context, it is interesting to note that Kovacsovics et al. (Fig. 4) found that DPG causes C1 activation through C1q adsorption, but that PI, as well as PC and PE, does not cause this activation [16]. Opposing some of these results, Marjan et al. found that PI activates complement via the classical pathway [18], a result found also by Chonn et al. [14]. On the other hand, Wassef et al. found that complement-mediated phagocytosis may be avoided by ganglioside GM1 and PI liposomes [17]. In summary, the surface charge seems to be of some importance for complement activation, but other features clearly play a role. The situation is further complicated by the occurrence of antibody-dependent [12,13] and antibody-independent [16,18] activation, and the connected associative adsorption, e.g., between IgG and C1q [85].

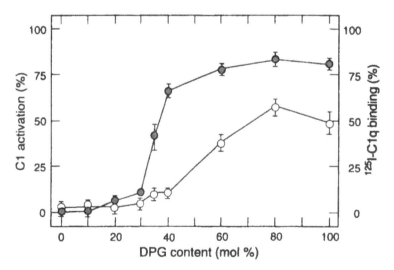

FIG. 4 C1q adsorption (open circles) and activation (filled circles) at PC:DPG vesicles as a function of the vesicle DPG content. (Data from Ref. 16.)

IV. SERUM PROTEIN ADSORPTION AT PEG-MODIFIED SURFACES

A. Controlled Circulation Time and Tissue Distribution Through Use of PEG-Containing Carriers

Although PEG-containing copolymers, and PEG-PPG-PEG block copolymers in particular [PPG being poly(propylene glycol)], are used extensively in order to stabilize colloidal systems in a host of pharmaceutical applications [19], a particularly interesting application area is parenteral administration of colloidal drug carriers. Examples of such colloidal systems stabilized with PEG-containing amphiphiles include emulsion droplets [36–43], liposomes [32–35], polymer particles [43–50], dispersed cubic phase [51], solid lipid nanoparticles [52,53], and solid drug particles [54]. As mentioned, numerous studies show that PEG-modified phospholipids are singularly efficient in reducing RES uptake and prolonging the circulation time of colloidal drug carriers [4,32–35,94,100–106]. This effect is most likely related to a very limited adsorption of a range of serum proteins at the carrier surface [4,6,7,85,112]. However, as shown in Fig. 5 for PC/cholesterol (CH) liposomes containing varying amounts of PEG-PE of different lengths, the PEG layer needs to be sufficiently dense and thick for advantageous effect. Although Fig. 5 refers to liposomes, analogous effects have also been obtained, e.g., for polystyrene-PEG particles by Dunn et al., who found that uptake of the particles by nonparenchymal liver cells in vitro decreased with an increasing PEG surface density of the particles, at the same time as in vivo studies in rats showed a prolonged circulation time for particles with a dense PEG coating [45]. Similar results were found by Verrecchia et al. [46] for poly(D,L-lactide)-PEG particles. Furthermore, modification of genetically produced proteins to be administered intravenously by covalently attaching PEG chains is an efficient way to reduce problems associated with an immunicity response and a consecutive rapid clearance of the protein from bloodstream circulation [55].

For emulsions and liposomes PEG-modified through adsorption of water-soluble PEG-containing block copolymers, several investigators have observed a decreased macrophage uptake, a prolongation of blood circulation time, and/or a more even biodistribution in vivo [47,48,86,107–111]. Just to mention one example, Illum et al. investigated the effect of Poloxamine 908 on the circulation time of emulsion droplets administered intravenously to rabbits and again found the general features outlined above for this system (Fig. 6) [109]. Analogous results have been found for these systems by Tan et al. [108] and Davis and Hansrani [86].

An application where adsorbed and stabilizing PEG-PPG-PEG block copolymers is of central importance is that of artificial blood, where these copolymers are used to stabilize perfluorochemical emulsions [39–42]. Although several different emulsifiers have been employed in artificial blood, PEG-PPG-PEG block copolymers were the first to be used commercially and are still frequently used in practical systems and in model investigations.

Furthermore, the use of PEG modification of liposomes has dramatically increased the (potential) usefulness of this type of formulation, since it reduces adverse effects related to short circulation time, uneven tissue distribution, storage stability problems, etc. For example, Doxil, a liposomal formulation for intravenous administration of doxorubicin for the treatment of Kaposi's sarcoma, is based on PEG-distearoyl phosphatidylethanolamine and has been declared approvable by the FDA

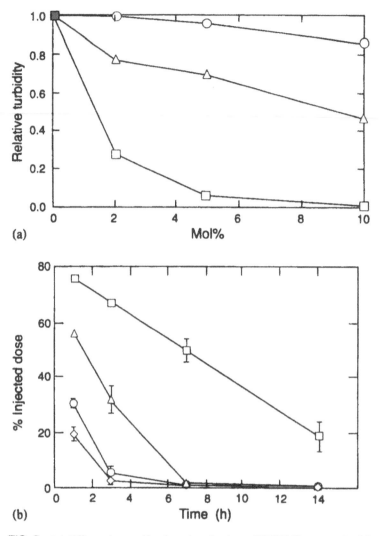

FIG. 5 (a) Effects (normalized to that for bare PC/CH liposomes) of increasing concentrations of PEO$_{750}$-PE (circles), PEO$_{2000}$-PE (triangles), and PEO$_{5000}$-PE (squares) on steptavidin-induced agglutination of liposomes containing biotin-cap-PE. Shown also are (b) blood clearance and (c) uptake in liver and spleen of intravenously injected ^{111}In-labeled liposomes in male Balb/c mice. [Symbols as in (a).] Diamonds show results obtained for the uncoated liposomes. (Data from Ref. 102.)

[34]. Several excellent reviews of recent developments within this field are currently available [32–34].

B. Mechanisms of Protein Rejection

Before discussing the effects of chain length, chain density, anchoring, etc., on the performance of PEG-based colloidal drug carriers in vitro and in vivo and the dependence of the protein-rejecting properties of PEG coatings in particular on these

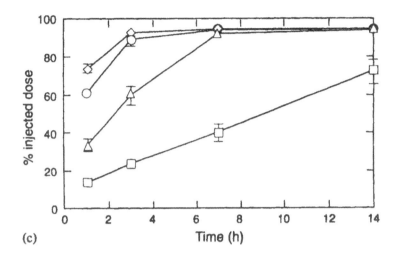

(c)

parameters, it is instructive to consider the mechanism of the protein rejection by these polymers. Therefore, let us consider the interaction between a protein and a surface as the protein approaches the surface. As discussed in more depth elsewhere in this volume, the long-range protein–surface interaction is typically dominated by electrostatic interactions, whereas at shorter separations other types of interaction (e.g., van der Waals, hydrophobic, hydrogen bonding) contribute significantly [97].

In the presence of a PEG layer, the protein–surface interaction is modified in a number of ways. First, at the conditions typically used in biomedical systems, PEG experiences fairly good solvency conditions. At such conditions, the osmotic equilibrium between chain entropy and the polymer–solvent interaction favors a high

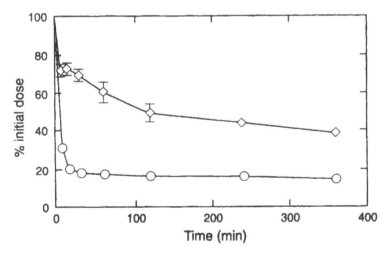

FIG. 6 Blood clearance of ^{123}I-labeled emulsions stabilized by egg lecithin (circles) and Poloxamine 908 (diamonds) intravenously administered to rabbit. (Data from Ref. 109.)

degree of hydration; i.e., the PEG chains are swollen and therefore dilute on average. Also within a layer containing (nonadsorbing) PEG chains attached to the surface through, e.g., an adsorbing block or a covalent coupling, these chains experience an analogous osmotic equilibrium and at good solvency conditions strive to maximize their hydration [96]. Consequently, there is an osmotic penalty associated with increasing the interfacial polymer concentration. (Naturally, this is one of the main reasons for an adsorbed amount saturation.) Similarly, when two PEG-coated particles approach each other, there is typically a repulsive interaction between the particles due to the increased polymer concentration in the region between the particles. This interaction is commonly referred to as steric repulsion [96,113]. Analogously, when a protein approaches a PEG-coated surface there will be a repulsion between the layer and the protein, since the interfacial polymer concentration must increase for the protein to be able to adsorb directly at the interface (in the absence of exchange) [114,115].

Second, the PEG layer dramatically reduces attractive components to the protein–surface interaction. Since PEG is uncharged and should tend to reduce the underlying surface charge density, e.g., through image charge interactions [97,116], electrostatic attractive interactions between the protein, on one hand, and the surface and the PEG layer, on the other, are virtually absent [117]. Furthermore, since the PEG layer water content is generally quite high (typically about 90%) [118], the van der Waals interaction between the PEG layer and the protein is generally small [97,114]. With some exceptions, such as the hydroxyl-mediated binding of C3 and its subsequent cleavage to C3b and C3a and the resulting complement activation [12,13], the PEG layer therefore does not provide any attractive components, or at least only weak ones, to the protein–"substrate" interaction. Considering this, and the simultaneous steric repulsion, it is hardly surprising that PEG coatings are very efficient as protein rejectors, a fact which by now is widely utilized in biomedical applications such as drug delivery, implants, solid-phase diagnostics, extracorporeal therapy, biosensors, affinity chromatography, etc. [119]. Note, however, that the only requirements for a steric stabilization and an effective elimination of attractive protein–surface interactions is that the polymer be flexible, polar, highly hydrated, and uncharged. Naturally, there are numerous polymers with these features (see Refs. 77, 78, 105, and 120), although PEG is definitely one of the most efficient. Furthermore, the limited toxicity of PEG makes it advantageous for many biomedical applications [19].

C. Different Types of PEG Modification

There are numerous ways to localize PEG chains at a surface. One of the most frequently used in intravenous drug delivery is to include a PEG-modified phospholipid (e.g., PEG-PE) in the liposome or emulsion droplet [4,32–35,94,100–106]. This technique has the advantage that the PEG chains are firmly anchored at the colloid surface, thus avoiding colloid destabilization and opsonization enhancement through collision-induced desorption or desorption on dilution. However, the technique is less straightforward than postadsorption of a PEG-containing extensively water-soluble compound (e.g., PEG-containing copolymers, PEG-containing fatty acid esters, etc.) after formation of the colloidal drug carrier [47,48,86,107–111]. On the other hand, the latter approach suffers from the risk of desorption on dilution (i.e., parenteral administration), which would be detrimental to the performance of this type of carrier. Yet another method of interest primarily for other types of biomedical applications, but also, e.g., for surface modification of solid nanoparticles and bio-degradable polymer spheres for parenteral drug delivery, is to attach PEG through a covalent link in one end of the PEG chains. Similarly to the PEG-PE modification, the latter has the advantage of resulting in an effectively irreversible PEG attachment to the carrier surface. On the other hand, the approach is even more complex than the use of PEG-PE and, therefore, often less interesting from a practical point of view. The main area of interest for covalent PEG grafting within intravenous drug delivery is the PEGylation of (protein) drugs to be administered parenterally [55–57].

D. Effects of Chain Length

As indicated, it has been found that the positive effects of PEG coatings for the circulation time and tissue distribution of colloidal drug carriers depend on the PEG molecular weight. More precisely, the longer the PEG chains at a given chain density, the longer is the circulation time and the more even is the tissue distribution (Fig. 5). In parallel to this, it has been found that the protein rejection of PEG coatings depends on the PEG chain length [108,112,121–123]. More precisely, it is generally found that once a certain chain length has been reached, the PEG layers formed effectively eliminate or dramatically reduce protein adsorption, an effect which above this is essentially independent of the PEG chain length. As an example, Table 3 in Chapter 27 shows the adsorption of Fgn at PEG-modified surfaces versus the PEG molecular weight. As can be seen from this table, the protein adsorption is essentially eliminated at a PEG weight of about 1500, whereafter little molecular-weight dependence is observed.

Although there seems to be a consensus about the general features outlined above, there is still some discussion regarding the minimum PEG chain length where efficient protein adsorption reduction is observed. Thus, while Tan et al., Bergström et al., Norman et al., and Mori et al. found the minimum chain length to be between 35 and 100 ethylene oxide units [108,112,121,122], it is interesting to note that Prime and Whitesides concluded that already at one or a few ethylene oxide units, a pronounced reduction in protein adsorption occurs, provided that the density of the ethylene oxide groups is sufficiently high. However, longer-chain PEGs were found to be protein resistant at a lower interfacial mole fraction [123]. These issues are discussed also by Holmberg and Quash in Chapter 27.

The general features outlined may be understood by a comparison to findings of steric stabilization of colloidal dispersions [113]. Also in the latter case, a repulsive interaction is observed at good solvency conditions for the stabilizing polymer. As long as the degree of polymerization of the stabilizing chains is sufficiently high, a good steric stabilization is found, and the destabilization occurs at the θ-conditions for the interfacial polymer layer. However, for polymers shorter than a minimal chain length, flocculation is observed already at significantly better than θ-conditions. Excluding the trivial case of insufficient anchoring (see below), the reason for this is that at sufficiently low-molecular-weight chains, the particles may approach each other without experiencing a steric repulsion to a distance of separation where attractive interactions, and notably the van der Waals interaction, may dominate, thus providing a mechanism for flocculation. Similarly, the van der Waals interaction between the protein and the underlying substrate may "shine through" the steric interaction due to the PEG chains, particularly for large proteins and for drug carriers with a high refractive index [113–115].

In the light of this, the findings by Prime and Whitesides [123] are particularly interesting. Thus, for one or a few ethylene oxide units only, clearly the van der Waals interactions are of a considerable magnitude at a protein–surface separation corresponding to the PEG layer thickness. The finding of an efficient protein rejection at these short PEG "chains" must therefore mean that a repulsive interaction outweighs the attractive interactions. Once more drawing the parallel to steric stabilization and indeed to polymer solutions, the osmotic repulsive interaction due to an elevated polymer concentration at good solvency conditions increases with the volume fraction in the overlap region [113]. Therefore, the higher the polymer chain density in the interaction region, the larger is the steric repulsion. Considering the very high "chain" densities used by Prime and Whitesides, the finding of an efficient steric stabilization already at very short PEG chains is not surprising. (Note that although the argument here has been based on chain molecules, and therefore obviously breaks down at sufficiently short chains, the essence of the argument should still be valid if the repulsive contribution is due to other essentially entropic interactions, such as hydration or protrusion forces.)

E. Effects of Chain Density

At a given chain length, it has been found that the circulation time and tissue distribution depend on the interfacial chain density; more specifically, the lower the chain density, the more pronounced the uptake of the drug carrier (see also Fig. 5). As discussed, the protein-rejecting capacity should increase with an increasing interfacial chain density, which has been found for both adsorbed and covalently attached PEG-containing layers (Fig. 7) [114,115,118,123–125]. As indicated in Fig. 7, the protein-rejecting capacity of a PEG coating with a fixed PEG chain length is essentially independent of the nature of the underlying surface, at least as long as the PEG-containing coating is firmly attached, but instead largely determined by the chain density. As shown in Tables 2 and 3, if the polymer coating is protein rejecting for one protein, it is generally protein rejecting also for others. Note, however, that a parameter of major importance for the protein-rejecting capacity of polymer layers is the average mesh size compared to the size of the protein. Thus, if the protein is sufficiently small it may be able to "slip through" the (dilute) polymer layer without disturbing the interfacial layer structure, thereby largely avoiding steric repulsive effects. As an example, the protein-rejecting capacity of a poly(ethylene imine)–poly(ethylene glycol) (PEI-PEG) coating is shown in Fig. 8 for one large protein (Fgn, approximate dimension 45 × 6 nm) and one small protein (insulin, approximate dimension 3.5 × 5 nm) [124]. The figure shows that insulin is sufficiently small to penetrate the adsorbed layer, at least to some extent, whereas Fgn is not. Consequently, the PEI-PEG coating is efficient in reducing Fgn adsorption but less so for insulin. (This effect is illustrated also in Fig. 9.) This is particularly important to consider when working with covalent immobilization of PEG chains. Due to the PEG-PEG effectively repulsive interaction at good solvency conditions, and the absence of a molecular domain counteracting this (e.g., a hydrophobic block in PEG-

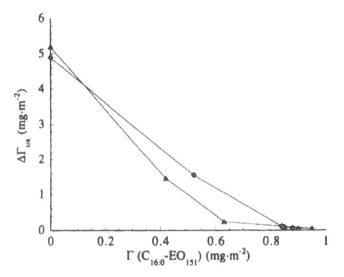

FIG. 7 Adsorbed amount of Fgn at $C_{16:0}$-EO_{151} fatty acid esters preadsorbed at methylated silica (diamonds) and PA (triangles) versus the preadsorbed amount. (Data from Ref. 118.)

containing block copolymers, phospholipids, or surfactants), the grafting yield is generally quite low at these solvency conditions. More precisely, the chain density rarely exceeds the mushroom-to-brush transition [96], which means that the average distance between two PEG chains is about twice the radius of gyration. Particularly for high-molecular-weight polymers, the grafting density may be very low in absolute terms, which makes protein rejection for small proteins increasingly difficult (Figs. 8 and 9).

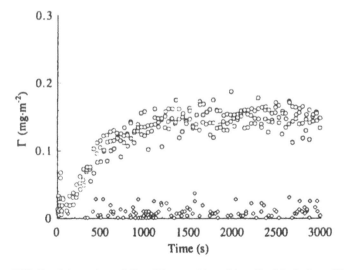

FIG. 8 Adsorption of Fgn (diamonds) and insulin (circles) at silica modified with PEI-PEG. (Data from Ref. 124.)

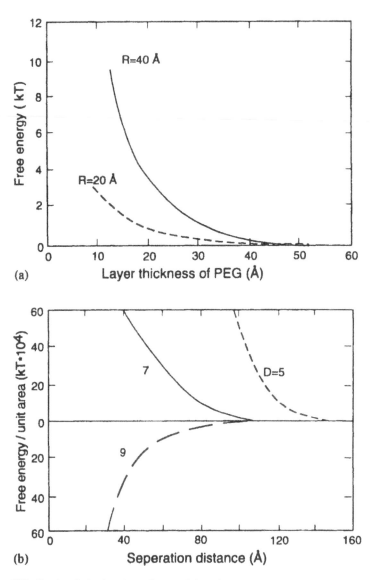

FIG. 9 (a) Calculated steric repulsion free energy versus the PEG layer thickness for a protein of radius 40 Å (solid line) and 20 Å (dashed line) at a fixed PEG chain length and grafting density of 120 monomer units and 1 chain per 17 Å, respectively. (b) Combined steric repulsion and hydrophobic attraction free energy per unit area (Å²) versus the protein-layer separation for a grafting density of 5 Å (short-dashed line), 7 Å (solid line), and 9 Å (long-dashed line) at a PEG chain length of 120. (Data from Refs. 114 and 115.)

F. Effects of Anchoring

For a polymer coating to be efficient, e.g., regarding steric stabilization or protein rejection, it has to be firmly attached to the interface. For example, it is generally found that it is more difficult to sterically stabilize an emulsion than a dispersion consisting of solid particles [113], which is due to lateral diffusion of the interfacial polymer layer on compression during collision. Ways to circumvent effects of this type are to increase the interfacial crowding (i.e., the adsorbed amount), to preclude lateral mass transport through entanglement (branched polymers), to work below the chain-melting temperature (in the case of phospholipids), etc.

An equally important consideration is the desorption of the adsorbed polymer molecules due to dilution. Thus, on intravenous administration, the colloidal drug formulation is extensively diluted. If the equilibrium between the interfacial and bulk solution polymer molecules is located at a sufficiently high bulk solution concentration, extensive desorption will be the result soon after the injection, and the protein-rejecting capacity will be lost (cf. Fig. 7). If an adsorbing PEG-containing layer is to be used, it is therefore essential to determine both the adsorption isotherm and, most importantly, the degree of desorption on dilution. Fortunately, polymers tend to adsorb in a high-affinity isotherm and display an effectively irreversible adsorption with respect to dilution. (For a more extensive discussion on this, see Chapter 1.) The origin of the latter is a combination of the high-affinity isotherm, the fact that polymers tend to be both polydisperse and heterogeneous, and the preferential adsorption of larger and/or more hydrophobic molecules, as well as the exceedingly slow kinetics for interfacial conformational changes, exchange, and desorption typically displayed by polymer systems. Therefore, although the initial desorption on dilution may be quite significant, a total desorption is rarely observed within the time frame relevant for parenteral administration. Consequently, at least for some systems, a sufficiently large amount remains at the surface in order to result in an efficient protein rejection (Tables 2 and 3) and a consecutive long circulation time (Fig. 6).

Another issue related to the anchoring of a polymer layer to a colloidal drug carrier is interfacial exchange. Thus, the adsorbed PEG-containing molecules may be displaced, e.g., by an adsorbing protein. Needless to say, this will eliminate any protective effects of the layer. The risk for such a displacement is particularly pronounced for low-molecular-weight PEG-containing polymers and/or PEG-containing copolymers with long PEG chains and short anchoring blocks at surfaces not strongly adsorbing PEG. For example, Muller and Malmsten studied the adsorption of Fgn at a hydrophobic surface modified with a PEG-poly(lactide) copolymer. At conditions where the hydrophobic lactide block was intact, the anchoring was good and the Fgn adsorption negligible. After hydrolysis of the lactide block, however, the anchoring was found to be very weak, and the Fgn adsorption was essentially identical to that at the bare hydrophobic surface (unpublished results). Thus, in the latter case little or no protein-rejecting effect exists. (Analogous effects have been observed also for polysaccharide systems [120].)

G. Mechanisms and Effects of Solvency

Both the steric repulsion and the protein–polymer layer interaction depend on the solvency of the polymer, as briefly outlined above. Thus, at poor solvency conditions, an interfacial polymer layer will collapse on itself (constant adsorbed amount) [126,127], which has at least two consequences. First, since the polymer strives to increase its concentration, the steric repulsion is largely absent at poor solvency conditions. Indeed, for the polymer solution and the sterically stabilized cases, poor solvency conditions result in a macroscopic phase separation and in a transition from steric repulsion to an effectively attractive colloid–colloid interaction and a consecutive flocculation, respectively [113]. Second, since the interfacial layer collapses, its water content decreases drastically [126], which increases the magnitude of the van der Waals attractive interaction between the protein and the polymer layer [97,114,115]. Consequently, the protein-rejecting properties of a polymer layer are largely eliminated at poor solvency conditions [114,115,127,128]. For example, Tiberg et al. investigated the collapse of a PEG-PPG–containing graft copolymer layer, achieved through a worsening of the solvency conditions through a temperature increase, as well as the effects of this process on the protein-rejecting properties of this layer [127]. As the solvency deteriorates, the adsorption of IgG was found to increase as a direct consequence of the collapse of the interfacial polymer layer. This behavior is very interesting since it may be used to control the protein–polymer interaction, which is of major importance when covalently attaching proteins at protein-rejecting surfaces and, hence, for, say, targeting of colloidal drug carriers.

H.   PEG-Containing Carriers and Targeting

Due to the protein-rejecting properties of PEG-modified drug carriers and the subsequent long circulation time in the bloodstream, these have great potential for targeting to specific tissues, types of cells, etc. Thus, if a biospecific molecule, e.g., a suitable antibody (fragment), can be covalently attached to such a carrier, it could be expected to circulate for a long time until it is taken up by a tissue or cell with a localized antigen. As an example of this type of system, Kabanov et al. observed specific targeting of fluorescein isothiocyanate solubilized in PEG-PPG-PEG block copolymer micelles to the brain when the copolymer was conjugated with antibodies to the antigen of brain glial cells (α_2-glycoprotein) [28,29]. Furthermore, incorporation of haloperidol into such micelles was found to result in a drastically increased therapeutic effect.

Similar effects could be expected for other types of PEG-modified colloidal drug carriers. Indeed, a similar approach has been used in solid-phase diagnostics and extracorporeal therapy. The latter application in particular is quite similar to that of antibody-based targeting of a PEG-modified colloidal drug carrier to a localized antigen. Issues related to solid-phase diagnostics and extracorporeal therapy are discussed more extensively by Holmberg and Quash in Chapter 27.

V.   CONCLUDING REMARKS

It seems clear from numerous studies that there is a pronounced correlation between circulation time and tissue distribution of intravenously administered colloidal drug carriers, on one hand, and the adsorption at the carrier surface of a number of key serum proteins, on the other. The latter seem to include, e.g., IgG, Fgn, Fnk, C3, and C1q. It was found that the adsorption from complex media such as serum or plasma could be understood to some extent from considerations of the adsorption of HSA and the proteins mentioned above, as well as interfacial exchange phenomena. By controlling the surface properties of the drug carrier, the serum protein adsorption, and, hence, the biological response to intravenously administered colloidal drug carriers, may be controlled to some extent. However, further knowledge particularly on the role of complement activation and immune response is probably essential for an increased understanding of these processes and for further facilitating the directed targeting of drugs to certain cells or tissues.

ACKNOWLEDGMENTS

This work was financed by Astra Arcus, Biacore, Nycomed Imaging, Pharmacia & Upjohn, and the Swedish National Board for Industrial and Technical Development (NUTEK).

REFERENCES

1. H. M. Patel, Serum opsonins and liposomes: their interaction and opsonophagocytosis. Crit. Rev. Ther. Drug Carrier Syst. 9:39–90, 1992.
2. Y. Tabata and Y. Ikada, Phagocytosis of polymer microspheres by macrophages. Adv. Polym. Sci. 94:107–141, 1990.
3. T. Eldem and P. Speiser, Endocytosis and intracellular drug delivery. Acta Pharm. Technol. 35:109–115, 1989.
4. S. C. Semple and A. Chonn, Liposome–blood protein interaction in relation to liposome clearance. J. Liposome Res. 6:33–60, 1996.
5. J. H. Senior, Fate and behaviour of liposomes in vivo: a review of controlling factors. Crit. Rev. Ther. Drug Carrier Syst. 3:123–193, 1987.
6. M. Malmsten, Ellipsometry studies of protein adsorption at lipid surfaces. J. Colloid Interface Sci. 168:247–254, 1994.
7. M. Malmsten, Protein adsorption at phospholipid surfaces. J. Colloid Interface Sci. 172:106–115, 1995.
8. A. Chonn and P. R. Cullis, Ganglioside GM1 and hydrophilic polymers increase liposome circulation times by inhibiting the association of blood proteins. J. Liposome Res. 2:397–410, 1992.
9. A. Chonn, S. C. Semple, and P. R. Cullis, Association of blood proteins with large unilamellar liposomes in vivo. J. Biol. Chem. 267:18759–18765, 1992.
10. A. Chonn, S. C. Semple, and P. R. Cullis, β_2-Glycoprotein I is a major protein associated with very rapidly cleared liposomes in vivo, suggesting a significant role in the immune clearance of "non-self" particles. J. Biol. Chem. 270:25845–25849, 1995.
11. A. Chonn, S. C. Semple, and P. R. Cullis, Separation of large unilamellar liposomes from blood components by a spin column procedure: towards identifying plasma proteins which mediate liposome clearance in vivo. Biochim. Biophys. Acta 1070:215–222, 1991.
12. W. E. Paul (ed.), Fundamental Immunology. 3rd ed. Raven Press, New York, 1993.
13. R. B. Sim (ed.), Activators and Inhibitors of Complement. Kluwer Academic, Dordrecht, 1993.
14. A. Chonn, P. R. Cullis, and D. V. Devine, The role of surface charge in the activation of the classical and alternative pathways of complement by liposomes. J. Immunol. 146:4234–4241, 1991.
15. K. Funato, R. Yoda, and H. Kiwada, Contribution of complement system on destabilization of liposomes composed of hydrogenated egg phosphatidylcholine in rat fresh plasma. Biochem. Biophys. Acta 1103:198–204, 1992.

16. T. Kovacsovics, J. Tschopp, A. Kress, and H. Isliker, Antibody-independent activation of C1, the first component of complement, by cardiolipin. J. Immunol. *135*:2695–2700, 1985.

17. N. M. Wassef, G. R. Matyas, and C. R. Alving, Complement-dependent phagocytosis of liposomes by macrophages: suppressive effects of "stealth" lipids. Biochem. Biophys. Res. Comm. *176*:866–874, 1991.

18. J. Marjan, Z. Xie, and D. V. Devine, Liposome-induced activation of the classical complement pathway does not require immunoglobulin. Biochim. Biophys. Acta *1192*: 35–44, 1994.

19. M. Malmsten, Block copolymers in pharmaceutics. In: *Amphiphilic Block Copolymers*: *Self-Assembly and Applications* B. Lindman and P. Alexandridis, eds. Elsevier, 2000.

20. G. S. Kwon and K. Kataoka, Block copolymer micelles as long-circulating drug vehicles. Adv. Drug Delivery Rev. *16*:295–309, 1995.

21. M. Yokoyama, Block copolymers as drug carriers. Crit. Rev. Ther. Drug Carrier Syst. *9*:213–248, 1992.

22. M. Yokoyama, T. Okano, Y. Sakurai, H. Ekimoto, C. Shibazaki, and K. Kataoka, Toxicity and antitumor activity against solid tumors of micelle-forming polymeric anticancer drug and its extremely long circulation in blood. Cancer Res. *51*:3229–3236, 1991.

23. G. S. Kwon, M. Yokoyama, T. Okano, Y. Sakurai, and K. Kataoka, Biodistribution of micelle-forming polymer-drug conjugates. Pharm. Res. *10*:970–974, 1993.

24. M. Yokoyama, M. Miyauchi, N. Yamada, T. Okano, Y. Sakurai, K. Kataoka, and S. Inoue, Polymer micelles as novel drug carrier: Adriamycin-conjugated poly(ethylene glycol)–poly(aspartic acid) block copolymer. J. Controlled Release *11*:269–278, 1990.

25. K. Kataoka, G. S. Kwon, M. Yokoyama, T. Okano, and Y. Sakurai, Block copolymer micelles as vehicles for drug delivery. J. Controlled Release *24*:119–132, 1993.

26. S. B. La, T. Okano, and K. Kataoka, Preparation and characterization of the micelle-forming polymeric drug Indomethacin-incorporated poly(ethylene oxide)–poly(β-benzyl-L-aspartate) block copolymer micelles. J. Pharm. Sci. *85*:85–90, 1996.

27. G. Kwon, S. Suwa, M. Yokoyama, T. Okano, Y. Sakurai, and K. Kataoka, Enhanced tumor accumulation and prolonged circulation times of micelle-forming poly(ethylene oxide–aspartate) block copolymer–adriamycin conjugates. J. Controlled Release *29*: 17–23, 1994.

28. A. V. Kabanov, E. V. Batrakova, N. S. Melik-Nubarov, N. A. Fedoseev, T. Y. Dorodnich, V. Y. Alakhov, V. P. Chekhonin, I. R. Nazarova, and V. A. Kabanov, A new class of drug carriers: micelles of poly(oxyethylene)–poly(oxypropylene) block copolymers as microcontainers for drug targeting from blood in brain. J. Controlled Release *22*: 141–158, 1992.

29. A. V. Kabanov, V. P. Chekhonin, V. Y. Alakhov, E. V. Batrakova, A. S. Lebedev, N. S. Melik-Nubarov, S. A. Arzhakov, A. V. Levashov, G. V. Morozov, E. S. Severin, and V. A. Kabanov, The neuroleptic activity of haloperidol increases after its solubilization in surfactant micelles—micelles as microcontainers for drug targeting. FEBS Lett. *258*: 343–345, 1989.

30. A. Rolland, J. O'Mullane, P. Goddard, L. Brookman, and K. Petrak, New macromolecular carriers for drugs. I. Preparation and characterization of poly(oxyethylene-b–isoprene-b–oxyethylene) block copolymer aggregates. J. Appl. Polym. Sci. *44*:1195–1203, 1992.

31. K. J. Zhu, L. Xiangzhou, and Y. Shilin, Preparation, characterization, and properties of polylactide (PLA)–poly(ethylene glycol) (PEG) copolymers: a potential drug carrier. J. Appl. Polym. Sci. *39*:1–9, 1990.

32. M. N. Jones, The surface properties of phospholipid liposome systems and their characterization. Adv. Colloid Interface Sci. *54*:93–128, 1995.

33. D. Lasic and F. Martin (eds.), *Stealth Liposomes*. CRC Press, Boca Raton, 1995.

34. G. Gregoriadis, Engineering liposomes for drug delivery: progress and problems. TIBTECH *13*:527–537, 1995.

35. D. Needham, K. Hristova, T. J. McIntosh, M. Dewhirst, N. Wu, and D. D. Lasic, Polymer-grafted liposomes: physical basis for the "stealth" property. J. Liposome Res. 2:411–430, 1992.

36. P. Becher (ed.), *Encyclopedia of Emulsion Technology*: Vol. 1. *Basic theory*. Marcel Dekker, New York, 1983.

37. P. Becher (ed.), *Encyclopedia of Emulsion Technology*: Vol. 2. *Applications*. Marcel Dekker, New York, 1985.

38. I. D. A. Johnston (ed.), *Current Perspectives in the Use of Lipid Emulsions*. Marcel Dekker, New York, 1983.

39. P. Schneider, Artificial blood substitutes. Transfus. Sci. *13*:357–370, 1992.

40. T. F. Zuck and J. G. Riess, Current status of injectable oxygen carriers. Crit. Rev. Clin. Lab. Sci. *31*:295–324, 1994.

41. G. M. Vercellotti, D. E. Hammerschmidt, P. R. Craddock, and H. S. Jacob, Activation of plasma complement by perfluorocarbon artificial blood: probable mechanism of adverse pulmonary reactions in treated patients and rationale for corticosteroid prophylaxis. Blood *59*:1299–1304, 1982.

42. D. A. Ingram, M. B. Forman, and J. J. Murray, Activation of complement by Fluosol attributable to the pluronic detergent micelle structure. J. Cardiovasc. Pharm. *22*:456–461, 1993.

43. H. A. Lieberman, M. M. Rieger, and G. S. Banker (eds.), *Pharmaceutical Dosage Forms*: *Disperse Systems*, Vol. 2. Marcel Dekker, New York, 1989.

44. A. G. A. Coombes, P. D. Scholes, M. C. Davies, L. Illum, and S. S. Davis, Resorbable polymeric microspheres for drug delivery—production and simultaneous surface modification using PEO-PPO surfactants. Biomaterials *15*:673–680, 1994.

45. S. E. Dunn, A. Brindley, S. S. Davis, M. C. Davies, and L. Illum, Polystyrene-poly(ethylene glycol) (PS-PEG2000) particles as model systems for site specific drug delivery. 2. The effect of PEG surface density on the in vitro cell interaction and in vivo biodistribution. Pharm. Res. *11*:1016–1022, 1994.

46. T. Verrecchia, G. Splenlehauer, D. V. Bazile, A. Murry-Brelier, Y. Archimbaud, and M. Veillard, Non-stealth (poly(lactic acid/albumin)) and stealth (poly(lactic acid–polyethylene glycol)) nanoparticles as injectable drug carriers. J. Controlled Release *36*:49–61, 1995.

47. S. S. Davis and L. Illum, Polymeric microspheres as drug carriers. Biomaterials *9*:111–115, 1988.

48. L. Illum, S. S. Davis, R. H. Müller, E. Mak, and P. West, The organ distribution and circulation time of intravenously injected colloidal carriers sterically stabilized with a blockcopolymer—Poloxamine 908. Life Sci. *40*:367–374, 1987.

49. P. J. Tarcha (ed.), *Polymers for Controlled Drug Delivery*. CRC Press, Boca Raton, 1991.

50. E. C. Lavelle, S. Sharif, N. W. Thomas, J. Holland, and S. S. Davis, The importance of gastrointestinal uptake of particles in the design of oral delivery systems. Adv. Drug Delivery Rev. *18*:5–22, 1995.

51. S. Engström, Cubic phases as drug delivery systems. Polym. Preps. *31*:157–158, 1990.

52. R. H. Müller, W. Mehnert, J.-S. Lucks, C. Schwarz, A. zur Mühlen, H. Weyhers, C. Freitas, and D. Rühl, Solid lipid nanoparticles (SLN)—an alternative colloidal carrier system for controlled drug delivery. Eur. J. Pharm. Biopharm. *41*:62–69, 1995.

53. B. Sjöström and B. Bergenståhl, Preparation of submicron drug particles in lecithin-stabilized o/w emulsions: I. Model studies of the preparation of cholesteryl acetate. Int. J. Pharm. *84*:107–116, 1992.

54. R. Gust, G. Bernhardt, T. Spruss, R. Krauser, M. Koch, H. Schönenberger, K.-H. Bauer, S. Schertl, and Z. Lu, Development of a parenterally administrable hydrosol preparation of "the third generation platinum complex" [(±)-1,2-bis(4-fluorophenyl)ethylenediamine]dichloroplatinum(II). Part 1. Preparation and studies on the stability and antitumor activity. Arch. Pharm. *328*:645–653, 1995.

55. C. Delgado, G. E. Francis, and D. Fisher, The uses and properties of PEG-linked proteins. Crit. Rev. Ter. Drug Carrier Syst. *9*:249–304, 1992.

56. K. E. Jensen-Pippo, K. L. Whitcomb, R. B. DePrince, L. Ralph, and A. D. Habberfield, Enteral bioavailability of human granulocyte colony stimulating factor conjugated with poly(ethylene glycol). Pharm. Res. *13*:102–107, 1996.

57. C. Delgado, M. Malmsten, and J. M. Van Alstine, Analytical partitioning of poly(ethylene glycol)–modified proteins. J. Chromatogr. B *692*:263–272, 1997.

58. S. E. Friberg and K. Larsson (eds.), *Food Emulsions*, 3rd ed. Marcel Dekker, New York, 1997.

59. D. G. Cornell, D. L. Patterson, and N. Hoban, The interaction of phospholipids in monolayers with bovine serum albumin and α-lactalbumin adsorbed from solution. J. Colloid Interface Sci. *140*:428–435, 1990.

60. J. A. Ibdah, K. E. Krebs, and M. C. Phillips, The surface properties of apolipoproteins A-I and A-II at the lipid/water interface. Biochim. Biophys. Acta *1004*:300–308, 1989.

61. R. B. Weinberg, J. A. Ibdah, and M. C. Phillips, Adsorption of apolipoprotein A-IV to phospholipid monolayers spread at the air/water interface. J. Biol. Chem. *267*:8977–8983, 1992.

62. S. L. Law, W. Y. Lo, S. H. Pai, G. W. Teh, and F. Y. Kou, The adsorption on bovine serum albumin by liposomes. Int. J. Pharm. *32*:237–241, 1986.

63. J. Senior, J. A. Waters, and G. Gregoriadis, Antibody-coated liposomes—the role of non-specific antibody adsorption. FEBS Lett. *196*:54–58, 1986.

64. M. Malmsten and B. Lassen, Competitive protein adsorption at phospholipid surfaces. Colloid Surf. B *4*:173–184, 1995.

65. P. A. Cuypers, J. W. Corsel, M. P. Janssen, J. M. M. Kop, W. T. Hermens, and H. C. Hemker, The adsorption of prothrombin to phosphatidylserine multilayers quantified by ellipsometry. J. Biol. Chem. *258*:2426–2431, 1983.

66. H. A. M. Andree, W. T. Hermens, and G. M. Willems, Testing protein adsorption models by off-null ellipsometry: determination of binding constants from a single adsorption curve. Colloids Surf. A *78*:133–141, 1993.

67. P. L. A. Giesen, G. M. Willems, H. C. Hemker, and W. T. Hermens, Membrane-mediated assembly of the prothrombinase complex. J. Biol. Chem. *266*:18720–18725, 1991.

68. P. L. A. Giesen, G. M. Willems, H. C. Hemker, M. C. A. Stuart, and W. T. Hermens, Monitoring of unbound protein in vesicle suspension with off-null ellipsometry. Biochim. Biophys. Acta *1147*:125–131, 1993.

69. J. W. Corsel, G. M. Willems, J. M. M. Kop, P. A. Cuypers, and W. T. Hermens, The role of intrinsic binding rate and transport rate in the adsorption of prothrombin, albumin, and fibrinogen at phospholipid bilayers. J. Colloid Interface Sci. *111*:544–554, 1986.

70. J. M. M. Kop, P. A. Cuypers, T. Lindhout, H. C. Hemker, and W. T. Hermens, The adsorption of prothrombin to phospholipid monolayers quantitated by ellipsometry. J. Biol. Chem. *259*:13993–13998, 1984.

71. M. Malmsten, to be published.

72. S. L. Cooper, C. H. Bamford, and T. Tsuruta (eds.), *Polymer Biomaterials in Solution, as Interfaces and as Solids*. VSP, Utrecht, 1995.

73. M. Szycher (ed.), *Biocompatible Polymers, Metals, and Composites*, Technomic, Lancaster, 1983.

74. W. Norde, Adsorption of proteins from solution at the solid–liquid interface. Adv. Colloid Interface Sci. *25*:267–340, 1986.
75. C. A. Haynes and W. Norde, Globular proteins at solid/liquid interfaces. Colloids Surf. B *2*:517–566, 1994.
76. M. Malmsten, Ellipsometry studies of the effects of surface hydrophobicity on protein adsorption. Colloids Surf. B *3*:297–308, 1995.
77. D. Chapman, Biomembranes and new hemocompatible materials. Langmuir *9*:39–45, 1993.
78. K. Ishihara, N. P. Ziats, B. P. Tierney, N. Nakabayashi, and J. M. Anderson, Protein adsorption from human plasma is reduced on phospholipid polymers. J. Biomed. Mater. Res. *25*:1397–1407, 1991.
79. T. A. Horbett and J. L. Brash (eds.), *Proteins at Interfaces. II. Fundamentals and Applications.* American Chemical Society, Washington, 1995.
80. B. Lassen and M. Malmsten, Competitive protein adsorption at plasma polymer surfaces. J. Colloid Interface Sci. *186*:9–16, 1997.
81. M. Malmsten, D. Muller, and B. Lassen, Sequential adsorption of human serum albumin (HSA), immunoglobulin G (IgG), and fibrinogen (Fgn) at HMDSO plasma polymer surfaces. J. Colloid Interface Sci. *193*:88–95, 1997.
82. M. Malmsten, B. Lassen, K. Holmberg, V. Thomas, and G. Quash, Effects of hydrophilization and immobilization on the interfacial behaviour of immunoglobulins. J. Colloid Interface Sci. *177*:70–78, 1966.
83. M. Malmsten, Ellipsometry studies of fibronectin adsorption. Colloids Surf. B *3*:371–381, 1995.
84. M. Malmsten, B. Bergenståhl, M. Masquelier, M. Pålsson, and C. Peterson, Adsorption of apolipoprotein B at phospholipid model surfaces. J. Colloid Interface Sci. *172*:485–493, 1995.
85. M. Malmsten, B. Lassen, J. M. Van Alstine, and U. R. Nilsson, Adsorption of complement proteins C3 and C1q. J. Colloid Interface Sci. *178*:123–134, 1996.
86. S. S. Davis and P. Hansrani, The influence of emulsifying agents on the phagocytosis of lipid emulsions by macrophages. Int. J. Pharm. *23*:69–77, 1985.
87. J. Senior, J. C. W. Crawley, and G. Gregoriadis, Tissue distribution of liposomes exhibiting long half-lives in the circulation after intravenous injection. Biochim. Biophys. Acta *839*:1–8, 1985.
88. R. L. Juliano and D. Stamp, The effect of particle size and charge on the clearance rates of liposomes and liposome encapsulated drugs. Biochem. Biophys. Res. Comm. *63*:651–658, 1975.
89. D. Papahadjopoulos, Fate of liposomes in vivo: a brief introductory review. J. Liposome Res. *6*:3–17, 1996.
90. K.-D. Lee, K. Hong, and D. Papahadjopoulos, Recognition of liposomes by cells: in vitro binding and endocytosis mediated by specific lipid headgroups and surface charge density. Biochim. Biophys. Acta *1103*:185–197, 1992.
91. T. M. Allen and A. Chonn, Large unilamellar liposomes with low uptake into the reticuloendothelial system. FEBS Lett. *223*:42–46, 1987.
92. A. Gabizon and D. Papahadjopoulos, Liposome formulations with prolonged circulation time in blood and enhanced uptake by tumors. Proc. Natl. Acad. Sci. USA *85*: 6949–6953, 1988.
93. T. M. Allen, C. Hansen, and J. Ruthledge, Liposomes with prolonged circulation times: factors affecting uptake by reticuloendothelial and other tissues. Biochim. Biophys. Acta *981*:27–35, 1989.
94. T. M. Allen, The use of glycolipids and hydrophilic polymers in avoiding rapid uptake of liposomes by the mononuclear phagocyte system. Adv. Drug Delivery Rev. *13*:285–309, 1994.

95. M. Malmsten, B. Lassen, J. Westin, C.-G. Gölander, R. Larsson, and U. R. Nilsson, Adsorption of complement protein C3 at polymer surfaces. J. Colloid Interface Sci. *179*:163–172, 1996.

96. G. J. Fleer, M. A. Cohen Stuart, J. M. H. M. Scheutjens, T. Cosgrove, and B. Vincent, *Polymers at interfaces*. Chapman & Hall, London, 1993.

97. J. N. Israelachvili, *Intermolecular and Surface Forces*, 2nd ed. Academic Press, London, 1992.

98. M. Malmsten, B. Bergenståhl, L. Nyberg, and G. Odham, Sphingomyelin from milk —characterization of liquid crystalline, liposome and emulsion properties. J. Am. Oil Chem. Soc. *71*:1021–1026, 1994.

99. V. P. Trochilin, V. R. Berdichevsky, A. A. Barsukov, and V. N. Smirnov, Coating liposomes with protein decreases their capture by macrophages. FEBS Lett. *111*:184–188, 1980.

100. T. M. Allen, C. Hansen, F. Martin, C. Redemann, and A. Yau-Young, Liposomes containing synthetic lipid derivatives of poly(ethylene glycol) show prolonged circulation half-lives in vivo. Biochim. Biophys. Acta *1066*:29–36, 1991.

101. A. L. Klibanov, K. Maruyama, V. P. Torchilin, and L. Huang, Amphipatic polyethyleneglycols effectively prolong the circulation time of liposomes. FEBS Lett. *268*: 235–237, 1990.

102. A. Mori, A. L. Klibanov, V. P. Torchilin, and L. Huang, Influence of the steric barrier activity of amphiphatic poly(ethylene glycol) and ganglioside GM1 on the circulation time of liposomes and on the target binding of immunoliposomes in vivo. FEBS Lett. *284*:263–266, 1991.

103. G. Blume and G. Cevc, Liposomes for the sustained drug release in vivo. Biochim. Biophys. Acta *1029*:91–97, 1990.

104. K. Hristova and D. Needham, The influence of polymer-grafted lipids on the physical properties of lipid bilayers: a theoretical study. J. Colloid Interface Sci. *168*:302–314, 1994.

105. V. P. Torchilin and V. S. Trubetskoi, Polymers on the surface of nanocarriers: modulation of carrier properties and biodistribution. Polym. Sci. *36*:1585–1598, 1994.

106. D. Kripotin, K. Hong, N. Mullah, D. Papahadjopoulos, and S. Zalipsky, Liposomes with detachable polymer coating: destabilization and fusion of dioleoylphosphatidylethanolamine vesicles triggered by cleavage of surface-grafted poly(ethylene glycol). FEBS Lett. *388*:115–118, 1996.

107. S. Stolnik, L. Illum, and S. S. Davis, Long circulating microparticulate drug carriers. Adv. Drug Del. Rev. *16*:195–214, 1995.

108. J. S. Tan, D. E. Butterfield, C. L. Voycheck, K. D. Caldwell, and J. T. Li, Surface modification of nanoparticles by PEO/PPO block copolymers to minimize interactions with blood components and prolong blood circulation in rats. Biomaterials *14*:823–833, 1993.

109. L. Illum, P. West, C. Washington, and S. S. Davis, The effect of stabilising agents on the organ distribution of lipid emulsions. Int. J. Pharm. *54*:41–49, 1989.

110. M. C. Woodle, M. S. Newman, and F. J. Martin, Liposome leakage and blood circulation: comparison of adsorbed block copolymers with covalent attachment of PEG. Int. J. Pharm. *88*:327–334, 1992.

111. C. J. H. Porter, S. M. Moghimi, M. C. Davies, S. S. Davis, and L. Illum, Differences in the molecular weight profile of poloxamer 407 affect its ability to redirect intravenously administered colloids to the bone marrow. Int. J. Pharm. *83*:273–276, 1992.

112. S. W. Kim and J. Feijen, Surface modification of polymers for improved blood compatibility. CRC Crit. Rev. Biocomp. *1*:229–260, 1986.

113. D. H. Napper, *Polymeric Stabilization of Colloidal Dispersions*. Academic Press, London, 1983.

114. S. I. Jeon, J. H. Lee, J. D. Andrade, and P. G. de Gennes, Protein–surface interaction in the presence of polyethylene oxide. 1. Simplified theory. J. Colloid Interface Sci. *142*:149–158, 1991.

115. S. I. Jeon and J. D. Andrade, Protein-surface interaction in the presence of polyethylene oxide. 2. Effect of protein size. J. Colloid Interface Sci. *142*:159–166, 1991.

116. I. Pezron, E. Pezron, P. M. Claesson, and M. Malmsten, Temperature-dependent forces between hydrophilic mica surfaces coated with ethyl (hydroxyethyl)cellulose. Langmuir 7:2248–2252, 1991.

117. N. L. Burns, J. M. Van Alstine, and J. M. Harris, Poly(ethylene glycol) grafted to quartz: analysis in terms of a site-dissociation model of electroosmotic fluid flow. Langmuir *11*:2768–2776, 1995.

118. M. Malmsten and J. M. Van Alstine, Adsorption of poly(ethylene glycol) amphiphiles to form coatings which inhibit protein adsorption. J. Colloid Interface Sci. *177*:502–512, 1996.

119. J. M. Harris (ed.), *Poly(ethylene glycol) Chemistry—Biotechnical and Biomedical Applications*. Plenum Press, New York, 1992.

120. E. Österberg, K. Bergström, K. Holmberg, J. A. Riggs, J. M. Van Alstine, T. P. Schuman, N. L. Burns, and J. M. Harris, Comparison of polysaccharide and poly(ethylene glycol) coatings for reduction of protein adsorption on polystyrene surfaces. Colloids Surf. A 77:159–169, 1993.

121. M. E. Norman, P. Williams, and L. Illum, Human serum albumin as a probe for surface conditioning (opsonization) of block copolymer-coated microspheres. Biomaterials *13*: 841–849, 1992.

122. K. Bergström, E. Österberg, K. Holmberg, A. S. Hoffman, T. P. Schuman, A. Kozlowski, and J. M. Harris, Effects of branching and molecular weight of surface-bound poly(ethylene oxide) on protein rejection. J. Biomater. Sci. Polym. Ed. 6:123–132, 1994.

123. K. L. Prime and G. M. Whitesides, Adsorption of proteins onto surfaces containing end-attached oligo(ethylene oxide): a model system using self-assembled monolayers. J. Am. Chem. Soc. *115*:10714–10721, 1993.

124. M. Malmsten and J. M. Van Alstine, Effects of chain density on the protein rejecting capacity of poly(ethylene glycol). J. Colloid Interface Sci., in press.

125. C. G. P. H. Schroen, M. A. Cohen Stuart, K. van der Voort Maarschalk, A. van der Padt, and K. van't Riet, Influence of preadsorbed block copolymers on protein adsorption. Surface properties, layer thickness, and surface coverage. Langmuir *11*:3068–3074, 1995.

126. M. Malmsten, P. M. Claesson, E. Pezron, and I. Pezron, Temperature-dependent forces between hydrophobic surfaces coated with ethyl(hydroxyethyl)–cellulose. Langmuir 6: 1572–1578, 1990.

127. F. Tiberg, C. Brink, M. Hellsten, and K. Holmberg, Immobilization of protein to surface-grafted PEO/PPO block copolymers. Colloid Polym. Sci. 270:1188–1193, 1992.

128. K. Bergström and K. Holmberg, Microemulsions as reaction media for immobilization of proteins to hydrophilized surfaces. Colloids Surf. 63:273–280, 1992.

27

Control of Protein Adsorption in Solid-Phase Diagnostics and Therapeutics

KRISTER HOLMBERG   Chalmers University of Technology, Göteborg, Sweden

GERARD QUASH   Laboratory of Immunochemistry, Faculty of Medicine Lyon-Sud, Oullins, France

I. INTRODUCTION

Proteins bind to a variety of molecules in vivo and the interaction, which is often highly specific, may initiate biologically important events. The interaction is often an interfacial process in which one of the interactive moieties is located at a cell surface. In several medical and biotechnological techniques the in vivo process is being transferred to an ex vivo or in vitro situation, utilizing the same type of specific interactions between proteins in solution and surface-bound molecules which may be antigens, antibodies, haptens, affinity ligands, etc. In all these techniques there is a potential problem with nonspecific adsorption of the soluble proteins. If nonspecific adsorption competes favorably with the specific interactions, a low accuracy, i.e., a low signal-to-noise ratio, in analytical procedures and a poor efficiency in therapeutic applications will be obtained.

The problem of competing nonspecific adsorption (sometimes referred to as passive adsorption) can be illustrated with an example from solid-phase immunoassay [1]. To detect 10 ng/mL or less of antigen-specific IgG in human serum, which is not an unusual analytical task, the procedure used must select one IgG molecule out of a million since serum contains approximately 10 mg/mL of IgG (assuming similar concentrations of specific and nonspecific binding sites). This selection must be based on a difference in idiotypic binding. Any nonspecific binding of serum IgGs to the surface-bound antigens, or adsorption to the underlying surface, with binding equilibrium constants of the order of 10^{-6} times those of the specific antigen–antibody binding may then also be detected. From this example it can be understood that methods to minimize all types of nonspecific interactions between the surface and the target molecule in the sample are desirable in order to reduce the background noise and, thus, bring down the percentage of false positive and false negative answers in the assays.

Also the adsorption of the analyte molecule, an antigen or an antibody, at the solid support can cause problems in solid-phase immunoassays. The standard procedure used in these assays is to coat a plastic plate with antibody or antigen by simply allowing the protein to adsorb at the surface, as will be described in detail in the following section. The adsorbed molecule will then capture the target molecule in the sample of body fluid by immunospecific binding. Two different problems may arise. First, adsorption of the analyte at the plastic surface may not be strong enough. The antigen–antibody complex may desorb, which may lead to false-negative answers in the test. This is a particular problem with hydrophilic analyte molecules, such as polysaccharide antigens, which do not adsorb strongly at hydrophobic surfaces.

The second problem related to adsorption of the analyte is that of too strong interaction between the bound protein and the surface. As is discussed in several other chapters of this book, protein adsorption is a dynamic event long after the initial bonding to the surface has taken place. The adsorbed protein, be it antibody or antigen, will gradually change its conformation, a process which often involves unfolding in order to maximize the interaction with the surface. Such a conformational change may eventually lead to loss of the biological activity, e.g., loss of antigen-binding capacity of an antibody. Such strong analyte–surface interaction is obviously unwanted since it leads to short shelf-life of the tests.

This chapter deals with protein adsorption related to solid-phase diagnostics and to therapeutics. The problems associated with too strong or too weak protein adsorption will be discussed in some detail, and strategies to minimize the problems will be presented. The principles discussed, although illustrated only with results from solid-phase immunoassays and therapeutic apheresis, are general and should be of relevance also to related areas which rely on the activity of surface-bound biomolecules. Examples of such other areas are biosensor technology, affinity chromatography, implants coated with biomolecules in order to promote tissue integration, and bioorganic synthesis using immobilized enzymes [2].

II. PRINCIPLES OF SOLID-PHASE DIAGNOSTICS

Solid-phase diagnostics uses a solid support to which an immunoactive species has been bound to detect the corresponding immunoreactive partner present in a mixture of other biological components (e.g., in serum, urine, milk etc.). The literature abounds with examples of antigens and antibodies adsorbed at insoluble supports such as polystyrene latex or nylon spheres, polyvinylchloride or polystyrene microtitre plates or tubes, nitrocellulose or nylon membranes, etc.

In one application of this procedure, an immobilized antigen (Ag1) on reacting with its corresponding antibody (Ab1) in a biological fluid results in the fixation of Ab1 to the support. After washing away all unreacted constituents, the amount of Ab1 present on the support is measured with the help of a second antibody (Ab2), specific for Ab1 and labeled with an enzymatic, colorimetric, or fluorescent marker. In some instances Ab2 can be replaced by a ligand with affinity for immunoglobulins such as Protein A, itself labeled with an enzymatic, colorimetric, or fluorescent marker. In other instances Ab2 is not tagged with a directly measurable marker but with biotin, whose presence can subsequently be revealed with labeled avidin or streptavidin. In all instances the determination of enzymatic activity, color, or fluorescence intensity provides a measure of the amount of Ab1 which has reacted with Ag1.

An alternative procedure is to adsorb an antibody (Ab1) directly at the support or immobilize it indirectly by allowing it to react with either a second antibody (Ab2), directed against Ab1, or with protein A, each of which had been previously adsorbed at the support. In this immunocapture method, the amount of antigen (Ag1) present in the biological fluid is measured by its ability to compete with pure Ag1 prelabeled with an enzymatic, colorimetric, or fluorescent marker (Ag1*). The more Ag1 present in the biological fluid, the smaller the amount of labeled Ag1* which will be retained by the immobilized Ab1. A comparison with a standard curve established with known amounts of unlabeled Ag1 permits the amount of Ag1 present in the biological fluid to be determined.

From the examples given above, it should be clear that the measurement of the amount of antibody or antigen in samples of biological fluids needs to be free of interference by other constituents present in the samples. Unfortunately this is seldom the case when an antibody or an antigen is adsorbed at an insoluble support. Indeed, sites which remain exposed after adsorption of the immunoactive species have to be "blocked" with proteins, e.g., bovine serum albumin or casein (usually in the form of skimmed milk), to prevent the nonspecific adsorption of other immunoglobulins or other antigens with no specificity for the immobilized immunoreactive partner. But, as mentioned earlier, protein adsorption is a dynamic process, which means that the addition of a blocking protein can bring about desorption of the Ag1 or Ab1 initially adsorbed. Such competitive adsorption, which has been verified experimentally [3], will severely decrease the accuracy of the assay.

The ultimate way of preventing desorption of the bound immunoreactive species is to immobilize the Ag1 or Ab1 to the support by covalent bonds. Microtiter plates with reactive surface groups are commercially available for this purpose. However, such an approach does not resolve the problem of the nonspecific adsorption of serum proteins due to the presence of residual hydrophobic sites on the support to which irrelevant proteins attach [4]. Such highly adsorbing sites should preferably be eliminated, e.g., by some kind of surface modification to transform the hydrophobic support into a hydrophilic one. Procedures for hydrophilizing microtiter plates and for covalent coupling of antibodies and antigens to such plates will be discussed in Section IV.

III. PRINCIPLES OF THERAPEUTIC APHERESIS

In therapeutic apheresis a body fluid, such as whole blood or plasma, is passed through some kind of device via an extracorporeal loop. A biologically active substance is immobilized to the inner surfaces of the device and exerts its action in immobilized state on components in the body fluid. Therapeutic apheresis is primarily intended for removal of harmful components in the blood, which may be antigens, circulating immune complexes, or other species [5–20]. Other applications of the procedure, such as stimulation of the immune system, have also been investigated [2,21].

The attractive feature of therapeutic apheresis is that injection of toxic substances to the patient can be avoided. The use of an immobilized ligand is particularly beneficial when the molecule has a very low half-life in vivo or is itself immunogenic. In this second context there exists a non-negligible risk of the patient developing antibodies to macromolecular nonhuman ligands as exemplified by the presence of human and anti-mouse antibodies in the serum of some patients receiving mouse monoclonal antibodies for diagnostic or therapeutic purposes [22]. For apheresis to be efficient and to present no risk to the patient, the ligands must be covalently bound to the insoluble support in order to prevent them from leaching out when in contact with the body fluid. Were this to occur, not only would the efficacy of the apheretic device decrease, but there would also arise the risk of formation of circulating immune complexes.

Therapeutic apheresis should preferably be performed on whole blood, not on plasma since plasmapheresis is costly and requires hospitalization under close clinical supervision. Treatment of whole blood without extensive clotting problems puts heavy demands on the design of the device. The device surface needs to be nonfouling, i.e., have a minimal tendency to adsorb plasma proteins. Such nonfouling surfaces can be obtained for instance by the hydrophilization technique discussed in Section IV.

Although therapeutic apheresis has been defined and studied experimentally for almost three decades [23,24], the method seems still not to be in clinical use on a routine basis. There are several reasons for the slow progress:

1. The first devices used had an insufficient biocompatibility. Device biocompatibility is mainly governed by the physicochemical properties of the surface exposed to the body fluid but is also dependent on the hydrodynamics of the fluid. Thus, the engineering of the extracorporeal device is of utmost importance.
2. There has been a lack of monospecific ligands, conforming to the human pharmacopoeia.
3. The methods of immobilization have been crude and have often caused considerable reduction in biological activity.

Research during the last decade has improved the situation. First, hydrophilic membranes well suited for the purpose are now available in the form of hollow fibers or flat surfaces. The hollow fiber devices have an extremely large inner surface, an important aspect for the efficiency of the process. The devices developed for the dialysis of patients with end-stage renal disease are of particular interest. Such cartridges, which are made from modified cellulose or synthetic polymers, have been designed to fulfill criteria such as hemocompatibility and innocuity, and they should have minimal complement-activating activity, all of which are relevant criteria also for a device for therapeutic apheresis. Second, a large number of ligands of high purity are now commercially available thanks to the development in genetic engineering and fermentation technology. Third, advances have been made in immobilization technology which provide a high degree of specificity in the coupling procedure and which give a considerable improvement of the biological activity of the bound biomolecule.

The macromolecules to be removed from a biological fluid may either be free in plasma or an integral part of the membrane of circulating cells. Relevant examples of published studies in the area are given below.

1. *Macromolecules free in plasma.* These may be
    a. *Antibodies (Ab),* such as
        - To DNA in the sera of patients suffering from systemic lupus erythematosus (SLE) [5,6]
        - To xenoantigens implicated in graft rejection [7,8]
        - To human leucocyte antigen (HLA) [9,10]
    b. *Antigens (Ag),* including
        - Tumour necrosis factor (TNFα), implicated in rheumatoid arthritis and in the clinical manifestations of other autoimmune diseases [11,12]
        - Cytokines, e.g., interleukin 6 (IL6), implicated in septic shock [13]
        - Endotoxins, implicated in perturbing the response of the inflammatory cascade [14–16]
    c. *Circulating Ag-Ab immune complexes (CICs),* e.g., DNA–anti-DNA complexes in the sera of SLE patients [17]
2. *Macromolecules as part of cell membranes.* The interaction of the Ag on circulating cells with the corresponding immobilized Ab should permit these cells to be selectively removed from whole blood. Examples include receptors present on the cell membrane of lymphocytes [18] and cell membrane antigens with specificity for cancer cells [19,20].

IV. SURFACE HYDROPHILIZATION AND IMMOBILIZATION

A. Protein Adsorption

Many proteins have a strong affinity for surfaces, and adsorption is governed both by the nature of the protein and by the character of the surface. The degree to which a specific protein adsorbs at a solid is governed by the nature of the surface. Even if proteins are a heterogeneous class of biomolecules with widely varying physicochemical characteristics, some general observations regarding the strength of interaction between proteins and surfaces can be made.

The majority of proteins adsorb in higher amounts on hydrophobic than on hydrophilic surfaces, as has been demonstrated by several groups. For instance, Malmsten et al. studied adsorption of a range of proteins on different surfaces including silica and methylated silica. Most proteins gave considerably higher adsorption at the methylated than at the nonmethylated silica, although examples of the opposite behavior were also found [25,26]. Surfaces with a hydrophobicity gradient have been used to investigate the effect of surface free energy on protein adsorption. For glass surfaces progressively hydrophobized with dichlorodimethylsilane, the adsorbed amount of fibrinogen increased from around 2 to around 5 mg/m^2 [27]. The strong tendency for adsorption at hydrophobic surfaces is due to hydrophobic interactions between the solid surface and hydrophobic domains in the protein. Such interactions can now be at least semiquantitatively measured by the surface force technique [28].

Electrostatic attraction forces are of importance for protein adsorption at charged surfaces. Since most proteins have a net negative charge at neutral pH, adsorption at positively charged surfaces can be expected to be particularly strong, as has indeed been found to be the case [29]. However, the relation between matching charges on the protein and on the surface, on the one hand, and adsorbed amount, on the other, is not straightforward. There are examples of proteins adsorbing strongly at a surface carrying the same net charge. Sometimes the adsorbed amount is even higher at a surface with the same charge than at an otherwise similar surface of opposite charge [30,31]. Evidently, the net charge of the protein is not always decisive. In order to fully understand the interactions involved in the adsorption process, one must be able to account for the individual interactions between charged groups on the protein and on the solid surface.

Most "real" surfaces contain both hydrophobic domains and charged or chargeable groups. (Also formally uncharged materials, such as polyolefins or polystyrene, invariably contain surface charges, as revealed by electrokinetic measurements. The net charge is usually negative and originates from spontaneous oxidation on exposure to air.) It seems clear that proteins normally use a combination of hydrophobic interactions and electrostatic attractions in the nonspecific binding. On most surfaces the hydrophobic interactions seem to be the dominating attractive force. Apart from hydrophobic interactions and electrostatic attractions, protein adsorption is influenced by factors such as degree of conformational freedom of the protein, ability of the surface to rearrange in order to expose different groups, etc. It is beyond the scope of this chapter to deal with these issues.

An efficient way to minimize both hydrophobic interactions and electrostatic attraction forces between a solid surface and a biopolymer in solution is to graft a dense layer of a hydrophilic, uncharged polymer to the surface. Poly(ethylene glycols) (PEGs) of molecular weights of 1500–6000 and polysaccharides such as dextran or cellulose derivatives have been found useful for the purpose. PEG grafting, in particular, has been the topic of many publications, with the first papers appearing in the early 1980s [32–35].

If such protein-resistant surfaces are to be used for solid-phase immunoassays, therapeutic apheresis, and similar applications, one can obviously not rely on simple adsorption of the analyte protein at the surface. The analyte, an antibody or an antigen, must be covalently coupled to reactive groups on the grafted polymer chains. Methods of grafting hydrophilic polymer chains, as well as techniques for covalent attachment of proteins to these polymers, will be discussed below.

B. Poly(Ethylene Glycol)–Grafted Surfaces

Following the pioneering work by the groups of Nagaoka, Merrill, Andrade, and others, grafting of poly(ethylene glycol) to surfaces has attracted much attention as a way to minimize biofouling [32–34]. It is by now a well-established fact that a dense and firmly attached PEG layer at the surface is a useful route for obtaining low protein adsorption and low cell adhesion characteristics. It has been shown that a PEG coating gives a marked suppression of plasma proteins and platelets at hydrophobic surfaces, leading to reduced risk of thrombus formation, as demonstrated both in vitro and in vivo [36,37]. (It should be noted, though, that a critique of the antithrombogenicity of PEG has been published [38].)

The grafted PEG layer is strongly hydrophilic, yet devoid of any charged groups. The polyoxyethylene moieties bind water strongly; thus the chains are highly hydrated. Van der Waals interactions between this strongly hydrated polymer layer and a protein approaching the surface are small [39]. Provided the grafted layer is thick enough, both electrostatic and hydrophobic attraction forces between the surface and proteins in solution should consequently be minimal. The PEG chains, when exposed to water, are highly mobile and oriented away from the surface. The conformational freedom of the grafted chains would be severely restricted by a close approach by a large molecule or a particle [40]. The resulting repulsive interaction between surface and dissolved molecule or particle can be seen as an example of steric stabilization and is analogous to stabilization of dispersed systems by poly-oxyethylene-based surfactants [41]. PEG modification of the surface can be made by adsorption of a surface active PEG derivative, by formation of a self-assembled monolayer that contains PEG, or by covalent coupling of a PEG with reactive end groups. The three approaches will be discussed in the following.

1. Adsorption of a Surface Active PEG Derivative

Block copolymers of ethylene oxide (EO) and propylene oxide (PO), often referred to as Pluronics, a BASF trade name, are the PEG derivatives most often used for surface modification by adsorption. PO is sometimes replaced by butylene oxide (BO) as hydrophobic monomer. The most common type of copolymer is the triblock EO-PO-EO. The central polyoxypropylene block is hydrophobic and has a strong driving force for adsorption at hydrophobic surfaces, leaving the polyoxyethylene segments oriented into the aqueous phase.

In the use of EO-PO block copolymers as hydrophilizing agents to prevent protein adsorption, a compromise will have to be made between driving force for adsorption and hydrophilicity of the modified surface. The smaller the EO-to-PO ratio in the copolymer, the stronger the driving force for adsorption and the better the packing at the interface. However, surface treatment with copolymers having a large polyoxypropylene block surrounded by smaller polyoxyethylene segments is likely to result in a surface which is not entirely hydrophilic since the polyoxypropylene portion may show through the relatively thin polyoxyethylene chain layer. Thus, proteins may be attracted by hydrophobic interactions to such a surface [42].

Ethylene oxide–propylene oxide block copolymers, like other oxyethylene-containing polymers and surfactants, show a reverse temperature-dependent phase behavior and a lower consolute temperature. The lower temperature phase boundary is often referred to as the cloud point (CP). The EO-to-PO ratio is the main governing factor of the CP; the higher the ratio, the higher the CP. Also, polymer molecular weight and cosolutes, such as electrolytes and alcohols, affect CP. By varying these parameters, phase separation temperatures (in water) ranging from less than 0 to more than 100°C can be achieved (see Section IV.C.1).

In principle, an EO-PO copolymer is hydrophilic and water soluble below, and hydrophobic and water insoluble above, the CP. When attached to a surface, the copolymer can be expected to impart protein repellency only when it is hydrophilic; i.e., below the CP. The following experiment with an ethoxylated cellulose ether containing hydrophobic side chains illustrates the behavior. The cellulose derivative was adsorbed from aqueous solution at hydrophobized mica surfaces [43]. By allowing two such modified surfaces to approach each other and measuring the forces that appear as the adsorbed polymer layers start to interact, one obtains information about the conformation of the adsorbed layers and the type of interaction, attractive or repulsive, that arises between the surfaces. It was found that as the temperature was raised toward the CP, the adsorbed polymer layers became more sense. The net force acting between the surfaces was repulsive, not only below the CP, but also some 5–10°C above the CP. Net repulsion also above the CP is an example of the so-called enhanced steric stabilization and has been experienced with several amphiphilic, water-soluble polymers [44]. Enhanced steric stabilization can be explained by an orientation of hydrophilic groups toward the solution, which lowers the local Flory–Huggins χ parameter in the interaction region. An additional contribution to the enhanced repulsive force is the high polymer concentration in the adsorbed layer at higher temperatures. This is likely to result in poor interpenetration of the polymer layers on compression and, consequently, in volume restriction [45].

Various types of EO-PO block copolymers have been used for preparation of nonfouling surfaces [42,46,47]. In general, the effect obtained in terms of hydrophilization is good, but the relatively small block copolymers are not always strongly adsorbed at the surface. On exposure to biological fluids they may be replaced by proteins with a high affinity for the surface. This is illustrated by an experiment in which a linear triblock polymer of EO-PO-EO type, as well as branched copolymers also with a central polyoxypropylene block, were used [42]. All copolymers gave efficient hydrophilization as judged from contact angle determination (complete wetting by water). However, none of the copolymers was very effective in preventing fibrinogen adsorption. As can be seen from Fig. 1, bars 2, 3, and 4, the effectiveness seemed to increase with increasing degree of branching. The relatively poor performance of the copolymers in this test is likely to be due to protein-induced desorption of the hydrophilizing agent. Tests with radiolabeled material showed that a substantial portion of the block copolymers desorbed from the surface during incubation with the dilute fibrinogen solution. As can be seen from Table 1, there is an inverse correlation between protein rejection efficiency and amount of radiolabeled copolymer in solution.

Other surface active PEG derivatives besides EO-PO block copolymers have been used to prepare protein-resistant surfaces by simple adsorption. Good results on rejection of a range of proteins have been obtained with fatty acid monoesters of a very long chain PEG (around 150 EO). However, this PEG derivative also seems to desorb in the presence of some proteins, viz., human serum albumin and the complement factor C3 [48].

2. Self-Assembled Monolayers that Contain PEG

Prime and Whitesides in a 1991 paper demonstrated that also very short PEG chains when packed in a self-assembled monolayer were efficient in preventing protein adsorption [49]. Using oligooxyethylene-terminated alkanethiols together with unfunctionalized alkanethiols and allowing self-assembly to occur on a noble metal surface, very ordered and tight packing of the oligooxyethylene chains were achieved. The nonfouling ability of the surfaces was monitored by FTIR and ESCA. Figure 2 shows the oligooxyethylene derivatives used in the experiments. The protein-rejecting efficiency varied between the derivatives and also differed depending on the choice of noble metal, gold or silver, used as substrate. Later investigations have led to the conclusion that the ability to prevent proteins to adsorb is related to the conformation of the short oligooxyethylene segment. The best correlation was obtained with compound B of Fig. 2. This oligooxyethylene derivative gave a com-

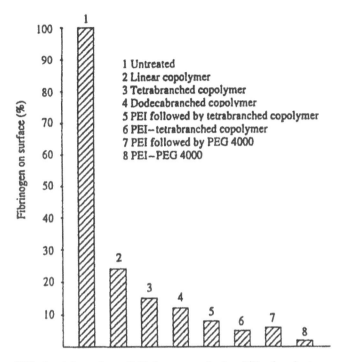

FIG. 1 Adsorption of fibrinogen at hydrophilized polystyrene surfaces. Surface treatment was done either by adsorption of linear or branched EO-PO block copolymers at unmodified polystyrene or by using PEI as anchoring polymer for PEG or EO-PO block copolymer. The latter procedure used oxidized polystyrene and was done either by the one-step or the two-step route illustrated in Fig. 4. (From Ref. 42.)

TABLE 1 Radiolabeled Copolymer in Solution After 30-min Incubation of Copolymer-Treated Polystyrene Plates with 0.1% Aqueous Solution of Human Fibrinogen

Hydrophilizing agent	Radioactivity (%)	Protein adsorption (%)
Adsorbed copolymer	9	15
PEI followed by copolymer	1	8
PEI-copolymer adduct	<1	5

The copolymer is a tetrabranched EO-PO block copolymer. The values are given as percent of total amount (surface bound + dissolved) of radiolabeled material. Also given are fibrinogen adsorption values as percent of adsorption at untreated polystyrene.
Source: Ref. 42.

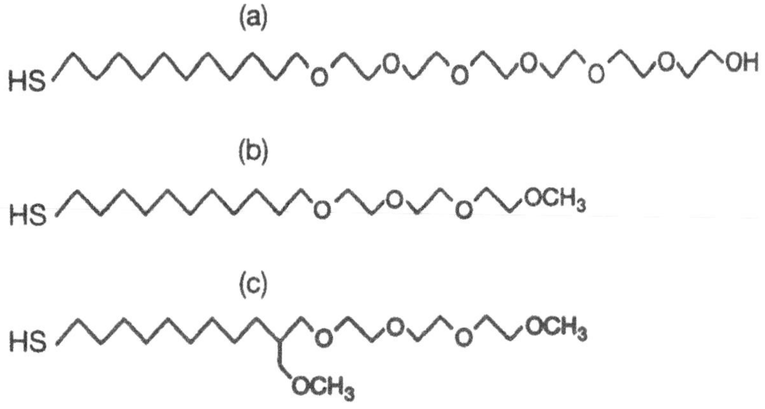

(a)

HS [chemical structure] OH

(b)

HS [chemical structure] OCH₃

(c)

HS [chemical structure] OCH₃
OCH₃

FIG. 2 Thiolated PEG derivatives used in Ref. 49 to form self-assembled monolayers.

pletely protein-resistant coating on gold but not on silver. The chains can pack in a helical or an all-trans structure, as shown in Fig. 3 [50]. The helical oligooxyethylene segment has a cross-sectional area that is larger than the unit cell dimension of unfunctionalized alkanethiols on silver but matches perfectly the unit cell dimension on gold (see Fig. 3). Thus, the helical structure can be accommodated on a planar gold surface but not on a planar silver surface. On the latter surface the molecule must adapt the all-trans conformation. The helical conformation is more hydrated, however, and thus more water repellent.

A serious limitation of the self-assembled monolayer method is that it seems to work well only on noble metal surfaces. An attempt to make protein-resistant surfaces by allowing an oligooxyethylene-terminated alkyltrichlorosilane to self-assemble onto silica showed moderate success. Considerably more protein adsorbed on these surfaces than on gold surfaces treated with the corresponding oligooxyethylene-terminated alkanethiols [51].

3. Attachment of PEG Chains by Covalent Bonds

The normal plastics used for solid-phase immunoassays and therapeutic applications do not contain reactive surface groups to which PEG can be covalently linked. Direct attachment of PEG chains to the polymeric surface requires some kind of activation step to create anchoring functional groups at the surface. Both wet chemical and dry methods (radiation techniques, plasma discharge, etc.) have been investigated for this purpose [52–54]. The strategy for grafting of PEG will then be either to have electrophilic groups at the surface and nucleophilic end groups on the PEG molecule, or vice versa. However, the result in terms of protein rejection is not very good for these two routes compared to the PEG-PEI route, which is discussed below [52]. Since the surface modification step is a rather complicated one with considerable difficulties in reproducibility, probably due to differences in surface characteristics between plate batches, the approach involving direct attachment of PEG chains to the plastic surface seems less attractive.

An alternative way of irreversibly attaching PEG chains to the plastic surface is to use a high-molecular-weight polymer which adsorbs strongly at the plastic

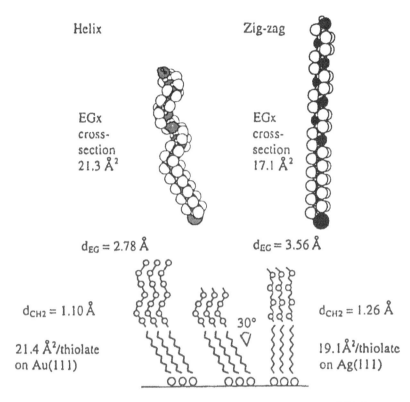

FIG. 3 Molecular cross-sectional areas per oxyethylene group (EG_x) for compound B of Fig. 2. Packing in a helical and in an all-trans conformation is shown. The unit cell dimensions for alkanethiols on gold and silver are also shown. (From Ref. 50.)

surface and which contains reactive groups for coupling to PEG or PEG derivatives. This approach gives several options with regard to choice of both anchoring polymer and PEG derivative. Much attention has been put on the use of branched high-molecular-weight poly(ethylene imine) (PEI) as anchoring polymer. This polymer contains both primary, secondary, and tertiary amino groups and PEG chains containing an electrophilic group, such as epoxide or succinimidyl carbonate, can be grafted in aqueous solution. The PEI adsorbs so strongly at negatively charged surfaces, even after PEG grafting, that protein-induced desorption seems not to be a problem. Most plastic surfaces carry a negative net charge from spontaneous surface oxidation in air. The concentration of negative surface charges can easily be increased, e.g., by oxidation with $KMnO_4/H_2SO_4$, which produces carboxylate, sulfate, and sulfonate groups [55]. In practice, PEG attached via PEI to such surfaces can be regarded as covalently bound to the surface [52].

Grafting of PEG to PEI may either be made prior to the adsorption step or after PEI has adsorbed. In the former route the graft copolymer, i.e., the PEG-PEI adduct, is the adsorbing species. The two routes are illustrated in Fig. 4.

Ellipsometry has been used to monitor the two surface modification routes. It was found that the grafted layer was much thicker when applied by the one-step route, i.e., using the preformed PEG-PEI adduct, than by the two-step procedure

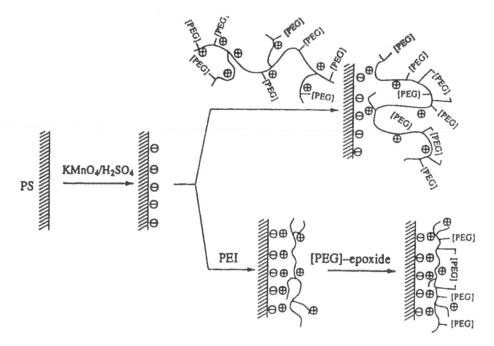

FIG. 4 Schematic of the PEG-grafted surface made either by adsorption of a PEI-PEG adduct or by consecutive surface treatments by PEI and PEG epoxide. Ellipsometry indicates an arrangement of PEI in loops on the surface in the former case and a flat adsorption of PEI in the latter case. (From Ref. 56.)

when PEI is first adsorbed and PEG grafting is made in a second step. Most likely, the one-step route gives a pronounced loops-and-trains conformation of the PEG-grafted PEI. (The schematic drawing of Fig. 4 is meant to illustrate the difference in conformation of the PEG-PEI adduct obtained by the two routes.)

The ellipsometry data are shown in Fig. 5 and Table 2 [56]. As can be seen from the table, the coating obtained from adsorption of PEI followed by reaction with PEG epoxide is rather thin, 4.9 nm, out of which PEI accounts for 4.0 nm. The small contribution by PEG to the layer thickness is an indication that the chains lie rather flat on the surface, which in turn implies that the PEG grafting is not very dense. (A fully extended PEG 4000 should be able to reach well over 20 nm.) The coating obtained from the preformed PEG-PEI adduct gives a very thick layer (50 nm), as measured by ellipsometry. A tentative explanation of this behavior is that adsorption of the PEI backbone of the PEG-PEI adduct in flat configuration on the polystyrene surface is prevented by unfavorable interaction between grafted PEG chains. Flat adsorption of PEI on the surface would require all PEG chains to be oriented away from the surface, toward the bulk water phase. Steric repulsion (steric stabilization) puts a limit to the packing of the PEG chains. The PEG-PEI copolymer can avoid unfavorable packing arrangement by forming loops above the surface. Hence, the large dimensions of the surface-bound PEG-PEI adduct obtained by the one-step route are likely to be due to two effects:

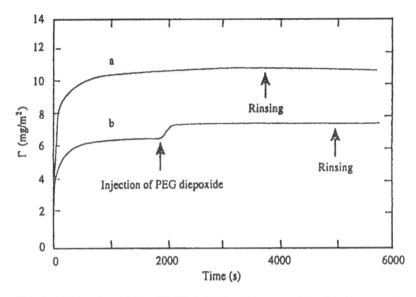

FIG. 5 Adsorption of the PEI-PEG 4000 adduct on oxidized polystyrene (curve a) and adsorption of PEI followed by treatment with PEG 4000 epoxide (curve b). (From Ref. 56.)

1. A dense packing of PEG chains, which leads to a high degree of chain elongation
2. A loops-and-trains arrangement of the PEI backbone on the surface

As shown in Table 1, adsorbing preformed PEG-PEI adduct to the surface is more efficient than the postgrafting route in preparing a fibrinogen-rejecting surface [42]. Similar results have been obtained with other proteins, such as IgG and albumin [56]. Table 1 also shows that there is practically no radioactive leakage from the coatings with PEI. Evidently, the PEG-PEI adduct desorbs to a negligible extent also

TABLE 2 Ellipsometry Measurements on a Series of Oxidized Polystyrene Samples

Sample	Γ^a (mg/m^2)	d (nm)
PEI on polystyrene	6.5	4.0
PEI followed by PEG 4000	7.5	4.9
PEI-PEG 4000	11	50
PEI-tetrabranched copolymer[b]	6.8	23

[a]Γ is total amount of adsorbed polymer and d is thickness of adsorbed layer.
[b]Tetrabranched copolymer stands for the same EO-PO copolymer as in Table 1.
Source: Ref. 56.

after incubation in a solution of fibrinogen, which is known to have a strong driving force for adsorption.

One may anticipate some kind of correlation between PEG grafting density and nonfouling properties of the surface. To date, conclusive evidence to support this hypothesis is lacking. No systematic study of correlation between PEG chain packing and protein repellency seems to have been performed. Ellipsometry can quantify the amount of surface-bound PEG. Provided the molecular weight of the PEG chains is known, a rough number of packing density, e.g., expressed as number of nm^2 per grafted PEG chain, can then be obtained. The two-step procedure, i.e., adsorption of PEI followed by PEG grafting, applied to an oxidized silicon wafer has given a value of 1.5 nm^2 per molecule for PEG 1900 [57]. Ellipsometry studies on a PEG 8000 derivative adsorbed at a hydrophobic surface have given values in the range of 10 nm^2 per chain [48].

Besides density of grafted PEG chains, the length of the chains, i.e., the PEG molecular weight, is a decisive parameter of protein repellency. Table 3 shows fibrinogen adsorption data related to two test series with a range of linear and tetra-branched PEGs of varying molecular weight [42]. Protein adsorption was measured by ELISA, which provides relative values rather than absolute amounts of protein on the surface. It has previously been shown that 100% fibrinogen adsorption under the exact conditions used for the experiments presented in Table 3 corresponds to adsorption of 100 ng/cm^2 [58].

As can be seen from the table, reduction in molecular weight of linear PEG below 1500 is accompanied by an increase in fibrinogen adsorption. This effect

TABLE 3 Effects of Linear (L) and Branched (B) PEGs Bound to Polystyrene on Fibrinogen Adsorption[a]

	Fibrinogen adsorption (%)	
PEG	Study 1 (Ref. 58)	Study 2 (Ref. 59)
Untreated	100 ± 11	100 ± 6
L-19,000	5 ± 1	—
L-4000	3 ± 2	5 ± 2
L-1500	5 ± 1	9 ± 1
L-1000	—	17 ± 3
L-500	—	52 ± 5
L-250	—	100 ± 1
B-6000	6 ± 5	—
B-1700	11 ± 3	14 ± 10
B-1000[b]	—	46 ± 17

[a]Surface grafting was made by the two-step route using PEI as anchoring polymer. All measurements are in triplicate. Error limits are standard deviations.
[b]Smaller B-PEGs could not be employed due to insolubility in water.

becomes especially pronounced below a molecular weight of 1000, with the 250-g/ mol material showing no protein rejection ability at all. The same general trend is found with the series of branched PEGs.

Several factors may contribute to the observed increase in fibrinogen adsorption for linear PEGs of low molecular weight. First, the small PEGs may be ineffective in hiding the exposed surface from approaching protein; i.e., the PEG layer produced by small PEGs may be so thin that the protein can sense the surface through the PEG layer. Alternatively, PEG packing density could be lower for the small PEGs so that the protein is able to find an exposed, unprotected portion of the surface. Evidence against this mechanism is that ESCA spectra for all the PEG-coated surfaces, regardless of the PEG molecular weight, are virtually identical [59]. In every case the ratio of —C—O— to —C—C— carbon in the C$1s$ peak is approximately 8 (within experimental error).

A second contributing factor derives from the theory that PEGs reject proteins because of the unfavorable entropy change that results upon compression of these conformationally random, heavily hydrated chains by protein adsorption [39,57,58,60]. Adsorption could either be by interpenetration to provide contact with the surface or it could be directly on top of the bound PEG layer. The events are visualized in Fig. 6. Although the chains are randomly oriented, there is considerable ordering of water around these chains, with 2–3 water molecules closely associated with each oxyethylene unit [61]. The negative entropy change upon PEG compression can be compensated by a positive entropy change associated with loss of water of hydration from the PEG chain, although there would be a large endothermic enthalpy change opposing this loss. Thus, if water is lost, protein adsorption is enthalpically unfavorable, and if water is not lost, protein adsorption is entropically unfavorable; in either case, the overall free energy change is positive and unfavorable [59]. This thermodynamic explanation of reduction in protein adsorption by the PEG coating can be related to the observed loss of effectiveness with low-molecular-weight PEGs. With the short chains the unfavorable entropy effect may not be large enough to overcome attractive forces involved in the adsorption process.

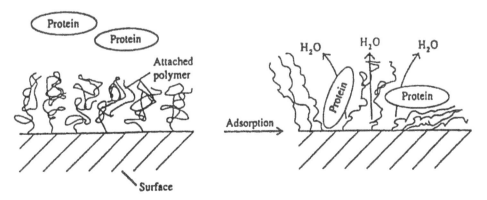

FIG. 6 Compression of PEG layer by protein adsorption. A negative entropy change accompanies compression of random, surface-bound PEG chains. (From Ref. 59.)

C. Polysaccharide-Grafted Surfaces

Polysaccharides, being hydrophilic and uncharged polymers, constitute an alternative to PEG as protein-rejecting polymer attached to a surface. The effect exerted by polysaccharides is well known. Both dextran, a polyglucose containing only 1,6-glucosidic linkages, and a nonionic cellulose ether, ethylhydroxyethylcellulose (EHEC), are commercially used to render surfaces protein resistant. In a comparative study it was found that dextran, when attached side-on to a surface, was equally effective as PEG in reducing fibrinogen adsorption at a hydrophobic surface [62]. The side-on attachment of dextran was made by first oxidizing the polysaccharide with periodate to generate aldehyde groups along the chain and subsequently coupling the oxidized dextran to a PEI-treated surface by reductive amination using sodium cyanoborohydride as reducing agent. End-on attachment of dextran was performed by direct coupling of the polysaccharide via its reducing end to the PEI-treated surface, again using sodium cyanoborohydride as reducing agent. The end-on treatment gave much inferior results in terms of protein repellency. This is interesting because the side-on attachment gives a thinner layer of hydrophilic polymer but better coverage of the surface, as compared with the end-on attachment. Evidently, similar to the PEG treatments discussed above, a good surface coverage is a prerequisite of a good protein rejection capability. A polysaccharide has also been successfully immobilized to a hydrophobic surface via poly(ethylene imine) [63]. The polysaccharide, EHEC, was first exposed to mild periodate oxidation to generate aldehyde groups and then coupled to the PEI by reductive amination. The polysaccharide–PEI adduct was subsequently adsorbed at the hydrophobic surface. The treatment was almost as efficient in terms of protein rejection as PEG treatment via the corresponding PEG-PEI adduct. In more recent work it was found that polysaccharide chains attached by radiofrequency glow discharge did not give as good protein rejection as the corresponding treatment with PEG as hydrophilic plasma grafted polymer. It was claimed that the relative inefficiency of the polysaccharide coating was due to inadequate surface coverage [64,65].

In an attempt to use the self-assembly approach based on the strong thiol–gold interaction, dextran was thiolated and the thiol-functional polymer was subsequently brought in contact with a gold surface [66]. The modified surfaces were found to be very protein resistant. The procedure for the thiol modification is shown in Fig. 7, which also depicts the arrangement of the polysaccharide on the surface. It was found that the thiolated polysaccharide readily self-assembled on silver and gold. The surface coverage, which is crucial to protein rejection, increased with the concentration of thiol groups. Increased dextran molecular weight gave a reduction in surface coverage, probably due to the larger molecules having reduced conformational freedom, therefore allowing relatively less thiol groups to be available for binding to the noble metal surface. The layer thickness seemed not be important for protein rejecting efficiency, an observation in line with previous findings with dextran immobilized to surfaces, as discussed above in relation to Ref. 62.

D. Immobilization of Proteins to Poly(Ethylene Glycol)–Grafted Surfaces

It seems natural that normal methods of immobilization of proteins to solid surfaces meet difficulties with surfaces grafted with uncharged, hydrophilic polymers, such

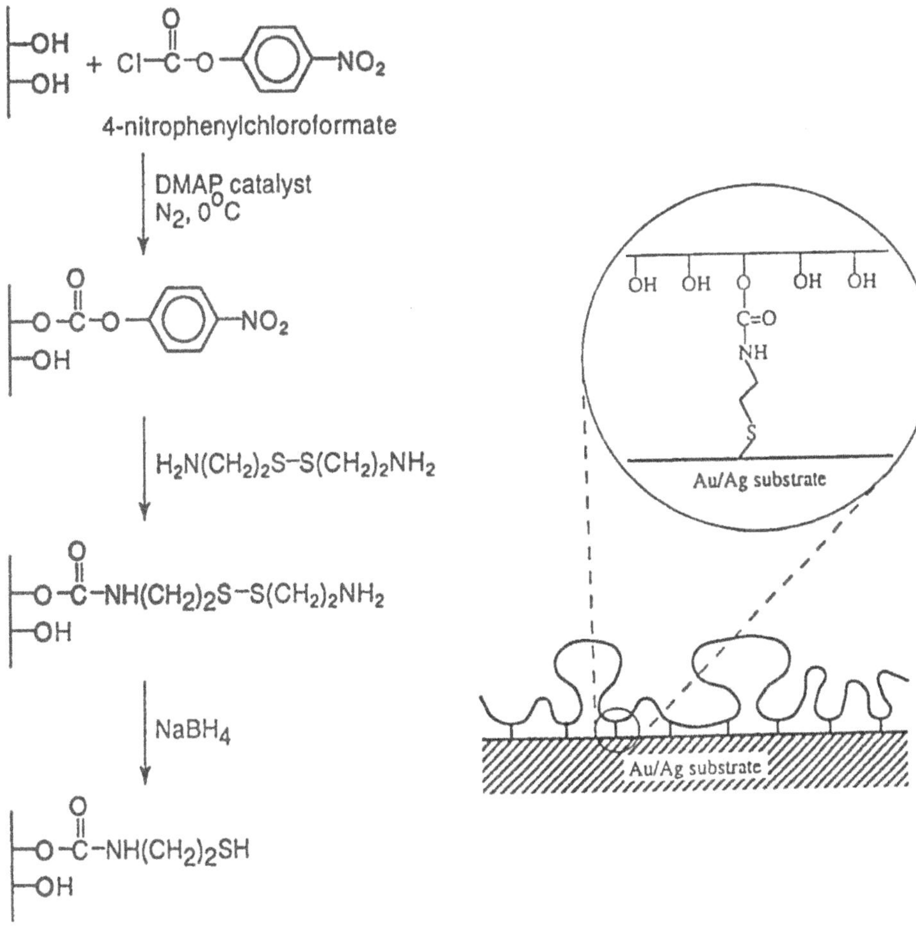

FIG. 7 Left: Reaction scheme showing the synthesis of thiol-functional dextran. Right: Arrangement of the thiol-functional dextran on a surface. (From Ref. 65.)

as PEG or polysaccharides. Such surfaces are deliberately made protein resistant; thus, the immobilization yield tends to be very low, even when highly reactive functional groups are attached to the polymer chains. The main approach taken to perform covalent coupling at a reasonable yield is to perform the immobilization in a reaction medium of poor solvency for the grafted polymer chains. On worsening the solvency, the free energy involved in separating polymer chains from solution decreases, which means that the interaction exerted by the polymer layer becomes less repulsive. At about theta conditions for the polymer, the repulsion is switched into an attractive interaction [66,67].

Two immobilization routes, both using the concept of a reaction medium of poor solvency for grafted polymer chains, are described below. Since the majority of papers dealing with the topic relate to PEG-grafted surfaces, the methods will be illustrated with this polymer. The second route—use of microemulsions as reaction medium—can be applied to polysaccharide-grafted surfaces as well [4]. Another method used for immobilization to polysaccharide-grafted surfaces, and in particular to surfaces grafted with dextran, is to first introduce carboxyl groups into the polymer, e.g., by reaction with monochloroacetate, and then immobilize at a pH of around 6, at which the carboxyl groups are dissociated and many proteins have a net positive charge, giving an attractive electrostatic interaction between the surface-bound polysaccharide and the approaching protein. Introduction of charges into the grafted polymer layer tends to make the surface less protein resistant, however, so this method is not ideal for prevention of nonspecific adsorption [63].

1. Use of the Cloud Point Concept

The reverse solubility–temperature relationship of PEG and PEG derivatives mentioned above can be taken advantage of in the immobilization of proteins, provided that one end of the polymer chain is attached to the surface and the other contains a free reactive group. The cloud point of an EO-PO block copolymer varies with the EO-to-PO ratio, as is shown in Table 4, and is also influenced by cosolutes such as electrolytes, alcohols, etc. Typically an EO-PO block copolymer with a CP in the immobilization medium around 40°C should be used. A surface grafted with such a copolymer is hydrophilic at room temperature, i.e., around 20°C below the cloud point. At the cloud point, where the immobilization is carried out, the polymer layer is relatively hydrophobic and the energy barrier for a protein to approach reactive terminal groups on the copolymer chains is small. After the reaction is completed the temperature is reduced and adsorbed proteins will desorb since the grafted polymer chains have again become hydrophilic. Presumably, covalently bound protein molecules will strive away from the surface. That the grafted layer contracts above the cloud point and expands below it can be shown by ellipsometry [68].

The procedure gives a high loading of protein on the surface. The drawback of the method, however, is that at 20°C below the cloud point the copolymer layer is not entirely hydrophilic. It has been shown that the protein repellancy of a surface grafted with such EO-PO block copolymers is not as good as that of a surface grafted with normal PEG [2].

2. Use of Microemulsions as Reaction Medium

A microemulsion is defined as a system of oil, water, and surfactant which is a single, optically isotropic and thermodynamically stable solution [69]. Microemul-

TABLE 4   Cloud Points of 1% Aqueous
Solutions of a Series of EO-PO Block
Copolymers with Varying Compositions

% EO in copolymer	Cloud point (°C)
10	23
20	33
40	40
80	>100

sions based on surfactant, aliphatic hydrocarbon, and a small amount of water have attracted interest as medium for enzymatic reactions [70]. The enzyme resides in the water droplets surrounded by a surfactant monolayer. Such microemulsions are also useful as medium for immobilization of proteins to PEG-grafted surfaces. The microemulsion is a nonsolvent (worse than theta solvent) for the PEG chains; thus the grafted polymer chains will be compressed and the surface will not be protein repelling. After covalent coupling of the protein to reactive groups at the PEG chains, the surface is washed with water whereby the PEG layer regains its hydrated and protein-rejecting character [2,71]. The majority of antibodies and antigens seem not to lose their biological activity after exposure to such microemulsions for the time required for immobilization (typically 1 h at 40°C). The principle of the microemulsion route of protein immobilization is shown in Fig. 8.

The concept of using a microemulsion as medium for protein immobilization is attractive since, unlike the cloud point route described above, the method works with normal PEG chains on the surface. Thus, the protein will become coupled to a surface layer which has maximum hydrophilicity and protein repellancy. Transformation of the PEG layer into a protein-resistant surface when contacted with water after exposure to the microemulsion is not instantaneous, however. This is probably due to a slow reorientation of the PEG chains from a compact entangled structure to an extended conformation. After soaking in water overnight complete hydration seems to have occurred, as judged from protein adsorption studies [2].

An interesting aspect of immobilization in a microemulsion of low water content is that the coupling pattern may be different from that in water. Since the microemulsion is a medium of low polarity, proteins will tend to adjust their conformation so as to expose the more hydrophobic amino acids and bury the hydrophilic residues in the inner regions, i.e., adopt a conformation opposite to that in water. This is likely to favor immobilization via the more hydrophobic amino acid residues. Furthermore, the relative reactivity between nucleophilic groups varies between water and media of low polarity. Water favors reaction with uncharged nucleophiles, such as amino groups, since water can stabilize the dipolar transition state of such reactions. Reactions with charged nucleophiles, such as deprotonated thiol and phenolic hydroxyl groups, on the other hand, are favored by unpolar solvents [72]. Taken together, microemulsions should, relative to water, favor coupling to Cys and Tyr and disfavor coupling to Lys [2]. This has been found to be the case in model experiments [73].

E. Site-Specific Immobilization

In the majority of works related to protein immobilization to grafted PEG chains, epoxides have constituted the reactive end group of the PEG. The oxirane ring is

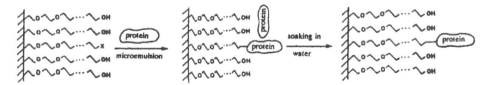

FIG. 8 Immobilization from a microemulsion. (From Ref. 2.)

readily opened by nucleophilic groups such as Lys–amino groups and Cys–thiol groups. These nucleophilic amino acid residues are randomly distributed along the polypeptide chain and the immobilization may be referred to as random coupling. Since reaction occurs without regiospecificity, there is the danger that lysines or cysteines implicated in the biological activity of the protein may be perturbed or that coupling via these amino acid side chains may provoke a change in conformation with repercussion on the active site.

Many proteins are in fact glycoproteins, i.e., they contain carbohydrate side chains. In glycoproteins, the carbohydrate moieties are linked to the polypeptide backbone via O glycosidic bonds formed with OH groups of serine or threonine and/ or via N glycosidic bonds formed with the β-carboxamide groups of asparagine. In IgG antibodies the carbohydrate segments are mainly located in the $C_{\gamma 2}$ region in the F_c part. This means that coupling such an antibody via its sugar residues is a way to minimize interaction with the antigen-binding F_{ab} part of the molecule.

Further, the majority of secreted glycoproteins have sialic acid as their terminal sugar preceded by galactose. These two sugars are readily oxidized by periodate under mild conditions with the formation of aldehyde groups, which can be used to anchor the glycoprotein to an insoluble support bearing an amine functionality at the end of a side arm. The Schiff base so formed is unstable to hydrolysis and needs to be reduced to a NH—CH$_2$ bond. Sodium cyanoborohydride is the reducing agent of choice for this.

Treatment with cyanoborohydride may also cause a reduction of disulfide bonds into thiols. For proteins such as immunoglobulins, in which disulfide bonds contribute to the secondary and tertiary structure, this reduction may bring about inactivation. A way to eliminate this problem is to make use of a hydrazide as the amine function on the side arms of the support [74,75]. The Schiff base formed from reaction between an aldehyde and a hydrazide is stable to hydrolysis; hence no reduction is needed to form a stable linkage. Hydrazide groups can easily be obtained from epoxide groups by reaction with an excess of a dihydrazide, e.g., adipic dihydrazide.

The procedure described above is an example of a site-specific immobilization of antibodies. It has been verified experimentally that an antibody immobilized in this way retains its activity better than an antibody bound by random coupling [4].

V. SOLID-PHASE DIAGNOSTICS ON HYDROPHILIZED PLATES

Before developing a diagnostic assay on hydrophilized plates the first question which must be asked is whether the hydrophilization procedure has in fact eliminated the problem of nonspecific adsorption. This can be addressed using fibrinogen, which is notorious for its propensity to adhere to hydrophobic surfaces, or a serum which has undergone several cycles of freezing and thawing. In this latter case some serum proteins, e.g., immunoglobulins, are known to adsorb nonspecifically to the support. Therefore a measure of adhering immunoglobulins during freeze-and-thaw cycles provides a good index of the efficiency of hydrophilization. Such a procedure has been described in the literature for microtiter plates hydrophilized with a nonionic cellulose ether [4]. Hydrophilized and nonhydrophilized plates were put into contact with a rabbit serum which had been frozen and thawed. Whereas the hydrophilized plate remained virtually free of immunoglobulins, the untreated plate adsorbed rabbit IgG in an amount that increased with the number of freeze and thaw cycles [4]. Similar results have been obtained on PEG-modified microtiter plates [76].

As mentioned in Section IV, proteins can be immobilized either by random coupling, normally by allowing nucleophilic groups of the polypeptide to react with electrophilic groups at the support surface, or by some kind of site-specific coupling. In Section IV.E a site-specific coupling procedure involving the carbohydrate side chains of glycoproteins was described [74,75]. This immobilization route is particularly suitable for glycosylated antibodies since the bonds formed will not involve the F_{ab} region, which is the antigen-recognizing part of the molecule.

A comparative study performed to determine the advantages of covalent coupling over adsorption using enzyme-linked immunoassay together with ellipsometry and total internal reflection fluorescence to analyze the amount and type of protein at the surface has been performed [3]. The study confirmed that hydrophilized plates containing reactive epoxide groups give very low nonspecific adsorption. It was also demonstrated that site-specific coupling, although resulting in less antibody bound per unit surface than in random coupling, results in a bound antibody with greater reactivity than that observed on random coupling. The immobilization routes used for random and site-specific coupling are shown in Fig. 9.

Hydrophilized and functionalized plates have also been used for measurement of levels of circulating immune complexes in the sera of viraemic patients. The detection was based on a dissociation and reassociation procedure in the presence of

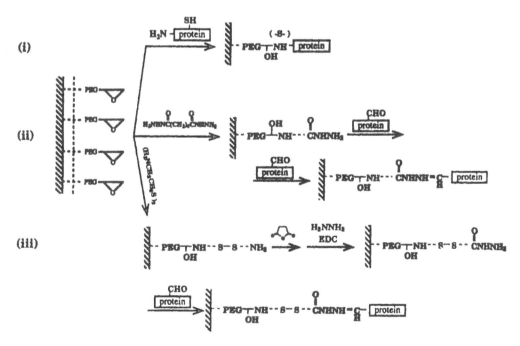

FIG. 9 Schematic of different routes of antibody immobilization to a surface grafted with PEG epoxide. Route i shows random coupling; routes ii and iii illustrate site-specific coupling. (From Ref. 3.)

the covalently bound viral antigen HTLV1. Dissociation was brought about by treatment of the immune complex in acid, and subsequent exposure to neutral pH caused the complex to reform [4].

The covalent coupling of the HTLV1 antigen permitted the dissociation and reassociation to be carried out under the conditions of high ionic strength and low pH (pH 2.25) needed to obtain high accuracy in the test. When a conventional procedure was used involving physical adsorption of the antigen, these conditions resulted in desorption of the antigen from the polystyrene support [4].

VI. NONSPECIFIC THERAPEUTIC APHERESIS: REMOVAL OF LOW-MOLECULAR-WEIGHT PROTEINS

As mentioned in Section III the cartridges developed for dialysis of patients with renal disease are well suited as devices for therapeutic apheresis. Of all the devices which exist, those made of an acrylonitrile/methallyl sulfonate copolymer (AN69XT) manufactured by Hospal, Gambro Renal Products were chosen for hydrophilization and functionalization as they already possess SO_3^- groups which obviate the need of performing the oxidation step to generate negative charges (see Section IV.B.3). It should be recalled that in dialysis low-molecular-weight metabolites (molecular weight below 1000) are removed by filtration, whereas in therapeutic apheresis the toxins to be eliminated have a molecular weight above 1000 and the molecular weight variation can be large. One example of a low-molecular-weight toxic protein is $\beta2$ microglobulin ($\beta2$-M) of molecular weight 11,800 and an isoelectric point of 5.7. Its level in plasma is correlated with the onset of β amyloid disease, and $\beta2$-M itself has been proven to be a major constituent of the amyloid deposits developing in patients undergoing hemodialysis [77–79].

Other examples of low-molecular-weight molecules whose levels increase in inflammatory states such as rheumatoid arthritis are the proinflammatory cytokines IL1 (molecular weight of 17,000), TNFα (molecular weight of 17,500), and IL6 (molecular weight of 26,000), the latter being responsible for the acute phase response [80]. This acute phase response is also one of the iatrogenic consequences of passing blood over nonbiocompatible membranes [81]. It leads to an increase in the synthesis and plasma levels of C-reactive protein serum anyloid A, $\alpha1$-proteinase inhibitor, $\alpha1$-acid glycoprotein, $\alpha1$-antichymotrypsin, fibrinogen, haptoglobulin, $\alpha2$-antiplasmin, C1-inactivator, and C2 of complement [80].

In view of this wide diversity of acute phase proteins, the removal of which would require the development of a specific cartridge for each protein, existing dialysis membranes have been first modified to try to reduce the initial protein–surface interactions which trigger off the cute phase response. Such an approach has been used with success to reduce the activation of kininogens by coating the inner surface of synthetic membranes made of AN69XT with poly(ethylene imine) [82]. Unequivocal evidence for the stability of PEI binding was obtained by the finding that <5% tritium-labeled PEI was leached from the cartridge through which was passed human plasma containing 333 U heparin/L [83]. However, kininogen activation is not the only deleterious effect of passing blood over insoluble supports. An increase in the circulating levels of acidic $\beta2$-M and basic anaphylatoxins C3a and C5a has also been described [81]. Accordingly, the removal of low-molecular-weight proteins from bovine plasma was investigated on different dialysis membranes using tritium-labeled β2-M (isoelectric point of 5.7) directly involved in β-amyloidosis and tritium-labeled lysozyme (isoelectric point of 10.8) as a model for the basic proteins C3a and C5a.

A. Removal by Filtration

It was found that tritium-labeled β2 microglobulin (β2-M) was removed by filtration with over 75% efficiency on hydroporous membranes such as AN69XT (acrylonitrile/ methallyl sulfonate copolymer), AN69ST surface-coated with PEI, and PS (poly-sulfone), whereas 95% was removed by adsorption on PMMA-BK (polymethyl-methacrylate-BK) [85]. β2-M was not removed by cuprophan, which has a cut-off limit of 1 kDa. It should be noted, though, that the adsorption of β2-M to PMMA-BK was *not* selective since this membrane adsorbed 35% of the proteins present in plasma. With AN69XT, AN69ST, and PS, protein adsorption was considerably reduced, amounting to less than 5%. Hence, for preventing amyloidosis due to β2-M accumulation, hydroporous membranes with ≈30 kDa cutoff are efficient and relatively selective.

B. Removal by Ionic Interaction

As regards tritium-labeled lysozyme it was filtered through PS but adsorbed strongly to AN69XT and AN69-ST in spite of the neutralization of the negative surface charges by PEI. In fact, the adsorption to the two types of AN69 membranes takes place with similar kinetics and is faster than the filtration kinetics through PS. This could be due to the fact that the AN69 PEI membrane has retained some of its negatively charged sulfonate groups of the copolymer since the PEI used for coating had a molecular weight above 100,000 which precludes it from penetrating into the pores. Hence a molecule like lysozyme can be expected to diffuse in a similar way into the pores of AN69XT, AN69ST, and PS membranes, but in the pores of the two AN69 membranes there is an additional interaction with sulfonate groups at the membrane surface. Observations similar to that made with lysozyme have been made for cytochrome C with an isoelectric point of 10.6 [85] and for the basic anaphylatoxins C3 and C5a [86]. These results suggest that nonspecific removal of low-molecular-weight proteins with acidic or basic isoelectric points can be achieved by filtration on neutral hydroporous membranes and by ionic interaction on negatively charged hydroporous membranes. Such cartridges are presently undergoing clinical trials.

VII. SPECIFIC THERAPEUTIC APHERESIS: REMOVAL OF HIGH-MOLECULAR-WEIGHT PROTEINS ON HYDROPHILIZED CARTRIDGES

For the removal of high-molecular-weight components (molecular weights above 50,000) implicated in a pathological condition such as autoantibodies IgG (MW 150 kDa) it is evident that filtration on hydromacroporous membranes could bring about more problems than it solves due to the concomitant loss of other essential molecules with similar or lower molecular weights such as albumin, etc.

To achieve selective antibody removal, a procedure has been developed based on the affinity of the antibody for its corresponding antigen. For the removal of antigens, the same principle can be applied, but the immobilized partner is in this case the corresponding antibody. In vitro examples of both procedures are given below.

A. Removal of Immunoglobulins According to Their Species Specificity

An AN69 XT cartridge in the form of hollow fibers was coated with partially succinylated PEI of molecular weight above 100,000 so as to have sufficient COOH groups available for protein coupling and enough NH_2 groups remaining for interaction with SO_3^- groups on the fibers [83]. After activation of the COOH groups with N-hydroxysuccinimide in the presence of a water-soluble carbodiimide a solution of rabbit IgG was circulated through the cartridge for 1 h in a closed circuit. Random amide coupling was adopted for this procedure because the cartridge was destined for removal of anti rabbit antibodies, which are known to react with both the F_{ab} and the F_c portions of IgG. Hence, coupling of the rabbit IgG by its carbohydrate residues as in site-specific immobilization (see Section IV.E) was purposely not used in order to preserve the antigenicity of the carbohydrate chains. In control cartridges the N-hydroxysuccinimide-activated COOH groups were reacted with ethanolamine in order to terminate with a hydrophilic OH group. Through each cartridge was passed 500 mL of a sheep serum spiked with 1 mg of goat IgG anti rabbit IgG labeled with fluorescein (IgG-FITC). After extensive washing and elution at pH 2.3 it was found that 300 μg of IgG-FITC were retained on rabbit IgG cartridges compared to 60 μg on control cartridges, corresponding to a fivefold increase.

The example given above provides evidence for the removal of antibody based on species specificity. This is *not* encountered in clinical practice but could have wide-ranging application in the biotechnology industry if the immobilized partner were IgG specific for a particular species. One example of this is the on-line isolation and purification of mouse or humanized monoclonal antibodies present in reactors containing culture medium with calf serum.

B. Removal of Antibodies of a Defined Specificity

Dialysis membranes in the form of hollow-fiber ready-to-use cartridges were exposed to an aqueous solution of a copolymer of poly(ethylene imine) and poly(ethylene glycol) having oxirane rings at the free ends of the grafted PEG chains [87]. As discussed in Section IV.B, the PEI backbone binds irreversibly to a slightly negatively charged plastic surface and the PEG chains, which carry the anchoring groups, are exposed to the aqueous phase. The oxirane rings were opened at pH 8.5 to generate a diol to which lipoic acid was attached by esterification, thus leaving the dithiolane ring intact. A diol-functionalized cartridge without lipoic acid and a nonhydrophilized cartridge, also without lipoic acid, were used as controls.

Through each of the three cartridges were circulated 200 mL plasma taken from a patient with acute final-stage systemic lupus erythematosus for 16 h at room temperature. After passage of the plasma, the cartridges were washed extensively until the protein content of the washing fluid was negligible. Bound proteins were then eluted in the following amounts:

    100 mg from the hydrophilized cartridge with lipoic acid
    20 mg from the hydrophilized reference cartridge
    80 mg from the nonhydrophilized reference cartridge

More importantly, the PAGE pattern of eluted proteins showed after Comassie blue-staining the presence of at least nine different proteins in the eluate from the non-hydrophilized cartridge, whereas from the two hydrophilized cartridges there were only three proteins, corresponding to IgG heavy chain (γ), IgG light chain (L), and IgM heavy chain (μ) (Fig. 10). This provides direct evidence that hydrophilization reduces nonspecific adsorption. Further, the intensity of staining of the μ and L chains was greater in the eluate derived from the hydrophilized cartridge with bound lipoic acid than in the eluate from the hydrophilized cartridge without lipoic acid. This is in accordance with the quantitative results given above, i.e., that there is a fivefold increase in the amount of antibody bound to the hydrophilized cartridge with lipoic acid compared to the control without lipoic acid. These results imply, though, that the hydrophilized cartridge without lipoic acid is not completely inert since a small amount of adsorption still takes place. In order to obtain further evidence for the retention of specific antibodies by the bound lipoic acid, the plasma flow-through fraction from this cartridge was passed through another cartridge with bound lipoic acid. No decrease in the total IgG titer of the flow-through fraction was obtained, indicating that no additional binding occurred.

As the antibodies eluted from the cartridge with bound lipoic acid reacted strongly in an immunoassay with DNA, these results provide one line of evidence

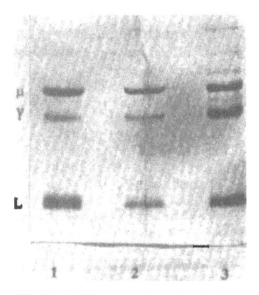

FIG. 10 PAGE pattern of proteins eluted from cartridges after passage of 200 mL of human plasma from a SLE patient. (1) Hydrophilized cartridge with bound lipoic acid, (2) hydrophilized cartridge without lipoic acid, (3) nonhydrophilized cartridge without lipoic acid.

TABLE 5 Elimination of Interleukin 6 (IL6) from Serum After Passage Through Cartridge Loaded with Anti-IL6 and Control Cartridge Loaded with Irrelevant Antibody

	Before passage	After passage through test cartridge	After passage through control cartridge
Concentration (pg/mL)	11,200	2512	9620
% retained	0	78	14

Source: Ref. 83.

that an apheretic device with bound lipoic acid can be used to remove circulating antibodies to DNA from plasma.

C. Removal of Antigens

An AN69 minidialyser made up of 170 hollow fibers with an internal surface area of 200 cm^2 was used for antigen removal [83]. The cartridge was first hydrophilized as described above. The oxirane rings were then opened with adipic dihydrazide to give hydrazide groups at the free ends of the grafted PEG chains. Rabbit IgG anti-IL6 was subjected to mild periodate oxidation and then coupled via the aldehyde groups generated in the carbohydrate side chains to the hydrazide groups of the grafted surface layer. A control with an irrelevant antibody was prepared in the same way. For actual coupling both IgGs at 1 μg/mL were circulated through the cartridge for 1 h. Using tritium-labeled IgG it was found after exhaustive washing that 11% of the circulated IgG was bound. This is an example of the site-specific immobilization discussed in the Section IV E.

The results obtained when 3 mL of serum from a septicaemic patient (11.2 ng IL6/mL) were passed over the cartridge are given in Table 5. Comparing the amount of IL6 eliminated by the two devices, it is evident that the cartridge loaded with anti-IL6 removes this cytokine with a high degree of specificity.

To try to increase the efficiency of IgG binding on hydrophilized and functionalized surfaces work is presently going on to modify the site-specific coupling procedure so as to have a polyvalent ready-to-use cartridge whose antigen binding capacity can be easily adapted by the end user according to the protein to be removed. The results of this work will be published separately [87].

VIII. CONCLUSIONS

Both solid-phase diagnostics and therapeutic apheresis require high specificity between an immunoactive species bound to a solid support and its immunoreactive partner present in the sample fluid. In both techniques it is essential that the immobilization procedure does not lead to a substantial decrease in biological activity and that nonspecific adsorption of other species in the sample is minimized. An effective way of keeping nonspecific adsorption of proteins and other biomolecules at a low level is to graft a dense layer of a hydrophilic, uncharged polymer, such as poly(ethylene glycol), to the solid surface. The grafted layer must contain reactive groups for covalent coupling of antibodies and antigens. Methods to perform immobilization to such hydrophilized surfaces are discussed, and the advantage of coupling antibodies in a site-specific manner is emphasized. Examples are given from solid-phase immunoassays and therapeutic apheresis.

REFERENCES

1. H. C. Graves, The effect of surface charge on non-specific binding of rabbit immunoglobulin G in solid-phase immunoassays. J. Immunol. Methods *111*:157–166, 1988.
2. K. Holmberg, K. Bergström, and M. B. Stark, Immobilization of proteins via PEG chains. In *Poly(Ethylene Glycol) Chemistry: Biotechnical and Biomedical Applications* (J. M. Harris, ed.). New York: Plenum Press, 1992, pp. 303–324.
3. M. Malmsten, B. Lassen, K. Holmberg, V. Thomas, and G. Quash, Effects of hydrophilization and immobilization on the interfacial behavior of immunoglobulins. J. Colloid Interface Sci. *177*:70–78, 1996.
4. V. Thomas, K. Bergström, G. Quash, and K. Holmberg, Hydrophilized and functionalized microtiter plates for the site-specific coupling of antigens and antibodies: application to the diagnosis of viral cardiac and autoimmune diseases. Colloids Surf. *77*:125–139, 1993.
5. Y. Taniguchi, N. Yorioka, S. Okushin, H. Oda, K. Usui, and M. Yamakido, Usefulness of immunoadsorption therapy for systemic lupus erythematosus associated with transverse myelitis. A case report. Int. J. Artif. Organs *18*:799–801, 1995.
6. F. Hiepe, K. Wolbart, W. Schossler, A. Speer, T. Montag, F. Mielke, and E. Apostoloff, Development of a DNA-adsorbent for the specific removal of anti-DNA autoantibodies in systemic lupus erythematosus (SLE). Biomater. Artif. Cells Artif. Organs *18*:683–688, 1990.
7. W. L. Kupin, K. K. Venkat, H. Hayashi, M. F. Mozes, H. K. Oh, and R. Watt, Removal of lymphcytotoxic antibodies by pretransplant immunoadsorption therapy in highly sensitized renal transplant recipients. Transplantation *51*:324–329, 1991.
8. I. Nathan, M. Aharon, G. Reisenfeld, and A. Dvilansky, A novel agarose acrobeads protein A column for selective immunoadsorbance of whole blood: performance, specificity and safety. Biomater. Artif. Cells Immobilization Biotechnol. *20*:23–30, 1992.
9. S. Gil-Vernet, J. M. Grino, J. Martorell, A. M. Castelao, D. Seron, C. Diaz, E. Andrés, L. Gonzalez-Castellanos, and J. Alsina, Anti-HLA antibody removal by immunoadsorption. Transplant Proc. *22*:1904–1905, 1990.
10. A. Palmer, D. Taube, K. Welsh, M. Bewick, P. Gjorstrup, and M. Thick, Removal of anti-HLA antibodies by extracorporeal immunoadsorption to enable renal transplantation. Lancet *1(8628)*:10–12, 1989.
11. M. J. Elliott, R. N. Maini, M. Feldmann, A. Long-Fox, P. Charles, P. Katsikis, F. M. Brennan, J. Walker, H. Bijl, J. Ghrayeb, and J. N. Woody, Treatment of rheumatoid arthritis with chimeric monoclonal antibodies to tumor necrosis factor alpha. Arthritis Rheum. *36*:1681–1690, 1993.
12. F. M. Brennan, R. N. Maini, and M. Feldmann. TNF-alpha—a pivotal role in rheumatoid arthritis? Br. J. Rheumatol. *31*:293–298, 1992.
13. P. C. Heinrich, J. V. Castell, and T. Andus, Interleukin-6 and the acute phase response. Biochem. J. *265*:621–636, 1990.
14. J. D. Baumgartner, M. P. Glauser, J. A. McCutchan, E. J. Ziegler, G. van Melle, M. R. Klauber, M. Vogt, E. Muehlen, R. Luethy, and R. Chiolero, Prevention of gram-negative

shock and death in surgical patients by antibody to endotoxin core glycolipid. Lancet
2(8446):59–63, 1985.

15. S. J. van Deventer, H. R. Buller, J. W. ten Cate, A. Sturk, and W. Pauw, Endotoxaemia:
an early predictor of septicaemia in febrile patients. Lancet 1(8586):605–609, 1988.

16. E. J. Ziegler, J. A. McCutchan, J. Fierer, M. P. Glauser, J. C. Sadoff, H. Douglas, A. I.
Braude, and N. Engl, Treatment of gram-negative bacteremia and shock with human
antiserum to a mutant Escherichia coli. N. Engl. J. Med. 307:1225–1230, 1982.

17. J. Traeger, M. Laville, P. F. Serres, M. Cronenberger, M. Thomas, M. J. Rey, and C.
Bourgeat, A new device for specific extracorporeal immunoadsorption of anti-DNA antibodies. In vitro and in vivo results. Ann. Med. Intern. 143(suppl. 1):9–12, 1992.

18. R. M. Lemoli, A. Fortuna, M. R. Motta, S. Rizzi, V. Giudice, A. Nannetti, G. Martinelli,
M. Cavo, M. Amabile, S. Mangianti, M. Fogli, R. Conte, and S. Tura, Concomitant
mobilization of plasma cells and hematopoietic progenitors into peripheral blood of
multiple myeloma patients: positive selection and transplantation of enriched CD34+
cells to remove circulating tumor cells. Blood 87:1625–1634, 1996.

19. J. E. Hardingham, D. Kotasek, B. Farmer, R. N. Butler, J. X. Mi, R. E. Sage, and A.
Dobrovic, Immunobead-PCR: a technique for the detection of circulating tumor cells
using immunomagnetic beads and the polymerase chain reaction. Cancer Res. 53:3455–
3458, 1993.

20. L. S. Wong, W. J. Bateman, A. G. Morris, and I. A. Fraser, Detection of circulating
tumor cells with the magnetic activated cell sorter. Br. J. Surg. 82:1333–1337, 1995.

21. H. Hydén and K. Holmberg, A new procedure to immobilize polyriboinosinic–polyri-
bocytidylic acid (Poly I:C) on the surface of polymethylmethacrylate and the use of
Poly I:C in clinical hemoperfusion. Arzneimittel-Forsch. 36:120–123, 1986.

22. S. Halpein, An overview of radioimmunoimagine. In Diagnosis of Colorectal and Ovarian Carcinoma: Application of Immunoscintigraphic Technology (R. T. Maguire and D.
Van Nostrand, eds.). New York: Marcel Dekker, 1992, pp. 1–22.

23. H. Hydén, L. E. Gelin, S. Larsson, and A. Saarne, A new specific chemotherapy: a pilot
study with an extracorporeal chamber. Rev. Surg. 31:305–320, 1974.

24. S. D. Saal and B. R. Gordon, Extracorporeal modification of plasma and whole blood.
Prog. Clin. Biol. Res. 106:375–384, 1982.

25. M. Malmsten, B. Bergenståhl, M. Masquelier, M. Pålsson, and C. Peterson, Adsorption
of apolipoprotein B at phospholipid model surfaces. J. Colloid Interface Sci. 172:485–
493, 1995.

26. M. Malmsten, B. Lassen, J. M. Van Alstine, and U. R. Nilsson, Adsorption of complement proteins C3 and C1q. J. Colloid Interface Sci. 178:123–134, 1996.

27. S. Welin-Klintström, A. Askendal, and H. Elwing. Surfactant and protein interactions
on wettability gradient surfaces. J. Colloid Interface Sci. 158:188–194, 1993.

28. P. M. Claesson, E. Blomberg, J. C. Fröberg, T. Nylander, and T. Arnebrant, Protein
interactions at solid surfaces. Adv. Colloid Interface Sci. 57:161–227, 1995.

29. B. Lassen and M. Malmsten, Competitive protein adsorption at plasma polymer surfaces.
J. Colloid Interface Sci. 186:9–16, 1997.

30. N. L. Burns, K. Holmberg, and C. Brink, Influence of surface charge on protein adsorption at an amphoteric surface: effects of varying acid to base ratio. J. Colloid Interface Sci. 178:116–122, 1996.

31. M. A. Bos, Z. Shervani, A. C. I. Anusien, M. Giesbers, W. Norde, and J. M. Kleijn,
Influence of the electric potential of the interface on the adsorption of proteins. Colloids
Surf. B 3:91–100, 1994.

32. S. Nagaoka, Y. Mori, H. Takiuchi, K. Yokota, H. Tanzawa, and S. Nishiumi, Interaction
between blood components and hydrogels with poloxyethylene chains. Polym. Prepr.
24:67–68, 1983.

33. E. W. Merrill and E. W. Salzman, Polyethylene oxide as a biomaterial. J. Am. Soc. Artif. Internal Organs 6:60–72, 1983.

34. J. H. Lee and J. D. Andrade, Surface properties of aqueous PEO/PPO block copolymer surfactants. In *Polymer Surface Dynamics* (J. D. Andrade, ed.). New York: Plenum Press, 1988, pp. 119–136.

35. S. W. Kim and J. Feijen, Surface modification of polymers for improved blood compatibility. Crit. Rev. Biocompat. 1:229–260, 1986.

36. S. Nagaoka, Hydrated dynamic surface. Trans. Am. Soc. Internal Organs 10:76–78, 1988.

37. S. Nagaoka and A. Nakao, Clinical application of antithrombogenic hydrogel with long poly(ethylene oxide) chains. Biomaterials 11:119–121, 1990.

38. G. R. Llanos and M. V. Sefton, Does polyethylene oxide possess a low thrombogenecity? J. Biomater. Sci. Polym. Ed. 4:381–400, 1993.

39. S. I. Jeon, J. H. Lee, J. D. Andrade, and P. G. de Gennes. Protein–surface interactions in the presence of polyethylene oxide. J. Colloid Interface Sci. 142:149–158, 1991.

40. D. W. J. Osmond, B. Vincent, and F. A. Waite, Steric stabilization: a reappraisal of current theory. Colloid Polym. Sci. 253:676–682, 1975.

41. T. F. Tadros, Polymer adsorption and dispersion stability. In *The Effect of Polymers on Dispersion Properties* (T. F. Tadros, ed.). New York: Academic Press, 1982, pp. 1–38.

42. K. Holmberg and J. M. Harris, PEG grafting as a way to prevent protein adsorption and bacterial adherence. In *Mittal Festschrift* (W. J. van Ooij and H. R. Anderson, Jr, eds.). Utrecht, The Netherlands: VSP, 1998, pp. 443–460.

43. M. Malmsten, P. M. Claesson, E. Pezron, and I. Pezron, Temperature-dependent forces between hydropohobic surfaces coated with ethyl(hydroxyethyl)cellulose. Langmuir 6:1572–1578, 1990.

44. J. W. Dobbie, R. Evans, D. V. Gibson, J. B. Smitham, and D. H. Napper, Enhanced steric stabilization. J. Colloid Interface Sci. 45:557–565, 1973.

45. D. H. Napper, *Polymeric Stabilization of Colloidal Dispersions*. London: Academic Press, 1983, pp. 198–215.

46. J. T. Li, K. D. Caldwell, and N. Rapoport, Surface properties of pluronic-coated polymeric colloids. Langmuir 10:4475–4482, 1994.

47. S. M. O'Connor, S. H. Gehrke, and G. S. Retzinger. Langmuir 15:2580–2588, 1999.

48. M. Malmsten and J. M. Van Alstine, Adsorption of polyethylene glycol amphiphiles to form coatings which inhibit protein adsorption. J. Colloid Interface Sci. 177:502–512, 1996.

49. K. L. Prime and G. M. Whitesides, Self-assembled organic monolayers: model systems for studying adsorption of proteins at surfaces. Science 252:1164–1167, 1991.

50. P. Harder, M. Grunze, R. Dahint, G. M. Whitesides, and P. E. Laibinis, Molecular conformation in oligo(ethylene glycol)–terminated self-assembled monolayers on gold and silver surfaces determines their ability to resist protein adsorption. J. Phys. Chem. B 102:426–436, 1998.

51. S.-W. Lee and P. E. Laibinis, Protein-resistant coatings for glass and metal oxide surfaces derived from oligo(ethylene glycol)–terminated alkyltrichlorosilanes. Biomaterials 19:1669–1675, 1998.

52. K. Holmberg, F. Tiberg, M. Malmsten, and C. Brink, Grafting with hydrophilic polymer chains to prepare protein-resistant surfaces. Colloids Surf. A 123–124:297–306, 1997.

53. K. Nilsson Ekdahl, B. Nilsson, C. G. Gölander, H. Elwing, B. Lassen, and U. R. Nilsson, Complement activation on radiofrequency, plasma modified polystyrene surfaces. J. Colloid Interface Sci. 158:121–128, 1993.

54. M. Malmsten, J. Å. Johansson, N. L. Burns, and H. K. Yasuda, Protein adsorption at n-butane plasma polymer surfaces. Colloids Surf. B 6:191–199, 1996.

55. C. G. Gölander and J. C. Eriksson, ESCA studies of the adsorption of polyethyleneimine and glutaraldehyde–reacted polyethyleneimine on polyethylene and mica surfaces. J. Colloid Interface Sci. *119*:38–48, 1987.

56. C. Brink, E. Österberg, K. Holmberg, and F. Tiberg, Using poly(ethylene imine) to graft poly(ethylene glycol) or polysaccharide to polystyrene. Colloids Surf. *66*:149–156, 1992.

57. C. G. Gölander, J. N. Herron, K. Lim, P. Claesson, P. Stenius, and J. D. Andrade, Properties of immobilized PEG films and the interaction with proteins: experiments and modelling. In *Poly(Ethylene Glycol) Chemistry: Biotechnical and Biomedical Applications* (J. M. Harris, ed.). New York: Plenum Press, 1992, pp. 221–246.

58. K. Bergström, K. Holmberg, A. Safranj, A. S. Hoffman, M. J. Edgell, A. Kozlowski, B. A. Hovanes, and J. M. Harris, Reduction of fibrinogen adsorption on PEG-coated polystyrene surfaces. J. Biomed. Mater. Res. *26*:779–790, 1992.

59. K. Bergström, E. Österberg, K. Holmberg, A. S. Hoffman, T. P. Schuman, A. Kozlowski, and J. M. Harris, Effects of branching and molecular weight of surface-bound polyethylene oxide on protein rejection. J. Biomater. Sci. Polym. Ed. *6*:123–132, 1994.

60. W. R. Gombotz, W. Guanghui, T. A. Horbett, and A. S. Hoffman, Protein adsorption to and elution from polyether surfaces. In *Poly(Ethylene Glycol) Chemistry: Biotechnical and Biomedical Applications* (J. M. Harris, ed.). New York: Plenum Press, 1992, pp. 247–262.

61. K. P. Antonsen and A. S. Hoffman, Water structure of PEG solutions by differential scanning calorimetry measurements. In *Poly(Ethylene Glycol) Chemistry: Biotechnical and Biomedical Applications* (J. M. Harris, ed.). New York: Plenum Press, 1992, pp. 15–28.

62. E. Österberg, K. Bergström, K. Holmberg, T. P. Schuman, J. A. Riggs, N. L. Burns, J. M. Van Alstine, and J. M. Harris, Protein-rejecting ability of surface-bound dextran in end-on and side-on configurations. Comparison to PEG. J. Biomed. Mater. Res. *29*: 741–747, 1995.

63. S. L. McArthur, K. M. McLean, P. Kingshott, H. A. W. StJohn, R. C. Chatelier, and H. J. Griesser, Effect of polysaccharide structure on protein adsorption. Colloids Surf. B *17*:37–48, 2000.

64. P. Kingshott and H. J. Griesser, Surfaces that resist bioadhesion. Curr. Opin. Solid State Mater. Sci. *4*:403–412, 1999.

65. R. A. Frazier, G. Matthijs, M. C. Davies, C. J. Roberts, E. Schacht, and S. J. B. Tendler, Characterization of protein-resistant dextran monolayers. Biomaterials *21*:957–966, 2000.

66. G. J. Fleer, M. A. Cohen Stuart, J. M. H. M. Scheutjens, T. Cosgrove, and B. Vincent, *Polymers at Interfaces*. London: Chapman & Hall, 1993, pp. 391–393, 449–453.

67. M. Malmsten, F. Tiberg, B. Lindman, and K. Holmberg, Effects of solvency on the interfacial behaviour in aqueous non-ionic polymer systems. Colloids Surf. A *77*:91–100, 1993.

68. F. Tiberg, C. Brink, M. Hellsten, and K. Holmberg, Immobilization of protein to surface-grafted PEO/PPO block copolymers. Colloid Polym. Sci. *270*:1188–1193, 1992.

69. I. Danielsson and B. Lindman, The definition of microemulsion. Colloids Surf. *3*:391–392, 1981.

70. K. Holmberg, Enzymatic reactions in microemulsions. In *Microemulsions: Fundamental and Applied Aspects* (P. Kumar and K. L. Mittal, eds.). New York: Marcel Dekker, 1999, pp. 713–742.

71. K. Bergström and K. Holmberg, Microemulsions as reaction media for immobilization of proteins to hydrophilized surfaces. Colloids Surf. *63*:273–280, 1992.

72. J. March, *Advanced Organic Chemistry*, 3rd ed. New York: Wiley, 1985, pp. 316–317.

73. K. Holmberg and M. B. Stark, Effect of medium polarity on the reactivity of amino acids in immobilization to silica. Colloids Surf. *47*:211–217, 1990.

74. G. A. Quash, V. Thomas, G. Ogier, S. El Alaoui, J. G. Delcros, H. Ripoll, A. M. Roch, S. Legastelois, R. Gilbert, and J. P. Ripoll, Diagnostic and therapeutic procedures with haptens and glycoproteins (antigens and antibodies) coupled covalently by specific sites to insoluble supports. In *Covalently Modified Antigens and Antibodies in Diagnosis and Therapy* (G. A. Quash and J. D. Rockwell, eds.). New York: Marcel Dekker, 1989, pp. 155–186.

75. G. Quash, A. M. Roch, A. Niveleau, J. Grange, T. Keolouangkhot, and J. Huppert, The preparation of latex particles with covalently bound polyamines, IgG and measles agglutinins and their use in visual agglutination tests. J. Immunol. Methods 22:165–174, 1978.

76. K. Bergström. Unpublished work.

77. J. M. Vandenbroucke, M. Jadoul, B. Maldague, J. P. Huaux, H. Noel, and C. Van Ypersele de Stribou, Possible role of dialysis membrane characteristics in amyloid osteoarthropathy. Lancet *1(8491)*:1240–1211, 1986.

78. S. Shaldon, J. Floege, C. Granolleras, G. Deschodt, G. Baudin, and K. M. Koch, High-flux synthetic versus cellulosic membranes for removal of beta 2-microglobulin during HD, HDF or HF. Kidney Int. *35*:1–10, 1989.

79. D. Hauglustaine, M. Waer, P. Michielsen, J. Goebels, and M. Vandeputte, Haemodialysis membranes, serum beta 2-microglobulin, and dialysis amyloidosis. Lancet *1(8491)*: 1211–1212, 1986.

80. P. Heinrich, J. Castell, and T. Andus, Interleukin-6 and the acute phase response. Biochem. J. *265*:621–636, 1990.

81. S. Jorstad, L. C. Smeby, T. Balstad, and T. E. Wideroe, Generation and removal of anaphylatoxins during hemofiltration with five different membranes. Blood Purif. *6*:325–335, 1988.

82. M. Thomas, P. Valette, A. L. Mausset, and P. Déjardin, High molecular weight molecular kininogen adsorption on hemodialysis membranes. Influence of pH and relationship with contact phase activation of blood plasma. Influence of pre-treatment with polyethylenimine. Int. J. Artif. Organs *23*:20–26, 2000.

83. G. Quash, N. Moachon, K. Holmberg, M. Malmsten, and M. Thomas, Fonctionnalisation de surfaces et application en immunoépuration. 10ème journées Europharmat. Revue de l'Association des Pharmaciens du Sud-Ouest *26*:73–79, 2001.

84. N. Moachon, C. Boullanger, S. Fraud, E. Vial, M. Thomas, and G. Quash, Influence of the charge of low molecular weight proteins on their efficacy of filtration and/or adsorption on dialysis membranes with different intrinsic properties. Biomaterials (in press).

85. P. Valette, M. Thomas, and P. Déjardin, Adsorption of low molecular weight proteins to hemodialysis membranes. Experimental results and simulations. Biomaterials *20*:1621–1634, 1999.

86. A. Kandus, R. Ponikvar, J. Drinovec, S. Kladnik, and P. Ivanovich, Anaphylatoxins C3a and C5a adsorption on acrylonitrile membrane of hollow-fiber and plate dialyser—an in vivo study. Int. J. Artif. Organs *13*:176–180, 1990.

87. S. Legalois, V. Thomas, G. Quash, M. P. Métais, J. Tebib, A. Morevia, and J. C. Monier, Naturally occurring antibodies reacting with lipoic acid: screening method, characterization and biochemical interest. J. Immunol. Methods *171*:111–119, 1994.

88. A. Nilsson, M. Nydén, K. Holmberg, N. Moachon, and G. Quash (to be published).

28
Protein Interfacial Behavior in Microfabricated Analysis Systems and Microarrays

HELENE DÉRAND   Gyros AB, Uppsala, Sweden

MARTIN MALMSTEN   Institute for Surface Chemistry and Royal Institute of Technology, Stockholm, Sweden

I. INTRODUCTION

Developments over the last decade in particular have made it possible to perform chemical and biochemical analyses in microfabricated devices. The concept of micro total analysis system (μTAS), or "lab-on-a-chip," was first introduced by Manz et al. some 10 years ago [1], and an increasing number of applications has been discussed and explored since then. Generally, such analysis systems consist of micrometer-sized reservoirs and flow channels fabricated in a planar substrate. For the traditionally open systems, such as micro plates, evaporation becomes a severe problem when working in nanoliter scale. Potentially, miniaturized systems offer a number of advantages, the first one being the inherently small scale itself. This implies that very small quantities of analytes and reagents are required, provided, of course, that the actual analysis and detection methods provide the necessary sensitivity. Also, the small dimensions in such systems yield short diffusion lengths, which translate into faster mass transport, and hence also in faster reaction times, which may be of significant interest, e.g., in clinical assays. Second, microanalysis systems generally refer to integrated systems, capable of several processing unit operations in sequence within the device, thereby reducing the need for intermediate robotic steps. These systems are also well suited for automation, e.g., of sample handling, analysis, and data processing. Altogether, μTAS offers interesting opportunities in high throughput applications. A particularly interesting area in this context is development of (bio)pharmaceuticals where thousands of substances are frequently screened for biological activity prior to even selecting a candidate drug for clinical trials.

Naturally, there are also numerous difficulties with microanalysis systems. In particular, these include larger handling errors with small volumes, variations in reaction conditions, and limitations in the dynamic range. Also, the small quantities used put high demands on the sensitivity of the detection methods used and may limit the analysis detection limit and increase the analysis uncertainty. Despite the numerous difficulties, a number of investigations have demonstrated that both microfluid-based approaches, relying on the movement of fluids in microchannels for analysis, and microarrays, using reagents immobilized on a substrate, allow high throughput, small sample volumes and ease of automation. In particular, this has been the case for DNA-based analysis methods. However, while the advances concerning DNA and other nucleic acids have been substantial, the use of μTAS for proteins and peptides is much less developed. Since much of the current interest among pharmaceutical companies in particular lie with protein and peptide drugs, this can be expected to change in the coming years. Such a development will require substantial efforts, however, e.g., since proteins are notoriously surface active, and frequently change their biological activity when adsorbed at an interface. Furthermore, signal amplification through polymerase chain reaction (PCR), which has proven to be so useful for DNA-based analysis, naturally cannot be used for proteins and peptides, which therefore puts higher demands on detection methods, particularly for low abundance proteins.

Despite difficulties with proteins and peptides in μTAS, e.g., related to structural and functional diversity, surface-induced deactivation, as well as sensitivity to drying, just to mention a few, progress has indeed already been made. The purpose of the present chapter is therefore to review some of these and to discuss some general aspects of protein interfacial behavior in microanalysis systems and microarrays.

II. ISSUES TO CONSIDER RELATED TO THE SMALL SCALE

A. Surface Area–to–Bulk Volume Ratio

There are a number of fundamental differences between "normal," or "macroworld," analysis systems, on the one hand, and "microanalysis" systems, on the other. For example, the surface area–to–bulk volume ratio may be one or even several orders of magnitude larger in the microworld compared to the macroworld. With surface-active compounds, such as proteins, the accumulation at the interface becomes more important with an increasing surface area. With the exception of phase-separating protein systems, where macroscopic interfacial films may form, protein adsorption is generally limited to a fraction of a dense monolayer, sometimes reaching up to a couple of monolayers for associating but non–phase-separating protein systems. This means that for one-phase protein systems, the total amount of protein which can be adsorbed increases with the surface area. If the surface area is large in relation to the bulk volume and/or the initial protein concentration in solution is low, the protein solution concentration after equilibration with the surface may be very low.

Not unexpectedly, such protein capture by the surface may cause significant problems for μTAS. The most obvious of these is the "loss of material" to the interface in itself, which may preclude detections based on proteins present in solution. There are, however, also other potential problems related to the large surface area in microsystems. For example, numerous proteins which adsorb at interfaces undergo conformational changes on adsorption, which ultimately may lead to loss of their biological activity [2,3]. If such proteins are released to the solution after being adsorbed for some time, they may not regain their native conformation. If the detection method is based on recognition of the native protein, e.g., through the use of antibodies as in the case of immunoassays, the detection will only work if the antibody recognizes a part of the protein which has been left intact after adsorption.

Furthermore, protein mixtures usually display competitive adsorption, where the different (surface-active) proteins compete for surface area. When a protein mixture is exposed to the surface, the initial adsorption will be dominated by the protein diffusing fastest to the surface, i.e., the smallest protein and/or the protein occurring in the highest concentration. Later on in the adsorption process, larger proteins and/or proteins occurring in low bulk concentrations arrive at the surface and may or may not replace the initially adsorbed protein. Particularly for more complex protein systems, such as blood serum/plasma, a rich competitive adsorption may be observed for certain surfaces, and the adsorbed layer being dominated by different proteins at different incubation times. As an illustration of this, Fig. 1 shows schematically the adsorption pattern on exposure of a hydrophilic surface (e.g., glass) to plasma. As can be seen, the initial adsorption is dominated by human serum albumin (HSA), which is a relatively small serum protein occurring in high concentrations. Later on, this protein is replaced by immunoglobulins (e.g., IgG), fibrinogen, fibronectin, and finally high-molecular-weight kininogen (HMWK) [4]. Thus, simply by incubating for different times, the surface protein composition will change, as will the bulk protein composition if the surface area is sufficiently large relative to the bulk volume. Furthermore, the "biological activity" of the coated surface will change over time. For example, many cells attach to surfaces through adhesion proteins such as fibrinogen or fibronectin. Thus, at intermediate incubation times for a system as described in Fig. 1, where fibrinogen and fibronectin dominate the adsorbed layer, cell adhesion can be substantial. HSA, on the other hand, is used as a blocking agent to prevent cell adhesion, and little or no adhesion will therefore be displayed at short incubation times, when HSA dominates the adsorbed layer. Hence, if the analysis is dependent on or affected by the biological activity of surface bound proteins, interfacial exchange processes must be considered.

Interestingly, the exchange processes depend on the nature of the surface, and it has been found, e.g., that hydrophobic surfaces, such as hydrophobized silica and glass, display less exchange than hydrophilic ones, such as silica and glass [5,6]. The use of hydrophobic surfaces or surface coatings may therefore be a way to avoid a time-dependent biological function of the adsorbed protein layer. Another possi-

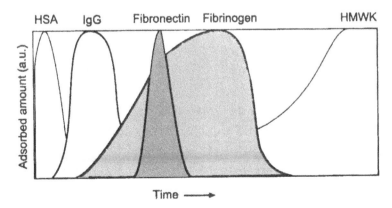

FIG. 1 Schematic illustration of the successive adsorption of various blood proteins as a function of incubation of a hydrophilic surface with serum, plasma, or blood. (From Ref. 4.)

bility to avoid such problems is naturally to avoid protein adsorption all together. Naturally, surface-induced effects such as those described above occur also in macrosystems. However, due to the smaller surface area–to–bulk volume ratio in the latter systems, such effects are quantitatively less important. For microsystems, on the other hand, the surfaces may dominate the system and require these matters to be addressed, e.g., through different types of surface modifications to eliminate or at least reduce unspecific protein adsorption.

B. Diffusion Length

Another fundamental difference between microsystems and macrosystems is the difference in diffusion length. Thus, for microfluidic systems based on microchannels, the radial dimension may be of the order 10–100 μm. For a typical protein, this distance is covered through self-diffusion in the order of seconds (Fig. 2) [7]. Since the mean displacement scales with $t^{1/2}$, this means that a factor of 100 larger dimension corresponds to a factor 10 longer diffusion time. Furthermore, due to the small dimensions, concentration gradients may also be very steep, which also contributes to a faster diffusion. The faster mass transport in microsystems offers potential advantages in the form of shorter reaction times, which is beneficial for, e.g., immunoassays and enzymatic reactions. In a similar way, dissipation of thermal energy may be very fast in microdimensional systems, although dependent on the thermal conduction properties of the materials used. Such fast thermal conduction may be of interest, e.g., in PCR and enzymatic assays.

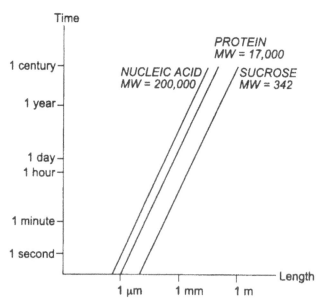

FIG. 2 Diffusion time–length relation for some important molecules of different molecular weight (MW) in water. (From Ref. 7.)

III. MICROFLUIDIC SYSTEMS

A. Fluid Control in Microfluidic Systems

Microfluidic systems, by which it is possible to move very small volumes of liquid in a controlled manner, open the possibility to miniaturize many powerful and well-developed macroscale techniques, ranging from sample preparation to reactions to separation to analysis. However, handling volumes in the nanoliter scale is in many aspects different from the corresponding handling at the macroscale. For example, low Reynolds numbers result in laminar flow and as a consequence mixing mainly occurs through diffusion. Precise control and movement of fluids on-chip constitutes a significant challenge in microfluidic systems. To date, electroosmotic flow is by far the most commonly used technique for controlling flow in microfluidic devices, but also other techniques, such as pressure- and centrifugation-driven displacement, have been explored. Recent developments and references to earlier work can be found in reviews on analytical microdevices [8–10].

B. Electroosmotic Flow

In a majority of the miniaturized systems described in literature today, the liquid flow is generated by electrokinetic pumping [8]. When an electric field is applied over a microchannel containing charged groups on its walls, the counterions move and an electroosmotic fluid flow (EOF) is obtained in the channel. Through controlling the applied potential, the EOF can be varied. Within the channel network, cross-intersections and T-intersections are used for valving and dispensing fluids in a reproducible manner. Through the use of such intersections, flow in different directions may be generated. Thus, liquid control can be achieved without using any moving parts, such as valves, which is a significant advantage.

Apart from the electric field strength, the EOF depends on the surface charge density and distribution, pH, and ionic strength. For example, the occurrence of a surface charge is required to generate an EOF in the first place, and the lower the surface charge density, the lower the sensitivity to the applied voltage. Naturally, any nonuniformity in charge density will degrade the performance of devices based on EOF. For protein analysis with µTAS based on EOF, also the surface activity of the proteins and its consequences should be considered. As discussed briefly above, nonspecific adsorption is likely to occur, and to significantly affect the performance of microanalysis systems, not the least due to the large surface area–to–bulk volume ratio in such systems. For EOF-based systems, this may pose problems also relating to fluid control, since the effective surface charge is changed due to the protein adsorption, and hence the fluid control deteriorates or is lost. As an illustration of this, Fig. 3 shows the effect of protein adsorption on the effective electrokinetic mobility of a (protein-coated) polystyrene substrate, as well as the EOF for a few plastic materials in the presence and absence of adsorbed protein [11,12]. As can be seen, significant changes to the electrokinetic mobility may result from the protein adsorption, and in some cases charge reversal may even occur.

Considering the above, it is not surprising that the critical dependence of EOF on surface chemistry has been rather widely recognized as a problem associated with microfluidic pumping based on this principle. However, although there seems to be an awareness of this issue, as judged from the introductions to numerous papers in

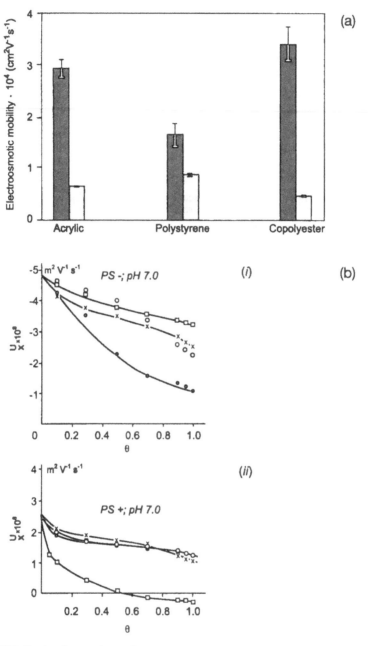

FIG. 3 (a) Comparison of electroosmotic mobility in the native (solid) versus the antibody-adsorbed (shaded) plastic. (From Ref. 11.) (b) Electrokinetic mobility as a function of surface coverage (fraction of saturation adsorption; θ) for lysozyme (filled circles), ribonuclease (open circles), myoglobin (crosses), and α-lactalbumin (squares) at negatively (i) and positively (ii) charged polystyrene at pH 7.0. (From Ref. 12.)

the field, only few reports discuss the effect of protein adsorption on the EOF rate in the microchannels [11,13,14]. Again, one way to control the adverse effects of protein adsorption is simply to reduce, or preferably eliminate, the protein adsorption itself. Such protein-rejecting coatings generally consist of effectively uncharged moieties and may be formed by water-soluble polymers [e.g., poly(ethylene oxide) (PEO) derivatives or polysaccharides]; PEO- or sugar-containing self-assembling monolayers; or zwitterionic phospholipids, just to mention a few. Although such coatings may reduce or even solve problems associated with nonspecific protein adsorption, they simultaneously also result in a reduced EOF, which may be detrimental for fluid transport. As an illustration of the effect of such coatings on EOF, Fig. 4 shows the EOF of amine-functionalized quartz grafted with PEO of different molecular weight [15]. As can be seen, with increasing polymer molecular weight, the underlying charges become increasingly screened, and for sufficiently long PEOs the initial surface charges are no longer observable, and essentially zero electroosmotic mobility is found. This, in turn, may require higher voltages than desired for practical or safety reasons, or make EOF for fluid control impossible. Clearly, therefore, coatings designed for EOF-based systems require some care in order to reach an acceptable level of nonspecific adsorption and simultaneously a sufficient EOF.

Naturally, since most charges titrate, pH may also affect the charge density, hence also the EOF (Fig. 4) and the fluid control. Furthermore, due to screening of electrostatic interactions, EOF becomes less sensitive at high ionic strengths and, more importantly, the EOF changes on changing the ionic strength. Taken together, this means that the sensitivity of EOF to changes in solution composition (pH and electrolyte concentration, but also presence of organic cosolvents) must be taken into account if this mechanism for fluid control is to be used to manipulate widely different liquids one after the other.

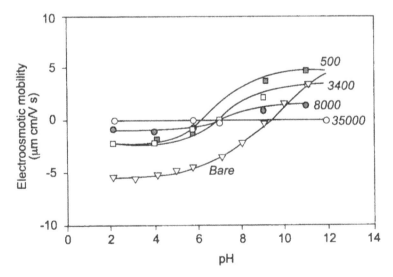

FIG. 4 Effect of pH on the charging of amine-functionalized quartz, and of coating the quartz with uncharged PEO of different molecular weight. Note that surface modification with high-molecular-weight PEO in particular results in the charges of the amine-functional surface being effectively masked. (From Ref. 15.)

Finally, since it is necessary at all times to keep the entire channel network filled with liquid, EOF-based systems tend to be rather sensitive to the occurrence of air bubbles, which may cause problems, particularly if hydrophobic surfaces and/ or elevated temperatures are used.

Despite the problems associated with volume control based on EOF, this approach has nevertheless been used successfully in a number of different contexts. During the first few years, the microstructures investigated were fabricated solely using silicon or glass. Glass has many advantages in relation to EOF-based μTAS, such as optical transparency, generation of high EOFs, and favorable clean surface characteristics after fabrication by etching. However, these microstructures are expensive and best suited for multiple-use devices. Unfortunately, when proteins and other biomolecules are handled, cross-contamination between runs due to adsorption/ desorption may become a severe problem, which strongly argues in favor of disposable devices. The issue of disposable EOF devices has, however, been at least partly resolved through the introduction of microfluidic networks in poly(dimethylsiloxane) (PDMS) [16], as well as convenient procedures to fabricate chips in PDMS [17,18]. The main advantages of the latter material include low cost, ease of handling, and optical transparency. Other polymer materials that have been used for EOF-based microfluidic devices include poly(methyl methacrylate) [13,19–21], polycarbonate [22,23], polystyrene [13,23], and poly(ethylene terephthatalate) (PET) [23,24]. Since native plastic substrates generally display different EOF mobilities, and since they can also be modified in various ways to control the EOF [20–22,25–28], disposable EOF-based systems may indeed be prepared from a range of materials.

C. Other Methods for Fluid Control

Pressure-driven methods, which predominate at the macroscale, do not suffer from any of the problems associated with EOF. However, they are not conveniently available for microscale fluid control due to connectors and valves required. Particularly when large numbers of analyses are to be run in parallel the devices may become very complex. Still many investigators in the field employ this method for demonstrating different aspects of miniaturization. Both discrete volumes and continuous systems can be used and the flow rates regulated by an external pressure generating system. However, accurate control of flow rates in the order of nL/s with incompressible liquids is extremely demanding by this approach, since a proper control of very small pressure differences is essential for controllable and reproducible results. Nevertheless, an equipment for precise flow control on a chip has recently been demonstrated by carrying out dye mixing and enzymatic assay titration using pressure-driven flow [29,30].

The use of centrifugal force to move liquids is yet another approach which has been exploited in relation to μTAS [31–35]. No connectors to the external world are necessary in such systems, and multiple microfluidic structures can be run in parallel. Compared to EOF, both pressure-based methods and methods based on centrifugal force are relatively independent on physiochemical properties such as surface charge density, ionic strength, pH, conductivity, and the presence of components, e.g., proteins, adsorbing to the walls. Discrete volumes of liquid are transported through the channel network and stopped at desired positions using valves. Different strategies have been demonstrated for creating such valves. One such approach is "burst valves," relying on the liquid being pinned at a junction [31,32]. Another is "hydrophobic valves" [33,34], relying on the resistance generated in a capillary as the surface properties change from hydrophilic to hydrophobic at a certain position. A third option that has been demonstrated is the use of "double U bends" to create a resistance for flow [33,35].

Particularly the use of hydrophobic valves may suffer from some problems related to protein adsorption. Thus, if protein solutions are (repeatedly) passed over this type of valve, protein adsorption can be expected to cause a reduced surface hydrophobicity [36], and therefore also in a deteriorating function. Also, protein adsorption at hydrophobic surfaces may be quite strong, precluding their removal, e.g., through use of surfactants (see, e.g., the Chapter 12).

IV. SEPARATION OF PROTEINS USING CAPILLARY ELECTROPHORESIS

One of the earliest examples of the use of microfabricated devices was the successful integration of capillary electrophoresis (CE) on a glass chip [37]. Excellent separation efficiencies were obtained since the small dimensions result in effective heat dissipation. Furthermore, high electric fields could be used, resulting in short run times and thus reduced diffusion and, consequently, improved separation. Numerous papers have followed in the evolution of this technology, where the separation element plays a central role. One notable breakthrough in the field was the demonstration of parallel separation of DNA. Starting with 12 channels [38], later reporting 96 [39], Mathies's group has now demonstrated a spectacular chip consisting of 384 channels (Fig. 5) [40,41].

Capillary electrophoresis, with its various complementary modes of operation, is also a powerful analytical technique for the separation of proteins. However, the separation performance of CE can be impaired by the adsorption of proteins at the surfaces of the channel walls. Such protein adsorption still limits many CE applications in routine biochemical analysis. Several different aspects can be recognized in this context. First, the adsorption leads to broad, asymmetrical peaks, thereby causing decreased efficiency and sensitivity. Low recovery of the separated analytes, sometimes due to irreversible protein adsorption, is another aspect. Furthermore, the magnitude of EOF becomes unpredictable, resulting in poor reproducibility of the migration times. Generally, shielding of the proteins from the capillary walls through surface modification is therefore necessary for optimal performance during separation [42–44].

A simple way to reduce protein adsorption which has been attempted is to run the separations with uncoated capillaries, but at high ionic strength [45]. However, effective removal of joule heating is necessary for successful use of this concept. Furthermore, while high electrolyte concentrations may limit the adsorption of some proteins, it may also result in an *increased* protein adsorption. In fact, increasing the concentration of at least certain salts has been found to be an efficient way to cause protein precipitation and crystallization [46]. This approach should therefore be used with some care. Alternatively, the separation may be carried out at extreme pH values, e.g., for minimizing or maximizing the ionization of the surface silanol groups and the charges of the proteins. Since many proteins have indeed been found

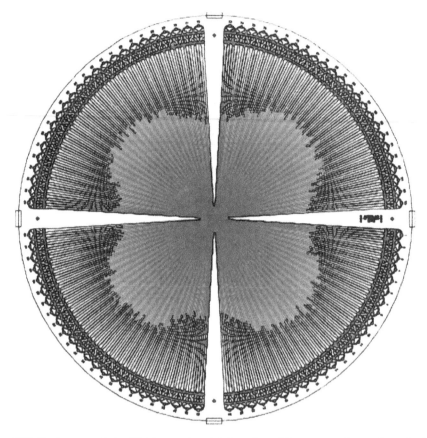

FIG. 5 Layout of the 384-lane microfabricated capillary array electrophoresis device with sample and cathode wells arrayed about the outer edge of the wafer. Each cathode reservoir is shared by four adjacent lanes. (From Ref. 41.)

to display maximum adsorption close to their isoelectric point [47], such an approach could in principle be feasible. However, this strategy limits the working pH range and, more importantly, extreme pH values may cause the protein to denature. In fact, the latter may counteract the intention of the approach, since denaturation-induced solubility reduction may actually result in an *increased* interfacial protein deposition.

Organic compounds included in the running buffer are frequently referred to as dynamic coatings in the CE literature. Cationic buffer additives [48], neutral polymers [49,50], and surfactants [51–53] have all been used as dynamic coatings. Advantages of dynamically coated capillaries include the versatility and simplicity of preparation and the ability to modulate the electrophoretic mobility. A major shortcoming associated with traditional dynamic coatings, however, is that they are present in the background electrolyte and often interfere with the separation or detection scheme. Furthermore, minor alterations to the electrophoretic buffer composition may drastically alter the coating performance, e.g., due to competitive adsorption and displacement of the dynamic coating. Therefore, such coatings frequently display a lack of robustness. For such coatings to be successful, the component forming the coating should adsorb with a high adsorbed amount, display fast adsorption kinetics as well as a high affinity adsorption isotherm (i.e., giving high adsorbed amounts already at low bulk concentrations), show no desorption on dilution, and present little displacement in the presence of proteins and other components present in the buffer. Also, care should be used when employing surfactants together with proteins, since these may cause the latter to denature (see, e.g., Refs. 54 and 55).

Many of the coatings which can be added to the running buffer can also be used for precoating of the capillaries. Particularly for high-molecular-weight coatings, e.g., those formed by polymers, this allows a longer adsorption time, which is generally beneficial for stable coating formation due to the slow adsorption frequently displayed by such systems. However, also with this type of coating, there may be a gradual loss of coating over time due to desorption or displacement. Therefore, in order to reach maximum stability and robustness, covalent immobilization of the coatings should preferably be employed. Not surprisingly, the latter approach has received considerable interest among the surface deactivation methods. It should be noted, however, that also covalently bound coatings may display limited stability if they contain hydrolytically unstable groups. If such groups hydrolyze, e.g., accelerated by high or low pH, the whole or parts of the coating can be detached, which naturally deteriorates the performance of the coating with regard to prevention of protein adsorption [56]. For glass substrates, also the hydrolytic instability for the siloxane bonds at high pH values effectively destroys all surface modifications.

As discussed above, however, the surface coating may affect the EOF considerably (Fig. 4), which should also be considered in the design of surface coatings for CE applications. Although nonionic coatings are without doubt the most interesting ones for reducing unspecific protein adsorption, also charged coatings have been attempted, e.g., in order to obtain a constant and finite, or a switchable EOF [57].

Separation of proteins using CE integrated into a microchip has been demonstrated by different groups. One recent example of this is the separation of low-density (LDL) and high-density lipoproteins (HDL). This is of interest since these two proteins are key for determining risk for atherosclerosis, high blood pressure, and several other cardiovascular disorders. Analysis of these components, among a few other lipoproteins, therefore forms the base for cardiac risk assessment. The CE separation in this investigation was performed for both fused silica capillaries and on chip-based capillaries. Using the same buffer and channel coating, similar results were obtained, as seen in Fig. 6 [58]. The only major difference between the two approaches is the analysis time, which is much shorter in the microdevice (40 s as compared to 15 min). The peak efficiency of LDL was later improved by adding sodium dodecyl sulfate (SDS) to the sample in a related study [52]. Chip-based methods have also been used to separate mixtures of model proteins [59], as well as antigen–antibody complexes from antigens in immunological assays [60–62]. Furthermore, a combination of pressure-driven flow and electrophoretic separation in this context was recently demonstrated [63].

V. IMMUNOASSAYS

Exploitation of microfluidic technology to design immunoassays is still in early stages. However, the modest requirement of material needed to perform assays in

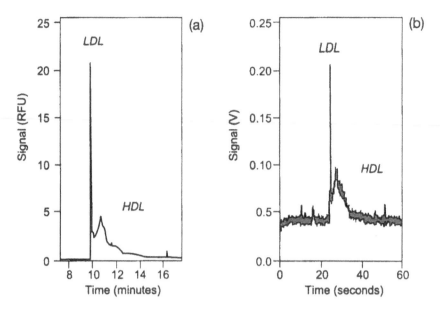

FIG. 6 Capillary electrophoresis of HDL/LDL fractions in (a) fused silica columns (separation conditions: capillary length 50 cm, field strength 333 V/cm, plug length 6 mm, column diameter 75 μm) and (b) glass chip (separation conditions: capillary length 4.2 cm, field strength 714 V/cm, plug length 200 μm, channel dimensions 20 μm deep by 50 μm wide). (From Ref. 58.)

microfluidic systems, the possibility to run numerous samples in parallel, thereby increasing the throughput, and the shorter diffusion lengths, potentially resulting in faster responses, again facilitating an increased throughput, all suggest that immunoassays may be a prime target for μTAS applications. Not surprisingly, therefore, several groups are developing technologies for assays both in solution (so-called homogeneous assays) and assays requiring a surface (so-called heterogeneous assays). While the early work in this area focused on homogeneous assays, demonstrating the ability to separate and quantify reactants and products on a chip [64–66], recent work has been focused also on heterogeneous immunoassays [14,67–73].

As an example of studies on homogeneous assays in microsystems, Chiem et al. used a microfluidic device etched in a glass substrate and performed an analysis of an anti-theophylline antibody and theophylline (Th) in serum [61,74]. Fluorescently labeled Th was mixed with a sample containing unlabeled Th, and the two were then allowed to compete for a limited amount of antibody. The reaction products from the free solution competitive assay were separated within 40 s. A calibration curve was obtained between 2.5 and 40 μg/mL, which includes the therapeutically useful range. In another study by the same group, the affinity constant of a monoclonal antibody to fluorescently labeled BSA was measured in diluted mouse ascites fluid using an identical chip layout [60]. In both of these studies a buffer combined with a detergent was used to reduce problems with adsorption of the components onto the channel walls. The same group also recently published a study where estradiol/antiestradiol was used in a competitive immunoassay [75]. Figure 7

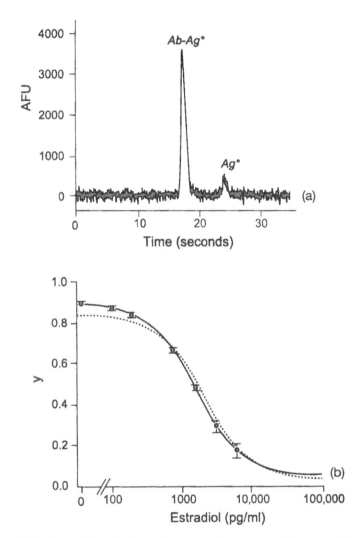

FIG. 7 (a) Typical electropherogram for a competitive estradiol assay, obtained in 20 mM borate, 20 mM NaCl, 0.01% Tween 20, pH 8.2 buffer solution. (b) Calibration curve obtained for the estradiol assay for 2.56 nM antibody and 3.125 nM antigen. Shown are theoretical predictions (dotted line), experimental results (circles), and a numerical fit to the latter (solid line), of the response y, defined as the ratio of the bound antigen to the total antigen. (From Ref. 75.)

shows reproducible results obtained in the latter study and illustrates currently available quality of data generated by this approach.

Generation of quantitative data based on fluorescence detection generally requires a calibration curve. If samples and standards are to be run simultaneously, a multichannel system is necessary. Indeed, the possibility to run such parallel analyses is one of the advantages of microfluidic systems in relation to immunoassays. As an illustration of this it can be noted that six independent mixing, reaction, and separation manifolds were recently demonstrated on one chip [62]. All channels were

operated at the same time, and a scanned fluorescence detection system was used to
monitor the separation. In analogy to the discussion above, the authors stressed that
designing an electrokinetically driven multichannel device constituted a significant
challenge. In particular, uniform behavior between channels must be achieved. The
surface charge and purity were inferred to have a crucial influence on the performance, and careful pretreatment of the surfaces with NaOH prior to analysis was
needed. Nevertheless, simultaneous calibration and analysis of ovalbumin and estradiol could be run within 30 s, with a detection limit in any single channel of 4.3
nM.

In heterogeneous assays the capturing antibody is attached to a solid phase
where it ideally should bind its antigen, thus allowing a location of the latter. An
advantage with this configuration is that unbound analyte antigen can be washed
away in a convenient manner, thereby potentially lowering the detection limits. The
possibility to perform rapid heterogeneous immunochemical reactions was demonstrated in a miniaturized device where the capturing agent was immobilized by adsorption direct to the channel walls [67]. Realizing that the potential advantage of
short diffusion distances can only be utilized if the rate-determining step for the
overall immunochemical reaction is diffusion controlled rather than controlled by
the inherent forward reaction rate, these authors studied the concentration and time
dependence of an immunoassay based on polyclonal anti–staphylococcal enterotoxin
B. The immunoassay kinetics of the sandwich assay was found to be in good agreement with numerical simulations of protein transport mainly governed by diffusion
control, and the overall reaction rate thus enhanced in the microsystem (Fig. 8).

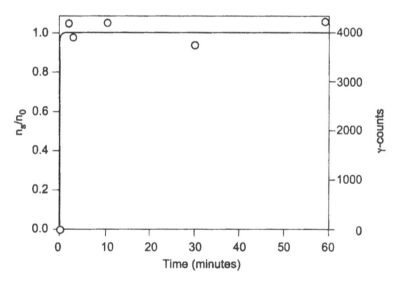

FIG. 8 Fick's second-law calculation of the fraction (n_s/n_0) of staphylococcal enterotoxin B
(SEB) molecules reaching the microchannel surface as a function of the incubation time (solid
line), as well as experimental results on the adsorption kinetics of SEB in a photoablated
microchannel (circles). The latter results were obtained by incubating the channel with 3.12
nM of anti-SEB for 1 h followed by incubation of 36 nM SEB for 1 to 60 min and subsequent
binding of a secondary radiolabeled anti-SEB. (From Ref. 67.)

Quantitatively, the equilibrium of the reaction could be achieved within about 1 min in the microchannels.

Yang et al. recently performed immunoassays based on supported lipid bilayers (SLBs) [68]. In this study, a buffer solution of small unilamellar egg phosphatidyl-choline (egg PC) vesicles was injected to PDMS microchannels placed on a planar glass substrate. Supported bilayers were assumed to be formed spontaneously on the channel surfaces through vesicle fusion. The binding of fluorescent labeled bivalent anti-dinitrophenyl (anti-DNP) antibodies to phospholipid bilayer containing dinitro-phenyl haptens was monitored using total internal reflection fluorescence microscopy (TIRM). SLBs from egg PC are known to resist nonspecific adsorption of proteins (see Chapter 21), and TIRM control experiments showed that IgG adsorption was suppressed by at least two orders of magnitude in comparison with uncoated chan-nels. Using the SLB surface modification, a linear array of 12 microchannels in parallel allowed an entire binding curve to be obtained in one single experiment (Fig. 9). The apparent binding constant for the DNP/anti-DNP system was found to be 1.8 μM, which is in agreement with data previously reported in literature [76,77]. A total volume of ~2 μL was used, while in total ~35 μg (230 pmol) was used to perform a single binding constant measurement.

Employing a slightly different approach for heterogeneous immunoassays in a microfluidic setup, Sato et al. used beads embedded inside a microchip to determine carcinoembryonic antigen (CEA) [69,70]. The chip used for this was composed of three quartz glass plates, where the middle carried the channel structure. Before performing the assay, the inner walls were blocked through casein adsorption. Poly-styrene beads precoated with anti-CEA were then introduced to the chip and packed against a shallow region (10 μm) in the channel system. In a stepwise reaction a serum sample containing CEA and then the second antibody conjugated with col-loidal gold were injected successively. The resulting complex, fixed to the bead surface, was quantified under a thermal lens microscope. Including washing steps and a detection procedure, this microchip-based diagnosis system required only 35 min to determine the CEA concentration. This can be compared with the analysis time for a conventional method performed with a microtiter plate which took more than 30 h. Furthermore, a detection limit at least one order of magnitude lower than that obtained with conventional ELISA was reached. Similarly, an IgA/anti-IgA re-action was compared between a microtiter plate assay and an integrated assay. As can be seen in Fig. 10, also in this case a significantly faster reaction was obtained for the integrated assay.

While the examples given above have demonstrated that miniaturized devices do indeed lead to the expected kinetic benefits of working in microscale, heteroge-neous immunoassays have more recently been more extensively integrated with mi-crofluidics. For example, heterogeneous bioassays have been performed in electro-kinetically driven microfluidic chips [14]. The reaction chamber had picoliter dimensions and was integrated into a network of microchannels etched in glass. In this study, protein A was first adsorbed to the reaction chamber walls, whereafter fluorescently labeled rabbit IgG (rIgG), serving as a sample, was added. In order to reach sufficient signal intensity, the rIgG was eluted into the detection zone using a buffer containing glycine at pH 2.0 (Fig. 11). Using this approach, a dose-response curve for rIgG was obtained for concentrations down to 50 nM with an incubation time of 200 s. Furthermore, the flexibility of the chip layout allowed a competitive

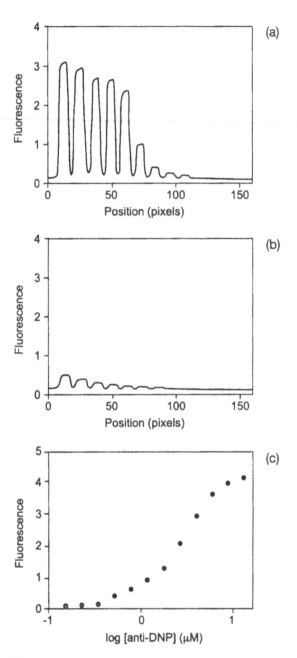

FIG. 9 (a) Line scan of fluorescence intensity across bilayer-coated parallel microchannels. The intensity was governed by binding of a fluorescence-labeled antibody to dinitrophenyl (DNP) present in the lipid membrane. (b) Similar to (a) but without any DNP-labeled lipids (i.e., background signal). (c) Point-by-point plot of the intensity in each channel in (a) after subtracting the background intensity of (b). (From Ref. 68.)

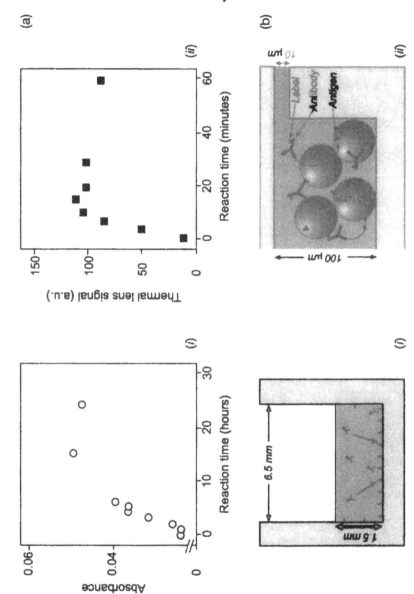

FIG. 10 (a) Time course of antigen–antibody reaction in a conventional microtiter plate assay (i) with an integrated immunoassay (ii). (b) Schematic illustration of the antigen–antibody reaction at a microtiter plate (i) and in the microchip (ii). (From Ref. 70.)

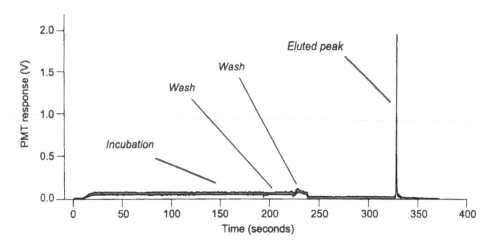

FIG. 11 Photomultiplier (PMT) response for adsorption of Cy5-rIgG (at 1.5 μM) on silanized glass, followed by double washing by 25 mM tris-HCl (pH 7.5) and elution by 25 mM glycine-HCl (pH 2.0). Note that a well-defined and controlled elution of the captured antibody complex could be achieved for this system. (From Ref. 14.)

assay to be accomplished using both a combined sample/tracer incubation and sequential addition of these solutions.

The same group has also demonstrated the use of PDMS microchannels for an immunoreaction model consisting of immobilized goat anti-human IgG and fluorescently labeled human IgG as target [71]. Due to problems with significant adsorption at the PDMS surface, a passivation method was developed based on consecutive deposition of three layers of biopolymers, i.e., biotinylated IgG, neutravidin, and biotinylated dextran. An immunoreactor exhibiting low nonspecific binding could be constructed by exchanging the dextran with, e.g., a biotinylated antibody, at desired locations in the microchannels.

The use of centrifugal force for fluid control has also been demonstrated in connection to immunoassays [72,73]. Specifically, a sandwich myoglobin assay was carried out with a bead-based method in which an antimyoglobin antibody was immobilized on agarose beads or on phenyldextran-coated polystyrene particles and then packed in 10-nL columns on a CD. Samples containing different concentrations of myoglobin in crude mixtures were processed simultaneously in blocks of 12 parallel structures, and bound myoglobin detected using an Alexa-labeled monoclonal antibody. Under these conditions the integrated fluorescence intensity generated a dose-response curve from 1–1020 nM myoglobin (Fig. 12) [78]. Twenty-four samples were *simultaneously* assayed in the CD within 22 min.

While immunoassays in microfluidic systems are only in their infancy, and the demands on data accuracy and reliability rather moderate, macroscale immunoassays are of course a mature technology, used extensively in numerous contexts, particularly in diagnostics. Although the overall success of immunoassays in the latter context cannot be disputed, it is also well known that particularly heterogeneous assays suffer from numerous shortcomings and difficulties. These may lead to either over-estimation or underestimation of the antigen or antibody assayed. For example, when

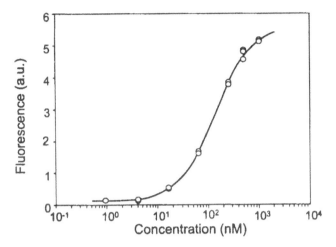

FIG. 12 Dose-response curve of myoglobin in fetal calf serum for a sandwich assay performed on a packed bed. Quantification was obtained by measuring fluorescent signals using a LIF detector by scanning the columns in situ while rotating the CD. (From Ref. 78.)

an immunoassay employs antibodies to recognize surface-bound antigens (e.g., proteins), immune-pair formation may not occur, e.g., if the antigen orientation at the surface is wrong, if the antigen is buried in a multilayer, if the antigen has undergone interfacial conformational changes, or if the surface layer is dense and antibody–antigen pair formation is precluded for steric reasons. All these scenarios lead to underestimation of the antigen concentration. Alternatively, if a sandwich immunoassay is used, unspecific adsorption of the labeled detection-step antibody may cause overestimation or false-positive answers. In fact, also in commercially available immunoassays, such false-positive answers continue to cause considerable problems, despite extensive use of different types of "blocking agents," such as diluted serum, casein, or albumin, the role of which is to adsorb at the base surface spots and thereby reduce unspecific adsorption of the detection-step antibody. Naturally, there are also other origins of false-positive answers in immunoassays, e.g., cross-reactivity between antibodies. In such cases, surface modifications of any type may not be successful, and other approaches are required.

Due to the large surface area–to–bulk volume ratio, the problems associated with nonspecific adsorption of proteins in microchannels also occur in relation to immunoassays. In fact, it could be expected to be even more pronounced than found for macroscale wells. Just to mention one example, this has been discussed for an enzyme-linked immunoassay of IgM in a microsystem [79]. In this investigation, conventional blocking cocktail used in ELISA assays based on polystyrene microtiter plate failed to block the nonspecific adsorption of the analyte, a secondary antibody. However, adding serum to the blocking solution seemed to improve the signal-to-noise ratio, and sensitivity down to 17 nM was obtained.

While the use of blocking agents may sometimes result in a satisfactory low unspecific adsorption of the detection step antibody, this is far from always the case. In such instances, different surface modifications must be used. Such coatings should display no or very low unspecific protein adsorption, and at the same time allow for covalent immobilization of the antibody/antigen. The use of such coatings is discussed elsewhere in this volume (Chapters 6 and 27).

VI. ENZYME REACTIONS

A. Enzymatic Assays

Analogously to immunoassays, advantages relating to throughput, speed, and sample requirements can be expected if enzymatic reactions are carried out in microscale. Also analogously to immunoassay applications in microsystems, precautions must most likely be taken to avoid surface-induced enzyme inactivation or inhibition if enzymatic reactions are to be carried out in nanoliter scale volumes. As discussed above and elsewhere in this volume, proteins frequently undergo conformational changes on adsorption at an interface. While such surface-induced conformational changes may in some cases be advantageous for enzymatic activity (e.g., lipases adsorbing at hydrophobic surfaces), the opposite is generally observed, i.e., enzymes tend to lose their activity when adsorbed at an interface [2,3]. As an illustration of this, Fig. 13 shows results on the immobilization of trypsin at glycidyl methacrylate beads. As can be seen, immobilization at this polymer surface causes the specific activity to decrease by some 90%. If, on the other hand, the trypsin is immobilized on the corresponding dextrane-coated glycidyl methacrylate beads, a considerably higher enzymatic activity can be retained. As with immunoassays, therefore, surface modification in order to reduce unspecific protein binding is crucial for successful enzymatic assays.

   Enzymatic assays have been demonstrated in microfluidic systems using different mechanisms for fluid control. For example, in an early study the β-galactosidase catalyzed hydrolysis of resorufin β-D-galactopyranoside was performed in an electrokinetically controlled microchip [80]. Using precise control of substrate, enzyme, and inhibitor concentrations, kinetic constants were obtained that agreed well with those obtained with conventional macroscale systems. It was, however, found that unspecific protein adsorption to the channel walls resulted in an increase of the background signal, and at low concentrations as much as 50% of the total signal was due to nonspecific adsorption. The assay reproducibility was also affected negatively by the adsorption. Nevertheless, the potential benefit of carrying out enzymatic assays in microformat was clearly demonstrated, as the total assay was performed in only 20 min, simultaneously reducing the reagent amounts by four orders of magnitude.

   Using a pressure-driven system, a fluorogenic assay for T-cell phosphatase was recently demonstrated on a glass chip (Fig. 14) [30]. Furthermore, enzymatic assays using kinases (Fig. 15) [81] and alkaline phosphatase [31] have been performed in parallel in microfluidic devices utilizing centrifugal force as the driving mechanism, thereby demonstrating the usefulness of these principles for fluid control also in the context of enzymatic assays.

B. Tryptic Digestion

An interesting area of enzymatic activity in relation to microsystem is tryptic digestion as a precursor to protein structure analysis. At present, 2D-gel electrophoresis

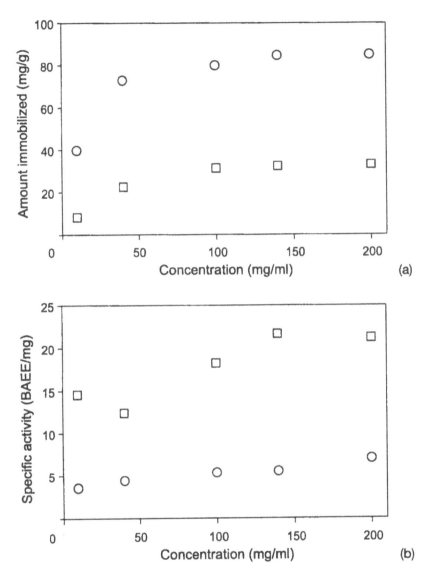

FIG. 13 Effects of trypsin concentration on its immobilization at unmodified (circles) and dextran-coated (squares) glycidyl methacrylate–based beads regarding the amount immobilized (a) and the specific activity displayed (b). The specific activity of trypsin in aqueous solution was 50 ± 5 BAEE/mg trypsin. (From Ref. 3.)

is the most powerful and extensively used separation method for proteins. Protein identification can then be carried out through peptide mass analysis using matrix-assisted laser desorption ionization–time of flight (MALDI-TOF) mass spectrometry (MS) techniques after cleavage of the protein. Generally, the enzymatic proteolysis requires times ranging from an hour to overnight, depending on reaction conditions and the concentration of the enzyme. Here, microsystems may offer real advantages relating to the faster reaction times in microsystems. Indeed, the feasibility of mi-

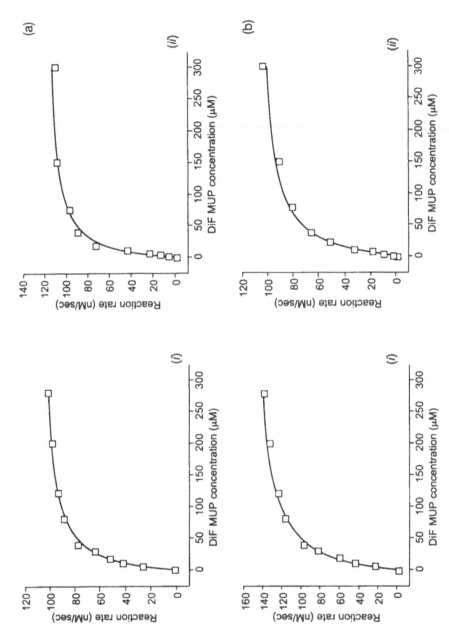

FIG. 14 Substrate titration curve obtained for T-cell phosphatase on-chip (i) and plate experiments (ii) with an aqueous buffer (a) and in a 50% glycerol/50% buffer mixture (b). Note that good agreement was obtained for the buffer, whereas the difference observed in the glycerol/buffer mixture was rather substantial. (From Ref. 30.)

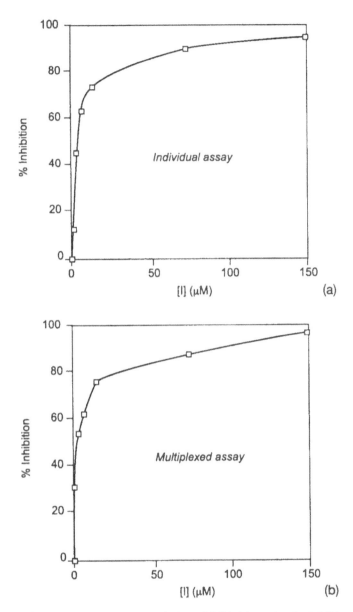

FIG. 15 Comparison of inhibition of PKA-kinase by the inhibitor PKI-(6-22) in multiplexed (i.e., multiple analytes in the same assay) and individual assays. The assay protocol and analysis procedure were essentially the same in the two cases. Note that the multiplexed assay gives results comparable to those obtained for the individual assay. (From Ref. 81.)

crosystems for this type of application has been demonstrated. For example, tryptic digestion of oxidized insulin has been performed in only 15 min under stopped flow conditions in a glass microchip device, as compared to the overnight digestion frequently employed in macrosystems [82], clearly demonstrating the increased effective reaction rate due to the small scale (Fig. 16) [83].

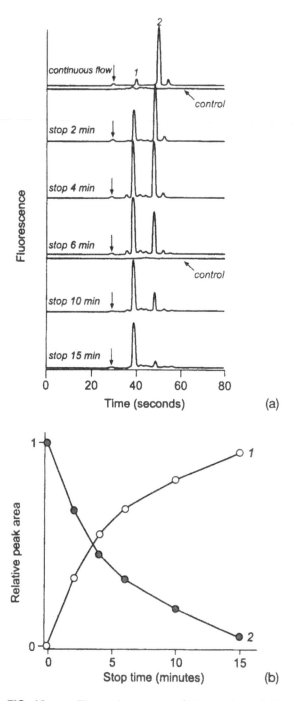

FIG. 16 (a) Electropherograms of the products following on-chip hydrolysis of oxidized insulin B-chain by bovine trypsin. A control run without insulin B-chains is shown for comparison. The arrows indicate the migration time of benzylamine, which was added as an electroosmotic mobility marker. All curves are plotted on the same scale with an offset for clarity. (b) Relative areas of the product (1, open circles) and intact β-chains (2, filled circles) as a function of the reaction time. Note the fast hydrolysis displayed. (From Ref. 82.)

Many proteins are unfortunately only available in very small amounts and often in the form of diluted solutions. Low abundance proteins are therefore particularly challenging to digest and analyze, and today it is the protein digestion step that limits MS analysis for protein identification. If high concentration of enzyme is used to increase the proteolysis rate, on the other hand, autoproteolysis often gives rise to fragments that interfere with the sample peptides in the MALDI spectrum. One approach to obtain fast digestion and simultaneously avoid or at least reduce auto-proteolysis is to immobilize trypsin on a solid phase [84–89]. It is important to note, however, that the protein either should be covalently immobilized at the surface, preferably through a spacer in order to retain as much as possible of the enzymatic activity (Fig. 13), or adsorbed at a hydrophobic surface. If, on the other hand, the enzyme is physically adsorbed at a hydrophilic surface such as silica or glass, au-toproteolysis may cause the enzyme adsorption to be only transient (Fig. 17) [90].

Using the approach of immobilizing trypsin in order to avoid autoproteolys, 10 fmol carbamidomethylated BSA could be successfully digested and analyzed by MS [84]. In an alternative strategy, proteins can be concentrated by adsorption onto reversed phase beads, washed, and then enzymatically digested while bound to the particles [91,92]. Using this procedure, proteins in concentrations of only 100 nM have been analyzed routinely (Fig. 18). Integrated devices here offer the potential benefit of reducing sample losses during sample preparation.

C. DNA Polymerase

Polymerase chain reactions using DNA polymerase is one of the most frequently investigated enzymatic systems in microscale devices [93–98]. Due to efficient heat transfer in the small scale, rapid heating/cooling cycles can be obtained, thus sig-nificantly reducing the processing time. However, inactivation of the DNA polymer-ase due to surface interactions has proven be significant, analogously to trypsin discussed above, and it is therefore important to "passivate" the surfaces prior to PCR reaction [99,100]. Such "passivation" has involved, e.g., silanization and sub-sequent treatment of different types. By the use of surface modification, successful amplification of DNA has been demonstrated in nanoliter volumes with a temperature cycle time of 30 s [96], clearly illustrating the benefits of chip devices.

VII. MICROARRAYS

Microarrays for genomic analysis is advancing rapidly and several commercial prod-ucts are available. These arrays can contain thousands of different probe reagents in spots or wells a few micrometers in size, placed on a solid support. The function of genes, however, is described by the activity of their translated proteins. Protein mi-croarrays (or protein chips) could therefore, in a similar way, become a versatile and powerful tool for proteomics, potentially allowing thousands of proteins to be studied simultaneously. These chips could be used not only to study protein–protein inter-actions, but also protein–small molecule interactions and enzyme–substrate reac-tions. They may also allow differential profiling, such as distinguishing the proteins of a healthy cell from those of a diseased cell. Identification of therapeutic markers and targets is another plausible application area.

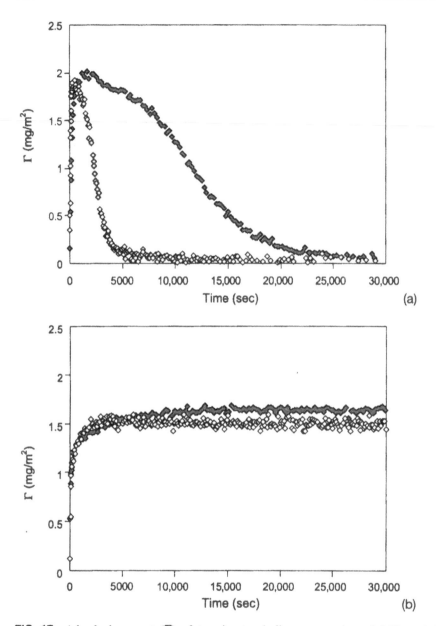

FIG. 17 Adsorbed amount (Γ) of trypsin at a bulk concentration of 0.01 mg/mL on (a) hydrophilic silica and (b) hydrophobic methylated silica at 22°C (filled symbols) and 34°C (open symbols). Note the autoproteolysis at silica and the absence of such effects at methylated silica. (From Ref. 90.)

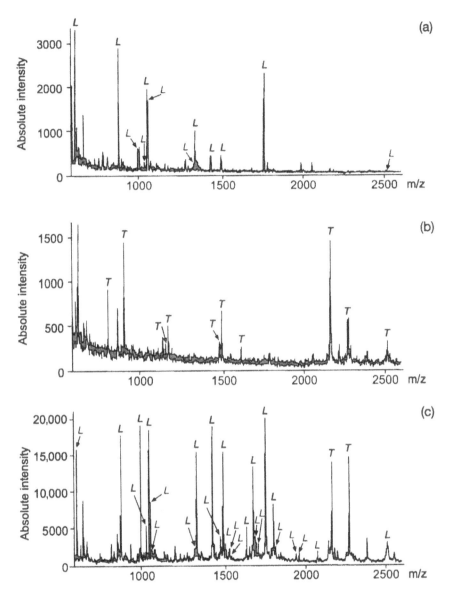

FIG. 18 MALDI spectra obtained from the tryptic digestion of (a) 500 n*M* lysozyme in solution, (b) 100 n*M* lysozyme in solution, and (c) 100 n*M* lysozyme that was concentrated and digested on porous beads. L denotes peptides from lysozyme, while T denotes peptides from trypsin. Note (1) that a lysozyme concentration of the order of 500 n*M* is required for obtaining sufficient amounts of lysozyme-originating peptides for MS analysis, while spectra at lower initial lysozyme concentrations are dominated by trypsin peptides, and (2) that bead immobilization improves the situation and in this case allows lysozyme to be fragmented and analyzed also at a concentration of 100 n*M*.

A functioning protein microarray consists of several different components: the chip itself including the surface, protein capture probes, e.g., antibodies, and a procedure to attach them onto the surface, a system to deliver the sample liquid to the probes, a detection system to read the chip and software programs to interpret the results. Depending on the purpose of the protein chip, the requested information may diverge in characteristics. For example, the applications of target discovery, on one hand, and diagnostic tests, on the other, most likely require different microarray setups.

While DNA arrays generally involve flat surfaces crowded with probe reagents, the protein arrays are likely to appear in different formats. Spots or wells patterned on a surface is of course one approach [101–110]. Another variant involves microfluidic networks (μFN), an array of channels (generally produced in PDMS), which are sealed against a flat surface [101,111–113]. The probes are immobilized in lines and the μFN is then rotated 90°. Samples can then be flushed across the different probes. The capturing probes can also be attached to individually labeled particles, which are then added to the sample. Since the beads are in the 5-μm range, this "3D array" is much denser than a typical spotted array, which has a distance of 100–200 μm from center to center of each probe [114,115]. Parallel processing of capture columns in nanoliter scale is yet another attractive format [72,73], enabling a large number of samples to be analyzed simultaneously with their calibration curves, thus being a suitable technique for quantitative determination in parallel.

A key question for protein arrays is the "level of quantification" required for a specific application. In some cases "presence" or "absence" may constitute a sufficient information level, while in other contexts the amount relative to that in another sample may be required, and in yet other cases absolute quantification is required. Unfortunately, no detection method available today is suitable for both identification and quantification of proteins in microanalysis systems. Particularly quantification may become a challenge in these systems due to the small size of the spots and the small amount of protein present in one such spot. A protein monolayer at a flat surface contains in the range of 1–10 mg protein/m^2, while the protein molecular weights generally lie between 10^4 and 10^6 g/mol. If a microarray spot has a diameter of 200 μm, a brief calculation yields protein amounts in the low fmol or sub-fmol range.

Apart from aspects relating to signal strength and detection level, the spot size in protein arrays could affect the performance of the latter also in other ways. In particular, analogously to the faster diffusion in microanalysis systems discussed above, one could expect diffusion-limited mass transport to be faster the smaller the spot size. Hence, the smaller the spot size for diffusion-limited systems, the faster the assay kinetics (Fig. 19a) [116]. Naturally, the spot size is of importance only when the difference in diffusion over the length scale variations is appreciable, and hence minor or no effects of the spot size can be expected, e.g., for flow systems where mass transport through diffusion is rather limited (Fig. 19b) [117], for rather minor spot size variations, or for high reactant concentrations.

Clearly, analysis of low abundance proteins will require detection methods having very high sensitivity. Furthermore, the dynamic concentration range of expressed proteins in biological samples is estimated to be at least seven orders of magnitude [118]. Fluorescence-based methods reach sensitivities in the sub-attomol range, and are today the most commonly used techniques. Quantification can also

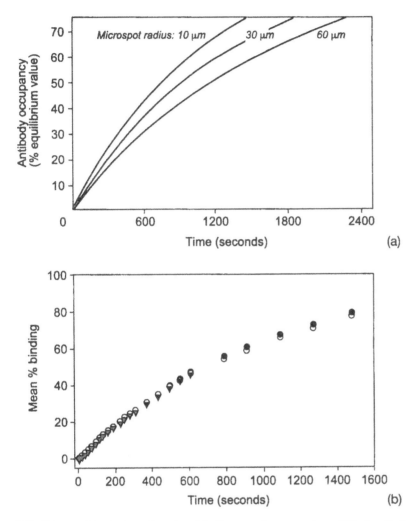

FIG. 19 (a) Simulation of analyte binding to microspots in a diffusion-limited system as a function of the microspot size. (From Ref. 116.) (b) Effect of spot size in a flow system. Binding of Cy5-labeled mouse IgG at PDMS patterned surfaces coated with goat anti-mouse IgG. The spot size was 260 μm (filled circles), 470 μm (open circles), 1145 μm (filled triangles). (From Ref. 117.)

be carried out using e.g., surface plasmon resonance (SPR), ellipsometry, reflectometry, or some other reflection-based technique [119,120] (see Chapters 21 and 22), having the advantage of not requiring any labels or genetically engineered tags that may interfere with the protein, but on the other hand providing a slightly poor sensitivity.

Mass spectrometry technology is widely viewed as the cornerstone proteomics tool for identification of proteins. Characterization is performed direct on proteins or via peptides using either MALDI-TOF technology or electrospray ionization MS. There is a strong focus today on developing mass spectrometers that provide even higher throughput and automation and more powerful and precise protein analysis. Consequently, MS systems with integrated high throughput microanalysis systems offer opportunities here. Unfortunately, MS does not provide a good tool for quantification and is therefore mainly used for identification. (For a more detailed discussion on the use of MS for studies of protein adsorption, see Chapter 24.)

Whereas ordered arrays of peptides have been used for some time now as a successful strategy for parallel protein analysis, characterization of more complex interactions between proteins is a newer field. Several approaches to protein microarrays and protein chips have been described recently and their progress discussed in recent reviews [121–123]. Glass slides are most commonly used in this context since they have the advantage that they can be used with standard scanners used for DNA chips [102]. Other variants of protein chips include filters [103] or aluminum surfaces [104]. Another strategy is "gel pad" slides with the purpose of increasing the immobilization capacity, thereby providing a good dynamic range for quantitative comparison. In such slides, polyacrylamide is photopolymerized on silane-treated glass surfaces to form gel-pads (as small as 10×10 μm) separated by a hydrophobic area [106,107]. Such gels have been reported to provide >100 times greater capacity for immobilization compared to a conventional glass support. In a similar manner, thin films of agarose coated on glass slides have been suggested as a support for protein arrays [105].

Micro- or nanowells engineered in either silicon chips or of silicone elastomer sheets is another format which has been attempted for protein microarrays. In one such approach, PDMS microwells were placed on top of a microscope slide and activated with a silane cross-linker [124]. Protein kinases were then covalently attached to the wells and kinase assays were performed successfully. In a somewhat different approach, nanowells were fabricated in silicon and then coated with nitrocellulose to improve the performance of the array [108]. Recombinant antibody fragments, used as capturing agents, were deposited into the nanovial chips using a dispenser. The antibody–antigen interactions were then analyzed with mass spectrometry.

Altogether, a variety of concepts of protein microarrays have been demonstrated, thereby indicating the possibilities of the field. However, most studies have been carried out using pure antibody–antigen solutions of various concentrations, while fewer studies have involved crude biological samples. Protein arrays are dependent on the characteristics of the capturing probes. Ideally, the capturing probes should be capable of binding their target protein with high efficiency and with sufficient specificity. Thus, they must be able to recognize an individual protein moiety from a complex mixture of proteins and discriminate all protein precursors and posttranslationally modified forms of proteins. Naturally, this makes the use of protein arrays together with multicomponent biological samples significantly more demanding than that of single protein solutions.

In a recent attempt to at least start to address the use of protein microarrays in complex biological samples, microarrays were constructed by printing microscopic spots of either antibodies or antigens into glass surface covered with poly(L-lysine) [109]. A comparative fluorescence assay was used to test the specificity, sensitivity, and accuracy of the array by analyzing the performance of 115 antibody–antigen pairs. Some of the antibody–antigen pairs allowed detection below 1 ng/mL, indicating sensitivities sufficient for measurements of many clinically important proteins in patient blood samples. The performance of the array was evaluated by adding varying concentration of fetal calf serum. It was found that the background fluorescence was about tenfold higher at the higher protein concentration, and the spots became much less clear against the high background. Higher overall protein concentrations also significantly reduced the precision of the measurements of specific proteins. Not surprisingly, therefore, the authors suggested better passivation of the array surface as one strategy for lowering the detection limits of the system.

While proteins can also be immobilized by physisorption, they adsorb with different efficiency when spotted onto solid supports, resulting in lack of compositional control of the formed array. Furthermore, proteins have a tendency to unfold upon adsorption, resulting in impaired function. For example, physical adsorption of antibodies for immunoassays has been shown to result in only limited surface activity of the immobilized probes (<20%) [67,125,126]. Again, an attractive strategy is to immobilize the capturing probes to a protein-resistant surface and thereby retain high protein activity [110] (see also Chapters 6 and 27). Clearly, it is desired to use immobilization techniques with capacity to yield high densities of the capturing probes in order to increase the sensitivity of the array, although the proteins there may also be an optimal packing density of the capturing probes depending on the size of either the probe or the analyte. Furthermore, site-specific immobilization may be useful for preserving the binding activity.

VIII. CONCLUSIONS

Following the lead of the developments within DNA arrays, where commercial products are already available, protein arrays are currently the focus of considerable attention, both in academic research and industrial product development. So far, no protein array products have been commercialized, but this is bound to change within the next year or so. As discussed in the present chapter, several challenges still exist, notably related to the surface activity of proteins and peptides, the functional diversity of such compounds, difficulties in analysis of low abundance components, etc. However, considerable progress has been made in a number of important areas, notably relating to fluid control, where significant advances have been made regarding both material (e.g., disposable plastics for EOF-based devices) and design. Furthermore, the potential of microfluidic systems has been clearly demonstrated for a number of interesting application areas. For such systems to obtain widespread use, however, further developments are required, e.g., in relation to data quality (particularly important for applications such as immunoassays, where quantitative information is often required), capacity to handle also complex biological samples (such as blood, serum, or plasma), and throughput (of importance, e.g., in drug development work). Furthermore, since proteins display widely different behavior, e.g., regarding biological function and solubility, at different conditions, e.g., pH, ionic strength, and presence of divalent cations, successful development of protein chips most likely will require careful probe selection and segmentation. Also, since the level of integration between different processes within one device—and between sample handling, separation, analysis, and data handling—is still generally low, considerable emphasis will have to be placed on true integration of such systems. Notably within the context of the present volume, protein interfacial behavior is of paramount interest, and further work will have to be directed to surface modification in order to facilitate sufficient performance. Here, much can most likely be learned from the extensive work done in other applications areas on protein interfacial behavior, although size-dependent effects may also occur.

Nevertheless, it seems clear that the drive toward miniaturized system will continue to grow, that many of the remaining challenges will be overcome, and that the practical importance of such systems will increase drastically within the coming years.

ACKNOWLEDGMENTS

Rolf Ehrnström, Per Andersson, Mats Inganäs, and Susanne Wallenborg are gratefully acknowledged for fruitful discussions. This work was financed by the Institute for Research and Competence Holding AB (IRECO), Sweden, and KK-stiftelsen, Sweden (MM).

REFERENCES

1. A. Manz, N. Graber, and H. M. Widmer, Miniaturized total chemical analysis systems: a novel concept for chemical sensing. Sens. Actuators $B1$:244–248, 1990.
2. M. Malmsten, Formation of adsorbed protein layers. J. Colloid Interface Sci. 207:186–199, 1998.
3. M. Malmsten and A. Larsson, Immobilization of trypsin on porous glycidyl methacrylate beads: effects of polymer hydrophilization. Colloids Surf. B 18:277–284, 2000.
4. L. Vroman, The importance of surfaces in contact with phase reactions. Sem. Thromb. Hemostatis 13:79–85, 1987.
5. M. Malmsten and B. Lassen, Competitive adsorption at hydrophobic surfaces from binary protein systems. J. Colloid Interface Sci. 166:490–498, 1994.
6. M. Malmsten, D. Mueller, and B. Lassen, Sequential adsorption of human serum albumin (HSA), immunoglobulin G (IgG), and fibrinogen (Fgn) at HDMSO plasma polymer surfaces. J. Colloid Interface Sci. 193:88–95, 1997.
7. F. W. Went, The size of man. Am. Scientist 56:400–413, 1968.
8. G. J. M. Bruin, Recent developments in electrokinetically driven analysis on microfabricated devices. Electrophoresis 21:3931–3951, 2000.
9. M. Krishnan, V. Namasivayam, R. Lin, R. Pal, and M. A. Burns, Microfabricated reaction and separation systems. Curr. Opin. Biotechnol. 12:92–98, 2001.
10. G. H. W. Sanders and A. Manz, Chip-based microsystems for genomic and proteomic analysis. Trends Anal. Chem. 19:364–378, 2000.
11. L. E. Locascio, C. E. Perso, and C. S. Lee, Measurement of the EOF in plastic imprinted microfluidic devices and the effect of protein adsorption on flow rate. J. Chromatogr. A 857:275–284, 1999.
12. T. Arai and W. Norde, The behavior of some model proteins at solid–liquid interfaces. 1. Adsorption from single protein solutions. Colloids Surf. 51:1–15, 1990.
13. U. J. Meyer, D. Trau, G. Key, M. Meusel, and F. Spener, Comparison of different types of immunaffinity reactors in an electrochemical flow injection immunoanalysis system developed for residue analysis. Biocat. Biotransform. 17:103–124, 1999.
14. A. Dodge, K. Fluri, E. Verpoorte, and N. F. de Rooij, Electrokinetically driven microfluidic chips with surface-modified chambers for heterogenous immunoassays. Anal. Chem. 73:3400–3409, 2001.

15. N. Burns, J. M. Van Alstine, and J. M. Harris, Poly(ethylene glycol) grafted to quartz: analysis in terms of a site-dissociation model of electroosmotic fluid flow. Langmuir 11:2768–2776, 1995.

16. C. S. Effenhauser, G. J. M. Bruin, and A. Paulus, Integrated chip-based capillary electrophoresis. Electrophoresis 18:2203–2213, 1997.

17. D. C. Duffy, J. A. Schueller, S. T. Brittain, and G. M. Whitesides, Rapid prototyping of microfluidic switches in poly(dimethyl siloxane) and their actuation by electroosmotic flow. J. Micromech. Microeng. 9:211–217, 1999.

18. D. C. Duffy, J. C. McDonald, J. A. Scheuller, and G. M. Whitesides, Rapid prototyping of microfluidic systems in poly(dimethyl siloxane). Anal. Chem. 70:4974–4984, 1998.

19. D. Ross, T. J. Johnson, and L. Locascio, Imaging of electroosmotic flow in plastic microchannels. Anal. Chem. 73:2509–2515, 2001.

20. T. J. Johnson, D. Ross, M. Gaitan, and L. Locascio, Laser modification of preformed polymer microchannels: application to reduce band broadening around turns subject to electrokinetic flow. Anal. Chem. 73:3656–3661, 2001.

21. A. C. Henry, T. J. Tutt, M. Galloway, Y. Y. Davidson, C. S. McWhorter, S. A. Soper, and R. L. McCarley, Surface modification of poly(methyl methacrylate) used in the fabrication of microanalytical devices. Anal. Chem. 72:5331–5337, 2000.

22. Y. Liu, D. Ganser, A. Scheider, R. Liu, P. Grodzinski, and N. Kroutchinina, Microfabricated polycarbonate CE devices for DNA analysis. Anal. Chem. 73:4196–4201, 2001.

23. M. A. Roberts, J. S. Rossier, P. Bercier, and H. H. Girault, UV laser machined polymer substrates for the development of microdiagnostic systems. Anal. Chem. 69:2035–2042, 1997.

24. J. S. Rossier, A. Schwarz, F. Reymond, R. Ferrigno, F. Bianchi, and H. H. Girault, Microchannel networks for electrophoretic separations. Electrophoresis 20:727–731, 1999.

25. F. Bianchi, Y. Chevolot, H. J. Mathieu, and H. H. Girault, Photomodification of polymeric microchannels induced by static and dynamic eximer ablation: effect on the electroosmotic flow. Anal. Chem. 73:3845–3853, 2001.

26. S. L. R. Barker, D. Ross, M. J. Tarlov, M. Gaitan, and L. E. Locasio, Control of flow direction in microfluidic devices with polyelectrolyte multilayers. Anal. Chem. 72:5925–5929, 2000.

27. Y. Liu, J. C. Fanguy, J. M. Bledsoe, and C. S. Henry, Dynamic coating using polyelectrolyte multilayers for chemical control of electroosmotic flow in capillary electrophoresis microchips. Anal. Chem. 72:5939–5944, 2000.

28. S. L. R. Barker, M. J. Tarlov, H. Canavan, J. J. Hickman, and L. Locascio, Plastic microfluidic devices modified with polyelectrolyte multilayers. Anal. Chem. 72:4899–4903, 2000.

29. R.-L. Chien and J. W. Parce, Muliport flow-control system for lab-on-a-chip microfluidic devices. Fresenius J. Anal. Chem. 371:106–111, 2001.

30. M. Kerby and R.-L. Chien, A fluorogenic assay using pressure driven flow on a microchip. Electrophoresis 22:3916–3923, 2001.

31. D. C. Duffy, H. L. Gillis, J. L. Norman, N. F. Sheppard, and G. Kellogg, Microfabricated centrifugal microfluidic systems: characterization and multiple enzymatic assays. Anal. Chem. 71:4669–4678, 1999.

32. R. D. Johnson, I. H. A. Badr, G. Barrett, S. Lai, Y. Lu, M. J. Madou, and L. G. Bachas, Development of a fully integrated analysis system for ions based on ion-selective optodes and centrifugal microfluidics. Anal. Chem. 73:3940–3946, 2001.

33. G. Ekstrand, C. Holmquist, A. Edman Örlefors, B. Hellman, A. Larsson, and P. Andersson, Microfluidics in a rotating CD. In Micro Total Analysis System (A. van den Berg, W. Olthuis, and P. Bergveld, eds.). The Netherlands: Enschede, 2000, pp. 311–314.

34. N. Thomas, A. Ocklind, I. Blikstad, S. Griffiths, M. Kendrick, H. Derand, G. Ekstrand, C. Ellström, A. Larsson, and P. Andersson, Integrated cell-based assays in micro-fabricated disposable CD devices. In *Micro Total Analysis Systems* (A. van den Berg, W. Olthuis, and P. Bergveld, eds.). The Netherlands: Enschede, 2000, pp. 249–252.

35. A. Eckersten, A. Edman Örlefors, C. Elleström, K. Erickson, E. Löfman, A. Eriksson, S. Eriksson, A. Jorsback, N. Tooke, H. Derand, G. Ekstrand, J. Engström, A. K. Honerud, A. Aksberg, H. Hedsten, L. Rosengren, M. Stjernström, T. Hultman, and P. Anderson, High-throughput SNP scoring in a disposable microfabricated CD device. In *Micro Total Analysis System* (A. van der Berg, W. Olthuis, and P. Bergveld, eds.). Dordrecht, 2000, pp. 521–524.

36. T. Nylander and F. Tiberg, Wetting of β-casein layers adsorbed at the solid–aqueous interface. Coll. Surf. B: Biointerface *15*:253, 1999.

37. D. J. Harrison, K. Fluri, K. Seiler, Z. Fau, C. S. Effenhauser, and A. Manz, Micro-machining a miniaturized capillary electrophoresis-based chemical analysis system on a chip. Science *261*:895–897, 1993.

38. A. T. Woolley, G. F. Sensabaugh, and R. A. Mathies, High-speed DNA genotyping using microfabricated capillary array electrophoresis chips. Anal. Chem. *69*:2181–2186, 1997.

39. Y. Shi, P. C. Simpson, J. R. Scherer, D. Wexter, C. Skibola, M. T. Smith, and R. A. Mathies, Radial capillary array electrophoresis microplate and scanner for high-performance nucleic adic analysis. Anal. Chem. *71*:5354–5361, 1999.

40. C. A. Emrich, I. L. Medintz, H. Tian, L. Berti, and R. A. Mathies, Ultra-high throughput genetic analysis using microfabricated capillary array electrophoresis devices. In *Micro Total Analysis Systems* (J. M. Ramsey and A. van den Berg, eds.). Monterey, 2001, pp. 13–15.

41. I. L. Medintz, B. M. Paegel, R. G. Blazej, C. A. Emrich, L. Berti, J. R. Scherer, and R. A. Mathies, High-performance genetic analysis using microfabricated capillary array electrophoresis microplates. Electrophoresis *22*:3845–3856, 2001.

42. M.-C. Millot and C. Vidal-Madjar, Overview of the surface modification techniques for the capillary electrophoresis of proteins. Adv. Chromatogr. *40*:427–467, 2000.

43. I. Rodriquez and S. F. Y. Li, Surface deactivation in protein and peptide analysis by capillary electrophoresis. Anal. Chim. Acta *383*:1–26, 1999.

44. S. N. Krylov and N. J. Dovichi, Capillary electrophoresis for the analysis of biopolymers. Anal. Chem. *72*:111R–128R, 2000.

45. I. Recio, E. Molina, M. Ramos, and M. de Frutos, Quantitative analysis of major whey proteins by capillary electrophoresis using uncoated capillaries. Electrophoresis *16*: 654–658, 1995.

46. A. Asanov, L. DeLucas, P. Oldham, and W. Wilson, Interfacial aggregation of bovine serum albumin related to crystallizations conditions studied by total internal reflection fluorescence. J. Colloid Interface Sci. *196*:62–73, 1997.

47. C. Hayes and W. Norde, Globular proteins at solid/liquid interfaces. Colloids Surf. B *2*:517–566, 1994.

48. D. Corradini and G. Cannarsa, Capillary electrophoresis analysis of proteins in bare fused silica capillaries. LC-GC *14*:326–332, 1996.

49. T. M. McNerney, S. K. Watson, J. H. Sim, and R. L. Bridenbaugh, Separation of recombinant human growth hormone from *Escherichia coli* cell pellet extract by cap-illary zone electrophoresis. J. Chromatogr. *744*:223–229, 1996.

50. M. Gilges, M. H. Kleemiss, and G. Shomberg, Capillary zone electrophoresis separa-tions of basic and acidic proteins using poly(vinyl alcohol) coatings in fused silica capillaries. Anal. Chem. *66*:2038–2046, 1994.

51. Å. Emmer, M. Jansson, and J. Roeraade, Separation of pig liver esterase isoenzymes and subunits by capillary zone electrophoresis in the presence of fluorinated surfactants. J. Chromatogr. *672*:231–236, 1994.

52. T. Shibata, L. Ceriotti, B. H. Weiller, M. A. Roberts, E. Verpoorte, and N. F. de Rooij, Low density lipoprotein (LDL) analysis in a chip-based system. In *Micro Total Analysis Systems* (J. M. Ramsey and A. van den Berg, eds.). Monterey, 2001, pp. 513–514.

53. K. K. C. Yeung and C. A. Lucy, Suppression of electroosmotic flow and prevention of wall adsorption in capillary zone electrophoresis using zwitterionic surfactants. Anal. Chem. *69*:3435–3441, 1997.

54. K. P. Ananthapadmanaban, *Interactions of Surfactants with Polymers and Proteins*. Boca Raton: CRC Press, 1993.

55. D. Patel, W. Ritschel, P. Chalasani, and S. Rao, Biological activity of insulin in microemulsion in mice. J. Pharm. Sci. *80*:613–614, 1991.

56. D. Muller, M. Malmsten, and T. S. C. Booth, Adsorption of diblock copolymers of poly(ethylene oxide) and polylactide at hydrophobized silica from aqueous solution. J. Colloid Interface Sci. *228*:326–334, 2000.

57. Z. El Rassi and W. Nashabeh, Carbohydrate analysis. In *High Performance Liquid Chromatography and Capillary Electrophoresis* (Z. El Rassi, ed.). Amsterdam: Elsevier, 1995, p. 267.

58. B. H. Weiller, T. Shibata, L. Ceriotti, M. A. Roberts, D. Rein, J. B. German, J. Lichtenberg, E. Verpoorte, and N. F. de Rooij, Development of a new, chip-based method for lipoprotein analysis by capillary electrophoresis. In *Micro Total Analysis Systems* (J. M. Ramsey and A. van den Berg, eds.). Monterey, 2001, pp. 426–428.

59. Y. Liu, R. S. Foote, C. T. Culbertson, S. C. Jacobsom, R. S. Ramsey, and J. M. Ramsey, Electrophoretic separation of proteins on microchip. J. Microcolumn Sep. *12*:407–411, 2000.

60. N. H. Chiem and D. J. Harrison, Monoclonal antibody binding affinity determined by microchip-based capillary electrophoresis. Electrophoresis *19*:3040–3044, 1998.

61. N. Chiem and D. J. Harrison, Microchip-based capillary electrophoresis for immunoassays: analysis of monoclonal antibodies and theophylline. Analytical Chemistry *69*: 373–378, 1997.

62. S. B. Cheng, C. D. Skinner, J. Taylor, S. Attiya, W. E. Lee, G. Picelli, and D. J. Harrison, Development of a multichannel microfluidic analysis system employing affinity capillary electrophoresis for immunoassay. Anal. Chem. *73*:1472–1479, 2001.

63. L. Bousse, S. Mouradian, W. Ausserer, and R. Dubrow, A lab-on-a-chip system for automated protein sizing. In *Micro Total Analysis Systems* (J. M. Ramsey and A. van den Berg, eds.). Monterey, 2001, pp. 45–47.

64. L. B. Koutny, D. Schmalzing, T. A. Taylor, and M. Fuchs, Microchip electrophoretic immunoassay for serum cortisol. Anal. Chem. *68*:18–22, 1996.

65. F. von Heeren, E. Verpoorte, A. Manz, and W. Thormann, Micellar electrokinetic chromatography separations and analyses of biological samples on a cyclic planar microstructure. Anal. Chem. *68*:2044–2053, 1996.

66. D. J. Harrison, K. Fluri, N. Chiem, T. Tang, and Z. Fau, In *Transducers '95*. Proceedings of the 8th international conference on solid-state sensors and actuators and eurosensors. Stockholm, Sweden, 1995, pp. 752–755.

67. J. S. Rossier, G. Gokulrangean, H. H. Girault, S. Svojanovsky, and G. S. Wilson, Characterization of protein adsorption and immunosorption kinetics in photoablated polymer microchannels. Langmuir *16*:8489–8494, 2000.

68. T. Yang, S. Jung, H. Mao, and P. S. Cremer, Fabrication of phospholipid bilayer–coated microchannels for on-chip immunoassays. Anal. Chem. *73*:165–169, 2001.

69. K. Sato, M. Tokeshi, H. Kimura, and T. Kitamori, Determination of carcinoembryonic antigen in human sera by integrated bead-bed immunoassay in a microchip for cancer diagnosis. Anal. Chem. *73*:1213–1218, 2001.

70. K. Sato, M. Tokeshi, T. Odake, H. Kimura, T. Ooi, M. Nakao, and T. Kitamori, Integration of an immunosorbent assay system: analysis of secretory human immunoglobulin A on polystyrene beads in a microchip. Anal. Chem. *72*:1144–1147, 2000.

71. V. Linder, E. Verpoorte, W. Thormann, N. F. de Rooij, and H. Sigrist, Surface biopassivation of replicated poly(dimethyl siloxane) microfluidic channels and application for heterogeneous immunoreaction with on-chip fluorescence detection. Anal. Chem. 73: 4181–4189, 2001.

72. A. Eckersten, J. Khoshnoodi, G. Ekstrand, M. Inganäs, H. Derand, J. Engström, M. Ljungström, and P. Andersson, Quantifying sub-femtomole amounts of protein: a novel microfluidic approach to process multiple samples in parallel. Proteomics, 2002 (submitted).

73. M. Inganäs, G. Ekstrand, J. Engström, A. Eckersten, H. Dérand, and P. Andersson, Quantitative bio-affinity assays at nanoliter scale, parallel analysis of crude protein mixtures. In *Micro Total Analysis Systems* (J. M. Ramsey and A. van den Berg, eds.). Monterey, 2001, pp. 91–92.

74. N. Chiem and D. J. Harrison, Microchip systems for immunoassay: an integrated immunoreactor with electrophoretic separation for serum theophylline determination. Clin. Chem. 44:591–598, 1998.

75. J. Taylor, G. Picelli, and D. J. Harrison, An evaluation of the detection limits possible for competitive capillary electrophoretic immunoassays. Electrophoresis 22:3699–3708, 2001.

76. M. L. Pisarchick and N. L. Thompson, Binding of a monoclonal antibody and its F_{ab} fragment to supported phospholipid monolayers measured by total internal reflection fluorescence microscopy. Biophys. J. 58:1235–1249, 1990.

77. M. Mammen, F. A. Gomez, and G. M. Whitesides, Determination of the binding of ligands containing the N-2,4-dinitrophenyl group to bivalent monoclonal rat anti-DNP antibody using affinity capillary electrophoresis. Anal. Chem. 67:3526–3535, 1995.

78. A. Eckersten, J. Khoshnoodi, G. Ekstrand, M. Inganäs, H. Derand, J. Engström, M. Ljungström, and P. Andersson. Unpublished data.

79. E. Eteshola and D. Leckband, Development and characterization of an ELISA assay in PDMS microfluidic channels. Sens. Actuators B B72:129–133, 2001.

80. A. G. Hadd, D. E. Raymond, J. W. Halliwell, S. C. Jacobson, and J. M. Ramsey, Microchip device for performing enzyme assays. Anal. Chem. 69:3407–3412, 1997.

81. Q. Xue, A. Wainright, and S. Ganggakhedkar, Multiplexed enzyme assays in capillary electrophoretic single-use microfluidic devices. Electrophoresis 22:4000–4007, 2001.

82. N. Gottschlich, T. Culbertson, T. E. McKnight, S. C. Jagcobson, and J. M. Ramsey, Integrated microchip-device for the digestion, separation and postcolumn labeling of proteins and peptides. J. Chromatogr. B 745:243–249, 2000.

83. Q. Xue, Y. M. Dunayevskiy, F. Foret, and B. L. Karger, Integrated multichannel microchip electrospray ionization mass spectrometry: analysis of peptides from on-chip tryptic digestion of melittin. Rapid Commun. Mass Sprectrom. 11:1253–1256, 1997.

84. J. Gobom, E. Nordhoff, R. Ekman, and P. Roepstorff, Rapid micro-scale proteolysis of proteins for MALDI-MS peptide mapping using immobilized trypsin. Int. J. Mass Spectrom. 169/170:153–163, 1997.

85. L. Licklider, W. G. Kuhr, M. P. Lacey, T. Keough, M. P. Purdon, and R. Takigiku, On-line microreactors/capillary electrophoresis/mass spectrometry for the analysis of proteins and peptides. Anal. Chem. 67:4170–4177, 1995.

86. D. Dogruel, P. M. Williams, and R. J. Nelson, Rapid tryptic mapping using enzymatically active mass spectrometer probes. Anal. Chem. 67:4343–4348, 1995.

87. S. Ekström, P. Önnerfjord, J. Nilsson, M. Bengtsson, T. Laurell, and G. Marko-Varga, Integrated microanalytical technology enabling rapid and automated protein identification. Anal. Chem. 72:286–293, 2000.

88. C. Wang, R. D. Oleschuk, F. Ouchen, J. Li, P. Thibault, and D. J. Harrison, Integration of immobilized trypsin bead beds for protein digestion within a microfluidic chip incorporating capillary electrophoresis separations and an electrospray mass spectrometry interface. Rapid Commun. Mass Spectrom. 14:1377–1383, 2000.

89. Y. Jiang and C. S. Lee, On-line coupling of micro-enzyme reactor with micro-membrane chromatography for protein digestion, peptide separation and protein identification using electrospray ionization mass spectrometry. J. Chromatogr. A 924:315–322, 2001.

90. I. Hahn Berg, D. Muller, T. Arnebrant, and M. Malmsten, Ellipsometry and TIRF studies of enzymatic degradation of interfacial proteinasceous layers. Langmuir 17:1641–1652, 2001.

91. M.-I. Aguilar, D. J. Clayton, P. Holt, V. Kronina, and R. I. Boysen, RP-HPLC binding domains of proteins. Anal. Chem. 70:5010–5018, 1998.

92. A. Doucette, D. Craft, and L. Li, Protein concentration and enzyme digestion on microbeads for MALDI-TOF peptide mass mapping of proteins from dilute solutions. Anal. Chem. 72:3355–3362, 2000.

93. V. I. Furdui and D. J. Harrison, Immunomagnetic separation of rare cells on chip for DNA assay sample preparation. In Micro Total Analysis Systems (J. M. Ramsey and A. van den Berg, eds.). Monterey, 2001, pp. 289–290.

94. C. F. Chou, R. Changrani, P. Roberts, D. Sadler, S. Lin, A. Mulholland, N. Swami, R. Terbrueggen, and F. Zenhausern, A miniaturized cyclic PCR device. In Micro Total Analysis Systems (J. M. Ramsey and A. van den Berg, eds.). Monterey, 2001, pp. 151–152.

95. H. Nagai, Y. Murakami, S. Wakida, E. Niki, and E. Tamiya, High throughput single cell PCR on a silicon microchamber array. In Micro Total Analysis Systems (J. M. Ramsey and A. van den Berg, eds.). Monterey, 2001, pp. 268–270.

96. E. T. Lagally and R. A. Mathies, Integrated PCR-CE system for DNA analysis to the single molecule limit. In Micro Total Analysis Systems (J. M. Ramsey and A. van den Berg, eds.). Monterey, 2001, pp. 117–118.

97. A. T. Woolley, D. Hadley, P. Landre, A. J. deMello, R. A. Mathies, and M. A. Northrup, Functional integration of PCR amplification and capillary electrophoresis in a microfabricated DNA analysis device. Anal. Chem. 68:4081–4086, 1996.

98. L. C. Waters, S. C. Jacobson, N. Kroutchinina, J. Khandurina, R. S. Foote, and J. M. Ramsey, Microchip device for cell lysis, multiplex PCR amplification, and electrophoretic sizing. Anal. Chem. 70:158–162, 1998.

99. M. A. Shoffner, J. Cheng, G. Hvichia, L. J. Kricka, and P. Wilding, Chip PCR. I: Surface passivation of microfabricated silicon-glass chips for PCR. Nucl. Acids Res. 24:375–379, 1996.

100. J. Cheng, M. A. Shoffner, G. Hvichia, L. J. Kricka, and P. Wilding, Chip PCR. II: Investigation of different PCR amplification systems in microfabricated silicon-glass chips. Nucl. Acids Res. 24:381–385, 1996.

101. D. Juncker, A. Bernard, I. Caelen, H. Schmid, A. Papra, B. Michel, N. F. de Rooij, and E. Delamarche, Microfluidic networks for patterning biomolecules and performing bioassays. In Micro Total Analysis Systems (J. M. Ramsey and A. van den Berg, eds.). Monterey, 2001, pp. 429–431.

102. G. Macbeath and S. L. Achreiber, Printing proteins as microarrays for high-throughput function determination. Science 289:1760–1763, 2000.

103. R. M. T. de Wildt, C. R. Mundy, B. D. Gorick, and I. A. Tomlinson, Antibody arrays for high throughput screening of antibody antigen interactions. Nature Biotech. 18:989–994, 2000.

104. N. V. Avseenko, T. Y. Morozova, F. I. Ataullakhanov, and V. N. Morozov, Immobilization of proteins in immunochemical microarrays fabricated by electrospray deposition. Anal. Chem. 73:6047–6052, 2001.

105. V. Afanassiev, V. Hanemann, and S. Wölfl, Preparation of DNA and protein micro arrays on glass slides coated with an agarose film. Nucl. Acids Res. 28:e66 i–v, 2000.

106. D. Guschin, G. Yershov, A. Zaslavsky, A. Gemmell, V. Shick, D. Proudnikov, P. Arenkov, and A. Mirzabekov, Manual manufacturing of oligonucleotide, DNA, and protein microchips. Anal. Biochem. 250:203–211, 1997.

107. P. Arenkov, A. Kukhtin, A. Gemmell, S. Voloschuk, V. Chupeeva, and A. Mirzabekov, Protein microchips: use for immunoassay and enzmatic reactions. Anal. Biochem. 278: 123–131, 2000.

108. C. A. K. Borrebaeck, S. Ekström, A. C. Malmborg Hager, J. Nilsson, T. Laurell, and G. Marko-Varga, Protein chips on recombinant antibody fragments: a highly sensitive approach as detected by mass spectrometry. BioTechniques 30:1126–1131, 2001.

109. B. B. Haab, M. J. Dunham, and P. O. Brown, Protein microarrays for highly parallel detection and quantitation of specific proteins and antibodies in complex solutions. Genome Biology 2:4.1–13, 2001.

110. L. A. Ruiz-Taylor, T. L. Martin, F. G. Zaugg, K. Witte, P. Indermuhle, S. Nock, and P. Wagner, Monolayers of derivatized poly(L-lysine)–grafted poly(etylene glycol) on metal oxides as a class of biomolecular interfaces. PNAS 98:852–857, 2001.

111. E. Delamarche, Patterned delivery of immunoglobulins to surfaces using microfluidic networks. Science 276:779–781, 1997.

112. A. Papra, A. Bernard, D. Juncker, N. B. Larsen, B. Michel, and E. Delamarche, Microfluidic networks made of poly(dimethylsiloxane), Si, and Au coated with PEG for patterning proteins onto surfaces. Langmuir 17:4090–4095, 2001.

113. A. Bernard, B. Michel, and E. Delamarche, Micromosaic immunoassays. Anal. Chem. 73:8–12, 2001.

114. J. A. Ferguson, F. J. Steemers, and D. R. Walt, High-density fiber-optic DNA random microsphere array. Anal. Chem. 72:5618–5624, 2000.

115. K. L. Michael, L. C. Taylor, S. L. Schultz, and D. R. Walt, Randomly ordered addressable high-density optical sensors arrays. Anal. Chem. 70:1242–1248, 1998.

116. R. Ekins. Ligand assays: from electrophoresis to miniaturized microarrays. Clin. Chem. 44(9):2015–2030, 1998.

117. K. E. Sapsford, Z. Liron, Y. S. Shubin, and F. S. Ligler, Kinetics of antigen binding to arrays of antibodies in different sized spots. Anal. Chem. 73:5518–5524, 2001.

118. G. L. Corthals, V. C. Wasinger, D. F. Hochstrasser, and J.-C. Sanchez, The dynamic range of protein expression: a challenge for proteomic research. Electrophoresis 21: 1104–1115, 2000.

119. R. W. Nelson, D. Nedelkov, and K. A. Tubbs, Biosensor chip mass spectrometry: a chip based proteomics approach. Electrophoresis 21:1155–1163, 2000.

120. C. P. Sönksen, E. Nordhoff, Ö. Jansson, M. Malmqvist, and P. Roepstorff, Combining MALDI mass spectrometry and biomolecular interaction analysis using a biomolecular interaction analysis instrument. Anal. Chem. 70:2731–2736, 1998.

121. H. Zhu and M. Snyder, Protein arrays and microarrays. Curr. Opin. Chem. Biol. 5:40–45, 2001.

122. R. E. Jenkins and S. R. Pennington, Arrays for protein expression profiling: towards a viable alternative to two-dimensional gel electrophoresis? Proteomics 1:13–29, 2001.

123. G. Walter, K. Bussow, D. Cahill, A. Lueking, and H. Lehrach, Protein arrays for gene expression and molecular interaction screening. Curr. Opin. Microbiol. 3:298–302, 2000.

124. H. Zhu, J. F. Klemic, S. Chang, P. Bertone, A. Casamayor, K. G. Klemic, D. Smith, M. Gerstein, M. A. Reed, and M. Snyder, Analysis of yeast protein kinases using protein chips. Nature Genetics 26:283–289, 2000.

125. J. N. Lin, J. D. Andrade, and I. N. Chang, The influence of adsorption of native and modified antibodies on their activity. J. Immunol. Methods 125:67–77, 1989.

126. J. N. Lin, I. N. Chang, J. D. Andrade, J. N. Herron, and D. A. Christensen, Comparison of site-specific coupling chemistry for antibody immobilization on different solid supports. J. Chromatogr. 542:41–54, 1991.

29
Protein Adsorption in the Oral Environment

THOMAS ARNEBRANT   Institute for Surface Chemistry, Stockholm, and Malmö University, Malmö, Sweden

I. INTRODUCTION

Saliva is a complex secretion containing proteinaceous, carbohydrate, as well as lipidic components. The composition varies among individuals but typical values are 2, 3–4, and 0.1 g/L, respectively [1,2]. For a thorough review on the composition of saliva and variations therein see, e.g., Refs. 1 and 3. The protein fraction includes more than 40 different components ranging in molecular weight from short peptides to assemblies with molecular weights in excess of 10^6 Da [4].

The presence of proteins and their inherent surface activity leads to a rapid coverage of interfaces by a proteinaceous film and formation of the so-called acquired pellicle, which is the acellular organic film that is deposited on the tooth surface after eruption or cleaning [5]. The role of this film is manifold, for example, it protects the enamel from acid attack and dissolution of the enamel as well as crystallization of calcium phosphate salts. In addition, it has a lubricating function and modulates bacterial attachment. The last function connects to the formation of dental plaque, which is an accumulation of microorganisms, mainly bacteria, forming a biofilm on oral interfaces. Proteins may also monitor bulk association/aggregation phenomena involving self-association, protein–protein interactions, and protein-mediated aggregation, so-called agglutination, of, e.g., bacteria.

The adsorption and adhesion sequences involved in pellicle and plaque formation on teeth, mucosa, and restorative materials as well as protective mechanisms relying on agglutination or barrier functions, originate in specific or nonspecific interactions that in turn depend on protein–surface or protein–protein interactions. Through such effects protein–surface activity will be a prerequisite for the adequate performance of the saliva in its protective role, including the mentioned functions of the pellicle, but may also be important, e.g., in emulsification of food components. Furthermore protein interfacial behavior will be affected by compositional changes in saliva not only in the proteinaceous components but also in, for example, polar lipids and carbohydrates as well as in the ionic environment. It is well known that subtle changes in a system of this complexity may give significant effects on adsorption and consequently influence all the above mentioned phenomena.

From a clinical perspective, the formation of dental plaque can in later stages give rise to so-called plaque-related diseases such as caries and periodontal disease.

811

Furthermore, as mentioned above, the protective role of saliva depends among other things on agglutination in solution and on keeping the mucosal barrier intact where associated defense proteins may play an important role. The latter function is also related to lubrication of oral surfaces, which involves both solid (teeth, restorative materials) surfaces and the mucosa and is a prerequisite to avoid abrasion and protect teeth and tissue from wear and damage. An understanding of protein adsorption in these applications will be an important piece of information needed to resolve the complex phenomena involved and to be able to develop more efficient antiplaque compounds, salivary substitutes, and dental care products for the consumer market.

With respect to the relevance of protein adsorption in dental applications, implant integration and biocompatibility of course are important areas. In these cases, however, material–tissue interactions are of main importance, and measures are taken to avoid exposure to saliva. As such interactions are comprehensively treated in the biomaterials literature the topic is left out of the present chapter.

The purpose of the present contribution is to outline the role of salivary protein adsorption in the oral environment through a review of the literature. Emphasis has been given to the relatively sparse data on physicochemical behavior and an attempt is made to relate these to the vast body of biochemical information available.

II. SALIVARY PROTEINS

Human whole saliva (HWS) contains the secretions from the major and minor salivary glands along with crevicular fluid and bacterial and cellular components. The most well-defined and characterized proteins are those secreted from the major salivary glands [6]. The reason is that the presence of bacteria and desquamated cells and their respective extracellular enzymes complicates the biochemical characterization of the proteins in the whole secretion. Furthermore, the secretions from the minor glands are only available in small quantities. The major glands, the parotid and sublingual-submandibular, are known to have different profiles with respect to protein content and composition. Human parotid saliva (HPS) is a serous secretion rich in, e.g., phosphoproteins like proline-rich proteins (PRPs) (see below), whereas the human submandibular-sublingual saliva (HSMSLS) has a higher content of mucuous proteins [1,3,7–9]. The molecular weights and isoelectric pHs (where available) of the major proteins found in saliva are illustrated in Fig. 1a.

Proteins may be divided into classes or families where differences within each may depend on transcriptional, translational, or post-translational modification (see, e.g., Ref. 10). From a functional perspective the salivary proteins may be classified according to different roles in the oral cavity where many of them are suggested to perform a multitude of roles [6,10]. Figure 1b gives an illustration of the main components and their functions. The main classes of protein in saliva are α-amylase, lactoferrin, lactoperoxidase, lysozyme, secretory IgA, cystatins, albumin, complement proteins, acidic proline-rich proteins, statherin, histatins, as well as glycosylated proline-rich glycoproteins and mucous glycoproteins [4,6,10,11].

Upon formation of the pellicle differences in concentration, diffusion rates and surface affinity of the proteins, selective adsorption may be anticipated and has also been demonstrated experimentally (cf. [8,12]). Furthermore, exchange reactions receiving substantial attention in research on blood contact materials [13–16] may be

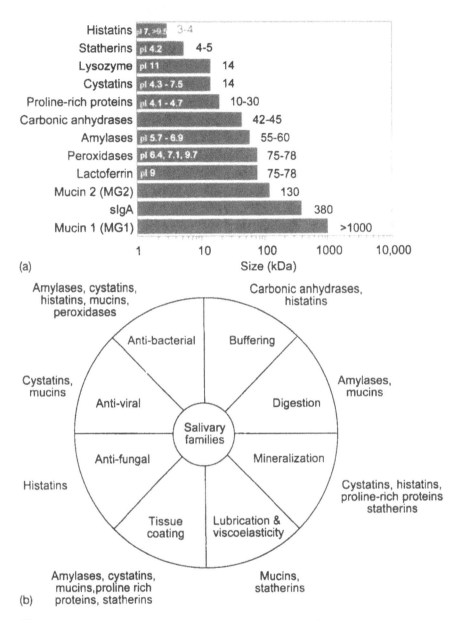

FIG. 1 (a) Approximate molecular weights of the major human salivary proteins and glycoproteins. The isoelectric pH (pI) is also inserted where data are available. Note that values for the latter vary among different fractions. (Adapted from Ref. 243.) (b) Functional properties of the major salivary protein classes. Note the multifunctional nature of the components. (From Ref. 243.)

expected to occur and be of importance. Although differences are reported between in vitro and in vivo pellicles (see details below), the major pellicle proteins identified are secretory IgA (sIgA), acidic proline-rich proteins, statherin, histatins, cystatins, high-molecular-weight mucin (MG-1), lactoferrin, lysozyme, amylase, albumin, and peroxidase [17,18], but IgM, IgG, and C3/C3c have also been detected [17].

Proteins that are found in the initial adsorbed layer are denoted "pellicle precursor proteins" by some authors (cf. [8,17,19]). These are mainly relatively low-molecular-weight proteins/peptides such as statherin and acidic PRPs.

The properties of the major proteins are shortly summarized below. As the flexibility of the structure is known to strongly affect the adsorption behavior of polymers (Chapter 1) and proteins (Chapter 2), a division into groups representing the globular and flexible structures, respectively, has been done.

A. Globular

1. Secretory IgA

Secretory IgA is a member of the adaptive immune response. The molecule is made up from two 4-chain units of IgA, one secretory unit of 70 kD and one joining (J) chain of 15 kD [20].

2. Lysozyme

Muramidase, or lysozyme, is a protein with an antibacterial effect [21]. It is also known to possess a globular structure with a high degree of conformational stability (see Refs. 22 and 23 and references therein). Cleaves $\beta(1\text{-}4)$ glycosidic bonds between muramic acid and N-acetylglucosamine residues in the peptidoglycan in the bacterial cell wall [21].

3. Lactoferrin

Lactoferrin is an iron-binding bactericidal protein first identified in milk. The protein binds 2 atoms of iron per mol with simultaneous binding of 2 mol bicarbonate. One suggested mechanism for the bactericidal effect is that it deprives the microorganisms of their essential nutrient, but other not-completely-understood mechanisms may be present. Only the apo-form of the enzyme possesses the bactericidal activity [4].

4. Lactoperoxidase

Lactoperoxidase and thiocyanate together with hydrogen peroxide produced by microbials yield oxidized thiocyanate ion derivatives toxic to many microorganisms. They exist in three active subfractions with isoelectric points: 6.4, 7.1, and 9.7 [4,18].

5. α-Amylase

α-Amylase catalyzes the hydrolysis of $\alpha(1\text{-}4)$ glycosidic linkages between glucose residues of polysaccharides like starch. α-Amylase has a digestive function and has also been demonstrated to have antimicrobial activity against a few specific organisms [4]. It is present in virtually every mucosal fluid in the body [20].

6. Albumin

Albumin is the most abundant protein in blood serum/plasma. Presence in saliva reflects passive contribution possibly due to epithelial inflammation, detected in healthy individuals only in very minor amounts [20]. It binds to hydroxyapatite (HA) and has been reported to take part in the formation of the pellicle [24].

7. Cystatins

Cystatins are cystein proteinase inhibitors and are present in all mucosal secretions. They are reported to bind to HA and also to inhibit HA crystal growth [25].

B. Flexible

1. Proline-Rich Proteins

The acidic PRPs are proteins found only in saliva [20]. The aPRPs are encoded by the *PRH1* and *PRH2* gene loci, and contain five common allelic variants; PRP-1, PRP-2, Db, PIF, and Pa deviating by single amino acid substitutions (PRP-2, PIF, Pa) and a 21 amino acid internal repeat (Db) [26] (see Fig. 2).

Acidic PRPs are secreted as large and truncated PRPs due to post-translational proteolysis. In saliva, they are subject to endogenous and bacterial proteolysis into low-molecular-weight peptides [27].

The acidic PRP molecule is highly asymmetrical with a 30-residue N-terminal phosphorylated domain (2 mol of serine phosphate per mol) interacting with HA surfaces and calcium, a proline-rich middle portion, and a C-terminal portion reported

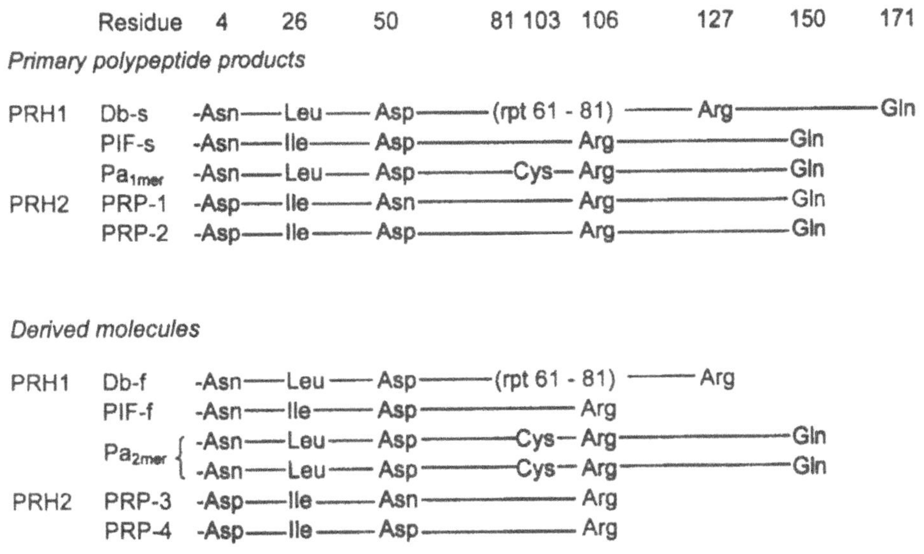

FIG. 2   Variants of acidic proline-rich proteins (PRPs). The PRP-1 and PRP-2 are encoded by PRH2 gene loci and Pa_{1mer}, Db-s and PIF-s by the PRH1. Post-translational cleavage results in the shorter PRP-3 and PRP-4 as well as Db-f, PIF-f. The Pa_{1mer} dimerizes to the Pa_{2mer} by a disulfide linkage at position 103. PRP-1 and PRP-2 differ by single amino acid substitutions at position 50. Db contains a 21 residue tandem repeat at residues 61–81. (From Ref. 26.)

to mediate bacterial adhesion. PRPs are potent inhibitors of spontaneous precipitation of calcium phosphate salts from saliva as well as of secondary crystal growth [28,29].

The molecular weights are 16 and 11 kD for the long (PRP-1 and -2) and short (PRP-3 and -4) PRPs and their isoelectric points (Ips) are 4.7 and 4.1, respectively [25,26]. Interestingly, salivary PRPs bear a structural resemblance to caseins in milk having a flexible structure and a highly segregated distribution of charged amino acids [30,31].

2. Statherin

Statherin is a 43 amino acid polypeptide rich in tyrosine. The molecule displays a highly negatively charged N-terminus interacting with calcium and HA surfaces [25] and a C-terminal domain containing hydrophobic amino acids [32]. Statherin is a potent inhibitor of spontaneous precipitation of calcium phosphate salts from saliva as well as of secondary crystal growth [28,29]. It is reported to contribute to the lubrication of oral surfaces [33].

3. Histatins

These proteins are exclusively found in saliva and belong to a family of neutral and basic histidine-rich peptides mainly in HPS but also to some extent in HSMSLS. Histatins are a family of polypeptides where histatins 1, 3, and 5 are the major ones and contain 38, 32, and 24 residues, respectively. Histatin 1 is neutral (Ip 7), whereas 3 and 5 are highly basic with Ip >9.5 [34]. Two histatins have been shown to originate from different gene loci, while the others are supposed to be formed by post-translational proteolysis. Involved on the formation of the (in vitro) acquired pellicle and in mineralization dynamics [25,35].

4. Glycosylated PRPs

Proline-rich glycoprotein (PRG) is secreted by the parotid gland and is a 39 kDa protein containing 40% carbohydrate. It contains a single peptide chain of 231 amino acids and 6 triantennary N-linked units evenly distributed along the chain. 75% of the amino acids are proline, glycine, and glutamine. Secondary structure is 70% random coil and 30% β-turn [11].

5. Mucins

Two major mucin populations in saliva, MG-1 (>1000 kD) and MG-2 (120–150 kD) have been identified [36,37] and are secreted by the submandibular-sublingual glands. MG-1 is composed of 78% carbohydrate, 15% protein, and 7% sulfate [38]. This large mucin has an oligomeric structure where mucin monomers are joined by disulfide bonds (Fig. 3).

The peptide chain contains heavily glycosylated hydrophilic domains flanked by less glycosylated regions that contain hydrophobic "patches" as evidenced by binding of the hydrophobic fluorescent probes (1-anilino-8-naphthalenesulfonate and N-phenyl-1-naphtylamine) [38]. These high-molecular-mass mucins bind strongly to HA in vitro [39] and participate in pellicle formation in vivo [40,41]. Recently MG-1 has been shown to be the product of the MUC5B gene [42–44]. MG-2, on the other hand, is composed of 70% carbohydrate and approximately 30% protein and

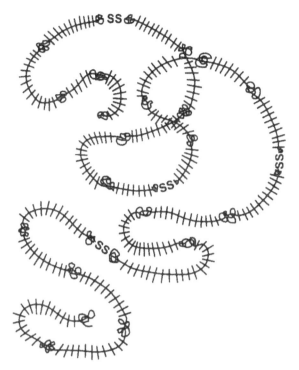

FIG. 3 Structure of a high-molecular-weight mucin MUC5B (MG-1). The structure is oligomeric with mucin subunits linked by disulfide bonds. The peptide chain contains heavily glycosylated hydrophilic domains flanked by regions containing hydrophobic domains. (For details, see Refs. 38 and 157.)

was previously described as a single monomeric molecule [36], but has recently been found to self-associate and thus form larger assemblies [45]. MG-2 interacts with oral bacterial and thereby has a function in bacterial clearance; for a review see Ref. 46.

C. Protein Association Phenomena and Structure Formation in Saliva

Protein self-association is well known to occur due to surface heterogeneity in protein structure as in the case of insulin [47]. More relevant in this context is lysozyme, which forms dimers and higher oligomers at higher concentration [21] (see also Ref. 23 and references therein). In analogy with the behavior of surfactants, amphiphilic proteins like caseins also exhibit this behavior, and β-casein has been demonstrated to have a "critical association concentration" at approximately 0.5 mg/mL [48,49]. Furthermore, the self-association of caseins is Ca^{2+} dependent and the basis for the formation of the so-called casein micelles in milk [50]. The caseins are also involved in the regulation of free calcium and inhibit crystal growth in the secretory cells [31]. As caseins possess structural resemblance to flexible salivary proteins like PRPs (see above) an analogy is obvious, and it is not surprising that such association phenomena are reported to take place also for salivary proteins [51,52].

Mucins are known to interact with lower-molecular-weight proteins and complex formation between MG-1 and amylase, PRPs, statherin, and histatins have been reported [53]. Furthermore, mapping of the binding domains for the interaction with histatin and statherin has been carried out [54]. A complexing between sIgA and MG-2 has also been demonstrated [55]. In a study by Wickström and coworkers sIgA, lysozyme, and lactoferrin were found to associate with MUC5B mucin (MG-1), and these results were interpreted as indicating a mucin matrix decorated with protective factors [56].

A complex between proline-rich glycoprotein and albumin has been reported to form and play a role in lubrication of oral tissues [57].

Yao and coworkers studied the cross-linking of pellicle precursor proteins like PRPs, statherin, and histatins and showed that complexes could form between PRP-1 and statherin [19] and that such cross-linking may take place by oral transglutaminase in vivo [58].

Salivary proteins also interact with lipids [2]. In parotid saliva the main part of the lipid exists as lipoproteins, whereas in submandibular saliva most of the lipids are associated with the high- and low-molecular-weight glycoprotein. It is indicated that the glycolipids interact with the glycosylated regions of the glycoproteins and that the phospholipids interact with the nonglycosylated parts. Furthermore, salivary mucins also contain fatty acids that are ester linked to nonglycosylated regions of the peptide chain [2].

As a result of the interactions discussed in the previous paragraph, possibilities for structural organization exist on several levels and supramolecular structures have been reported to form. In experiments with instantly frozen samples of human saliva investigated by TEM and light microcopy, Glantz and coworkers demonstrated the appearance of a filamentous network that was more clearly developed in HPS than in HSMSLS [59–62]. In a subsequent study it was reported that hydrophilic particles were accommodated in this network but that hydrophobic ones were less easily dispersed and present in the denser areas of the structures [63].

In a series of papers by Rykke and coworkers [51,52,64–66], they used photon correlation spectroscopy (PCS), SEM, TEM, cryo-TEM, and zeta potential measurements to demonstrate and characterize "micelle-like" aggregates in HPS and also that similar structures were observed on enamel after exposure in the oral cavity [64]. The studies show the presence of aggregates in the size range of 100–450 nm that were slowly growing in size with time, a process that could be inhibited by addition of pyrophosphate indicating a Ca^{2+} dependence [51]. It was also demonstrated that the structures could be enzymatically degraded, resulting in a reduction in size [67]. The effects of the enzymes investigated decreased in the order trypsin, pronase E, rennin. Furthermore, the aggregates were found to have a zeta potential of −9 mV at physiological conditions [66]. The similarities with casein micelles in milk are discussed in Ref. 52.

III.   ORAL INTERFACES

Interfaces in the oral cavity are of solid–liquid or air–liquid types. The latter are represented, e.g., by gas bubbles present in saliva and stabilized by adsorption of salivary protein or surfactants from dental care products such as toothpaste. The author has found no data on frequency or size distribution of bubbles in the oral cavity. It is clear, however, that the three-phase contact when bubbles travel along a solid–liquid interface will affect adsorption and adhesion. Langmuir–Blodgett deposition, which involves transfer of monolayers from air–liquid to solid–liquid interfaces, is well known and so is flotation of particles, and it has been recognized that bacterial adhesion experiments ought to be carried out in situ to avoid effects by moving three-phase boundaries [68,69]. Recently such phenomena have attracted interest in relation to adhesion and detachment of particles [70] and bacteria [71], due to the passage of an air–water interface, a process expected to occur at instances when a dried tooth surface is wetted by saliva or when bubbles are passing over the tooth surfaces. In such a process, the interfacial tension at the bubble–solution interface will effect the particle detachment [70]. In this connection, the magnitude and rate of surface tension reduction by salivary proteins is expected to play an important role. Analyses of salivary samples by the drop-volume method reveal that surface tension of whole saliva decays over a period of approximately 20 min and reaches a plateau at about 43 Mn/m [72]. In comparison with values for other protein systems this value is comparatively low, indicating a good packing efficiency in the adsorbed layer. It has also been demonstrated that the presence of clinically relevant concentrations of oral bacteria in the bulk phase did not significantly affect the magnitude and rate of surface tension decay [73].

The enamel surface consists of a mineral component, hydroxyapatite, and an organic matrix consisting of amelogenins, amelin, enamelin, enzymes, and serum-derived components such as albumin [74]. The former is present as crystallites of an approximately hexagonal shape (and packing) occupying approximately 86% of the volume (96–97% by weight) [75]. Crystallites are embedded in the organic phase and in turn arranged in prisms of a size of a few micrometers. Surface enamel may be aprismatic or prismatic, where crystals are aligned in parallel in the former but show abrupt changes in orientation at prism boundaries in the latter. Using the technique of AFM, erosion of enamel by acidic food constituents was recently investigated and it was found that erosion took place primarily in the aprismatic parts [76]. Using the same technique a study on the interaction between amelogenin and albumin and HA crystals was recently performed [74] and a preferential arrangement of adsorbed proteins, possibly correlated with charge domains at the crystal surface [77], was observed. Furthermore, a preferential adsorption of amelogenin to specific crystal faces was suggested. The surface composition of HA is dominated by its chemical constituents, that is, calcium, phosphate, and hydroxyl groups. Based on the composition [formula $CA_{10}(PO_4)_6(OH)_2$] the surface fractions have been calculated to be 16.5, 78.4, and 5.1% of the respective components. In aqueous solutions at neutral pH the surface is expected to be dominated by HPO_4^{2-} and $H_2PO_4^-$. Figure 4 illustrates the surface composition of enamel in an aqueous environment.

Arends [75] reported the points of zero charge for enamel and hydroxy- and fluoroapatites to be 4.4–5.0, 7.5–7.6, and 4–6, respectively. Zeta potentials for HA will depend on the ionic composition of the medium but reported values are conflicting [78]. To give some examples, zeta potentials of enamel and apatites were measured at different liquid-to-solid ratios by Arends [79]. At low solid-to-liquid ratios values were reported to be −9, −14–(−26), and −8 mV for enamel and hydroxy- and fluoroapatites in doubly distilled water. In the same study potentials were obtained at high solid-to-liquid ratios, representative of in vivo conditions, and

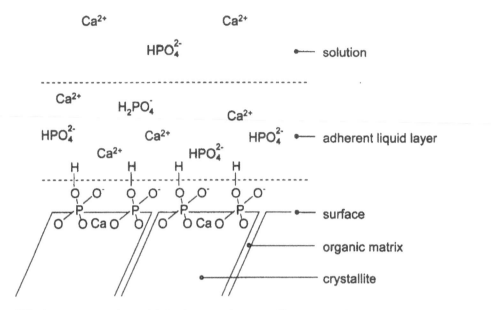

FIG. 4 Enamel surface with hydroxyapatite crystallites surrounded by the organic matrix. Note the dominance of ionizable surface phosphate groups. (From Ref. 75; for details, see also Refs. 74 and 76.)

positive values, 7–18 mV for HA and 16 mV for enamel, were reported. Neiders et al. reported a value of -10 mV for enamel in Hanks's buffer [80]. For pure HA, Olsson et al. reported a value of -7 mV [81] in 1 mM potassium phosphate buffer containing 50 mM KCl, 1 mM CaCl$_2$, and 0.1 mM MgCl$_2$. In 1 mM NaCl, HA was found to be slightly positively charged at pH 7.0 as determined by electroosmosis [82]. Furthermore, Arends reported that exposure to submandibular saliva and hence a buildup of a pellicle shifted the potentials for enamel and HA to lower (more negative) values by about 20 mV [79], whereas an increase in potential rendering the surfaces essentially uncharged after exposure to HPS was reported by Olsson [81].

The surface energies reported are 77 $\pm$ 10 to 87 $\pm$ 6 mJ/m^2 for enamel [83, 84] and 95 mJ/m^2 for fluoroapatite [85]. From extensive contact angle measurements on HA, fluoroapatite and enamel Glantz [86] reported total wettability for HA and fluoroapatite and values of the critical surface tension for wetting of 46.1 mN/m for human enamel and 45.1 mN/m for human dentine.

IV. PELLICLE

The acellular organic film which is deposited on the tooth surface after eruption or cleaning [5,40] is denoted the acquired pellicle. Several stages in growth can be distinguished, but in essence a bacteria-free film is inferred from the definition. During plaque formation the adhesion of microorganisms is preceded by the formation of this proteinaceous layer. By some authors this is also referred to as formation of a conditioning film [87]. It should be pointed out that adhesion may take place on clean substrates but that the presence of the pellicle has an effect on the initial colonization (see, for example, Ref. 88) and influences following adhesion sequences and hence has a modulating function. Furthermore, as mentioned earlier an important role of the pellicle is its protective function, including barrier and lubricating functions [58,89–93]. From an esthetic point of view discoloring of the pellicle may take place due to precipitation of chromogens from the diet, maillard reactions, and formation of pigmented metal sulfides and hence understanding of interaction between salivary and food components are important from this aspect [94].

A consensus view is that the initial film formation takes place within seconds or minutes both in vivo on model surfaces (cf. [95,96]) and enamel [97,98] and in vitro on HA and other model surfaces [82,99–101]. In the initial stage, the main increase in mass is reported to level out after 1–2 h in vivo [41,95,98] and in vitro [96,99]. Furthermore, the thickness of the adsorbed film in vitro has been reported to be 10–20 nm in this initial phase [95,96,102]; see Fig. 5 which illustrates the development of pellicle with time on silica surfaces exposed intraorally in 30 individuals [96].

The majority of the studies of salivary pellicles found in the literature are carried out with the aim of obtaining compositional information from a biochemical perspective, and the methodology applied is often not ideally suited to gain information on the physicochemical nature of these processes, e.g., surfaces are not defined with respect to charge and wettability; neither is high-resolution kinetic information available. In the following we first focus on the physicochemical properties of the adsorbed layers and then proceed to the compositional and biochemical properties and discuss their relation.

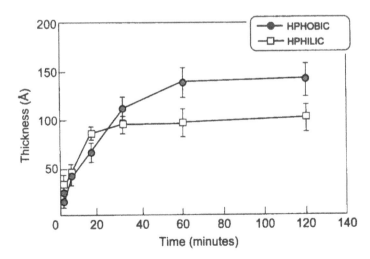

FIG. 5 Thickness of adsorbed films at hydrophilic and hydrophobic surfaces versus time of intraoral exposure. Error bars indicate standard deviations ($n = 30$). (From Ref. 96.)

A. Adsorption from Salivary Solutions—
Physicochemical Properties

In order to assess physicochemical properties common surface chemical methodology such as Fourier transform internal reflection infrared spectroscopy (FTIR), in situ ellipsometry, techniques for the measurements of surface forces, surface tension, contact angles, and zeta potentials have been applied. These enable determination of adsorption rates, thicknesses, and densities of adsorbed layers and their wettabilities. IR also provides some compositional information. In addition, studies have been made in order to gain detailed information on interactions between adsorbed layers including the influence by concentration, composition, surface properties, and ionic environment.

Pioneering work using germanium prisms exposed intraorally was carried out by Baier and Glantz [95], who demonstrated fast rates of adsorption, starting as early as seconds of exposure and reaching a plateau in film thickness of 150–200 Å after about 2 h. Furthermore, the adsorbed film on germanium exposed intraorally was found to contain mainly proteinaceous material evident from their IR spectra (amide I and II bands). Early work by Ericson and Arwin [103] was carried out where the buildup of in vitro pellicle from HPS onto hydrophilic silica and polished enamel surfaces was recorded in situ, and results were found to agree qualitatively at the two surfaces but the higher surface roughness on enamel precluded the recording of kinetics with a similar resolution as on silica.

Vassilakos and coworkers studied the rates of adsorption from HWS [99] and demonstrated that the main part of the in vitro pellicle on silica and hydrophobized silica was formed within minutes at as low concentrations of saliva as 1% (approximately 20 μg/mL protein) (see Fig. 6a).

By analysis of initial adsorption rates based on a model for mass transport by diffusion through a stagnant layer [104,105] Lindh and coworkers [106] showed that initial adsorption from whole (HWS) and glandular saliva (HPS and HSMSLS) onto negatively charged silica appeared to be mass transport controlled and involve proteins with a diffusion rate corresponding to those of PRPs and statherin, whereas adsorption to hydrophobic surfaces involved shorter peptides (Fig. 6b). Furthermore, this appeared to be more pronounced for HWS which is in agreement with its relatively higher content of proteolytically derived peptide fragments. Jensen et al. recently reported similar observations for adsorption onto synthetic HA [9].

Vassilakos et al. [99] investigated the concentration dependence of adsorption to model hydrophilic and hydrophobic surfaces. The adsorption "isotherms" depicted in Fig. 7a and b show that the amount adsorbed increases with concentration of saliva in the range 1–100% corresponding to approximate protein concentrations of 20 μg/mL to 2.0 mg/mL.

Later Lindh et al. showed that interindividual variations in adsorption from HWS [101], HPS, and HSMSLS [106] were statistically insignificant. Furthermore the low concentration region of the isotherms was investigated in more detail and revealed that concentration and also that adsorption started to decrease significantly at concentrations of HWS below 0.5% (10 μg/mL) (Fig. 8).

In a later publication, adsorption from HWS was compared to that from HPS and HSMSLS and differences in behavior were found; for example, higher amounts adsorbed were found for HSMSLS than HPS [106], which may be expected due to

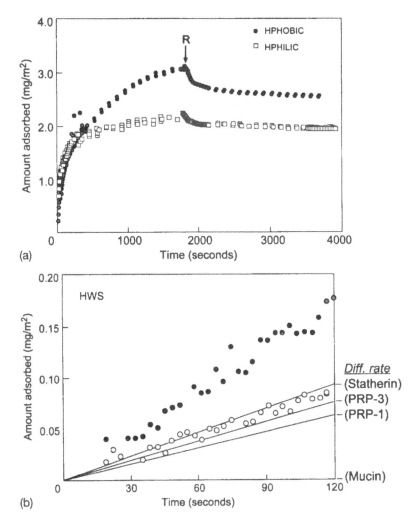

FIG. 6 (a) Adsorption to hydrophilic and hydrophobic surfaces from 1% human whole saliva (HWS) versus time. R indicates rinsing that was performed through a continuous flow at a rate of 20 mL/min for 5 min. (From Ref. 99.) (b) Initial adsorption from 0.05% human whole saliva (HWS) to hydrophilic (O) and hydrophobic (●) surfaces versus time. Solid lines indicate theoretical maximum adsorption rates for statherin, PRP-1, PRP-3, and intact MUC5B mucin. These were calculated assuming a mass transport controlled adsorption and that the concentration of each protein equalled the total protein concentration in the salivary sample. Note that the line for the mucin coincides with the x-axis on this time scale and under these conditions. (From Ref. 106; for details, see also Refs. 108 and 122.)

the higher content of mucins in the former and the dominance of small phosphoproteins in the latter. It is clear, however, that in the initial binding phase involving the first seconds and minutes mucins do not play an important role [106] but are found at the surface in later stages (see below and Refs. 107 and 108).

In spite of the complexity of the studied secretions a few conclusions may be drawn from the information available on rates and amounts. It is for example clear

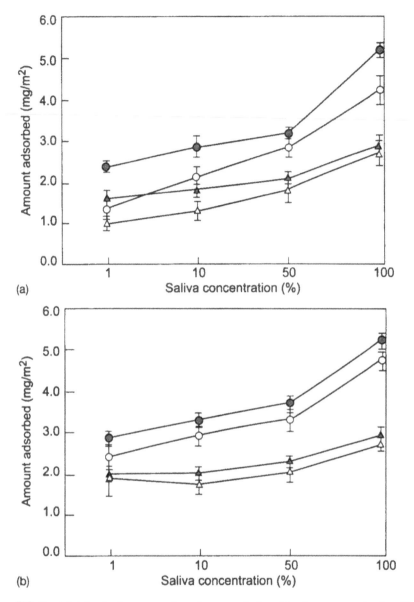

FIG. 7   (a) Adsorbed amounts from human whole saliva in water on hydrophilic and hydrophobic surfaces versus saliva concentration. Amounts were recorded before and after rinsing. Symbols are hydrophilic surfaces prior to rinse (○) and after rinse (△), and hydrophobic surfaces prior to rinse (●) and after rinse (▲). (From Ref. 99.) (b) Corresponding measurements in aqueous 0.15 M NaCl.

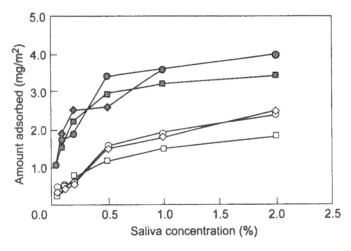

FIG. 8 Adsorbed amounts on hydrophilic (open symbols) and hydrophobic (filled symbols) surfaces from human whole saliva in 10 mM sodium phosphate buffer pH 7.0 containing 50 mM NaCl versus saliva concentration. Measurements were carried out with saliva from three donors as indicated by squares, circles, and diamonds, respectively. (From Ref. 101.)

that the initial adsorption, as expected, will be mass transport controlled and proceeds with a rate indicating the involvement of low-molecular-weight phosphoproteins such as PRPs and statherin, whereas mucins are expected to appear after longer times. Considering the amount adsorbed they correspond to approximate monolayer coverage but mixed composition and dynamic effects are expected. The adsorption isotherms indicate a high affinity binding with a plateau reached in the range of 10^{-5} g/mL protein in bulk solution.

Model surfaces with defined surface properties have proven useful in order to resolve mechanisms underlying adsorption of proteins in blood [13,109], milk [110,111], tears [112,113], and other biological fluids. Their use is mandatory, e.g., in order to resolve different contributions to the protein-surface interaction.

Well-defined, frequently used such model systems are silica and silica hydrophobized by reaction with alkylchlorosilanes. These have been characterized with respect to surface composition, wettability, and charge density [114,115]. In studies of salivary adsorption such surfaces have been frequently used [82,96,99–101,103,106,108,116–124] along with Ge prisms [95,125–129] modified to the desired wettability and HA [8,9,25,28,82,130–134]. A recent study of adsorption from HWS to HA discs was carried out by Hahn Berg et al. using in situ ellipsometry, and results revealed adsorption characteristics similar to those previously observed for adsorption to silica [82].

Studies of adsorption from HWS have demonstrated a higher affinity for hydrophobic surfaces than hydrophilic (Fig. 8) [101], and also higher amounts adsorbed (see Figs. 6 and 7) [99–101,135]. Also in the case of adsorption from HPS and from HSMSL corresponding effects by the surface wettability are observed. Obviously hydrophobic interactions constitute a strong driving force for adsorption in the case of salivary proteins. This is an observation made for many protein systems and has for example elegantly been demonstrated by the use of surfaces with a gradient in wettability [136].

In the case of adsorption from HWS to silica and hydrophobic silica the influence by the presence of monovalent electrolyte was investigated (see Fig. 7a and b) [99]. It could be concluded that adsorption to both types of surfaces increased in the presence of 0.15 M NaCl compared to in water. Furthermore, the effects were most pronounced at lower saliva concentrations (1 and 10%). A likely interpretation would be that the presence of salt reduces repulsive intermolecular electrostatic interactions between adsorbed molecules resulting in more efficient packing. The effect may be expected to be important as the major part of the pellicle proteins carry a negative charge at physiological conditions [4,18,130]. In measurements of interactions between adsorbed salivary films it was observed that the range of the repulsive interaction between salivary layers decreased slightly when the concentration of NaCl was increased from 1 to 80 mM, and this effect was assigned to screening of electrostatic segment–segment interactions [102].

Influence by pH in the range from 3 to 9 on the adsorption characteristics of HWS on silica and methylated silica has been found to be statistically insignificant [99]. In these studies only amounts bound were recorded and therefore possible changes in composition are not taken into account. In a study of adsorption at the air–saliva interface from whole saliva and parotid and submandibular secretions, Holterman and coworkers showed a correlation between pH of the secretion and ellipsometrically determined thickness indicating a thicker film at lower bulk pH [137]. A higher amount adsorbed may be anticipated as the main fraction of the pellicle proteins are acidic (see above), and a lowering in pH will bring these proteins closer to their isoelectric point. The trend was most pronounced for submandibular saliva. It is not clear why such effects are not observed for adsorption at solid–liquid interfaces. It should be noted, however, that the study of adsorption at the air–water interface involved long adsorption times and resulted in very thick adsorbed layers, more or less resembling a gel phase. In such a system intermolecular interactions may be expected to dominate.

The behavior of pellicle components with respect to self-exchange, exchange by other proteins, by surfactants, as well as by desorption upon rinsing is important with respect to functional aspects as well as analysis. For example, the first issues are relevant to consecutive adsorption and adhesion and the latter to the question of adequateness of elution methods. In protein adsorption typically the main fraction of the adsorbate tends to be apparently irreversibly attached with respect to dilution [138,139]. In many cases, however, a bi- or multiphasic behavior is observed where fractions of the adsorbate are reversibly adsorbed with respect to rinsing, exposure to salt, pH changes, or addition of surfactants. Such behavior is often taken as an indication of multiple binding states upon adsorption [140]. Considering the complex nature of the salivary secretion, a multiphasic nature of adsorption may be anticipated and is indeed what is observed (see Fig. 9). The figure illustrates the presence of a reversibly attached fraction desorbing upon rinsing and reappearing on saliva addition.

In many systems the fraction of the adsorbate that desorbs upon buffer rinsing is reported to be higher for hydrophilic surfaces than for hydrophobic ones [141,142]. For saliva the opposite behavior is observed, indicated by the presence of a larger reversibly adsorbed portion of the layer on hydrophobic surfaces [100]. The reason

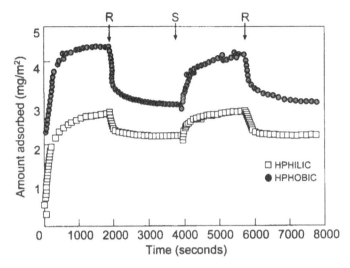

FIG. 9 Adsorbed amounts on hydrophilic and hydrophobic surfaces from 50% centrifuged human whole saliva in water versus time. Rinsing took place after 30 min of adsorption and saliva was readded after another 30 min. (From Ref. 100.)

for this is unclear, but it might be speculated that its presence may be of relevance for further interaction of the protein layer, e.g., in consecutive interaction with cells. It has been shown that hydrophobic surfaces perform comparatively well in exposure in vivo with respect to accumulation of dental plaque [86,143,144].

Interactions between surfactants and adsorbed protein layers, i.e., elution, have been used as an indicator of protein binding strength to the surface [145]. In dental applications these interactions are important as surfactants are usually present in toothpaste and mouth rinse formulations, and furthermore SDS elution has been used as one technique to elute adsorbed pellicles in vivo [146] and in vitro [147]. Surfactant-adsorbed protein interactions have been studied in vitro by ellipsometry in a number of papers with respect to whole saliva [82,117], glandular saliva [118,121], and salivary fractions [119]. It was found that in all instances the major parts of the pellicles were removed and that the fraction elutable of course depends on the surface type and type of surfactant. Hahn Berg and coworkers recently showed that while a HWS pellicle was completely removed by SDS from silica (Fig. 10a), the removal was substantially less efficient on HA (Fig. 10b). The difference may be ascribed to the different potentials of the surfaces; Si possesses a strong negative potential [115], whereas HA was found to be slightly positively charged or neutral at the experimental conditions [82].

Knowledge of interactions between adsorbed salivary films is of course highly relevant for oral adhesion, friction, and wear. In a surface force experiment the force acting between surfaces is measured versus their separation, and from the features of the force curves conclusions may be drawn regarding the extension of, e.g., adsorbed layers as well as the nature of the acting forces (Chapter 7) [22]. Interactions between adsorbed salivary films on mica surfaces have been studied using the interferometric surface force apparatus (SFA) by Nylander and coworkers [102], who measured the forces between mica surfaces in HWS as a function of time [102]. The

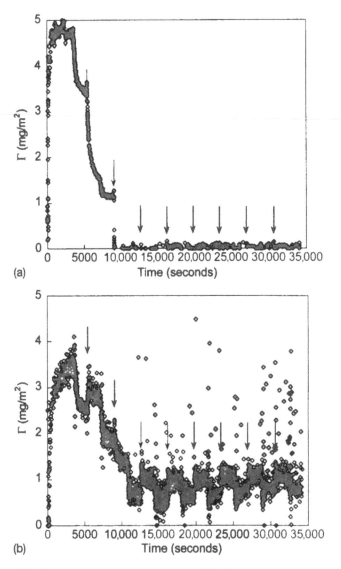

FIG. 10 (a) Sequential addition of SDS (0.4, 0.9, 1.3, 1.7, 2.2, 2.6, 3.0, and 3.5 cmc, additions indicated with arrows) to a salivary pellicle formed on Si/SiO$_2$. The figure shows adsorbed amounts versus time. An in vitro pellicle was allowed to form for 1 h from 10% human whole saliva and rinsing was performed prior to addition of SDS. (From Ref. 82.) (b) Sequential addition of SDS (0.4, 0.9, 1.3, 1.7, 2.2, 2.6, 3.0, and 3.5 cmc, additions indicated with arrows) to a salivary pellicle (10% saliva) formed on HA. Conditions as in (a).

studies revealed that adsorption from 10% HWS took place in a time-dependent fashion (Fig. 11a), showing a gradual increase and a plateau after about 4 h of adsorption (Fig. 11b).

In Fig. 11b the extension of the adsorbed layer at a load of 1 mN/m is plotted versus adsorption time and it is clear that the layer extends about 160 Å into the solution. This value is in good agreement with observations made by other methods

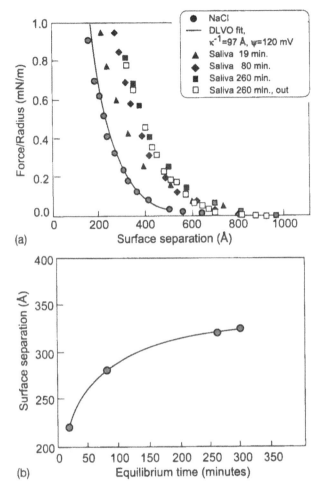

FIG. 11 (a) The normalized force versus distance between mica surfaces in 10% solution of human whole saliva. The figure shows the curves in pure 1 mM NaCl and in 10% saliva after different adsorption times. The curves on approach is presented except after 260 min of adsorption where both approach and decompression curves are given. The mica surfaces bearing the adsorbed salivary layer were approached up to a maximum repulsive force of 1 mN/m before they were separated. The solid line represents the DLVO-fit using a Debye length, κ^{-1}, of 97 Å and a surface potential, Ψ, of 120 mV. (From Ref. 102.) (b) The mica–mica separation at an applied force of 1 mN/m versus the time after addition of 10% saliva in 1 mM NaCl. (From Ref. 102.)

[95,96]. Note that the slow adsorption rates may be ascribed to mass transport which is limited to diffusion only in the case of the SFA, whereas convective transport plays a major role in the experimental setup used in the previously discussed ellipsometric measurements and may be expected to do so also in the in vivo situation. Interactions between adsorbed layers were found to be repulsive at all conditions and could not be described by ordinary electric double layer and van der Waals forces (DLVO) theory. Furthermore, interactions were demonstrated to be only mar-

ginally dependent on ionic strength, as explored by increasing the concentration of monovalent electrolyte. A more detailed analysis of the decay lengths of the interaction (see Fig. 12b) reveals that they are much longer than expected from electric double layer interactions. Taken together the features point to an interaction dominated by steric contributions and resembling that measured between adsorbed layers of mucins [148,149]. Considering the protective role of the adsorbed film its ability to relax after compression is highly relevant, and it was observed that this ability was dependent on the time of compression and that after long contact times (30 min) relaxation still took place but there was a less strong repulsion between the adsorbed layers.

In a subsequent study it was demonstrated that preadsorption of PRP-1 (discussed in more detail below) did not affect the main interaction between salivary layers, showing a predominance by the much longer range interaction exhibited by the layer adsorbed from whole saliva (Figs. 12a and b). Comparing Figs. 11 and 12 gives at hand that already at the first time point (20 min adsorption) the thickness of the layer adsorbed from whole saliva exceeds that of the pure PRP-1 layer, indicating a mixed layer formation and a possible presence of mucous glycoproteins at this point. Interestingly, adsorption of MG-1 to HA was previously reported not to be affected by the presence of cystein containing phosphoproteins [39].

B. Behavior of Isolated Protein Fractions

In order to elucidate compositional aspects of adsorbed layers data on adsorption from fractionated saliva and isolates are first reviewed. As a second step the biochemical analysis of in vitro and in vivo derived pellicles will be described.

Adsorption from fractionated saliva to silica and hydrophobic silica was studied by Vassilakos et al. [120]. It was clearly demonstrated that fractions with molecular weights >760–460 and 14–4.5 kD, respectively, possessed a high surface activity. These fractions are expected to contain high-molecular-weight glycoproteins and proline-rich proteins, statherin, cystatins, and histatins, respectively (see above). Upon adsorption to hydrophobic surfaces both fractions showed extensive adsorption and the lower-molecular-weight fraction was also found to dominate the adsorption at air–liquid interfaces as deduced from surface tension measurements. On pure silica, on the other hand, the highest adsorption was observed from the high-molecular-weight fraction.

In order to obtain affinities for HA, adsorption of many proteinaceous salivary components has been studied. This has been done for amino acids [132], polymeric amino acids [150], salivary proteins [151], and glycoproteins [152]. More detailed studies on the salivary phosphoproteins have been carried out by Moreno, Hay, and coworkers [133,134]. In the first of these studies, adsorption of statherin and PRP-3 to HA, fluoroapatite (FA), and fluorohydroxyapatites was investigated by a solution depletion technique. It was found that the adsorption affinity for these materials was considerably higher for PRP-3 than for statherin. On the other hand higher numbers of statherin molecules could be bound as expected due its smaller size. Furthermore, they found that the adsorption affinity increased with the fluoride content in the mineral, which was attributed to a decrease in the surface energy of the adsorbent. In a following study the adsorption of PRP-1, PRP-2, PRP-3, and PRP-4 as well as the 30 residue N-terminal portion of these proteins [134] was investigated. The ad-

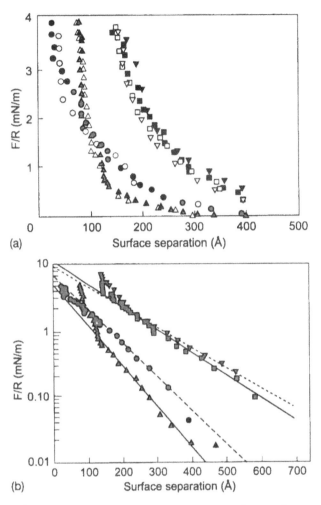

FIG. 12 (a) The normalized force versus distance between mica surfaces 20 h after the addition of 17 μg/mL (1.05×10^{-6} M) PRP-1 in 1 mM NaCl ($\bullet$) and in 1 mM CaCl$_2$ ($\blacktriangle$) as well as the curve recorded 20 h after the subsequent addition of 10% saliva to a layer of PRP-1 preadsorbed from 1 mM CaCl$_2$ solution ($\blacksquare$) and in pure 10% saliva ($\blacktriangledown$). The figure shows the force curves on a linear scale, and curves on compression and decompression are indicated by filled and unfilled symbols, respectively. (From Ref. 154.) (b) The force curves on compression are shown on a logarithmic scale. A linear fit was performed on the long-range part of the force and the resulting lines are inserted. An electrostatic repulsive force is roughly proportional to $e^{-\kappa D}$, where κ^{-1} is the Debye screening length. The resulting slopes for the linear fits correspond to the following values of κ^{-1}: 86 Å (lower dashed line) for PRP-1 in 1 mM NaCl, 72 Å (lower full line) for PRP-1 in 1 mM CaCl$_2$, 129 Å (upper full line) for saliva added after PRP-1 preadsorption, and 143 Å (upper dashed line) for pure saliva solution. (From Ref. 154.)

sorption was found to be endothermic and hence driven by increase in entropy. PRP-1 and PRP-3 along with the N-terminal peptide showed higher affinity for HA than did PRP-2 and PRP-4, indicating an influence by the residue at position 50 being aspargine in the case of PRP-1 and -3 and aspartic acid in PRP-2 and -4. An unexpected finding was that a lower number of short PRPs were accommodated on the surface at saturation. As for statherin, mentioned above, the adsorption affinity was found to be higher for FA than HA, again ascribed to differences in surface energies. Their data showed that adsorption of PRP-1 and PRP-4 to HA reached plateaus at concentrations above approximately 15 μg/mL. Assuming a content of 30% PRP in HWS this means that these proteins will be able to saturate the surface at a saliva dilution of 40 times. Their work on salivary protein adsorption onto Ca apatites is summarized in Ref. 153. In a paper by Johnsson [25] data for adsorption to HA of salivary proteins are compiled (Fig. 13).

Binding data were evaluated according to a Langmuir model, and binding constants are reported. They concluded that the highest affinities for HA were observed for statherin and the PRPs. Interestingly, the N-terminal portion of PRP-1 (PRP1-T1) was found to show higher binding affinity than intact PRPs, whereas the dephosphorylated peptide showed a binding constant about 10% of the phosphoserine-containing one. In line with this, it is suggested that the high-affinity binding to the HA surface takes place through interactions between negatively charged peptide sequences and positive surface sites. Figure 13 illustrates the differences in affinity and amounts and shows that adsorption appears to reach plateaus at concentrations below 50 μg/mL for all proteins investigated and that the amounts are in the range approximately corresponding to monolayers.

In the same paper the adsorption is correlated to effects on HA crystal growth, where PRPs and cystatins were shown to be most efficient. In a recent study the nucleation of Ca phosphate at a HA surface was monitored in situ and found to be

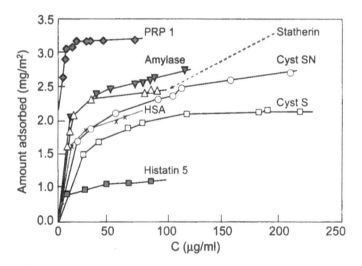

FIG. 13 Adsorption to hydroxyapatite versus concentration of various salivary proteins. (Compiled in Ref. 25; PRP-1 from Ref. 134; amylase from Ref. 244; statherin and cystatin S and SN from Ref. 245; HSA from Ref. 246; and histatin 5 from Richardson, unpublished.)

slowed down in the presence of a salivary pellicle [82]. In the same study it was also shown that the inhibitory effect increased with pellicle age (Fig. 14).

Adsorption of PRP-1, PRP-3, and statherin onto hydrophilic silica and hydrophobic model surfaces was recently studied in situ by ellipsometry [122]. In this study adsorption was found to be controlled by mass transport at concentrations up to the highest investigated (17 μg/mL). Furthermore, it was found that, except for PRP-1, much lower amounts were adsorbed on negatively charged hydrophilic surfaces as compared to hydrophobic ones. The concentration dependence of adsorption agrees well with previously observed behavior on HA, but maximum amounts adsorbed were 0.5–0.7 mg/m^2, which is less than half a monolayer side-on (assuming the area of a truncated β-casein with molecular weights corresponding to PRPs). Crucial importance of interactions involving the binding of phosphoserines to surface Ca^{2+} may be one reason for the differences with respect to the behavior on HA. In addition, uncertainty may exist in the determination of the accessible surface area for HA particles.

In a study of PRP-1 adsorption to mica and measurement of the interactions between adsorbed layers (see Figs. 12a and b) only weak adsorption was observed in 1 mM NaCl, and it was demonstrated that formation of a stable adsorbed layer relied on Ca^{2+} [154]. In the presence of Ca^{2+} the data indicated a biphasic structure with a dense inner structure and an outer compressible one. The thickness of the fully developed layer was approximately 60 Å and could be compressed to approximately 40 Å at high load. It may be tempting to assign the inner and outer fractions of the adsorbed layer to the N- and C-terminal portions of the protein as in the orientation previously suggested for adsorbed layers on apatitic surfaces. The adsorption and interaction behaviors resemble those of β-casein which is structurally related and has been characterized by surface force measurements [155] and by ellipsometric and neutron reflectivity measurements [156].

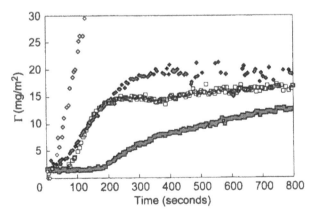

FIG. 14 The figure shows precipitation/crystallization from 10 mM sodium phosphate buffer containing 1 mM CaCl$_2$. Adsorbed amount versus time upon simultaneous addition of 10% saliva and 1 mM Ca^{2+} (open diamonds), upon addition of 1 mM Ca^{2+} after formation of a 1 h pellicle (filled diamonds), upon addition of 1 mM Ca^{2+} after formation of a 1 h pellicle and rinsing (open squares), and upon addition of 1 mM Ca^{2+} after formation of a 1 h pellicle subjected to rinsing and then aged for 12 h (filled squares). (From Ref. 82.)

Due to its importance for protection [56,157], role in bacterial adhesion [6,158], and also relevance for mucoadhesion [159] adsorption of mucins has been studied with respect to a variety of surfaces and by applying different techniques [39,108,148,149,159–162]. Lindh et al. investigated the adsorption of purified MUC5B mucin (MG-1) to hydrophilic and hydrophobic surfaces and compared its adsorption behavior with commercial bovine submaxillary mucin (BSM) [108]. The adsorption affinity of both mucins was found to be highest for hydrophobic surfaces. Adsorbed amounts of pure MUC5B were found to be 1.3 and 2.7 mg/m² at hydrophilic and hydrophobic surfaces, respectively, at a solution concentration of 0.1 mg/mL. Furthermore, adsorption of the MUC5B and the commercial preparation was found to agree well at low concentration, whereas at higher concentrations the commercial sample adsorbed less to the hydrophobic and more to the hydrophilic surface compared to the MUC5B, which may be ascribed to possible presence of proteinaceous impurities. It was also found that adsorption kinetics were slow and that adsorption was mass transport limited at low concentrations. Tabak et al. [39] investigated the adsorption of MG-1 and MG-2 to HA surfaces and evaluated the data in terms of a Langmuir model. A higher affinity for HA by MG-1 was reported, and furthermore it was found that the adsorption of MG-1 was inhibited by salivary glycolipids but was unaffected by a cysteine-containing salivary glycoprotein. Surface force measurements on BSM adsorbed on mica [148] and gastric mucins on hydrophobized mica [149] reveal that long-range steric repulsion dominate the interactions between such layers. This observation is relevant for the barrier function of mucous layers, and the resemblance to the interaction between salivary films mentioned above indicates that this function may be successfully fulfilled at oral interfaces.

C. Aspects of Composition and Development

It was early shown by histological staining that the pellicle was of proteinaceous nature. Furthermore, these observations were supported by the finding that the pellicle was removed upon exposure to proteolytic enzymes [163]. This was much later demonstrated in a real-time in situ experiment showing the degradation of an in vitro pellicle on silica by a mixture of proteolytic enzymes [123]. In an early report by Armstrong the composition of the pellicle was reported to be 46% amino acids, about 3% hexosamines, and a total of 14% carbohydrates [164].

The knowledge on the content and composition of lipids is less comprehensive, but cholesterol; cholesteryl esters; tri-, di-, and monoglycerides; phospholipids such as phosphatidylcholine, phosphatidyletanolamine, and sphingomyelin along with glycolipids are reported to be present [165–167]. Even though the lipid content in saliva is low (see above) the pellicle may contain substantial amounts and at 2 h pellicles are reported to contain 22–23% lipids, mainly glycolipids but also neutral lipids and phospholipids [165,167]. Carbohydrates found are glucose, glucosamine, galactose, and mannose [168]. A high content of glucose has been observed, while the proteins bound to the HA contain mainly galactose, mannose, and fucose, but only minor amounts of glucose, indicating that its presence in pellicle does not originate from pellicle proteins [130].

In an early investigation of adsorption from HWS to HA and enamel powders in vitro as well as on freshly extracted teeth, Hay [12] demonstrated that the fraction of the adsorbed salivary proteins that was eluted in 0.2 M phosphate buffer showed a selective binding in both cases and displayed similar electrophoretic mobilities. Furthermore, only minor amounts of sialoproteins were detected on the extracted teeth. In later papers the same author showed that proline-rich proteins, sIgA, amylase, and albumin could be eluted from HA powders exposed to HPS by ethylene diamine tetraacetate (EDTA) and that the highest affinity components were acidic peptides with a high content in glutamic acid, tyrosine, aspartate, and histidine, respectively [130]. By experiments involving varying HA-to-saliva ratios, Hay further showed that the initial adsorption to HA was dominated by a tyrosine-rich peptide (statherin) [8] followed by histidine-rich peptides (histatins) [131], and then by proline-rich proteins [132]. Embery demonstrated that low-molecular-weight phosphoproteins played an important part in the initial in vivo pellicle formation but gradually disappeared with pellicle maturation [40]. Work on homo- and heteropolyaminoacid adsorption to HA has demonstrated the importance of phosphate ester groups [169] and dicarboxylic amino acids [150] for high-affinity binding. Studies of the initial adsorption kinetics upon in vitro pellicle formation from glandular saliva to model surfaces also support the finding of initial binding of low-molecular-weight peptides/proteins of the size corresponding to statherin and PRPs to model surfaces (see Fig. 6b above) [106,122].

As indicated above mucins are believed to be incorporated much later in the pellicle, as shown by, e.g., Busscher et al. [107], who demonstrated the selective adsorption of mucins from HWS in vitro and showed that they were absent within the first hour but appeared after 2 h and continued to increase in concentration during the next 16 h. From experimental observations in vitro and simple mass transport analysis it was also shown that mucins were not involved in the initial adsorption from HSMSLS on hydrophobic and hydrophilic model surfaces [108].

The amino acid composition in acquired pellicles has been analyzed in a number of studies. One of the first was carried out by Sonju and Rolla in 1973 [41]. Data from many authors were compiled by Lendenmann [17] and show good agreement between different reports irrespective of the pellicle age, location of sampling area, presence of plaque, etc. Amino acid compositions alone are, however, of limited value in the prediction of the protein composition. Among the observations are preferential accumulation of hydrophobic amino acids in pellicle as demonstrated by Al-Hashimi et al. [170] and a lower content of proline than in the whole secretion [17,18].

By elution by sodium dodecyl sulfate (SDS) soaked foam sponges and subsequent electrophoresis and immunoblotting techniques Carlén et al. have investigated pellicles formed in vitro on dental materials [147] and in vivo and in vitro from HWS, HPS, and HSMSLS [146]. From the latter study it was concluded that pellicle compositions were reflecting the composition of saliva prevailing in the part of the mouth where the pellicles were formed, an observation that may be pertinent to the subsequent adhesion of bacteria. Furthermore, albumin was observed clearly in vivo but not in vitro. By using a combination of amino acid analysis, electrophoresis, gel filtration, and HPLC, Jensen et al. studied the protein composition on pellicles formed in vitro on HA [9] from HWS, HPS, and HSMSLS. Amylase, acidic and glycosylated proline-rich proteins, statherins, and histatins were found in the HPS pellicle, and by the use of cationic electrophoresis histatin 3 and 5 were identified. The HSMSLS pellicle was identical except for the presence of cystatins and absence of glycosylated PRPs. The HWS pellicle, on the other hand, was shown to contain mainly amylase, acidic PRPs, cystatins, and proteolytically derived peptides. In a study by Lamkin et al. the proteolytic degradation in HWS was further investigated and it was concluded that degradation was an ordered and consistent process and that the proteolytically derived fragments amounted to approximately 5% of the protein adsorbed to HA [171]. Yao and coworkers applied amino acid analysis, MALDI-TOF, and LC-MS to compare in vivo pellicles eluted by a swabbing procedure and in vitro HA pellicles from HWS [18]. They found that ammonium bicarbonate along with SDS were most efficient in removing pellicle components, removing 70 and 80%, respectively. In the in vitro pellicles lactoferrin, albumin, amylase A and B, PRP (Db, long, and short), lysozyme, cystatin SN, statherin, and peroxidase were identified. In vivo, on the other hand, the PRPs were absent and in addition low-molecular-weight peptides were observed in accordance with previous studies. It was argued that the differences were due to the different surface properties of enamel and HA and the presence of proteolytic enzymes in the in vivo case.

By use of SEM, TEM, and immunological techniques the presence of PRPs, histatins, and statherin were shown to be integral parts of the in vitro pellicle on HA and bovine enamel [35]. TEM was also used by Busscher and coworkers in a study of buildup of in vitro pellicle from reconstituted lyophilized HWS on enamel in a flow cell system [172]. They observed that the characteristics of the enamel surface disappeared after seconds and that within 10 s of salivary exposure three or four distinct homogeneous films were deposited on top of each other and an uneven knotted structure developed. This heterogeneous pattern was observed for at least 2 h. TEM was also used by Hannig to investigate the ultrastructure of salivary pellicles formed after 6 h on different dental materials in vivo. He found that pellicles on test pieces exposed on the buccal side ranged from 500–1000 nm and displayed a globular and heterogeneous structure, whereas on the lingual side a granular pellicle with a thickness of about 100 nm was formed [173]. In a subsequent study it was shown that the in vivo pellicle was initiated by a quick formation of a dense layer followed by attachment of an outer loosely arranged layer. At 24 h the pellicle was also shown to be influenced by the oral environment rather than properties of different dental materials [97].

In view of the significance of the function of each of the pellicle proteins, for example, the reported function of the precursor proteins in mineralization, lubrication, and bacterial modulation, their presence at the enamel surface may be of vital importance at specific and different time points. The heterogeneous protein composition of saliva, including proteins at varying concentrations with a wide range of molecular weights and structures and hence diffusion rates, means that exchange phenomena should be anticipated. Dynamic adsorption behavior featuring exchange reactions are known to occur in the case of synthetic polymer mixtures or polydisperse samples (Chapter 1). For proteins they have been demonstrated in the field of plasma protein interactions with blood contact surfaces [13,16,174,175]. In the latter system, the sequences involving a progression from low-molecular-weight proteins at high concentration followed by proteins with a higher affinity and molecular weight have been demonstrated.

In saliva analogous behavior may take place, which is supported by the reported affinity constants for HA binding [25] which indicate that, e.g., mucins should have a higher affinity for HA than phosphoproteins and thereby have the potential to replace the initially bound statherin and PRPs. The fact that this takes place in reality is supported by the experimental observations reviewed above. It is hence clear that at least in vivo the pellicle develops from essentially a monolayer of adsorbed salivary proteins formed within seconds to much thicker structures within the time frame of hours. Based on the compositional data presented above it may be reasonable to assume that the sequence in the case of saliva will initially involve the low-molecular-weight phosphoproteins and after this globular proteins adsorb as do large glycoproteins (agglutinins, mucins), thereby forming a complex mixed layer. It may be speculated that the content of large glycoproteins increases with time during this phase. Finally, after long times in contact with saliva, aggregates from solution such as the micelles described by Rykke et al. may attach. A mechanism along similar lines has previously been proposed by Embery [176].

V. PLAQUE FORMATION

One definition of dental plaque is "the nonmineralized microbial accumulation that adheres tenaciously to tooth surface, restorations, and prosthetic appliances, shows structural organization with predominance of filamentous forms, is composed of an organic matrix derived from salivary glycoproteins and extracellular microbial products, and cannot be removed by rinsing or water spray" [177]. Dental plaque also fulfils the criterion of being a typical biofilm [178,179].

Plaque is associated with the major dental disorders caries and periodontal diseases. A caries lesion is the result of demineralization of dental hard tissue caused by organic acids produced by fermentation of carbohydrates like sucrose, glucose, and fructose by oral bacteria. For enamel it occurs below a critical pH of about 5.5 and starts a few minutes after carbohydrate consumption and lasts for approximately 30 min. For dentin the critical pH is as high as pH 6.7. The presence of plaque imposes diffusion restrictions resulting in local pH drops and increased duration. Caries is associated with bacteria such as *Streptococci*, *Lactobacilli*, and *Actinomyces* strains.

Periodontal diseases include gingivitis and periodontitis, both caused by accumulation of plaque. Gingivitis is a reversible state characterized by a swollen gingiva, increase in pocket depth, and bleeding, whereas periodontitis also involves alveolar bone resorption and detachment of periodontal ligaments, which results in irreversible tissue damage. Associated bacteria are gram negative and mainly anaerobic, e.g., *Porphyromonas*, *Bacteroides*, *Fusobactrium*, *Capnocytophaga*, and *Spirochetes*, which produce extracellular enzymes and toxins that destroy connective tissue and contribute to the inflammatory response. For up-to-date reviews on the structure, composition, and buildup mechanisms of plaque biofilms, see Refs. 178, 180, and 181.

The initiation of plaque formation takes place with the adhesion of microorganisms, predominantly bacteria, onto oral interfaces such as the surfaces of teeth or restorative materials. This occurs rapidly and bacteria have been observed at germanium prisms exposed intraorally after 2 h [95] and are reported to be present at enamel surfaces after 4 h of in vivo exposure [182]. The primary colonizers are *Streptococcus*, *Actinomyces*, and *Veillonella* strains.

Adhesion may take place to clean or pellicle-coated surfaces, and due to the fast rate of protein deposition the latter is the normal. Protein adsorption will monitor adhesion of plaque bacteria through specific and nonspecific means (see below), and it has been shown that the initial adhesion strength depends on whether adhesion takes place on a clean surface or via a protein film [88].

Plaque composition is known to be different at different loci, for example, in supra and subgingival plaque, and furthermore a sequential pattern in the colonization has been reported to be present. Mechanisms behind the adhesion may be of specific and nonspecific nature, where the latter may be approached from a colloid chemical perspective [183,184]. The specific route involves adhesin−receptor (ligand on oral surfaces) interactions, possibly including conformational changes, and high-affinity binding. For reviews see Refs. 181 and 185. Both will have to be considered, as illustrated in Fig. 15, but their relative importance may vary [186].

The nonspecific adhesion stemming from general considerations of electric double layer and van der Waals forces (DLVO theory) may be predicted from data on available charge densities, sizes, and properties of the surrounding medium [183]. Along this route methods have been developed to quantify surface properties of oral interfaces [128,187,188] and bacteria [189−191].

The role of surface energies in determining adhesion strengths in this context was early recognized by Zisman [192], and later Baier, Shafrin, and Zisman [193] discussed the relevance of the critical surface tension in order to predict outcomes of biological adhesion events. This concept was employed by Glantz, who showed that the maximum amount of adhered plaque to tooth surfaces was dependent on the surface free energy of the solid as quantified by contact angle measurements [86]. Baier and Glantz [128,143,187,194] later showed that a range in critical surface tension between 20−30 mN/m exhibited low biological adhesion and was hence denoted "bioabhesive."

Regarding specific interactions it has been demonstrated that the adhesion of oral bacteria is mediated by interactions between adhesins on the bacterium surface or on fimbriae [195] and salivary components such as mucins [158], salivary agglu-

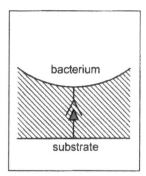
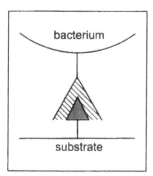

FIG. 15 Nonspecific (left) and specific (right) interactions upon bacterial adhesion. The former may be van der Waals and electrostatic double layer forces originating from the entire interacting surfaces. The latter, specific interactions may contain electrostatic, van der Waals, and hydrogen bond components, but take place between highly localized chemical groups. (From Ref. 247.)

tinins [176,196], statherin, and proline-rich proteins [197,198]. The affinity for the latter by *Streptococcus* and *Actinomyces* strains was early demonstrated by Gibbons [197–200].

An interesting feature of microorganisms is their ability to produce biosurfactants as defense against other colonizing species. This was demonstrated experimentally for *Streptococcus mitis*, which produces a biosurfactant that is able to substantially affect the adhesion of *Streptococcus mutans* [201]. The surfactant was found to be a glycolipid that was capable of reducing surface tension of aqueous solutions to 30–40 mN/m. Considering the presence of polar lipids in the pellicle, including reported interactions between MG-1 and glycolipids [39], the role of surfactants and polar lipids in oral adhesive events should not be neglected.

As outlined above the dental plaque may be considered a biofilm in the sense that it is a dynamic vital ecological unit that comprises a variety of bacterial accumulations that grow on an inert or biologically active surface. The plaque is subject to development and adaptation to changes in environmental conditions through changes in the relation between the more than 500 species found in a plaque [181]. It is clear that after the initial adhesion of the primary colonizers, which takes place in a sequential fashion, these species multiply and when the available surface is filled growth will continue normal to the surface in "towers" (see Ref. 180 and references therein). After approximately the third day filamentous bacteria are reported to be present in the predominantly coccoid plaque and corncob structures are detectable [180]. Figure 16 shows a hypothetical picture of the initial stages in bacterial adhesion and how these may affect the composition and structure of the biofilm. After about a week filamentous bacteria penetrate the coccoid plaque and starts to replace it, and after about 2 weeks columnar microbial colonies are replaced by a dense mat of filamentous colonies. In the process of coadhesion of bacteria the same reasoning

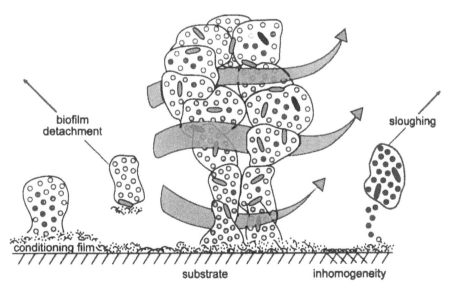

FIG. 16 Hypothetical picture illustrating initial bacterial adhesion and its role in determining biofilm growth. (Adapted from Ref. 247.)

as above would apply involving nonspecific and specific mechanisms. As in initial adhesion, adhesin-receptor mechanisms have been suggested to be involved [181]. Out of the currently identified models for biofilm structure—that is, water channel, heterogeneous mosaic, and dense biofilm model—the latter was suggested to conform with oral biofilms by Wimpenny and Colosanti [202].

Important components in biofilms in general, including plaque, are extracellular polysaccharides (EPS), which are important for the integrity of the biofilm and may comprise 50–95% of their dry weight [203]. The EPS produced differs between species and by accessibility of nutrient in oral biofilms; for example, glucans, fructans, and levans are common.

An interesting feature of bacteria in biofilms is their ability to adapt to the surroundings in the sense that a different phenotype may be expressed [179]. One consequence of this is their tolerance to antibiotics and antimicrobial substances, which may be a factor of 1000 higher [204]. Furthermore, the topic of intercell signalling by low-molecular-weight substances like homoserine lactone within communities of microorganisms, as demonstrated by Davies and coworkers [205,206], has received considerable attention. Furthermore, it has been shown that protein expression is different in planktonic and biofilm states. For recent reviews on oral biofilms, see, e.g., Refs. 178, 207, and 208.

In the literature a multitude of model systems have been developed and presented for studies of initial biofilm buildup [126,127,129], microbial adhesion [68,209], controlled biofilm growth [210–213], and plaque growth [214]. An important aspect in this respect is the possibility to study adhesion processes in situ without the complication of passing air–liquid interfaces.

VI. ROUTES FOR MODULATION OF PLAQUE GROWTH

Reduction in plaque and measures to minimize consequences of the presence of dental plaque are of course of prime importance in the dental field. Several strategies have been identified to combat plaque-related diseases (for reviews, see, e.g., Refs. 215–217). Some possible routes are summarized below:

1. Increasing the resistance of the teeth to decay
2. Modifying the surface of the teeth in order to reduce bacterial adhesion
3. Interfering with production of extracellular material
4. Using bactericidal compounds
5. Using antibiotics

Route 1 mainly involves treatment with fluorides, which from a surface chemical perspective also is known to affect the surface energy and hence adhesion when incorporated in enamel [86]. Routes 2–4, on the other hand, involve some interesting and challenging connections to the interfacial behavior of proteins. As in the field of blood contact surfaces effects of surface wettability have been addressed, and it has been shown experimentally that hydrophobic surfaces perform well with respect to accumulation of dental plaque in vivo [86,144]. Rolla and coworkers reported on the use of polydimethylsiloxane to reduce protein adsorption in vitro and in vivo [218]. Olsson et al. reported on a permanent hydrophobization by covalently attached silicone polymer, which was shown to reduce adsorption of model proteins and the amounts of stainable pellicle and plaque accumulated in vivo compared to untreated controls [219].

The potential of adhesion-reducing hydrophilizing compounds, of low molecular weight as well as polymeric nature, have been explored extensively over the last decades. Effects by surfactants in inhibiting bacterial adhesion have been investigated in a number of papers. Rolla and coworkers reported on the use of SDS [220], pyrophosphate [221], and phosphonates [222]. The effects of polyalkylene oxide derivatives on the binding of *Streptococcus mutans* in vitro was investigated by Olsson and coworkers, and it was found that both phosphates and phosphonates were able to reduce binding to pure HA but that the effect was limited on HA exposed to saliva [223]. Furthermore, efficient reduction in adhesion of *Streptococcus mutans* was accomplished in vitro by binary mixtures of surfactants involving alkyl phosphates aimed at binding to the HA surface and nonionic ethoxylates [81].

Coating of surfaces with hydrophilic noncharged polymers is an efficient route for reduction of protein adsorption provided that the layers are densely packed (Chapter 27). This concept was explored by Olsson and coworkers and compared to the previous approaches [135,224]. The potential of ethyl hydroxyethyl cellulose (EHEC) in unmodified and phosphated form was investigated, and the polymers were found to be effective adhesion reducers in absence of saliva. After preexposure of HA to saliva, however, effects were strongly reduced and neither approach gave significant plaque reductions in a small clinical trial [225]. Polyethylene imine (PEI)–polyethylene oxide (PEG) copolymers are often used to avoid nonspecific protein adsorption (Chapter 27). These polymers were shown to reduce protein adsorption and pellicle buildup almost completely, but showed much higher adhesion of *Streptococcus mutans* and in vivo plaque than hydrophobic reference surfaces prepared by plasma polymerization of hexamethyldisiloxane [135]. The surface modification work by Olsson, Holmberg, and coworkers is summarized in Ref. 226.

Other means of modulation include binding of proteins in order to passivate the surfaces or direct subsequent binding of bacteria to harmless flora. One alternative that has been explored is the use of the 30 residue anchoring peptide of PRP-1 as a high-affinity coating for HA. Furthermore beneficial properties with respect to caries protection have been reported from milk and milk components. Early studies focused on the casein phosphopeptides such as the caseinoglycomacropeptide [227]. It was later shown that proteose-peptone fractions 3 and 5, which are proteolytically derived peptides from β-casein, were found to have a high binding affinity for dental enamel and possess a strong reduction in the demineralization rate [228]. This finding is interesting to relate to the proteolytical degradation of proline-rich proteins and presence of derived peptides in HWS. Johansson has found that statherin-specific antibodies were able to block statherin-induced binding of *Candida albicans* to HA and to reduce it on buccal epithelial cells [229]. Furthermore, Edgerton showed that histatin 5 adsorption to modified PMMA [230] increased its anticandidal activity.

Concepts found in route 3 include interference with or modification of the exopolymer producing enzyme glycosyltransferase (GTF). This may be accomplished through inhibitors, substate analogs, and antibodies [215,231,232].

Another concept with interesting features is the antiplaque compound delmopinol, a surfactant having a low antimicrobial profile which is reported to influence several items mentioned in routes 2 and 3. For example, it has been shown to influence in vitro pellicles [118,233], wettability of tooth surfaces [233], surface potential and flocculation of oral bacteria [234], production of extracellular polysaccharides [214,235], as well as in vitro plaque cohesion [214].

Route 4 includes use of antibacterial compounds such as chlorhexidine (CHX), cetylpyridinium chloride (CPC), and triclosan, which have been widely used in order to control oral biofilms and many studies have been carried out showing their effects on plaque formation and related diseases (cf. [178]). CHX and CPC have been formulated in a two-phase formulation based on 15% vegetable and essential oils and an aqueous phase containing CPC or CPC plus CHX, and the concept was shown to have a high capacity in desorbing adhered *Acinetobacter calcoaceticus* RAG-1 from polystyrene in vitro [236]. CPC and CHX are cationic surface active compounds which means that effects on pellicles may be anticipated and have been demonstrated for CHX in vitro [237].

Triclosan, which is an oil-soluble bisphenol, has been solubilized in micelles of Tween 80 and sodium dodecyl sulfate, and it has been demonstrated that only the SDS formulation showed significant clinical effects [238,239]. This finding might be related to the efficiency of pellicle displacement reported for this surfactant [82,119]. Furthermore, triclosan was also used in combination with silicone oil [240] with positive results on bacterial adhesion and plaque reduction in vivo.

In routes 4 and 5, an attractive feature in antiplaque treatment is to increase the substantivity of the respective agent in the oral cavity, and therefore formulations of antibacterial compounds and antibiotics exhibiting sustained release properties are currently under development. Devices include various varnishes, strips, and gels to be inserted or applied on teeth or in gingival pockets (for a review, see Ref. 217). To mention an example, one interesting approach is a lipid-based formulation of metronidazole based on monoolein and triglyceride oil, where a dilution-induced phase transition from L_2 (reversed micellar) to H_2 (reversed hexagonal) is utilized in order to obtain in situ gelling properties.

VII. SALIVARY SUBSTITUTES

An important area in the dental field is the development of saliva substitutes. The use of salivary substitutes is necessary for reduced salivary flow rates encountered as a consequence of systemic diseases like Sjögrens's syndrome or caused by treatment of tumors by radio- or chemotherapy. Furthermore, many drugs have such negative side effects.

Traditionally, highly viscous polysaccharides like carboxymethylcellulose (CMC) have been used in aqueous solutions in this application [93]. However, an increasing interest is directed toward the use of molecules originating in or resembling those performing the lubricating roles in saliva. Therefore potential types of molecules are high-molecular-weight glycoproteins, like mucins, which are responsible for the viscoelastic properties of saliva [57]. Rheological characterization of potential components in saliva substitutes such as polysaccharides [93], submaxillary mucins [241], mucin and albumin alone and mixed [92]. Lubrication has also been measured using a "tooth on disc" setup, and the lubricating properties for saliva and selected salivary substitutes were determined [89,91]. It was found that little correlation existed between viscosity and lubricating efficiency. It was also observed that a CMC-based product was found not to be effective in lubricating oral hard interfaces, whereas a mucin-based product was a better lubricant, resembling the properties of human saliva [89]. Thin film (boundary) lubrication was reported for commercial salivary substitutes containing CMC and mucin, respectively, whereas a glycerol-based product showed hydrodynamic lubrication [91]. Furthermore, on a molar basis lubricating efficiency of salivary molecules was found to decrease in the order MG1 > MG2 > nonglycosylated α-amylases $\approx$ glycosylated α-amylases. In a paper by Reeh et al., enamel–enamel lubrication was assessed and lubricating properties of a number of CMC-, glycerol-, and mucin-based salivary substitutes reported [242]. In a recent paper comparing properties of commercially available saliva substitutes and saliva, the importance of adsorption and properties of adsorbed films for lubrication and duration (substantivity) were discussed [90]. Future development in this area may be expected to involve products replacing not only the lubricating properties of saliva, but also for example the ability to combat bacteria-mediated diseases [243].

ACKNOWLEDGMENTS

The Institute for Research and Competence Holding (IRECO) and the Foundation for Knowledge and Competence Development (KK-stiftelsen) are acknowledged for financial support.

REFERENCES

1. D. B. Ferguson, The salivary glands and their secretions. In *Oral Bioscience* (D. B. Ferguson, ed.). New York: Churchill Livingstone, 1999, pp. 117–150.
2. B. L. Slomiany, V. L. N. Murty, and A. Slomiany, Salivary lipids in health and disease. Prog. Lipid Res. *24*:311–324, 1985.
3. G. N. Jenkins, *The Physiology and Biochemistry of the Mouth*, 4th ed. Oxford: Blackwell Scientific, 1978.
4. D. I. Hay, Human glandular salivary proteins. In *CRC Handbook of Experimental Aspects of Oral Biochemistry* (E. P. Lazzari, ed.). Boca Raton: CRC Press, 1983, pp. 319–355.
5. C. Dawes, G. N. Jenkins, and C. H. Tonge, The nomenclature of the integuments of the enamel surface of teeth. Br. Dent. J. *115*:65–68, 1963.
6. M. J. Levine, P. C. Jones, R. E. Loomis, M. S. Reddy, I. Al-Hashimi, E. J. Bergey, S. D. Bradway, R. E. Cohen, and L. A. Tabak, Functions of human saliva and salivary mucins: an overview. In *Oral, Mucosal Diseases: Biology, Etiology and Therapy*. In *Proceeding of 2nd Dows Symposium*, October 6–9, 1985. University of Iowa, Iowa City, Iowa (I. C. Mackenzie, C. A. Squier, and E. Dabelsteen, eds.). Copenhagen: Laegeforeningens Forlag, 1987, pp. 24–27.
7. C. Dawes, Factors influencing salivary flow rate and composition. In *Saliva and Oral Health*, 2nd ed. (W. M. Edgar and D. M. O'Mullane, eds.). London: British Dental Journal, 1996, pp. 27–41.
8. D. I. Hay, The interaction of human parotid salivary proteins with hydroxyapatite. Arch. Oral Biol. *18*:1517–1529, 1973.
9. J. L. Jensen, M. S. Lamkin, and F. G. Oppenheim, Adsorption of human salivary proteins to hydroxyapatite: a comparison between whole saliva and glandular salivary secretions. J. Dent. Res. *71*:1569–1576, 1992.

10. M. J. Levine, Salivary macromolecules: a structure/function synopsis. In *Saliva as a Diagnostic Fluid* (D. Malamud and L. A. Tabak, eds.). New York: Annals of the New York Academy of Sciences, Vol. 694, 1993, pp. 11–16.

11. M. J. Levine, M. S. Reddy, L. A. Tabak, R. E. Loomis, E. J. Bergey, P. C. Jones, R. E. Cohen, M. W. Stinson, and I. Al-Hashimi, Structural aspects of salivary glycoproteins. J. Dent. Res. *66*:436–441, 1987.

12. D. I. Hay, The adsorption of salivary proteins by hydroxyapatite and enamel. Arch. Oral Biol. *12*:937–946, 1967.

13. L. Vroman and A. L. Adams, Identification of rapid changes in plasma–solid interfaces. J. Biomed. Mater. Res. *3*:43, 1969.

14. L. Vroman, A. L. Adams, M. Klings, G. C. Fischer, P. C. Munoz, and R. P. Solensky, Reactions of formed elements of blood with plasma proteins at interfaces. Ann. NY Acad. Sci. *288*:65–75, 1977.

15. L. Vroman and A. L. Adams, Why plasma proteins interact at interfaces. In *Proteins at Interfaces—Physicochemical and Biochemical Studies* (J. L. Brash and T. A. Horbett, eds.). Washington, DC: American Chemical Society, 1987, pp. 154–164.

16. J. L. Brash, Protein adsorption at the solid–solution interface in relation to blood–material interaction. In *Proteins at Interfaces—Physicochemical and Biochemical Studies* (J. L. Brash and T. A. Horbett, eds.). American Chemical Society, ACS Symposium Series, Vol. 343, 1987, pp. 490–506.

17. U. Lendenmann, J. Grogan, and F. G. Oppenheim, Saliva and dental pellicle—a review. Adv. Dent. Res. *14*:22–28, 2000.

18. Y. Yao, J. Grogan, M. Zehnder, U. Lendenmann, B. Nam, Z. Wu, C. E. Costello, and F. G. Oppenheim, Compositional analysis of acquired enamel pellicle by mass spectrometry. Arch. Oral Biol. *46*:293–303, 2001.

19. Y. Yao, M. S. Lamkin, and F. G. Oppenheim, Pellicle precursor proteins: acidic proline-rich proteins, statherin, and histatins, and their crosslinking reaction by oral transglutaminase. J. Dent. Res. *78*:1696–1703, 1999.

20. L. C. P. M. Schenkels, E. C. I. Veerman, and A. V. N. Nieuw Amerongen, Biochemical composition of human saliva in relation to other mucosal fluids. Crit. Rev. Oral Biol. Med. *6*:161–175, 1995.

21. T. Imoto, L. N. Johnson, A. C. T. North, D. C. Phillips, and J. A. Rupley, Vertebrate lysozymes. In *The Enzyme* (P. D. Boyer, ed.). Academic Press, 1972, pp. 665–868.

22. P. M. Claesson, E. Blomberg, J. C. Fröberg, T. Nylander, and T. Arnebrant, Protein interactions at solid surfaces. Adv. Colloid Interface Sci. *57*:161–227, 1995.

23. M. C. Wahlgren, T. Arnebrant, and I. Lundström, Adsorption of lysozyme to hydrophilic silicon oxide surfaces. J. Colloid Interface Sci. *175*:506–514, 1995.

24. K. H. Eggen and G. Rolla, Further information on the composition of the acquired enamel pellicle. In *Cariology Today* (B. Guggenheim, ed.). Karger, 1984, pp. 109–118.

25. M. Johnsson, M. J. Levine, and G. H. Nancollas, Hydroxyapatite binding domains in salivary proteins. Crit. Rev. Oral Biol. Med. *4*:371–378, 1993.

26. D. I. Hay, J. M. Ahern, S. K. Schuckebier, and D. H. Schlesinger, Human salivary acidic proline-rich protein polymorphisms and biosynthesis studied by high-performance liquid chomatography. J. Dent. Res. *73*:1717–1726, 1994.

27. T. Li, P. Bratt, A. P. Jonsson, M. Ryberg, I. Johansson, W. J. Griffiths, T. Bergman, and N. Strömberg, Possible release of an ArgGlyArgProGln Pentapeptide with innate immunity properties from acidic proline-rich proteins by proteolytic activity in commensal *Streptococcus* and *Actinomyces* species. Infect. Immun. *68*:5425–5429, 2000.

28. E. C. Moreno, K. Varughese, and D. I. Hay, Effect of human salivary proteins on the precipitation kinetics of calcium phosphate. Calcified Tissue Int. *28*:7–16, 1979.

29. D. I. Hay and E. C. Moreno, Statherin and the proline-rich proteins. In *Human Saliva: Clinical Chemistry and Microbiology* (J. O. Tenovuo, ed.). CRC Press, 1989, pp. 131–150.

30. W. N. Eigel, J. E. Butler, C. A. Ernstrom, H. M. Farrell, V. R. Harwalkar, R. Jenness, and R. Whitney, Nomenclature of proteins in cow's milk. J. Dairy Sci. 67:1599–1631, 1984.
31. C. Holt and L. Sawyer, Primary and secondary structure of the caseins in relation to their biological role. Protein Eng. 2:251, 1988.
32. D. H. Schlesinger and D. I. Hay, Complete covalent structure of statherin, a tyrosine-rich acidic peptide which inhibits calcium phosphate precipitation from human parotid saliva. J. Biol. Chem. 252:1689–1695, 1977.
33. W. H. Douglas, E. S. Reeh, N. Ramasubbu, P. A. Raj, K. K. Bhandary, and M. J. Levine, Statherin: a major boundary lubricant of human saliva. Biochem. Biophys. Res. Commun. 180:91–97, 1991.
34. F. G. Oppenheim, F. M. McMillian, S. M. Levitz, R. D. Diamond, G. D. Offner, and R. F. Troxler, Histatins, a novel family of histidine-rich proteins in human parotid secretion. Isolation, characterisation, primary structure and fungistatic effects on *Candida albicans*. J. Biol. Chem. 263:7472–7477, 1988.
35. P. Schüpbach, F. G. Oppenheim, U. Lendenmann, M. S. Lamkin, Y. Yao, and B. Guggenheim, Electron-microscopic demonstration of proline-rich proteins, statherin, and histatins in acquired enamel pellicles in vitro. Eur. J. Oral Sci. 109:60–68, 2001.
36. A. Prakobphol, M. J. Levine, L. A. Tabak, and M. S. Reddy, Purification of a low-molecular-weight, mucin-type glycoprotein from human submandibular-sublingual saliva. Carbohydr. Res. 108:111–122, 1982.
37. L. A. Tabak, Structure and function of human salivary mucins. Crit. Rev. Oral Biol. Med. 1:229–234, 1990.
38. R. E. Loomis, A. Prakobphol, M. J. Levine, M. S. Reddy, and P. C. Jones, Biochemical and biophysical comparison of two mucins from human submandibular-sublingual saliva. Arch. Biochem. Biophys. 258:452–464, 1987.
39. L. A. Tabak, M. J. Levine, J. K. Jane, R. A. Brian, R. A. Cohen, L. D. Monte, S. Zawacki, G. H. Nancollas, A. Slomiany, and B. L. Slomiany, Adsorption of human salivary mucins to hydroxyapatite. Arch. Oral Biol. 30:423–427, 1985.
40. G. Embery, T. G. Heaney, and J. B. Stanbury, Studies on the organic polyanionic constituents of human acquired dental pellicle. Arch. Oral Biol. 31:623–625, 1986.
41. T. Sonju and G. Rolla, Chemical analysis of the acquired pellicle formed in two hours on cleaned human teeth in vivo. Rate of formation and amino acid analysis. Caries Res. 7:30–38, 1973.
42. P. A. Nielsen, E. P. Bennet, H. H. Wandell, M. H. Therkildsen, J. Hannibal, and H. Clausen, Identification of a major human high molecular weight salivary mucin (MG1) as tracheobronchial mucin MUC5B. Glycobiology 7:413–419, 1997.
43. R. F. Troxler, I. Iontcheva, F. G. Oppenheim, D. P. Nunes, and G. D. Offner, Molecular characterization of a major molecular weight mucin from human sublingual gland. Glycobiology 7:965–973, 1997.
44. D. J. Thornton, N. Kahn, R. Mehrotra, E. Veerman, N. H. Packer, and J. K. Sheehan, Salivary mucin MG1 is comprised almost entirely of different glycosylated forms of the MUC5B gene product. Glycobiology 9:293–302, 1999.
45. R. Mehrotra, D. J. Thornton, and J. K. Sheehan, Isolation and physical characterization of the MUC7 (MG2) mucin from saliva: evidence for self-association. Biochem. J. 334:415–422, 1998.
46. L. A. Tabak, In defence of the oral cavity: structure, biosynthesis, and function of salivary mucins. Annu. Rev. Physiol. 57:547–564, 1995.
47. J. Brange, B. Skelbaek-Pedersen, L. Langkjaer, U. Damgaard, H. Ege, S. Havelund, L. G. Heding, K. H. Jørgensen, J. Lykkeberg, J. Markussen, et al. (eds.), *Galenics of Insulin. The Physico-Chemical and Pharmaceutical Aspects of Insulin and Insulin Preparations.* Berlin: Springer-Verlag, 1987.

48. D. G. Schmidt and T. A. J. Payens, The evaluation of positive and negative contributions to the second viral coefficient of some milk proteins. J. Colloid Interface Sci. *39*: 655–662, 1972.

49. T. A. J. Payens and B. W. Markwijk, Biochim. Biophys. Acta *71*:517, 1963.

50. P. Walstra and R. Jenness, *Dairy Chemistry and Physics*. New York: John Wiley and Sons, 1984.

51. M. Rykke, G. Smistad, G. Rölla, and J. Karlsen, Micelle-like structures in human saliva. Colloids Surf. B: Biointerfaces *4*:33–44, 1995.

52. M. Rykke, D. G. Devold, G. Smistad, and G. E. Vegarud, Comparison of bovine casein micelles and human salivary micelle-like structures. Int. Dairy J. *9*:365–366, 1999.

53. I. Iontcheva, F. G. Oppenheim, and R. F. Troxler, Human salivary mucin MG1 selectively forms heterotypic complexes with amylase, proline-rich proteins, statherin, and histatins. J. Dent. Res. *76*:734–743, 1997.

54. I. Iontcheva, F. G. Oppenheim, G. D. Offner, and R. F. Troxler, Molecular mapping of statherin- and histatin-binding domains in human salivary mucin MG1 (MUC5B) by the yeast two-hybrid system. J. Dent. Res. *79*:732–739, 2000.

55. A. R. Biesbrock, M. S. Reddy, and M. J. Levine, Interaction of a salivary mucin-secretory immunoglobulin A complex with mucosal antigens. Infect. Immun. *59*:3492–3497, 1991.

56. C. Wickstrom, C. Christersson, J. R. Davies, and I. Carlstedt, Macromolecular organization of saliva: identification of 'insoluble' MUC5B assemblies and non-mucin proteins in the gel phase. Biochem. J. *351*:421–428, 2000.

57. M. N. Hatton, R. E. Looomis, M. I. Levine, and L. A. Tabak, Masticatory lubrication. The role of carbohydrate in the lubricating property of a salivary glycoprotein–albumin complex. Biochem. J. *230*:817–820, 1985.

58. Y. Yao, M. S. Lamkin, and F. G. Oppenheim, Pellicle precursor protein crosslinking: characterisation of an adduct between acidic proline-rich protein (prp-1) and statherin generated by transglutaminase. J. Dent. Res. *79*:930–938, 2000.

59. P.-O. Glantz, S. E. Friberg, S. M. Wirth, and R. E. Baier, Thin-section transmission electron microscopy of human saliva. Acta Odont. Scand. *47*:111–116, 1989.

60. P.-O. Glantz, J. R. Natiella, C. D. Vaughan, A. E. Meyer, and R. E. Baier, Structural studies of human saliva. Acta Odont. Scand. *47*:17–24, 1989.

61. P.-O. Glantz, S. M. Wirth, R. E. Baier, and J. E. Wirth, Electron microscopic studies of human mixed saliva. Acta Odont. Scand. *47*:7–15, 1989.

62. P. O. Glantz, M. A. Meenaghan, K. Hyun, and S. Wirth, On the presence and localization of epidermal and nerve growth factors in human whole saliva. Acta Odont. Scand. *47*:287–292, 1989.

63. P.-O. Glantz, S. E. Friberg, C. Christersson, and R. E. Baier, Surface and colloid chemical aspects of saliva particle interactions. J. Oral Rehab. *22*:585–588, 1995.

64. G. Rolla and M. Rykke, Evidence for presence of micelle-like protein globules in human saliva. Colloids Surf. B: Biointerfaces *3*:117–182, 1994.

65. M. Rykke, A. Young, G. Rolla, D. G. Devold, and G. Smistad, Transmission electron microscopy of human saliva. Colloids Surf. B: Biointerfaces *9*:257–267, 1997.

66. M. Rykke, A. Young, G. Smistad, G. Rolla, and J. Karlsen, Zeta potentials of human salivary micelle-like structures. Colloids Surf. B: Biointerfaces *6*:51–56, 1996.

67. A. Young, G. Smistad, J. Karlsen, G. Rolla, and M. Rykke, The effect of some proteolytic enzymes on human salivary micelle-like structures. Colloids Surf. B: Biointerfaces *8*:189–198, 1997.

68. J. Sjollema, H. J. Busscher, and A. H. Weerkamp, Real time enumeration of adhering microorganisms in a parallel plate flow cell using automated image analysis. J. Microbiological Methods *9*:73–78, 1989.

69. W. G. Pitt, M. O. McBride, A. J. Barton, and D. Sagers, Air–water interface displaces adsorbed bacteria. Biomaterials *14*:605–608, 1993.

70. C. Gomez-Suarez, H. C. van der Mei, and H. J. Busscher, Air bubble–induced detachment of polystyrene particles with different sizes from collector surfaces in a parallel plate flow chamber. Colloids Surf. A *186*:211–219, 2001.

71. C. Gomez-Suarez, H. J. Busscher, and H. C. van der Mei, Analysis of bacterial detachment from substratum surfaces by the passage of air/liquid interfaces. Appl. Environ. Microbiol. *67*:2531–2537, 2001.

72. J. Sefton, T. Arnebrant, and P.-O. Glantz, Initial studies on the behaviour of salivary proteins at liquid/air interfaces. Acta Odont. Scand. *50*:221–226, 1992.

73. E. Adamczyk, T. Arnebrant, and P.-O. Glantz, Time-dependent interfacial tension of whole saliva and saliva–bacteria mixes. Acta Odont. Scand. *55*:384–389, 1998.

74. M. L. Wallwork, J. Kirkham, J. Zhang, D. A. Smith, S. J. Brookes, R. C. Shore, S. R. Wood, O. Ryu, and C. Robinson, Binding of matrix proteins to developing enamel crystals: an atomic force microscopy study. Langmuir *17*:2508–2513, 2001.

75. J. Arends and W. L. Jongebloed, The enamel substrate—characteristics of the enamel surface. Swed. Dent. J. *1*:215–224, 1977.

76. M. Finke, K. L. Jandt, and D. M. Parker, The early stages of native enamel dissolution studied with atomic force microscopy. J. Colloid Interface Sci. *232*:156–164, 2000.

77. J. Kirkham, J. Zhang, S. J. Brookes, R. C. Shore, S. R. Wood, D. A. Smith, M. L. Wallwork, O. Ryu, and C. Robinson, Evidence for charge domains on developing enamel crystal surfaces. J. Dent. Res. *79*:1943–1947, 2000.

78. S. Chandler and D. W. Fuerstenau, In *Adsorption On and Surface Chemistry of Hydroxy Apatite* (D. N. Misra, ed.). New York: Plenum Press, 1984, p. 29.

79. J. Arends, Zeta potentials of enamel and apatites. J. Dent. *7*:246–253, 1979.

80. M. E. Neiders, L. Weiss, and T. L. Cudney, An electrokinetic characterization of human tooth surfaces. Arch. Oral Biol. *15*:135–151, 1970.

81. J. Olsson, A. Carlén, and K. Holmberg, Inhibition of *Streptococcus mutans* adherence to hydroxy apatite with combination of alkyl phosphates and nonionic surfactants. Caries Res. *25*:51–57, 1991.

82. I. C. Hahn Berg, U. M. Elofsson, A. M. M. Joiner, and T. Arnebrant, Salivary protein adsorption onto hydroxyapatite and SDS-mediated elution studied by in situ ellipsometry. Biofouling *17*:173–187, 2001.

83. A. H. Weerkamp, H. M. Uyen, and H. J. Busscher, Effect of zeta potential and surface energy on bacterial adhesion to uncoated and saliva-coated human enamel and dentin. J. Dent. Res. *67*:1483–1487, 1988.

84. H. J. Busscher, D. H. Retief, and J. Arends, Relationship between surface free energies of dental resins and bond strengths to etched enamel. Dent. Mater. *3*:60–63, 1987.

85. M. Aning, D. O. Welch, and B. S. H. Royce, The surface energy of fluorapatite. Phys. Lett. *37A*:253–254, 1971.

86. P.-O. Glantz, On wettability and adhesiveness. Odontol. Rev. *20*:1–132, 1969.

87. R. E. Baier and R. C. Dutton, Initial events in interactions of blood with a foreign surface. J. Biomed. Mater. Res. *3*:191, 1969.

88. H. J. Busscher, J. Engelberg, and H. C. van der Mei, Initial microbial adhesion is a determinant for the strength of biofilm adhesion. FEMS Microbiol. Lett. *128*:229–234, 1995.

89. M. N. Hatton, M. J. Levine, J. E. Margarone, and A. Aguirre, Lubrication and viscosity features of human saliva and commercially available saliva substitutes. J. Oral Maxillofac. Surg. *45*:496–499, 1987.

90. C. Christersson, L. Lindh, and T. Arnebrant, Film-forming properties and viscosities of saliva substitutes and human whole saliva. Eur. J. Oral Sci. *108*:418–425, 2000.

91. A. Aquirre, B. Mendoza, M. S. Reddy, F. A. Scannapieco, M. J. Levine, and M. N. Hatton, Lubrication of selected salivary molecules and artificial salivas. Dysphagia *4*: 95–100, 1989.

92. J. Mellema, H. J. Holterman, H. A. Waterman, and C. Blom, Rheological aspects of mucin-containing solutions and saliva substitutes. Biorheology *29*:231–249, 1992.

93. W. A. van der Reijden, E. Veerman, and A. V. N. Nieuw Amerongen, Rheological properties of commercially available polysaccharides with potential use in saliva substitutes. Biorheology *31*:631–642, 1994.

94. A. Joiner, N. M. Jones, and S. J. Raven, Investigation of factors influencing stain formation utilising an in situ model. Adv. Dent. Res. *9*:471–476, 1995.

95. R. E. Baier and P.-O. Glantz, Characterization of oral in vivo films formed on different types of solid surfaces. Acta Odont. Scand. *36*:289–301, 1978.

96. N. Vassilakos, P.-O. Glantz, and T. Arnebrant, Reflectometry—a simple optical method for investigations of salivary film formation in the oral cavity. Scand. J. Dent. Res. *101*:339–343, 1993.

97. M. Hannig, Ultrastructural investigation of pellicle morphogenesis at two different intraoral sites during a 24 h period. Clin. Oral Invest. *3*:88–95, 1999.

98. K. K. Skjørland, M. Rykke, and T. Sønju, Rate of pellicle formation in vivo. Acta Odont. Scand. *53*:358–362, 1995.

99. N. Vassilakos, T. Arnebrant, and P.-O. Glantz, Adsorption of whole saliva onto hydrophilic and hydrophobic solid surfaces. The influence of concentration, ionic strength and pH. Scand. J. Dent. Res. *100*:346–353, 1992.

100. N. Vassilakos, T. Arnebrant, and P.-O. Glantz, An in vitro study of salivary film formation at solid/liquid interfaces. Scand. J. Dent. Res. *101*:133–137, 1993.

101. L. Lindh, T. Arnebrant, P.-E. Isberg, and P.-O. Glantz, Concentration dependence of adsorption from human whole resting saliva at solid/liquid interfaces—an ellipsometric study. Biofouling *14*:189–196, 1999.

102. T. Nylander, T. Arnebrant, and P.-O. Glantz, Interactions between salivary films adsorbed on mica surfaces. Colloids Surf. A: Physicochem. Eng. Aspects *129–130*:339–344, 1997.

103. T. Ericson, K. M. Pruitt, H. Arwin, and I. Lundström, Ellipsometric studies of film formation on tooth enamel and hydrophilic silicon surface. Acta Odont. Scand. *40*: 197–201, 1982.

104. H. J. Trurnit, Studies of enzyme systems at solid–liquid interface. II. The kinetics of adsorption and reaction. Arch. Biochem. Biophys. *51*:176–199, 1954.

105. J. M. M. Kop, J. W. Corsel, M. P. Janssen, P. A. Cuypers, and W. T. Hermens, Ellipsometric measurements of the association of prothrombine with phospholipid monolayers. J. Physique *C10*:491–493, 1983.

106. L. Lindh, P.-O. Glantz, P.-E. Isberg, and T. Arnebrant, An in vitro study of initial adsorption from human parotid and submandibular/sublingual resting saliva at solid/liquid interfaces. Biofouling *17*:227–239, 2001.

107. H. J. Busscher, M. van der Kuijl, J. Haker, R. Kalicharan, H. C. van der Mei, E. C. I. Veerman, and A. V. Nieuw Amerongen, Selective adsorption of salivary mucins by immunogold labeling. J. Dent. Res. *71*:601, 1992.

108. L. Lindh, P.-O. Glantz, I. Carlstedt, C. Wickstrom, and T. Arnebrant, Adsorption of MUC5B and the role of mucins in early salivary film formation. Colloid Surf. B: Biointerfaces 2001 (in press).

109. L. Vroman and A. L. Adams, Adsorption of proteins out of plasma and solution in narrow spaces. J. Colloid Interface Sci. *111*:391–402, 1986.

110. T. Nylander and N. M. Wahlgren, Competitive and sequential adsorption of β-casein and β-lactoglobulin on hydrophobic surfaces and the interfacial structure of β-casein. J. Colloid Interface Sci. *162*:151–162, 1994.

111. E. Dickinson, B. S. Murray, and G. Stainsby, Protein adsorption at the air–water and oil–water interfaces. In *Advances in Food Emulsions and Foams* (E. Dickinson and G. Stainsby, eds.). London: Elsevier Applied Science, 1988, pp. 123–162.

112. D. Horsley, J. Herron, V. Hlady, and J. D. Andrade, Human hen lysozyme adsorption: a comparative study using total internal reflection fluorescence spectroscopy and molecular graphics. In *Proteins at Interfaces—Physicochemical and Biochemical Studies* (J. L. Brash and T. A. Horbett, eds.). American Chemical Society, 1987, pp. 290–305. ACS Symposium Series, Vol. 343.

113. J. L. Bohnert, T. A. Horbett, B. D. Ratner, and F. H. Royce, Adsorption of proteins from artificial tear solution to contact lens materials. Invest. Ophtamol. Visual Sci. *29*: 362–373, 1988.

114. U. Jönsson, B. Ivarsson, I. Lundström, and L. Berghem, Adsorption behavior of fibronectin on well-characterized silica surfaces. J. Colloid Interface Sci. *90*:148–163, 1982.

115. M. Malmsten, N. Burns, and A. Veide, Electrostatic and hydrophobic effects of oligopeptide insertions on protein adsorption. J. Colloid Interface Sci. *204*:104–111, 1998.

116. N. Vassilakos, Some biophysical aspects of salivary film formation. Studies of salivary adsorption at solid/liquid and air/liquid interfaces. Ph.D. thesis. Lund University, Lund, Sweden, 1992.

117. N. Vassilakos, T. Arnebrant, and P.-O. Glantz, Interaction of anionic and cationic surfactants with salivary pellicles formed at solid surfaces in vitro. Biofouling *5*:277–286, 1992.

118. N. Vassilakos, T. Arnebrant, and J. Rundegren, In vitro interactions of delmopinol hydrochloride with salivary films adsorbed at solid/liquid interfaces. Caries Res. *27*: 176–182, 1993.

119. N. Vassilakos, T. Arnebrant, J. Rundegren, and P.-O. Glantz, In vitro interactions of anionic and cationic surfactants with salivary fractions on well defined solid surfaces. Acta Odont. Scand. *50*:179–188, 1992.

120. N. Vassilakos, J. Rundegren, T. Arnebrant, and P.-O. Glantz, Adsorption from salivary fractions at solid/liquid and air/liquid interfaces. Arch. Oral Biol. *37*:549–557, 1992.

121. T. Arnebrant and T. Simonsson, The effect of ionic surfactants on salivary proteins adsorbed on silica surfaces. Acta Odont. Scand. *49*:281–288, 1991.

122. L. Lindh, P.-O. Glantz, N. Strömberg, and T. Arnebrant, On the adsorption of human acidic proline rich proteins (PRP-1 and PRP-3) and statherin at solid/liquid interfaces. Biofouling *18*:87–94, 2002.

123. I. C. Hahn Berg, S. M. M. Kalfas, and T. Arnebrant, Proteolytic degradation of oral biofilms in vitro and in vivo: potential of proteases originating from *Euphausia superba* for plaque control. Eur. J. Oral Sci. *109*:316–324, 2001.

124. T. Ericson and H. Arwin, Molecular basis of saliva-mediated aggregation. In *Molecular Basis of Oral Microbial Adhesion* (S. E. Mergenhagen and B. Rosan, eds.). Washington, DC: American Society for Microbiology, 1985, pp. 144–150.

125. R. E. Baier, A. E. Meyer, D. M. Dombroski, U. Nassar, J. M. Merrick, D. I. Hay, and M. P. Olivieri, Surface expression of preferential bacterial binding sites by adsorbed salivary molecules. Surfaces in Biomaterials *1*:11–22, 1994.

126. C. Christersson and R. Dunford, Salivary film formation on defined solid surfaces in the absence and presence of microorganisms. Biofouling *3*:237–250, 1991.

127. C. E. Christersson, On salivary film formation and bacterial retention to solids. A methodological and experimental in vitro study. Thesis, Lund University, Lund, Sweden, 1991.

128. P.-O. Glantz and R. E. Baier, Recent studies of nonspecific aspects of intraoral adhesion. J. Adhesion *20*:227–244, 1986.

129. P.-O. Glantz, R. E. Baier, and C. E. Christersson, Biochemical and physiological considerations for modelling biofilms in the oral cavity. Dent. Mater. *12*:208–214, 1996.

130. D. I. Hay, Some observations on human saliva proteins and their role in the formation of the acquired enamel pellicle. J. Dent. Res. 48:806–810, 1969.

131. D. I. Hay, Arch. Oral Biol. 20:553–558, 1975.

132. D. I. Hay and E. C. Moreno, Differential adsorption and chemical affinities of proteins for apatitic surfaces. J. Dent. Res. (Special Issue B) 58:930–942, 1979.

133. E. C. Moreno, M. Kresak, and D. I. Hay, Adsorption of two human parotid salivary macromolecules on hydroxy-, fluorhydroxy-, and fluorapatites. Arch. Oral Biol. 23:525–533, 1978.

134. E. C. Moreno, M. Kresak, and D. I. Hay, Adsorption thermodynamics of acidic proline-rich human salivary proteins onto calcium apatites. J. Biol. Chem. 257:2981–2989, 1982.

135. B. Lassen, K. Holmberg, C. Brink, A. Carlén, and J. Olsson, Binding of salivary proteins and oral bacteria to hydrophobic and hydrophilic surfaces in vivo and in vitro. Colloid Polym. Sci. 272:1143–1150, 1994.

136. H. Elwing, S. Welin, A. Askendal, U. Nilsson, and I. Lundström, A wettability gradient method for studies of macromolecular interactions at the liquid/solid interface. J. Colloid Interface Sci. 119:203–209, 1987.

137. H. J. Holterman, E. J. Gravenmade, H. A. Waterman, J. Mellema, and C. Blom, An ellipsometric study of protein adsorption at the saliva–air interface. J. Colloid Interface Sci. 128:523–532, 1989.

138. W. Norde, Adsorption of proteins from solutions at the solid liquid interface. Adv. Colloid Interface Sci. 25:267–340, 1986.

139. W. Norde, F. MacRitchie, G. Nowicka, and J. Lyklema, Protein adsorption at solid–liquid interfaces: reversibility and conformational aspects. J. Colloid Interface Sci. 112:447–456, 1986.

140. T. A. Horbett and J. L. Brash, Proteins at interfaces: current issues and future prospects. In Proteins at Interfaces—Physicochemical and Biochemical Studies (J. L. Brash and T. A. Horbett, eds.). American Chemical Society, 1987, pp. 1–37. ACS Symposium Series, Vol. 343.

141. M. Wahlgren and T. Arnebrant, Adsorption of β-lactoglobulin into silica, methylated silica and polysulphone. J. Colloid Interface Sci. 136:259–265, 1990.

142. M. Wahlgren and T. Arnebrant, Protein adsorption to solid surfaces. TIBTECH 9:201–208, 1991.

143. R. E. Baier, A. E. Meyer, J. R. Natiella, R. R. Natiella, and J. M. Carter, Surface properties determine bioadhesive outcomes: methods and results. J. Biomed. Mater. Res. 18:337–355, 1984.

144. M. Quirynen, M. Marechal, H. J. Busscher, A. H. Weerkamp, J. Arends, P. L. Darius, and D. vanSteenberghe, The influence of surface free-energy on planimetric plaque growth in man. J. Dent. Res. 68:796–799, 1989.

145. J. L. Bohnert and T. A. Horbett, Changes in adsorbed fibrinogen and albumin interaction with polymers indicated by decreases in detergent elutability. J. Colloid Interface Sci. 111:363–377, 1986.

146. A. Carlén, A.-C. Börjesson, K. Nikdel, and J. Olsson, Composition of pellicles formed in vivo on tooth surfaces in different parts of the dentition, and in vitro on hydroxy-apatite. Caries Res. 32:447–455, 1998.

147. A. Carlén, K. Nikdel, A. Wennerberg, K. Holmberg, and J. Olsson, Surface characteristics and in vitro biofilm formation on glass ionomer and composite resin. Biomaterials 22:481–487, 2001.

148. E. Perez and J. E. Proust, Forces between mica surfaces covered with adsorbed mucin across aqueous solution. J. Colloid Interface Sci. 118:182–191, 1987.

149. M. Malmsten, E. Blomberg, P. M. Claesson, I. Carlstedt, and I. Ljusegren, Mucin layers on hydrophobic surfaces studied with ellipsometry and surface force measurements. J. Colloid Interface Sci. 151:579–590, 1992.

150. G. Bernardi and T. Kawasaki, Chromatography of polypeptides and proteins on hydroxy apatite columns. Biochim. Biophys. Acta *160*:301–310, 1968.

151. T. Ericson, Adsorption to hydroxyapatite of proteins and conjugated proteins from human saliva. Caries Res. *1*:52–58, 1967.

152. T. Ericson, Salivary glycoproteins: composition and adsorption to hydroxyapatite in relation to the formation of dental pellicles and calculus. Acta Odont. Scand. *26*:3–21, 1968.

153. E. C. Moreno, M. Kresak, and D. I. Hay, Adsorption of salivary proteins onto Ca apatites. Biofouling *4*:3–24, 1991.

154. T. Nylander, T. Arnebrant, R. E. Baier, and P.-O. Glantz, Interactions between layers of salivary acidic proline rich protein 1 (PRP 1) adsorbed on mica surfaces. Prog. Colloid Polym. Sci. *108*:34–39, 1998.

155. T. Nylander and N. M. Wahlgren, Forces between adsorbed layers of β-casein. Langmuir *13*:6219–6225, 1997.

156. T. Nylander, F. Tiberg, T.-J. Su, and R. K. Thomas, β-Casein adsorption at the hydrophobized silicon oxide–aqueous solution interface and the effect of added electrolyte. Biomacromolecules *2*:278–287, 2001.

157. C. Wickström, J. R. Davies, G. V. Eriksen, E. C. I. Veerman, and I. Carlstedt, MUC5B is a major gel-forming, oligomeric mucin from human salivary gland, respiratory tract and endocervix: identification of glycoforms and C-terminal cleavage. Biochem. J. *334*: 685–693, 1998.

158. M. J. Levine, L. A. Tabak, M. Reddy, and I. R. Mandel, Nature of salivary pellicles in microbial adherence: role of salivary mucins. In *Molecular Basis of Oral Microbial Adhesion* (S. E. Mergenhagen and B. Rosan, eds.). Washington, DC: IRL Press, 1985, pp. 125–130.

159. M. Malmsten, I. Ljusegren, and I. Carlstedt, Ellipsometry studies of the mucoadhesion of cellulose derivates. Colloids Surf. B: Biointerfaces *2*:463–470, 1994.

160. L. Shi and K. D. Caldwell, Mucin adsorption to hydrophobic surfaces. J. Colloid Interface Sci. *224*:372–381, 2000.

161. J. E. Proust, A. Baszkin, E. Perez, and M. M. Boissonnade, Bovine submaxillary mucin (BSM) adsorption at solid/liquid interfaces and surface forces. Colloids Surf. *10*:43–52, 1984.

162. A. V. N. Amerongen, C. H. Oderkerk, and E. C. I. Veerman, Interaction of human salivary mucins with hydroxyapatite. J. Biol. Buccale *17*:85–92, 1989.

163. A. H. Meckel, The nature and importance of organic deposits on dental enamel. Caries Res. *2*:104–114, 1968.

164. W. G. Armstrong, The composition of organic films formed on human teeth. Caries Res. *1*:89–103, 1967.

165. B. L. Slomiany, V. L. N. Murty, E. Zdebska, A. Slomiany, K. Gwozdzinski, and I. D. Mandel, Tooth surface lipids and their role in protection of dental enamel against lactic acid diffusion in man. Arch. Oral Biol. *31*:187–191, 1986.

166. B. Larsson, G. Olivecrona, and T. Ericson, Lipids in human saliva. Arch. Oral Biol. *41*:105–110, 1996.

167. V. L. N. Murty, B. L. Slomiany, W. Laszewicz, K. Petropoulou, and I. D. Mandel, Lipids of developing dental plaque in caries-resistant and caries-susceptable adult people. Arch. Oral Biol. *30*:171–175, 1985.

168. T. Sonju, Investigation of some glycoproteins and their possible role in pellicle formation. Norske Tannlaegeforenings Tidene *85*:393–403, 1975.

169. G. Bernardi and W. H. Cook, Biochim. Biophys. Acta *44*:96–105, 1960.

170. I. Al-Hashimi and M. J. Levine, Characterization of in vivo salivary-derived enamel pellicle. Arch. Oral Biol. *34*:289–295, 1989.

171. M. S. Lamkin, D. Migliari, Y. Yao, R. F. Troxler, and F. G. Oppenheim, New in vitro model for the acquired enamel pellicle: pellicles formed from whole saliva show inter-subject consistency in protein composition and proteolytic fragmentation patterns. J. Dent. Res. *80*:385–388, 2001.

172. H. J. Busscher, H. M. W. Uyen, I. Stokroos, and W. L. Jongebloed, A transmission electron microscopy study of the adsorption patterns of early developing artificial pel-licles on human enamel. Arch. Oral Biol. *34*:803–810, 1989.

173. M. Hannig, Transmission electron microscopic study of in vivo pellicle formation on dental materials. Eur. J. Oral Sci. *105*:422–433, 1997.

174. T. A. Horbett, Adsorption to biomaterials from protein mixtures. In *Proteins at Inter-faces—Physicochemical and Biochemical Studies* (J. L. Brash and T. A. Horbett, eds.). American Chemical Society, 1987, pp. 239–260. ACS Symposium Series, Vol. 343.

175. T. A. Horbett and J. L. Brash, Proteins at interfaces II. Fundamentals and applications. ACS Symposium, 1995.

176. G. Embery, S. D. Hogg, T. G. Heaney, J. B. Stanbury, and D. R. J. Green, Some considerations on dental pellicle formation and early bacterial colonisation: the role of high and low molecular weight proteins of the major and minor salivary glands. In *Bacterial Adhesion and Preventive Dentistry* (J. M. ten Cate, ed.). Proceedings of a workshop: Bacterial adhesion to biosurfaces with special emphasis on dental problems, November 28–December 1, 1983 Paterswolde, The Netherlands. Oxford: IRL Press, 1984, pp. 73–84.

177. W. A. Nolte, *Oral Microbiology*, 2nd ed. St. Louis: Mosby, 1973.

178. H. J. Busscher and L. V. Evans (eds.), *Oral Biofilms and Plaque Control*. Amsterdam: Harwood Academic, 1998.

179. J. W. Costerton and Z. Lewandowski, The biofilm lifestyle. In *Biofilms on Oral Sur-faces: Implications for Health and Disease*. 14th international conference on oral bi-ology, Monterey, CA (M. J. Novak, ed.). International American Associations for Dental Research, 1997, pp. 192–196.

180. M. A. Listgarten, The structure of dental plaque. Periodontology *5*:52–65, 2000.

181. B. Rosan and R. J. Lamont, Dental plaque formation. Microbes and Infection *2*:1599–1607, 2000.

182. T. Lie, Pellicle formation on hydroxyapatite splints attached to the human dentition: morphologic confirmation of the concept of adsorption. Arch. Oral Biol. *20*:739–742, 1975.

183. W. Norde and Lyklema, Protein adsorption and bacterial adhesion to solid surfaces: a colloid chemical approach. Colloids Surf. *38*:1–13, 1989.

184. H. J. Busscher and A. H. Weerkamp, Specific and non-specific interactions in bacterial adhesion to solid surfaces. FEMS Microbiol. Rev. *46*:165–173, 1987.

185. R. J. Gibbons, D. I. Hay, W. C. I. Childs, and G. Davis, Role of cryptic receptors (cryptitopes) in bacterial adhesion to oral surfaces. Arch. Oral Biol. *35(Suppl)*:107s–114s, 1990.

186. P.-O. Glantz, T. Arnebrant, T. Nylander, and R. E. Baier, Bioadhesion—a phenomenon with multiple dimensions. Acta Odont. Scand. *57*:238–241, 1999.

187. P. Glantz, R. Baier, R. Attström, and A. Meyer, Comparative clinical wettability of teeth and intraoral mucosa. J. Adhesion Sci. Technol. *5*:401–408, 1991.

188. M. D. Jendresen, Studies on intraoral adhesion. An in-vivo study of clinical adhesive-ness of teeth and selected dental materials. Ph.D. thesis, Lund University, Lund, Swe-den, 1980.

189. H. C. van der Mei, A. J. Leonard, A. H. Weerkamp, P. G. Rouxhet, and H. J. Busscher, Properties of oral *Streptococci* relevant for adherence: zeta potential surface free energy and elemental composition. Colloids Surf. *32*:297–305, 1988.

190. H. C. van der Mei, A. H. Weerkamp, and H. J. Busscher, A comparison of various methods to determine hydrophobic properties of streptococcal cell surfaces. J. Microbiol. Methods 6:277–287, 1987.

191. G. Westergren and J. Olsson, Hydrophobicity and adherence of oral *Streptococci* after repeated subculture in vitro. Infect. Immun. 40:432–435, 1983.

192. W. A. Zisman, Relation of the equilibrium contact angle to liquid and solid constitution. In *Contact Angles, Wettability and Adhesion* (R. Gould, ed.). American Chemical Society, 1964, pp. 1–51. ACS Adv. Chem. Ser., Vol. 43.

193. R. E. Baier, E. G. Schafrin, and W. A. Zisman, Science 162:1360, 1968.

194. R. E. Baier and A. E. Meyer, Surface energetics and biological adhesion. In *International Symposium on Physicochemical Aspects of Polymer Surfaces* (K. L. Mittal, ed.). New York: Plenum Press, 1981, pp. 895–909.

195. R. J. Gibbons, D. I. Hay, J. O. Cisar, and W. B. Clark, Adsorbed salivary proline-rich protein 1 and statherin: receptors for type 1 fimbriae of *Actenomyces viscosus* T14V-J1 on apatitic surfaces. Infect. Immun. 56:2990–2993, 1998.

196. A. Carlén, P. Bratt, C. Stenudd, J. Olsson, and N. Strömberg, Agglutinin and acidic proline-rich protein receptor patterns may modulate bacterial adherence and colonization on tooth surfaces. J. Dent. Res. 77:81–90, 1998.

197. R. J. Gibbons and D. I. Hay, Human salivary acidic proline-rich proteins and statherin promote the attachment of *Actinomyces viscosus* LY7 to apatitic surfaces. Infect. Immun. 56:439–445, 1988.

198. R. Gibbons and D. I. Hay (eds.), *Adsorbed Salivary Proline-Rich Proteins as Bacterial Receptors on Apatitic Surfaces.* New York: Springer-Verlag, 1989.

199. R. J. Gibbons and D. I. Hay, Adsorbed salivary acidic proline rich proteins contribute to the adhesion of *Streptococcus mutans* JBP to apatitic surfaces. J. Dent. Res. 68:1303–1307, 1989.

200. R. J. Gibbons, D. I. Hay, and D. H. Schlesinger, Delineation of a segment of adsorbed salivary acidic proline-rich proteins which promotes adhesion of *Streptococcus gordonii* to apatitic surfaces. Infect. Immun. 59:2948–2954, 1991.

201. C. G. van Hoogmoed, M. van der Kuijl-Booij, H. C. van der Mei, and H. J. Busscher, Inhibition of *Streptococcus mutans* NS adhesion to glass with and without a salivary conditioning film by biosurfactant-releasing *Streptococcus mitis* strains. Appl. Environ. Microbiol. 66:659–663, 2000.

202. J. W. T. Wimpenny and R. Colasanti, A unifying hypothesis for the structure of microbial biofilms based on cellular automaton models. FEMS Microbiol. Ecology 22:1–16, 1997.

203. I. W. Sutherland, Biofilm matrix polymers. In *Dental Plaque Revisited—Oral Biofilms in Health and Disease* (H. N. Newman and M. Wilson, eds.). Eastman Dental Institute, University College London and BioLine, Cardiff School of Biosciences, 1999.

204. D. J. Hassett, J. G. Elkins, J.-F. Ma, and T. R. McDermott, *Pseudomonas aeruginosa* biofilm sensitivity to biocides: use of hydrogen peroxide as model antimicrobial agent for examining resistance mechanisms. Method Enzymol. 310:599–608, 1999.

205. D. G. Davies, M. R. Parsek, J. P. Pearson, B. H. Iglewski, J. W. Costerton, and E. P. Greenberg, The involvement of cell-to-cell signals in the development of a bacterial biofilm. Science 280:295–298, 1998.

206. R. Kolter and R. Losick, One for all and all for one. Science 280:226–227, 1998.

207. J. W. Costerton, D. E. Lewandowski, D. R. Caldwell, D. R. Korber, and H. M. Lappin-Scott, Microbial biofilms. Annu. Rev. Microbiol. 49:711–745, 1995.

208. J. W. Costerton, Biofilms 2000. In *Biofilms 2000* (J. W. Costerton, ed.). Washington, DC: American Society for Microbiology, 2000.

209. N. Vassilakos, S. Kalfas, T. Arnebrant, and Rundegren, A simple flow cell system to evaluate in vitro bacterial adhesion on solids. Colloids Surf. B: Biointerfaces 1:341–347, 1993.

210. A. Kharazmi, B. Giwercman, and N. Hiby, Robbins device in biofilm research. Method Enzymol. *310*:207–215, 1999.

211. M. S. Zinn, R. D. Kirkegaard, R. J. Palmer, and D. C. White, Laminar flow chamber for continuous monitoring of biofilm formation and succession. Method Enzymol. *310*: 224–232, 1999.

212. D. G. Ahearn, R. N. Borazjani, R. B. Simmons, and M. M. Gabriel, Primary adhesion of *Pseudomonas aeruginosa* to inanimate surfaces including biomaterials. Method Enzymol. *310*:551–557, 1999.

213. J. Molin and S. Molin, CASE: complex adaptive system ecology. Adv. Microbial Ecology *15*:27–80, 1997.

214. J. Rundegren, T. Simonsson, L. Peterson, and E. Hansson, Effect of delmopinol on the cohesion of glucan-containing plaque formed by *Streptococcus mutans* in a flow cell system. J. Dent. Res. *71*:1792–1796, 1992.

215. I. D. Mandel, Caries prevention—current strategies, new directions. JADA *127*:1477–1488, 1996.

216. A. Mombelli, Antibiotics in periodontal therapy. In *Clinical Periodontology and Implant Dentistry* (J. Lindhe, ed.). Copenhagen: Munksgaard, 1999, pp. 488–507.

217. D. Steinberg and M. Friedman, Dental drug-delivery devices: local and sustained release applications. Crit. Rev. Therapeutic Drug Carrier Systems *16*:425–459, 1999.

218. M. Rykke and G. Rolla, Effect of silicone oil on protein adsorption to hydroxy apatite in vitro and pellicle formation in vivo. Scand. J. Dent. Res. *98*:401–411, 1990.

219. J. Olsson, Y. van der Heijde, and K. Holmberg, Plaque formation in vivo and bacterial attachment in vitro on permanently hydrophobic and hydrophilic surfaces. Caries Res. *26*:428–433, 1992.

220. M. Rykke, G. Rolla, and T. Sönju, Effect of sodium lauryl sulfate on protein adsorption to hydroxyapatite in vitro and on pellicle formation in vivo. Scand. J. Dent. Res. *98*: 135–143, 1990.

221. M. Rykke, G. Rolla, and T. Sönju, Effect of pyrophosphate on protein adsorption to hydroxyapatite in vitro and on pellicle formation in vivo. Scand. J. Dent. Res. *96*:517–522, 1988.

222. M. Rykke and G. Rolla, Effect of two organic phosphonates on protein adsorption in vitro and on pellicle formation in vivo. Scand. J. Dent. Res. *98*:486–496, 1990.

223. J. Olsson, A. Carlén, and K. Holmberg, Inhibition of *Streptococcus mutans* adherence by means of surface modification. J. Dent. Res. *69*:1586–1592, 1990.

224. J. Olsson, M. Hellsten, and K. Holmberg, Surface modification of hydroxyapatite to avoid bacterial adhesion. Colloid Polym. Sci. *269*, 1991.

225. J. Olsson, A. Carlén, and K. Holmberg, Modulation of bacterial binding to salivary pellicle by treatment with hydrophilising compounds. Arch. Oral Biol. *35*:1378–1405, 1990.

226. K. Holmberg and J. Olsson, Surface modification of hydroxy apatite for dental plaque inhibition. In *Desk Reference of Functional Polymers* (R. Arshady, ed.). American Chemical Society, 1997.

227. P. Schupbach, J. R. Neeser, M. Golliard, M. Rouvet, and B. Guggenheim, Incorporation of caseinoglycomacropeptide and caseinophosphopeptide into the salivary pellicle inhibits adherence of mutans *Streptococci*. J. Dent. Res. *75*:1779–1788, 1996.

228. T. H. Grenby, A. T. Andrews, M. Mistry, and R. J. H. Williams, Dental caries—protective agents in milk and milk products: investigations in vitro. J. Dent. *29*:83–92, 2001.

229. I. Johansson, P. Bratt, D. I. Hay, S. Schluckebier, and N. Strömberg, Adhesion of *Candida albicans*, but not *Candida krusei*, to salivary statherin and mimicking host molecules. Oral Microbiol. Immunol. *15*:112–118, 2000.

230. M. Edgerton, P. A. Raj, and M. J. Levine, Surface-modified poly(methyl methacrylate) enhances adsorption and retains anticandidal activities of salivary histatin 5. J. Biomed. Mater. Res. 29:1277–1286, 1995.

231. D. J. Smith, M. A. Taubman, W. F. King, S. Eida, J. R. Powell, and J. Eastcott, Immunological characteristics of a synthetic peptide associated with a catalytic domain of mutans streptococcal glucosyltransferase. Infect. Immun. 62:5470–5476, 1994.

232. A. M. Vacca Smith and W. H. Bowen, In situ studies of pellicle formation on hydroxyapatite discs. Arch. Oral Biol. 45:277–291, 2000.

233. T. Simonsson, T. Arnebrant, and L. Petersson, The effect of delmopinol on salivary pellicles. The wettability of tooth surfaces in vivo and bacterial cell surfaces in vitro. Biofouling 3:251–260, 1991.

234. T. Simonsson, T. Arnebrant, L. Petersson, and E. Bondesson Hvid, Influence of delmopinol on bacterial ζ-potentials and on the colloidal stability of bacterial suspensions. Acta Odont. Scand. 49:311–316, 1991.

235. J. Rundegren, T. Arnebrant, and L. Lindahl, Effect of whole saliva on the rheological behaviour of extracellular water-soluble glucan produced by Streptococcus mutans. Scand. J. Dent. Res. 99:484–488, 1991.

236. S. Goldberg and M. Rosenberg, Bacterial desorption by commercial mouthwashes vs two-phase oil:water formulations. Biofouling 3:193–198, 1991.

237. L. B. Freitas, N. Vassilakos, and T. Arnebrant, Interactions of chlorhexidine with salivary films adsorbed at solid/liquid and air/liquid interfaces. J. Periodontal Res. 28:92–97, 1993.

238. S. M. Waaler, G. Rolla, K. K. Skjörland, and B. Ögaard, Effects of oral rinsing with triclosan and sodium lauryl sulphate on dental plaque formation. Scand. J. Dent. Res. 101:192–195, 1993.

239. S. M. Waaler, G. Rolla, and V. Kjaerheim, Triclosan containing mouthwashes—does the nature of the solvent influence their clinical effect. Scand. J. Dent. Res. 102:46–49, 1994.

240. G. Rolla, J. E. Ellingsen, and D. Gaare, Polydimetylsiloxane as a tooth surface–bound carrier of triclosan: a new concept in chemical plaque inhibition. Adv. Dent. Res. 8:272–277, 1994.

241. C. M. McCullagh, L. M. Soby, A. M. Jamieson, and J. Blackwell, Viscoelastic behaviour of fractionated ovine submaxillary mucins. Biopolymers 32:1665–1674, 1992.

242. E. S. Reeh, W. H. Douglas, and M. J. Levine, Lubrication of saliva substitutes at enamel–enamel contacts in an artificial mouth. J. Prosthetic Dent. 75:649–656, 1996.

243. M. J. Levine, Development of artificial salivas. Crit. Rev. Oral Biol. Med. 4:279–286, 1993.

244. M. C. F. Johnsson, F. A. Richardson, F. A. Scannapieco, M. J. Levine, and G. H. Nancollas, The influence of salivary proteins on the growth, aggregation and surface properties of hydroxyapatite particles. In Material Synthesis Utilising Biological Processes (P. C. Rieke, P. D. Calvert, and M. Alper, eds.). 1990, pp. 81–86.

245. M. Johnsson, F. A. Richardson, E. J. Bergey, M. J. Levine, and G. H. Nancollas, The effects of human salivary cystatins and statherin on hydroxyapatite crystallisation. Arch. Oral Biol. 36:631–636, 1991.

246. V. Hlady and H. Füredi-Milhofer, Adsorption of human serum albumin on precipitated hydroxyapatite. J. Colloid Interface Sci. 69:460–468, 1979.

247. H. J. Busscher and H. C. van der Mei, Physico-chemical interactions in initial microbial adhesion and relevance for biofilm formation. Adv. Dent. Res. 11:24–32, 1997.

30

Adsorbed Biopolymers: Behavior in Food Applications

DAVID S. HORNE   Hannah Research Institute, Ayr, Scotland

J. M. RODRIGUEZ PATINO   University of Seville, Seville, Spain

I. INTRODUCTION

The protein film adsorbed at the oil–water or air–water interface is the source of many of the unique properties of food emulsions and foams, particularly their stability and interactions, which translate into the shelf-life and textural properties so desired by manufacturers and appreciated by consumers. This is the intuitively attractive picture that the reviews and textbooks would have us believe but the reality is that, despite all the research effort put into studying those interactions, we still do not have a reliable predictor of shelf-life in emulsion products. As for texture, we still have no quantitative instrumental measures of creaminess, mouth-feel, or even perceived thickness. Partly this arises because the interfacial layer in food emulsions is compositionally and structurally complex, but this facile explanation conceals an underlying lack of ability to formulate a realistic interaction potential between emulsion droplets or foam bubbles. That notwithstanding, considerable progress has been made in the last 10 to 15 years in extending our knowledge of the interactions between and within adsorbed protein films, much of it on proteins of interest to the food manufacturer, and much of it reviewed in extenso by Dickinson [1–6].

The problems posed by the complexity of food systems should not, however, be underestimated. Food dispersions are complicated multicomponent systems containing many emulsifiers which may show surface activity by themselves (proteins and low-molecular-weight surfactants) or by association with other components (proteins, low-molecular-weight surfactants, and polysaccharides, to name a few examples). In addition, food dispersions contain many other organic (ethanol, sugars, etc.) and inorganic (salts) reagents which may interact with emulsifiers in different complex fashion depending on pH, temperature, processing history, etc., all of which intensify the problems of the manufacturer who is attempting to control stability, shelf-life, or product texture. For up-to-date reviews, readers are directed to recent references [7–13].

Manufacturers employ two types of emulsifier or foaming agents in foods [1]. These are low-molecular-weight surfactants (mainly mono- and diglycerides, phospholipids, etc.) and macromolecules (proteins and certain hydrocolloids). In the main, the proteins are those of milk and eggs. Because early manufacturing practices tended to mimic processes developed initially in the kitchen, recipes have not strayed far from those early listings of ingredients. Since these recipes often employed whole milk or egg, proteins and lipids frequently coexist in manufactured emulsions and foams. Sometimes they are dissociated from one another, but in other cases, where they are deliberately introduced to develop specific functions in the finished product, they are found associated [14].

Lipids stabilize the dispersed droplets or bubbles by formation of a densely packed but much less rigid monomolecular layer, which is stabilized by dynamic processes (i.e., Gibbs–Marangoni effect). Polar lipids adsorb strongly to fluid–fluid interfaces giving close molecular packing at the interface to produce low surface and interfacial tensions [15,16]. In contrast, proteins act as polymeric emulsifiers with multiple anchoring sites at the interface that, together with the unfolding process of the adsorbing protein molecule [17–19], stabilize the interfacial layer kinetically. This behavior contributes significantly to the interfacial rheological properties and immobilizes proteins in the adsorbed layer.

Proteins and lipids have an important physical property in common, their amphiphilic nature. This property provides the potential for association, adsorption, and reorientation at fluid–fluid interfaces, depending on the properties of the components and the protein–lipid ratio [15,16,20,21]. However, more important in some products is the effect of the small molecule emulsifiers in destabilizing the emulsion [22,23]. In the formulation of ice cream the small molecule emulsifier (typically, mono- and diglycerides) is added to break the adsorbed layer of protein and allow the adsorption of fat to the surface of the air bubble. Thus, an important action of the small molecule emulsifiers is to promote the displacement of proteins (mainly caseins) from the interface. Competitive adsorption and/or displacement between lipids and proteins at fluid–fluid interfaces have been studied in detail in several investigations. For further information concerning the interfacial characteristics of food emulsifiers (proteins and lipids), the reader is referred to recent reviews [6,7,13,14,24–31].

This chapter will concentrate on the interfacial behavior of milk proteins. Our emphasis will be on the interface as a three-dimensional dynamic entity. Interdroplet interactions will be dominated by the dimension normal to the interface, but within the interface we will concern ourselves with the role of lateral interactions and bonding between neighboring molecules. We will consider protein structure and morphology at the interface, relaxation phenomena, and interfacial rheology at oil–water and air–water interfaces, all of this behavior relevant to the formation and stability of food emulsions and foams.

Conveniently the milk proteins subdivide into two categories, the disordered flexible caseins and the compact globular whey proteins. As a first approximation, a casein monomer may be regarded as a complex linear polymer that adsorbs to an oil or air interface to give an entangled monolayer of flexible chains having some regions of more hydrophobic character in direct contact with the surface (trains) and others protruding into the aqueous phase (loops and tails) [32]. The distribution of hydrophobic residues along the casein polypeptide chain is not uniform but clustered in a fashion specific to the various family members, α_{s1}-, α_{s2}-, β-, and κ-casein. This gives to each of the caseins some individuality, which is further enhanced by their differing content of phosphoseryl residues.

In milk, however, the caseins exist not as individual molecules but as strongly aggregated polydisperse complex particles containing all four members of the casein family. These aggregates, known as casein micelles, are held together by a combination of hydrophobic bonding between caseins and cross-linking through calcium phosphate microcrystals bonding to the phosphoseryl clusters [33,34]. Milk owes its low fluidity to the existence of these micelles, and many related food technology uses rely on their destabilization, as in the manufacture of cheese and yogurt. Precipitating the casein micelles by reducing milk pH to 4.6 and redispersing the precipitate in NaOH at pH 7.0 produces sodium caseinate. Sodium caseinate, which contains all of the caseins in the proportions found in milk, is widely used in the food industry as an emulsifier. Much of the relevant recent research work, however, has involved two major individual caseins, α_{s1}- and β-casein, which together comprise 75% of total bovine milk casein.

In contrast to the interfacial behavior of the caseins, research to be summarized herein has demonstrated that a closely packed globular protein network is perhaps better modelled as a dense two-dimensional network of strongly interacting particles with rheological and structural properties more akin to those of a concentrated heat-set globular protein gel. Again, while there are industrial preparations of whey proteins prepared from acid or cheese wheys [whey protein isolates (WPI) and whey protein concentrates (WPC)], the majority of research has been carried out on the major constituent protein at 75% of bovine whey protein, β-lactoglobulin.

II. FILM STRUCTURE AND MORPHOLOGY

A. β-Casein

In producing an emulsion, the adsorbing protein molecule is the one which happens to be in close proximity to the interface at the time it is created by high energy shearing. The protein layer effectively forms a macromolecular barrier at the oil–water interface to protect the freshly formed droplets against recoalescence [35]. While early studies concentrated on layer composition and competitive adsorption and displacement, experimental and theoretical work in recent years has emphasized the dynamic structure of the interfacial protein layer and its response to its environment.

The structure of adsorbed layers of β-casein at various solid and liquid interfaces has been investigated experimentally by the techniques of ellipsometry [36–39], small angle x-ray scattering [40], dynamic light scattering [41,42], and neutron reflectivity [43–49]. Ellipsometry and dynamic light scattering provide only gross measurements of layer thickness, whereas neutron reflectivity has emerged as a technique capable of providing a more detailed density profile of the adsorbed layer on an atomic scale in a direction normal to the interface. Data analysis is still restricted to the testing and fitting of appropriate models for the adsorbed film. Although a discrete two- or multilayer model, depending on the substrate, fits the neutron reflectivity data from most studies of adsorbed β-casein fairly well, a power-law model has also been employed to describe the volume fraction profile at the air–water interface [47].

Statistical modelling of the adsorption of a model β-casein-like polymer at a hydrophobic planar surface has also been carried out using the self-consistent field (scf) theory of Scheutjens and Fleer [50–53]. While this approach outputs a numerical density distribution, it does not lend itself readily to manipulative data fitting.

Only gross comparisons can therefore be made of its predictions with empirical observations. So many are the correlations with parameters drawn from all kinds of experiments that confidence in the conformations predicted from this modelling must be high.

The scf calculations predict that an adsorbed β-casein monolayer is produced at bulk protein concentrations below an ionic strength dependent threshold [52]. Above this threshold, there is multilayer condensation of self-associating protein onto the surface. Figure 1a shows the calculated segment density profile $\rho(z)$ at 10 mM ionic strength and bulk volume fraction of model protein (5×10^{-6}) for three different pH values. Each curve shows a smooth decrease in segment density moving away from the surface. Regardless of pH, the dense inner layer is about 1 nm thick with a volume fraction, ϕ, approaching 0.9. Volume fraction decreases to $\sim$0.01 by $z = 10$ nm. As pH is reduced, additional adsorbed material appears around the middle of the profile ($z \sim 2$ nm) with a distinct shoulder appearing at pH 5.5. In Fig. 1b, we present for comparison volume fraction profiles inverted from neutron reflectivity data, measured at 20 mM ionic strength and bulk protein concentration 5×10^{-3} wt% [46]. These too show an increase in adsorbed protein around the midrange of the profile at pH 6.0 and a doubling of the thickness of the inner layer at pH 5.5, broadly confirming the predicted behavior.

Replotting the scf data (Fig. 2a) as logarithm of the segment density as a function of distance, z, shows that the low density region ($\phi \sim 0.01$) continues out to around 15 nm before dropping abruptly by several orders of magnitude to the bulk solution volume fraction. This knee we take as defining the hydrodynamic thickness of the adsorbed layer. Inspection of this tail region shows that this hydrodynamic layer thickness is not a strong function of pH, with the calculated plots all lying very close to one another. Dynamic light scattering measurements of the hydrodynamic thickness of β-casein layers adsorbed onto polystyrene latices [42] confirm this behavior with pH (Fig. 2b), showing no major collapse of the β-casein layer at the reduced pH.

Detailed analysis of the distribution of individual segment types can also be extracted from the scf calculations. This has shown that the most hydrophilic residues, especially the carboxyl and phosphoserine residues, reside predominantly in the outer layer regions. In particular, the N-terminal amino residue lies mainly well away from the surface. All of this is consistent with the concept of a long sterically stabilizing tail that is composed of the N-terminus region of the β-casein molecule (Fig. 3).

Self-consistent field calculations performed for the fully dephosphorylated model of β-casein show that the loss of the phosphate residues produces an increase in surface coverage of the dephosphorylated analog, consistent with the much lower solubility of dephosphorylated β-casein. The additional adsorbed material is located in the inner layers at distances <11 nm, but the full hydrodynamic thickness of the dephosphorylated protein is significantly lower than that for the original β-casein model polyelectrolyte (Fig. 4a). Again dynamic light scattering measurements of the hydrodynamic thickness of protein layers adsorbed onto polystyrene latices confirm that the dephosphorylated protein layer is significantly thinner (Fig. 4b) [42].

This behavior indicates the importance of the charged phosphoserine residues in β-casein in maintaining a thick steric stabilizing monolayer. The phosphoserine residues are, however, mobile, as confirmed by ^{31}P nuclear magnetic resonance [54],

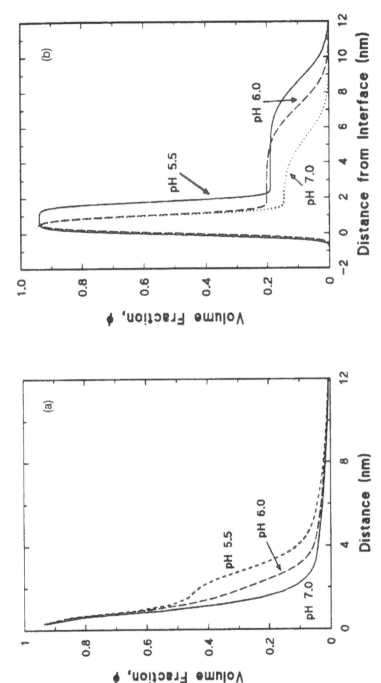

FIG. 1 Comparison of theoretical predictions and experimental measurements for adsorbed β-casein density profiles on adsorption to a hydrophobic interface at varying pH values. (a) Theoretical predictions from scf calculations using a model β-casein polymer. (b) Density profiles obtained by inverting neutron reflectivity data measured at air–water interface with 5×10^{-3} wt% protein in air–contrast matched water.

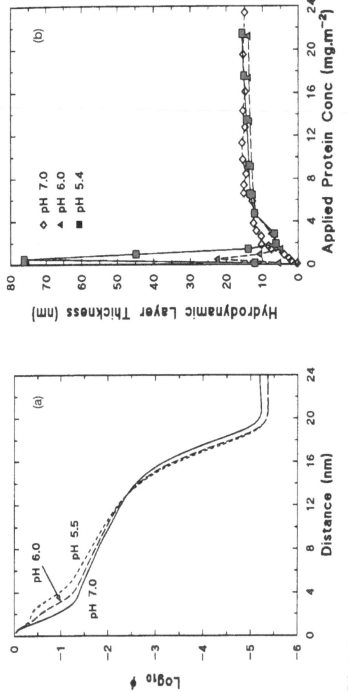

FIG. 2 Comparison of theoretical predictions and experimental measurements for adsorbed β-casein density profiles on hydrophobic surfaces. (a) Theoretical predictions from scf calculations, using a model β-casein polymer. Semi-log plots show low density behavior as a function of pH at distances remote from surface. (b) Experimental measurements of hydrodynamic layer thickness of β-casein adsorbed onto polystyrene latex particles at various pH values. Dynamic light scattering results are plotted as a function of bulk protein concentration relative to available particle surface area.

Hydrophobic interface

FIG. 3 Schematic diagrams depicting average conformations predicted by scf calculations for casein adsorbed at a planar hydrophobic interface. Bars are used to denote hydrophobic regions (B) of proteins and do not imply rigidity. Phosphoserine clusters (P) are located in loop (α_{S1}-casein) or tail (β-casein) regions.

and layer thickness is responsive to background buffer conditions, as evidenced by dynamic light scattering measurements in solutions of varying ionic strength and ionic calcium level [42]. The phosphoserine-rich cluster can be thought of as a negatively charged blob repelled from the polystyrene surface, but restrained from moving into bulk solution by a stretched spring region anchored strongly to the surface [55]. Neutralization of the charge by specific binding of ionic calcium or screening by salt addition causes the spring to relax back to give a thinner layer [42]. Leaver and Dalgleish [56] have also shown that the N-terminal tail is much more accessible to proteolytic digestion by trypsin than is the rest of the adsorbed molecule and that loss of the tail leads to a marked reduction in layer thickness.

Not all groups agree on the suggestion that β-casein adsorbs at a hydrophobic interface as a monolayer with spatial segregation of the molecule, as depicted above. Russev et al. [57] interpreted their observations on the kinetics of adsorption as the formation of a thick inner layer, 1.8 nm thick, of high density followed by a second layer of molecules. This second layer was much thicker at 5.4 nm and more diffuse. Without going into the details of the experiment and its results, it seems to these authors that the observations could equally well have been explained by the scenario depicted in Fig. 5. In the early stages of adsorption, the few molecules are relatively flat and signal is determined by the mainly thicker, denser hydrophobic segment. As the surface fills up in the middle stages, the tails are pushed more into the outer regions, giving detectable signal from this quarter and demanding a two-layer fit to the data. Finally in the later stages with a fully saturated surface, we have full complement in both layers but it should be emphasized that each apparent layer is still only part of the monolayer of β-casein molecules. Multilayer formation is not necessary to achieve the fits they require for their time-dependent ellipsometry signal.

If the scf calculations predict β-casein to adsorb in the tail–train conformation depicted in Fig. 3, similar calculations carried out on the α_{S1}-casein analog show it adopting a quite different structure, with a hydrophilic central loop containing the phosphoseryl clusters extending into the aqueous phase, flanked by two hydrophobic trains on either side (Fig. 3). It should be emphasized that these simple illustrations are statistical depictions of average behavior, composites of many coexisting molecular conformations.

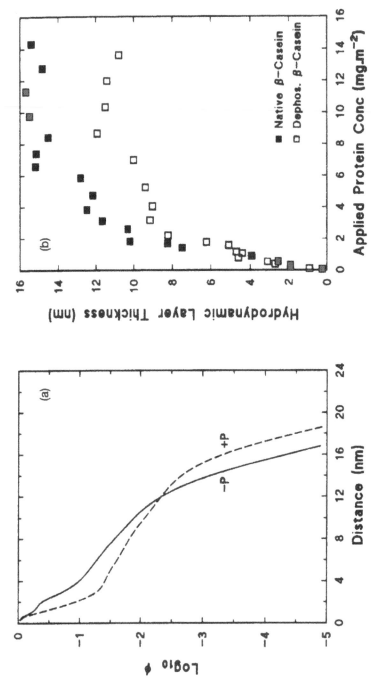

FIG. 4 Comparison of theoretical predictions and experimental measurements of influence of phosphorylation on β-casein conformation at hydrophobic interfaces. (a) Theoretical predictions from scf calculations of density profiles using model β-casein polymer. Plots show low density regions of profile at distances remote from interface. +P, native β-casein; −P, dephosphorylated analog. (b) Experimental measurements of hydrodynamic layer thickness of native and dephosphorylated β-casein adsorbed onto polystyrene latex particles.

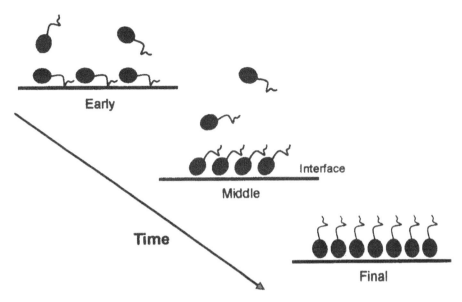

FIG. 5 Snapshot depictions of adsorption of β-casein onto a hydrophobic interface, showing buildup of monolayer as a function of elapsed time and revealing formation of a detectable second diffuse layer at a later point in proceedings.

B. Adsorbed β-Lactoglobulin

The differences in solution structure between β-casein, a flexible protein lacking secondary structure, and β-lactoglobulin, a globular protein with a compact conformation, are reflected in their adsorption behavior. Neutron reflectivity measurements at the air–water interface have shown that β-lactoglobulin adsorbs to give a monolayer that is thinner and of lower surface coverage than that of β-casein [46]. Layer thickness is of the order of the diameter of the native monomer in solution (3.6 nm) or perhaps slightly less on a solid hydrophobized surface [49]. These results are in broad agreement with small angle x-ray scattering [58].

In their neutron reflectivity study, Atkinson et al. [46] found β-lactoglobulin surface coverage to increase with time, the rate of change increasing as pH was lowered toward the protein isoelectric point. Initially a two-layer fit provided the best fit at higher pH, but as time went on the single-layer model superseded this, with an overall thinner layer. The initially adsorbed monomer deformed, rearranged, and unfolded with time, as its hydrophobic side chains moved to the nonaqueous region of the interface. Confirmation of this behavior comes from FTIR spectroscopy, which indicates some loss of intramolecular β-sheet structure on adsorption at the oil–water interface [59]. Some native structure, however, is retained in the emulsified state, as demonstrated by immunochemical analysis [60].

Immediately following adsorption, the β-lactoglobulin film can therefore be regarded as a close-packed monolayer of deformable particles [61]. On a hydrophilic solid–liquid interface, no further changes have been observed [62]. On hydrophobic surfaces, more rapid and more extensive rearrangements occur, converting the film into a two-dimensional gel-like layer [63,64], as a result of the increasing number of nonbonded intermolecular interactions and slow covalent cross-linking through interchange of disulfide bonds [65]. In consequence, the β-lactoglobulin layer shows some surface shear rheological properties characteristic of this gel-like nature [66].

C. Surface Studies

New techniques for microscopic observation and characterization of monolayers at the air–water interface have been used advantageously to clarify the structural characteristics of amphiphilic substances at fluid–fluid interfaces. Since its introduction [67,68], Brewster angle microscopy (BAM) has been used preferentially for the study of monolayers, as it reveals phase domains and heterogeneity in thin films without the use of any probe that may disturb the local environment of the probe and thereby cause artifacts. BAM, coupled with surface pressure (π)–area (A) isotherms, has been used to visualize and determine structural characteristics of protein monolayers at the air–water interface [11].

Whether obtained in a Wilhelmy trough [69] or in a Langmuir trough [70–72], results derived from π-A isotherms support the structures postulated above for β-casein, caseinate, and WPI films spread on similar subphase at pH 5 and 7. The forms of π-A isotherms (Fig. 6) confirm that β-casein and caseinate monolayers at the air–water interface adopt two different structures (Fig. 7) and a collapse phase [44,73–77]. The tail–train structure for β-casein (a, top) was observed at surface pressures lower than ca. 15 and 10 mN/m at pH 5 and 7, respectively. At higher surface pressures, and up to the equilibrium surface pressure, indicated in Fig. 7 by the vertical arrow, amino acid segments are extended into the underlying aqueous solution and adopt the form of loops and tails (b, top). However, WPI retains elements of the native structure, not fully unfolded at the interface (Fig. 7). Thus, most amino acid residues in WPI adopt a loop conformation at the air–water interface. This loop conformation becomes more condensed at higher surface pressures and is displaced toward the bulk phase at the collapse point. These data are in agreement with those deduced for globular proteins [74,75].

Results of BAM (in particular the relative reflectivity) as a function of surface pressure (Fig. 8) obtained with β-casein, caseinate, and WPI monolayers clearly show the same structural characteristics as those deduced from the π-A isotherms (Fig. 6). The domains that residues of protein molecules adopt at the air–water interface appeared to be of uniform reflectivity, suggesting homogeneity in thickness and film isotropy [71,72]. The relative reflectivity (Fig. 8) increases with the surface pressure and is a maximum at the collapse point at the highest surface pressure. At surface pressures lower than 17 mN/m, the relative reflectivity is independent of the protein, but at the collapse point the relative reflectivity is higher for β-casein and caseinate than for WPI. That is, at higher surface pressures ($\pi > 17$ mN/m), and especially at the collapse point, the thickness of the β-casein or caseinate monolayer is greater than that for WPI. These results are in good agreement with those obtained by atomic force microscopy [78]. Data in Fig. 8 show that a doubling in the thickness is produced at the highest surface pressure relative to the equilibrium surface pressure (π_e), the relative reflectivity increasing by about a factor of 4. At π_e, a saturation of the monolayer takes place. For β-casein at pH 7, the monolayer coverage saturation (Fig. 6) agrees well with literature data [79].

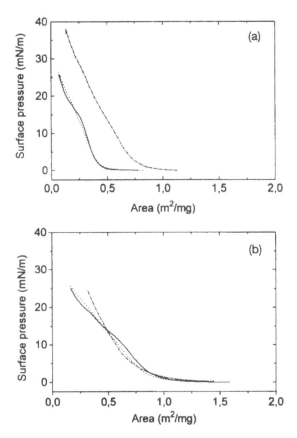

FIG. 6 Surface pressure–area isotherms for β-casein (——), caseinate ($\cdots$), and WPI monolayers (— $\cdot$ —) spread on aqueous solutions at pH 5 (a) and pH 7 (b) at 20°C.

At pH 5, the surface density of disordered proteins (β-casein and caseinate) was higher than that for the globular WPI (Fig. 6a). However, at pH 7 (Fig. 6b) the structural characteristics of globular (WPI) and disordered (β-casein and caseinate) proteins were essentially the same, especially at surface pressures lower than 17 mN/m. These differences are a consequence of the more compact packing of disordered residues in β-casein and caseinate on an acidic subphase close to the isoelectric point [71]. The fact that the WPI monolayer structure did not depend on the pH (Fig. 6) is consistent with the view that the protein components of WPI (which is mainly β-lactoglobulin) retains elements of the native structure, not fully unfolded, at the air–water interface.

III. KINETICS OF ADSORPTION OF PROTEINS AT FLUID INTERFACES

The rate of protein adsorption at fluid–fluid interfaces plays an important role in the formation and stabilization of food dispersions [1,80,81]. In fact, during the formation of a dispersed system the protein must be adsorbed at the interface to prevent

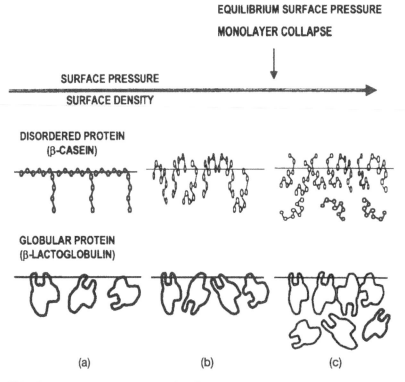

FIG. 7 Proposed structures of disordered proteins (β-casein) and globular proteins (β-lactoglobulin, BSA, etc.) at the air–water interface as a function of surface pressure or surface density. Disordered proteins adopt trains of amino acid segments at the interface (a, top), looping of the amino acid segments into the underlying aqueous solution (b, top), and finally (c, top) the collapse and multilayer formation. Globular proteins adopt loop conformation at the interface (a, bottom), which is more condensed at higher surface pressures (b, bottom), and then the collapse and multilayer formation (c, bottom). The monolayer collapse takes place at a surface pressure close to the equilibrium surface pressure.

the recoalescence of the initially formed bubbles or droplets. In addition, during the protein adsorption the surface or interfacial tension of fluid–fluid interfaces decreases, an important factor both in optimizing the energy required in the emulsification or foaming process [82] and in achieving smaller droplet and bubble size—which is an important factor for the stability of the dispersed system [1]. On the other hand, emulsification and foaming involve interfacial deformation, and the response of the adsorbed layer to such deformations is crucial for understanding the role of proteins in food systems [83].

The decrease in surface tension by proteins follows a series of different processes [84–87]. First, the protein has to move from the bulk phase to the subsurface (a layer immediately adjacent to the fluid interface) by diffusion and/or convection. This step is followed by the adsorption and unfolding of the protein at the interface. Third, the adsorbed protein segments rearrange at the fluid interface, a slow process caused by reorganization of the amino acid segments previously adsorbed on the interface. Adsorption of proteins is therefore a complex process, involving possibly

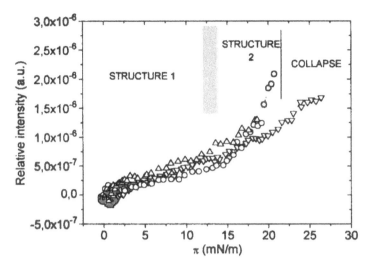

FIG. 8 Relative intensity as a function of surface pressure for β-casein (O), caseinate ($\triangle$), and WPI spread films (∇) at the air–water interface at 20°C and pH 5; I: 0.05 M.

several conformational changes that may be either reversible or irreversible and, in addition, may be time dependent [88,89].

When direct measurements of surface concentration, $\Gamma(\theta)$, are not possible, the kinetics of adsorption can be monitored by measuring the changes in surface pressure, π, with time, for which a range of techniques have been developed [90,91]. For the long-term stages of protein adsorption, which are controlled by unfolding at the interface and configurational rearrangements of the adsorbed molecules, the Wilhelmy plate is an adequate method. However, the diffusion step is too fast to be detected with this technique [17,18,92,93] and, for measurements at short adsorption time, an automatic drop tensiometer has been employed [19,94]. The same tensiometer has been used for the analysis of WPI adsorption at the oil–water interface [28,95].

Some authors are critical of the use of surface pressure measurements following adsorption [83,96–98], as it implicitly assumes that the surface equation of state is the same for spread and adsorbed protein films. However, it was recently shown by Rodriguez Niño and Rodriguez Patino [99] that the agreement between spread and adsorbed BSA monolayers is generally good, implying that the structures of the monolayers formed in the two different ways must be identical, at least for adsorption from low bulk protein concentrations. The same agreement between the adsorbed and spread isotherms has been observed for BSA [83,100,101] and for other proteins [83,102]. The good agreement between spread and adsorbed π-A isotherms provides support for the relevance of surface pressure measurements to adsorption studies [83,103].

A. Protein Adsorption at the Oil–Water Interface

For WPI adsorption at the oil–water interface we have observed that interfacial pressure, π, and surface dilatational modulus, E, increase; and the phase angle, ϕ, decreases with time, θ (Fig. 9). Normalization into a single master curve of E versus π data (Fig. 10) reflects the interfacial behavior of WPI adsorbed films for different protein concentrations, at different adsorption times, and under different processing conditions and suggests that the interfacial behavior of WPI films is mainly due to the amount of adsorbed protein.

The rate of WPI adsorption at the oil–water interface increases with protein concentration in solution (Fig. 9). The kinetics at short adsorption time are controlled by the diffusion of the protein toward the interface, in agreement with the Ward and Tordai model [104]. However, at long-term adsorption, a first-order model is a sat-

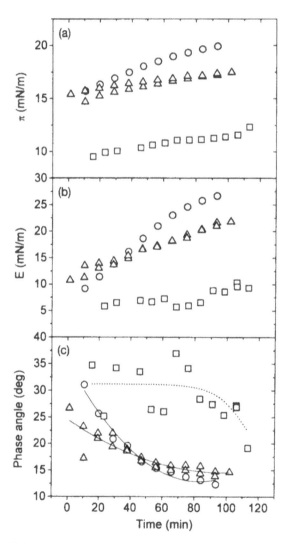

FIG. 9 Time-dependent surface pressure (a), surface dilatational modulus (b), and phase angle for WPI adsorbed films at the oil–water interface (c) at pH 5 and at 20°C; I: 0.05 M; frequency: 100 MHz; amplitude: 15%. Protein concentration in the drop bulk phase (%wt/wt): 10^{-1} (○), 10^{-2} (△), and 10^{-5} (□).

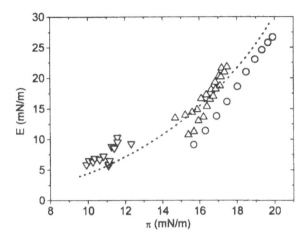

FIG. 10 Surface dilatational modulus as a function of surface pressure for WPI adsorbed films at the oil–water interface at pH 5 and 20°C; I: 0.05 M; frequency: 100 MHz; amplitude: 15%. Protein concentration in the drop bulk phase (%wt/wt): 10^{-1} (○), 10^{-2} (△), and 10^{-5} (▽).

isfactory mathematical description of the rheokinetic data, which show exponential changes in surface pressure or dilatational modulus with time [28,95]. These phenomena can be related to the protein unfolding and/or protein–protein interactions as a function of protein concentration in solution.

In practice, the plot of ln π or ln E versus time usually deviates from linearity. The initial slope is taken to correspond to a first-order rate constant of unfolding, while the second, later, slope is taken to correspond to a first-order rate constant of rearrangement, occurring among a more-or-less constant number of adsorbed molecules.

Transient surface dynamic properties of WPI adsorbed films also depend on the WPI concentration in the bulk phase [105]. It was observed that the rate of surface tension or surface dilatational modulus change over time increased when the WPI concentration in the bulk phase was increased. At high concentrations, the surface activity and surface dilatational modulus were high. Protein adsorption at the interface was therefore facilitated at higher protein concentrations in the bulk phase. However, the measured interfacial tension or surface dilatational modulus continued to increase with time, even at long-term adsorption. That is, a steady-state value of the surface dilatational modulus was not attained even at the higher adsorption time studied here. Thus, the study of WPI adsorption is very time consuming, especially at the lower protein concentrations in the bulk phase.

Over the adsorption period studied here, the film behaved, from a rheological point of view, as viscoelastic with a phase angle greater than zero. This phase angle decreased with time, with more marked time dependence at higher protein concentrations in the bulk phase. Such behavior is consistent with the existence of higher-order protein–protein interactions, which are thought to be due to a higher protein concentration at the interface [95,105], because of both the longer adsorption time and the higher protein concentration in the bulk phase.

In these experiments we have noted that the interfacial pressure and its rate of change for WPI adsorbed film in the presence of convection are the same as in the absence of convection [95]. As reported by Damodaran [81], the interfacial adsorption of proteins is not only dependent on diffusion to the interface, but the interfacial energy must be large enough to overcome an activation energy barrier for protein penetration and rearrangement into fluid–fluid interfaces. Thus, the surface forces, both of shear and dilation, may provide a means of altering protein interactions at the air–water interface [18,106,107] and, probably, at the oil–water interface as well. However, this effect was not observed in this study (at a frequency of 100 MHz and at an amplitude of compression/expansion cycle of 15%).

The time dependence of interfacial pressure for heat-treated and native WPI was also studied [95]. The results obtained at the oil–water interface indicate a greater time dependence in the interfacial pressure and surface dilatational modulus. This may be associated with the fact that for heat-treated WPI, the level of protein unfolding is already maximal, although some protein aggregation could also be taking place [24,108].

The unfolding of the protein upon adsorption, especially for heat-treated protein, increases the accessibility of the sulfydryl group and the potential for formation of intermolecular disulfide cross-links which are responsible for the high E value of heat-treated adsorbed WPI films [95]. As observed by Das and Kinsella [109], the surface hydrophobicity—a measure of alteration of the native structure of a protein —of β-lactoglobulin increases with heating at 80°C just as does the amount of protein adsorbed on emulsion droplets with the formation of multilayers. Dickinson and Hong [110] observed similar behavior for heat-treated β-lactoglobulin at 70°C in relation to time-dependent surface shear viscosity and interfacial surface coverage in emulsion droplets. The effect of gelation on the viscoelastic characteristics of adsorbed protein films at the oil–water interface is of theoretical and practical importance and will be considered more fully later.

B. Protein Adsorption at the Air–Water Interface

Figure 11 shows that the adsorption kinetics of BSA from water and aqueous solutions of ethanol and sucrose at short adsorption time [99], up to approximately 2 s, are controlled by the diffusion of the protein toward the interface, since the surface pressure is lower than about 5 mN/m, in agreement with the Ward and Tordai model (Table 1). The presence of sucrose in the aqueous phase increased the rate of BSA diffusion toward the interface, but the opposite was observed for aqueous solutions of ethanol, especially at higher concentrations of this reagent in the bulk phase. The presence of ethanol in the bulk phase apparently introduces an energy barrier for the BSA diffusion toward the interface. This could be attributable to a competition with previously adsorbed ethanol molecules for the penetration of the protein into the interface. In addition, if ethanol causes denaturation and/or aggregation of the protein in the bulk phase [111], the diffusion of the protein toward the interface could be diminished. Thus, the causes of the higher rate of BSA diffusion from aqueous solutions of sucrose, in comparison with that observed for water, must be different in aqueous ethanol solutions. Since protein molecules are preferentially hydrated in the presence of sucrose [112,113], it is possible that sucrose limits protein unfolding in the bulk phase [18] and reduces protein–protein interactions in the bulk phase

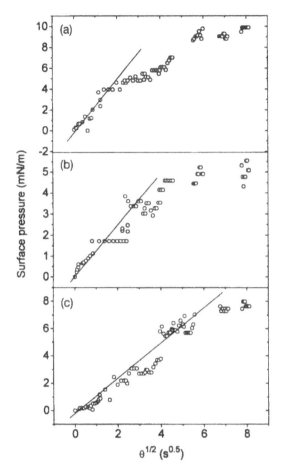

FIG. 11 Rate of BSA adsorption at the air–aqueous phase interface at 20°C and pH 7; I: 0.05 M. Protein concentration: 0.1%wt/wt. Aqueous phase composition: water (a), ethanol 1.0 M (b), and sucrose 0.5 M (c).

and at the interface [17,18,92]. Both of these phenomena may increase the rate of BSA diffusion toward the interface.

At longer adsorption time, in the period after that affected by the diffusion, an energy barrier exists against the BSA adsorption. This could be attributed to the penetration, unfolding, and rearrangements of the protein at the interface [99]. Clearly, the kinetics of adsorption of proteins at interfaces are highly complex, especially in the presence of typical food solutes such as ethanol and sucrose in the aqueous phase.

Simple unfolding at the interface with configurational rearrangements of adsorbed protein molecules produces changes in surface pressure which can be represented by a single exponential with time. However, when an energy barrier exists, the rate of protein penetration into the surface film will be rate limiting. Ward and Regan [114] have used a modified form of the Ward and Tordai equation to monitor this process:

TABLE 1 Characteristic Parameters for the Diffusion of BSA at the Air–Aqueous Phase Interface at 20°C

System	Slope π versus $\theta^{1/2}$ ($mN \cdot m^{-1} \cdot s^{-1/2}$)	LR
BSA $(1.10^{-3}\%)$–water	0.066	0.907
BSA $(7.5.10^{-3}\%)$–water	0.55	0.956
BSA $(5.10^{-3}\%)$–water	0.53	0.971
BSA $(1.10^{-2}\%)$–water	0.64	0.897
BSA $(7.5.10^{-2}\%)$–water	1.23	0.942
BSA $(5.10^{-2}\%)$–water	0.97	0.928
BSA $(1.10^{-1}\%)$–water	1.89	0.922
BSA $(7.5.10^{-1}\%)$–water	7.83	0.943
BSA $(5.10^{-1}\%)$–water	5.38	0.961
BSA (1%)–water	10.2	0.935
BSA (1%)–ethanol 0.5 M	9.92	0.921
BSA (1%)–ethanol 1 M	8.90	0.983
BSA $(1.10^{-1}\%)$–ethanol 0.1 M	1.42	0.947
BSA $(1.10^{-1}\%)$–ethanol 0.25 M	1.38	0.909
BSA $(1.10^{-1}\%)$–ethanol 0.5 M	1.36	0.948
BSA $(1.10^{-2}\%)$–ethanol 0.5 M	0.44	0.965
BSA $(1.10^{-1}\%)$–ethanol 0.75 M	1.28	0.960
BSA $(1.10^{-1}\%)$–ethanol 1 M	0.92	0.961
BSA $(1.10^{-2}\%)$–ethanol 1 M	0.34	0.917
BSA $(1.10^{-1}\%)$–ethanol 1.5 M	0.74	0.926
BSA $(1.10^{-1}\%)$–ethanol 2 M	0.66	0.939
BSA $(1.10^{-1}\%)$–sucrose 0.1 M	2.16	0.977
BSA $(1.10^{-1}\%)$–sucrose 0.25 M	2.24	0.952
BSA (1%)–sucrose 0.5 M	10.9	0.912
BSA $(1.10^{-1}\%)$–sucrose 0.5 M	2.36	0.940
BSA $(1.10^{-2}\%)$–sucrose 0.5 M	0.89	0.936
BSA $(1.10^{-1}\%)$–sucrose 0.75 M	2.41	0.964
BSA $(1.10^{-1}\%)$–sucrose 1 M	2.37	0.911
BSA $(1.10^{-1}\%)$–ethanol 1 M + sucrose 0.5 M	1.38	0.909

$$\ln\left(\frac{d\pi}{d\theta}\right) = \ln(k'\nu C_o) - \frac{\pi\Delta A}{KT} \tag{1}$$

where k' is the rate constant of adsorption, K is the Boltzman constant, ΔA is the molecular area required for the molecule to adsorb at the interface, and ν is the number of adsorbing groups per protein molecule. A plot of $\ln(d\pi/d\theta)$ versus π must be linear.

The long-term adsorption of BSA at the air–water interface, as an example, is given in Fig. 12. Different temperatures, protein concentrations in the bulk phase, and aqueous phase composition (aqueous solutions of ethanol and sucrose) give similar results [92,93]. We find, for all experiments of BSA adsorption, two or more linear regions in the semi-log plot of $\ln[(\pi_\infty - \pi_\theta)/(\pi_\infty - \pi_0)]$ versus θ or in the plot of $\ln(d\pi/d\theta)$ versus π [Eq. (1)].

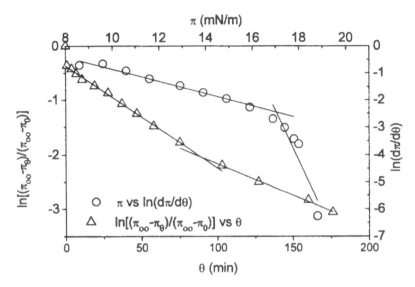

FIG. 12   Rate of BSA adsorption at the air–water interface at 20°C. Protein concentration: 10^{-2}%wt/wt. The subphase was phosphate buffer (pH 7; I: 0.05 M).

To summarize the effect of temperature, protein concentration, and aqueous phase composition on the time dependence of surface pressure during BSA adsorption from the bulk phase, the following conclusions were drawn [11,92,93] from the $\ln[(\pi_\infty - \pi_\theta)/(\pi_\infty - \pi_0)]$ versus θ plots. The rate of BSA adsorption at the interface increased with both BSA concentration in the aqueous phase and temperature. With ethanol in the subphase the existence of an induction period was observed, which could be associated with the competitive adsorption of BSA on an ethanol film—BSA has a higher affinity than ethanol for the interface due to its higher hydrophobicity. This phenomenon could also reflect the existence of BSA–solute interactions in the aqueous phase or at the interface. Sucrose has no affinity for the interface but instead a strong cohesive interaction with water molecules. Protein molecules are preferentially hydrated in the presence of sucrose, limiting the protein unfolding and protein–protein interactions and, consequently, the rate of BSA adsorption was observed to increase when sucrose was present in the bulk phase.

IV.   COMPARISON OF ADSORBED AND SPREAD PROTEIN FILMS AT EQUILIBRIUM

A.   Adsorbed Protein Films at the Oil–Water Interface

The effect of protein concentration on the equilibrium surface pressure is shown in Fig. 13 [28,72]. This adsorption isotherm was deduced from the data shown in Fig. 9 by extrapolating the plot of interfacial tension versus $\theta^{-1/2}$ to $1/\theta^{1/2} = 0$ [115,116]. As expected, the surface pressure (Fig. 13) and the surface dilatational modulus [28] increased markedly as the amount of emulsifier in the bulk phase increased. For WPI adsorbed films this continued until the protein concentration in the bulk phase approached a critical value as indicated by the achievement of a plateau value with no

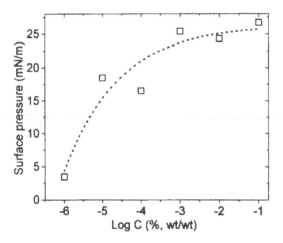

FIG. 13   Adsorption isotherm for WPI at the oil–water interface at 20°C and pH 5; I: 0.05 *M*.

further change as the concentration was increased. The break point in the curve of surface pressure as a function of bulk concentration suggests the existence of a critical concentration, at which the interactions between WPI molecules lead to a constant surface activity until the interface is saturated with protein. The concentration of protein at the break point represents a critical concentration that determines the existence of protein–lipid interactions at the oil–water interface [72].

B.   Adsorbed and Spread Protein Films at the Air–Water Interface

Equilibrium spreading pressure of proteins (β-casein, caseinate, and WPI) at the air–water interface in the temperature range between 5 and 40°C is shown in Fig. 14 [28,117]. The magnitude of π_e was dependent on the protein and the temperature. It can be seen that π_e for β-casein and caseinate was not affected by temperature over

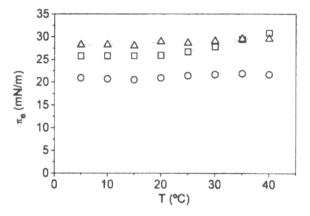

FIG. 14   Temperature dependence of equilibrium surface pressure (π_e) for spread monolayers of β-casein (○), caseinate (△), and WPI (▽) on air–water.

the range plotted. However, π_e for WPI increased slightly with temperature and more markedly at temperatures higher than 25°C. This phenomenon may be related to the temperature dependence on WPI spread films [105].

Figure 15 shows adsorption isotherms for proteins on water at 20°C [28,117]. The protein concentration dependence of surface pressure showed classical sigmoidal behavior. At low protein concentrations, the initial solutions caused only a small increase in the surface pressure. Thereafter, the surface pressure increased with protein concentration and then tended to a plateau. The beginning of this plateau occurred over the protein concentration range from $10^{-3}-1\%$wt/wt. Some differences existed between proteins. β-Casein showed significant surface activity at protein

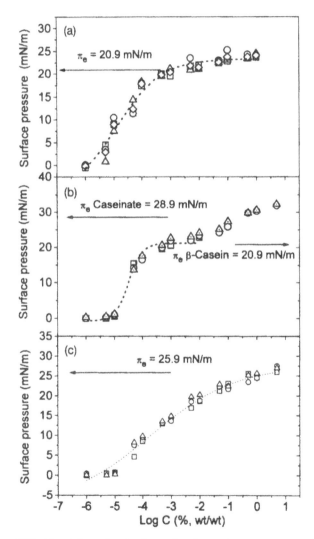

FIG. 15 Adsorption isotherm for β-casein (a), caseinate (b), and WPI (c) on buffered water at 20°C and pH 7; I: 0.05 M. Different symbols are for repetitive experiments. The equilibrium spreading pressures for proteins are indicated by the arrows.

concentrations in the bulk phase of $10^{-6}\%$, and surface pressure increased with protein concentration up to 1%wt/wt. On the other hand, the value of π_e for spread film was lower than that for the surface pressure at the plateau for an adsorbed film. Caseinate showed significant surface activity at concentrations in the bulk phase of $10^{-5}\%$wt/wt. This concentration was one order of magnitude higher than for β-casein. That is, at low concentrations, the surface activity of the proteins comprising caseinate gave lower surface activity than β-casein alone. However, at the higher concentrations, the surface activity of the proteins forming caseinate gave higher surface activity than β-casein alone, behavior similar to that observed with spread films (Fig. 14). These results suggest that the individual casein components in caseinate adsorb independently to the air–water interface, with few interactions between them. Finally, WPI showed significant surface activity at concentrations in the bulk phase of $5 \times 10^{-5}\%$wt/wt. The surface pressure increased with protein concentration and was tending to a plateau at the maximum protein concentration employed in the bulk phase (5%wt/wt). The value of π_e for spread WPI film was lower than that of the surface pressure of adsorbed film at the maximum protein concentration in the bulk phase (5%wt/wt).

The behavior of the adsorbed protein films (Fig. 15) can be interpreted in terms of monolayer coverage (see also Fig. 7). At the lower protein concentrations, as the surface pressure is close to zero, the adsorbed protein residues may be considered as a two-dimensional ideal gas. Proteins at higher concentrations, but lower than that of the plateau, form a monolayer of irreversibly adsorbed molecules. As the plateau is attained, the monolayer is saturated by protein that is irreversibly adsorbed. At higher protein concentrations, the protein molecules form multilayers beneath the primary monolayer, but these structures do not contribute significantly to surface pressure [75]. The presence of multilayers at the maximum protein concentration in the bulk phase has been deduced from Brewster angle microscopy (Fig. 8) and surface dilatational rheology (Fig. 17). Finally, differences observed between surface pressure at the plateau for adsorbed protein and equilibrium spreading pressure should be associated with a different rearrangement of residues when the protein is either adsorbed or spread on the interface at the highest surface density [105].

V.   RELAXATION AND VISCOELASTIC BEHAVIOR

A.   Long-Term Relaxation Phenomena of Spread Protein Films at the Air–Water Interface

Nonequilibrium processes occurring in systems containing fluid–fluid interfaces with a surfactant present are of great practical significance and include important technological operations such as emulsification and foaming. Two experimental approaches can be used for the analysis of long-term relaxation phenomena in emulsifier monolayers. In the first, the surface pressure (π) is kept constant, and the area (A) is measured as a function of time. This relaxation experiment is the usual, preferred method, and is capable of being interpreted kinetically [118]. In the second approach, area is kept constant (at the collapse), and the decrease in surface pressure is monitored as a function of time. Information on the various relaxation paths detailed in Fig. 16 can be derived from the data.

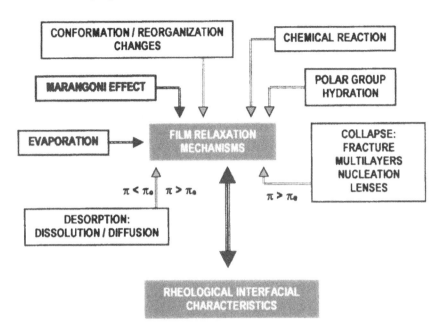

FIG. 16 Relaxation mechanisms in spread monolayers at the air–water interface, where π and π_e are the surface pressure and the equilibrium surface pressure, respectively.

Protein monolayers behave differently from typical lipids under the same experimental conditions. For protein monolayers an attempt was made to interpret the relaxation as a first-order monolayer molecular loss or collapse (Fig. 16). However, whether obtained from area measurement or surface pressure decays, fits of the experimental data at surfaces pressures lower than and higher than the equilibrium surface pressure (π_e) require two exponential decays [28,119,120]. The relaxation of protein monolayer therefore is not a simple process.

At surface pressures lower than π_e, the relaxation rate and the amplitude of the area relaxation depend on the surface pressure. The relaxation rate (quantified by means of the relaxation time, τ, inverse rate constant) is higher at the highest surface pressure. The amplitude of the area relaxation increased with surface pressure. This is a reversible process. In fact, at surface pressures lower than π_e no relaxation phenomena were observed in the π-A isotherm [119,120]. Nor was there any hysteresis during continuous compression–expansion cycles in this region.

Protein monolayer stability was also tested under the most adverse conditions, at the maximum interfacial density (at constant collapse area). Under these conditions, the relaxation phenomena in protein monolayers are controlled predominantly by the collapse mechanism [119,120]. At an increased relaxation time, the surface pressure tends to a plateau that is practically coincident with the value of π_e. As suggested by Graham and Phillips [75], the formation of multilayers of protein molecules under collapse conditions is more likely. In summary, the relaxation in relative molecular area at $\pi < \pi_e$ and in surface pressure at $\pi > \pi_e$—which is mainly limited to the first 50 min—should be attributed to processes related to monolayer organization/reordering and collapse, respectively (see Fig. 16).

B. Surface Dilatational Characteristics of Spread Protein Films at the Air–Water Interface

There are many experimental devices for measuring dilatational rheology [80,91,121–125]. Convenient techniques for measuring surface dilatational rheology are derived from the longitudinal wave method [126]. Trough methods are preferable for insoluble films at the air–water interface. In these methods the area change applied to the film may be oscillatory, a step change, or a continuous expansion–compression. The mechanically generated longitudinal or capillary wave—such as that obtained for the movement of the barriers in a Langmuir trough containing the film—produces a response of surface tension which is monitored by a probe (i.e., a Wilhelmy plate) some distance away from the barrier [127,128].

To obtain surface rheological parameters—such as surface dilatational modulus (E), elastic (E_d) and viscous (E_v) components, and loss angle tangent—a modified Wilhelmy-type film balance (KSV 3000) has been employed [69,129]. In this method the surface is subjected to small periodic sinusoidal compressions and expansions by means of two oscillating barriers at a given frequency (ω) and amplitude ($\Delta A/A$), and the response of the surface pressure is monitored (π). Surface pressure was directly measured by means of two roughened platinum plates that can be situated anywhere on the surface between the two barriers. With this device, an isotropic dilatational deformation of the surface, without interference of shear, can be achieved, as demonstrated by the results derived for monoglyceride spread mono-layers at the air–water interface [129]. In fact, the sinusoidal response in surface pressure due to sinusoidal area deformation was both the same in the two barriers and independent of the position of the barriers along the length of the film balance.

The surface dilatational modulus for WPI monolayers (Fig. 17) increased with increasing surface pressure up to the collapse point [69]. This increase is a result of an increase in the interactions between the monolayer molecules, as deduced from π-A isotherms (Fig. 6), BAM images [72], and monolayer thickness (Fig. 8). How-

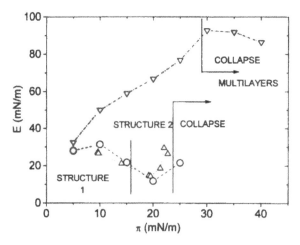

FIG. 17 Surface dilatational modulus as a function of surface pressure for β-casein ($\bigcirc$), caseinate ($\triangle$), and WPI spread films ($\triangledown$) at the air–water interface at 20°C and pH 5; I: 0.05 M.

ever, for the more disordered protein (β-casein and caseinate) the E-π dependence is more complex. E increased to a maximum with surface pressure for structure 1, but decreased with surface pressure and passed to a minimum with structure 2. Finally, E increased up to the collapse point. The same inflexion in the π-Γ curve was observed in the range of 10–20 mN/m and was attributed to the transition from an all-train configuration to a train-and-loop conformation of the β-casein molecule [130]. The results with protein monolayers indicate that the dilatational modulus is not only determined by the structure of protein molecules, but the internal nature of the spread protein molecule also plays an important role. In fact, for the more ordered β-lactoglobulin molecules in WPI the surface dilatational modulus is higher than that for β-casein or caseinate molecules (Fig. 17) with disordered structure at the same surface pressures.

Changes in surface dilatational properties for β-casein monolayer (as an example) as a function of frequency of oscillation over a range of 1 to 300 mHz, at a representative surface pressure (20 mN/m), are illustrated in Fig. 18 [69]. It can be seen that (1) the dilatational modulus increased with the frequency, (2) the dilatational modulus and its elastic component are essentially the same at frequencies lower than 50 mHz. However, significant differences between both rheological parameters were observed at frequencies higher than 50 mHz, mainly due to the decrease of the elastic component at increasing frequencies. (3) The value of the viscous component increased with the frequency and exceeded that of the elastic component at higher frequencies ($\omega > 200$ mHz). From these results it can be concluded that β-casein monolayers present rheological behavior in dilatational conditions that is essentially elastic at low frequencies ($\omega < 50$) and viscoelastic at higher frequencies ($\omega > 50$). As a consequence of the viscoelastic behavior, the tangent of the loss angle increased with frequency (Fig. 18). This behavior was observed with caseinate and WPI [69] and with BSA under dilatational deformation in a ring trough [131]. The

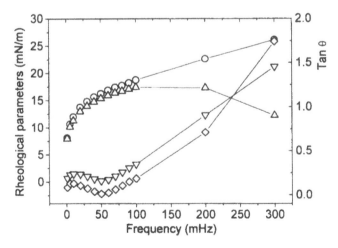

FIG. 18 Frequency dependence of surface rheological parameters—surface dilatational modulus, E ($\bigcirc$); surface dilatational elasticity, E_d ($\triangle$); surface dilatational viscosity, E_v ($\triangledown$); and loss angle tangent ($\Diamond$)—for β-casein spread films at the air–water interface. Surface pressure: 20 mN/m; temperature: 20°C; pH 5; I: 0.05 M; amplitude 5%.

frequency dependence of E was analyzed with the help of data derived from long-term relaxation phenomena discussed in the previous section [69].

VI. GELATION OF ADSORBED PROTEIN FILMS AT THE OIL–WATER INTERFACE

The stability and physicochemical properties of oil-in-water emulsions stabilized by WPI—protein surface coverage, surface shear viscosity, and stability [110,132–134] —are particularly sensitive to their thermal history [132]. Even though the phenomenon is not well understood [135,136], the interfacial gelation of globular proteins may possess great technological importance, because many protein-stabilized emulsions undergo some degree of thermal exposure during their processing, storage, or usage. We have therefore analyzed the viscoelastic characteristics of WPI heat-induced gels at the oil–water interface as a function of temperature and heating conditions [105,137].

Heat-induced interfacial aggregation of WPI, previously adsorbed at the oil–water interface, was studied by measuring interfacial dynamic characteristics (interfacial tension and surface dilatational properties) in an automatic drop tensiometer [105,137]. Dynamic rheological measurements in which sinusoidal oscillating stress or strain is applied to the sample are the preferred methods for characterizing viscoelastic foods [138,139]. These dynamic measurements allow coagulation and gelation to be monitored since the induced deformations are usually so small that their effect on structure is negligible.

Figure 19 exemplifies the time-dependent interfacial tension and surface dilational properties of WPI adsorbed films on the oil–water interface as a consequence of thermal treatment for concentrations of protein in the bulk phase of 10^{-1}%wt/wt. Overall, WPI adsorbed films behaved qualitatively in a similar manner after similar heat treatment, no matter what the protein concentration in the bulk phase, over the range 10^{-1} and 10^{-5}% wt/wt. Briefly, E decreased during heating, passed through a minimum, and then increased as the heating progressed and tended to a plateau value just at the end of the heating period before the isothermal treatment. During the isothermal treatment (at 40, 60, and 80°C), E tended to increase to a plateau, especially during the first period at 40°C. The E value at the plateau decreased after thermal treatment at 60 and 80°C due to the effect of temperature on rheological parameters [137]. The surface activity was increased with the heat treatment because the conformational changes of molecules during heating may include further unfolding, reorganization, and aggregation of the molecules to bring more hydrophobic segments from the interior of the molecule to the oil–water interface.

The rate of thermally induced changes in WPI adsorbed films at the oil–water interface increased with protein concentration in solution. Data in Fig. 19, especially the time dependence of E and phase angle, can be used in a quantitative kinetic analysis of the gelation process at the interface. If the interfacial gelation follows consecutive first-order steps, as for protein in solution [140], a first-order kinetics equation can be used to monitor the time dependence of E during heating. By plotting the data as ln E versus t it is possible to identify two linear regression regions [105], which account for the aggregation and/or cross-linking steps occurring consecutively and/or concurrently. The rate constants increased significantly as the protein concen-

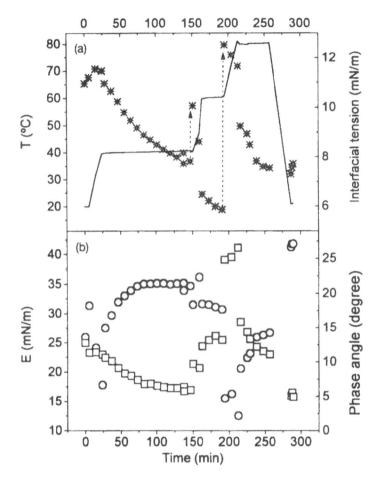

FIG. 19   (a) Time evolution of temperature (———), interfacial tension (*), and (b) surface rheological properties—surface dilational modulus, E (○) and phase angle (□)—for WPI adsorbed film during the isothermal treatment at 80°C. Protein concentration in solution: $10^{-1}\%$; pH 5; I: 0.05 M; frequency of oscillation: 100 mHz; amplitude of sinusoidal oscillation: 15%.

tration in solution increased. That is, conformational changes and protein aggregation increase with protein–protein interactions, which increase with protein concentration in solution. Thus it can be concluded that the gelling time, quantified by the rate constants for E–time dependence, is a useful parameter for detecting transition points in heat-treated protein films at the oil–water interface, just as for gelation in solution [141,142].

An important characteristic of WPI interfacial gelation, both from a theoretical and practical point of view, is the low protein concentration in solution necessary for interfacial gelation in comparison with that necessary for WPI gelation in solution. In this study we have observed the existence of significant changes in interfacial dynamic properties associated with WPI gelation in adsorbed films at the oil–water interface at protein concentration in solution as low as $10^{-5}\%$wt/wt, even at temperatures of 40°C. These values are far below those between 1 and 2.5% required for gelation in solution under similar conditions of pH [138,139,143,144], but it must be remembered that the local protein concentration at the interface is far above its concentration in solution. As observed in the experimental data (Fig. 6) the interfacial tension decreases steadily with heat treatment, which means that the interfacial activity of WPI films increases in the same way. At each protein concentration in solution, the minimum interfacial tension coincided with the maximum E value. The trend observed in interfacial characteristics (interfacial tension and viscoelastic characteristics) of WPI adsorbed films with the heat treatment is similar to the influence of temperature on the interactions and stability of WPI stabilized oil-in-water emulsions [65,132,134,145,146].

VII. INTERACTIONS AND STABILITY

A. Interactions Between Protein-Coated Surfaces

The scf approach described earlier has also been used to estimate the interaction energy between a pair of parallel hydrophobic surfaces in the presence of adsorbing β-casein or α_{s1}-casein [53]. Protein, solvent, and small ions were introduced into the gap between parallel hydrophobic surfaces and the system allowed to reach full thermodynamic equilibrium. Closing the gap and permitting equilibrium adsorption to be attained allows the interaction between the coated surfaces to be calculated as the difference from that calculated at infinite separation. It was found that the calculated interaction energy remained positive and repulsive for all separations for β-casein irrespective of pH or ionic strength.

Direct experimental measurements of interaction forces between β-casein-coated surfaces are obtainable as a function of surface separation by using the interferometric surface force apparatus [147]. On hydrophobized mica, a long-range repulsive electrostatic force was observed between β-casein layers. This could be represented by a DLVO-type potential with the plane of charge and origin of the van der Waals forces placed at the onset of the steric wall determined as 4 nm thick. As the surface separation was decreased, this long-range repulsion was overcome by an attractive force at a surface separation of about 25 nm. This caused the protein-covered surfaces to slide into contact, the force required to separate the adhering surfaces exceeding $2.4 \text{ mN} \cdot \text{m}^{-1}$.

The scf calculations show that reducing pH toward the protein isoelectric point reduces the strength and range of the interlayer repulsion. This implies a substantial electrostatic contribution to the calculated interaction energy, but the energy remains positive at all separations. There is no attraction corresponding to that seen with the surface force apparatus. The origin of this discrepancy remains obscure. On the one hand, the calculations report the result of an equilibrium being achieved on adsorbing to hydrophobic surfaces at the separation of interest. This is not conceptually identical to preadsorbing the protein at infinite separation and then closing the gap. Nor does the computed interaction include any contribution from ubiquitous van der Waals forces. It only includes that part of the overall interaction due to the casein chains. On the other hand, Nylander and Wahlgren [147] also performed measurements with β-casein preadsorbed onto hydrophilic surfaces. On bringing these surfaces together, the inward jump was no longer evident. The force was entirely repulsive and the same force was observed on decompression as on compression. The exponential decay again suggested double-layer forces contributed to this repulsive interaction. When the β-casein solution was replaced with a 1 mM NaCl buffer, a substantial change was observed in the force–distance profile. Not only was the magnitude of the double-layer force reduced, but also an attractive force became apparent at surface separations below 30 nm. Also the force curves measured for β-casein adsorbed onto the hydrophobic surface could be superimposed on that obtained with the hydrophilic counterpart provided it was shifted 4.7 nm outward. Nylander and Wahlgren [147] suggested that β-casein forms a bilayer on the hydrophilic mica; the first arrivals adsorbing flat onto the surface, thus exposing hydrophobic surfaces outward. This allows molecules forming the second layer to act as if they were adsorbing onto a hydrophobic surface and to orient in such a way that their hydrophilic parts protrude into solution. The final result is that, from the perspective of the bulk solution, the exposed interfaces look similar on the two surfaces.

B.   Comparison of α_{s1}-Casein and β-Casein Behavior

The scf calculations predict differences in interaction behavior for the two proteins β-casein and α_{s1}-casein. This main difference in the predicted interaction behavior occurs at higher ionic strengths. While the potential energy remains repulsive for β-casein at all separations, the theory predicts for α_{s1}-casein a potential well at pH 5.5 which deepens as ionic strength is increased in the range 50–200 mM. The underlying physical mechanism thought to be the origin of this attraction is the possible bridging of the gap by the α_{s1}-casein molecules with their two sticky ends. This bridging effect computes into the interaction as an attractive contribution, entropic in origin in thermodynamic terms. This predicted interaction behavior for the two proteins is in broad qualitative agreement with what is actually observed in model oil–water emulsions [148–150]. Those prepared with β-casein as sole emulsifier are stable toward NaCl addition (>2 M), whereas those prepared with α_{s1}-casein become extensively flocculated in 0.1–0.2 M NaCl. Casanova and Dickinson [149] have also shown that β-casein in mixed model films of α_{s1}-casein and β-casein has a strongly protective effect in overcoming the tendency of α_{s1}-casein emulsions to flocculate at these low ionic strengths.

While this agreement is gratifying, it may be illusory. We still have no reliable information on the form of the total interaction potential from which we might hope to quantitatively predict stability behavior. In keeping with the spirit of colloidal stability theory, we would imagine this interaction potential to be a balance of repulsive and attractive forces, perhaps as depicted in Fig. 20. It would be naive, however, to present this as a simple summation of van der Waals attraction, steric, and electrostatic repulsion terms. According to Nylander and Wahlgren [147], the long-range repulsion is electrostatic in origin, but in overcoming this barrier they entered an attractive well of unknown depth. The dynamic light scattering data of Brooksbank et al. [42] showed the steric stabilizing layer to be thinned before it was overcome and aggregation ensued. The position as well as the height of the steric stabilizing barrier are thus under electrostatic control. This coupling of steric and electrostatic influences lies at the heart of the representational problem and remains a challenge for the theoreticians.

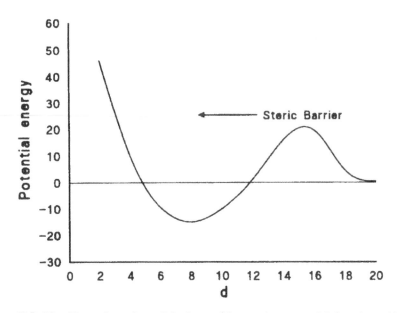

FIG. 20 Illustration of possible form of interaction potential function with steric barrier to be overcome before entry into an attractive potential well. Position as well as height of barrier could be functions of electrostatic repulsion, with barrier moving inward as electrostatic repulsion was shielded or neutralized. Shifts could be largely independent of barrier height (or at least couple with height), suggesting that simply additive steric and electrostatic terms may not be appropriate for prediction of an overall potential energy function.

C. Interactions Between β-Lactoglobulin-Coated Surfaces

Because the interfacial films are relatively thin, it is unlikely that β-lactoglobulin layers stabilize emulsions by the same steric stabilization mechanism invoked for β-casein. Ismailova, Yamployskaya, and Tulovskaya [151] revived Rehbinder's concept and related coalescence stability directly to the mechanical properties of the adsorbed protein layer. The Rehbinder hypothesis was developed further, showing that it was not simply the magnitude of the surface elastic modulus that was important, but rather the stress at which the film was ruptured. It is also likely that electrostatic interactions play a part in ensuring the stability of β-lactoglobulin-stabilized emulsions, since such emulsion droplets are highly susceptible to flocculation by calcium ions even though the native whey protein is not itself precipitated by ionic calcium in aqueous solution [152–154].

VIII. COMPETITIVE DISPLACEMENT

Milk proteins saturate the fluid interface at lower concentrations than small molecule surfactants [155], and hence the protein component dominates in mixture situations where the surfactant concentration is low. On the other hand, at high surfactant concentrations, more efficient packing of the surfactant molecules leads to lower interfacial tension from the surfactants, so the protein is displaced from the interface. The detailed structure and composition of the mixed layer depends on the balance of all interactions between the components both at the interface and in the bulk solution [14]. Nonionic water-soluble surfactants are generally more effective than oil-soluble surfactants at displacing proteins from the oil–water interface. For an anionic surfactant like sodium dodecyl sulfate, a much higher concentration of surfactant is required for displacement because of its propensity for complex formation with the protein both in solution and at the interface. All of these competitive adsorption effects which can occur in real systems have implications for the stability and shelf-life of milk protein–based emulsion products.

A. Chemistry

Protein displacement is readily quantified by determining the protein content of the serum phase of the emulsion in the presence of added surfactant. Hydrocarbon oils are frequently used in model emulsion systems, and there is some evidence to suggest that the hydrophobic phase has little influence on the conformational structure that the casein adopts at the interface [57]. The susceptibility of the proteins to displacement by surfactant is, however, determined by the oil phase. For example, Leaver et al. [156] found that Tween 20 readily displaced β-casein from fresh tetradecane emulsions but was less efficient from soya oil. Moreover, the displacement ability declined as the β-casein soya oil emulsions aged or following a heating regime of 80°C for 30 min, whereas neither aging nor heating had any effect on β-casein-stabilized tetradecane emulsions. Susceptibility to displacement was also found to be pH dependent and a function of the casein type involved [157]. Reversed phase high-pressure liquid chromatography analysis showed significant changes in peak shape and position in those proteins displaced from aged soya oil emulsions and relatively unchanged peaks from tetradecane emulsions over the same time scale. Tryptic digestion of the displaced proteins also showed modification of the primary structure, which increased with emulsion age. Gas chromatography–mass spectrometry of steam distillates from the emulsions showed the presence of a variety of aldehydes in soya oil emulsions that were not present in the original oil or in the tetradecane emulsions [158]. Several of these volatile aldehydes were identified as enals (α,β-unsaturated aldehydes) which react readily with lysine side chains, providing them with a hydrophobic tail which assists in anchoring the protein into the oil phase, thereby decreasing its susceptibility to displacement.

B. Physical Methods

Neutron reflectivity has also been used to study competitive displacement of β-lactoglobulin and β-casein from the air–water interface by the nonionic surfactant, hexaoxyethylene n-dodecyl ether ($C_{12}E_6$). The experiments for β-lactoglobulin took advantage of the differing neutron scattering cross-sections for hydrogen and deuterium in hydrogenated and partially deuterated surfactant to determine the amounts of surfactant and protein adsorbed and, importantly, the location of the surfactant [159]. The data are shown in Fig. 21 as a function of molar surfactant to protein ratio, R. In the presence of deuterated surfactant, we see the total amount of reflective material at the interface (i.e., protein + surfactant). In the presence of hydrogenated surfactant, we see only the protein displacement profile. Fitting the profiles with error functions, complementary for the protein and regular for the surfactant, we can

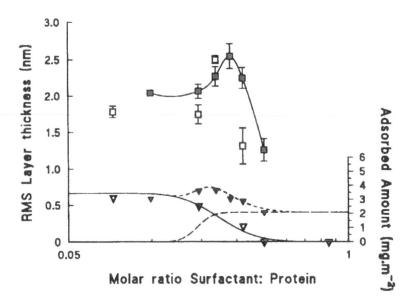

FIG. 21 Behavior of parameters derived from neutron reflectivity measurements during displacement of β-lactoglobulin by surfactant $C_{12}E_6$. Plots show adsorbed layer thickness and adsorbed amounts, derived from Guinier plots, as function of surfactant/protein molar ratio. Open symbols: with hydrogenated $C_{12}E_6$; closed symbols: with partially deuterated $C_{12}E_6$. Solid line through solid squares is spline fit to guide the eye. Solid line through open inverted triangles is error function fit to protein displacement curve. Long dashed line is error function fit for adsorbed surfactant. Summed with displacement error function (short dashed line through solid triangles), it gives data points for total adsorbed amount in presence of deuterated (reflecting) surfactant.

reproduce the total adsorbed amount as their summation. An interesting feature here is that at $R \approx 0.3$, when the amount of surfactant has almost saturated, there is apparently still a substantial amount of β-lactoglobulin at the interface. The rms thickness of the adsorbed films, calculated by Guinier analysis of the reflectivity data, is also shown in Fig. 21. This shows an increase in the thickness of the layer, both in the presence of deuterated and hydrogenated surfactant at a point at which surfactant adsorption had almost reached its plateau value, but at which it was estimated that some 60% of the original protein was still in the interfacial film. It was suggested that to produce these peaks in the layer thickness plots, the surfactant had formed an inner layer adjacent to the air–water interface, pushing off the protein. This location was deduced from the similar increase in thickness seen in the presence of the hydrogenated surfactant. Had this "invisible" surfactant layered on top of the protein, there would have been no measurable increase in thickness in its presence. On the high surfactant concentration side of the peak, the rms thickness becomes that for the deuterated surfactant layer alone.

The displacement behavior of β-lactoglobulin by nonionic surfactants has recently been visualized using atomic force microscopy at both air–water and oil–water interfaces [78,160]. These studies clearly show a demixing phenomenon and have led to the proposal of an "orogenic model" for protein displacement, where the surfactant domains exert a lateral surface pressure which compresses, buckles, and collapses the β-lactoglobulin protein regions prior to their displacement. They thus complement the neutron reflectivity measurements described above. Further support comes from a computer simulation study by Wijmans and Dickinson [161], who used a Brownian dynamics approach to model the competitive displacement process. Their model takes a bonded monolayer and introduces more strongly adsorbing, nonbonded particles. Initially the displacer particles fill up the gaps in the original bonded monolayer network. As the surface concentration of the displacer particles increases, the network "holes" grow, the strands become thinner, and some network particles are pushed away from the surface. Part of the bonded network remains pinned to the surface, but large parts of the film buckle out into the bulk solution, forming a relatively thick layer. Eventually the surface becomes saturated with displacer particles and the protein network detaches into the solution. The simulations emphasize the importance of the rheology of the protein network to the process and highlight the role of repulsive forces between protein and surfactant in promoting the phase separation.

C. Competitive Adsorption and Displacement in Protein–Lipid Mixed Systems

Competitive adsorption of proteins and lipids at fluid interfaces can affect the stability of food dispersions [2]. Thus knowledge of the lipids, proteins, and their mixtures at fluid–fluid interfaces is a key factor for the formation and stability of food dispersions (emulsions and foams). However, more important in some products is the effect of the small molecules in destabilizing the emulsion [21]. In the formulation of ice cream, the small molecule emulsifier is added to break the adsorbed layer of protein and allow the adsorption of fat to the surface of the air bubble. Thus, an important action of the small molecule emulsifiers is to promote the displacement of caseins from the interface.

There have been many studies of protein–lipid interactions, particularly between proteins and soluble lipids, in relation to the formation and stability of food emulsions and foams, but much less is known about the details of protein-insoluble lipid interaction [14,29].

1. Protein–Lipid Interactions at the Air–Water Interface at the Equilibrium

Surface tension data have been reported for the milk protein systems (β-casein, caseinate, and WPI) and their mixtures with three food-permitted insoluble lipids (monopalmitin, monoolein, and monolaurin) at equilibrium [117]. These experiments mimicked the behavior of emulsifiers in food emulsions in which an oil-soluble lipid (monopalmitin, monoolein, or monolaurin) diffuses to the interface to interact with a protein film which has adsorbed from the aqueous bulk phase. The effect of the protein/lipid ratio on the surface activity of mixed protein–monoglyceride systems is shown in Fig. 22. In these experiments, monoglyceride spread on a previously adsorbed protein film was maintained constant. Therefore, the variation in protein/lipid ratio is due to the protein added to the bulk phase in the range 5 to 1 $\times$

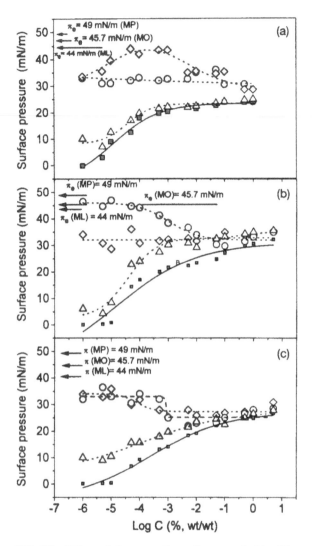

FIG. 22 Effect of the spreading of monopalmitin (○), monoolein (◊), and monolaurin (△) on a film of β-casein (a), caseinate (b), and WPI (c) previously adsorbed on the air–water interface. Temperature 20°C. Lipid superficial density (molecule.nm^{-2}): monopalmitin, 12; monoolein, 9.7; and monolaurin, 9.5. The arrows indicate the equilibrium spreading pressure, π_e, for monopalmitin [π_e (MP)], monoolein [π_e (MO)], and monolaurin [π_e (ML)].

10^{-5}%wt/wt. The monoglyceride density spread on the interface was higher than that required for the monolayer collapse, as was deduced from the π-A isotherm [162]. The protein concentration dependence on surface pressure for protein–monoglyceride mixed systems showed a sigmoidal behavior. However, the surface activity of the mixed systems depended on the protein/monoglyceride ratio and the monoglyceride spread on the interface (Fig. 22).

For all protein–monoglyceride mixed films, the surface pressure values approach those of pure protein films at higher relative protein concentrations in the mixed systems, as the protein saturated the monolayer. However, significant differences were observed at lower protein/monoglyceride ratios, especially for protein–monolaurin mixed films. In fact, for protein–monolaurin mixed films, the surface pressure followed the same dependence on protein concentration as pure protein.

For protein–monopalmitin and protein–monoolein mixed films, the protein/lipid ratio dependence on surface pressure was more complex. At lower relative protein concentrations, the surface pressure tended to the equilibrium surface pressure (π_e) of the pure monopalmitin or monoolein, which is indicated in Fig. 22 by the arrows. However, the level of surface pressure at the minimum protein/monoglyceride ratio depended on the interfacial composition. At the intermediate range of protein–monoglyceride concentrations, significant further reduction in the surface pressure was observed. The effect resulted in an inflection in the surface pressure curve in the intermediate region.

The general behavior described earlier for protein–monopalmitin and protein–monoolein mixed films at the air–water interface (Fig. 22) is essentially the same as that observed with other monoglyceride–protein mixed films [15]. Thus it appears that at higher protein relative concentrations in protein–monoglyceride mixed films the protein determines the surface activity, but at lower relative protein concentrations in the mixture monoglyceride (monopalmitin or monoolein) determines the surface activity of protein–monoglyceride mixed films. As the monoglyceride (monopalmitin or monoolein) surface densities spread here, the protein is removed and a collapsed monopalmitin or monoolein film saturates the interface. In the intermediate region, the surface activity is determined by the existence of protein and monoglyceride at the interface. However, what is more difficult to establish is the degree of interactions between film-forming components in the mixed film.

Protein–monolaurin mixed films behave differently. In fact, the protein, at every protein/monolaurin ratio, determined the surface activity of protein–monolaurin mixed films. This phenomenon may be associated with the instability of monolaurin monolayers at the air–water interface. In fact, from relaxation experiments we have observed [163] a significant monolaurin monolayer molecular loss, which is more pronounced as the surface pressure increases, with a maximum at the surface density utilized in this study (at the collapse point). These results support the hypothesis that a reduced level of interaction exists between monolaurin and protein at the air–water interface.

The extent of removal of protein by a surfactant is influenced by factors affecting the binding strength of a protein to a surface. Thus the displacement of protein by surfactant has been found to decrease in conditions favoring conformational changes. The ease of displacement, however, is not only influenced by protein properties, but also by the type of surfactant [14] and the aqueous composition [15,16]. In particular, as we have demonstrated here, the distribution of monoglyceride and proteins at the air–water interface is influenced by the spreading of monoglyceride on a previously adsorbed protein, especially when protein concentration in the bulk phase is higher than that required for complete coverage, at the plateau region (Fig. 22). Thus the way in which proteins and monoglycerides are spread or adsorbed on the interface may have a role in determining the interfacial characteristics of the mixed film.

2. Protein–Lipid Interactions at the Air–Water Interface at Steady State

From the π-A isotherms it is clear that some degree of interaction exists between monoglycerides and proteins in spread mixed films, which becomes more pronounced as the monolayer is compressed at the highest surface pressures [19,94,164,165]. Recent experiments (including measuring the π-A isotherm coupled with Brewster angle microscopy and relative thickness of spread monolayers) have shown for the first time [19,94,164,165] that in monoglyceride–protein mixed films, islands of protein and monoglyceride exist at the air–water interface on a microscopic level, but with few interactions between them, depending on the surface pressure (Fig. 23). At surface pressures lower than those for protein collapse a mixed monolayer of monoglyceride and protein may exist (Fig. 23a). However, at surface pressures higher than that for protein collapse, the mixed monolayers are essentially dominated by monoglyceride molecules. At higher surface pressures, collapsed pro-

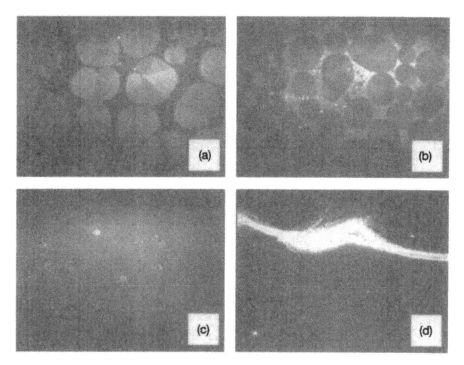

FIG. 23 Visualization of monopalmitin–WPI mixed monolayers by Brewster angle microscopy at 20°C and pH 5.0; *I*: 0.05 *M*. (a) At surface pressures lower than that for WPI collapse (π ≅ 31 mN/m) there exists coexistence between small circular liquid-condensed monopalmitin domains and a homogeneous phase of liquid-expanded monopalmitin and WPI domains. (b) At higher surface pressures, near and after the WPI collapse (25 < π < 37 mN/m), the squeezing out of WPI by monopalmitin can be distinguished in a region with LC domains of monopalmitin (dark area) over a sublayer of collapsed protein (bright area). (c) At the lipid collapse, the monopalmitin domains were so closely packed that the monolayer morphology acquired a high homogeneity. (d) In the region of highest surface pressure different regions of collapsed WPI were observed on the interface. Temperature 20°C. The horizontal direction of the image corresponds to 630 μm, and the vertical direction corresponds to 470 μm.

tein residues may be displaced from the interface by monoglyceride molecules (Fig. 23b). However, monoglyceride molecules are unable to completely displace protein residues from the air–water interface even at the highest surface pressure, at the collapse pressure of the monoglyceride (Figs. 23c and d). Over the overall range of existence of the mixed film the monolayer presents some heterogeneity due to the fact that domains of monoglyceride and protein residues are present during the monolayer compression, giving interfacial regions with different relative film thickness.

Interactions, miscibility, and displacement of proteins by monoglycerides from the air–water interface depend on the particular protein–monoglyceride system. Different proteins show different interfacial morphology, confirming the importance of protein secondary structure in determining the mechanism of interfacial interactions. On the other hand, the lower surface activity of monoolein justifies the idea that this lipid has a lower capacity than monopalmitin for protein displacement. In fact, monoolein requires higher surface pressures than monopalmitin for protein displacement from the air–water interface [165]. The prevalence of monoglyceride increases with its concentration in the mixture and at higher surface pressures. In summary, on a microscopic level, the distribution of lipids and proteins in mixed spread films at the air–water interface is heterogeneous and depends on the surface pressure and the lipid–protein ratio in the mixed film.

Until recently the exact mechanism for protein displacement was unclear. Recent studies [78,160,166] have shown that the displacement of milk proteins from the air–water interface by water-soluble surfactants (Tween 20 and SDS) involves a novel orogenic mechanism. Such a model provides an explanation for the displacement of protein by surfactants, but differences exist between the behavior of adsorbable water-soluble surfactants and spread water-insoluble monoglycerides. The results suggest that for spread monoglycerides [19,94,164,165] the first stage of the orogenic mechanism, which occurs at surface pressures lower than the equilibrium surface pressure of the protein, involves a displacement front of monoglyceride domains instead of the adsorption of water-soluble surfactant molecules at defects in the protein network. The second stage, which occurs at surface pressures near to and above the equilibrium surface pressure of the protein, involves a buckling of the monolayer and reordering of the molecules as the protein film gets thicker in response to the decreasing surface coverage. Finally, at sufficiently high surface pressures the protein network begins to fail, freeing proteins which then desorb from the interface [78,160,166]. But, for spread monoglyceride monolayers, the protein displacement is not total even at the highest surface pressure, at the collapse point of the mixed film. The orogenic displacement mechanism is a consequence of the low level of interaction between proteins and monoglycerides at the air–water interface [19,94,164,165].

3. Protein–Lipid Interactions at the Oil–Water Interface Under Dynamic Conditions

The existence of WPI–monoglyceride (monopalmitin and monoolein) interactions at the interface has been shown in measurements of interfacial tension and surface dilatational properties [105]. Systematic experimental studies of surface dynamic properties, as a function of time and at long-term adsorption, for protein–monoglyceride mixed films at the oil–water interface were carried out in an automated drop tensiometer. The dynamic behavior of protein + monoglyceride mixed films was found to depend on the adsorption time, the lipid and the protein/lipid ratio, all in a rather complicated manner. The protein determined the interfacial characteristics of the mixed film since, at WPI $\geq 10^{-2}$%wt/wt, the protein saturated the film (Fig. 13), no matter what the concentration of the monoglyceride. However, a competitive adsorption of the emulsifier (WPI and monoglycerides) does exist, as the concentration of protein in the bulk phase is far lower than that for interfacial saturation. This critical concentration determines the existence of protein–lipid interactions at the oil–water interface. That is, monopalmitin and, especially, monoolein in the oil bulk phase are unable to displace totally the adsorbed WPI film. In fact, the protein dominated the interfacial characteristics of WPI–monoolein mixed films throughout the whole range of protein/lipid ratios employed.

In these experiments, the existence of protein–monoglyceride interactions at the air–water interface has been demonstrated by different complementary techniques, including tensiometry under static and dynamic conditions, Langmuir- and Wilhelmy-type surface film balances, and Brewster angle microscopy. Surface tension measurement is an easier complementary experimental technique for providing information regarding the interfacial characteristics of pure protein and lipid films and the existence of protein–lipid interactions at the interface. However, surface pressure–concentration experiments are not sufficient to allow a full picture of the nature of protein–monoglyceride interactions at the interface. From the results derived from the different techniques, it can be concluded that the interfacial characteristics of mixed emulsifiers at fluid interfaces depend on the nature of the interface (either air–water or oil–water) and, in a complex fashion, on the way by which these emulsifiers are adsorbed to the interface (either by cooperative or competitive adsorption/spreading of the film-forming components).

On the other hand, lowering of the surface (interfacial) tension by emulsifiers (proteins and lipids) is only a first step in the preparation of stable food emulsions and foams. A low surface (interfacial) tension facilitates breaking up the oil phase into smaller droplets. However, dispersion requires rapid and substantial stretching of bubbles or drops, and consequently the surface (interfacial) tension may be far from equilibrium. Thus, dynamic properties of adsorbed emulsifier layers are also important due to their stabilizing function during emulsification.

IX. CONCLUDING REMARKS

In this chapter, we have analyzed the structure, morphology, adsorption, interactions, and dynamic properties of food dairy proteins (β-casein, caseinate, and WPI) at the air–water and oil–water interfaces. The summary includes an assessment of information derived from a variety of chemical and physical techniques. Combined surface chemistry (Langmuir- and Wilhelmy-type film balances and dynamic tensiometry) and microscopy (Brewster angle microscopy, BAM) techniques have been used to analyze the surface characteristics of proteins, with structural information included from neutron reflectivity, dynamic light scattering, and enzyme accessibility measurements. The results demonstrate that protein type and temperature affect the interfacial characteristics. The nature of emulsifier interactions at the interface has an important role on their physicochemical characteristics, including their role in conferring stability on emulsions and foams. Important functional differences have been demonstrated between globular (WPI) and disordered (β-casein and caseinate) proteins. The efficacy of the caseins as steric stabilizing agents of emulsions has been considered and the current state of theories regarding those interactions has been summarized.

ACKNOWLEDGMENTS

DSH acknowledges the support of the Scottish Executive Environment and Rural Affairs Department and BBSRC (UK) in this work. The research was also supported by the European Community through Grant FAIR-CT96-1216 (DSH, JMRP), by CICYT through Grant ALI97-1274-CE, and by DGICYT through Grant PB97-0734 (both JMRP).

REFERENCES

1. E. Dickinson, *An Introduction to Food Colloids*. Oxford: Oxford University Press, 1992.
2. E. Dickinson. J. Dairy Sci. *80*:2607–2619, 1997.
3. E. Dickinson. J. Chem. Soc. Faraday Trans. *94*:1657–1669, 1998.
4. E. Dickinson. In *Modern Aspects of Emulsion Science* (B. P. Binks, ed.). Cambridge: Royal Society of Chemistry, 1998, pp. 145–174.
5. E. Dickinson. Coll. Surf. B: Biointerfaces *15*:161–176, 1999.
6. E. Dickinson. Coll. Surf. B: Biointerfaces *20*:197–210, 2001.
7. S. Damodaran and A. Paraf. *Food Proteins and their Applications*, New York: Marcel Dekker, 1997.
8. S. E. Friberg and K. Larsson, *Food Emulsions*, 3rd ed. New York: Marcel Dekker, 1997.
9. R. Hartel and G. R. Hasenhuette, *Food Emulsifiers and Their Applications*. New York: Chapman and Hall, 1997.
10. F. D. Gunstone and F. B. Padley, *Lipid Technologies and Applications*. New York: Marcel Dekker, 1997.
11. R. E. McDonald and D. B. Min, *Food Lipids and Health*. New York: Marcel Dekker, 1996.
12. J. M. Rodríguez Patino and M. R. Rodríguez Niño. Colloids Surf. B *15*:235–252, 1999.
13. J. Sjöblom, *Emulsions and Emulsion Stability*. New York: Marcel Dekker, 1996.
14. M. Bos, T. Nylander, T. Arnebrandt, and D. C. Clark. In *Food Emulsions and Their Applications* (G. L. Hasenhuette and R. W. Hartel, eds.). New York: Chapman & Hall, 1997, pp. 95–146.
15. M. R. Rodríguez Niño and J. M. Rodríguez Patino. J. Am. Oil Chem. Soc. *75*:1233–1240, 1998.
16. M. R. Rodríguez Niño and J. M. Rodríguez Patino. J. Am. Oil Chem. Soc. *75*:1241–1248, 1998.
17. M. R. Rodríguez Niño, P. J. Wilde, D. C. Clark, F. A. Husband, and J. M. Rodríguez Patino. J. Agric. Food Chem. *45*:3010–3015, 1997.
18. M. R. Rodríguez Niño, P. J. Wilde, D. C. Clark, and J. M. Rodríguez Patino. J. Agric. Food Chem. *45*:3016–3021, 1997.
19. J. M. Rodíguez Patino, S. C. Carrera, and M. R. Rodríguez Niño. J. Agric. Food Chem. *47*:4998–5008, 1999.

20. M. R. Rodríguez Niño, P. J. Wilde, D. C. Clark, and J. M. Rodríguez Patino. J. Agric. Food Chem. 46:2177–2184, 1998.
21. M. R. Rodríguez Niño, P. J. Wilde, D. C. Clark, and J. M. Rodríguez Patino. Langmuir 14:2160–2166, 1998.
22. K. G. Berger. In *Food Emulsions* (S. E. Friberg and K. Larsson, eds.). New York: Marcel Dekker, 1997, pp. 423–490.
23. H. D. Goff and W. K. Jordan. J. Dairy Sci. 72:18–29, 1989.
24. P. Cayot and D. Lorient. In *Food Proteins and Their Applications* (S. Damodaran, A. Paraf, eds.). New York: Marcel Dekker, 1997, pp. 225–256.
25. D. G. Dalgleish. In *Emulsions and Emulsion Stability* (J. Sjöblom, ed.). New York: Marcel Dekker, 1996, pp. 287–325.
26. D. G. Dalgleish. In *Food Proteins and Their Applications* (S. Damodaran and A. Paraf, eds.). New York: Marcel Dekker, 1997, pp. 199–223.
27. T. Nylander and B. Ericsson. In *Food Emulsions* (S. E. Friberg and K. Larsson, eds.). New York: Marcel Dekker, 1997, pp. 189–233.
28. M. R. Rodríguez Niño, S. C. Carrera, J. M. Rodríguez Patino, F. M. Cejudo, and N. J. M. García. Chem. Eng. Commun. 2000 (in press).
29. R. Miller, V. F. Fainerman, A. V. Makievski, J. Krägel, D. O. Grigoriev, V. K. Kazakov, and O. V. Sinyachenko. Adv. Colloid Interface Sci. 86:39–82, 2000.
30. D. Mobius and R. Miller, *Studies of Interface Science*, Vol 7. Amsterdam: Elsevier, 2001.
31. M. Malmsten, *Biopolymers at Interfaces*. New York: Marcel Dekker, 1998.
32. D. S. Horne and J. Leaver. Food Hydrocolloids 9:91–95, 1995.
33. D. S. Horne. In *Encyclopaedia of Dairy Sciences* (H. Singh, ed.). New York: Academic Press (in press).
34. D. S. Horne. Int. Dairy J. 8:171–177, 1998.
35. P. Walstra. In *Gums and Stabilizers for the Food Industry*, Vol. 4 (G. O. Phillips, D. J. Wedlock, and P. A. Williams, eds.). Oxford: IRL Press, 1988, pp. 323–336.
36. J. Benjamins, J. A. DeFeijter, M. J. A. Evans, D. E. Graham, and M. C. Phillips. Faraday Discuss. Chem. Soc. 59:218–229, 1975.
37. B. W. Morrissey and C. C. Han. J. Colloid Interface Sci. 65:423–431, 1978.
38. T. Nylander and M. Wahlgren. J. Colloid Interface Sci. 162:151–162, 1994.
39. T. Kull, T. Nylander, F. Tiberg, and N. M. Wahlgren. Langmuir 13:5141–5147, 1997.
40. A. R. Mackie, J. Mingins, and A. N. North. J. Chem. Soc. Faraday Trans. 87:3043–3049, 1991.
41. D. G. Dalgleish. Colloids Surf. 46:141–155, 1990.
42. D. V. Brooksbank, C. M. Davidson, D. S. Horne, and L. Leaver. J. Chem. Soc. Faraday Trans. 89:3419–3425, 1993.
43. J. Penfold, R. M. Richardson, A. Zarbakhsh, J. R. P. Webster, D. G. Bucknall, A. R. Rennie, R. A. L. Jones, T. Cosgrove, R. K. Thomas, J. S. Higgins, P. D. I. Fletcher, E. Dickinson, S. J. Roser, I. A. McLure, A. R. Hillman, R. W. Richards, E. J. Staples, A. N. Burgess, E. A. Simister, and J. W. White. J. Chem. Soc. Faraday Trans. 93: 3899–3917, 1997.
44. E. Dickinson, D. S. Horne, J. S. Phipps, and R. M. Richardson. Langmuir 9:242–248, 1993.
45. G. Fragneto, R. K. Thomas, A. R. Rennie, and J. Penfold. Science 267:657–660, 1995.
46. P. J. Atkinson, E. Dickinson, D. S. Horne, and R. M. Richardson. J. Chem. Soc. Faraday Trans. 91:2847–2854, 1995.
47. B. Harzbollah, V. Angié-Béghin, R. Douillard, and L. Bosio. Int. J. Biol. Macromol. 23:73–84, 1998.
48. T. Nylander, F. Tiberg, T. J. Su, J. R. Lu, and R. K. Thomas. Biomacromolecules 2: 278–287, 2001.

49. G. Fragneto, T. J. Su, J. R. Lu, R. K. Thomas, and A. R. Rennie. Phys. Chem. Chem. Phys. 2:5214–5221, 2000.
50. M. A. Cohen Stuart, G. J. Fleer, J. Lyklema, W. Norde, and J. M. H. M. Scheutjens. Adv. Colloid Interface Sci. 34:477–535, 1991.
51. G. J. Fleer, M. A. Cohen Stuart, J. M. H. M. Scheutjens, T. Cosgrove, and B. Vincent, Polymers at Interfaces. London: Chapman & Hall, 1993.
52. F. A. M. Leermakers, P. J. Atkinson, E. Dickinson, and D. S. Horne. J. Colloid Interface Sci. 178:681–693, 1996.
53. E. Dickinson, D. S. Horne, V. J. Pinfield, and F. A. M. Leermakers. J. Chem. Soc. Faraday Trans. 93:425–432, 1997.
54. L. C. Ter Beek, M. Ketelaars, D. C. McCain, P. E. A. Smulders, P. Walstra, and M. A. Hemminga. Biophys. J. 70:2396–2402, 1996.
55. D. S. Horne and J. Leaver. Food Hydrocolloids 9:91–95, 1995.
56. J. Leaver and D. G. Dalgleish. Biochim. Biophys. Acta 1041:217–222, 1990.
57. S. C. Russev, T. V. Arguirov, and T. D. Gurkov. Colloids Surf. B: Biointerfaces 19: 89–100, 2000.
58. A. R. Mackie, J. Mingins, R. Dann, and A. N. North. In Food Polymers, Gels and Colloids (E. Dickinson, ed.). Cambridge: Royal Society of Chemistry, 1991, pp. 96–112.
59. Y. Fang and D. G. Dalgleish. J. Colloid Interface Sci. 196:292–298, 1997.
60. M. Shimizu. In Food Macromolecules and Colloids (E. Dickinson and D. Lorient, eds.). Cambridge: Royal Society of Chemistry, 1995, pp. 34–42.
61. J. A. DeFeijter, J. Benjamins, and M. Tamboer. J. Colloid Interface Sci. 90:289–292, 1982.
62. R. J. Marsh, R. A. L. Jones, and J. Penfold. J. Colloid Interface Sci. 218:347–349, 1999.
63. F. J. G. Boersboom, A. E. DeGroot-Mostert, A. Prins, and T. Van Vliet. Netherlands Milk Dairy J. 50:183–198, 1996.
64. J. T. Petlov, T. D. Gurkov, B. E. Campbell, and R. P. Borwankar. Langmuir 16:3703–3711, 1996.
65. E. Dickinson and Y. Matsumura. Int. J. Biol. Macromol. 13:26–30, 1991.
66. E. Dickinson, S. Rolfe, and D. G. Dalgleish. Food Hydrocolloids 3:193–203, 1989.
67. S. Hénon and J. Meunier. Rev. Sci. Instrum. 62:936–939, 1991.
68. D. Hönig and D. Möbius. J. Phys. Chem. 95:4590–4592, 1991.
69. J. M. Rodríguez Patino, S. C. Carrera, M. R. Rodríguez Niño, and M. Cejudo. J. Colloid Interface Sci. 242:141–151, 2001.
70. M. R. Rodríguez Niño, S. C. Carrera, and J. M. Rodríguez Patino. Colloids Surf. B 12:161–173, 1999.
71. J. M. Rodríguez Patino, S. C. Carrera, and M. R. Rodríguez Niño. Food Hydrocolloids 13:401–408, 1999.
72. J. M. Rodríguez Patino, S. C. Carrera, M. R. Rodríguez Niño, and F. M. Cejudo. In Food Colloids 2000: Fundamentals of Formulations (E. Dickinson and R. Miller, eds.). Cambridge: Royal Society of Chemistry, 2001, pp. 22–35.
73. J. V. Boyd, J. R. Mitchell, L. Irons, P. R. Musselwhite, and P. Sherman. J. Colloid Interface Sci. 45:478–486, 1973.
74. M. C. Phillips, M. T. A. Evans, D. E. Graham, and D. Oldani. Colloid Polym. Sci. 253:424–427, 1975.
75. D. E. Graham and M. C. Phillips. J. Colloid Interface Sci. 70:427–439, 1979.
76. J. R. Hunter, P. K. Kilpatrick, and R. J. Carbonell. J. Colloid Interface Sci. 142:429–447, 1991.
77. C. S. Gau, H. Yu, and G. Zografi. J. Colloid Interface Sci. 162:214–221, 1994.

78. A. R. Mackie, A. P. Gunning, P. J. Wilde, and V. J. Morris. J. Colloid Interface Sci. *210*:157–166, 1999.
79. J. Benjamins, J. S. de Feijter, M. T. A. Evans, D. E. Graham, and M. C. Phillips. Faraday Discuss. *59*:218–229, 1975.
80. B. S. Murray and E. Dickinson. Food Sci. Technol. Int. *2*:131–145, 1996.
81. S. Damodaran. Adv. Food Nutr. Res. *34*:1–79, 1990.
82. P. Walstra. Chem. Eng. Sci. *48*:333–349, 1993.
83. J. Benjamins, Static and dynamic properties of proteins adsorbed at liquid interfaces. Ph.D. Dissertation, Wageningen University, Wageningen, The Netherlands, 2000.
84. D. E. Graham and M. C. Phillips. J. Colloid Interface Sci. *70*:403–414, 1979.
85. F. MacRitchie. Adv. Protein Chem. *32*:283–326, 1978.
86. E. Tornberg. J. Colloid Interface Sci. *64*:391–402, 1978.
87. E. Tornberg. J. Sci. Food Agric. *29*:762–776, 1978.
88. A. V. Makievski, V. B. Fainerman, M. Bree, R. Wüstneck, J. Krägel, and R. Miller. J. Phys. Chem. B *102*:417–425, 1998.
89. F. MacRitchie, *Chemistry at Interfaces*. San Diego: Academic Press, 1990, pp. 156–185.
90. A. V. Adamson, *Physical Chemistry of Surfaces*, 5th ed. New York: Wiley, 1990.
91. S. S. Dukhin, G. Kretzschmar, and R. Miller, *Dynamics of Adsorption at Liquid Interfaces. Theory, Experiment, Application*. Amsterdam: Elsevier, 1995.
92. J. M. Rodríguez Patino and M. R. Rodríguez Niño. Colloids Surf. A *103*:91–108, 1995.
93. J. M. Rodríguez Patino and M. R. Rodríguez Niño. In *Food Macromolecules and Colloids* (E. Dickinson and D. Lorient, eds.). London: Royal Society of Chemistry, 1995, pp. 103–108.
94. J. M. Rodríguez Patino, S. C. Carrera, and M. R. Rodríguez Niño. Langmuir *15*:4777–4788, 1999.
95. J. M. Rodríguez Patino, M. R. Rodríguez Niño, and S. C. Carrera. J. Agric. Food Chem. *47*:2241–2248, 1999.
96. G. González and F. MacRitchie. J. Colloid Interface Sci. *32*:55–61, 1970.
97. F. MacRitchie. Colloids Surf. *41*:25–34, 1989.
98. P. Joos, M. van Uffelend, and G. Serrien. J. Colloid Interface Sci. *152*:521–533, 1992.
99. M. R. Rodríguez Niño and J. M. Rodríguez Patino. Ind. Eng. Chem. 2002 (in press).
100. D. E. Graham and M. C. Phillips. J. Colloid Interface Sci. *70*:415–426, 1979.
101. S. Damodaran and K. B. Song. Biochim. Biophys. Acta *954*: 253–264, 1988.
102. B. S. Murray, M. Faergemand, M. Trotereau, and A. Ventura. In *Food Emulsions and Foams: Interfaces, Interactions, and Stability* (E. Dickinson and J. M. Rodríguez Patino, eds.). Cambridge: Royal Society of Chemistry, 1999, pp. 223–235.
103. M. Paulsson and P. Dejmek. J. Colloid Interface Sci. *150*:394–403, 1992.
104. A. F. H. Ward and L. Tordai. J. Chem. Phys. *14*:453–461, 1946.
105. J. M. Rodríguez Patino, M. R. Rodríguez Niño, C. Carrera, J. M. Navarro, G. Rodriguez, and M. Cejudo. Colloids Surf. B *21*:87–99, 2001.
106. L. G. Phillips, S. E. Hawks, and J. B. German. J. Agric. Food. Chem. *43*:613–619, 1995.
107. A. Prins. In *Advances in Food Emulsions and Foams* (E. Dickinson and G. Stainsby, eds.). New York: Elsevier Applied Science, 1988, pp. 91–122.
108. J. N. de Wit and G. A. M. Swinkels. Biochim. Biophys. Acta *624*:40–50, 1980.
109. K. P. Das and J. E. Kinsella. J. Colloid Interface Sci. *139*:551–560, 1990.
110. E. Dickinson and S.-T. Hong, J. Agric. Food Chem. *42*:1602–1606, 1994.
111. M. Ahmed and E. Dickinson. Colloids Surf. *47*:353–365, 1990.
112. J. H. Crowe, L. M. Crowe, J. F. Carpenter, and C. A. Wistrom. Biochim. J. *242*:1–10, 1987.
113. J. C. Lee and S. N. Timasheff. J. Biol. Chem. *256*:7193–7201, 1981.

114. A. J. I. Ward and L. H. Regan. J. Colloid Interface Sci. *78*:389–395, 1980.
115. P. Chen, R. Prokop, S. S. Susnar, and A. W. Newman, Chapter 8. *Proteins at Liquid Interfaces*, Amsterdam: Elsevier, 1998.
116. R. Miller and G. Kretzschmar. Adv. Colloid Interface Sci. *37*:97–121, 1991.
117. M. R. Rodríguez Niño, C. Carrera, M. Cejudo, and J. M. Rodríguez Patino. J. Am. Oil Chem. Soc. *78*:873–879, 2001.
118. G. L. Gaines, *Insoluble Monolayers at Liquid–Gas Interface*. New York: Wiley, 1996.
119. J. M. Rodríguez Patino, M. R. Rodríguez Niño, and C. Carrera. AIChE J. 2001 (submitted).
120. J. M. Rodríguez Patino, M. R. Rodríguez Niño, and C. Carrera. AIChE J. 2001 (submitted).
121. B. S. Murray. In *Proteins at Liquid Interfaces* (D. Möbius and R. Miller, eds.). Amsterdam: Elsevier, 1998, pp. 179–220.
122. G. Kretzschmar and R. Miller. Adv. Colloid Interface Sci. *36*:65–124, 1991.
123. A. Prins. In *New Physico-Chemical Techniques for the Characterization of Complex Food Systems* (E. Dickinson, ed.). Glasgow: Blackie Academic, 1995, pp. 214–239.
124. E. I. Franses, O. A. Basaran, and C. H. Chag. Curr. Opin. Colloid Interface Sci. *1*: 296–312, 1996.
125. M. Bos and T. van Vliet. Adv. Colloid Interface Sci. *91*:437–471, 2001.
126. J. Lucassen and M. van den Tempel. Chem. Eng. Sci. *27*:1283–1291, 1972.
127. L. Ting, D. T. Wasan, and K. Miyano. J. Colloid Interface Sci. *107*:345–354, 1985.
128. A. Bonfillon and D. Langevin. Langmuir *9*:2172–2177, 1993.
129. J. M. Rodríguez Patino, S. C. Carrera, M. R. Rodríguez Niño, and M. Cejudo. Langmuir *17*:4003–4013, 2001.
130. E. H. Lucassen-Reynders and J. Benjamins. In *Food Emulsions and Foams: Interfaces, Interactions and Stability* (E. Dickinson and J. M. Rodríguez Patino, eds.). Cambridge: Royal Society of Chemistry, 1999, pp. 195–206.
131. M. R. Rodríguez Niño, P. J. Wilde, D. C. Clark, and J. M. Rodríguez Patino. Ind. Eng. Chem. Res. *35*:4449–4456, 1996.
132. K. Demetriades, J. N. Coupland, and D. J. McClements. J. Food Sci. *62*:462–467, 1997.
133. E. Dickinson and D. J. McClements, *Advances in Food Colloids*. London: Blackie Academic, 1995.
134. F. J. Monahan, D. J. McClements, and J. B. German. J. Food Sci. *61*:504–509, 1996.
135. A. Ball and R. A. L. Jones. Langmuir *11*:3542–3548, 1995.
136. R. J. Green, I. Hopkinson, and R. A. L. Jones. In *Food Emulsions and Foams: Interfaces, Interactions and Stability* (E. Dickinson and J. M. Rodríguez Patino, eds.). Cambridge: The Royal Society of Chemistry, 1999, pp. 285–295.
137. J. M. Rodríguez Patino, M. R. Rodríguez Niño, and C. Carrera. J. Agric. Food Chem. *47*:3640–3648, 1999.
138. G. R. Ziegler and E. A. Foegeding. Adv. Food Nutr. Res. *34*:203–298, 1990.
139. D. Oakenfull, J. Pearce, and R. W. Burley. In *Food Proteins and Their Applications* (S. Damodaran and A. Paraf, eds.). New York: Marcel Dekker, 1997, pp. 111–142.
140. J. I. Boye, C.-Y. Ma, and V. R. Harwalkar. In *Food Proteins and Their Applications* (S. Damodaran and A. Paraf, eds.). New York: Marcel Dekker, 1997, pp. 25–56.
141. T. Beveridge, L. Jones, and M. A. Tung. J. Agric. Food Chem. *32*:307–313, 1984.
142. J. E. Kinsella, D. J. Rector, and L. G. Phillips. In *Protein Structure–Function Relationships in Food* (R. Y. Yada, R. L. Jackman, and J. L. Smith, eds.). London: Blackie Academic, 1994.
143. M. Verheul, S. P. F. Roefs, J. Mellema, and K. G. deKruif. Langmuir *14*:2263–2268, 1998.

144. M. Verheul, S. P. F. Roefs, and K. G. deKruif. J. Agric. Food Chem. *46*:896–903, 1998.
145. J. A. Hunt and D. G. Dalgleish. J. Food Sci. *42*:2131–2135, 1994.
146. J. A. Hunt and D. G. Dalgleish. J. Food Sci. *50*:1120–1123, 1995.
147. T. Nylander and M. N. Wahlgren. Langmuir *13*:6219–6225, 1997.
148. E. Dickinson, R. H. Whyman, and D. G. Dalgleish. In *Food Emulsions and Foams* (E. Dickinson, ed.). London: Royal Society of Chemistry, 1987, pp. 40–51.
149. H. Casanova and E. Dickinson. J. Agric. Food Chem. *46*:72–76, 1998.
150. E. Dickinson, M. G. Semenova, and A. S. Antipova. Food Hydrocolloids *12*:227–235, 1998.
151. V. N. Ismailova, G. P. Yamployskaya, and B. D. Tulovskaya. Colloids Surf. A Physicochem. Eng. Aspects *160*:89–106, 1999.
152. S. O. Agboola and D. G. Dalgleish. J. Food Sci. *60*:399–404, 1995.
153. S. O. Agboola and D. G. Dalgleish. Lebensm. Wiss. Technol. *29*:425–432, 1996.
154. A. Kulmyrzaev, R. Chenamai, and D. J. McClements. Food Res. Int. *33*:15–20, 2000.
155. E. Dickinson and C. M. Woskett. In *Food Colloids* (R. D. Bee, P. Richmond, and J. Mingins, eds.). Cambridge: Royal Society of Chemistry, 1989, pp. 74–96.
156. J. Leaver, D. S. Horne, and A. J. R. Law. Int. Dairy J. *9*:319–322, 1999.
157. L. Leaver, A. J. R. Law, and D. S. Horne. In *Food Emulsions and Foams: Interfaces, Interactions and Stability* (E. Dickinson and J. M. Rodriguez Patino, eds.). Cambridge: Royal Society of Chemistry, 1999, pp. 258–268.
158. J. Leaver, A. J. R. Law, and E. Y. Brechany. J. Colloid Interface Sci. *210*:207–214, 1999.
159. D. S. Horne, P. J. Atkinson, E. Dickinson, V. J. Pinfield, and R. M. Richardson. Int. Dairy J. *8*:73–77, 1998.
160. A. R. Mackie, A. P. Gunning, P. J. Wilde, and V. J. Morris. Langmuir *16*:2243–2247, 2000.
161. C. M. Wijmans and E. Dickinson. Langmuir *15*:8344–8348, 1999.
162. J. M. Rodríguez Patino, S. C. Carrera, and M. R. Rodríguez Niño. Langmuir *15*:2484–2492, 1999.
163. S. C. Carrera, M. R. Rodríguez Niño, and J. M. Rodríguez Patino. Colloids Surf. B *12*:175–192, 1999.
164. J. M. Rodríguez Patino, M. R. Rodríguez Niño, S. C. Carrera, and M. Cejudo. J. Colloid Interface Sci. *240*:113–126, 2001.
165. J. M. Rodríguez Patino, M. R. Rodríguez Niño, S. C. Carrera, and M. Cejudo. Langmuir 2001 (in press).
166. A. R. Mackie, A. P. Gunning, M. J. Ridout, P. J. Wilde, and J. M. Rodríguez Patino. Biomacromolecules *2*:1001–1006, 2001.

Index

Acinetobacter calcoaceticus, 842
Activation, 405
Adenosin (*see* Nucleotides)
Adhesion proteins, 403
ADP (*see* Nucleotides)
Adriamycin, 712
Adsorption
  at ultrafiltration membranes, 678
  effect of pH, 17
  effect of surface charge density, 16
  from intraoral exposure, 821
  from plasma and serum, 721
  Langmuir model, 87, 132
  of alanine oligopeptides, 59
  of amino acids, 48
  of amylase, 832
  of antibodies, 151–155
  of anti-HSA, 151
  of apolipoprotein B, 719
  of β-casein, 274–277, 573–574, 859–869
  of β-lactoglobulin, 271, 322–323, 328, 331
  of BSA, 548–549, 873–875
  of C1q, 718, 722
  of C3, 718
  of carboxymethyl cellulose, 18
  of carboxypullulan, 15
  of cyclic peptides, 62
  of cytochrome b562, 593–596
  of cytochrome c, 421
  of DNA, 557–559, 599
  of ferritin, 423–424
  of fibrinogen, 554, 562–563, 591–592, 716, 718–719, 729, 749

[Adsorption]
  of fibronectin, 718
  of gelatin, 570
  of glycine oligopeptides, 60
  of histatin, 832
  of HSA, 429–432, 716, 718
  of human whole saliva, 823–830
  of IgG, 457–459, 548, 561–562, 716, 718
  of insulin, 262–267, 729
  of isoleucin oligopeptides, 107
  of krillase, 573–574
  of lipase, 565
  of lysozyme, 546, 617–638
  of MUC5b, 834
  of nucleic acid bases, 62–64
  of nucleosides, 65
  of nucleotides, 65–68
  of oligopeptides, 107, 109
  of peptides, 56–62
  of polyelectrolyte, 14–18
  of polymers, 3–18
  of prothrombin, 720
  of PRP, 832
  of RNase, 225, 564–565
  of salivary proteins, 812–843
  of savinase, 12–13, 563–564
  of SDS, 328
  of sodium caseinate, 275
  of spheres, 89
  of staphylococcal enterotoxin B, 786
  of statherin, 832
  of trypsin, 573–574, 793, 798
  of tryptophan oligopeptides, 107

[Adsorption]
  of Z, 105–110
  rate coefficient, 208–209
  to discrete receptor sites, 215–216
AFM, 481–483
  of multilayer systems, 355–356
  of polyelectrolyte shells, 386
Albumin (see HSA and BSA)
Alcohol dehydrogenase, 375
Alginate, 357
α-amino acids (see Amino acids)
α-amylase, 814
α-helix, 455, 475, 477, 505–506
α-lactalbumin
  apolar surface fraction, 38
  denaturation enthalpy of, 525–528
  effect of adsorption on electroosmotic
    mobility, 778
  effect on membrane flux, 685
  isoelectric point of, 300
  protein titration curve, 35
  thermal denaturation of, 29
Amide I, 452, 455
Amide II, 452, 467–489
Amino acids
  adsorption of, 48–56
  partitioning, 49
Amylase, 813, 832
Antibiotin, 154
Antibodies (see also IgG and IgA)
Antibodies
  biotinylation of, 121
  fluorophore conjugation of, 122
  immobilization of, 124–127
  orientation of, 154
  oxidation of, 120
  packing and orientation, 149–155
  photoaffinity labeling of, 123–124
  radiolabeling of, 121–122
Antifluorescein, 154
Anti-HSA, 151
Anti-IgG, 352
Apo-α-lactalbumin, 304
Apolar surface fraction, 38
Apolipoprotein B, 719
ATP (see Nucleotides)
ATR (see FTIR)
Autoantibodies IgG, 763
Autolysis, 571–573, 798
Autoproteolysis (see Autolysis)
Avidin, 143–145, 569

Bacteriorhodopsin, 375
β-casein
  adsorption of, 274–282, 573–574, 859–
    869, 876–884, 890
  hydrodynamic layer thickness of, 862, 864
  schematic adsorbed conformation, 863
  volume fraction profile of, 861, 864
β-D-galactopyranoside, 792
β-lactoglobulin, 271–274, 281–285, 865–
  867
  displacement by $C_{12}E_5$, 888
  effect of CTAB, 323
  effect of SDS, 322–323
  effect on membrane flux, 684–685
  removal by SDS, 331
β-sheet, 455, 475, 477, 487
Biomineralization, 361
Biotin, 143–145, 569, 598
Blood clearence, 713, 719, 725
Bovine pancreas ribonuclease (see RNase)
Bovine serum albumin (see BSA)
Brewster angle microscopy, 892
BSA
  absorbance of, 474–475
  adsorption at the air–water interface, 873–
    875
  adsorption kinetics of, 548–549
  aggregation in microfiltration, 674–675
  amide I, II, and III bands of, 475
  complexation with polyelectrolytes, 347
  dimension of, 499
  effect of dimers on adsorption, 110
  effect of metal chelators, 697
  effect on membrane flux, 681–685
  effect on membrane sieving, 680
  FRAPP result on, 231
  labile hydrogen atoms in, 613
  molecular weight of, 499
  surface diffusion of, 239–244, 249–250,
    252
  zeta potential of, 677

$C_{12}E_5$, 327
$C_{12}E_6$, 888
C1q, 718, 722
C3
  activation, 726
  adsorption of, 718
Candida albicans, 841
Capillary electrophoresis, 781–783
Carbonic anhydrase, 375, 813
Carboxymethyl cellulose, 18, 842

Carboxypullulan, 15
Catalase, 375
CD, 501–502
  of IgG, 457–459
  of lysozyme, 98–99, 504–506
  of Z, 106
Cells
  interactions with adhesion proteins, 405–408
  interactions with polyelectrolyte multilayers, 356–361
  stratified layers of, 552–553
Cellulose, 185–186,551–552
Cetyltrimethylammonium bromide (*see* CTAB)
Charge compensation, 16
Chitosan, 185, 187, 357
Chymotrypsinogen A, 76–77
Circular dichroism (*see* CD)
Cleaning food process equipment, 334
Cloud point, 758
cmc, 326
Coil, 1–2
Collagenase, 251–252
Competitive adsorption, 295–310
Complement activation, 566–567, 722, 726
Concavalin, 375
Conformation entropy, 26
Conformation of polymer at interface, 24
Coulomb interaction (*see* Electrostatic interaction)
Critical angle for total reflection, 436
Critical displacer concentration, 11
Critical micellization concentration (*see* cmc)
CTAB, 557–559
  effect on β-lactoglobulin adsorption, 323
Cubosomes, 714
Cystatins, 813, 815
Cytochrome b562, 593–596
Cytochrome c
  adsorption of, 421
  apolar surface fraction, 38
  isoelectric point of, 375
  molecular weight of, 375

Debye length, 15, 33, 169
Dextran, 357–358, 680, 687
Diaphorase, 375
Differential scanning calorimetry (*see* DSC)
Diffusion time–length relation, 776
Dipole moment, 77

Dispersion interaction (*see* van der Waals interaction)
DLVO, 688
DNA, 557–559, 596–601
DNA polymerase, 797
DSC, 503, 509–511, 518–535

Egg white lysozyme (*see* Lysozyme)
Electroosmotic flow, 777–780
Electrosorption, 16
Electrostatic interaction, 26–27, 29, 74–78, 169–170
ELISA, 155, 554
Ellipsometry, 539–574
Emulsion droplets, 714
Enamel, 819–820
Enzymatic assays, 792
Enzyme nanoreactors, 376–381
ESCA (*see* XPS)
Evanescent wave, 436
Evanescent wave immunosensor, 130
Exchange kinetics
  of apo-α-lactalbumin, 304
  of fibrinogen, 304
  of HSA, 305
  of IgG, 304
  of lysozyme, 302, 305
Excluded volume, 2
Exclusion zone, 210–212
Extracorporeal therapy (*see* Therapeutic apheresis)

Ferritin, 226
  adsorption of, 423–424
  surface diffusion of, 250, 252
Fibrinogen
  adsorbed layer formation, 562–563
  adsorption, 398, 554, 562–563, 716, 718–719, 729, 749
  amide II band of, 472
  antibody binding to, 403–404
  elutability by surfactants, 326
  exchange kinetics of, 304
  in competitive adsorption, 567–568
  in polyelectrolyte multilayers, 352
  isoelectric point of, 308
  layer thickness of, 591–592
  secondary structure, 353
  surface diffusion of, 251–252
Fibronectin
  adsorption of, 718
  antibody binding to, 403–404

[Fibronectin]
   secondary structure of, 487
Fick's law, 147, 786
Flory–Huggins parameter, 2
Fluorescence, 442, 502–503
   of HCA, 507–508
   of lysozyme, 504
   quenching, 234–236
   recovery after (pattern) photobleaching
      (see FRA(P)P)
Fluoroimmunoassays, 128–131
FRAP, 227–253
FTIR, 451–459, 467–489
   and blood proteins, 478
   and multiple protein solutions, 480–481
   and orientation, 479–480
   frequency precision in, 478
   quantification of secondary structure, 477
   relation to x-ray crystallography, 476–477
   structural correlations, 477

Galactolipids, 180
Galactose, 166
Ganglioside, 180–183
Gaussian coil, 2
Gelatin, 570, 572
Generalized ballistic deposition, 211
Glucoamylase, 371–381
Glucose, 166
Glucose oxidase, 371–381
   catalytic activity of, 377
   isoelectric point of, 375
   molecular weight of, 375
   storage stability of, 378
   thermal stability of, 379
Glucoside linkage, 184
Glycolipids, 180–183
Glycoproteins, 189–192
Glycosylated PRP, 816
Gouy-Stern model, 31
Gradient surface, 332–333, 550

Hamaker constant, 37, 80, 168, 547
HCA, 101–102, 500, 506–511
HDL, 330, 783–784
Helmholtz equation, 417
Hemicellululose (see Xylan)
Hemoglobin
   effect on membrane flux, 685
   isoelectric point of, 375
   molecular weight of, 375
Heparin, 358

High density lipoprotein (see HDL)
Histatin, 813, 816, 832
Histone f3, 375
HMWK, 307, 309
HSA
   adsorption at the air–water interface, 429–432
   adsorption of, 716, 718
   adsorption on polyelectrolyte multilayers, 351
   charge compensation of, 35
   diffusion on and in polyelectrolyte multi-layers, 354
   displacement by IgG and fibrinogen, 307–308
   effect of dimers on adsorption, 110
   exchange kinetics of, 305
   function in oral cavity, 815
   in competitive adsorption, 567–568
   in polyelectrolyte multilayers, 349–350
   isoelectric point of, 349
Human carbonic anhydrase (see HCA)
Human growth hormone, 441, 450
Human serum albumin (see HSA)
Human whole saliva, 823–830
Humicola lanuginosa lipase (see Lipase)
Hydrodynamic layer thickness, 9–10
Hydrogen bonds, 28, 171–172
Hydrophobic interaction, 26, 29, 171
Hydrophobicity gradient surface (see Gradi-ent surface)
Hydroxyapatite, 362, 819–820, 832–833

Ideal chain, 2
IgA, 566, 814
IgG
   adsorbed layer formation, 561–562
   adsorption, 457–459, 548, 561–562, 716, 718
   amide II band of, 472
   binding of anti-IgG, 557
   CD results of, 457–459
   conformational changes, 457
   exchange kinetics of, 304
   fluorescence of, 447–448
   in competitive adsorption, 567–568
   in microarrays, 801
   interaction with complement system, 566–567
   isoelectric point of, 375
   molecular weight of, 375
   structure of, 118

[IgG]
    thermogram of, 532–533
    TOF-SIMS spectra of, 652
IL1, 762
IL6, 762, 766
Immobilization
    by physical adsorption, 133–136
    of acid-pretreated antibodies, 154
    to hydrogel films, 136–143
    to PEG films, 136–143
    to SAMs, 143–145
Immunoassays, 127–156, 742–743, 760–
        762, 783–792
Infrared spectroscopy (see FTIR)
Inositol, 167
Insulin
    adsorption of, 261–271, 729
    electropherograms of, 796
Integrins, 394–395
Interaction
    between β-casein, 278–282
    between cellulose, 185–186
    between chitosan, 185, 197
    between galactolipids, 180
    between ganglioside, 180–183
    between insulin, 267–271
    between lysozyme, 100, 692–695
    between mucin, 191–192
    between octyl-β-glucoside, 176–180
    between proteoheparan sulfate, 189–190
    between PRP, 831
    between saliva surfaces, 829
    between xylan, 187–188
IR (see FTIR)
Isoelectric point
    of α-lactalbumin, 300
    of alcohol dehydrogenase, 375
    of amylases, 813
    of bacterorhodopsin, 375
    of BSA, 677
    of carbonic anhydrase, 375, 813
    of catalase, 375
    of concavalin, 375
    of cystatins, 813
    of cytochrome c, 375
    of diaphorase, 375
    of fibrinogen, 308
    of glucoamylase, 375
    of glucose oxidase, 375
    of hemoglobin, 375
    of histatins, 813
    of histone f3, 375

[Isoelectric point]
    of HSA, 349
    of IgG, 375
    of lactoferrin, 813
    of lactoperoxidase, 814
    of lysozyme, 300, 349, 375, 813
    of mucin, 813
    of myoglobin, 375
    of pepsin, 375
    of peroxidase, 375, 813
    of photosynthetic RC, 375
    of PRP, 813
    of RNase, 349
    of sIgA, 813
    of statherins, 813

Jamming limit, 88

Kretschmann configuration, 586
Krillase, 571–572

Lab-on-a-chip, 773
Lactoferrin, 813–814
Lactoperoxidase, 814
Langmuir isotherm, 87, 132, 151, 211
Langmuir monolayers, 415
Layer-by-layer adsorption, 368
LDL, 783–784
Lectin, 524, 527, 529–531
Lennard–Jones potential, 79
Lifshitz interactions (see van der Waals in-
        teractions)
Lipase, 104–105, 565
Liposomes, 713–714
Lysozyme
    adsorption at metal surface, 546
    adsorption of, 617–638
    α-helix content of, 99
    apolar surface fraction, 38
    CD of, 504–506
    conformational changes of, 503–506
    denaturation enthalpy of, 525–528
    dimension of, 500
    dipole moment of, 77
    effect of adsorption on electroosmotic mo-
        bility, 778
    effect of charges on adsorption, 104
    effect of structural stability of, 97–100
    effect on membrane flux, 684–685
    electrostatic interaction energy of, 77
    exchange kinetics of, 302, 305
    fluorescence of, 504

[Lysozyme]
  function of, 814
  homogenous exchange of, 303, 306
  in polyelectrolyte multilayers, 349–
    350
  interaction between, 100, 692–695
  interaction with surfactants, 329–330
  isoelectric point of, 300, 349, 375, 813
  labile hydrogen atoms in, 614
  MALDI-MS spectra of, 658
  membrane filtration of, 672
  molecular weight of, 375, 500, 813
  removal by $C_{12}E_5$, 327
  thermogram of, 522, 534
  MALDI spectra of trypsin, 799
  MALDI-MS, 656–664

Mannose, 166
MARCKS, 213
Memory function, 207–208
Metal chelators, 697
Micelles, 712–713
Microarrays, 797–803
Microemulsions, 758–759
Microfiltration membranes, 676–677
Molecular potency, 393, 396, 402–408
Molecular weight
  of alcohol dehydrogenase, 375
  of amylases, 813
  of autoantibodies IgG, 763
  of bacterorhodopsin, 375
  of BSA, 499
  of carbonic anhydrase, 375, 813
  of catalase, 375
  of concavalin, 375
  of cystatins, 813
  of cytochrome c, 375
  of diaphorase, 375
  of glucoamylase, 375
  of glucose oxidase, 375
  of hemoglobin, 375
  of histatins, 813
  of histone f3, 375
  of HCA, 500
  of IgG, 375
  of IL1, 762
  of IL6, 762
  of lactoferrin, 813
  of lysozyme, 375, 500, 813
  of mucin, 813
  of myoglobin, 375
  of pepsin, 375

[Molecular weight]
  of peroxidase, 375, 813
  of photosynthetic RC, 375
  of PRP, 813
  of sIgA, 813
  of statherins, 813
  of $TNF\alpha$, 762
Molten globule state, 498–499
Monolayer relaxation, 879
μTAS, 773
MUC5b, 834
Mucin, 191–192, 813, 816–817
Myoglobin
  effect of adsorption on electroosmotic
    mobility, 778
  effect on membrane flux, 685
  isoelectric point of, 375
  molecular weight of, 375

Neutron reflection
  exchange of labile hydrogens, 612–
    615
  principles of, 610–612
Nucleic acid bases, 62–64
Nucleosides, 65
Nucleotides, 65–68
Null ellipsometry, 540

Octyl-β-glucoside, 176–180
Oligopeptides, 107, 109
OSBP, 601–605
Ovalbumin, 149, 685
Overshoot, 13
Oxysterol binding protein (see OSBP)

PAGE, 765
PEG
  adsorption of a surface active PEG deriva-
    tive, 747–748
  antibody immobilization to, 136–143
  at the air–water interface, 427–432
  attachment by covalent bonds, 750–755
  blood interaction with, 400–401
  different types of PEG modification,
    726–727
  effect of anchoring, 730–731
  effect of chain density, 554, 728–730
  effect of chain length, 727–728
  effect of solvency, 731–732
  effect on electroosmotic mobility, 779
  fibrinogen adsorption to, 398–400
  immobilization to, 756–760
  in drug carriers, 723–724

[PEG]
  in drug targeting, 732
  mechanisms of protein rejection, 397–398, 724–726
  self-assembled monolayers containing PEG, 748–750
  site-specific immobilization, 759–760
  temperature-dependent contraction, 548
  use of microemulsions as reaction medium, 758–759
  use of the cloud point concept, 758
PEG-grafted surfaces, 746–747
Pellicle, 820–821, 828
PEO (*see* PEG)
Pepsin
  effect on membrane flux, 684–685
  isoelectric point of, 375
  molecular weight of, 375
Peptides, 56–62
Peroxidase, 375, 813
Phospholipids, 552, 715–716, 717–722
Photochemical modification, 699
Photosynthetic RC, 375
PKA-kinase, 795
Plaque formation, 837–840
Plasma, 721
Plasma clearence, 713, 719
Plasma modification, 551, 699
Poly(ethylene glycol) (*see* PEG)
Poly(ethylene oxide) (*see* PEG)
Polyampholyte, 14–18
Polydispersity, 11
Polyelectrolyte, 14–18
Polyion capsules, 385
Polylysine, 357
Polymer membranes, 671–703
Polypeptide chain, 22, 27
Polysaccharides, 183–188, 756
Proline-rich proteins (*see* PRP)
Protein-lipid interactions, 715–722, 889–894
Proteoheparan sulfate, 103, 189–190
Prothrombin, 720
PRP, 813, 815–816, 830–833
Pyranoses, 166

QCM, 371
Quartz crystal microbalance (*see* QCM)

Radioimmunoassays, 127–128
Random coil, 475
Random sequential adsorption (*see* RSA)

Reflectometry, 539–574
Refractive index increment, 542
Relative affinity, 395
RES, 711–712
Reticuloendothelial system (*see* RES)
RGD, 402, 486–488
Ribonuclease (*see* RNase)
RNase
  adsorption of, 564–565
  apolar surface fraction, 38
  effect of adsorption on electroosmotic mobility, 778
  effect on membrane flux, 685
  in polyelectrolyte multilayers, 349–350
  isoelectric point of, 349
  relative quantum yield, 233–234
  structure of, 23
  surface diffusion of, 250, 252
RSA, 88, 211, 222

Salivary proteins, 812–843
SAM, 143–145, 402, 406–408, 484–486, 550
Savinase, 12–13, 563–564
Scanning angle reflectometry (*see* Reflectometry)
Schitzophyllan, 183
SDS
  competitive adsorption with β-lactoglobulin, 328
  effect on β-lactoglobulin adsorption, 322–323
  effect on lysozyme adsorption, 330
  removal of β-lactoglobulin, 331
  removal of pellicle by, 828
Secondary structure (*see* α-Helix and β-Sheet)
Self-assembled monolayers (*see* SAM)
Self-avoiding walk, 2
Self-consistent field theory, 6
Semidilute solution, 2
Sequential adsorption, 300
SERS, 439
Serum, 721
sIgA, 813
Single-stranded binding protein (*see* SSB)
Sodium caseinate, 275
Sodium dodecyl sulfate (*see* SDS)
Solid phase diagnostics, 742–743, 760–762
Solvation forces, 81
Solvency, 2–3
Solvent quality (*see* Solvency)

SPR, 420, 429–432, 583–605
Spreading, 11–14
SSB, 596–601
Staphylococcal enterotoxin B, 786
Statherin, 813, 816, 832
Stern-Vollmer equation, 153, 451
Streptavidin, 532, 598
Streptococcus mitis, 839
Streptococcus mutans, 839, 841
Substrate activation (*see* Activation)
Sugar-based surfactants, 176–180
Surface diffusion
    mechanism, 246–248
    of BSA, 239–244, 249–250, 252
    of collagenase, 251–252
    of ferritin, 250, 252
    of fibrinogen, 251–252
    of RNase, 250, 252
Surface enhanced Raman scattering (*see*
    SERS)
Surface force apparatus, 173–174
Surface plasmon resonance (*see* SPR)
Surface plasmons, 585
Surface pressure, 418
Surface selection rule, 437

T4 lysozyme (*see* Lysozyme)
T-cell phosphatase, 792, 794
Therapeutic apheresis, 743–745, 762–766
Thermal transition, 520

Thermogram, 519
Thin-film balance, 175
TIRF, 298–299, 445–451, 570
TNFα, 762
TOF-SIMS, 649–656
Total internal reflection fluorescence spec-
    troscopy (*see* TIRF)
Trapping, 14
Triclosan, 842
Trypsin, 572–573, 792–798
Tryptic digestion, 792–798

Ultrafiltration membranes, 677–681
Urease, 384
UV modification, 699

van der Waals interacions, 27–29, 36–37,
    78–81, 167–169
van't Hoff equation, 521
Vroman effect, 296, 717, 775

WPI, 870–872, 876–884

Xanthan, 183
XPS, 643–649
X-ray photoelectron spectroscopy (*see* XPS)
Xylan, 187–188

Z, 105–110